New Video Resources

Author in Action videos are available for every learning objective of every section. The authors are all active teachers, and these Camtasia Studio® videos were recorded in the classroom as they presented the concepts to their own students.

- **Short, manageable video clips** cover all the important points of each learning objective, and are perfect for studying or reviewing on the students' schedule.

- **Icons** in the text and eText alert students to when a video is available.

- The **Author in Action videos** are now captioned in English and Spanish.

Solving Linear Equations

Example:
Solve the linear equation $3x - 9 = -24$.

① Isolate the term involving the variable.

$$3x - 9 = -24$$
$$3x - 9 + 9 = -24 + 9$$
$$3x = -15$$

② Get the coefficient on the variable to equal 1.

$$\frac{1}{3} \cdot 3x = \frac{1}{3} \cdot (-15)$$
$$x = -5$$

③ Check.

$$3(-5) - 9 \overset{?}{=} -24$$
$$-24 = -24 \checkmark$$
$$\{-5\}$$

Course:	Name:
Instructor:	Section:

Section 2.2 Video Guide
Linear Equations: Using the Properties Together

Objectives:
1. Use the Addition and Multiplication Properties of Equality to Solve Linear Equations
2. Combine Like Terms and Use the Distributive Property to Solve Linear Equations
3. Solve a Linear Equation with the Variable on Both Sides of the Equation
4. Use Linear Equations to Solve Problems

Section 2.2 – Objective 1: Use the Addition and Multiplication Properties of Equality to Solve Linear Equations
Video Length – 6:27

We will now solve linear equations where we need to use both the Addition Property of Equality and the Multiplication Property of Equality.

1. **Example:** Solve the linear equation $3x - 9 = -24$.

Write the steps in words	Show the steps with math
Step 1	
Step 2	
Step 3	

Final answer: _____

2. **Example:** Solve the linear equation $\frac{5}{4}x + 2 = 17$.

Final answer: _____

NEW! The **Video Notebook** is a note-taking tool that students use in conjunction with the "Author in Action" videos.

- A **Video Guide** for every section organizes the content by learning objective.

- **Ample space** is provided to allow students to write down important definitions and procedures, as well as show their work on examples, while watching the video.

- The **unbound, three-hole-punched format** allows students to insert additional class notes or homework, helping them build a course notebook and develop good study skills for future classes.

MyMathLab® = Your Resource for Success

In the lab, at home,...

- Access videos, PowerPoint® slides, and animations.
- Complete assigned homework and quizzes.
- Learn from your own personalized Study Plan.
- Print out the Video Notebook for additional practice.
- Explore even more tools for success.

...and on the go.

 Download the free MyDashBoard App to see instructor announcements and check your results on your Apple® or Android™ device. MyMathLab log-in required.

 Download the free Pearson eText App to access the full eText on your Apple® or Android™ device. MyMathLab log-in required.

 Use your Chapter Test as a study tool! Chapter Test Prep Videos show step-by-step solutions to all Chapter Test exercises. Access these videos in MyMathLab or by scanning the code.

Scan the code or go to: www.youtube.com/SullivanInter3e

Don't Miss Out! Log In Today.

MyMathLab delivers proven results in helping individual students succeed. It provides engaging experiences that personalize, stimulate, and measure learning for each student. And, it comes from a trusted partner with educational expertise and an eye on the future.

To learn more about how MyMathLab combines proven learning applications with powerful assessment, visit **www.mymathlab.com**

VIDEOS • POWERPOINT SLIDES • ANIMATIONS • HOMEWORK • QUIZZES • PERSONALIZED STUDY PLAN • TOOLS FOR SUCCESS

Michael Sullivan III • Katherine R. Struve

Intermediate Algebra

Second Custom Edition for Pasadena City College

Taken from:
Intermediate Algebra, Third Edition
by Michael Sullivan III and Katherine R. Struve

Strategies for Success for the College Math Student, Second Edition
by Lynn Marecek and MaryAnne Anthony-Smith

Cover Art: Courtesy of Matthew Henes.

Taken from:

Intermediate Algebra, Third Edition
by Michael Sullivan III and Katherine R. Struve
Copyright © 2014, 2010, 2007 by Pearson Education, Inc.
Boston, Massachusetts 02116

Strategies for Success Study Skills for the College Math Student, Second Edition
by Lynn Marecek and MaryAnne Anthony-Smith
Copyright © 2014, 2012 Pearson Education, Inc.
Boston, Massachusetts 02116

This special edition published in cooperation with Pearson Learning Solutions.

All trademarks, service marks, registered trademarks, and registered service marks are the property of their respective owners and are used herein for identification purposes only.

Pearson Learning Solutions, 501 Boylston Street, Suite 900, Boston, MA 02116
A Pearson Education Company
www.pearsoned.com

Printed in the United States of America

1 2 3 4 5 6 7 8 9 10 V011 18 17 16 15 14

000200010271914630

DS

ISBN 10: 1-269-92684-5
ISBN 13: 978-1-269-92684-3

Welcome to Pasadena City College!!!

The key to being successful in a math course is to take responsibility of your learning. You need to do some math every day and you need to find the answers to your questions right away. Do not wait until the day before a test to start studying. You should begin practicing problems a minimum of 3 days before the exam. There are many places on campus you can receive assistance with your math work.

1. See your instructor in his/her office.
    My teacher's office is in _____
    His/her office hours are _____

2. Work in the Math Resource Center's Homework Lab in R407.
    Their hours are   Monday _____
                      Tuesday _____
                      Wednesday _____
                      Thursday _____
                      Friday _____

3. Visit the Learning Assistance Center in D300.
    Their hours are   Monday _____
                      Tuesday_____
                      Wednesday _____
                      Thursday _____
                      Friday _____

4. Start a study group with your classmates. Teaching your classmates is the best way to learn math!
                      Name _____
                      number _____
                      e-mail _____
                      Name _____
                      number _____
                      e-mail _____

5. Call the PCC Homework Help Line 626-585-7056.
    Their hours are   Monday _____
                      Tuesday_____
                      Wednesday _____
                      Thursday _____
                      Friday _____

Time Management

MANAGING YOUR TIME

When deciding how many units to take, it is very important to plan time for reading, studying, and preparing for those classes. No time is given "in class" for study. You are expected to be ready for each class before the class begins.

Recommended formula to determine expected amount of **TOTAL TIME** required for success in a class is as follows:

Schedule two (2) hours per instruction hour per week for studying.

Example:

In-Class Time	=	$5\frac{1}{2}$ Hours per Week
Study Time	=	11 Hours per Week

(2 Studying Hours per Instruction Hour $\times$ $5\frac{1}{2}$ Instruction Hours = 11 Hours)

TOTAL TIME:
10 STUDYING HOURS PER WEEK FOR A 5-HOUR CLASS

MANAGING WORK AND SCHOOL:
RECOMMENDED COMBINATIONS

WORK	SCHOOL
15 Hours per Week	9-12 Units
25 Hours per Week	6-9 Units
40 Hours per Week	3-6 Units

DO NOT OVERLOAD YOURSELF WHEN YOU REGISTER FOR CLASSES!

Time Management Assessment

Rate yourself on this assessment using the following scale:

3 = true most of the time
2 = true some of the time
1 = true almost none of the time

_____ 1. I am usually on time.
_____ 2. I never seem to have enough time.
_____ 3. I am an effective time manager.
_____ 4. A time schedule makes me feel restricted.
_____ 5. I maintain a weekly time schedule (short-range planner).
_____ 6. College students really don't need to manage their time.
_____ 7. I maintain a monthly or semester plan (long-range planner).
_____ 8. I often feel like I'm wasting time.
_____ 9. I enjoy the structure provided by a time schedule.
_____ 10. Procrastination is my biggest time management problem.
_____ 11. I know how to set priorities.
_____ 12. There simply aren't sufficient hours in a day.
_____ 13. I keep my time commitments.
_____ 14. I don't manage my time because I don't have goals.
_____ 15. I use a daily "to do" list.
_____ 16. I allow my friends to distract me from important activities.
_____ 17. I maintain a study schedule.

Total score for odd numbered items. _____

Total score for even numbered items. _____

High ratings on the odd numbered items generally mean that you have a fairly good grasp of time management and are probably doing a pretty good job of managing your time.

High ratings on the even numbered items generally mean that you have difficulty with time management. If your total score on the even numbered items is higher than the odd items, you may want to discuss this with a counselor or enroll in Counseling 11 or 12 class at Pasadena City College.

7 Day
Anti-Procrastination Plan

MONDAY: *MAKE IT MEANINGFUL!*
List all the benefits of completing a task. Look at it in relation to your goals. Be specific about the rewards for getting it done, including how you will feel when the task is complete.

TUESDAY: *TAKE IT APART!*
Break big jobs into a series of small ones you can do in 15 minutes or less. If a long reading assignment intimidates you, divide it into two-page or three-page sections. Make a list of sections and cross off as you complete them so you can see your progress.

WEDNESDAY: *WRITE AN INTENTION STATEMENT!*
Write an Intention Statement on a 3×5 card. Carry it with you or post it in your study area where you can see it often. Example: *"I intend to reward myself with chocolate, guilt free."*

THURSDAY: *TELL EVERYONE!*
Announce publicly your intention to get it done. Include anyone who will ask whether you've completed it or who will suggest ways to get it done.

SATURDAY: *SETTLE IT NOW!*
Do it now! The minute you notice yourself procrastinating, plunge into the task. Image yourself wanting to swim but the swimming pool is unheated. Gradual immersion would be slow torture. It's often less painful to dive in and just start swimming.

SUNDAY: *SAY NO!*
When you keep pushing a task into the low-priority category, re-examine the purpose for doing it at all. If you realize you really don't intend to do something, quit telling yourself that you will.

P.S.
In some cases, procrastination is positive. Consider the following possibilities:

1. PROCRASTINATE DELIBERATELY.
You might discover that if you can choose to procrastinate, you can also choose not to procrastinate.

2. OBSERVE YOUR PROCRASTINATION.
Instead of doing something about your procrastination, look carefully at the process and its consequences. Avoid judgments. Be a scientist and record facts. See if it is keeping you from what you want. Seeing its cost may help you kick the habit.

Daily Schedule   Mathematician _____

	Monday	Tuesday	Wednesday	Thursday	Friday	Saturday	Sunday
06:00-07:00							
07:00-08:00							
08:00-09:00							
09:00-10:00							
10:00-11:00							
11:00-12:00							
12:00-13:00							
13:00-14:00							
14:00-15:00							
15:00-16:00							
16:00-17:00							
17:00-18:00							
18:00-19:00							
19:00-20:00							
20:00-21:00							
21:00-22:00							
22:00-23:00							
23:00-24:00							
00:00-01:00							
01:00-02:00							
02:00-03:00							
03:00-04:00							
04:00-05:00							
05:00-06:00							

Tips for preparing for each class meeting:

Use the questions below to read and study your assigned sections of the textbook and properly prepare yourself for the next class meeting.

1. **What is new in this section?**
   (a) New terminologies: List all new definitions.

   (b) New methodologies: Give two or more examples.

   (c) New types of problems: Give two or more examples.

2. **What mathematical concepts do you need to review in order to learn this section well? List all the required mathematical concepts and the sections where they can be found.**
   (1)

   (2)

3. **Is there anything in this section that you find difficult to understand?**

powered by CourseCompass™ and MathXL®

MyMathLab

Student Instructions for Registration and Login for Pasadena City College

Before you go online to register be sure you have:

- A valid **e-mail address**

- The MyMathLab **Course ID** from your instructor

- Pasadena City College zip code - **91106**

- A **student access code**, which should have come packaged with your textbook
  Sample: MMLST-TAROK-THOLE-PICON-SHRIK-PRAWN

If you are using a computer <u>off campus</u> (at home or work) you will need to have administrative access so you can install the necessary plug-ins, and be connected to the internet. *(If you're not sure your computer meets the system requirements, go to www.mymathlab.com/system.html, visit the MyMathLab Installation Wizard in your course, or contact tech support at 1-800-677-6337).*

If you are using a computer <u>on campus</u> the necessary plug-ins may already be installed in the lab. If you aren't sure, check with the lab coordinator or your instructor.

Instructions:
1. Launch Internet Explorer. *If you are off campus and using AOL to connect to the Internet you will need to minimize AOL and then launch Internet Explorer.*

NOTE: Your MyMathLab course may be available on both PC and Mac platforms, using Internet Explorer, Firefox, and Safari. Check with your instructor for more information.

2. If necessary, clear the address field and enter www.coursecompass.com.

3. Click the **Register** button for students.

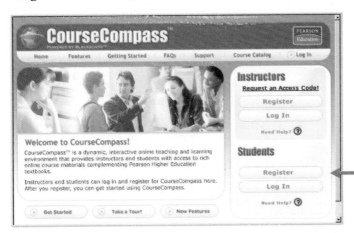

4. Confirm that you have the information necessary to continue and click **Next** on the "Before you Start..." page.

5. On the Product Selection page, enter the **course ID** your instructor gave you and click **Find Course**.

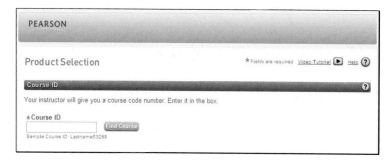

Your course information will appear. *Be sure to check that your instructor's course information is correct. If not, contact your instructor to verify the correct **course ID***.

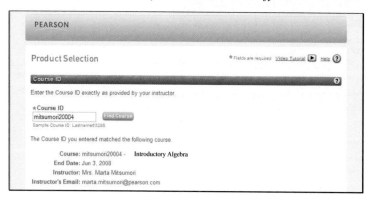

6. If your student access code came packaged with your textbook, select **Access Code**. If you do not have an access code and want to purchase access to your course with a credit card, select **Buy Now**.

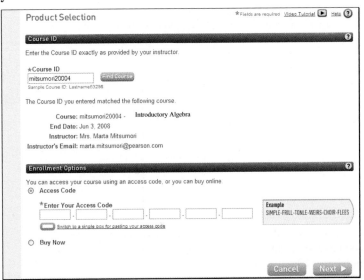

7. Type in your student access code using the tab key to move from one box to the next. Type letters in either all uppercase or all lowercase. Click **Next** after you have entered your code. *If you are buying access with a credit card, the on-screen instructions will guide you through the purchase process.*

8. Please read all information on the License Agreement and Privacy Policy page. Click on **Accept** if you agree to the terms of use.

9. On the Access Information Screen, you'll be asked whether you already have a Pearson Education Account.

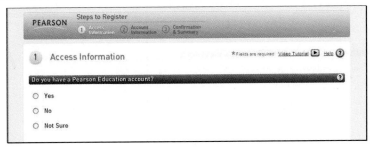

10. If you have registered for other Pearson online products and already have a login name and password, select **Yes**. Boxes will appear for you to enter your login information.

If you select Yes, you will see a Welcome Back screen, where you may be prompted to update your password and make changes to your account information as needed.

11. If this is the first time you have registered for a Pearson online product, select **No**. Boxes will appear for you to enter your desired **login name** and **password**. You may want to use your email address as your login name. If you do not use your email address, be prepared with a second login name choice if the one you first selected is already in use. Your login name must be at least 4 characters and cannot be the same as your password.

12. If you aren't sure whether you have a Pearson account or not, select **Not Sure**. Enter your email address and click **Search**. If you have an account, your login information will be sent to your email address within a few moments. Change your selection to Yes, and enter your login name and password as directed.

13. On the Account Information page, enter your first and last name and email address. Re-type your email address to make sure it is correct.

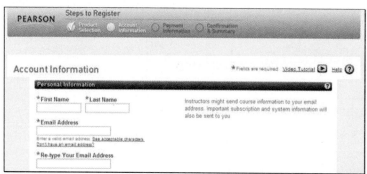

14. In the School Location section, select **United States** from the School Country drop-down menu (NOTE - United States appears at the top of this list). Enter your **school** zip code, and then select your school from the drop-down list. <u>**The zip code for PCC is 91106.**</u>

*If your school is not listed, scroll to the bottom of the drop-down list and select **Other**. Then enter your school name and city and select the state.*

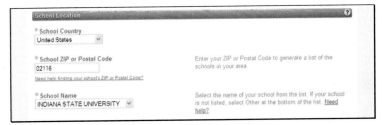

15. Select a security question and answer to ensure the privacy of your account. Then click **Next.**

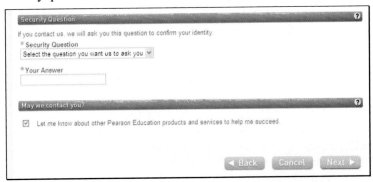

16. When your registration process is complete you will see a confirmation screen.

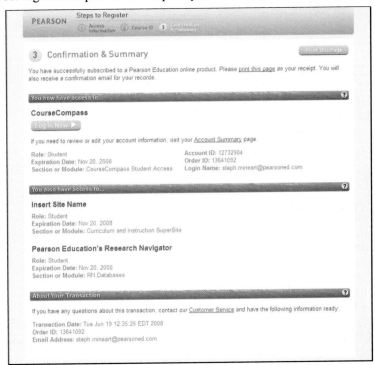

Congratulations – You have successfully registered for your MyMathLab course!

17. Print the Confirmation & Summary page so you will have a record of your login name and the email address used for your account. Be sure to keep the email confirmation sent to you.

18. Click **Log in Now** to reach your My CourseCompass page. You will see your MyMathLab course title in the **Courses** box on the left.

19. Click your course title to begin exploring MyMathLab!

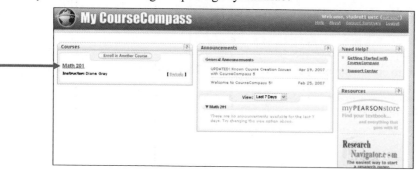

20. **Important!** If you are using a computer <u>off campus</u> and are logging into MyMathLab for the first time, you must click the **MyMathLab Installation Wizard** link now.

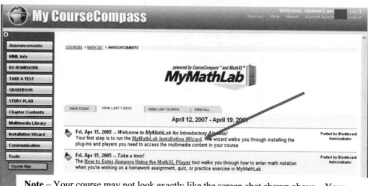

Note – Your course may not look exactly like the screen shot shown above – Your instructor may have specially customized the course navigation buttons and content.

21. The InstallationWizard (or Browser Check – depending on your course) detects and then helps you install the plug-ins and players you need to access the math exercises and multimedia content in your MyMathLab course. Follow the screen instructions to complete this process. After completing the installation process and closing the wizard you will be back on the Welcome page of your course.

22. Click the **How to Enter Answers Using the MathXL Player** link to learn how to enter answers when working within MyMathLab.

23. If you need help, contact the student technical support team at **1-800-677-6337**, or visit **www.mymathlab.com** and click the **Support** tab for more information.

24. When returning to work in MyMathLab, go to **www.coursecompass.com**. Click on the Log In button for students and log in using your login name and password.

25. **Important!** You will access your MyMathLab course from this site, so be sure to **bookmark this page!**

Taken from: *Strategies for Success for the College Math Student,* Second Edition,
by Lynn Marecek and MaryAnne Anthony-Smith

Table of Contents

Preface to the Instructor

Many students arrive at their first college math class without a clue about what it takes to be successful in college. They may have no role models of successful college students in their communities. They may not be aware of the support services the college offers them. Many students are burdened with job and family responsibilities making demands on their time. Some may expect that their course grade will be based on attendance rather than performance. They may not realize that one hour spent in class usually means at least two hours work outside of class. Even if our students can do some mathematics, their weak study skills hamper their overall success.

Strategies for Success are study skills activities specific to fostering success in college mathematics. They force students to take a pro-active approach to determine explicitly what they can do to become successful math students. By using *Strategies for Success*, students develop effective study skills to help them succeed in college. *Strategies for Success* take little class time and need few directions from the teacher, yet they produce big rewards in changed student behavior.

In this workbook, you will find 44 *Strategies for Success* activities. *Strategies for Success: Study Skills for the College Math Students, 2nd edition* includes several new activities that focus on specific study skills needed by students doing their homework exercises on the computer, such as in online, hybrid, redesigned, or other non-traditional classes. These new activities are also beneficial to students in traditional classrooms.

A complete instructor's manual is available through the Pearson Higher Education Instructor Resources website. In it, you will find a "To the Instructor" page for each activity with our suggestions about when and how to use that specific activity in your class. In addition, the *Strategies for Success* activities are available in electronic format through MyMathLab. Since students may be working independently and asynchronously in MyMathLab, a brief video provides an introduction to each activity, much as an instructor might do in a classroom setting.

The following suggestions address some general questions about implementation:

- **I've never taught study skills! How can I feel confident about integrating these worksheets into my curriculum?** For those of us who are used to teaching math, thinking about teaching study skills may make us uncomfortable. As with any new endeavor, we advise you to start slowly. We encourage you to try just one or two activities at first. Then try to use one new activity a month, and, as you build confidence, try one a week and eventually you'll be using the activities in almost every class meeting. By rolling them in gradually, as you feel comfortable, you'll find that they will become a natural part of your teaching repertoire.

- **How do I find class time to do the *Strategies for Success* activities?** Students derive great benefit from time spent helping them develop effective study skills, yet teachers are reluctant to give up time they use 'doing the math'. We felt the same way, but recognized that our students could not master the course content without effective study skills. *Strategies for Success* activities were designed to require very little class time, and students usually 'get the point' without much teacher input. Each activity can be used in several ways—for individual work, group work, large group discussion—and so you can be creative in how you use it in your class. We have students discuss a worksheet in small groups while we take attendance. We introduce a worksheet during the final few minutes of class and then assign it for homework. We sometimes sandwich an activity around a scheduled class break. We often plan to use an activity on no particular day, but have it ready for that day when there are few questions on the homework, or we finish more quickly than anticipated. Sometimes we even just assign a worksheet as homework, have students turn it in at the next class, and so no classtime is used at all. We are confident that once you have tried a few *Strategies for Success* activities with your students, you'll see the benefit and find ways to fit them in!

- **Do students do the worksheets on their own?** Many of the *Strategies for Success* activities can be done individually or in small groups. Some can be started individually and completed in small groups, and vice versa. Some lend themselves to being done individually as homework outside of class. Whether students do the activity in groups or individually, in class or out, they always benefit from a teacher-led wrap-up with the whole class. The "To the Instructor" page for each activity gives suggestions for how to implement it. There is no one best way to use the worksheets—your creativity and teaching style will determine how they work best for you.

- **How do I grade the *Strategies for Success* worksheets?** In order for students to take the activities seriously, we recommend that you assign some credit to completed worksheets. We usually give homework or classwork credit. Points are awarded for thoughtful completion of the assignment, since there are no right or wrong answers. Grading the worksheets takes very little of your time; yet reading students' responses gives you valuable insights into their lives. When used within MyMathLab, most of the *Strategies for Success* activities include open-ended responses and are graded for completion. Many of the activities also include worksheets for the student to print, fill out, and turn in.

- **Can I split the longer worksheets into smaller parts?** Yes, many of the worksheets can easily be split. Whenever you see the symbol ⌋ at the right-hand margin, it indicates a place where we suggest a split would be appropriate.

We are happy to share our *Strategies for Success* activities with you, and eager to get feedback from you as you use them in your classes!

Lynn Marecek   MaryAnne Anthony-Smith

Preface to the Student:

Congratulations! You are taking a college math class, one course on your path towards your educational goal. You have taken the initiative and enrolled in math, knowing that you need to succeed in this class in order to reach your dream.

Now that you are in a math class, you need to do more than sit back and wish for success. Become pro-active in setting yourself up to reach your dream! Identify what successful study habits you used in your earlier school years and vow to continue using them in college. Think about some areas where you could improve and resolve to work on them. Find out what resources your college provides to help you succeed in your math class and take advantage of them.

We, the authors, have seen thousands of students at different stages of their college math careers. We know that many students are unsuccessful at college math, not because they can't do the math, but due to weak study skills. We have noticed what behaviors and habits are common to successful students, and we want to help you recognize and develop those habits, too.

Strategies for Success will guide you in determining specifically what you can do to become a successful math student. By using *Strategies for Success* diligently, you will develop effective study skills, which you can use in your other college classes, too. You'll see that *Strategies for Success* take just a little of your time, but they produce big rewards in changed behavior.

We are honored to accompany you along this part of your educational path and wish you success in reaching your dream.

Lynn Marecek   MaryAnne Anthony-Smith

Taken from: *Intermediate Algebra,* Third Edition, by Michael Sullivan III and Katherine R. Struve

Contents

CHAPTER 6 **Radicals and Rational Exponents 472**

CHAPTER 7 **Quadratic Equations and Functions 541**

CHAPTER 8 **Exponential and Logarithmic Functions 623**

To my children , Michael, Kevin, and Marissa.
—Michael Sullivan

To Ruth and Ed Struve, for loving hospitality.
—Katherine R. Struve

About the Authors

With training in mathematics, statistics, and economics, Michael Sullivan, III has a varied teaching background that includes 23 years of instruction in both high school and college-level mathematics. He is currently a full-time professor of mathematics at Joliet Junior College. Michael has numerous textbooks in publication, including an Introductory Statistics series and a Precalculus series, which he writes with his father, Michael Sullivan.

Michael believes that his experiences writing texts for college-level math and statistics courses give him a unique perspective as to where students are headed once they leave the developmental mathematics tract. This experience is reflected in the philosophy and presentation of his developmental text series. When not in the classroom or writing, Michael enjoys spending time with his three children, Michael, Kevin, and Marissa, and playing golf. Now that his two sons are getting older, he has the opportunity to do both at the same time!

Kathy Struve has been a classroom teacher for nearly 35 years, first at the high school level and, for the past 20 years, at Columbus State Community College. Kathy embraces classroom diversity: diversity of age, learning styles, and previous learning success. She is aware of the challenges of teaching mathematics at a large, urban community college, where students have varied mathematics backgrounds and may enter college with a high level of mathematics anxiety.

Kathy served as Lead Instructor of the Developmental Algebra sequence at Columbus State, where she developed curriculum, conducted workshops, and provided leadership to adjunct faculty in the mathematics department. She embraces the use of technology in instruction, and teaches web and hybrid classes in addition to traditional face-to-face classes. She is always looking for ways to more fully involve students in the learning process. In her spare time Kathy enjoys spending time with her two adult daughters, her two granddaughters, and biking, hiking, and traveling with her husband.

Preface

We would like to thank the reviewers, class testers, and users of the second editions of *Intermediate Algebra* who helped to make the book an overwhelming success. Their thoughtful comments and suggestions provided strong guidance for improvements to the third edition that we believe will enhance this solid, student-friendly text.

Intermediate Algebra is a gateway course to other college-level mathematics courses. The goal of the course is to provide students with the mathematical skills that are prerequisites for courses such as College Algebra, Elementary Statistics, Liberal-Arts Math, and Mathematics for Teachers. In addition, Intermediate Algebra must expose students to a variety of mathematical concepts that build on one another and that range from the basics such as linear equations to sophisticated concepts such as exponential functions.

Of particular importance in this course are rigor and mathematical thinking. It is imperative that the coverage be sufficiently rigorous to teach students how to study math successfully. At the same time, the course must develop students' ability to think mathematically. The rigor in the course exists both in the material presented (such as a more thorough development of functions) as well as in the array of problems and examples. As a result, it should be clear to students that this course is not simply a rehash of Elementary Algebra.

Most students have seen at least some of the content of this course at some point in their high school careers or in other college coursework. For some students, success at studying and facility with math concepts did not develop during their previous contact with the material, and they need a fresh start. In addition, the number of nontraditional students who have lost some of their math skills over the course of time continues to grow, especially at community colleges. Nontraditional students often are highly motivated to succeed because of their life experiences, yet they may be rusty at the business of "going to school." For nontraditional students, this course refreshes and reinforces their study skills as well as their mathematical skills.

To address the many needs and the diversity of today's Intermediate Algebra students, we have been guided by the following ideas as broad goals for this text:

- Provide the student with a strong conceptual foundation in mathematics through a clear, comprehensive presentation of topics with a special emphasis on functions.
- Present a variety of innovative pedagogical features, tools, study tips, and easy-to use aids to help students see the value of the text as an important resource and guide that will increase their success in the course.
- Provide comprehensive exercise sets with paired exercises that build problem-solving skills, show a variety of applications of mathematics, and reinforce mathematical concepts for students.
- Streamline the Intermediate Algebra course through the strategic placement of topics that will provide instructors with the flexibility to review material, as needed, instead of reteaching it.

New to the Third Edition

- The popular Author in Action videos have been edited so that the length of each video is under 12 minutes. Therefore, some objectives will have more than one lecture video. Students will be alerted to the availability of a video with the ▶ icon. The videos are available in MyMathLab, the Multimedia Textbook (in MyMathLab), or on DVD. Now captioned in English and Spanish.
- A new video notebook will be available with the third edition—ideal for online, emporium/redesign courses, or inverted classrooms. This video notebook assists students in taking thorough, organized, and understandable notes as they watch the Author in Action videos by asking students to complete definitions, procedures, and examples based on the content of the videos.

- The Quick Check exercises now have more fill-in-the-blank and True/False questions to assess the student's understanding of vocabulary and formulas. The end-of-section exercises went through a thorough review. The goal of the review was to be sure that all the various problem types are covered within the exercise set. The result is a comprehensive set of exercises that slowly build in level of difficulty.
- The text went through an extensive review of the exposition with the goal of reducing word count and making the reading more accessible to students. Now, the text provides concise explanations with a page design that is friendly and inviting.

Organizational Changes

- Domain of a function was moved from Section 2.3 to Section 2.2.
- Inverse variation was moved from Section 2.7 to Section 5.7. This allows students to see the relation between inverse variation and rational expressions more clearly.
- The section on quadratic inequalities was renamed polynomial inequalities. This section now contains both quadratic and higher-order inequalities.

Build a Strong Foundation Through a Functions Approach

The approach that we take in Intermediate Algebra is that the function is the overriding theme of the text. The reason for this stress on functions is twofold. First, Intermediate Algebra is not a terminal course but rather a gateway to the future, and functions form the basis for much study in mathematics. The introduction of functions helps make the "jump" from Intermediate Algebra to College Algebra less severe because students feel more comfortable with functions and function notation. Second, today's students like to learn in context so that they can see the relevancy of the material. The function provides a great way to present the usefulness of the material we are teaching.

Develop an Effective Text for Use In and Out of the Classroom

Given the hectic lives led by most students, coupled with the anxiety and trepidation with which they approach this course, an outstanding developmental mathematics text must provide pedagogical support that makes the text valuable to students as they study and do assignments. Pedagogy must be presented within a framework that teaches students how to study math; pedagogical devices must also address what students see as the "mystery" of mathematics—and solve that mystery.

To encourage students and to clarify the material, we developed a set of pedagogical features that help students develop good study skills, garner an understanding of the connections between topics, and work smarter in the process. The pedagogy used is based upon the more than 70 years of classroom teaching experience that the authors bring to this text.

Examples are often the determining factor in how valuable a textbook is to a student. Students look to examples to provide them with guidance and instruction when they need it most—the times when they are away from the instructor and the classroom. We have developed several example formats in an attempt to provide superior guidance and instruction for students. The formats include:

Innovative Sullivan/Struve Examples

The innovative *Sullivan/Struve Example* has a two-column format in which annotations are provided to the **left** of the algebra, rather than the right, as is the practice in most texts. Because we read from **left to right,** placing the annotation on the left will make more sense to the student. It becomes clear that the annotation describes what we are about to do instead of what was just done. The annotations may be thought of as the teacher's voice offering clarification immediately before writing the next step in the solution on the board. Consider the following:

EXAMPLE 6   **Solving Linear Inequalities**

Solve the inequality: $x - 4 \geq 5x + 12$

Solution

$$x - 4 \geq 5x + 12$$

Add 4 to both sides: $x - 4 + 4 \geq 5x + 12 + 4$

$$x \geq 5x + 16$$

Subtract 5x from both sides: $x - 5x \geq 5x + 16 - 5x$

$$-4x \geq 16$$

Divide both sides by −4. Don't forget to
change the direction of the inequality: $\dfrac{-4x}{-4} \leq \dfrac{16}{-4}$

$$x \leq -4$$

Figure 20

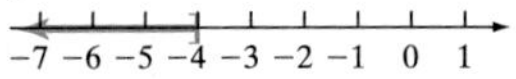

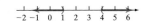
$-7 \ -6 \ -5 \ -4 \ -3 \ -2 \ -1 \ \ 0 \ \ 1$

The solution using set-builder notation is $\{x \mid x \leq -4\}$. The solution using interval
notation is $(-\infty, -4]$. See Figure 20 for the graph of the solution set.

Quick ✓

*In Problems 20–22, solve each linear inequality. Write the solution using set-builder
notation and interval notation. Graph the solution set.*

20. $3x + 1 > x - 5$ **21.** $-2x + 1 \leq 3x + 11$

22. $-5x + 12 < x - 3$

Showcase Examples

Showcase Examples are used strategically to introduce key topics or important problem
solving techniques. These examples provide "how-to" instruction by offering a guided,
step-by-step approach to solving a problem. Students can then immediately see how
each of the steps is employed. We remind students that the *Showcase Example* is meant
to provide "how-to" instruction by including the words "how to" in the example title.
The *Showcase Example* has a three-column format in which the left column describes
a step, the middle column provides a brief annotation, as needed, to explain the step,
and the right column presents the algebra. With this format, students can see each
step in the problem-solving process in context so that the steps make more sense. This
approach is more effective than simply stating each step in the text.

EXAMPLE 9   **How to Solve an Inequality Involving** $>$

Solve the inequality $|2x - 5| > 3$. Express the solution set in set-builder notation and interval notation. Graph the
solution set.

Step-by-Step Solution

Step 1: The inequality is in the form $|u| > a$,
where $u = 2x - 5$ and $a = 3$. Rewrite the
inequality as a compound inequality that
does not involve absolute value.

$$|2x - 5| > 3$$

$|u| > a$ means $u < -a$ or $u > a$: $2x - 5 < -3$ or $2x - 5 > 3$

Step 2: Solve each inequality separately.

	$2x - 5 < -3$		$2x - 5 > 3$
Add 5 to both sides:	$2x < 2$	Add 5 to both sides:	$2x > 8$
Divide both sides by 2:	$x < 1$	Divide both sides by 2:	$x > 4$

Step 3: Find the union of the solution sets
of each inequality.

The solution set is $\{x \mid x < 1 \text{ or } x > 4\}$ or, using interval notation,
$(-\infty, 1) \cup (4, \infty)$. See Figure 46 for the graph of the solution set.

Figure 46
$-2 \ -1 \ \ 0 \ \ 1 \ \ 2 \ \ 3 \ \ 4 \ \ 5 \ \ 6$

Quick ✓

34. $|x - 9| > 6$ is equivalent to $x - 9 > 6$ or $x - 9$ ___ -6.

*In Problems 35–40, solve each inequality. Express the solution set in set-builder notation
and interval notation. Graph the solution set.*

35. $|x + 3| > 4$ **36.** $|4x - 3| \geq 5$

37. $|-3x + 2| > 7$ **38.** $|2x + 5| - 2 > -2$

39. $|6x - 5| \geq 0$ **40.** $|2x + 1| > -3$

Work Smart

$|u| > a$

CANNOT be written as

$-a > u > a$

Quick Check Exercises

Placed at the conclusion of most examples, the *Quick Check* exercises provide students with an opportunity for immediate reinforcement. By working the problems that mirror the example just presented, students get instant feedback and gain confidence in their understanding of the concept. All *Quick Check* exercises answers are provided in the back of the text. The *Quick Check* exercises should be assigned as homework to encourage students to read, consult, and use the text regularly.

Superior Exercise Sets: Paired with Purpose

Students learn algebra by doing algebra. The superior end-of-section exercise sets in this text provide students with ample practice of both procedures and concepts. The exercises are paired and present problem types with every possible derivative. The exercises also present a gradual increase in difficulty level. The early, basic exercises keep the student's focus on as few "levels of understanding" as possible. The later or higher-numbered exercises are "multi-task" (or Mixed Practice) exercises where students are required to utilize multiple skills, concepts, or problem-solving techniques.

Throughout the textbook, the exercise sets are grouped into nine categories— some of which appear only as needed:

1. **Are You Ready? . . .** problems are located at the opening of the section. They are problems that deal with prerequisite material for the section along with page references so students may remediate, if necessary. Answers to the Ready?… problems appear as a footnote on the page.

2. **Quick Check** exercises, which provide the impetus to get students into the text, follow most examples and are numbered sequentially as the first problems in each section exercise set. By doing these problems as homework and the first exercises attempted, the student is directed into the material in the section. If a student gets stuck, he or she will learn that the example immediately preceding the Quick Check exercise illustrates the concepts needed to solve the problem.

3. **Building Skills** exercises are drill problems that develop the student's understanding of the procedures and skills in working with the methods presented in the section. These exercises can be linked back to a single objective in the section. Notice that the Building Skills problems begin the numbering scheme where the Quick Checks leave off. For example, if the last Quick Check exercise is Problem 20, then we begin the Building Skills exercises with Problem 21. This serves as a reminder that Quick Check exercises should be assigned as homework.

4. **Mixed Practice** exercises are also drill problems, but they offer a comprehensive assessment of the skills learned in the section by asking problems that relate to more than one concept or objective. In addition, we may present problems from previous sections so students must first recognize the type of problem and then employ the appropriate technique to solve the problem.

5. **Applying the Concepts** exercises are problems that allow students to see the relevance of the material learned within the section. Problems in this category either are situational problems that use material learned in the section to solve "real-world" problems or are problems that ask a series of questions to enhance a student's conceptual understanding of the mathematics presented in the section.

6. **Extending the Concepts** exercises can be thought of as problems that go beyond the basics. Within this block of exercises an instructor will find a variety of problems to sharpen students' critical-thinking skills.

7. **Explaining the Concepts** problems require students to think about the big picture concepts of the section and express these ideas in their own words. It is our belief that students need to improve their ability to communicate complicated ideas both orally and in writing. When they are able to explain mathematical methods or

concepts to another individual, they have truly mastered the ideas. These problems can serve as a basis for classroom discussion or can be used as writing assignments.

8. Starting with Chapter 5, we provide **Synthesis Review** exercises to help students grasp the "big picture" of algebra—once they have a sufficient conceptual foundation to build upon from their work in Chapters R through 4. Synthesis Review exercises ask students to perform a single operation (adding, solving, and so on) on several objects (polynomials, rational expressions, and so on). The student is then asked to discuss the similarities and differences in performing the same operation on the different objects.

9. Finally, we include coverage of the **graphing calculator.** Instructors' philosophies about the use of graphing devices vary considerably. Because instructors disagree about the value of this tool, we have made an effort to make graphing technology entirely optional. When appropriate, technology exercises are included at the close of a section's exercise set.

Problem Icons In addition to the carefully structured categories of exercises, selected problems are flagged with icons to denote that:

-   Problems whose number is green have complete worked-out solutions found in MyMathLab.
-   △These problems focus on geometry concepts.

Hallmark Features of Sullivan/Struve

Streamlining Intermediate Algebra: *Getting Ready for Chapter … Review Sections*

To maintain the pace of the course, we created several *Getting Ready* sections that review material taught in Elementary Algebra courses. The *Getting Ready* sections are designed to allow students to brush up on topics and skills as needed before beginning the chapters in the Intermediate Algebra text where the skills will be used or further developed. These optional, yet integrated, sections provide the student with timely review. They also streamline the Intermediate Algebra course by providing the instructors with the flexibility to decide if the *Getting Ready* sections should be covered in their entirety, briefly reviewed, or skipped, depending upon the needs of their students. *Getting Ready* review sections have been placed before Chapters 4, 5, and 6 in the text.

Quick Check Exercises: Encourage Study Skills that Lead to Independent Learning

What is one of the overarching goals of an education? We believe it is to learn to solve problems independently. In particular, we would like to see students develop the ability to pick up a text or manual and teach themselves the skills they need. In our mathematics classes, however, we are often frustrated because students rarely read the text and often struggle to understand the concepts independently.

To encourage students to use the text more effectively and to help them achieve greater success in the course, we have structured the exercises in the third edition of our text differently from other mathematics textbooks. The aim of this new structure is to get students "into the text" in order to increase their ability and confidence to work any math problem—particularly when they are away from the classroom and an instructor who can help.

Each section's exercise set begins with *Quick Check* exercises. The *Quick Checks* are consecutively numbered. The end-of-section exercises begin their numbering scheme based on where the *Quick Checks* end. For example:

-   Section 1.1: *Quick Checks* end at Problem 30, so the end-of-section exercise set starts with Problem 31 (see page 57).

- Section 1.2: *Quick Checks* end at Problem 30, so the end-of-section exercise set starts with Problem 31 (see page 69).

The *Quick Checks* follow most examples and provide the platform for students to get "into the text." By integrating these exercises into the exercise set, we direct students to the instructional material in that section. Our hope is that students will then become more aware of the instructional value of the text and will be more likely to succeed when studying away from the classroom and the instructor.

Answer annotations to Quick Checks and exercises have been placed directly next to each problem in the Annotated Instructor's Edition to make it easier for instructors to create assignments.

We have used the same background color for the Quick Checks and the exercise sets to reinforce the connection between them visually. The colored background will also make the Quick Checks easier to find on the page.

Answers to Selected Exercises at the back of the text integrate the answers to *every* Quick Check exercise with the answers to *every odd* problem from the section exercise sets.

Study Skills and Student Success

We have included study skills and student success as regular themes throughout this text starting with *Section R.1, Success in Mathematics*. In addition to this dedicated section that covers many of the basics that are essential to success in any math course, we have included several recurring study aids that appear in the margin. These features were designed to anticipate the student's needs and to provide immediate help—as if the teacher were looking over his or her shoulder. These margin features include *In Words; Work Smart;* and *Work Smart: Study Skills*.

Section R.1: *Success in Mathematics* focuses the student on basic study skills, including what to do during the first week of the semester; what to do before, during, and after class; how to use the text effectively; and how to prepare for an exam.

In Words helps to address the difficulty that students have in reading mathematically precise definitions and theorems by explaining them in plain English.

Work Smart provides "tricks of the trade" hints, tips, reminders, and alerts. It also identifies some common errors to avoid and helps students work more efficiently.

Work Smart: Study Skills reminds students of study skills that will help them to succeed at various points in the course. Attention to these practices will help them to become better, more proficient learners.

Test Preparation and Student Success

The Chapter Tests in this text and the companion Chapter Test Prep Videos have been designed to help students make the most of their valuable study time.

Chapter Test In preparation for their classroom test, students should take the practice test to make sure they understand the key topics in the chapter. The exercises in the Chapter Tests have been crafted to reflect the level and types of exercises a student is likely to see on a classroom test.

Chapter Test Prep Video The Chapter Test Prep Videos provide students with help at the critical juncture when they are studying for a test. The videos present step-by-step solutions to the exact exercises found in each of the book's Chapter Tests. Easy video navigation allows students instant access to the worked-out solutions to the exercises they want to study or review. These videos are available in MyMathLab or the video lecture series DVD.

How It All Fits Together: The Big Picture

Another important role of the pedagogy in this text is to help students see and understand the connection between the mathematical topics being presented. Several section-opening and margin features help to reinforce connections:

The Big Picture: Putting It Together (Chapter Opener) This feature is based on how we start each chapter in the classroom—with a quick sketch of what we plan to cover.

Before tackling a chapter, we tie concepts and techniques together by summarizing material covered previously and then relate these ideas to material we are about to discuss. It is important for students to understand that content truly builds from one chapter to the next. We find that students need to be reminded that the familiar operations of addition, subtraction, multiplication, and division are being applied to different or more complex objects.

Are You Ready for This Section? As part of this building process, we think it is important to remind students of specific material that they will need from earlier in the course to be successful within a given section. The *Are You Ready?* feature that begins each section not only provides a list of prerequisite skills that a student should understand before tackling the content of a new section but also offers a short quiz to test students' preparedness. Answers to the quiz are provided in a footnote on the same page, and a cross-reference to the material in the text is provided so that the student can remediate when necessary.

Putting the Concepts Together (Mid-Chapter Review) Each chapter has a group of exercises at the appropriate point in the chapter, entitled *Putting the Concepts Together*. These exercises serve as a review—synthesizing material introduced up to that point in the chapter. The exercises in these mid-chapter reviews are carefully chosen to assist students in seeing the "big picture."

Synthesis Review Exercises Starting with Chapter 5, we provide Synthesis Review exercises to help students grasp the "big picture" of algebra—once they have a sufficient conceptual foundation to build upon from their work in Chapters R through 4.

Cumulative Review Learning algebra is a building process, and building involves considerable reinforcement. The Cumulative Review exercises at the end of each odd-numbered chapter, starting with Chapter 1, help students reinforce and solidify their knowledge by revisiting concepts and using them in context. This way, studying for the final exam should be fairly easy. Cumulative Reviews for each even-numbered chapter can be found on the Instructor's Resource Center. Answers to all cumulative review problems appear in the back of the text.

In Closing

When we started writing this textbook, we discussed what improvement we could make in coverage; in staples such as examples and problems; and in any pedagogical features that we found truly useful. After writing and rewriting, and reading many thoughtful reviews from instructors, we focused on the following features of the text to set it apart.

- **Functions** are introduced early and revisited often throughout the course. This integration helps prepare students for the quantitative courses that they will take after Intermediate Algebra.

- The **innovative *Sullivan/Struve Examples*** and ***Showcase Examples*** provide students with superior guidance and instruction when they need it most—when they are away from the instructor and the classroom. Each of the margin features ***In Words, Work Smart,*** and ***Work Smart: Study Skills*** are designed to improve study skills, make the textbook easier to navigate, and increase student success.

- **Exercise Sets: Paired with Purpose**—The exercise sets are structured to assess student understanding of vocabulary, concepts, drill, problem solving, and applications. The exercise sets are graded in difficulty level to build confidence and to enhance students' mathematical thinking. The ***Quick Check*** exercises provide students with immediate reinforcement and instant feedback to determine their understanding of the concepts presented in the examples. ***Putting the Concepts Together*** and ***Synthesis Review*** help students see the big picture and provide a structure for learning each new concept and skill in the course.

- The text is written to streamline Intermediate Algebra (and distinguish it from Elementary Algebra) through a single-chapter presentation of linear equations and inequalities along with the strategic placement of ***Getting Ready*** review sections that provide instructors with the flexibility to review material instead of reteaching it.

Student and Instructor Resources

STUDENT RESOURCES

Available for purchase at MyPearsonStore.com

Student's Solutions Manual
(ISBN: 0321881354 / 9780321881359)
Complete worked solutions to the odd-numbered problems in the end-of-section exercises and all of the Quick Checks and end-of-chapter exercises.

***Do the Math* Workbook**
(ISBN: 0321881362 / 9780321881366)
A collection of 5-minute Warm-Up exercises, Guided Practice exercises, and *Do the Math* exercises for each section in the text. These exercises are designed for students to show their work in homework, during class, or in a lab setting.

Author in Action Videos on DVD for Intermediate *Algebra*, 3/e
(ISBN: 0321881575 / 9780321881571)
Keyed to each section in the text, objective specific mini-lectures provide a 12-minute review of the key concepts. Videos include optional subtitles in English and Spanish. The Chapter Test Prep videos are included in the package.

Video Notebook
(ISBN: 0321880358/9780321880352)
Ideal for online, emporium based/redesign courses or inverted classrooms. Students learn to take organized notes as they watch the Author in Action videos. Students are asked to complete definitions, procedures, and examples based on the videos.

INSTRUCTOR RESOURCES

Available through your Pearson representative.

Annotated Instructor's Edition
(ISBN: 0321880439 / 9780321880437)

Instructor's Solutions Manual
(Available for download from the IRC)

Instructor's Resource Manual
(Available for download from the IRC)
Includes Mini-Lectures (one-page lesson plans at-a-glance) for each section in the text that feature key examples and Teaching Tips on how students respond to the material. Printed Test Forms (free response and multiple choice), Additional Exercises.

PowerPoint Lecture Slides
(Available for download from the IRC)

TestGen® for *Intermediate Algebra*, 3/e
(Available for download from the IRC)

Online Resources

MyMathLab® (access code required)

MathXL® (access code required)

Acknowledgments

Textbooks are written by authors but evolve through the efforts of many people. We would like to extend our thanks to the following individuals for their important contributions to the project. From Pearson: Mary Beckwith, whose editorial expertise has been an invaluable asset; Michelle Renda for her innovative marketing ideas; Chris Hoag for her support and encouragement; Heather Scott and the design team for the attractive and functional design; Rose Kernan, Patty Bergin and Lauren Morse for keeping a watchful eye and managing countless production details and finally, the Pearson Arts & Sciences sales team for their confidence and support of our books.

We would also like to thank John Bialas, Peggy Irish, Teri Lovelace, Peter Wilk, and Cindy Trimble for their attention to details and consistency in accuracy checking the text and answer sections. A huge thanks goes out to Val Villegas, who created the outstanding classroom video notebooks that accompany the text. We offer many thanks to all the instructors from across the country who participated in reviewer conferences and focus groups, reviewed or class-tested some aspect of the manuscript, and taught from the previous editions. Their insights and ideas form the backbone of this text. Hundreds of instructors contributed their time, energy, and ideas to help us shape this text. We will attempt to thank them all here. We apologize for any omissions.

The following individuals, many of whom reviewed or class-tested the first edition, provided direction and guidance in shaping the third edition.

Marwan Abu–Sawwa, *Florida Community College—Jacksonville*
MaryAnne Anthony, *Santa Ana College*
Darla Aguilar, *Pima State University*
Grant Alexander, *Joliet Junior College*
Philip Anderson, *South Plains College*
Mary Lou Baker, *Columbia State Community College*
Bill Bales, *Rogers State*
Tony Barcellos, *American River College*
John Beachy, *Northern Illinois University*
Donna Beatty, *Ventura College*
David Bell, *Florida Community College—Jacksonville*
Sandy Berry, *Hinds Community College*
John Bialas, *Joliet Junior College*
Linda Blanco, *Joliet Junior College*
Kevin Bodden, *Lewis and Clark College*
Rebecca Bonk, *Joliet Junior College*
Cherie Bowers, *Santa Ana College*
Becky Bradshaw, *Lake Superior College*
Lori Braselton, *Georgia Southern University*
Tim Britt, *Jackson State Community College*
Beverly Broomell, *Suffolk Community College*
Joanne Brunner, *Joliet Junior College*
Hien Bui, *Hillsborough Community College—Dale Mabry*
Connie Buller, *Metropolitan Community College*
Annette Burden, *Youngstown State University*
James Butterbach, *Joliet Junior College*
Marc Campbell, *Daytona Beach Community College*

Elena Catoiu, *Joliet Junior College*
Nancy Chell, *Anne Arundel Community College*
John F. Close, *Salt Lake Community College*
Bobbi Cook, *Indian River Community College*
Carlos Corona, *San Antonio College*
Faye Dang, *Joliet Junior College*
Shirley Davis, *South Plains College*
Vivian Dennis-Monzingo, *Eastfield College*
Alvio Dominguez, *Miami Dade Community College—Wolfson*
Karen Driskell, *South Plains College*
Thomas Drucker, *University of Wisconsin—Whitewater*
Brenda Dugas, *McNeese State University*
Doug Dunbar, *Okaloosa-Walton Junior College*
Laura Dyer, *Southwestern Illinois State University*
Bill Echols, *Houston Community College—Northwest*
Laura Egner, *Joliet Junior College*
Jason Eltrevoog, *Joliet Junior College*
Nancy Eschen, *Florida Community College—Jacksonville*
Mike Everett, *Santa Ana College*
Phil Everett, *Ohio State University*
Scott Fallstrom, *Shoreline Community College*
Betsy Farber, *Bucks County Community College*
Fitzroy Farqharson, *Valencia Community College—West*
Jacqueline Fowler, *South Plains College*
Dorothy French, *Community College of Philadelphia*

Randy Gallaher, *Lewis and Clark College*
Sanford Geraci, *Broward Community College*
Donna Gerken, *Miami Dade Community College—Kendall*
Adrienne Goldstein, *Miami Dade Community College—Kendall*
Marion Graziano, *Montgomery County Community College*
Susan Grody, *Broward Community College*
Tom Grogan, *Cincinnati State University*
Barbara Grover, *Salt Lake Community College*
Shawna Haider, *Salt Lake Community College*
Margaret Harris, *Milwaukee Area Technical College*
Teresa Hasenauer, *Indian River Community College*
Mary Henderson, *Okaloosa-Walton Junior College*
Celeste Hernandez, *Richland College*
Paul Hernandez, *Palo Alto College*
Pete Herrera, *Southwestern College*
Bob Hervey, *Hillsborough Community College—Dale Mabry*
Teresa Hodge, *Broward Community College*
Sandee House, *Georgia Perimeter College*
Becky Hubiak, *Tidewater Community College—Virginia Beach*
Sally Jackman, *Richland College*
John Jarvis, *Utah Valley State College*
Nancy Johnson, *Broward Community College*
Steven Kahn, *Anne Arundel Community College*

Linda Kass, *Bergen Community College*
Donna Katula, *Joliet Junior College*
Mohammed Kazemi, *University of North Carolina—Charlotte*
Doreen Kelly, *Mesa Community College*
Mike Kirby, *Tidewater Community College—Virginia Beach*
Keith Kuchar, *College of Dupage*
Carla Kulinsky, *Salt Lake Community College*
Julie Labbiento, *Leigh Carbon Community College*
Kathy Lavelle, *Westchester Community College*
Deanna Li, *North Seattle Community College*
Brian Macon, *Valencia Community College—West*
Lynn Marecek, *Santa Ana College*
Jim Matovina, *Community College of Southern Nevada*
Jean McArthur, *Joliet Junior College*
Michael McComas, *Marshall University*
Mikal McDowell, *Cedar Valley College*
Lee McEwen, *Ohio State University*
David McGuire, *Joliet Junior College*
Angela McNulty, *Joliet Junior College*
Debbie McQueen, *Fullerton College*
Judy Meckley, *Joliet Junior College*
Lynette Meslinsky, *Erie Community College—City Campus*
Kausha Miller, *Lexington Community College*
Chris Mizell, *Okaloosa Walton Junior College*
Jim Moore, *Madison Area Technical College*
Ronald Moore, *Florida Community College—Jacksonville*
Elizabeth Morrison, *Valencia Community College—West*
Roya Namavar, *Rogers State*
Hossein Navid-Tabrizi, *Houston Community College*

Carol Nessmith, *Georgia Southern University*
Kim Neuburger, *Portland Community College*
Larry Newberry, *Glendale Community College*
Elsie Newman, *Owens Community College*
Charlotte Newsome, *Tidewater Community College*
Charles Odion, *Houston Community College*
Viann Olson, *Rochester Community and Technical College*
Linda Padilla, *Joliet Junior College*
Carol Perry, *Marshall Community and Technical College*
Faith Peters, *Miami Dade Community College—Wolfson*
Dr. Eugenia Peterson, *Richard J. Daley College*
Jean Pierre-Victor, *Richard J. Daley College*
Philip Pina, *Florida Atlantic University*
Carol Poos, *Southwestern Illinois University*
Elise Price, *Tarrant County Community College*
R.B. Pruitt, *South Plains College*
William Radulovich, *Florida Community College—Jacksonville*
Pavlov Rameau, *Miami Dade Community College—Wolfson*
David Ray, *University of Tennessee—Martin*
Nancy Ressler, *Oakton Community College*
Michael Reynolds, *Valencia Community College—West*
George Rhys, *College of the Canyons*
Jorge Romero, *Hillsborough Community College—Dale Mabry*
David Ruffato, *Joliet Junior College*
Carol Rychly, *Augusta State University*

David Santos, *Community College of Philadelphia*
Togba Sapolucia, *Houston Community College*
Doug Smith, *Tarrant Community College*
Catherine J.W. Snyder, *Alfred State College*
Gisela Spieler-Persad, *Rio Hondo College*
Raju Sriram, *Okaloosa-Walton Junior College*
Patrick Stevens, *Joliet Junior College*
Bryan Stewart, *Tarrant Community College*
Jennifer Strehler, *Oakton Community College*
Elizabeth Suco, *Miami Dade Community College—Wolfson*
Katalin Szucs, *East Carolina University*
KD Taylor, *Utah Valley State College*
Mary Ann Teel, *University of North Texas*
Suzanne Topp, *Salt Lake Community College*
Suzanne Trabucco, *Nassau Community College*
Jo Tucker, *Tarrant Community College*
Bob Tuskey, *Joliet Junior College*
Mary Vachon, *San Joaquin Delta College*
Carol Walker, *Hinds Community College*
Kim Ward, *Eastern Connecticut State University*
Richard Watkins, *Tidewater Community College*
Natalie Weaver, *Daytona Beach Community College*
Darren Wiberg, *Utah Valley State College*
Rachel Wieland, *Bergen Community College*
Christine Wilson, *Western Virginia University*
Brad Wind, *Miami Dade Community College—North*
Roberta Yellott, *McNeese State University*
Steve Zuro, *Joliet Junior College*
Jon Anderson, *Utah Valley University*
Darren Wiberg, *Utah Valley University*

Additional Acknowledgments

We also would like to extend thanks to our colleagues at Joliet Junior College and Columbus State Community College, who provided encouragement, support, and the teaching environment where the ideas and teaching philosophies in this text were developed.

Michael Sullivan, III
Katherine R. Struve

Strategies for Success

Syllabus Search

Name_____

By doing this Syllabus Search you will be making the information in your course syllabus more meaningful and useful to you.

- **General Information**

    1) My instructor's name is _____

    2) I can contact my instructor by phone: _____ email:_____

    3) My instructor's office is located in _____

    4) My instructor's office hours are _____

    5) Matching my schedule with my instructor's office hours, the times that I will be able to meet

    with my instructor are _____

    6) The website address for this class is _____

    7) The required textbook for this class is titled _____

    and I can buy it on campus at _____

    8) For this class I need a (circle one) scientific/graphing calculator such as a _____

    9) Other materials I need are _____

    10) The attendance policy is _____

    _____

    11) The cheating policy is _____

    _____

    12) If my cell phone goes off in class, I _____

- **Course Grading Policy**

    13) I plan to earn a(n) A/B/C_____ in this course.

    14) The **grading scale** will be: A=_____ B=_____ C=_____ D=_____ F=_____

    15) My course grade will be based on my scores on:

    _____ homework   _____quizzes   _____tests

    _____final examination   _____class work   _____participation

    _____ other:_____   _____ other:_____   _____ other:_____

16) When is **homework** due? _____ How do I turn it in? _____

17) Is late homework accepted? _____ If so, is there a penalty? _____

18) Each homework assignment is worth _____ points and all homework is worth _____

points total for the course.

19) There will/will not (circle one) be **quizzes** in this class. If so, each quiz is worth _____

points and all quizzes are worth _____ points total for the course.

20) This class has _____ **tests** that are scheduled on _____

21) Each test is worth _____ points and all tests together contribute _____ points towards

my course grade.

22) The **makeup test** policy is _____

23) The **Final Exam** is scheduled on _____ and is worth _____ points.

24) Other work that will contribute to my grade: _____

25) Questions for my instructor about the grading policies:

- **Resources for this Course**

26) If I need help in this course, I can use the following resources:

1. _____

2. _____

3. _____

4. _____

27) If I need a tutor for this course, I can call _____

or go to_____

28) If I need accommodation due to a disability I need to _____

_____

29) If I need to contact a classmate from this class I would call or email

1. _____phone: _____email:_____

2. _____phone: _____email:_____

3. _____phone: _____email:_____

30) A good time for me to meet with a study group is_____

Strategies for Success

Notebook Preparation   **Name**_____

Good organization is a study skill that is essential for success in mathematics. Some people seem to be naturally organized, while other people are not. But it is possible to learn to be organized. Preparing a math notebook is a good way to develop this important skill.

1) You will **need**:
   - a **three-ring binder** (with rings at least 1")
   - **dividers**
   - **notebook paper** and **graph paper**
   - **a hole-puncher** (most useful if it fits right in your binder)

2) To **prepare** your notebook, **label** the tabs for your dividers:
   - Course information
   - Notes
   - Assignments
   - Tests
   - Vocabulary
   - Paper
   - Graph paper
   - If your instructor wants any other labels, list them here:_____

3) To **assemble** your notebook,
   - put the labeled dividers into the notebook.
   - file your papers into the proper section.

When you have **finished**, have your instructor sign you off below.

Date: _____   Instructor: _____

Strategies for Success

Reading the Textbook Part I
Forming Good Habits

Name_____

Have you ever thought about how you read a math textbook? It's different from reading a novel or a magazine. Most people don't read math textbooks for relaxation or entertainment! They read math textbooks to learn how to do math. The book speaks to you, like a teacher does in class, showing and explaining how to do math.

In this activity, you will reflect on your past practices when reading math texts up to now and identify one or more new behaviors you will try so that you will read your textbook more effectively.

1) The table below lists some behaviors that may help you read your math book effectively. Think about how you usually read your math book. Then check the appropriate column next to each behavior.

When I read my math book, …	Yes	No, but I know I should	No, I never thought of it
a) I sit with an alert, but comfortable, posture.			
b) I am prepared to do some math--I have a pencil in my hand. If I cannot write in my textbook, I have some paper, too.			
c) I read every single word.			
d) I look at all diagrams, graphs, and pictures carefully.			
e) I underline important words and ideas.			
f) I work each step of the examples on paper.			
g) I make sure I understand all the math steps. If I don't understand how one step follows the step before, I put a question mark and get help as soon as possible.			

2) Did you check 'Yes' for all the behaviors? _____ If so, you have formed good habits! If not, what will you do to make sure you read your math book more effectively? Which of the behaviors from the table will you try next?

Strategies for Success

Reading the Textbook Part II
Reading your Math Book Effectively

Name_____

In this activity you will practice reading your math book effectively.
Read the first two pages of one section of your math textbook. (Your instructor may assign a specific section.) Then answer the following questions:

1) What section are you reading? _____ What is the title of the section?

2) Find the section objectives.

   a) What are the section objectives?

   b) Why are they listed at the start of the section?

3) Read the first paragraph.

   a) What does the first paragraph tell you?

   b) List any words you don't understand.

4) Continue reading the first page.

   a) List any words you don't understand.

   b) What can you do to find out what these words mean?

5) Look for pictures, graphs, or diagrams.

   a) Are there any pictures, graphs, or diagrams? ____yes ____no

   b) What purpose do they serve?

6) Continue reading until you get to Example 1.

    a) What is Example 1 titled?

    b) How does the title relate to the section objectives?

    c) What do the directions say?

    d) Restate the directions in your own words.

    e) Copy all the math steps in Example 1 into the table below. To the right of each line, use your own words to explain what math you did in that step. If you do not understand how one step follows the line before, try to pinpoint exactly what you don't understand and write it down next to that step. How can you get help to understand that step?

Copy the math steps in Example 1	Explain the math in your own words

7) Now look at the Exercises at the end of this section.

    a) Which Exercises are like Example 1?

    b) Do you feel ready to work those exercises? _____yes _____no   Why or why not?

Strategies for Success

Taking Notes-Part I   **Name**_____
Notes from Reading the Text

Taking notes while you read your textbook can save you valuable study time before a test or exam. If you have good notes, you won't have to re-read the book. Your notes will be your own personal study guide! How do you take notes when you read your math textbook? Can you take notes from an e-book the same way you do for a print textbook?

1) The tables below list some ways successful students take notes when reading math textbooks. Think about what you usually do to take notes when reading. Choose the appropriate table and then fill in the checklist. Use Table I if you use an e-book. Use Table II if you use a print textbook for this course.

Table I – Use this checklist if you are using an electronic textbook (e-book).

When I take notes from my math e-book,…	Always	Sometimes	Never
a) I take notes on paper so that I can use them without going back to the e-book.			
b) I label my notes with the section number and title at the top of the page to help me keep organized.			
c) I use the 'highlight' feature of the e-book so it is easy to find important concepts and ideas.			
d) I write important concepts and ideas to remind me of the topics covered.			
e) I copy the definitions and formulas so that I can refer to them when doing the exercises.			
f) I list the steps of important procedures.			
g) I work examples to make sure I understand all the steps.			
h) I label examples with the section number and problem number so I know where they came from.			
i) I note places where I don't fully understand so I can ask my instructor.			
j) I use the 'pushpin' feature of the e-book to note place I want to return to.			
k) I list page numbers of portions I want to re-read to help me find them easily.			
l) I make comments to remind me of how new concepts relate to concepts I have already learned.			

Table II – Use this checklist if you are using a print (hardcopy) textbook.

When I take notes from my math textbook,…	Always	Sometimes	Never
a) I label my notes with the section number and title at the top of the page to help me keep organized.			
b) I highlight or underline important concepts and ideas to make them easy to find.			
c) I list the steps of important procedures.			
d) I work examples on a separate piece of paper to make sure I understand all the steps.			
e) I label examples with the section number and problem number so I know where they came from.			
f) I mark places where I don't fully understand so I can ask my instructor.			
g) I make comments to remind me of how new concepts relate to concepts I have already learned.			

2) What will you do to make sure you take notes more effectively? Which of the behaviors from the table will you try next?

3) Use the form on the next page to record your notes from one section of your math textbook. In the left column write down
  - key ideas
  - concepts
  - definitions
  - formulas
  - examples

In the right column, write any comments and questions you may have about the reading.

Strategies for Success

Notes from Reading the Text   **Name**_____

Section: _____   Title: _____

Key ideas, concepts, definitions, formulas, and examples	My comments and questions

Strategies for Success

Math Autobiography   **Name**_____

We all arrived in this class by different paths. Each of us has had many experiences that have influenced our attitudes and beliefs about math and our abilities to do math. This exercise will help you reflect on the past and begin to focus on the future.

1) Write your math autobiography—your life story with math. In your autobiography you should:
   a) discuss your present attitude about math.
   b) relate any specific experiences you have had that may have influenced your attitude about math. Think back to your earliest memories and then trace your story forward to today. (These may or may not be experiences in school.)
   c) discuss your fears and concerns about this course.
   d) describe your strengths and relate how they will help you as you progress through this course.

Strategies for Success

On Time and Ready to Go!   **Name**_____

What does it take to get you to class on time? Do you come to class directly from home or from work, or are you already on campus?

In this activity, you'll think about what you need to do to make sure you regularly arrive in class on time and ready to go.

1) My math class meets on_____ from _____to _____.

2) To be in class on time regularly I have to:

      1.

      2.

      3.

3) To get to class on time, I may need to adjust my schedule by doing the following:

4) Every class day I have to bring the following tools:

      1.

      2.

      3.

5) To be sure I have these tools every day in class, I should:

6) In order to be ready for the next session of this class, I have to:

      1.

      2.

      3.

7) If I need help with the course material, I will:

      1.

      2.

      3.

8) To devote enough time to succeed in this class, I will make these adjustments to my life:

      1.

      2.

      3.

Strategies for Success

Test Preparation Skills

Name_____

How do you prepare for a test? Have you ever just 'shown up' for a test and then were disappointed by the results?

Successful test preparation requires a strategy and a plan. If you make a plan and carry it out, not only will you be better prepared, but also you will feel more confident and less anxious about the test.

Strategies for careful test preparation

- **Start your test preparation early**, at least several days before the test. Successful test prep involves several steps and you need sufficient time to complete each one.

- Check that you have **completed every homework assignment** that the test will cover. Not completing every assignment causes holes in your body of knowledge.

- Check that **every problem is understood and done with integrity**. Integrity means that you did not copy from the student solution manual or another student and that you re-did any problems for which you got help to guarantee that you can do them yourself!

- **Review your class notes**. Pay particular attention to areas you had marked for further study.

- **Review the Chapter Summary** in your textbook to make sure you understand all the key concepts. Go back to any section where you need more practice and work some of the exercises.

- Go to each section and **reread the section objectives**. For each objective, **choose a representative problem** that best typifies this objective. Write this problem on a 3x5 card, being sure to list the section and problem number where you found it. Write the answer on the back of the card. Put the 3x5 cards together to create your own practice test.

- Work the **practice test** you created. Check your answers with those on the backs of the cards. Go back and review the objectives of any you got wrong.

- **Work out the Chapter Review and/or the Chapter Test**. Do this in a 'test' setting, if possible.

- **Use all available resources** to get help on topics you did not understand.

1) Use this checklist to analyze how you prepared for your last test and to design a plan to prepare for your next test.

Strategy	My prep for last test	Will do before next test
a) Test prep started several days before the test		
b) Every homework assignment completed		
c) Every problem understood		
d) Every problem completed with integrity		
e) Class notes reviewed		
f) Chapter Summary reviewed		
g) One problem chosen for each objective		
h) My Practice Test worked		
i) Chapter Review/Chapter Test worked		
j) Resources for help used		

2) To be better prepared for the next test I plan to (choose one):

_____ continue what I've been doing

_____ make a few changes to my test prep strategies

_____ make major changes to my test prep strategies

3) List the resources available to you to support your test preparation.

Strategies for Success

Test Stress Reduction   **Name**_____

Do you stress out when taking a test? Many people do. Do you know that stress may arise when you feel you are in a situation that is out of your control? You can reduce your test stress by **taking control** of your success with some strategies that are easy to incorporate into your test prep routine.

Prepare yourself mathematically for the test so you will have confidence in your ability to succeed. If you feel prepared and confident, you will believe you can do well. These positive thoughts will carry over to your actions on the test. Lack of preparation causes students to be nervous, feel overwhelmed, get discouraged, and 'blank out'. **Take care of your body** so you are in good shape to do your best on the test. Your brain needs fuel and rest in order to work effectively. **Plan ahead** to make sure you arrive at your test on time and with all needed materials.

- **Take control by being prepared mathematically.**
    o   Start your test preparation several days before the test.
    o   Complete all your homework assignments.
    o   Make sure you understand every problem and get help if there are any you don't understand.
    o   Review your class notes.
    o   Review the chapter summary.
    o   Create a practice test for yourself and then work it.
    o   Work the exercises in the Chapter Review and/or Chapter Test.

1) To be prepared mathematically I will:

- **Take control by taking care of your body.**

    o   Maintain your **exercise routine**. Exercise helps reduce stress and improves circulation to all of your body, including your brain.

    o   Get a **good night's sleep.** Your body becomes refreshed as you sleep. Rest will help you think more clearly during the test. You will not do your best if you stay up all night cramming.

    o   **Eat properly and maintain good nutrition**. Keep your body strong to better handle the stress of a test.

    o   **Dress for your success.** Dress in a way that makes you feel confident and comfortable. Some students like to dress up a bit for tests and others prefer to wear their favorite jeans. Choose what works for you

2) To take care of my body I will:

⌐

- **Take control by planning ahead.**
  - o **Plan your transportation** so that you arrive early and relaxed.
  - o Make sure that you have all the **required materials** packed and ready to go.
    - o Pencils, erasers, highlighter
    - o Calculator
    - o Scantron, Blue Book or other materials required by your teacher
    - o Any assignment that you need to turn in
  - o Pack **personal items** that add to your comfort such as tissues, water, a jacket or sweatshirt.

3) To plan ahead I will:

⌐

Strategies for Success

Test Taking Skills   **Name**_____

Have you ever thought about how your behavior before or during a test contribute to your success on the test?

1) For each statement in the list of test taking strategies below, check Always (A), Sometimes (S), or Never (N)

A   S   N

Before the test

___ ___ ___   a) I arrive on time or even early so I feel calm and ready.

___ ___ ___   b) I set out the required materials so I feel prepared.

___ ___ ___   c) If a problem in the rest of my life may interfere with my test performance, I write it down on a card and put it away until after the test.

___ ___ ___   d) I ignore others in the room so I won't pick up their negativity or anxiety. I am prepared and confident.

___ ___ ___   e) I check my inner voice. I turn any negative thoughts into positive statements. "I am prepared; I've done what I can; I am ready to succeed; I can do math!"

___ ___ ___   f) I use the restroom. Most teachers do not allow exit/re-entry during the test.

___ ___ ___   other: _____

Taking the test

___ ___ ___   g) I do a "data dump" as soon as I get the test. Then I no longer need to think about remembering the facts/formulas.

___ ___ ___   h) I scan the test, reading all problems before I begin any work.

___ ___ ___   i) I read directions carefully. I circle, underline or highlight key words and directions.

___ ___ ___   j) I note easy problems and do them first to build my confidence and ensure those points.

___ ___ ___   k) If I can't do a problem immediately, I write down anything I can think of such as formulas, pictures, etc., then I move on and return to it later. The solution may come to me as I work on the other problems.

___ ___ ___   l) If I do not know how to do something, I try to relate it to something I do know.

___ ___ ___   m) I show all my work. I write all steps, reasoning, and supporting evidence. This is really helpful if my teacher awards partial credit.

A   S   N

___ ___ ___ n) I check my work.

___ ___ ___ o) I check answers. I make sure word problems have reasonable answers.

___ ___ ___ p) I ignore others. I remember that those done early may be turning in a blank test.

___ ___ ___ q) I pace myself.

___ ___ ___ r) I do not turn in my test early. I use the time to go over my work carefully.

___ ___ ___ other: _____

Reducing stress during the test

___ ___ ___ s) I check my inner voice. I turn any negative thoughts into positive statements.

___ ___ ___ t) I imagine and visualize that I am in my favorite pleasant relaxing situation.

___ ___ ___ u) I take mental breaks.

___ ___ ___ v) I do stress reducing exercises.

___ ___ ___ w) I do deep breathing.

___ ___ ___ x) I do muscle tensing and relaxing.

___ ___ ___ other: _____

Look at your checklist.

2) Are there any techniques that you use regularly that are not on the checklist? Add them in the checklist.

3) Identify the skills for which you checked 'Sometimes' or 'Never.' List three skills that you will try during the next test.

        1.

        2.

        3.

Strategies for Success

Post Test Check up

Name_____

Seeing your graded test gives you an opportunity to reflect on how you earned your grade. Use this activity to analyze how you prepared for your last test

1) I attended every class since the last test.   Yes_____ No_____

2) I am satisfied with the quality of my lecture notes.   Yes_____ No_____

3) I used the following test prep strategies:

Strategy	Yes	No
a) Test prep started several days before the test		
b) Every homework assignment completed		
c) Every problem understood		
d) Every problem completed with integrity		
e) Class notes reviewed		
f) Chapter Summary reviewed		
g) One problem chosen for each objective		
h) My Practice Test worked		
i) Chapter Review/ Chapter Test worked		
j) Resources for help used		

4) I also prepared for the test by…

5) Fill in the side of the table that applies to how you feel about your grade on the last test.

I am happy	I am unhappy
a) The study skills and strategies that worked for me and that I plan to continue are….	a) In order to do better on my next test, I need to change…
b) The area that I need to improve is…	b) I will effect this change by…
	c) The study skills and strategies that worked for me and that I plan to continue are….

6) Fill in the table below to commit to completing the loop on this assessment process:

I will...	Yes	No
a) correct every problem that I missed on the test.		
b) make sure I can do each problem on my own and understand it completely. • This will fill in my gaps in the knowledge tested and help me as I progress through the course. • This will also help prepare me for the Final Exam.		
c) rework, on a separate sheet of paper, all the problems where I missed even one point. Then I will staple that paper with all the corrections to the test. This will give me a good Final Exam Study Guide for this unit.		

Strategies for Success

Test Analysis   **Name**_____

Have you ever thought of your graded test as a learning experience? There is a lot you can learn about yourself, your study habits, and your test-taking skills by examining your graded test after you get it back. Did you do as well as you thought you could? Or is there room for improvement? You may think, "the test was too hard" or "the teacher didn't give us enough time", but, chances are, your instructor has been giving a similar test under similar conditions to many students before you. So let's see what **you** can do to earn a higher score on your next test.

Look at your graded test and analyze whether each point loss was due to your having been **unprepared** for that problem, a **concept error,** or a **careless error.**

- Being **unprepared** for a problem means you didn't know how to do the problem because you hadn't done the homework that would have prepared you for it.

- A **concept error** is one where you really didn't understand the concept behind the problem. No matter how much time was available for a problem like this, you wouldn't have been able to do it because you didn't know how to approach it.

- A **careless error** is one where you understood the problem and knew how to solve it, but you made a mistake that could have been avoided. Maybe you copied the problem or your handwriting incorrectly, made a relatively minor mistake in calculation, or some similar error.

1) On a separate piece of paper, make a chart like the one below with one line for each problem on the test. Put the **number of points you missed** on each problem under the correct heading and then find the total in each column.

Problem	unprepared	concept error	careless error
1			
2			
3			
4			
5			
6			
…			
	Total points	Total points	Total points

2) In which column did you have the most missed points?

3) What does this tell you about yourself?

4) What can you learn from this exercise?

Being Unprepared

Consider the points lost because you were **unprepared.**

Why did you take the test without being fully prepared? Oftentimes, activities and responsibilities in life interfere with good intentions about being diligent in attending class, reading the textbook, and doing all the assignments. It may be time to:

- **re-examine your weekly schedule** and make sure you are devoting a sufficient amount of time to this class. Lay out a time management grid of your schedule making sure to schedule your math study time.

- **re-commit yourself to succeeding in this class.** Think about your college and career goals and remind yourself of how this course helps you get one step closer to achieving them.

5) List two steps you will take to remedy being unprepared.

    1.

    2.

Concept Errors

Now consider the **concept error** point loss.

A high total in this column tells you that you didn't understand the concepts very well. As you do your work day-to-day you might think you "get it", but you don't always make sure that you completely understand each problem in the homework. You may understand a math concept for the two hours you're working on the homework problems, but forget it by the next day.

- **Review earlier sections**. Regularly review earlier sections, instead of saving all the review for test time.

- **Get the help you need immediately!** Math concepts build on each other. Each new idea is based on many previous concepts. Make sure you get the help you need immediately, as soon as you find yourself beginning to feel lost, so that the confusion doesn't compound itself – otherwise it can become like a snowball, getting bigger and bigger as it rolls through the snow.

If your total loss due to concept errors is fairly large, find out where you can get the help you need. Your school has places available just for you to get help with your math.

6) List two places you can go to get help with your math:

    1.

    2.

A high concept error total is cause for concern and must be addressed immediately to guarantee success!

Careless Errors

Next, look at the **careless error** point loss.

Careless errors are often caused by hurrying during a test or by lack of concentration due to test anxiety or over-confidence. So here are some strategies that have worked for other math students:

- **Do the easiest problems first.** When you first start on a test, look it over from beginning to end and note which problems will be easiest for you. Do all those problems first, to ensure that you don't leave an easy problem blank, just because it is at the end of the test. Finishing the problems you find easy will help build your confidence! Then go through the rest of the test from beginning to end.

- **Work carefully and neatly**. As you do each problem try to focus on one step at a time.

- **Review each problem to look for careless errors** when you finish the test. Find and correct common careless errors like arithmetic mistakes and sign errors before you turn in your test.

- Whenever possible **check the problem**.

A lot of points can be gained by slowing down and being careful!

7) What are two things you will do next time to prevent careless errors?

    1.

    2.

8) Now take half of your 'careless' point total and add it back to your test total.
    a) What could your test grade have been? _____
    b) Would that have changed your A/B/C grade? _____

Strategies for Success

Know Where You Stand

Name_____

Do you know your current grade in this course? Successful students continually monitor their grade and understand how each assignment, quiz, and test contributes to their course grades.

Accessing your recorded grade

1) Do you know your current grade in this course?   _____ yes _____ no

2) Is your current grade for this course available online?   _____ yes _____ no

3) How often do you have access to your updated grade?
    _____ 24/7 _____ weekly _____ at instructor's choice _____ other: _____

4) Describe step-by-step the process you would follow to access your grade, as if you are telling your classmate how to do it.

Tracking your grades

5) Check the components of your grade record that you can access online and then indicate what your current grade (or score) is for each of those components.

Grade component	My current grade (or score) is:
_____a) Attendance	a)
_____b) Homework	b)
_____c) Quizzes	c)
_____d) Tests	d)
_____e) Notes/Notebook	e)
_____f) Participation	f)
_____g) Other: _____	g)
_____h) Overall current grade	h)

6) How much does each grade component that you checked above contribute to your course grade? Write this information in the chart below.

Grade component	Percentage of Final Grade

Analyzing your grade

7) What is your current overall grade in this class? _____

8) How can you improve your overall grade? Refer to the different components that contribute to your grade and describe specifically what behaviors you can change to improve your grade

Strategies for Success

Successful Student Behavior   **Name**_____

Being successful in a math class is about more than the math. Successful students often exhibit similar behaviors. Do you see yourself as a successful student? Do you practice the behaviors of successful students?

1) Name the person who you see as the best role model of a student. Why is this person a role model to you?

2) What are some behaviors you think a successful student should exhibit?
   List 4 of them here.

   1.

   2.

   3.

   4.

3) From this list, which do you already do on a regular basis?

4) Which behavior do you think you could try next?

Strategies for Success

Email Etiquette

Name_____

Emails and texts messages are the primary way many people communicate. Have you ever considered the impression others get of you from your online communication?

Your email address

1) Do you have more than one email address? _____ yes ____ no

2) Which email address do you send from when you communicate with your professor?

3) Does your email address include your name? Does that name match your name on the professor's roster? Why is that helpful to your professor?

4) Do you think your email address presents a professional image of yourself as a successful college student? Why or why not?

Emailing your instructor

5) Think about writing an email or text to your instructor. The chart shows greetings that some students have used. Check the greetings on the chart that are appropriate.

Greeting	Always appropriate	May be appropriate	Not appropriate
a) Dear Professor _____			
b) Professor			
c) Dr./Mr./Mrs./Ms _____			
d) Hello			
e) Hi			
f) Hey			
g) Yo			
h) Hey Professor			
i) Hello Professor			
j) Hi Professor			
k) Good morning/afternoon/evening			

6) Describe what you do to make sure your instructor knows your email is from one of his or her students.

7) What might you write in the subject line of an email to your professor to identify it as coming from a student?

8) What device do you use to email your instructor? ___computer ___tablet ___phone ___other

9) Think about communicating with your instructor by email or text message. The list shows items often included in student messages. Check the items that are appropriate. Add to the list any other items you include.

	In an Email	In a Text	Not needed	Not Appropriate
a) Proper greeting				
b) Your name				
c) Class enrolled				
d) Complete sentences				
e) Good grammar				
f) Slang and abbreviations				
g) Appropriate capitalizations				
h) Emoticons				
i) Closing				
j)				
k)				

Submitting assignments via email

10) How does your instructor prefer for you submit assignments via email?
_____ within the email body
_____ as attachments to emails
_____ other: _____

11) Describe what you do to make sure your instructor knows your email contains a class assignment.

Organizing your emails

12) Do you create a folder in your inbox for the emails from this course? Or do have another way of organizing your inbox so that you can easily identify emails from this course? Describe what method you use to organize your emails for this course.

Strategies for Success

Textbook Tour

Name_____

By now, you are a few weeks into your math course. Are you familiar with all the features of your textbook? Do you realize each feature has a unique purpose?

1) Match each feature with its purpose.

Textbook feature	Purpose
___ 1. Preface	a) Lists answers to the exercises so I can check my work and correct my mistakes
___ 2. Objectives	b) Exercises I work so that I practice and master a concept
___ 3. Example	c) Summarizes the steps of a certain procedure
___ 4. Procedure box	d) Explains and shows me how to work a specific type of problem
___ 5. Definition box	e) Helps me review the chapter by summarizing key concepts and terms, and practicing typical exercises of each section
___ 6. Margin note	f) A sample test I can use to assess my readiness for my class exam
___ 7. Exercises	g) Lists topics I will learn and master in this section
___ 8. Chapter Review	h) Gives me the meaning of an important term or concept
___ 9. Practice Test	i) Gives me tips and cautions me about common mistakes
___ 10. Answers to Exercises	j) Explains the author's philosophy and introduces features of the book

2) List any other features of your textbook and identify the purpose of each:

Textbook feature	Purpose

Strategies for Success

Time Management Part I   **Name**_____
Managing Your Weekly Schedule

Have you ever heard the saying "If you want something done, ask a busy person to do it"? People who have many demands on their schedule and manage to accomplish a lot are usually very organized. They use their time wisely. Time management is a skill you can learn, and it will help you become a more successful student.

A weekly schedule showing all your regular activities is a useful tool to help you manage your time and commitments. Once you create your schedule on paper, you'll be able to look at it, know that all your commitments are accounted for, and see what times are available for other things. Then you can easily match your free hours to your instructors' office hours, plan study group sessions, and set up regular meetings with a tutor. You can see when you can schedule things that occasionally come up, like counseling and doctor's appointments. And you can see how much time you have for fun activities.

Start by making a chart showing all 168 hours of the week – that's 24 hours per day for 7 days. You may want to use the one on the next page, or make one like it on a separate piece of paper.

1) First show all the classes you are taking this term, making sure to block out the number of hours for each class meeting. Also show the time it takes you to get to school and return home.

2) Many students work at jobs, in addition to taking classes. Do you have a job? If so, mark your typical weekly job schedule in the chart. Don't forget to include commuting time!

3) Now think about what activities you do every day, other than school and work. Your basic needs like sleeping, eating, bathing, exercising, etc. all take time. If you are responsible for cooking meals for your family or caring for young children, you know those tasks take time, too. Show all your usual daily activities in the chart.

4) Where does your study time fit in? The guideline for college students to do all the reading, homework, and studying required for their classes is to count 2 hours outside of class for each hour in class. For each of your courses, multiply the total number of hours each week you are in class by 2, then block out and label that many hours for studying for that course on your weekly chart. Try to schedule as many hours as possible in your college's math center and library, where you will find help nearby if you need it. Keep in mind that it is more effective to study in several small sessions instead of a couple of 'marathon' sessions.

5) Last, you may schedule time to spend with friends, going out, relaxing, leisure reading, playing video games, or watching tv.

	Monday	Tuesday	Wednesday	Thursday	Friday	Saturday	Sunday
1 am							
2							
3							
4							
5							
6							
7							
8							
9							
10							
11am							
noon							
1 pm							
2							
3							
4							
5							
6							
7							
8							
9							
10							
11 pm							
12 am							

6) a) Now that you have completed a schedule of your typical week, take a good look at it. Do you have spare time, or are you 'overscheduled'?

b) Is this a feasible schedule for you? Will you be able to meet all your commitments without overstressing?

c) What changes can you make to your schedule to make it work better for you?

Strategies for Success

Time Management Part II   **Name_____**
Managing Your Schedule for the Term

Being a successful college student takes a lot of time. You have to attend classes, do the required reading and homework, turn in assignments and projects, and take exams. And then there's the rest of your life—your family, your friends, your job, your health--making demands on your time, too. Schedule conflicts may tempt you to miss classes, skip readings, do assignments at the last minute or even late, or be unprepared for exams. In general, poor time management can cause stress, which prevents you from doing your best.

To be successful in college, you have to be a good planner. When you plan ahead and use your time wisely, you keep the chance of scheduling conflicts to a minimum. Making a schedule of the entire term can help you to become a good planner. The schedule will show you when readings, homework, assignments, and projects are due and when you will have exams. You should also record important events in the rest of your life that you know will take place this term. After you make the schedule, you'll be able to refer to it to see how early you should plan to start on a project, when you should start studying for an exam, or when you can schedule a doctor's appointment.

Use the charts on the next pages to make a schedule of this semester (or quarter). Use one chart for each month.

1) What month did this term start? Write that month name on the top of the first chart. Then write the names of all the other months in this term at the top of the other charts.

2) Fill in the dates of each month, making sure to write them on the right days of the week.

3) Check your college's academic calendar for any holidays or breaks this term, then note them on your calendar.

4) Record all due dates for readings, homework, assignments, and projects, and all scheduled tests.

5) Write Final Exam on the dates your final exams are scheduled.

6) Do you have any appointments already scheduled with your counselor? Do you know the date you can register for next term? Is there a deadline for scholarship or financial aid applications? If so, write those on your calendar.

7) Now record any big events from the rest of your life that you know will take place this term (weddings, births, celebrations, job training, business trips, etc.).

After you have completed making your schedule for this term, look over every month.

8) a) Do you see any potential schedule conflicts? _____ If so, list them here.

39

b) Is there a way for you to reschedule to avoid the conflicts?

c) For those conflicts you can't resolve by rescheduling, make a plan to minimize the impact of the conflicts. For example, if you will be attending a wedding the weekend before your math test, you could resolve this conflict with your study time by preparing for the exam several days earlier.

List all your schedule conflicts and describe a plan to minimize the impact of each

Schedule conflict	I can resolve the conflict by...

9) Looking at your schedule for the term, what dates do you have available for routine doctor's appointments, dental checkups, etc.?

Month: _____

Sunday	Monday	Tuesday	Wednesday	Thursday	Friday	Saturday

Month: _____

Sunday	Monday	Tuesday	Wednesday	Thursday	Friday	Saturday

Month: _____

Sunday	Monday	Tuesday	Wednesday	Thursday	Friday	Saturday

Month: _____

Sunday	Monday	Tuesday	Wednesday	Thursday	Friday	Saturday

Month: _____

Sunday	Monday	Tuesday	Wednesday	Thursday	Friday	Saturday

Strategies for Success

Homework Skills

Name_____

Why do homework?

1) Why do math teachers assign homework?

2) Who benefits when you do homework?

3) How does practice help you improve a skill?

4) What are the advantages of doing homework?

5) What are the disadvantages of not doing homework?

6) How can your graded homework be useful to you?

Looking at Our Homework

Work with a partner or small group of classmates. Trade your last math homework paper with someone in your group.

7) Looking at your classmate's homework paper,

list 3 good things you notice:

1.

2.

3.

list 3 things that could be improved:

1.

2.

3.

8) Share your results with your group. Is there one common area in which you all could improve?

Doing Homework

Answer each of the following questions individually and then discuss your answers with your group members.

9) When do you usually do your math homework?

10) Where do you usually do your math homework?

11) What is going on around you when you do your math homework?

12) What would be the best environment for you to do your homework?

13) What do you do if you get stuck on a homework problem?

14) When you finish a homework assignment, what does the paper look like?

15) How do you feel when you finish a homework assignment?

Homework Practices

16) This table lists several practices that students use when doing math homework. Check the practices that you usually do. In the next column, check all the practices that will help you succeed in this class.

Homework Practice	I usually do	Will help me succeed!
a) Do homework where there is help available		
b) Write name on top of paper		
c) List assignment at top of page		
d) Identify each problem by number		
e) Keep problems in order on paper		
f) Write neatly and legibly		
g) Show all work-not just answers		
h) Refer to similar examples in text		
i) Check answers in back of book		
j) Attempt to correct wrong answers		
k) Highlight problems on which I need extra help		
l) Redo a problem on my own, if I received help with it		
m) Save homework to review before test		
n) Feel proud of my work		

Homework grading policy

17) How much of your course grade is based on homework?

18) How often is your homework assigned?

19) When is your homework due?

20) How much does each assignment count toward your homework total?

21) Does your teacher require you to do your homework in a specific format? If so, what is it?

22) Where do you turn in your homework?

23) Describe the scoring system used by your teacher to evaluate your homework.

24) Will your teacher accept late homework? Is there a penalty?

Strategies for Success

Taking Notes—Part II
Notes from Online Exercises

Name_____

When doing online homework exercises, it is usually necessary to do some work on paper before entering the final answer on the computer. Are the papers from your online work neat and organized? Sometimes you may want to refer to one exercise when you do a later exercise, and reviewing a chapter's homework exercises is always good test preparation strategy. Do you keep notes from your online exercises in a way that is helps you be successful in your class?

1) Think about how you do your online exercises, then fill in the chart below.

When I do online exercises, I...	Always	Sometimes	Never
a) copy the problem onto my paper so that I can refer to it later.			
b) label each problem with section and problem number so that I know where it came from.			
c) note which section objective the exercise relates to so that I know how the exercise fits into the big picture.			
d) write all my work neatly so that I can review it when I do a similar exercise or when I study for the test.			
e) file my homework papers in a folder or notebook so that I know where to find them.			
f) keep my homework papers for each chapter (or unit) for future reference, at least until I have finished the test for that chapter.			

2) Use the **Exercise Log** on the next page to organize your notes as you do your online exercises. You may need several copies of the Exercise Log to record the whole assignment.

You may find the **Exercise Log for Word Problems** useful when you do Word Problems.

Strategies for Success

Notes from Online Exercises

Name_____

Exercise Log

Name:		Date:
Section:_____		
Exercise Number	Explain your steps in words.	Do the math.

Strategies for Success

Exercise Log

Name_____

Exercise Log for Word Problems

Section _____ Exercise _____	Name
Step 1 Look at the objectives listed at the start of this section. Which objective does this exercise address?	
Step 2 Copy the exercise.	
Step 3 What are you asked to do?	

Step 4 Work the exercise. Be sure to show all your steps. Explain each step in words in the left column and do the math in the right column.	
Explain your steps in words.	**Do the math.**

Strategies for Success

Mid-term Check-up Part I
Making the Grade

Name_____

Midterm is a good time for you to pause for a while and reflect on your goals for this class, your current course grade, and the study habits you have used up to now.

1) My goal is to have a grade of _____ for my final course grade.

2) My grade in this class right now is _____.

3) I feel   proud / ok / disappointed with my class grade, because:
     (circle one)

4) The one study skill that has helped me most so far was _____,
    because

5) Two other study skills I used that have also been helpful were:

    1.

    2.

6) In order to ensure I meet my grade goal, I need to improve my math study habits. Three specific strategies I will use are:

    1.

    2.

    3.

Strategies for Success

Mid-term Check-up-Part II   **Name**_____
Evaluating Your Study Habits

1) Evaluate your study habits by completing the following checklist.

 So far in this class I have:

 a) Been absent
  _____ never   _____ 1 or 2 times   _____ 3 or more times

 b) Arrived in class on time
  _____ always   _____ usually   _____ rarely

 c) Brought my text, notebook, and calculator to class
  _____ always   _____ usually   _____ rarely

 d) Paid close attention and taken good notes in class
  _____ always   _____ usually   _____ rarely

 e) Organized my papers in my notebook the way my teacher recommends
  _____ always   _____ usually   _____ rarely

 f) Scheduled time for homework
  _____every day   _____ 2-3 times/week   _____ once a week

 g) Re-read or re-copied my class notes before doing the homework
  _____ always   _____ usually   _____ rarely

 h) Completed each homework assignment before the due date
  _____ always   _____ usually   _____ rarely

 i) Reviewed topics and/or problems that gave me trouble
  _____ always   _____ usually   _____ rarely

 j) Studied with a friend or study group
  _____ always   _____ sometimes   _____ never

 k) Used my instructor's office hours
  _____ often   _____ 1 or 2 times   _____ never

 l) Used the Math Center or Tutoring Center
  _____ every week   _____ 1 or 2 times   _____ never

2) I will improve my chances of success in this class by taking the following steps. I will:

Strategies for Success

Attendance   **Name**_____

Have you ever thought about how your attendance affects your success in this class?

1) How many absences from this class do you have so far this term?

2) Why is regular class attendance important?

3) My attendance has affected my performance in this class in these two ways:

    1.

    2.

4) If I miss a class, I know I have missed important information.

To get back on track, I:	Always	Sometimes	Never
a) check the course syllabus to see what I missed			
b) read the textbook			
c) try to find a way to turn in any assignments that are due			
d) contact a classmate to find out what was covered and get the new assignment			
e) get the notes from a classmate			
f) do the homework that was assigned, even if I can no longer turn it in			
g) go to my school's math tutoring center			
h) go to my instructor's office hour			
i) get help from a friend or family member			
j) other:			

5) a) When I miss class, it is usually because:

    b) To prevent this from happening again, I will:

6) Two things I can do to improve my attendance are:

    1.

    2.

Strategies for Success

Study Group   **Name**_____

Do you have a study group to help you with this class? A study group is two or more students who have set a regular time to get together to work on their coursework. Research has shown that participating in a study group is an important factor that contributes to successful course completion. A study group may choose to meet at a place convenient to all its members, such as the college math center, a library, a local coffee shop – anywhere that has tables and chairs. Or, the group may prefer to have virtual meetings by setting up an online chat room or conference phone call.

1) How can you benefit from being in a study group? Several possibilities are listed in the chart.

Benefit	I know this will help me succeed	I hadn't thought of this
a) I can share and compare notes with other people in my class.		
b) I have colleagues to work with on practice problems and homework.		
c) I have colleagues to ask for help with concepts I don't understand.		
d) I have people to call if I have to be absent from class.		
e) I can improve my understanding by explaining concepts to others.		
f) I can review and quiz with my study group before a test.		
g) I make friends with my classmates.		
h) I can get support and encouragement from my study group.		
i) Other:		

2) How can you form a study group? A few ideas are listed below – check which ones you can do. Then list two other ways to form a study group.

_____ ask people who usually sit near me in class

_____ ask classmates I see in the math center

_____ write a note on the board of my classroom

_____ post a comment on my class website

_____

_____

Strategies for Success

Help!   **Name**_____

Sometime the thought of asking for help with your coursework can be intimidating. Knowing where help is available for your class is the first step in getting the help you need. There are usually several sources of help, so try different ones and see which works best for you. Wherever you go to get help, if you organize and prepare your questions before you go, you will get the maximum benefit from the help session.

- **Getting help**
  1) The following places or people provide resources for me to get the help I need to be successful in my math class.

Source	Yes	No
a) Teacher		
b) Math Center with Computers		
c) Drop-in Tutoring Center		
d) Tutoring Center by appointment		
e) Supplemental Instruction Workshop		
f) Classmates		
g) Study group		
h) Online class discussion board		
i) On-line chat room		
j) Private tutor		
k) Family		
l) Friends		

- **Getting help from my teacher**
  o **Office hours**
  2) My teacher has scheduled office hours: _____ online _____ in person, at: _____

  3) My teacher's office hours are _____

  4) From those hours, the times that best fit into my schedule are _____

  o **Email**
  5) My teacher's email address is _____

  6) I can generally expect a response within_____ hours/days (circle one)

  o **Telephone**
  7) My teacher's phone number is _____

  8) My teacher has said I can generally expect a response within _____hours/days (circle one)

    o   **Within the classroom management system such as Blackboard, MyMathLab, etc.**

9) Describe how you would contact your teacher using the classroom management system (such as Blackboard, MyMathLab, etc.) for your course.

    o   **Other**

10) Describe other ways of contacting your instructor.

- **Getting help in the math center with computers**

  11) The math center where I can work on a computer is located _____

  12) The hours the center is open that fit in my schedule are _____

  13) The best way to get help in the math computer center is _____

- **Getting help in the drop-in tutoring center**

  14) The drop-in tutoring center is located_____

  15) The hours the center is open that fit in my schedule are _____

  16) The best way to get help in the drop-in tutoring center is _____

- **Getting help by appointment at a tutoring center**

  17) To get help in the tutoring center, I need to make an appointment. I can make the appointment:

      in person, by going to the center located at _____

      by calling the telephone number _____

      through the website with url: _____

- **Getting help in a supplemental instruction workshop**

  18) I can find out about supplemental instruction workshops by _____

  19) The location for the workshops is _____

  20) To enroll in a workshop I need to _____

- **Getting help from a discussion board**

  21) List the steps you would use to access the class discussion board.

  22) The best way for me to post a question on my class discussion board is _____

  23) List ways the discussion board might be more helpful to you than some of the other resources.

- **Getting help from an online chat room**
  24) List the steps you would use to access the online chat room.

  25) The best way for me to post a question on the online chat room is _____
  26) List ways the online chat room would be more helpful to you than some of the other resources.

- **Getting help from classmates**
  27) To get help from a classmate I _____

  28) The best way for my classmate to help me is to _____

- **Getting help from a study group**
  29) I can form a study group by _____

  30) Good days and times for me to meet with a study group are _____

- **Getting help from a private tutor**
  31) To find a private tutor I _____

  32) I will check the tutor's qualifications by

  _____

  33) I will agree on a price for the tutoring before we begin. _____yes _____no

  34) Good days and times for me to meet with a private tutor are _____

  35) Meeting with a private tutor is best done in a public place, to ensure safety. To create a good academic environment, we need to meet where we are free of distractions and have a table and chairs.
  A good place for me to meet with a private tutor is _____

- **Getting help from family**
  36) The family member who is best able to help me is _____

  37) Good days and times for me to meet with this family member are _____

- **Getting help from a friend**
  38) The friend who is best able to help me is _____

  39) Good days and times for me to meet with this friend are _____

  40) A good place for me to meet with this friend is _____

Strategies for Success

Preparing to Get Help

Name_____

Have you ever said to your teacher, "I don't understand anything. I just don't get it!"? While your exclamation is an effective way of expressing your frustration, it doesn't give the teacher enough information to focus attention on what specifically you need to have clarified. When you organize your work, identify your areas of concern, and prepare your questions before meeting with your teacher or tutor, the help session will be much more useful.

Preparing to get help

1) How do you prepare to get help? For each statement below, check Always (A), Sometimes (S), or Never (N)

<u>A   S   N </u>

___ ___ ___ a) I make a list of the questions I want to ask and put the list somewhere easy to find.

___ ___ ___ b) I make a list of the problems I need help with and put the list somewhere easy to find.

___ ___ ___ c) I try to analyze the differences between problems I can do and those I need help with.

___ ___ ___ d) For each exercise that I have questions on, I work out the problems I have questions on as far as I can go, showing all my work.

___ ___ ___ e) I make sure my homework pages are labeled and in order.

___ ___ ___ f) I make sure my notes are organized.

___ ___ ___ g) I highlight questions I have in my notes so they are easy to find.

___ ___ ___ h) I mark my notes with a post-it to show the page I have questions on.

___ ___ ___ i) I note passages in the text that I don't understand and bookmark the pages.

___ ___ ___ j) I find and highlight questions I had written in my reading log.

___ ___ ___ k) I turn off my cell phone so I will not be distracted.

___ ___ ___ l) Other_____

2) If you checked 'Never' for any of the strategies listed above, which do you think you will try the next time you need to get help?

3) Ask your teacher what you should do to prepare for a meeting with her to get help. What does she recommend?

Knowing good help when you find it!

4) When you need help with your math, you want someone to help you whose methods will help you understand the math so that you can do it yourself later. Sometimes a peer or even a tutor does not use good teaching techniques that will help you learn. Which of the following are characteristic of a quality tutoring experience?

The tutor...	Yes	No
a) asks to see my work on the problem to find out where I got stuck.		
b) diagnoses my weakness and difficulties.		
c) laughs at my mistakes.		
d) models good problem solving strategies so I learn to use them too.		
e) asks to see my class notes.		
f) asks what method my teacher demonstrated.		
g) just gives me the answer.		
h) explains the concepts until I understand them.		
i) leads me through some examples.		
j) has me try problems on my own.		
k) does all the work for me.		
l) shows me 'shortcuts' but doesn't explain the math behind them.		
m) makes sure I understand a problem by asking me to explain how I did it.		
n) builds my confidence.		

5) Which of the characteristics in the chart above is most helpful to you?

Strategies for Success

Goals

Name_____

What are your goals and dreams? To achieve your goals and make your dreams become reality, you must recognize the barriers that may arise and learn to go over, under, or around those barriers. With a firm plan, and a good solid backup plan, you can reach your goal!

1) What is your **short-term** educational goal?

2) Name 3 things you must do to achieve it.

    1.

    2.

    3.

3) In the chart below, list 3 potential barriers to achieving your short–term goal and for each barrier, name something specific you can do to overcome or go around that barrier.

Potential barrier to achieving my short-term goal	My plan to overcome it

Reaching your ultimate educational goal may take years, but it is important to keep that goal in sight. Every step you take in your education brings you closer to your goal!

4) What is your **long-term** educational goal?

5) List at least three steps you must take to achieve this goal.

    1.

    2.

    3.

6) In the chart below, list 3 potential barriers to achieving your long-term goal and for each barrier, name something specific you can do to overcome or go around that barrier.

Potential barrier to achieving my long-term goal	My plan to overcome it

Strategies for Success

Thoughts in Charge!

Name_____

Do you know that your thoughts can affect your emotions, body sensations and behaviors? Thoughts, emotions, body sensations, and behaviors are interrelated. Each one of them influences the other three. (Adapted from Ooten, C. (2003). *Managing the Mean Math Blues.* Upper Saddle River, NJ: Pearson Education, Inc.)

These "interrelationship" charts use arrows to show how thoughts, emotions, body sensations and behaviors are all related to each other. Read these two interrelationship charts, starting at the top left with the "Thoughts." Notice the effects of a negative thought on your emotions, body sensations and behaviors.

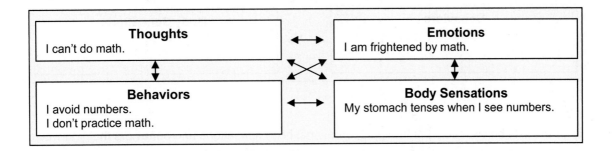

As you read the next chart, notice the effects of neutral thoughts.

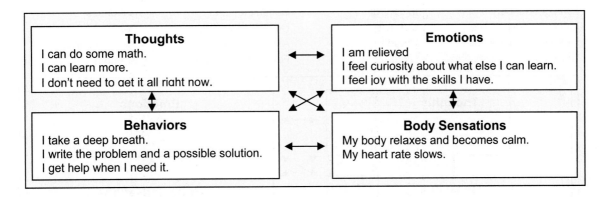

1) What is your reaction to the two charts above?

2) In the lists below, circle the thoughts, emotions, body sensations and behaviors that you have experienced. Then use them to create an "interrelationship" chart.

Thoughts	Emotions	Behaviors	Body Sensations
I will fail.	Frustrated	Avoid math	Palpitations
I will never do well.	Embarrassed	Blame the teacher	Sweating
I am incompetent.	Helpless	Tune out in class	Stomach ache
Math teachers hate me.	Anxious	Waste time	Hives
I am helpless.	Panicky	Avoid homework	Crying

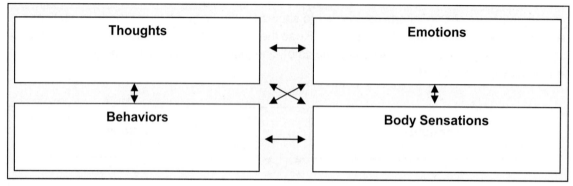

3) In the lists below, circle the thoughts, emotions, body sensations and behaviors that you would like to experience. Then use them to create an "interrelationship" chart.

Thoughts	Emotions	Behaviors	Body Sensations
Practice helps.	Excited	Come to class prepared	Relaxation
I can get support.	In control	Consult teacher	Calmness
I have learned before.	Capable	Do my homework	Peace
I intend to understand.	Proud	Ask questions	Steady heartbeat
Understanding takes time.	Calm	Form a study group	Strength

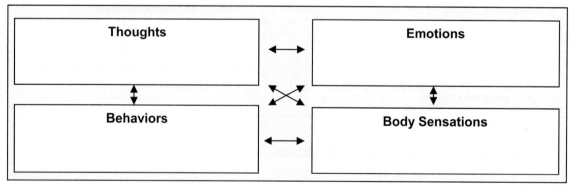

4) Look at the "interrelationship" chart you created in #3. Do you think you can turn the thoughts, emotions, body sensations and behaviors in the "interrelationship" chart into reality? Why or why not?

Strategies for Success

Neutralize Negative Thoughts

Name_____

Recognizing and acknowledging your negative thoughts about mathematics is the first step towards changing your negative emotions, body sensations and behaviors. This activity will help you recognize negative math thoughts, so you can then consciously intervene to neutralize the negativity they create. The next activity, *Intervention Strategies for Negative Thoughts,* will show you some interventions strategies you can use. (Adapted from Ooten, C. (2003). *Managing the Mean Math Blues.* Upper Saddle River, NJ: Pearson Education, Inc.)

1) Read each thought below and identify whether it is a negative statement or a neutral statement.

Negative   Neutral

_____   _____ a) I will never understand math.

_____   _____ b) I feel dumb and stupid.

_____   _____ c) Math is out to get me.

_____   _____ d) This has happened to me before. I have worked through it.

_____   _____ e) Math problems contain tricks meant to stump me.

_____   _____ f) I cannot do math.

_____   _____ g) Maybe I need to ask some questions, or do some examples again.

_____   _____ h) Everyone understands what to do except for me.

_____   _____ i) Because I don't understand this, I will never be able to do math.

_____   _____ j) The learning process is challenging. There must be something I don't quite understand here.

_____   _____ k) I feel like an idiot.

_____   _____ l) I have many resources to assist me—the book, notes, examples, the instructor, friends and tutors.

_____   _____ m) Just because these few problems are difficult doesn't mean all the rest will be difficult too. This is an opportunity for me to figure out what I misunderstood and correct it!

_____   _____ n) I should understand this.

_____   _____ o) I will take notes or ask a question so I can clarify this concept.

Negative   Neutral

_____ _____ p) Math is a process and it has a way of being harder, then easier, then harder, then easier.

_____ _____ q) The material is new to me. I am not expected to understand it all immediately. I have many resources to help me understand this.

_____ _____ r) The teacher will be upset with me if I ask questions about this.

_____ _____ s) Establishing a relationship with my teacher might make asking questions during or after class easier.

_____ _____ t) I can act positively by taking a deep breath and congratulating myself for being courageous enough to put myself in this class.

2) Choose two of the statements that you identified as negative thoughts above and rewrite each of them as a more neutral and useful, but still true, statement. For example:

Negative thought: "I hate fractions!"

More neutral and useful statement: "I don't completely understand fractions, but I can learn to work with them."

1. Negative thought: _____

   More neutral and useful statement: _____

   _____

2. Negative thought: _____

   More neutral and useful statement: _____

   _____

⅃Ｊ

Strategies for Success

Intervention Strategies for Negative Thoughts   Name_____

Here are some intervention strategies for negative math thoughts. When you catch your brain forming one of these negative thoughts, try an intervention strategy. With practice, you will be able to change the negative thoughts into alternative thoughts that will lead to more productive emotions, body sensations and behaviors. (Adapted from Ooten, C. (2003). *Managing the Mean Math Blues.* Upper Saddle River, NJ: Pearson Education, Inc.)

Strategy #1: Examine the Evidence.
- What is the evidence that your negative thought is really true?
- What would you do differently if this thought were false?

1) Suppose your negative thought is "I'm sure I will fail this class."

    a) What evidence could you check to see if you really will fail the class?

    b) If you are not failing the class, how might your behavior change?

Strategy #2: Get a Different Perspective.
- Tell yourself what you would tell a close friend who has this thought.

2) Suppose your brother said, "I am too stupid to do math." What would you tell him to convince him he is not too stupid to do math?

Strategy #3: Do Something Differently.
- Identify a behavior that contributes to your negative math thought. Behave in a new way to get a different result.

3) When learning math, just like when learning sports, music, or a foreign language, if you don't practice -- by doing homework-- you can't expect to become proficient. Not doing homework is a behavior that contributes to not being able to do math.

    a) How can you change this behavior to get a better result?

    b) Name one of your behaviors that contributes to your negative thoughts about math.

    c) How can you change that behavior to get a better result?

Strategy #4: Change the Wording.
- Restate a negative thought in a way that it becomes neutral or positive. Add the words "right now," or "yet."

For example:
- Change "I can't do math" to "Right now I am unable to do these math problems."
- Change "I don't understand" to "I don't understand yet."
- Change "I'm not prepared" to "I am not prepared yet."

4) Change the wording of two of your negative thoughts.

1. Change _____

   to _____

2. Change _____

   to _____

Strategy #5: Act "As If."
- Act as if you have the trait you lack or already are the kind of person you would like to become.

5) If you want to be a successful math student, think about how good math students act. What behaviors could you do to act "as if" you were a successful math student?

Strategies for Success

Can You Hear Me Now???

Name_____

Have you ever been in a movie theater and heard someone's cell phone ring? Or maybe someone in the theater was carrying on a cell phone conversation while you were trying to concentrate on the movie? When someone in the audience takes a phone call during a movie, it disrupts the theater environment. Similarly, cell phone use during class is disruptive to the academic environment.

1) The table below lists several statements related to the use of cell phones during class. Indicate whether you agree or disagree with each statement.

	Agree	Disagree	Neutral
a) My classmates have the right to have an academic environment that supports their learning.			
b) A cell phone is 'on' even when it is in silent mode.			
c) It is rude to have my cell phone on during class.			
d) If my cell phone rings during class it is distracting to my classmates and teacher as it disrupts the academic environment.			
e) If my cell phone rings during class it interrupts the flow of the class.			
f) Even if my cell phone is on silent, its vibrations are noticed by those sitting near me.			
g) Taking phone calls during class is disrespectful of the teacher who put great effort into preparing the lesson.			
h) If I leave class to answer a call, I have distracted my classmates, lost my concentration, and missed out on the material covered while I was gone.			
i) Texting in class puts my thoughts and focus somewhere other than on the math being covered.			
j) My cell phone has voicemail.			
k) There are very few situations in my daily life that cannot wait two hours for my reply.			
l) If I have an urgent situation going on in my life that may require me to be contacted immediately, I can talk to my instructor about discretely and silently keeping my cell phone on during class.			

2) List 3 reasons why students should not text during class:

1.

2.

3.

Strategies for Success

A Gift to Yourself

Name_____

Choosing to get an education is a precious gift you give to yourself and to your future. Do everything you can to protect that gift and cherish it as much as possible. View each class meeting as irreplaceable and the knowledge you gain there as a critical piece of your education. Give your full focus every class meeting. Take full advantage of your gift of time! Each class and each assignment make a contribution to the end result—your education.

1) Consider each statement in the table and indicate whether you agree or disagree.

	Agree	Disagree	Neutral
a) I want to succeed in college.			
b) I want this course to give me a solid foundation for my next math class.			
c) Every topic in this course is important to my success in the next math course.			
d) I want to get as much from each class meeting as possible.			
e) If I earn an A in this course I will be better prepared for my next math course than if I earn a C.			
f) I am less efficient when I multitask than when I focus on a single job.			
g) I cannot give 100% of my attention to two things at the same time.			
h) I put away my cell phone, books, and homework from other classes when I am in my math class.			

2) I show my appreciation for my gift of time in this class by demonstrating these 3 behaviors:

    1.

    2.

    3.

Strategies for Success

Math Plan   **Name**_____

Because math skills in each course build upon the skills from the previous course, it is best to take your math classes in sequence, without taking a break of one or more terms between them. Now is the time to start thinking about your next math class!

1) To reach my college or career goal, I need to take math up through _____.

2) The math class I should take next semester is _____.

3) The successful student behaviors that helped me the most in this course have been:

   1.

   2.

4) The study strategies that helped me the most in this course have been:

   1.

   2.

5) In order to increase my chance of success in math classes, I will improve my study skills by:

Strategies for Success

The End is in Sight!   **Name**_____

Crunch time is coming! You are now more than half-way through with this course and the end of the term is in sight. But you need to make sure you don't lose your momentum or get distracted by your other duties and responsibilities. The effort you put into completing your math class successfully will really pay off. As a marathon runner might say, "it's time to sprint towards the finish line!" What work do you have left to do for this course?

1) There are _____ weeks left until the end of this course.

2) I have _____ math tests scheduled between now and the final exam. They are scheduled on the following dates_____.

3) Other math assignments that will need to be done before the final exam are:

4) I have usually spent _____ hours per week, not counting class hours, for this class.

5) In order to stay focused and not jeopardize my success in this class, the number of hours I spend each week for this class should

_____stay the same.

_____increase, and I will make the following adjustments to my weekly schedule:

Strategies for Success

Excuses! Excuses!   **Name_____**

Math teachers hear all sorts of excuses from their students for not fully completing class assignments. Many students use excuses to attempt to justify their failure to plan ahead or maintain good study habits. Let's look at some of the most common excuses. For each excuse listed below:

a) explain why this is not a valid excuse.

b) describe what a successful student would do in this situation.

1) "I don't have my homework because I was absent"

a)

b)

2) "I don't have my homework because I don't have a book."

a)

b)

3) "I did not finish the assignment. I forgot about it!"

a)

b)

4) "I left my homework paper at home."

a)

b)

5) "I didn't do the chapter review/chapter test assignment. Can I do it after I take the test?"

a)

b)

6) "I was absent—did you do anything important?" *(This is the favorite of many math teachers.)*

a)

b)

Strategies for Success

Excuses! Excuses!   **Name_____**
Online Homework Version

Math teachers hear all sorts of excuses from their students for not fully completing their online math assignments. Let's look at some of the most common excuses. For each excuse listed below:

> a) explain why this is not a valid excuse.
>
> b) describe what a successful student would do in this situation.

1) "I tried to complete the assignment, but the computer wouldn't let me access it!!"

> a)
>
> b)

2) "I don't have my homework done because my internet access is unreliable."

> a)
>
> b)

3) "I can't do any more online homework because my room-mate moved out and cancelled the internet."

> a)
>
> b)

4) ""I was stuck on one problem and I couldn't go on because it took you too long to answer my email."

> a)
>
> b)

5) "I missed class, so you have to extend my due dates."

> a)
>
> b)

6) "I didn't have time to do all my homework."

     a)

     b)

7) "I got shut out of the assignment I was working on when the time passed the deadline."

     a)

     b)

8) "I kept typing in the right answer but the computer marked me wrong."

     a)

     b)

Strategies for Success

Support from Family and Friends   Name_____

When you were in elementary school, did your family help you prepare for the first day of school? Maybe they took you shopping for new clothes, a new pair of shoes, and a notebook? As a college student, you still need the support of your family and friends to help you succeed. But the support you need from them may not be as obvious as when you were younger. Analyze your own situation to identify what you need to be a successful student. Then be pro-active in letting your family and friends know how to help you!

1) The table below lists several ways you may inform your family and friends of your needs as a student. How do these apply to you?

To inform your family and friends of your needs as a student, you might say…	I have said this	I didn't, but I wanted to	I had no idea!	Does not apply to me
a) I need a quiet place to study.				
b) I need to have fewer chores at home so I have time to study.				
c) I need to work fewer hours so I have more time to study.				
d) I can't go to the movies/party/dance because I have to study.				
e) I can't miss class to go to the doctor/dentist.				
f) I can't miss class to babysit.				
g) I can't miss class to run errands for my family.				
g) I can't miss class to go to a funeral.				

2) I have also asked my family and friends to help me succeed in college by …

3) As a result of this exercise, I now realize it would be a good idea to ask my family and friends to help me succeed in college by making the following requests:

⅃

83

Strategies for Success

Stay on Campus--Stay on Task!   **Name**_____

Your overall time commitment to college includes study and homework time. Most colleges have places where students can study and do homework. Staying on campus to study can help you succeed by keeping your family and social life separate from school. Scheduling study time on campus also makes it easier for you to use campus support services, like tutoring, counseling, and your instructor's office hours. And when you complete your homework at school, you leave campus with the satisfaction of knowing that you are free!

1) I stay on campus after my classes are done. _____yes _____no

2) Two places on campus where I can do my math homework are:

      1.

      2.

3) The hours I can do my math on campus are:

Monday	_____	Friday	_____
Tuesday	_____	Saturday	_____
Wednesday	_____	Sunday	_____
Thursday	_____		

4) Staying on campus to do my math will help me because:

Strategies for Success

Final Exam Prep Part I **Name**_____
Get the Facts and Get Organized.

Final exam time is very stressful for both students and faculty. So why do we have final exams anyway??? During the course you learned the material in chunks—sections and chapters. It is now time to pull it all together and firm up the concepts before you head to the next course.

The more prepared you are, the less stressed and more confident you will feel. All the test taking strategies from earlier in this course still apply to preparing for the final exam, but there are some additional things to do and consider.

Get the facts: I need to find out all I can about the final exam so I can be prepared, knowledgeable and ready to go!

1) My final exam is:

   Day_____ Date_____ Time_____ Location_____

2) My final exam will cover the material contained in:

   Whole book_____   or   Chapters ___to___   or   Other_____

3) Thinking about the types of questions, number of questions, time limit, etc, I know the format of the final will be:

4) The final exam affects my course grade. The amount the final exam counts toward my course grade is:

Get organized: By getting organized ahead of time, I will reduce stress and feel more confident on the test day!

5) I need to bring the following materials to the final:

   _____Student ID   _____Pencils
   _____Eraser   _____Calculator
   _____Scantron   _____Blue Book
   _____Review assignments that must be turned in
   _____Anything else due the day of the final exam: _____.
   _____Other:_____

Strategies for Success

Final Exam Prep Part II　　　　　　　　Name_____
Make a Study Plan

Most students say they will "study" for the final, but what exactly does that mean? You need a specific study plan to review the material covered and to ensure your success. It has been shown that cycling through the material 3 times reinforces learning.

Three strategies that will help you review the course material are:

1. **Review each Chapter**
   o Chapter Summary – read it carefully. If there are words or concepts you don't remember, you should reread the referenced sections.
   o Chapter Review - do all the odd problems on paper and if you need help, go to the referenced sections for more practice.
   o Chapter Practice Test - do every problem to insure mastery

2. **Redo your old Tests** - put pencil to paper and redo every problem. Then check your answers with your test corrections.

3. **Cumulative Reviews** – do at least every fourth problems in each review. This will be most like the final exam and the time you spend here will pay off!

1) Create and order your own Study Plan. Use a separate sheet of paper. You may want to follow this example:

1. Review each chapter

_____Chapter 1
   o Review the Chapter Summary
   o Chapter Review
   o Chapter Practice Test
_____Chapter 2
   o Review the Chapter Summary
   o Chapter Review
   o Chapter Practice Test
_____Chapter 3
   o Review the Chapter Summary
   o Chapter Review
   o Chapter Practice Test
_____Chapter 4
   o Review the Chapter Summary
   o Chapter Review
   o Chapter Practice Test
   etc.

2. Redo your class tests

- _____ Redo Test 1
- _____ Redo Test 2
- _____ Redo Test 3
- _____ Redo Test 4

etc.

3. Do Cumulative Reviews in the textbook

- _____Do Cumulative Review on Chapters 1-2
- _____Do Cumulative Review on Chapters 1-3
- _____Do Cumulative Review on Chapters 1-4

etc.

2) Obviously, you cannot do all that preparation the night before the final! Look at your study plan and decide how many weeks before the final you need to start to get it all done.

a) I will start my Final Exam preparation on_____ (date) which is _____ weeks before the final exam.

b) My first step in my final exam preparation will be to:

c) After I have completed that, then my next step will be to:

Strategies for Success

Final Exam Prep Part III   **Name**_____
Make a Time Management Plan

Success on a final exam involves more than just showing up the day of the test. Laying out a time management plan will help you get organized and make sure you allow enough time for preparation. Once you see everything you need to do to prepare for the exam, you will know what adjustments you may need to make in order to have enough time to study.

1) Make a time management plan a few weeks before the final exam to show what you need to do when.

    Create a chart like the one shown below and include:
- Classes
- Major assignments that are due
- Final exams
- Work schedule
- Blocks of time to prepare for finals
  - Show when you will study each chapter
  - Show preparation time for math as well as all your other classes
- All other scheduled activities

	Monday	Tuesday	Wednesday	Thursday	Friday	Saturday	Sunday
2 weeks before final							
1 week before final							
Week of final							

2) Now that you have your time management plan, look at it carefully. You may need to make some adjustments to your usual schedule in order to have enough time to prepare for your final exam. The table below lists some strategies students use to find more study time. For each strategy, check the response that applies to you.

In order to have enough time to prepare for my final exams, …	Yes, this would help!	No, this won't work for me.	Does not apply to me
a) I will adjust my weekly schedule so I will have more time to study and prepare for my final exam.			
b) I will adjust my work schedule for the next few weeks, so I will work fewer hours.			
c) I will ask my family to cover my usual chores. (I may need to promise I'll do more after the term ends!)			
d) I will prepare my household to run as smoothly as possible while I concentrate on my studies.			
e) I will postpone family celebrations and holiday preparations until the finals are over. Or I will prepare really early—I don't want to add to final exam stress.			
f) I will restrict my social obligations until after finals are finished.			
g) I will postpone all non-essential personal appointments until after exams.			
h) I will modify my child care arrangements for the next few weeks to give me more time to prepare.			
i) I plan to celebrate afterwards!			
j) I will _____			

Strategies for Success

Grade Check Up

Name_____

You are quickly approaching the end of this class! Do you know what your course grade is right now? What score will you need to earn on the final exam in order to pass the class?

1) Looking at my course syllabus,
    a) the grading scale for this course is:

        A=_____ B=_____ C=_____ D=_____ F=_____

    b) Are these percents or cumulative points?

2) My goal for this class is to earn a(n) _____A _____B _____C.

    a) To achieve this goal, I need _____ points/percent.

    b) Right now, I have _____ points/percent in this class.

    c) Fill in the chart below.

      Points needed:

      Minus points I have now:

      Equals points needed to achieve my goal:

    d) Therefore, I need _____ points/percent more in order to achieve my goal for a final grade.

3) The final exam contributes _____ points/percent to my course grade.

4) a) I still have the following assignments to turn in:

    b) They are worth _____ points/percent towards my course grade.

5) Reality Check: is my goal attainable? _____ yes _____no If not, what is a more realistic goal at this point?

6) Three things I will do to make sure I do well enough on the final exam to help achieve my goal are:

1.

2.

3.

Strategies for Success

Look Back, Look Forward   Name_____

As you finish up this term, it is good to take some time to reflect back on the study skills you used. At the end of the year many people reflect on their past behavior and make resolutions for the new year. Similarly, now is a good time for you to identify what study skills worked for you this past term and resolve to continue your successful strategies and adopt some new good behaviors in your next term in college.

1) List 3 study skills that helped you to succeed this term:

     1.

     2.

     3.

2) Were there any of your study skills this term that could use improvement? List them here:

3) For each study skill you listed in #2, identify how you can change to be more successful in your next math class:

Strategies for Success

Reward Yourself!   **Name**_____

Congratulations! You have nearly achieved your goal of completing this class successfully! It's now time to savor your accomplishment and appreciate what you have achieved.

Reflect on the following.

1) The most important thing I have learned in this class is…

> (This may or may not be a math concept.)

2) I am most proud of…

3) Fill in this table:

The hardest thing I had to overcome was…	To overcome this barrier I ……

Celebrate your accomplishment! You sacrificed a lot to achieve your goal of completing this class. Think of a specific thing you can do to reward yourself after the final exam. It doesn't have to be costly - just something you give to yourself to acknowledge your success. Be creative and ask yourself what would make you feel special and honored. Take time to reward your success!

4) To reward my accomplishment I will:

Taken from: *Intermediate Algebra,* Third Edition,
by Michael Sullivan III and Katherine R. Struve

R Real Numbers and Algebraic Expressions

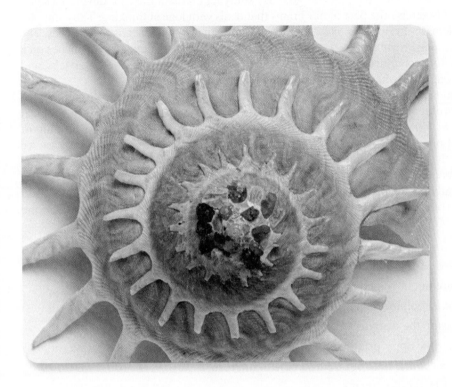

The image to the left of a sunburst carrier shell fossil (*Stellaria solaris*) demonstrates how a specific sequence of numbers, called the Fibonacci sequence, occurs frequently in nature. Problem 142 in Section R.3 explores the Fibonacci sequence further.

The Big Picture: Putting It Together

As the "R" in the title implies, this chapter is a review. The purpose of this chapter is to help you recall mathematical concepts that you learned in earlier courses. The topics in this chapter are important building blocks that will help you succeed in this course.

 Your instructor may or may not decide to cover this chapter, depending on the course syllabus. Regardless, as you proceed through the book, references will be made to Chapter R so that you can use it as a "just-in-time" review.

R.1 Success in Mathematics

Objectives

1. What to Do the First Week of the Semester
2. What to Do Before, During, and After Class
3. How to Use the Text Effectively
4. How to Prepare for an Exam

Let's start by having a discussion about the big picture goals of the course and how this text can help you be successful at mathematics. Our first "big picture" goal is to develop algebraic skills and gain an appreciation for the power of algebra and mathematics. But there is also a second "big picture" goal. By studying mathematics, we develop a sense of logic and exercise the part of our brains that deals with logical thinking. The examples and problems in this text are like the crunches we do in a gym to exercise our bodies. The goal of running or walking is to get from point A to point B, so doing fifty crunches on a mat does not accomplish that goal, but crunches do make our upper bodies, backs, and hearts stronger for when we need to run or walk.

Logical thinking can assist us in solving difficult everyday problems, and solving algebra problems "builds the muscles" in the part of our brain that performs logical thinking. So, when you are studying algebra and getting frustrated with the amount of work that needs to be done, and you say, "My brain hurts," remember the phrase we all use in the gym, "No pain, no gain."

Another phrase to keep in mind is "Success breeds success." Mathematics is everywhere. You already are successful at doing some everyday mathematics. With practice, you can take your initial successes and become even more successful. Have you ever done any of the following everyday activities?

- Compare the price per ounce of different sizes of jars of peanut butter or jam.
- Leave a tip at a restaurant.
- Figure out how many calories your bowl of breakfast cereal provides.
- Take an opinion survey along with many other people.
- Measure the distances between cities as you plan a vacation.
- Order the appropriate number of gallons of paint to cover the walls of a room.
- Buy a car and take out a car loan with interest.
- Double a cookie recipe.
- Exchange American dollars for Canadian dollars.
- Fill up a basketball or soccer ball with air (balls are spheres, after all).
- Coach a Little League team (scores, statistics, catching, and throwing all involve math).
- Check the percentages of saturated and unsaturated fats in a chocolate bar.

You may do five or ten mathematical activities in a single day! The everyday mathematics that you already know is the foundation for your success in this course.

1 What to Do the First Week of the Semester

The first week of the semester gives you the opportunity to prepare for a successful course. Here are the things you should do:

1. **Pick a good seat.** Choose a seat that gives you a good view of the room. Sit close enough to the front so you can easily see the board and hear the professor.

2. **Read the syllabus to learn about your instructor and the course.** Take note of your instructor's name, office location, e-mail address, telephone number, and office hours. Pay attention to any additional help such as tutoring centers, videos in the library, software, and online tutorials. Be sure you fully understand all of the instructor's policies for the class, including those on absences, missed exams or quizzes, and homework. Ask questions.

3. **Learn the names of some of your classmates and exchange contact information.** One of the best ways to learn math is through group study sessions. Try to create time each week to study with your classmates. Knowing how to get in contact with classmates is also useful, because you can obtain the assignment for the day if you ever miss class.

4. Budget your time. Most students have a tendency to "bite off more than they can chew." To help with time management, consider the following general rule: Plan on studying *at least* two hours outside of class for each hour in class. So, if you enrolled in a four-hour math class, you should set aside at least eight hours each week to study for the course. You will also need to set aside time for other courses. Consider your work schedule and personal life when creating your time budget.

② What to Do Before, During, and After Class

Now that the semester is under way, we present the following ideas for what to do before, during, and after each class meeting. These suggestions may sound overwhelming, but we guarantee that by following them, you will be successful in mathematics (and other courses). Also, you will find that following this plan will make studying for exams much easier.

Before Class Begins

1. Read the section or sections that will be covered in the upcoming class meeting. Watch the video lectures that accompany the text.

2. Based on your reading, write down a list of questions. Your questions will probably be answered through the lecture, but if not, you can then ask any questions that are not answered completely.

3. Arrive early and make sure you are mentally prepared for class. Your mind should be alert and ready to concentrate for the entire class. (Invest in a cup of coffee, and eat lots of protein for breakfast!)

During Class

1. Stay alert. Do not doze off or daydream during class. Understanding the lecture will be very difficult when you "return to class."

2. Take thorough notes. It is normal not to understand certain topics the first time you hear them in a lecture. However, this does not mean you should throw your hands up in despair. Rather, continue to take class notes.

3. Do not be afraid to ask questions. In fact, instructors love questions, for two reasons. First, if one student has a question, other students are likely to have the same question. Second, by asking questions, you teach the teacher what topics cause difficulty.

After Class

1. Reread (and possibly rewrite) your class notes. In our experience as students, we were amazed how often our confusion during class disappeared after studying our in-class notes after class.

2. Reread the section. This is an especially important step. Once you have heard the lecture, the section will make more sense and you will understand much more.

3. Do your homework. **Homework is not optional.** There is an old Chinese proverb that says,

> I hear ... and I forget
> I see ... and I remember
> I do ... and I understand

This proverb applies to any situation in life in which you want to succeed. Would a pianist expect to be the best if she didn't practice? The only way you are going to learn algebra is by doing algebra.

4. When you get a problem wrong, try to figure out why you got the problem wrong. If you can't discover your error, be sure to ask for help.

5. If you have questions, visit your professor during office hours. You can also ask someone in your study group or go to the tutoring center on campus, if available.

Learning Is a Building Process

Learning is the art of making connections among thousands of neurons (specialized cells) in the brain. Memory is the ability to reactivate these neural networks—it is a conversation among neurons.

Math isn't a mystery. You already know some math. But you do have to practice what you know and expand your knowledge. Why? The brain contains thousands of neurons. Through repeated practice, signals in the brain travel faster. The cells "fire" more quickly, and connections are made faster and with less effort. Practice allows us to retrieve concepts and facts at test time. Remember those crunches, which are a way of making your body more robust and nimble—learning does the same to your brain.

Have We Mentioned Asking Questions?

To move information from short-term memory to long-term memory, we need to think about the information, comprehend its meaning, and ask questions about it.

❸ How to Use the Text Effectively

When we sat down to write this text, we knew from our teaching experience that students typically do not read their math books. We decided to accept the steps students usually go through:

1. Attend the lecture and watch the instructor do some problems on the board. Perhaps work some problems in class.
2. Go home and work on the homework assignment.
3. After each problem, check the answer in the back of the text. If right, move on, but if wrong, go back and see where the solution went wrong.
4. If the mistake cannot be identified, go to the class notes or try to find a similar example in the text. With a little luck, the student can determine where the solution went wrong in the problem.
5. If Step 4 fails, mark the problem and ask about it in the next class meeting, which leads us back to Step 1.

With this model in mind, we started to develop this text so that there is more than one way to extract the information you need from it.

All of the features in the text are there to help you succeed. These features are based on techniques we use in class. The paragraphs that follow outline the features that appear, an explanation of the purpose of each feature, and how each can be used to help you succeed in this course.

Are You Ready for This Section? Warming Up

Each section, beginning with Section 1.1, starts with a short "readiness quiz." This quiz asks questions about material that was presented earlier in the course and is needed for the upcoming section. Take the readiness quiz to be sure you understand the material that the new section is based on. Answers to the quiz appear as footnotes on the page where the quiz appears. Check your answers. If you get a problem wrong, or if you don't know how to do a problem, go back to the section listed and review the material.

Objectives: A "Road Map" Through the Course

To the left of the readiness quiz is a list of objectives to be covered in the section. If you follow the objectives, you will get a good idea of the section's "big picture"—the important concepts, techniques, and procedures that it introduces.

The objectives are numbered. (See the numbered headline at the beginning of this section.) When we begin discussing a particular objective within the section, the objective number appears along with the stated objective.

Examples: Where to Look for Information

Examples are meant to provide you with guidance and instruction when you are away from the instructor and the classroom. With this in mind, we have developed two special example formats.

Step-by-Step Examples have a three-column format. The left column describes a step, the middle column briefly explains the step, and the right column presents the algebra. Thus, the left and middle columns can be thought of as your instructor's voice during a lecture. *Step-by-Step Examples* introduce key topics or important problem-solving strategies. They provide easy-to-understand, practical instructions by including the words "how to" in the examples' title.

Annotated Examples have a two-column format with explanations to the left of the algebra. The explanation clearly describes what operations we are about to perform and in the order in which we will do it. Again, annotations are like your instructor's voice as he or she writes each step of the solution on the board.

Authors in Action: Lecture Videos to Help You Learn

Every objective has one or more classroom lecture video of the authors teaching their students marked by a ▶ icon. These "live" classroom lectures can be used to supplement your instructor's presentations and your reading of the text. They can be found in the Multimedia Library of MyMathLab or on a DVD video series.

In Words: Math in Everyday Language

Have you ever been given a math definition in class and said, "What in the world does that mean?" We have heard that from our students. So we added the "In Words" feature, which restates mathematical definitions in everyday language. This margin feature will help you understand the language of mathematics better.

Work Smart

These "tricks of the trade" that appear in the margin can help you solve problems. They also show alternative problem-solving approaches. There is more than one way to solve a math problem!

Work Smart: Study Skills

These margin notes highlight the study skills required for success in this and other mathematics courses.

Exercises: A Unique Numbering Scheme

As teachers, we know that students typically jump right to the exercises after attending class. This means they tend to skip all of the examples and explanations of concepts in the section. To help you use the book most effectively to learn the math, we have structured the exercises differently from other texts you have used. Our structure is designed to encourage the reading of the book while boosting your confidence and ability to work any mathematical problem. Thus the exercises in each section are broken into as many as seven parts. Each exercise set will have some, or all, of the following exercise types.

1. Quick Checks
2. Building Skills
3. Mixed Practice
4. Applying the Concepts
5. Extending the Concepts
6. Explaining the Concepts
7. Synthesis Review
8. The Graphing Calculator

1. **Quick Checks: Learning to Ride a Bicycle with Training Wheels** Do you remember when you were first learning to ride a bicycle? Training wheels were placed on the bicycle to assist you in learning balance. The Quick Checks are like exercises with training wheels. These exercises appear right after the example or examples that illustrate the concept being taught. So if you get stuck on a Quick Check problem, you can simply consult the example immediately preceding it, rather than searching through the text. The Quick Check exercises also verify your understanding of new vocabulary. See Quick ✓ on page 19 in Section R.3.

2. **Building Skills: Learning to Ride a Bicycle with Assistance** Once you felt ready to ride without training wheels, you probably had an adult follow closely behind you, holding the bicycle for balance and building your confidence. The Building Skills problems serve a similar purpose. They are linked to the objectives within the section, so the directions for the problem indicate which objective is being developed. As a result, you know exactly which objective (but not exactly which example) to consult if you get stuck. See page 31 in Section R.3.

3. **Mixed Practice: Now You Are Ready to Ride!** After mastering training wheels and learning to balance with assistance, you are ready to ride alone. This stage corresponds to the Mixed Practice exercises. These exercises include problems that develop your ability to see the big picture of mathematics. They are not linked to a particular objective, and they require you to determine the appropriate approach to solving a problem on your own. See page 31 in Section R.3.

4. **Applying the Concepts: Where Will I Ever Use This Stuff?** The Applying the Concepts exercises not only illustrate the application of mathematics in your life, but also provide problems that test your conceptual understanding of the mathematics. See page 32 in Section R.3.

5. **Extending the Concepts: Stretching Your Mind** Sometimes we need to be challenged. These exercises extend your skills to a new level and provide further insight into where mathematics can be used. See page 32 in Section R.3.

6. **Explaining the Concepts: Verbalize Your Understanding** These problems require you to express the section's big-picture concepts in your own words. Students need to improve their ability to communicate complicated ideas (both oral and written). If you truly understand the material in the section, you should be able to articulate the concepts clearly. See page 32 in Section R.3.

7. **Synthesis Review: Seeing the Forest for the Trees** The Synthesis Review exercises, which begin in Chapter 5, can help you grasp the "big picture" of algebra. These exercises ask you to perform a single operation (adding, solving, and so on) on several objects (polynomials, rational expressions, and so on). You are then asked to discuss the similarities and differences in performing the same operation on the different objects. See page 409 in Section 5.1.

8. **The Graphing Calculator** The graphing calculator is a great tool for verifying answers and for helping to visualize results. These exercises illustrate how the graphing calculator can be incorporated into the material of the section. See page 33 in Section R.3.

Chapter Review

The chapter review is arranged by section. For each section, we state key concepts, key terms, and objectives. We also list the examples, page numbers from the text that illustrate each objective, and the review exercises that assess your understanding of each objective. If you get a problem wrong, use this feature to determine where to look in the book to help you work the problem.

Chapter Test

We have included a chapter test. Once you think you are prepared for the exam, take the chapter test. If you do well on the chapter test, chances are you will do well on your in-class exam. Be sure to take the test under the same conditions that you will face in class. If you are unsure how to solve a problem in the chapter test, watch the Chapter Test Prep Videos, available in MyMathLab and the lecture video DVD, which shows an instructor solving each chapter test problem.

Cumulative Review: Reinforcing Your Knowledge

The building process of learning algebra involves a lot of reinforcement. For this reason, we provide cumulative reviews at the end of every odd-numbered chapter starting with Chapter 1. Do these cumulative reviews after each chapter test, so that you are always refreshing your memory—making those neurons do their calisthenics. This way, studying for the final exam should be fairly easy.

④ How to Prepare for an Exam

The following steps are time-tested suggestions to help you prepare for an exam.

Step 1: **Revisit your homework and the chapter review problems** About one week before your exam, start to redo your homework assignments. If you don't understand a topic, seek out help. Work the problems in the chapter review as well. The problems are linked to the section objectives. If you get a problem wrong, identify the objective and the examples that illustrate the objective. Then review this material and try the problem in the chapter review again. If you still get the problem wrong, seek out help.

Step 2: **Test yourself** A day or two before the exam, take the chapter test under test conditions. Be sure to check your answers. If you get any problems wrong, determine why and remedy the situation.

Step 3: **View the Chapter Test Prep Videos** These videos show step-by-step solutions to the problems found in each of the book's chapter tests. Follow the worked-out solutions to any of the exercises on the chapter test that you want to study or review.

Work Smart: Study Skills

Do not "cram" for an exam by pulling an "all-nighter."

Step 4: **Follow these rules as you train** Be sure to arrive early at the location of the exam. Prepare your mind for the exam. Be sure you are well rested. Don't try to pull "all-nighters." If you need to study all night for an exam, then your time management is poor, and you should rethink how you are using your time or whether you have allocated enough time to the course.

R.1 Exercises

1. Why do you want to be successful in mathematics? Are your goals positive or negative? If you stated your goal negatively ("Just get me out of this course!"), can you restate it positively?

2. Name three activities in your daily life that involve the use of math (for instance, playing cards, operating your computer, or reading a credit card bill).

3. What is your instructor's name?

4. What are your instructor's office hours? Where is your instructor's office?

5. What is your instructor's e-mail address?

6. Does your class have a website? Do you know how to access it? What information is located on the website?

7. Are there tutors available for this course? If so, where are they located? When are they available?

8. Name two other students in your class. What is their contact information? When can you meet with them to study?

9. List some of the things that you should do before class begins.

10. List some of the things you should do during class.

11. List some of the things you should do after class.

12. What is the point of the Chinese proverb on page 3?

13. What is the "readiness quiz"? How should it be used?

14. Name three features that appear in the margins. What is the purpose of each of them?

15. Name the categories of exercises that appear in this book.

16. How should the chapter review material be used?

17. How should the chapter test be used?

18. What are the Chapter Test Prep Videos?

19. List the four steps that should be followed when preparing for an exam. Can you think of other methods of preparing for an exam that have worked for you?

20. How is mathematics like doing crunches at the gym?

21. Use the chart to help manage your time. Be sure to fill in the time allocated to various activities in your life, including school, work, and leisure.

	Monday	Tuesday	Wednesday	Thursday	Friday	Saturday	Sunday
7 am							
8 am							
9 am							
10 am							
11 am							
Noon							
1 pm							
2 pm							
3 pm							
4 pm							
5 pm							
6 pm							
7 pm							
8 pm							
9 pm							

R.2 Sets and Classification of Numbers

Objectives

1. Use Set Notation
2. Classify Numbers
3. Approximate Decimals by Rounding or Truncating
4. Plot Points on the Real Number Line
5. Use Inequalities to Order Real Numbers

1 Use Set Notation

A **set** is a well-defined collection of objects. "Well-defined" means that there is a rule for determining whether a given object is in the set. For example, the students enrolled in Intermediate Algebra at your college is a set. The collection of numbers 0, 1, 2, 3, 4, 5, 6, 7, 8, and 9 may also be identified as a set. If we let D represent this set of numbers, then we can write

$$D = \{0, 1, 2, 3, 4, 5, 6, 7, 8, 9\}$$

This notation uses braces $\{\ \}$ to enclose the objects, or **elements,** in the set. This method of representing a set is called the **roster method.**

EXAMPLE 1   **Using the Roster Method**

Write the set that represents the vowels.

Solution

The vowels are a, e, i, o, and u, so we write

$$V = \{a, e, i, o, u\}$$

Another way to denote a set is to use **set-builder notation.** The numbers in the set $D = \{0, 1, 2, 3, 4, 5, 6, 7, 8, 9\}$ are called digits. Using set-builder notation, the set D of digits can be written as

$$D = \{x \mid x \text{ is a digit}\}$$

In algebra, we use letters such as x, y, a, b, and c to represent numbers. When the letter can be any number in a set of numbers, it is called a **variable.** In the set D, the letter x can represent any digit, so x is a variable that can take on the value 0, 1, 2, 3, 4, 5, 6, 7, 8, or 9.

EXAMPLE 2 **Using Set-Builder Notation**

Use set-builder notation to represent the following sets.

(a) The set of all even digits

(b) The set of all odd digits

Solution

(a) We will let E represent the set of all even digits, so that

$$E = \{x \mid x \text{ is an even digit}\}$$

(b) We will let O represent the set of all odd digits, so that

$$O = \{x \mid x \text{ is an odd digit}\}$$

Quick ✓

1. A ___ is a well-defined collection of objects.

2. The objects in a set are called _____.

In Problems 3 and 4, use set-builder notation and the roster method to represent each set.

3. The set of all digits less than 5

4. The set of all digits greater than or equal to 6

When we list the elements in a set, we never list an element more than once. For example, we do not write $\{1, 2, 3, 2\}$; we write $\{1, 2, 3\}$. Also, the order in which the elements are listed does not matter. For example, $\{2, 3\}$ and $\{3, 2\}$ represent the same set.

We now introduce more notation for describing sets.

Set Notation

- If two sets A and B have the same elements, then we say that A **equals** B and write $A = B$.

- If every element of a set A is also an element of a set B, then we say that A is a **subset** of B and write $A \subseteq B$.

- If $A \subseteq B$ and $A \neq B$, then we say that A is a **proper subset** of B and write $A \subset B$. Put another way, A is a proper subset of B if all elements in A are also in B and there are elements in B that are not in A.

- If a set A has no elements, it is called the **empty set,** or **null set,** and is denoted by the symbol $\varnothing$ or $\{\ \}$. The empty set is a subset of every set; that is $\varnothing \subseteq A$ for any set A.

When working with sets, we usually designate a **universal set,** which is the set of all elements of interest to us. For instance, in Example 2, we were interested in the set of all digits, so the universal set is the set of all digits.

It is often helpful to draw pictures of sets because the pictures help us visualize relations among sets. We call pictures of sets **Venn diagrams,** in honor of John Venn (1834–1923). In Venn diagrams we represent sets as circles enclosed in a rectangle. The rectangle represents the universal set. For example, if $A = \{1, 2\}$ $B = \{1, 2, 3, 4, 5\}$, and the universal set is $U = \{0, 1, 2, 3, 4, 5, 6, 7, 8, 9\}$, then $A \subset B$. Figure 1 illustrates the relation between sets A and B in a Venn diagram.

Figure 1
Venn diagram with $A \subset B$.

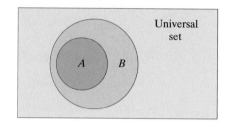

EXAMPLE 3

Using Set Notation

Let $A = \{0, 1, 2, 3, 4, 5\}$, $B = \{3, 4, 5\}$, $C = \{5, 4, 3\}$, and $D = \{3, 4, 5, 6\}$. Write True or False for each statement.

(a) $B \subseteq A$ **(b)** $D \subseteq A$ **(c)** $B = C$

(d) $C = D$ **(e)** $B \subset C$ **(f)** $\varnothing \subseteq C$

Solution

(a) The statement $B \subseteq A$ is True because all the elements that are in B are also elements in A.

(b) The statement $D \subseteq A$ is False because there is an element that is in D, 6, that is not in A.

(c) The statement $B = C$ is True because sets B and C have the same elements.

(d) The statement $C = D$ is False because there is an element in D, 6, that is not in C.

(e) In order for B to be a proper subset of C, it must be the case that all the elements in B are also elements in C. In addition, there must be at least one element in C that is not in B. Because $B = C$, the statement $B \subset C$ is False. Note, however, that $B \subseteq C$ is True and $B \subset D$ is also True.

(f) The statement $\varnothing \subseteq C$ is True because the empty set is a subset of every set. ●

Quick ✓

5. *True or False* The order in which elements are listed in a set does not matter.

6. If every element of a set A is also an element of a set B, then we say that A is a _____ of B and write $A \subseteq B$.

7. *True or False* If a set has no elements, it is called the empty set and is denoted $\{\varnothing\}$.

In Problems 8–11, let $A = \{a, b, c, d, e, f, g\}$, $B = \{a, b, c\}$, $C = \{c, d\}$, and $D = \{c, b, a\}$. Write True or False for each statement. Be sure to justify your answer.

8. $B \subseteq A$ **9.** $B = C$ **10.** $B \subset D$ **11.** $\varnothing \subseteq A$

We use the symbol $\in$ (which is read "is an element of") to denote that a particular element is in a set. For example, $7 \in \{1, 3, 5, 7, 9\}$ means "7 is an element of the set $\{1, 3, 5, 7, 9\}$." If an element is not in a set, we use the symbol $\notin$ (which is read "is not an element of"). For example, we write "b is not a vowel" as $b \notin \{a, e, i, o, u\}$.

EXAMPLE 4 **Using Set Notation**

Write True or False for each statement.

(a) $3 \in \{x \mid x \text{ is a digit}\}$

(b) $h \notin \{a, e, i, o, u\}$

(c) $\frac{1}{2} \in \left\{x \mid x = \frac{p}{q}, \text{where } p \text{ and } q \text{ are digits}, q \neq 0\right\}$

Solution

(a) The statement $3 \in \{x \mid x \text{ is a digit}\}$ is True because 3 is a digit.

(b) The statement $h \notin \{a, e, i, o, u\}$ is True because h is not an element of the set $\{a, e, i, o, u\}$.

Work Smart

Using correct notation is important:
$3 \subset \{1, 2, 3\}$ is incorrect.
$3 \in \{1, 2, 3\}$ is correct.

(c) The statement $\frac{1}{2} \in \left\{x \mid x = \frac{p}{q}, \text{where } p \text{ and } q \text{ are digits}, q \neq 0\right\}$ is True because $\frac{1}{2}$ is of the form $\frac{p}{q}$, where $p = 1$ and $q = 2$. •

Quick ✓

In Problems 12–14, answer True or False for each statement.

12. $5 \in \{0, 1, 2, 3, 4, 5\}$

13. Michigan $\notin$ {Illinois, Indiana, Michigan, Wisconsin}

14. $\frac{8}{3} \in \left\{x \mid x = \frac{p}{q}, \text{where } p \text{ and } q \text{ are digits}, q \neq 0\right\}$

Set Notation

$A = B$	means	all the elements in set A are also elements in set B, and all the elements in set B are also elements in set A.
$A \subseteq B$	means	all the elements in set A are also elements in set B.
$A \subset B$	means	all the elements in set A are also elements in set B, but there is at least one element in set B that is not in set A.
$\varnothing$ or $\{ \ \}$	means	the empty set. The set has no elements.
$5 \in A$	means	5 is an element in set A.
$5 \notin A$	means	5 is not an element in set A.

▷ ❷ **Classify Numbers**

We discussed sets because it is helpful to classify the various kinds of numbers that we deal with as sets.

Definition

The **natural numbers,** or **counting numbers,** are the numbers in the set
$\mathbb{N} = \{1, 2, 3, \dots\}$.

The three dots in the definition above, called an *ellipsis,* indicate that the pattern continues indefinitely. The counting numbers are often used to count things. For example, we can count the cars that arrive at a McDonald's drive-through, or we can count the letters in the alphabet.

Suppose you had $7 and purchased a drink and a hot dog at a baseball game for $7. Can we use the counting numbers to describe the amount of money you have left? No! We need a new number system to describe the remaining amount. We need the *whole number system.*

> **Definition**
>
> The **whole numbers** are the numbers in the set $W = \{0, 1, 2, 3, \dots\}$.

We can see that the whole numbers consist of the set of counting numbers together with the number 0, so $\mathbb{N} \subset W$.

Now suppose you have a balance of $100 in your checking account and you write a check for $120. Can the whole numbers be used to describe your new balance? No! You need the *integers*.

> **Definition**
>
> The **integers** are the numbers in the set $\mathbb{Z} = \{\dots, -3, -2, -1, 0, 1, 2, 3, \dots\}$.

In Words

The set of natural numbers is a subset of the set of whole numbers. The set of whole numbers is a subset of the set of integers.

Notice that the set of natural numbers is a proper subset of the set of whole numbers, and the set of whole numbers is a proper subset of the set of integers. That is, $\mathbb{N} \subset W$ and $W \subset \mathbb{Z}$. As we expand a number system, we are able to discuss new, and usually more complicated, problems. For example, the whole numbers allow us to discuss the absence of something because they include zero, but the counting numbers do not. The integers allow us to deal with problems involving both negative and positive quantities, such as profit (positive counting numbers) and loss (negative counting numbers).

To represent a portion of a dollar or a portion of a whole pie, we enlarge our number system to include *rational numbers*.

In Words

A rational number is a number that can be expressed as a fraction, where the numerator is any integer and the denominator is a nonzero integer.

> **Definition**
>
> A **rational number** is a number that can be expressed as a quotient $\dfrac{p}{q}$ of two integers. The integer p is the **numerator,** and the integer q, which cannot be 0, is the **denominator.** The set of rational numbers are the numbers
>
> $$\mathbb{Q} = \left\{ x \,\middle|\, x = \frac{p}{q}, \text{ where } p \text{ and } q \text{ are integers, } q \neq 0 \right\}.$$

Examples of rational numbers are $\dfrac{3}{4}, \dfrac{4}{3}, \dfrac{0}{6}, -\dfrac{4}{5}$, and $\dfrac{33}{5}$. Because $\dfrac{p}{1} = p$ for any integer p, it follows that the set of integers is a proper subset of the set of rational numbers ($\mathbb{Z} \subset \mathbb{Q}$). For example, 5 is a rational number because it can be written as $\dfrac{5}{1}$, but more specifically, it is an integer. More specifically than that, it is a counting number.

We can also represent rational numbers as *decimals*. The **decimal** representation of a rational number is found by carrying out the division indicated. For example,

Work Smart

To write $\dfrac{2}{11}$ as a decimal, write $11\overline{)2}$ and carry out the division as follows:

```
      0.181
11)2.000
     11
     ──
      90
      88
      ──
       20
       11
       ──
        9
```

and so on.

$$\frac{4}{5} = 0.8 \qquad \frac{7}{2} = 3.5 \qquad -\frac{2}{3} = -0.6666\dots = -0.\overline{6} \qquad \frac{2}{11} = 0.181818\dots = 0.\overline{18}$$

Notice the line above the 6 in $-0.\overline{6}$. This **repeat bar** represents the fact that the pattern continues. Similarly, $0.\overline{18}$ means the block of numbers 18 will continue indefinitely to the right of the decimal point.

Every rational number may be represented by a decimal that either **terminates** (as in the case of $\dfrac{4}{5} = 0.8$ and $\dfrac{7}{2} = 3.5$) or is **nonterminating** with a repeating block of decimals (as in the case of $-\dfrac{2}{3} = -0.\overline{6}$ and $\dfrac{2}{11} = 0.\overline{18}$).

Work Smart

Although it is true that $\sqrt{2}$ is an irrational number, not all numbers involving the $\sqrt{}$ symbol are irrational. For example, $\sqrt{4} = 2$, a positive integer.

What if a decimal neither terminates nor has a block of digits that repeat? These decimals represent a set of numbers called *irrational numbers*. Every **irrational number** may be represented by a decimal that neither terminates nor repeats. This means that irrational numbers cannot be written in the form $\dfrac{p}{q}$, where p and q are integers and $q \neq 0$. An example of an irrational number is π, whose value is approximately 3.14159265359. Another example of an irrational number is $\sqrt{2}$. This is the number whose square is 2. Its value is approximately 1.414213562.

Definition

Together, the set of rational numbers and the set of irrational numbers form the set of **real numbers.** The set of real numbers is denoted using the symbol $\mathbb{R}$.

Figure 2 shows the relationships among the various types of numbers. The universal set in Figure 2 is the set of all real numbers. It is extremely important to be able to distinguish among the various number systems—in particular, it is important to know the difference between a rational number and an irrational number. Notice that if a number is rational, it cannot be irrational, and vice versa.

Figure 2

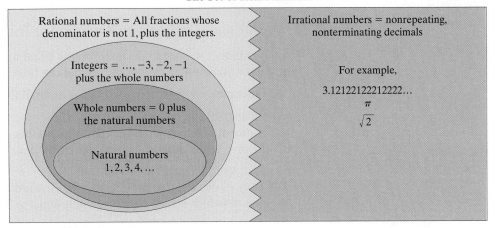

The Set of Real Numbers

EXAMPLE 5   **Classifying Numbers in a Set**

List the numbers in the set

$$\left\{ 3, -\frac{3}{5}, -1, 0, \sqrt{2}, 5.\overline{94}, 4.2122122212222\ldots, -\frac{8}{2}, \frac{\pi}{2} \right\}$$

that are

(a) Natural numbers   (b) Whole numbers   (c) Integers

(d) Rational numbers   (e) Irrational numbers   (f) Real numbers

Solution

(a) 3 is the only natural number.

(b) 0 and 3 are the whole numbers.

(c) $3, -1, 0,$ and $-\dfrac{8}{2}$ are the integers $\left(-\dfrac{8}{2} \text{ is an integer because it simplifies to } -4 \right)$.

(d) $3, -\dfrac{3}{5}, -1, 0, 5.\overline{94},$ and $-\dfrac{8}{2}$ are the rational numbers.

(e) $\sqrt{2}, 4.2122122212222\ldots,$ and $\dfrac{\pi}{2}$ are the irrational numbers. Note that $4.2122122212222\ldots$ is irrational because it has infinitely many nonrepeating digits.

(f) All of the numbers are real numbers.

Quick ✓

15. The _____ numbers are numbers in the set $\{0, 1, 2, 3, 4, \dots\}$.

16. The numbers in the set $\left\{ x \middle| x = \dfrac{p}{q}, \text{where } p \text{ and } q \text{ are integers, } q \neq 0 \right\}$ are called _____ numbers.

17. The set of _____ numbers have decimal representations that neither terminate nor repeat.

18. *True or False* The rational numbers are a subset of the set of irrational numbers.

19. *True or False* If a number is expressed as a decimal, then it is rational.

20. *True or False* Every rational number is a real number.

21. *True or False* If a number is rational, it cannot be irrational.

In Problems 22–27, list the numbers in the set

$$\left\{ \frac{7}{3}, -9, 10, 4.\overline{56}, 5.7377377737777\dots, \frac{0}{3}, \pi, -\frac{4}{7}, \frac{12}{4}, \sqrt{11} \right\} \text{ that are}$$

22. Natural numbers

23. Whole numbers

24. Integers

25. Rational numbers

26. Irrational numbers

27. Real numbers

▶ ❸ Approximate Decimals by Rounding or Truncating

Every number written in decimal form is a real number that may either be rational or irrational. In addition, every real number can be represented by a decimal. For example, the rational number $\dfrac{3}{4}$ can be written in decimal form as 0.75. The rational number $\dfrac{2}{3}$ is equivalent to $0.666\dots$ or $0.\overline{6}$ as a decimal.

Irrational numbers have decimals that neither terminate nor repeat. The irrational numbers $\sqrt{2}$ and π have decimal representations that begin as follows:

$$\sqrt{2} = 1.414213\dots \qquad \pi = 3.14159\dots$$

We have to use approximations to write irrational numbers as decimals. We use the symbol $\approx$ (which is read "approximately equal to") to do so. For example,

$$\sqrt{2} \approx 1.4142 \qquad \pi \approx 3.1416$$

In approximating decimals, we either *round* or *truncate* to a given number of decimal places. The number of decimal places determines the location of the *final digit* in the decimal approximation.

Definition

Truncation: Drop all the digits that follow the specified final digit in the decimal.
Rounding: Identify the specified final digit in the decimal. If the next digit is 5 or more, add 1 to the final digit; if the next digit is 4 or less, leave the final digit as it is. Then truncate all digits to the right of the final digit.

(EXAMPLE 6) **Approximating a Decimal by Truncating and by Rounding**

Approximate 13.9463 to two decimal places by

(a) Truncating   (b) Rounding

Solution

We want to approximate the decimal to two decimal places, so the final digit is 4:

13.9463

(a) To truncate, remove all digits to the right of the final digit, 4, to get 13.94.

(b) To round to two decimal places, find the digit to the right of the final digit. It is 6. Because 6 is 5 or more, add 1 to the final digit 4 and then truncate everything to the right. We get 13.95.

Quick ✓

In Problems 28 and 29, approximate each number by (a) truncating and (b) rounding to the indicated number of decimal places.

28. 5.694392; three decimal places

29. −4.9369102; two decimal places

The Graphing Calculator: Does Your Calculator Truncate or Round?

Calculators are capable of displaying only a certain number of decimal places. Most scientific calculators display eight digits, and most graphing calculators display ten to twelve digits. When a number has more digits than the calculator can display, the calculator will either round or truncate.

　To see whether your calculator rounds or truncates, divide 2 by 3. How many digits do you see? Is the last digit a 6 or a 7? If it is a 6, your calculator truncates; if it is a 7, your calculator rounds. Figure 3 shows the result on a TI-84 Plus graphing calculator. Does the calculator shown in Figure 3 round or truncate? Because the last digit displayed is a 7, it rounds.

Figure 3

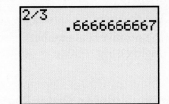

▶ ④ **Plot Points on the Real Number Line**

Figure 4
The real number line.

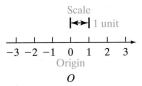

The real numbers can be represented by points on a line called the **real number line.** Every real number corresponds to a point on the line, and each point on the line has a unique real number associated with it.

　To construct a real number line, pick a point on a line somewhere in the center, and label it O. This point, called the **origin,** corresponds to the real number 0. See Figure 4. The point 1 unit to the right of O corresponds to the real number 1. The distance between 0 and 1 determines the **scale** of the number line. For example, the point associated with the number 2 is twice as far from O as 1 is. Notice that an arrowhead on the right end of the line indicates the direction in which the numbers increase. Points to the left of the origin correspond to the real numbers −1, −2, and so on.

In Words
We can think of the real number line as the graph of the set of all real numbers.

Definition

The real number associated with a point P is called the **coordinate** of P. The **real number line** is the set of all points that have been assigned coordinates.

EXAMPLE 7   **Plotting Points on the Real Number Line**

On the real number line, plot the points with coordinates 0, 5, −1, 1.5, $-\dfrac{1}{2}$.

Solution

We draw a real number line with a scale of 1 and then plot the points. See Figure 5.

Figure 5

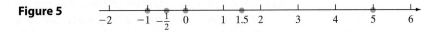

Quick ✓

30. On the real number line, plot the points with coordinates $0, 3, -2, \frac{1}{2}$, and 3.5.

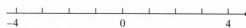

The real number line consists of three classes (or categories) of real numbers, as shown in Figure 6.

Figure 6

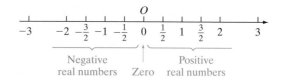

- The **negative real numbers** are the coordinates of points to the left of the origin O.
- The real number **zero** is the coordinate of the origin O.
- The **positive real numbers** are the coordinates of points to the right of the origin O.

Figure 7

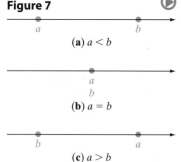

▶ ⑤ Use Inequalities to Order Real Numbers

Given two numbers (points) a and b, a must be to the left of b, the same as b, or to the right of b. See Figure 7.

If a is to the left of b, we say "a is less than b" and write $a < b$. If a is at the same location as b, then we say that "a is equal to b" and write $a = b$. If a is to the right of b, we say "a is greater than b" and write $a > b$. If a is either less than or equal to b, we write $a \leq b$. Similarly, $a \geq b$ means that a is either greater than or equal to b. Collectively, the symbols $<, >, \leq$, and $\geq$ are called **inequality symbols.**

Note that $a < b$ and $b > a$ mean the same thing, so we can write $2 < 3$ (2 is to the left of 3) or $3 > 2$ (3 is to the right of 2).

EXAMPLE 8   **Using Inequality Symbols**

(a) $2 < 5$ because the coordinate 2 lies to the left of the coordinate 5 on the real number line.

(b) $-1 > -3$ because -1 lies to the right of -3 on the real number line.

(c) $3.5 \leq \frac{7}{2}$ because $3.5 = \frac{7}{2}$.

(d) $\frac{5}{6} > \frac{4}{5}$ because $\frac{5}{6} = 0.8\overline{3}$ and $\frac{4}{5} = 0.8$, so $\frac{5}{6}$ lies to the right of the $\frac{4}{5}$ on the real number line. ●

Quick ✓

In Problems 31–36, replace the question mark by $<, >$ *or =, whichever is correct.*

31. $3\ ?\ 6$

32. $-3\ ?\ -2$

33. $\frac{2}{3}\ ?\ \frac{1}{2}$

34. $\frac{5}{7}\ ?\ 0.7$

35. $\frac{2}{3}\ ?\ \frac{10}{15}$

36. $\pi\ ?\ 3.14$

Notice in Example 8 that the inequality symbol always points to the smaller number. Inequalities of the form $a < b$ or $b > a$ are **strict inequalities,** whereas inequalities of the form $a \leq b$ or $b \geq a$ are **nonstrict** (or **weak**) **inequalities.**

Based upon the discussion so far, we conclude that

$a > 0$ is equivalent to a is positive

$a < 0$ is equivalent to a is negative

We read $a > 0$ as "a is positive" or "a is greater than 0." If $a \geq 0$, then either $a > 0$ or $a = 0$, and we may read this as "a is nonnegative" or "a is greater than or equal to 0."

Work Smart: Study Skills

Selected problems in exercise sets are identified by a green color. For extra help, worked solutions to these problems are in MyMathLab or on the book's DVD Lecture Series.

R.2 Exercises   MyMathLab®   Exercise numbers in green have complete video solutions in MyMathLab.

Problems 1–36, are the Quick ✓s that follow the EXAMPLES.

Building Skills

In Problems 37–42, write each set using the roster method. See Objective 1.

37. $\{x \mid x$ is a whole number less than $6\}$

38. $\{x \mid x$ is a natural number less than $4\}$

39. $\{x \mid x$ is an integer between -3 and $5\}$

40. $\{x \mid x$ is an integer between -4 and $6\}$

41. $\{x \mid x$ is a natural number less than $1\}$

42. $\{x \mid x$ is a whole number less than $0\}$

In Problems 43–50, let $A = \{1, 3, 5, 7, 9\}$, $B = \{2, 4, 6, 8\}$, $C = \{1, 2, 3, 4, 5, 6, 7, 8, 9\}$, and $D = \{8, 6, 4, 2\}$. Write True or False for each statement. Then justify your answer. See Objective 1.

43. $B \subseteq C$ 44. $A \subseteq C$ 45. $B \subset D$ 46. $A \subset C$

47. $B = D$ 48. $B \subseteq D$ 49. $\varnothing \subset C$ 50. $\varnothing \subset B$

In Problems 51–54, fill in the blank with the appropriate symbol, $\in$ or $\notin$. See Objective 1.

51. $\dfrac{1}{2}$ ___ $\{x \mid x$ is an integer$\}$

52. $4.\overline{5}$ ___ $\{x \mid x$ is a rational number$\}$

53. π ___ $\{x \mid x$ is a real number$\}$

54. 0 ___ $\{x \mid x$ is a natural number$\}$

In Problems 55–58, list the numbers in each set that are (a) Natural numbers, (b) Integers, (c) Rational numbers, (d) Irrational numbers, (e) Real numbers. See Objective 2.

55. $A = \left\{ -5, 4, \dfrac{4}{3}, -\dfrac{7}{5}, 5.\overline{1}, \pi \right\}$

56. $B = \left\{ 13, 0, -4.5656\ldots, 2.43, \sqrt{2} \right\}$

57. $C = \left\{ 100, -5.423, \dfrac{8}{7}, \sqrt{2} + 4, -64 \right\}$

58. $D = \left\{ 15, -\dfrac{6}{1}, 7.3, \sqrt{2} + \pi \right\}$

In Problems 59–62, approximate each number (a) by truncating and (b) by rounding to the indicated number of decimal places. See Objective 3.

59. 19.93483; 4 decimal places

60. -93.432101; 2 decimal places

61. 0.06345; 1 decimal place

62. 9.9999; 2 decimal places

In Problems 63 and 64, plot the points whose coordinates are given on the real number line. See Objective 4.

63. $2, 0, -3, 1.5, -\dfrac{3}{2}$ 64. $4, -5, 2.5, \dfrac{5}{3}, -\dfrac{1}{2}$

In Problems 65–70, replace the question mark by $<$, $>$, or $=$, whichever is correct. See Objective 5.

65. $-5 \,?\, -3$ 66. $4 \,?\, 2$ 67. $\dfrac{3}{2} \,?\, 1.5$

68. $\dfrac{2}{3} \,?\, \dfrac{2}{5}$ 69. $\dfrac{1}{3} \,?\, 0.3$ 70. $-\dfrac{8}{3} \,?\, -\dfrac{8}{5}$

Applying the Concepts

71. Death Valley Death Valley in California is the lowest point in the United States, with an elevation that is 282 feet below sea level. Express this elevation as an integer. (SOURCE: *Information Please Almanac*)

72. Dead Sea The Dead Sea, Israel–Jordan, is the lowest point in the world, with an elevation that is 1349 feet below sea level. Express this elevation as an integer. (SOURCE: *Information Please Almanac*)

73. Best Buy Best Buy lost $3.37 per share of common stock in 2012. Express this loss as a rational number. (SOURCE: *Yahoo!Finance*)

74. Cisco Systems Cisco Systems stock lost $0.37 in a recent trading day. Express this loss as a rational number. (SOURCE: *Yahoo!Finance*)

75. Golf In the game of golf, your score is often given in relation to par. For example, if par is 72 and a player shoots a 66, then he is 6 under par. Express this score as an integer.

76. It's a Little Chilly! The normal high temperature in Las Vegas, Nevada, on January 20 is 60°F. On January 20, 2012, the temperature was 6°F below normal. Express the departure from normal as an integer. (SOURCE: *USA Today*)

Extending the Concepts

77. Research the history of the set of irrational numbers. Your research should concentrate on the Greek cult called the Pythagoreans. Write a report on your findings.

78. Research the origins of the number 0. Is there any single person who can claim its discovery?

79. The first known computation of the decimal approximation of the number π is attributed to Archimedes around 200 B.C. Research Archimedes and find out his approximation.

80. The irrational number e is attributed to Euler. Research Euler and find out the decimal approximation of e.

Explaining the Concepts

81. Are there any real numbers that are both rational and irrational? Are there any real numbers that are neither? Explain your reasoning.

82. Explain why the sum of a rational number and an irrational number must be irrational.

83. Explain what a set is. Give an example of a set.

84. Explain why it is impossible to list the set of rational numbers using the roster method.

85. Explain the difference between a subset and a proper subset.

86. Describe the difference between 0.45 and $0.\overline{45}$. Are both rational?

87. Explain the circumstances under which rounding and truncating will both result in the same decimal approximation.

88. Is there a positive real number "closest" to 0?

The Graphing Calculator

89. Use your calculator to express $\frac{8}{7}$ rounded to three decimal places. Express $\frac{8}{7}$ truncated to three decimal places.

90. Use your calculator to express $\frac{19}{7}$ rounded to four decimal places. Express $\frac{19}{7}$ truncated to four decimal places.

R.3 Operations on Signed Numbers; Properties of Real Numbers

Objectives

① Compute the Absolute Value of a Real Number

② Add and Subtract Signed Numbers

③ Multiply and Divide Signed Numbers

④ Perform Operations on Fractions

⑤ Know the Associative and Distributive Properties

The symbols used in algebra for the operations of addition, subtraction, multiplication, and division and the words used to describe the results of these operations are shown in Table 1 on the following page.

In algebra, we avoid using the multiplication sign × (to avoid confusion with the often-used *x*) and the division sign ÷ (to avoid confusion with +). Also, two expressions placed next to each other without an operation symbol, as in *ab* or $(a)(b)$, are understood to be **factors** that are to be multiplied.

We also do not use mixed numbers in algebra. When mixed numbers are used, addition is understood—for example, $2\frac{3}{4}$ means $2 + \frac{3}{4}$. But in algebra, the absence of an operation

Table 1

Operation	Symbol	Result of Operation
Addition	$a + b$	**Sum:** a plus b
Subtraction	$a - b$	**Difference:** a minus b b is subtracted from a
Multiplication	$a \cdot b, (a) \cdot b, a \cdot (b), (a) \cdot (b),$ $ab, (a)b, a(b), (a)(b)$	**Product:** a times b
Division	a/b or $\dfrac{a}{b}$	**Quotient:** a divided by b

Work Smart

Do not use mixed numbers in algebra.

symbol between two terms is taken to mean multiplication. To avoid any confusion, $2\dfrac{3}{4}$ is written as 2.75 or as $\dfrac{11}{4}$.

The symbol $=$, which is called an **equal sign** and read as "equals" or "is," is used to express the idea that the expression on the left side of the equal sign is equivalent to the expression on the right.

▶ ❶ Compute the Absolute Value of a Real Number

Let's use the real number line to describe the concept of *absolute value*.

Work Smart

The absolute value of a number can never be negative because it represents a distance.

Definition

The **absolute value** of a number a, written $|a|$, is the distance from 0 to a on the real number line.

Figure 8

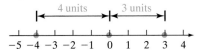

This definition of absolute value is sometimes called the geometric definition because it provides a geometric interpretation. For example, because the distance from 0 to 3 on the real number line is 3, the absolute value of 3, $|3|$, is 3. Because the distance from 0 to -4 on the real number line is 4, $|-4| = 4$. See Figure 8.

EXAMPLE 1   **Computing Absolute Value**

(a) $|8| = 8$ because the distance from 0 to 8 on the real number line is 8.

(b) $|-5| = 5$ because the distance from 0 to -5 on the real number line is 5.

(c) $|0| = 0$ because the distance from 0 to 0 on the real number line is 0.

Quick ✓

1. In the expression $a \cdot b$, the expressions a and b are called _____.

2. *True or False* The absolute value of a number is always positive.

In Problems 3 and 4, evaluate each expression.

3. $|6|$ 4. $|-10|$

❷ Add and Subtract Signed Numbers

▶ The following rules are used to add two real numbers.

In Words

To add two real numbers that have the same sign, add their absolute values and then keep the sign of the original numbers. To add two real numbers with different signs, subtract the absolute value of the smaller number from the absolute value of the larger number. The sign of the sum will be the sign of the number whose absolute value is larger.

Adding Two Nonzero Real Numbers

The approach to adding two real numbers depends on the signs of the two numbers.
1. *Both Positive:* Add the numbers.
2. *Both Negative:* Add the absolute values of the numbers. The sum will be negative.
3. *One Positive, One Negative:* Determine the absolute value of each number. Subtract the smaller absolute value from the larger absolute value.
   - If the larger absolute value was originally the positive number, then the sum is positive.
   - If the larger absolute value was originally the negative number, then the sum is negative.
   - If the two absolute values are equal, then the sum is 0.

EXAMPLE 2   **Adding Two Real Numbers**

Perform the indicated operation.

    (a) $-11 + (-3)$ **(b)** $9.3 + (-6.4)$

Solution

  (a) Both numbers are negative, so we first find their absolute values: $|-11| = 11$ and $|-3| = 3$. We now add the absolute values: $11 + 3 = 14$. Because both numbers to be added are negative, the sum must be negative. So

$$-11 + (-3) = -14$$

  (b) One number is positive and the other is negative. We find the absolute value of each number: $|9.3| = 9.3$ and $|-6.4| = 6.4$. We subtract the smaller absolute value from the larger absolute value and get $9.3 - 6.4 = 2.9$. The larger absolute value was originally positive, so the sum is positive:

$$9.3 + (-6.4) = 2.9$$

Work Smart

Another way to add two real numbers is to use a number line. In Example 2(a), we place a point at -11 on the real number line. Then we move three places to the left (since we are adding -3). See Figure 9. We end up at the point whose coordinate is -14, so we know that $-11 + (-3) = -14$.

Figure 9

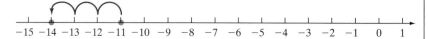

Quick ✓

In Problems 5–9, perform the indicated operation.

  5. $18 + (-6)$ **6.** $-21 + 10$ **7.** $-5.4 + (-1.2)$

  8. $-6.5 + 4.3$ **9.** $-9 + 9$

What do you notice about the sum in Problem 9 from the Quick Check above? The result of Problem 9 is true in general.

Work Smart

The additive inverse of a, $-a$, is sometimes called the *negative* of a or the *opposite* of a. Be careful when using these terms because they suggest that the additive inverse is a negative number, which may not be true! For example, the additive inverse of -3 is 3, a positive number.

Additive Inverse Property

For any real number a, there is a real number $-a$, called the **additive inverse,** or **opposite,** of a, having the following property:

$$a + (-a) = -a + a = 0$$

For example, the opposite of 3 is -3, so $-3 + 3 = 0$. We say that -3 and 3 are additive inverses, or opposites. The opposite of -42 is 42, and the opposite of $-\dfrac{5}{8}$ is $\dfrac{5}{8}$, so we have the following property:

Double Negative Property

For any real number a,

$$-(-a) = a$$

EXAMPLE 3 | **Finding an Additive Inverse**

 (a) The additive inverse of 6 is -6 because $6 + (-6) = 0$.

 (b) The additive inverse of -10 is $-(-10) = 10$ because $-10 + 10 = 0$. ●

Quick ✓

10. For any real number a, there is a real number $-a$, called the _____ _____, or _____, of a such that $a + (-a) = -a + a = $ __.

11. For any real number a, $-(-a) = $ __.

In Problems 12–16, determine the additive inverse of the given real number.

12. 5   **13.** $\dfrac{4}{5}$ **14.** -12 **15.** $-\dfrac{5}{3}$ **16.** 0

The real number 0 is the only number that can be added to any real number a and result in the same number a. We call this the *Identity Property of Addition.*

Identity Property of Addition

For any real number a,

$$0 + a = a + 0 = a$$

That is, the sum of any number and 0 is that number. We call 0 the **additive identity.**

We will use the Identity Property of Addition throughout the entire course (and future courses) to create a new expression that is equivalent to a previous expression. For example, $a = a + 0 = a + 6 + (-6)$ since $6 + (-6) = 0$.

Addition is also *commutative.* That is to say, we get the same result whether we compute $a + b$ or $b + a$.

In Words

We get the same results whether we compute $a + b$ or $b + a$.

Commutative Property of Addition

If a and b are real numbers, then

$$a + b = b + a \quad ,$$

Now we can present a more formal definition of absolute value.

In Words

The absolute value of a number greater than or equal to 0 is the number itself. The absolute value of a number less than zero is the additive inverse of the number.

Definition

The **absolute value** of a real number a, denoted by the symbol $|a|$, is defined by the rules

$$|a| = a \quad \text{if } a \geq 0 \qquad \text{and} \qquad |a| = -a \quad \text{if } a < 0$$

This definition of absolute value is sometimes called the algebraic definition. For example $|12| = 12$ and $|-13| = -(-13) = 13$.

We can use the additive inverse to define subtraction between two real numbers.

In Words

To subtract b from a, add the "opposite" of b to a.

Definition

If a and b are real numbers, then the **difference** $a - b$, read "a minus b" or "a less b," is defined as

$$a - b = a + (-b)$$

Based on this definition, we see that subtracting b from a is really just adding the additive inverse of b to a.

EXAMPLE 4 **Working with Differences**

Evaluate each expression.

    (a) $10 - 4$ **(b)** $-7.3 - (-4.2)$

Solution

    (a) $10 - 4 = 10 + (-4) = 6$

    (b) Notice that we are subtracting a negative number. Subtracting -4.2 is the same as adding 4.2, so

$$-7.3 - (-4.2) = -7.3 + 4.2$$
$$= -3.1$$

Quick ✓

In Problems 17–22, evaluate each expression.

17. $6 - 2$ **18.** $4 - 13$ **19.** $-3 - 8$

20. $12.5 - 3.4$ **21.** $-8.5 - (-3.4)$ **22.** $-6.9 - 9.2$

❸ Multiply and Divide Signed Numbers

From arithmetic, we know that multiplication is repeated addition. For example, $3 \cdot 5$ is equivalent to adding 5 three times. That is,

$$3 \cdot 5 = \underbrace{5 + 5 + 5}_{\text{Add 5 three times}} = 15$$

Also, because

$$5 \cdot 3 = 3 + 3 + 3 + 3 + 3 = 15$$

we see that multiplication of two real numbers a and b is commutative, just like addition.

Commutative Property of Multiplication

If a and b are real numbers, then

$$a \cdot b = b \cdot a$$

When we multiply real numbers, we need to follow rules for determining the sign of the product:

Rules of Signs for Multiplying Two Real Numbers

1. The product of two positive real numbers is positive.

2. The product of one positive real number and one negative real number is negative.

3. The product of two negative real numbers is positive.

EXAMPLE 5 **Multiplying Signed Numbers**

(a) $3(-5) = -(3 \cdot 5) = -15$ **(b)** $-7 \cdot 3 = -(7 \cdot 3) = -21$

(c) $(-9) \cdot (-4) = 36$ **(d)** $-1.5 \cdot (2.6) = -3.9$

Quick ✓

23. *True or False* The product of two negative real numbers is positive.

24. *True or False* Addition and multiplication are commutative.

In Problems 25–29, multiply.

25. $-6 \cdot (8)$ **26.** $12 \cdot (-5)$ **27.** $4 \cdot 14$

28. $-7 \cdot (-15)$ **29.** $-1.9 \cdot (-2.7)$

Work Smart

In Example 5(a), $3 \cdot (-5)$ means to add -5 three times. That is, $3 \cdot (-5)$ means

$$-5 + (-5) + (-5)$$

which is -15. In Example 5(b),

$$-7 \cdot 3 = 3 \cdot (-7)$$
$$= -7 + (-7) + (-7)$$
$$= -21$$

Do you see why a positive times a negative is negative?

The real number 1 has an interesting property. Recall that the expression $3 \cdot 5$ is equivalent to $5 + 5 + 5$. Therefore, $1 \cdot 5$ means to add 5 one time, so, $1 \cdot 5 = 5$. This result is true in general.

Identity Property of Multiplication

For any real number a,

$$a \cdot 1 = 1 \cdot a = a$$

We call 1 the **multiplicative identity**.

In Words

The product of any number and 1 is that number.

The multiplicative identity lets us create equivalent expressions. For example, the expressions $\frac{4}{5}$ and $\frac{4}{5} \cdot \frac{3}{3}$ are equivalent because $\frac{3}{3} = 1$.

For each *nonzero* real number a, there is a real number $\frac{1}{a}$, called the *multiplicative inverse* of a, having the following property:

In Words

When reciprocals are multiplied, their product is 1.

Multiplicative Inverse Property

For each *nonzero* real number a, there is a real number $\frac{1}{a}$, called the **multiplicative inverse** or **reciprocal** of a, having the following property:

$$a \cdot \frac{1}{a} = \frac{1}{a} \cdot a = 1 \qquad a \neq 0$$

EXAMPLE 6   **Finding the Multiplicative Inverse or Reciprocal**

(a) The multiplicative inverse or reciprocal of 5 is $\dfrac{1}{5}$.

(b) The multiplicative inverse or reciprocal of -4 is $-\dfrac{1}{4}$.

(c) The multiplicative inverse of $\dfrac{2}{3}$ is $\dfrac{3}{2}$ because $\dfrac{2}{3} \cdot \dfrac{3}{2} = 1$.

Quick ✓

30. The additive inverse of a, $-a$, is also called the _____ of a. The multiplicative inverse of a, $\dfrac{1}{a}$, is also called the _____ of a.

In Problems 31–34, find the multiplicative inverse or reciprocal of the given real number.

31. 10   **32.** -8 **33.** $\dfrac{2}{5}$ **34.** $-\dfrac{1}{5}$

We now use the idea behind the multiplicative inverse to define division of real numbers.

In Words

Quotients are rational numbers.

Definition

If a is a real number and b is a nonzero real number, the **quotient** $\dfrac{a}{b}$, read as "a divided by b" or "the ratio of a to b," is defined as

$$\frac{a}{b} = a \cdot \frac{1}{b} \qquad \text{if } b \neq 0$$

For example, $\dfrac{5}{8} = 5 \cdot \dfrac{1}{8}$. Because division of real numbers can be represented as multiplication, the same rules of signs that apply to multiplication also apply to division.

Rules of Signs for Dividing Two Real Numbers

1. The quotient of two positive numbers is positive.
2. The quotient of one positive real number and one negative real number is negative.
3. The quotient of two negative real numbers is positive.

Put another way, if a and b are real numbers and $b \neq 0$, then

$$-\frac{a}{b} = \frac{-a}{b} = \frac{a}{-b} \qquad \text{and} \qquad \frac{-a}{-b} = \frac{a}{b}$$

We now introduce additional properties of the numbers 0 and 1.

Multiplication by Zero

For any real number a, the product of a and 0 is always 0; that is,

$$a \cdot 0 = 0 \cdot a = 0$$

Division Properties

For any nonzero real number a,

$$\frac{0}{a} = 0 \qquad \frac{a}{a} = 1 \qquad \frac{a}{0} \text{ is undefined}$$

Perhaps you are wondering what $\frac{0}{0}$ equals. The answer, which may surprise you, is that the value of $\frac{0}{0}$ cannot be determined! Why? If $\frac{0}{0} = n$, then $0 \cdot n = 0$. But $0 \cdot n = 0$ for any real number n. For this reason, we say that $\frac{0}{0}$ is **indeterminate.**

④ Perform Operations on Fractions

▶ We use the following property to write a fraction in *lowest terms*. A fraction is said to be in **lowest terms** if the numerator and the denominator share no common factor other than 1.

Reduction Property

If a, b, and c are real numbers, then

$$\frac{ac}{bc} = \frac{a}{b} \quad \text{if } b \neq 0, c \neq 0$$

EXAMPLE 7   **Using the Reduction Property**

(a)
$$\frac{5 \cdot 3}{5 \cdot 4} = \frac{\cancel{5} \cdot 3}{\cancel{5} \cdot 4}$$

Divide out the 5's: $= \dfrac{3}{4}$

(b) Factor the numerator and denominator and divide out any common factors.

$$\frac{18}{12} = \frac{\cancel{6} \cdot 3}{\cancel{6} \cdot 2}$$

Divide out the 6's: $= \dfrac{3}{2}$

Quick ✓

35. A fraction is written in _____ _____ if the numerator and the denominator share no common factor other than 1.

In Problems 36–39, use the Reduction Property to simplify each expression.

36. $\dfrac{2 \cdot 6}{2 \cdot 5}$ **37.** $\dfrac{33}{24}$ **38.** $-\dfrac{24}{20}$ **39.** $\dfrac{4}{9}$

We now have all the tools we need to perform arithmetic operations on rational numbers.

In Words

To multiply two rational numbers, multiply the numerators and then multiply the denominators. To divide two rational numbers, multiply the rational number in the numerator by the reciprocal of the rational number in the denominator. To add two rational numbers, write the rational numbers over a common denominator and then add the numerators.

Arithmetic of Rational Numbers

$$\frac{a}{b} \cdot \frac{c}{d} = \frac{ac}{bd} \qquad\qquad \text{if } b \neq 0, d \neq 0$$

$$\frac{a}{b} \div \frac{c}{d} = \frac{\dfrac{a}{b}}{\dfrac{c}{d}} = \frac{a}{b} \cdot \frac{d}{c} = \frac{ad}{bc} \qquad \text{if } b \neq 0, c \neq 0, d \neq 0$$

$$\frac{a}{c} + \frac{b}{c} = \frac{a+b}{c} \qquad\qquad \text{if } c \neq 0$$

$$\frac{a}{b} + \frac{c}{d} = \frac{ad}{bd} + \frac{bc}{bd} = \frac{ad+bc}{bd} \qquad \text{if } b \neq 0, d \neq 0$$

EXAMPLE 8 **Multiplying and Dividing Rational Numbers**

Perform the indicated operation. Express your answer in lowest terms.

▶ **(a)** $\dfrac{8}{3} \cdot \dfrac{15}{4}$ ▶ **(b)** $-\dfrac{3}{5} \div \dfrac{6}{7}$

Solution

(a) Multiply the numerators and then multiply the denominators using $\dfrac{a}{b} \cdot \dfrac{c}{d} = \dfrac{ac}{bd}$.

$$\frac{8}{3} \cdot \frac{15}{4} = \frac{8 \cdot 15}{3 \cdot 4}$$

Factor the numerator: $\qquad = \dfrac{4 \cdot 2 \cdot 5 \cdot 3}{3 \cdot 4}$

Divide out like factors: $\qquad = \dfrac{\cancel{4} \cdot 2 \cdot 5 \cdot \cancel{3}}{\cancel{3} \cdot \cancel{4}}$

$$= \frac{2 \cdot 5}{1}$$

$$= \frac{10}{1}$$

$$= 10$$

Work Smart

Notice that when all the factors in the denominator of Example 8(a) divide out, the denominator is 1.

(b) Rewrite the division problem as an equivalent multiplication problem using $\dfrac{a}{b} \div \dfrac{c}{d} = \dfrac{a}{b} \cdot \dfrac{d}{c}$.

$$-\frac{3}{5} \div \frac{6}{7} = -\frac{3}{5} \cdot \frac{7}{6}$$

$$= -\frac{3 \cdot 7}{5 \cdot 3 \cdot 2}$$

$$= -\frac{7}{10}$$

Quick ✓

In Problems 40–43, perform the indicated operation. Express each answer in lowest terms.

40. $\dfrac{5}{3} \cdot \dfrac{12}{25}$

41. $\dfrac{2}{3} \cdot \left(-\dfrac{5}{4} \right)$

42. $\dfrac{4}{3} \div \dfrac{8}{3}$

43. $\dfrac{\dfrac{10}{3}}{\dfrac{5}{12}}$

▶ EXAMPLE 9 **Adding and Subtracting Rational Numbers with the Same Denominator**

Perform the indicated operation. Express your answer in lowest terms.

(a) $\dfrac{5}{12} + \dfrac{11}{12}$ **(b)** $\dfrac{4}{15} - \dfrac{7}{15}$

Solution

(a) Because the denominators are the same, we add the numerators and write the sum over the common denominator.

$$\frac{5}{12} + \frac{11}{12} = \frac{5 + 11}{12}$$

$$= \frac{16}{12}$$

Factor the numerator and denominator. Divide out like factors:
$$= \frac{4 \cdot 4}{4 \cdot 3}$$

$$= \frac{4}{3}$$

(b) Because the denominators are the same, we subtract the numerators and write the difference over the common denominator.

$$\frac{4}{15} - \frac{7}{15} = \frac{4 - 7}{15}$$

$$= \frac{-3}{15}$$

$$= \frac{-1 \cdot 3}{3 \cdot 5}$$

$$= -\frac{1}{5}$$

Quick ✓

In Problems 44–47, perform the indicated operation. Express your answer in lowest terms.

44. $\dfrac{3}{11} + \dfrac{2}{11}$ **45.** $\dfrac{8}{15} - \dfrac{13}{15}$ **46.** $\dfrac{3}{7} + \dfrac{11}{7}$ **47.** $\dfrac{8}{5} - \dfrac{3}{5}$

▶ To add two rational numbers with different denominators, we first find the **least common denominator (LCD)**-the smallest number that all the denominators have as a common multiple. The next example illustrates the idea.

EXAMPLE 10   **How to Find the Least Common Denominator**

Find the least common denominator of the rational numbers $\frac{8}{15}$ and $\frac{5}{12}$. Then rewrite each rational number with the least common denominator.

Step-by-Step Solution

Step 1: Factor each denominator as the product of prime factors.

$$15 = 5\cdot 3 \qquad 12 = 4\cdot 3$$
$$= 2\cdot 2\cdot 3$$

Step 2: Write the factor(s) that both denominators have in common. Then write the remaining factors.

The common factor is 3. The remaining factors are 5, 2, and 2.

Step 3: Multiply the factors listed in Step 2. The product is the least common denominator (LCD).

$$LCD = 3\cdot 5\cdot 2\cdot 2 = 60$$

Step 4: Rewrite $\frac{8}{15}$ and $\frac{5}{12}$ with a denominator of 60.

Multiply $\frac{8}{15}$ by $\frac{4}{4}$. Multiply $\frac{5}{12}$ by $\frac{5}{5}$.

$$\frac{8}{15} = \frac{8}{15}\cdot\frac{4}{4} = \frac{8\cdot 4}{15\cdot 4} = \frac{32}{60}$$
$$\frac{5}{12} = \frac{5}{12}\cdot\frac{5}{5} = \frac{5\cdot 5}{12\cdot 5} = \frac{25}{60}$$

Quick ✓

48. The ____ _____ _____ is the smallest number that each denominator has as a common multiple.

In Problems 49 and 50, find the least common denominator (LCD) of each pair of rational numbers. Then rewrite each rational number with the LCD.

49. $\frac{3}{20}$ and $\frac{2}{15}$

50. $\frac{5}{18}$ and $-\frac{1}{45}$

EXAMPLE 11   **Adding Rational Numbers Using the Least Common Denominator**

Perform the indicated operation.

(a) $\frac{8}{15} + \frac{5}{12}$ **(b)** $-\frac{3}{10} - \frac{2}{15}$

Solution

(a) From Example 10, we know that the LCD of $\frac{8}{15}$ and $\frac{5}{12}$ is 60. Rewrite each fraction with a denominator of 60:

$$\frac{8}{15} + \frac{5}{12} = \frac{8}{15}\cdot\frac{4}{4} + \frac{5}{12}\cdot\frac{5}{5}$$
$$= \frac{32}{60} + \frac{25}{60}$$

$\frac{a}{c} + \frac{b}{c} = \frac{a+b}{c}$: $= \frac{32+25}{60}$

$$= \frac{57}{60}$$
$$= \frac{19}{20}$$

(b) First, write the subtraction problem as an equivalent addition problem using $a - b = a + (-b)$:

$$-\frac{3}{10} - \frac{2}{15} = \frac{-3}{10} + \left(\frac{-2}{15}\right)$$

$$\text{LCD} = 30: \quad = \frac{-3}{10} \cdot \frac{3}{3} + \left(\frac{-2}{15} \cdot \frac{2}{2}\right)$$

$$= \frac{-9}{30} + \left(\frac{-4}{30}\right)$$

$$= \frac{-9 + (-4)}{30}$$

$$= \frac{-13}{30}$$

$$\frac{-a}{b} = -\frac{a}{b}: \quad = -\frac{13}{30}$$

Quick ✓

In Problems 51–54, perform the indicated operation. Express your answer in lowest terms.

51. $\dfrac{3}{20} + \dfrac{2}{15}$

52. $\dfrac{5}{14} - \dfrac{11}{21}$

53. $-\dfrac{4}{25} - \dfrac{7}{30}$

54. $-\dfrac{5}{18} - \dfrac{1}{45}$

⑤ Know the Associative and Distributive Properties

Example 12 illustrates that the order in which we add or multiply three real numbers does not affect the final result. This property is called the *Associative Property*.

▶ EXAMPLE 12   Illustrating the Associative Property

(a) $17 + (3 + 5) = 17 + 8 = 25$
$(17 + 3) + 5 = 20 + 5 = 25$
$17 + (3 + 5) = (17 + 3) + 5$

(b) $13 \cdot (2 \cdot 5) = 13 \cdot 10 = 130$
$(13 \cdot 2) \cdot 5 = 26 \cdot 5 = 130$
$13 \cdot (2 \cdot 5) = (13 \cdot 2) \cdot 5$

Associative Properties of Addition and Multiplication

If a, b, and c are real numbers, then

$$a + (b + c) = (a + b) + c = a + b + c$$
$$a \cdot (b \cdot c) = (a \cdot b) \cdot c = a \cdot b \cdot c$$

▶ The next property will be used throughout the course and in future courses.

The Distributive Property

If a, b, and c are real numbers, then

$$a \cdot (b + c) = a \cdot b + a \cdot c$$
$$(a + b) \cdot c = a \cdot c + b \cdot c$$

(**EXAMPLE 13**) **Using the Distributive Property**

Use the Distributive Property to remove the parentheses.

(a) $2(x + 3)$ (b) $-3(2y + 1)$ (c) $(z - 4) \cdot 3$ (d) $\dfrac{1}{2}(4x - 10)$

Solution

(a) $2(x + 3) = 2 \cdot x + 2 \cdot 3 = 2x + 6$

(b) $-3(2y + 1) = -3 \cdot 2y + (-3) \cdot 1 = -6y - 3$

(c) $(z - 4) \cdot 3 = z \cdot 3 - 4 \cdot 3 = 3z - 12$

(d) $\dfrac{1}{2}(4x - 10) = \dfrac{1}{2} \cdot 4x - \dfrac{1}{2} \cdot 10 = \dfrac{4x}{2} - \dfrac{10}{2} = 2x - 5$ ●

When parentheses are preceded by a minus sign, such as in $-(3x + 7)$, we can use the Distributive Property as shown below.

$$-(3x + 7) = -1(3x + 7) = -1 \cdot 3x + -1 \cdot 7 = -3x - 7$$

We can generalize this property.

In Words

The opposite of a sum is the sum of the opposites.

The Opposite of a Sum

For any real numbers a and b,

$$-(a + b) = -a + (-b) = -a - b$$

Quick ✓

55. The Distributive Property states that $a \cdot (b + c) = \underline{} + \underline{}$.

56. $-(a + b) = \underline{}$

In Problems 57–62, use the Distributive Property to remove the parentheses.

57. $5(x + 3)$ **58.** $-6(x + 1)$

59. $-4(z - 8)$ **60.** $\dfrac{1}{3}(6x + 9)$

61. $-(11p + 8)$ **62.** $-(-9 - 4t)$

Summary **Properties of the Real Number System**

If a, b, and c are real numbers,

- Identity Properties $a + 0 = 0 + a = a; \quad 1 \cdot a = a \cdot 1 = a$

- Inverse Properties $a + (-a) = (-a) + a = 0; \quad a \cdot \dfrac{1}{a} = \dfrac{1}{a} \cdot a = 1 \, (a \neq 0)$

- Double Negative Property $-(-a) = a$
- Commutative Properties $a + b = b + a; \quad a \cdot b = b \cdot a$
- Multiplication Property of 0 $a \cdot 0 = 0 \cdot a = 0$
- Division Properties $\dfrac{0}{a} = 0, a \neq 0; \quad \dfrac{a}{a} = 1, a \neq 0;$

                                      $\dfrac{a}{0}$ is undefined, $a \neq 0$

- Reduction Property $\dfrac{ac}{bc} = \dfrac{a}{b}, b \neq 0, c \neq 0$

- Associative Properties $(a + b) + c = a + (b + c); \quad (ab)c = a(bc)$
- Distributive Property $a(b + c) = ab + ac$
- Opposite of a Sum $-(a + b) = -a - b$

R.3 Exercises MyMathLab® Exercise numbers in green have complete video solutions in MyMathLab.

*Problems **1–62,** are the Quick ✓ s that follow the EXAMPLES.*

Building Skills

In Problems 63–66, evaluate each expression. See Objective 1.

63. $\left|\dfrac{2}{3}\right|$ **64.** $\left|\dfrac{5}{6}\right|$ **65.** $\left|-\dfrac{8}{3}\right|$ **66.** $\left|-\dfrac{7}{2}\right|$

In Problems 67–72, perform the indicated operation. See Objective 2.

67. $-13 + 4$ **68.** $-6 + 10$ **69.** $12 - 5$

70. $9 + (-3)$ **71.** $4.3 - 6.8$

72. $-8.2 - 4.5$

In Problems 73–78, perform the indicated operation. See Objective 3.

73. $4 \cdot (-8)$ **74.** $-5 \cdot (-15)$

75. $-6 \cdot (-14)$ **76.** $7 \cdot (-15)$

77. $-4.3 \cdot (8.5)$ **78.** $-10.4 \cdot (-0.6)$

In Problems 79–82, use the Reduction Property to simplify each expression. See Objective 4.

79. $\dfrac{28}{12}$ **80.** $\dfrac{25}{40}$ **81.** $\dfrac{30}{25}$ **82.** $\dfrac{40}{16}$

In Problems 83–94, perform the indicated operation. See Objective 4.

83. $\dfrac{3}{4} \cdot \dfrac{20}{9}$ **84.** $\dfrac{2}{5} \cdot \dfrac{15}{8}$

85. $\dfrac{2}{3} \cdot \left(-\dfrac{9}{14}\right)$ **86.** $-\dfrac{7}{3} \cdot \left(-\dfrac{12}{35}\right)$

87. $\dfrac{7}{4} \div \dfrac{21}{8}$ **88.** $-\dfrac{10}{3} \div \dfrac{15}{21}$

89. $\dfrac{\frac{2}{5}}{\frac{8}{25}}$ **90.** $\dfrac{\frac{18}{7}}{\frac{3}{14}}$

91. $-\dfrac{5}{12} + \dfrac{7}{12}$ **92.** $\dfrac{8}{5} - \dfrac{18}{5}$

93. $\dfrac{7}{15} + \dfrac{9}{20}$ **94.** $\dfrac{3}{8} - \dfrac{7}{18}$

In Problems 95–104, use the Distributive Property to remove the parentheses. See Objective 5.

95. $2(x + 4)$ **96.** $3(y - 5)$

97. $-3(z - 2)$ **98.** $-5(x - 4)$

99. $(x - 10) \cdot 3$ **100.** $(3x + y) \cdot 2$

101. $\dfrac{3}{4}(8x - 12)$ **102.** $-\dfrac{2}{3}(3x + 15)$

103. $-(5z + 17)$ **104.** $-(11 - 8k)$

Mixed Practice

In Problems 105–126, perform the indicated operation. Express all rational numbers in lowest terms.

105. $|13 - 16|$ **106.** $|-4| + 12$

107. $\dfrac{51}{42} - \left(\dfrac{-8}{35}\right)$ **108.** $-\dfrac{18}{45} - \dfrac{23}{24}$

109. $5 \div \dfrac{15}{4}$ **110.** $\dfrac{2}{3} \div 8$

111. $|6.2 - 9.5|$ **112.** $|-5.4 + 10.5|$

113. $-|-8 \cdot (4)|$ **114.** $-|-5 \cdot 9|$

115. $\left|-\dfrac{1}{2} - \dfrac{4}{5}\right|$ **116.** $\left|\dfrac{4}{3} - \dfrac{8}{7}\right|$

117. $\left|-\dfrac{5}{6} - \dfrac{3}{10}\right|$ **118.** $\left|\dfrac{4}{15} - \dfrac{1}{6}\right|$

119. $\dfrac{21}{32} + (-5)$ **120.** $-\dfrac{10}{21} + 6$

121. $\dfrac{\frac{2}{3}}{6}$ **122.** $\dfrac{\frac{20}{5}}{4}$

123. $\dfrac{18}{0}$ **124.** $-\dfrac{7}{0}$

125. $\dfrac{0}{20}$ **126.** $\dfrac{0}{5}$

In Problems 127–134, state the property that is being illustrated.

127. $5 \cdot 3 = 3 \cdot 5$

128. $9 + (-9) = 0$

129. $5 \cdot \dfrac{1}{5} = 1$

130. $\dfrac{a}{a} = 1, a \neq 0$

131. $\dfrac{42}{10} = \dfrac{21}{5}$

132. $3(x - 4) = 3x - 12$

133. $3 + (4 + 5) = (3 + 4) + 5$

134. $\dfrac{0}{6} = 0$

Applying the Concepts

135. Age of Presidents The youngest president at the time of inauguration was Theodore Roosevelt (42 years of age). The oldest president at the time of inauguration was Ronald Reagan (69 years of age). What is the difference in age at the time of inauguration between the oldest and youngest presidents?

136. Life Expectancy In South Korea, the life expectancy for females born in 1950 was 49 years of age. The life expectancy for females born in 2012 was 81 years of age. Compute the difference in life expectancy between 1950 and 2012.

137. Football The Chicago Bears obtained the following yardages for each of the first three plays of a game: 4, −3, 8. How many total yards did they gain for the first three plays? If 10 yards are required for a first down, did the Bears obtain a first down?

138. Balancing a Checkbook At the beginning of the month, Paul had $400 in his checking account. During the month he wrote four checks for $20, $45, $60, and $105. He also made a deposit in the amount of $150. What is Paul's balance at the end of the month?

139. Peaks and Valleys In the United States, the highest elevation is Mount McKinley in Alaska (20,320 feet above sea level); the lowest elevation is Death Valley in California (282 feet below sea level). What is the difference between the highest and lowest elevations?

140. More Peaks and Valleys In Louisiana, the highest elevation is Driskill Mountain (535 feet above sea level); the lowest elevation is New Orleans (8 feet below sea level). What is the difference between the highest and lowest elevations?

Extending the Concepts

141. Illustrate why the product of two positive numbers is positive. Illustrate why the product of a positive number and a negative number is negative.

142. The Fibonacci Sequence The Fibonacci sequence is a famous sequence of numbers that were discovered by Leonardo Fibonacci of Pisa. The numbers in the sequence are 1, 1, 2, 3, 5, 8, 13, 21, 34, 55, . . . , where each term after the second term is the sum of the two preceding terms.

(a) Compute the ratio of consecutive terms in the sequence. That is, compute $\frac{1}{1}, \frac{2}{1}, \frac{3}{2}, \frac{5}{3}$, and so on.

(b) What number does the ratio approach? This number is called the **golden ratio** and has application in many different areas.

(c) Research Fibonacci numbers and cite three different applications.

Problems 143–148 use the following definition.

If P and Q are two points on a real number line with coordinates a and b, respectively, the **distance between P and Q**, denoted by $d(P, Q)$, is

$$d(P, Q) = |b - a|$$

Since $|b - a| = |a - b|$, it follows that $d(P, Q) = d(Q, P)$.

143. Plot the points $P = -4$ and $Q = 10$ on the real number line, and then find $d(P, Q)$.

144. Plot the points $P = -2$ and $Q = 6$ on the real number line, and then find $d(P, Q)$.

145. Plot the points $P = -3.2$ and $Q = 7.2$ on the real number line, and then find $d(P, Q)$.

146. Plot the points $P = -9.3$ and $Q = 1.6$ on the real number line, and then find $d(P, Q)$.

147. Plot the points $P = -\frac{10}{3}$ and $Q = \frac{6}{5}$ on the real number line, and then find $d(P, Q)$.

148. Plot the points $P = -\frac{7}{5}$ and $Q = 6$ on the real number line, and then find $d(P, Q)$.

149. Why Is the Product of Two Negatives Positive? In this problem, we use the Distributive Property to illustrate why the product of two negative real numbers is positive.

(a) Express the product of any real number a and 0.

(b) Use the Additive Inverse Property to write 0 from part (a) as $b + (-b)$.

(c) Use the Distributive Property to distribute the a into the expression in part (b).

(d) Suppose that $a < 0$ and $b > 0$. What can be said about the product ab? Now, what must be true regarding the product $a(-b)$ in order for the sum to be zero?

150. Math for the Future Find the set of all ratios $\frac{p}{q}$ such that $p \in A$ and $q \in B$.

$$A = \{-9, -3, -1, 1, 3, 9\}, B = \{-4, -2, -1, 1, 2, 4\}$$

Explaining the Concepts

151. Explain why 0 does not have a multiplicative inverse.

152. Why does $2(4 \cdot 5)$ not equal $(2 \cdot 4) \cdot (2 \cdot 5)$?

153. Explain the flaw in the following reasoning:

$$\frac{4 + 3}{2 + 5} = \frac{2 \cdot 2 + 3}{2 + 5} = \frac{2 \cdot 2 + 3}{2 + 5} = \frac{2 + 3}{1 + 5} = \frac{5}{6}$$

154. Is subtraction commutative? Support your conclusion with an example.

155. Is subtraction associative? Support your conclusion with an example.

156. Is division commutative? Support your conclusion with an example.

157. Is division associative? Support your conclusion with an example.

158. Explain why $\frac{1}{4} + \frac{1}{8} = \frac{3}{8}$.

The Graphing Calculator

Many calculators have the ability to compute absolute value. Figure 10 shows the results of Example 1 using a TI-84 Plus graphing calculator. Calculators have the ability to add, subtract, multiply, and divide rational numbers and write the solution in lowest terms! Figure 11 shows the results of Examples 8(b) and 11(b).

Figure 10

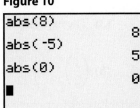

Figure 11

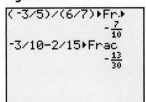

In Problems 159–168, use a calculator to perform the indicated operation. Express all rational numbers in lowest terms.

159. $5.4 - 9.2$

160. $-2.9 + (-6.3)$

161. $-3(6.4)$

162. $-5.4(-4.8)$

163. $\frac{3}{10} - \frac{4}{15}$

164. $-\frac{3}{8} + \frac{1}{10}$

165. $|4.5(-3.2)|$

166. $|-3.65| \cdot (5.4)$

167. $\frac{6}{5} \cdot \frac{45}{12}$

168. $-\frac{3}{4} \div \frac{9}{20}$

R.4 Order of Operations

Objectives

① Evaluate Real Numbers with Exponents

② Use the Order of Operations to Evaluate Expressions

▶ ① Evaluate Real Numbers with Exponents

Integer exponents indicate repeated multiplication. For example,

$$3^4 = 3 \cdot 3 \cdot 3 \cdot 3 = 81$$

In the expression 3^4, 3 is the **base** and 4 is the **exponent** or **power.** The power tells us the number of times to use the base as a factor.

In Words

The notation a^n means to multiply a by itself n times.

Definition

If a is a real number and n is a positive integer, the expression a^n represents the product of n factors of a. That is,

$$a^n = \underbrace{a \cdot a \cdot \cdots \cdot a}_{n \text{ factors}}$$

Thus, we have

$$a^1 = a$$
$$a^2 = a \cdot a$$
$$a^3 = a \cdot a \cdot a$$

and so on.

We read a^n as "a raised to the power n" or as "a to the nth power." We usually read a^2 as "a squared" and a^3 as "a cubed."

EXAMPLE 1   **Evaluating Expressions Containing Exponents**

Evaluate each expression.

(a) 2^3 **(b)** $\left(\frac{1}{2}\right)^2$ **(c)** 5^4

Solution

(a) The expression 2^3 means that we should use the base, 2, as a factor 3 times, so
$$2^3 = 2 \cdot 2 \cdot 2 = 8$$
(b) $\left(\dfrac{1}{2}\right)^2 = \dfrac{1}{2} \cdot \dfrac{1}{2} = \dfrac{1}{4}$ (c) $5^4 = 5 \cdot 5 \cdot 5 \cdot 5 = 625$

EXAMPLE 2   **Evaluating Expressions Containing Exponents**

Evaluate each expression.

(a) $(-3)^4$ (b) $(-3)^5$ (c) -3^4 (d) $-(-3)^4$

Solution

(a) $(-3)^4 = (-3) \cdot (-3) \cdot (-3) \cdot (-3) = 81$
(b) $(-3)^5 = (-3) \cdot (-3) \cdot (-3) \cdot (-3) \cdot (-3) = -243$
(c) The expression -3^4 means the additive inverse, or opposite, of 3^4.
     That is, $-3^4 = -(3) \cdot (3) \cdot (3) \cdot (3) = -81$.
(d) $-(-3)^4 = -(-3) \cdot (-3) \cdot (-3) \cdot (-3) = -81$

Work Smart

The expressions $(-3)^4$ and -3^4 are different. $(-3)^4$ means to use (-3) as a factor four times, while -3^4 means to determine the additive inverse of 3^4. That is,

$(-3)^4 = (-3)(-3)(-3)(-3) = 81$

and

$-3^4 = -(3)(3)(3)(3) = -81$

The results of Examples 2(a) and 2(b) suggest that raising a negative number to a positive, even integer exponent yields a positive number, while raising a negative number to a positive, odd integer exponent yields a negative number. This conclusion is correct in general.

Quick ✓

1. In the expression 5^6, the number 5 is the ____ and 6 is the _____ or ____.

2. *True or False* $-7^4 = (-7) \cdot (-7) \cdot (-7) \cdot (-7)$

In Problems 3–8, evaluate each expression.

3. 4^3 **4.** $(-7)^2$ **5.** $(-10)^3$ **6.** $\left(\dfrac{2}{3}\right)^3$ **7.** -8^2 **8.** $-(-5)^3$

❷ Use the Order of Operations to Evaluate Expressions

To evaluate an expression containing both multiplication and addition, such as $2 + 3 \cdot 7$, we agree to the following.

> If the two operations of addition and multiplication separate three numbers, always multiply first, then add.

Therefore, $2 + 3 \cdot 7 = 2 + 21 = 23$.

▶ EXAMPLE 3   **Finding the Value of an Expression**

Evaluate each expression.

(a) $3 + 4 \cdot 5$ (b) $8 \cdot 2 + 1$

Solution

(a) Don't forget! Multiply before adding.
$$3 + 4 \cdot 5 = 3 + 20$$
$$= 23$$

(b) $8 \cdot 2 + 1 = 16 + 1$
$$= 17$$

Quick ✓

In Problems 9 and 10, evaluate each expression.

9. $5 \cdot 2 + 6$

10. $3 \cdot 2 + 5 \cdot 6$

To add two numbers before multiplying use parentheses. For example, in $(2 + 3) \cdot 7$ we would add 2 and 3 first, the multiply the sum by 7.

▶ **EXAMPLE 4** **Finding the Value of an Expression**

(a) $(6 + 2) \cdot 4 = 8 \cdot 4$
$$= 32$$

(b) $(8 - 3) \cdot (7 + 3) = 5 \cdot 10$
$$= 50$$ ●

Quick ✓

In Problems 11–14, evaluate each expression.

11. $4 \cdot (5 + 3)$

12. $8 \cdot (9 - 3)$

13. $(12 - 4) \cdot (18 - 13)$

14. $(4 + 9) \cdot (6 - 4)$

Work Smart

Never use the Reduction Property across addition.

This is incorrect:

$$\frac{4 + 5}{6 + 12} = \frac{\cancel{2} \cdot 2 + 5}{\cancel{2} \cdot 3 + 12}$$
$$= \frac{2 + 5}{3 + 12} = \frac{7}{15}$$

This is correct:

$$\frac{4 + 5}{6 + 12} = \frac{9}{18}$$
$$= \frac{1}{2}$$

In expressions like $\dfrac{4 + 5}{6 + 12}$, it is understood that the division bar acts like parentheses. This means we evaluate the expressions in the numerator and denominator FIRST, and then divide. That is,

$$\frac{4 + 5}{6 + 12} = \frac{(4 + 5)}{(6 + 12)} = \frac{9}{18} = \frac{\cancel{9} \cdot 1}{\cancel{9} \cdot 2} = \frac{1}{2}$$

This rule will be extremely important throughout the course.

In addition to parentheses, grouping symbols include brackets [ ], braces { }, and absolute value symbols | |.

If one set of grouping symbols is embedded within a second set of grouping symbols, as in $2 + [3 + 2(9 - 4)]$, we simplify within the innermost grouping symbols first.

▶ **EXAMPLE 5** **Finding the Value of an Expression**

Evaluate each expression.

(a) $\dfrac{3 + 7}{6 + 4 \cdot 6}$

(b) $9 \cdot [8 + 4(10 - 7)]$

Solution

(a) $\dfrac{3 + 7}{6 + 4 \cdot 6} = \dfrac{3 + 7}{6 + 24}$
$$= \frac{10}{30}$$
$$= \frac{1}{3}$$

(b) Evaluate the expression in the innermost grouping symbols first.

$$9 \cdot [8 + 4(10 - 7)] = 9 \cdot [8 + 4(3)]$$
$$= 9 \cdot [8 + 12]$$
$$= 9 \cdot 20$$
$$= 180$$ ●

Work Smart

When one set of parentheses is embedded within another set of parentheses, simplify the expression in the innermost parentheses first.

Quick ✔

In Problems 15–18, evaluate each expression.

15. $\dfrac{3 + 7}{4 + 9}$

16. $1 - 4 + 8 \cdot 2 + 5$

17. $25 \cdot [2(8 - 3) - 8]$

18. $\dfrac{3 + 5}{2 \cdot (9 - 4)}$

▶ When, in the order of operations, do we evaluate exponents? In the expression $3 \cdot 2^4$, do we multiply 3 and 2 first and then raise this product to the 4th power to get 1296, or do we first raise 2 to the 4th power to get 16 and multiply this by 3 to obtain 48? Because $2^4 = 2 \cdot 2 \cdot 2 \cdot 2$, we have $3 \cdot 2^4 = 3 \cdot 2 \cdot 2 \cdot 2 \cdot 2 = 48$, which implies that we evaluate exponents before multiplication.

> **Order of Operations**
>
> **1.** Evaluate expressions within **parentheses** first. When an expression has more than one set of grouping symbols, begin with the innermost grouping symbols and work outward.
>
> **2.** Evaluate expressions containing **exponents,** working from left to right.
>
> **3.** Perform **multiplication** and **division,** working from left to right.
>
> **4.** Perform **addition** and **subtraction,** working from left to right.

Quick ✔

19. The order of operations is (1) _____, (2) _____, (3) _____ and _____, (4) _____ and _____.

In Problems 20–24, evaluate each expression.

20. $6 + 5 \cdot 2$

21. $(3 + 9) \cdot 4$

22. $\dfrac{7 + 5}{4 + 10}$

23. $4 + [(8 - 3) \cdot 2]$

24. $4 \cdot (6 - 2) - 9$

EXAMPLE 6 **Evaluating Expressions Using the Order of Operations**

Evaluate each expression.

(a) $6 + 4 \cdot 3^2$

(b) $\dfrac{1}{4} \cdot 2^3 + 9 \cdot 4 - 2 \cdot 5^2$

Solution

(a) Evaluate the exponent first:

$$6 + 4 \cdot 3^2 = 6 + 4 \cdot 9$$
$$\text{Multiply:} = 6 + 36$$
$$= 42$$

(b) $\dfrac{1}{4} \cdot 2^3 + 9 \cdot 4 - 2 \cdot 5^2 = \dfrac{1}{4} \cdot 8 + 9 \cdot 4 - 2 \cdot 25$

$$\text{Multiply:} = 2 + 36 - 50$$
$$\text{Add from left to right:} = 38 - 50$$
$$= -12$$

●

EXAMPLE 7   **Evaluating Expressions Using the Order of Operations**

Evaluate each expression.

(a) $\dfrac{2 + 3 \cdot 4^2}{5 \cdot (6 - 2)}$

(b) $-2 \cdot \left| -(-2)^3 - 4 \cdot (10 - 7)^2 \right|$

Solution

(a)
$$\frac{2 + 3 \cdot 4^2}{5 \cdot (6 - 2)} = \frac{2 + 3 \cdot 16}{5 \cdot 4}$$
$$= \frac{2 + 48}{20}$$
$$= \frac{50}{20}$$
$$= \frac{5 \cdot 10}{2 \cdot 10}$$

Divide out the common factor, 10: $= \dfrac{5}{2}$

(b) The absolute value bars form a grouping symbol, so we simplify within the absolute value bars first.

Evaluate inner parentheses first
↓

$$-2 \cdot \left| -(-2)^3 - 4 \cdot (10 - 7)^2 \right| = -2 \cdot \left| -(-2)^3 - 4 \cdot (3)^2 \right|$$

Evaluate exponents: $= -2 \cdot \left| -(-8) - 4 \cdot 9 \right|$
$$= -2 \cdot \left| 8 - 36 \right|$$
$$= -2 \cdot \left| -28 \right|$$
$$= -2 \cdot 28$$
$$= -56$$

Quick ✓

In Problems 25–30, evaluate each expression.

25. $-8 + 2 \cdot 5^2$

26. $5 \cdot 3 - 3 \cdot 2^3$

27. $5 \cdot (10 - 8)^2$

28. $3 \cdot (-2)^2 + 6 \cdot 3 - 3 \cdot 4^2$

29. $\dfrac{4 + 6^2}{2 \cdot 3 + 2}$

30. $-3 \cdot \left| 7^2 - 2 \cdot (8 - 5)^3 \right|$

R.4 Exercises

Exercise numbers in green
have complete video solutions
in MyMathLab.

*Problems **1–30** are the Quick ✓s that follow the EXAMPLES.*

Building Skills
In Problems 31–38, evaluate each expression. See Objective 1.

31. $(-3)^4$

32. $(-4)^3$

33. -5^4

34. -2^4

35. $-(-2)^3$

36. $-(-7)^3$

37. $-\left(\dfrac{2}{3}\right)^2$

38. $-\left(\dfrac{3}{4}\right)^4$

Mixed Practice
In Problems 39–74, evaluate each expression.

39. $4^2 - 3^2$

40. $(-3)^2 + (-2)^2$

41. $3 \cdot 2 + 9$

42. $-5 \cdot 3 + 12$

43. $4 + 2 \cdot (6 - 2)$

44. $2 + 5 \cdot (8 - 5)$

45. $-2[10 - (3 - 7)]$

46. $3[15 - (7 - 3)]$

47. $\dfrac{5 - (-7)}{4}$

48. $\dfrac{12 - 4}{-2}$

49. $\left| 3 \cdot 2 - 4 \cdot 5 \right|$

50. $\left| 6 \cdot 2 - 5 \cdot 3 \right|$

51. $12 \cdot \dfrac{2}{3} - 5 \cdot 2$

52. $15 \cdot \dfrac{3}{5} + 4 \cdot 3$

53. $3 \cdot [2 + 3 \cdot (1 + 5)]$

54. $2 \cdot [25 - 2 \cdot (10 - 4)]$

55. $(3^2 - 3) \cdot (3 - (-3)^3)$

56. $-2 \cdot (5 - 2) - (-5)^2$

57. $|3 \cdot (6 - 3^2)|$

58. $-2 \cdot (4 + |2 \cdot 3 - 5^2|)$

59. $|4 \cdot [2 \cdot 5 + (-3) \cdot 4]|$

60. $|6 \cdot (3 \cdot 2 - 10)|$

61. $\dfrac{2 \cdot 5 + 15}{2^2 + 3 \cdot 2}$

62. $\dfrac{2 \cdot 3^2}{4^2 - 4}$

63. $\dfrac{2 \cdot (4 + 8)}{3 + 3^2}$

64. $\dfrac{3 \cdot (5 + 2^2)}{2 \cdot 3^3}$

65. $\dfrac{6 \cdot [12 - 3 \cdot (5 - 2)]}{5 \cdot [21 - 2 \cdot (4 + 5)]}$

66. $\dfrac{4 \cdot [3 + 2 \cdot (8 - 6)]}{5 \cdot [14 - 2 \cdot (2 + 3)]}$

67. $\left(\dfrac{2}{3}\right)^2 \cdot \left(\dfrac{1 + 2^3}{2^3 - 2}\right)$

68. $\left(\dfrac{3^2}{29 - 3 \cdot 2^3}\right) \cdot \dfrac{5}{4 + 5}$

69. $\dfrac{3^3 - 2^4 \cdot 3}{4 \cdot (3^2 - 2 \cdot 3)}$

70. $2^5 + 4 \cdot (-5) + 4^2$

71. $\dfrac{2 \cdot 4 - 5}{4^2 + (-2)^3} + \dfrac{3^2}{2^3}$

72. $\dfrac{5 \cdot (37 - 6^2)}{6 \cdot 2 - 3^2} + \dfrac{7 \cdot 2 - 4^2}{5 + 4}$

73. $\dfrac{\dfrac{2 \cdot 3 + 3^2}{2 \cdot 5 - 8} + \dfrac{4}{3}}{\dfrac{7 \cdot (5 - 3)}{14 - 2^3}}$

74. $\dfrac{\dfrac{4}{4^2 - 1} - \dfrac{3}{5 \cdot (7 - 5)}}{\dfrac{-(-2)^2}{4 \cdot 7 + 2}}$

Applying the Concepts

In Problems 75–78, insert parentheses in order to make each statement true.

75. $3 \cdot 7 - 2 = 15$

76. $-2 \cdot 3 - 5 = 4$

77. $3 + 5 \cdot 6 - 3 = 18$

78. $3 + 5 \cdot 6 - 3 = 24$

△ **79. Geometry** The surface area of a closed right circular cylinder whose radius is 5 inches and height is 12 inches is given approximately by $2 \cdot 3.1416 \cdot 5^2 + 2 \cdot 3.1416 \cdot 5 \cdot 12$. Evaluate this expression, rounding your answer to two decimal places.

△ **80. Geometry** The surface area of a sphere whose radius is 3 centimeters is given approximately by $4 \cdot 3.1416 \cdot 3^2$. Evaluate this expression, rounding your answer to two decimal places.

81. Hitting a Golf Ball The height (in feet) of a golf ball hit with an initial speed of 100 feet per second after 3 seconds is given by $-16 \cdot 3^2 + 50 \cdot 3$. Evaluate this expression in order to determine the height of the golf ball after 3 seconds.

82. Horsepower The horsepower rating of an engine is $\dfrac{10^2 \cdot 8}{2.5}$. Evaluate this expression in order to determine the horsepower rating of the engine.

Math for the Future *Problems 83 and 84 show some computations required in statistics. Completely simplify each expression.*

83. $\dfrac{105 + 80 + 115 + 95 + 105}{5}$

84. $\dfrac{(105 - 100)^2 + (80 - 100)^2 + (115 - 100)^2 + (95 - 100)^2 + (105 - 100)^2}{4}$

Explaining the Concepts

85. Explain why $\dfrac{2 + 7}{2 + 9} \neq \dfrac{7}{9}$.

86. Develop an example that illustrates why we perform multiplication before addition when there are no grouping symbols.

87. Develop an example that illustrates why we evaluate exponents before multiplication when there are no grouping symbols.

88. Explain the difference between -4^3 and $(-4)^3$.

The Graphing Calculator

Calculators can be used to evaluate exponential expressions. Figures 12(a) and (b) shows the results of Examples 2(a), (b), and (c) using a TI-84 Plus graphing calculator.

Figure 12

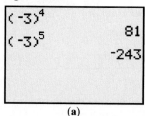
(a)

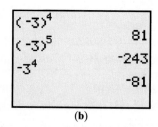
(b)

Graphing calculators know the order of operations. Figure 13(a) on the following page shows the result of Example 6(b), and Figure 13(b) the result of Example 7(a), using a TI-84 Plus graphing calculator. Be careful with the placement of parentheses when using the calculator!

Figure 13

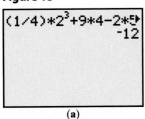

| (a) | (b) |

In Problems 89–96, use a calculator to evaluate each expression. When necessary, express answers rounded to two decimal places.

89. $\dfrac{4}{5} - \left(\dfrac{2}{3}\right)^2$

90. $3 - \left(\dfrac{6}{5}\right)^3$

91. $\dfrac{4^2 + 1}{13}$

92. $\dfrac{3^2 - 2^3}{1 + 3 \cdot 2}$

93. $\dfrac{3.5^3}{1.3^2} - 6.2^3$

94. $2.3^4 \cdot \dfrac{4}{11} - (3.7)^2 \cdot \dfrac{8}{3}$

95. $4.3[9.3^2 - 4(34.2 + 18.5)]$

96. $6.3^2 + 4.2^2$

R.5 Algebraic Expressions

Objectives

① Translate English Expressions into Mathematical Language

② Evaluate Algebraic Expressions

③ Simplify Algebraic Expressions by Combining Like Terms

④ Determine the Domain of a Variable

In Words
In algebra, letters of the alphabet are used to represent numbers.

We mentioned earlier that if a letter is used to represent *any* number from a given set of numbers, it is called a **variable**. A **constant** is either a fixed number, such as 5 or $\sqrt{2}$, or a letter that represents a fixed (possibly unspecified) number. For example, in Einstein's Theory of Relativity, $E = mc^2$, E and m are variables that represent total energy and mass, respectively, while c is a constant that represents the speed of light (299,792,458 meters per second).

An **algebraic expression** is any combination of variables, constants, grouping symbols such as parentheses ( ) and brackets [ ], and mathematical operations such as addition, subtraction, multiplication, division, and exponents. The following are examples of algebraic expressions.

$$3x + 4 \qquad 2y^2 - 5y - 12z \qquad \frac{5v^4 - v}{4 - v}$$

▶ ① **Translate English Expressions into Mathematical Language**

One of the neat features of mathematics is that we can translate various English phrases into a few math symbols. Table 2 lists some English words and phrases and their corresponding math symbols.

Work Smart

An algebraic expression is not the same as an algebraic statement. An expression might be $x + 3$ while a statement might be $x + 3 = 7$. Do you see the difference?

Table 2 Math Symbols and the Words They Represent			
Add (+)	Subtract (−)	Multiply (·)	Divide (/)
sum	difference	product	quotient
plus	minus	times	divided by
more than	subtracted from	of	per
exceeds by	less	twice	ratio
in excess of	less than		
added to	decreased by		
increased by			

EXAMPLE 1 **Writing English Phrases Using Math Symbols**

Express each English phrase as an algebraic expression.

(a) The sum of 3 and 8

(b) The quotient of 50 and some number y

(c) The number 11 subtracted from a number z

(d) Twice the sum of a number x and 5

Solution

(a) A sum implies the $+$ symbol, so "The sum of 3 and 8" is represented mathematically as $3 + 8$.

(b) "The quotient of 50 and some number y" is represented mathematically as $\dfrac{50}{y}$.

(c) "The number 11 subtracted from a number z" is represented mathematically as $z - 11$.

(d) "Twice the sum of a number x and 5" is represented as $2(x + 5)$. Notice that the example said "twice the sum." (The English phrase that leads to $2x + 5$ might be "the sum of twice a number and 5." Do you see the difference?) ●

Quick ✓

1. A _____ is a letter used to represent any number from a given set of numbers.

2. A _____ is either a fixed number or a letter used to represent a fixed (possibly unspecified) number.

In Problems 3–8, express each English phrase using an algebraic expression.

3. The sum of 3 and 11

4. The product of 6 and 7

5. The quotient of y and 4

6. The difference of 3 and z

7. Twice the difference of x and 3

8. The difference of twice a number x and 3

▶ ❷ **Evaluate Algebraic Expressions**

To **evaluate an algebraic expression,** substitute a numerical value for each variable in the expression, and simplify the result.

EXAMPLE 2 **Evaluating an Algebraic Expression**

Evaluate each expression for the given value of the variable.

(a) $4x + 3$ for $x = 5$ (b) $\dfrac{-z^2 + 4z}{z + 1}$ for $z = 9$ (c) $|10y - 8|$ for $y = \dfrac{1}{2}$

Solution

(a) We substitute 5 for x in the expression $4x + 3$.

$$4(5) + 3 = 20 + 3 = 23$$

(b) We substitute 9 for z in the expression $\dfrac{-z^2 + 4z}{z + 1}$.

$$\frac{-(9)^2 + 4 \cdot 9}{9 + 1} = \frac{-81 + 36}{10} = \frac{-45}{10} = \frac{-9 \cdot 5}{5 \cdot 2} = \frac{-9}{2} = -\frac{9}{2}$$

(c) We substitute $\dfrac{1}{2}$ for y in the expression $|10y - 8|$.

$$\left| 10 \cdot \frac{1}{2} - 8 \right| = |5 - 8| = |-3| = 3$$ ●

Quick ✓

9. To _____ an algebraic expression, substitute a numerical value for each variable in the expression, and simplify the result.

In Problems 10–13, evaluate each expression for the given value of the variable.

10. $-5x + 3$ for $x = 2$

11. $y^2 - 6y + 1$ for $y = -4$

12. $\dfrac{w + 8}{3w}$ for $w = 4$

13. $|4x - 5|$ for $x = \dfrac{1}{2}$

EXAMPLE 3 **Evaluating an Algebraic Expression**

Figure 14

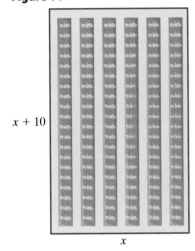

$x + 10$

x

The algebraic expression $2x + 2(x + 10)$ represents the perimeter of a rectangular field whose length is 10 yards more than its width. See Figure 14. Evaluate the algebraic expression for $x = 4, 8,$ and 10.

Solution
We evaluate the algebraic expression $2x + 2(x + 10)$ for each value of x.

$x = 4$: $2(4) + 2(4 + 10) = 8 + 2(14) = 8 + 28 = 36$ yards

$x = 8$: $2(8) + 2(8 + 10) = 16 + 2(18) = 16 + 36 = 52$ yards

$x = 10$: $2(10) + 2(10 + 10) = 20 + 2(20) = 20 + 40 = 60$ yards ●

Quick ✓

14. The algebraic expression $80x$ represents the number of Japanese yen that you could purchase for x dollars. Evaluate the algebraic expression for $x = 100$, 1000, and $10,000$ dollars. (source: *Yahoo! Finance*)

15. The algebraic expression $\dfrac{5}{9}(x - 32)$ represents the temperature in degrees Celsius that is equivalent to x degrees Fahrenheit. Determine the equivalent temperature for $x = 32, 86,$ and 212 degrees Fahrenheit.

❸ Simplify Algebraic Expressions by Combining Like Terms

One way to simplify algebraic expressions is to combine *like terms*. A **term** is a number or the product of a number and one or more variables raised to a power. In algebraic expressions, terms are separated by addition signs. See Table 3, which lists the terms in various algebraic expressions.

Table 3

Algebraic Expression	Terms
$5x + 4$	$5x, 4$
$7x^2 - 8x + 3 = 7x^2 + (-8x) + 3$	$7x^2, -8x, 3$
$3x^2 + 7y^2$	$3x^2, 7y^2$

Notice that in the algebraic expression $7x^2 - 8x + 3$, we first rewrite it with only addition signs as $7x^2 + (-8x) + 3$ to identify the terms.

Terms that have the same variable(s) and the same exponent(s) on the variable(s) are **like terms**. For example, $3x$ and $8x$ are like terms because both terms have the variable x raised to the 1st power. Also, $-4x^3y$ and $10x^3y$ are like terms because both have the variable x raised to the 3rd power, along with y raised to the 1st power. The numerical factor of the term is the **coefficient**. For example, the coefficient of $3x$ is 3; the coefficient of $-4x^3y$ is -4. Terms that have no number as a factor, such as xy, have the coefficient one (1) because $xy = 1 \cdot xy$. The coefficient of the term $-z$ is negative one (-1) because $-z = -1 \cdot z$. If a term consists of just a constant, the coefficient is the number itself. For example, the coefficient of the term 19 is 19.

We combine like terms by using the Distributive Property "in reverse."

▶ **EXAMPLE 4** **Combining Like Terms**

Simplify each algebraic expression by combining like terms.

(a) $5x + 3x$ **(b)** $z + 7z - 5$

Solution

(a) $5x + 3x = (5 + 3)x = 8x$

(b) Since $z = 1 \cdot z$ because of the multiplicative identity, we have

$$z + 7z - 5 = 1z + 7z - 5$$
$$= (1 + 7)z - 5$$
$$= 8z - 5$$

Quick ✓

16. Terms that have the same variable(s) and the same exponent(s) on the variables are called ___ _____.

17. The coefficient of the term $-mn$ is ___.

In Problems 18–21, simplify each expression by combining like terms.

18. $4x - 9x$ **19.** $-2x^2 + 13x^2$

20. $-5x - 3x + 6 - 3$ **21.** $6x - 10x - 4y + 12y$

Sometimes we must rearrange terms using the Commutative Property of Addition to combine like terms.

EXAMPLE 5 **Combining Like Terms Using the Commutative and Distributive Properties**

Simplify each algebraic expression.

(a) $13y^2 + 8 - 4y^2 + 3$ **(b)** $12z + 5 + 5z - 8z - 2$

Solution

(a) Use the Commutative Property to rearrange the terms.

$$13y^2 + 8 - 4y^2 + 3 = 13y^2 - 4y^2 + 8 + 3$$

Use the Distributive Property "in reverse": $\quad = (13 - 4)y^2 + 8 + 3$

Combine like terms: $\quad = 9y^2 + 11$

(b) We again rearrange terms before using the Distributive Property.

$$12z + 5 + 5z - 8z - 2 = 12z + 5z - 8z + 5 - 2$$

Use the Distributive Property "in reverse": $\quad = (12 + 5 - 8)z + 5 - 2$

Combine like terms: $\quad = 9z + 3$

Quick ✓

In Problems 22–24, simplify each expression by combining like terms.

22. $10y - 3 + 5y + 2$ **23.** $0.5x^2 + 1.3 + 1.8x^2 - 0.4$

24. $4z + 6 - 8z - 3 - 2z$

We also may need to remove parentheses using the Distributive Property before collecting like terms. This is consistent with the Rule for Order of Operations since expressions must be simplified by multiplying before adding or subtracting.

▶ **EXAMPLE 6**   **Combining Like Terms Using the Distributive Property**

Simplify each algebraic expression.

(a) $5(x - 3) - 2x$ **(b)** $x + 4 - 2(x + 3)$

Solution

(a) First, use the Distributive Property to remove the parentheses.

$$5(x - 3) - 2x = 5x - 15 - 2x$$

Rearrange terms: $= 5x - 2x - 15$

Combine like terms: $= 3x - 15$

(b) Distribute -2 to remove the parentheses.

$$x + 4 - 2(x + 3) = x + 4 - 2x - 6$$

Rearrange terms: $= x - 2x + 4 - 6$

Combine like terms: $= -x - 2$

●

EXAMPLE 7   **Combining Like Terms Using the Distributive Property**

Simplify each algebraic expression.

(a) $5(x - 3) - (7x - 4)$ **(b)** $\dfrac{1}{2}(4x + 3) + \dfrac{8x - 3}{5}$

Solution

(a) Recall that when grouping symbols are preceded by a minus sign, we multiply the terms in the grouping symbol by -1.

$$5(x - 3) - (7x - 4) = 5(x - 3) - 1(7x - 4)$$

Use the Distributive Property: $= 5x - 15 - 7x + 4$

Rearrange terms: $= 5x - 7x - 15 + 4$

Combine like terms: $= -2x - 11$

(b) Use the fact that $\dfrac{a}{b} = \dfrac{1}{b} \cdot a$, so $\dfrac{8x - 3}{5} = \dfrac{1}{5}(8x - 3)$.

$$\frac{1}{2}(4x + 3) + \frac{8x - 3}{5} = \frac{1}{2}(4x + 3) + \frac{1}{5}(8x - 3)$$

Use the Distributive Property: $= \dfrac{1}{2} \cdot 4x + \dfrac{1}{2} \cdot 3 + \dfrac{1}{5} \cdot 8x + \dfrac{1}{5} \cdot (-3)$

Simplify: $= 2x + \dfrac{3}{2} + \dfrac{8}{5}x - \dfrac{3}{5}$

Rearrange terms; rewrite fractions with least common denominator: $= \dfrac{10}{5}x + \dfrac{8}{5}x + \dfrac{15}{10} - \dfrac{6}{10}$

Combine like terms: $= \dfrac{18}{5}x + \dfrac{9}{10}$

●

Quick ✓

In Problems 25–30, simplify each expression by combining like terms.

25. $3(x - 2) + x$

26. $5(y + 3) - 10y - 4$

27. $3(z + 4) - 2(3z + 1)$

28. $-4(x - 2) - (2x + 4)$

29. $\dfrac{1}{2}(6x + 4) - \dfrac{15x + 5}{5}$

30. $\dfrac{5x - 1}{3} + \dfrac{5x + 9}{2}$

▶ ④ Determine the Domain of a Variable

In some algebraic expressions, the variable may be allowed to take on values from only a certain set of numbers. For example, in the expression $\dfrac{1}{x}$, the variable x cannot take on the value 0, because this would cause division by 0, which is not defined.

> **Definition**
>
> The set of values that a variable may assume is called the **domain of the variable.**

EXAMPLE 8 **Determining the Domain of a Variable**

Determine which of the following numbers are in the domain of the variable x for the expression $\dfrac{4}{x + 3}$.

(a) $x = 3$ **(b)** $x = 0$ **(c)** $x = -3$

Solution

We need to determine whether the value of the variable causes division by 0. That is, we need to determine whether the value of the variable causes $x + 3$ to equal 0. If it does, we exclude it from the domain.

(a) When $x = 3$, the denominator is $x + 3 = 3 + 3 = 6$, so 3 is in the domain of the variable.

(b) When $x = 0$, the denominator is $x + 3 = 0 + 3 = 3$, so 0 is in the domain.

(c) When $x = -3$, the denominator $x + 3 = -3 + 3 = 0$, so -3 is NOT in the domain. ●

Quick ✓

31. The set of values that a variable may assume is called the _____ of the variable.

In Problems 32–34, determine which of the following numbers are in the domain of the variable.

(a) $x = 2$ **(b)** $x = 0$ **(c)** $x = 4$ **(d)** $x = -3$

32. $\dfrac{2}{x - 4}$

33. $\dfrac{x}{x + 3}$

34. $\dfrac{x + 3}{x^2 + x - 6}$

R.5 Exercises MyMathLab®

Exercise numbers in green
have complete video solutions
in MyMathLab.

Problems **1–34,** are the Quick ✓s that follow the EXAMPLES.

Building Skills

In Problems 35–46, express each English phrase using an algebraic
expression. See Objective 1.

35. The sum of 5 and a number x

36. The difference of 10 and a number y

37. The product of 4 and a number z

38. The ratio of a number x and 5

39. A number y decreased by 7

40. A number z increased by 30

41. Twice the sum of a number t and 4

42. The sum of a number x and 5 divided by 10

43. Three less than five times a number x

44. Three times a number z increased by the quotient
of z and 8

45. The quotient of some number y and 3 increased by
the product of 6 and some number x

46. Twice some number x decreased by the ratio of a
number y and 3

In Problems 47–62, evaluate each expression for the given value of
the variable. See Objective 2.

47. $4x + 3$ for $x = 2$

48. $-5x + 1$ for $x = -3$

49. $x^2 + 5x - 3$ for $x = -2$

50. $y^2 - 4y + 5$ for $y = 3$

51. $4 - z^2$ for $z = -5$

52. $3 + z - 2z^2$ for $z = -4$

53. $\dfrac{2w}{w^2 + 2w + 1}$ for $w = 3$

54. $\dfrac{4z + 3}{z^2 - 4}$ for $z = 3$

55. $\dfrac{v^2 + 2v + 1}{v^2 + 3v + 2}$ for $v = 5$

56. $\dfrac{2x^2 + 5x + 2}{x^2 + 5x + 6}$ for $x = 3$

57. $|5x - 4|$ for $x = -5$

58. $|x^2 - 6x + 1|$ for $x = 2$

59. $(x + 2y)^2$ for $x = 3, y = -4$

60. $(a - 5b)^2$ for $a = 1, b = 3$

61. $\dfrac{(x + 2)^2}{|4x - 10|}$ for $x = 1$

62. $\dfrac{|3 - 5z|}{(z - 4)^2}$ for $z = 4$

In Problems 63–92, simplify each expression by combining like
terms. See Objective 3.

63. $3x - 2x$

64. $5y + 2y$

65. $-4z - 2z + 3$

66. $8x - 9x + 1$

67. $13z + 2 - 14z - 7$

68. $-10x + 6 + 4x - x + 1$

69. $\dfrac{3}{4}x + \dfrac{1}{6}x$

70. $\dfrac{3}{10}y + \dfrac{4}{15}y$

71. $2x + 3x^2 - 5x + x^2$

72. $-x - 3x^2 + 4x - x^2$

73. $-1.3x - 3.4 + 2.9x + 3.4$

74. $2.5y - 1.8 - 1.4y + 0.4$

75. $3x - 2 - x + 3 - 5x$

76. $10y + 3 - 2y + 6 + y$

77. $-2(5x - 4) - (4x + 1)$

78. $3(2y + 5) - 6(y + 2)$

79. $5(z + 2) - 6z$

80. $\dfrac{1}{2}(20x - 14) + \dfrac{1}{3}(6x + 9)$

81. $\dfrac{2}{5}(5x - 10) + \dfrac{1}{4}(8x + 4)$

82. $4(w + 2) + 3(4w + 3)$

83. $2(v - 3) + 5(2v - 1)$

84. $-4(w - 3) - (2w + 1)$

85. $\dfrac{3}{5}(10x + 4) - \dfrac{8x + 3}{2}$

86. $\dfrac{4}{3}(5y + 1) - \dfrac{2}{5}(3y - 4)$

87. $\dfrac{5}{6}\left(\dfrac{3}{10}x - \dfrac{2}{5}\right) + \dfrac{2}{3}\left(\dfrac{1}{6}x + \dfrac{1}{2}\right)$

88. $\frac{1}{4}\left(\frac{2}{3}x - \frac{1}{2}\right) + \frac{1}{10}\left(\frac{5}{2}x - \frac{15}{4}\right)$

89. $4.3(1.2x - 2.3) + 9.3x - 5.6$

90. $0.4(2.9x - 1.6) - 2.7(0.3x + 6.2)$

91. $6.2(x - 1.4) - 5.4(3.2x - 0.6)$

92. $9.3(0.2x - 0.8) + 3.8(1.3x + 6.3)$

In Problems 93–98, determine which of the following numbers are in the domain of the variable. See Objective 4.

(a) $x = 5$ **(b)** $x = -1$ **(c)** $x = -4$ **(d)** $x = 0$

93. $\frac{3}{x - 5}$

94. $\frac{7}{x + 1}$

95. $\frac{x + 1}{x + 5}$

96. $\frac{x + 4}{x - 1}$

97. $\frac{x}{x^2 - 5x}$

98. $\frac{x + 1}{x^2 + 5x + 4}$

Applying the Concepts

△ **99. Volume of a Cube** The algebraic expression s^3 represents the volume of a cube whose sides are length s. See the figure. Evaluate the algebraic expression for $s = 1, 2, 3,$ and 4 inches.

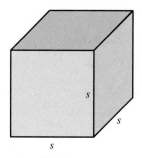

△**100. Area of a Triangle** The algebraic expression $\frac{1}{2}h(h + 2)$ represents the area of a triangle whose base is 2 centimeters longer than its height h. See the figure. Evaluate the algebraic expression for $h = 2, 5,$ and 10 centimeters.

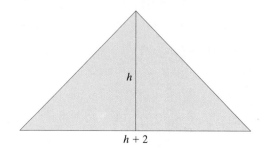

101. Projectile Motion The algebraic expression $-16t^2 + 75t$ represents the height (in feet) of a golf ball hit at an angle of 30° to the horizontal after t seconds.

  (a) Evaluate the algebraic expression for $t = 0, 1, 2, 3,$ and 4 seconds.

  (b) Use the results of part (a) to describe what happens to the golf ball as time passes.

102. Cost of Production The algebraic expression $30x + 1000$ represents the cost of manufacturing x watches in a day. Evaluate the algebraic expression for $x = 20, 30,$ and 40 watches.

103. How Old Is Tony? Suppose that Bob is x years of age. Write a mathematical expression for the following: "Tony is 5 years older than Bob." Evaluate the expression if Bob is $x = 13$ years of age.

104. Getting a Discount Suppose the regular price of a computer is p dollars. Write a mathematical expression for the following: "I'll give you $50 off regular price." Evaluate the expression if the regular price of the computer is $890.

105. Getting a Big Discount Suppose the regular price of a computer is p dollars. Write a mathematical expression for the following: "I'll give you half off regular price." Evaluate the expression if the regular price of the computer is $900.

106. How Old Is Marissa's Mother? Suppose that Marissa is x years of age. Write a mathematical expression for the following: The age of "Marissa's mother is twice the sum of Marissa's age and 3." How old is Marissa's mother if Marissa is $x = 18$ years of age?

Math for the Future *For Problems 107 and 108, evaluate the expression* $\frac{X - \mu}{\sigma}$ *for the given values.*

107. $X = 120, \mu = 100, \sigma = 15$

108. $X = 40, \mu = 50, \sigma = 10$

Extending the Concepts

In Problems 109–114, write an English phrase that would translate into the given mathematical expression.

109. $2z - 5$ **110.** $5x + 3$ **111.** $2(z - 5)$

112. $5(x + 3)$ **113.** $\frac{z + 3}{2}$ **114.** $\frac{t}{3} - 2t$

Explaining the Concepts

115. Explain the difference between a variable and a constant.

116. Explain the difference between a term and a factor.

117. Explain what like terms are. Explain how the Distributive Property is used to combine like terms.

118. Explain the difference between the direction "evaluate the algebraic expression" and "simplify the algebraic expression."

The Graphing Calculator

A graphing calculator can be used to evaluate an algebraic expression. Figure 15 shows the results of Example 2 using a TI-84 Plus graphing calculator.

Figure 15

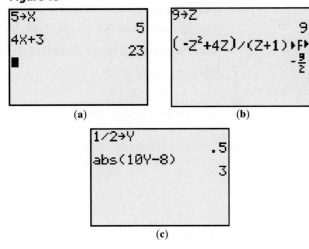

(a)

(b)

(c)

In Problems 119–126, use a graphing calculator to evaluate each algebraic expression for the given values of the variable.

119. $-4x + 3$ for **(a)** $x = 0$ **(b)** $x = -3$

120. $-5x + 9$ for **(a)** $x = 4$ **(b)** $x = -3$

121. $4x^2 - 8x + 3$ for **(a)** $x = 5$ **(b)** $x = -2$

122. $-9x^2 + x - 5$ for **(a)** $x = 3$ **(b)** $x = -4$

123. $\dfrac{3z - 1}{z^2 + 1}$ for **(a)** $z = -2$ **(b)** $z = 8$

124. $\dfrac{2y^2 + 5}{3y - 1}$ for **(a)** $y = 3$ **(b)** $y = -8$

125. $|-9x + 5|$ for **(a)** $x = 8$ **(b)** $x = -3$

126. $|-3x^2 + 5x - 2|$ for **(a)** $x = 6$ **(b)** $x = -4$

127. Use your graphing calculator to evaluate $\dfrac{2x + 1}{x - 5}$ when $x = 5$. What result does the calculator display? Why?

128. Use your graphing calculator to evaluate $\dfrac{x + 2}{(x + 2)(x - 2)}$ when $x = -2$. What result does your calculator display? Why?

1 Linear Equations and Inequalities

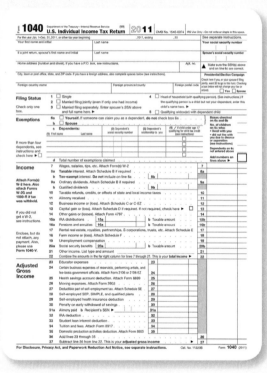

Did you know that the amount of income tax an individual or a couple pays to the federal government can be found by solving a linear equation? See Problems 103 and 104 in Section 1.1.

The Big Picture: Putting It Together

In Chapter R, we reviewed skills learned in earlier courses. You will use these skills throughout the text.

We now begin our discussion of *algebra*. The word "algebra" is derived from the Arabic word *al-jabr*. The word is part of the title of a ninth-century work, "Hisâb al-jabr w'al-muqâbalah," written during the golden age of Islamic science and mathematics by Muhammad ibn Mûqâ al-Khowârizmî.

The word *al-jabr* means "restoration," a reference to the fact that if a number is added to one side of an equation, then it must also be added to the other side in order to "restore" the equality. The title of the work "Hisâb al-jabr w'al-muqâbalah" means "the science of restoring and canceling." Today, algebra means a great deal more.

Chapter 1 consists of Part I, a review of linear equations and inequalities in one variable, and Part II, a review of linear equations and inequalities in two variables.

Outline

Part I: Linear Equations and Inequalities in One Variable

1.1 Linear Equations in One Variable

Objectives

1 Determine Whether a Number Is a Solution to an Equation

2 Solve Linear Equations

3 Determine Whether an Equation Is a Conditional Equation, an Identity, or a Contradiction

Are You Ready for This Section?

Before getting started, take this readiness quiz. If you get a problem wrong, go back to the section cited and review the material.

R1. Determine the additive inverse of 5. [Section R.3, p. 21]

R2. Determine the multiplicative inverse of -3. [Section R.3, pp. 23–24]

R3. Use the Reduction Property to simplify $\frac{1}{5} \cdot 5x$. [Section R.3, p. 25]

R4. Find the Least Common Denominator of $\frac{3}{8}$ and $\frac{5}{12}$. [Section R.3, pp. 27–28]

R5. Use the Distributive Property to remove the parentheses: $6(z - 2)$ [Section R.3, pp. 29–30]

R6. What is the coefficient of $-4x$? [Section R.5, pp. 41–42]

R7. Simplify by combining like terms: $4(y - 2) - y + 5$ [Section R.5, pp. 41–44]

R8. Evaluate the expression $-5(x + 3) - 8$ when $x = -2$. [Section R.5, pp. 40–41]

R9. Is $x = 3$ in the domain of $\frac{2}{x + 3}$? Is $x = -3$ in the domain? [Section R.5, p. 44]

▶ 1 Determine Whether a Number Is a Solution to an Equation

An **equation in one variable** is a statement made up of two expressions that are equal, where at least one of the expressions contains the variable. The expressions are called the **sides** of the equation. Examples of equations in one variable are

$$2y + 5 = 0 \qquad 4x + 5 = -2x + 10 \qquad \frac{3}{z + 2} = 9$$

In this section, we will concentrate on solving *linear equations in one variable.*

In Words

In the equation $2y + 5 = 0$, the expression $2y + 5$ is the *left side* of the equation and 0 is the *right side*. In this equation, only the left side contains the variable, y. In the equation $4x + 5 = -2x + 10$, the expression $4x + 5$ is the *left side* of the equation and $-2x + 10$ is the *right side*. In this equation, both sides have expressions that contain the variable, x.

Definition

A **linear equation in one variable** is an equation with one unknown that is written to the first power. Linear equations in one variable can be written in the form

$$ax + b = 0$$

where a and b are real numbers and $a \neq 0$.

The following are all examples of linear equations in one variable because they can be written in the form $ax + b = 0$ with a little algebraic manipulation.

$$4x - 3 = 12 \qquad \frac{2}{3}y + \frac{1}{5} = \frac{2}{15} \qquad -0.73p + 1.23 = 1.34p + 8.05$$

Because an equation is a statement, it can be either true or false, depending on the value of the variable. Any value of the variable that results in a true statement is called a **solution** of the equation. If a value of the variable results in a true statement, we say that the value **satisfies** the equation. To determine whether a number satisfies an equation, we replace the variable with the number. If the left side of the equation equals the right side, then the statement is true and the number is a solution.

Ready? ... Answers **R1.** -5
R2. $-\frac{1}{3}$ **R3.** x **R4.** 24 **R5.** $6z - 12$
R6. -4 **R7.** $3y - 3$ **R8.** -13 **R9.** Yes; No

(**EXAMPLE 1**) **Determining Whether a Number Is a Solution to a Linear Equation**

Determine whether the following numbers are solutions to the equation

$$3(x - 1) = -2x + 12$$

(**a**) $x = 5$ (**b**) $x = 3$

Solution

(**a**) Let $x = 5$ in the equation and simplify.

$$3(x - 1) = -2x + 12$$
$$3(5 - 1) \stackrel{?}{=} -2(5) + 12$$

Simplify: $\qquad\qquad 3(4) \stackrel{?}{=} -10 + 12$

$$12 \neq 2$$

> **In Words**
> The symbol $\stackrel{?}{=}$ is used to indicate that we are unsure whether the left side of the equation equals the right side of the equation.

Because the left side of the equation does not equal the right side of the equation, we do not have a true statement. Therefore, $x = 5$ is not a solution.

(**b**) $\qquad\qquad\qquad\qquad 3(x - 1) = -2x + 12$

Let $x = 3$: $\qquad 3(3 - 1) \stackrel{?}{=} -2(3) + 12$

Simplify: $\qquad\qquad 3(2) \stackrel{?}{=} -6 + 12$

$$6 = 6 \quad \text{True}$$

Because the left and right sides of the equation are equal, we have a true statement. Therefore, $x = 3$ is a solution to the equation. ●

Quick ✓

1. The equation $3x + 5 = 2x - 3$ is a _____ equation in one variable. The expressions $3x + 5$ and $2x - 3$ are called _____ of the equation.

2. The _____ of a linear equation is the value or values of the variable that satisfy the equation.

In Problems 3–5, determine which of the given numbers are solutions to the equation.

3. $-5x + 3 = -2; x = -2, x = 1, x = 3$

4. $3x + 2 = 2x - 5; x = 0, x = 6, x = -7$

5. $-3(z + 2) = 4z + 1; z = -3, z = -1, z = 2$

Work Smart

The directions "solve", "simplify", and "evaluate" are different! We *solve* equations. We *simplify* algebraic expressions to form equivalent algebraic expressions. We *evaluate* algebraic expressions to find the value of the expression for a specific value of the variable.

❷ Solve Linear Equations

To **solve an equation** means to find ALL the solutions of the equation. The set of all solutions to the equation is called the **solution set** of the equation.

One method for solving equations uses a series of *equivalent equations* beginning with the original equation and ending with a solution.

> **Definition**
> Two or more equations with the same solutions are **equivalent equations.**

But how do we find equivalent equations? One way is to use the *Addition Property of Equality.*

In Words
The Addition Property says that whatever you add to one side of an equation, you must also add to the other side.

Addition Property of Equality

The **Addition Property of Equality** states that for real numbers a, b, and c,

$$\text{if } a = b, \quad \text{then} \quad a + c = b + c$$

For example, if $x = 3$, then $x + 2 = 3 + 2$ (we added 2 to both sides of the equation). Because $a - b$ is equivalent to $a + (-b)$, the Addition Property can be used to add a real number to each side of an equation or to subtract a real number from each side of an equation. You will use this handy property a great deal in algebra.

We can also find an equivalent equation by using the *Multiplication Property of Equality*.

In Words
The Multiplication Property says that whenever you multiply one side of an equation by a nonzero expression, you must also multiply the other side by the same nonzero expression.

Multiplication Property of Equality

The **Multiplication Property of Equality** states that for real numbers a, b, and c where $c \neq 0$,

$$\text{if } a = b, \quad \text{then} \quad ac = bc$$

For example, if $5x = 30$, then $\dfrac{1}{5} \cdot 5x = \dfrac{1}{5} \cdot 30$. Also, because $\dfrac{a}{b} = a \cdot \dfrac{1}{b}$, the Multiplicative Property of Equality says that we may divide both sides of an equation by the same nonzero number and the result will be an equivalent equation. Thus if $5x = 30$, then it is also true that $\dfrac{5x}{5} = \dfrac{30}{5}$. So the Multiplication Property can be used to multiply or divide each side of the equation by some nonzero number.

EXAMPLE 2 **Using the Addition and Multiplication Properties to Solve a Linear Equation**

Solve the linear equation: $\dfrac{1}{3}x - 2 = 4$

Solution

Work Smart

The number in front of the variable expression is the coefficient. For example, the coefficient in the expression $2x$ is 2.

The goal in solving any linear equation is to get the variable by itself with a coefficient of 1—that is, to **isolate the variable.**

$$\frac{1}{3}x - 2 = 4$$

Addition Property of Equality; add 2 to both sides: $\quad \left(\dfrac{1}{3}x - 2\right) + 2 = 4 + 2$

Simplify: $\quad \dfrac{1}{3}x = 6$

Multiplication Property of Equality; multiply both sides by 3: $\quad 3\left(\dfrac{1}{3}x\right) = 3 \cdot 6$

Simplify: $\quad x = 18$

Check

$$\frac{1}{3}x - 2 = 4$$

Let $x = 18$ in the original equation: $\quad \dfrac{1}{3}(18) - 2 \overset{?}{=} 4$

$$6 - 2 \overset{?}{=} 4$$

$$4 = 4 \quad \text{True}$$

The solution set is $\{18\}$.

6. *True or False* The Addition Property of Equality says that whatever you add to one side of an equation you must also add to the other side.

7. To _____ the variable means to get the variable by itself with a coefficient of 1.

In Problems 8–10, solve each equation and verify your solution.

8. $3x + 8 = 17$　　　　**9.** $-4a - 7 = 1$　　　　**10.** $5y + 1 = 2$

Often, we must combine like terms or use the Distributive Property to eliminate parentheses before we can use the Addition or Multiplication Properties. As usual, our goal is to isolate the variable.

▶ **EXAMPLE 3**　**Solving a Linear Equation by Combining Like Terms**

Solve the linear equation: $3y - 2 + 5y = 2y + 5 + 4y + 3$

Solution

$$3y - 2 + 5y = 2y + 5 + 4y + 3$$

Combine like terms:　　　　$8y - 2 = 6y + 8$

Subtract 6y from both sides:　　$8y - 2 - 6y = 6y + 8 - 6y$

$$2y - 2 = 8$$

Add 2 to both sides:　　$2y - 2 + 2 = 8 + 2$

$$2y = 10$$

Divide both sides by 2:　　$\dfrac{2y}{2} = \dfrac{10}{2}$

$$y = 5$$

Check　　　　$3y - 2 + 5y = 2y + 5 + 4y + 3$

Let y = 5 in the original equation:　$3(5) - 2 + 5(5) \stackrel{?}{=} 2(5) + 5 + 4(5) + 3$

$$15 - 2 + 25 \stackrel{?}{=} 10 + 5 + 20 + 3$$

$$38 = 38 \quad \text{True}$$

The solution set is $\{5\}$.

In Problems 11–13, solve each linear equation and verify your solution.

11. $2x + 3 + 5x + 1 = 4x + 10$

12. $4b + 3 - b - 8 - 5b = 2b - 1 - b - 1$

13. $2w + 8 - 7w + 1 = 3w - 1 + 2w - 5$

▶ **EXAMPLE 4**　**Solving a Linear Equation Using the Distributive Property**

Solve the linear equation: $4(x + 3) = x - 3(x - 2)$

Solution

Use the Distributive Property　　$4(x + 3) = x - 3(x - 2)$
to remove parentheses:　　　$4x + 12 = x - 3x + 6$

Combine like terms:　　$4x + 12 = -2x + 6$

Add 2x to both sides:　　$4x + 12 + 2x = -2x + 6 + 2x$

$$6x + 12 = 6$$

Subtract 12 from both sides: $\quad 6x + 12 - 12 = 6 - 12$

$$6x = -6$$

Divide both sides by 6: $\qquad \dfrac{6x}{6} = \dfrac{-6}{6}$

$$x = -1$$

Check

Let $x = -1$ in the original equation:

$$4(x + 3) = x - 3(x - 2)$$

$$4(-1 + 3) \overset{?}{=} -1 - 3(-1 - 2)$$

$$4(2) \overset{?}{=} -1 - 3(-3)$$

$$8 \overset{?}{=} -1 + 9$$

$$8 = 8 \quad \text{True}$$

Because $x = -1$ satisfies the equation, the solution of the equation is -1, or the solution set is $\{-1\}$.

Quick ✓

In Problems 14–17, solve each linear equation. Be sure to verify your solution.

14. $4(x - 1) = 12$

15. $-2(x - 4) - 6 = 3(x + 6) + 4$

16. $4(x + 3) - 8x = 3(x + 2) + x$

17. $5(x - 3) + 3(x + 3) = 2x - 3$

We now summarize the steps for solving a linear equation. Not every step is always necessary.

Solving a Linear Equation

Step 1: Remove any parentheses using the Distributive Property.

Step 2: Combine like terms on each side of the equation.

Step 3: Use the Addition Property of Equality to get the terms with the variable on one side of the equation and the constants on the other side.

Step 4: Use the Multiplication Property of Equality to get the coefficient of the variable to equal 1.

Step 5: Check the solution to verify that it satisfies the original equation.

▶ **Linear Equations with Fractions or Decimals**

To rewrite a linear equation that contains fractions as an equivalent equation without fractions, multiply both sides of the equation by the Least Common Denominator (LCD) of all the fractions in the equation.

EXAMPLE 5 **How to Solve a Linear Equation That Contains Fractions**

Solve the linear equation: $\dfrac{y + 1}{4} + \dfrac{y - 2}{10} = \dfrac{y + 7}{20}$

Step-by-Step Solution

Before we follow the summary steps, rewrite the equation without fractions by multiplying both sides by the LCD, which is 20.

$$20 \cdot \left(\dfrac{y + 1}{4} + \dfrac{y - 2}{10} \right) = 20 \cdot \left(\dfrac{y + 7}{20} \right)$$

Now follow Steps 1–5 for solving a linear equation.

Step 1: Remove all parentheses using the Distributive Property.

$$20 \cdot \left(\frac{y+1}{4} + \frac{y-2}{10} \right) = 20 \cdot \left(\frac{y+7}{20} \right)$$

Use the Distributive Property: $\quad 20 \cdot \dfrac{y+1}{4} + 20 \cdot \dfrac{y-2}{10} = 20 \cdot \dfrac{y+7}{20}$

Divide out common factors: $\quad 5(y+1) + 2(y-2) = y+7$

Use the Distributive Property: $\quad 5y + 5 + 2y - 4 = y + 7$

Step 2: Combine like terms on each side of the equation.

$$7y + 1 = y + 7$$

Step 3: Use the Addition Property of Equality to get the terms with the variable on one side of the equation and the constants on the other side.

Subtract y from both sides: $\quad 7y + 1 - y = y + 7 - y$

$$6y + 1 = 7$$

Subtract 1 from both sides: $\quad 6y + 1 - 1 = 7 - 1$

$$6y = 6$$

Step 4: Use the Multiplication Property of Equality to get the coefficient on the variable to equal 1.

Divide both sides by 6: $\quad \dfrac{6y}{6} = \dfrac{6}{6}$

$$y = 1$$

Step 5: Check: Verify the solution.

$$\frac{y+1}{4} + \frac{y-2}{10} = \frac{y+7}{20}$$

Let $y = 1$ in the original equation:

$$\frac{1+1}{4} + \frac{1-2}{10} \overset{?}{=} \frac{1+7}{20}$$

$$\frac{2}{4} + \frac{-1}{10} \overset{?}{=} \frac{8}{20}$$

Rewrite each rational number with LCD = 20:

$$\frac{2}{4} \cdot \frac{5}{5} + \frac{-1}{10} \cdot \frac{2}{2} \overset{?}{=} \frac{8}{20}$$

$$\frac{10}{20} + \frac{-2}{20} \overset{?}{=} \frac{8}{20}$$

$$\frac{8}{20} = \frac{8}{20} \quad \text{True}$$

The solution of the equation is 1, or the solution set is $\{1\}$.

Quick ✓

18. To solve a linear equation containing fractions, we can multiply each side of the equation by the ____ _____ _____ to clear the fractions.

In Problems 19–22, solve each linear equation and verify your solution.

19. $\dfrac{3y}{2} + \dfrac{y}{6} = \dfrac{10}{3}$

20. $\dfrac{3x}{4} - \dfrac{5}{12} = \dfrac{5x}{6}$

21. $\dfrac{x+2}{6} + 2 = \dfrac{5}{3}$

22. $\dfrac{4x+3}{9} - \dfrac{2x+1}{2} = \dfrac{1}{6}$

To rewrite a linear equation that contains decimals as an equivalent equation without decimals, multiply both sides of the equation by a power of 10 that removes the decimals.

For example, multiplying $0.7 \left(\text{or } \dfrac{7}{10}\right)$ by 10 "eliminates" the decimal, and multiplying $0.03 \left(\text{or } \dfrac{3}{100}\right)$ by 100 "eliminates" the decimal.

EXAMPLE 6   **Solving a Linear Equation That Contains Decimals**

Solve the linear equation: $0.5x - 0.4 = 0.3x + 0.2$

Solution

Multiplying both sides of the equation by 10 will "eliminate" the decimal because each of the decimals is written to the tenths position.

$$10(0.5x - 0.4) = 10(0.3x + 0.2)$$

Use the Distributive Property: $\quad 10(0.5x) - 10(0.4) = 10(0.3x) + 10(0.2)$

$$5x - 4 = 3x + 2$$

Subtract 3x from both sides: $\quad 5x - 4 - 3x = 3x + 2 - 3x$

$$2x - 4 = 2$$

Add 4 to both sides: $\quad 2x - 4 + 4 = 2 + 4$

$$2x = 6$$

Divide both sides by 2: $\quad \dfrac{2x}{2} = \dfrac{6}{2}$

$$x = 3$$

Check $0.5x - 0.4 = 0.3x + 0.2$

Let $x = 3$ in the original equation: $\quad 0.5(3) - 0.4 \overset{?}{=} 0.3(3) + 0.2$

$$1.5 - 0.4 \overset{?}{=} 0.9 + 0.2$$

$$1.1 = 1.1 \quad \text{True}$$

Because $x = 3$ satisfies the equation, the solution of the equation is 3, or the solution set is $\{3\}$. ●

Quick ✓

In Problems 23 and 24, solve each linear equation. Be sure to verify your solution.

23. $0.07x - 1.3 = 0.05x - 1.1$ **24.** $0.4(y + 3) = 0.5(y - 4)$

▶ ❸ **Determine Whether an Equation Is a Conditional Equation, an Identity, or a Contradiction**

Each linear equation we have studied so far has had one solution. Linear equations may have one solution, no solution, or infinitely many solutions. We can classify equations based on the number of their solutions.

The equations that we have solved so far are called *conditional equations.*

Definition

A **conditional equation** is an equation that is true for some values of the variable and false for other values of the variable.

For example, the equation $x + 7 = 10$ is a conditional equation because it is true when $x = 3$ and false for every other real number x.

Definition

An equation that is false for every value of the variable is a **contradiction.**

For example, the equation $3x + 8 = 3x + 6$ is a contradiction because it is false for any value of x. We recognize contradictions by creating equivalent equations. For example, if we subtract $3x$ from both sides of $3x + 8 = 3x + 6$, we get $8 = 6$, which is clearly false. Contradictions have no solution, so the solution set is the empty set and is written as $\varnothing$ or $\{\ \}$.

Definition

An equation that is satisfied for every value of the variable for which both sides of the equation are defined is called an **identity.**

For example, $2x + 3 + x + 8 = 3x + 11$ is an identity because any real number x satisfies the equation. We recognize identities by creating equivalent equations. For example, if we combine like terms in the equation $2x + 3 + x + 8 = 3x + 11$, we get $3x + 11 = 3x + 11$, which is true no matter what value of x we choose. Therefore, the solution set of linear equations that are identities is the set of all real numbers, which we express as either $\{x \mid x \text{ is any real number}\}$ or $\mathbb{R}$.

EXAMPLE 7 **Classifying a Linear Equation**

Solve the linear equation $3(x + 3) - 6x = 5(x + 1) - 8x$. Then state whether it is an identity, a contradiction, or a conditional equation.

Solution

As with any linear equation, our goal is to isolate the variable.

$$3(x + 3) - 6x = 5(x + 1) - 8x$$

Use the Distributive Property: $\qquad 3x + 9 - 6x = 5x + 5 - 8x$

Combine like terms: $\qquad -3x + 9 = -3x + 5$

Add $3x$ to both sides: $\qquad -3x + 9 + 3x = -3x + 5 + 3x$

$$9 = 5$$

Work Smart

In the solution to Example 7, we obtained the equation

$$-3x + 9 = -3x + 5$$

You may recognize at this point that the equation is a contradiction and express the solution set as $\varnothing$ or $\{\ \}$.

Because the last statement, $9 = 5$, is false, the equation is a contradiction. The original equation is a contradiction and has no solution. The solution set is $\varnothing$ or $\{\ \}$.

EXAMPLE 8 **Classifying a Linear Equation**

Solve the linear equation $-4(x - 2) + 3(4x + 2) = 2(4x + 7)$. Then state whether the equation is an identity, a contradiction, or a conditional equation.

Solution

$$-4(x - 2) + 3(4x + 2) = 2(4x + 7)$$

Use the Distributive Property: $\qquad -4x + 8 + 12x + 6 = 8x + 14$

Combine like terms: $\qquad 8x + 14 = 8x + 14$

The equation $8x + 14 = 8x + 14$ is true for all real numbers x. So the original equation is an identity, and its solution set is $\{x \mid x \text{ is any real number}\}$ or $\mathbb{R}$.

If we continued to solve the equation in Example 8, we could write

$$8x + 14 - 8x = 8x + 14 - 8x$$
$$14 = 14$$

The statement $14 = 14$ is true for all real numbers x, so the solution set of the original equation is all real numbers.

Quick ✓

25. A(n) _____ _____ is an equation that is true for some values of the variable and false for others.

26. A(n) _____ is an equation that is false for every value of the variable. A(n) _____ is an equation that is satisfied by every allowed choice of the variable.

In Problems 27–30, solve the equation and state whether it is an identity, a contradiction, or a conditional equation.

27. $4(x + 2) = 4x + 2$

28. $3(x - 2) = 2x - 6 + x$

29. $-4x + 2 + x + 1 = -4(x + 2) + 11$

30. $-3(z + 1) + 2(z - 3) = z + 6 - 2z - 15$

1.1 Exercises MyMathLab® PRACTICE

Exercise numbers in green have complete video solutions in MyMathLab.

*Problems **1–30** are the* Quick ✓ *s that follow the* EXAMPLES.

Building Skills

In Problems 31–36, determine which of the numbers are solutions to the given equation. See Objective 1.

31. $8x - 10 = 6; x = -2, x = 1, x = 2$

32. $-4x - 3 = -15; x = -2, x = 1, x = 3$

33. $5m - 3 = -3m + 5; m = -2, m = 1, m = 3$

34. $6x + 1 = -2x + 9; x = -2, x = 1, x = 4$

35. $4(x - 1) = 3x + 1; x = -1, x = 2, x = 5$

36. $3(t + 1) - t = 4t + 9; t = -3, t = -1, t = 2$

In Problems 37–56, solve each linear equation, and be sure to verify your solution. See Objective 2.

37. $3x + 1 = 7$

38. $8x - 6 = 18$

39. $4z + 3 = 2$

40. $-6x - 5 = 13$

41. $-3w + 2w + 5 = -4$

42. $-7t - 3 + 5t = 11$

43. $5x + 2 - 2x + 3 = 7x + 2 - x + 5$

44. $-6x + 2 + 2x + 9 + x = 5x + 10 - 6x + 11$

45. $3(x + 2) = -6$

46. $4(z - 2) = 12$

47. $\dfrac{4y}{5} - \dfrac{14}{15} = \dfrac{y}{3}$

48. $\dfrac{3x}{2} + \dfrac{x}{6} = -\dfrac{5}{3}$

49. $\dfrac{4x + 3}{9} - \dfrac{2x + 1}{2} = \dfrac{1}{6}$

50. $\dfrac{2x + 1}{3} - \dfrac{6x - 1}{4} = -\dfrac{5}{12}$

51. $\dfrac{y}{10} + 6 = \dfrac{y}{4} + 12$

52. $\dfrac{n}{5} + 3 = \dfrac{n}{2} + 6$

53. $0.5x - 3.2 = -1.7$

54. $0.3z + 0.8 = -0.1$

55. $0.14x + 2.23 = 0.09x + 1.98$

56. $0.12y - 5.26 = 0.05y + 1.25$

In Problems 57–72, solve the equation. Identify each equation as an identity, a contradiction, or a conditional equation. Be sure to verify your solution. See Objective 3.

57. $4(x + 1) = 4x$

58. $5(s + 3) = 3s + 2s$

59. $4m + 1 - 6m = 2(m + 3) - 4m$

60. $10(x - 1) - 4x = 2x - 1 + 4(x + 1)$

61. $2(y + 1) - 3(y - 2) = 5y + 8 - 6y$

62. $8(w + 2) - 3w = 7(w + 2) + 2(1 - w)$

63. $\dfrac{x}{4} + \dfrac{3x}{10} = -\dfrac{33}{20}$

64. $\dfrac{z-2}{4} + \dfrac{2z-3}{6} = 7$

65. $3p - \dfrac{p}{4} = \dfrac{11p}{4} + 1$

66. $\dfrac{r}{2} + 2(r-1) = \dfrac{5r}{2} + 4$

67. $\dfrac{2x+1}{2} - \dfrac{x+1}{5} = \dfrac{23}{10}$

68. $\dfrac{3x+1}{4} - \dfrac{7x-4}{2} = \dfrac{26}{3}$

69. $0.4(z+1) - 0.7z = -0.1z + 0.7 - 0.2z - 0.3$

70. $0.9(z-3) - 0.2(z-5) = 0.4(z+1) + 0.3z - 2.1$

71. $\dfrac{1}{3}(2x-3) + 2 = \dfrac{5}{6}(x+3) - \dfrac{11}{12}$

72. $\dfrac{4}{5}(y-4) + 3 = \dfrac{2}{3}(y+1) + \dfrac{4}{15}$

Mixed Practice

In Problems 73–90, solve each linear equation and verify the solution. State whether the equation is an identity, a contradiction, or a conditional equation.

73. $7y - 8 = -7$ **74.** $4y + 5 = 7$

75. $4a + 3 - 2a + 4 = 5a - 7 + a$

76. $-5x + 5 + 3x + 7 = 5x - 6 + x + 12$

77. $4(p+3) = 3(p-2) + p + 18$

78. $7(x+2) = 5(x-2) + 2(x+12)$

79. $4b - 3(b+1) - b = 5(b-1) - 5b$

80. $13z - 8(z+1) = 2(z-3) + 3z$

81. $8(4x+6) = 11-(x+7)$

82. $5(2x+3) = 9 - (x+5)$

83. $\dfrac{m+1}{4} + \dfrac{5}{6} = \dfrac{2m-1}{12}$

84. $\dfrac{z-4}{6} - \dfrac{2z+1}{9} = \dfrac{1}{3}$

85. $0.3x - 1.3 = 0.5x - 0.7$

86. $-0.8y + 0.3 = 0.2y - 3.7$

87. $-0.8(x+1) = 0.2(x+4)$

88. $0.5(x+3) = 0.2(x-6)$

89. $\dfrac{1}{4}(x-4) + 3 = \dfrac{1}{3}(2x+6) - \dfrac{5}{6}$

90. $\dfrac{1}{5}(2a-5) - 4 = \dfrac{1}{2}(a+4) - \dfrac{7}{10}$

Applying the Concepts

91. Find a such that the solution set of $ax + 3 = 15$ is $\{-3\}$.

92. Find a such that the solution set of $ax + 6 = 20$ is $\{7\}$.

93. Find a such that the solution set of $a(x-1) = 3(x-1)$ is the set of all real numbers.

94. Find a such that the solution set of $ax + 3 = 2x + 3(x+1)$ is the set of all real numbers.

In Section R.5, we introduced the domain of a variable. The domain of a variable is the set of all values that a variable may assume. Recall that division by zero is not defined, so any value of the variable that results in division by zero must be excluded from the domain. In Problems 95–100, determine which values of the variable must be excluded from the domain.

95. $\dfrac{5}{2x+1}$ **96.** $\dfrac{-3}{5x+8}$

97. $\dfrac{3x+7}{4x-3}$ **98.** $\dfrac{2x}{3x+1}$

99. $\dfrac{6x-2}{3(x+1)-6}$ **100.** $\dfrac{-2x+7}{4(x-3)+2}$

101. Interest Suppose you have a credit card debt of $2000. Last month, the bank charged you $25 interest on the debt. The solution to the equation $25 = \dfrac{2000}{12} \cdot r$ represents the annual interest rate r on the credit card. Find the annual interest rate on the credit card.

102. How Much Do I Make? Last week, before taxes, you earned $539 after working 26 hours at your regular hourly rate and 6 hours at time-and-a-half. The solution to the equation $26x + 9x = 539$ represents your regular hourly rate, x. Determine your regular hourly rate.

103. Paying Your Taxes You are single and have just determined that you paid $4412.50 in federal income tax in 2011. The solution to the equation $4412.5 = 0.15(I - 8500) + 850$ represents your annual adjusted income I for 2011. Determine your annual adjusted income for 2011. (SOURCE: *Internal Revenue Service*)

104. Paying Your Taxes You are married and have just determined that you paid $11,037.50 in federal income tax in 2011. The solution to the equation $11,037.5 = 0.25(I - 69,000) + 9500$ represents your annual adjusted income I for 2011. Determine your annual adjusted income for 2011. (SOURCE: *Internal Revenue Service*)

Explaining the Concepts

105. Explain the difference between $4(x + 1) - 2$ and $4(x + 1) = 2$. In general, what is the difference between an algebraic expression and an equation?

106. Explain the difference between the directions "solve" and "simplify."

107. Make up a linear equation that has one solution. Make up a linear equation that has no solution. Make up a linear equation that is an identity. Comment on the differences and similarities in making up these equations.

1.2 An Introduction to Problem Solving

Objectives

1 Translate English Sentences into Mathematical Statements

2 Model and Solve Direct Translation Problems

3 Model and Solve Mixture Problems

4 Model and Solve Uniform Motion Problems

Are You Ready for This Section?

Before getting started, take the following readiness quiz. If you get a problem wrong, go back to the section cited and review the material.

In Problems R1–R6, express each English phrase as an algebraic expression. [Section R.5, pp. 39–40]

R1. The sum of 12 and a number z

R2. The product of 4 and x

R3. A number y decreased by 87

R4. The quotient of z and 12

R5. Four times the sum of x and 7

R6. The sum of four times a number x and 7

1 Translate English Sentences into Mathematical Statements

Work Smart

We learned in the last section that an equation is a statement in which two algebraic expressions, separated by an equal sign, are equal.

In English, a complete sentence must contain a subject and a verb, so expressions or "phrases" are not complete sentences. For example, the expression "5 more than a number x" does not contain a verb and therefore is not a complete sentence. The statement "5 more than a number x is 18" is a complete sentence because it contains a subject and a verb, so we can translate it into a mathematical statement. Mathematical statements are represented symbolically through equations. In English, statements can be true or false. Mathematical statements can be true or false as well.

You may want to look back at page 39 in Section R.5 for a review of key words that translate into mathematical symbols. Table 1 provides a summary of words that typically translate into an equal sign.

Table 1 Words That Translate into an Equal Sign

is	yields	are	equals	is equivalent to
was	gives	results in	is equal to	

Let's translate some English sentences into mathematical statements.

▶ EXAMPLE 1 **Translating English Sentences into Mathematical Statements**

Translate each sentence into a mathematical statement. Do not solve the equation.

(a) Five more than a number x is 20.

(b) Four times the sum of a number z and 3 is 15.

(c) The difference of x and 5 equals the quotient of x and 2.

Solution

(a) 5 more than a number x ... is ... 20

$$x + 5 \qquad = 20$$

(b) The expression "Four times the sum of" tells us to find the sum first and then multiply this result by 4.

Four times sum of a number z and 3 ... is ... 15

$$4(z + 3) \qquad = 15$$

(c) The difference of x and 5 ... equals ... the quotient of x and 2

$$x - 5 \qquad = \qquad \frac{x}{2}$$

Work Smart

The English statement "The sum of four times a number z and 3 is 15" would be expressed mathematically as $4z + 3 = 15$. Do you see the difference between this statement and the one in Example 1(b)?

Quick ✓

1. Mathematical statements are represented symbolically as _____.

In Problems 2–6, translate each English statement into a mathematical statement. Do not solve the equation.

2. The sum of x and 7 results in 12.

3. The product of 3 and y is equal to 21.

4. Two times the sum of n and 3 is 5.

5. The difference of x and 10 equals the quotient of x and 2.

6. The sum of two times n and 3 is 5.

▶ An Introduction to Problem Solving and Mathematical Models

Every day we encounter various types of problems that must be solved. **Problem solving** is the ability to use information, tools, and our own skills to achieve a goal. For example, suppose Kevin wants a glass of water, but he is too short to reach the sink. Kevin has a problem. To solve the problem, he finds a step stool and pulls it over to the sink. He uses the step stool to climb on the counter, opens the kitchen cabinet, and pulls out a cup. He then crawls along the counter top, turns on the faucet, and proceeds to fill his cup with water. Problem solved!

Of course, Kevin could solve the problem in other ways. Just as there are various ways to solve life's everyday problems, there are many ways to solve problems using mathematics. However, regardless of the approach, there are always some common aspects in solving any problem. For example, regardless of how Kevin ultimately ends up with his cup of water, someone must get a cup from the cabinet and someone must turn on the faucet.

One of the purposes of learning algebra is to be able to solve certain types of problems. To solve these problems, we translate the verbal description of the problem into equations that we can solve. The process of turning a verbal description into a mathematical equation is **mathematical modeling.** We call the equation that is developed the **mathematical model.**

Not all models are mathematical. In general, a **model** is a way of using graphs, pictures, equations, or even verbal descriptions to represent a real-life situation. Because the world is a complex place, we need to simplify information when we develop a model. For example, a map is a model of our road system. Maps don't show all the details of the system (such as trees, buildings, or potholes), but they do a good job of describing how to get from point A to point B. Mathematical models are similar in that we often make assumptions regarding our world in order to make the mathematics less complicated.

Every problem is unique in some way. However, because many problems are similar, we can categorize problems. In this text, we will solve five categories of problems.

Five Categories of Problems

1. **Direct Translation**—problems in which we translate from English directly into an equation by using key words in the verbal description
2. **Mixture**—problems where two or more quantities are combined in some fashion
3. **Geometry**—problems where the unknown quantities are related through geometric formulas
4. **Uniform Motion**—problems where an object travels at a constant speed
5. **Work Problems**—problems where two or more entities join forces to complete a job

In this section, we will concentrate on direct translation problems, mixture problems, and uniform motion problems. However, the following guidelines will help you solve any category of problem.

Work Smart

When you solve a problem by making a mathematical model, check your work to make sure you have the right answer. Typically, errors can happen in two ways.

- One type of error occurs if you correctly translate the problem into a model but then make an error solving the equation. This type of error is usually easy to find.

- However, if you misinterpret the problem and develop an incorrect model, then the solution you obtain may still satisfy your model, but it probably will not be the correct solution to the original problem. We can check for this type of error by determining whether the solution is reasonable. Does your answer make sense? Always be sure that you are answering the question that is being asked.

Solving Problems With Mathematical Models

Step 1: Identify What You Are Looking For Read the problem very carefully, perhaps two or three times. Identify the type of problem and the information you wish to learn. It is fairly typical for the last sentence in the problem to indicate what you need to solve for.

Step 2: Give Names to the Unknowns Assign variables to the unknown quantities. Choose a variable that is representative of the unknown quantity. For example, use t for time.

Step 3: Translate the Problem into Mathematics Read the problem again. This time, after each sentence is read, determine whether the sentence can be translated into a mathematical statement or expression in terms of the variables identified in Step 2. It is often helpful to create a table, chart, or figure. Combine the mathematical statements or expressions into an equation that can be solved.

Step 4: Solve the Equation(s) Found in Step 3 Solve the equation for the variable.

Step 5: Check the Reasonableness of Your Answer Check your answer to be sure that it makes sense. If it does not, go back and try again.

Step 6: Answer the Question Write your answer in a complete sentence.

---

❷ Model and Solve Direct Translation Problems

Let's solve some **direct translation** problems, which can be set up by reading the problem and translating the verbal description into an equation.

▶ **EXAMPLE 2**   **Finding Consecutive Integers**

The sum of three consecutive odd integers results in 45. Find the integers.

Solution

Step 1: Identify This is a direct translation problem. We are looking for three odd integers. The odd integers are 1, 3, 5, and so on.

Step 2: Name If we let x represent the first odd integer, then $x + 2$ is the next odd integer, and $x + 4$ is the third odd integer.

Step 3: Translate Since we know that their sum is 45, we have

$$\underbrace{x}_{\text{First Integer}} \quad + \quad \underbrace{x + 2}_{\text{Second Integer}} \quad + \quad \underbrace{x + 4}_{\text{Third Integer}} \quad = 45 \qquad \text{The Model}$$

Step 4: Solve Now we can solve the equation.

$$x + x + 2 + x + 4 = 45$$

Combine like terms: $\qquad 3x + 6 = 45$

Subtract 6 from both sides: $\qquad 3x = 39$

Divide both sides by 3: $\qquad x = 13$

Step 5: Check Since x represents the first odd integer, the remaining two odd integers are $13 + 2 = 15$ and $13 + 4 = 17$. It is always a good idea to make sure your answer is reasonable. Since $13 + 15 + 17 = 45$, we know we have the right answer!

Step 6: Answer the Question The three consecutive odd integers are 13, 15, and 17.

Quick ✓

7. *True or False* If n represents the first of two consecutive odd integers, $n + 1$ represents the next consecutive odd integer.

In Problems 8 and 9, translate each English statement into an equation and solve the equation.

8. The sum of three consecutive even integers is 60. Find the integers.

9. The sum of three consecutive integers is 78. Find the integers.

▶ (**EXAMPLE 3**)   **Finding Hourly Wage**

Before taxes, Marissa earned $725 one week after working 52 hours. Her employer pays time-and-a-half for all hours worked in excess of 40 hours. What is Marissa's hourly wage?

Solution

Step 1: Identify In this direct translation problem, we want to find Marissa's hourly wage.

Step 2: Name Let w represent Marissa's hourly wage.

Step 3: Translate We know that Marissa earned $725 by working 40 hours at her regular wage and 12 hours at 1.5 times her regular wage. If her regular wage is w dollars, then her overtime wage is $1.5w$ dollars. Therefore, her total salary is

$$\underset{\text{Regular Earnings}}{\underbrace{40w}} \quad + \quad \underset{\text{Overtime Earnings}}{\underbrace{12(1.5w)}} \quad = \quad \underset{\text{Total Earnings}}{\underbrace{725}} \qquad \text{The Model}$$

Step 4: Solve

$$40w + 12(1.5w) = 725$$

Simplify: $\qquad 40w + 18w = 725$

Combine like terms: $\qquad 58w = 725$

Divide both sides by 58: $\qquad \dfrac{58w}{58} = \dfrac{725}{58}$

$$w = 12.50$$

Step 5: Check If Marissa's hourly wage is $12.50, then for the first 40 hours she earned $12.50(40) = 500 and for the next 12 hours she earned $1.5($12.50)(12) = 225. Her total salary was $500 + $225 = 725. This checks with the information presented in the problem.

Step 6: Answer the Question Marissa's hourly wage is $12.50.

Quick ✓

10. Before taxes, Melody earned $735 one week after working 46 hours. Her employer pays time-and-a-half for all hours worked in excess of 40 hours. What is Melody's hourly wage?

11. Before taxes, Jim earned $564 one week after working 30 hours at his regular wage, 6 hours at time-and-a-half on Saturday, and 4 hours at double-time on Sunday. What is Jim's hourly wage?

(EXAMPLE 4) **Choosing a Long-Distance Carrier**

MCI has a long-distance phone plan that charges $5.95 a month plus $0.05 per minute of usage. AT&T has a long-distance phone plan that charges $3.95 a month plus $0.07 per minute of usage. For how many minutes of long-distance calls will the costs of the two plans be the same? (SOURCE: *MCI and Sprint*)

Solution

Step 1: Identify This is a direct translation problem. We are looking for the number of minutes for which the two plans cost the same.

Step 2: Name Let m represent the number of long-distance minutes used in the month.

Step 3: Translate The monthly cost for MCI is $5.95 + $0.05 for each minute used. So if one minute is used, the cost is $5.95 + 0.05(1) = 6.00$ dollars. For 2 minutes, the cost is $5.95 + 0.05(2) = 6.05$ dollars. In general, for m minutes, the cost is $5.95 + 0.05m$ dollars. Similar logic reveals that the monthly cost of AT&T's plan is $3.95 + 0.07m$ dollars. To find the number of minutes for which the cost for the two plans will be the same, we need to solve

$$\text{Cost for MCI} = \text{Cost for AT\&T}$$
$$5.95 + 0.05m = 3.95 + 0.07m \qquad \text{The Model}$$

Step 4: Solve

$$5.95 + 0.05m = 3.95 + 0.07m$$

Subtract 3.95 from both sides: $2 + 0.05m = 0.07m$

Subtract 0.05m from both sides: $2 = 0.02m$

Divide both sides by 0.02: $100 = m$

Step 5: Check For $m = 100$ minutes, the cost of MCI's plan will be $5.95 + 0.05(100) = 10.95, and the cost of AT&T's plan will be $3.95 + 0.07(100) = 10.95. They are the same!

Step 6: Answer the Question The cost of the two plans will be the same if 100 minutes are used.

Quick ✓

12. You need to rent a moving truck. EZ-Rental charges $35 per day plus $0.15 per mile. Do It Yourself Rental charges $20 per day plus $0.25 per mile. For how many miles will the cost of renting be the same?

13. You need a new cell phone for emergencies only. Company A charges $12 per month plus $0.10 per minute, and Company B charges $0.15 per minute with no monthly service charge. For how many minutes will the monthly cost be the same?

There are many types of direct translation problems. One type of direct translation problem is a "percent problem." Typically, these problems involve discounts or mark-ups that businesses use in determining their prices.

Percent means *divided by* 100 or *per hundred*. The symbol % denotes percent, so 45% means 45 out of 100 or $\dfrac{45}{100}$ or 0.45. In applications involving percents, we often see the word "of," as in 20% of 60. Remember that the word "of" translates into multiplication in mathematics, so 20% of 60 means

$$20\% \cdot 60 \quad \text{or} \quad 0.20 \cdot 60$$

When dealing with percents and the price of goods, it is helpful to remember the following:

$$\text{Original Price} - \text{Discount} = \text{Sale Price}$$

$$\text{Wholesale Price} + \text{Markup} = \text{Selling Price}$$

▶ (EXAMPLE 5) **Finding Discounted Price**

Everything in your favorite clothing store is marked at a discount of 40% off. If the sale price of a suit is $144, what was its original price?

Solution

Step 1: Identify This is a direct translation problem involving percents. We wish to find the original price of the suit.

Step 2: Name Let p represent the original price.

Step 3: Translate The original price minus the discount is the sale price, $144, so

$$p - \text{discount} = 144$$

The discount was 40% off of the original price, so the discount is $0.40p$. Substitute this expression into the equation $p - \text{discount} = 144$:

$$p - 0.40p = 144 \quad \text{The Model}$$

Step 4: Solve
$$p - 0.40p = 144$$

Combine like terms: $\quad 0.60p = 144$

Divide both sides by 0.60: $\quad \dfrac{0.60p}{0.60} = \dfrac{144}{0.60}$

Simplify: $\quad p = 240$

Work Smart

You could eliminate the decimals by multiplying both sides of the equation by 10.

Step 5: Check If p, the original price of the suit, was $240, then the discount would be $0.4($240$) = $96. Subtracting $96 from the original price gives $144, the sale price. This agrees with the information in the problem.

Step 6: Answer the Question The original price of the suit was $240. ●

Quick ✓

14. Percent means "divided by ___."

15. What is 40% of 100?

16. 8 is 5% of what number?

17. 15 is what percent of 20?

18. Suppose you have just entered your favorite clothing store and find that everything in the store is marked at a discount of 30% off. If the sale price of a shirt is $21, what was its original price?

19. A Milex Tune-Up automotive facility marks up its parts 35%. Suppose that Milex charges its customers $1.62 for each spark plug it installs. What is Milex's cost for each spark plug?

Interest is money paid for the use of money. The total amount borrowed is called the **principal.** The principal can be in the form of a loan (an individual borrows from the bank) or a deposit (the bank borrows from the individual). The **rate of interest,** expressed as a percent, is the amount charged for the use of the principal for a given period of time, usually for a year (that is, on a per annum basis).

> **Simple Interest Formula**
>
> If a principal of P dollars is borrowed for a period of t years at an annual interest rate r, expressed as a decimal, then the interest I charged is
>
> $$I = Prt$$
>
> Interest charged according to this formula is called **simple interest.**

▶ **EXAMPLE 6** **Computing Credit Card Interest**

Yolanda has a credit card balance of $2800. Each month, the credit card company charges 14% annual simple interest on any outstanding balances. How much interest will Yolanda be charged on this loan after one month? What is her credit card balance after one month? Round answers to the nearest penny.

Solution

We want to find the interest I charged on the loan. Because the interest rate is given as an annual rate we must express the length of time the money is borrowed in years. One month is $\frac{1}{12}$ of a year, so $t = \frac{1}{12}$. The outstanding balance, or principal, is $P = \$2800$. The annual interest rate is $r = 14\% = 0.14$. Substituting into $I = Prt$, we obtain

$$I = (\$2800)\,(0.14)\left(\frac{1}{12}\right) = \$32.67$$

Yolanda will owe the amount borrowed, $2800, plus accrued interest, $32.67, for a total of $2800 + $32.67 = $2832.67. ●

> **Quick** ✓
>
> **20.** _____ is money paid for the use of money. The total amount borrowed is called the _____.
>
> **21.** Suppose that Dave has a car loan of $6500. The bank charges 6% annual simple interest. What is the interest charge on Dave's car loan after 1 month?
>
> **22.** Suppose that you have $1400 in a savings account. The bank pays 1.5% annual simple interest. What would be the interest paid after 6 months? What is the balance in the account?

❸ **Model and Solve Mixture Problems**

In **mixture problems,** two or more items are combined to form a third item. The different types of mixture problems all rely on the same basic idea:

Portion from Item A + Portion from Item B = Whole or Total

One type of mixture problem is the so-called interest problem.

▶ **EXAMPLE 7** **Financial Planning**

Sergio has $15,000 to invest. His goal is to earn 9%, or $1350, annually. His financial advisor recommends investing some of the money in corporate bonds that pay 12% and the rest in government-backed Treasury bonds paying 4%. How much should Sergio place in each investment in order to achieve his goal?

Solution

Step 1: Identify This is a mixture problem involving simple interest. Sergio needs to know how much to invest into corporate bonds and how much to invest into Treasury bonds in order to earn $1350 in interest.

Step 2: Name Let b represent the amount invested in corporate bonds; then $15,000 - b$ represents the amount invested in Treasury bonds.

Step 3: Translate We organize the given information in Table 2.

Table 2

	Principal $	Rate %	Time Yr	Interest $
Corporate Bond	b	0.12	1	$0.12\,b$
Treasury Bond	$15,000 - b$	0.04	1	$0.04(15,000 - b)$
Total	$15,000$	0.09	1	$0.09(15,000) = \$1350$

Sergio wants to earn 9% each year on his principal of $15,000. The total interest is the sum of the interest from the corporate bonds and the Treasury bonds.

$$\text{Interest from corporate bonds} + \text{Interest from Treasury bonds} = \$1350$$

$$0.12b + 0.04(15,000 - b) = 1350 \quad \text{The Model}$$

Step 4: Solve
$$0.12b + 0.04(15,000 - b) = 1350$$

Use the Distributive Property: $\quad 0.12b + 600 - 0.04b = 1350$

Combine like terms: $\quad 0.08b + 600 = 1350$

Subtract 600 from both sides: $\quad 0.08b = 750$

Divide both sides by 0.08: $\quad b = 9375$

Step 5: Check If Sergio invests $9375 in corporate bonds and $15,000 - \$9375 = \$5625$ in Treasury bonds, the simple interest earned each year on the corporate bonds is $(\$9375)(0.12)(1) = \$1125$, and that on the Treasury bonds is $(\$5625)(0.04)(1) = \$225$. The total interest earned is $\$1125 + \$225 = \$1350$, which is the amount he wanted to earn.

Step 6: Answer the Question Sergio should invest $9375 in corporate bonds and $5625 in Treasury bonds.

Quick ✓

23. Sophia has recently retired and requires an extra $5400 per year in income. She has $90,000 to invest and can invest in either an Aaa-rated bond that pays 5% per annum or a B-rated bond paying 9% per annum. How much should Sophia place in each investment in order to achieve her goal?

24. Steve has $25,000 to invest and wishes to earn an overall annual rate of return of 8%. His financial advisor recommends that he invest some of the money in a 5-year CD paying 4% per annum and the rest in a corporate bond paying 9% per annum. How much should Steve place in each investment in order to achieve his goal?

Mixing two quantities together creates a new blend. For example, a chef might mix buckwheat flour with wheat flour to make buckwheat pancakes. Or a coffee shop might mix two different types of coffee to create a new coffee blend.

(EXAMPLE 8) **Blending Coffees**

A coffee shop manager wishes to form a new blend of coffee. She wants to mix distinctive-tasting Sumatra beans that sell for $12 per pound with milder Brazilian beans that sell for $8 per pound to get 50 pounds of the new blend. The new blend will sell for $9 per pound, and the revenue from selling the new blend and that from selling the beans separately will be the same. See Figure 2. How many pounds of the Sumatra beans and how many pounds of Brazilian beans will be required?

Figure 2

| Sumatra | Brazilian | Blend |
| $12 per pound | $8 per pound | $9 per pound |

Solution

Step 1: Identify This is a blend problem. We want to know the number of pounds of Sumatra beans and the number of pounds of Brazilian beans that are required in the new blend.

Step 2: Name Let s represent the required number of pounds of Sumatra beans. Thus $50 - s$ will be the required number of pounds of Brazilian beans.

Step 3: Translate The revenue from selling the new blend and that from selling the beans separately will be the same. This means that if the blend contains one pound of Sumatra and one pound of Brazilian, we should collect $12(1) + \$8(1) = \$20$, which is what we would collect if we sold the beans separately.

We set up Table 3.

Table 3

	Price $/pound	Number of Pounds	Revenue
Sumatra	12	s	$12s$
Brazilian	8	$50 - s$	$8(50 - s)$
Blend	9	50	$9(50) = 450$

In general, if the blend contains s pounds of Sumatra beans, we should collect $\$12s$. If it contains $50 - s$ pounds of Brazilian beans, we should collect $\$8(50 - s)$. All 50 pounds of the blend, selling at $9 per pound, will sell for $9(50) = \$450$.

$$\left(\begin{array}{c}\text{Price per pound}\\\text{of Sumatra}\end{array}\right)\left(\begin{array}{c}\text{Pounds of}\\\text{Sumatra}\end{array}\right) + \left(\begin{array}{c}\text{Price per pound}\\\text{of Brazilian}\end{array}\right)\left(\begin{array}{c}\text{Pounds of}\\\text{Brazilian}\end{array}\right) = \left(\begin{array}{c}\text{Price per pound}\\\text{of blend}\end{array}\right)\left(\begin{array}{c}\text{Pounds of}\\\text{blend}\end{array}\right)$$

$$\underbrace{12 \quad \cdot \quad s}_{\text{Revenue from Sumatra}} + \underbrace{8 \quad \cdot \quad (50 - s)}_{\text{Revenue from Brazilian}} = \underbrace{\$9 \quad \cdot \quad 50}_{\text{Revenue from Mixture}}$$

We have the equation

$$12s + 8(50 - s) = 450 \quad \text{The Model}$$

Step 4: Solve   Solve the equation for s.

Use the Distributive Property to remove the parentheses:	$12s + 8(50 - s) = 450$
	$12s + 400 - 8s = 450$
Combine like terms:	$4s + 400 = 450$
Subtract 400 from both sides:	$4s = 50$
Divide both sides by 4:	$s = 12.5$

Step 5: Check   If the manager mixes 12.5 pounds of Sumatra beans with $50 - 12.5 = 37.5$ pounds of Brazilian beans, the total revenue will be $\$12(12.5) + \$8(37.5) = \$150 + \$300 = \$450$. This equals the revenue from the blend and checks with the given information.

Step 6: Answer the Question   The manager should mix 12.5 pounds of Sumatra beans with 37.5 pounds of Brazilian beans.

Quick ✓

25. Suppose you want to blend two teas to obtain 10 pounds of a new blend selling for $3.50 per pound. Tea A sells for $4.00 per pound, and Tea B sells for $2.75 per pound. Assuming that the revenue from selling the new blend and that from selling the tea separately will be the same, determine the number of pounds of each tea required in the blend.

26. "We're Nuts!" sells cashews for $6.00 per pound and peanuts for $1.50 per pound. The manager has decided to make a trail mix. She wants the mix to sell for $3.00 per pound, and there should be no loss in revenue from selling the trail mix rather than selling the nuts alone. How many pounds of cashews and peanuts are required to create 30 pounds of trail mix?

▶ ❹ Model and Solve Uniform Motion Problems

Objects that move at a constant velocity are in **uniform motion.** We can interpret an object's average velocity as its constant velocity. For example, a car traveling at an average velocity of 40 miles per hour is in uniform motion.

> **In Words**
> The uniform motion formula states that distance equals rate times time.

Uniform Motion Formula

If an object moves at an average speed r, the distance d covered in time t is given by the formula

$$d = rt$$

Note that **rate** is another term for speed.

(EXAMPLE 9)   **Uniform Motion**

Roger and Bill decide to have a 10-mile race. Roger can run at an average speed of 12 miles per hour, and Bill can run at an average speed of 10 miles per hour. To "even things up," Roger agrees to give Bill a head start of 0.15 hour. When will Roger catch up to Bill?

Solution

Step 1: Identify   This is a uniform motion problem. We need the number of hours it will take for Roger to catch up to Bill if each man runs at his average speed.

Step 2: Name   Let t represent the number of hours it takes for Roger to catch up to Bill. Since Bill has a head start of 0.15 hour, he will have been running for $t + 0.15$ hours.

Step 3: Translate Figure 3 illustrates the situation. We will set up Table 4.

Figure 3

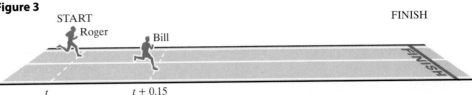

Table 4			
	Rate, mph	Time, hours	Distance, miles
Roger	12	t	$12t$
Bill	10	$t + 0.15$	$10(t + 0.15)$

Roger will travel $12t$ miles since his average speed is 12 miles per hour and he runs for t hours. Bill will travel $10(t + 0.15)$ miles, since his average speed is 10 miles per hour and he runs for $t + 0.15$ hours. Because each travels the same distance, we set up the following model:

$$\text{Distance Roger runs} = \text{Distance Bill runs}$$

$$12t = 10(t + 0.15) \quad \text{The Model}$$

Step 4: Solve

$$12t = 10(t + 0.15)$$

Use the Distributive Property:
$$12t = 10t + 1.5$$

Subtract 10t from both sides:
$$2t = 1.5$$

Divide both sides by 2:
$$t = \frac{1.5}{2} = 0.75$$

Step 5: Check It should take Roger 0.75 hour to catch up to Bill. After 0.75 hour, Roger will have traveled $12(0.75) = 9$ miles, and Bill will have traveled $10(0.75 + 0.15) = 9$ miles. It checks!

Step 6: Answer the Question It will take Roger 0.75 hour, or 45 minutes, to catch up to Bill. ●

Quick ✓

27. Objects that move at a constant velocity are said to be in _____ _____.

28. Another term for speed is ____.

29. A Chevrolet Cruze left Omaha, Nebraska, traveling at an average velocity of 40 miles per hour. Two hours later, a Dodge Charger left Omaha, Nebraska, traveling on the same road at an average velocity of 60 miles per hour. When will the Charger catch up to the Cruze? How far will each car have traveled?

30. A train leaves Union Station traveling at an average velocity of 50 miles per hour. Four hours later, a helicopter follows along the train tracks traveling at an average velocity of 90 miles per hour. When will the helicopter catch up to the train? How far will each have traveled?

1.2 Exercises MyMathLab® PRACTICE

Exercise numbers in green have complete video solutions in MyMathLab.

Problems 1–30 are the Quick ✓ s that follow the EXAMPLES.

Building Skills

31. What is 25% of 40?

32. What is 150% of 70?

33. 12 is 30% of what?

34. 50 is 90% of what?

35. 30 is what percent of 80?

36. 90 is what percent of 120?

In Problems 37–46, translate each English statement into a mathematical statement. Then solve the equation. See Objective 1.

37. The sum of a number x and 12 is 20.

38. The difference between 10 and a number z is 6.

39. Twice the sum of y and 3 is 16.

40. The sum of two times y and 3 is 16.

41. The difference between w and 22 equals three times w.

42. The sum of x and 4 results in twice x.

43. Four times a number x is equivalent to the sum of two times x and 14.

44. Five times a number x is equivalent to the difference of three times x and 10.

45. 80% of a number is equivalent to the sum of the number and 5.

46. 40% of a number equals the difference between the number and 10.

Applying the Concepts

47. Number Sense Grant is thinking of two numbers. He says that one of the numbers is twice the other number and the sum of the numbers is 39. What are the numbers?

48. Number Sense Pattie is thinking of two numbers. She says that one of the numbers is 8 more than the other number and the sum of the numbers is 56. What are the numbers?

49. Consecutive Integers The sum of three consecutive integers is 75. Find the integers.

50. Consecutive Integers The sum of four consecutive odd integers is 104. Find the integers.

51. Computing Grades Going into the final exam, which counts as two grades, Kendra has test scores of 84, 78, 64, and 88. What score does Kendra need on the final exam in order to have an average of 80?

52. Computing Grades Going into the final exam, which counts as three grades, Mark has test scores of 65, 79, 83, and 68. What score does Mark need on the final exam in order to have an average of 70?

53. Comparing Printers Jacob is trying to decide between two laser printers, one manufactured by Hewlett-Packard, the other by Brother. Both have similar features and warranties, so price is the determining factor. The Hewlett-Packard costs $180 and printing costs are approximately $0.03 per page. The Brother costs $230 and printing costs are approximately $0.01 per page. How many pages need to be printed for the two printers to cost the same?

54. Comparing Job Offers Maria has just been offered two sales jobs. The first job offer is a base monthly salary of $2500 plus a commission of 3% of total sales. The second job offer is a base monthly salary of $1500 plus a commission of 3.5% of total sales. For what level of monthly sales are the salaries offered by these two jobs equivalent?

55. Finance An inheritance of $800,000 is to be divided among Avery, Connor, and Olivia in the following manner: Olivia is to receive $^3/_4$ of what Connor gets, while Avery gets $^1/_4$ of what Connor gets. How much does each receive?

56. Sharing the Cost of a Pizza Judy and Linda agree to share the cost of a $21 pizza based on how much each ate. If Judy ate $^3/_4$ of the amount that Linda ate, how much should each pay?

57. Sales Tax In the state of Connecticut there is a state sales tax of 6.35% on all goods purchased. If Jan buys a television for $599, what will be the final bill, including state sales tax?

58. Sales Tax In Austin, Texas, there is a sales tax of 8.25% on all goods purchased. If Megan buys a sofa for $450, what will be the final bill, including the sales tax?

59. Markups A new 2012 Honda Accord has a list price of $23,950. Suppose that the dealer mark-up on this car is 15%. What is the dealer's cost?

60. Markups Suppose that the price of a new Intermediate Algebra text is $95. The bookstore has a policy of marking texts up 30%. What is the cost of the text to the bookstore?

61. Discount Pricing Suppose that you just received an e-mail alert indicating that 32-gigabyte flash drives have just been discounted by 60%. If the sale price of the flash drive is now $27.55, what was the original price?

62. Discount Pricing Suppose that you just received an e-mail alert from Kohls indicating that the fall line of clothing has just been discounted by 30% and knit polo shirts are now $28. What was the original price of a polo shirt?

63. Cars A Mazda 6s weighs 136 pounds more than a Nissan Altima. A Honda Accord EX weighs 119 pounds more than a Nissan Altima. The total weight of all three cars is 9834 pounds. How much does each car weigh? (SOURCE: *Road and Track magazine*)

64. Cars The Honda Accord EX has 1.9 cubic feet less cargo space than a Mazda 6s. Together the cars have 31.3 cubic feet of cargo space. How much cargo space does each car have? (SOURCE: *Road and Track magazine*)

65. Finance A total of $20,000 is going to be split between Adam and Krissy with Adam receiving $3000 less than Krissy. How much will each get?

66. Finance A total of $40,000 is going to be invested in stocks and bonds. A financial advisor recommends that $6000 more be invested in stocks than in bonds. How much is invested in stocks? How much is invested in bonds?

67. Investments Suppose that your long-lost Aunt Sara has left you an unexpected inheritance of $24,000. You have decided to invest the money rather than spending it on frivolous purchases. Your financial advisor has recommended that you diversify by placing some of the money in stocks and some in bonds. Based upon current market conditions, she has recommended that the amount in bonds equal three-fifths of the amount invested in stocks. How much should be invested in stocks? How much should be invested in bonds?

68. Investments Jack and Diane have $60,000 to invest. Their financial advisor has recommended that they diversify by placing some of the money in stocks and some in bonds. Based upon current market conditions, he has recommended that the amount in bonds equal two-thirds of the amount invested in stocks. How much should be invested in stocks? How much should be invested in bonds?

69. Simple Interest Elena has a credit card balance of $2500. The credit card company charges 14% per annum simple interest. What is the interest charge on Elena's credit card after 1 month?

70. Simple Interest Faye has a home equity loan of $70,000. The bank charges Faye 6% per annum simple interest. What is the interest charge on Faye's loan after 1 month?

71. Banking A bank has loaned out $500,000, part of it at 6% per annum and the rest of it at 11% per annum. If the bank receives $43,750 in interest each year, how much was loaned at 6%?

72. Banking Patrick is a loan officer at a bank. He has $2,000,000 to lend out and has two loan programs. His home equity loan is currently priced at 6% per annum, and his unsecured personal loan is priced at 14%. The bank president wants Patrick to earn a rate of return of 12% on the $2,000,000 available. How much should Patrick lend out at 6%?

73. Investments Pedro wants to invest his $25,000 bonus check. His investment advisor has recommended that he put some of the money in a bond fund that yields 5% per annum and the rest in a stock fund that yields 9% per annum. If Pedro wants to earn $1875 each year from his investments, how much should he place in each investment?

74. Investments Johnny is a shrewd 8-year-old. For Christmas, his grandparents gave him $10,000. Johnny decides to invest some of the money in a savings account that pays 2% per annum and the rest in a stock fund paying 10% per annum. Johnny wants his investments

to yield 7% per annum. How much should Johnny put into each account?

75. Making Coffee Suppose that you want to blend two coffees in order to obtain 50 pounds of a new blend. If x represents the number of pounds of coffee A, write an algebraic expression that represents the number of pounds of coffee B.

76. Candy "Sweet Tooth!" candy store sells chocolate-covered almonds for $6.50 per pound and chocolate-covered peanuts for $4.00 per pound. The manager decides to make a bridge mix that combines the almonds with the peanuts. She wants the bridge mix to sell for $5.00 per pound, and there should be no loss in revenue from selling the bridge mix rather than the almonds and peanuts alone. How many pounds of chocolate-covered almonds and how many pounds of chocolate-covered peanuts are required to create 50 pounds of bridge mix?

77. Coins Bobby has been saving quarters and dimes. He opened up his piggy bank and found that it contained 47 coins worth $9.50. Determine how many dimes and quarters were in the piggy bank.

78. More Coins Diana has been saving nickels and dimes. She opened up her piggy bank and found that it contained 48 coins worth $4.50. Determine how many nickels and dimes were in the piggy bank.

79. Gold The purity of gold is measured in karats, with pure gold being 24 karats. Other purities of gold are expressed as proportional parts of pure gold. For example, 18-karat gold is $\frac{18}{24}$, or 75%, pure gold; 12-karat gold is $\frac{12}{24}$, or 50%, pure gold; and so on. How much pure gold should be mixed with 12-karat gold to obtain 72 grams of 18-karat gold?

80. Antifreeze The cooling system of a car has a capacity of 15 liters. The system is currently filled with a mixture that is 40% antifreeze. How much of this mixture should be drained and replaced with pure antifreeze so that the system is filled with a solution that is 50% antifreeze?

81. A Biathlon Suppose that you have entered a 62-mile biathlon that consists of a run and a bicycle race. During your run, your average velocity is 8 miles per hour, and during your bicycle race, your average velocity is 20 miles per hour. You finish the race in 4 hours. What is the distance of the run? What is the distance of the bicycle race?

82. A Biathlon Suppose that you have entered a 15-mile biathlon that consists of a run and swim. During your run, your average velocity is 7 miles per hour, and during your swim, your average velocity is 2 miles per hour. You finish the race in 2.5 hours. What is the length of the swim? What is the distance of the run?

83. Collision Course Two cars that are traveling toward each other are 455 miles apart. One car is traveling 10 miles per hour faster than the other car. The cars pass after 3.5 hours. How fast is each car traveling?

84. Collision Course Two planes that are traveling toward each other are 720 miles apart. One plane is traveling 40 miles per hour faster than the other. The planes pass after 0.75 hour. How fast is each plane traveling?

85. Boats Two boats leave a port at the same time, one traveling north and the other traveling south. The northbound boat travels at 12 miles per hour (mph) faster than the southbound boat. If the southbound boat is traveling at 25 mph, how long will it be before they are 155 miles apart?

86. Cyclists Two cyclists leave a city at the same time, one going east and the other going west. The westbound cyclist bikes at 3 mph faster than the eastbound cyclist. If after 6 hours they are 162 miles apart, how fast is each cyclist riding?

87. Walking At 10:00 A.M. two people leave their homes that are 15 miles apart and begin walking toward each other. If one person walks at a rate that is 2 mph faster than the other and they meet after 1.5 hours, how fast was each person walking?

88. Trains At 9:00 A.M., two trains are 715 miles apart traveling toward each other on parallel tracks. If one train is traveling at a rate that is 10 miles per hour (mph) faster than the other train and they meet after 5.5 hours, how fast is each train traveling?

Extending the Concepts

89. Computing Average Speed On a recent trip to Florida, we averaged 50 miles per hour. On the return trip we averaged 60 miles per hour. What do you think the average speed of the trip to Florida and back was?

Defend your position. Algebraically, determine the average speed of the trip to Florida and back.

90. Discount Pricing Suppose that you are the manager of a clothing store and have just purchased 100 shirts for $15 each. After 1 month of selling the shirts at the regular price, you plan to have a sale giving 30% off the original selling price. However, you want to make a profit of $6 on each shirt at the sale price. What should you price the shirts at initially to ensure this?

91. Critical Thinking Make up an applied problem that would result in the equation $x - 0.05x = 60$.

92. Critical Thinking Make up an applied problem that would result in the equation $10 + 0.14x = 50$.

93. Uniform Motion Suppose that you are walking along the side of train tracks at 2 miles per hour. A train that is traveling 20 miles per hour in the same direction requires 1 minute to pass you. How long is the train?

Explaining the Concepts

94. Consider the phrase "… the revenue from selling the new blend and that from selling the beans separately will be the same," which appeared in Example 8. Explain what this means.

95. How is mathematical modeling related to problem solving?

96. Why do we make assumptions when creating mathematical models?

97. Name the different categories of problems presented in this section. Name two different kinds of mixture problems.

98. Think of two models that you use in your everyday life. Are any of them mathematical models? Describe your models to members of your class.

1.3 Using Formulas to Solve Problems

Objectives

➊ Solve for a Variable in a Formula

➋ Use Formulas to Solve Problems

Are You Ready for This Section?

Before getting started, take this readiness quiz. If you get a problem wrong, go back to the section cited and review the material.

In Problems R1 and R2, (a) round, and (b) truncate each decimal to the indicated number of places. [Section R.2, pp. 14–15]

R1. 3.00343; three decimal places

R2. 14.957; two decimal places

A known relation that exists between two or more variables can be used to solve certain types of problems. A **formula** is an equation that describes how two or more variables are related.

EXAMPLE 1 **Answering Questions with Formulas**

Work Smart

Think of a formula as another kind of mathematical model.

The information in Table 5 gives descriptions, in words, of known relations and the corresponding formulas.

Table 5

Verbal Description	Formula
Geometry: The area *A* of a rectangle is the product of its length *l* and width *w*.	$A = lw$
Physics: The energy of an object in motion, kinetic energy *K*, is one-half the product of the mass *m* and the square of the velocity *v*.	$K = \dfrac{1}{2}mv^2$
Grading: The final grade, *G*, in your Intermediate Algebra class is the average of your four exam grades, *A*, *B*, *C*, and *D*.	$G = \dfrac{A + B + C + D}{4}$
Finance: The future value of money, *A*, is the product of the present value *P* and the sum of 1 and the annual interest rate *r*.	$A = P(1 + r)$

Quick ✓

1. A _____ is an equation that describes how two or more variables are related.

In Problems 2–5, translate the verbal description into a mathematical formula.

2. The area *A* of a circle is the product of the number π and the square of its radius *r*.

3. The volume *V* of a right circular cylinder is the product of the number π, the square of its radius *r*, and its height *h*.

4. The daily cost *C* of manufacturing computers is $175 times the number of computers manufactured *x*, plus $7000.

5. The distance *s* that an object free-falls is one-half the product of acceleration due to gravity *g* and the square of time *t*.

▶ **1** **Solve for a Variable in a Formula**

To "solve for a variable" means to isolate the variable on one side of the equation, with all the other variables and constants, if any, on the other side, by forming equivalent equations. For example, the formula for the area of a rectangle, $A = lw$, is solved for *A* because *A* is by itself on one side of the equation. The steps we follow when solving a formula for a certain variable are identical to those we follow when solving an equation.

EXAMPLE 2 **Solving for a Variable in a Formula**

The volume *V* of a cone is given by the formula $V = \dfrac{1}{3}\pi r^2 h$, where *r* is the radius and *h* is the height of the cone. See Figure 4 on the following page.

(a) Solve the formula for *h*.

(b) Use the result from part (a) to find the height of a cone if its volume is 50π cubic feet and its radius is 5 feet.

Solution

(a) To solve the formula for *h*, we need to isolate *h* on one side of the equation.

$$V = \frac{1}{3}\pi r^2 h$$

Figure 4

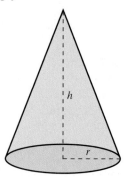

Multiply both sides by 3:	$3 \cdot V = \left(\dfrac{1}{3}\pi r^2 h\right) \cdot 3$
Divide out common factors:	$3V = \pi r^2 h$
Divide both sides by πr^2:	$\dfrac{3V}{\pi r^2} = \dfrac{\pi r^2 h}{\pi r^2}$
Divide out common factors:	$\dfrac{3V}{\pi r^2} = h$
If $a = b$, then $b = a$:	$h = \dfrac{3V}{\pi r^2}$

(b) Substituting $V = 50\pi$ ft^3 and $r = 5$ ft into $h = \dfrac{3V}{\pi r^2}$, we obtain

$$h = \frac{3(50\pi \text{ ft}^3)}{\pi(5 \text{ ft})^2}$$

$$h = \frac{150\pi \text{ ft}^3}{25\pi \text{ ft}^2}$$

$$h = 6 \text{ feet}$$

Work Smart

Solving for a variable is just like solving an equation with one unknown. When solving for a variable, treat all the other variables as constants.

Geometry Formulas

Let's review a few common terms from geometry.

Definitions

The **perimeter** is the sum of the lengths of all the sides of a figure.

The **area** is the amount of space enclosed by a two-dimensional figure, measured in square units.

The **surface area** of a solid is the sum of the areas of the surfaces of a three-dimensional figure.

The **volume** is the amount of space occupied by a three-dimensional figure, measured in units cubed.

The **radius** r of a circle is the line segment that extends from the center of the circle to any point on the circle.

The **diameter** of a circle is any line segment that extends from one point on the circle through the center to a second point on the circle. The length of a diameter is two times the length of the radius, $d = 2r$.

In circles, we use the term **circumference** to mean the perimeter.

Work Smart

When working with formulas, keep track of the units. Verify that the units in your answer are reasonable.

Formulas from geometry are useful in solving many types of problems. We list some of these formulas in Table 6.

Table 6

Plane Figures	Formulas	Plane Figures	Formulas
Square	**Area:** $A = s^2$ **Perimeter:** $P = 4s$	Trapezoid	**Area:** $A = \dfrac{1}{2}h(B + b)$ **Perimeter:** $P = a + b + c + B$
Rectangle	**Area:** $A = lw$ **Perimeter:** $P = 2l + 2w$	Parallelogram	**Area:** $A = bh$ **Perimeter:** $P = 2a + 2b$

Table 6 (*continued*)

Solids	Formulas	Solids	Formulas
Triangle	**Area:** $A = \dfrac{1}{2}bh$ **Perimeter:** $P = a + b + c$	Circle	**Area:** $A = \pi r^2$ **Circumference:** $C = 2\pi r$ $\qquad\qquad\quad = \pi d$
Cube	**Volume:** $V = s^3$ **Surface Area:** $S = 6s^2$	Right Circular Cylinder	**Volume:** $V = \pi r^2 h$ **Surface Area:** $S = 2\pi r^2 + 2\pi rh$
Rectangular Solid	**Volume:** $V = lwh$ **Surface Area:** $S = 2lw + 2lh + 2wh$	Cone	**Volume:** $V = \dfrac{1}{3}\pi r^2 h$
Sphere	**Volume:** $V = \dfrac{4}{3}\pi r^3$ **Surface Area:** $S = 4\pi r^2$		

Quick ✓

6. The _____ is the amount of space enclosed by a two-dimensional figure and is measured in square units, whereas _____ is the amount of space occupied by a three-dimensional figure and is measured in cubic units.

7. *True or False* The perimeter of a rectangle 6 feet long and 4 feet wide is 24 square feet.

8. The area A of a triangle is given by the formula $A = \dfrac{1}{2}bh$, where b is the base of the triangle and h is the height.

   (a) Solve the formula for h.

   (b) Find the height of the triangle whose area is 10 square inches and whose base is 4 inches.

9. The perimeter P of a parallelogram is given by the formula $P = 2a + 2b$, where a is the length of one side of the parallelogram and b is the length of the adjacent side.

   (a) Solve the formula for b.

   (b) Find the length of one side of a parallelogram whose perimeter is 60 cm and length of the other side is 20 cm.

  (EXAMPLE 3)   **Solving for a Variable in a Formula**

The formula $Y = C + bY + I + G + N$ is a model used in economics to describe the total income of an economy. In the model, Y is income, C is consumption, I is investment in capital, G is government spending, N is net exports, and b is a constant. Solve the formula for Y.

Solution

We want to get all terms with Y on the same side of the equal sign.

$$Y = C + bY + I + G + N$$

Subtract bY from both sides: $Y - bY = C + bY + I + G + N - bY$

Combine like terms: $Y - bY = C + I + G + N$

Use the Distributive Property "in reverse" to isolate Y: $Y(1 - b) = C + I + G + N$

Divide both sides by $1 - b$: $\dfrac{Y(1 - b)}{1 - b} = \dfrac{C + I + G + N}{1 - b}$

Simplify: $Y = \dfrac{C + I + G + N}{1 - b}$

Quick ✓

In Problems 10–13, solve for the indicated variable.

10. $I = Prt$ for P **11.** $Ax + By = C$ for y

12. $2xh - 4x = 3h - 3$ for h **13.** $S = na + (n - 1)d$ for n

❷ Use Formulas to Solve Problems

Formulas are often used to solve certain types of word problems. We will use the problem-solving steps from page 61.

▶ (**EXAMPLE 4**)   **Finding the Perimeter of a Window**

The perimeter of a rectangular picture window is 466 inches. The length of the window is 55 inches more than the width. See Figure 5. Find the dimensions of the window.

Solution

Step 1: Identify This is a geometry problem that requires the formula for the perimeter of a rectangle. We want to determine the dimensions of the rectangular window. That is, we want to find the window's length and width.

Step 2: Name Let w represent the width. The length is 55 inches more than the width, so $l = w + 55$.

Figure 5

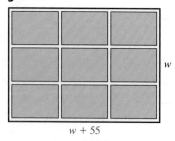

w

$w + 55$

Step 3: Translate The perimeter of a rectangle is $P = 2l + 2w$. We substitute the known values into this formula.

$$P = 2l + 2w$$

$P = 466; l = w + 55$: $466 = 2(w + 55) + 2w$ The Model

Step 4: Solve $466 = 2(w + 55) + 2w$

Distribute the 2: $466 = 2w + 110 + 2w$

Combine like terms: $466 = 4w + 110$

Subtract 110 from both sides: $356 = 4w$

Divide both sides by 4: $89 = w$

Step 5: Check If the width of the window is 89 inches, then the length is $89 + 55 = 144$ inches. The perimeter of the window is $2(144) + 2(89) = 466$ inches. It checks!

Step 6: Answer the Question The window's width is 89 inches and its length is 144 inches.

There is an interesting side note to the result of Example 4. If we compute the ratio of the window's length to its width, we get $\frac{144}{89} \approx 1.618$. Rectangles whose dimensions form this ratio are **golden rectangles.** Golden rectangles are said to have dimensions that are "pleasing to the eye." The golden rectangle was first constructed by the Greek philosopher Pythagoras in the sixth century B.C. These rectangles are used in architecture (the Parthenon) and in art (the Mona Lisa). See Figure 6.

Figure 6

Dagli Orti (A)/Picture Desk, Inc./
Kobal Collection

Quick ✓

14. The perimeter of a rectangular pool is 180 feet. If the length of the pool is to be 10 feet more than the width, find the dimensions of the pool.

15. The opening of a rectangular bookcase has a perimeter of 224 inches. If the height of the bookcase is 32 inches more than the width, determine the dimensions of the opening of the bookcase.

▶ (EXAMPLE 5) **Constructing a Soup Can**

A can of Campbell's soup has a surface area of 46.5 square inches. See Figure 7. The surface area S of a right circular cylinder is $S = 2\pi r^2 + 2\pi rh$, where r is the radius of the can and h is its height. Find the height of a can of Campbell's soup if its radius is 1.375 inches. Round your answer to two decimal places.

Figure 7

$r = 1.375''$

Solution

Step 1: Identify This is a geometry problem that requires the formula for the surface area of a cylinder. We want to find the height of the can of soup.

Step 2: Name Let h represent the height of the can of soup.

Step 3: Translate We know that the surface area S of the can is 46.5 square inches. The radius is 1.375 inches. Substitute these values into the formula for the surface area of the can.

$$S = 2\pi r^2 + 2\pi rh$$

$S = 46.5;\ r = 1.375:$ $46.5 = 2\pi (1.375)^2 + 2\pi (1.375)h$ The Model

Work Smart

Round-off error occurs when decimals are continually rounded during the course of solving a problem. The more times we round, the more inaccurate the results may be. So do not do any rounding until the last step.

Step 4: Solve Solve for h. To avoid round-off error, do not compute any of the values until the last calculation.

$$46.5 = 2\pi (1.375)^2 + 2\pi (1.375)h$$

Subtract $2\pi (1.375)^2$ from both sides: $46.5 - 2\pi (1.375)^2 = 2\pi (1.375)h$

Divide both sides by $2\pi (1.375)$: $\dfrac{46.5 - 2\pi (1.375)^2}{2\pi (1.375)} = h$

Figure 8

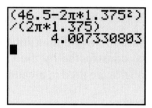

We will use a calculator to evaluate this expression. Figure 8 shows the output from a TI-84 Plus graphing calculator.

Thus $h = 4.01$ inches, rounded to two decimal places.

Step 5: Check The surface area S of the can with height 4.01 inches is

$$S = 2\pi r^2 + 2\pi rh$$
$$= 2\pi (1.375)^2 + 2\pi (1.375)(4.01)$$
$$\approx 46.5 \text{ square inches.}$$

Step 6: Answer the Question The height of the can is 4.01 inches, rounded to two decimal places.

Quick ✓

16. A can of peaches has a surface area of 51.8 square inches. The surface area S of a right circular cylinder is $S = 2\pi r^2 + 2\pi rh$, where r is the radius of the can and h is the height of the can. Find the height of a can of peaches if its radius is 1.5 inches. Round your answer to two decimal places.

1.3 Exercises MyMathLab® Math XP PRACTICE Exercise numbers in green have complete video solutions in MyMathLab.

*Problems **1–16** are the Quick ✓ s that follow the EXAMPLES.*

Building Skills

In Problems 17–20, translate the verbal description into a mathematical formula.

17. Force F equals the product of mass m and acceleration a.

18. The area A of a triangle is one-half the product of its base b and its height h.

19. The volume V of a sphere is four-thirds the product of the number π and the cube of its radius r.

20. The revenue R of selling computers is $800 times the number of computers sold x.

In Problems 21–32, solve the formula for the indicated variable. See Objective 1.

21. Uniform Motion Solve $d = rt$ for r.

22. Direct Variation Solve $y = kx$ for k.

23. Algebra Solve $y - y_1 = m(x - x_1)$ for m.

24. Algebra Solve $y = mx + b$ for m.

25. Statistics Solve $Z = \dfrac{x - \mu}{\sigma}$ for x.

26. Statistics Solve $E = \dfrac{Z \cdot \sigma}{\sqrt{n}}$ for $\sqrt{n}$.

27. Newton's Law of Gravitation Solve $F = G\,\dfrac{m_1 m_2}{r^2}$ for m_1.

28. Sequences Solve $S - rS = a - ar^5$ for S.

29. Finance Solve $A = P + Prt$ for P.

30. Bernoulli's Equation Solve $p + \dfrac{1}{2}\rho v^2 + \rho gy = a$ for ρ.

31. Temperature Conversion Solve $C = \dfrac{5}{9}(F - 32)$ for F.

32. Trapezoid Solve $A = \dfrac{1}{2}h(B + b)$ for b.

In Problems 33–40, solve for y.

33. $2x + y = 13$

34. $-4x + y = 12$

35. $9x - 3y = 15$

36. $4x + 2y = 20$

37. $4x + 3y = 13$

38. $5x - 6y = 18$

39. $\dfrac{1}{2}x + \dfrac{1}{6}y = 2$

40. $\dfrac{2}{3}x - \dfrac{5}{2}y = 5$

Applying the Concepts

41. Cylinders The volume V of a right circular cylinder is given by the formula $V = \pi r^2 h$, where r is the radius and h is the height.

(a) Solve the formula for h.

(b) Find the height of a right circular cylinder whose volume is 32π cubic inches and whose radius is 2 inches.

42. Cylinders The surface area S of a right circular cylinder is given by the formula $S = 2\pi rh + 2\pi r^2$, where r is the radius and h is the height.

(a) Solve the formula for h.

(b) Determine the height of a right circular cylinder whose surface area is 72π square centimeters and whose radius is 4 centimeters.

43. Maximum Heart Rate The model $M = -0.711A + 206.3$ was developed by Londeree and Moeschberger to determine the maximum heart rate M of an individual who is age A. (SOURCE: Londeree and Moeschberger, "Effect of Age and Other Factors on HR max," *Research Quarterly for Exercise and Sport,* 53(4), 297–304)

(a) Solve the model for A.

(b) According to this model, what is the age of an individual whose maximum heart rate is 160?

44. Maximum Heart Rate The model $M = -0.85A + 217$ was developed by Miller to determine the maximum heart rate M of an individual who is age A. (SOURCE: Miller et al., "Predicting Max HR," *Medicine and Science in Sports and Exercise,* 25(9), 1077–1081)

(a) Solve the model for A.

(b) According to this model, what is the age of an individual whose maximum heart rate is 160?

45. Finance The formula $A = P(1 + r)^t$ can be used to relate the future value A of a deposit of P dollars in an account that earns an annual interest rate r (expressed as a decimal) after t years.

(a) Solve the formula for P.

(b) How much would you have to deposit today in order to have $5000 in 5 years in a bank account that pays 4% annual interest?

46. Federal Income Taxes For a single filer with an annual adjusted income in 2011 over $35,350$ but less than $85,650$, the federal income tax T for an annual adjusted income I is found using the formula $T = 0.25(I - 35{,}350) + 4867.5$.

(a) Solve the formula for I.

(b) Determine the adjusted income of a single filer whose tax bill is $14,780.

47. Supplementary Angles Two angles are **supplementary** if the sum of the measures of the angles is 180°. If one angle is 30° more than its supplement, find the measures of the two angles.

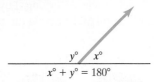

$x° + y° = 180°$

48. Supplementary Angles See Problem 47. If one angle is twice the measure of its supplement, find the measures of the two angles.

49. Complementary Angles Two angles are **complementary** if the sum of the measures of the angles is 90°. If one angle is 10° more than 3 times its complement, find the measures of the two angles.

$x° + y° = 90°$

50. Complementary Angles See Problem 49. If one angle is 30° less than twice its complement, find the measures of the two angles.

51. Dimensions of a Window The perimeter of a rectangular window is 26 feet. The width of the window is 3 feet more than the length. What are the dimensions of the window?

52. Dimensions of a Window The perimeter of a rectangular window is 120 inches. The length of the window is twice its width. What are the dimensions of the window?

53. Art An artist wants to place a piece of round stained glass into a square that is made of copper wire. See the figure. If the perimeter of the square is 40 inches, what is the area of the largest circular piece of stained glass that can fit into the copper square?

54. Art An artist wants to place two circular pieces of stained glass into a rectangle that is made of copper wire. See the figure. The perimeter of the rectangle is 36 centimeters, and the length is twice the width. Find the area of each circle, assuming they are to be as large as possible and still fit inside the rectangle.

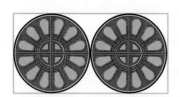

55. Angles in a Triangle The sum of the measures of the interior angles in a triangle is 180°. The measure of the second angle is 15° more than the measure of the first angle. The measure of the third angle is 45° more than the measure of the first angle. Find the measures of the interior angles in the triangle.

$x° + y° + z° = 180°$

56. Angles in a Triangle See Problem 55. The measure of the second angle is 3 times the measure of the first angle. The measure of the third angle is 20° more than the measure of the first angle. Find the measures of the interior angles in the triangle.

57. Designing a Patio Suppose that you wish to build a rectangular cement patio that is to have a perimeter of 80 feet. The length is to be 5 feet more than the width.

(a) Find the dimensions of the patio.

(b) If the building code requires that the cement be 4 inches deep, how much cement do you have to purchase?

58. Designing a Foundation You have just purchased a circular gazebo whose diameter is 12 feet. Before you have the gazebo delivered, you must lay a cement foundation to place the gazebo on.

(a) Find the area of the base of the gazebo.

(b) If the building code requires that the cement be 4 inches deep, how much cement do you have to purchase?

Extending the Concepts

59. Critical Thinking Suppose that you have just purchased a swimming pool whose diameter is 25 feet. You decide to build around the pool a deck that is 3 feet wide.

(a) What is the exact area of the deck?

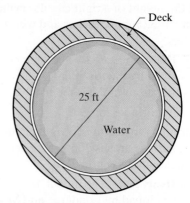

(b) The building code requires that the pool must be enclosed in a fence. How much fence is needed to encircle the pool and deck?

(c) If the fence costs $25 per linear foot to install, how much will the fence cost?

60. Critical Thinking Suppose that you wish to install a window with the dimensions given in the figure.

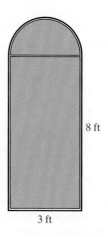

(a) What is the area of the opening of the window?

(b) What is the perimeter of the window?

(c) If glass costs $8.25 per square foot, what is the cost of the glass for the window?

61. If the radius of a circle is doubled, does the area double? Explain. If the length of the side of a cube is doubled, what happens to the volume?

1.4 Linear Inequalities in One Variable

Objectives

1 Represent Inequalities Using the Real Number Line and Interval Notation

2 Understand the Properties of Inequalities

3 Solve Linear Inequalities

4 Solve Problems Involving Linear Inequalities

Are You Ready for This Section?

Before getting started, take the following readiness quiz. If you get a problem wrong, go back to the section cited and review the material.

In Problems R1–R4, replace the question mark by $<$, $>$, or $=$ to make the statement true. [Section R.2, pp. 16–17]

R1. $3 \; ? \; 6$

R2. $-3 \; ? \; -6$

R3. $\dfrac{1}{2} \; ? \; 0.5$

R4. $\dfrac{2}{3} \; ? \; \dfrac{3}{5}$

R5. *True or False* The inequality $\geq$ is called a strict inequality. [Section R.2, p. 17]

R6. Use set-builder notation to represent the set of all digits that are divisible by 3. [Section R.2, pp. 8–9]

▶ An **inequality in one variable** is a statement involving two expressions, at least one containing the variable, separated by the inequality symbol $<$, $\leq$, $>$, or $\geq$. To **solve an inequality** means to find all values of the variable for which the statement is true. These values are called **solutions** of the inequality. The set of all solutions is called the **solution set.**

> ### Definition
>
> A **linear inequality in one variable** is an inequality that can be written in the form
>
> $$ax + b < c \quad \text{or} \quad ax + b \leq c \quad \text{or} \quad ax + b > c \quad \text{or} \quad ax + b \geq c$$
>
> where a, b, and c are real numbers and $a \neq 0$.

For example, the following are all linear inequalities involving one variable:

$$x - 4 > 9 \qquad 5x - 1 \leq 14 \qquad 8z < 0 \qquad 5x - 1 \geq 3x + 8$$

Before we discuss methods for solving linear inequalities, we introduce two new ways to represent the solution set: *interval notation* and graphing the solution set on a real number line.

1 Represent Inequalities Using the Real Number Line and Interval Notation

▶ Suppose that a and b are two real numbers and $a < b$. The notation

$$a < x < b$$

Work Smart

Remember that the inequalities $<$ and $>$ are called strict inequalities, while $\leq$ and $\geq$ are called nonstrict inequalities.

means that x is a number *between* a and b. So the expression $a < x < b$ is equivalent to the two inequalities $a < x$ and $x < b$. Similarly, the expression $a \leq x \leq b$ is equivalent to the two inequalities $a \leq x$ and $x \leq b$. We define $a \leq x < b$ and $a < x \leq b$ similarly. Expressions such as $-2 < x < 5$ and $x \geq 5$ are in **inequality notation.**

Even though the expression $3 \geq x \geq 2$ is technically correct, we prefer that the numbers in an inequality go from smaller values to larger values. So we write $3 \geq x \geq 2$ as $2 \leq x \leq 3$.

A statement such as $3 \leq x \leq 1$ is false because there is no number x for which $3 \leq x$ and $x \leq 1$. We also never mix inequalities directions, as in $2 \leq x \geq 3$.

Now let's see how to use **interval notation** to represent the solution set of an inequality.

Ready?...Answers **R1.** $<$ **R2.** $>$
R3. $=$ **R4.** $>$ **R5.** False
R6. $\{\, x \mid x$ is a digit that is divisible by 3 $\}$

Definitions

Let a and b represent two real numbers with $a < b$.

A **closed interval,** denoted by $[a, b]$, consists of all real numbers x for which $a \leq x \leq b$.

An **open interval,** denoted by (a, b), consists of all real numbers x for which $a < x < b$.

The **half-open,** or **half-closed, intervals** are $(a, b]$, consisting of all real numbers x for which $a < x \leq b$, and $[a, b)$, consisting of all real numbers x for which $a \leq x < b$.

In each of these definitions, a is the **left endpoint** and b is the **right endpoint** of the interval.

The symbol ∞ (which is read "infinity") is not a real number but a notational device that indicates unboundedness in the positive direction. In other words, the symbol ∞ means the inequality has no right endpoint. The symbol $-\infty$ (which is read "minus infinity" or "negative infinity") is also not a number but a notational device that indicates unboundedness in the negative direction. The symbol $-\infty$ means the inequality has no left endpoint. The symbols ∞ and $-\infty$ allow us to define five other kinds of intervals.

Work Smart

The symbols ∞ and $-\infty$ are never included as endpoints because they are not real numbers. Therefore, we use parentheses when $-\infty$ or ∞ is an endpoint.

Interval Notation

$[a, \infty)$	consists of all real numbers x where $x \geq a$
(a, ∞)	consists of all real numbers x where $x > a$
$(-\infty, a]$	consists of all real numbers x where $x \leq a$
$(-\infty, a)$	consists of all real numbers x where $x < a$
$(-\infty, \infty)$	consists of all real numbers x (that is, $-\infty < x < \infty$)

Another way to represent inequalities is to draw a graph on the real number line. The inequality $x > 3$ or the interval $(3, \infty)$ consists of all numbers x that lie to the right of 3 on the real number line. We graph these values by shading the real number line to the right of 3. We use a *parenthesis* on the endpoint to indicate that 3 is not included in the set. See Figure 9.

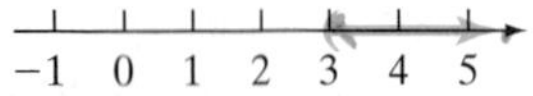

To graph the inequality $x \geq 3$ or the interval $[3, \infty)$, we shade to the right of 3 but use a *bracket* on the endpoint to indicate that 3 is included in the set. See Figure 10.

Table 7 summarizes interval notation, inequality notation, and their graphs.

Figure 9
$x > 3$

Figure 10
$x \geq 3$

Table 7

Interval Notation	Inequality Notation	Graph
The open interval (a, b)	$\{x \mid a < x < b\}$	
The closed interval $[a, b]$	$\{x \mid a \leq x \leq b\}$	
The half-open interval $[a, b)$	$\{x \mid a \leq x < b\}$	
The half-open interval $(a, b]$	$\{x \mid a < x \leq b\}$	
The interval $[a, \infty)$	$\{x \mid x \geq a\}$	
The interval (a, ∞)	$\{x \mid x > a\}$	
The interval $(-\infty, a]$	$\{x \mid x \leq a\}$	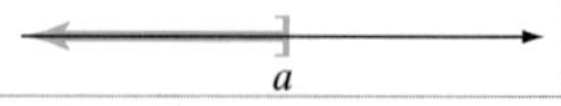
The interval $(-\infty, a)$	$\{x \mid x < a\}$	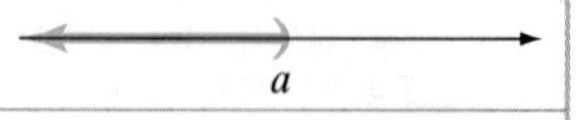
The interval $(-\infty, \infty)$	$\{x \mid x \text{ is a real number}\}$	

EXAMPLE 1 **Using Interval Notation and Graphing Inequalities**

Write each inequality using interval notation. Graph the inequality.

(a) $-2 \leq x \leq 4$ (b) $1 < x \leq 5$

Solution

(a) $-2 \leq x \leq 4$ describes all numbers x between -2 and 4, inclusive. In interval notation, we write $[-2, 4]$. To graph $-2 \leq x \leq 4$, we place brackets at -2 and 4 and shade in between. See Figure 11.

Figure 11

(b) $1 < x \leq 5$ describes all numbers x greater than 1 and less than or equal to 5. In interval notation, we write $(1, 5]$. To graph $1 < x \leq 5$, we place a parenthesis at 1 and a bracket at 5 and shade in between. See Figure 12.

Figure 12

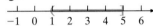

EXAMPLE 2 **Using Interval Notation and Graphing Inequalities**

Write each inequality using interval notation. Graph the inequality.

(a) $x < 2$ (b) $x \geq -3$

Solution

(a) $x < 2$ describes all numbers x less than 2. In interval notation, we write $(-\infty, 2)$. To graph $x < 2$, we place a parenthesis at 2 and then shade to the left. See Figure 13.

Figure 13

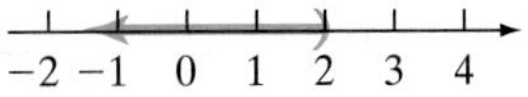

(b) $x \geq -3$ describes all numbers x greater than or equal to -3. In interval notation, we write $[-3, \infty)$. To graph $x \geq -3$, we place a bracket at -3 and then shade to the right. See Figure 14.

Figure 14

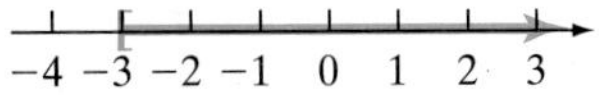

Quick ✓

1. A(n) _____ _____, denoted $[a, b]$, consists of all real numbers x for which $a \leq x \leq b$.

2. In the interval (a, b), a is called the _____ _____ and b is called the _____ _____ of the interval.

3. *True or False* The inequality $x \leq -11$ is written $[-11, -\infty)$ in interval notation.

In Problems 4–7, write each inequality in interval notation. Graph the inequality.

4. $-3 \leq x \leq 2$ 5. $3 \leq x < 6$

6. $x \leq 3$ 7. $\dfrac{1}{2} < x < \dfrac{7}{2}$

▶ (**EXAMPLE 3**) **Using Inequality Notation and Graphing Inequalities**

Write each interval in inequality notation involving x. Graph the inequality.

(a) $[-2, 4)$ **(b)** $(1, 5)$

Solution

(a) The interval $[-2, 4)$ consists of all numbers x for which $-2 \le x < 4$. See Figure 15 for the graph.

(b) The interval $(1, 5)$ consists of all numbers x for which $1 < x < 5$. See Figure 16 for the graph.

Figure 15

Figure 16

(**EXAMPLE 4**) **Using Inequality Notation and Graphing Inequalities**

Write each interval in inequality notation involving x. Graph the inequality.

(a) $\left[\dfrac{3}{2}, \infty\right)$ **(b)** $(-\infty, 1)$

Solution

(a) The interval $\left[\dfrac{3}{2}, \infty\right)$ consists of all numbers x for which $x \ge \dfrac{3}{2}$. See Figure 17 for the graph.

(b) The interval $(-\infty, 1)$ consists of all numbers x for which $x < 1$. See Figure 18 for the graph.

Figure 17

Figure 18

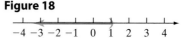

Quick ✓

In Problems 8–11, write each interval as an inequality. Graph the inequality.

8. $(0, 5]$ **9.** $(-6, 0)$

10. $(5, \infty)$ **11.** $\left(-\infty, \dfrac{8}{3}\right]$

▶ ❷ **Understand the Properties of Inequalities**

Consider the inequality $2 < 5$. Because $2 < 5$, if we add 4 to both sides of the inequality, we have $2 + 4 < 5 + 4$ or $6 < 9$. Also, since $3 > -1$, then $3 + (-2) > -1 + (-2)$ or $1 > -3$ so, adding the same quantity to both sides of an inequality does not change the direction of the inequality.

In Words

The Addition Property states that the direction of the inequality does not change when the same quantity is added to each side of the inequality.

Addition Property of Inequalities

For real numbers a, b, and c,

$$\text{If} \quad a < b, \quad \text{then} \quad a + c < b + c$$
$$\text{If} \quad a > b, \quad \text{then} \quad a + c > b + c$$

Because $-b$ is equivalent to $a + (-b)$, the Addition Property of Inequalities can also be used to subtract a real number from each side of an inequality without changing the direction of the inequality.

What happens when we multiply both sides by a nonzero constant? Let's multiply both sides of $3 < 5$ by 2. The left side becomes 6 and the right side becomes 10. Certainly $6 < 10$, so the direction of the inequality did not change.

Let's now multiply both sides of the inequality $3 < 5$ by -2. The left side becomes -6 and the right side becomes -10. Because $-6 > -10$, the direction of the inequality is reversed.

These results are true in general and lead us to the next properties.

> **In Words**
>
> The Multiplication Property states that if both sides of an inequality are multiplied by a positive real number, the direction of the inequality is unchanged. If both sides of an inequality are multiplied by a negative real number, the direction of the inequality is reversed.

Multiplication Properties of Inequality

Let a, b, and c be real numbers.

$$\text{If } a < b \text{ and } c > 0, \text{ then } ac < bc$$
$$\text{If } a > b \text{ and } c > 0, \text{ then } ac > bc$$
$$\text{If } a < b \text{ and } c < 0, \text{ then } ac > bc$$
$$\text{If } a > b \text{ and } c < 0, \text{ then } ac < bc$$

Because the quotient $\dfrac{a}{b}$ is equivalent to $a \cdot \dfrac{1}{b}$, the Multiplication Properties of Inequality can also be used to divide both sides of an inequality by a nonzero real number.

Quick ✓

In Problems 12–16, write the inequality that results from following the directions. Then state the inequality property you have illustrated.

12. Add 5 to each side of the inequality $4 < 7$.

13. Subtract 3 from each side of the inequality $x + 3 > -6$.

14. Multiply both sides of the inequality $2 < 8$ by $\dfrac{1}{2}$.

15. Divide both sides of the inequality $-6 < 9$ by -3.

16. Divide both sides of the inequality $5x < 30$ by 5.

❸ Solve Linear Inequalities

Two inequalities that have exactly the same solution set are called **equivalent inequalities.** To solve linear inequalities, we use a process similar to the one used to solve linear equations: We create a series of equivalent inequalities to arrive at a solution.

Although not essential, it is easier to read an inequality when the variable is on the left side and the constant is on the right. We can rewrite inequalities using the fact that

> **In Words**
>
> If the sides of an inequality are interchanged, the direction of the inequality reverses.

$$a < x \quad \text{is equivalent to} \quad x > a$$
$$\text{and}$$
$$a > x \quad \text{is equivalent to} \quad x < a$$

(**EXAMPLE 5**)   **How to Solve a Linear Inequality**

Solve the inequality $3x - 2 > 13$. Graph the solution set.

Step-by-Step Solution

The goal in solving any linear inequality is to get the variable by itself with a coefficient of 1.

Step 1: Isolate the term containing the variable.	Add 2 to both sides (Addition Property): $\quad 3x - 2 > 13$ $3x - 2 + 2 > 13 + 2$ $3x > 15$

(continued)

Step 2: *Get a coefficient of 1 on the variable.*

Divide both sides by 3
(Multiplication Property):

$$\frac{3x}{3} > \frac{15}{3}$$

$$x > 5$$

Figure 19

The solution using set-builder notation is $\{x \mid x > 5\}$. The solution using interval notation is $(5, \infty)$. Figure 19 shows the graph of the solution set.

Quick ✓

In Problems 17–19, solve each linear inequality. Write the solution using set-builder notation and interval notation. Graph the solution set.

17. $x + 3 > 5$

18. $\frac{1}{3}x \le 2$

19. $-2x + 1 \le 13$

EXAMPLE 6   **Solving Linear Inequalities**

Solve the inequality: $x - 4 \ge 5x + 12$

Solution

$$x - 4 \ge 5x + 12$$

Add 4 to both sides: $x - 4 + 4 \ge 5x + 12 + 4$

$$x \ge 5x + 16$$

Subtract 5x from both sides: $x - 5x \ge 5x + 16 - 5x$

$$-4x \ge 16$$

Divide both sides by −4. Don't forget to change the direction of the inequality: $\dfrac{-4x}{-4} \le \dfrac{16}{-4}$

$$x \le -4$$

Figure 20

The solution using set-builder notation is $\{x \mid x \le -4\}$. The solution using interval notation is $(-\infty, -4]$. See Figure 20 for the graph of the solution set.

Quick ✓

In Problems 20–22, solve each linear inequality. Write the solution using set-builder notation and interval notation. Graph the solution set.

20. $3x + 1 > x - 5$

21. $-2x + 1 \le 3x + 11$

22. $-5x + 12 < x - 3$

▶ EXAMPLE 7   **Solving Linear Inequalities**

Solve the inequality: $3(x - 1) + 2x < 6x + 3$

Solution

$$3(x - 1) + 2x < 6x + 3$$

Distribute the 3: $3x - 3 + 2x < 6x + 3$

Combine like terms: $5x - 3 < 6x + 3$

Add 3 to both sides: $5x - 3 + 3 < 6x + 3 + 3$

$$5x < 6x + 6$$

Subtract 6x from both sides: $5x - 6x < 6x + 6 - 6x$

$$-x < 6$$

Multiply both sides of the inequality by −1: $(-1)(-x) > (-1)6$

$$x > -6$$

Figure 21

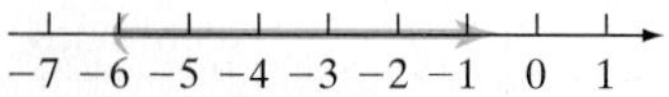

The solution using set-builder notation is $\{x \mid x > -6\}$. The solution using interval notation is $(-6, \infty)$. See Figure 21 for the graph of the solution set.

Quick ✓

In Problems 23–25, solve each linear inequality. Write the solution using set-builder notation and interval notation. Graph the solution set.

23. $4(x - 2) < 3x - 4$

24. $-2(x + 1) \geq 4(x + 3)$

25. $7 - 2(x + 1) \leq 3(x - 5)$

(**EXAMPLE 8**) **Solving Linear Inequalities Involving Fractions**

Solve the inequality: $\dfrac{2x + 1}{3} > \dfrac{x - 2}{2}$

Solution

We clear the linear inequality of fractions by multiplying both sides of the inequality by 6, the least common denominator.

$$6 \cdot \left(\frac{2x + 1}{3}\right) > 6 \cdot \left(\frac{x - 2}{2}\right)$$

$$2(2x + 1) > 3(x - 2)$$

Distribute: $\qquad 4x + 2 > 3x - 6$

Subtract 2 from both sides: $\qquad 4x + 2 - 2 > 3x - 6 - 2$

$$4x > 3x - 8$$

Subtract 3x from both sides: $\qquad 4x - 3x > 3x - 8 - 3x$

$$x > -8$$

Figure 22

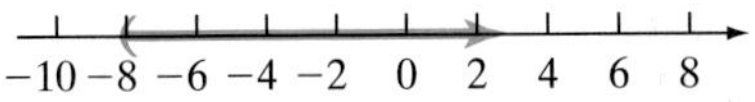

The solution using set-builder notation is $\{x \mid x > -8\}$. The solution using interval notation is $(-8, \infty)$. See Figure 22 for the graph of the solution set. ●

Quick ✓

In Problems 26–28, solve each linear inequality. Write the solution using set-builder notation and interval notation. Graph the solution set.

26. $\dfrac{3x + 1}{5} \geq 2$ $\qquad$ **27.** $\dfrac{2}{5}x + \dfrac{3}{10} < \dfrac{1}{2}$

28. $\dfrac{1}{2}(x + 3) > \dfrac{1}{3}(x - 4)$

▶ ❹ **Solve Problems Involving Linear Inequalities**

One of the first things you need to do to solve a word problem is look for key words. Certain phrases can give you a clue about the type of inequality problem you have. Some of these phrases are shown in Table 8.

Table 8			
Phrase	**Inequality**	**Phrase**	**Inequality**
At least	$\geq$	No more than	$\leq$
No less than	$\geq$	At most	$\leq$
More than	$>$	Fewer than	$<$
Greater than	$>$	Less than	$<$

To solve applications involving linear inequalities, we use the problem-solving steps from Section 1.2 on page 61.

EXAMPLE 9 | **Comparing Credit Cards**

US Bank offers several credit card options. You qualify for the SKYPASS Visa Signature card, which charges an annual fee of $80 plus 12.99% annual interest on outstanding balances, and for the SKYPASS Visa Classic card, which has an annual fee of $50 plus 14.24% annual interest on outstanding balances. What annual balances will result in the Signature card costing less than the Classic card? (SOURCE: US Bank)

Solution

Step 1: Identify We want to know the credit card balance for which the Signature credit card costs less than the Classic card. The phrase "costs less" implies that this is an inequality problem.

Step 2: Name Let b represent the credit card balance on each card.

Step 3: Translate Because each card charges an annual fee plus simple interest, the cost for each card will be "annual fee + interest."

Recall from Section 1.2 the simple interest formula is

$$I = Prt$$

where I is the interest charged, P is the balance on the credit card, r is the annual interest rate, and t is time.

In this problem, we let b represent the credit card balance. The annual interest rate r is 0.1299 (for the Signature card) or 0.1424 (for the Classic card). Because we are discussing annual cost, we let $t = 1$.

Since we want to know what balance results in Signature card costing less than the Classic card, we have the following inequality:

$$\text{Signature Card Cost} < \text{Classic Card Cost}$$

Annual Fee for Signature	Interest Charged for Signature	Annual Fee for Classic	Interest Charged for Classic	
80 +	$0.1299b$ <	50 +	$0.1424b$	The Model

Step 4: Solve Solve the inequality for b. $\qquad\qquad 80 + 0.1299b < 50 + 0.1424b$

Subtract 80 from both sides: $\qquad\qquad\qquad 0.1299b < -30 + 0.1424b$

Subtract $0.1424b$ from both sides: $\qquad\qquad -0.0125b < -30$

Divide both sides by -0.0125 and reverse the inequality symbol: $\qquad\qquad b > 2400$

Step 5: Check If the balance is $2400, then the annual cost for the Signature card is $80 + 0.1299(2400) = \$391.76$, and the annual cost for the Classic card is $50 + 0.1424(2400) = \$391.76$. What about balances greater than $2400, say $2500? The annual cost for the Signature card is $80 + 0.1299(2500) = \$404.75$. The annual cost for the Classic card is $50 + 0.1424(2500) = \$406.00$.

Step 6: Answer the Question The Signature card costs less than the Classic card if the annual balance is greater than $2400.

Quick ✓

29. You are choosing between two credit cards. Bank A's card has an annual fee of $24 and charges 9.95% simple interest. Bank B's card has no annual fee but charges 14.95% simple interest. For what annual balance will the Bank A card cost less than the Bank B card?

30. Suppose the daily revenue from selling x boxes of candy is given by the equation $R = 12x$. The daily cost of operating the store and making the candy is given by the equation $C = 8x + 96$. For how many boxes of candy will revenue exceed costs? That is, solve $R > C$.

1.4 Exercises   MyMathLab®   MathXP PRACTICE

Exercise numbers in green are complete video solutions in MyMathLab

Problems **1–30** *are the* Quick ✓ s *that follow the* EXAMPLES.

Building Skills

In Problems 31–38, write each inequality using interval notation. Graph the inequality. See Objective 1.

31. $2 \leq x \leq 10$ **32.** $1 < x < 7$

33. $-4 \leq x < 0$ **34.** $-8 < x \leq 1$

35. $x \geq 6$ **36.** $x < 0$

37. $x < \dfrac{3}{2}$ **38.** $x \geq -\dfrac{5}{2}$

In Problems 39–46, write each interval as an inequality involving x. Graph each inequality. See Objective 1.

39. $(1, 8)$ **40.** $[-2, 3]$

41. $(-5, 1]$ **42.** $[1, 4)$

43. $(-\infty, 5)$ **44.** $(2, \infty)$

45. $[3, \infty)$ **46.** $(-\infty, 8]$

In Problems 47–54, fill in the blank with the correct inequality symbol. State which property of inequalities is being utilized. See Objective 2.

47. If $x - 3 < 7$, then x __ 10.

48. If $2x - 5 > 6$, then $2x$ __ 11.

49. If $\dfrac{1}{3}x > 5$, then x __ 15.

50. If $4x > 36$, then x __ 9.

51. If $2x + 5 \leq 9$, then $2x$ __ 4.

52. If $\dfrac{2}{5}x + 6 \leq 8$, then $\dfrac{2}{5}x$ __ 2.

53. If $-2x \geq 10$, then x __ -5.

54. If $-6x < 30$, then x __ -5.

In Problems 55–92, solve each linear inequality. Write the solution using set-builder notation and interval notation. Graph the solution set. See Objective 3.

55. $x - 4 \leq 2$ **56.** $x + 6 < 9$

57. $6x < 24$ **58.** $4x \geq 20$

59. $-7x < 21$ **60.** $-8x > 32$

61. $\dfrac{4}{15}x > \dfrac{8}{5}$ **62.** $\dfrac{3}{8}x < \dfrac{9}{16}$

63. $3x + 2 > 11$ **64.** $5x - 4 \leq 16$

65. $-3x + 1 > 13$

66. $-6x - 5 < 13$

67. $6x + 5 \leq 3x + 2$

68. $8x + 3 \geq 5x - 9$

69. $-3x + 1 < 2x + 11$

70. $3x + 4 \geq 5x - 8$

71. $3(x - 3) < 2(x + 4)$

72. $3(x - 2) + 5 > 4(x + 1) + x$

73. $4(x + 1) - 2x \geq 5(x - 2) + 2$

74. $-3(x + 4) + 5x < 4(x + 3) - 14$

75. $0.5x + 4 \leq 0.2x - 5$

76. $2.3x - 1.2 > 1.8x + 0.4$

77. $\dfrac{3x + 1}{4} < \dfrac{1}{2}$

78. $\dfrac{2x - 3}{3} > \dfrac{4}{3}$

79. $\dfrac{1}{2}(x - 4) > \dfrac{3}{4}(2x + 1)$

80. $\dfrac{1}{3}(3x + 5) < \dfrac{1}{6}(x + 4)$

81. $\dfrac{3}{5} - x > \dfrac{5}{3}$

82. $\dfrac{2}{3} - \dfrac{5}{6}x > 2$

83. $-5(x - 3) \geq 3[4 - (x + 4)]$

84. $-3(2x + 1) \leq 2[3x - 2(x - 5)]$

85. $4(3x - 1) - 5(x + 4) \geq 3[2 - (x + 3)] - 6x$

86. $7(x + 2) - 4(2x + 3) < -2[5x - 2(x + 3)] + 7x$

87. $\dfrac{2}{3}(4x - 1) - \dfrac{4}{9}(x - 4) > \dfrac{5}{12}(2x + 3)$

88. $\dfrac{5}{6}(3x - 2) - \dfrac{2}{3}(4x - 1) < -\dfrac{2}{9}(2x + 5)$

89. $\dfrac{4x - 3}{3} < 3$

90. $\dfrac{2}{5}x + \dfrac{3}{10} < \dfrac{1}{2}$

91. $\dfrac{2}{3}x < \dfrac{1}{4}(2x + 3)$

92. $\dfrac{x}{12} \geq \dfrac{x}{2} - \dfrac{2x + 1}{4}$

Mixed Practice

In Problems 93–106, solve each linear inequality. Write the solution using set-builder notation and interval notation. Graph the solution set.

93. $y + 8 > -7$

94. $y - 5 \geq 7$

95. $-3a \leq -21$

96. $-5x < 30$

97. $13x - 5 > 10x - 6$

98. $4x + 3 \geq -6x - 2$

99. $4(x + 2) \leq 3(x - 2)$

100. $5(y + 7) < 6(y + 4)$

101. $3(4 - 3x) > 6 - 5x$

102. $2(5 - x) - 3 \leq 4 - 5x$

103. $2[4 - 3(x + 1)] \leq -4x + 8$

104. $3[1 + 2(x - 4)] \geq 3x + 3$

105. $\dfrac{x}{2} + \dfrac{3}{4} \geq \dfrac{3}{8}$

106. $\dfrac{b}{3} + \dfrac{5}{6} < \dfrac{11}{12}$

Applying the Concepts

107. Find the set of all x such that the sum of twice x and 5 is at least 13.

108. Find the set of all x such that the difference between 3 times x and 2 is less than 7.

109. Find the set of all z such that the product of 4 and z minus 3 is no more than 9.

110. Find the set of all y such that the sum of twice y and 3 is greater than 13.

111. Computing Grades In order to earn an A in Mr. Ruffatto's Intermediate Algebra course, Jackie must earn at least 540 points. Thus far, Jackie has earned 90, 83, 95, and 90 points on her four exams. The final exam, which counts as 200 points, is rapidly approaching. How many points does Jackie need to earn on the final to earn an A in Mr. Ruffatto's class?

112. Computing Grades In order to earn an A in Mrs. Padilla's Intermediate Algebra course, Mark must obtain an average score of at least 90. On his first four exams, Mark scored 94, 83, 88, and 92. The final exam counts as two test scores. What score does Mark need on the final to earn an A in Mrs. Padilla's class?

113. McDonald's Suppose that you have ordered one medium order of French fries and one 16-ounce triple-thick chocolate shake from McDonald's. The fries have 19 grams of fat, and the shake has 21 grams of fat. Each McDonald's hamburger has 9 grams of fat. How many hamburgers can you order and still keep the total fat content of the meal to no more than 67 grams? (SOURCE: *McDonald's Corporation*)

114. Burger King Suppose that you have ordered one medium onion ring and one 16-ounce chocolate shake from Burger King. The onion rings have 21 grams of fat and the shake has 21 grams of fat. Each cheeseburger has 13 grams of fat. How many cheeseburgers can you order and still keep the total fat content of the meal to no more than 81 grams? (SOURCE: *Burger King*)

115. Payload Restrictions An Airbus A320 has a maximum payload of 45,686 pounds. Suppose that on a flight from Chicago to New Orleans, United Airlines sold out the flight at 179 seats. Assuming that the average passenger weighs 150 pounds, determine the weight of the luggage and other cargo that the plane can carry. (SOURCE: *Airbus*)

116. Moving Trucks A 15-foot moving truck from Budget costs $39.95 per day plus $0.65 per mile. If your budget allows for you to spend at most $125.75, what is the number of miles you can drive? (SOURCE: *Budget*)

117. Health Benefits The average monthly benefit B, in dollars, for individuals on disability is given by the equation $B = 19.25t + 585.72$, where t is the number of years since 1990. In what year will the monthly benefit B exceed 1000? That is, solve $B > 1000$.

118. Health Expenditures Total private health expenditures H, in billions of dollars, are given by the equation $H = 26t + 411$, where t is the number of years since 1990. In what year will total private health expenditures exceed 1 trillion (1000 billion)? That is, solve $26t + 411 > 1000$.

119. Commissions Susan sells computer systems. Her annual base salary is $34,000. She also earns a commission of 1.2% on the sale price of all computer systems that she sells. For what value of the computer systems sold will Susan's annual salary be at least $100,000?

120. Commissions Al sells used cars for a Chevy dealer. His annual base salary is $24,300. He also earns a commission of 3% on the sale price of the cars that he sells. For what value of the cars sold will Al's annual salary be more than $60,000? If used cars at this particular dealership sell for an average of $15,000, how many cars does Al have to sell to meet his salary goal?

121. Supply and Demand The quantity demanded of custom monogrammed shirts is given by the equation $D = 1000 - 20p$, where p is the price of a shirt. The quantity supplied of custom monogrammed shirts is given by the equation $S = -200 + 10p$. For what prices will quantity supplied exceed quantity demanded, thereby resulting in a surplus of shirts? That is, solve $S > D$.

122. Supply and Demand The quantity supplied of digital cameras is given by the equation $S = -2800 + 13p$, where p is the price of a camera. The quantity demanded of digital cameras is given by the equation $D = 1800 - 12p$. For what prices will quantity demanded exceed quantity supplied, thereby resulting in a shortage of cameras? That is, solve $D > S$.

Extending the Concepts

123. Solve the linear inequality
$3(x + 2) + 2x > 5(x + 1)$.

124. Solve the linear inequality
$-3(x - 2) + 7x > 2(2x + 5)$.

Explaining the Concepts

125. Write a brief paragraph that explains the circumstances under which the direction, or sense, of an inequality changes.

126. Explain why the inequality $5 < x < 1$ is false.

127. Explain why we never mix inequalities as in $4 < x > 7$.

Putting the Concepts Together (Sections 1.1–1.4)

We designed these problems so that you can review Sections 1.1 through 1.4 and show your mastery of the concepts. Take time to work these problems before proceeding with the next section. The answers are at the back of the text on page AN-4.

1. Determine which, if any, of the following are solutions to

$$5(2x - 3) + 1 = 2x - 6$$

(a) $x = -3$ **(b)** $x = 1$

In Problems 2 and 3, solve the equation.

2. $3(2x - 1) + 6 = 5x - 2$

3. $\dfrac{7}{3}x + \dfrac{4}{5} = \dfrac{5x + 12}{15}$

Determine whether the equation is an identity, a contradiction, or a conditional equation.

4. $5 - 2(x + 1) + 4x = 6(x + 1) - (3 + 4x)$

In Problems 5 and 6, translate the English statement into a mathematical statement. Do not solve the equation.

5. The difference of a number and 3 is two more than half the number.

6. The quotient of a number and 2 is less than the number increased by 5.

7. Mixture Two acid solutions are available to a chemist. One is a 20% nitric acid solution, and the other is a 40% nitric acid solution. How much of each type of solution should be mixed together to form 16 liters of a 35% nitric acid solution?

8. Travel Two cars leave from the same location and travel in opposite directions along a straight road. One car travels 30 miles per hour, while the other travels at 45 miles per hour. How long will it take the two cars to be 255 miles apart?

9. Solve $3x - 2y = 4$ for y.

10. Solve the formula $A = P + Prt$ for r.

11. The volume of a right circular cylinder is given by the formula $V = \pi r^2 h$, where r is the radius of the cylinder and h is the height of the cylinder.

(a) Solve the formula for h.

(b) Use the result from part (a) to find the height of a right circular cylinder with volume $V = 294\pi$ inches^3 and radius $r = 7$ inches.

12. Write the following inequalities in interval notation. Graph the inequality.

(a) $x > -3$

(b) $2 < x \leq 5$

13. Write the interval in inequality notation involving x. Graph the inequality.

(a) $(-\infty, -1.5]$

(b) $(-3, 1]$

In Problems 14–16, solve the inequality and graph the solution set on a real number line.

14. $2x + 3 \leq 4x - 9$

15. $-3 > 3x - (x + 5)$

16. $x - 9 \leq x + 3(2 - x)$

17. Birthday Party A recreational center offers a children's birthday party for $75 plus $5 for each child. How many children can Logan invite to his birthday party if the budget for the party is no more than $125?

Part II: Linear Equations and Inequalities in Two Variables

1.5 Rectangular Coordinates and Graphs of Equations

Objectives

1 Plot Points in the Rectangular Coordinate System

2 Determine Whether an Ordered Pair Is a Point on the Graph of an Equation

3 Graph an Equation Using the Point-Plotting Method

4 Identify the Intercepts from the Graph of an Equation

5 Interpret Graphs

Are You Ready for This Section?

Before getting started, take this readiness quiz. If you get a problem wrong, go back to the section cited and review the material.

R1. Plot the following points on the real number line: $-2, 4, 0, \dfrac{1}{2}$. [Section R.2, pp. 15–16]

R2. Determine which of the following are solutions to the equation $3x - 5(x + 2) = 4$.

(a) $x = 0$ (b) $x = -3$ (c) $x = -7$ [Section 1.1, pp. 49–50]

R3. Evaluate the expression $2x^2 - 3x + 1$ for the given values of the variable.

(a) $x = 0$ (b) $x = 2$ (c) $x = -3$ [Section R.5, pp. 40–41]

R4. Solve the equation $3x + 2y = 8$ for y. [Section 1.3, pp. 73–74]

R5. Evaluate $|-4|$. [Section R.3, p. 19]

▶ **1** Plot Points in the Rectangular Coordinate System

In Section R.2, when we located a point on the real number line, we were working in one dimension. Now we will work in two dimensions.

We begin by drawing two real number lines—one horizontal and one vertical—that intersect at right (90°) angles. We call the horizontal line the **x-axis** and the vertical line the **y-axis**. The point where the **axes** (plural of "axis") intersect is called the **origin, O**. See Figure 23.

Figure 23

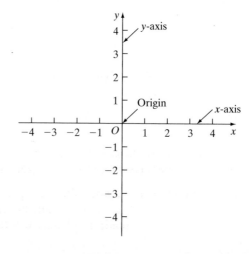

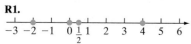

The origin O has a value of 0 on each axis. On the x-axis, points to the right of O are positive real numbers, and points to the left are negative. On the y-axis, points above O are positive real numbers, and points below are negative. In Figure 23 we label the x-axis "x" and the y-axis "y". An arrow at the end of each axis denotes the positive direction.

The coordinate system in Figure 23 is called a **rectangular** or **Cartesian coordinate system,** named after René Descartes (1596–1650), a French mathematician, philosopher, and theologian. The plane formed by the x-axis and the y-axis is often called the **xy-plane,** and the x-axis and y-axis are called the **coordinate axes.**

We can represent any point P in the rectangular coordinate system by using an **ordered pair** (x, y) of real numbers. If $x > 0$, then P is x units to the right of the y-axis; if $x < 0$, then P is $|x|$ units to the left of the y-axis. If $y > 0$, then P is y units above the x-axis; if $y < 0$, then P is $|y|$ units below the x axis. The ordered pair (x, y) is also called the **coordinates** of P.

The origin O has coordinates $(0, 0)$. Any point on the x-axis has coordinates of the form $(x, 0)$, and any point on the y-axis has coordinates of the form $(0, y)$.

If point P has coordinates (x, y), then x is the **x-coordinate,** or **abscissa,** of P, and y is the **y-coordinate,** or **ordinate,** of P.

Look back at Figure 23. Notice that the x- and y-axes divide the plane into four separate regions or **quadrants.** In quadrant I, both the x- and y-coordinate are positive; in quadrant II, x is negative and y is positive; in quadrant III, both x and y are negative; and in quadrant IV, x is positive and y is negative. Points on the coordinate axes do not belong to a quadrant. See Figure 24.

Figure 24

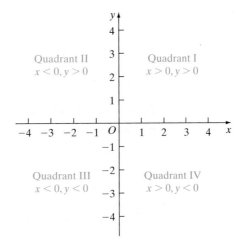

EXAMPLE 1   **Plotting Points in the Rectangular Coordinate System and Determining the Quadrant in Which the Point Lies**

Plot the points in the xy-plane. State which quadrant each point is in.

(a) $A(3, 2)$ (b) $B(-2, 4)$ (c) $C(-1, -3)$
(d) $D(3, -4)$ (e) $E(-2, 0)$

Solution

First, draw a rectangular, or Cartesian, coordinate system. See Figure 25(a) on the next page. Now plot the points.

(a) To plot point $A(3, 2)$, begin at the origin O, travel 3 units to the right, and then travel 2 units up. Label the point A. See Figure 25(b). Point A is in quadrant I because both x and y are positive.

Figure 25

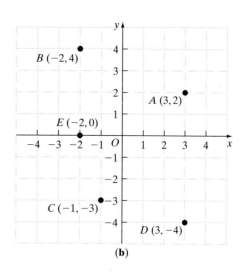

(a)   (b)

(b) To plot point $B(-2, 4)$, begin at the origin O, and travel 2 units to the left and then 4 units up. Label the point B. See Figure 25(b). Point B is in quadrant II.

(c) See Figure 25(b). Point C is in quadrant III.

(d) See Figure 25(b). Point D is in quadrant IV.

(e) See Figure 25(b). Point E is not in a quadrant because it lies on the x-axis.

Quick ✓

1. The point where the x-axis and the y-axis intersect in the Cartesian coordinate system is called the _____.

2. *True or False* If a point lies in quadrant III of the Cartesian coordinate system, then both x and y are negative.

In Problems 3 and 4, plot each point in the xy-plane. Tell in which quadrant or on which coordinate axis each point lies.

3. **(a)** $A(5, 2)$
   (b) $B(4, -2)$
   (c) $C(0, -3)$
   (d) $D(-4, -3)$

4. **(a)** $A(-3, 2)$
   (b) $B(-4, 0)$
   (c) $C(3, -2)$
   (d) $D(6, 1)$

▶ ❷ Determine Whether an Ordered Pair Is a Point on the Graph of an Equation

Recall from Section 1.1 that the solution of a linear equation must be a single value of the variable, the empty set, or all real numbers. Now we will learn a method for representing the solution to an *equation in two variables*.

> **Definition**
>
> An **equation in two variables,** say x and y, is a statement in which the algebraic expressions involving x and y are equal. The expressions are called **sides** of the equation.

Since an equation is a statement, it may be true or false, depending on the values of the variables. Any values of the variable that make the equation true **satisfy** the equation.

For example, the following are all equations in two variables.

$$x^2 = y + 2 \qquad 3x + 2y = 6 \qquad y = -4x + 5$$

The first equation is satisfied when $x = 3$ and $y = 7$ since $3^2 = 7 + 2$. It is also satisfied when $x = -2$ and $y = 2$. In fact, infinitely many choices of x and y satisfy this equation. However, some choices of x and y do not satisfy it. For example $x = 3$ and $y = 4$ does not satisfy the equation because $3^2 \neq 4 + 2$.

If two values of x and y, respectively, satisfy an equation, then the point whose coordinates are (x, y) is on the *graph of the equation*. For the remainder of the course, we will say "the point (x, y)" rather than "the point whose coordinates are (x, y)" for the sake of brevity.

In Words

The graph of an equation is a geometric way of representing the set of all points that make the equation a true statement.

> **Definition**
>
> The **graph of an equation in two variables** x and y is the set of all points whose coordinates, (x, y), in the xy-plane satisfy the equation.

EXAMPLE 2 **Determining Whether a Point Is on the Graph of an Equation**

Determine whether the following points are on the graph of $3x - y = 6$.

(a) $(2, 0)$ (b) $(1, -2)$ (c) $\left(\dfrac{1}{2}, -\dfrac{9}{2}\right)$

Solution

(a) For $(2, 0)$, check to see whether $x = 2, y = 0$ satisfies the equation $3x - y = 6$.

$$3x - y = 6$$
$$\text{Let } x = 2, y = 0: \quad 3(2) - 0 \overset{?}{=} 6$$
$$6 = 6 \quad \text{True}$$

The statement is true, so $(2, 0)$ is on the graph of the equation.

(b) For $(1, -2)$, we have

$$3x - y = 6$$
$$\text{Let } x = 1, y = -2: \quad 3(1) - (-2) \overset{?}{=} 6$$
$$3 + 2 \overset{?}{=} 6$$
$$5 = 6 \quad \text{False}$$

The statement $5 = 6$ is false, so the point $(1, -2)$ is not on the graph.

(c) For $\left(\dfrac{1}{2}, -\dfrac{9}{2}\right)$, we have

$$3x - y = 6$$
$$\text{Let } x = \frac{1}{2}, y = -\frac{9}{2}: \quad 3\left(\frac{1}{2}\right) - \left(-\frac{9}{2}\right) \overset{?}{=} 6$$
$$\frac{3}{2} + \frac{9}{2} \overset{?}{=} 6$$
$$\frac{12}{2} \overset{?}{=} 6$$
$$6 = 6 \quad \text{True}$$

The statement is true, so the point $\left(\dfrac{1}{2}, -\dfrac{9}{2}\right)$ is on the graph.   ●

Quick ✓

5. *True or False* The graph of an equation in two variables x and y is the set of all points whose coordinates, (x, y), in the xy-plane satisfy the equation.

6. Determine whether the following points are on the graph of $2x - 4y = 12$.

(a) $(2, -3)$ (b) $(2, -2)$ (c) $\left(\dfrac{3}{2}, -\dfrac{9}{4}\right)$

7. Determine whether the following points are on the graph of $y = x^2 + 3$.

(a) $(1, 4)$ (b) $(-2, -1)$ (c) $(-3, 12)$

▶ ❸ Graph an Equation Using the Point-Plotting Method

One of the simplest ways to graph an equation is the **point-plotting method.** With this method, we choose values for one of the variables and use the equation to find the corresponding values of the other variable. It does not matter whether we choose values of x and use the equation to find the corresponding values of y or choose y and find x. We let convenience determine which variable to begin with.

(EXAMPLE 3) **How to Graph an Equation by Plotting Points**

Graph the equation $y = -2x + 4$ by plotting points.

Step-by-Step Solution

Step 1: We want to find points (x, y) that satisfy the equation. To determine these points, choose values of x (do you see why?) and use the equation to determine the corresponding values of y. See Table 9.

Table 9

x	$y = -2x + 4$	(x, y)
-3	$y = -2(-3) + 4 = 10$	$(-3, 10)$
-2	$y = -2(-2) + 4 = 8$	$(-2, 8)$
-1	$y = -2(-1) + 4 = 6$	$(-1, 6)$
0	$y = -2(0) + 4 = 4$	$(0, 4)$
1	$y = -2(1) + 4 = 2$	$(1, 2)$
2	$y = -2(2) + 4 = 0$	$(2, 0)$
3	$y = -2(3) + 4 = -2$	$(3, -2)$

Step 2: Plot the ordered pairs listed in the third column of Table 9 as shown in Figure 26(a). Now connect the points to obtain the graph of the equation (*a line*), as shown in Figure 26(b).

Figure 26

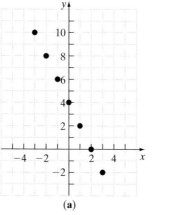

(a)

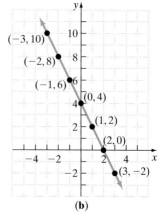

(b)

The graph of the equation in Figure 26(b) does not show all of the equation's points such as the point (8, –12). We use arrows on the ends of the graph to indicate that the pattern continues. It is important to show enough of the graph so that the rest of the graph is obvious. Such a graph is called a **complete graph.**

(EXAMPLE 4) **Graphing an Equation by Plotting Points**

Graph the equation $y = x^2$ by plotting points.

Solution

Table 10 shows several points on the graph.

Table 10

x	$y = x^2$	(x, y)
−4	$y = (-4)^2 = 16$	$(-4, 16)$
−3	$y = (-3)^2 = 9$	$(-3, 9)$
−2	$y = (-2)^2 = 4$	$(-2, 4)$
−1	$y = (-1)^2 = 1$	$(-1, 1)$
0	$y = (0)^2 = 0$	$(0, 0)$
1	$y = (1)^2 = 1$	$(1, 1)$
2	$y = (2)^2 = 4$	$(2, 4)$
3	$y = (3)^2 = 9$	$(3, 9)$
4	$y = (4)^2 = 16$	$(4, 16)$

In Figure 27(a), we plot the ordered pairs listed in Table 10. In Figure 27(b), we connect the points in a smooth curve.

Figure 27

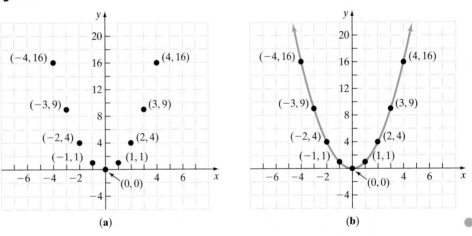

(a)   (b)

Work Smart

Note that in Figure 27, we use different scales on the *x*- and *y*-axes.

Two questions that you might be asking yourself right now are "How do I know how many points are sufficient?" and "How do I know which *x*-values (or *y*-values) I should choose in order to obtain points on the graph?" Often, the type of equation we wish to graph indicates the number of points that are necessary. For example, we will learn in the next section that if the equation is of the form $y = mx + b$, then its graph is a line, and only two points are required to obtain the graph (as in Example 3). Other times, more points are required. At this stage in your math career, you will need to plot quite a few points to obtain a complete graph. However, as your experience and knowledge grow, you will become more efficient in obtaining complete graphs.

Work Smart

Experience will play a huge role in determining which *x*-values to choose in creating a table of values. For the time being, start by choosing values of *x* around $x = 0$, as in Table 10.

Quick ✓

In Problems 8–10, graph each equation using the point-plotting method.

8. $y = 3x + 1$ **9.** $2x + 3y = 8$ **10.** $y = x^2 + 3$

EXAMPLE 5 **Graphing the Equation $x = y^2$**

Graph the equation $x = y^2$ by plotting points.

Solution

Because the equation is solved for *x*, we will choose values of *y* and use the equation to find the corresponding values of *x*. See Table 11 on the following page. Plot the ordered pairs listed in Table 11 and connect the points in a smooth curve. See Figure 28.

Table 11

y	$x = y^2$	(x, y)
-3	$x = (-3)^2 = 9$	$(9, -3)$
-2	$x = (-2)^2 = 4$	$(4, -2)$
-1	$x = (-1)^2 = 1$	$(1, -1)$
0	$x = 0^2 = 0$	$(0, 0)$
1	$x = 1^2 = 1$	$(1, 1)$
2	$x = 2^2 = 4$	$(4, 2)$
3	$x = 3^2 = 9$	$(9, 3)$

Figure 28

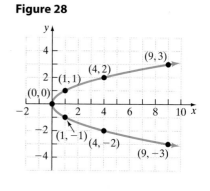

Quick ✓

In Problems 11 and 12, graph each equation using the point-plotting method.

11. $x = y^2 + 2$ **12.** $x = (y - 1)^2$

▶ ④ **Identify the Intercepts from the Graph of an Equation**

Work Smart

In order for a graph to be complete, all of its intercepts must be displayed.

A complete graph needs to include the *intercepts* of the graph.

Figure 29

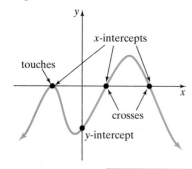

Definition

The **intercepts** are the coordinates of the points, if any, where a graph crosses or touches the coordinate axes. An **x-intercept** is the coordinates of a point where the graph crosses or touches the x-axis. A **y-intercept** is the coordinates of a point where the graph crosses or touches the y-axis.

See Figure 29 for an illustration. Notice that an x-intercept exists when $y = 0$ and a y-intercept exists when $x = 0$.

EXAMPLE 6 **Finding Intercepts from a Graph**

Find the intercepts of the graph shown in Figure 30. What are the x-intercepts? What are the y-intercepts?

Figure 30

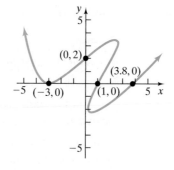

Solution

The intercepts of the graph are the points $(-3, 0)$, $(0, 2)$, $(1, 0)$, and $(3.8, 0)$. The x intercepts are $(-3, 0)$, $(1, 0)$, and $(3.8, 0)$; the y-intercept is $(0, 2)$.

Quick ✓

13. The points, if any, at which a graph crosses or touches a coordinate axis are called _____.

14. *True or False* An x-intercept exists when $x = 0$.

15. Find the intercepts of the graph shown in the figure. What are the x-intercepts? What are the y-intercepts?

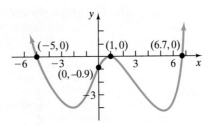

▷ ⑤ **Interpret Graphs**

Graphs help us to visualize a relationship between two variables or quantities. A graph is a "picture" that illustrates the relationship between two variables. Visualizing allows us to see important information and draw conclusions about the relationship between the two variables.

(EXAMPLE 7) **Interpreting a Graph**

The graph in Figure 31 shows the profit P for selling x gallons of gasoline in an hour at a gas station. The vertical axis represents the profit, and the horizontal axis represents the number of gallons of gasoline sold.

Figure 31

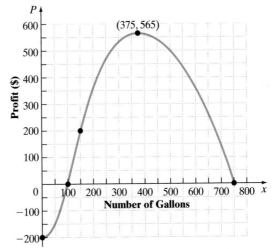

(a) What is the profit if 150 gallons of gasoline are sold?

(b) How many gallons of gasoline are sold when profit is highest? What is the highest profit?

(c) Identify and interpret the intercepts.

Solution

(a) Locate the point on the graph with x-coordinate 150. The y-coordinate at this point is 200. The profit from selling 150 gallons of gasoline is $200.

(b) The profit is highest when 375 gallons of gasoline are sold. The highest profit is $565.

(c) The intercepts are $(0, -200)$, $(100, 0)$, and $(750, 0)$. For $(0, -200)$: If 0 gallons of gasoline are sold, the profit is $-\$200$. The negative profit is due to the fact that the company has $\$0$ in revenue but has hourly expenses of $\$200$. For $(100, 0)$: When 100 gallons of gasoline are sold, there is $\$0$ profit. The company sells just enough gas to pay its bills. For $(750, 0)$: If 750 gallons of gasoline are sold, the profit is $\$0$. The 750 gallons sold represents the maximum number of gallons that the station can pump and still break even.

Quick ✓

16. The graph shown represents the cost C (in thousands of dollars) of refining x gallons of gasoline per hour (in thousands). The vertical axis represents the cost, and the horizontal axis represents the number of gallons of gasoline refined.

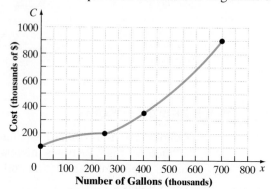

(a) What is the cost of refining 250 thousand gallons of gasoline per hour?

(b) What is the cost of refining 400 thousand gallons of gasoline per hour?

(c) In the context of the problem, explain the meaning of the graph ending at 700 thousand gallons of gasoline.

(d) Identify and interpret the intercept.

1.5 Exercises MyMathLab®
PRACTICE

Exercise numbers in green have complete video solutions in MyMathLab.

*Problems **1–16** are the* Quick ✓ *s that follow the* EXAMPLES.

Building Skills

17. Determine the coordinates of each of the points plotted. Tell in which quadrant or on what coordinate axis each point lies. See Objective 1.

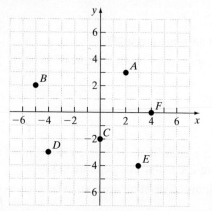

18. Determine the coordinates of each of the points plotted. Tell in which quadrant or on what coordinate axis each point lies. See Objective 1.

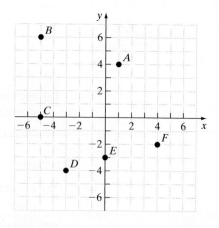

In Problems 19 and 20, plot each point in the xy-plane. Tell in which quadrant or on what coordinate axis each point lies. See Objective 1.

19. $A(3, 5)$

$B(-2, -6)$

$C(5, 0)$

$D(1, -6)$

$E(0, 3)$

$F(-4, 1)$

20. $A(-3, 1)$

$B(-6, 0)$

$C(2, -5)$

$D(-6, -2)$

$E(1, 2)$

$F(0, -5)$

In Problems 21–26, determine whether the given points are on the graph of the equation. See Objective 2.

21. $2x + 5y = 12$
(a) $(1, 2)$
(b) $(-2, 3)$
(c) $(-4, 4)$
(d) $\left(-\dfrac{3}{2}, 3\right)$

22. $-4x + 3y = 18$
(a) $(1, 7)$
(b) $(0, 6)$
(c) $(-3, 10)$
(d) $\left(\dfrac{3}{2}, 4\right)$

23. $y = -2x^2 + 3x - 1$
(a) $(-2, -15)$
(b) $(3, 10)$
(c) $(0, 1)$
(d) $(2, -3)$

24. $y = x^3 - 3x$
(a) $(2, 2)$
(b) $(3, 8)$
(c) $(-3, -18)$
(d) $(0, 0)$

25. $y = |x - 3|$
(a) $(1, 4)$
(b) $(4, 1)$
(c) $(-6, 9)$
(d) $(0, 3)$

26. $x^2 + y^2 = 1$
(a) $(0, 1)$
(b) $(1, 1)$
(c) $\left(\dfrac{1}{2}, \dfrac{1}{2}\right)$
(d) $\left(\dfrac{\sqrt{3}}{2}, \dfrac{1}{2}\right)$

In Problems 27–54, graph each equation by plotting points. See Objective 3.

27. $y = 4x$

28. $y = 2x$

29. $y = -\dfrac{1}{2}x$

30. $y = -\dfrac{1}{3}x$

31. $y = x + 3$

32. $y = x - 2$

33. $y = -3x + 1$

34. $y = -4x + 2$

35. $y = \dfrac{1}{2}x - 4$

36. $y = -\dfrac{1}{2}x + 2$

37. $2x + y = 7$

38. $3x + y = 9$

39. $y = -x^2$

40. $y = x^2 - 2$

41. $y = 2x^2 - 8$

42. $y = -2x^2 + 8$

43. $y = |x|$

44. $y = |x| - 2$

45. $y = |x - 1|$

46. $y = -|x|$

47. $y = x^3$

48. $y = -x^3$

49. $y = x^3 + 1$

50. $y = x^3 - 2$

51. $x^2 - y = 4$

52. $x^2 + y = 5$

53. $x = y^2 - 1$

54. $x = y^2 + 2$

In Problems 55–58, the graph of an equation is given. List the intercepts of the graph. See Objective 4.

55.

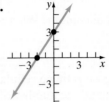

56.

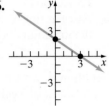

57.

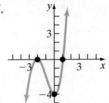

58.

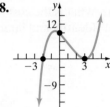

Applying the Concepts

59. If $(a, 4)$ is a point on the graph of $y = 4x - 3$, what is a?

60. If $(a, -2)$ is a point on the graph of $y = -3x + 5$, what is a?

61. If $(3, b)$ is a point on the graph of $y = x^2 - 2x + 1$, what is b?

62. If $(-2, b)$ is a point on the graph of $y = -2x^2 + 3x + 1$, what is b?

63. Area of a Window Bob Villa wishes to put a new window in his home. He wants the perimeter of the window to be 100 feet. The graph below shows the relation between the width, x, of the opening and the area of the opening.

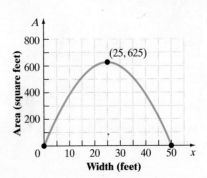

(continued)

(a) What is the area of the opening if the width is 10 feet?

(b) What is the width of the opening in order for area to be a maximum? What is the maximum area of the opening?

(c) Identify and interpret the intercepts.

64. Projectile Motion The graph to the right shows the height, in feet, of a ball thrown straight up with an initial speed of 80 feet per second from an initial height of 96 feet after t seconds.

(a) What is the height of the object after 1.5 seconds?

(b) At what time is the height a maximum? What is the maximum height?

(c) Identify and interpret the intercepts.

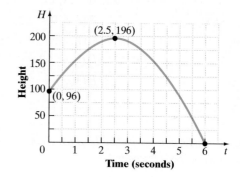

65. Cell Phones We all struggle with selecting a cellular phone provider. The graph below shows the relation between the monthly cost of a cellular phone and the number of minutes used, m.

SOURCE: *Sprint.com*

(a) What is the cost of talking for 200 minutes in a month? 500 minutes?

(b) What is the cost of talking 8000 minutes in a month?

(c) Identify and interpret the intercept.

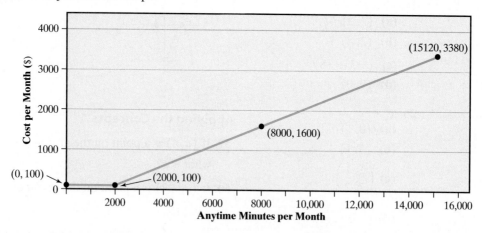

66. Wind Chill It is 10° Celsius outside. The wind is calm but then gusts up to 20 meters per second. You feel the chill go right through your bones. The following graph shows the relation between the wind chill temperature (in degrees Celsius) and wind speed (in meters per second).

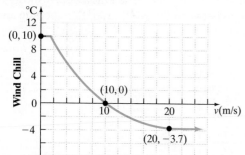

(a) What is the wind chill if the wind is blowing 4 meters per second?

(b) What is the wind chill if the wind is blowing 20 meters per second?

(c) Identify and interpret the intercepts.

67. Plot the points $(4, 0)$, $(4, 2)$, $(4, -3)$, and $(4, -6)$. Describe the set of all points of the form $(4, y)$ where y is a real number.

68. Plot the points $(4, 2)$, $(1, 2)$, $(0, 2)$, and $(-3, 2)$. Describe the set of all points of the form $(x, 2)$ where x is a real number.

Extending the Concepts.

69. Draw a graph of an equation that contains two x-intercepts, $(-2, 0)$ and $(3, 0)$. At the x-intercept $(-2, 0)$ the graph crosses the x-axis; at the x-intercept $(3, 0)$, the graph touches the x-axis. Compare your graph with those of your classmates. How are they similar? How are they different?

70. Draw a graph that contains the points $(-3, -1)$, $(-1, 1)$, $(0, 3)$, and $(1, 5)$. Compare your graph with those of your classmates. How many of the graphs are straight lines? How many are "curved"?

71. Make up an equation that is satisfied by the points $(2, 0)$, $(4, 0)$, and $(1, 0)$. Compare your equation with those of your classmates. How are they similar? How are they different?

72. Make up an equation that contains the points $(0, 3)$, $(1, 3)$, and $(-4, 3)$. Compare your equation with those of your classmates. How many are the same?

Explaining the Concepts

73. Explain what is meant by a complete graph.

74. Explain what the graph of an equation represents.

75. What is the point-plotting method for graphing an equation?

76. What is the y-coordinate of a point that is an x-intercept? What is the x-coordinate of a point that is a y-intercept?

The Graphing Calculator

Just as we have graphed equations using point plotting, the graphing calculator also graphs equations by plotting points. Figure 32 shows the graph of $y = x^2$, and Table 12 shows points on the graph of $y = x^2$, using a TI-84 Plus graphing calculator.

Figure 32

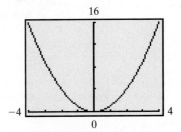

Table 12

X	Y1
-3	9
-2	4
-1	1
0	0
1	1
2	4
3	9

$Y_1 \boxminus X^2$

In Problems 77–84, use a graphing calculator to draw a complete graph of each equation. Use the TABLE feature to assist in selecting an appropriate viewing window.

77. $y = 3x - 9$

78. $y = -5x + 8$

79. $y = -x^2 + 8$

80. $y = 2x^2 - 4$

81. $y + 2x^2 = 13$

82. $y - x^2 = -15$

83. $y = x^3 - 6x + 1$

84. $y = -x^3 + 3x$

1.6 Linear Equations in Two Variables

Objectives

1 Graph Linear Equations Using Point Plotting

2 Graph Linear Equations Using Intercepts

3 Graph Vertical and Horizontal Lines

4 Find the Slope of a Line Given Two Points

5 Interpret Slope as an Average Rate of Change

6 Graph a Line Given a Point and Its Slope

7 Use the Point-Slope Form of a Line

8 Identify the Slope and y-Intercept of a Line from Its Equation

9 Find the Equation of a Line Given Two Points

Are You Ready for This Section?

Before getting started, take this readiness quiz. If you get a problem wrong, go back to the section cited and review the material.

R1. Solve: $3x + 12 = 0$ [Section 1.1, pp. 50–52]

R2. Solve: $2x + 5 = 13$ [Section 1.1, pp. 50–52]

R3. Solve for y: $3x - 2y = 10$ [Section 1.3, pp. 73–74]

R4. Evaluate: $\dfrac{5 - 2}{-2 - 4}$ [Section R.4, pp. 35–36]

R5. Evaluate: $\dfrac{-7 - 2}{-5 - (-2)}$ [Section R.4, pp. 35–36]

R6. Distribute: $-2(x + 3)$ [Section R.3, pp. 29–30]

▶ 1 Graph Linear Equations Using Point Plotting

In Section 1.5, we discussed how to graph any equation using the point-plotting method. Remember, the graph of an equation is the set of all ordered pairs (x, y) that satisfy the equation.

We are now going to learn methods for graphing a specific type of equation called a *linear equation in two variables*.

> **Definition**
>
> A **linear equation in two variables** is an equation of the form
> $$Ax + By = C$$
> where A, B, and C are real numbers. A and B cannot both be 0.

A linear equation written in the form $Ax + By = C$ is in **standard form.***
Some examples of linear equations in standard form are

$$3x - 4y = 9 \qquad \frac{1}{2}x + \frac{2}{3}y = 4 \qquad 3x = 9 \qquad -2y = 5$$

The graph of a linear equation is a **line.** Let's graph a linear equation using the point-plotting method.

EXAMPLE 1 **Graphing a Linear Equation Using the Point-Plotting Method**

Graph the linear equation: $4x + 2y = 6$

Solution

To graph the linear equation, choose various values for x and then use the equation to find the corresponding values of y. For this equation, we will let $x = -2, -1, 0,$ and 1.

$x = -2$:	$4(-2) + 2y = 6$		$x = -1$:	$4(-1) + 2y = 6$
	$-8 + 2y = 6$			$-4 + 2y = 6$
Add 8 to both sides:	$2y = 14$		Add 4 to both sides:	$2y = 10$
Divide both sides by 2:	$y = 7$		Divide both sides by 2:	$y = 5$
$(-2, 7)$ is on the graph.			$(-1, 5)$ is on the graph.	

Work Smart

In Example 1, we chose to pick x-values and use the equation to find the corresponding y-values; however, we could also have chosen y-values and used the equation to find the corresponding x-values.

$x = 0$:	$4(0) + 2y = 6$		$x = 1$:	$4(1) + 2y = 6$
	$0 + 2y = 6$			$4 + 2y = 6$
	$2y = 6$		Subtract 4 from both sides:	$2y = 2$
Divide both sides by 2:	$y = 3$		Divide both sides by 2:	$y = 1$
$(0, 3)$ is on the graph.			$(1, 1)$ is on the graph.	

Table 13 shows the points that are on the graph of $4x + 2y = 6$. We plot these ordered pairs in the Cartesian plane and connect the points in a straight line. See Figure 33.

Table 13

x	y	(x, y)
-2	7	$(-2, 7)$
-1	5	$(-1, 5)$
0	3	$(0, 3)$
1	1	$(1, 1)$

Figure 33

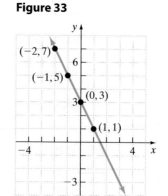

*Some texts refer to this as the general form.

Work Smart

We recommend that you find at least three points when graphing linear equations to be sure your graph is correct. Also, remember that a complete graph is a graph that shows all the interesting features of the graph, such as its intercepts.

Example 1 shows that we need only two points to obtain a complete graph of a linear equation. To guard against error, however, you should plot at least three points.

Quick ✓

1. A(n) _____ is an equation of the form $Ax + By = C$, where A, B, and C are real numbers. A and B cannot both be 0.

2. The graph of a linear equation is a ____.

In Problems 3–5, graph each linear equation using the point-plotting method.

3. $y = 2x - 3$ **4.** $\dfrac{1}{2}x + y = 2$ **5.** $-6x + 3y = 12$

❷ Graph Linear Equations Using Intercepts

In Section 1.5, we said a complete graph should display any intercepts. The intercepts of the graph of an equation are the points, if any, where the graph crosses or touches the coordinate axes. See Figure 34. Now we will explain how to find an equation's intercepts algebraically. We use the fact that at an x-intercept, the value of y is 0, and at a y-intercept, the value of x is 0.

Figure 34

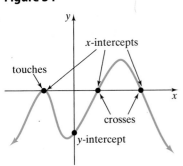

Finding Intercepts

- To find the x-intercept(s), if any, of the graph of an equation, let $y = 0$ in the equation and solve for x.
- To find the y-intercept(s), if any, of the graph of an equation, let $x = 0$ in the equation and solve for y.

The procedure given can be used to find the intercepts of any type of equation. Let's use this procedure to find the intercepts of a linear equation.

EXAMPLE 2

Graphing a Linear Equation by Finding Its Intercepts

Graph the linear equation $3x + 2y = 12$ by finding its intercepts.

Solution

To find the x-intercept, let $y = 0$ and solve the equation $3x + 2y = 12$ for x.

Let $y = 0$: $3x + 2(0) = 12$
$3x + 0 = 12$
$3x = 12$
Divide both sides by 3: $x = 4$

The x-intercept is $(4, 0)$.

The find the y-intercept, let $x = 0$ and solve the equation $3x + 2y = 12$ for y.

Let $x = 0$: $3(0) + 2y = 12$
$0 + 2y = 12$
$2y = 12$
Divide both sides by 2: $y = 6$

The y-intercept is $(0, 6)$.

We find one additional point on the graph by letting $x = 2$ (or any other value of x besides 0 or 4), and find y to be 3. Table 14 shows points on the graph. We plot the points and connect them in a straight line, obtaining the graph in Figure 35.

Figure 35

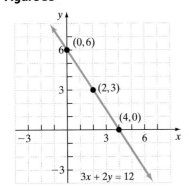

Table 14		
x	y	(x, y)
0	6	(0, 6)
4	0	(4, 0)
2	3	(2, 3)

Work Smart

Linear equations in one variable have no solution, one solution, or infinitely many solutions. Because the procedure for finding intercepts of linear equations in two variables results in a linear equation in one variable, linear equations can have no x-intercepts, one x-intercept, or infinitely many x-intercepts. The same applies to y-intercepts.

Quick ✓

6. *True or False*   To find the x-intercept(s), if any, of the graph of an equation, let $y = 0$ in the equation and solve for x.

In Problems 7 and 8, graph each linear equation by finding its intercepts.

7. $x + y = 4$ **8.** $4x - 5y = 20$

▶ (**EXAMPLE 3**)   **Graphing a Linear Equation by Finding Its Intercepts**

Graph the linear equation $x + 3y = 0$ by finding its intercepts.

Solution

x-intercept:

$$\text{Let } y = 0: \quad x + 3(0) = 0$$
$$x + 0 = 0$$
$$x = 0$$

The x-intercept is $(0, 0)$.

y-intercept:

$$\text{Let } x = 0: \quad 0 + 3y = 0$$
$$3y = 0$$
$$\text{Divide both sides by 3:} \quad y = 0$$

The y-intercept is $(0, 0)$.

Work Smart

We chose $x = -3$ and $x = 3$ to avoid fractions. This makes plotting the points easier.

Because both the x- and y-intercepts are $(0, 0)$ we will find *two* additional points on the graph. By letting $x = 3$, we find that $y = -1$. By letting $x = -3$, we find that $y = 1$. Plot the points $(0, 0)$, $(-3, 1)$, and $(3, -1)$, connect them in a straight line, and obtain the graph in Figure 36.

Figure 36

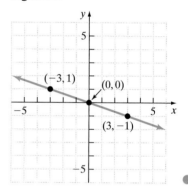

Example 3 shows that any equation of the form $Ax + By = 0$, where $A \neq 0$ and $B \neq 0$, has only one intercept at $(0, 0)$. Therefore, to graph equations of this form, we find two additional points on the graph.

Quick ✓

In Problem 9, graph the equation by finding its intercepts.

9. $3x - 2y = 0$

▶ ❸ **Graph Vertical and Horizontal Lines**

We have said that in the equation of a line, $Ax + By = C$, A and B cannot both be zero. But what if $A = 0$ or $B = 0$?

(**EXAMPLE 4**)   **Graphing a Vertical Line**

Graph the equation $x = 3$ using the point-plotting method.

Solution

Because the equation $x = 3$ can be written as $1x + 0y = 3$, the graph is a line. For the equation $x = 3$, it should be clear that no matter what value of y we choose, the corresponding value of x is going to be 3. Therefore, the points $(3, -2)$, $(3, -1)$, $(3, 0)$, $(3, 1)$, and $(3, 2)$ are all on the line. See Figure 37.

Figure 37
$x = 3$

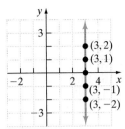

Based on the results of Example 4, we can write the equation of a vertical line:

> **Equation of a Vertical Line**
>
> A **vertical line** is given by an equation of the form
> $$x = a$$
> where $(a, 0)$ is the x-intercept.

Now let's look at equations that lead to graphs that are horizontal lines.

EXAMPLE 5 **Graphing a Horizontal Line**

Graph the equation $y = -2$ using the point-plotting method.

Solution

Because the equation $y = -2$ can be written as $0x + 1y = -2$, we know the graph is a line. For the equation $y = -2$, it should be clear that no matter what value of x we choose, the corresponding value of y is going to be -2. Therefore, the points $(-2, -2), (-1, -2), (0, -2), (1, -2),$ and $(2, -2)$ are all on the line. See Figure 38. ●

Based on the results of Example 5, we can generalize:

Figure 38
$y = -2$

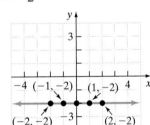

> **Equation of a Horizontal Line**
>
> A **horizontal line** is given by an equation of the form
> $$y = b$$
> where $(0, b)$ is the y-intercept.

Quick ✓

10. The equation of a vertical line is of the form _ = a, where $(a, 0)$ is the x-intercept. The equation of a horizontal line is the form _ = b, where $(0, b)$ is the y-intercept.

In Problems 11–13, graph each equation.

11. $x = 5$ **12.** $y = -4$ **13.** $-3x + 4 = 1$

▶ ❹ Find the Slope of a Line Given Two Points

Consider the staircase shown in Figure 39(a). If we draw a line at the edge of each step (in blue), we can see that each step has exactly the same horizontal **run** and the same vertical **rise**. We call the ratio of the rise to the run the *slope* of the line. Thus slope is a numerical measure of the steepness of the line. For example, the staircase in Figure 39(a) has a run of 7 inches and a rise of 6 inches, so the line's slope is $\dfrac{\text{Rise}}{\text{Run}} = \dfrac{6 \text{ inches}}{7 \text{ inches}} = \dfrac{6}{7}$. In Figure 39(b), the slope of the line is $\dfrac{\text{Rise}}{\text{Run}} = \dfrac{6 \text{ inches}}{10 \text{ inches}} = \dfrac{6}{10}$. And in Figure 39(c), the line's slope is $\dfrac{\text{Rise}}{\text{Run}} = \dfrac{6 \text{ inches}}{4 \text{ inches}} = \dfrac{6}{4}$.

Figure 39

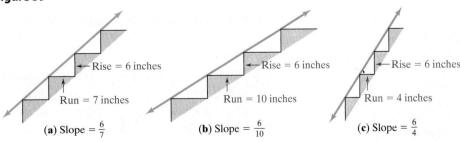

(a) Slope $= \frac{6}{7}$ (b) Slope $= \frac{6}{10}$ (c) Slope $= \frac{6}{4}$

We can define the slope of a line using rectangular coordinates.

Definition

Let $P = (x_1, y_1)$ and $Q = (x_2, y_2)$ be two distinct points. If $x_1 \neq x_2$, the **slope** m of the nonvertical line L containing P and Q is defined by the formula

$$m = \frac{y_2 - y_1}{x_2 - x_1}, \quad x_1 \neq x_2$$

If $x_1 = x_2$, then L is a vertical line and the slope m of L is **undefined** (since this results in division by 0).

The accepted symbol for the slope of a line is m. It comes from the French word *monter*, which means "to ascend or climb." Figure 40(a) illustrates the slope of a nonvertical line; Figure 40(b) shows a vertical line.

Figure 40

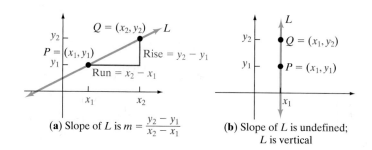

(a) Slope of L is $m = \frac{y_2 - y_1}{x_2 - x_1}$ (b) Slope of L is undefined; L is vertical

Notice that the slope m of a nonvertical line may be viewed as

$$m = \frac{y_2 - y_1}{x_2 - x_1} = \frac{\text{rise}}{\text{run}}$$

We can also write the slope m of a nonvertical line as

$$m = \frac{y_2 - y_1}{x_2 - x_1} = \frac{\text{Change in } y}{\text{Change in } x} = \frac{\Delta y}{\Delta x}$$

The symbol Δ is the Greek letter delta, which comes from the Greek word for "difference," *diaphora*. In mathematics, we read Δ as "change in." Thus we read $\frac{\Delta y}{\Delta x}$ as "change in y divided by change in x." The slope m of a nonvertical line measures the amount that y changes (the vertical change) as x changes from x_1 to x_2 (the horizontal change).

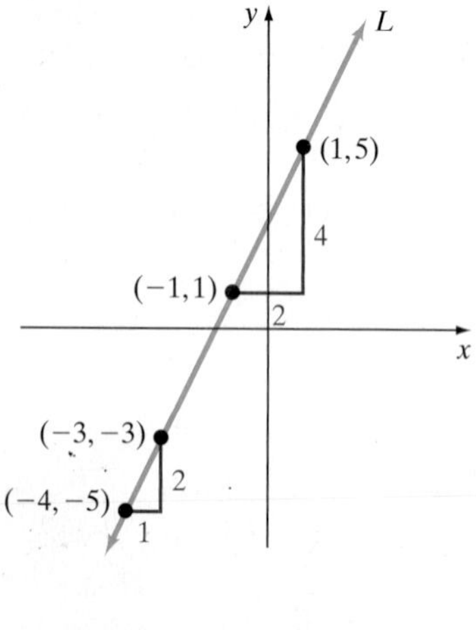

Comments Regarding the Slope of a Nonvertical Line

1. Any two different points on the line can be used to compute the slope of the line shown in Figure 41. The slope m of the line L is given by

$$m = \frac{-5 - (-3)}{-4 - (-3)} = \frac{-2}{-1} = 2 \quad \text{or} \quad m = \frac{5 - 1}{1 - (-1)} = \frac{4}{2} = 2$$

This result is due to the fact that the two triangles in Figure 41 are similar (the measure of the angles is the same in both triangles). Therefore, the ratios of the sides are proportional.

2. The slope of a line may be computed from $P = (x_1, y_1)$ to $Q = (x_2, y_2)$ or from Q to P because

$$m = \frac{y_2 - y_1}{x_2 - x_1} = \frac{-(y_1 - y_2)}{-(x_1 - x_2)} = \frac{y_1 - y_2}{x_1 - x_2}$$

In Words
Slope is the change in y divided by the change in x.

Figure 41

(graph showing line L through points (1,5), (−1,1), (−3,−3), (−4,−5) with slope triangles labeled 4, 2 and 2, 1)

Work Smart
It doesn't matter whether we compute the slope of the line from point P to Q or from point Q to P.

EXAMPLE 6 **Finding and Interpreting the Slope of a Line**

Find and interpret the slope of the line containing the points $(3, 6)$ and $(-2, 2)$.

Solution

Plot the points $P = (x_1, y_1) = (3, 6)$ and $Q = (x_2, y_2) = (-2, 2)$ and draw a line through them, as shown in Figure 42. The slope of the line is

$$m = \frac{y_2 - y_1}{x_2 - x_1} = \frac{\Delta y}{\Delta x} = \frac{2 - 6}{-2 - 3} = \frac{-4}{-5} = \frac{4}{5}$$

We could also compute the slope as

$$m = \frac{y_1 - y_2}{x_1 - x_2} = \frac{6 - 2}{3 - (-2)} = \frac{4}{5}$$

We interpret a slope of $\frac{4}{5}$ as follows: For every 5-unit increase in x, y increases by 4 units. Or for every 5-unit decrease in x, y decreases by 4 units.

Figure 42

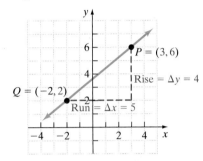

Quick ✓

14. If a line is vertical, then its slope is _____.

15. On a line, for every 10-foot run there is a 4-foot rise. The slope of the line is _.

16. *True or False* If $P = (x_1, y_1)$ and $Q = (x_2, y_2)$ are two distinct points with $x_1 \neq x_2$, the slope m of the line L containing P and Q is defined by the formula

$$m = \frac{x_2 - x_1}{y_2 - y_1}, \quad y_1 \neq y_2.$$

17. *True or False* If the slope of a line is $\frac{1}{2}$, then if x increases by 2, y increases by 1.

In Problems 18–21, find and interpret the slope of the line containing the points.

18. $(0, 3); (3, 12)$

19. $(-1, 3); (3, -4)$

20. $(3, 2); (-3, 2)$

21. $(-2, 4); (-2, -1)$

▶ **EXAMPLE 7** **Finding Slopes of Different Lines Each of Which Contains (3, 5)**

Find the slope of the lines $L_1, L_2, L_3,$ and L_4 containing the following pairs of points. Graph the lines in the Cartesian plane.

$L_1: P(3, 5), Q_1(5, 8)$ $L_2: P(3, 5), Q_2(6, 5)$

$L_3: P(3, 5), Q_3(5, \ 2)$ $L_4: P(3, 5), Q_4(3, 0)$

Solution

Let m_1, m_2, m_3, and m_4 denote the slopes of the lines L_1, L_2, L_3, and L_4, respectively. Then

$$m_1 = \frac{8-5}{5-3} = \frac{3}{2} \qquad\qquad m_2 = \frac{5-5}{6-3} = \frac{0}{3} = 0$$

$$m_3 = \frac{-2-5}{5-3} = \frac{-7}{2} = -\frac{7}{2} \qquad m_4 = \frac{0-5}{3-3} = \frac{-5}{0} \quad \text{undefined}$$

The four lines are shown in Figure 43.

Figure 43

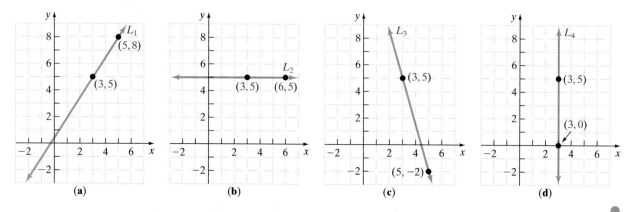

(a)   (b)   (c)   (d)

Based on Figure 43, we have the following properties of slope:

Properties of Slope

- When the slope of a line is positive, the line slants upward from left to right, as shown by L_1 in Figure 43(a).
- When the slope of a line is zero, the line is horizontal, as shown by L_2 in Figure 43(b).
- When the slope of a line is negative, the line slants downward from left to right, as shown by L_3 in Figure 43(c).
- When the slope of a line is undefined, the line is vertical, as shown by L_4 in Figure 43(d).

Quick ✓

22. *True or False* If the slope of a line is negative, then the line slants downward from left to right.

23. A horizontal line has slope = _.

24. Find the slope of the lines L_1, L_2, L_3, and L_4 containing the following pairs of points. Graph all four lines in the same Cartesian plane.

$L_1: P(1,3), Q_1(6,4)$ $L_2: P(1,3), Q_2(1,8)$

$L_3: P(1,3), Q_3(-3,7)$ $L_4: P(1,3), Q_4(-4,3)$

▶ ⑤ **Interpret Slope as an Average Rate of Change**

The slope m of a nonvertical line measures the amount that y changes as x changes from x_1 to x_2. The slope of a line is also called the **average rate of change** of y with respect to x.

In applications, we are often asked how the change in one variable will affect some other variable. For example, if your income increases by \$1000, how much will your spending (on average) change? Or if the speed of your car increases by 10 miles per hour, how much (on average) will your car's gas mileage change?

EXAMPLE 8 **Slope as an Average Rate of Change**

A strain of *E. coli* Beu 397-recA441 is placed into a Petri dish at 30° Celsius and allowed to grow. The data shown in Table 15 are collected. The population is measured in grams and the time in hours. The population growth is shown in Figure 44.

Table 15

Time (hours), x	Population (grams), y
0	0.09
1	0.12
2	0.16
3	0.22
4	0.29
5	0.39

SOURCE: *Dr. Polly Lavery, Joliet Junior College*

Figure 44

Growth of *E. coli*

(a) Compute and interpret the average rate of change in the population between hours 0 and 1.

(b) Compute and interpret the average rate of change in the population between hours 3 and 4.

(c) Based on your results to parts (a) and (b), do you think that the population grows linearly? Why?

Solution

(a) To find the average rate of change, compute the slope of the line between the points $(0, 0.09)$ and $(1, 0.12)$.

$$m = \text{average rate of change} = \frac{0.12 - 0.09}{1 - 0} = 0.03 \text{ gram per hour}$$

The population of *E. coli* grew at the rate of 0.03 gram per hour between hours 0 and 1.

(b) The slope of the line between the points $(3, 0.22)$ and $(4, 0.29)$ is

$$m = \text{average rate of change} = \frac{0.29 - 0.22}{4 - 3} = 0.07 \text{ gram per hour}$$

The population growth rate was 0.07 gram per hour between hours 3 and 4.

(c) The population is not growing linearly, because the average rate of change (slope) is not constant.

Quick ✓

25. The data to the left represent the total revenue that would be received from selling *x* bicycles at Gibson's Bicycle Shop.

(a) Plot the ordered pairs (x, y) on a graph and connect the points with straight lines.

(b) Compute and interpret the average rate of change in the revenue between 0 and 25 bicycles sold.

(c) Compute and interpret the average rate of change in the revenue between 102 and 150 bicycles sold.

(d) Based on your results to parts (a), (b), and (c), do you think the revenue grows linearly? Why?

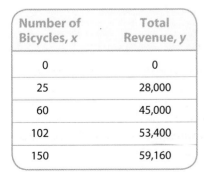

Number of Bicycles, x	Total Revenue, y
0	0
25	28,000
60	45,000
102	53,400
150	59,160

▶ ⑥ Graph a Line Given a Point and Its Slope

One reason we care about slope is because it can be used to help us graph lines.

$\boxed{\text{EXAMPLE 9}}$ **Graphing a Line Given a Point and Its Slope**

Draw a graph of the line that contains the point $(1, 2)$ and has a slope of

(a) 3   **(b)** $-\dfrac{3}{2}$

Solution

(a) Because the slope $= \dfrac{\text{Rise}}{\text{Run}} = \dfrac{\Delta y}{\Delta x}$, we know that $3 = \dfrac{3}{1} = \dfrac{\Delta y}{\Delta x}$. Thus as x increases by 1 unit, y increases by 3 units. Start at $(1, 2)$ and move 1 unit to the right and then 3 units up, and arrive at the point $(2, 5)$. Our graph is the line through the points $(1, 2)$ and $(2, 5)$. See Figure 45.

(b) Because the slope $= \dfrac{\text{Rise}}{\text{Run}} = \dfrac{\Delta y}{\Delta x}$, we know that $-\dfrac{3}{2} = \dfrac{-3}{2} = \dfrac{\Delta y}{\Delta x}$. Thus as x increases by 2 units, y decreases by 3 units. Start at $(1, 2)$ and move 2 units to the right and then 3 units down, and arrive at the point $(3, -1)$. Our graph is the line through the points $(1, 2)$ and $(3, -1)$. See Figure 46.

It is perfectly acceptable to set $\dfrac{\Delta y}{\Delta x} = -\dfrac{3}{2} = \dfrac{3}{-2}$ and move left 2 units from $(1, 2)$ and then up 3 units. We would then end up at $(-1, 5)$, which is also on the graph of the line in Figure 46. ●

Figure 45

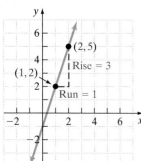

Quick ✓

26. Draw a graph of the line that contains the point $(-1, 3)$ and has a slope of

(a) $\dfrac{1}{3}$ **(b)** -4 **(c)** 0

▶ ⑦ Use the Point-Slope Form of a Line

Another reason we care about slope is that we can use it to help us find the equation of a line. Suppose L is a nonvertical line that has slope m and contains the point (x_1, y_1). See Figure 47.

Figure 46

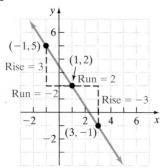

Figure 47

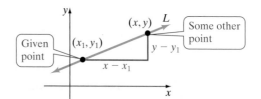

For any other point (x, y) on L, we know from the formula for the slope of a line that

$$m = \dfrac{y - y_1}{x - x_1}$$

Multiplying both sides by $x - x_1$, we can rewrite this expression as

$$y - y_1 = m(x - x_1)$$

Point-Slope Form of an Equation of a Line

An equation of a nonvertical line with slope m that contains the point (x_1, y_1) is

Slope
↓

$$y - y_1 = m(x - x_1)$$
↑   ↑

Given Point

EXAMPLE 10   **Using the Point-Slope Form of an Equation of a Line**

Find the equation of a line that has slope 3 and contains the point $(-2, 5)$. Graph the line.

Solution

Figure 48

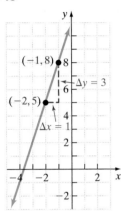

Because we are given the slope and a point on the line, we use the point-slope form of a line with $m = 3$ and $(x_1, y_1) = (-2, 5)$.

$$y - y_1 = m(x - x_1)$$
$$m = 3, x_1 = -2, y_1 = 5: \quad y - 5 = 3[x - (-2)]$$
$$y - 5 = 3(x + 2)$$

See Figure 48 for a graph of the line.   ●

Quick ✓

27. An equation in the form $y - y_1 = m(x - x_1)$ is said to be in _____ form.

In Problems 28–31, find an equation of the line with the given properties. Graph the line.

28. $m = 2$, $(x_1, y_1) = (3, 5)$

29. $m = -4$, $(x_1, y_1) = (-2, 3)$

30. $m = \dfrac{1}{3}$, $(x_1, y_1) = (3, -4)$

31. $m = 0$, $(x_1, y_1) = (4, -2)$

▶ **❽ Identify the Slope and y-Intercept of a Line from Its Equation**

Let's solve the equation in Example 10 for y:

$$y - 5 = 3(x + 2)$$
$$\text{Distribute the 3:} \quad y - 5 = 3x + 6$$
$$\text{Add 5 to both sides:} \quad y = 3x + 11$$

The coefficient of x, 3, is the slope of the line; the y-intercept of the line is $(0, 11)$ since $y = 3(0) + 11 = 11$. An equation written in the form $y = mx + b$, is in *slope-intercept form*.

Slope-Intercept Form of an Equation of a Line

An equation of a line L with slope m and y-intercept $(0, b)$ is

$$y = mx + b$$

EXAMPLE 11   **Finding the Slope and y-Intercept of a Line from Its Equation**

Write the equation $x - 3y = 9$ in slope-intercept form. Find the slope m and y-intercept $(0, b)$ of the line. Graph the line.

Solution

To put the equation in slope-intercept form, $y = mx + b$, solve the equation for y.

$$x - 3y = 9$$

Subtract x from both sides: $\qquad -3y = -x + 9$

Divide both sides by -3: $\qquad y = \dfrac{-x + 9}{-3}$

Divide -3 into both terms in the numerator: $\qquad y = \dfrac{1}{3}x - 3$

Since $y = \dfrac{1}{3}x - 3$ is in the form $y = mx + b$, the coefficient of x, $\dfrac{1}{3}$, is the slope, and $(0, -3)$ is the y-intercept.

Graph the line by plotting a point at $(0, -3)$. Then use the slope to find an additional point by moving right 3 units and up 1 unit from the point $(0, -3)$ to the point $(3, -2)$. Draw a line through the two points and obtain the graph shown in Figure 49.

Figure 49

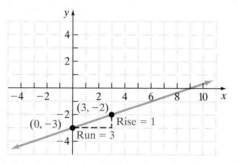

Quick ✓

32. An equation in the form $y = mx + b$ is said to be in _____ form.

In Problems 33–36, find the slope and y-intercept of each line. Graph the line.

33. $3x - y = 2$ 34. $6x + 2y = 8$

35. $3x - 2y = 7$ 36. $7x + 3y = 0$

▶ ❾ Find the Equation of a Line Given Two Points

We know that two points are all that we need to graph a line. If we are given two points, we can find an equation of the line through the points by first finding the slope of the line and then using the point-slope form of a line.

EXAMPLE 12   **How to Find an Equation of a Line from Two Points**

Find the equation of a line through the points $(-1, 4)$ and $(2, -5)$. If possible, write the equation in slope-intercept form. Graph the line.

Step-by-Step Solution

Step 1: Find the slope of the line containing the points.

Let $(x_1, y_1) = (-1, 4)$ and $(x_2, y_2) = (2, -5)$.
Substitute these values into the formula for the slope of a line.

$$m = \frac{y_2 - y_1}{x_2 - x_1} = \frac{-5 - 4}{2 - (-1)} = \frac{-9}{3} = -3$$

Step 2: Use the point-slope form of a line to find the equation.

With $m = -3$, $x_1 = -1$, and $y_1 = 4$, we have

$$y - y_1 = m(x - x_1)$$
$$y - 4 = -3[x - (-1)]$$
$$y - 4 = -3(x + 1)$$

Step 3: Solve the equation for y.

Distribute the -3: $y - 4 = -3x - 3$

Add 4 to both sides: $\qquad y = -3x + 1$

Work Smart

In Step 2 of Example 12, we chose to use $x_1 = -1$ and $y_1 = 4$, but we could also have used $x_1 = 2$ and $y_1 = -5$. Choose the values of x and y that make the algebra easiest.

The slope of the line is -3 and the y-intercept is $(0, 1)$. See Figure 50 for the graph.

Figure 50

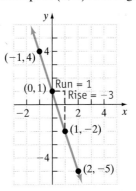

Quick ✓

In Problems 37–39, find the equation of the line containing the given points. If possible, write the equation in slope-intercept form. Graph the line.

37. $(1, 3)$; $(4, 9)$ **38.** $(-2, 4)$; $(2, 2)$

39. $(-4, 6)$; $(3, 6)$

(EXAMPLE 13) **Finding an Equation of a Line from Two Points**

Find the equation of a line through the points $(-3, 2)$ and $(-3, -2)$. If possible, write the equation in slope-intercept form. Graph the line.

Solution

Let $(x_1, y_1) = (-3, 2)$ and $(x_2, y_2) = (-3, -2)$. Substitute these values into the slope formula.

$$m = \frac{y_2 - y_1}{x_2 - x_1} = \frac{-2 - 2}{-3 - (-3)} = \frac{-4}{0} \quad \text{undefined}$$

Work Smart

If you plot the points first, it will be clear that the line through the points is vertical.

The slope is undefined, so the line is vertical. The equation of the line is $x = -3$. See Figure 51 for the graph.

Quick ✓

40. Find an equation of the line containing the points $(3, 2)$ and $(3, -4)$. If possible, write the answer in slope-intercept form. Graph the line.

Figure 51

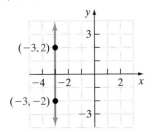

This section dealt with two types of problems:

1. Given an equation, classify and graph it. (See Examples 1–5.)
2. Given a graph, or information about a graph, find its equation. (See Examples 10, 12, and 13.)

Work Smart: Study Skills

To determine which equation of a line to use, ask yourself "What information do I know?"

    If you know the slope and a point that isn't the y-intercept, then use the point-slope form. If you know the slope and the y-intercept, use the slope-intercept form. If you know two points, use the slope formula with the point-slope form. If the slope is undefined, the line is vertical.

Summary   **Equations of Lines**

Form of Line	Formula	Comments
Horizontal Line	$y = b$	Graph is a horizontal line (slope is 0) with y-intercept $(0, b)$.
Vertical Line	$x = a$	Graph is a vertical line (undefined slope) with x-intercept $(a, 0)$.
Point-slope	$y - y_1 = m(x - x_1)$	Useful for finding the equation of a line given a point and a slope or two points.
Slope-intercept	$y = mx + b$	Useful for quickly determining the slope and y-intercept of the line.
Standard	$Ax + By = C$	Useful for finding the x- and y-intercepts.

1.6 Exercises MyMathLab® PRACTICE

Exercise numbers in green
have complete video solutions
in MyMathLab.

*Problems **1–40** are the* Quick ✓ *s that follow the* EXAMPLES.

Building Skills

In Problems 41–48, graph each linear equation by plotting points.
See Objective 1.

41. $x - 2y = 6$ **42.** $2x - y = -8$

43. $3x + 2y = 12$ **44.** $-5x + y = 10$

45. $\frac{2}{3}x + y = 6$ **46.** $2x - \frac{3}{2}y = 10$

47. $5x - 3y = 6$ **48.** $-7x + 3y = 9$

In Problems 49–58, graph each linear equation by finding its
intercepts. See Objective 2.

49. $3x + y = 6$ **50.** $-2x + y = 4$

51. $5x - 3y = 15$ **52.** $-4x + 3y = 24$

53. $\frac{1}{3}x - \frac{1}{2}y = 1$ **54.** $\frac{1}{4}x + \frac{1}{5}y = 2$

55. $2x + y = 0$ **56.** $4x + 3y = 0$

57. $\frac{2}{3}x - \frac{1}{2}y = 0$ **58.** $-\frac{3}{2}x + \frac{3}{4}y = 0$

In Problems 59–64, graph each linear equation. See Objective 3.

59. $x = -5$ **60.** $x = 5$

61. $y = 1$ **62.** $y = 6$

63. $2y + 8 = -6$ **64.** $3y + 20 = -10$

In Problems 65–68, (a) find the slope of the line and (b) interpret
the slope. See Objective 4.

65.   **66.**

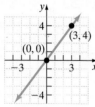

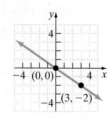

67.   **68.**

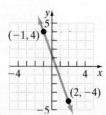

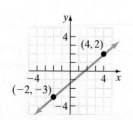

In Problems 69–80, plot each pair of points and graph the line
containing them. Determine the slope of the line. See Objective 4.

69. $(0, 0)$; $(1, 5)$

70. $(0, 0)$; $(-2, 5)$

71. $(-2, 3)$; $(1, -6)$

72. $(3, -1)$; $(-2, 11)$

73. $(-2, 3)$; $(3, 7)$

74. $(1, -4)$; $(-1, 3)$

75. $(-3, 2)$; $(4, 2)$

76. $(-3, 1)$; $(2, 1)$

77. $(10, 2)$; $(10, -3)$

78. $(4, 1)$; $(4, -3)$

79. $\left(\frac{1}{2}, \frac{5}{3}\right)$; $\left(\frac{9}{4}, \frac{11}{6}\right)$

80. $\left(\frac{7}{3}, \frac{5}{2}\right)$; $\left(\frac{13}{9}, \frac{13}{4}\right)$

In Problems 81–90, graph the line that contains the given point and
has slope m. Do not find an equation of the line. See Objective 6.

81. $m = 3$; $(1, 2)$ **82.** $m = 2$; $(-1, 4)$

83. $m = -2$; $(-3, 1)$ **84.** $m = -4$; $(-1, 5)$

85. $m = \frac{1}{3}$; $(-3, 4)$ **86.** $m = \frac{4}{3}$; $(-2, -5)$

87. $m = -\frac{3}{2}$; $(2, 5)$ **88.** $m = -\frac{1}{2}$; $(3, 3)$

89. $m = 0$; $(1, 2)$ **90.** m is undefined; $(-5, 2)$

In Problems 91 and 92, the slope and a point on a line are given.
Use the information to find three additional points on the line.
Answers may vary.

91. $m = \frac{5}{2}$; $(-2, 3)$ **92.** $m = -\frac{2}{3}$; $(1, -3)$

In Problems 93–96, find an equation of the line. Express your
answer in slope-intercept form. See Objective 7.

93.   **94.**

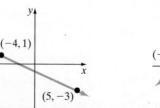

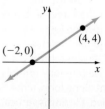

95.

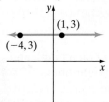

96.

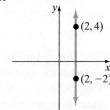

In Problems 97–106, find an equation of the line that has the given slope and contains the given point. Express your answer in slope-intercept form, if possible. See Objective 7.

97. $m = 2$; $(0, 0)$

98. $m = -1$; $(0, 0)$

99. $m = -3$; $(-1, 1)$

100. $m = 4$; $(2, -1)$

101. $m = \dfrac{4}{3}$; $(3, 2)$

102. $m = \dfrac{1}{2}$; $(2, 1)$

103. $m = -\dfrac{5}{4}$; $(-2, 4)$

104. $m = -\dfrac{4}{3}$; $(1, -3)$

105. m undefined; $(6, 1)$

106. $m = 0$; $(3, -2)$

In Problems 107–118, find an equation of the line that contains the given points. Express your answer in slope-intercept form, if possible. See Objective 9.

107. $(0, 0)$; $(5, 7)$

108. $(0, 0)$; $(4, -3)$

109. $(3, 2)$; $(4, 7)$

110. $(1, 3)$; $(3, 7)$

111. $(-2, 1)$; $(5, -2)$

112. $(-3, 1)$; $(1, 6)$

113. $(-1, -3)$; $(-1, 5)$

114. $(-3, -4)$; $(1, -4)$

115. $(1, 3)$; $(-3, -7)$

116. $(-5, 1)$; $(1, -1)$

117. $(2, 4)$; $(-4, 4)$

118. $(3, 1)$; $(3, -4)$

In Problems 119–130, find the slope and y-intercept of each line. Graph the line. See Objective 8.

119. $y = 2x - 1$

120. $y = 3x + 2$

121. $y = -4x$

122. $y = -7x$

123. $2x + y = 3$

124. $-3x + y = 1$

125. $4x + 2y = 8$

126. $3x + 6y = 12$

127. $x - 4y - 2 = 0$

128. $2x - 5y - 10 = 0$

129. $x = 3$

130. $y = -4$

Applying the Concepts

131. Find an equation for the x-axis.

132. Find an equation for the y-axis.

133. **Maximum Heart Rate** The data below represent the maximum number of heartbeats that a healthy individual should have during a 15-second interval of time while exercising for different ages.

(a) Plot the ordered pairs (x, y) on a graph and connect the points with straight lines.

(b) Compute and interpret the average rate of change in the maximum number of heartbeats between 20 and 30 years of age.

(c) Compute and interpret the average rate of change in the maximum number of heartbeats between 50 and 60 years of age.

(d) Based on your results for parts (a), (b), and (c), do you think that the maximum number of heartbeats is linearly related to age? Why?

Age, x	Maximum Number of Heartbeats, y
20	50
30	47.5
40	45
50	42.5
60	40
70	37.5

SOURCE: *American Heart Association*

134. Raisins The following data represent the weight (in grams) of a box of raisins and the number of raisins in the box.

(a) Plot the ordered pairs (x, y) on a graph and connect the points with straight lines.
(b) Compute and interpret the average rate of change in the number of raisins between 42.3 and 42.5 grams.
(c) Compute and interpret the average rate of change in the number of raisins between 42.7 and 42.8 grams.
(d) Based on your results for parts (a), (b), and (c), do you think that the number of raisins is linearly related to weight? Why?

Weight (in grams), x	Number of Raisins, y
42.3	82
42.5	86
42.6	89
42.7	91
42.8	93

SOURCE: *Jennifer Maxwell, student at Joliet Junior College*

135. Average Income An individual's income varies with age. The following data show the average income of individuals of different ages in the United States for 2010.

(a) Plot the ordered pairs (x, y) on a graph and connect the points with straight lines.
(b) Compute and interpret the average rate of change in average income between 20 and 30 years of age.
(c) Compute and interpret the average rate of change in average income between 50 and 60 years of age.
(d) Based on your results for parts (a), (b), and (c), do you think that average income is linearly related to age? Why?

Age, x	Average Income, y
20	$10,036
30	$31,914
40	$42,224
50	$44,731
60	$41,296
70	$25,877

SOURCE: *Statistical Abstract, 2012*

136. U.S. Population The following data represent the population of the United States between 1930 and 2000.

(a) Plot the ordered pairs (x, y) on a graph and connect the points with straight lines.
(b) Compute and interpret the average rate of change in population between 1930 and 1940.
(c) Compute and interpret the average rate of change in population between 2000 and 2010.
(d) Based on your results for parts (a), (b), and (c), do you think that population is linearly related to the year? Why?

Year, x	Population, y
1930	123,202,624
1940	132,164,569
1950	151,325,798
1960	179,323,175
1970	203,302,031
1980	226,542,203
1990	248,709,873
2000	281,421,906
2010	313,131,065

SOURCE: *U.S. Census Bureau*

137. Measuring Temperature The relationship between Celsius (°C) and Fahrenheit (°F) degrees for measuring temperature is linear. Find an equation relating °C and °F if 0°C corresponds to 32°F and 100°C corresponds to 212°F. Use the equation to find the Celsius measure of 60°F.

138. Building Codes As a result of the Americans with Disabilities Act (ADA, 1990), the building code states that access ramps must have a slope not steeper than $\frac{1}{12}$. Interpret what this result means.

139. Which of the following equations might have the graph shown? (More than one answer is possible.)

(a) $y = 3x - 1$
(b) $y = -2x + 3$
(c) $y = 2x + 3$
(d) $3x - 2y = 4$
(e) $-3x + 2y = -4$

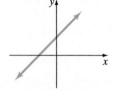

140. Which of the following equations might have the graph shown? (More than one answer is possible.)

(a) $y = 2x - 5$
(b) $y = -x + 2$
(c) $y = -\frac{2}{3}x - 3$
(d) $4x + 3y = -5$
(e) $-2x + y = -4$

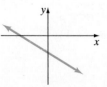

Explaining the Concepts

141. Name the five forms of equations of lines given in this section.

142. What type of line has one x-intercept but no y-intercept?

143. What type of line has one y-intercept but no x-intercept?

144. What type of line has one x-intercept and one y-intercept?

145. Are there any lines that have no intercepts? Explain your answer.

146. Exploration Graph $y = 2x$, $y = 2x + 3$, $y = 2x + 7$, and $y = 2x - 4$ in the same Cartesian plane. What pattern do you observe? In general, describe the graph of $y = 2x + b$.

147. Exploration Graph $y = \frac{1}{2}x$, $y = x$, and $y = 2x$ in the same Cartesian plane. What pattern do you observe? In general, describe the graph of $y = ax$ with $a > 0$.

148. Exploration Graph $y = -\frac{1}{2}x$, $y = -x$, and $y = -2x$ in the same Cartesian plane. What pattern do you observe? In general, describe the graph of $y = ax$ with $a < 0$.

The Graphing Calculator

149. To see the role that the slope m plays in the graph of a linear equation $y = mx + b$, graph the following lines on the same screen.

$$Y_1 = 0x + 2 \qquad Y_2 = \frac{1}{2}x + 2$$
$$Y_3 = 2x + 2 \qquad Y_4 = 6x + 2$$

State some general conclusions about the graph of $y = mx + b$ for $m \geq 0$. Now graph

$$Y_1 = -\frac{1}{2}x + 2 \quad Y_2 = -2x + 2 \quad Y_3 = -6x + 2$$

State some general conclusions about the graph of $y = mx + b$ for $m < 0$.

150. To see the role that the y-intercept $(0, b)$ plays in the graph of a linear equation $y = mx + b$, graph the following lines on the same screen.

$$Y_1 = 2x \qquad Y_2 = 2x + 2$$
$$Y_3 = 2x + 5 \qquad Y_4 = 2x - 4$$

State some general conclusions about the graph of $y = 2x + b$.

1.7 Parallel and Perpendicular Lines

Objectives

1 Define Parallel Lines

2 Find Equations of Parallel Lines

3 Define Perpendicular Lines

4 Find Equations of Perpendicular Lines

Are You Ready for This Section?

Before getting started, take this readiness quiz. If you get a problem wrong, go back to the section cited and review the material.

R1. Determine the reciprocal of 3.　　　　[Section R.3, pp. 23–24]

R2. Determine the reciprocal of $-\frac{3}{5}$.　　　　[Section R.3, pp. 23–24]

▶ **1** Define Parallel Lines

Work Smart

The words "if and only if" given in the definition mean that there are two statements being made:
(1) If two nonvertical lines are parallel, then their slopes are equal and they have different y-intercepts.
(2) If two nonvertical lines have equal slopes and different y-intercepts, then they are parallel.

When two lines (in the Cartesian plane) do not intersect (that is, they have no points in common), they are said to be *parallel*.

> **Parallel Lines**
>
> Two nonvertical lines are **parallel** if and only if their slopes are equal and they have different y-intercepts. Vertical lines are parallel if they have different x-intercepts.

Ready?...Answers　**R1.** $\frac{1}{3}$

R2. $-\frac{5}{3}$

Figure 52(a) on the next page shows nonvertical parallel lines. Figure 52(b) on the next page shows vertical parallel lines.

Figure 52
Parallel lines

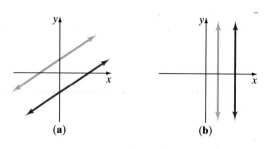

<table><tr><td>(a)</td><td>(b)</td></tr></table>

EXAMPLE 1   **Determining Whether Two Lines Are Parallel**

Determine whether the given lines are parallel.

(a) $L_1: 4x + y = 8$

$L_2: 6x + 2y = 12$

(b) $L_1: -3x + 2y = 6$

$L_2: \quad 6x - 4y = 8$

Solution

Find the slope and y-intercept of each line by writing their equations in slope-intercept form. If the slopes are the same but the y-intercepts are different, then the lines are parallel.

(a)   Solve L_1 for y: $4x + y = 8$

Subtract $4x$
from both sides: $y = -4x + 8$

The slope of L_1 is -4 and the y-intercept is $(0, 8)$.

Solve L_2 for y: $6x + 2y = 12$

Subtract $6x$ from
both sides: $2y = -6x + 12$

Divide both sides
by 2: $y = \dfrac{-6x + 12}{2}$

Divide each term in
the numerator by 2: $y = -3x + 6$

The slope of L_2 is -3 and the y-intercept is $(0, 6)$.

Because the lines have different slopes, they are not parallel.

(b)   Solve L_1 for y: $-3x + 2y = 6$

Add $3x$ to both sides: $2y = 3x + 6$

Divide both sides by 2: $y = \dfrac{3x + 6}{2}$

Divide each term in
the numerator by 2: $y = \dfrac{3}{2}x + 3$

The slope of L_1 is $\dfrac{3}{2}$ and the y-intercept is $(0, 3)$.

Solve L_2 for y: $6x - 4y = 8$

Subtract $6x$ from
both sides: $-4y = -6x + 8$

Divide both sides
by -4: $y = \dfrac{-6x + 8}{-4}$

Divide each term in
the numerator by -4: $y = \dfrac{3}{2}x - 2$

The slope of L_2 is $\dfrac{3}{2}$ and the y-intercept is $(0, -2)$.

The lines have the same slope but different y-intercepts, so they are parallel.

Quick ✓

1. Two lines are parallel if and only if they have the same _____ and different

_____ .

In Problems 2–4, determine whether the two lines are parallel.

2. $L_1: y = 3x + 1$

$L_2: y = -3x - 3$

3. $L_1: 6x + 3y = 3$

$L_2: -8x - 4y = 12$

4. $L_1: -3x + 5y = 10$

$L_2: 6x + 10y = 10$

❷ Find Equations of Parallel Lines

Let's now discuss how to find the equation of a line that is parallel to a given line.

EXAMPLE 2 **How to Find the Equation of a Line Parallel to a Given Line**

Find an equation for the line that is parallel to $4x + 2y = 2$ and contains the point $(-2, 3)$. Graph the lines.

Step-by-Step Solution

Step 1: Find the slope of the given line by putting the equation in slope-intercept form.

$$4x + 2y = 2$$

Subtract 4x from both sides: $\quad 2y = -4x + 2$

Divide both sides by 2: $\quad y = -2x + 1$

The slope of the line is -2.

Step 2: Use the point-slope form of a line with the given point and the slope found in Step 1 to find the equation of the parallel line.

$$y - y_1 = m(x - x_1)$$

$m = -2, x_1 = -2, y_1 = 3$: $\quad y - 3 = -2[x - (-2)]$

Step 3: Put the equation in slope-intercept form by solving for y.

$$y - 3 = -2(x + 2)$$

Distribute the -2: $\quad y - 3 = -2x - 4$

Add 3 to both sides: $\quad y = -2x - 1$

The line parallel to $4x + 2y = 2$ that contains $(-2, 3)$ is $y = -2x - 1$. Notice that the slopes of the two lines are the same, but the y-intercepts are different. Figure 53 shows the graph of the parallel lines.

Figure 53

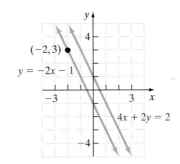

Quick ✓

In Problems 5 and 6, find the equation of the line that contains the given point and is parallel to the given line. Write the line in slope-intercept form. Graph the lines.

5. $(5, 8); y = 3x + 1$ **6.** $(-2, 4); 3x + 2y = 10$

▶ ❸ Define Perpendicular Lines

Figure 54
Perpendicular lines

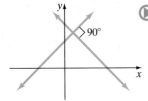

Two lines that intersect at a right angle (90°), are **perpendicular.** See Figure 54.

We use the slopes of the lines to determine whether two lines are perpendicular.

Work Smart
If $m_1 \cdot m_2 = -1$, then $m_1 = \dfrac{-1}{m_2}$.

Perpendicular Lines

Two nonvertical lines are **perpendicular** if and only if the product of their slopes is -1. Alternatively, two nonvertical lines are perpendicular if their slopes are negative reciprocals of each other. Any vertical line is perpendicular to any horizontal line.

EXAMPLE 3   Finding the Slope of a Line Perpendicular to a Given Line

Find the slope of any line that is perpendicular to a line whose slope is $\frac{5}{4}$.

Solution

The given line's slope is $\frac{5}{4}$. Any line whose slope is $-\frac{4}{5}$ will be perpendicular to the given line, because $\frac{5}{4} \cdot \left(-\frac{4}{5} \right) = -1$.

Quick ✓

7. Two lines are perpendicular if and only if the product of their slopes is ___.

8. Find the slope of any line perpendicular to a line whose slope is -3.

EXAMPLE 4   Determining Whether Two Lines Are Perpendicular

Determine whether the given lines are perpendicular.

(a) $L_1: y = 4x + 1$
     $L_2: y = -4x - 3$

(b) $L_1: y = \frac{2}{3}x - 5$
     $L_2: y = -\frac{3}{2}x + 2$

Solution

(a) The slope of L_1 is $m_1 = 4$. The slope of L_2 is $m_2 = -4$. Because the product of the slopes, $m_1 m_2 = 4 \cdot (-4) = -16 \neq -1$, the lines are not perpendicular.

(b) The slope of L_1 is $m_1 = \frac{2}{3}$. The slope of L_2 is $m_2 = -\frac{3}{2}$. The product of the slopes is $m_1 \cdot m_2 = \frac{2}{3} \cdot \left(-\frac{3}{2} \right) = -1$, so the lines are perpendicular.

Quick ✓

In Problems 9–11, determine whether the given lines are perpendicular.

9. $L_1: y = 5x - 3$
   $L_2: y = -\frac{1}{5}x - 4$

10. $L_1: 4x - y = 3$
    $L_2: x - 4y = 2$

11. $L_1: 2y + 4 = 0$
    $L_2: 3x - 6 = 0$

▶ ❹ Find Equations of Perpendicular Lines

We now find the equation of a line perpendicular to a second line.

EXAMPLE 5   How to Find the Equation of a Line Perpendicular to a Given Line

Find an equation of the line that is perpendicular to the line $2x + 5y = 10$ and contains the point $(4, -1)$. Write the equation in slope-intercept form. Graph the two lines.

Step-by-Step Solution

Step 1: Find the slope of the given line by putting the equation in slope-intercept form.

$$2x + 5y = 10$$

Subtract $2x$ from both sides: $\quad 5y = -2x + 10$

Divide both sides by 5: $\quad y = -\dfrac{2}{5}x + 2$

The slope of the line is $-\dfrac{2}{5}$.

Step 2: Find the slope of the perpendicular line.

The slope of the perpendicular line is the negative reciprocal of $-\dfrac{2}{5}$, which is $\dfrac{5}{2}$.

Step 3: Use the point-slope form of a line with the given point and the slope found in Step 2 to find the equation of the perpendicular line.

$$y - y_1 = m(x - x_1)$$

$m = \dfrac{5}{2}, x_1 = 4, y_1 = -1: \quad y - (-1) = \dfrac{5}{2}(x - 4)$

Step 4: Put the equation in slope-intercept form by solving for y.

$$y + 1 = \dfrac{5}{2}(x - 4)$$

Distribute the $\dfrac{5}{2}$: $\quad y + 1 = \dfrac{5}{2}x - 10$

Subtract 1 from both sides: $\quad y = \dfrac{5}{2}x - 11$

The equation of the line perpendicular to $2x + 5y = 10$ through $(4, -1)$ is $y = \dfrac{5}{2}x - 11$. See Figure 55.

Figure 55

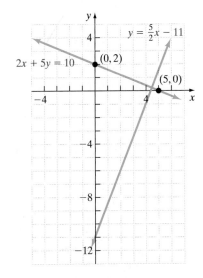

In Problems 12–14, find the equation of the line that contains the given point and is perpendicular to the given line. Write the line in slope-intercept form. Graph the lines.

12. $(-4, 2); y = 2x + 1$

13. $(-3, -4); 3x - 4y = 8$

14. $(3, -2); x = 3$

1.7 Exercises MyMathLab® MathXL PRACTICE

Problems **1–14** are the Quick ✓s that follow the EXAMPLES.

Building Skills

In Problems 15–18, a slope of a line is given. Determine (a) the slope of any line parallel to the line whose slope is given and (b) the slope of any line perpendicular to the line whose slope is given. See Objectives 1 and 3.

15. $m = 5$

16. $m = -\dfrac{8}{5}$

17. $m = -\dfrac{5}{6}$

18. $m = 0$

In Problems 19–26, determine whether the given linear equations are parallel, perpendicular, or neither. See Objectives 1 and 3.

19. $y = 5x + 4$
$y = 5x - 7$

20. $y = 3x - 1$
$y = -\dfrac{1}{3}x - 5$

21. $8x + y = 12$
$2x - 8y = 3$

22. $-3x - y = 3$
$6x + 2y = 9$

23. $-4x + 2y = 12$
$x + 2y = 6$

24. $10x - 3y = 5$
$5x + 6y = 3$

25. $-x + \dfrac{1}{3}y = \dfrac{1}{3}$
$x - \dfrac{1}{3}y = \dfrac{5}{3}$

26. $\dfrac{1}{2}x - \dfrac{3}{2}y = 3$
$2x + \dfrac{2}{3}y = 1$

In Problems 27–32, find an equation of the line L. Express your answer in slope-intercept form. See Objectives 2 and 4.

27.

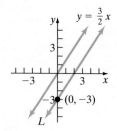

L is parallel to $y = \dfrac{3}{2}x$

28.

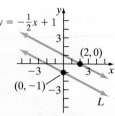

L is parallel to $y = -\dfrac{1}{2}x + 1$

29.

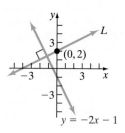

L is perpendicular to $y = -2x - 1$

30.

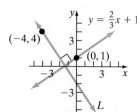

L is perpendicular to $y = \dfrac{2}{3}x + 1$

31.

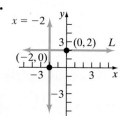

L is perpendicular to $x = -2$

32.

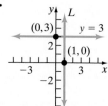

L is perpendicular to $y = 3$

In Problems 33–46, find an equation of the line with the given properties. Express your answer in slope-intercept form. Graph the lines. See Objectives 2 and 4.

33. Parallel to $y = 2x + 3$ through the point $(3, 1)$

34. Parallel to $y = -3x + 1$ through the point $(2, 5)$

35. Perpendicular to $y = -2x + 1$ through the point $(2, 3)$

36. Perpendicular to $y = 4x + 3$ through the point $(4, 1)$

37. Parallel to $y = 1$ through the point $(-1, -3)$

38. Parallel to $x = -2$ through the point $(2, 5)$

39. Perpendicular to $x = 1$ through the point $(1, 3)$

40. Perpendicular to $y = 8$ through the point $(2, -4)$

41. Parallel to $3x - y = 2$ through the point $(1, 5)$

42. Parallel to $2x + y = 5$ through the point $(-4, 3)$

43. Perpendicular to $4x + 3y - 1 = 0$ through the point $(-4, 1)$

44. Perpendicular to $-2x + 5y - 3 = 0$ through the point $(2, -3)$

45. Parallel to $5x + 2y = 1$ through the point $(-2, -3)$

46. Perpendicular to $3x + y = 1$ through the point $(3, -1)$

Mixed Practice

In Problems 47–52, two points on L_1 and two points on L_2 are given. Plot the points in the Cartesian plane and draw a line through the points. Compute the slope of the line containing these points and determine whether the lines are parallel, perpendicular, or neither.

47. L_1: $(1, 2)$; $(6, 5)$
   L_2: $(-2, 3)$; $(1, -2)$

48. L_1: $(1, 1)$; $(4, 3)$
   L_2: $(-1, 3)$; $(3, -3)$

49. L_1: $(-2, 4)$; $(1, -3)$
   L_2: $(-1, -6)$; $(-4, 1)$

50. L_1: $(-3, 0)$; $(0, 2)$
   L_2: $(4, 8)$; $(2, 5)$

51. L_1: $(0, 5)$; $(1, 3)$
   L_2: $(-4, 6)$; $(0, 14)$

52. L_1: $(1, -3)$; $(5, -4)$
   L_2: $(0, 4)$; $(8, 2)$

Applying the Concepts

△ **53. Geometry** Given the points $A = (1, 1)$, $B = (4, 3)$, and $C = (2, 6)$,

(a) Plot the points in a Cartesian plane. Connect the points to form a triangle.
(b) Verify that the triangle is a right triangle by showing that the line segment $\overline{AB}$ is perpendicular to the line segment $\overline{BC}$ and therefore forms a right angle.

△ **54. Geometry** Given the points $A = (-2, -2)$, $B = (3, 1)$, and $C = (-5, 3)$,

(a) Plot the points in a Cartesian plane. Connect the points to form a triangle.
(b) Verify that the triangle is a right triangle by showing that the line segment $\overline{AB}$ is perpendicular to the line segment $\overline{AC}$ and therefore forms a right angle.

△ **55. Geometry** In geometry, we learn that a **parallelogram** is a quadrilateral in which both pairs of opposite sides are parallel. Given the points $A = (2, 2)$, $B = (7, 3)$, $C = (8, 6)$, and $D = (3, 5)$,

(a) Plot the points in a Cartesian plane. Connect the points to form a quadrilateral.
(b) Verify that the quadrilateral is a parallelogram by showing that the opposite sides are parallel.

△ **56. Geometry** In geometry, we learn that a parallelogram is a quadrilateral in which both pairs of opposite sides are parallel. Given the points $A = (-2, -1)$, $B = (4, 1)$, $C = (5, 5)$, and $D = (-1, 3)$,

(a) Plot the points in a Cartesian plane. Connect the points to form a quadrilateral.
(b) Verify that the quadrilateral is a parallelogram by showing that the opposite sides are parallel.

Extending the Concepts

57. Find A so that $Ax + 4y = 12$ is perpendicular to $4x + y = 3$.

58. Find B so that $-6x + By = 3$ is perpendicular to $2x - 3y = 8$.

59. The figure shows the graph of two parallel lines. Which of the following pairs of equations might have such a graph?

(a) $y = x + 3$
   $y = -x - 1$
(b) $y = 2x + 3$
   $y = 2x + 1$
(c) $x - 2y = 4$
   $x - 2y = -3$
(d) $-2x + y = 5$
   $-2x + y = 2$
(e) $x - y = 3$
   $3x - 3y = 9$

60. The figure shows the graph of two perpendicular lines. Which of the following pairs of equations might have such a graph?

(a) $y = 3x + 4$
   $y = -\dfrac{1}{3}x - 2$
(b) $-2x + y = 3$
   $x + 2y = 1$
(c) $2x + 3y = -2$
   $3x - 2y = 5$
(d) $3x + 4y = 5$
   $-3x + 4y = -2$
(e) $x - 2y = 6$
   $2y + x = 2$

Explaining the Concepts

61. If two nonvertical lines have the same x-intercept but different y-intercepts, can they be parallel? Explain your answer.

62. Why don't we say that a horizontal line is perpendicular to a vertical line if they have slopes that are negative reciprocals of each other?

1.8 Linear Inequalities in Two Variables

Objectives

1. Determine Whether an Ordered Pair Is a Solution to a Linear Inequality
2. Graph Linear Inequalities
3. Solve Problems Involving Linear Inequalities

Are You Ready for This Section?

Before getting started, take this readiness quiz. If you get a problem wrong, go back to the section cited and review the material.

R1. Determine whether $x = 4$ satisfies the inequality $3x + 1 \geq 7$. [Section 1.4, pp. 85–87]

R2. Solve the inequality: $-4x - 3 > 9$ [Section 1.4, pp. 85–87]

R3. *True or False* The inequality $5 > 3$ is a strict inequality. [Section R.2, p. 17]

In Section 1.4, we solved inequalities in one variable. In this section, we discuss linear inequalities in two variables.

1 Determine Whether an Ordered Pair Is a Solution to a Linear Inequality

A linear inequality in two variables is an inequality in one of the forms

$$Ax + By < C \qquad Ax + By > C \qquad Ax + By \leq C \qquad Ax + By \geq C$$

where A, B, and C are real numbers and A and B are not both zero.

If we replace the inequality symbol with an equal sign, we obtain the equation of a line, $Ax + By = C$. The line separates the xy-plane into two regions called **half-planes.** See Figure 56.

Figure 56

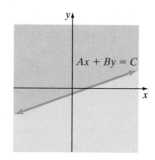

A linear inequality in two variables x and y is **satisfied** by an ordered pair (a, b) if a true statement results when x is replaced by a and y is replaced by b.

EXAMPLE 1 **Determining Whether an Ordered Pair Is a Solution to a Linear Inequality in Two Variables**

Determine which of the following ordered pairs are solutions to the linear inequality $3x + y < 7$.

(a) $(3, 2)$ **(b)** $(-3, 1)$ **(c)** $(1, -2)$

Solution

(a) Let $x = 3$ and $y = 2$ in the inequality. If a true statement results, then $(3, 2)$ is a solution to the inequality.

$$3x + y < 7$$
$$x = 3, y = 2: \quad 3(3) + 2 \stackrel{?}{<} 7$$
$$9 + 2 \stackrel{?}{<} 7$$
$$11 \stackrel{?}{<} 7 \quad \text{False}$$

The statement is false, so $(3, 2)$ is not a solution to the inequality.

(b) Let $x = -3$ and $y = 1$ in the inequality. If a true statement results, then $(-3, 1)$ is a solution to the inequality.

$$3x + y < 7$$

$$x = -3, y = 1: \quad 3(-3) + 1 \overset{?}{<} 7$$

$$-9 + 1 \overset{?}{<} 7$$

$$-8 \overset{?}{<} 7 \quad \text{True}$$

The statement is true, so $(-3, 1)$ is a solution to the inequality.

(c) Let $x = 1$ and $y = -2$ in the inequality.

$$3x + y < 7$$

$$x = 1, y = -2: \quad 3(1) + (-2) \overset{?}{<} 7$$

$$3 - 2 \overset{?}{<} 7$$

$$1 \overset{?}{<} 7 \quad \text{True}$$

The statement is true, so $(1, -2)$ is a solution to the inequality. ●

Quick ✓

1. If we replace the inequality symbol in $Ax + By > C$ with an equal sign, we obtain the equation of a line, $Ax + By = C$. The line separates the xy-plane into two regions called _____.

2. Determine which of the following ordered pairs are solutions to the linear inequality $-2x + 3y \geq 3$.

   (a) $(4, 1)$ **(b)** $(-1, 2)$ **(c)** $(2, 3)$ **(d)** $(0, 1)$

▶ ❷ Graph Linear Inequalities

A **graph of a linear inequality in two variables** x and y consists of all points (x, y) whose coordinates satisfy the inequality.

The graph of any linear inequality in two variables may be obtained by graphing the equation corresponding to the inequality, using dashes if the inequality is strict ($<$ or $>$) and a solid line if the inequality is nonstrict ($\leq$ or $\geq$). This graph will separate the xy-plane into two half-planes. In each half-plane, either all points satisfy the inequality or no points satisfy the inequality. So the use of a single test point is all that is required to obtain the graph of a linear inequality in two variables.

EXAMPLE 2   **How to Graph a Linear Inequality in Two Variables**

Graph the linear inequality: $3x + y < 7$

Step-by-Step Solution

Step 1: Replace the inequality symbol with an equal sign and graph the corresponding line. If the inequality is strict ($<$ or $>$), graph the line as a dashed line. If the inequality is nonstrict ($\leq$ or $\geq$), graph the line as a solid line.

Replace $<$ with $=$ to obtain $3x + y = 7$. Graph the line $3x + y = 7$ by first writing it in slope-intercept form: $y = -3x + 7$. The y-intercept of the line is $(0, 7)$ and the slope is -3, so plot the point $(0, 7)$. Use slope $= -\dfrac{3}{1} = \dfrac{\text{rise}}{\text{run}}$ to count right 1 unit and, down 3 units and plot the point $(1, 4)$. Graph the line through $(0, 7)$ and $(1, 4)$, using a dashed line because the inequality is strict. See Figure 57 (a) on the next page.

(continued)

Step 2: Select any test point not on the line—the origin $(0, 0)$ is often a good choice—and determine whether the point satisfies the inequality. If it does, shade the half-plane containing that point; otherwise, shade the other half-plane.

$$3x + y < 7$$
Test Point: $(0, 0)$: $3(0) + 0 \overset{?}{<} 7$
$$0 \overset{?}{<} 7 \quad \text{True}$$

Because 0 is less than 7, the point $(0, 0)$ satisfies the inequality. Therefore, shade the half-plane containing the point $(0, 0)$. See Figure 57(b). The shaded region represents the solutions to the linear inequality.

Figure 57

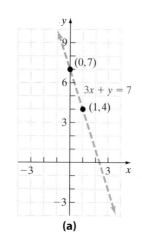

(a)

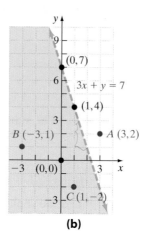

(b)

Only one test point is needed to obtain the graph of the inequality. Why? Consider the inequality $3x + y < 7$ presented in Examples 1 and 2. Notice that $A(3, 2)$ does not satisfy the inequality, while $B(-3, 1)$ and $C(1, -2)$ satisfy the inequality. Notice that point A is not in the shaded region of Figure 57(b), while points B and C are in the shaded region. So if a point does not satisfy the inequality, then none of the points in the half-plane containing that point satisfy the inequality. If a point does satisfy the inequality, then all the points in the half-plane containing the point satisfy the inequality.

Graphing a Linear Inequality in Two Variables

Step 1: Replace the inequality symbol with an equal sign, and graph the resulting equation. If the inequality is strict ($<$ or $>$), use a dashed line; if the inequality is nonstrict ($\leq$ or $\geq$) use a solid line. The graph separates the xy-plane into two half-planes.

Step 2: Select a test point P that is not on the line.

    (a) If the coordinates of P satisfy the inequality, then shade the half-plane containing P.

    (b) If the coordinates of P do not satisfy the inequality, then shade the half-plane that does not contain P.

Work Smart

An alternative to using test points is to solve the inequality for y. If the inequality is of the form $y >$ or $y \geq$, shade above the line. If the inequality is of the form $y <$ or $y \leq$, shade below the line.

Quick ✓

3. *True or False* The graph of a linear inequality is a line.

4. *True or False* In a graph of a linear inequality in two variables with a strict inequality, the line separating the two half-planes should be dashed.

In Problems 5 and 6, graph each linear inequality.

5. $y < -2x + 3$

6. $6x - 3y \leq 15$

EXAMPLE 3   **Graphing a Linear Inequality in Two Variables**

Graph the linear inequality: $y \geq \dfrac{1}{2}x$

Solution

We graph the corresponding line $y = \dfrac{1}{2}x$. The line is solid because the inequality is nonstrict. See Figure 58(a).

Now, select any test point not on the line. The line contains the origin, so we decide to use $(2, 0)$ as the test point.

Work Smart

Do not use (0, 0) as a test point for equations of the form $Ax + By = 0$ because the graph of this equation contains the origin.

$$y \geq \frac{1}{2}x$$

$$\text{Test Point: } (2, 0): \quad 0 \overset{?}{\geq} \frac{1}{2} \cdot 2$$

$$0 \overset{?}{\geq} 1 \quad \text{False}$$

Because 0 is not greater than 1, the point $(2, 0)$ does not satisfy the inequality. Therefore, shade the half-plane that does not contain $(2, 0)$. The shaded region in Figure 58(b) represents the solutions to the linear inequality.

Work Smart

Notice the inequality is of the form $y \geq$, so we shade above the line.

Figure 58

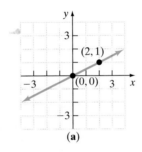   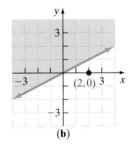

(a)   (b)

Quick ✓
7. Graph the linear inequality: $2x + y < 0$

▶ ❸ **Solve Problems Involving Linear Inequalities**

Linear inequalities involving two variables can be used to solve problems in areas such as manufacturing, sales, or (as in the next example) nutrition.

EXAMPLE 4   **Saturated Fat Intake**

Randy really enjoys Wendy's Junior Cheeseburgers and Biggie French Fries. However, he knows that his intake of saturated fat during lunch should not exceed 16 grams. Each Junior Cheeseburger contains 6 grams of saturated fat, and each order of Biggie Fries contains 3 grams of saturated fat. (SOURCE: *wendys.com*)

(a) Write a linear inequality that describes all the combinations of Junior Cheeseburgers and Biggie Fries that Randy can order.

(b) Can Randy eat two Junior Cheeseburgers and one order of Biggie Fries during lunch and stay within his allotment of saturated fat?

(c) Can Randy eat three Junior Cheeseburgers and one order of Biggie Fries during lunch and stay within his allotment of saturated fat?

Solution

(a) We are going to use the first three steps in the problem-solving strategy given in Section 1.2 on page 61.

Step 1: Identify We want to determine the number of Junior Cheeseburgers and Biggie Fries Randy can eat while not exceeding 16 grams of saturated fat.

Step 2: Name the Unknowns Let x represent the number of Junior Cheeseburgers that Randy eats, and let y represent the number of Biggie Fries Randy eats.

Step 3: Translate One Junior Cheeseburger has 6 grams of saturated fat. Two cheeseburgers have $6(2) = 12$ grams. In general, x Junior Cheeseburgers have $6x$ grams of saturated fat. Similar logic for the Biggie Fries tells us that y Biggie Fries have $3y$ grams of saturated fat. The words "cannot exceed' imply a $\leq$ inequality. Therefore, a linear inequality that describes the combinations Randy can have is

$$6x + 3y \leq 16 \quad \text{The Model}$$

(b) Letting $x = 2$ and $y = 1$, we obtain

$$6(2) + 3(1) \overset{?}{\leq} 16$$
$$15 \overset{?}{\leq} 16 \quad \text{True}$$

Because the inequality is true, Randy can eat two Junior Cheeseburgers and one order of Biggie Fries and remain within 16-grams of saturated fat.

(c) Letting $x = 3$ and $y = 1$, we obtain

$$6(3) + 3(1) \overset{?}{\leq} 16$$
$$21 \overset{?}{\leq} 16 \quad \text{False}$$

Because the inequality is false, Randy cannot eat three Junior Cheeseburgers and one order of Biggie Fries and remain within his limit.

Quick ✓

8. Avery is on a diet that requires that he consume no more than 900 calories for lunch. He really enjoys Wendy's Homestyle Chicken filet and Frosties. Each chicken filet has 560 calories, and each Frosty has 300 calories.

  (a) Write a linear inequality that describes Avery's options for eating at Wendy's.

  (b) Can Avery eat one Chicken Homestyle filet and one Frosty and stay within his allotment of calories?

  (c) Can Avery eat two Homestyle Chicken filets and one Frosty and stay within his allotment of calories?

1.8 Exercises MyMathLab®   PRACTICE

Exercise numbers in green have complete video solutions in MyMathLab.

Problems **1–8** are the Quick ✓ s that follow the EXAMPLES.

Building Skills

In Problems 9–12, determine whether the given points are solutions to the linear inequality. See Objective 1.

9. $x + 3y < 6$
  (a) $(0, 1)$
  (b) $(-2, 4)$
  (c) $(8, -1)$

10. $2x + y > -3$
  (a) $(2, -1)$
  (b) $(1, -3)$
  (c) $(-5, 4)$

11. $-3x + 4y \geq 12$
  (a) $(-4, 2)$
  (b) $(0, 2)$
  (c) $(0, 3)$

12. $2x - 5y \leq 2$
  (a) $(1, 0)$
  (b) $(3, 0)$
  (c) $(1, 2)$

In Problems 13–32, graph each inequality. See Objective 2.

13. $y > 3$

14. $y < -2$

15. $x \geq -2$

16. $x < 7$

17. $y < 5x$

18. $y \geq \dfrac{2}{3}x$

19. $y > 2x + 3$

20. $y < -3x + 1$

21. $y \leq \dfrac{1}{2}x - 5$

22. $y \geq -\dfrac{4}{3}x + 5$

23. $3x + y \leq 4$

24. $-4x + y \geq -5$

25. $2x + 5y \leq -10$

26. $3x + 4y \geq 12$

27. $-4x + 6y > 24$

28. $-5x + 3y < 30$

29. $\dfrac{x}{2} + \dfrac{y}{3} < 1$

30. $\dfrac{x}{3} - \dfrac{y}{4} \le 1$

31. $\dfrac{2}{3}x - \dfrac{3}{2}y \ge 2$

32. $\dfrac{5}{4}x - \dfrac{3}{5}y \le 2$

Applying the Concepts

33. Nutrition Sammy goes to McDonald's for lunch. He is on a diet that requires that he consume no more than 125 grams of carbohydrates for lunch. Sammy enjoys McDonald's Filet-o-Fish sandwiches and French fries. Each Filet-o-Fish has 39 grams of carbohydrates, and each small order of French fries has 29 grams of carbohydrates. (SOURCE: *McDonald's Corp.*)

   (a) Write a linear inequality that describes Sammy's options for eating at McDonald's.

   (b) Can Sammy eat two Filet-o-Fish and one order of fries and meet his carbohydrate requirement?

   (c) Can Sammy eat three Filet-o-Fish and one order of fries and meet his carbohydrate requirement?

34. Salesperson Juanita sells two different computer models. For each Model A computer sold she makes $45, and for each Model B computer sold she makes $65. Juanita set a monthly goal of earning at least $4000.

   (a) Write a linear inequality that describes Juanita's options for making her sales goal.

   (b) Will Juanita make her sales goal if she sells 50 Model A and 28 Model B computers?

   (c) Will Juanita make her sales goal if she sells 41 Model A and 33 Model B computers?

35. Production Planning Acme Switch Company is a small manufacturing firm that makes two different styles of microwave switches. Switch A requires 2 hours to assemble, and Switch B requires 1.5 hours to assemble. Suppose that there are at most 80 hours of assembly time available each week.

   (a) Write a linear inequality that describes Acme's options for making the microwave switches.

   (b) Can Acme manufacture 24 of Switch A and 41 of Switch B in a week?

   (c) Can Acme manufacture 16 of Switch A and 45 of Switch B in a week?

36. Budget Constraints Johnny can spend no more than $3.00 that he got from his grandparents. He goes to the candy store and wants to buy gummy bears that cost $0.10 each and suckers that cost $0.25 each.

   (a) Write a linear inequality that describes Johnny's options for buying candy.

   (b) Can Johnny buy 18 gummy bears and 5 suckers?

   (c) Can Johnny buy 19 gummy bears and 4 suckers?

Extending the Concepts

In Problems 37–40, determine the linear inequality whose graph is given.

37.
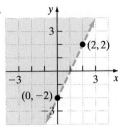

38.
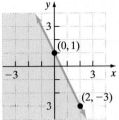

39.
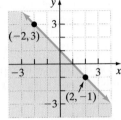

40.
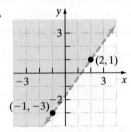

Explaining the Concepts

41. If an ordered pair does not satisfy a linear inequality, then we shade the side opposite the point. Explain why this works.

42. Explain why we cannot use a test point that lies on the line separating the two half-planes when determining the solutions to a linear inequality involving two variables.

The Graphing Calculator

Graphing calculators can be used to graph linear inequalities in two variables. Figure 59 shows the result of Example 2 using a TI-84 Plus graphing calculator. Consult your owner's manual for specific keystrokes.

Figure 59

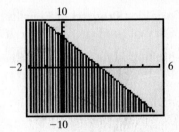

In Problems 43–54, graph the inequalities using a graphing calculator.

43. $y > 3$

44. $y < -2$

45. $y < 5x$

46. $y \ge \dfrac{2}{3}x$

47. $y > 2x + 3$

48. $y < -3x + 1$

49. $y \le \dfrac{1}{2}x - 5$

50. $y \ge -\dfrac{4}{3}x + 5$

51. $3x + y \le 4$

52. $-4x + y \ge -5$

53. $2x + 5y \le -10$

54. $3x + 4y \ge 12$

Chapter 1 Activity: Pass the Paper

Focus: Solving equations and inequalities

Time: 15 minutes

Group size: 4

Materials needed: One blank piece of notebook paper per group member

To the right are two equations and two inequalities. In this activity you will work together to solve these problems by following the procedure below. Be sure to read through the entire procedure together before beginning the activity so all group members will understand the procedure.

Procedure

1. Write one of the problems shown at the top of your paper. Be sure that each group member chooses a different problem.

2. Two lines below the original problem, write out the first step for solving the problem.

3. Fold the top of the paper down to cover the original problem, leaving your first step visible.

4. Pass your paper to a different group member. You might want to arrange your seats so that the papers can be passed around in a circle.

5. Continue solving the problems one step at a time, covering the step above yours and passing the paper to the next group member, until all problems are solved.

6. As a group, discuss the solutions and decide whether they are correct. If any solutions are incorrect, solve them correctly together.

Problems

(a) $4 - (x - 3) = -10 + 5(x + 1)$

(b) $3(x - 2) + 8 = 5x - 2(x - 1)$

(c) $3(x - 4) - 5x > 2x + 12$

(d) $4(x + 1) - 2x \geq 5x - 11$

Chapter 1 Review

Section 1.1   Linear Equations in One Variable

KEY CONCEPTS

- **Linear Equation in One Variable**
  An equation equivalent to one of the form $ax + b = 0$, where a and b are real numbers with $a \neq 0$.

- **Addition Property of Equality**
  For real numbers a, b, and c, if $a = b$, then $a + c = b + c$.

- **Multiplication Property of Equality**
  For real numbers a, b, and c where $c \neq 0$, if $a = b$, then $ac = bc$.

KEY TERMS

Equation in one variable
Sides of the equation
Solution
Satisfies
Solve an equation
Solution set
Equivalent equations
Conditional equation
Contradiction
Identity

You Should Be Able To...	EXAMPLE	Review Exercises
1 Determine whether a number is a solution to an equation (p. 49)	Example 1	1–4
2 Solve linear equations (p. 50)	Examples 2 through 6	5–18
3 Determine whether an equation is a conditional equation, an identity, or a contradiction (p. 55)	Examples 7 and 8	5–14

In Problems 1–4, determine which of the numbers, if any, are solutions to the given equation.

1. $3x - 4 = 6 + x$; $x = 5$, $x = 6$

2. $-1 - 4x = 2(3 - 2x) - 7$; $x = -2$, $x = -1$

3. $4y - (1 - y) + 5 = -6 - 2(3y - 5) - 2y$; $y = -2$, $y = 0$

4. $\dfrac{w - 7}{3} - \dfrac{w}{4} = -\dfrac{7}{6}$; $w = -14$, $w = 7$

In Problems 5–14, solve the linear equation. State whether the equation is an identity, a contradiction, or a conditional equation.

5. $2w + 9 = 15$

6. $-4 = 8 - 3y$

7. $2x + 5x - 1 = 20$

8. $7x + 5 - 8x = 13$

9. $-2(x - 4) = 8 - 2x$

10. $3(2r + 1) - 5 = 9(r - 1) - 3r$

11. $\dfrac{2y + 3}{4} - \dfrac{y}{2} = 5$

12. $\dfrac{x}{3} + \dfrac{2x}{5} = \dfrac{x - 20}{15}$

13. $0.2(x - 6) + 1.75 = 4.25 + 0.1(3x + 10)$

14. $2.1w - 3(2.4 - 0.2w) = 0.9(3w - 5) - 2.7$

In Problems 15 and 16, determine which values of the variable must be excluded from the domain.

15. $\dfrac{8}{2x + 3}$

16. $\dfrac{6x - 5}{6(x - 1) + 3}$

17. State Income Tax A resident of Missouri completes her state tax return and determines that she paid $2370 in state income tax in 2011. The solution to the equation $2370 = 0.06(x - 9000) + 315$ represents her Missouri taxable income x in 2011. Solve the equation to determine her Missouri taxable income. (SOURCE: *Missouri Department of Revenue*)

18. Movie Club The DVD club to which you belong offers unlimited DVDs at $10 off the regular price if you buy 1 at the regular price. You purchase 5 DVDs through this offer and spend $69.75 (not including tax and shipping). The solution to the equation $x + 4(x - 10) = 69.75$ represents the regular club price x for a DVD. Solve the equation to determine the regular club price for a DVD.

Section 1.2 An Introduction to Problem Solving

KEY CONCEPTS

- **Simple Interest Formula**
  $I = Prt$, where I is interest, P is principal, r is the per annum interest rate expressed as a decimal, and t is time in years

- **Uniform Motion Formula**
  $d = rt$, where d is distance, r is average speed, and t is time

KEY TERMS

Problem solving
Mathematical modeling
Modeling process
Mathematical model
Direct translation
Interest

Principal
Rate of interest
Simple interest
Mixture problems
Uniform motion

You Should Be Able To...	EXAMPLE	Review Exercises
① Translate English sentences into mathematical statements (p. 59)	Example 1	19–22
② Model and solve direct translation problems (p. 61)	Examples 2 through 6	23–28
③ Model and solve mixture problems (p. 65)	Examples 7 and 8	29–32
④ Model and solve uniform motion problems (p. 68)	Example 9	33–34

In Problems 19–22, translate each of the following English statements into a mathematical statement. Do not solve the equation.

19. The sum of three times a number and 7 is 22.

20. The difference of a number and 3 is equivalent to the quotient of the number and 2.

21. 20% of a number equals the difference of the number and 12.

22. The product of 6 and a number is the same as 4 less than twice the number.

For Problems 23 and 24, translate each English statement into a mathematical statement. Then solve the equation.

23. Shawn is 8 years older than Payton and the sum of their ages is 18. What are their ages?

24. The sum of five consecutive odd integers is 125. Find the integers.

25. Computing Grades Logan is in an elementary statistics course and has test scores of 85, 81, 84, and 77. If the final exam counts the same as two tests, what score does Logan need on the final to have an average of 80?

26. Home Equity Loans On February 23, 2012, Bank of America offered a home equity line of credit at a rate of 5.49% annual simple interest. If Cherie has such a credit line with a balance of $15,000, how much interest will she accrue at the end of 1 month?

27. Discounted Price Suppose that REI sells a 0° sleeping bag at the discounted price of $94.50. If this price represents a discount of 30% off the original selling price, find the original price.

28. Minimum Wage On July 24, 2009, the federal minimum wage was increased 10.7% to $7.25. Determine the federal minimum wage prior to July 24, 2009.

29. Making a Mixture CoffeeAM sells chocolate-covered blueberries for $10.95 per pound and chocolate-covered strawberries for $13.95 per pound. The company wants to create a mix of the two that will sell for $12.95 per pound with no loss in revenue. How many pounds of each treat should be used to make 12 pounds of the mix?

30. A Sports Mix The Candy Depot sells baseball gumballs for $3.50 per pound and soccer gumballs for $4.50 per pound. The company wants to sell a "sports mix" that sells for $3.75 per pound with no loss in revenue. How many pounds of each gumball type should be included to make 10 pounds of the mix?

31. Investments Angie received an $8000 bonus and wants to invest the money. She can invest part of the money at 8% simple interest with a moderate risk and the rest at 18% simple interest with a high risk. She wants an overall annual return of 12% but does not want to risk losing any more than necessary. How much should she invest at each rate to reach her goal?

32. Antifreeze A 2012 Chevrolet Malibu has an engine coolant system capacity of 7.5 quarts. If the system is currently filled with a mixture that is 30% antifreeze, how much of this mixture should be drained and replaced with pure antifreeze so that the system is filled with a mixture that is 50% antifreeze?

33. Road Trip On a 300-mile trip to Chicago, Josh drove part of the time at 60 miles per hour and the remainder of the trip at 70 miles per hour. If the total trip took 4.5 hours, for how many miles did Josh drive at a rate of 60 miles per hour?

34. Uniform Motion An F15 Strike Eagle near New York City and an F14 Tomcat near San Diego are about 2200 miles apart and traveling toward each other. The F15 is traveling 200 miles per hour faster than the F14, and the planes pass each other after 50 minutes. How fast is each plane traveling?

Section 1.3 Using Formulas to Solve Problems

KEY CONCEPTS

- **Geometry Formulas (see pages 74–75)**

KEY TERMS

Formula
Golden rectangle
Supplementary angles
Complementary angles

You Should Be Able To...	EXAMPLE	Review Exercises
① Solve for a variable in a formula (p. 73)	Examples 2 and 3	35–44
② Use formulas to solve problems (p. 76)	Examples 4 and 5	45–52

In Problems 35–40, solve for the indicated variable.

35. Solve $y = \dfrac{k}{x}$ for x.

36. Solve $F = \dfrac{9}{5}C + 32$ for C.

37. Solve $P = 2L + 2W$ for W.

38. Solve $\rho = m_1 v_1 + m_2 v_2$ for m_2.

39. Solve $PV = nRT$ for T.

40. Solve $S = 2LW + 2LH + 2WH$ for W.

In Problems 41–44, solve for y.

41. $3x + 4y = 2$

42. $-5x + 4y = 10$

43. $48x - 12y = 60$

44. $\dfrac{2}{5}x + \dfrac{1}{3}y = 8$

45. Temperature Conversions To convert temperatures from Fahrenheit to Celsius, we can use the formula

$C = \dfrac{5}{9}(F - 32)$. If the melting point for platinum is 3221.6°F, convert this temperature to degrees Celsius.

46. Angles in a Triangle The measure of each congruent angle in an isosceles triangle is 30 degrees larger than the measure of the remaining angle. Determine the measures of all three angles.

47. Window Dimensions The perimeter of a rectangular window is 76 feet. The window is 8 feet longer than it is wide. Find the dimensions of the window.

48. Long-Distance Phone Calls A long-distance telephone company charges a monthly fee of $2.95 and a per-minute charge of $0.04. The monthly cost for long distance is given by $C = 2.95 + 0.04x$, where x is the number of minutes used.

(a) Solve the equation for x.

(b) How many full minutes can Debbie use in one month on this plan if she does not want to spend more than $20 on long-distance calls in one month?

49. Concrete Rick has 80 cubic feet of concrete to pour for his new patio. If the patio is rectangular and Rick wants it to be 12 feet by 18 feet, how thick will the patio be if he uses all the concrete?

50. Right Circular Cones The lateral surface area for a frustum of a right circular cone is given by

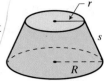

$$A = \pi s (R + r)$$

where s is the slant height of the frustum, R is the radius of the base, and r is the radius of the top.

(a) Solve the equation for r.

(b) If the frustum of a right circular cone has a lateral surface area of 10π square feet, a slant height of 2 feet, and a base whose radius is 3 feet, what is the radius of the top of the frustum?

51. Heating Bills The winter energy charge C for Illinois Power residential service is computed using the formula $C = 7.48 + 0.08674x$ for x kilowatt-hours (kwh).

(a) Solve the equation for x.

(b) How many kwh were used if the winter energy charge is $206.03? Round to the nearest whole number.

52. Supplementary and Complementary Angles The supplement of an angle and the complement of the same angle sum to 150°. What is the measure of the angle?

Section 1.4   Linear Inequalities in One Variable

KEY CONCEPTS

KEY TERMS

- **Linear Inequality in One Variable**
  An inequality of the form $ax + b < c$, $ax + b \le c$, $ax + b > c$, or $ax + b \ge c$, where a, b, and c are real numbers with $a \ne 0$.

- **Interval Notation versus Inequality Notation**

Solve an inequality
Solutions
Solution set
Interval notation
Closed interval
Open interval
Half-open or half-closed
   interval
Left endpoint
Right endpoint
Equivalent inequalities

INTERVAL NOTATION	INEQUALITY NOTATION	GRAPH
The open interval (a, b)	$\{x \mid a < x < b\}$	
The closed interval $[a, b]$	$\{x \mid a \le x \le b\}$	
The half-open interval $[a, b)$	$\{x \mid a \le x < b\}$	
The half-open interval $(a, b]$	$\{x \mid a < x \le b\}$	
The interval $[a, \infty)$	$\{x \mid x \ge a\}$	
The interval (a, ∞)	$\{x \mid x > a\}$	
The interval $(-\infty, a]$	$\{x \mid x \le a\}$	
The interval $(-\infty, a)$	$\{x \mid x < a\}$	
The interval $(-\infty, \infty)$	$\{x \mid x \text{ is a real number}\}$	

- **Addition Property of Inequalities**
  For real numbers a, b, and c

  If $a < b$, then $a + c < b + c$.

  If $a > b$, then $a + b > b + c$.

- **Multiplication Properties of Inequalities**
  For real numbers a, b, and c

  If $a < b$ and $c > 0$, then $ac < bc$.

  If $a > b$ and $c > 0$, then $ac > bc$.

  If $a < b$ and $c < 0$, then $ac > bc$.

  If $a > b$ and $c < 0$, then $ac < bc$.

You Should Be Able To...	EXAMPLE	Review Exercises
① Represent inequalities using the real number line and interval notation (p. 81)	Examples 1 through 4	53–56
② Understand the properties of inequalities (p. 84)		57–58
③ Solve linear inequalities (p. 85)	Examples 5 through 8	59–68
④ Solve problems involving linear inequalities (p. 87)	Example 9	69–72

In Problems 53 and 54, write each inequality using interval notation and graph the inequality.

53. $2 < x \leq 7$

54. $x > -2$

In Problems 55 and 56, write each interval as an inequality involving x and graph the inequality.

55. $(-\infty, 4]$

56. $[-1, 3)$

In Problems 57 and 58, use the Addition Property and/or the Multiplication Property to find a and b.

57. If $5 \leq x \leq 9$, then $a \leq 2x - 3 \leq b$.

58. If $-2 < x < 0$, then $a < 3x + 5 < b$.

In Problems 59–68, solve each linear inequality. Express your solution using set-builder notation and interval notation. Graph the solution set.

59. $3x + 12 \leq 0$

60. $2 < 1 - 3x$

61. $-7 \leq 3(h + 1) - 8$

62. $-7x - 8 < -22$

63. $3(p - 2) + (5 - p) > 2 - (p - 3)$

64. $2(x + 1) + 1 > 2(x - 2)$

65. $5(x - 1) - 7x > 2(2 - x)$

66. $0.03x + 0.10 > 0.52 - 0.07x$

67. $-\dfrac{4}{9}w + \dfrac{7}{12} < \dfrac{5}{36}$

68. $\dfrac{2}{5}y - 20 > \dfrac{2}{3}y + 12$

69. Octoberfest The German Club plans to rent a hall for its annual Octoberfest banquet. The hall costs $150 to rent plus $7.50 for each person who attends. If the club does not want to spend more than $600 for the event, how many people can attend the banquet?

70. Car Rentals A Ford Taurus at Enterprise Rent-a-Car rents for $43.46 per day. You receive 150 free miles per day but are charged $0.25 per mile for any additional miles. How many miles can you drive per day, on average, and not exceed your daily budget of $60.00?

71. Fund Raising A middle school band sells $1 candy bars at a carnival to raise money for new instruments. The band pays $50.00 to rent a booth and must pay the candy company $0.60 for each bar sold. How many bars must the band sell to make a profit?

72. Movie Club A DVD club offers unlimited DVDs for $9.95 if you purchase one for $24.95. How many DVDs can you purchase without spending more than $72.00?

Section 1.5 Rectangular Coordinates and Graphs of Equations

KEY CONCEPTS

- **Graph of an Equation in Two Variables**
  The set of all ordered pairs (x, y) in the xy-plane that satisfy the equation

- **Intercepts**
  The points, if any, where a graph crosses or touches the coordinate axes

KEY TERMS

x-axis	Quadrants
y-axis	Equation in two
Origin	variables
Rectangular or Cartesian	Sides
coordinate system	Satisfy
xy-plane	Graph of an equation in
Coordinate axes	two variables
Ordered pair	Point-plotting method
Coordinates	Complete graph
x-coordinate	Intercept
y-coordinate	x-intercept
Abscissa	y-intercept
Ordinate	

You Should Be Able To...	EXAMPLE	Review Exercises
1 Plot points in the rectangular coordinate system (p. 92)	Example 1	73, 74
2 Determine whether an ordered pair is a point on the graph of an equation (p. 94)	Example 2	75, 76
3 Graph an equation using the point-plotting method (p. 96)	Examples 3 through 5	77–82
4 Identify the intercepts from the graph of an equation (p. 98)	Example 6	83
5 Interpret graphs (p. 99)	Example 7	84

In Problems 73 and 74, plot each point in the same xy-plane. Tell in which quadrant or on what coordinate axis each point lies.

73. $A\,(2, -4)$
$B\,(-1, -3)$
$C\,(0, 4)$
$D\,(-5, 1)$
$E\,(1, 0)$
$F\,(4, 3)$

74. $A\,(3, 0)$
$B\,(1, 5)$
$C\,(-3, -5)$
$D\,(-1, 4)$
$E\,(5, -2)$
$F\,(0, -5)$

In Problems 75 and 76, determine whether the given points are on the graph of the equation.

75. $3x - 2y = 7$
(a) $(3, 1)$
(b) $(2, -1)$
(c) $(4, 0)$
(d) $\left(\dfrac{1}{3}, -3\right)$

76. $y = 2x^2 - 3x + 2$
(a) $(-1, 3)$
(b) $(1, 1)$
(c) $(-2, 16)$
(d) $\left(\dfrac{1}{2}, \dfrac{3}{2}\right)$

In Problems 77–82, graph each equation by plotting points.

77. $y = x + 2$
78. $2x + y = 3$
79. $y = -x^2 + 4$
80. $y = |x + 2| - 1$
81. $y = x^3 + 2$
82. $x = y^2 + 1$

In Problem 83, the graph of an equation is given. List the intercepts of the graph.

83.

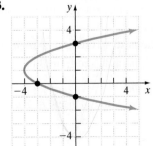

84. Cell Phones A cellular phone company offers a plan for $40 per month for 3000 minutes with additional minutes costing $0.05 per minute. The graph below shows the monthly cost, in dollars, when x minutes are used.

(a) If you talk for 2250 minutes in a month, how much is your monthly bill?

(b) Use the graph to estimate your monthly bill if you talk for 12 thousand minutes.

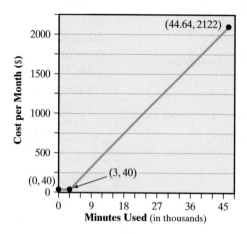

Section 1.6   Linear Equations in Two Variables

KEY CONCEPTS

- **Standard Form of a Line**
  $Ax + By = C$, where A, B, and C are real numbers. A and B are not both 0.

- **Finding Intercepts**
  x-intercept(s): Let $y = 0$ in the equation and solve for x
  y-intercept(s): Let $x = 0$ in the equation and solve for y

- **Equation of a Vertical Line**
  $x = a$, where $(a, 0)$ is the x-intercept

- **Equation of a Horizontal Line**
  $y = b$, where $(0, b)$ is the y-intercept

- **Slope of a Line**
  Let $P = (x_1, y_1)$ and $Q = (x_2, y_2)$ be two distinct points. The slope m of the nonvertical line L containing P and Q is defined by the formula

  $$m = \frac{y_2 - y_1}{x_2 - x_1}, \quad x_1 \neq x_2$$

  If $x_1 = x_2$, then L is a vertical line and the slope m of L is undefined (since this results in division by 0).

- When the slope of a line is positive, the line slants upward from left to right.
- When the slope of a line is negative, the line slants downward from left to right.
- When the slope of a line is zero, the line is horizontal.
- When the slope of a line is undefined, the line is vertical.

- **Point-slope Form of a Line**
  An equation of a nonvertical line of slope m that contains the point (x_1, y_1) is $y - y_1 = m(x - x_1)$.

- **Slope-intercept Form of a Line**
  An equation of a nonvertical line with slope m and y-intercept $(0, b)$ is $y = mx + b$.

KEY TERMS

Linear equation
Standard form
Line
Vertical line
Horizontal line
Run
Rise
Slope
Undefined slope
Average rate of change

You Should Be Able To...	EXAMPLE	Review Exercises
❶ Graph linear equations using point plotting (p. 103)	Example 1	85–88
❷ Graph linear equations using intercepts (p. 105)	Examples 2 and 3	89–92
❸ Graph vertical and horizontal lines (p. 106)	Examples 4 and 5	93–95
❹ Find the slope of a line given two points (p. 107)	Examples 6 and 7	96–99
❺ Interpret slope as an average rate of change (p. 110)	Example 8	100
❻ Graph a line given a point and its slope (p. 112)	Example 9	101–104
❼ Use the point-slope form of a line (p. 112)	Example 10	105–108
❽ Identify the slope and y-intercept of a line from its equation (p. 113)	Example 11	113–114
❾ Find the equation of a line given two points (p. 114)	Examples 12 and 13	109–112

In Problems 85–88, graph each linear equation by plotting points.

85. $x + y = 7$

86. $x - y = -4$

87. $5x - 2y = 6$

88. $-3x + 2y = 8$

In Problems 89–92, graph each linear equation by finding its intercepts.

89. $5x + 3y = 30$

90. $4x + 3y = 0$

91. $\frac{3}{4}x - \frac{1}{2}y = 1$

92. $4x + y = 8$

In Problems 93–95, graph each linear equation.

93. $x = 4$

94. $y = -8$

95. $3x + 5 = -1$

In Problems 96 and 97, (a) find the slope of the line and (b) interpret the slope.

96.

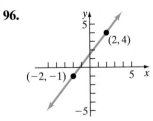

97.

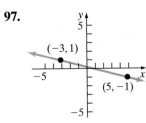

In Problems 98 and 99, plot each pair of points and determine the slope of the line containing them. Graph the line.

98. $(-1, 5)$; $(2, -1)$

99. $(4, 5)$; $(0, -1)$

100. Illinois's Population The following data represent the population of Illinois between 1940 and 2010.

Year, x	Population, y
1940	7,897,241
1950	8,712,176
1960	10,081,158
1970	11,110,285
1980	11,427,409
1990	11,430,602
2000	12,419,293
2010	12,830,632

SOURCE: *U.S. Census Bureau*

(a) Plot the ordered pairs (x, y) on a graph and connect the points with straight lines.

(b) Compute and interpret the average rate of change in population between 1940 and 1950.

(c) Compute and interpret the average rate of change in population between 1980 and 1990.

(d) Compute and interpret the average rate of change in population between 2000 and 2010.

(e) Based on the results to parts (a), (b), (c), and (d), do you think that population is linearly related to the year? Why?

In Problems 101–104, graph the line that contains the point P and has slope m. Do not find the equation of the line.

101. $m = 4; P(-1, -5)$

102. $m = -\dfrac{2}{3}; P(3, 2)$

103. m is undefined; $P(2, -4)$

104. $m = 0; P(-3, 1)$

In Problems 105 and 106, find an equation of the line. Express your answer in either slope-intercept or standard form, whichever you prefer.

105.

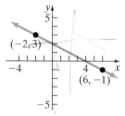

106.

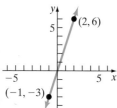

In Problems 107 and 108, find an equation of the line that has the given slope and contains the given point. Express your answer in either slope-intercept or standard form, whichever you prefer.

107. $m = -1; (3, 2)$

108. $m = \dfrac{3}{5}; (-10, -4)$

In Problems 109–112, find an equation of the line containing the given points. Express your answer in either slope-intercept or standard form, whichever you prefer.

109. $(6, 2); (-3, 5)$

110. $(-2, 3); (4, 3)$

111. $(4, -1); (1, -7)$

112. $(-1, 2); (8, -1)$

In Problems 113–114, find the slope and y-intercept of each line. Graph the line.

113. $y = 4x - 6$

114. $2x + 3y = 12$

Section 1.7 Parallel and Perpendicular Lines

KEY CONCEPTS

- **Parallel Lines**
  Two lines are parallel if they have the same slope but different *y*-intercepts.

- **Slopes of Perpendicular Lines**
  Two lines are perpendicular if the product of their slopes is -1. Alternatively, two lines are perpendicular if their slopes are negative reciprocals of each other.

KEY TERMS

Parallel lines
Perpendicular lines

You Should Be Able To...	EXAMPLE	Review Exercises
❶ Define parallel lines (p. 119)	Example 1	115, 117–120
❷ Find equations of parallel lines (p. 121)	Example 2	121–123
❸ Define perpendicular lines (p. 121)	Examples 3 and 4	116, 117–120
❹ Find equations of perpendicular lines (p. 122)	Example 5	124–126

In Problems 115–116, the slope of a line L is $m = -\dfrac{3}{8}$.

115. Determine the slope of a line that is parallel to *L*.

116. Determine the slope of a line that is perpendicular to *L*.

In Problems 117–120, determine whether the given pairs of linear equations are parallel, perpendicular, or neither.

117. $x - 3y = 9$
$9x + 3y = -3$

118. $6x - 8y = 16$
$3x + 4y = 28$

119. $\begin{aligned} 2x - y &= 3 \\ -6x + 3y &= 0 \end{aligned}$

120. $x = 2$
$\quad\;\; y = 2$

In Problems 121–126, find an equation of the line with the given properties. Express your answer in slope-intercept form, if possible. Graph the lines.

121. Parallel to $y = -2x - 5$ through $(1, 2)$

122. Parallel to $5x - 2y = 8$ through $(4, 3)$

123. Parallel to $x = -3$ through $(1, -4)$

124. Perpendicular to $y = 3x + 7$ through $(6, 2)$

125. Perpendicular to $3x + 4y = 6$ through $(-3, -2)$

126. Perpendicular to $x = 2$ through $(5, -4)$

Section 1.8 Linear Inequalities in Two Variables

KEY CONCEPT

- Linear inequalities in two variables are inequalities in one of the forms

$$Ax + By < C \qquad Ax + By > C \qquad Ax + By \le C \qquad Ax + By \ge C$$

where A, B, and C are real numbers and A and B are not both zero.

KEY TERMS

Half-planes
Satisfied
Graph of a linear inequality
in two variables

You Should Be Able To...	EXAMPLE	Review Exercises
❶ Determine whether an ordered pair is a solution to a linear inequality (p. 126)	Example 1	127–128
❷ Graph linear inequalities (p. 127)	Examples 2 and 3	129–134
❸ Solve problems involving linear inequalities (p. 129)	Example 4	135–136

In Problems 127 and 128, determine whether the given points are solutions to the linear inequality.

127. $5x + 3y \le 15$
  (a) $(4, -2)$
  (b) $(-6, 15)$
  (c) $(5, -1)$

128. $x - 2y > -4$
  (a) $(2, 3)$
  (b) $(5, -2)$
  (c) $(-1, 3)$

In Problems 129–134, graph each inequality.

129. $y < 3x - 2$

130. $2x - 4y \le 12$

131. $3x + 4y > 20$

132. $y \ge 5$

133. $2x + 3y < 0$

134. $x > -8$

135. Entertainment Budget Ethan's entertainment budget permits him to spend a maximum of $60 per month on movie tickets and MP3 music downloads. Movie tickets cost on average $7.50 each. MP3 music downloads average $2.00 each.

(a) Write a linear inequality that describes Ethan's options for spending the $60 maximum budget.

(b) Can Ethan buy 5 movie tickets and 15 music downloads?

(c) Can Ethan buy 6 movie tickets and 5 music downloads?

136. Fund Raising For a fund-raiser, the Math Club agrees to sell candy bars and candles. The club's profit will be 50¢ for each candy bar it sells and $2.00 for each candle it sells. The club needs to earn at least $1000 in order to pay for an upcoming field trip.

(a) Write a linear inequality that describes the combination of candy bars and candles that must be sold.

(b) Will selling 500 candy bars and 350 candles earn enough for the trip?

(c) Will selling 600 candy bars and 400 candles earn enough for the trip?

Chapter 1 Test *Step-by-step test solutions are found on the Chapter Test Prep Videos available in* MyMathLab® *or on* YouTube.

1. Determine which, if any, of the following are solutions to $3(x - 7) + 5 = x - 4$.
   (a) $x = 6$
   (b) $x = -2$

2. Write the following inequalities in interval notation, and graph them on a real number line.
   (a) $x > -4$
   (b) $3 < x \le 7$

In Problems 3 and 4, translate the English statement into a mathematical statement. Do not attempt to solve.

3. Three times a number, decreased by 8, is 4 more than the number.

4. Two-thirds of a number, increased by twice the difference of the number and 5, is more than 7.

In Problems 5 and 6, solve the equation. Determine whether the equation is an identity, a contradiction, or a conditional equation.

5. $5x - (x - 2) = 6 + 2x$

6. $7 + x - 3 = 3(x + 1) - 2x$

In Problems 7–9, solve the inequality and graph the solution set on a real number line.

7. $x + 2 \le 3x - 4$

8. $4x + 7 > 2x - 3(x - 2)$

9. $-x + 4 \le x + 3$

10. Solve $7x + 4y = 3$ for y.

11. **Computer Sales** Glen works as a computer salesman and earns $400 weekly plus 8% commission on his weekly sales. If he wants to make at least $750 in a week, how much must his sales be?

12. **Party Costs** A recreational center offers a children's birthday party for $75 plus $5 for each child. How many children were at Payton's birthday party if the total cost for the party was $145?

13. **Sandbox** Rick is building a rectangular sandbox for his daughter. He wants the length of the sandbox to be 2 feet more than the width, and he has 20 feet of lumber to build the frame. Find the dimensions of the sandbox.

14. **Mixture** Two acid solutions are available to a chemist. One is a 10% nitric acid solution, and the other is a 40% nitric acid solution. How much of each type of solution should be mixed together to form 12 liters of a 20% nitric acid solution?

15. **Ironman Race** The last leg of the Ironman competition is a 26.2-mile run. Contestant A runs at a constant rate of 8 miles per hour. If Contestant B starts the run 30 minutes after Contestant A and runs at a constant rate of 10 miles per hour, how long will it take Contestant B to catch up to Contestant A?

16. Plot the following ordered pairs in the same xy-plane. Tell in which quadrant or on what coordinate axis each point lies.
    $A(3, -4), B(0, 2), C(3, 0), D(2, 1), E(-1, -4), F(-3, 5)$

17. Determine whether the ordered pair is a point on the graph of the equation $y = 3x^2 + x - 5$.
    (a) $(-2, 4)$
    (b) $(-1, -3)$
    (c) $(2, 9)$

In Problems 18 and 19, graph the equations by plotting points.

18. $y = 4x - 1$

19. $y = 4x^2$

20. Identify the intercepts from the graph below.

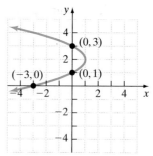

21. The following graph represents the speed of a car over time.

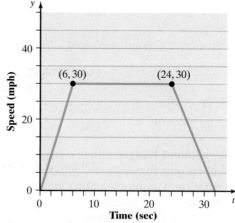

   (a) What is the speed of the car at 6 seconds?
   (b) Identify and interpret the intercepts.

In Problems 22–26, graph each linear equation, using any appropriate method.

22. $x - y = 8$

23. $3x + 5y = 0$

24. $3x + 2y = 12$

25. $\frac{3}{2}x - \frac{1}{4}y = 1$

26. $x = -7$

27. Find and interpret the slope of the line containing the points $(5, -2)$ and $(-1, 6)$.

28. Draw a graph of the line that contains the point $(2, -4)$ and has a slope of $-\frac{3}{5}$. Do not find the equation of the line.

29. Determine whether the graphs of the following pair of linear equations are parallel, perpendicular, or neither.

$$8x - 2y = 1$$
$$x + 4y = -2$$

In Problems 30–33, find the equation of the line with the given properties. Express your answer in either slope-intercept or standard form, whichever you prefer.

30. Through the point $(-3, 1)$ and having a slope of 4

31. Through the points $(6, 1)$ and $(-3, 7)$

32. Parallel to $x - 5y = 15$ and through the point $(10, -1)$

33. Perpendicular to $3x - y = 4$ and through the point $(6, 2)$

34. Determine whether the given points are solutions to the linear inequality $3x - y > 10$.

(a) $(3, -1)$ **(b)** $(4, 5)$ **(c)** $(5, 3)$

In Problems 35 and 36, graph each linear inequality.

35. $y \leq -2x + 1$

36. $5x - 2y < 0$

37. Area of a Circle The following data show the relationship between the diameter of a circle and the area of that circle.

Diameter (feet), x	Area (square feet), y
1	0.79
3	7.07
6	28.27
7	38.48
8	50.27
10	78.54
13	132.73

(a) Plot the ordered pairs (x, y) on a graph and connect the points with straight lines.

(b) Compute and interpret the average rate of change in area between diameter lengths of 1 and 3 feet.

(c) Compute and interpret the average rate of change in area between diameter lengths of 10 and 13 feet.

(d) Based upon the results to parts (a), (b), and (c), do you think that the area of the circle is linearly related to the diameter? Why?

Cumulative Review Chapters R–1

1. Approximate each number by (i) truncating and (ii) rounding to the indicated number of decimal places.

(a) 27.2357; three decimal places.

(b) 1.0729; one decimal place.

2. Plot the points $-4, -\frac{5}{2}, 0,$ and $\frac{7}{2}$ on a real number line.

In Problems 3–8, evaluate the expressions.

3. $-|-14|$

4. $-3 + 4 - 7$

5. $\frac{-3(12)}{-6}$

6. $(-3)^4$

7. $5 - 2(1 - 4)^3 + 5 \cdot 3$

8. $\frac{2}{3} + \frac{1}{2} - \frac{1}{4}$

9. Evaluate $3x^2 + 2x - 7$ when $x = 2$.

10. Simplify: $4a^2 - 6a + a^2 - 12 + 2a - 1$

11. Determine whether the given values are in the domain of x for the expression $\dfrac{x + 3}{x^2 + x - 2}$.

(a) $x = -2$ **(b)** $x = 0$

12. Use the Distributive Property to remove parentheses and then simplify: $3(x + 2) - 4(2x - 1) + 8$

13. Determine whether $x = 3$ is a solution to the equation $x - (2x + 3) = 5x - 1$.

In Problems 14 and 15, solve the equation.

14. $4x - 3 = 2(3x - 2) - 7$

15. $\dfrac{x + 1}{3} = x - 4$

16. Solve $2x - 5y = 6$ for y.

In Problems 17 and 18, solve the inequality and graph the solution set on a real number line.

17. $\dfrac{x + 3}{2} \leq \dfrac{3x - 1}{4}$

18. $5(x - 3) \geq 7(x - 4) + 3$

19. Plot the following ordered pairs in the same Cartesian plane.

$$A(-3, 0), \quad B(4, -2), \quad C(1, 5),$$
$$D(0, 3), \quad E(-4, -5), \quad F(-5, 2)$$

In Problems 20 and 21, graph the linear equation using the method you prefer.

20. $y = -\dfrac{1}{2}x + 4$ **21.** $4x - 5y = 15$

In Problems 22 and 23, find the equation of the line with the given properties. Express your answer in either slope-intercept or standard form, whichever you prefer.

22. Through the points $(3, -2)$ and $(-6, 10)$

23. Parallel to $y = -3x + 10$ and through the point $(-5, 7)$

24. Graph $x - 3y > 12$.

25. Computing Grades Shawn really wants an A in his geometry class. His four exam scores are 94, 95, 90, and 97. The final exam is worth two exam scores. To have an A, his average must be at least 93. For what range of scores on the final exam will Shawn be able to earn an A in the course?

26. Body Mass Index The body mass index (BMI) of a person who is 62 inches tall and weighs x pounds is given by $0.2x - 2$. A person with a BMI of 30 or more is considered obese. For what weights would a person 62 inches tall be considered obese?

27. Supplementary Angles Two angles are supplementary. The measure of the larger angle is 15 degrees more than twice the measure of the smaller angle. Find the angle measures.

28. Cylinders Max has 100 square inches of aluminum with which to make a closed cylinder. If the radius of the cylinder must be 2 inches, how tall will the cylinder be? (Round to the nearest hundredth of an inch.)

29. Consecutive Integers Find three consecutive even integers such that the sum of the first two is 22 more than the third.

2 Relations, Functions, and More Inequalities

Your cell phone bill depends on the cost of the data plan you choose. The revenue a Brownie troop earns from selling Girl Scout cookies depends on the number of boxes of cookies they sell. The area of a circle depends on the length of its radius. These three relations are examples of *functions*. Did you know that your weekly earnings are a function of the number of hours you work? See Problem 87 on page 162.

The Big Picture: Putting It Together

In Chapter 1 we studied linear equations in two variables. We graphed the solution of a linear equation in two variables using the rectangular coordinate system. The rectangular coordinate system connects algebra and geometry. Before the introduction of the rectangular coordinate system, algebra and geometry were thought to be separate subjects.

We begin this chapter with the concept of a function, probably the single most important concept of algebra. We then discuss compound inequalities, absolute value equations and inequalities, and the graphs of solutions to these equations and inequalities.

2.1 Relations

Objectives

1. Understand Relations
2. Find the Domain and the Range of a Relation
3. Graph a Relation Defined by an Equation

Are You Ready for This Section?

Before getting started, take this readiness quiz. If you get a problem wrong, go back to the section cited and review the material.

R1. Write the inequality $-4 \leq x \leq 4$ in interval notation. [Section 1.4, pp. 81–84]

R2. Write the interval $[2, \infty)$ using an inequality. [Section 1.4, pp. 81–84]

R3. Plot the ordered pairs $(-2, 4)$, $(3, -1)$, $(0, 5)$, and $(4, 0)$ in the rectangular coordinate system. [Section 1.5, pp. 92–94]

R4. Graph the equation: $2x + 5y = 10$ [Section 1.6, pp. 103–106]

R5. Graph the equation $y = x^2 - 3$ by plotting points. [Section 1.5, pp. 96–98]

▶ 1 Understand Relations

When the value of one variable is related to the value of a second variable, we have a *relation*. For example, an individual's level of education is related to annual income. Engine size is related to gas mileage.

> **Definition**
>
> When the elements in one set are associated with elements in a second set, we have a **relation.** If x and y are two elements in these sets, and if a relation exists between x and y, then we say that x **corresponds** to y or that y **depends on** x, and we write $x \rightarrow y$. We may also write a relation where y depends on x as an ordered pair (x, y).

EXAMPLE 1   **Illustrating a Relation**

The data presented in Figure 1 show a correspondence between states and senators in 2012 for randomly selected senators. We might name the relation "is represented in the U.S. Senate by." Thus we would say "Indiana is represented by Dan Coats." In Figure 1, we use **mapping** to represent the relation by drawing an arrow from an element in the set "state" to an element in the set "senator."

Figure 1

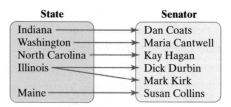

We could also represent this relation by using ordered pairs in the form (state, senator) as follows:

{ (Indiana, Dan Coats), (Washington, Maria Cantwell),
(North Carolina, Kay Hagan), (Illinois, Dick Durbin),
(Illinois, Mark Kirk), (Maine, Susan Collins) }

Quick ✓

1. If a relation exists between x and y, then we say that x _____ to y or that y _____ on x, and we write $x \rightarrow y$.

Ready?...Answers **R1.** $[-4, 4]$
R2. $x \geq 2$
R3.  **R4.**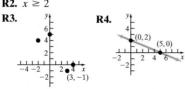

R5.

2. Use the map to represent the relation as a set of ordered pairs.

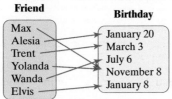

3. Use the set of ordered pairs to represent the relation as a map.

$$\{ (1, 3), (5, 4), (8, 4), (10, 13) \}$$

❷ Find the Domain and the Range of a Relation

▷ In a relation we say that y depends on x and can write the relation as a set of ordered pairs (x, y). We can think of the set of all x as the **inputs** of the relation. The set of all y can be thought of as the **outputs** of the relation. We use this interpretation of a relation to define *domain* and *range*.

> **Definition**
>
> The **domain** of a relation is the set of all inputs of the relation. The **range** is the set of all ouputs of the relation.

EXAMPLE 2 **Finding the Domain and the Range of a Relation**

Find the domain and the range of the relation presented in Figure 1 from Example 1.

Solution

The domain of the relation is the set of all inputs, therefore, the domain of the relation is

{ Indiana, Washington, North Carolina, Illinois, Maine }

The range of the relation is the set of all outputs. Therefore, the range of the relation is

{ Dan Coats, Maria Cantwell, Kay Hagan, Dick Durbin, Mark Kirk, Susan Collins }

Notice that we did not list Illinois twice in the domain. The domain and the range are sets, and we never list elements in a set more than once. Also, it does not matter in what order we list the elements in the domain or range.

Work Smart

Never list elements in the domain or range more than once.

Quick ✓

4. The _____ of a relation is the set of all inputs of the relation. The _____ is the set of all outputs of the relation.

5. State the domain and the range of the relation.

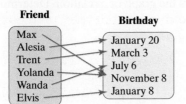

6. State the domain and the range of the relation.

$$\{ (1, 3), (5, 4), (8, 4), (10, 13) \}$$

Relations can also be represented by plotting a set of ordered pairs. The set of all x-coordinates represents the domain of the relation, and the set of all y-coordinates represents the range of the relation.

EXAMPLE 3 **Finding the Domain and the Range of a Relation**

Figure 2 shows the graph of a relation. Identify the domain and the range of the relation.

Figure 2

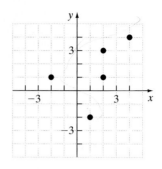

Solution

Work Smart

Write the points as ordered pairs to assist in finding the domain and range.

The ordered pairs in the graph are $(-2, 1)$, $(1, -2)$, $(2, 1)$, $(2, 3)$, and $(4, 4)$, so the domain is the set of all x-coordinates: $\{-2, 1, 2, 4\}$. The range is the set of all y-coordinates: $\{-2, 1, 3, 4\}$.

Quick ✓

7. Identify the domain and the range of the relation shown in the figure.

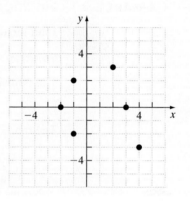

▶ A third way to define a relation is by a graph. Remember that the graph of an equation is the set of all ordered pairs (x, y) that make the equation a true statement. If a graph exists at some ordered pair (x, y), then the x-coordinate is in the domain and the y-coordinate is in the range.

EXAMPLE 4 **Identifying the Domain and the Range of a Relation from Its Graph**

Figure 3 shows the graph of a relation. Determine the domain and the range of the relation.

Figure 3

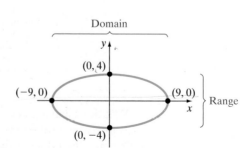

Work Smart

You can think of the domain as the shadow created by the graph on the x-axis by vertical beams of light. The range can be thought of as the shadow created by the graph on the y-axis by horizontal beams of light.

Solution

The domain of the relation consists of all x-coordinates at which the graph exists. The graph exists at all x-values between -9 and 9, inclusive. Therefore, the domain is $\{x \mid -9 \le x \le 9\}$ or, using interval notation, $[-9, 9]$.

The range of the relation consists of all y-coordinates at which the graph exists. The graph exists at all y-values between -4 and 4, inclusive. Therefore, the range is $\{y \mid -4 \le y \le 4\}$ or, using interval notation, $[-4, 4]$.

Quick ✓

8. *True or False* If the graph of a relation does not exist at $x = 3$, then 3 is not in the domain of the relation.

9. *True or False* The range of a relation is always the set of all real numbers.

In Problems 10 and 11, identify the domain and range of the relation from its graph.

10.

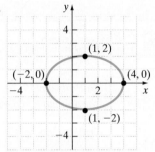

11.

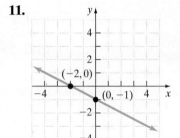

Work Smart

We can define relations by
1. Mapping
2. Sets of ordered pairs
3. Graphs
4. Equations

▶ ③ Graph a Relation Defined by an Equation

Another way to define a relation (instead of a map, a set of ordered pairs, or a graph) is to use equations such as $x + y = 4$ or $x = y^2$. When relations are defined by equations, we typically graph the relation in order to visualize how the variables are related. As we saw in Example 4, the graph of the relation also helps us identify its domain and range.

(EXAMPLE 5) **Relations Defined by Equations**

Graph the relation $y = -x^2 + 4$. Find its domain and range using the graph.

Solution

The relation says to take the input x, square it, multiply this result by -1, and then add 4 to get the output y. We use the point-plotting method to graph the relation. Table 1 shows some points on the graph. Figure 4 shows a graph of the relation.

Table 1

x	$y = -x^2 + 4$	(x, y)
-3	$-(-3)^2 + 4 = -5$	$(-3, -5)$
-2	$-(-2)^2 + 4 = 0$	$(-2, 0)$
-1	3	$(-1, 3)$
0	4	$(0, 4)$
1	3	$(1, 3)$
2	0	$(2, 0)$
3	-5	$(3, -5)$

Figure 4

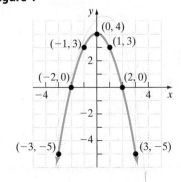

The graph extends indefinitely to the left and to the right (that is, the graph exists for all x-values). Therefore, the domain of the relation is the set of all real numbers, or $\{x \mid x \text{ is any real number}\}$, or, using interval notation, $(-\infty, \infty)$. Notice that there are no y-values greater than 4, but the graph exists everywhere for y-values less than or equal to 4. The range of the relation is $\{y \mid y \leq 4\}$ or, using interval notation, $(-\infty, 4]$.

Quick ✓

In Problems 12–14, graph each relation. Use the graph to identify the domain and range.

12. $y = 3x - 8$ **13.** $y = x^2 - 8$ **14.** $x = y^2 + 1$

2.1 Exercises

MyMathLab® Math XL PRACTICE

Exercise numbers in green have complete video solutions in MyMathLab.

Problems **1–14** are the Quick ✓s that follow the EXAMPLES.

Building Skills

In Problems 15–18, write each relation as a set of ordered pairs. Then identify the domain and the range of the relation. See Objectives 1 and 2.

15.

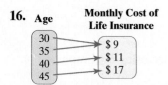

Newspaper	Daily Circulation (in millions)
USA Today	1.83
Wall Street Journal	2.12
New York Times	0.92
Los Angeles Times	0.61
Washington Post	0.55

SOURCE: *Information Please Almanac*

16.

Age	Monthly Cost of Life Insurance
30	$9
35	$11
40	$17
45	

SOURCE: *wholesale insurance.net*

17.

Level of Education	Average Annual Income
Less than 9th Grade	$13,992
9th – 12th Grade — No Diploma	$14,460
High School Graduate	$23,520
Associate's Degree	$36,012
Bachelor's Degree	$51,108

SOURCE: *United States Census Bureau*

18.

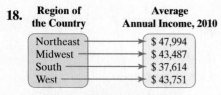

Region of the Country	Average Annual Income, 2010
Northeast	$47,994
Midwest	$43,487
South	$37,614
West	$43,751

SOURCE: *United States Census Bureau*

In Problems 19–24, write each relation as a map. Then identify the domain and the range of the relation. See Objectives 1 and 2.

19. $\{(-3, 4), (-2, 6), (-1, 8), (0, 10), (1, 12)\}$

20. $\{(-2, 6), (-1, 3), (0, 0), (1, -3), (2, 6)\}$

21. $\{(-2, 4), (-1, 2), (0, 0), (1, 2), (2, 4)\}$

22. $\{(-2, -8), (-1, -1), (0, 0), (1, 1), (2, 8)\}$

23. $\{(0, -4), (-1, -1), (-2, 0), (-1, 1), (0, 4)\}$

24. $\{(-3, 0), (0, 3), (3, 0), (0, -3)\}$

In Problems 25–32, identify the domain and the range of the relation from the graph. See Objective 2.

25.

26.

27.

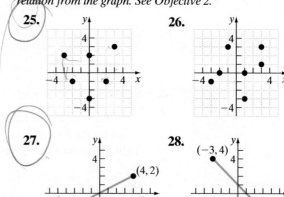

28.

29.

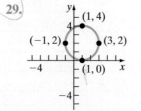

30.

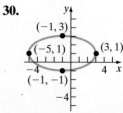

31.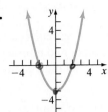

32.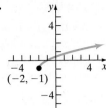

In Problems 33–54, use the graph of the relation obtained in Problems 33–54 from Section 1.5 to identify the domain and the range of the relation. See Objective 3.

33. $y = -3x + 1$

34. $y = -4x + 2$

35. $y = \dfrac{1}{2}x - 4$

36. $y = -\dfrac{1}{2}x + 2$

37. $2x + y = 7$

38. $3x + y = 9$

39. $y = -x^2$

40. $y = x^2 - 2$

41. $y = 2x^2 - 8$

42. $y = -2x^2 + 8$

43. $y = |x|$

44. $y = |x| - 2$

45. $y = |x - 1|$

46. $y = -|x|$

47. $y = x^3$

48. $y = -x^3$

49. $y = x^3 + 1$

50. $y = x^3 - 2$

51. $x^2 - y = 4$

52. $x^2 + y = 5$

53. $x = y^2 - 1$

54. $x = y^2 + 2$

Applying the Concepts

55. Area of a Window Bob Villa wishes to put a new window in his home. He wants the perimeter of the window to be 100 feet. The graph shows the relation between the width, x, of the opening and the area of the opening.

(a) Determine the domain and the range of the relation.

(b) Explain why the domain obtained in part (a) is reasonable.

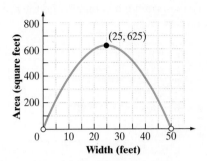

56. Projectile Motion The graph below shows the height, in feet, of a ball thrown straight up with an initial speed of 80 feet per second from an initial height of 96 feet after t seconds. Determine the domain and the range of the relation.

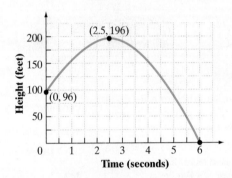

57. Cell Phones The graph below shows the relation between the monthly cost, C, of a cellular phone and the number of anytime minutes used, m.

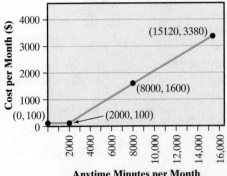

(a) Determine the domain and the range of the relation.

(b) If anytime minutes are from 7:00 A.M. to 7:00 P.M. Monday through Friday, explain why the domain obtained in part (a) is reasonable, assuming there are 21 nonweekend days per month.

58. Wind Chill It is 10° Celsius outside. The wind is calm but then gusts up to 20 meters per second. You feel the chill go right through your bones. The graph below shows the relation between the wind chill temperature (in degrees Celsius) and wind speed (in meters per second). Determine the domain and the range of the relation.

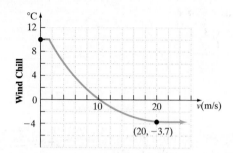

Extending the Concepts

59. Draw the graph of a relation whose domain is all real numbers, but whose range is a single real number. Compare your graph with those of your classmates. How are they similar?

60. Draw the graph of a relation whose domain is a single real number, but whose range is all real numbers. Compare your graph with those of your classmates. How are they similar?

Explaining the Concepts

61. Explain what a relation is. Be sure to include an explanation of domain and range.

62. State the four methods for describing a relation that are presented in this section. When is using ordered pairs most appropriate? When is using a graph most appropriate? Support your opinion.

2.2 An Introduction to Functions

Objectives

1. Determine Whether a Relation Expressed as a Map or Ordered Pairs Represents a Function
2. Determine Whether a Relation Expressed as an Equation Represents a Function
3. Determine Whether a Relation Expressed as a Graph Represents a Function
4. Find the Value of a Function
5. Find the Domain of a Function
6. Work with Applications of Functions

Are You Ready for This Section?

Before getting started, take this readiness quiz. If you get a problem wrong, go back to the section cited and review the material.

R1. Evaluate the expression $2x^2 - 5x$ for

    (a) $x = 1$ **(b)** $x = 4$ **(c)** $x = -3$ [Section R.5, pp. 40–41]

R2. Evaluate $\dfrac{3}{2x + 1}$ for $x = -\dfrac{1}{2}$.   [Section R.5, pp. 40–41]

R3. Express the inequality $x \le 5$ using interval notation.   [Section 1.4, pp. 81–84]

R4. Express the interval $(2, \infty)$ using set-builder notation. [Section 1.4, pp. 81–84]

▶ ① Determine Whether a Relation Expressed as a Map or Ordered Pairs Represents a Function

We now present one of the most important concepts in algebra—the *function*. A function is a special type of relation. To understand functions, let's revisit the relation in Example 1 from Section 2.1, shown again in Figure 5. In this correspondence between states and their senators, we named the relation "is represented in the U.S. Senate by."

Figure 5

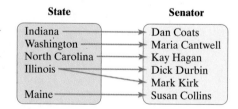

In this relation, if you were asked to name the senator who represents Illinois, you could respond "Dick Durbin" or "Mark Kirk." In other words, the input "state" does not correspond to a single output "senator."

Let's consider the relation in Figure 6(a), a correspondence between states and their populations. If asked for the population that corresponds to North Carolina, you could only respond "9535 thousand." In other words, each input "state" corresponds to exactly one output "population."

Figure 6(b) is a relation that shows a correspondence between animals and life expectancies. If asked to determine the life expectancy of a dog you could only respond "11 years." You would have the same answer about the life expectancy of a cat.

Figure 6

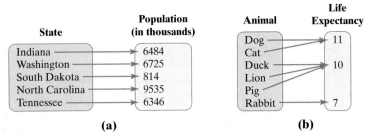

(a) (b)

What do the relations in Figures 6(a) and 6(b) have in common that is missing from the relation in Figure 5? In the relations in Figure 6(a) and 6(b), each input corresponds to only one output. In Figure 5, however, the input Illinois corresponds to two outputs—Dick Durbin and Mark Kirk. This leads to the definition of a *function*.

In Words
For a relation to be a function, each input may have only one output.

Definition

A **function** is a relation in which each input, or element in the domain of the relation, corresponds to exactly one output, or element in the range of the relation.

The idea behind functions is predictability. If an input is known, a function can be used to determine the output with 100% certainty, as we saw in Figures 6(a) and (b). With nonfunctions, we don't have this predictability (Figure 5).

EXAMPLE 1 **Determining Whether a Relation Represents a Function**

Determine whether the relation is a function. If so, state its domain and range.

(a) See Figure 7(a). The domain represents the length (mm) of the right humerus, and the range represents the length (mm) of the right tibia for each of five rats that had been sent to space.

(b) See Figure 7(b). The domain represents the weight of six pear-cut diamonds, and the range represents their price.

(c) See Figure 7(c). The domain represents the age of five males, and the range represents their HDL (or "good") cholesterol (mg/dL).

Figure 7

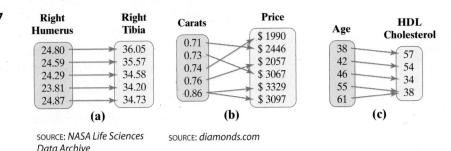

(a) (b) (c)

SOURCE: *NASA Life Sciences Data Archive*   SOURCE: *diamonds.com*

Solution

(a) The relation in Figure 7(a) is a function because each element in the domain corresponds to exactly one element in the range. The domain is { 24.80, 24.59, 24.29, 23.81, 24.87 }. The range is { 36.05, 35.57, 34.58, 34.20, 34.73 }.

(b) The relation in Figure 7(b) is not a function because an element in the domain, 0.86, corresponds to two elements in the range. A single price cannot be determined for the 0.86-carat diamonds.

(c) The relation in Figure 7(c) is a function because each element in the domain corresponds to exactly one element in the range. Notice that it is okay for more than one element in the function's domain to correspond to the same element in the range. The domain of the function is $\{38, 42, 46, 55, 61\}$. The range of the function is $\{57, 54, 34, 38\}$.

Quick ✓

1. A _____ is a relation in which each element in the domain of the relation corresponds to exactly one element in the range of the relation.

2. *True or False* Every relation is a function.

In Problems 3 and 4, determine whether the relation is a function. If so, state its domain and range.

3.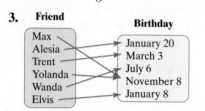

4.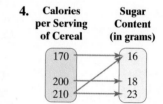

> **Work Smart**
>
> All functions are relations, but not all relations are functions!

We may also think of a function as a set of ordered pairs (x, y) in which no ordered pairs have the same first coordinate and different second coordinates.

(**EXAMPLE 2**)   **Determining Whether a Relation Represents a Function**

Determine whether the relation is a function. If so, state its domain and range.

(a) $\{(1, 3), (-1, 4), (0, 6), (2, 8)\}$

(b) $\{(-2, 6), (-1, 3), (0, 2), (1, 3), (2, 6)\}$

(c) $\{(0, 3), (1, 4), (4, 5), (9, 5), (4, 1)\}$

Solution

(a) This relation is a function because no ordered pairs have the same first coordinate and different second coordinates. The domain of the function is the set of all first coordinates, $\{-1, 0, 1, 2\}$. The range of the function is the set of all second coordinates, $\{3, 4, 6, 8\}$.

(b) This relation is a function because no ordered pairs have the same first coordinate and different second coordinates. The domain is, $\{-2, -1, 0, 1, 2\}$. The range is, $\{2, 3, 6\}$.

(c) This relation is not a function because two ordered pairs, $(4, 5)$ and $(4, 1)$, have the same first coordinate and different second coordinates.

> **Work Smart**
>
> In a function, two different inputs can correspond to the same output, but two different outputs cannot be the result of a single input.

In Example 2(b), notice that inputs -2 and 2 each correspond to output 6. This does not violate the definition of a function—two different first coordinates can have the same second coordinate. A violation of the definition occurs when two ordered pairs have the same first coordinate and different second coordinates, as in Example 2(c).

Quick ✓

In Problems 5 and 6, determine whether each relation is a function. If so, state its domain and range.

5. $\{(-3,3), (-2,2), (-1,1), (0,0), (1,1)\}$

6. $\{(-3,2), (-2,5), (-1,8), (-3,6)\}$

▶ ❷ Determine Whether a Relation Expressed as an Equation Represents a Function

We now know how to identify when a relation defined by a map or ordered pairs is a function. In Section 2.1, we expressed relations as equations. We will now address the circumstances under which equations are functions.

To determine whether an equation, where y depends on x, is a function, it is often easiest to solve the equation for y. If each value of x corresponds to exactly one value of y, the equation is a function; otherwise, it is not.

EXAMPLE 3 | Determining Whether an Equation Is a Function

Determine whether the equation $y = 3x + 5$ shows y as a function of x.

Solution

The rule for getting from x to y is to multiply x by 3 and then add 5. Since only one output y can result by performing these operations on any input x, the equation is a function. ●

EXAMPLE 4 | Determining Whether an Equation Is a Function

Determine whether the equation $y = \pm x^2$ shows y as a function of x.

Solution

> **In Words**
> The symbol $\pm$ is a shorthand device and is read "plus or minus." For example, ± 4 means "negative four or positive four."

Notice that for any single value of x (other than 0), two values of y result. For example, if $x = 2$, then $y = \pm 4$ (-4 or $+4$). Since a single x corresponds to more than one y, the equation is not a function. ●

Quick ✓

In Problems 7–9, determine whether each equation shows y as a function of x.

7. $y = -2x + 5$ **8.** $y = \pm 3x$ **9.** $y = x^2 + 5x$

▶ ❸ Determine Whether a Relation Expressed as a Graph Represents a Function

Remember that the graph of an equation is the set of all ordered pairs (x, y) that satisfy the equation. For a relation to be a function, each number x in the domain can correspond to only one number y in the range. This means that a graph is *not* a function if two points with the same x-coordinate have different y-coordinates. This leads to the following test.

> **Vertical Line Test**
>
> A set of points in the xy-plane is the graph of a function if and only if every vertical line intersects the graph in at most one point.

EXAMPLE 5   **Using the Vertical Line Test to Identify Graphs of Functions**

Which of the graphs in Figure 8 are graphs of functions?

Figure 8

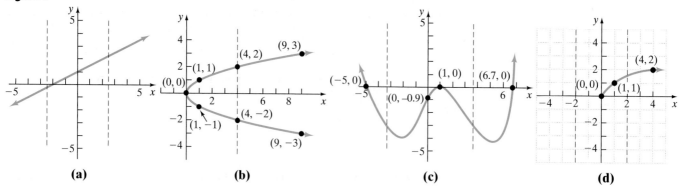

(a)   (b)   (c)   (d)

Solution

The graphs in Figures 8(a), (c), and (d) are functions because every vertical line intersects these graphs in at most one point. The graph in Figure 8(b) is not a function, because a vertical line intersects the graph in more than one point. ●

Does Example 5 show you why the vertical line test works? If a vertical line intersects the graph of an equation in two or more points, then the same *x*-coordinate corresponds to two or more different *y*-coordinates, and we have violated the definition of a function.

Quick ✓

10. *True or False* For a graph to be a function, any vertical line can intersect the graph in at most one point.

In Problems 11 and 12, use the vertical line test to determine whether the graph is a function.

11.

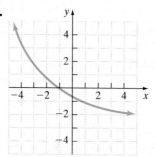

12.

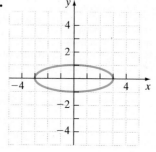

❹ Find the Value of a Function

▶ Functions are often denoted by letters such as f, F, g, G, and so on. If f is a function, we use special notation, called *function notation*, to represent the function. The following table shows some equations where y is a function of x. In each case, we denote the function by the letter f.

Equation in Two Variables	Function Notation
$y = \dfrac{1}{2}x + 5$	$f(x) = \dfrac{1}{2}x + 5$
$y = -3x + 1$	$f(x) = -3x + 1$
$y = 2x^2 - 3x$	$f(x) = 2x^2 - 3x$

Work Smart

Be careful with function notation. In the expression $y = f(x)$, y is the dependent variable, x is the independent variable, and f is the name given to a rule that relates the input x to the output y.

Notice that $f(x)$ replaces the variable y. If f is a function, then for each number x in its domain, the corresponding value in the range is denoted $f(x)$, which is read as "f of x" or as "f at x." Understand that $f(x)$ does not mean "f times x."

We call $f(x)$ the **value of the function f at the number x**; $f(x)$ is the number that results when the function is applied to x. Finding the value of a function parallels finding the value of y in an equation in two variables when y is a function of x. For example, "Given the equation $y = \dfrac{1}{2}x + 5$, find the value of y when $x = 4$" means the same as "Given the function $f(x) = \dfrac{1}{2}x + 5$, evaluate $f(4)$." In both cases, we replace the value of x with 4 and evaluate.

EXAMPLE 6 **Finding Values of a Function**

If $f(x) = \dfrac{3}{2}x - 6$ and $g(x) = x^2 + 6x$, evaluate:

(a) $f(10)$ **(b)** $f(-8)$ **(c)** $g(3)$ **(d)** $g(-2)$

Solution

(a) Wherever we see an x in the equation defining the function f, substitute 10.

$$f(10) = \frac{3}{2}(10) - 6$$
$$= 15 - 6$$
$$= 9$$

So $f(10) = 9$.

(b) Substitute -8 for x in the expression $\dfrac{3}{2}x - 6$ to get

$$f(-8) = \frac{3}{2}(-8) - 6$$
$$= -12 - 6$$
$$= -18$$

So $f(-8) = -18$.

(c) Substitute 3 for x in the equation defining the function g.

$$g(3) = (3)^2 + 6(3)$$
$$= 9 + 18$$
$$= 27$$

(d) Substitute -2 for x in the expression $x^2 + 6x$ to get

$$g(-2) = (-2)^2 + 6(-2)$$
$$= 4 + (-12)$$
$$= -8$$

Quick ✓

In Problems 13–16, let $f(x) = 3x + 2$ and $g(x) = -2x^2 + x + 3$ to evaluate each function.

13. $f(4)$ **14.** $f(-2)$ **15.** $g(-3)$ **16.** $g(1)$

For a function $y = f(x)$, the variable x is called the **independent variable,** because it can be assigned any of the numbers in the domain. The variable y is called the **dependent variable,** because its value depends on x.

The independent variable is also called the **argument** of the function. For example, if f is the function defined by $f(x) = x^2$, then f tells us to square the argument. Thus, $f(2)$ means to square 2, $f(a)$ means to square a, and $f(x + h)$ means to square the quantity $x + h$.

The notation $f(x)$ plays a dual role—it represents the rule for getting from the input to the output and it represents the output y of the function. For example, in Example 6(c), the rule for getting from the input to the output is given by $g(x) = x^2 + 6x$. In words, the function says to "take some input x, square it, and add the result to six times the input x." If the input is 3, then $g(3)$ represents the output, 27.

▶ **EXAMPLE 7** **Finding Values of a Function**

For the function $F(z) = 4z + 7$, evaluate:

(a) $F(z + 3)$ **(b)** $F(z) + F(3)$

Solution

(a) Wherever we see a z in the equation defining F, substitute $z + 3$ to get

$$F(z + 3) = 4(z + 3) + 7$$
$$= 4z + 12 + 7$$
$$= 4z + 19$$

(b) $F(z) + F(3) = \underbrace{4z + 7}_{F(z)} + \underbrace{4 \cdot 3 + 7}_{F(3)}$

$$= 4z + 7 + 12 + 7$$
$$= 4z + 26$$ ●

Quick ✓

17. In the function $H(q) = 2q^2 - 5q + 1$, H is called the _____ variable, and q is called the _____ variable or _____.

In Problems 18 and 19, let $f(x) = 2x - 5$ to evaluate each function.

18. $f(x - 2)$ **19.** $f(x) - f(2)$

Summary Important Facts About Functions

1. Each x in the domain has exactly one corresponding y in the range.

2. We use letters such as f to denote the function. It represents the rule we use to get from an x in the domain to $f(x)$ in the range.

3. If $y = f(x)$, then x is the independent variable or argument of f, and y is the dependent variable or the value of f at x.

▶ ⑤ **Find the Domain of a Function**

We always need to know which inputs make sense for the function we are working with.

In Words

The domain of a function is the set of all inputs for which the function gives an output that is a real number and makes sense.

Definition

When only the equation of a function is given, we agree that the **domain of f** is the set of real numbers x for which $f(x)$ is a real number.

When identifying the domain of a function, don't forget that division by zero is undefined. Exclude values of the variable that result in division by zero.

EXAMPLE 8   **Finding the Domain of a Function**

Find the domain of each of the following functions:

(a) $G(x) = x^2 + 1$ (b) $g(z) = \dfrac{z - 3}{z + 1}$

Solution

(a) The function G squares a number x and then adds 1 to the result. These operations can be performed on any real number, so the domain of G is the set of all real numbers, which we can express as $\{x \mid x \text{ is a real number}\}$ or, using interval notation, $(-\infty, \infty)$.

(b) The function g involves division. Since division by 0 is not defined, the denominator $z + 1$ cannot be 0. Therefore, z cannot equal -1. The domain of g is $\{z \mid z \neq -1\}$.   ●

Quick ✓

20. When only the equation of a function f is given, we agree that the _____ of f is the set of real numbers x for which $f(x)$ is a real number.

In Problems 21 and 22, find the domain of each function.

21. $f(x) = 3x^2 + 2$ **22.** $h(x) = \dfrac{x + 1}{x - 3}$

In an application, a function's domain may be restricted by physical or geometric considerations, in addition to mathematical restrictions. For example, the domain of $f(x) = x^2$ is the set of all real numbers. However, if f is used to find the area of a square given side x, then we restrict the domain of f to the positive real numbers, since the length of a side cannot be 0 or negative.

EXAMPLE 9   **Finding the Domain of a Function**

The number N of computers produced at a Dell Computers' manufacturing facility in one day after t hours is given by the function, $N(t) = 336t - 7t^2$. What is the domain of this function?

Solution

The independent variable is t, the number of hours in the day. Therefore the function's domain is $\{t \mid 0 \leq t \leq 24\}$, or the interval $[0, 24]$.   ●

Quick ✓

23. The function $A(r) = \pi r^2$ gives the area of a circle A as a function of the radius r. What is the domain of the function?

▷ ❻ **Work with Applications of Functions**

In practice, the symbols used for the independent and dependent variables should remind us what they represent. For example, in economics, we use C for cost and q for quantity, so that $C(q)$ represents the cost of manufacturing q units of a good. Here C is the dependent variable, q is the independent variable, and $C(q)$ is the rule that tells us how to get the output C from the input q.

EXAMPLE 10 **Life Cycle Hypothesis**

The Life-Cycle Hypothesis from economics was presented by Franco Modigliani in 1954. It states that income is a function of age. The function

$$I(a) = -55a^2 + 5119a - 54{,}448$$

represents the relation between average annual income I and age a.

(a) Identify the dependent and independent variables.

(b) Evaluate $I(20)$ and explain what $I(20)$ means.

Solution

(a) Because income depends on age, the dependent variable is income, I, and the independent variable is age, a.

(b) We let $a = 20$ in the function.

$$I(20) = -55(20)^2 + 5119(20) - 54{,}448 = 25{,}932$$

The average annual income of a 20-year-old is $25,932.

Quick ✓

24. In 2010, the *Deepwater Horizon* oil explosion spilled millions of gallons of oil in the Gulf of Mexico. The oil slick takes the shape of a circle. Suppose that the area A (in square miles) of the circle contaminated with oil can be determined using the function $A(t) = 0.25\pi t^2$, where t represents the number of days since the rig exploded.

(a) Identify the dependent and independent variables.

(b) Evaluate $A(30)$ and explain what $A(30)$ means.

2.2 Exercises MyMathLab® Math XP
PRACTICE

Exercise numbers in green have complete video solutions in MyMathLab.

*Problems **1–24** are the Quick ✓s that follow the EXAMPLES.*

Building Skills

In Problems 25–34, determine whether each relation represents a function. State the domain and the range of each relation. See Objective 1.

25.

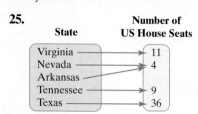

State — Number of US House Seats

26.

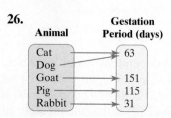

Animal — Gestation Period (days)

27.

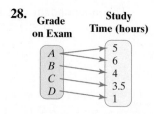

Horse-power — Top Speed

28.

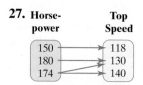

Grade on Exam — Study Time (hours)

29. $\{(0,3), (1,4), (2,5), (3,6)\}$

30. $\{(-1,4), (0,1), (1,-2), (2,-5)\}$

31. $\{(-3,5), (1,5), (4,5), (7,5)\}$

32. $\{(-2,3), (-2,1), (-2,-3), (-2,9)\}$

33. $\{(-10, 1), (-5, 4), (0, 3), (-5, 2)\}$

34. $\{(-5, 3), (-2, 1), (5, 1), (7, -3)\}$

In Problems 35–44, determine whether each equation shows y as a function of x. See Objective 2.

35. $y = 2x + 9$

36. $y = -6x + 3$

37. $2x + y = 10$

38. $6x - 3y = 12$

39. $y = \pm 5x$

40. $y = \pm 2x^2$

41. $y = x^2 + 2$

42. $y = x^3 - 3$

43. $x + y^2 = 10$

44. $y^2 = x$

In Problems 45–52, determine whether the graph is that of a function. See Objective 3.

45.

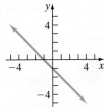

46.

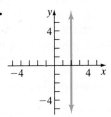

47.

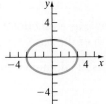

48.

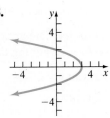

49.

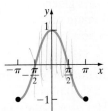

50.

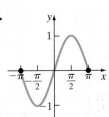

51.

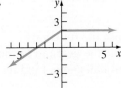

52.

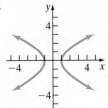

In Problems 53–60, find the indicated value of each function. See Objective 4.

    (a) $f(0)$ **(b)** $f(3)$ **(c)** $f(-2)$

53. $f(x) = 2x + 3$

54. $f(x) = 3x + 1$

55. $f(x) = -5x + 2$

56. $f(x) = -2x - 3$

57. $f(x) = x^2 - 3x$

58. $f(x) = 2x^2 + 5x$

59. $f(x) = -x^2 + x + 3$

60. $f(x) = -x^2 + 2x - 5$

In Problems 61–64, find the indicated value of each function. See Objective 4.

    (a) $f(-x)$ **(b)** $f(x + 2)$ **(c)** $f(2x)$

    (d) $-f(x)$ **(e)** $f(x + h)$

61. $f(x) = 2x - 5$

62. $f(x) = 4x + 3$

63. $f(x) = 7 - 5x$

64. $f(x) = 8 - 3x$

In Problems 65–72, find the value of each function. See Objective 4.

65. $f(x) = x^2 + 3; f(2)$

66. $f(x) = -2x^2 + x + 1; f(-3)$

67. $s(t) = -t^3 - 4t; s(-2)$

68. $g(h) = -h^2 + 5h - 1; g(4)$

69. $F(x) = |x - 2|; F(-3)$

70. $G(z) = 2|z + 5|; G(-6)$

71. $F(z) = \dfrac{z + 2}{z - 5}; F(4)$

72. $h(q) = \dfrac{3q^2}{q + 2}; h(2)$

In Problems 73–80, find the domain of each function. See Objective 5.

73. $f(x) = 4x + 7$

74. $G(x) = -8x + 3$

75. $F(z) = \dfrac{2z + 1}{z - 5}$

76. $H(x) = \dfrac{x + 5}{2x + 1}$

77. $f(x) = 3x^4 - 2x^2$

78. $s(t) = 2t^2 - 5t + 1$

79. $G(x) = \dfrac{3x - 5}{3x + 1}$

80. $H(q) = \dfrac{1}{6q + 5}$

Applying the Concepts

81. If $f(x) = 3x^2 - x + C$ and $f(3) = 18$, what is the value of C?

82. If $f(x) = -2x^2 + 5x + C$ and $f(-2) = -15$, what is the value of C?

83. If $f(x) = \dfrac{2x + 5}{x - A}$ and $f(0) = -1$, what is the value of A?

84. If $f(x) = \dfrac{-x + B}{x - 5}$ and $f(3) = -1$, what is the value of B?

△ **85. Geometry** Express the area A of a circle as a function of its radius, r. Determine the area of a circle whose radius is 4 inches. That is, find $A(4)$.

△ **86. Geometry** Express the area A of a triangle as a function of its height h, assuming that the length of the base is 8 centimeters. Determine the area of this triangle if its height is 5 centimeters. That is, find $A(5)$.

87. Salary Express the gross salary G of Jackie, who earns $15 per hour, as a function of the number of hours worked, h. Determine the gross salary of Jackie if she works 25 hours. That is, find $G(25)$.

88. Commissions Roberta is a commissioned salesperson. She earns a base weekly salary of $250 per week plus 15% of the sales price of items sold. Express her gross salary G as a function of the price p of items sold. Determine the weekly gross salary of Roberta if the value of items sold is $10,000. That is, find $G(10,000)$.

89. Population as a Function of Age The function $P(a) = 18.75a^2 - 5309.62a + 321{,}783.32$ represents the population (in thousands) of U.S. residents in 2010, P, that are a years of age or older.
SOURCE: *United States Census Bureau*

(a) Identify the dependent and independent variables.

(b) Evaluate $P(20)$. Provide a verbal explanation of the meaning of $P(20)$.

(c) Evaluate $P(0)$. Provide a verbal explanation of the meaning of $P(0)$.

90. Number of Rooms The function $N(r) = -1.33r^2 + 14.68r - 17.09$ represents the number of housing units (in millions), N, in 2010 that have r rooms, where $1 \le r \le 9$.
SOURCE: *United States Census Bureau*

(a) Identify the dependent and independent variables.

(b) Evaluate $N(3)$. Provide a verbal explanation of the meaning of $N(3)$.

(c) Why is it unreasonable to evaluate $N(0)$?

91. Revenue Function The function $R(p) = -p^2 + 200p$ represents the daily revenue R earned from selling MP3 players at p dollars for $0 \le p \le 200$.

(a) Identify the dependent and independent variables.

(b) Evaluate $R(50)$. Provide a verbal explanation of the meaning of $R(50)$.

(c) Evaluate $R(120)$. Provide a verbal explanation of the meaning of $R(120)$.

92. Average Trip Length The function $T(x) = 0.01x^2 - 0.12x + 8.89$ represents the average vehicle trip length T (in miles) x years since 1969.

(a) Identify the dependent and independent variables.

(b) Evaluate $T(35)$. Provide a verbal explanation of the meaning of $T(35)$.

(c) Evaluate $T(0)$. Provide a verbal explanation of the meaning of $T(0)$.

△ **93. Geometry** The volume V of a sphere as a function of its radius r is given by $V(r) = \dfrac{4}{3}\pi r^3$. What is the domain of this function?

△ **94. Geometry** The area A of a triangle as a function of its height h, assuming that the length of the base is 5 centimeters, is $A = \dfrac{5}{2}h$. What is the domain of the function?

95. Salary The gross salary G of Jackie as a function of the number of hours worked, h, is given by $G(h) = 22.5h$. What is the domain of the function if she can work up to 60 hours per week?

96. Commissions Roberta is a commissioned salesperson. She earns a base weekly salary of $350 plus 12% of the sales price of items sold. Her gross salary G as a function of the price p of items sold is given by $G(p) = 350 + 0.12p$. What is the domain of the function?

97. Demand for Hot Dogs Suppose the function $D(p) = 1200 - 10p$ represents the demand for hot dogs, whose price is p, at a baseball game. Find the domain of the function.

98. Revenue Function The function $R(p) = -p^2 + 200p$ represents the daily revenue earned from selling MP3 players at p dollars for $0 \le p \le 200$. Explain why any p greater than $200 is not in the domain of the function.

Extending the Concepts

99. Math for the Future: College Algebra A piecewise-defined function is a function defined by more than one equation. For example, the absolute value function $f(x) = |x|$ is actually defined by two equations: $f(x) = x$ if $x \geq 0$ and $f(x) = -x$ if $x < 0$. We can combine these equations into one expression as

$$f(x) = \begin{cases} x & x \geq 0 \\ -x & x < 0 \end{cases}$$

To evaluate $f(3)$, we recognize that $3 \geq 0$, so we use the rule $f(x) = x$ and obtain $f(3) = 3$. To evaluate $f(-4)$, we recognize that $-4 < 0$, so we use the rule $f(x) = -x$ and obtain $f(-4) = -(-4) = 4$.

(a) $f(x) = \begin{cases} x + 3 & x < 0 \\ -2x + 1 & x \geq 0 \end{cases}$

(i) Find $f(3)$.   **(ii)** Find $f(-2)$.   **(iii)** Find $f(0)$.

(b) $f(x) = \begin{cases} -3x + 1 & x < -2 \\ x^2 & x \geq -2 \end{cases}$

(i) Find $f(-4)$.   **(ii)** Find $f(2)$.   **(iii)** Find $f(-2)$.

100. Math for the Future: Calculus

(a) If $f(x) = 3x + 7$, find $\dfrac{f(x + h) - f(x)}{h}$.

(b) If $f(x) = -2x + 1$, find $\dfrac{f(x + h) - f(x)}{h}$.

Explaining the Concepts

101. Investigate when the use of function notation $y = f(x)$ first appeared. Start by researching Lejeune Dirichlet.

102. Are all relations functions? Are all functions relations? Explain your answers.

103. Explain what a function is. Be sure to include the terms *domain* and *range* in your explanation.

104. Explain why the vertical line test can be used to identify the graph of a function.

105. What are the four forms of a function presented in this section?

106. Explain why the terms *independent variable* for x and *dependent variable* for y make sense in the function $y = f(x)$.

The Graphing Calculator

Graphing calculators have the ability to evaluate any function you wish. The figure shows the results obtained in Example 6(c) and 6(d) using a TI-84 Plus graphing calculator.

 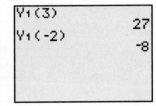

In Problems 107–114, use a graphing calculator to find the value of each function.

107. $f(x) = x^2 + 3; f(2)$

108. $f(x) = -2x^2 + x + 1; f(-3)$

109. $F(x) = |x - 2|; F(-3)$

110. $g(h) = \sqrt{2h + 1}; g(4)$

111. $H(x) = \sqrt{4x - 3}; H(7)$

112. $G(z) = 2|z + 5|; G(-6)$

113. $F(z) = \dfrac{z + 2}{z - 5}; F(4)$

114. $h(q) = \dfrac{3q^2}{q + 2}; h(2)$

2.3 Functions and Their Graphs

Objectives

1 Graph a Function

2 Obtain Information from the Graph of a Function

3 Know Properties and Graphs of Basic Functions

4 Interpret Graphs of Functions

Ready?...Answers   R1. $\{4\}$

R2.

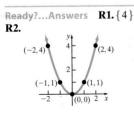

Are You Ready for This Section?

Before getting started, take this readiness quiz. If you get a problem wrong, go back to the section cited and review the material.

R1. Solve: $3x - 12 = 0$ [Section 1.1, pp. 50–52]

R2. Graph $y = x^2$ by point-plotting.   [Section 1.5, pp. 96–98]

▶ **1** Graph a Function

The graph of the linear equation $y = -2x + 4$ (Figure 9 on the next page) passes the vertical line test so, the equation is a function. We can write it as $f(x) = -2x + 4$. The graph of a function is the same as the graph of the equation that defines the function. The horizontal axis represents the independent variable and the vertical axis represents the dependent variable. When we graph functions, we label the vertical axis either by y or by the name of the function.

Figure 9

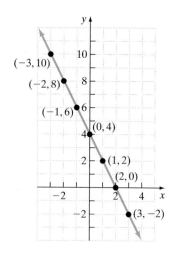

Definition

When a function is defined by an equation in x and y, the **graph of the function** is the set of *all* ordered pairs (x, y) such that $y = f(x)$.

So, if $f(3) = 8$, the point whose ordered pair is $(3, 8)$ is on the graph of $y = f(x)$.

EXAMPLE 1 **Graphing a Function**

Graph the function $f(x) = |x|$.

Solution

To graph $f(x) = |x|$, first determine some ordered pairs $(x, f(x)) = (x, y)$ such that $y = |x|$. See Table 2. Now plot the ordered pairs (x, y) from Table 2 and connect the points as shown in Figure 10.

Table 2

x	$f(x) = \|x\|$	$(x, f(x))$
-3	$\|-3\| = 3$	$(-3, 3)$
-2	$\|-2\| = 2$	$(-2, 2)$
-1	$\|-1\| = 1$	$(-1, 1)$
0	$\|0\| = 0$	$(0, 0)$
1	$\|1\| = 1$	$(1, 1)$
2	$\|2\| = 2$	$(2, 2)$
3	$\|3\| = 3$	$(3, 3)$

Figure 10
$f(x) = |x|$

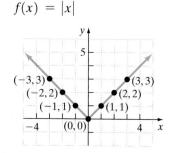

Quick ✓

1. When a function is defined by an equation in x and y, the _____ of the _____ is the set of all ordered pairs (x, y) such that $y = f(x)$.

2. If $f(4) = -7$, then the point whose ordered pair is (__ __) is on the graph of $y = f(x)$.

In Problems 3–5, graph each function.

3. $f(x) = -2x + 9$ 4. $f(x) = x^2 + 2$ 5. $f(x) = |x - 2|$

❷ Obtain Information from the Graph of a Function

▶ Remember, the domain of a function is the set of all inputs, and the range is the set of all outputs of the function. We can find the domain and the range of a function from its graph.

EXAMPLE 2 **Finding the Domain and Range of a Function from Its Graph**

Figure 11 shows the graph of a function.

(a) Determine the function's domain and range.

(b) Identify the intercepts.

Figure 11

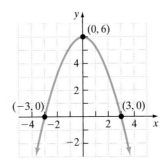

Solution

> **In Words**
> When the graph of a function is given, its domain may be viewed as the shadow created by the graph on the x-axis by vertical beams of light. Its range can be viewed as the shadow created by the graph on the y-axis by horizontal beams of light.

(a) The domain of the function consists of all of the graph's x-coordinates. Because the graph exists for all real numbers x, the domain is $\{x \mid x$ is any real number$\}$, or the interval, $(-\infty, \infty)$.

The range of the function consists of all of the graph's y-coordinates. Because the graph exists for all real numbers y less than or equal to 6, the range is $\{y \mid y \leq 6\}$, or the interval, $(-\infty, 6]$.

(b) The x-intercepts are $(-3, 0)$ and $(3, 0)$. The y-intercept is $(0, 6)$. ●

Quick ✓

6. Use the graph of the function to answer parts (a) and (b).

(a) Determine the domain and range of the function.

(b) Identify the intercepts.

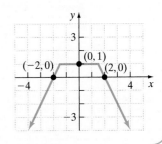

The next example illustrates how to obtain information about a function from its graph. Remember, if $(1, 5)$ is a point on the graph of f, then $f(1) = 5$.

EXAMPLE 3 **Obtaining Information from the Graph of a Function**

The Wonder Wheel is a Ferris wheel in Coney Island. See Figure 12 on the next page. Let f be the distance (in feet) above the ground of a person riding in a car on the Wonder Wheel as a function of time x (in minutes). Use the graph of f in Figure 13 to answer the following questions.

Figure 12

Figure 13

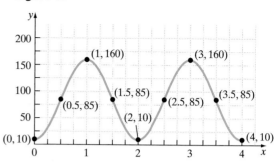

(a) Find $f(1.5)$ and $f(3)$. Interpret these values.

(b) What is the domain of f?

(c) What is the range of f?

(d) For what values of x does $f(x) = 85$? That is, solve $f(x) = 85$.

Solution

(a) Since $(1.5, 85)$ is on the graph of f, then $f(1.5) = 85$. After 1.5 minutes, a Wonder Wheel rider is 85 feet above the ground. Similarly, since $(3, 160)$ is on the graph, $f(3) = 160$. After 3 minutes, a Wonder Wheel rider is 160 feet above the ground.

(b) To determine the domain of f, we notice that the graph exists for each number x between 0 and 4, inclusive. Therefore, the domain of f is $\{x \,|\, 0 \le x \le 4\}$, or the interval $[0, 4]$.

(c) The points on the graph have y-coordinates between 10 and 160, inclusive. Therefore, the range of f is $\{y \,|\, 10 \le y \le 160\}$, or the interval $[10, 160]$.

(d) Since $(0.5, 85)$, $(1.5, 85)$, $(2.5, 85)$, and $(3.5, 85)$ are the only points on the graph for which $y = f(x) = 85$, the solution set to the equation $f(x) = 85$ is $\{0.5, 1.5, 2.5, 3.5\}$.

Quick ✓

7. If the point $(3, 8)$ is on the graph of a function f, then $f(\_) = \_$. If $g(-2) = 4$, then $(\_, \_)$ is a point on the graph of g.

8. Use the graph of $y = f(x)$ to answer the following questions.

   (a) Find $f(-3)$ and $f(1)$.

   (b) What is the domain of f?

   (c) What is the range of f?

   (d) Identify the intercepts.

   (e) For what value of x does $f(x) = 15$? That is, solve $f(x) = 15$.

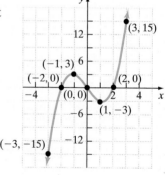

▶ **EXAMPLE 4** **Obtaining Information about the Graph of a Function**

Consider the function $f(x) = 2x - 5$.

(a) Is the point $(3, -1)$ on the graph of the function?

(b) If $x = 1$, what is $f(x)$? Based on this result, what point is on the graph of the function?

(c) If $f(x) = 3$, what is x? Based on this result, what point is on the graph of f?

Solution

(a) If $x = 3$, then
$$f(x) = 2x - 5$$
$$f(3) = 2(3) - 5 = 6 - 5 = 1$$
Since $f(3) = 1$, the point $(3, 1)$ is on the graph but $(3, -1)$ is not.

(b) If $x = 1$, then
$$f(1) = 2(1) - 5 = 2 - 5 = -3$$
Therefore, the point $(1, -3)$ is on the graph of f.

(c) If $f(x) = 3$, then

$$f(x) = 3$$
$$2x - 5 = 3$$
Add 5 to both sides: $\quad 2x = 8$
Divide both sides by 2: $\quad x = 4$

If $f(x) = 3$, then $x = 4$. Therefore, the point $(4, 3)$ is on the graph of f.

Work Smart: Study Skills
Do not confuse the directions "Find $f(3)$" with "If $f(x) = 3$, what is x?" Write down and study errors that you commonly make so that you can avoid them.

Quick ✓

9. Consider the function $f(x) = -3x + 7$.
   (a) Is the point $(-2, 1)$ on the graph of the function?
   (b) If $x = 3$, what is $f(x)$? Based on this result, what point is on the graph of the function?
   (c) If $f(x) = -8$, what is x? What point is on the graph of f?

▶ The Zero of a Function

If $f(r) = 0$ for some number r, then r is a **zero** of f. For example, if $f(x) = x^2 - 4$, then -2 and 2 are zeros of f because $f(-2) = 0$ and $f(2) = 0$. We can identify the zeros of a function from its graph by identifying the x-intercepts of the graph. Why? If $f(r) = 0$, then the point $(r, 0)$ is on the graph of f, and any point with coordinates $(r, 0)$ is an x-intercept of the graph.

EXAMPLE 5 **Finding the Zeros of a Function from Its Graph**

Find the zeros of the function f whose graph is shown in Figure 14.

Figure 14

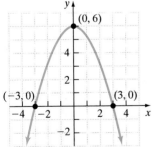

Solution

The x-intercepts of the graph are $(-3, 0)$ and $(3, 0)$. Therefore, the zeros of f are -3 and 3.

Quick ✓

In Problems 10–12, determine whether the value is a zero of the function.

10. $f(x) = 2x + 6; -3$
11. $g(x) = x^2 - 2x - 3; 1$
12. $h(z) = -z^3 + 4z; 2$
13. Find the zeros of the function f whose graph is shown.

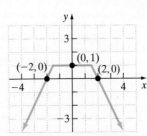

▶ ❸ Know Properties and Graphs of Basic Functions

In Table 3, we list a number of basic functions, their properties, and their graphs.

Table 3

Function	Properties	Graph
Linear Function $f(x) = mx + b$ m and b are real numbers	• Domain and range are all real numbers. • Graph is nonvertical line with slope $= m$ and y-intercept $= (0, b)$.	y $f(x) = mx + b, m > 0$ $(0, b)$ x
Identity Function (special type of linear function) $f(x) = x$	• Domain and range are all real numbers. • Graph is a line with slope of $m = 1$ and y-intercept $= (0, 0)$. • The line consists of all points for which the x-coordinate equals the y-coordinate.	y $f(x) = x$ $(1, 1)$, $(0, 0)$, $(-1, -1)$
Constant Function (special type of linear function) $f(x) = b$ b is a real number	• Domain is the set of all real numbers, and range is the set consisting of a single number b. • Graph is a horizontal line with slope $m = 0$ and y-intercept of $(0, b)$.	y $f(x) = b$ $(0, b)$ x
Square Function $f(x) = x^2$	• Domain is the set of all real numbers, and range is the set of nonnegative real numbers. • The intercept of the graph is $(0, 0)$.	y $f(x) = x^2$ $(-2, 4)$, $(2, 4)$, $(-1, 1)$, $(1, 1)$, $(0, 0)$
Cube Function $f(x) = x^3$	• Domain and range are the set of all real numbers. • The intercept of the graph is $(0, 0)$.	y $f(x) = x^3$ $(1, 1)$, $(0, 0)$, $(-1, -1)$
Absolute Value Function $f(x) = |x|$	• Domain is the set of all real numbers, and range is the set of nonnegative real numbers. • The intercept of the graph is $(0, 0)$. • If $x \geq 0$, then $f(x) = x$, and the graph of f is part of the line $y = x$; if $x < 0$, then $f(x) = -x$, and the graph of f is part of the line $y = -x$.	y $f(x) = |x|$ $(-2, 2)$, $(2, 2)$, $(-1, 1)$, $(1, 1)$, $(0, 0)$

❹ Interpret Graphs of Functions

We can use the graph of a function to give a visual description of many different scenarios. Consider the following example.

EXAMPLE 6 **Graphing a Verbal Description**

Maria decides to take a walk. She leaves her house and walks 3 blocks in 2 minutes at a constant speed. She realizes that she left her front door unlocked, so she runs home in 1 minute. It takes Maria 1 minute to find her keys and lock the door. She next runs 10 blocks in 3 minutes and then rests for 1 minute. She walks 4 more blocks in 10 minutes, and finally hitches a ride home with her neighbor, who happens to drive by, and gets home in 2 minutes. Draw a graph of Maria's distance from home (in blocks) as a function of time.

Solution

Because distance from home is a function of time, we draw a Cartesian plane with the horizontal axis representing the independent variable, time, and the vertical axis representing the dependent variable, distance from home.

The ordered pair $(0, 0)$ corresponds to starting the walk. The ordered pair $(2, 3)$ represents being 3 blocks from home after 2 minutes. We start the graph at $(0, 0)$ and then draw a straight line from $(0, 0)$ to $(2, 3)$. Next, we draw a straight line to $(3, 0)$, which represents the return trip home to lock the door. Draw a line segment from $(3, 0)$ to $(4, 0)$ to represent the time it takes to lock the door. Draw a line segment from $(4, 0)$ to $(7, 10)$, which represents the 10 block run in 3 minutes. Now draw a horizontal line from $(7, 10)$ to $(8, 10)$. This represents the resting period. Draw a line from $(8, 10)$ to $(18, 14)$ to represent the 4-block walk in 10 minutes. Finally, draw a line segment from $(18, 14)$ to $(20, 0)$ to represent the ride home. See Figure 15.

Figure 15

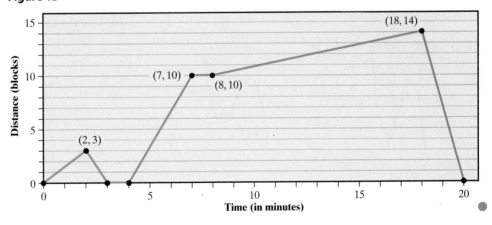

Quick ✓

14. Maria decides to take a walk. She leaves her house and walks 5 blocks in 5 minutes at a constant speed. She realizes that she left her front door unlocked, so she runs home in 2 minutes. It takes her 1 minute to find her keys and lock the door. She next jogs 8 blocks in 5 minutes, and then runs 3 blocks in 1 minute. After resting for 2 minutes she walks home in 10 minutes. Draw a graph of Maria's distance from home (in blocks) as a function of time (in minutes).

2.3 Exercises MyMathLab® MathXL PRACTICE

Exercise numbers in green
have complete video solutions
in MyMathLab.

Problems 1–14 are the Quick ✓s that follow the EXAMPLES.

Building Skills

In Problems 15–22, graph each function. See Objective 1.

15. $f(x) = 4x - 6$ **16.** $g(x) = -3x + 5$

17. $h(x) = x^2 - 2$ **18.** $F(x) = x^2 + 1$

19. $G(x) = |x - 1|$ **20.** $H(x) = |x + 1|$

21. $g(x) = x^3$ **22.** $h(x) = x^3 - 3$

In Problems 23–32, for each graph of a function, find (a) the domain and range, (b) the intercepts, if any, and (c) the zeros, if any. See Objective 2.

23.

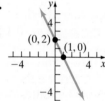

24.

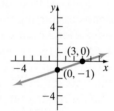

25.

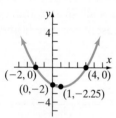

26.

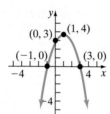

27.

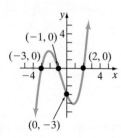

28.

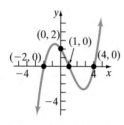

29.

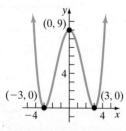

30.

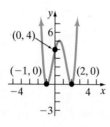

31.

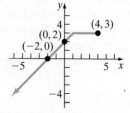

32.

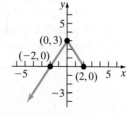

33. Use the graph of the function f shown to answer parts (a)–(l).

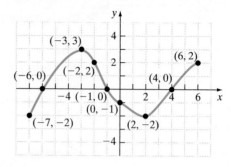

(a) Find $f(-7)$.

(b) Find $f(-3)$.

(c) Find $f(6)$.

(d) Is $f(2)$ positive or negative?

(e) For what numbers x is $f(x) = 0$?

(f) What is the domain of f?

(g) What is the range of f?

(h) What are the x-intercepts?

(i) What is the y-intercept?

(j) For what numbers x is $f(x) = -2$?

(k) For what number x is $f(x) = 3$?

(l) What are the zeros of f?

34. Use the graph of the function g shown to answer parts (a)–(l).

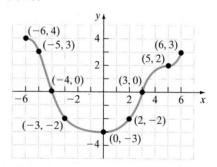

(a) Find $g(-3)$.

(b) Find $g(5)$.

(c) Find $g(6)$.

(d) Is $g(-5)$ positive or negative?

(e) For what numbers x is $g(x) = 0$?

(f) What is the domain of g?

(g) What is the range of g?

(h) What are the x-intercepts?

(i) What is the y-intercept?

(j) For what numbers x is $g(x) = -2$?

(k) For what number x is $g(x) = 3$?

(l) What are the zeros of g?

35. Use the table of values for the function F to answer questions (a)–(e).

x	$F(x)$
-4	0
-2	3
-1	5
0	2
3	-6

(a) What is $F(-2)$?   **(b)** What is $F(3)$?

(c) For what number(s) x is $F(x) = 5$?

(d) What is the x-intercept of the graph of F?

(e) What is the y-intercept of the graph of F?

36. Use the table of values for the function G to answer questions (a)–(e).

x	$G(x)$
-0	-3
-4	0
0	5
3	8
7	5

(a) What is $G(3)$?   **(b)** What is $G(7)$?

(c) For what number(s) x is $G(x) = 5$?

(d) What is the x-intercept of the graph of G?

(e) What is the y-intercept of the graph of G?

In Problems 37–40, answer the questions about the given function. See Objective 2.

37. $f(x) = 4x^2 - 9$

(a) Is the point $(2, 1)$ on the graph of the function?

(b) If $x = 3$, what is $f(x)$? What point is on the graph of the function?

(c) If $f(x) = 7$, what is x? What point is on the graph of f?

(d) Is 2 a zero of f?

38. $f(x) = 3x + 5$

(a) Is the point $(-2, 1)$ on the graph of the function?

(b) If $x = 4$, what is $f(x)$? What point is on the graph of the function?

(c) If $f(x) = -4$, what is x? What point is on the graph of f?

(d) Is -2 a zero of f?

39. $g(x) = -\dfrac{1}{2}x + 4$

(a) Is the point $(4, 2)$ on the graph of the function?

(b) If $x = 6$, what is $g(x)$? What point is on the graph of the function?

(c) If $g(x) = 10$, what is x? What point is on the graph of g?

(d) Is 8 a zero of g?

40. $H(x) = \dfrac{2}{3}x - 4$

(a) Is the point $(3, -2)$ on the graph of the function?

(b) If $x = 6$, what is $H(x)$? What point is on the graph of the function?

(c) If $H(x) = -4$, what is x? What point is on the graph of H?

(d) Is 6 a zero of H?

In Problems 41–46, match each graph to the function listed whose graph most resembles the one given. See Objective 3.

(a) Constant function   **(b)** Linear function
(c) Square function   **(d)** Cube function
(e) Absolute value function   **(f)** Identity function

41. 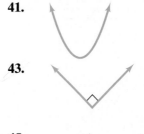   **42.**

43.    **44.**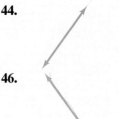

45.   **46.**

In Problems 47–50, sketch the graph of each function. Label at least three points. See Objective 3.

47. $f(x) = x^2$ **48.** $f(x) = x^3$

49. $f(x) = |x|$ **50.** $f(x) = 4$

Applying the Concepts

51. Match each of the following functions with the graph on the following page that best describes the situation.

(a) The distance from ground level of a person who is jumping on a trampoline as a function of time

(b) The cost of a telephone call as a function of time

(c) The height of a human as a function of time

(d) The revenue earned from selling cars as a function of price

(e) The book value of a machine that is depreciated by equal amounts each year as a function of the year

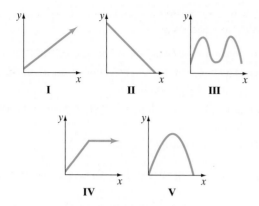

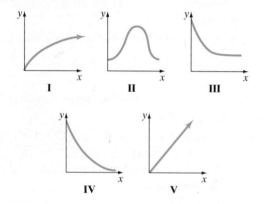

52. Match each of the following functions with the graph that best describes the situation.

(a) The average high temperature each day as a function of the day of the year

(b) The number of bacteria in a Petri dish as a function of time

(c) The distance that a person rides her bicycle at a constant speed as a function of time

(d) The temperature of a pizza after it is removed from the oven as a function of time

(e) The value of a car as a function of time

53. Pulse Rate Consider the following scenario: Zach starts jogging on a treadmill. His resting pulse rate is 70. As he continues to jog on the treadmill, his pulse increases at a constant rate until, after 10 minutes, his pulse is 120. He then starts jogging faster and his pulse increases at a constant rate for 2 minutes, at which time his pulse is up to 150. He then begins a cooling-off period for 7 minutes until his pulse backs down to 110. He then gets off the treadmill and his pulse returns to 70 after 12 minutes. Draw a graph of Zach's pulse as a function of time.

54. Altitude of an Airplane Suppose that a plane is flying from Chicago to New Orleans. The plane leaves the gate and taxis for 5 minutes. The plane takes off and gets up to 10,000 feet after 5 minutes. The plane continues to ascend at a constant rate until it reaches its cruising altitude of 35,000 feet after another 25 min es. For the next 80 minutes, the plane maintains a constant height of 35,000 feet. The plane then descends at a constant rate until it lands after 20 minutes.

It requires 5 minutes to taxi to the gate. Draw a graph of the height of the plane as a function of time.

55. Height of a Swing An 8-year-old girl gets on a swing and starts swinging for 10 minutes. Draw a graph that represents the height of the child from the ground as a function of time.

56. Temperature of Pizza Marissa is hungry and would like a pizza. Her mother pulls a frozen pizza out of the freezer and puts it in the oven. After 12 minutes the pizza is done, but Mom lets the pizza cool for 5 minutes before serving it to Marissa. Draw a graph that represents the temperature of the pizza as a function of time.

57. The graph below shows the weight of a person as a function of his age. Describe the weight of the individual over the course of his life.

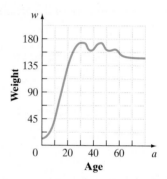

58. The following graph shows the depth of a lake (in feet) as a function of time (in days). Describe the depth of the lake over the course of the year.

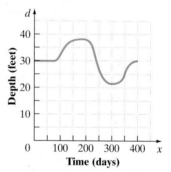

Extending the Concepts

59. Draw a graph of a function f with the following characteristics: x-intercepts: $(-4, 0)$, $(-1, 0)$, and $(2, 0)$; y-intercept: $(0, -2)$; $f(-3) = 7$ and $f(3) = 8$.

60. Draw a graph of a function f with the following characteristics: x-intercepts: $(-3, 0)$, $(2, 0)$, and $(5, 0)$; y-intercept: $(0, 3)$; $f(3) = -2$.

Explaining the Concepts

61. Using the definition of a function, explain why the graph of a function can have at most one y-intercept.

62. Explain what the domain of a function is. In your explanation, discuss how domains are determined in applications.

63. Explain what the range of a function is.

64. Explain the relationship between the x-intercepts of a function and its zeros.

Putting the Concepts Together (Sections 2.1–2.3)

We designed these problems so that you can review Sections 2.1–2.3 and show your mastery of the concepts. Take time to work these problems before proceeding with the next section. The answers are at the back of the text on page AN-13.

1. Explain why the following relation is a function. Then express the function as a set of ordered pairs.

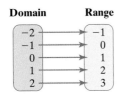

2. Determine which of the following relations represent functions.

   (a) $y = x^3 - 4x$ **(b)** $y = \pm 4x + 3$

3. Is the following relation a function? If so, state the domain and range.

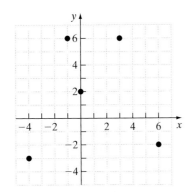

4. Explain why the relation whose graph is below is a function. If the name of the function is f, find $f(5)$.

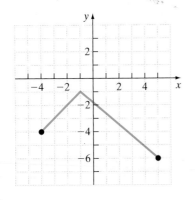

5. What is the zero of the function whose graph is shown?

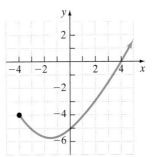

6. Let $f(x) = -5x + 3$ and $g(x) = -2x^2 + 5x - 1$. Find the value of each of the following.

   (a) $f(4)$ **(b)** $g(-3)$

   (c) $f(x) - f(4)$ **(d)** $f(x - 4)$

7. Find the domain of each of the following functions.

   (a) $G(h) = h^2 + 4$ **(b)** $F(w) = \dfrac{w - 4}{3w + 1}$

8. Graph $f(x) = |x| - 2$. Use the graph to state the domain and the range of f.

9. **Vertical Motion** The graph below shows the height, h, in feet, of a ball thrown straight up with an initial speed of 40 feet per second from an initial height of 80 feet after t seconds.

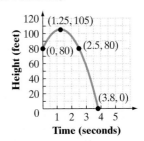

   (a) Find and interpret $h(2.5)$.

   (b) What is the domain of h?

   (c) What is the range of h?

   (d) For what value of t does $h(t) = 105$?

10. Consider the function $f(x) = 5x - 2$.

   (a) Is the point $(3, 12)$ on the graph of the function?

   (b) If $x = -2$, what is $f(x)$? What point is on the graph of the function?

   (c) If $f(x) = -22$, what is x? What point is on the graph of f?

   (d) Is $\dfrac{2}{5}$ a zero of f?

2.4 Linear Functions and Models

Objectives

1. Graph Linear Functions
2. Find the Zero of a Linear Function
3. Build Linear Models from Verbal Descriptions
4. Build Linear Models from Data

Are You Ready for This Section?

Before getting started, take this readiness quiz. If you get a problem wrong, go back to the section cited and review the material.

R1. Graph: $y = 2x - 3$ [Section 1.6, pp. 113–114]

R2. Graph: $\dfrac{1}{2}x + y = 2$ [Section 1.6, pp. 105–106]

R3. Graph: $y = -4$ [Section 1.6, pp. 106–107]

R4. Graph: $x = 5$ [Section 1.6, pp. 106–107]

R5. Find and interpret the slope of the line through $(-1, 3)$ and $(3, -4)$.   [Section 1.6, pp. 107–109]

R6. Find the equation of the line through $(1, 3)$ and $(4, 9)$.   [Section 1.6, pp. 114–115]

R7. Solve: $0.5(x - 40) + 100 = 84$ [Section 1.1, pp. 50–55]

R8. Solve: $4x + 20 \geq 32$ [Section 1.4, pp. 85–87]

▶ 1 Graph Linear Functions

Recall from Chapter 1 that a linear equation in two variables has the form $Ax + By = C$, where A, B, and C are real numbers, and A and B are not both zero.

Consider the graphs of the four lines in Figure 16. Notice that lines that rise from left to right have positive slope, and lines that fall from left to right have negative slope. Lines that have zero slope are horizontal lines, and lines that have undefined slope are vertical lines. Remember, just as we read a text from left to right, we also read graphs from left to right.

Figure 16

Positive Slope	Negative Slope	Zero Slope	Undefined Slope
$m > 0$	$m < 0$	$m = 0$	m is undefined
Line rises from left to right	Line falls from left to right	Horizontal Line	Vertical Line

In Figure 16, all the graphs except one pass the vertical line test for identifying graphs of functions. We conclude that **all linear equations except those of the form $x = a$, vertical lines, are functions.**

We can therefore write any linear equation that is in the form $Ax + By = C$ using function notation, provided $B \neq 0$, as follows:

$$Ax + By = C \qquad B \neq 0$$

Subtract Ax from both sides: $\quad By = -Ax + C$

Divide both sides by B: $\quad \dfrac{By}{B} = \dfrac{-Ax + C}{B}$

Simplify: $\quad y = -\dfrac{A}{B}x + \dfrac{C}{B}$

$$\qquad\qquad \updownarrow \qquad \updownarrow \qquad \updownarrow$$

$$f(x) = mx + b$$

Ready?...Answers

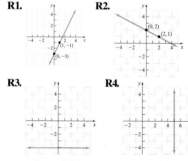

R1. **R2.**

R3. **R4.**

R5. $-\dfrac{7}{4}$; y decreases by 7 when x increases by 4. **R6.** $y = 2x + 1$ **R7.** $\{8\}$
R8. $\{x \,|\, x \geq 3\}$; $[3, \infty)$

This leads to the following definition:

> **Definition**
> A **linear function** is a function of the form
> $$f(x) = mx + b$$
> where m is the slope and $(0, b)$ is the y-intercept. The graph of a linear function is a line.

We can graph linear functions using the same techniques we used to graph linear equations written in slope-intercept form, $y = mx + b$ (See Section 1.6).

(EXAMPLE 1) **Graphing a Linear Function**

Graph the linear function: $f(x) = 3x - 5$

Solution

Since $f(x) = 3x - 5$ is in the form $f(x) = mx + b$, the y-intercept is $(0, -5)$, so we plot that point in the coordinate plane. Because the slope $m = 3 = \dfrac{3}{1} = \dfrac{\Delta y}{\Delta x} = \dfrac{\text{rise}}{\text{run}}$, we then go to the right 1 unit and up 3 units and end up at $(1, -2)$. The line through these two points is the graph of $f(x) = 3x - 5$, shown in Figure 17.

Figure 17

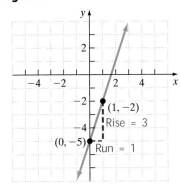

Quick ✓

1. For the graph of a linear function $f(x) = mx + b$, m is the _____ and $(0, b)$ is the _____.

2. The graph of a linear function is called a ___.

3. *True or False* All linear equations are functions.

4. For the linear function $G(x) = -2x + 3$, the slope is ___ and the y-intercept is ____.

In Problems 5–8, graph each linear function.

5. $f(x) = 2x - 3$ 6. $G(x) = -5x + 4$ 7. $h(x) = \dfrac{3}{2}x + 1$ 8. $f(x) = 4$

▶ ➋ **Find the Zero of a Linear Function**

In Section 2.3 we stated that if r is a zero of a function f, then $f(r) = 0$. To find the zero of any function f, we solve the equation $f(x) = 0$.

(EXAMPLE 2) **Finding the Zero of a Linear Function**

Find the zero of $f(x) = -4x + 12$.

Solution

We find the zero by solving $f(x) = 0$.

$$f(x) = 0$$
$$-4x + 12 = 0$$

Subtract 12 from both sides of the equation: $\qquad -4x = -12$

Divide both sides of the equation by -4: $\qquad x = 3$

Check: Since $f(3) = -4(3) + 12 = 0$, the zero of f is 3.

Quick ✓

In Problems 9–11, find the zero of each linear function.

9. $f(x) = 3x - 15$ **10.** $G(x) = \dfrac{1}{2}x + 4$ **11.** $F(p) = -\dfrac{2}{3}p + 8$

▶ **Applications of Linear Functions**

Linear functions have many applications. For example, the cost of cab fare, sales commissions, and the cost of breakfast as a function of the number of eggs ordered can be modeled by linear functions.

EXAMPLE 3 | **Sales Commissions**

Tony's weekly salary at Apple Chevrolet is 0.75% of his weekly sales plus $450, so $S(x) = 0.0075x + 450$ describes Tony's weekly salary S as a linear function of his weekly sales x.

   (a) What is the implied domain of the function?

   (b) If Tony sells cars worth a total of $50,000 one week, what is his salary?

   (c) If Tony earned $600 one week, what was the value of the cars that he sold?

   (d) Draw a graph of the function.

   (e) For what value of cars sold will Tony's weekly salary exceed $1200?

Solution

   (a) The independent variable is weekly sales, x. Because negative weekly sales do not make sense, the function's domain is $\{x \mid x \geq 0\}$ or, using interval notation, $[0, \infty)$.

   (b) If Tony's weekly sales are $x = \$50{,}000$, then he earns

$$S(50{,}000) = 0.0075(50{,}000) + 450$$
$$= \$825$$

Tony will earn $825 for the week, if he sells $50,000 worth of cars.

   (c) Here, we need to solve the equation $S(x) = 600$.

$$0.0075x + 450 = 600$$

Subtract 450 from both sides: $0.0075x = 150$

Divide both sides by 0.0075: $x = \$20{,}000$

If Tony earned $600, then he sold $20,000 worth of cars in a week.

   (d) Plot the independent variable, weekly sales, on the horizontal axis and the dependent variable, salary, on the vertical axis. Graph the equation by plotting points. Part (b) shows that (50,000, 825) is on the graph. Part (c) indicates that (20,000, 600) is on the graph. Why do we also know that (0, 450) is on the graph? Plot these points to get the graph in Figure 18.

Figure 18

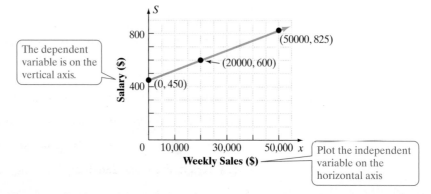

(e) Solve the inequality $S(x) > 1200$.

$$0.0075x + 450 > 1200$$

Subtract 450 from both sides: $\qquad 0.0075x > 750$

Divide both sides of the inequality by 0.0075: $\qquad x > 100,000$

If Tony sells more than $100,000 worth of cars for the week, his salary will exceed $1200. ●

Notice in Figure 18 that we graph the function only over its domain, $[0, \infty)$—that is, only in quadrant I. Also notice that we labeled the horizontal axis x for the independent variable, weekly sales, and the vertical axis S for the dependent variable, salary. For this reason, the intercept on the vertical axis is the S-intercept, not the y-intercept. We also indicated on the axes what x and S represent. Labeling your axes is always a good practice.

Quick ✓

12. The cost, C, of renting a 12-foot moving truck for a day is $40 plus $0.35 times the number of miles driven. The linear function $C(x) = 0.35x + 40$ describes the cost C of driving the truck x miles.

(a) What is the domain of this linear function?

(b) Determine the C-intercept of the graph of the linear function.

(c) What is the rental cost if the truck is driven 80 miles?

(d) How many miles was the truck driven if the rental cost is $85.50?

(e) Graph the linear function.

(f) How many miles can you drive if you can spend up to $127.50?

③ Build Linear Models from Verbal Descriptions

A linear function has the form $f(x) = mx + b$, where m is the slope of the linear function and $(0, b)$ is its y-intercept. In Section 1.6, we said that slope can be thought of as an average rate of change. Slope describes how much a dependent variable changes for a given change in the independent variable. For example, in the linear function $f(x) = 4x + 3$, the slope is $4 = \dfrac{4}{1} = \dfrac{\Delta y}{\Delta x}$, so the dependent variable y increases by 4 units for every 1-unit increase in x (the independent variable). When the average rate of change of a function is constant, we can use a linear function to model the situation. For example, if your phone company charges $0.05 per minute to talk regardless of the number of minutes on the phone, then we can use a linear function to model the cost of talking with slope $m = \dfrac{0.05 \text{ dollar}}{1 \text{ minute}}$.

▶ **EXAMPLE 4** **Cost Function**

In the linear cost function $C(x) = ax + b$, b represents the fixed costs of operating a business, and a represents the costs associated with manufacturing one additional item. Suppose that a bicycle manufacturer has daily fixed costs of $2000 and each bicycle costs $80 to manufacture.

(a) Write a linear function showing the cost to manufacture x bicycles in a day.

(b) What is the cost to manufacture 5 bicycles in a day?

(c) How many bicycles can be manufactured for $2800?

(d) Graph the linear function.

Figure 19

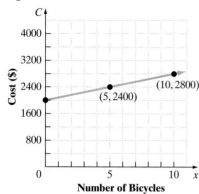

Number of Bicycles

Solution

(a) Because each bicycle costs $80 to manufacture, $a = 80$. The fixed costs are $2000, so $b = 2000$. Therefore, the cost function is

$$C(x) = 80x + 2000$$

(b) We evaluate the function for $x = 5$.

$$C(5) = 80(5) + 2000$$
$$= \$2400$$

It will cost $2400 to manufacture 5 bicycles.

(c) Solve $C(x) = 2800$.

$$C(x) = 2800$$
$$80x + 2000 = 2800$$

Subtract 2000 from both sides: $\quad 80x = 800$

Divide both sides by 80: $\quad x = 10$

Ten bicycles can be manufactured for $2800.

(d) Label the horizontal axis x and the vertical axis C. See Figure 19 for the graph of the cost function.

Quick ✓

13. Suppose the business presented in Example 4 must pay a tax of $1 per bicycle produced.

(a) Write a linear function that expresses the cost C of producing x bicycles in a day.

(b) What is the cost of manufacturing 5 bicycles in a day?

(c) How many bicycles can be made for $2810?

(d) Graph the linear function.

 EXAMPLE 5 **Straight-Line Depreciation**

The *book value* of an asset such as a building or piece of machinery is the value that the company uses to create its balance sheet. Some companies use *straight-line depreciation* so that the book value of the asset declines by a constant amount each year. The amount of the decline depends on the useful life that the company places on the asset. Suppose Pearson Publishing Company just purchased a new fleet of cars for its sales force at a cost of $29,400 per car. The company uses the straight-line depreciation method for 7 years.

(a) Write a linear function that expresses the book value V of each car as a function of its age, x.

(b) What is the domain of this linear function?

(c) What is the book value of each car after 3 years?

(d) When will the book value of each car be $12,600?

(e) Graph the linear function.

Solution

(a) We let the linear function $V(x) = mx + b$ represent the book value of each car after x years. The original value of the car is $29,400, so $V(0) = 29,400$. Thus the V-intercept of the function is $(0, 29,400)$. After 7 years, the book value of the car is $0. We use the ordered pairs $(0, 29,400)$ and $(7, 0)$ to find the slope, or amount of yearly (annual) depreciation, of V:

$$m = \frac{0 - 29,400}{7 - 0} = \frac{-29,400}{7} = -4200$$

So each car depreciates by $4200 per year. The linear function that represents the book value of each car after x years is

$$V(x) = -4200x + 29,400$$

(b) A car cannot have a negative age, so the age x must be greater than or equal to zero. In addition, each car is depreciated over 7 years, at which time $V(7) = 0$. Therefore, the domain of the function is $\{x \mid 0 \le x \le 7\}$, or $[0, 7]$ using interval notation.

(c) The book value of each car after $x = 3$ years is given by $V(3)$.

$$V(3) = -4200(3) + 29,400$$
$$= \$16,800$$

(d) To find when the book value is $12,600, solve the equation

$$V(x) = 12,600$$
$$-4200x + 29,400 = 12,600$$

Subtract 29,400 from both sides: $\qquad -4200x = -16,800$

Divide both sides by -4200: $\qquad x = 4$

Each car will have a book value of $12,600 in 4 years.

(e) Label the horizontal axis x and the vertical axis V. Since $V(0) = 29,400$, the point $(0, 29400)$ is on the graph. Since $V(7) = 0$, the point $(7, 0)$ is on the graph. To graph the function, we use these intercepts, along with the points $(3, 16800)$ and $(4, 12600)$ from parts (c) and (d). See Figure 20.

Figure 20

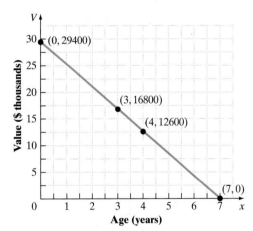

Quick ✓

14. Roberta's monthly payments for her new car are $250 per month. She estimates that maintenance and gas cost her $0.18 per mile.

(a) Write a linear function that expresses the monthly cost C of operating the car as a function of miles driven, x.

(b) What is the domain of this linear function?

(c) What is the monthly cost of driving 320 miles?

(d) How many miles can Roberta drive each month if she can afford the monthly cost to be $282.40?

(e) Graph the linear function.

▶ ❹ Build Linear Models from Data

If we have a set of data with more than two points, how can we tell whether the data (the two variables) are related linearly? There are some methods (beyond the scope of this course) for determining whether two variables are linearly related, but we can draw a picture of the data, a *scatter diagram*, and learn whether the variables *might* be linearly related.

Scatter Diagrams

The graph consisting of the ordered pairs that make up the relation in the Cartesian plane is called a **scatter diagram.**

(EXAMPLE 6) **Drawing a Scatter Diagram**

The on-base percentage for a baseball team is the percent of time that the team safely reaches base. Table 4 shows the number of runs scored and the on-base percentage for various teams in the 2011 season.

Table 4

Team	On-Base Percentage, x	Runs Scored, y	(x, y)
NY Yankees	34.3	867	(34.3, 867)
Los Angeles Angels	31.3	667	(31.3, 667)
Texas Rangers	34.0	855	(34.0, 855)
Toronto Blue Jays	31.7	743	(31.7, 743)
Minnesota Twins	30.6	619	(30.6, 619)
Oakland A's	31.1	645	(31.1, 645)
Kansas City Royals	32.9	730	(32.9, 730)
Baltimore Orioles	31.6	708	(31.6, 708)

SOURCE: *espn.com*

(a) Draw a scatter diagram of the data, treating on-base percentage as the independent variable.

(b) Describe what happens as the on-base percentage increases.

Figure 21

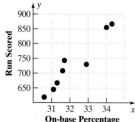

On-base Percentage

Solution

(a) To draw a scatter diagram, plot the ordered pairs listed in Table 4. See Figure 21.

(b) The scatter diagram reveals that as the on-base percentage increases, the number of runs scored also increases. While this relation is not perfectly linear (because the points don't all fall on a straight line), the *pattern* of the data is linear. ●

Quick ✓

15. The data listed below represent the total cholesterol (in mg/dL) and age of males.

Age, x	Total Cholesterol, y	Age, x	Total Cholesterol, y
25	180	38	239
25	195	48	204
28	186	51	243
32	180	62	228
32	197	65	269

(a) Draw a scatter diagram, treating age as the independent variable.

(b) Describe the relation between age and total cholesterol.

Recognizing the Type of Relation Between Two Variables

Scatter diagrams help us see the type of relation between two variables. In this section, we will focus distinguishing linear relations from nonlinear relations. See Figure 22.

Figure 22

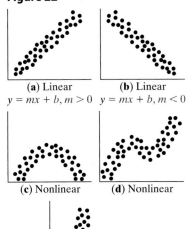

(a) Linear
$y = mx + b, m > 0$

(b) Linear
$y = mx + b, m < 0$

(c) Nonlinear

(d) Nonlinear

(e) Nonlinear

EXAMPLE 7 **Distinguishing Between Linear and Nonlinear Relations**

Determine whether each relation in Figure 23 is linear or nonlinear. If it is linear, state whether the slope is positive or negative.

Figure 23

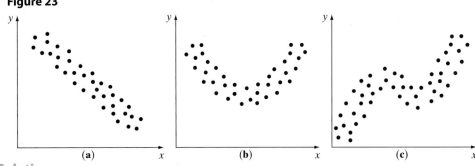

(a) (b) (c)

Solution

(a) Linear with negative slope   **(b)** Nonlinear   **(c)** Nonlinear   ●

Quick ✓

In Problems 16 and 17, determine whether each relation is linear or nonlinear. If it is linear, state whether the slope is positive or negative.

16.

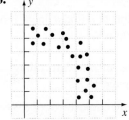

17.

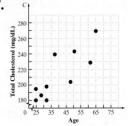

Fitting a Line to Data

Suppose a scatter diagram indicates a linear relationship, as in Figure 22(a) or (b). To find a linear equation for such data, we draw a line through two points on the scatter diagram. Then we determine the equation of that line by using the point-slope form of a line, $y - y_1 = m(x - x_1)$. To review using this formula, work Problem R6 in "*Are You Ready for This Section?*" on page 174, if you haven't already done so.

EXAMPLE 8 **Finding a Model for Linearly Related Data**

Using the data in Table 4 from Example 6,

(a) Select two points and find a linear function for the line connecting the points.

(b) Graph the line on the scatter diagram from Example 6(a).

(c) Use the function found in part (a) to predict the number of runs scored by a team whose on-base percentage is 33.5%.

(d) Interpret the slope. Does it make sense to interpret the *y*-intercept?

Solution

(a) Select two points—for example, (31.1, 645) and (34.3, 867). (You should select your own two points and work through the solution.) The slope of the line joining these points is

$$m = \frac{867 - 645}{34.3 - 31.1} = \frac{222}{3.2} = 69.375$$

Use the point-slope form to find the equation of the line that has slope 69.375 passing through (31.1, 645):

Point-slope form:	$y - y_1 = m(x - x_1)$
$m = 69.375; x_1 = 31.1, y_1 = 645:$	$y - 645 = 69.375(x - 31.1)$
Distribute 69.375:	$y - 645 = 69.375x - 2157.5625$
Add 645 to both sides:	$y = f(x) = 69.375x - 1512.5625$

Figure 24

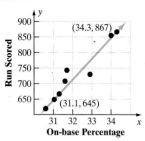

(b) Figure 24 shows the scatter diagram with the graph of the line found in part (a). We graphed the line through the two points selected in part (a).

(c) Evaluate $f(x) = 69.375x - 1512.5625$ at $x = 33.5$.

$$f(33.5) = 69.375(33.5) - 1512.5625$$
$$= 811.5$$

Round this to the nearest whole number. We predict that a team whose on-base percentage is 33.5% will score 812 runs.

(d) The slope of the linear function is 69.375, so if the on-base percentage increases by 1, then the number of runs scored will increase by about 69. The y-intercept, -1512.5625, represents the runs scored when the on-base percentage is 0. Since negative runs scored does not make sense and we have no observations near zero, interpreting the y-intercept does not make sense. ●

Quick ✓

18. Using the data from Quick Check Problem 15 on page 180:

(a) Select two points and find a linear model that describes the relation between the points.

(b) Graph the line on the scatter diagram obtained in Quick Check Problem 15 (page 180).

(c) Predict the total cholesterol of a 39-year-old male.

(d) Interpret the slope. Does it make sense to interpret the y-intercept?

2.4 Exercises MyMathLab®

PRACTICE

Exercise numbers in green have complete video solutions in MyMathLab.

Problems **1–18.** are the Quick ✓ s that follow the EXAMPLES.

Building Skills

For Problems 19–30, graph each linear function.
See Objective 1.

19. $F(x) = 5x - 2$ **20.** $F(x) = 4x + 1$

21. $G(x) = -3x + 7$ **22.** $G(x) = -2x + 5$

23. $H(x) = -2$ **24.** $P(x) = 5$

25. $f(x) = \frac{1}{2}x - 4$ **26.** $f(x) = \frac{1}{3}x - 3$

27. $F(x) = -\frac{5}{2}x + 5$ **28.** $P(x) = -\frac{3}{5}x - 1$

29. $G(x) = -\frac{3}{2}x$ **30.** $f(x) = \frac{4}{5}x$

In Problems 31–38, find the zero of the linear function.
See Objective 2.

31. $f(x) = 2x + 10$ **32.** $f(x) = 3x + 18$

33. $G(x) = -5x + 40$ **34.** $H(x) = -4x + 36$

35. $s(t) = \frac{1}{2}t - 3$ **36.** $p(q) = \frac{1}{4}q + 2$

37. $P(z) = -\frac{4}{3}z + 12$ **38.** $F(t) = -\frac{3}{2}t + 6$

In Problems 39–42, determine whether the scatter diagram indicates that a linear relation may exist between the two variables. If a linear relation does exist, indicate whether the slope is positive or negative. See Objective 4.

39.

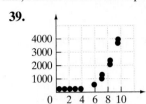

40.

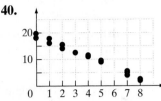

41.

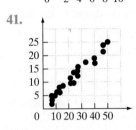

42.

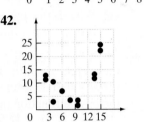

In Problems 43–46,

(a) *Draw a scatter diagram of the data.*

(b) *Select two points from the scatter diagram and find the equation of the line containing the points selected.**

(c) *Graph the line found in part (b) on the scatter diagram. See Objective 4.*

43.

x	2	4	5	8	9
y	1.4	1.8	2.1	2.3	2.6

44.

x	2	3	5	6	7
y	5.7	5.2	2.8	1.9	1.8

45.

x	1.2	1.8	2.3	3.5	4.1
y	8.4	7.0	7.3	4.5	2.4

46.

x	0	0.5	1.4	2.1	3.9
y	0.8	1.3	1.9	2.5	5.0

Mixed Practice

47. Suppose that $f(x) = 3x + 2$.

(a) What is the slope?

(b) What is the y-intercept?

(c) What is the zero of f?

(d) Solve $f(x) = 5$. What point is on the graph of f?

(e) Solve $f(x) \leq -1$.

(f) Graph f.

48. Suppose that $g(x) = 8x + 3$.

(a) What is the slope?

(b) What is the y-intercept?

(c) What is the zero of g?

(d) Solve $g(x) = 19$. What point is on the graph of g?

(e) Solve $g(x) > -5$.

(f) Graph g.

*Answers will vary.

49. Suppose that $f(x) = x - 5$ and $g(x) = -3x + 7$.

(a) Solve $f(x) = g(x)$. What is the value of f at the solution? What point is on the graph of f? What point is on the graph of g?

(b) Solve $f(x) > g(x)$.

(c) Graph f and g in the same Cartesian plane. Label the intersection point.

50. Suppose that $f(x) = \dfrac{4}{3}x + 5$ and $g(x) = \dfrac{1}{3}x + 1$.

(a) Solve $f(x) = g(x)$. What is the value of f at the solution? What point is on the graph of f? What point is on the graph of g?

(b) Solve $f(x) \leq g(x)$.

(c) Graph f and g in the same Cartesian plane. Label the intersection point.

51. Find a linear function f such that $f(2) = 6$ and $f(5) = 12$. What is $f(-2)$?

52. Find a linear function g such that $g(1) = 5$ and $g(5) = 17$. What is $g(-3)$?

53. Find a linear function h such that $h(3) = 7$ and $h(-1) = 14$. What is $h\left(\dfrac{1}{2}\right)$?

54. Find a linear function F such that $F(2) = 5$ and $F(-3) = 9$. What is $F\left(-\dfrac{3}{2}\right)$?

55. In parts (a)–(e), use the figure shown below.

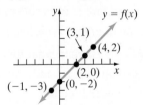

(a) Solve $f(x) = 1$.   (b) Solve $f(x) = -3$.

(c) What is $f(4)$?

(d) What are the intercepts of the function $y = f(x)$?

(e) Write the equation of the function whose graph is given in the form $f(x) = mx + b$.

56. In parts (a)–(e), use the figure shown below.

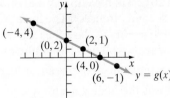

(a) Solve $g(x) = 1$.

(b) Solve $g(x) = -1$.

(c) What is $g(-4)$?

(d) What are the intercepts of the function $y = g(x)$?

(e) Write the equation of the function whose graph is given in the form $g(x) = mx + b$.

Applying the Concepts

57. Taxes The function $T(x) = 0.15(x - 10,850) + 870$ represents the federal income tax bill T of a single person whose adjusted gross income in 2011 was x dollars for income between $10,850 and $37,500, inclusive. SOURCE: *Internal Revenue Service*

  (a) What is the domain of this linear function?

  (b) What was a single filer's tax bill if adjusted gross income was $20,000?

  (c) Which variable is independent and which is dependent?

  (d) Graph the linear function over the domain specified in part (a).

  (e) What was a single filer's adjusted gross income if his or her tax bill was $2996.25?

58. Sales Commissions Tanya works for Pearson Education as a book representative. The linear function $I(s) = 0.01s + 20,000$ describes the annual income I of Tanya when she has total sales s.

  (a) What is the domain of this linear function?

  (b) What is $I(0)$? Explain what this result means.

  (c) What is Tanya's salary if she sells $500,000 in books for the year?

  (d) Graph the linear function.

  (e) At what level of sales will Tanya's income be $45,000?

59. Cab Fare The linear function $C(m) = 1.5m + 2$ describes the cab fare C for a ride of m miles.

  (a) What is the domain of this linear function?

  (b) What is $C(0)$? Explain what this result means.

  (c) What is cab fare for a 5-mile ride?

  (d) Graph the linear function.

  (e) How many miles can you ride in a cab if you have $13.25?

  (f) Over what range of miles can you ride if you can spend no more than $39.50?

60. Luxury Tax In 2002, Major League Baseball signed a labor agreement with the players. In this agreement, any team whose payroll exceeds $178 million in 2012 will have to pay a luxury tax of 40% (for second offenses). The linear function $T(p) = 0.4(p - 178)$ describes the luxury tax T of a team whose payroll was p (in millions).

  (a) What is the domain of this linear function?

  (b) What was the luxury tax for a team whose payroll was $190 million?

  (c) Graph the linear function.

  (d) What was the payroll of a team that paid a luxury tax of $12.8 million?

61. Health Costs The annual cost of health insurance H as a function of age a is given by the function

$H(a) = 22.8a - 117.5$ for $15 \le a \le 90$.

SOURCE: *Statistical Abstract*

  (a) What are the independent and dependent variables?

  (b) What is the domain of this linear function?

  (c) What is the health insurance premium of a 30-year-old?

  (d) Graph the linear function over its domain.

  (e) What is the age of an individual whose health insurance premium is $976.90?

62. Birth Rate A multiple birth is any birth with 2 or more children born. The birth rate is the number of births per 1000 women. The birth rate B of multiple births as a function of age a is given by the function $B(a) = 1.73a - 14.56$ for $15 \le a \le 44$.

SOURCE: *Centers for Disease Control*

  (a) What are the independent and dependent variables?

  (b) What is the domain of this linear function?

  (c) What is the multiple birth rate of women who are 22 years of age, according to the model?

  (d) Graph the linear function over its domain.

  (e) What is the age of women whose multiple birth rate is 49.45?

63. Phone Charges Sprint has a long-distance phone plan that charges a monthly fee of $5.95 plus $0.05 per minute. SOURCE: *Sprint.com*

  (a) Find a linear function that expresses the monthly bill B as a function of minutes used m.

  (b) What are the independent and dependent variables?

  (c) What is the domain of this linear function?

  (d) What is the monthly bill if 300 minutes are used for long-distance phone calls?

  (e) How many minutes were used for long distance if the long-distance phone bill was $17.95?

  (f) Graph the linear function.

  (g) Over what range of minutes can you talk each month if you don't want to spend more than $18.45?

64. RV Rental The weekly rental cost R of a class C 20-foot recreational vehicle is $129 plus $0.32 per mile, up to a maximum of 500 miles. SOURCE: *westernrv.com*

  (a) Find a linear function that expresses the cost R as a function of miles driven m.

  (b) What are the independent and dependent variables?

  (c) What is the domain of this linear function?

  (d) What is the rental cost if 360 miles are driven?

  (e) How many miles were driven if the rental cost is $275.56?

(f) Graph the linear function.

(g) Over what range of miles can you drive if you have a budget of $273?

65. Depreciation Suppose that a company has just purchased a new computer for $2700. The company chooses to depreciate the computer using the straight-line method over 3 years.

(a) Find a linear function that expresses the book value V of the computer as a function of its age x.

(b) What is the domain of this linear function?

(c) What is the book value of the computer after the first year?

(d) What are the intercepts of the graph of the linear function?

(e) When will the book value of the computer be $900?

(f) Graph the linear function.

66. Depreciation Suppose that a company just purchased a new machine for its manufacturing facility for $1,200,000. The company chooses to depreciate the machine using the straight-line method over 20 years.

(a) Find a linear function that expresses the book value V of the machine as a function of its age x.

(b) What is the domain of this linear function?

(c) What is the book value of the machine after three years?

(d) What are the intercepts of the graph of the linear function?

(e) When will the book value of the machine be $480,000?

(f) Graph the linear function.

67. Diamonds The relation between the cost of a diamond and its weight is linear. In looking at two diamonds, we find that one of the diamonds weighs 0.7 carat and costs $3543, while the other diamond weighs 0.8 carat and costs $4378. source: *diamonds.com*

(a) Find a linear function that relates the price of a diamond, C, to its weight, x, treating weight as the independent variable.

(b) Predict the price of a diamond that weighs 0.77 carat.

(c) Interpret the slope.

(d) If a diamond costs $5300, what do you think it should weigh?

68. Apartments In the North Chicago area, an 820-square-foot apartment rents for $1507 per month. A 970-square-foot apartment rents for $1660. Suppose that the relation between area and rent is linear. source: *apartments.com*

(a) Find a linear function that relates the rent of a North Chicago apartment, R, to its area, x, treating area as the independent variable.

(b) Predict the rent of a 900-square-foot apartment in North Chicago.

(c) Interpret the slope.

(d) If the rent of a North Chicago apartment is $1300 per month, how big would you expect it to be?

69. The Consumption Function A famous theory in economics developed by John Maynard Keynes states that personal consumption expenditures are a linear function of disposable income. An economist wishes to develop a model that relates income and consumption and obtains the following information from the United States Bureau of Economic Analysis. In 2006, personal disposable income was $9916 billion and personal consumption expenditures were $9323 billion. In 2010, personal disposable income was $11,380 billion and personal consumption expenditures were $10,349 billion.

(a) Find a linear function that relates personal consumption expenditures, C, to disposable income, x, treating disposable income as the independent variable.

(b) In 2007, personal disposable income was $9742 billion. Use this information to find personal consumption expenditures in 2007.

(c) Interpret the slope. In economics, this slope is called the **marginal propensity to consume.**

(d) If personal consumption expenditures were $9520 billion, what do you think that disposable income was?

70. Birth Weight According to the National Center for Health Statistics, the average birth weight of babies born to 22-year-old mothers is 3280 grams. The average birth weight of babies born to 32-year-old mothers is 3370 grams. Suppose that the relation between age of mother and birth weight is linear.

(a) Find a linear function that relates age of mother a to birth weight W, treating age of mother as the independent variable.

(b) Predict the birth weight of a baby born to a mother who is 30 years old.

(c) Interpret the slope.

(d) If a baby weighs 3310 grams, how old do you expect the mother to be?

71. Concrete As concrete cures, it gains strength. The data on the following page represent the 7-day and 28-day strength (in pounds per square inch) of a certain type of concrete.

7-day Strength, x	28-day Strength, y
2300	4070
3390	5220
2430	4640
2890	4620
3330	4850
2480	4120
3380	5020
2660	4890
2620	4190
3340	4630

(a) Draw a scatter diagram of the data, treating 7-day strength as the independent variable.

(b) What type of relation appears to exist between 7-day strength and 28-day strength?

(c) Select two points and find an equation of the line containing the points.

(d) Graph the line on the scatter diagram drawn in part (a).

(e) Predict the 28-day strength of a slab of concrete if its 7-day strength is 3000 psi.

(f) Interpret the slope of the line found in part (c).

72. Candy The following data represent the weight (in grams) of various candy bars and the corresponding number of calories.

Candy Bar	Weight, x	Calories, y
Hershey's Milk Chocolate	44.28	230
Nestle Crunch	44.84	230
Butterfinger	61.30	270
Baby Ruth	66.45	280
Almond Joy	47.33	220
Twix (with Caramel)	58.00	280
Snickers	61.12	280
Heath	39.52	210

SOURCE: *Megan Pocius, student at Joliet Junior College*

(a) Draw a scatter diagram of the data, treating weight as the independent variable.

(b) What type of relation appears to exist between the weight of a candy bar and the number of calories?

(c) Select two points and find an equation of the line containing the points.

(d) Graph the line on the scatter diagram drawn in part (a).

(e) Predict the number of calories in a candy bar that weighs 62.3 grams.

(f) Interpret the slope of the line found in part (c).

73. Raisins The following data represent the weight (in grams) of a box of raisins and the number of raisins in the box.

Weight, w	Number of Raisins, N
42.3	87
42.7	91
42.8	93
42.4	87
42.6	89
42.4	90
42.3	82
42.5	86
42.7	86
42.5	86

SOURCE: *Jennifer Maxwell, student at Joliet Junior College*

(a) Does the relation defined by the set of ordered pairs (w, N) represent a function?

(b) Draw a scatter diagram of the data, treating weight as the independent variable.

(c) Select two points and find the equation of the line containing the points.

(d) Graph the line on the scatter diagram drawn in part (b).

(e) Express the relationship found in part (c) using function notation.

(f) Predict the number of raisins in a box that weighs 42.5 grams.

(g) Interpret the slope of the line found in part (c).

74. Height versus Head Circumference The following data represent the height (in inches) and head circumference (in inches) of 9 randomly selected children.

Height, h	Head Circumference, C
25.25	16.4
25.75	16.9
25	16.9
27.75	17.6
26.50	17.3
27.00	17.5
26.75	17.3
26.75	17.5
27.5	17.5

SOURCE: *Denise Slucki, student at Joliet Junior College*

(a) Does the relation defined by the set of ordered pairs (h, C) represent a function?

(b) Draw a scatter diagram of the data, treating height as the independent variable.

(c) Select two points and find the equation of the line containing the points.

(d) Graph the line on the scatter diagram drawn in part (b).

(e) Express the relationship found in part (c) using function notation.

(f) Predict the head circumference of a child who is 26.5 inches tall.

(g) Interpret the slope of the line found in part (c).

Extending the Concepts

75. Math for the Future: Calculus The **average rate of change** of a function $y = f(x)$ from c to x is defined as

$$\text{Average rate of change} = \frac{\Delta y}{\Delta x} = \frac{f(x) - f(c)}{x - c}, \quad x \neq c$$

provided that c is in the domain of f. The average rate of change of a function is simply the slope of the line joining the points $(c, f(c))$ and $(x, f(x))$. We call the line joining these points a **secant line.** The slope of the secant line is

$$m_{\text{sec}} = \frac{f(x) - f(c)}{x - c}$$

The following figure illustrates the idea.

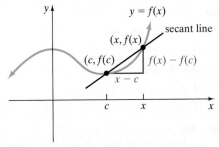

Below, we show the graph of the function $f(x) = 2x^2 - 4x + 1$.

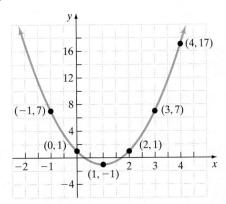

(a) On the graph of $f(x) = 2x^2 - 4x + 1$, draw a line through the points $(1, f(1))$ and $(x, f(x))$, where $x = 4$.

(b) Find the slope of the secant line through $(1, f(1))$ and $(x, f(x))$, where $x = 4$.

(c) Find the equation of the secant line through $(1, f(1))$ and $(x, f(x))$, where $x = 4$

(d) Repeat parts (a)–(c) for $x = 3, x = 2, x = 1.5$, and $x = 1.1$.

(e) What happens to the slope of the secant line as x gets closer to 1?

76. A strain of *E. coli* Beu 397-recA441 is placed into a Petri dish at 30° Celsius and allowed to grow. The population is estimated by means of an optical device in which the amount of light that passes through the Petri dish is measured. The data below are collected. Do you think that a linear function could be used to describe the relation between the two variables? Why or why not?

Time, x	Population, y
0	0.09
2.5	0.18
3.5	0.26
4.5	0.35
6	0.50

SOURCE: *Dr. Polly Lavery, Joliet Junior College*

The Graphing Calculator

The equation of the line obtained in Example 8 depends on the points selected, which will vary from person to person. So the line we found might be different from the line that you found. Although the line that we found in Example 8 fits the data well, there may be a line that "fits better." Do you think that your line fits the data better? Is there a line of *best fit*? As it turns out, there is a method for finding the line that best fits linearly related data (called the *line of best fit*).[*]

Graphing utilities can be used to draw scatter diagrams and find the line of best fit. Figure 25(a) shows a scatter diagram of the data presented in Table 4 from Example 6, drawn on a TI-84 Plus graphing calculator. Figure 25(b) shows the line of best fit from a TI-84 Plus graphing calculator.

The line of best fit is $y = 63.556x - 1316.466$.

Figure 25

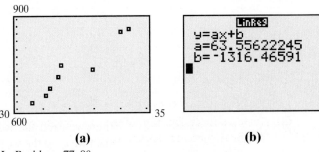

(a)   (b)

In Problems 77–80,

(a) *Draw a scatter diagram using a graphing calculator.*

(b) *Find the line of best fit using a graphing calculator for the data in the problem specified.*

77. Problem 71   **78.** Problem 72

79. Problem 73   **80.** Problem 74

---

[*]We shall not discuss in this book the underlying mathematics of lines of best fit. Books on elementary statistics discuss this topic.

2.5 Compound Inequalities

Objectives

1 Determine the Intersection or Union of Two Sets

2 Solve Compound Inequalities Involving "and"

3 Solve Compound Inequalities Involving "or"

4 Solve Problems Using Compound Inequalities

Are You Ready for This Section?

Before getting started, take the following readiness quiz. If you get a problem wrong, go back to the section cited and review the material.

R1. Use set-builder notation and interval notation to name the set of all real numbers x such that $-2 \le x \le 5$. [Section 1.4, pp. 81–84]

R2. Graph the inequality $x \ge 4$. [Section 1.4, pp. 81–84]

R3. Use interval notation to express the inequality shown in the graph. $\underset{-2\ -1\ \ 0\ \ 1\ \ 2\ \ 3\ \ 4}{\longmapsto}$ [Section 1.4, pp. 81–84]

R4. Solve: $2(x + 3) - 5x = 15$ [Section 1.1, pp. 50–53]

R5. Solve: $2x + 3 > 11$ [Section 1.4, pp. 85–87]

R6. Solve: $x + 8 \ge 4(x - 1) - x$ [Section 1.4, pp. 85–87]

▶ 1 Determine the Intersection or Union of Two Sets

Table 5 contains information about students in an Intermediate Algebra course. We can classify these people in various sets. For example, we can define set A as the set of all students whose age is less than 25. Then

$$A = \{\text{Grace, Sophia, Kevin, Jack, George, Teresa}\}$$

If we define set B as the set of all students who are female, then

$$B = \{\text{Grace, Sophia, Mary, Nancy, Teresa}\}$$

Now list all the students who are in set A and set B, that is, students who are under 25 years of age and female.

$$A \text{ and } B = \{\text{Grace, Sophia, Teresa}\}$$

Now list all the students who are in set A or in set B or in both sets.

$$A \text{ or } B = \{\text{Grace, Sophia, Kevin, Jack, George, Teresa, Mary, Nancy}\}$$

Figure 26 shows a Venn diagram illustrating the relations among A, B, A and B, and A or B. Notice that Grace, Sophia, and Teresa are in both A and B, while Robert is in neither A nor B.

Table 5

Student	Age	Gender
Grace	19	Female
Sophia	23	Female
Kevin	20	Male
Robert	32	Male
Jack	19	Male
Mary	35	Female
Nancy	40	Female
George	22	Male
Teresa	20	Female

Figure 26

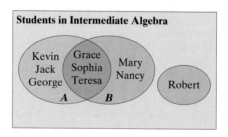

When we used the word "and," we listed elements common to both set A and set B. When we used the word "or," we listed elements in set A or in set B or in both sets. These results lead us to the following definitions.

Definitions

- The **intersection** of two sets A and B, denoted $A \cap B$, is the set of all elements in both set A and set B.
- The **union** of two sets A and B, denoted $A \cup B$, is the set of all elements in set A or in set B or in both set A and set B.
- The word **and** implies intersection, while the word **or** implies union.

Ready?...Answers

R1. $\{x \mid -2 \le x \le 5\}$; $[-2, 5]$

R2. $\underset{0\ \ 1\ \ 2\ \ 3\ \ 4\ \ 5\ \ 6}{\longmapsto}$

R3. $(-1, 3]$ **R4.** $\{-3\}$

R5. $\{x \mid x > 4\}$; $(4, \infty)$;
$\underset{2\ \ 3\ \ 4\ \ 5\ \ 6\ \ 7}{\longmapsto}$

R6. $\{x \mid x \le 6\}$; $(-\infty, 6]$;
$\underset{2\ \ 3\ \ 4\ \ 5\ \ 6\ \ 7}{\longmapsto}$

EXAMPLE 1 Finding the Intersection and Union of Sets

Let $A = \{1, 3, 5, 7, 9\}$ and let $B = \{1, 2, 3, 4, 5\}$. Find

(a) $A \cap B$ (b) $A \cup B$

Solution

(a) $A \cap B$ is the set of all elements that are in both A and B. So,

$$A \cap B = \{1, 3, 5\}$$

Work Smart

When finding the union of two sets, we list each element only once, even if it occurs in both sets.

(b) $A \cup B$ is the set of all elements that are in A or B, or both. So,

$$A \cup B = \{1, 2, 3, 4, 5, 7, 9\}$$

Quick ✓

1. The _____ of two sets A and B, denoted $A \cap B$, is the set of all elements that belong to both set A and set B.

2. The word ___ implies intersection. The word __ implies union.

3. *True or False* The intersection of two sets can be the empty set.

4. *True or False* The symbol for the union of two sets is $\cap$.

In Problems 5–10, let $A = \{1, 2, 3, 4, 5, 6\}$, $B = \{1, 3, 5, 7\}$, and $C = \{2, 4, 6, 8\}$.

5. Find $A \cap B$.   6. Find $A \cap C$.

7. Find $A \cup B$.   8. Find $A \cup C$.

9. Find $B \cap C$.   10. Find $B \cup C$.

EXAMPLE 2 Finding the Intersection and Union of Two Sets

Suppose $A = \{x \mid x \leq 5\}$, $B = \{x \mid x \geq 1\}$, and $C = \{x \mid x < -2\}$.

(a) Determine $A \cap B$. Graph the set on a real number line and write it in set-builder notation and interval notation.

(b) Determine $B \cup C$. Graph the set on a real number line and write it in set-builder notation and interval notation.

Solution

(a) $A \cap B$ is the set of all real numbers less than or equal to 5 and greater than or equal to 1. We can identify this set by determining where the graphs of the inequalities overlap. See Figure 27.

Figure 27

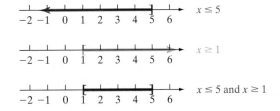

$A \cap B$ is $\{x \mid 1 \leq x \leq 5\}$ in set-builder notation, and $[1, 5]$ in interval notation.

(b) $B \cup C$ is the set of all real numbers greater than or equal to 1 or less than -2. See Figure 28. Therefore, $B \cup C$ is $\{x \mid x < -2 \text{ or } x \geq 1\}$ in set-builder notation, and $(-\infty, -2) \cup [1, \infty)$ in interval notation.

Figure 28

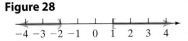

Work Smart

Throughout the text when a solution is written in set-builder notation, we use the word "or". When a solution is written in interval notation, we use the union symbol, U.

Quick ✓

Let $A = \{x | x > 2\}, B = \{x | x < 7\}, and\ C = \{x | x \leq -3\}.$

11. Determine $A \cap B$. Graph the set on a real number line and write it in set-builder notation and interval notation.

12. Determine $A \cup C$. Graph the set on a real number line and write it in set-builder notation and interval notation.

▶ ❷ Solve Compound Inequalities Involving "and"

A **compound inequality** is formed by joining two inequalities with the word "and" or "or." For example,

$$3x + 1 > 4 \quad \text{and} \quad 2x - 3 < 7$$
$$5x - 2 \leq 13 \quad \text{or} \quad 2x - 5 > 3$$

are of compound inequalities. To **solve a compound inequality** means to find all possible values of the variable such that the compound inequality results in a true statement. For example, the compound inequality

$$3x + 1 > 4 \quad \text{and} \quad 2x - 3 < 7$$

is true for $x = 2$ but false for $x = 0$.

Let's look at an example that illustrates how to solve compound inequalities involving the word "and."

(**EXAMPLE 3**)   **How to Solve a Compound Inequality Involving "and"**

Solve $3x + 2 > -7$ and $4x + 1 \leq 9$. Express your solution using set-builder and interval notation. Graph the solution set.

Step-by-Step Solution

Step 1: Solve each inequality separately.

	$3x + 2 > -7$		$4x + 1 \leq 9$
Subtract 2 from both sides:	$3x > -9$	Subtract 1 from both sides:	$4x \leq 8$
Divide both sides by 3:	$x > -3$	Divide both sides by 4:	$x \leq 2$

Step 2: Find the intersection of the solution sets, which will represent the solution set to the compound inequality.

To find the intersection of the two solution sets, we graph each inequality separately. See Figures 29(a) and (b).

Figure 29

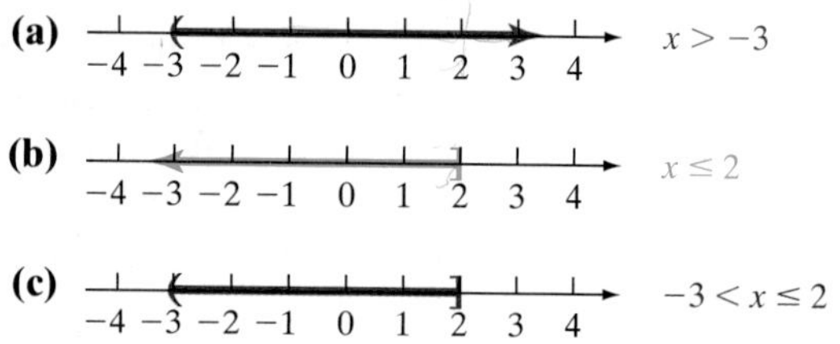

(a)

$x > -3$

(b)

$x \leq 2$

(c)

$-3 < x \leq 2$

The intersection of $x > -3$ and $x \leq 2$ is $-3 < x \leq 2$, so the solution set is $\{x | -3 < x \leq 2\}$ or, using interval notation, $(-3, 2]$. The graph of the solution set is shown in Figure 29(c).

The steps below summarize the procedure for solving compound inequalities involving "and."

Work Smart

The words "and" and "intersection" suggest overlap. When solving these types of problems, look for the overlap of the graphs.

Solving Compound Inequalities Involving "and"

Step 1: Solve each inequality separately.

Step 2: Find the INTERSECTION of the solution sets of the respective inequalities.

EXAMPLE 4   **Solving a Compound Inequality with "and"**

Solve $-2x + 5 > -1$ and $5x + 6 \le -4$. Express the solution using set-builder notation and interval notation. Graph the solution set.

Solution

Solve each inequality separately:

$$-2x + 5 > -1 \qquad\qquad\qquad 5x + 6 \le -4$$

Subtract 5 from both sides: $\quad -2x > -6 \quad$ Subtract 6 from both sides: $\quad 5x \le -10$

Divide both sides by -2 and

reverse the direction of the $\qquad\qquad x < 3 \qquad$ Divide both sides by 5: $\qquad x \le -2$

inequality!

The intersection of the solution sets is the solution set to the compound inequality. See Figures 30(a) and (b).

Figure 30   (a) ![number line with arrow x < 3 from -4 to 4] $x < 3$

(b) ![number line x ≤ -2] $x \le -2$

(c) ![number line x ≤ -2 and x < 3] $x \le -2$ and $x < 3$

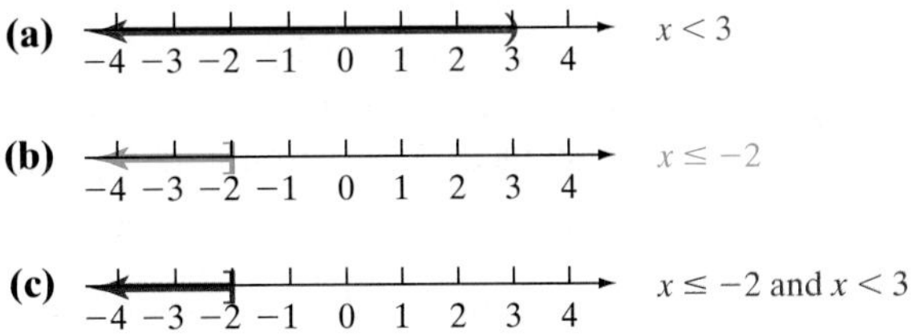

The intersection of $x < 3$ and $x \le -2$ is $x \le -2$. The solution set is $\{x \mid x \le -2\}$ or, using interval notation, $(-\infty, -2]$. The graph of the solution set is shown in Figure 30(c).

Quick ✓

In Problems 13–15, solve each compound inequality. Express the solution using set-builder notation and interval notation. Graph the solution set.

13. $2x + 1 \ge 5$ and $-3x + 2 < 5$

14. $4x - 5 < 7$ and $3x - 1 > -10$

15. $-8x + 3 < -5$ and $\dfrac{2}{3}x + 1 < 3$

EXAMPLE 5   **Solving a Compound Inequality with "and"**

Solve $x - 5 > -1$ and $2x - 3 \le -5$. Express the solution using set-builder and interval notation. Graph the solution set.

Solution

Solve each inequality separately:

$$x - 5 > -1 \qquad\qquad\qquad 2x - 3 \le -5$$

Add 5 to both sides: $\qquad\qquad x > 4 \qquad$ Add 3 to both sides: $\qquad 2x \le -2$

Divide both sides by 2: $\qquad x \le -1$

Figure 31

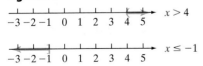

$x > 4$

$x \le -1$

The intersection of the solution sets is the solution set to the compound inequality. See Figure 31. The intersection is the empty set, so the solution set is $\{\ \}$ or $\varnothing$.

Quick ✓

In Problems 16 and 17, solve each compound inequality. Express the solution using set-builder notation and interval notation. Graph the solution set.

16. $3x - 5 < -8$ and $2x + 1 > 5$

17. $5x + 1 \le 6$ and $3x + 2 \ge 5$

Sometimes, we can combine "and" inequalities into a more streamlined notation.

> **Writing Inequalities Involving "and" Compactly**
>
> If $a < b$, then we can write
> $$a < x \quad \text{and} \quad x < b$$
> more compactly as
> $$a < x < b$$

For example, we can write
$$-3 < -4x + 1 \quad \text{and} \quad -4x + 1 < 13$$
as
$$-3 < -4x + 1 < 13$$

We solve such compound inequalities by isolating the variable "in the middle" with a coefficient of 1.

▶ **EXAMPLE 6** **Solving a Compound Inequality**

Solve $-3 < -4x + 1 < 13$. Express the solution using set-builder notation and interval notation. Graph the solution set.

Solution

Isolate the variable "in the middle" with a coefficient of 1:

$$-3 < -4x + 1 < 13$$

Subtract 1 from all three parts (Addition Property): $\quad -3 - 1 < -4x + 1 - 1 < 13 - 1$

$$-4 < -4x < 12$$

Divide all three parts by −4 and reverse the inequalities' direction. $\quad \dfrac{-4}{-4} > \dfrac{-4x}{-4} > \dfrac{12}{-4}$

$$1 > x > -3$$

If $b > x > a$, then $a < x < b$: $\quad -3 < x < 1$

The solution set is $\{x \mid -3 < x < 1\}$ or, in interval notation, $(-3, 1)$. Figure 32 shows the graph of the solution set.

Figure 33

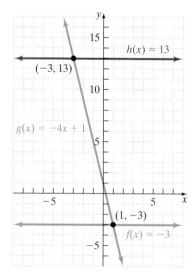

Figure 32

$$\begin{array}{c} \overset{|}{-5} \ \overset{|}{-4} \ \overset{(}{-3} \ \overset{|}{-2} \ \overset{|}{-1} \ \overset{|}{0} \ \overset{)}{1} \ \overset{|}{2} \ \overset{|}{3} \end{array}$$

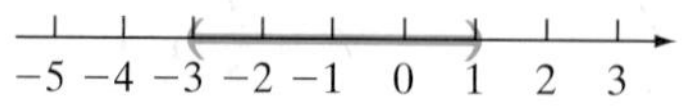

To visualize the results of Example 6, look at Figure 33, which shows the graph of $f(x) = -3$, $g(x) = -4x + 1$, and $h(x) = 13$. Note that the graph of $g(x) = -4x + 1$ is between the graphs of $f(x) = -3$ and $h(x) = 13$ for $-3 < x < 1$. Thus the solution set of $-3 < -4x + 1 < 13$ is $\{x \mid -3 < x < 1\}$.

Quick ✓

In Problems 18–20, solve each compound inequality. Express the solution using set-builder notation and interval notation. Graph the solution set.

18. $-2 < 3x + 1 < 10$

19. $0 < 4x - 5 \leq 3$

20. $3 \leq -2x - 1 \leq 11$

▶ ❸ **Solve Compound Inequalities Involving "or"**

The solution to compound inequalities involving the word "or" is the union of the solutions to each inequality.

EXAMPLE 7 How to Solve a Compound Inequality Involving "or"

Solve $3x - 5 < -2$ or $4 - 5x \leq -16$. Express the solution using set-builder notation and interval notation. Graph the solution set.

Step-by-Step Solution

Step 1: Solve each inequality separately.

$$3x - 5 < -2$$

Add 5 to each side: $\quad 3x < 3$

Divide both sides by 3: $\quad x < 1$

$$4 - 5x \leq -16$$

Subtract 4 from both sides: $\quad -5x \leq -20$

Divide both sides by -5;
Reverse the direction of the inequality: $\quad x \geq 4$

Step 2: Find the union of the solution sets, which will represent the solution set to the compound inequality.

The union of the two solution sets is $x < 1$ or $x \geq 4$. The solution set using set-builder notation is $\{x \mid x < 1 \text{ or } x \geq 4\}$. The solution set using interval notation is $(-\infty, 1) \cup [4, \infty)$. Figure 34 shows the graph of the solution set.

Figure 34

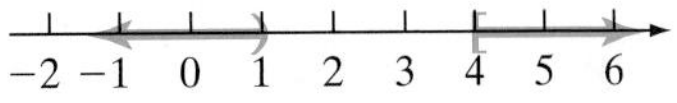

Steps for Solving Compound Inequalities Involving "or"

Step 1: Solve each inequality separately.

Step 2: Find the UNION of the solution sets of each inequality.

Quick ✓

In Problems 21–24, solve each compound inequality. Express the solution using set-builder notation and interval notation. Graph the solution set.

21. $x + 3 < 1$ or $x - 2 > 3$

22. $3x + 1 \leq 7$ or $2x - 3 > 9$

23. $2x - 3 \geq 1$ or $6x - 5 \geq 1$

24. $\dfrac{3}{4}(x + 4) < 6$ or $\dfrac{3}{2}(x + 1) > 15$

Work Smart

A common error is to write the solution $x < 1$ or $x > 4$ as $1 > x > 4$, which is incorrect. There are no real numbers that are less than 1 *and* greater than 4. Another common error is to "mix" symbols, as in $1 < x > 4$. This notation makes no sense!

EXAMPLE 8 Solving Compound Inequalities Involving "or"

Solve $\dfrac{1}{2}x - 1 < 1$ or $\dfrac{2x - 1}{3} \geq -1$. Express the solution using set-builder notation and interval notation. Graph the solution set.

Solution

Solve each inequality separately: $\quad \dfrac{1}{2}x - 1 < 1 \qquad\qquad \dfrac{2x - 1}{3} \geq -1$

Add 1 to each side: $\quad \dfrac{1}{2}x < 2 \quad$ Multiply both sides by 3: $\quad 2x - 1 \geq -3$

Multiply both sides by 2: $\quad x < 4 \qquad$ Add 1 to both sides: $\quad 2x \geq -2$

Divide both sides by 2: $\quad x \geq -1$

Find the union of these solution sets. The graphs in Figure 35 on the following page shows that the union of the two solution sets is the set of all real numbers.

Figure 35

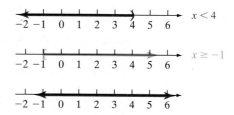

The solution set is $\{x \mid x \text{ is any real number}\}$ in set-builder notation, and $(-\infty, \infty)$ in interval notation.

Quick ✓

In Problems 25 and 26, solve each compound inequality. Express the solution using set-builder notation and interval notation. Graph the solution set.

25. $3x - 2 > -5$ or $2x - 5 \le 1$

26. $-5x - 2 \le 3$ or $7x - 9 > 5$

▶ ❹ Solve Problems Using Compound Inequalities

We now look at an application involving compound inequalities.

EXAMPLE 9   **Federal Income Taxes**

In 2012, married couples filing a joint federal tax return who were in the 25% tax bracket paid between $9735 and $27,735 in federal income taxes. These couples' federal income taxes equal $9735 plus 25% of their taxable income over $70,700. Find the range of taxable incomes into which a married couple must fall to have been in the 25% tax bracket. SOURCE: *Internal Revenue Service*

Solution

Step 1: Identify We want to find the range of taxable incomes for married couples in the 25% tax bracket. This direct translation problem involves an inequality.

Step 2: Name We let t represent the taxable income.

Step 3: Translate The federal tax bill equals $9735 plus 25% of the taxable income over $70,700. If the couple had taxable income equal to $71,700, their tax bill was $9735 plus 25% of $1000 ($1000 is the amount over $70,700). In general, if the couple has taxable income t, then their tax bill will be

$$\underbrace{\$9735}_{\$9735} \quad \underset{+}{\text{plus}} \quad \underbrace{25\%}_{0.25} \quad \underset{\cdot}{} \quad \underbrace{\text{of the amount over } \$70,700}_{(t - \$70,700)}$$

Because the tax bill was between $9735 and $27,735, we have

$$9735 \le 9735 + 0.25(t - 70,700) \le 27,735 \quad \text{The Model}$$

Step 4: Solve

$$9735 \le 9735 + 0.25(t - 70,700) \le 27,735$$

Distribute 0.25: $9735 \le 9735 + 0.25t - 17,675 \le 27,735$

Combine like terms: $9735 \le -7940 + 0.25t \le 27,735$

Add 7940 to all three parts: $17,675 \le 0.25t \le 35,675$

Divide all three parts by 0.25: $70,700 \le t \le 142,700$

Step 5: Check If a married couple had taxable income of $70,700, then their tax bill was $9735 + 0.25($70,700 - $70,700) = $9735. If a married couple had taxable income of $142,700, then their tax bill was $9735 + 0.25($142,700 - $70,700) = $27,735.

Step 6: Answer the Question A married couple who filed a federal joint income tax return with a tax bill between $9735 and $27,735 had taxable income between $70,700 and $142,700.

Work Smart

The word "range" tells us that an inequality is to be solved.

Quick ✓

27. In 2012, an individual filing a federal tax return whose income placed him or her in the 25% tax bracket paid federal income taxes between $4867.50 and $17,442.50. The individual had to pay federal income taxes equal to $4867.50 plus 25% of the amount over $35,350. Find the range of taxable income in order for an individual to have been in the 25% tax bracket. (SOURCE: *Internal Revenue Service*)

28. AT&T offers a long-distance phone plan that charges $2.00 per month plus $0.10 per minute. During the course of a year, Sophia's long-distance phone bill ranges from $6.50 to $26.50. What was the range of monthly minutes?

2.5 Exercises MyMathLab® Math XP PRACTICE

Exercise numbers in green have complete video solutions in MyMathLab.

Problems **1–28** *are the* Quick ✓ *s that follow the* EXAMPLES.

Building Skills

In Problems 29–34, use $A = \{4, 5, 6, 7, 8, 9\}$, $B = \{1, 5, 7, 9\}$, *and* $C = \{2, 3, 4, 6\}$ *to find each set. See Objective 1.*

29. $A \cup B$

30. $A \cup C$

31. $A \cap B$

32. $A \cap C$

33. $B \cap C$

34. $B \cup C$

In Problems 35–38, use the graph of the inequality to find each set. See Objective 1.

35. $A = \{x \mid x \le 5\}$; $B = \{x \mid x > -2\}$
Find (a) $A \cap B$ and (b) $A \cup B$.

```
 ─┼──┼──┼──┼──┼──┼──┼──┼──┼──┼─→
 -3 -2 -1  0  1  2  3  4  5  6
```

```
 ─┼──┼──┼──┼──┼──┼──┼──┼──┼──┼─→
 -3 -2 -1  0  1  2  3  4  5  6
```

36. $A = \{x \mid x \ge 4\}$; $B = \{x \mid x < 1\}$
Find (a) $A \cap B$ and (b) $A \cup B$.

```
 ─┼──┼──┼──┼──┼──┼──┼──┼─→
 -1  0  1  2  3  4  5  6
```

```
 ─┼──┼──┼──┼──┼──┼──┼──┼─→
 -1  0  1  2  3  4  5  6
```

37. $E = \{x \mid x > 3\}$; $F = \{x \mid x < -1\}$
Find (a) $E \cap F$ and (b) $E \cup F$.

```
 ─┼──┼──┼──┼──┼──┼──┼──┼─→
 -3 -2 -1  0  1  2  3  4
```

```
 ─┼──┼──┼──┼──┼──┼──┼──┼─→
 -3 -2 -1  0  1  2  3  4
```

38. $E = \{x \mid x \le 2\}$; $F = \{x \mid x \ge -2\}$
Find (a) $E \cap F$ and (b) $E \cup F$.

```
 ─┼──┼──┼──┼──┼──┼──┼─→
 -3 -2 -1  0  1  2  3
```

```
 ─┼──┼──┼──┼──┼──┼──┼─→
 -3 -2 -1  0  1  2  3
```

In Problems 39–42, use the graph to solve the compound inequality. Graph the solution set. See Objectives 2 and 3.

39. **(a)** $-5 \le 2x - 1 \le 3$

(b) $2x - 1 < -5$ or $2x - 1 > 3$

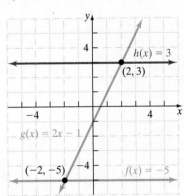

40. **(a)** $-1 \le \frac{1}{2}x + 1 \le 3$

(b) $\frac{1}{2}x + 1 < -1$ or $\frac{1}{2}x + 1 > 3$

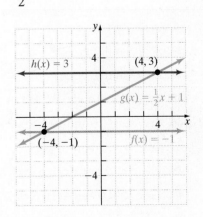

41. (a) $-4 < -\dfrac{5}{3}x + 1 < 6$

(b) $-\dfrac{5}{3}x + 1 \leq -4$ or $-\dfrac{5}{3}x + 1 \geq 6$

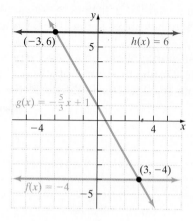

42. (a) $-3 < \dfrac{5}{4}x + 2 < 7$

(b) $\dfrac{5}{4}x + 2 \leq -3$ or $\dfrac{5}{4}x + 2 \geq 7$

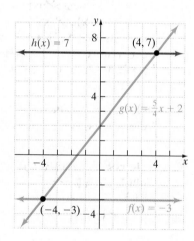

In Problems 43–66, solve each compound inequality. Graph the solution set. See Objective 2.

43. $x < 3$ and $x \geq -2$

44. $x \leq 5$ and $x > 0$

45. $4x - 4 < 0$ and $-5x + 1 \leq -9$

46. $6x - 2 \leq 10$ and $10x > -20$

47. $4x - 3 < 5$ and $-5x + 3 > 13$

48. $x - 3 \leq 2$ and $6x + 5 \geq -1$

49. $7x + 2 \geq 9$ and $4x + 3 \leq 7$

50. $-4x - 1 < 3$ and $-x - 2 > 3$

51. $-3 \leq 5x + 2 < 17$

52. $-10 < 6x + 8 \leq -4$

53. $-3 \leq 6x + 1 \leq 10$

54. $-12 < 7x + 2 \leq 6$

55. $3 \leq -5x + 7 < 12$

56. $-6 < -3x + 6 \leq 4$

57. $-1 \leq \dfrac{1}{2}x - 1 \leq 3$

58. $0 < \dfrac{3}{2}x - 3 \leq 3$

59. $3 \leq -2x - 1 \leq 11$

60. $-3 < -4x + 1 < 17$

61. $\dfrac{2}{3}x + \dfrac{1}{2} < \dfrac{5}{6}$ and $-\dfrac{1}{5}x + 1 < \dfrac{3}{10}$

62. $x - \dfrac{3}{2} \leq \dfrac{5}{4}$ and $-\dfrac{2}{3}x - \dfrac{2}{9} < \dfrac{8}{9}$

63. $-2 < \dfrac{3x + 1}{2} \leq 8$

64. $-4 \leq \dfrac{4x - 3}{3} < 3$

65. $-8 \leq -2(x + 1) < 6$

66. $-6 < -3(x - 2) < 15$

In Problems 67–80, solve each compound inequality. Graph the solution set. See Objective 3.

67. $x < -2$ or $x > 3$

68. $x < 0$ or $x \geq 6$

69. $x - 2 < -4$ or $x + 3 > 8$

70. $x + 3 \leq 5$ or $x - 2 \geq 3$

71. $6(x - 2) < 12$ or $4(x + 3) > 12$

72. $4x + 3 > -5$ or $8x - 5 < 3$

73. $-8x + 6x - 2 > 0$ or $5x > 3x + 8$

74. $3x \geq 7x + 8$ or $x < 4x - 9$

75. $2x + 5 \leq -1$ or $\dfrac{4}{3}x - 3 > 5$

76. $-\dfrac{4}{5}x - 5 > 3$ or $7x - 3 > 4$

77. $\dfrac{1}{2}x < 3$ or $\dfrac{3x - 1}{2} > 4$

78. $\dfrac{2}{3}x + 2 \leq 4$ or $\dfrac{5x - 3}{3} \geq 4$

79. $3(x - 1) + 5 < 2$ or $-2(x - 3) < 1$

80. $2(x + 1) - 5 \leq 4$ or $-(x + 3) \leq -2$

Mixed Practice

In Problems 81–94, solve each compound inequality. Graph the solution set.

81. $3a + 5 < 5$ and $-2a + 1 \leq 7$

82. $5x - 1 < 9$ and $5x > -20$

83. $5(x + 2) < 20$ or $4(x - 4) > -20$

84. $3(x + 7) < 24$ or $6(x - 4) > -30$

85. $-4 \leq 3x + 2 \leq 10$

86. $-8 \leq 5x - 3 \leq 4$

87. $2x + 7 < -13$ or $5x - 3 > 7$

88. $3x - 8 < -14$ or $4x - 5 > 7$

89. $5 < 3x - 1 < 14$

90. $-5 < 2x + 7 \leq 5$

91. $\dfrac{x}{3} \leq -1$ or $\dfrac{4x - 1}{2} > 7$

92. $\dfrac{x}{2} \leq -4$ or $\dfrac{2x - 1}{3} \geq 2$

93. $-3 \leq -2(x + 1) < 8$

94. $-15 < -3(x + 2) \leq 1$

Applying the Concepts

In Problems 95–100, use the Addition Property and/or Multiplication Properties to find a and b.

95. If $-3 < x < 4$, then $a < x + 4 < b$.

96. If $-2 < x < 3$, then $a < x - 3 < b$.

97. If $4 < x < 10$, then $a < 3x < b$.

98. If $2 < x < 12$, then $a < \dfrac{1}{2}x < b$.

99. If $-2 < x < 6$, then $a < 3x + 5 < b$.

100. If $-4 < x < 3$, then $a < 2x - 7 < b$.

101. Systolic Blood Pressure Blood pressure is measured using two numbers. One of the numbers measures systolic blood pressure. The systolic blood pressure represents the pressure while the heart is beating. In a healthy person, the systolic blood pressure should be greater than 90 and less than 140. If we let the variable x represent a person's systolic blood pressure, express the systolic blood pressure of a healthy person using a compound inequality.

102. Diastolic Blood Pressure Blood pressure is measured using two numbers. One of the numbers measures diastolic blood pressure. The diastolic blood pressure represents the pressure while the heart is resting between beats. In a healthy person, the diastolic blood pressure should be greater than 60 and less than 90. If we let the variable x represent a person's diastolic blood pressure, express the diastolic blood pressure of a healthy person using a compound inequality.

103. Computing Grades Joanna desperately wants to earn a B in her history class. Her current test scores are 74, 86, 77, and 89. Her final exam is worth two test scores. In order to earn a B, Joanna's average must lie between 80 and 89, inclusive. What range of scores can Joanna receive on the final and earn a B in the course?

104. Computing Grades Jack needs to earn a C in his sociology class. His current test scores are 67, 72, 81, and 75. His final exam is worth three test scores. In order to earn a C, Jack's average must lie between 70 and 79, inclusive. What range of scores can Jack receive on the final exam and earn a C in the course?

105. Federal Tax Withholding The percentage method of withholding for federal income tax (2012) states that a single person whose weekly wages, after subtracting withholding allowances, are over $721, but not over $1688, shall have $93.60 plus 25% of the excess over $721 withheld. Over what range does the amount withheld vary if the weekly wages vary from $800 to $900, inclusive? SOURCE: *Internal Revenue Service*

106. Federal Tax Withholding Rework Problem 105 if the weekly wages vary from $1000 to $1100, inclusive.

107. Commission Gerard had an offer for a medical equipment sales position that pays $2500 per month plus 1% of all sales. What total sales is required to earn between $3000 and $5000 per month?

108. Commission Juanita had a job offer to be an automobile sales position that pays $1500 per month plus 2.5% of all sales. What total sales is required to earn between $4000 and $6000 per month?

109. Electric Bill In North Carolina, Duke Energy charges $42.41 plus $0.092897 for each additional kilowatt hour (kwh) used during the months from July through October for usage in excess of 350 kwh. Suppose one homeowner's electric bill ranged from a low of $88.86 to a high of $137.16 during this time period. Over what range (in kwh) did the usage vary?

110. Electric Bill In North Carolina, Duke Energy charges $42.41 plus $0.084192 for each additional kilowatt hour (kwh) used during the months from November through June for usage in excess of 350 kwh. Suppose one homeowner's electric bill ranged from a low of $55.04 to a high of $89.56 during this time period. Over what range (in kwh) did the usage vary?

111. The Arithmetic Mean If $a < b$, show that
$$a < \frac{a+b}{2} < b.$$ We call $\frac{a+b}{2}$ the **arithmetic mean** of a and b.

△ **112. Identifying Triangles** A triangle has the property that the length of the longest side is greater than the difference of the other sides, and the length of the longest side is less than the sum of the other sides. That is, if a, b, and c are sides such that $a \le b \le c$, then $b - a < c < b + a$. Determine which of the following could be lengths of the sides of a triangle.

(a) 3, 4, 5   (b) 4, 7, 12

(c) 3, 3, 5   (d) 1, 9, 10

Extending the Concepts

113. Solve $2x + 1 \le 5x + 7 \le x - 5$.

114. Solve $x - 3 \le 3x + 1 \le x + 11$.

115. Solve $4x + 1 > 2(2x + 1)$. Provide an explanation that generalizes the result.

116. Solve $4x - 2 > (2x - 1)$. Provide an explanation that generalizes the result.

117. Consider the following analysis, assuming that $x < 2$.

$$5 > 2$$
$$5(x - 2) > 2(x - 2)$$
$$5x - 10 > 2x - 4$$
$$3x > 6$$
$$x > 2$$

How can it be that the final line in the analysis states that $x > 2$, when the original assumption stated that $x < 2$?

2.6 Absolute Value Equations and Inequalities

Objectives

1. Solve Absolute Value Equations
2. Solve Absolute Value Inequalities Involving $<$ or $\le$
3. Solve Absolute Value Inequalities Involving $>$ or $\ge$
4. Solve Applied Problems Involving Absolute Value Inequalities

Figure 36
$|-5| = 5$

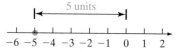

Are You Ready for This Section?

Before getting started, take this readiness quiz. If you get a problem wrong, go back to the section cited and review the material.

In Problems R1–R4, evaluate each expression.

R1. $|3|$ **R2.** $|-4|$ **R3.** $|-1.6|$ **R4.** $|0|$ [Section R.3, p. 19]

R5. Express the distance between the origin, 0, and 5 as an absolute value.   [Section R.3, p. 19]

R6. Express the distance between the origin, 0, and -8 as an absolute value.   [Section R.3, p. 19]

R7. Solve: $4x + 5 = -9$ [Section 1.1, pp. 50–52]

R8. Solve: $-2x + 1 > 5$ [Section 1.4, pp. 85–87]

Recall that the absolute value of a number is its distance from the origin on the real number line. For example, $|-5| = 5$ because the distance on the real number line from 0 to -5 is 5 units. See Figure 36.

▶ 1 Solve Absolute Value Equations

(**EXAMPLE 1**) **Solving an Absolute Value Equation**

Solve the equation $|x| = 4$.

Solution

We will present two geometric solutions and one algebraic solution.

Geometric Solution 1: The equation $|x| = 4$ asks, "Which real numbers x are 4 units from 0 on the number line?" Figure 37 shows the two numbers, -4 and 4. The solution set is $\{-4, 4\}$.

Ready?...Answers **R1.** 3   **R2.** 4
R3. 1.6   **R4.** 0   **R5.** $|5|$
R6. $|-8|$ **R7.** $\left\{-\dfrac{7}{2}\right\}$
R8. $\{x | x < -2\}; (-\infty, -2)$

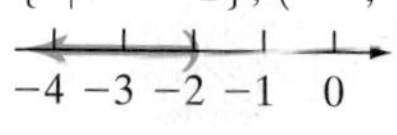

Figure 37

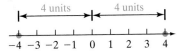

Figure 38

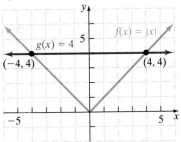

Geometric Solution 2: We can solve $|x| = 4$ by graphing $f(x) = |x|$ and $g(x) = 4$ on the same xy-plane, as in Figure 38. The x-coordinates of the points of intersection are -4 and 4, which are the solutions to the equation $f(x) = g(x)$. Again, the solution set is $\{-4, 4\}$.

Algebraic Solution: We know that $|a| = a$ if $a \geq 0$ and $|a| = -a$ if $a < 0$. Since we do not know whether x is positive or negative in the equation $|x| = 4$, we solve the problem for $x \geq 0$ or $x < 0$.

If $x \geq 0$		If $x < 0$
$|x| = 4$		$|x| = 4$
$|x| = x$ since $x \geq 0$: $x = 4$	$|x| = -x$ since $x < 0$: $-x = 4$	
	Multiply both sides by -1: $x = -4$	

The solution set is $\{-4, 4\}$.

Quick ✓

In Problems 1 and 2, solve the equation geometrically and algebraically.

1. $|x| = 7$ **2.** $|z| = 1$

The results of Example 1 and Quick Checks 1 and 2 lead us to the following result.

Figure 39

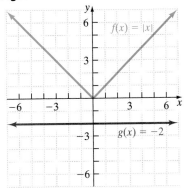

Equations Involving Absolute Value

If a is a positive real number and u is any algebraic expression, then

$$|u| = a \quad \text{is equivalent to} \quad u = a \ \text{ or } \ u = -a$$

Note: If $a = 0$, the equation $|u| = 0$ is equivalent to $u = 0$. If $a < 0$, the equation $|u| = a$ has no solution, as explained below.

In the equation $|u| = a$, we require that a be nonnegative (greater than or equal to 0). If a is negative, the equation has no real solution. To see why, consider the equation $|x| = -2$. Figure 39 shows the graph of $f(x) = |x|$ and $g(x) = -2$. Notice that the graphs do not intersect, which shows that the equation $|x| = -2$ has no real solution. The solution set is the empty set, $\varnothing$ or $\{\ \}$.

EXAMPLE 2 **How to Solve an Equation Involving Absolute Value**

Solve the equation: $|2x - 1| + 3 = 12$

Step-by-Step Solution

Step 1: Isolate the expression containing the absolute value.	$\begin{aligned} |2x - 1| + 3 &= 12 \\ \text{Subtract 3 from both sides:} \quad |2x - 1| &= 9 \end{aligned}$
Step 2: Rewrite the equation $|u| = a$ as: $u = a$ or $u = -a$, where u is the algebraic expression in the absolute value symbol.	In the equation $|2x - 1| = 9$, $u = 2x - 1$ and $a = 9$. $$2x - 1 = 9 \qquad \text{or} \qquad 2x - 1 = -9$$
Step 3: Solve each equation.	Add 1 to each side: $2x = 10$ Add 1 to each side: $2x = -8$ Divide both sides by 2: $x = 5$ Divide both sides by 2: $x = -4$

(continued)

Step 4: Check: *Verify each solution.*

$$|2x - 1| + 3 = 12$$

Let $x = 5$: $\quad |2(5) - 1| + 3 \overset{?}{=} 12$

$$|10 - 1| + 3 \overset{?}{=} 12$$

$$9 + 3 \overset{?}{=} 12$$

$$12 = 12 \text{ True}$$

$$|2x - 1| + 3 = 12$$

Let $x = -4$: $\quad |2(-4) - 1| + 3 \overset{?}{=} 12$

$$|-8 - 1| + 3 \overset{?}{=} 12$$

$$9 + 3 \overset{?}{=} 12$$

$$12 = 12 \text{ True}$$

Both solutions check, so the solution set is $\{-4, 5\}$. ●

Solving Absolute Value Equations with One Absolute Value

Step 1: Isolate the expression containing the absolute value.

Step 2: Rewrite the absolute value equation as two equations: $u = a$ and $u = -a$, where u is the expression in the absolute value symbol.

Step 3: Solve each equation.

Step 4: Verify your solution.

Quick ✓

3. $|u| = a$ is equivalent to $u = \underline{\quad}$ or $u = \underline{\quad}$.

4. $|2x + 3| = 5$ is equivalent to $2x + 3 = 5$ or $\underline{\qquad\qquad}$.

In Problems 5–8, solve each equation.

5. $|2x - 3| = 7$

6. $|3x - 2| + 3 = 10$

7. $|-5x + 2| - 2 = 5$

8. $3|x + 2| - 4 = 5$

▶ **EXAMPLE 3** **Solving an Equation Involving Absolute Value with No Solution**

Solve the equation: $|x + 5| + 7 = 5$

Solution

Work Smart

The equation $|u| = a$, where a is a negative real number, has no real solution. Why? See Figure 39.

$$|x + 5| + 7 = 5$$

Subtract 7 from both sides: $\quad |x + 5| = -2$

Since the absolute value of any real number is always nonnegative (greater than or equal to zero), the equation has no real solution. The solution set is $\{\ \}$ or $\emptyset$. ●

Quick ✓

9. *True or False* $|x| = -4$ has no real solution.

In Problems 10–12, solve each equation.

10. $|5x + 3| = -2$

11. $|2x + 5| + 7 = 3$

12. $|x + 1| + 3 = 3$

What if an absolute value equation has two absolute values, as in $|3x - 1| = |x + 5|$? The signs of the algebraic expressions in the absolute value symbol have four possibilities.

1. both algebraic expressions are positive,

2. both are negative,

3. the left is positive, and the right is negative, or

4. the left is negative and the right is positive.

To see how the solution works, we use the fact that $|a| = a$, if $a \geq 0$ and $|a| = -a$, if $a < 0$.

Thus, if $3x - 1 \geq 0$, then $|3x - 1| = 3x - 1$. However, if $3x - 1 < 0$, then $|3x - 1| = -(3x - 1)$. This leads us to a method for solving absolute value equations with two absolute values.

Case 1: Both Expressions Are Positive	Case 2: Both Expressions Are Negative	Case 3: The Expression on the Left Is Positive, and That on the Right Is Negative	Case 4: The Expression on the Left Is Negative, and That on the Right Is Positive
$|3x - 1| = |x + 5|$ $3x - 1 = x + 5$	$|3x - 1| = |x + 5|$ $-(3x - 1) = -(x + 5)$ $3x - 1 = x + 5$	$|3x - 1| = |x + 5|$ $3x - 1 = -(x + 5)$	$|3x - 1| = |x + 5|$ $-(3x - 1) = x + 5$

When both algebraic expressions are positive, or both negative, we end up with equivalent equations. Therefore, Case 1 and Case 2 result in equivalent equations. Also, if one side is positive and the other is negative, we end up with equivalent equations. Therefore, Case 3 and Case 4 result in equivalent equations. The four possibilities reduce to two possibilities.

> **Equations Involving Two Absolute Values**
>
> If u and v are any algebraic expression, then
>
> $$|u| = |v| \quad \text{is equivalent to} \quad u = v \quad \text{or} \quad u = -v$$

▶ **EXAMPLE 4** **Solving an Absolute Value Equation Involving Two Absolute Values**

Solve the equation: $|2x - 3| = |x + 6|$

Solution

The equation is in the form $|u| = |v|$, where $u = 2x - 3$ and $v = x + 6$. Rewrite the equation in the form $u = v$ or $u = -v$ and then solve each equation.

$u = v$: $2x - 3 = x + 6$ $\qquad\qquad$ $u = -v$: $2x - 3 = -(x + 6)$

Add 3 to both sides: $\qquad 2x = x + 9$ $\qquad\qquad$ Distribute the -1: $2x - 3 = -x - 6$

Subtract x from both sides: $\qquad x = 9$ $\qquad\qquad$ Add 3 to both sides: $\qquad 2x = -x - 3$

$\qquad\qquad\qquad\qquad\qquad\qquad\qquad\qquad\qquad\qquad$ Add x to both sides: $\qquad 3x = -3$

$\qquad\qquad\qquad\qquad\qquad\qquad\qquad\qquad\qquad\qquad$ Divide both sides by 3: $\qquad x = -1$

Check

$x = 9$: $\quad |2(9) - 3| \overset{?}{=} |9 + 6|$ $\qquad$ $x = -1$: $\quad |2(-1) - 3| \overset{?}{=} |-1 + 6|$

$\qquad\qquad |18 - 3| \overset{?}{=} |15|$ $\qquad\qquad\qquad\qquad |-2 - 3| \overset{?}{=} |5|$

$\qquad\qquad\quad |15| \overset{?}{=} 15$ $\qquad\qquad\qquad\qquad\qquad |-5| \overset{?}{=} 5$

$\qquad\qquad\qquad 15 = 15 \quad$ True $\qquad\qquad\qquad\qquad\quad 5 = 5 \quad$ True

Both solutions check, so the solution set is $\{-1, 9\}$.

Quick ✓

13. $|u| = |v|$ is equivalent to $\_ = \_$ or $\_ = \_$.

In Problems 14–17, solve each equation.

14. $|x - 3| = |2x + 5|$ $\qquad\qquad\qquad\qquad$ **15.** $|8z + 11| = |6z + 17|$

16. $|3 - 2y| = |4y + 3|$ $\qquad\qquad\qquad\qquad$ **17.** $|2x - 3| = |5 - 2x|$

▶ ② Solve Absolute Value Inequalities Involving < or ≤

The method for solving absolute value equations relies on the geometric interpretation of absolute value. That is, the absolute value of a real number x is its distance from the origin on the real number line. We also use the geometric interpretation of absolute value to solve absolute value inequalities.

EXAMPLE 5 Solving an Absolute Value Inequality

Solve the inequality $|x| < 4$. Graph the solution set.

Solution

The inequality $|x| < 4$ asks for all real numbers x that are less than 4 units from the origin on the real number line. See Figure 40. We can see from the figure that any number between -4 and 4 satisfies the inequality. The solution set is $\{x \mid -4 < x < 4\}$ or, using interval notation, $(-4, 4)$.

We could also visualize these results by graphing $f(x) = |x|$ and $g(x) = 4$. See Figure 41. To solve $f(x) < g(x)$, we look for all x-coordinates such that the graph of $f(x)$ is below the graph of $g(x)$. We can see that this is true for all x between -4 and 4, as we found above.

Figure 40

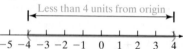

The results of Example 5 lead to the following.

Figure 41

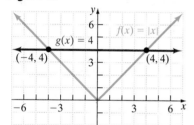

Inequalities of the Form < or ≤ Involving Absolute Value

If a is a positive real number and if u is an algebraic expression, then

$$|u| < a \quad \text{is equivalent to} \quad -a < u < a$$
$$|u| \le a \quad \text{is equivalent to} \quad -a \le u \le a$$

Note: If $a = 0$, $|u| < 0$ has no solution; $|u| \le 0$ is equivalent to $u = 0$. If $a < 0$, the inequality has no solution.

Quick ✓

18. If $a > 0$, then $|u| < a$ is equivalent to _____.

19. To solve $|3x + 4| < 10$, solve ___ $< 3x + 10 <$ ___.

In Problems 20 and 21, solve each inequality. Graph the solution set.

20. $|x| \le 5$

21. $|x| < \dfrac{3}{2}$

EXAMPLE 6 How to Solve an Absolute Value Inequality Involving ≤

Solve the inequality $|2x + 3| \le 5$. Express the solution set in set-builder notation and interval notation. Graph the solution set.

Step-by-Step Solution

Step 1: The inequality is in the form $|u| \le a$, where $u = 2x + 3$ and $a = 5$. Rewrite the inequality as a compound inequality that does not involve absolute value.

Use the fact that $|u| \le a$ means $-a \le u \le a$:

$$|2x + 3| \le 5$$
$$-5 \le 2x + 3 \le 5$$

Step 2: Solve the resulting compound inequality.

Subtract 3 from all three parts:

$$-5 - 3 \le 2x + 3 - 3 \le 5 - 3$$
$$-8 \le 2x \le 2$$

Divide all three parts by 2:

$$\frac{-8}{2} \le \frac{2x}{2} \le \frac{2}{2}$$
$$-4 \le x \le 1$$

The solution is $\{x \mid -4 \le x \le 1\}$ or, in interval notation, $[-4, 1]$. Figure 42 shows the graph of the solution set.

Figure 42

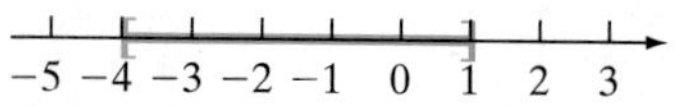

Work Smart

As a partial check of the solution of Example 6, we can check a number in the interval. Let's try $x = -3$.

$$|2(-3) + 3| \overset{?}{\leq} 5$$
$$|-6 + 3| \overset{?}{\leq} 5$$
$$|-3| \overset{?}{\leq} 5$$
$$3 \leq 5 \quad \text{True}$$

Quick ✓

In Problems 22–24, solve each inequality. Express the solution set in set-builder notation and interval notation. Graph the solution set.

22. $|x + 3| < 5$

23. $|2x - 3| \leq 7$

24. $|7x + 2| < -3$

EXAMPLE 7 **Solving an Absolute Value Inequality Involving $<$**

Solve the inequality $|-3x + 2| + 4 < 14$. Express the solution set in set-builder notation and interval notation. Graph the solution set.

Solution

First, isolate the absolute value.

$$|-3x + 2| + 4 < 14$$

Subtract 4 from both sides:
$$|-3x + 2| < 10$$

$|u| < a$ means $-a < u < a$:
$$-10 < -3x + 2 < 10$$

Subtract 2 from all three parts:
$$-10 - 2 < -3x + 2 - 2 < 10 - 2$$
$$-12 < -3x < 8$$

Divide all three parts by -3.
Reverse the direction of the inequalities.
$$\frac{-12}{-3} > \frac{-3x}{-3} > \frac{8}{-3}$$
$$4 > x > -\frac{8}{3}$$

$b > x > a$ is equivalent to $a < x < b$:
$$-\frac{8}{3} < x < 4$$

Figure 43

The solution is $\left\{ x \middle| -\dfrac{8}{3} < x < 4 \right\}$ using set-builder notation and $\left(-\dfrac{8}{3}, 4 \right)$ using interval notation. Figure 43 shows the graph of the solution set. ●

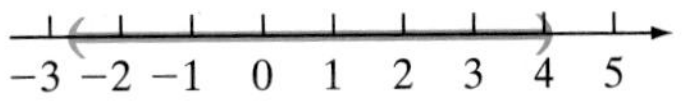

Quick ✓

25. *True or False* $|x + 1| + 2 < 7$ is equivalent to $-7 < x + 1 + 2 < 7$.

In Problems 26–29, solve each inequality. Express the solution set in set-builder notation and interval notation. Graph the solution set.

26. $|x| + 4 < 6$

27. $|x - 3| + 4 \leq 8$

28. $3|2x + 1| \leq 9$

29. $|-3x + 1| - 5 < 3$

▶ ❸ **Solve Absolute Value Inequalities Involving $>$ or $\geq$**

EXAMPLE 8 **Solving an Absolute Value Inequality Involving $>$**

Solve the inequality $|x| > 3$. Graph the solution set.

Solution

The inequality $|x| > 3$ asks for all real numbers x that are more than 3 units from the origin on the real number line. See Figure 44.

Figure 44   More than 3 units from origin, 0.   More than 3 units from origin, 0.

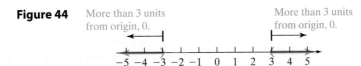

Figure 45

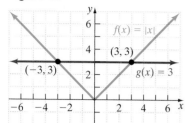

We can see that any number less than -3 or greater than 3 satisfies the inequality. The solution set is $\{x \mid x < -3 \text{ or } x > 3\}$ or, using interval notation, $(-\infty, -3) \cup (3, \infty)$.

We could also visualize these results by graphing $f(x) = |x|$ and $g(x) = 3$. See Figure 45. To solve $f(x) > g(x)$, we look for all x-coordinates such that the graph of $f(x)$ is above the graph of $g(x)$. We can see that this is true for all x less than -3 or all x greater than 3, as we found above.

Based on Example 8, we have the following results.

> **Inequalities of the Form $>$ or $\geq$ Involving Absolute Value**
>
> If a is a positive real number and u is an algebraic expression, then
>
> $$|u| > a \quad \text{is equivalent to} \quad u < -a \quad \text{or} \quad u > a$$
> $$|u| \geq a \quad \text{is equivalent to} \quad u \leq -a \quad \text{or} \quad u \geq a$$

Quick ✓

30. $|u| > a$ is equivalent to _____ or ____.

31. $|5x - 2| \geq 7$ is equivalent to $5x - 2 \leq$ ___ or $5x - 2 \geq$ ___.

In Problems 32 and 33, solve each inequality. Graph the solution set.

32. $|x| \geq 6$

33. $|x| > \dfrac{5}{2}$

EXAMPLE 9 **How to Solve an Inequality Involving $>$**

Solve the inequality $|2x - 5| > 3$. Express the solution set in set-builder notation and interval notation. Graph the solution set.

Step-by-Step Solution

Step 1: The inequality is in the form $|u| > a$, where $u = 2x - 5$ and $a = 3$. Rewrite the inequality as a compound inequality that does not involve absolute value.

$$|2x - 5| > 3$$

$|u| > a$ means $u < -a$ or $u > a$: $\quad 2x - 5 < -3 \text{ or } 2x - 5 > 3$

Step 2: Solve each inequality separately.

	$2x - 5 < -3$		$2x - 5 > 3$
Add 5 to both sides:	$2x < 2$	Add 5 to both sides:	$2x > 8$
Divide both sides by 2:	$x < 1$	Divide both sides by 2:	$x > 4$

Step 3: Find the union of the solution sets of each inequality.

The solution set is $\{x \mid x < 1 \text{ or } x > 4\}$ or, using interval notation, $(-\infty, 1) \cup (4, \infty)$. See Figure 46 for the graph of the solution set.

Figure 46

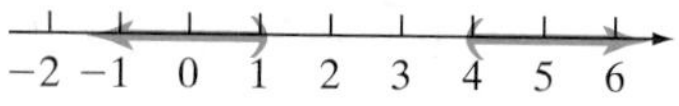

$\overset{}{\underset{-2\ -1\ \ 0\ \ 1\ \ 2\ \ 3\ \ 4\ \ 5\ \ 6}{\longleftarrow\!\!\!\longrightarrow}}$

Quick ✓

34. $|x - 9| > 6$ is equivalent to $x - 9 > 6$ or $x - 9$ ___ -6.

In Problems 35–40, solve each inequality. Express the solution set in set-builder notation and interval notation. Graph the solution set.

35. $|x + 3| > 4$

36. $|4x - 3| \geq 5$

37. $|-3x + 2| > 7$

38. $|2x + 5| - 2 > -2$

39. $|6x - 5| \geq 0$

40. $|2x + 1| > -3$

Work Smart

$$|u| > a$$

CANNOT be written as

$$-a > u > a$$

Summary Solving Absolute Value Equations and Inequalities

Absolute Value Form	Equation/Inequality Form	Example		
		Algebraic Solution	**Graphical Solution**	

Absolute Value Form	Equation/Inequality Form	Algebraic Solution	Graphical Solution
$\lvert u \rvert = a$	$u = a$ or $u = -a$	$\lvert x + 3 \rvert = 5$ $x + 3 = 5$ or $x + 3 = -5$ $x = 2$ or $x = -8$ Solution set: $\{-8, 2\}$	
$\lvert u \rvert = \lvert v \rvert$	$u = v$ or $u = -v$	$\lvert x - 2 \rvert = \lvert 2x \rvert$ $x - 2 = 2x$ or $x - 2 = -2x$ $-2 = x$ or $3x = 2$ $x = -2$ or $x = \dfrac{2}{3}$ Solution set: $\left\{ -2, \dfrac{2}{3} \right\}$	
$\lvert u \rvert < a$ $\lvert u \rvert \le a$	$-a < u < a$ $-a \le u \le a$	$\lvert x - 1 \rvert \le 4$ $-4 \le x - 1 \le 4$ $-3 \le x \le 5$ Solution set: $\{x \mid -3 \le x \le 5\}$ or $[-3, 5]$	
$\lvert u \rvert > a$ $\lvert u \rvert \ge a$	$u < -a$ or $u > a$ $u \le -a$ or $u \ge a$	$\lvert x + 1 \rvert > 3$ $x + 1 < -3$ or $x + 1 > 3$ $x < -4$ or $x > 2$ Solution set: $\{x \mid x < -4 \text{ or } x > 2\}$ or $(-\infty, -4) \cup (2, \infty)$	

4 Solve Applied Problems Involving Absolute Value Inequalities

You may have read phrases such as "margin of error" and "tolerance" in the newspaper, For example, according to a Gallup poll conducted in March 2012, 67% of those polled said that recent increases in gas prices have caused financial hardship for their household. The poll had a margin of error of 4%. The 67% is an estimate of the true percentage of Americans who have experienced financial hardship as a consequence of increases in gas prices. If we let p represent the true percentage of Americans who have experienced financial hardship due to the increase in gas prices, then we can represent the poll's margin of error mathematically as

$$\lvert p - 67 \rvert \le 4$$

As another example, the tolerance of a belt used in a pulley system whose width is 6 inches is $\dfrac{1}{16}$ inch. If x represents the actual width of the belt, then we can represent the acceptable belt widths as

$$\lvert x - 6 \rvert \le \dfrac{1}{16}$$

EXAMPLE 10   **Analyzing the Margin of Error in a Poll**

The inequality

$$\lvert p - 67 \rvert \le 4$$

represents the percentage of Americans who said they experienced financial hardship due to increase in gas prices. Solve the inequality and interpret the results.

Solution

$$|p - 67| \le 4$$

$|u| \le a$ means $-a \le u \le a$: $-4 \le p - 67 \le 4$

Add 67 to all three parts: $63 \le p \le 71$

The percentage of Americans who have experienced financial hardship due to increases in gas prices is between 63% and 71%, inclusive.

Quick ✓

41. The inequality $|x - 4| \le \dfrac{1}{32}$ represents the acceptable belt width x (in inches) for a belt that is manufactured for a pulley system. Determine the acceptable belt width.

42. In a poll conducted by ABC News, 9% of Americans stated that they have been shot at. The margin of error in the poll was 1.7%. If we let p represent the true percentage of people who have been shot at, we can represent the margin of error as

$$|p - 9| \le 1.7$$

Solve the inequality and interpret the results.

2.6 Exercises   MyMathLab®   MathXL PRACTICE

Exercise numbers in green have complete video solutions in MyMathLab.

*Problems **1–42** are the Quick ✓s that follow the EXAMPLES.*

Building Skills

In Problems 43–64, solve each absolute value equation. See Objective 1.

43. $|x| = 10$

44. $|z| = 9$

45. $|y - 3| = 4$

46. $|x + 3| = 5$

47. $|-3x + 5| = 8$

48. $|-4y + 3| = 9$

49. $|y| - 7 = -2$

50. $|x| + 3 = 5$

51. $|2x + 3| - 5 = 3$

52. $|3y + 1| - 5 = -3$

53. $-2|x - 3| + 10 = -4$

54. $3|y - 4| + 4 = 16$

55. $|-3x| - 5 = -5$

56. $|-2x| + 9 = 9$

57. $\left| \dfrac{3x - 1}{4} \right| = 2$

58. $\left| \dfrac{2x - 3}{5} \right| = 2$

59. $|3x + 2| = |2x - 5|$

60. $|5y - 2| = |4y + 7|$

61. $|8 - 3x| = |2x - 7|$

62. $|5x + 3| = |12 - 4x|$

63. $|4y - 7| = |9 - 4y|$

64. $|5x - 1| = |9 - 5x|$

In Problems 65–78, solve each absolute value inequality. Graph the solution set on a real number line. See Objective 2.

65. $|x| < 9$

66. $|x| \le \dfrac{5}{4}$

67. $|x - 4| \le 7$

68. $|y + 4| < 6$

69. $|3x + 1| < 8$ **70.** $|4x - 3| \le 9$

71. $|6x + 5| < -1$ **72.** $|4x + 3| \le 0$

73. $2|x - 3| + 3 < 9$ **74.** $3|y + 2| - 2 < 7$

75. $|2 - 5x| + 3 < 10$ **76.** $|-3x + 2| - 7 \le -2$

77. $|(2x - 3) - 1| < 0.01$ **78.** $|(3x + 2) - 8| < 0.01$

In Problems 79–90, solve each absolute value inequality. Graph the solution set on the real number line. See Objective 3.

79. $|y - 5| > 2$ **80.** $|x + 4| \ge 7$

81. $|-4x - 3| \ge 5$ **82.** $|-5y + 3| > 7$

83. $2|y| + 3 > 1$ **84.** $3|z| + 8 > 2$

85. $|-5x - 3| - 7 > 0$ **86.** $|-9x + 2| - 11 \ge 0$

87. $4|-2x + 1| > 4$ **88.** $3|8x + 3| \ge 9$

89. $|1 - 2x| \ge |-5|$ **90.** $|3 - 5x| > |-7|$

Mixed Practice

In Problems 91–94, use the graphs of the functions given to solve each problem.

91. $f(x) = |x|$, $g(x) = 5$

  (a) $f(x) = g(x)$
  (b) $f(x) \le g(x)$
  (c) $f(x) > g(x)$

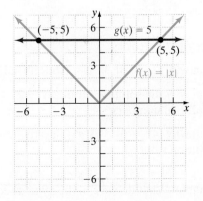

92. $f(x) = |x|$, $g(x) = 6$

  (a) $f(x) = g(x)$
  (b) $f(x) \le g(x)$
  (c) $f(x) > g(x)$

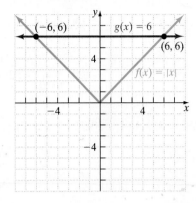

93. $f(x) = |x + 2|$, $g(x) = 3$

  (a) $f(x) = g(x)$
  (b) $f(x) < g(x)$
  (c) $f(x) \ge g(x)$

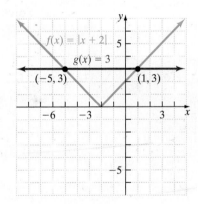

94. $f(x) = |2x|$, $g(x) = 10$

  (a) $f(x) = g(x)$
  (b) $f(x) < g(x)$
  (c) $f(x) \ge g(x)$

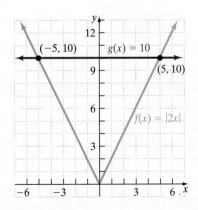

In Problems 95–114, solve each absolute value equation or inequality. For absolute value inequalities, graph the solution set on a real number line.

95. $|x| > 5$

96. $|x| \geq \dfrac{8}{3}$

97. $|2x + 5| = 3$

98. $|4x + 3| = 1$

99. $7|x| = 35$

100. $8|y| = 32$

101. $|5x + 2| \leq 8$

102. $|7y - 3| < 11$

103. $|-2x + 3| = -4$

104. $|3x - 4| = -9$

105. $|3x + 2| \geq 5$

106. $|5y + 3| > 2$

107. $|3x - 2| + 7 > 9$

108. $|4y + 3| - 8 \geq -3$

109. $|5x + 3| = |3x + 5|$

110. $|3z - 2| = |z + 6|$

111. $|4x + 7| + 6 < 5$

112. $|4x + 1| > 0$

113. $\left| \dfrac{x - 2}{4} \right| = \left| \dfrac{2x + 1}{6} \right|$

114. $\left| \dfrac{1}{2}x - 3 \right| = \left| \dfrac{2}{3}x + 1 \right|$

Applying the Concepts

115. Express the fact that x differs from 5 by less than 3 as an inequality involving absolute value. Solve for x.

116. Express the fact that x differs from -4 by less than 2 as an inequality involving absolute value. Solve for x.

117. Express the fact that twice x differs from -6 by more than 3 as an inequality involving absolute value. Solve for x.

118. Express the fact that twice x differs from 7 by more than 3 as an inequality involving absolute value. Solve for x.

119. Tolerance A certain rod in an internal combustion engine is supposed to be 5.7 inches long. The tolerance on the rod is 0.0005 inch. If x represents

the length of a rod, the acceptable lengths of a rod can be expressed as $|x - 5.7| \leq 0.0005$. Determine the acceptable lengths of the rod. SOURCE: *WiseCo Piston*

120. Tolerance A certain rod in an internal combustion engine is supposed to be 6.125 inches long. The tolerance on the rod is 0.0005 inch. If x represents the length of a rod, the acceptable lengths of a rod can be expressed as $|x - 6.125| \leq 0.0005$. Determine the acceptable lengths of the rod.

121. IQ Scores According to the Stanford-Binet IQ test, a normal IQ score is 100. It can be shown that anyone with an IQ x that satisfies the inequality $\left| \dfrac{x - 100}{15} \right| > 1.96$ has an unusual IQ score. Determine the IQ scores that would be considered unusual.

122. Gestation Period The length of human pregnancy is about 266 days. It can be shown that a mother whose gestation period x satisfies the inequality $\left| \dfrac{x - 266}{16} \right| > 1.96$ has an unusual length of pregnancy. Determine the length of pregnancy that would be considered unusual.

Extending the Concepts

In Problems 123–130, solve each equation.

123. $|x| - x = 5$

124. $|y| + y = 3$

125. $z + |-z| = 4$

126. $y - |-y| = 12$

127. $|4x + 1| = x - 2$

128. $|2x + 1| = x - 3$

129. $|x + 5| = -(x + 5)$

130. $|y - 4| = y - 4$

Explaining the Concepts

131. Explain why $|2x - 3| + 1 = 0$ has no solution.

132. Explain why the solution set of $|5x - 3| > -5$ is the set of all real numbers.

133. Explain why $|4x + 3| + 3 < 0$ has the empty set as the solution set.

134. Solve $|x - 5| = |5 - x|$. Explain why the result is reasonable. What do we call this type of equation?

Chapter 2 Activity: Shifting Discovery

Focus: Using graphing skills, discover the possible "rules" for graphing functions.

Time: 30–35 minutes

Group size: 4

Materials Needed: Graph paper (2–3 pieces)

Each member of the group needs to

1. Draw a coordinate plane and label the x-axis and y-axis.

2. By plotting points, graph the primary function: $f(x) = x^2$.

3. In the same coordinate plane, each group member graphs *one* of the following functions by plotting points. Be sure to graph the primary function and one of the functions (a)–(d) in the same coordinate plane.

   (a) $f(x) = x^2 + 3$ (b) $f(x) = (x - 3)^2$
   (c) $f(x) = x^2 - 3$ (d) $f(x) = (x + 3)^2$

As a group, discuss the following:

4. What shape are the graphs?

5. Each member of the group should share the difference between the graph of your primary function and the other function you chose.

6. As a group, can you develop rules for these differences?

With this possible rule in mind, each member of the group needs to

7. Draw a coordinate plane and label the x-axis, the y-axis, and -10 to 10 on each axis.

8. Graph the primary function by plotting points: $f(x) = |x|$.

9. In the same coordinate plane, each group member graphs *one* of the following functions by plotting points. Be sure to graph the primary function and one of the functions (a)–(d) on the same coordinate plane.

   (a) $f(x) = |x| + 4$ (b) $f(x) = |x - 4|$
   (c) $f(x) = |x| - 4$ (d) $f(x) = |x + 4|$

10. Did the rules you developed in Problem 6 hold true? Discuss.

Chapter 2 Review

Section 2.1   Relations

KEY CONCEPT

- **Relation**
  A correspondence between two variables x and y, where y depends on x. Relations can be represented through maps, sets of ordered pairs, equations, or graphs.

KEY TERMS

Relation	Inputs
Corresponds	Outputs
Depends on	Domain
Mapping	Range

You Should Be Able To...	EXAMPLE	Review Exercises
❶ Understand relations (p. 146)	Example 1	1–4
❷ Find the domain and the range of a relation (p. 147)	Examples 2 through 4	1–16
❸ Graph a relation defined by an equation (p. 149)	Example 5	9–14

In Problems 1 and 2, write each relation as a set of ordered pairs. Then identify the domain and range of the relation.

1.

U.S. Coin	Weight (g)
Penny	2.500
Nickel	5.000
Dime	2.268
Quarter	5.670
Half Dollar	11.340
Dollar	8.100

SOURCE: *U.S. Mint website*

2.

Pieces	Price of Lego Set
16	$ 12.99
28	$ 14.99
30	$ 24.99
59	$ 29.99
85	

SOURCE: *Lego website*

In Problems 3 and 4, write each relation as a map. Then identify the domain and range of the relation.

3. $\{(2, 7), (-4, 8), (3, 5), (6, -1), (-2, -9)\}$

4. $\{(3, 1), (3, 7), (5, 1), (-2, 8), (1, 4)\}$

In Problems 5–8, identify the domain and range of the relation from the graph.

5.
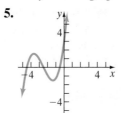

6.
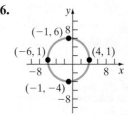

7.
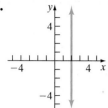

8.
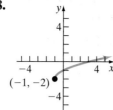

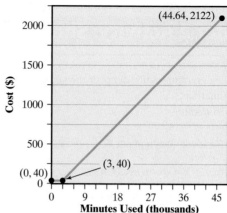

In Problems 9–14, graph the relation. Use the graph of the relation to identify the domain and range of the relation.

9. $y = x + 2$ **10.** $2x + y = 3$ **11.** $y = -x^2 + 4$

12. $y = |x + 2| - 1$ **13.** $y = x^3 + 2$ **14.** $x = y^2 + 1$

15. Cell Phones A cellular phone company offers a plan for $40 per month for 3000 minutes with additional minutes costing $0.05 per minute. The graph to the right shows the monthly cost, in dollars, when x minutes are used.

(a) What are the domain and range of relation?

(b) Explain why the domain obtained in part (a) is reasonable.

16. Vertical Motion The graph below shows the height, in feet, of a ball thrown straight up with an initial speed of 40 feet per second from an initial height of 96 feet after t seconds. What are the domain and the range of the relation?

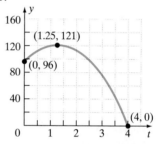

Section 2.2   An Introduction to Functions

KEY CONCEPTS

- **Functions**
  A special type of relation where any given input, x, corresponds to only one output y. Functions can be represented through maps, sets of ordered pairs, equations, or graphs.

- **Vertical Line Test**
  A set of points in the xy-plane is the graph of a function if and only if every vertical line intersects the graph in at most one point.

- **Domain of a Function**
  When only an equation of a function is given, the domain of the function is the set of real numbers x for which $f(x)$ is a real number. However, in applications, the domain of a function is the largest set of real numbers for which the output of the function is reasonable.

KEY TERMS

Function
Vertical Line Test
Value of f at the number x
Independent variable
Dependent variable
Argument
Domain of f
Graph of the function

You Should Be Able To...	EXAMPLE	**Review Exercises**
1 Determine whether a relation expressed as a map or ordered pairs represents a function (p. 152)	Examples 1 and 2	17, 18
2 Determine whether a relation expressed as an equation represents a function (p. 155)	Examples 3 and 4	19–22
3 Determine whether a relation expressed as a graph represents a function (p. 155)	Example 5	23–26
4 Find the value of a function (p. 156)	Examples 6 and 7	27–30
5 Find the domain of a function (p. 158)	Examples 8 and 9	31–34
6 Work with applications of functions (p. 159)	Example 10	35, 36

In Problems 17 and 18, determine whether the given relation represents a function. State the domain and the range of each relation.

17. (a) $\{(-1, -2), (-1, 3), (5, 0), (7, 2), (9, 4)\}$

(b)

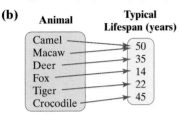

18. (a) $\{(-2, 4), (2, 3), (-3, -1), (5, 7), (4, 7)\}$

(b)

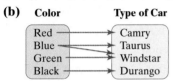

In Problems 19–22, determine whether each relation shows y as a function of x.

19. $3x - 5y = 18$

20. $x^2 + y^2 = 81$

21. $y = \pm 10x$

22. $y = x^2 - 14$

In Problems 23–26, determine whether the graph is that of a function.

23.

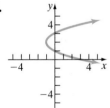

24.

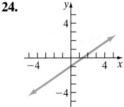

25.

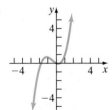

26.

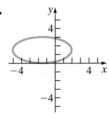

In Problems 27–30, find the indicated values for the given functions.

27. $f(x) = x^2 + 2x - 5$
  (a) $f(-2)$
  (b) $f(3)$

28. $g(z) = \dfrac{2z + 1}{z - 3}$
  (a) $g(0)$
  (b) $g(2)$

29. $F(x) = -2x + 7$
  (a) $F(5)$
  (b) $F(-x)$

30. $G(x) = 2x + 1$
  (a) $G(7)$
  (b) $G(x + h)$

For Problems 31–34, find the domain of each function.

31. $f(x) = -\dfrac{3}{2}x + 5$

32. $g(w) = \dfrac{w - 9}{2w + 5}$

33. $h(t) = \dfrac{t + 2}{t - 5}$

34. $G(t) = 3t^2 + 4t - 9$

35. Population Using census data from 1900 to 2010, the function $P(t) = 0.132t^2 - 4.782t + 43.947$ represents the population, P, of Orange County in Florida (in thousands) t years after 1900.

  (a) Identify the dependent and independent variables.

  (b) Evaluate $P(120)$ and explain what it represents.

  (c) Evaluate $P(-70)$ and explain what it represents. Is the result reasonable? Explain.

36. Education The function $P(a) = -0.0064a^2 + 0.6826a - 6.82$ represents the percent of the population a years of age with an advanced degree, where $a \geq 25$. SOURCE: *Current Population Survey*

  (a) Identify the dependent and independent variables.

  (b) Evaluate $P(30)$ and explain what it represents.

Section 2.3 Functions and Their Graphs

KEY CONCEPTS

- **Graph of a Function**
  The graph of a function, f, is the set of all ordered pairs $(x, f(x))$.

- **Zero of a Function**
  If $f(r) = 0$ for some number r, then r is a zero of f.

KEY TERMS

Zero of a function

You Should Be Able To...	EXAMPLE	Review Exercises
❶ Graph a function (p. 163)	Example 1	37–40
❷ Obtain information from the graph of a function (p. 165)	Examples 2 through 5	41–45, 47, 48
❸ Know properties and graphs of basic functions (p. 168)	p. 168	46
❹ Interpret graphs of functions (p. 169)	Example 6	49, 50

In Problems 37–40, graph each function.

37. $f(x) = 2x - 5$

38. $g(x) = x^2 - 3x + 2$

39. $h(x) = (x - 1)^3 - 3$

40. $f(x) = |x + 1| - 4$

In Problems 41–44, for each function whose graph is shown, find (a) the domain and the range, and (b) the intercepts, if any.

41.

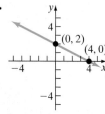

42.

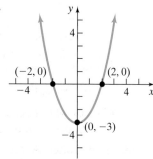

43.

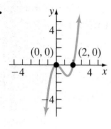

44.

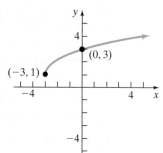

45. The graph of $y = f(x)$ is shown.

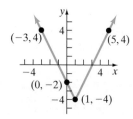

    (a) What is $f(-3)$?

    (b) For what value of x does $f(x) = -4$?

    (c) What are the zeros of f?

46. Graph each of the following functions.

    (a) $f(x) = x^2$ **(b)** $f(x) = x$

In Problems 47 and 48, answer the questions about the given function.

47. $h(x) = 2x - 7$

    (a) Is the point $(3, -1)$ on the graph of the function?

    (b) If $x = -2$, what is $h(x)$? What point is on the graph of the function?

    (c) If $h(x) = 4$, what is x? What point is on the graph of h?

48. $g(x) = \dfrac{3}{5}x + 4$

    (a) Is the point $(-5, 2)$ on the graph of the function?

    (b) If $x = 3$, what is $g(x)$? What point is on the graph of the function?

    (c) If $g(x) = -2$, what is x? What point is on the graph of g?

49. Travel by Train A Metrolink train leaves L.A. Union Station and travels 6 miles at a constant speed for 10 minutes, arriving at the Glendale station, where it waits 1 minute for passengers to board and depart. The train continues traveling at the same speed for 5 more minutes to reach downtown Burbank, which is 3 miles from Glendale. Sketch a graph that represents the distance of the train as a function of time until it reaches downtown Burbank.

50. Filling a Tub With the faucet running at a constant rate, it takes Angie 7 minutes to fill her bathtub. She turns off the faucet when the tub is full and realizes the water is too hot. She then opens the drain, letting water out at a constant rate that is half the rate of the faucet. After draining for 2 minutes, she stops the drain and turns on the faucet at the same rate as before (but at a cooler temperature) until the tub is full. Sketch a graph that represents the amount of water in the tub as a function of time.

Section 2.4 · Linear Functions and Models

KEY CONCEPT

- **Linear Function**
  A linear function is a function of the form $f(x) = mx + b$, where m is the slope and $(0, b)$ is the y-intercept of the graph.

KEY TERMS

Linear function
Line
Scatter diagram

You Should Be Able To...	EXAMPLE	Review Exercises
1 Graph linear functions (p. 174)	Example 1	51–54
2 Find the zero of a linear function (p. 175)	Example 2	51–54
3 Build linear models from verbal descriptions (p. 177)	Examples 4 and 5	57–60
4 Build linear models from data (p. 179)	Examples 6 through 8	61–64

In Problems 51–54, graph each linear function. Find the zero of each function.

51. $g(x) = 2x - 6$

52. $H(x) = -\dfrac{4}{3}x + 5$

53. $F(x) = -x - 3$

54. $f(x) = \dfrac{3}{4}x - 3$

55. Long Distance A phone company offers a plan for long-distance calls that charges $5.00 per month plus 7¢ per minute. The monthly long-distance cost C for talking x minutes is given by the linear function $C(x) = 0.07x + 5$.

(a) What is the domain of this linear function?

(b) What is the cost if a person made 235 minutes worth of long-distance calls during one month?

(c) Graph the linear function.

(d) In one month, how many minutes of long-distance can be purchased for $75?

56. Straight-Line Depreciation Using straight-line depreciation, the value V of a particular computer x years after purchase is given by the linear function $V(x) = 1800 - 360x$ for $0 \le x \le 5$.

(a) What are the independent and dependent variables?

(b) What is the domain of this linear function?

(c) What is the initial value of the computer?

(d) What is the value of the computer 2 years after purchase?

(e) Graph the linear function over its domain.

(f) After how long will the value of the computer be $0?

57. Credit Scores Your Fair Isaacs Corporation (FICO) credit score is used to determine your ability to get a loan. The higher your credit score, the better your credit history. Suppose a bank offers a 36-month auto loan at 9% for a FICO score of 675 and at 5% for a FICO score of 750.

(a) Assuming a linear relation between FICO credit score and auto loan rate, find a linear function that relates the FICO credit score rate to the auto loan rate (as a percentage), treating the FICO credit score as the independent variable.

(b) Predict the auto loan rate for a FICO credit score of 710, to the nearest whole percent.

(c) Interpret the slope.

(d) For what FICO credit score will a bank offer a 36-month auto loan at 6.5%?

58. Heart Rates According to the American Geriatric Society, the maximum recommended heart rate for a 20-year-old man under stress is 200 beats per minute. The maximum recommended heart rate for a 60-year-old man under stress is 160 beats per minute.

(a) Find a linear function that relates the maximum recommended heart rate for men to age.

(b) Predict the maximum recommended heart rate for a 45-year-old man under stress.

(c) Interpret the slope.

(d) For what age would the maximum recommended heart rate under stress be 168 beats per minute?

59. Car Rental The daily rental charge for a particular car is $35 plus $0.12 per mile.

(a) Find a linear function that expresses the rental cost C as a function of the miles driven m.

(b) What are the independent and dependent variables?

(c) What is the domain of this linear function?

(d) For a one-day rental, what is the rental cost if 124 miles are driven?

(e) For a one-day rental, how many miles were driven if the rental cost was $67.16?

(f) Graph the linear function.

60. Satellite Television Bill A satellite television company charges $33.99 per month for a 100-channel package, plus $3.50 for each pay-per-view movie watched that month.

(a) Find a linear function that expresses the monthly bill B as a function of x, the number of pay-per-view movies watched that month.

(b) What are the independent and dependent variables?

(c) What is the domain of this linear function?

(d) What is the monthly bill if 5 pay-per-view movies are watched that month?

(e) For one month, how many pay-per-view movies were watched if the bill was $58.49?

(f) Graph the linear function.

In Problems 61 and 62,

(a) *Draw a scatter diagram of the data.*

(b) *Select two points from the scatter diagram and find the equation of the line containing the points selected.*

(c) *Graph the line found in part (b) on the scatter diagram.*

61.

x	2	5	8	11	14
y	13.3	11.6	8.4	7.2	4.6

62.

x	0	0.4	1.5	2.3	4.2
y	0.6	1.1	1.3	1.8	3.0

63. The table below gives the number of calories and the total carbohydrates (in grams) for a one-cup serving of seven name-brand cereals (not including milk).

Cereal	Calories, x	Total Carbohydrates (in grams), y
Rice Krispies®	96	23.2
Life®	160	33.3
Lucky Charms®	120	25.0
Kellogg's Complete®	120	30.7
Wheaties®	110	24.0
Cheerios®	110	22.0
Honey Nut Chex®	160	34.7

SOURCE: *Quaker Oats, General Mills, and Kellogg*

(a) Draw a scatter diagram of the data, treating calories as the independent variable.

(b) What type of relation appears to exist between calories and total carbohydrates in a one-cup serving of cereal?

(c) Select two points and find an equation of the line containing the points.

(d) Graph the line on the scatter diagram drawn in part (a).

(e) Predict the total carbohydrates in a one-cup serving of cereal that has 140 calories.

(f) Interpret the slope of the line found in part (c).

64. Second-Day Delivery Costs The table below lists some selected prices charged by Federal Express for FedEx 2Day delivery, depending on the weight of the package.

Weight (in pounds), x	FedEx 2Day® Delivery Charge, y
1	$21.09
3	$22.46
6	$27.76
8	$31.75
9	$33.92
11	$37.79

SOURCE: *Federal Express Corporation*

(a) Draw a scatter diagram of the data, treating weight as the independent variable.

(b) What type of relation appears to exist between the weight of the package and the FedEx 2Day delivery charge?

(c) Select two points and find an equation of the line containing the points.

(d) Graph the line on the scatter diagram drawn in part (a).

(e) Predict the FedEx 2Day delivery charge for shipping a 5-pound package.

(f) Interpret the slope of the line found in part (c).

Section 2.5 Compound Inequalities

KEY CONCEPT

• If $a < b$, then we can write $a < x$ and $x < b$ as $a < x < b$.

KEY TERMS

Intersection
Union

Compound inequality
Solve a compound inequality

You Should Be Able To...	EXAMPLE	Review Exercises
1 Determine the intersection or union of two sets (p. 188)	Examples 1 and 2	65–70
2 Solve compound inequalities involving "and" (p. 190)	Examples 3 through 6	71, 72, 75, 76, 80
3 Solve compound inequalities involving "or" (p. 192)	Examples 7 and 8	73, 74, 77, 78, 79
4 Solve problems using compound inequalities (p. 194)	Example 9	81, 82

In Problems 65–68, use $A = \{2, 4, 6, 8\}$, $B = \{-1, 0, 1, 2, 3, 4\}$, and $C = \{1, 2, 3, 4\}$ to find each set.

65. $A \cup B$ **66.** $A \cap C$

67. $B \cap C$ **68.** $A \cup C$

In Problems 69 and 70, use the graph of the inequality to find each set.

69. $A = \{x \mid x \leq 4\}; B = \{x \mid x > 2\}$
Find **(a)** $A \cap B$ and **(b)** $A \cup B$.

70. $E = \{x \mid x \geq 3\}; F = \{x \mid x < -2\}$
Find **(a)** $E \cap F$ and **(b)** $E \cup F$.

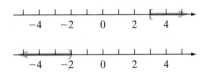

In Problems 71–80, solve each compound inequality. Graph the solution set.

71. $x < 4$ and $x + 3 > 2$

72. $3 < 2 - x < 7$

73. $x + 3 < 1$ or $x > 2$

74. $x + 6 \geq 10$ or $x \leq 0$

75. $3x + 2 \leq 5$ and $-4x + 2 \leq -10$

76. $1 \leq 2x + 5 < 13$

77. $x - 3 \leq -5$ or $2x + 1 > 7$

78. $3x + 4 > -2$ or $4 - 2x \geq -6$

79. $\dfrac{1}{3}x > 2$ or $\dfrac{2}{5}x < -4$

80. $x + \dfrac{3}{2} \geq 0$ and $-2x + \dfrac{3}{2} > \dfrac{1}{4}$

81. Heart Rates The normal heart rate for healthy adults between the ages of 21 and 60 should be between 70 and 75 beats per minute (inclusive). If we let x represent the heart rate of an adult between the ages of 21 and 60, express the normal range of values using a compound inequality.

82. Heating Bills For usage above 800 kilowatt hours, the non-space heat energy charge for American Electric Power residential service was $43.56 plus $0.038752 per kilowatt hour for usage over 800 kwh. During one winter, a customer's charge ranged from a low of $52.62 to a high of $88.22. Over what range of values did electrical usage vary (in kilowatt hours)? Express answers rounded to the nearest tenth of a kilowatt hour.

Section 2.6   Absolute Value Equations and Inequalities

KEY CONCEPTS

- **Equations Involving Absolute Value**
  If a is a positive real number and if u is any algebraic expression, then $|u| = a$ is equivalent to $u = a$ or $u = -a$.

- **Equations Involving Two Absolute Values**
  If u and v are any algebraic expression, then $|u| = |v|$ is equivalent to $u = v$ or $u = -v$.

- **Inequalities of the Form** $<$ **or** $\leq$ **Involving Absolute Value**
  If a is a positive real number and if u is any algebraic expression, then $|u| < a$ is equivalent to $-a < u < a$, and $|u| \leq a$ is equivalent to $-a \leq u \leq a$.

- **Inequalities of the Form** $>$ **or** $\geq$ **Involving Absolute Value**
  If a is a positive real number and if u is any algebraic expression, then $|u| > a$ is equivalent to $u < -a$ or $u > a$, and $|u| \geq a$ is equivalent to $u \leq -a$ or $u \geq a$.

You Should Be Able To...	EXAMPLE	Review Exercises
❶ Solve absolute value equations (p. 198)	Examples 1 through 4	83–88
❷ Solve absolute value inequalities involving $<$ or $\leq$ (p. 202)	Examples 5 through 7	89, 91, 94, 95
❸ Solve absolute value inequalities involving $>$ or $\geq$ (p. 203)	Examples 8 and 9	90, 92, 93, 96
❹ Solve applied problems involving absolute value inequalities (p. 205)	Example 10	97, 98

In Problems 83–88, solve the absolute value equation.

83. $|x| = 4$

84. $|3x - 5| = 4$

85. $|-y + 4| = 9$

86. $-3|x + 2| - 5 = -8$

87. $|2w - 7| = -3$

88. $|x + 3| = |3x - 1|$

In Problems 89–96, solve each absolute value inequality. Graph the solution set on a real number line.

89. $|x| < 2$

90. $|x| \geq \dfrac{7}{2}$

91. $|x + 2| \leq 3$

92. $|4x - 3| \geq 1$

93. $3|x| + 6 \geq 1$

94. $|7x + 5| + 4 < 3$

95. $|(x - 3) - 2| \leq 0.01$

96. $\left| \dfrac{2x - 3}{4} \right| > 1$

97. Tolerance The diameter of a certain ball bearing is required to be 0.503 inch. The tolerance on the bearing is 0.001 inch. If x represents the diameter of a bearing, the acceptable diameters of the bearing can be expressed as $|x - 0.503| \leq 0.001$. Determine the acceptable diameters of the bearing.

98. Tensile Strength The tensile strength of paper used to make grocery bags is about 40 lb/in.^{2} A paper grocery bag whose tensile strength satisfies the inequality $\left| \dfrac{x - 40}{2} \right| > 1.96$ has an unusual tensile strength. Determine the tensile strengths that would be considered unusual.

Chapter 2 Test *Step-by-step test solutions are found on the Chapter Test Prep Videos available in MyMathLab® or on You Tube.*

1. Write the relation as a map. Then identify the domain and the range of the relation.

$$\{ (2, 8), (5, -2), (7, 12), (-4, -7), (7, 3), (5, -1) \}$$

2. Identify the domain and range of the relation from the graph.

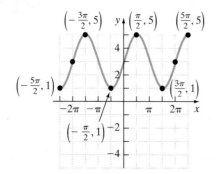

$\left(-\frac{3\pi}{2}, 5\right)$ $\left(\frac{\pi}{2}, 5\right)$ $\left(\frac{5\pi}{2}, 5\right)$

$\left(-\frac{5\pi}{2}, 1\right)$ $\left(-\frac{\pi}{2}, 1\right)$ $\left(\frac{3\pi}{2}, 1\right)$

3. Graph the relation $y = x^2 - 3$ by plotting points. Use the graph of the relation to identify the domain and range.

In Problems 4 and 5, determine whether each relation represents a function. Identify the domain and the range of each relation.

4.

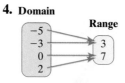

5.

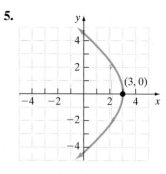

6. Does the equation $y = \pm 5x$ represent a function? Why or why not?

7. For $f(x) = -3x + 11$, find $f(x + h)$.

8. For $g(x) = 2x^2 + x - 1$, find the indicated values.

(a) $g(-2)$ **(b)** $g(0)$ **(c)** $g(3)$

9. Sketch the graph of $f(x) = x^2 + 3$.

10. Using data from 1996 to 2010, the function $P(x) = 0.24x + 4.34$ approximates the average movie ticket price (in dollars) x years after 1996.

SOURCE: *National Association of Theater Owners*

(a) Identify the dependent and independent variables.

(b) Evaluate $P(14)$ and explain what it represents.

(c) In what year will the average movie ticket be $10.00?

11. Find the domain of $f(x) = \dfrac{-15}{x + 2}$.

12. $h(x) = -5x + 12$

(a) Is the point $(2, 2)$ on the graph of the function?

(b) If $x = 3$, what is $h(x)$? What point is on the graph of the function?

(c) If $h(x) = 27$, what is x? What point is on the graph of h?

(d) What is the zero of h?

13. The following graph represents the speed of a car as a function of time.

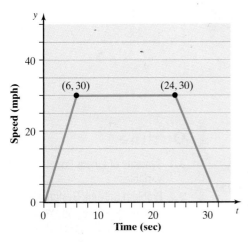

(a) When does the car stop accelerating?

(b) For how long does the car maintain a constant speed?

14. Crafts Fair Sales Henry plans to sell small wooden shelves at a crafts fair for $30 each. A booth at the fair costs $100 to rent. Henry estimates his expenses for producing the shelves to be $12 each, so his profit will be $18 per shelf.

(a) Write a function that expresses Henry's profit P as a function of the number of shelves sold x.

(b) What is the implied domain of this linear function?

(c) What is the profit if Henry sells 34 shelves?

(d) Graph the linear function.

(f) If Henry's profit is $764, how many shelves did he sell?

15. Shetland Pony Weights The table below lists the average weight of a Shetland pony, depending on the age of the pony.

Age (months), x	Average Weight (kilograms), y
3	60
6	95
12	140
18	170
24	185

SOURCE: *The Merck Veterinary Manual*

(a) Draw a scatter diagram of the data, treating age as the independent variable.

(b) What type of relation appears to exist between the age and the weight of a Shetland pony?

(c) Select two points and find an equation of the line containing the points.

(d) Graph the line on the scatter diagram drawn in part (a).

(e) Predict the weight of a 9-month-old Shetland pony.

(f) Interpret the slope of the line found in part (c).

16. Solve: $|2x + 5| - 3 = 0$

In Problems 17–20, solve each inequality and graph the solution set on a real number line.

17. $x + 2 < 8$ and $2x + 5 \geq 1$

18. $x > 4$ or $2(x - 1) + 3 < -2$

19. $2|x - 5| + 1 < 7$

20. $|-2x + 1| \geq 5$

3 Systems of Linear Equations and Inequalities

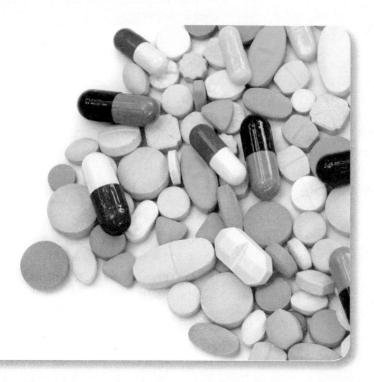

Finally! You've landed an offer for a sales position at your ideal job: pharmaceutical sales. You have been offered two salary options: an annual base salary of $15,000 plus a commission of 1% on all sales, or an annual base salary of $25,000 plus a commission of 0.75% on all sales. What annual sales are required for these options to result in the same annual earnings? See Problem 49 in Section 3.2.

The Big Picture: Putting It Together

Recall from the first part of Chapter 1 that linear equations in one variable can have no solution, one solution, or infinitely many solutions. In the second part of Chapter 1, we graphed linear equations and inequalities in two variables. We will use graphs of linear equations in this chapter to help us visualize results.

In this chapter, we will discuss solving *systems of equations:* two or more linear equations involving two or more variables. We will learn a variety of techniques for solving these systems. We will also learn that these systems can have no solution, one solution, or infinitely many solutions, just like linear equations in one variable. We conclude the chapter by looking at systems of linear inequalities.

Outline

3.1 Systems of Linear Equations in Two Variables

Objectives

1 Determine Whether an Ordered Pair Is a Solution of a System of Linear Equations

2 Solve a System of Two Linear Equations by Graphing

3 Solve a System of Two Linear Equations by Substitution

4 Solve a System of Two Linear Equations by Elimination

5 Identify Inconsistent Systems

6 Write the Solution of a System with Dependent Equations

Are You Ready for This Section?

Before getting started, take the following readiness quiz. If you get a problem wrong, go back to the section cited and review the material.

R1. Evaluate $2x - 3y$ for $x = 5, y = 4$.   [Section R.5, pp. 40–41]

R2. Determine whether the point $(4, -1)$ is on the graph of the equation $2x - 3y = 11$.   [Section 1.5, pp. 94–95]

R3. Graph: $y = 3x - 7$ [Section 1.6, p. 113–114]

R4. Find the equation of the line parallel to $y = -3x + 1$ containing the point $(2, 3)$.   [Section 1.7, p. 121]

R5. Determine the slope and y-intercept of $4x - 3y = 15$.   [Section 1.6, pp. 113–114]

R6. What is the additive inverse of 4?   [Section R.3, p. 21]

R7. Solve: $2x - 3(-3x + 1) = -36$ [Section 1.1, pp. 50–53]

Recall from Section 1.6 that an equation in two variables is linear if it can be written in the form $Ax + By = C$, where A, B, and C are real numbers and A and B are not both zero.

A **system of linear equations** is a grouping of two or more linear equations, each of which contains one or more variables.

EXAMPLE 1   **Examples of Systems of Linear Equations**

(a) $\begin{cases} 2x + y = 5 \\ x - 5y = -10 \end{cases}$ Two linear equations containing two variables, x and y

(b) $\begin{cases} x + 3y + z = 8 \\ 3x - y + 6z = 12 \\ -4x - y + 2z = -1 \end{cases}$ Three linear equations containing three variables, x, y, and z

We use a brace, as shown in the systems in Example 1, to remind us that we are dealing with a *system* of equations. In this section, we concentrate on systems of two linear equations containing two variables, such as the system in Example 1(a).

▶ **1** Determine Whether an Ordered Pair Is a Solution of a System of Linear Equations

A **solution** of a system of equations consists of values for the variables that are solutions of each equation of the system. We represent the solution of a system of two linear equations containing two unknowns as an ordered pair, (x, y).

EXAMPLE 2   **Determining Whether an Ordered Pair Is a Solution of a System of Linear Equations**

Determine whether the given ordered pairs are solutions of the system of equations.

$$\begin{cases} 2x + 3y = 9 \\ -5x - 3y = 0 \end{cases}$$

(a) $(6, -1)$ **(b)** $(-3, 5)$

Solution

To help organize our thoughts, we name $2x + 3y = 9$ equation (1) and $-5x - 3y = 0$ equation (2).

$$\begin{cases} 2x + 3y = 9 & (1) \\ -5x - 3y = 0 & (2) \end{cases}$$

(a) Let $x = 6$ and $y = -1$ in both equations (1) and (2). If both equations are true, then $(6, -1)$ is a solution.

Equation (1):

$$2x + 3y = 9$$
$$x = 6, y = -1: \quad 2(6) + 3(-1) \stackrel{?}{=} 9$$
$$12 - 3 \stackrel{?}{=} 9$$
$$9 = 9 \quad \text{True}$$

Equation (2):

$$-5x - 3y = 0$$
$$x = 6, y = -1: \quad -5(6) - 3(-1) \stackrel{?}{=} 0$$
$$-30 + 3 \stackrel{?}{=} 0$$
$$-27 = 0 \quad \text{False}$$

Work Smart

A solution of a system must satisfy all of the equations in the system.

Although $x = 6$, $y = -1$ satisfy equation (1), they do not satisfy equation (2); therefore, $(6, -1)$ is not a solution of the system of equations.

(b) Let $x = -3$ and $y = 5$ in both equations (1) and (2). If both equations are true, then $(-3, 5)$ is a solution.

Equation (1):

$$2x + 3y = 9$$
$$x = -3, y = 5: \quad 2(-3) + 3(5) \stackrel{?}{=} 9$$
$$-6 + 15 \stackrel{?}{=} 9$$
$$9 = 9 \quad \text{True}$$

Equation (2):

$$-5x - 3y = 0$$
$$x = -3, y = 5: \quad -5(-3) - 3(5) \stackrel{?}{=} 0$$
$$15 - 15 \stackrel{?}{=} 0$$
$$0 = 0 \quad \text{True}$$

Because $x = -3$, $y = 5$ satisfy both equations (1) and (2), the ordered pair $(-3, 5)$ is a solution of the system of equations. ●

Work Smart

It is a good idea to number the equations in a system so that it is easier to keep track of your work.

For the remainder of the chapter, we shall number each equation as we did in Example 1. When solving homework problems, you should do the same.

Quick ✓

1. A _____ __ ____ _____ is a grouping of two or more linear equations, each of which contains one or more variables.

2. A _____ of a system of equations consists of values of the variables that satisfy each equation of the system.

3. Which of the following points is a solution of the system of equations?

$$\begin{cases} 2x + 3y = 7 \\ 3x + y = -7 \end{cases}$$

(a) $(2, 1)$ **(b)** $(-4, 5)$ **(c)** $(-2, -1)$

Visualizing the Solutions in a System of Two Linear Equations

The graph of each equation in the system is a line, so a system of two linear equations containing two variables represents a pair of lines. The graphs of the two lines can appear in one of three ways:

1. INTERSECT: If the lines intersect, then the system of equations has one solution given by the point of intersection. The system is **consistent** and the equations are **independent.** See Figure 1(a).

2. PARALLEL: If the lines are parallel, then the system of equations has no solution because the lines never intersect. The system is **inconsistent** and the equations are **independent.** See Figure 1(b).

3. COINCIDENT: If the lines lie on top of each other (are coincident), then the system of equations has infinitely many solutions. The solution set is the set of all points on the line. The system is **consistent** and the equations are **dependent.** See Figure 1(c).

Figure 1

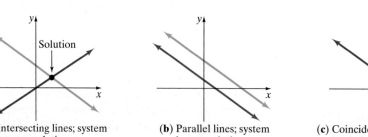

(a) Intersecting lines; system has one solution

(b) Parallel lines; system has no solution

(c) Coincident lines; system has infinitely many solutions

Quick ✓

4. If a system of equations has no solution, it is said to be _____.

5. If a system of equations has infinitely many solutions, the system is said to be _____ and the equations are _____.

6. *True or False* A system of two linear equations containing two variables always has at least one solution.

7. *True or False* When the lines in a system of equations are parallel, then the system is inconsistent and has no solution.

8. *True or False* The visual representation of a consistent system consisting of dependent equations is one line.

For now, we will concentrate on solving systems that have a single solution.

▶ ❷ Solve a System of Two Linear Equations by Graphing

The solution of a system of equations is the point or points of intersection, if any, of the two graphs, because both equations are satisfied at the point(s) of intersection.

(EXAMPLE 3) **Solving a System of Two Linear Equations Using Graphing**

Solve the following system by graphing: $\begin{cases} x + y = -1 \\ -2x + y = -7 \end{cases}$

Solution

First, name $x + y = -1$ equation (1) and name $-2x + y = -7$ equation (2).

$$\begin{cases} x + y = -1 & (1) \\ -2x + y = -7 & (2) \end{cases}$$

Figure 2

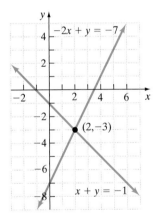

In order to graph each equation, put them in slope-intercept form. Equation (1) in slope-intercept form is $y = -x - 1$, which has slope -1 and y-intercept $(0, -1)$. Equation (2) in slope-intercept form is $y = 2x - 7$, which has slope 2 and y-intercept $(0, -7)$. Figure 2 shows their graphs. (Note that we could also have graphed the lines using the intercepts.) The lines appear to intersect at $(2, -3)$, so we believe that the ordered pair $(2, -3)$ is the solution to the system.

Check Let $x = 2$ and $y = -3$ in both equations in the system:

Equation (1): $\qquad x + y = -1 \qquad$ Equation (2): $\qquad -2x + y = -7$

$x = 2, y = -3$: $\quad 2 + (-3) \overset{?}{=} -1 \qquad x = 2, y = -3$: $\quad -2(2) + (-3) \overset{?}{=} -7$

$\qquad\qquad\qquad 2 - 3 \overset{?}{=} -1 \qquad\qquad\qquad\qquad\qquad -4 - 3 \overset{?}{=} -7$

$\qquad\qquad\qquad -1 = -1$ True $\qquad\qquad\qquad\qquad\qquad -7 = -7$ True

Both equations are true, so the solution is the ordered pair $(2, -3)$. ●

Quick ✓

In Problems 9 and 10, solve the system by graphing.

9. $\begin{cases} y = -3x + 10 \\ y = \quad 2x - 5 \end{cases}$

10. $\begin{cases} 2x + \quad y = -1 \\ -2x + 2y = 10 \end{cases}$

▶ ❸ Solve a System of Two Linear Equations by Substitution

Work Smart

Obtaining exact solutions using graphical methods can be difficult. Therefore, algebraic methods should be used.

Finding the exact point of intersection of two lines can be difficult if the x- and y-coordinates of that point are not integers. Therefore, we prefer to use algebraic, rather than graphical, methods for solving systems of equations. One algebraic method is *substitution*. The goal of the substitution method is to obtain a single linear equation involving a single unknown.

EXAMPLE 4 How to Solve a System of Two Equations by Substitution

Solve the following system by substitution: $\begin{cases} 3x + y = -9 & (1) \\ -2x + 3y = 17 & (2) \end{cases}$

Step-by-Step Solution

Step 1: Solve one of the equations for one of the unknowns.

It is easier to solve equation (1) for y since the coefficient of y is 1.

$$3x + y = -9$$

Subtract $3x$ from both sides: $\qquad y = -3x - 9$

Step 2: Substitute $-3x - 9$ for y in equation (2).

Equation (2): $\qquad -2x + 3y = 17$

$$-2x + 3(-3x - 9) = 17$$

Step 3: Solve the equation for x.

Distribute the 3: $\qquad -2x - 9x - 27 = 17$

Combine like terms: $\qquad -11x - 27 = 17$

Add 27 to both sides: $\qquad -11x = 44$

Divide both sides by -11: $\qquad x = -4$

Step 4: Substitute -4 for x in the equation from Step 1.

$$y = -3x + 9$$
$$y = -3(-4) - 9$$
$$y = 12 - 9$$
$$y = 3$$

Step 5: Check Verify that $x = -4$ and $y = 3$ satisfy each equation in the original system.

Equation (1): $\qquad 3x + y = -9 \qquad$ Equation (2): $\qquad -2x + 3y = 17$

$x = -4, y = 3$: $\quad 3(-4) + 3 \stackrel{?}{=} -9 \qquad\qquad -2(-4) + 3(3) \stackrel{?}{=} 17$

$\qquad\qquad -12 + 3 \stackrel{?}{=} -9 \qquad\qquad\qquad\qquad 8 + 9 \stackrel{?}{=} 17$

$\qquad\qquad -9 = -9 \quad$ True $\qquad\qquad\qquad\qquad 17 = 17 \quad$ True

Both equations are satisfied, so the solution is the ordered pair $(-4, 3)$.

Work Smart

When using substitution, solve for the variable whose coefficient is 1 or -1 in order to simplify the algebra.

> **Solving a System of Two Linear Equations by Substitution**
>
> **Step 1:** Solve one of the equations for one of the unknowns. Choose the equation that is easier to solve for a variable. Typically, this would be an equation that has a variable whose coefficient is 1 or -1.
>
> **Step 2:** Substitute the expression solved for in Step 1 into the *other* equation. The result will be a single linear equation in one unknown.
>
> **Step 3:** Solve the linear equation in one unknown found in Step 2.
>
> **Step 4:** Substitute the value of the variable found in Step 3 into one of the original equations to find the value of the other variable.
>
> **Step 5:** Check your answer by substituting the ordered pair into both of the original equations.

EXAMPLE 5 Solving a System of Two Equations by Substitution

Solve the following system by substitution: $\begin{cases} 2x - 3y = -6 & (1) \\ -8x + 3y = 3 & (2) \end{cases}$

Solution

We choose to solve equation (1) for x because it seems easiest.

Equation (1): $\quad 2x - 3y = -6$

Add $3y$ to both sides: $\qquad 2x = 3y - 6$

$$\text{Divide both sides by 2:} \qquad x = \frac{3y - 6}{2}$$

$$\text{Divide both terms in the numerator by 2:} \qquad x = \frac{3}{2}y - 3$$

Now substitute $\frac{3}{2}y - 3$ for x in equation (2) and then solve for y.

$$\text{Equation (2):} \qquad -8x + 3y = 3$$

$$-8\left(\frac{3}{2}y - 3\right) + 3y = 3$$

$$\text{Distribute the } -8: \qquad -8 \cdot \frac{3}{2}y - (-8) \cdot 3 + 3y = 3$$

$$\text{Multiply:} \qquad -12y + 24 + 3y = 3$$

$$\text{Combine like terms:} \qquad -9y + 24 = 3$$

$$\text{Subtract 24 from both sides:} \qquad -9y = -21$$

$$\text{Divide both sides by } -9: \qquad y = \frac{-21}{-9}$$

$$y = \frac{7}{3}$$

Now substitute $\frac{7}{3}$ for y in $x = \frac{3}{2}y - 3$ to find the value of x.

$$x = \frac{3}{2}\left(\frac{7}{3}\right) - 3$$

$$\text{Multiply:} \quad x = \frac{7}{2} - 3$$

$$x = \frac{1}{2}$$

Check Equation (1): $\quad 2x - 3y = -6 \qquad$ Equation (2): $\quad -8x + 3y = 3$

$$x = \frac{1}{2}, y = \frac{7}{3}: \quad 2\left(\frac{1}{2}\right) - 3\left(\frac{7}{3}\right) \overset{?}{=} -6 \qquad -8\left(\frac{1}{2}\right) + 3\left(\frac{7}{3}\right) \overset{?}{=} 3$$

$$1 - 7 \overset{?}{=} -6 \qquad\qquad -4 + 7 \overset{?}{=} 3$$

$$-6 = -6 \quad \text{True} \qquad\qquad 3 = 3 \quad \text{True}$$

Both equations are satisfied, so the solution is the ordered pair $\left(\frac{1}{2}, \frac{7}{3}\right)$.

Quick ✓

In Problems 11 and 12, solve the system using substitution.

11. $\begin{cases} y = -3x - 5 \\ 5x + 3y = 1 \end{cases}$ **12.** $\begin{cases} 2x + y = -2 \\ -3x - 2y = -2 \end{cases}$

▶ ④ Solve a System of Two Linear Equations by Elimination

Using substitution in Example 5 led to some complicated equations containing fractions. A second algebraic method for solving a system of linear equations is the *elimination method*. This method is usually preferred over the substitution method if substitution leads to fractions.

The basic idea in using elimination is to get the coefficients of one of the variables to be additive inverses, or opposites, such as 5 and −5, so that we can add the equations together and get a single linear equation involving one unknown.

Let's go over an example that shows how to solve a system of linear equations by elimination.

EXAMPLE 6 **How to Solve a System of Linear Equations by Elimination**

Solve: $\begin{cases} 5x + 2y = -5 & (1) \\ -2x - 4y = -14 & (2) \end{cases}$

Step-by-Step Solution

Step 1: *Our first goal is to get the coefficients on one of the variables to be additive inverses.*

We can make the coefficients of y additive inverses by multiplying equation (1) by 2.

$\begin{cases} 5x + 2y = -5 & (1) \\ -2x - 4y = -14 & (2) \end{cases}$

Multiply both sides of (1) by 2:

$\begin{cases} 2(5x + 2y) = 2(-5) & (1) \\ -2x - 4y = -14 & (2) \end{cases}$

Use the Distributive Property:

$\begin{cases} 10x + 4y = -10 & (1) \\ -2x - 4y = -14 & (2) \end{cases}$

Step 2: *Now add equations (1) and (2) to eliminate the variable y and then solve for x.*

$\begin{cases} 10x + 4y = -10 & (1) \\ -2x - 4y = -14 & (2) \end{cases}$

Add (1) and (2): $\qquad 8x \qquad = -24$

Divide both sides by 8: $\qquad x = -3$

The x-coordinate of the solution is -3.

Step 3: *Substitute -3 for x in either equation (1) or (2) and solve for y. We will use equation (1).*

Equation (1): $\qquad 5x + 2y = -5$

$x = -3$: $\qquad 5(-3) + 2y = -5$

$\qquad -15 + 2y = -5$

Add 15 to both sides: $\qquad 2y = 10$

Divide both sides by 2: $\qquad y = 5$

The y-coordinate of the solution is 5. The solution appears to be $(-3, 5)$.

Step 4: Check *Verify that $x = -3$ and $y = 5$ satisfy each equation in the original system.*

Equation (1): $\qquad 5x + 2y = -5$

$x = -3, y = 5$: $\quad 5(-3) + 2(5) \overset{?}{=} -5$

$-15 + 10 \overset{?}{=} -5$

$-5 = -5 \quad$ True

Equation (2): $\quad -2x - 4y = -14$

$-2(-3) - 4(5) \overset{?}{=} -14$

$6 - 20 \overset{?}{=} -14$

$-14 = -14 \quad$ True

The solution is the ordered pair $(-3, 5)$. ●

The following steps can be used to solve a system of linear equations by elimination.

Solving a System of Linear Equations by Elimination

Step 1: Write both equations of the system in standard form, $Ax + By = C$. If necessary, multiply both sides of one equation or both equations by a nonzero constant so that the coefficients of one of the variables are additive inverses.

Step 2: Add equations (1) and (2) to eliminate the variable whose coefficients are now additive inverses. Solve the resulting equation for the unknown.

Step 3: Substitute the value of the variable found in Step 2 into one of the original equations to find the value of the remaining variable.

Step 4: Check your answer by substituting the ordered pair in both of the original equations.

What allows us to add two equations and use the result to replace an equation? Remember, an equation is a statement that the left side equals the right side. When we add equation (2) to equation (1), we are adding the same quantity to both sides of equation (1).

EXAMPLE 7 **Solving a System of Linear Equations by Elimination**

Solve:
$$\begin{cases} \dfrac{5}{2}x + 2y = 5 & (1) \\[2mm] \dfrac{3}{2}x + \dfrac{3}{2}y = \dfrac{9}{4} & (2) \end{cases}$$

Solution

Because both equations have fractions, the first goal is to clear the fractions by multiplying both sides of equation (1) by 2 and both sides of equation (2) by 4.

$$\begin{cases} \dfrac{5}{2}x + 2y = 5 & (1) \\[2mm] \dfrac{3}{2}x + \dfrac{3}{2}y = \dfrac{9}{4} & (2) \end{cases}$$

Multiply both sides of (1) by 2:
$$\begin{cases} 2\left(\dfrac{5}{2}x + 2y\right) = 2 \cdot 5 & (1) \\[2mm] 4\left(\dfrac{3}{2}x + \dfrac{3}{2}y\right) = 4 \cdot \dfrac{9}{4} & (2) \end{cases}$$

Multiply both sides of (2) by 4:

Use the Distributive Property:
$$\begin{cases} 5x + 4y = 10 & (1) \\ 6x + 6y = 9 & (2) \end{cases}$$

Divide both sides of equation (2) by 3:
$$\begin{cases} 5x + 4y = 10 & (1) \\ 2x + 2y = 3 & (2) \end{cases}$$

Multiply both sides of equation (2) by -2:
$$\begin{cases} 5x + 4y = 10 & (1) \\ -4x - 4y = -6 & (2) \end{cases}$$

Add (1) and (2):
$$x = 4$$

Work Smart

If you are not afraid of fractions, you can eliminate x by multiplying both sides of equation (1) by -3 and both sides of equation (2) by 5. Try it!

Work Smart

Although it is not necessary to divide both sides of equation (2) by 3, it makes solving the problem easier. Do you see why?

Back-substitute 4 for x in the original equation (1).

Equation (1):
$$\dfrac{5}{2}x + 2y = 5$$

$x = 4$:
$$\dfrac{5}{2}(4) + 2y = 5$$

$$10 + 2y = 5$$

Subtract 10 from both sides:
$$2y = -5$$

Divide both sides by 2:
$$y = -\dfrac{5}{2}$$

We have that $x = 4$ and $y = -\dfrac{5}{2}$.

We leave the check to you. The solution is the ordered pair $\left(4, -\dfrac{5}{2}\right)$.

Quick ✓

13. The basic idea in using the elimination method is to get the coefficients of one of the variables to be _____ _____, such as 5 and −5.

In Problems 14–16, solve the system using elimination. **Note:** *Problem 14 is the system that we solved by using substitution in Example 5.*

14. $\begin{cases} 2x - 3y = -6 \\ -8x + 3y = 3 \end{cases}$

15. $\begin{cases} -2x + y = 4 \\ -5x + 3y = 7 \end{cases}$

16. $\begin{cases} -3x + 2y = 3 \\ 4x - 3y = -6 \end{cases}$

Work Smart

The figure below shows a consistent and independent system.

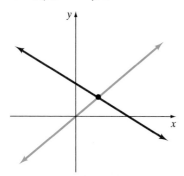

Which Method Should I Use?

When should you use each of the three methods for solving systems of two linear equations containing two unknowns? Suggestions are given below.

Method	Advantages/Disadvantages	When Should I Use It?
Graphical	Allows us to "see" the answer, but if the solutions are not integers, it can be difficult to determine the solution.	When a visual solution is required
Substitution	Method gives exact solutions. The algebra can be easy if one of the variables has a coefficient of 1.	If one of the coefficients of the variables is 1 or one of the variables is already solved for (as in $x =$ or $y =$)
Elimination	Method gives exact solutions. It is easy to use when none of the variables has a coefficient of 1.	If both equations are in standard form $(Ax + By = C)$

▶ ⑤ Identify Inconsistent Systems

Examples 2 through 7 dealt only with consistent and independent systems of equations—that is, systems with a single solution. Let's now consider systems that are inconsistent, which means the lines representing their equations are parallel, and systems that are consistent but have dependent equations, which means the lines representing their equations are coinciding (in other words, the same line).

EXAMPLE 8 | **An Inconsistent System**

Solve: $\begin{cases} 3x + 2y = 2 & (1) \\ -6x - 4y = 8 & (2) \end{cases}$

Solution

We will use the elimination method to solve this system because none of the variables has a coefficient of 1. Notice that we can make the coefficients of x additive inverses by multiplying equation (1) by 2.

$$\begin{cases} 3x + 2y = 2 & (1) \\ -6x - 4y = 8 & (2) \end{cases}$$

Multiply (1) by (2): $\begin{cases} 2(3x + 2y) = 2(2) & (1) \\ -6x - 4y = 8 & (2) \end{cases}$

Use the Distributive Property: $\begin{cases} 6x + 4y = 4 & (1) \\ -6x - 4y = 8 & (2) \end{cases}$

Add (1) and (2): $ 0 = 12$

Work Smart

When solving a system of equations, if you end up with a statement "0 = some nonzero constant," the system is inconsistent. Graphically, the lines representing the system are parallel.

Figure 3

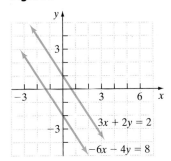

The equation $0 = 12$ is false. We conclude that the system has no solution, so the solution set is $\varnothing$ or $\{\ \}$. The system is inconsistent.

Figure 3 shows the graph of the lines from Example 8. Notice that the two lines both have slope $-\dfrac{3}{2}$. Equation (1) has a y-intercept of $(0, 1)$, and equation (2) has a y-intercept of $(0, -2)$. Therefore, the lines are parallel and do not intersect. This geometric statement is equivalent to the algebraic statement that the system has no solution.

Quick ✓

17. While solving a system of equations, a *false* statement, such as $-8 = 0$, results. This means that the solution of the system is ___ or ___.

In Problem 18, show that the system is inconsistent. Draw a graph to support your result.

18. $\begin{cases} -3x + y = 2 \\ 6x - 2y = 1 \end{cases}$

▶ ⑥ **Write the Solution of a System with Dependent Equations**

The system in the next example has infinitely many solutions.

(**EXAMPLE 9**) **Solving a System with Dependent Equations**

Solve: $\begin{cases} 3x + y = 1 & (1) \\ -6x - 2y = -2 & (2) \end{cases}$

Solution

We will use the substitution method, because solving equation (1) for y is straightforward.

$$\text{Equation (1):} \qquad 3x + y = 1$$
$$\text{Subtract } 3x \text{ from both sides:} \qquad y = -3x + 1$$

Substitute $-3x + 1$ for y in equation (2).

$$\text{Equation (2):} \qquad -6x - 2y = -2$$
$$-6x - 2(-3x + 1) = -2$$
$$\text{Distribute the } -2: \qquad -6x + 6x - 2 = -2$$
$$-2 = -2$$

Figure 4

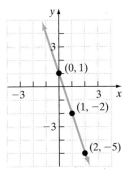

The equation $-2 = -2$ is true. This means that any values of x and y that satisfy $3x + y = 1$ or $-6x - 2y = -2$ are solutions. Some ordered pairs that satisfy both equations are $(0, 1)$, $(1, -2)$, and $(2, -5)$.

The system is consistent and its equations are dependent (the value of y that makes the equation true depends on the value of x), so there are infinitely many solutions. We write the solution in either of two equivalent ways:

$$\{(x, y) \mid 3x + y = 1\} \qquad \text{or} \qquad \{(x, y) \mid -6x - 2y = -2\}$$

In Words

The solution to a consistent system with dependent equations is "the set of all ordered pairs such that one of the equations in the system is true."

Figure 4 illustrates the system in Example 9. The graphs of the two equations are lines, and both have slope -3 and y-intercept $(0, 1)$. The lines are coincident.

Look back at the equations in Example 9. Notice that the terms in equation (2) are -2 times the terms in equation (1). This is another way to identify dependent equations when you have two equations with two unknowns.

Quick ✓

19. When solving a system of equations, a true statement, such as $11 = 11$, results. This means that the equations are _____.

In Problems 20–22, solve the system. Draw a graph to support your result.

20. $\begin{cases} -3x + 2y = 8 \\ 6x - 4y = -16 \end{cases}$ **21.** $\begin{cases} 2x - 3y = -16 \\ -3x + 2y = 19 \end{cases}$ **22.** $\begin{cases} \dfrac{1}{3}x + \dfrac{1}{5}y = 7 \\ \dfrac{1}{6}x - \dfrac{2}{5}y = -4 \end{cases}$

3.1 Exercises MyMathLab®

Exercise numbers in green have complete video solutions in MyMathLab.

PRACTICE

*Problems **1–22** are the Quick ✓s that follow the EXAMPLES.*

Building Skills

In Problems 23–26, determine whether the ordered pairs listed are solutions of the system of linear equations. See Objective 1.

23. $\begin{cases} 2x + y = 13 \\ -5x + 3y = 6 \end{cases}$
   (a) $(5, 3)$
   (b) $(3, 7)$

24. $\begin{cases} x - 2y = -11 \\ 3x + 2y = -1 \end{cases}$
   (a) $(-5, 3)$
   (b) $(-3, 4)$

25. $\begin{cases} 5x + 2y = 9 \\ -10x - 4y = -18 \end{cases}$
   (a) $(1, 2)$
   (b) $\left(2, -\dfrac{1}{2}\right)$

26. $\begin{cases} -3x + y = 5 \\ 6x - 2y = 6 \end{cases}$
   (a) $(-2, -1)$
   (b) $(2, 0)$

In Problems 27–30, use the graph of the system to determine whether the system is consistent or inconsistent. If consistent, indicate whether the equations are independent or dependent. See Objective 1.

27. $\begin{cases} x + y = 1 \\ x + 2y = 0 \end{cases}$ **28.** $\begin{cases} -2x + y = 4 \\ 2x + y = 0 \end{cases}$

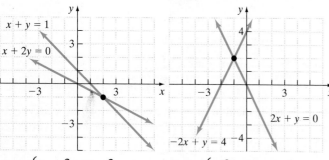

29. $\begin{cases} x - 2y = -2 \\ x - 2y = 2 \end{cases}$ **30.** $\begin{cases} 3x + y = 1 \\ -6x - 2y = -2 \end{cases}$

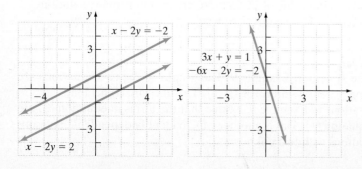

In Problems 31–34, solve the system of equations by graphing. See Objective 2.

31. $\begin{cases} y = 3x \\ y = -2x + 5 \end{cases}$ **32.** $\begin{cases} y = -2x + 4 \\ y = 2x - 4 \end{cases}$

33. $\begin{cases} 2x + y = 2 \\ x + 3y = -9 \end{cases}$ **34.** $\begin{cases} -x + 2y = -9 \\ 2x + y = -2 \end{cases}$

In Problems 35–42, solve the system of equations using substitution. See Objective 3.

35. $\begin{cases} y = -\dfrac{1}{2}x + 1 \\ y + 2x = 10 \end{cases}$ **36.** $\begin{cases} y + 3x = -4 \\ y = 4x + 17 \end{cases}$

37. $\begin{cases} x = \dfrac{2}{3}y \\ 3x - y = -3 \end{cases}$ **38.** $\begin{cases} y = \dfrac{1}{2}x \\ x - 4y = -4 \end{cases}$

39. $\begin{cases} 2x - 4y = 2 \\ x + 2y = 0 \end{cases}$ **40.** $\begin{cases} 3x + 2y = 0 \\ 6x + 2y = 5 \end{cases}$

41. $\begin{cases} x + y = 10{,}000 \\ 0.05x + 0.07y = 650 \end{cases}$

42. $\begin{cases} x + y = 5000 \\ 0.04x + 0.08y = 340 \end{cases}$

In Problems 43–50, solve the system of equations using elimination. See Objective 4.

43. $\begin{cases} x + y = -5 \\ -x + 2y = 14 \end{cases}$ **44.** $\begin{cases} x + y = -6 \\ -2x - y = 0 \end{cases}$

45. $\begin{cases} x + 2y = -5 \\ 3x + 3y = 9 \end{cases}$ **46.** $\begin{cases} -3x + 2y = -5 \\ 2x - y = 10 \end{cases}$

47. $\begin{cases} 2x + 5y = -3 \\ x + \dfrac{5}{4}y = -\dfrac{1}{2} \end{cases}$ **48.** $\begin{cases} x + 2y = -\dfrac{8}{3} \\ 3x - 3y = 5 \end{cases}$

49. $\begin{cases} 0.05x + 0.1y = 5.25 \\ 0.08x - 0.02y = 1.2 \end{cases}$

50. $\begin{cases} 0.04x + 0.06y = 2.1 \\ 0.06x - 0.03y = 0.15 \end{cases}$

In Problems 51–54, use either substitution or elimination to show that the system is inconsistent. Draw a graph to support your result. See Objective 5.

51. $\begin{cases} 3x + y = 1 \\ -6x - 2y = -4 \end{cases}$

52. $\begin{cases} -2x + 4y = 9 \\ x - 2y = -3 \end{cases}$

53. $\begin{cases} 5x - 2y = 2 \\ -10x + 4y = 3 \end{cases}$

54. $\begin{cases} 6x - 4y = 6 \\ -3x + 2y = 3 \end{cases}$

In Problems 55–60, use either substitution or elimination to show that the system is consistent and its equations are dependent. Solve the system and draw a graph to support your solution. See Objective 6.

55. $\begin{cases} y = \dfrac{1}{2}x + 1 \\ 2x - 4y = -4 \end{cases}$

56. $\begin{cases} y = -\dfrac{2}{3}x + 3 \\ 2x + 3y = 9 \end{cases}$

57. $\begin{cases} x + 3y = 6 \\ -\dfrac{x}{3} - y = -2 \end{cases}$

58. $\begin{cases} -4x + y = 8 \\ x - \dfrac{y}{4} = -2 \end{cases}$

59. $\begin{cases} \dfrac{1}{3}x - 2y = 6 \\ -\dfrac{1}{2}x + 3y = -9 \end{cases}$

60. $\begin{cases} \dfrac{5}{4}x - \dfrac{1}{2}y = 6 \\ -\dfrac{5}{3}x + \dfrac{2}{3}y = -8 \end{cases}$

Mixed Practice

In Problems 61–68, solve the system of equations using either substitution or elimination.

61. $\begin{cases} x + 3y = 0 \\ -2x + 4y = 30 \end{cases}$

62. $\begin{cases} 2x + y = -1 \\ -3x - 2y = 7 \end{cases}$

63. $\begin{cases} x = 5y - 3 \\ -3x + 15y = 9 \end{cases}$

64. $\begin{cases} y = \dfrac{1}{2}x + 2 \\ x - 2y = -4 \end{cases}$

65. $\begin{cases} 2x - 4y = 18 \\ 3x + 5y = -3 \end{cases}$

66. $\begin{cases} 12x + 45y = 0 \\ 8x + 6y = 24 \end{cases}$

67. $\begin{cases} \dfrac{5}{6}x - \dfrac{1}{3}y = -5 \\ -x + \dfrac{2}{5}y = 1 \end{cases}$

68. $\begin{cases} \dfrac{1}{3}x - \dfrac{1}{2}y = -5 \\ -\dfrac{4}{5}x + \dfrac{6}{5}y = 1 \end{cases}$

Applying the Concepts

In Problems 69–72, write each equation in the system of equations in slope-intercept form. Use the slope-intercept form to determine the number of solutions the system has.

69. $\begin{cases} 2x + y = -5 \\ 5x + 3y = 1 \end{cases}$

70. $\begin{cases} 4x - 2y = 8 \\ -10x + 5y = 5 \end{cases}$

71. $\begin{cases} 3x - 2y = -2 \\ -6x + 4y = 4 \end{cases}$

72. $\begin{cases} 2x - y = -5 \\ -4x + 3y = 9 \end{cases}$

△ **73. Parallelogram** Use the parallelogram shown to the right to answer parts (a) and (b).

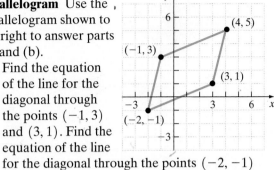

(a) Find the equation of the line for the diagonal through the points $(-1, 3)$ and $(3, 1)$. Find the equation of the line for the diagonal through the points $(-2, -1)$ and $(4, 5)$.

(b) Find the point of intersection of the diagonals.

△ **74. Rhombus** A rhombus is a parallelogram whose adjacent sides are congruent. Use the rhombus to the right to answer parts (a), (b), and (c).

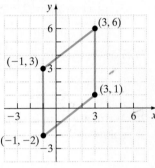

(a) Find the equation of the line for the diagonal through the points $(-1, 3)$ and $(3, 1)$. Find the equation of the line for the diagonal through the points $(-1, -2)$ and $(3, 6)$.

(b) Find the point of intersection of the diagonals.

(c) Compare the slopes of the diagonals. What can be said about the diagonals of a rhombus?

Extending the Concepts

75. Which of the following ordered pairs could be a solution to the system graphed below?

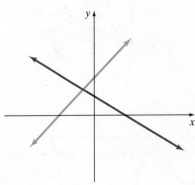

(a) $(2, 4)$ **(b)** $(-2, 0)$ **(c)** $(-3, 1)$

(d) $(5, -2)$ **(e)** $(-1, -3)$ **(f)** $(-1, 3)$

76. Which of the following systems of equations could have the graph below?

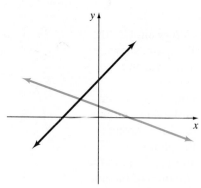

(a) $\begin{cases} 2x + 3y = 12 \\ 2x + y = -2 \end{cases}$ **(b)** $\begin{cases} 2x + 3y = 3 \\ -2x + y = 2 \end{cases}$

(c) $\begin{cases} 2x - 3y = 12 \\ x + 2y = 2 \end{cases}$

77. For the system $\begin{cases} Ax + 3By = 2 \\ -3Ax + By = -11 \end{cases}$, find A and B such that $x = 3$, $y = 1$ is a solution.

78. Write a system of equations that has $(3, 5)$ as a solution.

79. Write a system of equations that has $(-1, 4)$ as a solution.

△ **80. Centroid** The medians of a triangle are the line segments from each vertex to the midpoint of the opposite side. The centroid of a triangle is the point where the medians of the triangle intersect. Use the information given in the figure of the triangle to find its centroid.

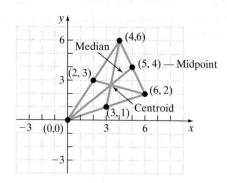

In Problems 81–84, solve each system using either substitution or elimination. Draw a graph to support your solution.

81. $\begin{cases} 3x + y = 5 \\ x + y = 3 \\ x + 3y = 7 \end{cases}$ **82.** $\begin{cases} x + y = -2 \\ -3x + 2y = 16 \\ 2x - 4y = -16 \end{cases}$

83. $\begin{cases} y = \dfrac{2}{3}x - 5 \\ 4x - 6y = 30 \\ x - 5y = 11 \end{cases}$

84. $\begin{cases} -4x + 3y = 33 \\ 3x - 4y = -37 \\ 2x - 3y = 15 \end{cases}$

Explaining the Concepts

85. In this section, we presented two algebraic methods for solving a system of linear equations. Are there any circumstances where one method is preferable to the other? What are these circumstances?

86. Describe geometrically the three possibilities for a solution of a system of two linear equations containing two variables.

87. The solution of a system of two linear equations in two unknowns is $x = 3$, $y = -2$. Where do the lines in the system intersect? Why?

88. In the process of solving a system of linear equations, what tips you off that the system is consistent and the equations are dependent? What tips you off that the system is inconsistent?

The Graphing Calculator

A graphing calculator can be used to approximate the point of intersection between two equations using its INTERSECT command. We illustrate this feature of the graphing calculator by doing Example 3. Start by graphing each equation in the system as shown in Figure 5(a). Then use the INTERSECT command and find that the lines intersect at $x = 2$, $y = -3$. See Figure 5(b). The solution is the ordered pair $(2, -3)$.

Figure 5

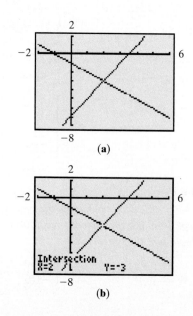

In Problems 89–96, use a graphing calculator to solve each system of equations. If necessary, express your solution rounded to two decimal places.

89. $\begin{cases} y = 3x - 1 \\ y = -2x + 5 \end{cases}$

90. $\begin{cases} y = \dfrac{3}{2}x - 4 \\ y = -\dfrac{1}{4}x + 3 \end{cases}$

91. $\begin{cases} 3x - y = -1 \\ -4x + y = -3 \end{cases}$

92. $\begin{cases} -6x - 2y = 4 \\ 5x + 3y = -2 \end{cases}$

93. $\begin{cases} 4x - 3y = 1 \\ -8x + 6y = -2 \end{cases}$

94. $\begin{cases} -2x + 5y = -2 \\ 4x - 10y = 1 \end{cases}$

95. $\begin{cases} 2x - 3y = 12 \\ 5x + y = -2 \end{cases}$

96. $\begin{cases} x - 3y = 21 \\ x + 6y = -2 \end{cases}$

3.2 Problem Solving: Systems of Two Linear Equations Containing Two Unknowns

Objectives

1 Model and Solve Direct Translation Problems

2 Model and Solve Geometry Problems

3 Model and Solve Mixture Problems

4 Model and Solve Uniform Motion Problems

5 Find the Intersection of Two Linear Functions

Are You Ready for This Section?

Before getting started, take the following readiness quiz. If you get a problem wrong, go back to the section cited and review the material.

R1. List the problem-solving steps given in Section 1.2. [Section 1.2, p. 61]

R2. A total of $25,000 is invested in stocks and bonds, with s representing the amount in stocks. Write an algebraic expression for the amount invested in bonds [Section 1.2, pp. 65–66]

R3. Suppose that you have a credit card balance of $3500 and the credit card company charges 12.5% annual interest on outstanding balances. How much interest will you have to pay after 1 month? [Section 1.2, pp. 64–65]

R4. Write a linear cost function if the fixed costs are $500 and the cost of producing each additional unit is $15 per unit. [Section 2.4, pp. 177–178]

In Sections 1.2 and 1.3, we modeled and solved problems using a single variable. In this section, we will model problems using two variables. The variables will be related through a system of equations. This approach is different from the problems in Section 1.2 because the Section 1.2 problems had a single unknown whose value was determined from a single equation.

▶ **1** Model and Solve Direct Translation Problems

Let's first look at examples of direct translation.

EXAMPLE 1 **What Are the Numbers?**

The sum of one number and three times another is −12. The difference of twice the first number and the second number is 32. What are the numbers?

Solution

Step 1: Identify We want to know the value of the two numbers.

Step 2: Name Let x represent the first number. Let y represent the second number.

Step 3: Translate "The sum of one number and three times the other is -12," can be represented by $x + 3y = -12$. We call this equation (1). "The difference of twice the first number and the second number is 32" can be represented by $2x - y = 32$. We call this equation (2). We combine the equations to form the following system:

$$\begin{cases} x + 3y = -12 & (1) \\ 2x - y = 32 & (2) \end{cases} \quad \text{The Model}$$

Step 4: Solve We will solve the system using the elimination method.

Eliminate y by multiplying equation (2) by 3:
$$\begin{cases} x + 3y = -12 & (1) \\ 3(2x - y) = 3(32) & (2) \end{cases}$$

Distribute:
$$\begin{cases} x + 3y = -12 \\ 6x - 3y = 96 \end{cases}$$

Add equations (1) and (2): $\quad 7x = 84$

Divide both sides by 7: $\quad x = 12$

Substitute 12 for x in equation (1) and solve for y.

$$x + 3y = -12$$
$$12 + 3y = -12$$

Subtract 12 from both sides: $\quad 3y = -24$

Divide both sides by 3: $\quad y = -8$

Step 5: Check If $x = 12$ (the first number) and $y = -8$ (the second number), then "the sum of one number and three times the other" is $12 + 3(-8) = 12 + (-24) = -12$, which we want. "The difference of twice the first number and the second number" is $2(12) - (-8) = 24 + 8 = 32$, which is correct.

Step 6: Answer The two numbers are 12 and -8.

Quick ✓

1. The sum of twice one number and three times another is 9. The difference of the numbers is 22. Find the numbers.

▶ (EXAMPLE 2) **Take Me Out to the Ball Game**

At a baseball game, Adrienne bought her family 4 hot dogs and 3 large Cokes for $22.25. Dave bought his family 5 hot dogs and 4 large Cokes for $28.50. How much does each hot dog cost? How much does each large Coke cost?

Solution

Step 1: Identify We want to know the cost of each hot dog and large Coke.

Step 2: Name Let h represent the cost of a hot dog. Let c represent the cost of a large Coke.

Step 3: Translate If one hot dog costs h dollars, then 4 hot dogs cost $4h$ dollars. Similarily 3 large Cokes cost $3c$ dollars. We add these costs together to get $22.25, Adrienne's purchase.

Adrienne's purchase: $\quad 4h + 3c = 22.25 \quad (1)$

Using the same notation, we describe Dave's purchase:

Dave's purchase: $\quad 5h + 4c = 28.50 \quad (2)$

We combine equations (1) and (2) to form the following system:

$$\begin{cases} 4h + 3c = 22.25 & (1) \\ 5h + 4c = 28.50 & (2) \end{cases} \quad \text{The Model}$$

Step 4: Solve

Eliminate h by multiplying
equation (1) by 5 and equation (2) by -4:

$$\begin{cases} 5(4h + 3c) = 5(22.25) & \text{(1)} \\ -4(5h + 4c) = -4(28.50) & \text{(2)} \end{cases}$$

Distribute:

$$\begin{cases} 20h + 15c = 111.25 & \text{(1)} \\ -20h - 16c = -114 & \text{(2)} \end{cases}$$

Add equations (1) and (2): $\qquad -c = -2.75$

Divide both sides by -1: $\qquad c = 2.75$

Substitute 2.75 for c in equation (1) and solve for h.

$$4h + 3(2.75) = 22.25$$

$$4h + 8.25 = 22.25$$

Subtract 8.25 from both sides: $\qquad 4h = 14$

Divide both sides by 4: $\qquad h = 3.50$

Step 5: Check If a hot dog costs \$3.50 and a Coke costs \$2.75, and Adrienne buys 4 hot dogs and 3 Cokes, then she'll spend $4(\$3.50) + 3(\$2.75) = \$14.00 + \$8.25 = \$22.25$. If Dave buys 5 hot dogs and 4 Cokes, then he'll spend $5(\$3.50) + 4(\$2.75) = \$17.50 + \$11.00 = \$28.50$. It checks.

Step 6: Answer Each hot dog costs \$3.50 and each large Coke costs \$2.75.

Quick ✓

2. At a fast-food joint, 4 cheeseburgers and 2 medium shakes cost \$10.10. At the same fast-food joint, 3 cheeseburgers and 3 medium shakes cost \$10.35. What is the cost of a cheeseburger? What is the cost of a medium shake?

❷ Model and Solve Geometry Problems

You'll need to use geometry formulas to solve many types of problems. The formula needed will depend on the problem. Remember that a formula is a model—and in geometry, formulas describe relationships and shapes.

▶ **EXAMPLE 3** Enclosing a Yard with a Fence

The Fitzgeralds just bought a dog, so they need to enclose a large area of their backyard with a fence. The perimeter of the rectangular enclosure is 180 feet. The length is 25 feet longer than the width. What are the dimensions of the enclosure?

Solution

Step 1: Identify We are looking for the length and the width of the enclosure.

Step 2: Name Let l represent the length and w represent the width of the enclosure.

Step 3: Translate The perimeter P of a rectangle is $P = 2w + 2l$, where l is the length and w is the width. Thus we know that

Equation (1): $\quad 2w + 2l = 180$

We also know that the length is 25 feet longer than the width, so

Equation (2): $\quad l = w + 25$

Equations (1) and (2) are combined to form the following system:

$$\begin{cases} 2w + 2l = 180 & \text{(1)} \\ l = w + 25 & \text{(2)} \end{cases}$$

Step 4: Solve We will use the method of substitution because equation (2) is solved for l.

Let $l = w + 25$ in equation (1):	$2w + 2(w + 25) = 180$
Distribute:	$2w + 2w + 50 = 180$
Combine like terms:	$4w + 50 = 180$
Subtract 50 from both sides:	$4w = 130$
Divide both sides by 4:	$w = 32.5$

Substitute 32.5 for w in $l = w + 25$.

$$l = 32.5 + 25$$
$$= 57.5$$

Step 5: Check If $l = 57.5$ and $w = 32.5$, then the perimeter is $2(57.5) + 2(32.5) = 180$ feet. The length, 57.5, is 25 feet more than the width, 32.5.

Step 6: Answer The width of the backyard enclosure is 32.5 feet, and the length is 57.5 feet. ●

Quick ✓

3. A rectangular field has a perimeter of 360 yards. The length of the field is twice the width. What is the length of the field? What is the width of the field?

▶ **EXAMPLE 4** **Solving a Geometry Problem**

Figure 6

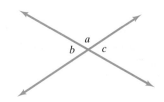

Work Smart

Two angles are supplementary if the sum of the angles is 180°. When two lines intersect, the nonadjacent angles are vertical.

Two intersecting lines are shown in Figure 6. Suppose the measure of angle a is $(x + 3y)°$, the measure of angle b is $(x + y)°$, and the measure of angle c is $(3x - 2y - 5)°$. Find x and y.

Solution

Step 1: Identify This geometry problem involves supplementary angles. Angle a and b are supplementary, which means that the sum of the measures of angles a and b equals 180°. In addition, angles b and c are vertical angles. Therefore, the measure of angle b equals the measure of angle c.

Step 2: Name The names of the unknowns, x and y, were given in the problem.

Step 3: Translate The measure of angle a plus the measure of angle b equals 180°, so

$$\underbrace{x + 3y}_{\text{Measure of Angle } a} + \underbrace{x + y}_{\text{Measure of Angle } b} = 180$$

This simplifies to

Equation (1): $2x + 4y = 180$

Also, the measure of angle b equals the measure of angle c, so

$$\underbrace{x + y}_{\text{Measure of Angle } b} = \underbrace{3x - 2y - 5}_{\text{Measure of Angle } c}$$

This simplifies to

Equation (2): $-2x + 3y = -5$

We form a system using equations (1) and (2).

$$\begin{cases} 2x + 4y = 180 & (1) \\ -2x + 3y = -5 & (2) \end{cases} \quad \text{The Model}$$

Step 4: Solve Use the elimination method by adding equations (1) and (2):

$$\begin{cases} 2x + 4y = 180 & (1) \\ -2x + 3y = -5 & (2) \end{cases}$$

Add: $\qquad\qquad 7y = 175$

Divide both sides by 7: $\qquad y = 25$

Let $y = 25$ in equation (1) and find that $x = 40$.

Step 5: Check If $x = 25$ and $y = 40$, then the measure of angle a is $(x + 3y)° = (40 + 3(25))° = 115°$. The measure of angle b is $(x + y)° = (40 + 25)° = 65°$. The measure of angle c is $(3x - 2y - 5)° = (3(40) - 2(25) - 5)° = 65°$. The sum of the measures of angle a and angle b is $115° + 65° = 180°$, so angle a and angle b are supplementary. In addition, we find that the measure of angle b equals the measure of angle c.

Step 6: Answer The value of x is 40 and the value of y is 25.

Quick ✓

4. Suppose that lines m and n are parallel. Line p transverses lines m and n. Suppose that the measure of angle a is $(x + 3y)°$, the measure of angle c is $(3x + y)°$, and the measure of angle e is $(5x + y)°$. Find x and y.

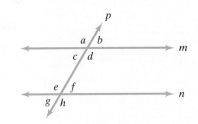

❸ Model and Solve Mixture Problems

In Section 1.2, we solved a variety of mixture problems by writing a single equation with a single unknown. An alternative approach to solving these problems is to write a system of equations. This is one of the beauties of modeling: Different models can be developed to solve the exact same problem! Let's revisit a mixture problem and solve it by developing a model that uses a system of equations.

▶ **EXAMPLE 5**   **Financial Planning**

Kevin has $15,000 to invest. His goal is to earn 9%, or $1350, annually. His financial advisor recommends investing some of the money in corporate bonds that pay 12% and the rest in government-backed Treasury notes paying 4%. How much should Kevin place in each investment in order to achieve his goal?

Solution

Step 1: Identify This mixture problem involves simple interest. Kevin needs to know how much to place in corporate bonds and how much to place in Treasury notes to earn $1350 in interest.

Step 2: Name Let c represent the amount invested in corporate bonds and t represent the amount invested in Treasury notes.

Step 3: Translate We organize the given information in Table 1.

Table 1

	Principal ($)	Rate (%)	Time (Yr)	Interest ($)
Corporate Bonds	c	0.12	1	0.12 c
Treasury Notes	t	0.04	1	0.04 t
Total	15,000	0.09	1	$0.09(15{,}000) = \$1350$

Kevin wants to earn 9%, or $1350, each year on his principal of $15,000. The total interest will be the sum of the interest from the corporate bonds and the Treasury notes.

Kevin wants to earn a total of $1350 per year, so since $t = 1$ year, we have

$$\text{Interest from corporate bonds} + \text{Interest from Treasury notes} = \$1350$$

$$\text{Equation (1):} \quad 0.12c + 0.04t = 1350$$

Since the total investment is $15,000, the sum of the amounts invested in corporate bonds and in Treasury notes must equal $15,000:

$$\text{Equation (2):} \quad c + t = 15,000$$

We use equations (1) and (2) to form a system of equations.

$$\begin{cases} 0.12c + 0.04t = 1350 & (1) \\ c + \quad t = 15,000 & (2) \end{cases} \text{The Model}$$

Step 4: Solve

Use the substitution method by first solving equation (2) for t:	$t = 15,000 - c$
Substitute $15,000 - c$ for t in equation (1):	$0.12c + 0.04(15,000 - c) = 1350$
Use the Distributive Property:	$0.12c + 600 - 0.04c = 1350$
Combine like terms:	$0.08c + 600 = 1350$
Subtract 600 from both sides:	$0.08c = 750$
Divide both sides by 0.08:	$c = 9375$

Since $c = 9375$, then $t = 15,000 - c = 15,000 - 9375 = 5625$.

Step 5: Check The simple interest earned each year on the corporate bonds is $(\$9375)(0.12)(1) = \$1125$. The simple interest earned each year on the Treasury notes is $(\$5625)(0.04)(1) = \$225$. The total interest earned is $\$1125 + \$225 = \$1350$, the amount Kevin wanted.

Step 6: Answer the Question Kevin will invest $9375 in corporate bonds and $5625 in Treasury notes. ●

Quick ✓
5. Maria has recently retired and requires an extra $7200 per year in income. She has $120,000 to invest and can invest in an Aaa-rated bond that pays 5% per annum or a B-rated bond paying 10% per annum. How much should be placed in each investment in order for Maria to achieve her goal?

⊙ (EXAMPLE 6) **Blending Coffees**

A coffee shop manager wishes to form a new blend of coffee. She wants to mix strong, distinctive Sumatra beans that sell for $12 per pound with milder Brazilian beans that sell for $8 per pound to get 50 pounds of the new blend. The new blend will sell for $9 per pound. The revenue from selling the new blend will be the same as that from selling the beans separately. How many pounds of the Sumatra and Brazilian beans are required?

Solution

Step 1: Identify This is a mixture problem. We want to know the number of pounds of Sumatra beans and the number of pounds of Brazilian beans required in the new blend.

Step 2: Name Let s represent the required number of pounds of Sumatra beans and b represent the required number of pounds of Brazilian beans.

Step 3: Translate There is to be no difference in revenue between selling the Sumatra and Brazilian separately and selling the blend. This means that if the blend contains one pound of Sumatra and one pound of Brazilian, the shop should collect $\$12(1) + \$8(1) = \$20$, which is the amount it would collect if it sold the beans separately.

Table 2 summarizes the information, and Figure 7 illustrates the idea.

Table 2

	Price ($/Pound)	Number of Pounds	Revenue
Sumatra	12	s	$12s$
Brazilian	8	b	$8b$
Blend	9	50	$9(50) = 450$

Figure 7

Sumatra	Brazilian	Blend
s pounds	b pounds	50 pounds
Sumatra	Brazilian	Blend
$12 per pound	$8 per pound	$9 per pound

$$\left(\begin{array}{c}\text{Price per pound}\\ \text{of Sumatra}\end{array}\right)\left(\begin{array}{c}\text{Pounds of}\\ \text{Sumatra}\end{array}\right) + \left(\begin{array}{c}\text{Price per pound}\\ \text{of Brazilian}\end{array}\right)\left(\begin{array}{c}\text{Pounds of}\\ \text{Brazilian}\end{array}\right) = \left(\begin{array}{c}\text{Price per pound}\\ \text{of blend}\end{array}\right)\left(\begin{array}{c}\text{Pounds of}\\ \text{blend}\end{array}\right)$$

$$\$12 \quad \cdot \quad s \quad + \quad \$8 \quad \cdot \quad b \quad = \quad \$9 \quad \cdot \quad 50$$

We have the equation

$$12s + 8b = 450 \quad (1)$$

The number of pounds of Sumatra beans plus the number of pounds of Brazilian beans should equal 50 pounds, so

$$s + b = 50 \quad (2)$$

We use equations (1) and (2) to form a system of equations.

$$\begin{cases} 12s + 8b = 450 & (1) \\ s + b = 50 & (2) \end{cases} \text{The Model}$$

Step 4: Solve

Use the substitution method by solving equation (2) for b:	$b = 50 - s$
Substitute $50 - s$ for b in equation (1):	$12s + 8(50 - s) = 450$
Use the Distributive Property:	$12s + 400 - 8s = 450$
Combine like terms:	$4s + 400 = 450$
Subtract 400 from both sides:	$4s = 50$
Divide both sides by 4:	$s = 12.5$

Since $s = 12.5$ pounds, then $b = 50 - s = 50 - 12.5 = 37.5$ pounds.

Step 5: Check If the manager mixes 12.5 pounds of Sumatra with 37.5 pounds of Brazilian, the total revenue will be $\$12(12.5) + \$8(37.5) = \$150 + \$300 = \$450$, which equals the revenue from the blend. This checks with the given information.

Step 6: Answer the Question The manager should mix 12.5 pounds of Sumatra beans with 37.5 pounds of Brazilian beans to make the blend.

Work Smart

Remember that mixtures can include interest (money), solids (nuts), liquids (chocolate milk), and even gases (Earth's atmosphere).

Quick ✓

6. "We're Nuts!" sells cashews for $7.00 per pound and peanuts for $2.50 per pound. The manager has decided to make a cashew and peanut trail mix. She wants the mix to sell for $4.00 per pound, and the revenue from selling the trail mix should be the same as that from selling the nuts alone. How many pounds of cashews and how many pounds of peanuts are required to create 30 pounds of trail mix?

▶ ❹ Model and Solve Uniform Motion Problems

Let's now look at uniform motion problems. Remember, these problems use the fact that distance equals rate times time $(d = rt)$.

EXAMPLE 7 | **Uniform Motion**

The airspeed of a plane is its speed through the air, assuming there is no wind. This speed contrasts with the plane's groundspeed—its speed relative to the ground. The ground-speed of an airplane is affected by the speed of the wind. Suppose that a Boeing 767 flying west from Washington, D.C., to San Francisco, a distance of 2400 miles, takes 6 hours. The trip east from San Francisco to Washington, D.C., takes 4 hours. Find the airspeed of the plane and the effect wind resistance has on the plane.

Solution

Step 1: Identify In this uniform motion problem, we want to find the airspeed of the plane.

Step 2: Name We will let a represent the airspeed of the plane and w represent the effect of wind resistance on the plane.

Step 3: Translate Going west, the plane is flying into the jet stream, so the plane is slowed down by the wind. Therefore, the groundspeed of the plane is $a - w$. Going east, the jet stream is helping the plane, so the groundspeed of the plane is $a + w$. We set up Table 3.

Table 3

	Distance	= Rate	· Time
Against Wind (West)	2400	$a - w$	6
With Wind (East)	2400	$a + w$	4

Going against the wind, we have the equation

Equation (1): $2400 = 6(a - w)$ or $2400 = 6a - 6w$

Going with the wind, we have the equation

Equation (2): $2400 = 4(a + w)$ or $2400 = 4a + 4w$

Now, we can form a system of two linear equations containing two unknowns.

$$\begin{cases} 6a - 6w = 2400 & (1) \\ 4a + 4w = 2400 & (2) \end{cases} \text{The Model}$$

Work Smart

In equation (1), we could have divided both sides by 6 rather than distributing. What could we have done with equation (2)?

Step 4: Solve Use the elimination method by first multiplying equation (1) by 2 and equation (2) by 3 and then adding equations (1) and (2).

$$\begin{cases} 2(6a - 6w) = 2(2400) & (1) \\ 3(4a + 4w) = 3(2400) & (2) \end{cases}$$

Work Smart

When checking your work, make sure that the airspeed of the plane is greater than the effect of wind resistance.

Distribute: $\begin{cases} 12a - 12w = 4800 & (1) \\ 12a + 12w = 7200 & (2) \end{cases}$

Add (1) and (2): $\quad 24a \quad\quad = 12{,}000$

Divide both sides by 24: $\quad\quad\quad a = 500$

We let $a = 500$ in equation (1) to find that $w = 100$.

Step 5: Check Flying west, the plane's groundspeed is $500 - 100 = 400$ miles per hour, so the plane will fly (400 miles per hour) (6 hours) = 2400 miles. Flying east, the plane's groundspeed is $500 + 100 = 600$ miles per hour, so the plane will fly 2400 miles in 4 hours.

Step 6: Answer The airspeed of the plane is 500 miles per hour. The effect of wind resistance on the plane is 100 miles per hour. ●

Figure 8

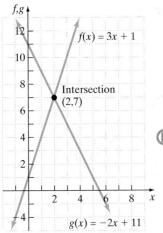

Quick ✓

7. Suppose that a plane flying 1200 miles west requires 4 hours and flying 1200 miles east requires 3 hours. Find the airspeed of the plane and the effect wind resistance has on the plane.

▶ ⑤ **Find the Intersection of Two Linear Functions**

We are often interested in knowing when two linear functions are equal. For example, we might want to know the value of x such that $f(x) = 3x + 1$ will equal $g(x) = -2x + 11$. This would require that we solve the equation $f(x) = g(x)$, or $3x + 1 = -2x + 11$. The solution to this equation is represented geometrically as the point of intersection of the graphs of the two functions, as shown in Figure 8.

Let's look at an application of intersecting functions from business.

⎛ **EXAMPLE 8** ⎞ **Break-Even Analysis**

Business is motivated by profit. A company's profit is the difference between revenues and cost. Companies need to understand how many units of their product they must manufacture and sell in order to be profitable. Suppose a gas grill company sells its entry-level grill for $130. The cost of manufacturing each grill is $80, and the fixed costs per month are $8500.

(a) Write revenue R as a function of the number of grills sold x.

(b) Write cost C as a function of the number of grills manufactured per month x.

(c) Graph the revenue function and cost function in the same Cartesian plane.

(d) The **break-even point** is the point where revenue equals cost. Find the break-even number of grills to be manufactured and sold. What is the revenue when this number of grills is sold? Label the break-even point on the graph drawn in part (c).

Solution

(a) If one grill is sold, the revenue is $130. If two grills are sold, the revenue is $130(2) = 260. If x grills are sold, the revenue is $130x$. The revenue function is $R(x) = 130x$.

(b) The cost function is $C(x) = ax + b$, where a is the cost of manufacturing each additional unit and b is the fixed cost. With $a = 80$ and $b = 8500$, we have $C(x) = 80x + 8500$.

(c) Figure 9 on the next page shows the graphs of the revenue and cost functions.

(d) To find the break-even point, we need to solve the equation $R(x) = C(x)$.

$$R(x) = C(x)$$
$$130x = 80x + 8500$$

Subtract 80x from both sides: $\quad 50x = 8500$

Divide both sides by 50: $\quad\quad x = 170$

Figure 9

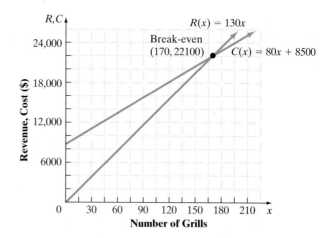

It appears that the company needs to sell 170 grills each month to break even. We verify this by determining the revenue and cost when $x = 170$.

$$R(170) = 130(170) = \$22{,}100$$

$$C(170) = 80(170) + 8500 = \$22{,}100$$

Since the revenue equals the cost when $x = 170$ grills are manufactured and sold, the break-even point is 170 grills per month. We label this point on the graph in Figure 9. Notice that if more than 170 grills are sold, the company will make a profit.

Quick ✓

8. Suppose that a nursery sells 8-foot Austrian Pine trees for \$230. The nursery can buy the trees from a tree farm for \$160. The fixed costs at the nursery amount to \$2100 per month.

  (a) Write revenue R as a function of the number of trees sold x.

  (b) Write cost C as a function of the number of trees purchased from the tree farm x.

  (c) Graph the revenue function and cost function in the same Cartesian plane.

  (d) Find the break-even number of trees to be sold. What is the revenue when this number of trees is sold? Label the break-even point on the graph drawn in part (c).

3.2 Exercises MyMathLab®   PRACTICE

Exercise numbers in green have complete video solutions in MyMathLab.

*Problems **1–8** are the Quick ✓s that follow the* EXAMPLES.

Building Skills

In Problems 9–14, solve each direct translation problem. See Objective 1.

9. The sum of two numbers is 18. The difference of the two numbers is −2. Find the numbers.

10. The sum of two numbers is 25. The difference of the two numbers is 3. Find the numbers.

11. Janice is thinking of two numbers. She says that two times the first number plus the second number is 47. In addition, the first number plus three times the second number is 81. Find the numbers.

12. Ashad is thinking of two numbers. He says that three times the first number minus the second number is 118. In addition, two times the first number plus the second number is 147. Find the numbers.

13. Rental Costs Gina rented a moon-walk for 5 hours at a total cost of \$235. Lori rented the same moon-walk for 3 hours at a total cost of \$165. The cost of renting is based upon a flat set-up fee plus a rental rate per hour. How much is the set-up fee? What is the hourly rental fee?

14. Making Change Johnny has \$6.75 in dimes and quarters. He has 8 more dimes than quarters. How many quarters does Johnny have? How many dimes does Johnny have?

In Problems 15–18, solve each geometry problem. See Objective 2.

△ **15. Perimeter** The perimeter of a rectangle is 120 meters. If the length of the rectangle is 20 meters more than the width, what are the dimensions of the rectangle?

△ **16. Perimeter** The perimeter of a rectangle is 260 centimeters. If the width of the rectangle is 15 centimeters less than the length, what are the dimensions of the rectangle?

△ **17. Angles** Two lines that are not parallel are shown in the figure. Suppose it is known that the measure of angle 1 is $(10x + 6y)°$, the measure of angle 2 is $(4y)°$, and the measure of angle 3 is $(7x + 2y)°$. Find x and y.

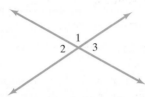

△ **18. Angles** Suppose that lines m and n are parallel. Line p transverses lines m and n. Suppose that it is known that the measure of angle 1 is $(2x + 3y)°$, the measure of angle 2 is $(4x)°$, and the measure of angle 3 is $(5x + y + 1)°$. Find x and y.

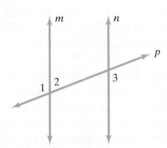

In Problems 19–22, solve each mixture problem. See Objective 3.

19. Investments Suppose that you received an unexpected inheritance of $36,000. You have decided to invest the money by placing some of the money in stocks and some in bonds. To diversify, you decided that five times the amount invested in bonds should equal three times the amount invested in stocks. How much should be invested in stocks? How much should be invested in bonds?

20. Investments Marge and Homer have $80,000 to invest. Their financial advisor has recommended that they diversify by placing some of the money in stocks and some in bonds. Based upon current market conditions, he has recommended that three times the amount invested in bonds should equal two times the amount invested in stocks. How much should be invested in stocks? How much should be invested in bonds?

21. Making Coffee Suppose that you want to blend two coffees in order to obtain a new blend. The blend will be made with the best Arabica beans that sell for $9.00 per pound and select African Robustas that sell for $11.50 per pound to obtain 100 pounds of the new blend. The new blend will sell for $10.00 per pound and there will be no difference in revenue from selling the new blend versus selling the beans separately. How many pounds of the Arabica and Robusta beans are required?

22. Candy A candy store sells chocolate-covered almonds for $6.50 per pound and chocolate-covered peanuts for $4.00 per pound. The manager decides to make a bridge mix that combines the almonds with the peanuts. She wants the bridge mix to sell for $6.00 per pound, and there should be no loss in revenue from selling the bridge mix versus the almonds and peanuts alone. How many pounds of chocolate-covered almonds and chocolate-covered peanuts are required to create 50 pounds of bridge mix?

In Problems 23–26, solve each uniform motion problem. See Objective 4.

23. Canoeing in the River Jonathon and Samantha paddle their canoe 26 miles downstream in 2 hours. After a picnic, they paddle back upstream. After 3 hours they have traveled only 9 miles back. Assuming that they paddle at a constant rate and the river's current is constant, find the speed at which Jonathon and Samantha can paddle in still water.

24. Against the Wind A Piper Arrow can fly 510 miles in 3 hours with a tailwind. Against this same wind, the plane can fly 390 miles in 3 hours. Find the airspeed of the plane. What is the impact of the wind on the plane?

25. A Car Ride An Infiniti G35 travels at an average speed of 50 miles per hour. A Lincoln Aviator travels at an average speed of 40 miles per hour. In the time it takes the Lincoln to travel a certain distance d, the Infiniti travels 100 miles farther. Find the distance that each car travels and the time of the trip.

26. Runners Enrique leaves his house and starts to run at an average speed of 6 miles per hour. Half an hour later, Enrique's younger (and faster) brother leaves the house to catch up to Enrique running at an average speed of 8 miles per hour. How long will it take for Enrique's brother to run half the distance that Enrique has run?

In Problems 27 and 28, let R represent a company's revenue, let C represent the company's cost, and let x represent the number of units produced and sold each day. See Objective 5.

(a) *Graph the revenue function and cost function in the same Cartesian plane.*

(b) *Find the company's break-even point; that is, find x so that*
$R(x) = C(x)$. Label this point on the graph drawn in part (a).

27. $R(x) = 12x$
$C(x) = 5.5x + 9880$

28. $R(x) = 16x$
$C(x) = 7x + 3645$

Applying the Concepts

△ **29. Triangle** An isosceles triangle is one in which two of the sides are the same length (congruent). The perimeter of an isosceles triangle is 35 cm. If the length of each of the congruent sides is 3 times the length of the third side, find the dimensions of the triangle.

△ **30. Trapezoid** A trapezoid is a quadrilateral in which two of the sides are parallel. A trapezoid is an isosceles trapezoid if the two non-parallel sides are congruent (the same length). See the figure. Suppose that the length of each of the congruent sides of an isosceles trapezoid is 12 cm. The perimeter of the trapezoid is 100 cm. Of the remaining two sides, one length is 14 cm shorter than two times the other. Find the length of each of the two remaining sides.

12 cm   12 cm

31. Banking A bank has loaned out $750,000, part of it at 5% per annum and the rest of it at 8% per annum. If the bank receives $52,500 in interest each year, how much was loaned at 5%?

32. One Bad Investment Horace invested $70,000 in two stocks. After 1 year, one of the stocks increased by 13%, while the other stock declined by 5%. His total gain for the year was $2800. How much was invested in the stock that earned 13%? How much was invested in the stock that lost 5%?

33. The sum of twice a first number and three times a second number is 81. If the second number is subtracted from three times the first number, the result is 17. Find the numbers.

34. The sum of four times a first number and a second number is 68. If the first number is decreased by twice the second number, the result is −1. Find the numbers.

35. Moving Walkways Moving walkways are common in airports. Drenell conducts an experiment in which he walks with a 126-foot walkway and then against the walkway. It takes Drenell 18 seconds to walk the 126-foot walkway going with the walkway and it takes him 63 seconds to walk the walkway going against the walkway. What is Drenell's normal walking speed?

36. Biking Suppose that Shannon bikes into the wind for 60 miles and it takes her 6 hours. After a long

rest, she returns (with the wind at her back) in 5 hours. Determine the speed at which Shannon can ride her bike in still air, and determine the effect that the wind had on her speed.

△ **37. Angles** Find the values of x and y based upon the values for the angles in the figure. The measure of angle 1 is $(4x + 2y)°$, the measure of angle 2 is $(10x + 5y)°$, the measure of angle 3 is $(7x + 5y)°$, and the measure of angle 4 is $(9x + 8y)°$.

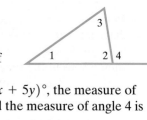

△ **38. Angles** Find the values of x and y based upon the values for the angles in the parallelogram. The measure of angle 1 is $(5x + 7y)°$, the measure of angle 2 is $(10x + 5y)°$, and the measure of angle 3 is $(15x - 9y)°$.

39. Pharmacy A doctor's prescription calls for a daily intake of liquid containing 40 mg of vitamin C and 30 mg of vitamin D. Your pharmacy stocks two liquids that can be used: Liquid A contains 20% vitamin C and 30% vitamin D; Liquid B contains 50% vitamin C and 25% vitamin D. How many milligrams of each liquid should be mixed to fill the prescription?

40. Pharmacy A doctor's prescription calls for the creation of pills that contain 10 units of vitamin B_{12} and 13 units of vitamin E. Your pharmacy stocks two powders that can be used to make these pills: Powder A contains 20% vitamin B_{12} and 40% vitamin E; Powder B contains 50% vitamin B_{12} and 30% vitamin E. How many units of each powder should be mixed in each pill?

41. Counting Calories Suppose that Kristin ate two McDonald's orders of chicken nuggets and drank one medium Coke, for a total of 770 calories. Kristin's friend Jack ate three orders of nuggets and drank two medium Cokes (Jack takes advantage of free refills), for a total of 1260 calories. How many calories are in an order of chicken nuggets? How many calories are in a medium Coke?

42. Carbs Yvette and José go to McDonald's for breakfast. Yvette ordered two sausage biscuits and one orange juice. The entire meal had 98 grams of carbohydrates. José ordered three sausage biscuits and two orange juices and his meal had 162 grams of carbohydrates. How many grams of carbohydrates are in a sausage biscuit? How many grams of carbohydrates are in an orange juice?

43. Making an Alloy A metallurgist has an alloy with 10% titanium and an alloy with 25% titanium. He needs 100 grams of an alloy with 16% titanium. How much of each alloy should be mixed to obtain the 100 grams of alloy with 16% titanium?

44. Hydrochloric Acid A chemist has a solution that is 20% hydrochloric acid (HCl) and a solution that is 50% HCl. How much of each solution should be mixed to obtain 300 cubic centimeters (cc) of a solution that is 30% HCl?

45. Supply and Demand Suppose that the quantity supplied S and the quantity demanded D of hot dogs at a baseball game are given by these functions:

$$S(p) = -2000 + 3000p$$
$$D(p) = 10{,}000 - 1000p$$

where p is the price. The **equilibrium price** of a market is defined as the price at which quantity supplied equals quantity demanded $(S = D)$.

(a) Graph each of the two functions in the same Cartesian plane.

(b) Find the equilibrium price for hot dogs at the baseball game. What is the equilibrium quantity? Label the equilibrium point on the graph drawn in part (a).

46. Supply and Demand See Problem 45. Suppose that the quantity supplied S and the quantity demanded D of baseball hats at a baseball game are given by the following functions:

$$S(p) = 9p - 17$$
$$D(p) = -35p + 995$$

where p is the price.

(a) Graph each of the two functions in the same Cartesian plane.

(b) Find the equilibrium price for baseball hats at the baseball game. What is the equilibrium quantity? Label the equilibrium point on the graph drawn in part (a).

47. College Grads The percent of adult males who hold a bachelor's degree M as a function of the year t is $M(t) = 0.25t + 14.4$, where t is the number of years since 1990. The percent of adult females who hold a bachelor's degree F as a function of the year t is $F(t) = 0.29t + 12$, where t is the number of years since 1990. (SOURCE: *Statistical Abstract*)

(a) Graph each of the two functions in the same Cartesian plane.

(b) Assuming that this linear trend continues, find the year in which the percent of male college grads and the percent of female college grads are equal. That is, solve $M(t) = F(t)$. What is the percent of college grads in this year? Label this point on the graph.

48. Weekly Earnings The average weekly earnings of 16- to 24-year-old males M as a function of the year t is $M(t) = 8.92t + 272.11$, where t is the number of years since 1990. The average weekly earnings of 16- to 24-year-old females F as a function of the year t is $F(t) = 9.44t + 244.92$, where t is the number of years since 1990.

(a) Graph each of the two functions in the same Cartesian plane.

(b) Assuming that this linear trend continues, find the year in which the average weekly earnings of 16- to 24-year-old males will equal the average weekly earnings of 16- to 24-year-old females. That is, solve $M(t) = F(t)$. What is the average weekly earnings in this year? Label this point on the graph.

49. Salary Suppose that you are offered a sales position for a pharmaceutical company. It offers you two salary options. Option A would pay you an annual base salary of \$15,000 plus a commission of 1% on sales. Option B would pay you an annual base salary of \$25,000 plus a commission of 0.75% on sales.

(a) Find a function for each salary option.

(b) Graph each function found in part (a) in the same Cartesian plane.

(c) Determine the annual sales required for the options to result in the same annual salary. What would the annual salary be? Label this point on the graph.

50. Salary Suppose you are offered a sales position for a medical equipment company. It offers you two salary options. Option A would pay you an annual base salary of \$27,000 plus a commission of 1.5% on sales. Option B would pay you an annual base salary of \$45,500 plus a commission of 1% on sales.

(a) Find a function for each salary option.

(b) Graph the functions found in part (a) in the same Cartesian plane.

(c) Determine the annual sales required for the options to result in the same annual salary. What would the annual salary be? Label this point on the graph.

51. Break-Even A wood craftsman makes children's desks. He sells the desks for \$60 each. His monthly fixed costs of operating the business are \$3500. Each desk costs \$35 in material.

(a) Find the revenue function R, treating the number of desks x as the independent variable.

(b) Find the cost function C, treating the number of desks x as the independent variable.

(c) Graph the revenue and cost functions in the same Cartesian plane.

(d) Find the break-even number of desks that must be manufactured and sold. What are the revenue and cost at this number of desks? Label the point on the graph drawn in part (c).

52. Break-Even Audra wants to establish a lemonade stand on her corner. Her father buys some lumber and other materials and makes Audra a lemonade stand for $40. Audra goes to the store and buys lemonade. She determines that each cup of lemonade will cost her $0.03 to make, but the actual cup will cost her an additional $0.07. She decides to sell the lemonade for $0.30 a cup.

(a) Find the revenue function R, treating the number of cups of lemonade x as the independent variable.

(b) Find the cost function C, treating the number of cups of lemonade x as the independent variable.

(c) Graph the revenue and cost functions in the same Cartesian plane.

(d) Find the break-even number of cups of lemonade that Audra must sell. What are the revenue and cost at this number of cups of lemonade? Label the point on the graph drawn in part (c).

Extending the Concepts

53. The Olympics The data in the following table represent the winning times in the 200-meter run in the finals of the Olympics.

(a) Draw a scatter diagram of the men's winning time, treating the year as the independent variable.

(b) On the same graph, draw a scatter diagram of the women's winning time, treating the year as the independent variable. Be sure to use a

Year	Men's Time (in seconds)	Women's Time (in seconds)
1968	19.83	22.50
1972	20.00	22.40
1976	20.23	22.37
1980	20.19	22.03
1984	19.80	21.81
1988	19.75	21.34
1992	19.73	21.72
1996	19.32	22.12
2000	20.09	21.84
2004	19.79	22.05
2008	19.30	21.93
2012	19.32	21.69

different plotting symbol to label the points (such as a □ and a ○).

(c) Draw a line through any two points for the men's winning time. Use these points to find the equation of the line.

(d) Draw a line through any two points for the women's winning time. Use these points to find the equation of the line.

(e) Use the equations found in parts (c) and (d) to find the year in which the winning time for men will equal the winning time for women. What is the winning time?

(f) Do you think that your answer to part (e) is reasonable? Why or why not?

(g) Use the lines from parts (c) and (d) to predict the results from the 2012 Olympic games. Compare your predictions to the actual men's and women's winning times.

3.3 Systems of Linear Equations in Three Variables

Objectives

1 Solve Systems of Three Linear Equations

2 Identify Inconsistent Systems

3 Write the Solution of a System with Dependent Equations

4 Model and Solve Problems Involving Three Linear Equations

Are You Ready for This Section?

Before getting started, take the following readiness quiz. If you get the problem wrong, go back to the section cited and review the material.

R1. Evaluate the expression $3x - 2y + 4z$ for $x = 1$, $y = -2$, and $z = 3$. [Section R.5, pp. 40–41]

1 Solve Systems of Three Linear Equations

▶ An example of a linear equation in three variables is $2x - y + z = 8$. An example of a system of three linear equations containing three variables, x, y, and z is

Ready ?...Answer **R1.** 19

$$\begin{cases} x + 3y + z = 8 \\ 3x - y + 6z = 12 \\ -4x - y + 2z = -1 \end{cases}$$

Systems of three linear equations with three variables have the same possible solutions as a system of two linear equations containing two variables:

1. **Exactly one solution**—A consistent system with independent equations

2. **No solution**—An inconsistent system

3. **Infinitely many solutions**—A consistent system with dependent equations

We can think of a system of three linear equations containing three variables as a geometry problem. The graph of each equation in a system of linear equations containing three variables is a plane in space. A system of three linear equations containing three variables represents three planes in space. Figure 10 illustrates some of the possibilities.

Recall that a **solution** to a system of equations consists of values for the variables that are solutions of each equation of the system. We write the solution to a system of three equations containing three unknowns as an **ordered triple** (x, y, z).

Figure 10

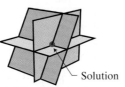 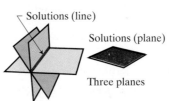 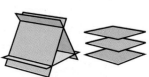

Solutions (line)

Solutions (plane)

Three planes

Solution

(**a**) Consistent system; one solution

(**b**) Consistent system; infinite number of solutions

(**c**) Inconsistent system; no solution

Quick ✓

1. If a system of equations has no solution, it is said to be _____. If a system of equations has infinitely many solutions, the system is said to be _____ and the equations are _____.

2. A _____ to a system of equations consists of values for the variables that are solutions of each equation of the system.

3. *True or False* A system of three linear equations containing three variables always has at least one solution.

4. *True or False* When the planes in a system of equations are parallel, the system is inconsistent and has no solution.

EXAMPLE 1 **Determining Whether Values Are a Solution of a System**

Determine which of the following ordered triples are solutions of the system.

$$\begin{cases} x + y + z = 0 \\ 2x - y + 3z = 17 \\ -3x + 2y - z = -21 \end{cases}$$

(**a**) $(1, 3, -4)$ (**b**) $(3, -5, 2)$

Solution

First, we name the system's equations (1), (2), and (3).

$$\begin{cases} x + y + z = 0 & (1) \\ 2x - y + 3z = 17 & (2) \\ -3x + 2y - z = -21 & (3) \end{cases}$$

(**a**) Let $x = 1$, $y = 3$, and $z = -4$ in equations (1), (2), and (3). If all three equations are true, then $(1, 3, -4)$ is a solution.

Equation (1): $x + y + z = 0$
$$1 + 3 + (-4) \overset{?}{=} 0$$
$$0 = 0 \text{ True}$$

Equation (2): $2x - y + 3z = 17$
$$2(1) - 3 + 3(-4) \overset{?}{=} 17$$
$$2 - 3 - 12 \overset{?}{=} 17$$
$$-13 = 17 \text{ False}$$

Equation (3): $-3x + 2y - z = -21$
$$-3(1) + 2(3) - (-4) \overset{?}{=} -21$$
$$-3 + 6 + 4 \overset{?}{=} -21$$
$$7 = -21 \text{ False}$$

Although this ordered triple satisfies (1), it does not satisfy (2) or (3). Therefore, the ordered triple $(1, 3, -4)$ is not a solution.

(b) Let $x = 3, y = -5$, and $z = 2$ in equations (1), (2), and (3).

Equation (1): $x + y + z = 0$
$$3 + (-5) + (2) \overset{?}{=} 0$$
$$0 = 0 \ \text{True}$$

Equation (2): $2x - y + 3z = 17$
$$2(3) - (-5) + 3(2) \overset{?}{=} 17$$
$$6 + 5 + 6 \overset{?}{=} 17$$
$$17 = 17 \ \text{True}$$

Equation (3): $-3x + 2y - z = -21$
$$-3(3) + 2(-5) - (2) \overset{?}{=} -21$$
$$-9 - 10 - 2 \overset{?}{=} -21$$
$$-21 = -21 \ \text{True}$$

Because the ordered triple $(3, -5, 2)$ satisfies all three equations, it is a solution to the system.   ●

Quick ✓

5. Determine which of the following ordered triples are solutions of the system.

$$\begin{cases} x + y + z = 3 \\ 3x + y - 2z = -23 \\ -2x - 3y + 2z = 17 \end{cases}$$

(a) $(3, 2, -2)$ **(b)** $(-4, 1, 6)$

Typically, to solve a system of three linear equations containing three variables, we use the elimination method. We eliminate one variable by multiplying equations by nonzero constants to get the coefficients of the variables to be additive inverses. We then add these equations to remove that variable. We can also interchange any two equations or multiply (or divide) each side of an equation by the same nonzero constant. Another approach is to replace an equation by the sum of that equation and a multiple of a second equation.

The first step in solving a system of three linear equations with three unknowns is to reduce the system to one with two linear equations with two unknowns. We then solve the smaller system using the methods of Section 3.1.

▶ (**EXAMPLE 2**)   **How to Solve a System of Three Linear Equations**

Use the method of elimination to solve the system:
$$\begin{cases} x + y - z = -1 & (1) \\ 2x - y + 2z = 8 & (2) \\ -3x + 2y + z = -9 & (3) \end{cases}$$

Step-by-Step Solution

Step 1: The first step is to eliminate the same variable from two of the equations. Notice that we can use equation (1) to eliminate the variable x from equations (2) and (3). We can do this by multiplying equation (1) by -2 and adding the result to equation (2). The resulting equation is equation (4). We also multiply equation (1) by 3 and add the result to equation (3). The resulting equation is equation (5).

$$\begin{array}{l} x + y - z = -1 \quad (1) \\ 2x - y + 2z = 8 \quad (2) \end{array}$$

Multiply (1) by -2:
$$\begin{array}{l} -2x - 2y + 2z = 2 \quad (1) \\ \underline{2x - y + 2z = 8} \quad (2) \\ \text{Add:} \quad -3y + 4z = 10 \end{array}$$

$$\begin{array}{l} x + y - z = -1 \quad (1) \\ -3x + 2y + z = -9 \quad (3) \end{array}$$

Multiply (1) by 3:
$$\begin{array}{l} 3x + 3y - 3z = -3 \quad (1) \\ \underline{-3x + 2y + z = -9} \quad (3) \\ \text{Add:} \quad 5y - 2z = -12 \end{array}$$

$$\begin{cases} x + y - z = -1 & (1) \\ -3y + 4z = 10 & (4) \\ 5y - 2z = -12 & (5) \end{cases}$$

Step 2: Treat equations (4) and (5) as a system of two equations with two variables. Eliminate z by multiplying equation (5) by 2 and then adding equations (4) and (5). The result is equation (6).

$$\begin{cases} -3y + 4z = 10 & (4) \\ 5y - 2z = -12 & (5) \end{cases}$$

Multiply (5) by 2:
$$\begin{array}{l} -3y + 4z = 10 \quad (4) \\ \underline{10y - 4z = -24} \quad (5) \\ \text{Add:} \quad 7y = -14 \quad (6) \end{array}$$

Step 3: Solve equation (6) for y.

$$7y = -14 \quad (6)$$

Divide both sides by 7: $\quad y = -2$

Step 4: Back-substitute -2 for y in equation (4) and solve for z.

$$-3y + 4z = 10 \quad (4)$$
$$-3(-2) + 4z = 10$$
$$6 + 4z = 10$$
$$4z = 4$$
$$z = 1$$

Step 5: Back-substitute -2 for y and 1 for z in equation (1) and solve for x.

$$x + y - z = -1 \quad (4)$$
$$x + (-2) - 1 = -1$$
$$x - 3 = -1$$
$$x = 2$$

The solution appears to be $(2, -2, 1)$.

Step 6: Check Verify that $(2, -2, 1)$ is the solution.

Equation (1): $\quad x + y - z = -1$

$x = 2, y = -2, z = 1$: $\quad 2 + (-2) - 1 \stackrel{?}{=} -1$

$$-1 = -1 \quad \text{True}$$

Equation (2): $\quad 2x - y + 2z = 8$

$$2(2) - (-2) + 2(1) \stackrel{?}{=} 8$$
$$4 + 2 + 2 \stackrel{?}{=} 8$$
$$8 = 8 \quad \text{True}$$

Equation (3): $\quad -3x + 2y + z = -9$

$$-3(2) + 2(-2) + 1 \stackrel{?}{=} -9$$
$$-6 - 4 + 1 \stackrel{?}{=} -9$$
$$-9 = -9 \quad \text{True}$$

The solution, $(2, -2, 1)$, checks.

Work Smart

In Example 2, we eliminated x from equations (2) and (3). We could also have eliminated y from equation (2) by adding equations (1) and (2). We could then eliminate y from equation (3) by multiplying equation (1) by -2 and adding the result to equation (3). Had we done this, we would have ended up with the following system of two equations with two unknowns:

$$\begin{cases} 3x + z = 7 \\ -5x + 3z = -7 \end{cases}$$

The solution of this system is $x = 2$ and $z = 1$. Substituting these values into equation (1) yields $y = -2$, so the solution to the system is $(2, -2, 1)$, which agrees with the solution we found in Example 2. There is more than one way to solve a problem!

Solving a System of Three Linear Equations Containing Three Unknowns by Elimination

Step 1: Select two of the equations and eliminate one of the variables from one of the equations. Select any two other equations and eliminate the *same variable* from one of the equations.

Step 2: You will have two equations with two unknowns. Solve this system using the techniques from Section 3.1.

Step 3: Use the two known values of the variables found in Step 2 to find the value of the third variable.

Step 4: Check your answer.

Work Smart

By eliminating a variable in two of the equations in Step 1, we are creating a system of two equations with two unknowns we can solve. Remember, in mathematics we often want to reduce a problem to one we already know how to solve.

Quick ✓

6. Use the elimination method to solve the system:

$$\begin{cases} x + y + z = -3 \\ 2x - 2y - z = -7 \\ -3x + y + 5z = 5 \end{cases}$$

(EXAMPLE 3) **Solving a System of Three Linear Equations**

Use the elimination method to solve the system: $\begin{cases} 4x \quad\;\; + \;\; z = 4 & (1) \\ 2x + 3y \quad\;\;\;\;\; = -4 & (2) \\ \quad\;\; 2y - 4z = -15 & (3) \end{cases}$

Solution

Eliminate z from equation (3) by multiplying equation (1) by 4 and adding the result to equation (3). The resulting equation is equation (4).

$$\begin{array}{l} 4x \quad\;\; + \;\; z = 4 \quad (1) \\ \quad\;\; 2y - 4z = -15 \quad (3) \end{array} \qquad \text{Multiply (1) by 4:} \quad \begin{array}{r} 16x \quad\quad + 4z = 16 \quad (1) \\ \underline{2y - 4z = -15} \quad (3) \\ 16x + 2y \quad\quad = 1 \end{array} \longrightarrow \begin{cases} 4x \quad\quad + z = 4 & (1) \\ 2x + 3y \quad\quad = -4 & (2) \\ 16x + 2y \quad\quad = 1 & (4) \end{cases}$$

Focus on the system containing equations (2) and (4). To eliminate the variable x, multiply equation (2) by -8 and then add equations (2) and (4). The result is equation (5).

$$\begin{cases} 2x + 3y = -4 & (2) \\ 16x + 2y = 1 & (4) \end{cases} \qquad \begin{array}{l} \text{Multiply (2) by } -8: \\ \\ \text{Add:} \end{array} \quad \begin{array}{r} -16x - 24y = 32 \quad (2) \\ \underline{16x + \;\; 2y = \;\; 1} \quad (4) \\ -22y = 33 \quad (5) \end{array}$$

Solve equation (5) for y by dividing both sides of the equation by -22.

$$\text{Equation (5):} \quad -22y = 33$$

$$y = \frac{33}{-22} = -\frac{3}{2}$$

Now substitute $-\dfrac{3}{2}$ for y in equation (2) and solve for x.

$$\text{Let } y = -\frac{3}{2} \text{ in Equation (2):} \quad 2x + 3y = -4$$

$$2x + 3\left(-\frac{3}{2}\right) = -4$$

$$2x - \frac{9}{2} = -4$$

$$2x = \frac{1}{2}$$

$$x = \frac{1}{4}$$

Now substitute $\dfrac{1}{4}$ for x in equation (1) and solve for z.

$$4x + z = 4 \quad (1)$$

$$4\left(\frac{1}{4}\right) + z = 4$$

$$1 + z = 4$$

$$z = 3$$

We leave the check to you. The solution is the ordered triple $\left(\dfrac{1}{4}, -\dfrac{3}{2}, 3\right)$.

Quick ✓

7. Use the elimination method to solve the system: $\begin{cases} 2x \quad\quad - 4z = -7 \\ x + 6y \quad\quad = 5 \\ \quad\; 2y - \;\; z = 2 \end{cases}$

⊙ ❷ Identify Inconsistent Systems

Examples 2 and 3 were consistent and independent systems resulting in a single solution. We now look at an inconsistent system.

EXAMPLE 4 **An Inconsistent System of Linear Equations**

Use the elimination method to solve the system: $\begin{cases} x + 2y - z = 4 & (1) \\ -2x + 3y + z = -4 & (2) \\ x + 9y - 2z = 1 & (3) \end{cases}$

Solution

Notice that we can use equation (1) to eliminate x from equations (2) and (3). We can do this by multiplying equation (1) by 2 and adding the result to equation (2). The resulting equation is equation (4). Next, multiply equation (1) by -1 and add the result to equation (3). The resulting equation is equation (5). Equations (4) and (5) form a system of two equations with two unknowns.

$$\begin{array}{llll} x + 2y - z = 4 & (1) & \text{Multiply by 2:} & 2x + 4y - 2z = 8 \quad (1) \\ -2x + 3y + z = -4 & (2) & & \underline{-2x + 3y + z = -4} \quad (2) \\ & & \text{Add:} & 7y - z = 4 \end{array}$$

Work Smart

These figures illustrate two different inconsistent systems geometrically.

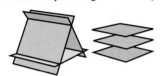

$$\begin{array}{llll} x + 2y - z = 4 & (1) & \text{Multiply by } -1: & -x - 2y + z = -4 \quad (1) \\ x + 9y - 2z = 1 & (3) & & \underline{x + 9y - 2z = 1} \quad (3) \\ & & \text{Add:} & 7y - z = -3 \longrightarrow \end{array}$$

$\begin{cases} 7y - z = 4 & (4) \\ 7y - z = -3 & (5) \end{cases}$

Now focus on the system containing equations (4) and (5). Multiply equation (4) by -1 and then add equations (4) and (5).

Work Smart

Whenever you end up with a false statement such as $0 = -7$, you have an inconsistent system.

$$\begin{array}{llll} 7y - z = 4 & (4) & \text{Multiply by } -1: \longrightarrow & -7y + z = -4 & (4) \\ 7y - z = -3 & (5) & & \underline{7y - z = -3} & (5) \\ & & \text{Add:} & 0 = -7 & (6) \quad \text{False} \end{array}$$

Equation (6) now states that $0 = -7$, which is a false statement. Therefore, the system is inconsistent. The solution set is $\varnothing$ or $\{\ \}$. ●

Quick ✓

8. Use the elimination method to solve the system: $\begin{cases} x - y + 2z = -7 \\ -2x + y - 3z = 5 \\ x - 2y + 3z = 2 \end{cases}$

⊙ ❸ Write the Solution of a System with Dependent Equations

Now we look at a system of dependent equations.

EXAMPLE 5 **Solving a System with Dependent Equations**

Work Smart

These figures illustrate two different dependent systems.

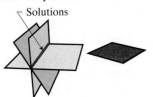

Solutions

Use the elimination method to solve the system: $\begin{cases} x - 3y - z = 4 & (1) \\ x - 2y + 2z = 5 & (2) \\ 2x - 5y + z = 9 & (3) \end{cases}$

Solution

We can use equation (1) to eliminate x from equations (2) and (3). We can do this by multiplying equation (1) by -1 and adding the result to equation (2). The resulting equation is equation (4). We also multiply equation (1) by -2 and add the result to equation (3). The resulting equation is equation (5).

$$\begin{array}{ll} x - 3y - z = 4 & (1) \\ x - 2y + 2z = 5 & (2) \end{array} \quad \text{Multiply by } -1:$$

$$\begin{array}{ll} -x + 3y + z = -4 & (1) \\ \underline{x - 2y + 2z = 5} & (2) \\ \text{Add:} \quad y + 3z = 1 \end{array}$$

$$\begin{array}{ll} x - 3y - z = 4 & (1) \\ 2x - 5y + z = 9 & (3) \end{array} \quad \text{Multiply by } -2:$$

$$\begin{array}{ll} -2x + 6y + 2z = -8 & (1) \\ \underline{2x - 5y + z = 9} & (3) \\ \text{Add:} \quad y + 3z = 1 \end{array} \quad \begin{cases} y + 3z = 1 & (4) \\ y + 3z = 1 & (5) \end{cases}$$

We now concentrate on the system containing equations (4) and (5).

$$\begin{array}{ll} y + 3z = 1 & (4) \\ y + 3z = 1 & (5) \end{array} \quad \text{Multiply by } -1:$$

$$\begin{array}{ll} -y - 3z = -1 & (4) \\ \underline{y + 3z = 1} & (5) \\ \text{Add:} \quad 0 = 0 & (6) \end{array}$$

The statement $0 = 0$ in equation (6) indicates that we may have a consistent system with dependent equations. We can show how the values for x and y *depend* on the value of z by letting z represent any real number. Then solve equation (4) for y and obtain y in terms of z.

Equation (4): $\quad y + 3z = 1$

Subtract $3z$ from both sides: $\quad y = -3z + 1$

Let $y = -3z + 1$ in equation (1) and solve for x in terms of z.

Equation (1): $\quad x - 3y - z = 4$

Let $y = -3z + 1$ in (1): $\quad x - 3(-3z + 1) - z = 4$

Distribute: $\quad x + 9z - 3 - z = 4$

Combine like terms: $\quad x + 8z - 3 = 4$

Subtract $8z$ from both sides; add 3 to both sides: $\quad x = -8z + 7$

The solution to the system is $\{(x, y, z) \,|\, x = -8z + 7, y = -3z + 1, z \text{ is any real number}\}$. To find specific solutions to the system, choose any value of z and use the equations $x = -8z + 7$ and $y = -3z + 1$ to determine x and y. Some specific solutions to the system are $(7, 1, 0)$, $(-1, -2, 1)$, and $(15, 4, -1)$. ●

Quick ✓

9. Use the elimination method to solve the system: $\begin{cases} x - y + 3z = 2 \\ -x + 2y - 5z = -3 \\ 2x - y + 4z = 3 \end{cases}$

Notice that in Examples 2 – 5, we did not always eliminate x first and y second. The order in which variables are eliminated from a system does not matter, and different approaches to solving the problem will lead to the right answer if done correctly. For example, in Example 4 we could have chosen to eliminate z from equations (1) and (3).

▶ ④ Model and Solve Problems Involving Three Linear Equations

Now let's look at problems involving systems of three equations containing three variables.

EXAMPLE 6 Production

A manufacturer makes three different models of swing sets. The Monkey takes 2 hours to cut the wood, 2 hours to stain, and 3 hours to assemble. The Gorilla takes 3 hours to cut the wood, 4 hours to stain, and 4 hours to assemble. The King Kong takes 4 hours to cut the wood, 5 hours to stain, and 5 hours to assemble. The company has 61 hours available to cut the wood, 73 hours available to stain, and 83 hours available to assemble each day. How many of each type of swing set can be manufactured each day?

Solution

Step 1: Identify We want to determine the number of Monkey, Gorilla, and King Kong swing sets that can be manufactured each day.

Step 2: Name Let m represent the number of Monkey swing sets, g represent the number of Gorilla swing sets, and k represent the number of King Kong swing sets.

Step 3: Translate We organize the given information in Table 4.

Table 4				
	Monkey	Gorilla	King Kong	Total Hours Available
Cut Wood	2	3	4	61
Stain	2	4	5	73
Assemble	3	4	5	83

Making m Monkey swing sets takes $2m$ hours to cut the wood. Similarly, it takes $3g$ and $4k$ hours to cut the wood for g Gorilla and k King Kong sets. Since 61 hours are available, we have

$$2m + 3g + 4k = 61 \quad \text{Equation (1)}$$

It takes $2m$ hours to stain m Monkey sets, $4g$ hours to stain g Gorilla sets, and $5k$ hours to stain k King Kong sets. Since 73 hours are available, we have

$$2m + 4g + 5k = 73 \quad \text{Equation (2)}$$

We need $3m$ hours to assemble m Monkey sets, $4g$ hours to assemble g Gorilla sets, and $5k$ hours to assemble k King Kong sets. Since 83 hours are available, we have

$$3m + 4g + 5k = 83 \quad \text{Equation (3)}$$

We combine equations (1), (2), and (3) to form the following system:

$$\begin{cases} 2m + 3g + 4k = 61 & (1) \\ 2m + 4g + 5k = 73 & (2) \quad \text{The Model} \\ 3m + 4g + 5k = 83 & (3) \end{cases}$$

Step 4: Solve The solution of this system of equations is $m = 10, g = 7$, and $k = 5$.

Step 5: Check Manufacturing 10 Monkeys, 7 Gorillas, and 5 King Kongs requires $2(10) + 3(7) + 4(5) = 61$ hours to cut the wood; $2(10) + 4(7) + 5(5) = 73$ hours to stain; and $3(10) + 4(7) + 5(5) = 83$ hours to assemble.

Step 6: Answer The company can manufacture 10 Monkey swing sets, 7 Gorilla swing sets, and 5 King Kong swing sets each day. ●

Quick ✓

10. The Mowing 'Em Down lawn mower company makes three styles of lawn mower. The 21-inch model requires 2 hours to mold, 3 hours for engine manufacturing, and 1 hour to assemble. The 24-inch model requires 3 hours to mold, 3 hours for engine manufacturing, and 1 hour to assemble. The 40-inch model requires 4 hours to mold, 4 hours for engine manufacturing, and 2 hours to assemble. The company has 81 hours available to mold, 95 hours available for engine manufacturing, and 35 hours available to assemble each day. How many of each type of mower can be manufactured each day?

3.3 Exercises MyMathLab® Exercise numbers in green have complete video solutions in MyMathLab.

PRACTICE

*Problems **1–10** are the Quick ✓s that follow the EXAMPLES.*

Building Skills

In Problems 11 and 12, determine whether the ordered triples listed are solutions of the system of linear equations. See Objective 1.

11. $\begin{cases} x + y + 2z = 6 \\ -2x - 3y + 5z = 1 \\ 2x + y + 3z = 5 \end{cases}$

    (a) $(6, 2, -1)$ **(b)** $(-3, 5, 2)$

12. $\begin{cases} 2x + y - 2z = 6 \\ -2x + y + 5z = 1 \\ 2x + 3y + z = 13 \end{cases}$

    (a) $(3, 2, 1)$ **(b)** $(10, -4, 5)$

In Problems 13–20, solve each system of three linear equations containing three unknowns. See Objective 1.

13. $\begin{cases} x + y + z = 5 \\ -2x - 3y + 2z = 8 \\ 3x - y - 2z = 3 \end{cases}$

14. $\begin{cases} x + 2y - z = 4 \\ 2x - y + 3z = 8 \\ -2x + 3y - 2z = 10 \end{cases}$

15. $\begin{cases} x - 3y + z = 13 \\ 3x + y - 4z = 13 \\ -4x - 4y + 2z = 0 \end{cases}$

16. $\begin{cases} x + 2y - 3z = -19 \\ 3x + 2y - z = -9 \\ -2x - y + 3z = 26 \end{cases}$

17. $\begin{cases} x - 4y + z = 5 \\ 4x + 2y + z = 2 \\ -4x + y - 3z = -8 \end{cases}$

18. $\begin{cases} 2x + 2y - z = -7 \\ x + 2y - 3z = -8 \\ 4x - 2y + z = -11 \end{cases}$

19. $\begin{cases} x - 3y = 12 \\ 2y - 3z = -9 \\ 2x + z = 7 \end{cases}$

20. $\begin{cases} 2x + z = -7 \\ 3y - 2z = 17 \\ -4x - y = 7 \end{cases}$

In Problems 21 and 22, show that each system of equations has no solution. See Objective 2.

21. $\begin{cases} x + y - 2z = 6 \\ -2x - 3y + z = 12 \\ -3x - 4y + 3z = 2 \end{cases}$

22. $\begin{cases} -x + 4y - z = 8 \\ 4x - y + 3z = 9 \\ 2x + 7y + z = 0 \end{cases}$

In Problems 23–26, solve each system with dependent equations. See Objective 3.

23. $\begin{cases} x + y + z = 4 \\ -2x - y + 2z = 6 \\ x + 2y + 5z = 18 \end{cases}$

24. $\begin{cases} x + 2y + z = 4 \\ -3x + y + 4z = -2 \\ -x + 5y + 6z = 6 \end{cases}$

25. $\begin{cases} x + 3z = 5 \\ -2x + y = 1 \\ y + 6z = 11 \end{cases}$

26. $\begin{cases} 2x - y = 2 \\ -x + 5z = 3 \\ -y + 10z = 8 \end{cases}$

Mixed Practice

In Problems 27–40, solve each system of equations.

27. $\begin{cases} 2x - y + 2z = 1 \\ -2x + 3y - 2z = 3 \\ 4x - y + 6z = 7 \end{cases}$

28. $\begin{cases} x - y + 3z = 2 \\ -2x + 3y - 8z = -1 \\ 2x - 2y + 4z = 7 \end{cases}$

29. $\begin{cases} x - y + z = 5 \\ -2x + y - z = 2 \\ x - 2y + 2z = 1 \end{cases}$

30. $\begin{cases} x - y + 2z = 3 \\ 2x + y - 2z = 1 \\ 4x - y + 2z = 0 \end{cases}$

31. $\begin{cases} 2y - z = -3 \\ -2x + 3y = 10 \\ 4x + 3z = -11 \end{cases}$

32. $\begin{cases} x - 3z = -3 \\ 3y + 4z = -5 \\ 3x - 2y = 6 \end{cases}$

33. $\begin{cases} x - 2y + z = 5 \\ -2x + y - z = 2 \\ x - 5y - 4z = 8 \end{cases}$

34. $\begin{cases} x + 2y - z = -4 \\ -2x + 4y - z = 6 \\ 2x + 2y + 3z = 1 \end{cases}$

35. $\begin{cases} x + 2y - z = 1 \\ 2x + 7y + 4z = 11 \\ x + 3y + z = 4 \end{cases}$

36. $\begin{cases} x + y - 2z = 3 \\ -2x - 3y + z = -7 \\ x + 2y + z = 4 \end{cases}$

37. $\begin{cases} x + y + z = 5 \\ 3x + 4y + z = 16 \\ -x - 4y + z = -6 \end{cases}$

38. $\begin{cases} x + y + z = 4 \\ 2x + 3y - z = 8 \\ x + y - z = 3 \end{cases}$

39. $\begin{cases} x + y + z = 3 \\ -x + \frac{1}{2}y + z = \frac{1}{2} \\ -x + 2y + 3z = 4 \end{cases}$

40. $\begin{cases} x + \frac{1}{2}y + \frac{1}{2}z = \frac{3}{2} \\ -x + 2y + 3z = 1 \\ 3x + 4y + 5z = 7 \end{cases}$

Applying the Concepts

41. Role Reversal Write a system of three linear equations containing three unknowns that has the solution $(2, -1, 3)$.

42. Role Reversal Write a system of three linear equations containing three unknowns that has the solution $(-4, 1, -3)$.

43. Curve Fitting The function $f(x) = ax^2 + bx + c$ is a quadratic function, where a, b, and c are constants.

(a) If $f(1) = 4$, then $4 = a(1)^2 + b(1) + c$ or $a + b + c = 4$. Find two additional linear equations if $f(-1) = -6$ and $f(2) = 3$.

(b) Use the three linear equations found in part (a) to determine a, b, and c. What is the quadratic function that contains the points $(-1, -6)$, $(1, 4)$, and $(2, 3)$?

44. Curve Fitting The function $f(x) = ax^2 + bx + c$ is a quadratic function, where a, b, and c are constants.

(a) If $f(-1) = 6$, then $6 = a(-1)^2 + b(-1) + c$ or $a - b + c = 6$. Find two additional linear equations if $f(1) = 2$ and $f(2) = 9$.

(b) Use the three linear equations found in part (a) to determine a, b, and c. What is the quadratic function that contains the points $(-1, 6)$, $(1, 2)$, and $(2, 9)$?

45. Electricity: Kirchhoff's Rules An application of Kirchhoff's Rule to the circuit shown results in the following system of equations:

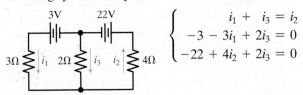

$\begin{cases} i_1 + i_3 = i_2 \\ -3 - 3i_1 + 2i_3 = 0 \\ -22 + 4i_2 + 2i_3 = 0 \end{cases}$

In the system circuit, V is the voltage, Ω is resistance, and i is the current. Find the currents i_1, i_2, and i_3.

46. Electricity: Kirchhoff's Rules An application of Kirchhoff's Rule to the circuit shown results in the following system of equations:

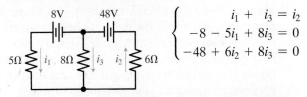

$\begin{cases} i_1 + i_3 = i_2 \\ -8 - 5i_1 + 8i_3 = 0 \\ -48 + 6i_2 + 8i_3 = 0 \end{cases}$

In the system circuit, V is the voltage, Ω is resistance, and i is the current. Find the currents i_1, i_2, and i_3.

47. Minor League Baseball In the Joliet Jackhammers baseball stadium, there are three types of seats available. Box seats are $9, reserved seats are $7, and lawn seats are $5. The stadium capacity is 4100. If all the seats are sold, the total revenue to the club is $28,400. If $\frac{1}{2}$ of the box seats are sold, $\frac{1}{2}$ of the reserved seats are sold, and all the lawn seats are sold, the total revenue is $18,300. How many are there of each kind of seat?

48. Theater Revenues A theater has 600 seats, divided into orchestra, main floor, and balcony seating. Orchestra seats sell for $80, main floor seats for $60, and balcony seats for $25. If all the seats are sold, the total revenue to the theater is $33,500. One evening, all the orchestra seats were sold, $\frac{3}{5}$ of the main seats were sold, and $\frac{4}{5}$ of the balcony seats were sold. The total revenue collected was $24,640. How many are there of each kind of seat?

49. Nutrition Nancy's dietitian wants her to consume 470 mg of sodium, 89 g of carbohydrates, and 20 g of protein for breakfast. This morning, Nancy wants to have Chex® cereal, 2% milk, and orange juice for breakfast. Each serving of Chex® cereal contains 220 mg of sodium, 26 g of carbohydrates, and 1 g of protein. Each serving of 2% milk contains 125 mg of sodium, 12 g of carbohydrates, and 8 g of protein. Each serving of orange juice contains 0 mg of sodium, 26 g of carbohydrates, and 2 g of protein. How many servings of each does Nancy need?

50. Nutrition Antonio is on a special diet that requires he consume 1325 calories, 172 grams of carbohydrates, and 63 grams of protein for lunch. He wishes to have a Broccoli and Cheese Baked Potato, Chicken BLT Salad, and a medium Coke. Each Broccoli and Cheese Baked Potato has 480 calories, 80 g of carbohydrates, and 9 g of protein. Each Chicken BLT Salad has 310 calories, 10 g of carbohydrates, and 33 g of protein. Each Coke has 140 calories, 37 g of carbohydrates, and 0 g of protein. How many servings of each does Antonio need?

51. Finance Sachi has $25,000 to invest. Her financial planner suggests that she diversify her investment into three investment categories: Treasury bills that yield 3% simple interest annually, municipal bonds that yield 5% simple interest annually, and corporate bonds that yield 9% simple interest annually. Sachi would like to earn $1210 per year in income. In addition, Sachi wants her investment in Treasury bills to be $7000 more than her investment in corporate bonds. How much should Sachi invest in each investment category?

52. Finance Delu has $15,000 to invest. She decides to place some of the money into a savings account paying 2% annual interest, some in Treasury bonds paying 5% annual interest, and some in a mutual fund paying 10% annual interest. Delu would like to earn $720 per year in income. In addition, Delu wants her investment in the savings account to be twice the amount in the mutual fund. How much should Delu invest in each investment category?

△ **53. Geometry** A circle is inscribed in $\triangle ABC$ as shown in the figure above. Suppose that $AB = 6$, $AC = 14$, and $BC = 12$. Find the length of $\overline{AM}$, $\overline{BN}$, and $\overline{CO}$. (*Hint:* $\overline{AM} \cong \overline{AO}$; $\overline{BM} \cong \overline{BN}$; $\overline{NC} \cong \overline{OC}$)

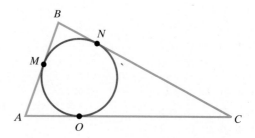

Extending the Concepts

In Problems 54–57, solve each system of equations.

54. $\begin{cases} \dfrac{1}{4}x + \dfrac{1}{4}y + \dfrac{1}{2}z = 6 \\[2mm] -\dfrac{1}{8}x + \dfrac{1}{2}y - \dfrac{1}{5}z = -5 \\[2mm] \dfrac{1}{2}x + \dfrac{1}{2}y - \dfrac{1}{2}z = -3 \end{cases}$

55. $\begin{cases} \dfrac{2}{5}x + \dfrac{1}{2}y - \dfrac{1}{3}z = 0 \\[2mm] \dfrac{3}{5}x - \dfrac{1}{4}y + \dfrac{1}{2}z = 10 \\[2mm] -\dfrac{1}{5}x + \dfrac{1}{4}y - \dfrac{1}{6}z = -4 \end{cases}$

56. $\begin{cases} x + y + z + w = 0 \\ 2x - 3y - z + w = -17 \\ 3x + y + 2z - w = 8 \\ -x + 2y - 3z + 2w = -7 \end{cases}$

57. $\begin{cases} x + y + z + w = 3 \\ -2x - y + 3z - w = -1 \\ 2x + 2y - 2z + w = 2 \\ -x + 2y - 3z + 2w = 12 \end{cases}$

Explaining the Concepts

58. Suppose that $(3, 2, 5)$ is the only solution of a system of three linear equations containing three variables. What does this mean geometrically?

59. Why is it necessary to eliminate the same variable in Step 1 of the box "Steps for Solving a System of Three Linear Equations Containing Three Unknowns by Elimination" (page 247)?

Putting the Concepts Together (Sections 3.1–3.3)

We designed these problems so that you can review Sections 3.1 through 3.3 and show your mastery of the concepts. Take time to work these problems before proceeding with the next section. The answers are at the back of the text on page AN-22.

1. Use the graph of the system of linear equations to determine the solution.

$$\begin{cases} 5x - 2y = -25 \\ x + y = 2 \end{cases}$$

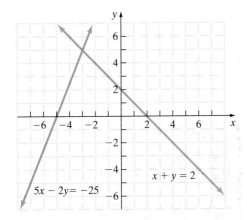

2. Solve the system of linear equations by graphing.

$$\begin{cases} 3x + y = 7 \\ -2x + 3y = -12 \end{cases}$$

In Problems 3–6, solve the system of linear equations using either substitution or elimination.

3. $\begin{cases} x = 2 - 3y \\ 3x + 10y = 5 \end{cases}$ **4.** $\begin{cases} 4x + 3y = -1 \\ 2x - y = 3 \end{cases}$

5. $\begin{cases} x + 3y = 8 \\ \dfrac{1}{5}x + \dfrac{1}{2}y = 1 \end{cases}$ **6.** $\begin{cases} 8x - 4y = 12 \\ -10x + 5y = -15 \end{cases}$

In Problems 7 and 8, solve each system of three linear equations containing three unknowns.

7. $\begin{cases} 2x + y + 3z = 10 \\ x - 2y + z = 10 \\ -4x + 3y + 2z = 5 \end{cases}$ **8.** $\begin{cases} x + 2y - 2z = 3 \\ x + 3y - 4z = 6 \\ 4x + 5y - 2z = 6 \end{cases}$

9. Museum Tickets A museum sells adult tickets for $9.00 and youth tickets for $5.00. On Saturday, the museum collected a total of $5925. It is also known that 825 people passed through the turnstiles. How many adult tickets were sold? How many youth tickets were sold?

10. Theater Seating A theater has 1200 seats that are divided into three sections: orchestra, mezzanine, and balcony. Seats sell for $65 each in the orchestra section, $48 each in the mezzanine section, and $35 each in the balcony section. There are 150 more balcony seats than mezzanine seats. If all the seats are sold, the theater will bring in $55,640 in revenue. Find the number of seats in each section of the theater.

3.4 Using Matrices to Solve Systems

Objectives

1 Write the Augmented Matrix of a System

2 Write the System from the Augmented Matrix

3 Perform Row Operations on a Matrix

4 Solve Systems Using Matrices

5 Solve Consistent Systems with Dependent Equations and Solve Inconsistent Systems

Are You Ready for This Section?

Before getting started, take the following readiness quiz. If you get a problem wrong, go back to the section cited and review the material.

R1. Determine the coefficients of the expression $4x - 2y + z$. [Section R.5, p. 41–42]

R2. Solve $x - 4y = 3$ for x. [Section 1.3, pp. 73–76]

R3. Evaluate $3x - 2y + z$ when $x = 1$, $y = -3$, and $z = 2$. [Section R.5, pp. 40–41]

We now introduce a way to solve systems of linear equations that streamlines the notation and makes working with systems more manageable.

But first, we need to introduce a new concept called the *matrix*.

Definition

A **matrix** is a rectangular array of numbers.

A matrix has rows and columns. The number of rows and columns determines the size of the matrix. For example, a matrix with 2 rows and 3 columns is called a "2 by 3 matrix," which is denoted "2 × 3 matrix." Below are some examples of matrices.

2 × 3 matrix
3 columns

2 rows $\begin{bmatrix} 3 & -1 & 4 \\ 8 & 0 & -5 \end{bmatrix}$

3 × 3 matrix
3 columns

3 rows $\begin{bmatrix} 2 & -8 & 12 \\ 0 & 7 & -2 \\ 5 & -2 & 1 \end{bmatrix}$

Work Smart

A spreadsheet such as Microsoft Excel is a matrix!

▶ ❶ Write the Augmented Matrix of a System

A system of linear equations may be represented in an *augmented matrix*. To write the **augmented matrix** of a system, write the terms containing the variables of each equation on the left side of the equal sign in the same order (x, y, and z, for example), and write the constants on the right side. A variable that does not appear in an equation has a coefficient of 0. For example, consider the following system of two linear equations containing two unknowns:

coefficients of x coefficients of y

$\begin{cases} a_1 x + b_1 y = c_1 \\ a_2 x + b_2 y = c_2 \end{cases}$ is written as an augmented matrix as $\begin{bmatrix} a_1 & b_1 & c_1 \\ a_2 & b_2 & c_2 \end{bmatrix}$ constants

Notice that the first column represents the coefficients of the variable x, the second column represents the coefficients of the variable y, the vertical bar represents the equal signs, and the constants are to the right of the vertical bar. The first row in the matrix is equation (1), and the second row in the matrix is equation (2).

EXAMPLE 1 **Writing the Augmented Matrix of a System of Linear Equations**

Write each system of linear equations as an augmented matrix.

(a) $\begin{cases} x - 3y = 2 \\ -2x + 5y = 7 \end{cases}$ (b) $\begin{cases} 2x + 3y - z = 1 \\ x - 2z + 1 = 0 \\ -4x - y + 3z = 5 \end{cases}$

Solution

(a) In the augmented matrix, we let the first column represent the coefficients of the variable x. The second column represents the coefficients of the variable y. The vertical line signifies the equal signs. The third column represents the constants to the right of the equal sign.

$\begin{array}{cc} x & y \end{array}$
$\begin{bmatrix} 1 & -3 & 2 \\ -2 & 5 & 7 \end{bmatrix}$ $\begin{array}{l} x - 3y = 2 \\ -2x + 5y = 7 \end{array}$

(b) We have to write the system with all the variables on the left side of the equal sign and the constants on the right. We also need to write 0 as the coefficient of any missing variable. Therefore, we rewrite the system

$\begin{cases} 2x + 3y - z = 1 \\ x - 2z + 1 = 0 \\ -4x - y + 3z = 5 \end{cases}$ as $\begin{cases} 2x + 3y - z = 1 \\ x + 0y - 2z = -1 \\ -4x - y + 3z = 5 \end{cases}$

The augmented matrix is

$\begin{bmatrix} 2 & 3 & -1 & 1 \\ 1 & 0 & -2 & -1 \\ -4 & -1 & 3 & 5 \end{bmatrix}$

Quick ✓

1. An m by n rectangular array of numbers is called a(n) _____.
2. The matrix used to represent a system of linear equations is called a(n) _____ matrix.
3. A 4×3 matrix has _ rows and _ columns.
4. *True or False* The augmented matrix of a system of two equations containing two unknowns has 2 rows and 2 columns.

In Problems 5 and 6, write the augmented matrix of each system of equations.

5. $\begin{cases} 3x - y = -10 \\ -5x + 2y = 0 \end{cases}$

6. $\begin{cases} x + 2y - 2z = 11 \\ -x - 2z = 4 \\ 4x - y + z - 3 = 0 \end{cases}$

▶ ❷ Write the System from the Augmented Matrix

EXAMPLE 2   **Writing the System from the Augmented Matrix**

Write the system of linear equations corresponding to each augmented matrix.

(a) $\left[\begin{array}{cc|c} 2 & 1 & -5 \\ -1 & 3 & 2 \end{array}\right]$ (b) $\left[\begin{array}{ccc|c} 1 & -3 & 2 & 5 \\ -2 & 0 & 4 & -3 \\ -1 & 4 & 1 & 0 \end{array}\right]$

Solution

(a) The augmented matrix has two rows, so it represents a system of two equations. There are two columns to the left of the vertical bar, so the system has two variables, which we can call x and y. The system of equations is

$$\begin{cases} 2x + y = -5 \\ -x + 3y = 2 \end{cases}$$

(b) The augmented matrix has three rows, so it represents a system of three equations. The three columns to the left of the vertical bar represent three variables, which we can call x, y, and z. The system of equations is

$$\begin{cases} x - 3y + 2z = 5 \\ -2x + 4z = -3 \\ -x + 4y + z = 0 \end{cases}$$

Quick ✓
In Problems 7 and 8, write the system of linear equations corresponding to the given augmented matrix.

7. $\left[\begin{array}{cc|c} 1 & -3 & 7 \\ -2 & 5 & -3 \end{array}\right]$

8. $\left[\begin{array}{ccc|c} 1 & -3 & 2 & 4 \\ 3 & 0 & -1 & -1 \\ -1 & 4 & 0 & 0 \end{array}\right]$

▶ ❸ Perform Row Operations on a Matrix

We use row operations on an augmented matrix to solve the corresponding system of equations.

> **Row Operations**
> 1. Interchange any two rows.
> 2. Multiply all entries in a row by a nonzero constant.
> 3. Replace a row by the sum of that row and a nonzero multiple of some other row.

These are the same operations that we performed on systems of equations in Section 3.3. The main reason for using a matrix to solve a system of equations is that its notation is more efficient and helps us organize the mathematics. To see how row operations work, consider the following augmented matrix.

$$\begin{bmatrix} 1 & 3 & | & -1 \\ -2 & 1 & | & 3 \end{bmatrix}$$

Suppose we want to apply a row operation that results in a matrix whose entry in row 2, column 1 is a 0. The current entry here is -2. The row operation to use is

Multiply each entry in row 1 by 2, and then add the result to the corresponding entry in row 2. Have this result replace the current row 2.

To streamline this a little, we introduce some notation. If we use R_2 to represent the new entries in row 2 and we use r_1 and r_2 to represent the original entries in rows 1 and 2, respectively, then we can express the row operation given above by

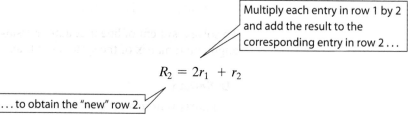

Multiply each entry in row 1 by 2 and add the result to the corresponding entry in row 2 . . .

$$R_2 = 2r_1 + r_2$$

. . . to obtain the "new" row 2.

We demonstrate this row operation below.

$$\begin{bmatrix} 1 & 3 & | & -1 \\ -2 & 1 & | & 3 \end{bmatrix} \xrightarrow{R_2 = 2r_1 + r_2} \begin{bmatrix} 1 & 3 & | & -1 \\ 2(1) + (-2) & 2(3) + 1 & | & 2(-1) + 3 \end{bmatrix} = \begin{bmatrix} 1 & 3 & | & -1 \\ 0 & 7 & | & 1 \end{bmatrix}$$

We now have a 0 in row 2, column 1, as desired. Notice the second row represents the equation $7y = 1$, which is easy to solve.

EXAMPLE 3) **Applying a Row Operation to an Augmented Matrix**

Apply the row operation $R_2 = -3r_1 + r_2$ to the augmented matrix

$$\begin{bmatrix} 1 & 2 & | & -3 \\ 3 & 4 & | & 5 \end{bmatrix}$$

Solution

The row operation $R_2 = -3r_1 + r_2$ tells us to multiply the entries in row 1 by -3 and then add the result to the entries in row 2. The result should replace the current row 2.

$$\begin{bmatrix} 1 & 2 & | & -3 \\ 3 & 4 & | & 5 \end{bmatrix} \xrightarrow{R_2 = -3r_1 + r_2} \begin{bmatrix} 1 & 2 & | & -3 \\ -3(1) + 3 & -3(2) + 4 & | & -3(-3) + 5 \end{bmatrix} = \begin{bmatrix} 1 & 2 & | & -3 \\ 0 & -2 & | & 14 \end{bmatrix}$$

●

Quick ✓

9. Apply the row operation $R_2 = 4r_1 + r_2$ to the augmented matrix

$$\begin{bmatrix} 1 & -2 & | & 5 \\ -4 & 5 & | & -11 \end{bmatrix}$$

EXAMPLE 4 **Finding a Particular Row Operation**

For the augmented matrix

$$\left[\begin{array}{cc|c} 1 & -3 & -12 \\ 0 & 1 & 5 \end{array}\right]$$

find and perform a row operation that will result in the entry in row 1, column 2 becoming a 0.

Solution

To get a 0 in row 1, column 2, we multiply row 2 by 3 and add the result to row 1. That is, apply the row operation $R_1 = 3r_2 + r_1$.

$$\left[\begin{array}{cc|c} 1 & -3 & -12 \\ 0 & 1 & 5 \end{array}\right] \xrightarrow{R_1 = 3r_2 + r_1} \left[\begin{array}{cc|c} 3(0)+1 & 3(1)+(-3) & 3(5)+(-12) \\ 0 & 1 & 5 \end{array}\right] = \left[\begin{array}{cc|c} 1 & 0 & 3 \\ 0 & 1 & 5 \end{array}\right]$$

●

Work Smart

If you want to change the entries in row 1, then you should multiply the entries in some other row by a "well-chosen" constant and then add this result to the entries in row 1.

Quick ✓

10. For the augmented matrix $\left[\begin{array}{cc|c} 1 & 5 & 13 \\ 0 & 1 & 2 \end{array}\right]$, find and perform a row operation that will result in the entry in row 1, column 2 becoming a 0.

❹ Solve Systems Using Matrices

▶ To solve a system of linear equations using matrices, we use row operations on the augmented matrix of the system to obtain a matrix that is in *row echelon form*.

Definition

A matrix is in **row echelon form** when

1. The entry in row 1, column 1 is a 1, and 0s appear below it.
2. The first nonzero entry in each row below the first row is a 1, 0s appear below it, and it appears to the right of the first nonzero entry in any row above.
3. Any rows that contain all 0s to the left of the vertical bar appear at the bottom.

For a system of two equations containing two variables with a single solution, the row echelon form of the augmented matrix is

$$\left[\begin{array}{cc|c} 1 & a & b \\ 0 & 1 & c \end{array}\right]$$

where a, b, and c are real numbers. For a system of three equations containing three variables with a single solution, the row echelon form is

$$\left[\begin{array}{ccc|c} 1 & a & b & d \\ 0 & 1 & c & e \\ 0 & 0 & 1 & f \end{array}\right]$$

where a, b, c, d, e, and f are real numbers.

The augmented matrix $\left[\begin{array}{cc|c} 1 & a & b \\ 0 & 1 & c \end{array}\right]$ is equivalent to the system $\begin{cases} x + ay = b \ (1) \\ y = c \ (2) \end{cases}$.

Because c is a known number, we know y. We can then back-substitute to find x. For example, the augmented matrix $\left[\begin{array}{cc|c} 1 & 3 & -5 \\ 0 & 1 & -3 \end{array}\right]$ is in row echelon form and is equivalent to the system $\begin{cases} x + 3y = -5 \ (1) \\ y = -3 \ (2) \end{cases}$. Equation (2) states that $y = -3$. Using back-substitution, we find that $x = 4$, so the solution is $(4, -3)$.

EXAMPLE 5 **How to Solve a System of Two Linear Equations Containing Two Variables Using Matrices**

Solve: $\begin{cases} 3x - 2y = -19 \\ x + 2y = 7 \end{cases}$

Step-by-Step Solution

Step 1: Write the augmented matrix of the system.	$\begin{bmatrix} 3 & -2 & \vert & -19 \\ 1 & 2 & \vert & 7 \end{bmatrix}$
Step 2: We want the entry in row 1, column 1 to be 1, so we interchange rows 1 and 2.	$\begin{bmatrix} 1 & 2 & \vert & 7 \\ 3 & -2 & \vert & -19 \end{bmatrix}$
Step 3: We want the entry in row 2, column 1 to be 0, so we use the row operation $R_2 = -3r_1 + r_2$. The entries in row 1 remain unchanged.	$R_2 = -3r_1 + r_2$: $\begin{bmatrix} 1 & 2 & \vert & 7 \\ 0 & -8 & \vert & -40 \end{bmatrix}$
Step 4: Now we want the entry in row 2, column 2 to be 1. This is accomplished by multiplying row 2 by $-\dfrac{1}{8}$. We use the row operation $R_2 = -\dfrac{1}{8}r_2$.	$R_2 = -\dfrac{1}{8}r_2$: $\begin{bmatrix} 1 & 2 & \vert & 7 \\ 0 & 1 & \vert & 5 \end{bmatrix}$
Step 5: Row 2 says that $y = 5$. Substitute this value into row 1, which is $x + 2y = 7$. Solve for x.	Let $y = 5$: $\begin{aligned} x + 2y &= 7 \\ x + 2(5) &= 7 \\ x + 10 &= 7 \\ x &= -3 \end{aligned}$
Step 6: Check: We let $x = -3$ and $y = 5$ in equations (1) and (2) to verify our solution.	Equation (1): $\begin{aligned} 3x - 2y &= -19 \\ x = -3, y = 5: \quad 3(-3) - 2(5) &\overset{?}{=} -19 \\ -9 - 10 &\overset{?}{=} -19 \\ -19 &= -19 \quad \text{True} \end{aligned}$ Equation (2): $\begin{aligned} x + 2y &= 7 \\ x = -3, y = 5: \quad -3 + 2(5) &\overset{?}{=} 7 \\ -3 + 10 &\overset{?}{=} 7 \\ 7 &= 7 \quad \text{True} \end{aligned}$

The solution is the ordered pair $(-3, 5)$.

Quick ✓

11. *True or False* The matrix $\begin{bmatrix} 1 & 3 & \vert & 2 \\ 0 & 1 & \vert & -5 \end{bmatrix}$ is in row echelon form.

12. Solve the following system using matrices: $\begin{cases} 2x - 4y = 20 \\ 3x + y = 16 \end{cases}$

▶ **EXAMPLE 6** **How to Solve a System of Three Linear Equations Containing Three Variables Using Matrices**

Solve: $\begin{cases} x + y + z = 3 \\ -2x - 3y + 2z = 13 \\ 4x + 5y + z = -3 \end{cases}$

Step-by-Step Solution

Step 1: Write the augmented matrix of the system.

$$\begin{bmatrix} 1 & 1 & 1 & \vert & 3 \\ -2 & -3 & 2 & \vert & 13 \\ 4 & 5 & 1 & \vert & -3 \end{bmatrix}$$

Step 2: We want the entry in row 1, column 1 to be 1. This is already done.

$$\begin{bmatrix} 1 & 1 & 1 & | & 3 \\ -2 & -3 & 2 & | & 13 \\ 4 & 5 & 1 & | & -3 \end{bmatrix}$$

Step 3: We want the entry in row 2, column 1 to be 0, so we use the row operation $R_2 = 2r_1 + r_2$. We also want the entry in row 3, column 1 to be 0, so we use the row operation $R_3 = -4r_1 + r_3$. The entries in row 1 remain unchanged.

$$\begin{array}{c} R_2 = 2r_1 + r_2: \\ R_3 = -4r_1 + r_3: \end{array} \begin{bmatrix} 1 & 1 & 1 & | & 3 \\ 0 & -1 & 4 & | & 19 \\ 0 & 1 & -3 & | & -15 \end{bmatrix}$$

Step 4: Now we want the entry in row 2, column 2 to be 1. This is accomplished by interchanging rows 2 and 3.

$$\begin{bmatrix} 1 & 1 & 1 & | & 3 \\ 0 & 1 & -3 & | & -15 \\ 0 & -1 & 4 & | & 19 \end{bmatrix}$$

Step 5: We need the entry in row 3, column 2 to be 0. We use the row operation $R_3 = r_2 + r_3$.

$$R_3 = r_2 + r_3: \begin{bmatrix} 1 & 1 & 1 & | & 3 \\ 0 & 1 & -3 & | & -15 \\ 0 & 0 & 1 & | & 4 \end{bmatrix}$$

Step 6: We want the entry in row 3, column 3 to be 1. This is already done.

$$\begin{bmatrix} 1 & 1 & 1 & | & 3 \\ 0 & 1 & -3 & | & -15 \\ 0 & 0 & 1 & | & 4 \end{bmatrix}$$

Step 7: The augmented matrix is in row echelon form. Write the corresponding system of equations and solve.

$$\begin{cases} x + y + z = 3 & (1) \\ y - 3z = -15 & (2) \\ z = 4 & (3) \end{cases}$$

Equation (3) says $z = 4$. Let $z = 4$ in equation (2) to find y.

$$\begin{array}{rl} \text{Equation (2):} & y - 3z = -15 \\ z = 4: & y - 3(4) = -15 \\ & y - 12 = -15 \\ & y = -3 \end{array}$$

Let $y = -3$ and $z = 4$ in equation (1) to find that $x = 2$.

Step 8: Check: We let $x = 2, y = -3$, and $z = 4$ in each equation in the original system to verify our solution.

We leave it to you to verify the solution, $(2, -3, 4)$

The solution is the ordered triple $(2, -3, 4)$.

Work Smart

Look at Step 4 in Example 6. Other ways to obtain a 1 in row 2, column 2 are to use the row operation $R_2 = 2r_3 + r_2$ or $R_2 = -r_2$.

Quick ✓

In Problems 13 and 14, solve each system of equations using matrices.

13. $\begin{cases} x - y + 2z = 7 \\ 2x - 2y + z = 11 \\ -3x + y - 3z = -14 \end{cases}$

14. $\begin{cases} 2x + 3y - 2z = 1 \\ 2x - 4z = -9 \\ 4x + 6y - z = 14 \end{cases}$

> **Solving a System of Linear Equations Using Matrices**
>
> **Step 1:** Write the augmented matrix of the system.
>
> **Step 2:** Perform row operations so the entry in row 1, column 1 is 1.
>
> **Step 3:** Perform row operations so all the entries below the 1 in row 1, column 1 are 0s.
>
> **Step 4:** Perform row operations so the entry in row 2, column 2 is 1. Be sure the entries in column 1 remain unchanged. If it is impossible to place a 1 in row 2, column 2, then use operations to obtain a 1 in row 2, column 3. (*Note:* If you get a row with all 0s, then place it in the last row of the matrix.)
>
> **Step 5:** Perform row operations so that the entry below the 1 in row 2, column 2, is 0.
>
> **Step 6:** Perform row operations so that the entry in row 3, column 3, is 1.
>
> **Step 7:** With the augmented matrix in row echelon form, write the corresponding system of equations. Back-substitute to find the values of the variables.
>
> **Step 8:** Check your answer.

▶ ❺ Solve Consistent Systems with Dependent Equations and Solve Inconsistent Systems

The matrix method for solving a system of linear equations also identifies systems that have infinitely many solutions (dependent equations) and systems with no solution (inconsistent systems).

EXAMPLE 7 **Solving a Consistent System with Dependent Equations Using Matrices**

$$\text{Solve: } \begin{cases} x + y + 3z = 3 \\ -2x - y - 8z = -5 \\ 3x + 2y + 11z = 8 \end{cases}$$

Solution

We will write the augmented matrix of the system and perform row operations to get the matrix into row echelon form.

$$\begin{bmatrix} 1 & 1 & 3 & | & 3 \\ -2 & -1 & -8 & | & -5 \\ 3 & 2 & 11 & | & 8 \end{bmatrix} \xrightarrow[\substack{R_2 = 2r_1 + r_2 \\ R_3 = -3r_1 + r_3}]{} \begin{bmatrix} 1 & 1 & 3 & | & 3 \\ 0 & 1 & -2 & | & 1 \\ 0 & -1 & 2 & | & -1 \end{bmatrix} \xrightarrow[R_3 = r_2 + r_3]{} \begin{bmatrix} 1 & 1 & 3 & | & 3 \\ 0 & 1 & -2 & | & 1 \\ 0 & 0 & 0 & | & 0 \end{bmatrix}$$

Notice that the last row is all 0s. The augmented matrix is in row echelon form. The corresponding system of equations is

$$\begin{cases} x + y + 3z = 3 & (1) \\ y - 2z = 1 & (2) \\ 0 = 0 & (3) \end{cases}$$

The statement $0 = 0$ in equation (3) indicates that the equations are dependent. If we let z represent any real number, then we can solve equation (2) for y in terms of z.

$$\text{Equation (2):} \quad y - 2z = 1$$
$$\text{Add } 2z \text{ to both sides:} \quad y = 2z + 1$$

Let $y = 2z + 1$ in equation (1) and solve for x in terms of z.

$$\text{Equation (1):} \qquad x + y + 3z = 3$$
$$\text{Let } y = 2z + 1: \ x + (2z + 1) + 3z = 3$$
$$\text{Combine like terms:} \qquad x + 5z + 1 = 3$$
$$\text{Subtract } 5z + 1 \text{ from both sides:} \qquad x = -5z + 2$$

The solution to the system is $\{(x, y, z) \mid x = -5z + 2, y = 2z + 1, z \text{ is any real number}\}$. To find specific solutions to the system, choose any value of z and use the equations $x = -5z + 2$ and $y = 2z + 1$ to determine x and y. Some specific solutions to the system are $(2, 1, 0)$, $(-3, 3, 1)$ and $(7, -1, -1)$.

If we evaluate the system at some of the specific solutions, we see that all of them satisfy the original system. This leads us to believe that our solution is correct. ●

Quick ✓

In Problems 15 and 16, solve each system of equations using matrices.

15. $\begin{cases} 2x + 5y = -6 \\ -6x - 15y = 18 \end{cases}$

16. $\begin{cases} x + y - 3z = 8 \\ 2x + 3y - 10z = 19 \\ -x - 2y + 7z = -11 \end{cases}$

EXAMPLE 8 **Solving an Inconsistent System Using Matrices**

Solve: $\begin{cases} 2x + y - z = 5 \\ -x + 2y + z = 1 \\ 3x + 4y - z = -2 \end{cases}$

Solution

Write the augmented matrix of the system and perform row operations to get the matrix into row echelon form.

$$\begin{bmatrix} 2 & 1 & -1 & | & 5 \\ -1 & 2 & 1 & | & 1 \\ 3 & 4 & -1 & | & -2 \end{bmatrix} \xrightarrow{R_1 = r_2 + r_1} \begin{bmatrix} 1 & 3 & 0 & | & 6 \\ -1 & 2 & 1 & | & 1 \\ 3 & 4 & -1 & | & -2 \end{bmatrix} \xrightarrow[R_3 = -3r_1 + r_3]{R_2 = r_1 + r_2} \begin{bmatrix} 1 & 3 & 0 & | & 6 \\ 0 & 5 & 1 & | & 7 \\ 0 & -5 & -1 & | & -20 \end{bmatrix}$$

$$\xrightarrow{R_2 = \frac{1}{5}r_2} \begin{bmatrix} 1 & 3 & 0 & | & 6 \\ 0 & 1 & \frac{1}{5} & | & \frac{7}{5} \\ 0 & -5 & -1 & | & -20 \end{bmatrix} \xrightarrow{R_3 = 5r_2 + r_3} \begin{bmatrix} 1 & 3 & 0 & | & 6 \\ 0 & 1 & \frac{1}{5} & | & \frac{7}{5} \\ 0 & 0 & 0 & | & -13 \end{bmatrix}$$

We want the entry in row 3, column 3 to be 1. This cannot be done (without changing the entries in row 3, column 1 or 2). The corresponding system of equations is

$$\begin{cases} x + 3y \qquad = 6 & (1) \\ \quad y + \dfrac{1}{5}z = \dfrac{7}{5} & (2) \\ \qquad\qquad 0 = -13 & (3) \end{cases}$$

The statement $0 = -13$ in equation (3) indicates that the system is inconsistent. The solution is $\varnothing$ or $\{\ \}$. ●

Quick ✓

In Problems 17 and 18, solve each system of equations using matrices.

17. $\begin{cases} -2x + 3y = 4 \\ 10x - 15y = 2 \end{cases}$

18. $\begin{cases} -x + 2y - z = 5 \\ 2x + y + 4z = 3 \\ 3x - y + 5z = 0 \end{cases}$

3.4 Exercises · MyMathLab® · PRACTICE

Exercise numbers in green are complete video solutions in MyMathLab

*Problems **1–18** are the* Quick ✓ *s that follow the* **EXAMPLES**.

Building Skills

In Problems 19–26, write the augmented matrix of the given system of equations. See Objective 1.

19. $\begin{cases} x - 3y = 2 \\ 2x + 5y = 1 \end{cases}$

20. $\begin{cases} -x + y = 6 \\ 5x - y = -3 \end{cases}$

21. $\begin{cases} x + y + z = 3 \\ 2x - y + 3z = 1 \\ -4x + 2y - 5z = -3 \end{cases}$

22. $\begin{cases} x + y - z = 2 \\ -2x + y - 4z = 13 \\ 3x - y - 2z = -4 \end{cases}$

23. $\begin{cases} -x + y - 2 = 0 \\ 5x + y + 5 = 0 \end{cases}$

24. $\begin{cases} 6x + 4y + 2 = 0 \\ -x - y + 1 = 0 \end{cases}$

25. $\begin{cases} x + z = 2 \\ 2x + y = 13 \\ x - y + 4z + 4 = 0 \end{cases}$

26. $\begin{cases} 2x + 7y = 1 \\ -x - 6z = 5 \\ 5x + 2y - 4z + 1 = 0 \end{cases}$

In Problems 27–32, write the system of linear equations corresponding to each augmented matrix. Use x, y; or x, y, z as variables. See Objective 2.

27. $\left[\begin{array}{cc|c} 2 & 5 & 3 \\ -4 & 1 & 10 \end{array}\right]$

28. $\left[\begin{array}{cc|c} -1 & 4 & 0 \\ 2 & -7 & -9 \end{array}\right]$

29. $\left[\begin{array}{ccc|c} 1 & 5 & -3 & 2 \\ 0 & 3 & -1 & -5 \\ 4 & 0 & 8 & 6 \end{array}\right]$

30. $\left[\begin{array}{ccc|c} -2 & 3 & 0 & 3 \\ 4 & 1 & -8 & -6 \\ 7 & 2 & -5 & 10 \end{array}\right]$

31. $\left[\begin{array}{ccc|c} 1 & -2 & 9 & 2 \\ 0 & 1 & -5 & 8 \\ 0 & 0 & 1 & \frac{4}{3} \end{array}\right]$

32. $\left[\begin{array}{ccc|c} 1 & 4 & -1 & \frac{2}{3} \\ 0 & 1 & 3 & 0 \\ 0 & 0 & 1 & 10 \end{array}\right]$

In Problems 33–38, perform each row operation on the given augmented matrix. See Objective 3.

33. $\left[\begin{array}{cc|c} 1 & -3 & 2 \\ -2 & 5 & 1 \end{array}\right]$

  (a) $R_2 = 2r_1 + r_2$ followed by

  (b) $R_2 = -r_2$

34. $\left[\begin{array}{cc|c} 1 & 5 & 7 \\ 3 & 11 & 13 \end{array}\right]$

  (a) $R_2 = -3r_1 + r_2$ followed by

  (b) $R_2 = -\dfrac{1}{4} r_2$

35. $\left[\begin{array}{ccc|c} 1 & 1 & -1 & 4 \\ 2 & 5 & 3 & -3 \\ -1 & -3 & 2 & 1 \end{array}\right]$

  (a) $R_2 = -2r_1 + r_2$ followed by

  (b) $R_3 = r_1 + r_3$

36. $\left[\begin{array}{ccc|c} 1 & -1 & 1 & 6 \\ -2 & 1 & -3 & 3 \\ 3 & 2 & -2 & -5 \end{array}\right]$

  (a) $R_2 = 2r_1 + r_2$ followed by

  (b) $R_3 = -3r_1 + r_3$

37. $\left[\begin{array}{ccc|c} 1 & 1 & 1 & 4 \\ 0 & 5 & 3 & -3 \\ 0 & -4 & 2 & 8 \end{array}\right]$

  (a) $R_2 = r_3 + r_2$ followed by

  (b) $R_3 = \dfrac{1}{2} r_3$

38. $\left[\begin{array}{ccc|c} 1 & -3 & 4 & 11 \\ 0 & 3 & 6 & -12 \\ 0 & -2 & -3 & 8 \end{array}\right]$

  (a) $R_2 = \dfrac{1}{3} r_2$ followed by

  (b) $R_3 = 2r_2 + r_3$

In Problems 39–44, solve each system of equations using matrices.
See Objective 4.

39. $\begin{cases} 2x + 3y = 1 \\ -x + 4y = -28 \end{cases}$

40. $\begin{cases} 3x + 5y = -1 \\ -2x - 3y = 2 \end{cases}$

41. $\begin{cases} x + 5y - 2z = 23 \\ -2x - 3y + 5z = -11 \\ 3x + 2y + z = 4 \end{cases}$

42. $\begin{cases} x - 4y + 2z = -2 \\ -3x + y - 3z = 12 \\ 2x + 3y - 4z = -20 \end{cases}$

43. $\begin{cases} 4x + 5y = 0 \\ -8x + 10y - 2z = -13 \\ 15y + 4z = -7 \end{cases}$

44. $\begin{cases} 2x + y - 3z = 4 \\ 2y + 6z = 18 \\ -4x - y = -14 \end{cases}$

In Problems 45–50, solve each system of equations using matrices.
See Objective 5.

45. $\begin{cases} x - 3y = 3 \\ -2x + 6y = -6 \end{cases}$

46. $\begin{cases} -4x + 2y = 4 \\ 6x - 3y = -6 \end{cases}$

47. $\begin{cases} x - 2y + z = 3 \\ -2x + 4y - z = -5 \\ -8x + 16y + z = -21 \end{cases}$

48. $\begin{cases} x - 3y + 2z = 4 \\ 3x + y + 4z = -5 \\ -x - 7y = 5 \end{cases}$

49. $\begin{cases} x + y - 2z = 4 \\ -4x + 3z = -4 \\ -2x + 2y - z = 4 \end{cases}$

50. $\begin{cases} -x + 3y + 3z = 0 \\ x + 5y = -2 \\ 8y + 3z = -2 \end{cases}$

Mixed Practice

In Problems 51–56, an augmented matrix of a system of linear equations is given. Write the system of equations corresponding to the given matrix. Use x, y; or x, y, z as variables. Determine whether the system is consistent with independent equations, consistent with dependent equations, or inconsistent. If it is consistent with independent equations, give the solution.

51. $\left[\begin{array}{cc|c} 1 & 4 & -5 \\ 0 & 1 & -2 \end{array}\right]$

52. $\left[\begin{array}{cc|c} 1 & -2 & 3 \\ 0 & 1 & -5 \end{array}\right]$

53. $\left[\begin{array}{ccc|c} 1 & 3 & -2 & 6 \\ 0 & 1 & 5 & -2 \\ 0 & 0 & 0 & 4 \end{array}\right]$

54. $\left[\begin{array}{ccc|c} 1 & -2 & -4 & 6 \\ 0 & 1 & -5 & -3 \\ 0 & 0 & 0 & 0 \end{array}\right]$

55. $\left[\begin{array}{ccc|c} 1 & -2 & -1 & 3 \\ 0 & 1 & -2 & -8 \\ 0 & 0 & 1 & 5 \end{array}\right]$

56. $\left[\begin{array}{ccc|c} 1 & 2 & -1 & -7 \\ 0 & 1 & 2 & -4 \\ 0 & 0 & 1 & -3 \end{array}\right]$

In Problems 57–78, solve each system of equations using matrices.

57. $\begin{cases} x - 3y = 18 \\ 2x + y = 1 \end{cases}$

58. $\begin{cases} x + 5y = 2 \\ -2x + 3y = 9 \end{cases}$

59. $\begin{cases} 2x + 4y = 10 \\ x + 2y = 3 \end{cases}$

60. $\begin{cases} 5x - 2y = 3 \\ -15x + 6y = -9 \end{cases}$

61. $\begin{cases} x - 6y = 8 \\ 2x + 8y = -9 \end{cases}$

62. $\begin{cases} 3x + 3y = -1 \\ 2x + y = 1 \end{cases}$

63. $\begin{cases} 4x - y = 8 \\ 2x - \dfrac{1}{2}y = 4 \end{cases}$

64. $\begin{cases} 5x - 2y = 10 \\ 2x - \dfrac{4}{5}y = 4 \end{cases}$

65. $\begin{cases} x + y + z = 0 \\ 2x - 3y + z = 19 \\ -3x + y - 2z = -15 \end{cases}$

66. $\begin{cases} x + y + z = 5 \\ 2x - y + 3z = -3 \\ -x + 2y - z = 10 \end{cases}$

67. $\begin{cases} 2x + y - z = 13 \\ -x - 3y + 2z = -14 \\ -3x + 2y - 3z = 3 \end{cases}$

68. $\begin{cases} 2x - y + 2z = 13 \\ -x + 2y - z = -14 \\ 3x + y - 2z = -13 \end{cases}$

69. $\begin{cases} 2x - y + 3z = 1 \\ -x + 3y + z = -4 \\ 3x + y + 7z = -2 \end{cases}$

70. $\begin{cases} -x + 2y + z = 1 \\ 2x - y + 3z = -3 \\ -x + 5y + 6z = 2 \end{cases}$

71. $\begin{cases} 3x + y - 4z = 0 \\ -2x - 3y + z = 5 \\ -x - 5y - 2z = 3 \end{cases}$

72. $\begin{cases} -x + 4y - 3z = 1 \\ 3x + y - z = -3 \\ x + 9y - 7z = -1 \end{cases}$

73. $\begin{cases} 2x - y + 3z = -1 \\ 3x + y - 4z = 3 \\ x + 7y - 2z = 2 \end{cases}$

74. $\begin{cases} x + y + z = 8 \\ 2x + 3y + z = 19 \\ 2x + 2y + 4z = 21 \end{cases}$

75. $\begin{cases} 3x + 5y + 2z = 6 \\ 10y - 2z = 5 \\ 6x + 4z = 8 \end{cases}$

76. $\begin{cases} 2x + y + 3z = 3 \\ 2x - 3y = 7 \\ 4y + 6z = -2 \end{cases}$

77. $\begin{cases} x - z = 3 \\ 2x + y = -3 \\ 2y - z = 7 \end{cases}$

78. $\begin{cases} x + 2y - z = 3 \\ y + z = 1 \\ x - 3z = 2 \end{cases}$

Applying the Concepts

79. **Curve Fitting** The function $f(x) = ax^2 + bx + c$ is a quadratic function, where a, b, and c are constants.

    (a) If $f(-1) = 6$, then $6 = a(-1)^2 + b(-1) + c$ or $a - b + c = 6$. Find two additional linear equations if $f(1) = 0$ and $f(2) = 3$.

    (b) Use the three linear equations found in part (a) to determine a, b, and c. What is the quadratic function that contains the points $(-1, 6)$, $(1, 0)$, and $(2, 3)$?

80. **Curve Fitting** The function $f(x) = ax^2 + bx + c$ is a quadratic function, where a, b, and c are constants.

    (a) If $f(-1) = -6$, then $-6 = a(-1)^2 + b(-1) + c$ or $a - b + c = -6$. Find two additional linear equations if $f(1) = 0$ and $f(2) = -3$.

    (b) Use the three linear equations found in part (a) to determine a, b, and c. What is the quadratic function that contains the points $(-1, -6)$, $(1, 0)$, and $(2, -3)$?

81. **Finance** Carissa has $20,000 to invest. Her financial planner suggests that she diversify her investment into three investment categories: Treasury bills that yield 4% simple interest, municipal bonds that yield 5% simple interest, and corporate bonds that yield 8% simple interest. Carissa would like to earn $1070 per year in income. In addition, she wants her investment in Treasury bills to be $3000 more than her investment in corporate bonds. How much should Carissa invest in each investment category?

82. **Finance** Marlon has $12,000 to invest. He decides to place some of the money into a savings account paying 2% interest, some in Treasury bonds paying 4% interest, and some in a mutual fund paying 9% interest. Marlon would like to earn $440 per year in income. In addition, Marlon wants his investment in the savings account to be $4000 more than the amount in Treasury bonds. How much should Marlon invest in each investment category?

Extending the Concepts

*Sometimes it is advantageous to write a matrix in **reduced row echelon form**. In this form, row operations are used to obtain entries that are 0 above and below the leading 1 in a row. Augmented matrices for systems in reduced row echelon form with two equations containing two variables and three equations containing three variables are shown below.*

$$\left[\begin{array}{cc|c} 1 & 0 & a \\ 0 & 1 & b \end{array}\right] \qquad \left[\begin{array}{ccc|c} 1 & 0 & 0 & a \\ 0 & 1 & 0 & b \\ 0 & 0 & 1 & c \end{array}\right]$$

The obvious advantage to writing an augmented matrix in reduced row echelon form is that the solution to the system is readily seen. In the system with two equations containing two variables, the solution is $x = a$ and $y = b$. In the system with three equations containing three variables, the solution is $x = a$, $y = b$, and $z = c$.

In Problems 83–86, solve the system of equations by writing the augmented matrix in reduced row echelon form.

83. $\begin{cases} 2x + y = 1 \\ -3x - 2y = -5 \end{cases}$

84. $\begin{cases} 2x + y = -1 \\ -3x - 2y = -3 \end{cases}$

85. $\begin{cases} x + y + z = 3 \\ 2x + y - 4z = 25 \\ -3x + 2y + z = 0 \end{cases}$

86. $\begin{cases} x + y + z = 3 \\ 3x - 2y + 2z = 38 \\ -2x \quad\;\; - 3z = -19 \end{cases}$

Explaining the Concepts

87. Write a paragraph that outlines the strategy for putting an augmented matrix in row echelon form.

88. When solving a system of linear equations using matrices, how do you know that the system is inconsistent?

89. What would be the next row operation on the given augmented matrix?

$$\left[\begin{array}{cc|c} 1 & 3 & 8 \\ 0 & 5 & 10 \end{array}\right]$$

90. What would you recommend as the next row operation on the given augmented matrix? Why?

$$\left[\begin{array}{ccc|c} 1 & 3 & -2 & 4 \\ 0 & 5 & 3 & 2 \\ 0 & -4 & 6 & -5 \end{array}\right]$$

The Graphing Calculator

Graphing calculators have the ability to solve systems of linear equations by writing the augmented matrix in row echelon form. We enter the augmented matrix into the graphing calculator and name it A. See Figure 11(a). Figure 11(b) shows the results of Example 6 using a TI-84 Plus graphing calculator. Since the entire matrix does not fit on the screen, we need to scroll right to see the rest of it. See Figure 11(c).

Figure 11

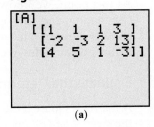 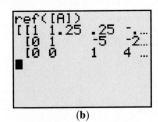

(a)   (b)

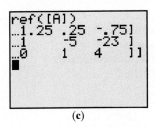

(c)

Notice that the row echelon form of the augmented matrix found using the graphing calculator differs from the row echelon form in the algebraic solution presented in Example 6, yet both matrices provide the same solution! This is because the two solutions use different row operations to obtain the row echelon form.

In Problems 91–94, solve each system of equations using a graphing calculator.

91. $\begin{cases} 2x + 3y = 1 \\ -3x - 4y = -3 \end{cases}$

92. $\begin{cases} 3x + 2y = 4 \\ -5x - 3y = -4 \end{cases}$

93. $\begin{cases} 2x + 3y - 2z = -12 \\ -3x + y + 2z = 0 \\ 4x + 3y - z = 3 \end{cases}$

94. $\begin{cases} 2x - 3y - 4z = 16 \\ -3x + y + 2z = -23 \\ 4x + 3y - z = 13 \end{cases}$

3.5 Determinants and Cramer's Rule

Objectives

1 Evaluate the Determinant of a 2 × 2 Matrix

2 Use Cramer's Rule to Solve a System of Two Equations

3 Evaluate the Determinant of a 3 × 3 Matrix

4 Use Cramer's Rule to Solve a System of Three Equations

Are You Ready for This Section?

Before getting started, take the following readiness quiz. If you get a problem wrong, go back to the section cited and review the material.

R1. Evaluate: $4 \cdot 2 - 3 \cdot (-3)$ [Section R.4, pp. 34–37]

R2. Simplify: $\dfrac{18}{6}$ [Section R.3, p. 25]

At this point, we know how to solve systems of linear equations using substitution, elimination, and row operations on augmented matrices. In this section we will learn another method. This method, which requires that the number of equations equal the number of variables, is called *Cramer's Rule*. It is based on the concept of a *determinant*.

▶ 1 Evaluate the Determinant of a 2 × 2 Matrix

A matrix is **square** if the number of rows equals the number of columns. We can compute the *determinant* of any square matrix.

Work Smart

The methods for solving systems presented in this section work only when the number of equations equals the number of variables.

> **Definition**
>
> Suppose a, b, c, and d are four real numbers. The **determinant** of a 2 × 2 matrix $\begin{bmatrix} a & b \\ c & d \end{bmatrix}$, denoted $\begin{vmatrix} a & b \\ c & d \end{vmatrix}$, is
>
> $$\begin{vmatrix} a & b \\ c & d \end{vmatrix} = ad - bc$$

Notice that the determinant of a matrix is a number, not a matrix.

EXAMPLE 1   **Evaluating a 2 × 2 Determinant**

Evaluate each determinant:

(a) $\begin{vmatrix} 5 & 2 \\ 3 & 4 \end{vmatrix}$ (b) $\begin{vmatrix} -1 & 3 \\ -4 & 5 \end{vmatrix}$

Solution

(a) $\begin{vmatrix} 5 & 2 \\ 3 & 4 \end{vmatrix} = 5(4) - 3(2)$

$= 20 - 6$

$= 14$

(b) $\begin{vmatrix} -1 & 3 \\ -4 & 5 \end{vmatrix} = -1(5) - (-4)(3)$

$= -5 - (-12)$

$= -5 + 12$

$= 7$

Quick ✓

1. $D = \begin{vmatrix} a & b \\ c & d \end{vmatrix} = $ _____.

2. A matrix is _____ if the number of rows and the number of columns are equal.

In Problems 3 and 4, evaluate each determinant.

3. $\begin{vmatrix} 5 & 3 \\ 4 & 6 \end{vmatrix}$ 4. $\begin{vmatrix} -2 & -5 \\ 1 & 7 \end{vmatrix}$

▶ ❷ Use Cramer's Rule to Solve a System of Two Equations

We can use 2×2 determinants to solve a system of two equations containing two variables.

Cramer's Rule for Two Equations Containing Two Variables

The solution to the system of equations

$$\begin{cases} ax + by = s & (1) \\ cx + dy = t & (2) \end{cases}$$

is given by

$$x = \frac{\begin{vmatrix} s & b \\ t & d \end{vmatrix}}{\begin{vmatrix} a & b \\ c & d \end{vmatrix}} = \frac{D_x}{D} \qquad y = \frac{\begin{vmatrix} a & s \\ c & t \end{vmatrix}}{\begin{vmatrix} a & b \\ c & d \end{vmatrix}} = \frac{D_y}{D}$$

where

$$D = \begin{vmatrix} a & b \\ c & d \end{vmatrix} = ad - bc \neq 0$$

Look very carefully at the pattern in Cramer's Rule. The denominator in the solution is the determinant of the coefficients of the variables.

$$\begin{cases} ax + by = s \\ cx + dy = t \end{cases} \qquad D = \begin{vmatrix} a & b \\ c & d \end{vmatrix}$$

In the solution for x, the numerator is the determinant formed by replacing the entries in the first column (the coefficients of x) in D by the system's constants. That is,

$$D_x = \begin{vmatrix} s & b \\ t & d \end{vmatrix}$$

In the solution for y, the numerator is the determinant formed by replacing the entries in the second column (the coefficients of y) in D by the constants. That is,

$$D_y = \begin{vmatrix} a & s \\ c & t \end{vmatrix}$$

EXAMPLE 2 **How to Solve a System of Two Linear Equations Containing Two Variables Using Cramer's Rule**

Use Cramer's Rule to solve the system: $\begin{cases} x + 2y = 0 \\ -2x - 8y = -9 \end{cases}$

Step-by-Step Solution

Step 1: Find the determinant of the coefficients of the variables, D.

$$D = \begin{vmatrix} 1 & 2 \\ -2 & -8 \end{vmatrix} = 1(-8) - (-2)(2) = -8 - (-4) = -8 + 4 = -4$$

Step 2: Since $D \neq 0$, find D_x by replacing the first column in D with the system's constants. Find D_y by replacing the second column in D with the constants.

$$D_x = \begin{vmatrix} 0 & 2 \\ -9 & -8 \end{vmatrix} = 0(-8) - (-9)(2) = 0 - (-18) = 18$$

$$D_y = \begin{vmatrix} 1 & 0 \\ -2 & -9 \end{vmatrix} = 1(-9) - (-2)(0) = -9 - 0 = -9$$

(continued)

Step 3: Find $x = \dfrac{D_x}{D}$ and $y = \dfrac{D_y}{D}$.

$$x = \frac{D_x}{D} = \frac{18}{-4} = -\frac{9}{2} \qquad y = \frac{D_y}{D} = \frac{-9}{-4} = \frac{9}{4}$$

Step 4: Check: $x = -\dfrac{9}{2}$ and $y = \dfrac{9}{4}$

We leave the check of the solution $\left(-\dfrac{9}{2}, \dfrac{9}{4}\right)$ to you. ●

If the determinant of the coefficients of the variables, D, is found to be 0, then the system of equations is either consistent with dependent equations, or inconsistent. To determine if the system has no solution or infinitely many solutions, solve the system using the methods of Section 3.1, 3.3, or 3.4.

Quick ✓

In Problems 5 and 6, use Cramer's Rule to solve the system, if possible.

5. $\begin{cases} 3x + 2y = 1 \\ -2x - y = 1 \end{cases}$ **6.** $\begin{cases} 4x - 2y = 8 \\ -6x + 3y = 3 \end{cases}$

▷ ❸ **Evaluate the Determinant of a 3 × 3 Matrix**

To use Cramer's Rule to solve a system of three equations containing three variables, we need to define a 3 × 3 determinant.

The **determinant of a 3 × 3 matrix** is symbolized by

$$\begin{vmatrix} a_{1,1} & a_{1,2} & a_{1,3} \\ a_{2,1} & a_{2,2} & a_{2,3} \\ a_{3,1} & a_{3,2} & a_{3,3} \end{vmatrix}$$

where $a_{1,1}, a_{1,2}, \dots, a_{3,3}$ are real numbers.

The subscript is used to identify the row and column of an entry. For example, $a_{2,3}$ is the entry in row 2, column 3.

Definition

The **value of the determinant of a 3 × 3 matrix** may be defined in terms of 2 × 2 determinants as follows:

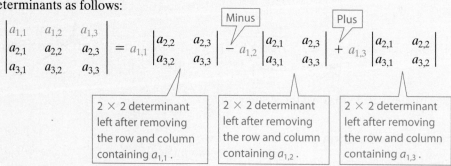

The 2 × 2 determinants shown in the definition above are called **minors** of the 3 × 3 determinant. Notice that once again we have reduced a problem to something we already know how to do—here we reduced the 3 × 3 determinant to three different 2 × 2 determinants.

The formula given in the definition above is easiest to remember by noting that each entry in row 1 is multiplied by the 2 × 2 determinant that remains after the row and column containing the entry have been removed.

EXAMPLE 3 **Evaluating a 3 × 3 Determinant**

Evaluate: $\begin{vmatrix} 3 & 2 & -4 \\ 1 & 7 & -3 \\ 0 & 2 & -5 \end{vmatrix}$

Solution

$$\begin{vmatrix} 3 & 2 & -4 \\ 1 & 7 & -3 \\ 0 & 2 & -5 \end{vmatrix} = 3\begin{vmatrix} 7 & -3 \\ 2 & -5 \end{vmatrix} - 2\begin{vmatrix} 1 & -3 \\ 0 & -5 \end{vmatrix} + (-4)\begin{vmatrix} 1 & 7 \\ 0 & 2 \end{vmatrix}$$

$$= 3(-35 - (-6)) - 2(-5 - 0) - 4(2 - 0)$$
$$= 3(-29) - 2(-5) - 4(2)$$
$$= -87 + 10 - 8$$
$$= -85$$

The definition shows one way to evaluate a 3 × 3 determinant—by *expanding* across row 1. In fact, the expansion can take place across any row or down any column. The terms to be added or subtracted consist of the row (or column) entry times the value of the 2 × 2 determinant that remains after removing the row and column containing the entry. There is only one glitch—the signs of the terms in the expansion change depending on the row or column that is expanded on. The signs of the terms obey the following scheme:

$$\begin{matrix} + & - & + \\ - & + & - \\ + & - & + \end{matrix}$$

For example, if we choose to expand down column 2, we obtain

$$\begin{vmatrix} a_{1,1} & a_{1,2} & a_{1,3} \\ a_{2,1} & a_{2,2} & a_{2,3} \\ a_{3,1} & a_{3,2} & a_{3,3} \end{vmatrix} = \overset{\text{Minus}}{-a_{1,2}}\begin{vmatrix} a_{2,1} & a_{2,3} \\ a_{3,1} & a_{3,3} \end{vmatrix} + \overset{\text{Plus}}{a_{2,2}}\begin{vmatrix} a_{1,1} & a_{1,3} \\ a_{3,1} & a_{3,3} \end{vmatrix} \overset{\text{Minus}}{- a_{3,2}}\begin{vmatrix} a_{1,1} & a_{1,3} \\ a_{2,1} & a_{2,3} \end{vmatrix}$$

EXAMPLE 4 **Evaluating a 3 × 3 Determinant**

Evaluate $\begin{vmatrix} 3 & 2 & -4 \\ 1 & 7 & -3 \\ 0 & 2 & -5 \end{vmatrix}$ (from Example 3), by expanding down column 1.

Solution

We choose to expand down column 1 because it has a 0 in it.

$$\begin{vmatrix} 3 & 2 & -4 \\ 1 & 7 & -3 \\ 0 & 2 & -5 \end{vmatrix} = 3\begin{vmatrix} 7 & -3 \\ 2 & -5 \end{vmatrix} - 1\begin{vmatrix} 2 & -4 \\ 2 & -5 \end{vmatrix} + 0\begin{vmatrix} 2 & -4 \\ 7 & -3 \end{vmatrix}$$

$$= 3(-35 - (-6)) - 1(-10 - (-8)) + 0(-6 - (-28))$$
$$= 3(-29) - 1(-2) + 0$$
$$= -87 + 2$$
$$= -85$$

Work Smart

Expand across the row, or down the column, with the most 0s.

Notice that the results of Example 3 and 4 are the same! However, the computation in Example 4 was a little easier because we expanded down the column that contains a 0.

To make the computation easier, expand across the row that contains the most 0s or down the column that contains the most 0s.

Quick ✓

7. Evaluate: $\begin{vmatrix} 2 & -3 & 5 \\ 0 & 4 & -1 \\ 3 & 8 & -7 \end{vmatrix}$

8. Evaluate: $\begin{vmatrix} 3 & 2 & 1 \\ 1 & 1 & -3 \\ -5 & -1 & -5 \end{vmatrix}$

▶ ④ Use Cramer's Rule to Solve a System of Three Equations

Cramer's Rule can also be applied to a system of three linear equations.

Cramer's Rule for Three Equations Containing Three Variables

For the system of three equations containing three variables

$$\begin{cases} a_1 x + b_1 y + c_1 z = d_1 \\ a_2 x + b_2 y + c_2 z = d_2 \\ a_3 x + b_3 y + c_3 z = d_3 \end{cases}$$

with

$$D = \begin{vmatrix} a_1 & b_1 & c_1 \\ a_2 & b_2 & c_2 \\ a_3 & b_3 & c_3 \end{vmatrix} \neq 0 \quad D_x = \begin{vmatrix} d_1 & b_1 & c_1 \\ d_2 & b_2 & c_2 \\ d_3 & b_3 & c_3 \end{vmatrix} \quad D_y = \begin{vmatrix} a_1 & d_1 & c_1 \\ a_2 & d_2 & c_2 \\ a_3 & d_3 & c_3 \end{vmatrix} \quad D_z = \begin{vmatrix} a_1 & b_1 & d_1 \\ a_2 & b_2 & d_2 \\ a_3 & b_3 & d_3 \end{vmatrix}$$

then

$$x = \frac{D_x}{D} \qquad y = \frac{D_y}{D} \qquad z = \frac{D_z}{D}$$

EXAMPLE 5 **How to Use Cramer's Rule**

Use Cramer's Rule to solve the following system:

$$\begin{cases} x - 2y + z = -9 \\ -3x + y - 2z = 5 \\ 4x + 3z = 1 \end{cases}$$

Step-by-Step Solution

Step 1: Find the determinant of the coefficients of the variables, D. We choose to expand down column 2. Do you see why?

$$D = \begin{vmatrix} 1 & -2 & 1 \\ -3 & 1 & -2 \\ 4 & 0 & 3 \end{vmatrix} = -(-2)\begin{vmatrix} -3 & -2 \\ 4 & 3 \end{vmatrix} + 1\begin{vmatrix} 1 & 1 \\ 4 & 3 \end{vmatrix} - 0\begin{vmatrix} 1 & 1 \\ -3 & -2 \end{vmatrix}$$

$$= 2[-3(3) - 4(-2)] + 1[1(3) - 4(1)] - 0$$

$$= 2(-1) + 1(-1) - 0$$

$$= -2 - 1$$

$$= -3$$

Step 2: Since $D \neq 0$, continue by determining D_x by replacing column 1 in D with the system's constants.

$$D_x = \begin{vmatrix} -9 & -2 & 1 \\ 5 & 1 & -2 \\ 1 & 0 & 3 \end{vmatrix} \overset{\underset{\text{Expand down column 2}}{\downarrow}}{=} -(-2)\begin{vmatrix} 5 & -2 \\ 1 & 3 \end{vmatrix} + 1\begin{vmatrix} -9 & 1 \\ 1 & 3 \end{vmatrix} - 0\begin{vmatrix} -9 & 1 \\ 5 & -2 \end{vmatrix}$$

$$= 2[\,15 - (-2)\,] + 1[\,-27 - 1\,] - 0$$
$$= 2(17) - 28$$
$$= 6$$

Find D_y by replacing column 2 in D with the system's constants.

$$D_y = \begin{vmatrix} 1 & -9 & 1 \\ -3 & 5 & -2 \\ 4 & 1 & 3 \end{vmatrix} \overset{\underset{\text{Expand down column 1}}{\downarrow}}{=} 1\begin{vmatrix} 5 & -2 \\ 1 & 3 \end{vmatrix} - (-3)\begin{vmatrix} -9 & 1 \\ 1 & 3 \end{vmatrix} + 4\begin{vmatrix} -9 & 1 \\ 5 & -2 \end{vmatrix}$$

$$= -15$$

Find D_z by replacing column 3 in D with the system's constants.

$$D_z = \begin{vmatrix} 1 & -2 & -9 \\ -3 & 1 & 5 \\ 4 & 0 & 1 \end{vmatrix} \overset{\underset{\text{Expand down column 2}}{\downarrow}}{=} -(-2)\begin{vmatrix} -3 & 5 \\ 4 & 1 \end{vmatrix} + 1\begin{vmatrix} 1 & -9 \\ 4 & 1 \end{vmatrix} - 0\begin{vmatrix} 1 & -9 \\ -3 & 5 \end{vmatrix}$$

$$= -9$$

Step 3: Find $x = \dfrac{D_x}{D}, y = \dfrac{D_y}{D},$ and $z = \dfrac{D_z}{D}.$

$$x = \frac{D_x}{D} = \frac{6}{-3} \qquad\qquad y = \frac{D_y}{D} = \frac{-15}{-3} \qquad\qquad z = \frac{D_z}{D} = \frac{-9}{-3}$$

$$= -2 \qquad\qquad\qquad = 5 \qquad\qquad\qquad = 3$$

Step 4: Check.

$$x - 2y + z = -9 \qquad\qquad -3x + y - 2z = 5 \qquad\qquad 4x + 3z = 1$$

$$-2 - 2(5) + 3 \overset{?}{=} -9 \qquad\qquad -3(-2) + 5 - 2(3) \overset{?}{=} 5 \qquad\qquad 4(-2) + 3(3) \overset{?}{=} 1$$

$$-2 - 10 + 3 \overset{?}{=} -9 \qquad\qquad 6 + 5 - 6 \overset{?}{=} 5 \qquad\qquad -8 + 9 \overset{?}{=} 1$$

$$-9 = -9 \quad \text{True} \qquad\qquad 5 = 5 \quad \text{True} \qquad\qquad 1 = 1 \quad \text{True}$$

The solution is the ordered triple $(-2, 5, 3)$.

Work Smart: Study Skills

At the end of Section 3.1, we summarized the methods that can be used to solve a system of linear equations in two variables. Make up a summary table of your own for solving systems of linear equations in three variables. When is elimination appropriate? When would you use matrices? When is Cramer's Rule best?

Quick ✓

9. Suppose that you wish to solve a system of equations using Cramer's Rule and find that $D = 4$, $D_x = 2$, $D_y = -8$, and $D_z = -4$. What is the solution of the system?

10. Use Cramer's Rule to solve the system

$$\begin{cases} x - y + 3z = -2 \\ 4x + 3y + z = 9 \\ -2x + 5z = 7 \end{cases}$$

In the case where $D = 0$, Cramer's Rule does not apply. To determine whether the system has no solution or infinitely many solutions, it is necessary to solve the system using the methods of Section 3.1, 3.3, or 3.4.

3.5 Exercises MyMathLab® PRACTICE

*Problems **1–10** are the Quick ✔s that follow the EXAMPLES.*

Building Skills

In Problems 11–14, find the value of each determinant. See Objective 1.

11. $\begin{vmatrix} 4 & 2 \\ 1 & 3 \end{vmatrix}$

12. $\begin{vmatrix} 5 & 3 \\ 2 & 4 \end{vmatrix}$

13. $\begin{vmatrix} -2 & -4 \\ 1 & 3 \end{vmatrix}$

14. $\begin{vmatrix} -8 & 5 \\ -4 & 3 \end{vmatrix}$

In Problems 15–22, solve each system of equations using Cramer's Rule, if applicable. See Objective 2.

15. $\begin{cases} x + y = -4 \\ x - y = -12 \end{cases}$

16. $\begin{cases} x + y = 6 \\ x - y = 4 \end{cases}$

17. $\begin{cases} 2x + 3y = 3 \\ -3x + y = -10 \end{cases}$

18. $\begin{cases} 2x + 4y = -6 \\ 3x + 2y = 7 \end{cases}$

19. $\begin{cases} 3x + 4y = 1 \\ -6x + 8y = 4 \end{cases}$

20. $\begin{cases} 4x - 2y = 9 \\ -8x - 2y = -11 \end{cases}$

21. $\begin{cases} 2x - 6y - 12 = 0 \\ 3x - 5y - 11 = 0 \end{cases}$

22. $\begin{cases} 3x - 6y - 2 = 0 \\ x + 2y - 4 = 0 \end{cases}$

In Problems 23–28, find the value of each determinant. See Objective 3.

23. $\begin{vmatrix} 2 & 0 & -1 \\ 3 & 8 & -3 \\ 1 & 5 & -2 \end{vmatrix}$

24. $\begin{vmatrix} -2 & 1 & 6 \\ -3 & 2 & 5 \\ 1 & 0 & -2 \end{vmatrix}$

25. $\begin{vmatrix} -3 & 2 & 3 \\ 0 & 5 & -2 \\ 1 & 4 & 8 \end{vmatrix}$

26. $\begin{vmatrix} 8 & 4 & -1 \\ 2 & -7 & 1 \\ 0 & 5 & -3 \end{vmatrix}$

27. $\begin{vmatrix} 0 & 2 & 1 \\ 1 & -6 & -4 \\ -3 & 4 & 5 \end{vmatrix}$

28. $\begin{vmatrix} -3 & 4 & -2 \\ 1 & -2 & 0 \\ 0 & 6 & 6 \end{vmatrix}$

In Problems 29–42, solve each system of equations using Cramer's Rule, if applicable. See Objective 4.

29. $\begin{cases} x - y + z = -4 \\ x + 2y - z = 1 \\ 2x + y + 2z = -5 \end{cases}$

30. $\begin{cases} x + y - z = 6 \\ x + 2y + z = 6 \\ -x - y + 2z = -7 \end{cases}$

31. $\begin{cases} x + y + z = 4 \\ 5x + 2y - 3z = 7 \\ 2x - y - z = 5 \end{cases}$

32. $\begin{cases} x + y + z = -3 \\ -2x - 3y - z = 1 \\ 2x - y - 3z = -5 \end{cases}$

33. $\begin{cases} 2x + y - z = 4 \\ -x + 2y + 2z = -6 \\ 5x + 5y - z = 6 \end{cases}$

34. $\begin{cases} -x + 2y - z = 2 \\ 2x + y + 2z = -6 \\ -x + 7y - z = 0 \end{cases}$

35. $\begin{cases} 3x + y + z = 5 \\ x + y - 3z = 9 \\ -5x - y - 5z = -10 \end{cases}$

36. $\begin{cases} x - 2y - z = 1 \\ 2x + 2y + z = 3 \\ 6x + 6y + 3z = 6 \end{cases}$

37. $\begin{cases} 2x + z = 27 \\ -x - 3y = 6 \\ x - 2y + z = 27 \end{cases}$

38. $\begin{cases} x + 2y + z = 0 \\ -x - 3y - 2z = 3 \\ 2x - 3z = 7 \end{cases}$

39. $\begin{cases} 5x + 3y = 2 \\ -10x + 3z = -3 \\ y - 2z = -9 \end{cases}$

40. $\begin{cases} 2x + 4y = 0 \\ -2x + z = -5 \\ -4y - 3z = 1 \end{cases}$

41. $\begin{cases} x + y + z = 3 \\ -y + 2z = 1 \\ -x + z = 0 \end{cases}$

42. $\begin{cases} 3y - z = -1 \\ x + 5y - z = -4 \\ -3x + 6y + 2z = 11 \end{cases}$

Mixed Practice

In Problems 43–46, solve for x.

43. $\begin{vmatrix} x & 3 \\ 1 & 2 \end{vmatrix} = 7$

44. $\begin{vmatrix} -2 & x \\ 3 & 4 \end{vmatrix} = 1$

45. $\begin{vmatrix} x & -1 & -2 \\ 1 & 0 & 4 \\ 3 & 2 & 5 \end{vmatrix} = 5$

46. $\begin{vmatrix} 2 & x & -1 \\ 3 & 5 & 0 \\ -4 & 1 & 2 \end{vmatrix} = 0$

Applying the Concepts

Problems 47–50, use the following result. Determinants can be used to find the area of a triangle. If $(a_1, b_1), (a_2, b_2)$, and (a_3, b_3) are the vertices of a triangle, the area of the triangle is $|D|$, where

$$D = \frac{1}{2} \begin{vmatrix} a_1 & a_2 & a_3 \\ b_1 & b_2 & b_3 \\ 1 & 1 & 1 \end{vmatrix}$$

△ **47. Geometry: Area of a Triangle** Given the points $A = (1, 1), B = (5, 1),$ and $C = (5, 6),$

  (a) Plot the points in the Cartesian plane and form triangle ABC.

  (b) Find the area of the triangle ABC.

△ **48. Geometry: Area of a Triangle** Given the points $A = (-1, -1), B = (3, 2),$ and $C = (0, 6),$

  (a) Plot the points in the Cartesian plane and form triangle ABC.

  (b) Find the area of the triangle ABC.

△ **49. Geometry: Area of a Parallelogram** Find the area of a parallelogram by doing the following:

(a) Plot the points $A = (2, 1)$, $B = (7, 2)$, $C = (8, 4)$, and $D = (3, 3)$ in the Cartesian plane.

(b) Form triangle ABC and find the area of triangle ABC.

(c) Find the area of the triangle ADC.

(d) Conclude that the diagonal of the parallelogram forms two triangles of equal area. Use this result to find the area of the parallelogram.

△ **50. Geometry: Area of a Parallelogram** Find the area of a parallelogram by doing the following:

(a) Plot the points $A = (-3, -2)$, $B = (3, 1)$, $C = (4, 4)$, and $D = (-2, 1)$ in the Cartesian plane.

(b) Form triangle ABC and find the area of triangle ABC.

(c) Find the area of the triangle ADC.

(d) Conclude that the diagonal of the parallelogram forms two triangles of equal area. Use this result to find the area of the parallelogram.

△ **51. Geometry: Equation of a Line** An equation of the line containing the two points (x_1, y_1) and (x_2, y_2) may be expressed as the determinant

$$\begin{vmatrix} x & y & 1 \\ x_1 & y_1 & 1 \\ x_2 & y_2 & 1 \end{vmatrix} = 0$$

(a) Find the equation of the line containing $(3, 2)$ and $(5, 1)$ using the determinant.

(b) Verify your result by using the slope formula and the point-slope formula.

△ **52. Geometry: Collinear Points** The distinct points (x_1, y_1), (x_2, y_2), and (x_3, y_3) are collinear if and only if

$$\begin{vmatrix} x_1 & y_1 & 1 \\ x_2 & y_2 & 1 \\ x_3 & y_3 & 1 \end{vmatrix} = 0$$

(a) Plot the points $(-3, -2)$, $(1, 2)$, and $(7, 8)$ in the Cartesian plane.

(b) Show that the points are collinear using the determinant.

(c) Show that the points are collinear using the idea of slopes.

Extending the Concepts

53. Evaluate the determinant $\begin{vmatrix} 3 & -2 \\ 1 & 4 \end{vmatrix}$. Now interchange rows 1 and 2 and recompute the determinant. What do you notice? Do you think that this result is true in general?

54. Evaluate the determinant $\begin{vmatrix} -3 & 1 \\ 6 & 5 \end{vmatrix}$. Multiply the entries in column 2 by 3 and recompute the determinant. What do you notice? Do you think that this result is true in general?

The Graphing Calculator

Graphing calculators have the ability to evaluate 2×2 determinants. First, enter the matrix into the graphing calculator and name it A. Then compute the determinant of matrix A. Figure 12 shows the results of Example 1(a) using a TI-84 Plus graphing calculator.

Figure 12

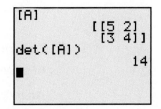

Because graphing calculators have the ability to compute determinants of matrices, they can be used to solve systems of equations using Cramer's Rule. Figure 13 shows the results of solving Example 2 using Cramer's Rule, where $A = D$, $B = D_x$, and $C = D_y$.

Figure 13

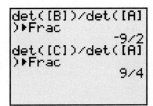

In Problems 55–60, use a graphing calculator to solve the system of equations.

55. $\begin{cases} x + y = -4 \\ x - y = -12 \end{cases}$

56. $\begin{cases} x + y = 6 \\ x - y = 4 \end{cases}$

57. $\begin{cases} 2x + 3y = 3 \\ -3x + y = -10 \end{cases}$

58. $\begin{cases} 2x + 4y = -6 \\ 3x + 2y = 7 \end{cases}$

59. $\begin{cases} x - y + z = -4 \\ x + 2y - z = 1 \\ 2x + y + 2z = -5 \end{cases}$

60. $\begin{cases} x + y - z = 6 \\ x + 2y + z = 6 \\ -x - y + 2z = -7 \end{cases}$

3.6 Systems of Linear Inequalities

Objectives

1 Determine Whether an Ordered Pair Is a Solution of a System of Linear Inequalities

2 Graph a System of Linear Inequalities

3 Solve Problems Involving Systems of Linear Inequalities

Are You Ready for This Section?

Before getting started, take the following readiness quiz. If you get a problem wrong, go back to the section cited and review the material.

R1. Determine whether $x = -2$ satisfies the inequality $-3x + 2 \geq 7$. [Section 1.4, pp. 85–87]

R2. Solve the inequality: $-2x + 1 > 7$ [Section 1.4, pp. 85–87]

R3. Graph the inequality: $3x + 2y > -6$ [Section 1.8, pp. 127–129]

In Section 1.8, we graphed a single linear inequality in two variables. In this section, we discuss how to graph a system of linear inequalities in two variables.

▶ **1** Determine Whether an Ordered Pair Is a Solution of a System of Linear Inequalities

An ordered pair **satisfies** a system of linear inequalities if it makes each inequality in the system a true statement.

EXAMPLE 1 **Determining Whether an Ordered Pair Is a Solution of a System of Linear Inequalities**

Which of the following points, if any, is a solution of the system of linear inequalities?

$$\begin{cases} 3x + y \leq 7 \\ 4x - 2y \leq 8 \end{cases}$$

(a) $(2, 4)$ **(b)** $(-3, 1)$

Solution

(a) Let $x = 2$ and $y = 4$ in each inequality in the system. If each inequality is true, then $(2, 4)$ is a solution of the system of inequalities.

$$3x + y \leq 7 \qquad\qquad 4x - 2y \leq 8$$
$$3(2) + 4 \overset{?}{\leq} 7 \qquad\qquad 4(2) - 2(4) \overset{?}{\leq} 8$$
$$6 + 4 \overset{?}{\leq} 7 \qquad\qquad 8 - 8 \overset{?}{\leq} 8$$
$$10 \leq 7 \quad \text{False} \qquad\qquad 0 \leq 8 \quad \text{True}$$

The inequality $3x + y \leq 7$ is not true when $x = 2$ and $y = 4$, so $(2, 4)$ is not a solution of the system.

(b) Let $x = -3$ and $y = 1$ in each inequality in the system. If each inequality is true, then $(-3, 1)$ is a solution of the system of inequalities.

$$3x + y \leq 7 \qquad\qquad 4x - 2y \leq 8$$
$$3(-3) + 1 \overset{?}{\leq} 7 \qquad 4(-3) - 2(1) \overset{?}{\leq} 8$$
$$-9 + 1 \overset{?}{\leq} 7 \qquad\qquad -12 - 2 \overset{?}{\leq} 8$$
$$-8 \leq 7 \quad \text{True} \qquad\qquad -14 \leq 8 \quad \text{True}$$

Both inequalities are true, so $(-3, 1)$ is a solution of the system. ●

Ready?...Answers **R1.** Yes
R2. $\{x \mid x < -3\}$ or $(-\infty, -3)$ **R3.**

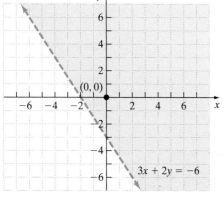

Quick ✓

1. An ordered pair _____ a system of linear inequalities if it makes each inequality in the system a true statement.

2. Determine which of the following ordered pairs is a solution of the system of linear inequalities.

$$\begin{cases} -4x + y < -5 \\ 2x - 5y < 10 \end{cases}$$

(a) $(1, 2)$ **(b)** $(3, 1)$ **(c)** $(1, -2)$

❷ Graph a System of Linear Inequalities

▶ The graph of a system of inequalities in two variables x and y is the set of all points (x, y) that simultaneously satisfy *each* of the inequalities in the system. To graph a system of linear inequalities, graph each linear inequality individually and then determine where, if at all, they intersect. A graph is the *only* way we show the solution of a system of linear inequalities.

EXAMPLE 2 **Graphing a System of Linear Inequalities**

Graph the system: $\begin{cases} 3x + y \leq 7 \\ 4x - 2y \leq 8 \end{cases}$

Solution

First, graph the inequality $3x + y \leq 7$. To graph $3x + y \leq 7$, first graph the solid boundary line $y = -3x + 7$. The test point $(0, 0)$ satisfies the inequality $3x + y \leq 7$, so shade below the boundary line. See Figure 14(a).

Then graph the inequality $4x - 2y \leq 8$ by first graphing the solid boundary line $4x - 2y = 8$ (or $y = 2x - 4$). The test point $(0, 0)$ satisfies the inequality $4x - 2y \leq 8$, so shade above the boundary line. See Figure 14(b).

Now combine the graphs in Figures 14(a) and (b). The intersection of the shaded regions is the solution of the system of linear inequalities. See Figure 14(c). Notice that the point $(-3, 1)$, which was verified to be a solution in Example 1(b), and the point $(0, 0)$ are in the shaded region.

Work Smart

To obtain the boundary line $y = 2x - 4$, solve the equation $4x - 2y = 8$ for y. Or, use intercepts.

Figure 14

Work Smart

Don't forget that we graph the boundary line as a solid line when the inequality is nonstrict ($\leq$ or $\geq$) and as a dashed line when the inequality is strict ($<$ or $>$).

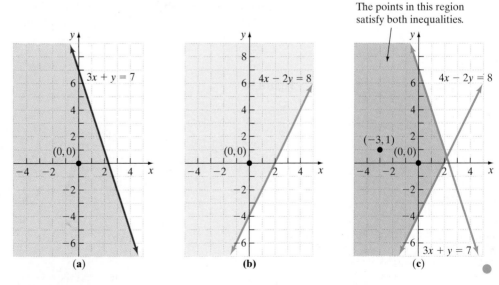

Have you noticed that if we write the inequality in the form $y >$ or $y \geq$, we shade above the line? If the inequality is of the form $y <$ or $y \leq$, shade below the line. For example, $3x + y \leq 7$ can be written as $y \leq -3x + 7$, so we shade below the boundary line. Instead of using a test point to determine shading, we will use this method.

Rather than using multiple graphs to obtain the solution set to the system of linear inequalities, we can use a single graph to determine the overlapping region.

EXAMPLE 3 **Graphing a System of Linear Inequalities**

Graph the system: $\begin{cases} -2x + y > 5 \\ 3x + 2y \geq -4 \end{cases}$

Solution

Graph the inequality $-2x + y > 5$ by adding $2x$ to both sides and obtaining $y > 2x + 5$. Graph $y = 2x + 5$ as a dashed line and shade above (since the inequality is of the form $y >$). To graph the inequality $3x + 2y \geq -4$, solve for y and get $y \geq -\dfrac{3}{2}x - 2$. Graph $y = -\dfrac{3}{2}x - 2$ as a solid line and shade above (since the inequality is of the form $y \geq$).

The region shaded purple in Figure 15 represents the solution. The points on the boundary line $y = -\dfrac{3}{2}x - 2$ are solutions of the system because the inequality is not strict. ●

Figure 15

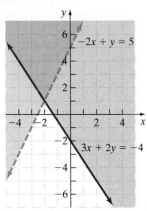

Quick ✓

In Problems 3 and 4, graph the system of linear inequalities.

3. $\begin{cases} 2x + y \leq 5 \\ -x + y \geq -4 \end{cases}$

4. $\begin{cases} 3x + y > -2 \\ 2x + 3y < 3 \end{cases}$

⊙ EXAMPLE 4 **Graphing a System of Linear Inequalities with No Solution**

Graph the system: $\begin{cases} 2x + y > 2 \\ 2x + y < -2 \end{cases}$

Solution

Graph the inequality $2x + y > 2$ (or $y > -2x + 2$) using a dashed line because the inequality is strict. In the same Cartesian plane, graph the inequality $2x + y < -2$ (or $y < -2x - 2$) using a dashed line. See Figure 16.

Figure 16

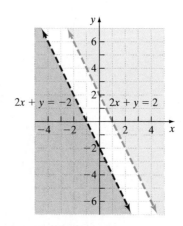

Because no overlapping region results, there are no points in the Cartesian plane that satisfy both inequalities. The system has no solution, so the solution set is $\varnothing$ or $\{\ \}$. ●

5. *True or False* A system of linear inequalities must always have a nonempty solution set.

Graph the system of linear inequalities.

6. $\begin{cases} -2x + 3y \geq 9 \\ 6x - 9y \geq 9 \end{cases}$

Often, we encounter systems with more than two linear inequalities.

▷ **EXAMPLE 5** | **Graphing a System of Four Linear Inequalities**

Graph the system: $\begin{cases} x + y \leq 5 \\ 2x + y \leq 7 \\ x \quad\quad \geq 0 \\ \quad\quad y \geq 0 \end{cases}$

Solution

Figure 17

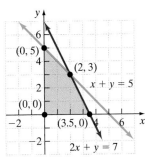

The two inequalities $x \geq 0$ and $y \geq 0$ require that the graph be in quadrant I. The inequality $x + y \leq 5$ requires that we shade below the line $x + y = 5$. The inequality $2x + y \leq 7$ requires that we shade below the line $2x + y = 7$. Figure 17 shows the graph.

Notice that in Figure 17 we have labeled specific points on the graph. These are points of intersection between two boundary lines in the system and are called **corner points.** The corner points in Figure 17 are (0, 0), (0, 5), (2, 3), and (3.5, 0). They are found by solving the following systems of equations formed by two boundary lines:

$\begin{cases} x = 0 \\ y = 0 \end{cases}$ $\begin{cases} x + y = 5 \\ x \quad\quad = 0 \end{cases}$ $\begin{cases} x + y = 5 \\ 2x + y = 7 \end{cases}$ $\begin{cases} 2x + y = 7 \\ y = 0 \end{cases}$

Corner point: (0, 0)   Corner point: (0, 5)   Corner point: (2, 3)   Corner point: (3.5, 0)

The graph of the system of linear inequalities in Example 5 is **bounded** because it can be contained within a circle. Look at the systems graphed in Figure 14 or Figure 15. They are **unbounded** because the solution set cannot be contained within a circle. Those graphs extend indefinitely.

7. The points of intersection between two boundary lines in a system of linear inequalities are called _____ _____.

8. If the solution set of a system of linear inequalities cannot be contained within a circle, then the system is _____.

Graph the system of linear inequalities. Tell whether the graph is bounded or unbounded, and label the corner points.

9. $\begin{cases} x + y \leq 6 \\ 2x + y \leq 10 \\ x \quad\quad \geq 0 \\ \quad\quad y \geq 0 \end{cases}$

▷ ❸ **Solve Problems Involving Systems of Linear Inequalities**

In problems that lead to systems of linear inequalities, we use the problem-solving strategy presented in Section 1.2.

EXAMPLE 6   **Breakfast**

Aman likes French toast sticks and orange juice for breakfast. He wants to eat no more than 500 calories and to consume no more than 425 mg of sodium. Each French toast stick (with syrup) has 90 calories and 100 mg of sodium. Each small orange juice has 140 calories and 25 mg of sodium. Write a system of linear inequalities that represents the possible combinations of French toast sticks and orange juice that Aman can consume. Graph the system.

Solution

Step 1: Identify We want to know how many French toast sticks and orange juice drinks Aman can consume while staying within his diet restrictions.

Step 2: Name Let x represent the number of French toast sticks and y represent the number of orange juice drinks that Aman consumes.

Step 3: Translate One French toast stick has 90 calories. Two sticks have 180 calories, so x French toast sticks have $90x$ calories. Similarly y orange juice drinks must contain $140y$ calories. The words "no more than" imply a $\leq$ inequality. A linear inequality that describes Aman's options while staying within his calorie restriction is

$$90x + 140y \leq 500$$

A linear inequality that describes Aman's options while staying within his sodium restriction is

$$100x + 25y \leq 425$$

Finally, Aman cannot consume negative quantities of French toast sticks or orange juice, so we have the **nonnegativity constraints**

$$x \geq 0 \quad \text{and} \quad y \geq 0$$

In Words

A nonnegativity constraint means that the value of the variable has to be greater than or equal to zero.

Combining these four linear inequalities, we obtain the system

$$\begin{cases} 90x + 140y \leq 500 \\ 100x + 25y \leq 425 \\ x \geq 0 \\ y \geq 0 \end{cases} \quad \text{The Model}$$

Step 4: Solve Figure 18 shows the graph of the system.

Figure 18

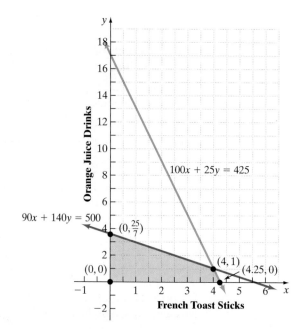

Step 5: Check Although it is not a check, we can gather evidence that our graph is correct by choosing a point in the shaded region and determining whether all the inequalities in the system are satisfied. For example, the combination of $x = 3$ French toast sticks and $y = 1$ orange juice drinks satisfies all four inequalities.

Quick ✓

10. Jack and Mary recently retired, and they have $25,000 to invest. Their financial advisor has recommended that they place at least $10,000 in Treasury notes yielding 5% and no more than $15,000 in corporate bonds yielding 8%. Write a system of linear inequalities that represents the possible combinations of investments. Graph the system.

3.6 Exercises

MyMathLab®
PRACTICE

Exercise numbers in green have complete video solutions in MyMathLab.

*Problems **1–10** are the Quick ✓s that follow the EXAMPLES.*

Building Skills

In Problems 11–16, determine which of the points, if any, satisfies the system. See Objective 1.

11. $\begin{cases} x + y \le 4 \\ 3x + y \ge 10 \end{cases}$
 (a) $(3, 2)$
 (b) $(5, -1)$

12. $\begin{cases} x + y \ge 2 \\ -3x + y \le 10 \end{cases}$
 (a) $(-3, 6)$
 (b) $(4, 1)$

13. $\begin{cases} 3x + 2y < 12 \\ -2x + 3y > 12 \end{cases}$
 (a) $(3, 2)$
 (b) $(1, 4)$

14. $\begin{cases} 5x + 2y < 10 \\ 4x - 3y < 24 \end{cases}$
 (a) $(1, 3)$
 (b) $(1, 1)$

15. $\begin{cases} x + y \le 8 \\ 3x + y \le 12 \\ x \ge 0 \\ y \ge 0 \end{cases}$
 (a) $(4, 2)$ (b) $(2, 5)$

16. $\begin{cases} x + y \ge 6 \\ 2x + y \ge 10 \\ x \ge 0 \\ y \ge 0 \end{cases}$
 (a) $(4, 2)$ (b) $(2, 5)$

In Problems 17–28, graph each system of linear inequalities. See Objective 2.

17. $\begin{cases} x + y \le 4 \\ 3x + y \ge 10 \end{cases}$

18. $\begin{cases} x + y \ge 2 \\ -3x + y \le 10 \end{cases}$

19. $\begin{cases} 3x + 2y < 12 \\ -2x + 3y > 12 \end{cases}$

20. $\begin{cases} 5x + 2y < 10 \\ 4x - 3y < 24 \end{cases}$

21. $\begin{cases} x - \dfrac{1}{2}y \le 3 \\ \dfrac{3}{2}x + y > 3 \end{cases}$

22. $\begin{cases} -x + \dfrac{1}{3}y < 3 \\ \dfrac{4}{3}x + y \ge 4 \end{cases}$

23. $\begin{cases} -2x + y < -8 \\ 2x - y > 8 \end{cases}$

24. $\begin{cases} 3x - 2y < -6 \\ -6x + 4y > 12 \end{cases}$

25. $\begin{cases} -5x + 3y < 12 \\ 5x - 3y < 9 \end{cases}$

26. $\begin{cases} 4x + 3y > -9 \\ -8x - 6y > 12 \end{cases}$

27. $\begin{cases} y \le 8 \\ x \ge 3 \end{cases}$

28. $\begin{cases} y \le 4 \\ x \ge -1 \end{cases}$

In Problems 29–36, graph each system of linear inequalities. Tell whether the graph is bounded or unbounded, and label the corner points. See Objective 2.

29. $\begin{cases} x + y \le 6 \\ 3x + y \le 12 \\ x \ge 0 \\ y \ge 0 \end{cases}$

30. $\begin{cases} x + y \le 9 \\ 3x + 2y \le 24 \\ x \ge 0 \\ y \ge 0 \end{cases}$

31. $\begin{cases} x + y \ge 8 \\ 4x + 2y \ge 28 \\ x \ge 0 \\ y \ge 0 \end{cases}$

32. $\begin{cases} x + y \ge 8 \\ x + 3y \ge 12 \\ x \ge 0 \\ y \ge 0 \end{cases}$

33. $\begin{cases} 2x + 3y \le 30 \\ 3x + 2y \le 25 \\ 5x + 2y \le 35 \\ x \ge 0 \\ y \ge 0 \end{cases}$

34. $\begin{cases} 2x + 3y \le 36 \\ x + y \le 14 \\ 3x + y \le 30 \\ x \ge 0 \\ y \ge 0 \end{cases}$

35. $\begin{cases} 2x + y \ge 13 \\ x + 2y \ge 11 \\ x \ge 4 \\ y \ge 0 \end{cases}$

36. $\begin{cases} 7x + 3y \ge 45 \\ 5x + 3y \ge 39 \\ x \ge 0 \\ y \ge 3 \end{cases}$

Applying the Concepts

37. Breakfast Daria is on a special diet that requires she consume at least 500 mg of potassium and 14 grams of protein for breakfast. One morning, Daria decides to have orange juice, which contains 450 mg of potassium and 2 g of protein per serving, and cereal (with 2% milk), which contains 50 mg of potassium and 6 g of protein per serving.

 (a) Let x denote the number of servings of orange juice and y denote the number of servings of cereal. Write a system of linear inequalities that represents the possible combinations of foods.

 (b) Graph the system and label the corner points.

38. Manufacturing Printers Bob's Printer Emporium manufactures two printers: an ink jet printer and a laser printer. Each ink-jet printer requires 2 hours for molding and 3 hours for assembly; each laser printer requires 3 hours for molding and 4 hours for assembly. There are a total of 80 hours per week available in the work schedule for molding and 120 hours per week available for assembly.
   (a) Using x to denote the number of ink-jet printers and y to denote the number of laser printers, write a system of linear inequalities that describes the possible numbers of each model printer that can be manufactured in a week (it is possible to manufacturer a portion of a printer).
   (b) Graph the system and label the corner points.

39. Financial Planning Jack and Mary recently retired, and they have $25,000 to invest. Their financial advisor has recommended that they place at least $5000 in Treasury notes yielding 5% and no more than $15,000 in corporate bonds yielding 8%. In addition, they want to earn at least $1400 each year.
   (a) Let x denote the amount invested in Treasury notes and y denote the amount invested in corporate bonds. Write a system of linear inequalities that represents the possible combinations of investments.
   (b) Graph the system and label the corner points.

40. Mixing Nuts You've Got to Be Nuts is a store that specializes in selling nuts. The owner finds that she has excess inventory of 100 pounds (1600 ounces) of cashews and 120 pounds (1920 ounces) of peanuts. She decides to make two types of 1-pound nut mixes from the excess inventory. A premium mix will contain 12 ounces of cashews and 4 ounces of peanuts and the standard mix will contain 6 ounces of cashews and 6 ounces of peanuts.
   (a) Use x to denote the number of premium mixes and y to denote the number of standard mixes. Write a system of linear inequalities that describes the possible numbers of each kind of mix.
   (b) Graph the system and label the corner points.

Extending the Concepts

In Problems 41–44, write a system of linear inequalities that has the given graph.

41.

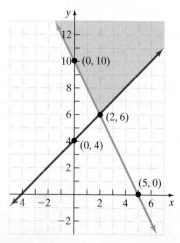

42.

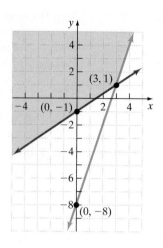

43.

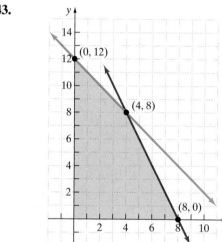

44.

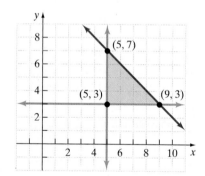

45. Math for the Future: Linear Programming A **linear programming problem** in two variables x and y consists of maximizing (or minimizing) a **linear objective function** $z = Ax + By$, where A and B are real numbers, not both 0, subject to certain constraints. The constraints form a system of linear inequalities. A typical linear programming problem is

$$\text{Maximize:} \qquad z = 2x + 3y$$
$$\text{subject to} \qquad x + y \le 8$$
$$2x + y \le 10$$
$$x \ge 0, y \ge 0$$

(a) The constraints form the system of linear

inequalities $\begin{cases} x + y \le 8 \\ 2x + y \le 10 \\ x \ge 0 \\ y \ge 0 \end{cases}$. Graph this system.

The shaded region is called the **feasible region.**

(b) The equation $z = 2x + 3y$ is the *objective function.* The goal is to make z as large as possible while (x, y) remains in the feasible region. Graph the objective function for $z = 0, z = 12, z = 18,$ and $z = 24$.

(c) What is the largest value of z that is possible while still remaining within the feasible region? What values of x and y correspond to this value?

Explaining the Concepts

46. Explain how to graph a system of linear inequalities in two variables.

47. If one inequality in a system of two linear inequalities contains a strict inequality and the other inequality contains a nonstrict inequality, is the corner point a solution to the system? Explain your answer.

48. Is it possible for a system of linear inequalities to have the entire Cartesian plane as the solution set? Explain your answer, and if your answer is yes, then provide an example.

49. Is it possible for a system of linear inequalities to have a straight line as the solution set? Explain your answer, and if your answer is yes, then provide an example.

50. What does it mean for a system of linear inequalities to be bounded? What does it mean for a system of linear inequalities to be unbounded?

The Graphing Calculator

Graphing calculators have the ability to graph systems of linear inequalities. Figure 19 shows the results of Example 3 using a TI-84 Plus graphing calculator.

Figure 19

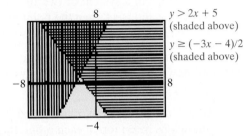

$y > 2x + 5$
(shaded above)
$y \ge (-3x - 4)/2$
(shaded above)

In Problems 51–56, graph each system of linear inequalities using a graphing calculator.

51. $\begin{cases} x + y \le 4 \\ 3x + y \ge 10 \end{cases}$

52. $\begin{cases} x + y \ge 2 \\ -3x + y \le 10 \end{cases}$

53. $\begin{cases} 3x + 2y < 12 \\ -2x + 3y > 12 \end{cases}$

54. $\begin{cases} -5x + 2y < 10 \\ 4x - 3y < 24 \end{cases}$

55. $\begin{cases} x + y \le 6 \\ 3x + y \le 12 \\ x \ge 0 \\ y \ge 0 \end{cases}$

56. $\begin{cases} x + y \le 9 \\ 3x + 2y \le 24 \\ x \ge 0 \\ y \ge 0 \end{cases}$

Chapter 3 Activity: Find the Numbers

Focus: Solving systems of equations

Time: 15 minutes

Group size: 2

Consider the following dialogue between two students.

Michele: Think of two numbers between 1 and 10, and don't tell me what they are.

Rafael: OK, I've thought of two numbers.

Michele: Now tell me the sum of the two numbers and the difference of the two numbers, and I'll tell you what your two numbers are.

Rafael: Their sum is 14 and their difference is 6.

Michele: Your numbers are 10 and 4.

Rafael: That's right! How did you do that?

Michele: I set up a system of equations using the sum and difference that you gave me, along with the variables x and y.

1. Each group member should set up and solve the system of equations described by Michele.

Discuss your results and be sure that you both arrive at the solutions 10 and 4.

2. Now each of you will think of two new numbers, and the other will try to find the numbers by solving a system of equations. But this time the system will be a bit trickier! Give each other the following information about your two numbers, and then figure out each other's numbers:

five more than twice the sum of the numbers
four less than three times the difference of the numbers

3. Would the systems in this activity work if negative numbers were used? Try it and see!

Chapter 3 Review

Section 3.1 Systems of Linear Equations in Two Variables

KEY CONCEPTS

- **Recognizing Solutions to Systems of Two Linear Equations with Two Unknowns**

  - If the lines in a system of two linear equations containing two unknowns intersect, then the point of intersection is the solution, and the system is consistent and its equations are independent.

  - If the lines in a system of two linear equations containing two unknowns are parallel, then the system has no solution, and the system is inconsistent.

  - If the lines in a system of two linear equations containing two unknowns lie on top of each other, then the system has infinitely many solutions. The solution set is the set of all points on the line, the system is consistent, and its equations are dependent.

KEY TERMS

System of linear equations
Solution
Consistent system and independent equations
Inconsistent
Consistent system and dependent equations

You Should Be Able To...	EXAMPLE	Review Exercises
1 Determine whether an ordered pair is a solution of a system of linear equations (p. 219)	Example 2	1–4
2 Solve a system of two linear equations by graphing (p. 221)	Example 3	5–10
3 Solve a system of two linear equations by substitution (p. 221)	Examples 4 and 5	11–14, 19–24
4 Solve a system of two linear equations by elimination (p. 223)	Examples 6 and 7	15–24
5 Identify inconsistent systems (p. 226)	Example 8	9, 23
6 Write the solution of a system with dependent equations (p. 227)	Example 9	12, 17

In Problems 1–4, determine whether the given values of the variables listed are solutions of the system of equations.

1. $\begin{cases} x + 3y = -2 \\ 2x - y = 10 \end{cases}$
  (a) $x = 3, y = -1$
  (b) $x = 4, y = -2$

2. $\begin{cases} -2x + 5y = 4 \\ 4x - 5y = -3 \end{cases}$
  (a) $x = \dfrac{1}{2}, y = 1$
  (b) $x = -1, y = \dfrac{1}{3}$

3. $\begin{cases} 6x - 5y = -12 \\ x - y = -3 \end{cases}$
  (a) $x = 3, y = 6$
  (b) $x = 1, y = 4$

4. $\begin{cases} 3x - y = 9 \\ 8x + 3y = 7 \end{cases}$
  (a) $x = 3, y = 0$
  (b) $x = 2, y = -3$

In Problems 5 and 6, use the graph of the system in the next column to determine the solution.

5. $\begin{cases} x + 4y = 12 \\ 2x - y = 6 \end{cases}$

6. $\begin{cases} 3x + 2y = 4 \\ x - 2y = 4 \end{cases}$

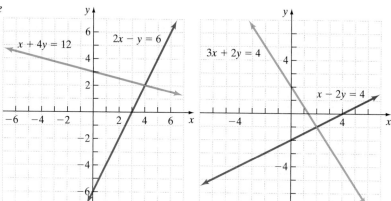

In Problems 7–10, solve each system of linear equations by graphing.

7. $\begin{cases} y = -3x + 1 \\ y = \dfrac{1}{2}x - 6 \end{cases}$

8. $\begin{cases} -2x + 3y = -9 \\ 3x + y = 8 \end{cases}$

9. $\begin{cases} y = -\dfrac{1}{3}x + 2 \\ x + 3y = -9 \end{cases}$

10. $\begin{cases} 2x - 3y = 0 \\ 2x - y = -4 \end{cases}$

In Problems 11–14, solve each system of linear equations using substitution.

11. $\begin{cases} y = -\dfrac{1}{4}x + 2 \\ y = 4x - 32 \end{cases}$ **12.** $\begin{cases} y = -\dfrac{3}{4}x + 2 \\ 3x + 4y = 8 \end{cases}$

13. $\begin{cases} y = 3x - 9 \\ 4x + 3y = -1 \end{cases}$ **14.** $\begin{cases} x - 2y = 7 \\ 3x - y = -4 \end{cases}$

In Problems 15–18, solve each system of linear equations using elimination.

15. $\begin{cases} 2x - y = 9 \\ 3x + y = 11 \end{cases}$ **16.** $\begin{cases} -x + 3y = 4 \\ 3x - 4y = -2 \end{cases}$

17. $\begin{cases} 2x - 4y = 8 \\ -3x + 6y = -12 \end{cases}$ **18.** $\begin{cases} 3x - 4y = -11 \\ 2x - 3y = -7 \end{cases}$

In Problems 19–24, solve each system of linear equations using either substitution or elimination.

19. $\begin{cases} x + y = -4 \\ 2x - 3y = 12 \end{cases}$ **20.** $\begin{cases} 5x - 3y = 2 \\ x + 2y = -10 \end{cases}$

21. $\begin{cases} 3x - 2y = 5 \\ 4x - 5y = 9 \end{cases}$ **22.** $\begin{cases} 12x + 20y = 21 \\ 3x - 2y = 0 \end{cases}$

23. $\begin{cases} 6x + 9y = -3 \\ 8x + 12y = 7 \end{cases}$ **24.** $\begin{cases} 6x + 11y = 2 \\ 5x + 8y = -3 \end{cases}$

Section 3.2 Problem Solving: Systems of Two Linear Equations Containing Two Unknowns

KEY TERM

Break-even point

You Should Be Able To...	EXAMPLE	Review Exercises
1 Model and solve direct translation problems (p. 231)	Examples 1 and 2	25–27
2 Model and solve geometry problems (p. 233)	Examples 3 and 4	28–30
3 Model and solve mixture problems (p. 235)	Examples 5 and 6	31–33
4 Model and solve uniform motion problems (p. 238)	Example 7	34–36
5 Find the intersection of two linear functions (p. 239)	Example 8	37–38

25. Numbers The sum of two numbers is 56. The difference of the two numbers is 14. Find the two numbers.

26. Phi Theta Kappa The local chapter of the Phi Theta Kappa International Honor Society for Two-Year Colleges will induct 73 new members this semester. There are 11 more female inductees than male inductees. How many males and how many females will be inducted?

27. Counting Calories Shawn and Randy ate lunch at a pizza buffet. Shawn ate 3 slices of pepperoni pizza and 5 slices of Italian sausage pizza for a total of 2600 calories. Randy ate 4 slices of pepperoni pizza and 2 slices of Italian sausage pizza for a total of 1880 calories. Assuming the pizza slices are exactly the same size, how many calories are in a slice of pepperoni pizza? How many calories are in a slice of Italian sausage pizza?

△ **28. Rectangle** The length of a rectangle is 5 inches less than twice the width. The perimeter of the rectangle is 68 inches. Find the dimensions of the rectangle.

△ **29. Angles** In the right triangle shown in the figure to the right, the measure of angle y is twice the measure of angle x. Find the measures of angles x and y.

30. Angles Two intersecting lines are shown in the figure to the right. Suppose that the measure of angle 1 is $(x + 4y)°$, the measure of angle 2 is $(8x + 5y)°$, and the measure of angle 3 is $(2x + y)°$. Find x and y.

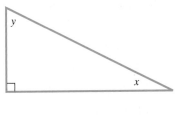

31. Coins Jerome's piggy bank contains a total of 40 nickels and quarters. The value of the coins in the bank is $4.40. How many nickels and how many quarters are in the bank?

32. Hydrochloric Acid Solution For an experiment, a chemist needs 12 liters of a solution that is 30% hydrochloric acid. However, she only has two solutions that are 25% hydrochloric acid and 40% hydrochloric acid. How many liters of each should she mix in order to obtain the needed solution?

33. Investments Verna invested part of a $10,000 inheritance in stocks paying 6.5% simple interest annually. She invested the rest in bonds paying 4.25% simple interest annually. If Verna will receive $582.50 in interest after one year, how much did she invest in stocks? How much did she invest in bonds?

34. Airplane Speed An airplane averaged 160 miles per hour with the wind and 112 miles per hour against the wind. Determine the speed of the plane and the effect of the wind resistance.

35. Catching Up A Pontiac Grand Am enters the Will Rogers Turnpike traveling at 50 miles per hour. One-half hour later, a Ford Mustang enters the turnpike at the same location and travels in the same direction at 80 miles per hour. How long will it take for the Mustang to catch up to the Grand Am? When it catches up, how far will the two cars have traveled on the turnpike?

36. Boat Speed A boat traveled 30 miles upstream in 1.5 hours. The return trip downstream took only 1 hour. Assuming that the boat travels at a constant rate and that the river's current is constant, find the speed of the boat in still water and the speed of the current.

37. High School Sports Participation The number of male students (in millions) who participate in high school sports can be estimated by the function $M(t) = 0.056t + 3.354$, where t is the number of years since 1990. The number of female students (in millions) who participate in high school sports can be estimated by $F(t) = 0.103t + 1.842$. (*SOURCE: National Federation of State High School Associations*)

(a) Graph each of the two functions in the same Cartesian plane.

(b) Assuming these linear trends continue, find the year in which participation by male and by female students will be the same.

38. Break-Even Karen bakes pies for a living. She sells pies for $15 each. Her monthly fixed costs are $1200. Each pie costs $2.50 in ingredients.

(a) Find the revenue function R, treating the number of pies x as the independent variable.

(b) Find the cost function C, treating the number of pies x as the independent variable.

(c) Graph the revenue and cost functions in the same Cartesian plane.

(d) Find the break-even number of pies that Karen must make and sell each month. What are the revenue and cost at this number of pies? Label the point on the graph in part (c).

Section 3.3 Systems of Linear Equations in Three Variables

KEY TERMS

Ordered triple   Exactly one solution   No solution   Infinitely many solutions

You Should Be Able To...	EXAMPLE	Review Exercises
1 Solve systems of three linear equations (p. 244)	Examples 1 through 3	39–46
2 Identify inconsistent systems (p. 249)	Example 4	42
3 Write the solution of a system with dependent equations (p. 249)	Example 5	43
4 Model and solve problems involving three linear equations (p. 250)	Example 6	47, 48

In Problems 39–46, solve each system of three linear equations containing three unknowns.

39. $\begin{cases} x + y - z = 1 \\ x - y + z = 7 \\ x + 2y + z = 1 \end{cases}$

40. $\begin{cases} 2x - 2y + z = -10 \\ 3x + y - 2z = 4 \\ 5x + 2y - 3z = 7 \end{cases}$

41. $\begin{cases} x + 2y = -1 \\ 3y + 4z = 7 \\ 2x - z = 6 \end{cases}$

42. $\begin{cases} x + 2y - 3z = -4 \\ x + y + 3z = 5 \\ 3x + 4y + 3z = 7 \end{cases}$

43. $\begin{cases} 3x + y - 2z = 6 \\ x + y - z = -2 \\ -x - 3y + 2z = 14 \end{cases}$

44. $\begin{cases} 9x - y + 2z = -5 \\ -3x - 4y + 4z = 3 \\ 15x + 3y - 2z = -10 \end{cases}$

45. $\begin{cases} 4x - 5y + 2z = -8 \\ 3x + 7y - 3z = 21 \\ 7x - 4y + 2z = -5 \end{cases}$

46. $\begin{cases} 3x - 2y + 5z = -7 \\ 4x + y + 3z = -2 \\ 2x - 3y + 7z = -4 \end{cases}$

△ **47. Angles** In the triangle shown to the right, the measure of angle z is twice the measure of angle y. The measure of angle y is $10°$ less than three times the measure of angle x. Find the measures of angles x, y, and z.

48. Counting Calories Mike, Clint, and Charlie decided to stop for a snack. Mike consumed 1000 calories by eating a cheeseburger, a medium order of fries, and a medium Coke. Clint consumed 1300 calories by eating two cheeseburgers, a medium order of fries, and a medium Coke. Charlie consumed 1590 calories by eating two cheeseburgers, a medium order of fries, and two medium Cokes. Find the number of calories each in a cheeseburger, in a medium order of fries, and in a medium Coke.

Section 3.4 Using Matrices to Solve Systems

KEY CONCEPTS

- **Augmented Matrix**

  $\begin{cases} a_1x + b_1y = c_1 \\ a_2x + b_2y = c_2 \end{cases}$ is written as an augmented matrix as $\begin{bmatrix} a_1 & b_1 & | & c_1 \\ a_2 & b_2 & | & c_2 \end{bmatrix}$

- **Row Operations**
  1. Interchange any two rows.
  2. Multiply all entries in a row by a nonzero constant.
  3. Replace a row by the sum of that row and a nonzero multiple of some other row.

KEY TERMS

Matrix
Augmented matrix
Row echelon form

You Should Be Able To...	EXAMPLE	Review Exercises
1 Write the augmented matrix of a system (p. 256)	Example 1	49–52
2 Write the system from the augmented matrix (p. 257)	Example 2	53–56
3 Perform row operations on a matrix (p. 257)	Examples 3 and 4	57–60
4 Solve systems using matrices (p. 259)	Examples 5 and 6	61–70
5 Solve consistent systems with dependent equations and solve inconsistent systems (p. 262)	Examples 7 and 8	64, 65, 68, 69

In Problems 49–52, write the augmented matrix of the given system of linear equations.

49. $\begin{cases} 3x + y = 7 \\ 2x + 5y = 9 \end{cases}$

50. $\begin{cases} x - 5y = 14 \\ -x + y = -3 \end{cases}$

51. $\begin{cases} 5x - y + 4z = 6 \\ -3x - 3z = -1 \\ x - 2y = 0 \end{cases}$

52. $\begin{cases} 8x - y + 3z = 14 \\ -3x + 5y - 6z = -18 \\ 7x - 4y + 5z = 21 \end{cases}$

In Problems 53–56, write the system of linear equations corresponding to the given augmented matrix.

53. $\begin{bmatrix} 1 & 2 & | & 12 \\ 0 & 3 & | & 15 \end{bmatrix}$

54. $\begin{bmatrix} 3 & -4 & | & -5 \\ -1 & 2 & | & 7 \end{bmatrix}$

55. $\begin{bmatrix} 1 & 3 & 4 & | & 20 \\ 0 & 1 & -2 & | & -16 \\ 0 & 0 & 1 & | & 7 \end{bmatrix}$

56. $\begin{bmatrix} -3 & 7 & 9 & | & 1 \\ 4 & 10 & 7 & | & 5 \\ 2 & -5 & -6 & | & -8 \end{bmatrix}$

In Problems 57–60, perform each row operation on the given augmented matrix.

57. $\begin{bmatrix} 1 & -5 & | & 22 \\ -2 & 9 & | & -40 \end{bmatrix}$

    (a) $R_2 = 2r_1 + r_2$ followed by

    (b) $R_2 = -r_2$

58. $\begin{bmatrix} 1 & -4 & | & 7 \\ 3 & -7 & | & 6 \end{bmatrix}$

(a) $R_2 = -3r_1 + r_2$ followed by

(b) $R_2 = \dfrac{1}{5}r_2$

59. $\begin{bmatrix} -1 & 2 & 1 & | & 1 \\ 2 & -1 & 3 & | & -3 \\ -1 & 5 & 6 & | & 2 \end{bmatrix}$

(a) $R_2 = 2r_1 + r_2$ followed by

(b) $R_3 = -r_1 + r_3$

60. $\begin{bmatrix} 1 & 3 & 4 & | & 4 \\ 0 & 5 & 10 & | & -15 \\ 0 & -4 & -7 & | & 7 \end{bmatrix}$

(a) $R_2 = \dfrac{1}{5}r_2$ followed by

(b) $R_3 = 4r_2 + r_3$

In Problems 61–70, solve each system of linear equations using matrices.

61. $\begin{cases} x + 2y = 1 \\ x + 3y = -2 \end{cases}$

62. $\begin{cases} 6x - 2y = -7 \\ 4x + 3y = 17 \end{cases}$

63. $\begin{cases} 3x + 2y = -10 \\ 2x - y = -9 \end{cases}$

64. $\begin{cases} -3x + 9y = 15 \\ 5x - 15y = 11 \end{cases}$

65. $\begin{cases} 4x - 2y = 6 \\ 6x - 3y = 9 \end{cases}$

66. $\begin{cases} x + 4y = 4 \\ 4x - 8y = 7 \end{cases}$

67. $\begin{cases} x + 2y - 2z = -11 \\ 2x + z = -6 \\ 5y - 3z = -7 \end{cases}$

68. $\begin{cases} 2x - 5y + 2z = 9 \\ -x + y - 2z = -2 \\ -x - 2y - 4z = 8 \end{cases}$

69. $\begin{cases} 2x - 7y + 11z = -5 \\ 4x - 2y + 6z = 2 \\ -2x + 19y - 27z = 17 \end{cases}$

70. $\begin{cases} 5x + 3y + 7z = 9 \\ 3x + 5y + 4z = 8 \\ x + 3y + 3z = 9 \end{cases}$

Section 3.5 Determinants and Cramer's Rule

KEY CONCEPTS

- **Evaluate a 2 × 2 Determinant**

$$\begin{vmatrix} a & b \\ c & d \end{vmatrix} = ad - bc$$

- **Evaluate a 3 × 3 Determinant**

$$\begin{vmatrix} a_{1,1} & a_{1,2} & a_{1,3} \\ a_{2,1} & a_{2,2} & a_{2,3} \\ a_{3,1} & a_{3,2} & a_{3,3} \end{vmatrix} = a_{1,1}\begin{vmatrix} a_{2,2} & a_{2,3} \\ a_{3,2} & a_{3,3} \end{vmatrix} - a_{1,2}\begin{vmatrix} a_{2,1} & a_{2,3} \\ a_{3,1} & a_{3,3} \end{vmatrix} + a_{1,3}\begin{vmatrix} a_{2,1} & a_{2,2} \\ a_{3,1} & a_{3,2} \end{vmatrix}$$

- **Cramer's Rule:**

The solution of the system of equations $\begin{cases} ax + by = s \\ cx + dy = t \end{cases}$ with $D = \begin{vmatrix} a & b \\ c & d \end{vmatrix} \neq 0$

is given by $x = \dfrac{\begin{vmatrix} s & b \\ t & d \end{vmatrix}}{\begin{vmatrix} a & b \\ c & d \end{vmatrix}} = \dfrac{D_x}{D}$ $y = \dfrac{\begin{vmatrix} a & s \\ c & t \end{vmatrix}}{\begin{vmatrix} a & b \\ c & d \end{vmatrix}} = \dfrac{D_y}{D}$

The solution of the system of equations $\begin{cases} a_1 x + b_1 y + c_1 z = d_1 \\ a_2 x + b_2 y + c_2 z = d_2 \\ a_3 x + b_3 y + c_3 z = d_3 \end{cases}$

with $D = \begin{vmatrix} a_1 & b_1 & c_1 \\ a_2 & b_2 & c_2 \\ a_3 & b_3 & c_3 \end{vmatrix} \neq 0$

KEY TERMS

Square matrix
2 × 2 determinant
3 × 3 determinant
Minor

(continued)

$$D_x = \begin{vmatrix} d_1 & b_1 & c_1 \\ d_2 & b_2 & c_2 \\ d_3 & b_3 & c_3 \end{vmatrix} \qquad D_y = \begin{vmatrix} a_1 & d_1 & c_1 \\ a_2 & d_2 & c_2 \\ a_3 & d_3 & c_3 \end{vmatrix} \qquad D_z = \begin{vmatrix} a_1 & b_1 & d_1 \\ a_2 & b_2 & d_2 \\ a_3 & b_3 & d_3 \end{vmatrix}$$

is given by

$$x = \frac{D_x}{D} \qquad y = \frac{D_y}{D} \qquad z = \frac{D_z}{D}$$

You Should Be Able To...	EXAMPLE	Review Exercises
1 Evaluate the determinant of a 2 × 2 matrix (p. 268)	Example 1	71–74
2 Use Cramer's Rule to solve a system of two equations (p. 269)	Example 2	79–84
3 Evaluate the determinant of a 3 × 3 matrix (p. 270)	Examples 3 and 4	75–78
4 Use Cramer's Rule to solve a system of three equations (p. 272)	Example 5	85–90

In Problems 71–78, find the value of each determinant.

71. $\begin{vmatrix} 3 & 4 \\ -1 & 2 \end{vmatrix}$

72. $\begin{vmatrix} 2 & -3 \\ -6 & 9 \end{vmatrix}$

73. $\begin{vmatrix} -5 & 7 \\ -4 & 6 \end{vmatrix}$

74. $\begin{vmatrix} -7 & 2 \\ 6 & -3 \end{vmatrix}$

75. $\begin{vmatrix} 5 & 0 & 1 \\ 2 & -3 & -1 \\ 3 & 6 & -4 \end{vmatrix}$

76. $\begin{vmatrix} 1 & -4 & 5 \\ 0 & 1 & -3 \\ 2 & -6 & 4 \end{vmatrix}$

77. $\begin{vmatrix} 3 & 0 & -1 \\ 2 & 6 & 7 \\ 2 & 5 & 4 \end{vmatrix}$

78. $\begin{vmatrix} 2 & 3 & 1 \\ 1 & -3 & -7 \\ -5 & 4 & 8 \end{vmatrix}$

In Problems 79–90, solve each system of linear equations using Cramer's Rule, if applicable.

79. $\begin{cases} x + 2y = -1 \\ 2x + 3y = 1 \end{cases}$

80. $\begin{cases} 4x + 9y = -13 \\ -5x + 6y = 22 \end{cases}$

81. $\begin{cases} 4x - y = -6 \\ 3x + 5y = 7 \end{cases}$

82. $\begin{cases} x - y = 2 \\ 2x + y = 5 \end{cases}$

83. $\begin{cases} 6x + 2y = 5 \\ 15x + 5y = 8 \end{cases}$

84. $\begin{cases} 12x + y = 6 \\ -6x + 7y = 15 \end{cases}$

85. $\begin{cases} x - y + 2z = 9 \\ 3x + 2y - 4z = 7 \\ 3y + 5z = -1 \end{cases}$

86. $\begin{cases} x - 3y - 3z = -5 \\ 7x + y - 2z = 24 \\ 6x - 5y - 4z = -9 \end{cases}$

87. $\begin{cases} x + y + z = 1 \\ 3x + 2y + 4z = -1 \\ 2x + 2y + 3z = 0 \end{cases}$

88. $\begin{cases} 4x - 3y + z = -6 \\ 4x - 2y + 3z = -3 \\ 8x - 5y - 2z = -12 \end{cases}$

89. $\begin{cases} x - y - 4z = 7 \\ 4x - 3y - 3z = 4 \\ 3x - 2y + z = -3 \end{cases}$

90. $\begin{cases} x + y = -3 \\ 2y - z = -1 \\ 5x + z = 1 \end{cases}$

Section 3.6 Systems of Linear Inequalities

KEY TERMS

Satisfies   Bounded

Corner points   Unbounded

You Should Be Able To...	EXAMPLE	Review Exercises
1 Determine whether an ordered pair is a solution of a system of linear inequalities (p. 276)	Example 1	91–94
2 Graph a system of linear inequalities (p. 277)	Examples 2 through 5	95–102
3 Solve problems involving systems of linear inequalities (p. 279)	Example 6	103, 104

In Problems 91–94, determine which of the ordered pairs, if any, satisfies the system of linear inequalities.

91. $\begin{cases} x - y > 2 \\ 2x + 3y > 8 \end{cases}$

(a) $(5, 2)$

(b) $(-3, 5)$

92. $\begin{cases} 7x + 3y \leq 21 \\ -2x + y \geq 5 \end{cases}$

(a) $(-1, 2)$

(b) $(-3, 4)$

93. $\begin{cases} x + 2y \leq 6 \\ 3x - y \geq 2 \\ x \geq 0 \\ y \geq 0 \end{cases}$

(a) $(2, 1)$

(b) $(1, 2)$

94. $\begin{cases} 3x - 2y \leq 12 \\ 2x + y \leq 15 \\ x \geq 0 \\ y \geq 0 \end{cases}$

(a) $(-2, 3)$

(b) $(4, 1)$

In Problems 95–100, graph each system of linear inequalities.

95. $\begin{cases} x + y \geq 7 \\ 2x - y \geq 5 \end{cases}$

96. $\begin{cases} 2x + 3y > 9 \\ x - 3y > -18 \end{cases}$

97. $\begin{cases} x - 4y > -4 \\ x + 2y \leq 8 \end{cases}$

98. $\begin{cases} x - y < -2 \\ 2x - 2y > 6 \end{cases}$

99. $\begin{cases} 2x - y \geq -2 \\ 2x - 3y \leq 6 \end{cases}$

100. $\begin{cases} x \geq 2 \\ y < -3 \end{cases}$

In Problems 101 and 102, graph each system of linear inequalities. Tell whether the graph is bounded or unbounded, and label the corner points.

101. $\begin{cases} 3x + 5y \leq 30 \\ 4x - 5y \geq 5 \\ x \leq 8 \\ y \geq 0 \end{cases}$

102. $\begin{cases} 3x + 2y \geq 10 \\ x + 2y \geq 6 \\ x \geq 0 \\ y \geq 0 \end{cases}$

103. Investments Options Anna has a maximum of $4000 to invest. One investment option pays 6% simple interest annually, while a second investment option pays 8% simple interest annually. Anna's financial advisor recommends that she invest at least $500 at 6% and no more than $2500 at 8%. Anna wants to earn at least $275 annually in interest.

(a) Let x denote the amount invested at 6% and y denote the amount invested at 8%. Write a system of linear inequalities that represents Anna's possible combinations of investments.

(b) Graph the system and label the corner points.

104. Purchasing Tulip Bulbs Jordan has a maximum of $144 to purchase tulip bulbs for her yard. Red tulip bulbs cost $6 per dozen, and yellow tulip bulbs cost $4 per dozen. Jordan wants at least 4 dozen more red tulip bulbs than yellow tulip bulbs.

(a) Let x denote the number of dozens of red tulip bulbs and y denote the number of dozens of yellow tulip bulbs. Write a system of linear inequalities that represents Jordan's possible combinations of purchases.

(b) Graph the system and label the corner points.

Chapter 3 Test

Step-by-step test solutions are found on the Chapter Test Prep Videos available in MyMathLab® or on YouTube.

1. Solve the system of linear equations by graphing.

$\begin{cases} 2x - y = 0 \\ 4x - 5y = 12 \end{cases}$

In Problems 2–5, solve the system of linear equations using either substitution or elimination.

2. $\begin{cases} 5x + 2y = -3 \\ y = 2x - 6 \end{cases}$

3. $\begin{cases} 9x + 3y = 1 \\ x - 2y = 4 \end{cases}$

4. $\begin{cases} 6x - 9y = 5 \\ 8x - 12y = 7 \end{cases}$

5. $\begin{cases} 2x + y = -4 \\ \dfrac{1}{3}x + \dfrac{1}{2}y = 2 \end{cases}$

In Problems 6 and 7, solve each system of three linear equations containing three unknowns.

6. $\begin{cases} x - 2y + 3z = 1 \\ x + y - 3z = 7 \\ 3x - 4y + 5z = 7 \end{cases}$

7. $\begin{cases} 2x + 4y + 3z = 5 \\ 3x - y + 2z = 8 \\ x + y + 2z = 0 \end{cases}$

In Problems 8 and 9, perform each row operation on the given augmented matrix.

8. $\begin{bmatrix} 1 & -3 & | & -2 \\ 2 & -4 & | & 8 \end{bmatrix}$

(a) $R_2 = -2r_1 + r_2$
followed by

(b) $R_2 = \dfrac{1}{2}r_2$

9. $\begin{bmatrix} 1 & -2 & 1 & | & -2 \\ 3 & -5 & 2 & | & 1 \\ 0 & -4 & 5 & | & -32 \end{bmatrix}$

(a) $R_2 = -3r_1 + r_2$
followed by

(b) $R_3 = 4r_2 + r_3$

In Problems 10 and 11, write the augmented matrix for each system of linear equations. Then use the matrix to solve the system.

10. $\begin{cases} x - 5y = 2 \\ 2x + y = 4 \end{cases}$

11. $\begin{cases} x + 2y + z = 3 \\ 4y + 3z = 5 \\ 2x + 3y = 1 \end{cases}$

In Problems 12 and 13, find the value of each determinant.

12. $\begin{vmatrix} 3 & -5 \\ 4 & -8 \end{vmatrix}$

13. $\begin{vmatrix} 0 & 1 & 2 \\ 3 & 3 & -1 \\ -2 & 1 & 2 \end{vmatrix}$

In Problems 14 and 15, solve each system of linear equations using Cramer's Rule, if applicable.

14. $\begin{cases} x - y = -2 \\ 5x + 3y = -8 \end{cases}$

15. $\begin{cases} x + y + z = -2 \\ x + y - 2z = 1 \\ 4x + 2y + 3z = -15 \end{cases}$

In Problems 16 and 17, graph each system of linear inequalities. Label the corner points.

16. $\begin{cases} 2x - y > -2 \\ x - 3y < 9 \end{cases}$

17. $\begin{cases} 3x + 2y \le 12 \\ x - 2y \ge -4 \\ x \ge 0 \\ y \ge 0 \end{cases}$

18. Grain Storage A grain-storage warehouse has a total of 50 bins. Some of the bins hold 25 tons of grain each, and the rest hold 20 tons each. How many of each type of bin are there if the capacity of the warehouse is 1160 tons?

△ **19. Triangle** In the triangle shown below, the measure of angle z is 10° larger than the measure of angle y. Five times the measure of angle x equals the sum of the measures of angles y and z. Find the measures of angles x, y, and z.

20. Shopping It's Margaret's lucky day. While shopping, she has happened onto a sale where designer blouses are being sold for $12 each and designer sweaters are being sold for $18 each. Unfortunately, Margaret has no more than $180 to spend. She also decides that she can buy at most 13 items.

(a) Let x denote the number of blouses and y denote the number of sweaters that Margaret will buy. Write a system of linear inequalities that represents the possible combinations of purchases.

(b) Graph the system and label the corner points.

Cumulative Review Chapters R–3

1. Consider the following set of numbers:

$$\left\{ -13, -\frac{7}{8}, 0, \frac{\pi}{2}, 2.7, 4\sqrt{2}, 11 \right\}.$$ List the numbers that are:

(a) Counting numbers

(b) Integers

(c) Rational numbers

(d) Irrational numbers

(e) Real numbers

2. Translate the following statement into a mathematical statement: Twice the sum of a number and four equals three times the number decreased by six.

3. Evaluate: $\dfrac{3 - 7(5)}{4}$

4. Evaluate $\dfrac{x^2 - 5x + 4}{x - 1}$ for $x = 7$.

5. Simplify: $x(x - 4) - 7(x + 5) + 2(x^2 + 4)$

6. Solve: $3|x + 7| - 4 = 20$

7. Solve: $\dfrac{x + 1}{5} \ge \dfrac{5x + 29}{10}$

8. Solve: $|2x - 7| > 3$

9. For $A = \{0, 2, 4, 6, 8, 10, 12\}$ and $B = \{0, 3, 6, 9, 12, 15\}$, find each of the following.

(a) $A \cap B$

(b) $A \cup B$

10. Graph: $y = 2|x| - 5$

11. Find the zero of $f(x) = \dfrac{2}{3}x - 4$.

12. Determine which, if any, of the following graphs are graphs of functions.

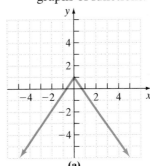

(a)

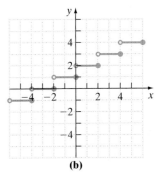

(b)

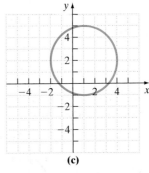

(c)

13. Determine the domain of $f(x) = \dfrac{2x + 7}{2x - 1}$.

14. For $g(x) = 3x^2 + 4x - 7$, find each of the following.

   (a) $g(-4)$

   (b) $g\left(\dfrac{2}{3}\right)$

   (c) $g(-x)$

15. Graph the linear equation $2x - 5y = 20$.

16. Determine whether the graphs of the following linear equations are parallel, perpendicular, or neither.

$$\begin{cases} 3x - 9y = 5 \\ 5x - 15y = -27 \end{cases}$$

17. Find the equation of the line passing through the points $(-6, 14)$ and $(9, -11)$. Write your answer in either slope-intercept or standard form, whichever you prefer.

18. The weight of a collection of quarters varies directly with the number of quarters in the collection. If 50 quarters weigh 281 grams, determine the weight of 35 quarters.

19. Using data from the U.S. Census Bureau for the years 1930 through 2010, the population (in thousands) of New Orleans, Louisiana, x years after 1930 is given by $P(x) = -0.123x^2 + 8.543x + 447.873$.

   (a) Identify the dependent and independent variables.

   (b) What is the domain of this function?

   (c) Find $P(60)$ and explain what this value represents.

   (d) Use the function to find the population of New Orleans in 2010.

20. Long Distance Earth Tones offers the Green Saver long-distance plan that charges \$4.95 per month plus 8¢ per minute.

   (a) Write the monthly long-distance cost C as a function of long-distance minutes talked x.

   (b) What is the implied domain of this linear function?

   (c) What is the cost when a person makes 645 minutes of long-distance calls during one month?

   (d) During one month, how many minutes of long-distance calls can be made for \$130.95?

In Problems 21 and 22, solve each system of equations.

21. $\begin{cases} y = \dfrac{1}{2}x + 9 \\ 5x + 4y = -20 \end{cases}$

22. $\begin{cases} 6x - 4y + 3z = 3 \\ 3x + 10y + z = -4 \\ 9x + 2y + 4z = 1 \end{cases}$

In Problems 23 and 24, find the value of each determinant.

23. $\begin{vmatrix} -4 & 2 \\ -5 & 3 \end{vmatrix}$

24. $\begin{vmatrix} 2 & 3 & 0 \\ -1 & 4 & 2 \\ 2 & -2 & -3 \end{vmatrix}$

25. Graph the following system of linear inequalities.

$$\begin{cases} x + 3y < 12 \\ x \geq 3 \end{cases}$$

Getting Ready for Chapter 4
Laws of Exponents and Scientific Notation

Objectives

1. Simplify Exponential Expressions Using the Product Rule
2. Simplify Exponential Expressions Using the Quotient Rule
3. Evaluate Exponential Expressions with a Zero or Negative Exponent
4. Simplify Exponential Expressions Using the Power Rule
5. Simplify Exponential Expressions Containing Products or Quotients
6. Simplify Exponential Expressions Using the Laws of Exponents
7. Convert Between Scientific Notation and Decimal Notation
8. Use Scientific Notation to Multiply and Divide

Work Smart

We read a^n as "a raised to the power n" or "a raised to the nth power." We usually read a^2 as "a squared" and a^3 as "a cubed."

In Words

When multiplying two exponential expressions with the same base, add the exponents. Then write the common base to the power of this sum.

A quick review of positive integer exponents is in order. Recall that if a is a real number and n is a positive integer, then the symbol a^n means that we should use a as a factor n times. That is,

$$a^n = \underbrace{a \cdot a \cdot \cdots \cdot a}_{n \text{ factors}}$$

For example,

$$4^3 = \underbrace{4 \cdot 4 \cdot 4}_{3 \text{ factors}}$$

In the notation a^n, we call a the **base** and n the **power** or **exponent.**

① Simplify Exponential Expressions Using the Product Rule

Several general rules can be discovered for simplifying expressions involving *positive* integer exponents. The first rule that we introduce is used when multiplying two exponential expressions that have the same base. Consider the following:

$$x^2 \cdot x^4 = \underbrace{(x \cdot x)}_{2 \text{ factors}} \underbrace{(x \cdot x \cdot x \cdot x)}_{4 \text{ factors}} = \underbrace{x \cdot x \cdot x \cdot x \cdot x \cdot x}_{6 \text{ factors}} = x^6$$

Same base — Same base — Sum of powers 2 and 4

Based on the above, we have the following result:

Product Rule for Exponents (Positive Integer Exponents)

If a is a real number, and m and n are positive integers, then
$$a^m \cdot a^n = a^{m+n}$$

EXAMPLE 1 **Using the Product Rule to Simplify Expressions Involving Exponents**

Simplify each expression. All answers should contain only positive integer exponents.

▶ **(a)** $2^2 \cdot 2^3$ ▶ **(b)** $3z^2 \cdot 4z^4$

Solution

(a) $2^2 \cdot 2^3 = 2^{2+3}$
$\qquad\qquad = 2^5$
$\qquad\qquad = 32$

(b) $3z^2 \cdot 4z^4 = 3 \cdot 4 \cdot z^2 \cdot z^4$
$\qquad\qquad\quad = 12z^{2+4}$
$\qquad\qquad\quad = 12z^6$

Quick ✓

1. In the notation a^n we call a the ____ and n the _____ or _____.

2. If a is a real number, and m and n are positive integers, then $a^m \cdot a^n =$ ____.

In Problems 3–7, simplify each expression. All answers should contain only positive integer exponents.

3. $5^2 \cdot 5$ 4. $(-3)^2 \cdot (-3)^3$ 5. $y^4 \cdot y^3$

6. $(5x^2) \cdot (-2x^5)$ 7. $6y^3 \cdot (-y^2)$

▶ ❷ Simplify Exponential Expressions Using the Quotient Rule

To find a general rule for the quotient of two exponential expressions with *positive* integer exponents, we use the Reduction Property to divide out common factors. Consider the following:

$$\frac{y^6}{y^2} = \frac{\overbrace{y \cdot y \cdot y \cdot y \cdot \cancel{y} \cdot \cancel{y}}^{6 \text{ factors}}}{\underbrace{\cancel{y} \cdot \cancel{y}}_{2 \text{ factors}}} = \underbrace{y \cdot y \cdot y \cdot y}_{4 \text{ factors}} = y^4 \quad \overset{\text{Difference of powers}}{\underset{\downarrow}{6 \text{ and } 2}}$$

We conclude from this result that

$$\frac{y^6}{y^2} = y^{6-2} = y^4$$

This result is true in general.

> **In Words**
>
> When dividing two exponential expressions with a common base, subtract the exponent in the denominator from the exponent in the numerator. Then write the common base to the power of this difference.

> **Quotient Rule for Exponents (Positive Integer Exponents)**
>
> If a is a real number and if m and n are positive integers, then
>
> $$\frac{a^m}{a^n} = a^{m-n} \quad \text{if } a \neq 0$$

EXAMPLE 2 **Using the Quotient Rule to Simplify Expressions Involving Exponents**

Simplify each expression. Answers should contain only positive integer exponents.

(a) $\dfrac{8^5}{8^3}$

(b) $\dfrac{27z^9}{12z^4}$

Solution

(a) $\dfrac{8^5}{8^3} = 8^{5-3}$

$\quad\quad = 8^2$

$\quad\quad = 64$

(b) $\dfrac{27z^9}{12z^4} = \dfrac{9 \cdot 3}{4 \cdot 3} z^{9-4}$

$\quad\quad\quad\quad = \dfrac{9}{4} z^5$ ●

Quick ✓

8. *True or False* To divide two exponential expressions having the same base, keep the base and subtract the exponents.

In Problems 9–12, simplify each expression. All answers should contain only positive integer exponents.

9. $\dfrac{5^6}{5^4}$

10. $\dfrac{y^8}{y^6}$

11. $\dfrac{16a^6}{10a^5}$

12. $\dfrac{-24b^5}{16b^3}$

❸ Evaluate Exponential Expressions with a Zero or Negative Exponent

Let's now extend the definition of exponential expressions to *all* integer exponents. That is, let's evaluate exponential expressions where the exponent can be a positive integer, zero, or a negative integer. We begin with raising a real number to the 0 power.

▶ **Zero-Exponent Rule**

If a is a nonzero real number, we define

$$a^0 = 1 \quad \text{if } a \neq 0$$

This rule is based on the Product Rule and the Identity Property of Multiplication. From the Product Rule for Exponents, we know

$$a^0 a^n = a^{0+n}$$
$$= a^n$$
$$= 1 \cdot a^n$$

From the Identity Property of Multiplication, it must be that $a^0 = 1$.

Suppose we want to simplify $\dfrac{z^3}{z^5}$. If we use the Quotient Rule for Exponents, we get

$$\frac{z^3}{z^5} = z^{3-5} = z^{-2}$$

We could also simplify this expression using the Reduction Property.

$$\frac{z^3}{z^5} = \frac{\cancel{z} \cdot \cancel{z} \cdot \cancel{z}}{\cancel{z} \cdot \cancel{z} \cdot \cancel{z} \cdot z \cdot z} = \frac{1}{z^2}$$

This implies that $z^{-2} = \dfrac{1}{z^2}$. This result suggests that we define a raised to a negative power as follows:

▶ **Negative-Exponent Rule**

If n is a positive integer and if a is a nonzero real number, then we define

$$a^{-n} = \frac{1}{a^n} \quad \text{or} \quad \frac{1}{a^{-n}} = a^n \quad \text{if } a \neq 0$$

EXAMPLE 3 **Evaluating Exponential Expressions Containing Integer Exponents**

Simplify each expression. All exponents should be positive integers.

(a) 3^{-4} (b) $\dfrac{1}{3^{-2}}$ (c) $5x^0$ (d) $4x^{-5}$

Solution

(a) $3^{-4} = \dfrac{1}{3^4}$ (b) $\dfrac{1}{3^{-2}} = 3^2$ (c) $5x^0 = 5 \cdot 1$ (d) $4x^{-5} = \dfrac{4}{x^5}$

$\quad\quad = \dfrac{1}{81}$ $\quad\quad = 9$ $\quad\quad = 5$

Quick ✓

13. $a^0 =$ ___, provided $a \neq$ _. **14.** $a^{-n} =$ ___, provided $a \neq$ _.

In Problems 15–20, simplify each expression. All exponents should be positive integers.

15. 5^{-3} **16.** $5z^{-7}$ **17.** $\dfrac{1}{x^{-4}}$ **18.** $\dfrac{5}{y^{-3}}$ **19.** -4^0 **20.** $(-10)^0$

▶ **EXAMPLE 4** **Evaluating Exponential Expressions Containing Integer Exponents**

Simplify each expression. All exponents should be positive integers.

(a) $\left(\dfrac{2}{3}\right)^{-3}$

(b) $\left(\dfrac{1}{7}\right)^{-2}$

Solution

(a) $\left(\dfrac{2}{3}\right)^{-3} = \dfrac{1}{\left(\dfrac{2}{3}\right)^3}$

$= \dfrac{1}{\dfrac{2}{3} \cdot \dfrac{2}{3} \cdot \dfrac{2}{3}}$

$= \dfrac{1}{\dfrac{8}{27}}$

$= \dfrac{27}{8}$

(b) $\left(\dfrac{1}{7}\right)^{-2} = \dfrac{1}{\left(\dfrac{1}{7}\right)^2}$

$= \dfrac{1}{\dfrac{1}{7} \cdot \dfrac{1}{7}}$

$= \dfrac{1}{\dfrac{1}{49}}$

$= 49$

The following shortcut is based on the results of Example 4:

In Words

To evaluate $\left(\dfrac{a}{b}\right)^{-n}$, determine the reciprocal of the base and then raise it to the nth power.

If a and b are real numbers and n is an integer, then

$$\left(\dfrac{a}{b}\right)^{-n} = \left(\dfrac{b}{a}\right)^{n} \qquad \text{if } a \neq 0, b \neq 0$$

Quick ✓

In Problems 21–24, simplify each expression. All exponents should be positive integers.

21. $\left(\dfrac{4}{3}\right)^{-2}$

22. $\left(-\dfrac{1}{4}\right)^{-3}$

23. $\left(\dfrac{3}{x}\right)^{-2}$

24. $\dfrac{5}{2^{-2}}$

Now that we have definitions for 0 as an exponent and negative exponents, we restate the Product Rule and Quotient Rule for Exponents assuming that the exponent is any integer (positive, negative, or zero).

Product Rule for Exponents

If a is a real number and m and n are integers, then

$$a^m \cdot a^n = a^{m+n}$$

If $m, n,$ or $m + n$ is 0 or negative, then a cannot be 0.

Quotient Rule for Exponents

If a is a real number and if m and n are integers, then

$$\dfrac{a^m}{a^n} = a^{m-n} \quad \text{if } a \neq 0$$

Notice that allowing the exponents to be any integer (not just any positive integer) requires that we include restrictions on the value of the base.

EXAMPLE 5 **Using the Product Rule to Simplify Expressions Containing Exponents**

Simplify each expression. All exponents should be positive integers.

(a) $(-3)^2(-3)^{-4}$

(b) $\frac{3}{4}y^5 \cdot \frac{20}{9}y^{-2}$

Solution

(a) $(-3)^2(-3)^{-4} = (-3)^{2+(-4)}$

$= (-3)^{-2}$

$= \frac{1}{(-3)^2}$

$= \frac{1}{9}$

(b) $\frac{3}{4}y^5 \cdot \frac{20}{9}y^{-2} = \frac{3}{4} \cdot \frac{20}{9}y^{5+(-2)}$

$= \frac{5}{3}y^3$

EXAMPLE 6 **Using the Quotient Rule to Simplify Expressions Containing Exponents**

Simplify each expression. All exponents should be positive integers.

(a) $\frac{w^{-2}}{w^{-5}}$

(b) $\frac{20a^3b}{4ab^4}$

Solution

(a) $\frac{w^{-2}}{w^{-5}} = w^{-2-(-5)}$

$= w^{-2+5}$

$= w^3$

(b) $\frac{20a^3b}{4ab^4} = 5a^{3-1}b^{1-4}$

$= 5a^2b^{-3}$

$a^{-n} = \frac{1}{a^n} \colon = \frac{5a^2}{b^3}$

Quick ✓

In Problems 25–30, simplify each expression. All exponents should be positive integers.

25. $6^3 \cdot 6^{-5}$

26. $\frac{10^{-3}}{10^{-5}}$

27. $(4x^2y^3) \cdot (5xy^{-4})$

28. $\left(\frac{3}{4}a^3b\right) \cdot \left(\frac{8}{9}a^{-2}b^3\right)$

29. $\frac{-24b^5}{16b^{-3}}$

30. $\frac{50s^2t}{15s^5t^{-4}}$

▶ ❹ **Simplify Exponential Expressions Using the Power Rule**

Another law of exponents applies when an exponential expression containing a power is itself raised to a power.

$$(3^2)^4 = 3^2 \cdot 3^2 \cdot 3^2 \cdot 3^2 = \underbrace{(3 \cdot 3)}_{2 \text{ factors}} \cdot \underbrace{(3 \cdot 3)}_{2 \text{ factors}} \cdot \underbrace{(3 \cdot 3)}_{2 \text{ factors}} \cdot \underbrace{(3 \cdot 3)}_{2 \text{ factors}} = 3^8$$

$$\underbrace{}_{4 \text{ factors}}$$

$$2 \cdot 4 = 8 \text{ factors}$$

We have the following result:

In Words
If an exponential expression contains a power raised to a power, keep the base and multiply the powers.

Power Rule for Exponential Expressions

If a is a real number and m and n are integers, then

$$(a^m)^n = a^{m \cdot n}$$

If m or n is 0 or negative, then a must not be 0.

EXAMPLE 7 **Using the Power Rule to Simplify Exponential Expressions**

Simplify each expression. All exponents should be positive integers.

(a) $(y^3)^5$ **(b)** $[(-3)^3]^2$ **(c)** $(6^3)^0$

Solution

(a) $(y^3)^5 = y^{3 \cdot 5}$ **(b)** $[(-3)^3]^2 = (-3)^{3 \cdot 2}$ **(c)** $(6^3)^0 = 6^{3 \cdot 0}$
$\qquad\qquad = y^{15}$ $= (-3)^6$ $= 6^0$
$\qquad\qquad\qquad\qquad\qquad\qquad\qquad\quad = 729$ $= 1$

Quick ✓

In Problems 31–36, simplify each expression. All exponents should be positive integers.

31. $(2^2)^3$ **32.** $(5^8)^0$ **33.** $[(-4)^3]^2$
34. $(a^3)^5$ **35.** $(z^3)^{-6}$ **36.** $(s^{-3})^{-7}$

❺ Simplify Exponential Expressions Containing Products or Quotients

▶ We have two additional laws of exponents. The first deals with raising a product to a power, and the second deals with raising a quotient to a power. Consider the following product to a power:

$$(x \cdot y)^3 = (x \cdot y) \cdot (x \cdot y) \cdot (x \cdot y)$$
$$= (x \cdot x \cdot x) \cdot (y \cdot y \cdot y)$$
$$= x^3 \cdot y^3$$

We have the following result:

Work Smart

Do not use this rule to try to simplify $(a + b)^2$ as $a^2 + b^2$ or $(a + b)^3$ as $a^3 + b^3$. For this rule to apply, the base must be the *product* of two numbers—not a sum.

> **Product-to-a-Power Rule**
>
> If a and b are real numbers and n is an integer, then
>
> $$(a \cdot b)^n = a^n \cdot b^n$$
>
> If n is 0 or negative, neither a nor b can be 0.

EXAMPLE 8 **Using the Product-to-a-Power Rule to Simplify Exponential Expressions**

Simplify each expression. All exponents should be positive integers.

(a) $(3z)^4$ **(b)** $(-5y^{-2})^{-3}$ **(c)** $(-4a^2)^{-2}$

Solution

(a) $(3z)^4 = 3^4 z^4$ **(b)** $(-5y^{-2})^{-3} = (-5)^{-3}(y^{-2})^{-3}$ **(c)** $(-4a^2)^{-2} = \dfrac{1}{(-4a^2)^2}$
$\qquad\qquad = 81z^4$ $= \dfrac{y^{-2(-3)}}{(-5)^3}$ $= \dfrac{1}{(-4)^2(a^2)^2}$
$\qquad\qquad\qquad\qquad\qquad\qquad\qquad = \dfrac{y^6}{-125}$ $= \dfrac{1}{16a^4}$
$\qquad\qquad\qquad\qquad\qquad\qquad\qquad = -\dfrac{y^6}{125}$

Quick ✓

In Problems 37–40, simplify each expression. All exponents should be positive integers.

37. $(5y)^3$ **38.** $(6y)^0$ **39.** $(3x^2)^4$ **40.** $(4a^3)^{-2}$

Now let's look at a quotient raised to a power:

$$\left(\frac{2}{3}\right)^4 = \left(\frac{2}{3}\right)\cdot\left(\frac{2}{3}\right)\cdot\left(\frac{2}{3}\right)\cdot\left(\frac{2}{3}\right) = \frac{2^4}{3^4}$$

We have the following result:

> ▶ **Quotient-to-a-Power Rule**
>
> If a and b are real numbers and n is an integer, then
>
> $$\left(\frac{a}{b}\right)^n = \frac{a^n}{b^n} \qquad \text{if } b \neq 0$$
>
> If n is negative or 0, then a cannot be 0.

EXAMPLE 9   **Using the Quotient-to-a-Power Rule to Simplify Exponential Expressions**

Simplify each expression. All exponents should be positive integers.

(a) $\left(\dfrac{w}{4}\right)^3$

(b) $\left(\dfrac{2x^2}{y^3}\right)^4$

Solution

(a) $\left(\dfrac{w}{4}\right)^3 = \dfrac{w^3}{4^3}$

$\phantom{(a)\left(\dfrac{w}{4}\right)^3} = \dfrac{w^3}{64}$

(b) $\left(\dfrac{2x^2}{y^3}\right)^4 = \dfrac{(2x^2)^4}{(y^3)^4}$

$\phantom{(b)\left(\dfrac{2x^2}{y^3}\right)^4} = \dfrac{2^4(x^2)^4}{(y^3)^4}$

$\phantom{(b)\left(\dfrac{2x^2}{y^3}\right)^4} = \dfrac{16x^{2\cdot4}}{y^{3\cdot4}}$

$\phantom{(b)\left(\dfrac{2x^2}{y^3}\right)^4} = \dfrac{16x^8}{y^{12}}$

Quick ✓

In Problems 41–44, simplify each expression. All exponents should be positive integers.

41. $\left(\dfrac{z}{3}\right)^4$ **42.** $\left(\dfrac{x}{2}\right)^{-5}$ **43.** $\left(\dfrac{x^2}{y^3}\right)^4$ **44.** $\left(\dfrac{3a^{-2}}{b^4}\right)^3$

⑥ Simplify Exponential Expressions Using the Laws of Exponents

We now summarize the Laws of Exponents.

> **The Laws of Exponents**
>
> If a and b are real numbers and if m and n are integers, then assuming the expression is defined,
>
> | | | |
> |---|---|---|
> | **Zero-Exponent Rule:** | $a^0 = 1$ | if $a \neq 0$ |
> | **Negative-Exponent Rule:** | $a^{-n} = \dfrac{1}{a^n}$ | if $a \neq 0$ |
> | **Product Rule:** | $a^m \cdot a^n = a^{m+n}$ | |
> | **Quotient Rule:** | $\dfrac{a^m}{a^n} = a^{m-n}$ | if $a \neq 0$ |

(continued)

The Laws of Exponents (*continued*)

Power Rule: $\qquad\qquad\qquad\qquad (a^m)^n = a^{m \cdot n}$

Product-to-a-Power Rule: $\qquad\quad (a \cdot b)^n = a^n \cdot b^n$

Quotient-to-a-Power Rule: $\qquad \left(\dfrac{a}{b}\right)^n = \dfrac{a^n}{b^n} \qquad$ if $b \neq 0$

Quotient-to-a-Negative-Power Rule: $\left(\dfrac{a}{b}\right)^{-n} = \left(\dfrac{b}{a}\right)^n \qquad$ if $a \neq 0, b \neq 0$

Now let's do some examples where we use one or more of the preceding rules.

EXAMPLE 10 Using the Laws of Exponents

Simplify each expression. All exponents should be positive integers. None of the variables is zero.

▶ **(a)** $\dfrac{a^3 b^{-1}}{(a^2 b)^3}$

▶ **(b)** $\left(\dfrac{3xy}{x^2 y^{-2}}\right)^2 \cdot \left(\dfrac{9x^2 y^{-3}}{x^3 y^2}\right)^{-1}$

Solution

(a)

$$\overset{\overset{\displaystyle (a \cdot b)^n = a^n \cdot b^n}{\downarrow}}{\dfrac{a^3 b^{-1}}{(a^2 b)^3} = \dfrac{a^3 b^{-1}}{(a^2)^3 b^3}}$$

$(a^m)^n = a^{m \cdot n}:\qquad = \dfrac{a^3 b^{-1}}{a^6 b^3}$

$\dfrac{a^m}{a^n} = a^{m-n}:\qquad = a^{3-6} b^{-1-3}$

$\qquad\qquad\qquad\quad = a^{-3} b^{-4}$

$a^{-n} = \dfrac{1}{a^n}:\qquad = \dfrac{1}{a^3 b^4}$

(b)

$$\left(\dfrac{3xy}{x^2 y^{-2}}\right)^2 \cdot \left(\dfrac{9x^2 y^{-3}}{x^3 y^2}\right)^{-1} \overset{\overset{\displaystyle \frac{a^m}{a^n} = a^{m-n}}{\downarrow}}{=} (3x^{1-2} y^{1-(-2)})^2 \cdot (9x^{2-3} y^{-3-2})^{-1}$$

$\qquad\qquad\qquad\qquad\qquad = (3x^{-1} y^3)^2 \cdot (9x^{-1} y^{-5})^{-1}$

$(a \cdot b)^n = a^n \cdot b^n:\qquad = 3^2 \cdot (x^{-1})^2 (y^3)^2 \cdot 9^{-1} \cdot (x^{-1})^{-1} (y^{-5})^{-1}$

$(a^m)^n = a^{m \cdot n}:\qquad = 9x^{-2} y^6 \cdot \dfrac{1}{9} \cdot xy^5$

$a^m a^n = a^{m+n}:\qquad = x^{-2+1} y^{6+5}$

$a^{-n} = \dfrac{1}{a^n}:\qquad = x^{-1} y^{11} = \dfrac{y^{11}}{x}$

Work Smart: Study Skills

Many different approaches may be taken to simplify exponential expressions. In Example 10(b), we could have used the Quotient-to-a-Power Rule first and then continued to simplify, for example. Try working a problem one way and then working it again a second way to see if you obtain the same answer.

Quick ✓

In Problems 45–47, simplify each expression. All exponents should be positive integers.

45. $\dfrac{(3x^2 y)^2}{12xy^{-2}}$

46. $(3ab^3)^3 \cdot (6a^2 b^2)^{-2}$

47. $\left(\dfrac{2x^2 y^{-1}}{x^{-2} y^2}\right)^2 \cdot \left(\dfrac{4x^3 y^2}{xy^{-2}}\right)^{-1}$

7 **Convert Between Scientific Notation and Decimal Notation**

Figure 1

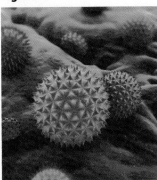

Measurements of physical quantities can range from very small to very large. For example, the mass of an electron (Figure 1) is approximately 0.00000000000000000000000000911 gram, and the mass of Earth (Figure 2) is about 5,980,000,000,000,000,000,000,000 kilograms. These numbers are difficult to write and difficult to read, so we use exponents to write them.

> **Definition**
>
> A number written as the product of a number a, where $1 \le |a| < 10$, and a power of 10 is written in **scientific notation.** That is, a number is written in scientific notation when it is of the form
>
> $$a \times 10^N$$
>
> where
>
> $$1 \le |a| < 10 \quad \text{and} \quad N \text{ is an integer}$$

Figure 2

For example, in scientific notation,

$$\text{Mass of an electron} = 9.11 \times 10^{-28} \text{ gram}$$

$$\text{Mass of Earth} = 5.98 \times 10^{24} \text{ kilograms}$$

EXAMPLE 11   **How to Convert from Decimal Notation to Scientific Notation**

Write each number in scientific notation.

(a) 94,873

(b) 0.042

Step-by-Step Solution

For a number to be in scientific notation, the decimal must be moved so there is a single nonzero digit to the left of the decimal point. All remaining digits must appear to the right of the decimal point.

(a)

Step 1: The decimal point in 94,873 follows the 3. Therefore, we will move the decimal point to the left $N = 4$ places until it is between the 9 and the 4. Do you see why?

$$9\,4\,8\,7\,3.$$

Step 2: The original number is greater than 1, so we write 94,873 in scientific notation as

$$9.4873 \times 10^4$$

(b)

Step 1: Because 0.042 is less than 1, we move the decimal point to the right $N = 2$ places until it is between the 4 and the 2.

$$0.0\,4\,2$$

Step 2: The original number is between 0 and 1, so we write 0.042 in scientific notation as

$$4.2 \times 10^{-2}$$

Converting a Decimal to Scientific Notation

To change a number from decimal to scientific notation,

Step 1: Count the number N of decimal places that the decimal point must be moved in order to arrive at a number a, where $1 \le |a| < 10$.

Step 2: If the absolute value of the original number is greater than or equal to 1, the scientific notation is $a \times 10^N$. If the absolute value of the original number is between 0 and 1, the scientific notation is $a \times 10^{-N}$.

▶ (EXAMPLE 12) **Converting a Negative Number from Decimal Notation to Scientific Notation**

Write $-520,000,000$ in scientific notation.

Work Smart

In Example 11(a), we move the decimal point four places to the left because, when the number is written in scientific notation, we are multiplying 9.4873 by $10^4 = 10,000$ —which, if simplified, would move the decimal point four places to the right.

Solution

Because the absolute value of the number is greater than 1, we move the decimal point to the left $N = 8$ places. Therefore,

$$-520,000,000 = -5.2 \times 10^8$$

Quick ✓

In Problems 48–51, write each number in scientific notation.

48. 532   **49.** $-1,230,000$ **50.** 0.034   **51.** -0.0000845

Now we are going to convert numbers from scientific notation to decimal notation.

Converting a Number from Scientific Notation to Decimal Notation

Determine the exponent on the number 10. If the exponent is negative, then move the decimal $|N|$ decimal places to the left. If the exponent is positive, then move the decimal N decimal places to the right.

(EXAMPLE 13) **Converting from Scientific Notation to Decimal Notation**

Write each number in decimal notation.

(a) 3.2×10^3 **(b)** 7.54×10^{-5}

Solution

(a) The exponent on 10 is 3 so we move the decimal point three places to the right.
$$3.2 \times 10^3 = 3.200 \times 10^3 = 3200$$

(b) The exponent on 10 is -5, so we move the decimal point five places to the left.
$$7.54 \times 10^{-5} = 000007.54 \times 10^{-5} = 0.0000754$$

Work Smart

We move the decimal point three places to the right in Example 13(a) because $10^3 = 1000$. Thus we are really multiplying 3.2 by 1000. We move the decimal point five places to the left in Example 13(b) because $10^{-5} = \dfrac{1}{10^5} = \dfrac{1}{100,000} = 0.00001$

Quick ✓

In Problems 52–55, write each number in decimal notation.

52. 5×10^2 **53.** 9.1×10^5 **54.** 1.8×10^{-4} **55.** 1×10^{-6}

▶ ⑧ Use Scientific Notation to Multiply and Divide

To multiply or divide two numbers written in scientific notation, we use two Laws of Exponents:

$$a^m \cdot a^n = a^{m+n} \quad \text{and} \quad \frac{a^m}{a^n} = a^{m-n}$$

We will use these laws where the base is 10.

EXAMPLE 14 **Multiplying Using Scientific Notation**

Perform the indicated operation. Express the answer in scientific notation.

    (a) $(5 \times 10^2) \cdot (3 \times 10^8)$

    (b) $(2.5 \times 10^{-3}) \cdot (4.3 \times 10^{-4})$

Solution

  (a) $\quad (5 \times 10^2) \cdot (3 \times 10^8) = (5 \cdot 3) \times (10^2 \cdot 10^8)$

$5 \cdot 3 = 15;$ use $a^m \cdot a^n = a^{m+n}$: $\quad = 15 \times 10^{10}$

$15 = 1.5 \times 10^1$: $\quad = (1.5 \times 10^1) \times 10^{10}$

$\quad = 1.5 \times 10^{11}$

  (b) $\quad (2.5 \times 10^{-3}) \cdot (4.3 \times 10^{-4}) = (2.5 \cdot 4.3) \times (10^{-3} \cdot 10^{-4})$

$2.5 \cdot 4.3 = 10.75;$ use $a^m \cdot a^n = a^{m+n}$: $\quad = 10.75 \times 10^{-7}$

$10.75 = 1.075 \times 10^1$: $\quad = (1.075 \times 10^1) \times 10^{-7}$

$\quad = 1.075 \times 10^{-6}$ ●

Quick ✓

In Problems 56–58, perform the indicated operation. Express the solution in scientific notation.

56. $(3 \times 10^3) \cdot (2 \times 10^5)$ **57.** $(2 \times 10^{-4}) \cdot (4 \times 10^{-7})$

58. $(6 \times 10^{-5}) \cdot (4 \times 10^8)$

EXAMPLE 15 **Dividing Using Scientific Notation**

Perform the indicated operation. Express the answer in scientific notation.

    (a) $\dfrac{8 \times 10^5}{2 \times 10^3}$ **(b)** $\dfrac{2.5 \times 10^5}{5 \times 10^{-3}}$

Solution

  (a) $\quad \dfrac{8 \times 10^5}{2 \times 10^3} = \dfrac{8}{2} \times \dfrac{10^5}{10^3}$

$\dfrac{a^m}{a^n} = a^{m-n}$: $\quad = 4 \times 10^2$

  (b) $\quad \dfrac{2.5 \times 10^5}{5 \times 10^{-3}} = \dfrac{2.5}{5} \times \dfrac{10^5}{10^{-3}}$

$\dfrac{a^m}{a^n} = a^{m-n}$: $\quad = 0.5 \times 10^{5-(-3)}$

$\quad = 0.5 \times 10^8$

$0.5 = 5 \times 10^{-1}$: $\quad = (5 \times 10^{-1}) \times 10^8$

$\quad = 5 \times 10^7$ ●

Quick ✓

In Problems 59–62, perform the indicated operation. Express the solution in scientific notation.

59. $\dfrac{6 \times 10^8}{3 \times 10^6}$ **60.** $\dfrac{6.8 \times 10^{-8}}{3.4 \times 10^{-5}}$ **61.** $\dfrac{4.8 \times 10^7}{9.6 \times 10^3}$ **62.** $\dfrac{3 \times 10^{-5}}{8 \times 10^7}$

EXAMPLE 16 **Using Scientific Notation to Multiply and Divide**

Perform the indicated operation. Express the answer in decimal notation.

(a) $(3,000,000) \cdot (90,000)$

(b) $\dfrac{0.00000075}{0.00015}$

Solution

In each of these problems, we will first write the numbers in scientific notation and then perform the indicated operation.

(a) $(3,000,000) \cdot (90,000) = (3 \times 10^6) \cdot (9 \times 10^4)$

$a^m \cdot a^n = a^{m+n}:\quad = 27 \times 10^{10}$

$= (2.7 \times 10^1) \times 10^{10}$

$= 2.7 \times 10^{11}$

$= 270,000,000,000$

(b) $\dfrac{0.00000075}{0.00015} = \dfrac{7.5 \times 10^{-7}}{1.5 \times 10^{-4}}$

$= \dfrac{7.5}{1.5} \times \dfrac{10^{-7}}{10^{-4}}$

$\dfrac{a^m}{a^n} = a^{m-n}:\quad = 5 \times 10^{-7-(-4)}$

$= 5 \times 10^{-3}$

$= 0.005$

Quick ✓

In Problems 63–66, perform the indicated operation. Express the solution in decimal notation.

63. $(8,000,000) \cdot (30,000)$

64. $\dfrac{0.000000012}{0.000004}$

65. $(25,000,000) \cdot (0.00003)$

66. $\dfrac{0.000039}{13,000,000}$

Getting Ready for Chapter 4 Exercises

MyMathLab®

*Problems **1–66** are the Quick ✓s that follow the EXAMPLES.*

Mixed Practice

In Problems 67–112, simplify each expression. All exponents should be positive integers.

67. -5^2

68. 5^{-2}

69. -5^{-2}

70. -5^0

71. $-8^2 \cdot 8^{-2}$

72. $\dfrac{8^7}{8^5} \cdot 8^{-2}$

73. $\left(\dfrac{4}{9}\right)^{-2}$

74. $\left(\dfrac{3}{4}\right)^{-3}$

75. $(-3)^2 \cdot (-3)^{-5}$

76. $(-4)^{-5} \cdot (-4)^3$

77. $\dfrac{(-4)^2}{(-4)^{-1}}$

78. $\dfrac{(-3)^3}{(-3)^{-2}}$

79. $\dfrac{2^3 \cdot 3^{-2}}{2^{-2} \cdot 3^{-4}}$

80. $\dfrac{3^{-2} \cdot 5^3}{3^2 \cdot 5}$

81. $(6x)^3 (6x)^{-3}$

82. $(5a^2)^5 (5a^2)^{-5}$

83. $(2s^{-2}t^4)(-5s^2t)$

84. $(6ab) \cdot (3a^3b^{-4})$

85. $\left(\dfrac{1}{4}xy\right) \cdot (20xy^{-2})$

86. $(3xy^3) \cdot \left(\dfrac{1}{9}x^2y\right)$

87. $\dfrac{36x^7y^3}{9x^5y^2}$

88. $\dfrac{25a^2b^3}{5ab^6}$

89. $\dfrac{21a^2b}{14a^3b^{-2}}$

90. $\dfrac{25x^{-2}y}{10xy^3}$

91. $(x^{-2})^4$

92. $(z^2)^{-6}$

93. $(3x^2y)^3$

94. $(5a^2b^{-1})^2$

95. $\left(\dfrac{z}{4}\right)^{-3}$

96. $\left(\dfrac{x}{y}\right)^{-8}$

97. $(3a^{-3})^{-2}$

98. $(2y^{-2})^{-4}$

99. $(-2a^2b^3)^{-4}$

100. $(-4a^{-2}b^2)^{-2}$

101. $\dfrac{2^3 \cdot xy^{-2}}{12(x^2)^{-2}y}$

102. $\dfrac{3^2 \cdot x^{-3}(y^2)^3}{15x^2y^8}$

103. $\left(\dfrac{15a^2b^3}{3a^{-4}b^5}\right)^{-2}$

104. $\left(\dfrac{15x^4y^7}{18x^{-3}y}\right)^{-1}$

105. $(4x^4y^{-2})^{-1} \cdot (2x^2y^{-1})^2$

106. $(9a^2b^{-4})^{-1} \cdot (3ab^{-2})^2$

107. $\dfrac{(-2)^2x^3(yz)^2}{-4xy^{-2}z}$

108. $\dfrac{(-3)^3a^3(ab)^{-2}}{9ab^4}$

109. $\dfrac{(3x^{-1}yz^2)^2}{(xy^{-2}z)^3}$

110. $\dfrac{(2ab^2c)^{-1}}{(a^{-1}b^3c^2)^{-2}}$

111. $\dfrac{(6a^3b^{-2})^{-1}}{(2a^{-2}b)^{-2}} \cdot \left(\dfrac{3ab^3}{2a^2b^{-3}}\right)^2$

112. $\left(\dfrac{a^{-3}b^{-1}}{2a^4b^{-2}}\right)^2 \cdot \dfrac{(4a^2b)^2}{(2a^{-2}b)^3}$

In Problems 113–122, perform the indicated operation. Express the solution in scientific notation.

113. $(-5.3 \times 10^{-4}) \cdot (2.8 \times 10^{-3})$

114. $(6.2 \times 10^3) \cdot (-3.8 \times 10^5)$

115. $(4 \times 10^6)^3$

116. $(5 \times 10^8)^2$

117. $\dfrac{5 \times 10^{-6}}{8 \times 10^{-4}}$

118. $\dfrac{1 \times 10^7}{5 \times 10^{-4}}$

119. $\dfrac{(4 \times 10^3) \cdot (6 \times 10^7)}{3 \times 10^4}$

120. $\dfrac{(3 \times 10^9) \cdot (8 \times 10^{-6})}{4 \times 10^4}$

121. $\dfrac{6.2 \times 10^{-3}}{(3.1 \times 10^4) \cdot (2 \times 10^{-7})}$

122. $\dfrac{1.5 \times 10^{13}}{(3 \times 10^4) \cdot (5 \times 10^8)}$

In Problems 123–128, perform the indicated operation by changing to scientific notation first. Express the solution in decimal notation.

123. $(4{,}000{,}000) \cdot (3{,}000{,}000)$

124. $(15{,}000) \cdot (3{,}000{,}000)$

125. $\dfrac{0.00008}{0.002}$

126. $\dfrac{0.00012}{0.0000002}$

127. $\dfrac{(0.000004) \cdot 1{,}600{,}000}{(0.0008) \cdot (0.002)}$

128. $\dfrac{(0.0001) \cdot (3{,}500{,}000)}{(0.0005) \cdot (1{,}400{,}000)}$

Applying the Concepts

△ **129. Cubes** Suppose the length of a side of a cube is x^2. Find the volume of the cube in terms of x.

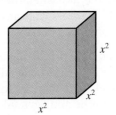

△ **130. Circles** The radius r of a circle is $\dfrac{d}{2}$, where d is the diameter. The area of a circle is given by the formula $A = \pi r^2$. Find the area of a circle in terms of its diameter d.

131. Diameter of a Plant Cell The diameter of a plant cell is 0.00001276 meter. Express this number in scientific notation.

132. Diameter of an Atom The diameter of an atom is about 0.0000000001 meter. Express this diameter as a number in scientific notation.

133. Diameter of Earth The diameter of Earth is 12,760,000 meters. Express this number in scientific notation.

134. National Debt As of March 2012, the debt of the United States federal government was \$15,500,000,000,000. Express this number in scientific notation.

135. A Nanometer One nanometer is 1×10^{-9} meter. Express this number in decimal notation.

136. A Light-Year One light-year is 1×10^{16} meters. Express this number in decimal notation.

137. **Rubik's Cube** The Rubik's Cube has 4×10^{19} possible states. Express this number in decimal notation.

138. **Hair Growth** Human hair grows at a rate of 1×10^{-8} mile per hour. Express this number in decimal notation. What is the growth rate in miles per year?

139. **Per Capita GDP** The per capita gross domestic product (GDP) is the GDP divided by the United States population. It represents the average output of each resident of the United States and is measured in dollars per person. In 2011, the gross domestic product of the United States was $\$1.53 \times 10^{13}$. In 2011, the United States population was 3.1×10^{8} people. Determine the per capita GDP of the United States in 2011. Express your answer in decimal notation rounded to the nearest dollar.

140. **Astronomy** The speed of light is 186,000 miles per second. How long does it take a beam of light to reach Earth from the Sun when the Sun is 93,000,000 miles from Earth? Express your answer in seconds, using scientific notation.

141. **Visits to Facebook** In April 2012, Facebook reported that it had 1.75×10^{6} visitors each day. How many visitors to Facebook were there in April 2012? Express the answer in scientific and decimal notation. (SOURCE: *Facebook*)

142. **Google Hits** Reports suggest that Google.com has 2.93×10^{8} hits per day. How many hits does Google.com get in a 30-day month? Express the answer in scientific and decimal notation.

143. **Population Density** Population density is the number of people living per some unit of land. In other words, Population Density $= \dfrac{\text{Population}}{\text{Area of Land}}$. The United States is 3.54×10^{6} square miles. In 2000, the United States population was 2.81×10^{8} people. In 2011, the United States population was 3.11×10^{8} people.

   (a) Determine the population density of the United States in 2000. Interpret the result.
   (b) Determine the population density of the United States in 2011. Interpret the result.
   (c) By how much did the population density increase? What does this result mean?

144. **Fossil Fuels** In 2011, the United States imported 4.2×10^{9} barrels of oil. The United States population was 3.1×10^{8} people in 2011.

   (a) Determine the per capita number of barrels of oil imported into the United States in 2011.

   (b) One barrel of oil contains 42 gallons. Determine the number of gallons of oil imported into the United States in 2011.
   (c) Use the result from part (b) to compute the per capita number of gallons of oil imported into the United States in 2011.

In Problems 145–148, simplify each algebraic expression by rewriting each factor with a common base. (Hint: Consider that $8 = 2^{3}$.)

145. $\dfrac{2^{x+3}}{4}$

146. $\dfrac{3^{2x}}{27}$

147. $3^{x} \cdot 27^{3x+1}$

148. $9^{-x} \cdot 3^{x+1}$

149. If $3^{x} = 5$, what does 3^{4x} equal?

150. If $4^{x} = 6$, what does 4^{5x} equal?

151. If $2^{x} = 7$, what does 2^{-4x} equal?

152. If $5^{x} = 3$, what does 5^{-3x} equal?

Explaining the Concepts

153. A friend of yours has a homework problem in which he must simplify $\left(x^{4}\right)^{3}$. He tells you that he thinks the answer is x^{7}. Is he right? If not, explain where he went wrong.

154. A friend of yours is convinced that x^{0} must equal 0. Write an explanation that details why $x^{0} = 1$. Include any restrictions that must be placed on x.

155. Explain why a cannot be 0 when m, n, or $m + n$ is negative or 0 in the expression a^{m+n}. Use examples to support your explanation.

156. Explain why a cannot be 0 when n is negative or 0 in the expression $\left(a^{m}\right)^{n}$. Use examples to support your explanation.

157. Explain why neither a nor b can be 0 when n is 0 or negative in the expression $(a \cdot b)^{n} = a^{n} \cdot b^{n}$.

158. Provide a justification for the Product Rule for Exponents.

159. Provide a justification for the Quotient Rule for Exponents.

160. Provide a justification for the Power Rule for Exponents.

161. Provide a justification for the Product-to-a-Power Rule for Exponents.

162. Explain how to convert a decimal number to scientific notation. Explain how to convert a number in scientific notation to decimal notation.

163. Explain the benefits of using scientific notation to multiply or divide very large or very small numbers.

4 Polynomials and Polynomial Functions

Did you know that your late 40s and early 50s are your so-called "peak earning years"? In fact, economists develop models that can be used to describe your earnings over your lifetime. Not surprisingly, your earnings will rise from your early 20s to your mid-40s. Then, from your mid-50s and beyond, your earnings start to decrease. See Problem 104 in Section 4.1.

The Big Picture: Putting It Together

In Chapter 2, we discussed functions and, in particular, linear functions. We learned about properties of linear functions and how to graph them.

We are now going to generalize our discussion of functions a little more. Linear functions belong to a class of functions called *polynomial functions.* In this chapter, we learn how to add, subtract, multiply, and divide polynomials and polynomial functions. We then learn about factoring techniques, which enables us to "undo" polynomial multiplication. You will find that the Distributive Property comes in handy in many of these operations. Finally, we learn how to use factoring to solve equations that have polynomial expressions.

Outline

4.1 Adding and Subtracting Polynomials

Objectives

1. Define Monomial and Determine the Coefficient and Degree of a Monomial
2. Define Polynomial and Determine the Degree of a Polynomial
3. Simplify Polynomials by Combining Like Terms
4. Evaluate Polynomial Functions
5. Add and Subtract Polynomial Functions

Are You Ready for This Section?

Before getting started, take the following readiness quiz. If you get a problem wrong, go back to the section cited and review the material.

R1. What is the coefficient of $-4x^5$?   [Section R.5, p. 41]

R2. Combine like terms: $5x^2 - 3x + 1 - 2x^2 - 6x + 3$ [Section R.5, pp. 41–43]

R3. Use the Distributive Property to remove the parentheses: $-4(x - 3)$ [Section R.3, pp. 29–30]

R4. Given $f(x) = -4x + 3$, find $f(3)$.   [Section 2.2, pp. 156–158]

▶ **1 Define Monomial and Determine the Coefficient and Degree of a Monomial**

Recall from Section R.5 that a *term* is a number or the product of a number and one or more variables raised to a power. The numerical factor of a term is the *coefficient*. For example, consider Table 1, where some algebraic expressions are given and their terms are identified.

Table 1

Algebraic Expression	Terms
$5x + 4$	$5x, 4$
$7x^2 - 8x + 3 = 7x^2 + (-8x) + 3$	$7x^2, -8x, 3$
$3x^2 + 7y^{-1}$	$3x^2, 7y^{-1}$

In this chapter, we study *polynomials*. Polynomials have terms that are *monomials*.

Work Smart

The nonnegative integers are $0, 1, 2, 3, \ldots$.

Definition

A **monomial in one variable** is the product of a constant and a variable raised to a nonnegative integer power. A monomial in one variable is of the form

$$ax^k$$

where a is a constant, x is a variable, and $k \geq 0$ is an integer. The constant a is the **coefficient** of the monomial. If $a \neq 0$, then k is the **degree** of the monomial.

What if $a = 0$? Since $0 = 0x = 0x^2 = 0x^3 = \ldots$, we cannot assign a degree to 0. Therefore, we say 0 has no degree.

EXAMPLE 1   **Identifying the Coefficient and Degree of Monomials**

	MONOMIAL	COEFFICIENT	DEGREE
(a)	$5x^3$	5	3
(b)	$-\dfrac{2}{3}x^6$	$-\dfrac{2}{3}$	6
(c)	$8 = 8x^0$	8	0
(d)	$x^2 = 1x^2$	1	2
(e)	$-x = -1 \cdot x$	-1	1
(f)	0	0	no degree

Now let's look at some expressions that are not monomials.

EXAMPLE 2   **Expressions That Are Not Monomials**

(a) $4x^{\frac{1}{2}}$ is not a monomial because the exponent of the variable x is $\dfrac{1}{2}$, and $\dfrac{1}{2}$ is not a nonnegative integer.

(b) $5x^{-3}$ is not a monomial because the exponent of the variable x is -3, and -3 is not a nonnegative integer.

Quick ✓

1. A _____ in one variable is the product of a number and a variable raised to a nonnegative integer power.

In Problems 2–5, determine whether the expression is a monomial. For those that are monomials, name the coefficient and give the degree.

2. $8x^5$ 3. $5x^{-2}$

4. 12 5. $x^{\frac{1}{3}}$

A monomial may contain more than one variable factor. For example, the monomial $ax^m y^n$ has two variables, x and y, where m and n are nonnegative integers. The **degree of the monomial** $ax^m y^n$ is the sum of the exponents, $m + n$.

EXAMPLE 3   **Monomials in More than One Variable**

(a) $-4x^3 y^4$ is a monomial in x and y of degree $3 + 4 = 7$. The coefficient is -4.

(b) $10ab^5$ is a monomial in a and b of degree $1 + 5 = 6$. The coefficient is 10.   ●

Quick ✓

6. The degree of a monomial in the form $ax^m y^n$ is _____.

In Problems 7–10, determine whether the expression is a monomial. For those that are monomials, determine the coefficient and degree.

7. $3x^5 y^2$ 8. $-2m^3 n$

9. $4ab^{\frac{1}{2}}$ 10. $-xy$

▶ ❷ Define Polynomial and Determine the Degree of a Polynomial

We begin with a definition.

Definition

A **polynomial** is a monomial or the sum of monomials.

A polynomial is in **standard form** if it is written with the terms in descending order according to degree. The **degree of a polynomial** is the highest degree of all the terms of the polynomial. Remember, the degree of a nonzero constant is 0, and the number 0 has no degree.

EXAMPLE 4   **Examples of Polynomials**

	POLYNOMIAL	DEGREE
(a)	$7x^3 - 2x^2 + 6x + 4$	3
(b)	$3 - 8x + x^2 = x^2 - 8x + 3$	2
(c)	$-7x^4 + 24$	4
(d)	$x^3 y^4 - 3x^3 y^2 + 2x^3 y$	7
(e)	$p^2 q - 8p^3 q^2 + 3 = -8p^3 q^2 + p^2 q + 3$	5
(f)	6	0
(g)	0	No Degree

  ●

EXAMPLE 5   **Is the Algebraic Expression a Polynomial?**

(a) $4x^{-2} - 5x + 1$ is not a polynomial because the exponent on the first term, -2, is negative.

(b) $\dfrac{4}{x^3}$ is not a polynomial because it can be written as $4x^{-3}$, and -3 is less than 0. Remember, the exponents on polynomials must be integers greater than or equal to 0.

(c) $\dfrac{8x^2 + 16}{2}$ is a polynomial of degree 2 because it can be written as $4x^2 + 8$ after dividing 2 into each term in the numerator.

(d) $\dfrac{3xy + 1}{xy - 2}$ is not a polynomial because it is the quotient of two polynomials, and the expression cannot be simplified to a polynomial.

Quick ✓

11. *True or False* The degree of the polynomial $2m^2n + 5mn^3 - \dfrac{2}{3}m^2n^3$ is 5.

In Problems 12–16, determine whether the algebraic expression is a polynomial. For those that are polynomials, determine the degree.

12. $-3x^3 + 7x^2 - x + 5$

13. $5z^{-1} + 3$

14. $\dfrac{x - 1}{x + 1}$

15. $\dfrac{3x^2 - 9x + 27}{3}$

16. $5p^3q - 8pq^2 + pq$

Certain polynomials have special names. A polynomial with exactly one term is a **monomial**; a polynomial that has two monomials that are not like terms is called a **binomial;** and a polynomial that contains three monomials that are not like terms is called a **trinomial.** So

$-14x$ is a polynomial   but more specifically $-14x$ is a monomial
$2x^3 - 5x$ is a polynomial   but more specifically $2x^3 - 5x$ is a binomial
$-x^3 - 4x + 11$ is a polynomial   but more specifically $-x^3 - 4x + 11$ is a trinomial
$3x^2 + 6xy - 2y^2$ is a polynomial   but more specifically $3x^2 + 6xy - 2y^2$ is a trinomial

Work Smart

The prefix "bi" means "two," as in bicycle. The prefix "tri" means "three," as in tricycle. The prefix "poly" means "many."

▶ ❸ Simplify Polynomials by Combining Like Terms

Work Smart

Remember, like terms have the same variable and the same exponent on the variable.

To simplify a polynomial means to perform all indicated operations such as addition, and combine like terms. **To add polynomials, combine their like terms.**

EXAMPLE 6 **Simplifying Polynomials: Addition**

Simplify: $(-4x^3 + 9x^2 + x - 3) + (2x^3 + 6x + 5)$

Solution

We can find the sum using horizontal or vertical addition.

Horizontal Addition: The idea here is to combine like terms.

$$(-4x^3 + 9x^2 + x - 3) + (2x^3 + 6x + 5)$$

Remove parentheses: $= -4x^3 + 9x^2 + x - 3 + 2x^3 + 6x + 5$

Rearrange terms: $= -4x^3 + 2x^3 + 9x^2 + x + 6x - 3 + 5$

Distributive Property: $= (-4 + 2)x^3 + 9x^2 + (1 + 6)x + (-3 + 5)$

Simplify: $= -2x^3 + 9x^2 + 7x + 2$

Vertical Addition: Here, we line up like terms in each polynomial vertically and then add the coefficients.

$$
\begin{array}{ccccc}
x^3 & x^2 & x^1 & x^0 \\
-4x^3 + & 9x^2 + & x & - 3 \\
2x^3 & & + 6x & + 5 \\
\hline
-2x^3 + & 9x^2 + & 7x & + 2
\end{array}
$$

Quick ✓

17. *True or False* $5y^3 + 7y^3 = 12y^6$

In Problems 18–20, simplify by adding the polynomials.

18. $(2x^2 - 3x + 1) + (4x^2 + 5x - 3)$

19. $(5w^4 - 3w^3 + w - 8) + (-2w^4 + w^3 - 7w^2 + 3)$

20. $\left(\dfrac{1}{2}x^2 - \dfrac{4}{3}x + 1\right) + \left(\dfrac{1}{4}x^2 + \dfrac{2}{3}x - 8\right)$

EXAMPLE 7 **Simplifying Polynomials in Two Variables: Addition**

Simplify: $(5a^2b - 3ab + 2ab^2) + (a^2b + 5ab - 4ab^2)$

Solution

Although you may add horizontally or vertically, we present only the horizontal format. The first step is to remove the parentheses.

$$
\begin{aligned}
(5a^2b - 3ab + 2ab^2) + (a^2b + 5ab - 4ab^2) &= 5a^2b - 3ab + 2ab^2 + a^2b + 5ab - 4ab^2 \\
\text{Rearrange terms: } &= 5a^2b + a^2b - 3ab + 5ab + 2ab^2 - 4ab^2 \\
\text{Distributive Property: } &= (5+1)a^2b + (-3+5)ab + (2-4)ab^2 \\
\text{Simplify: } &= 6a^2b + 2ab - 2ab^2
\end{aligned}
$$

Quick ✓

In Problem 21, simplify by adding the polynomials.

21. $(8x^2y + 2x^2y^2 - 7xy^2) + (-3x^2y + 5x^2y^2 + 3xy^2)$

We can subtract polynomials horizontally or vertically as well. Remember,

$$a - b = a + (-b)$$

Thus, to subtract one polynomial from another, we add the opposite of each term in the polynomial following the subtraction sign and then combine like terms.

EXAMPLE 8 **Simplifying Polynomials: Subtraction**

Simplify: $(5z^3 + 3z^2 - 3) - (-2z^3 + 7z^2 - z + 2)$

Solution

Horizontal Subtraction: Recall that $a - b = a + (-1) \cdot b$.

$$
\begin{aligned}
(5z^3 + 3z^2 - 3) - (-2z^3 + 7z^2 - z + 2) &= 5z^3 + 3z^2 - 3 + (-1)(-2z^3 + 7z^2 - z + 2) \\
\text{Distribute the } -1\text{: } &= 5z^3 + 3z^2 - 3 + 2z^3 - 7z^2 + z - 2 \\
\text{Rearrange terms: } &= 5z^3 + 2z^3 + 3z^2 - 7z^2 + z - 3 - 2 \\
\text{Combine like terms: } &= 7z^3 - 4z^2 + z - 5
\end{aligned}
$$

Vertical Subtraction: We line up like terms, change the sign of each coefficient of the second polynomial, and add.

Work Smart
Be careful with vertical subtraction: The sign of every term of the second polynomial must be changed. Vertical subtraction will be used when we divide polynomials.

z^3	z^2	z^1	z^0
$5z^3 +$	$3z^2$		$- 3$
$-(-2z^3 +$	$7z^2 -$	$z +$	$2)$

z^3	z^2	z^1	z^0
$5z^3 +$	$3z^2$		$- 3$
$+ 2z^3 -$	$7z^2 +$	$z -$	2
$7z^3 -$	$4z^2 +$	$z -$	5

Quick ✓

In Problems 22–24, simplify by subtracting the polynomials.

22. $(5x^3 - 6x^2 + x + 9) - (4x^3 + 10x^2 - 6x + 7)$

23. $(8y^3 - 5y^2 + 3y + 1) - (-3y^3 + 6y + 8)$

24. $(8x^2y + 2x^2y^2 - 7xy^2) - (-3x^2y + 5x^2y^2 + 3xy^2)$

▶ ④ Evaluate Polynomial Functions

Up to now, we have discussed only polynomial expressions, such as $4x^3 + x^2 - 7x + 1$. If we write $f(x) = 4x^3 + x^2 - 7x + 1$, then we have a *polynomial function.*

Definition

A **polynomial function** is a function whose rule is a polynomial. The domain of every polynomial function is the set of all real numbers. The **degree** of a polynomial function is the highest degree of all the variable terms.

Recall from Chapter 2 that a linear function is a function of the form $f(x) = mx + b$. Table 2 illustrates some specific types of polynomial functions, their functional forms, specific examples, and their degree.

Table 2

Polynomial	Functional Form	Examples	Degree
Linear	$f(x) = a_1x + a_0$, where a_1 is the slope and a_0 is the y-intercept	$f(x) = 4x + 5$ $f(x) = -6x$ $f(x) = 3$	1 1 0
Quadratic	$f(x) = a_2x^2 + a_1x + a_0, a_2 \neq 0$	$f(x) = 3x^2 - 5x + 1$	2
Cubic	$f(x) = a_3x^3 + a_2x^2 + a_1x + a_0, a_3 \neq 0$	$f(x) = -2x^3 + 4x - 1$	3

To **evaluate** a polynomial function, we substitute the value of the variable and simplify, just as we did in Section 2.2.

EXAMPLE 9   **Evaluating a Polynomial Function**

For the polynomial function $P(x) = 2x^3 - 5x^2 + x - 3$, find

(a) $P(3)$ **(b)** $P(-1)$

Solution

(a) Substitute 3 for x in the expression $2x^3 - 5x^2 + x - 3$ to get

$$P(3) = 2(3)^3 - 5(3)^2 + 3 - 3$$
$$= 2 \cdot 27 - 5 \cdot 9 + 3 - 3$$
$$= 54 - 45 + 3 - 3$$
$$= 9$$

(b) Substitute −1 for x: $P(-1) = 2(-1)^3 - 5(-1)^2 + (-1) - 3$

$$= 2 \cdot (-1) - 5 \cdot 1 + (-1) - 3$$
$$= -2 - 5 - 1 - 3$$
$$= -11$$

Quick ✓

25. Find the following values of the polynomial function $g(x) = -2x^3 + 7x + 1$.

(a) $g(0)$ (b) $g(2)$ (c) $g(-3)$

Polynomial functions can be used to model a variety of situations.

EXAMPLE 10 A Polynomial Model for Higher Education

The percentage of Americans with an advanced degree (more than a bachelor's degree) is a function of age. See Figure 1. The polynomial function

$$D(a) = -0.006a^2 + 0.683a - 6.82$$

approximates the percent D of Americans of age a who have an advanced degree (a degree higher than a bachelor's).

(a) Use the function to estimate the percent of 35-year-olds with an advanced degree.

(b) Use the function to estimate the percent of 65-year-olds with an advanced degree.

Figure 1

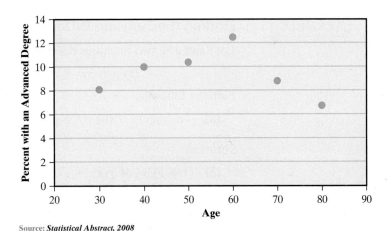

Source: *Statistical Abstract, 2008*

Solution

(a) The independent variable a represents the age, so $a = 35$. We evaluate $D(35)$.

$$D(a) = -0.006a^2 + 0.683a - 6.82$$
$$D(35) = -0.006(35)^2 + 0.683(35) - 6.82$$
$$= -7.35 + 23.905 - 6.82$$
$$= 9.735$$

We estimate that the percent of 35-year-olds with an advanced degree is 9.735%. This looks to be close to the pattern in the scatter diagram.

(b) We evaluate $D(65)$.

$$D(65) = -0.006(65)^2 + 0.683(65) - 6.82$$
$$= -25.35 + 44.395 - 6.82$$
$$= 12.225$$

We estimate that the percent of 65-year-olds with an advanced degree is 12.225%. This also looks to be close to the pattern in the scatter diagram. ●

Quick ✓

26. The polynomial function $B(a) = -1399a^2 + 70{,}573a - 495{,}702$ represents the number of first-time births to women who were a years of age in the United States in 2010.

(a) Use the function to predict the number of first-time births to 25-year-old women.

(b) Use the function to predict the number of first-time births to 40-year-old women.

▶ ❺ **Add and Subtract Polynomial Functions**

Functions, just like numbers, can be added, subtracted, multiplied, and divided. In this section, we concentrate on adding and subtracting polynomial functions.

> **Definition**
>
> If f and g are two functions,
>
> The **sum** $f + g$ is the function defined by
> $$(f + g)(x) = f(x) + g(x)$$
> The **difference** $f - g$ is the function defined by
> $$(f - g)(x) = f(x) - g(x)$$

(EXAMPLE 11) **Finding the Sum and Difference of Two Polynomial Functions**

Let f and g be two functions defined as

$$f(x) = 3x^2 - x + 1 \quad \text{and} \quad g(x) = -x^2 + 5x - 6$$

Find the following:

(a) $(f + g)(x)$ (b) $(f - g)(x)$ (c) $(f + g)(2)$ (d) $(f - g)(-1)$

Solution

(a) $(f + g)(x) = f(x) + g(x)$
$$= (3x^2 - x + 1) + (-x^2 + 5x - 6)$$
$$= 3x^2 - x + 1 - x^2 + 5x - 6$$
$$= 3x^2 - x^2 - x + 5x + 1 - 6$$
$$= 2x^2 + 4x - 5$$

(b) $(f - g)(x) = f(x) - g(x)$
$$= (3x^2 - x + 1) - (-x^2 + 5x - 6)$$
$$= 3x^2 - x + 1 + x^2 - 5x + 6$$
$$= 3x^2 + x^2 - x - 5x + 1 + 6$$
$$= 4x^2 - 6x + 7$$

(c) Because $(f + g)(x) = 2x^2 + 4x - 5$, we have
$$(f + g)(2) = 2(2)^2 + 4(2) - 5$$
$$= 8 + 8 - 5$$
$$= 11$$

We could also have evaluated $(f + g)(2)$ as follows:
$$(f + g)(2) = f(2) + g(2)$$
$$= 11 + 0$$
$$= 11$$

(d) Because $(f - g)(x) = 4x^2 - 6x + 7$, we have
$$(f - g)(-1) = 4(-1)^2 - 6(-1) + 7$$
$$= 4 + 6 + 7$$
$$= 17$$

Quick ✓

27. Let f and g be two functions defined as

$$f(x) = 3x^2 - x + 1 \quad \text{and} \quad g(x) = -x^2 + 5x - 6$$

Find the following:

(a) $(f + g)(x)$ (c) $(f + g)(1)$

(b) $(f - g)(x)$ (d) $(f - g)(-2)$

Let's look at an application of the difference of two functions.

EXAMPLE 12   **The Profit Function**

Profit is defined as total revenue minus total cost. The profit function of a company that manufactures and sells x units of a product is given by

$$P(x) = R(x) - C(x)$$

where P represents the company's profits

       R represents the company's revenue

       C represents the company's cost

(a) If a company sells a scientific calculator for \$12, its revenue function is $R(x) = 12x$. If the cost of each calculator manufactured is \$7 and fixed costs are \$1200 per week, its cost function is given by $C(x) = 7x + 1200$. Find the company's profit function, $P(x)$.

(b) Determine and interpret $P(800)$.

Solution

(a) The profit function is

$$\begin{aligned} P(x) &= R(x) - C(x) \\ &= 12x - (7x + 1200) \\ &= 12x - 7x - 1200 \\ &= 5x - 1200 \end{aligned}$$

(b) $\begin{aligned} P(800) &= 5(800) - 1200 \\ &= 4000 - 1200 \\ &= 2800 \end{aligned}$

If the company manufactures and sells 800 calculators in a week, its profit will be \$2800.

Quick ✓

28. The calculator company presented in Example 12 just gave its employees a raise that increases the cost of each calculator manufactured to \$8. In addition, it renewed its lease on the plant, so its weekly fixed costs increased to \$1250.

(a) Determine the new profit function.

(b) Determine and interpret $P(800)$.

4.1 Exercises MyMathLab® Exercise numbers in green have complete video solutions in MyMathLab.

*Problems **1–28** are the Quick✓ s that follow the EXAMPLES.*

Building Skills

In Problems 29–36, determine the coefficient and degree of each monomial. See Objective 1.

29. $3x^2$

30. $5x^4$

31. $-8x^2y^3$

32. $-12xy$

33. $\dfrac{4}{3}x^6$

34. $-\dfrac{5}{3}z^5$

35. 2

36. -7

In Problems 37–40, state why each of the following is not a polynomial. See Objective 2.

37. $2x^{-1} + 3x$

38. $6p^{-3} - p^{-2} + 3p^{-1}$

39. $\dfrac{4}{z-1}$

40. $\dfrac{x^2 + 2}{x}$

In Problems 41–56, determine whether the algebraic expression is a polynomial (Yes or No). If it is a polynomial, write the polynomial in standard form, determine the degree, and state whether it is a monomial, a binomial, or a trinomial. If it is a polynomial with more than three terms, say the expression is a polynomial. See Objective 2.

41. $5x^2 - 9x + 1$

42. $-3y^2 + 8y + 1$

43. $\dfrac{-20}{n}$

44. $\dfrac{1}{x}$

45. $3y^{\frac{1}{3}} + 2$

46. $8m - 4m^{\frac{1}{2}}$

47. $\dfrac{5}{8}$

48. -12

49. $5 - 8y + 2y^2$

50. $7 - 5p + 2p^2 - p^3$

51. $7x^{-1} + 4$

52. $4y^{-2} + 6y - 1$

53. $3x^2y^2 + 2xy^4 + 4$

54. $4mn^3 - 2m^2n^3 + mn^8$

55. $4pqr + 2p^2q + 3pq^{\frac{1}{4}}$

56. $-2xyz^2 + 7x^3z - 8y^{\frac{1}{2}}z$

In Problems 57–80, simplify each polynomial by adding or subtracting, as indicated. Express your answer as a single polynomial in standard form. See Objective 3.

57. $5z^3 + 8z^3$

58. $10y^4 - 6y^4$

59. $(x^2 + 5x + 1) + (3x^2 - 2x - 3)$

60. $(x^2 - 4x + 1) + (5x^2 + 2x + 7)$

61. $(6p^3 - p^2 + 3p - 4) + (2p^3 - 7p + 3)$

62. $(2w^3 - w^2 + 6w - 5) + (-3w^3 + 5w^2 + 9)$

63. $\begin{array}{r} 8x^3 + 4x^2 - 3x + 1 \\ + \ -x^3 - 2x^2 - 3x + 7 \\ \hline \end{array}$ **64.** $\begin{array}{r} -7n^3 + 2n^2 - 5n - 3 \\ + \ 2n^3 - 5n^2 + n + 1 \\ \hline \end{array}$

65. $(5x^2 + 9x + 4) - (3x^2 + 5x + 1)$

66. $(7y^2 + 9y + 12) - (4y^2 + 8y - 3)$

67. $(7s^2t^3 + st^2 - 5t - 8) - (4s^2t^3 + 5st^2 - 7)$

68. $(-2x^3y^3 + 7xy - 3) - (x^3y^3 + 5y^2 + xy - 3)$

69. $(3 - 5x + x^2) + (-2 + 3x - 5x^2)$

70. $(-3 - 5z + 3z^2) + (1 + 2z + z^2)$

71. $(6 - 2y + y^3) - (-2 + y^2 - 2y^3)$

72. $(8 - t^3) - (1 + 3t + 3t^2 + t^3)$

73. $\left(\dfrac{1}{4}x^2 + \dfrac{3}{2}x + 3\right) + \left(\dfrac{1}{2}x^2 - \dfrac{1}{4}x - 2\right)$

74. $\left(\dfrac{3}{4}y^3 - \dfrac{1}{8}y + \dfrac{2}{3}\right) + \left(\dfrac{1}{2}y^3 + \dfrac{5}{12}y - \dfrac{5}{6}\right)$

75. $(5x^2y^2 - 8x^2y + xy^2) + (3x^2y^2 + x^2y - 4xy^2)$

76. $(7a^3b + 9ab^2 - 4a^2b) + (-4a^3b + 3a^2b - 8ab^2)$

77. $(3x^2y + 7xy^2 + xy) - (2x^2y - 4xy^2 - xy)$

78. $(-5xy^2 + 3xy - 9y^2) - (5xy^2 + 7xy - 8y^2)$

79. $\begin{array}{r} 9a^3 + 2a^2 - 5a - 8 \\ - \ 5a^3 - 2a^2 + a - 6 \\ \hline \end{array}$ **80.** $\begin{array}{r} -11p^3 + 8p^2 - p - 7 \\ - \ -p^3 + 5p^2 + p + 1 \\ \hline \end{array}$

Applying the Concepts

81. Add $5x^3 - 5x + 3$ to $-4x^3 + x^2 - 2x + 1$.

82. Add $2x^3 - 3x^2 - 5x + 7$ to $x^3 + 3x^2 - 6x - 4$.

83. Subtract $4b^3 - b^2 + 3b - 1$ from $2b^3 + 5b^2 - b + 3$.

84. Subtract $2q^3 - 3q^2 + 7q - 2$ from $-5q^3 + q^2 + 2q - 1$.

In Problems 85–88, simplify each of the following.

85. $(2x^2 - 3x + 1) + (x^2 + 9) - (4x^2 - 2x - 5)$

86. $(y^2 + 6y - 2) + (4y^2 - 9) - (2y^2 - 7y - 10)$

87. $(4n - 3) + (n^3 - 9) - (2n^2 - 7n + 3)$

88. $(y^3 - 1) + (y^2 - 9) - (y^3 + 2y^2 - 7y + 3)$

In Problems 89–94, find the following values for each polynomial function. See Objective 4.

   (a) $f(0)$ **(b)** $f(2)$ **(c)** $f(-3)$

89. $f(x) = x^2 - 4x + 1$ **90.** $f(x) = x^2 + 5x - 3$

91. $f(x) = 2x^3 - 7x + 3$ **92.** $f(x) = -2x^3 + 3x - 1$

93. $f(x) = -x^3 + 3x^2 - 2x + 3$

94. $f(x) = -2x^3 + x^2 + 5x - 3$

In Problems 95–100, for the given functions f and g, find the following. See Objective 5.

   (a) $(f + g)(x)$ **(b)** $(f - g)(x)$
   (c) $(f + g)(2)$ **(d)** $(f - g)(1)$

95. $f(x) = 2x + 5; g(x) = -5x + 1$

96. $f(x) = 4x + 3; g(x) = 2x - 3$

97. $f(x) = x^2 - 5x + 3; g(x) = 2x^2 + 3$

98. $f(x) = 3x^2 + x + 2; g(x) = x^2 - 3x - 1$

99. $f(x) = x^3 + 6x^2 + 12x + 2; g(x) = x^3 - 8$

100. $f(x) = 8x^3 + 1; g(x) = x^3 + 3x^2 + 3x + 1$

△ **101. Area** A rectangle has one corner on the graph of $y = 6 - 2x$, another at the origin, a third on the positive *y*-axis, and the fourth on the positive *x*-axis (see the figure).

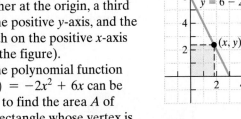

The polynomial function $A(x) = -2x^2 + 6x$ can be used to find the area A of the rectangle whose vertex is at (x, y).

  (a) Find the area of the rectangle whose vertex is at $(2, 2)$.

  (b) Find the area of the rectangle whose vertex is at $(1, 4)$.

  (c) How can the coordinates of the vertex be used to find the area? What is the area?

△ **102. Area** A rectangle has one corner on the graph of $y = 10 - 2x$, another at the origin, a third on the positive *y*-axis, and the fourth on the positive *x*-axis (see the figure).

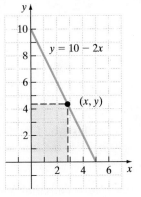

The polynomial function $A(x) = -2x^2 + 10x$ can be used to find the area A of the rectangle whose vertex is at (x, y).

  (a) Find the area of the rectangle whose vertex is at $(3, 4)$.

  (b) Find the area of the rectangle whose vertex is at $(4, 2)$.

---

103. Price per Square Foot of a New Home The bar graph shown represents the average price per square foot for a new single-family home in the United States for the years 1995–2010.

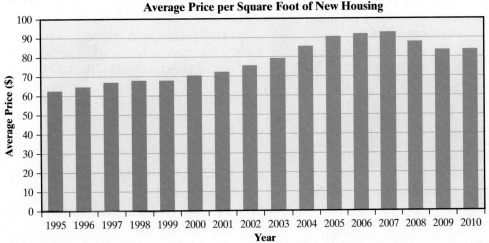

SOURCE: *United States Census Bureau*

The polynomial function $P(t) = -0.043t^3 + 0.879t^2 - 2.184t + 65.100$ can be used to approximate the average price per square foot P, where t is the number of years since 1995.

(a) Use the function to estimate the average price per square foot in 1995.

(b) Use the function to predict the average price per square foot in 2015. Does this result seem reasonable?

104. Income The bar graph shown represents the average per-capita income for residents of the United States by age in 2009.

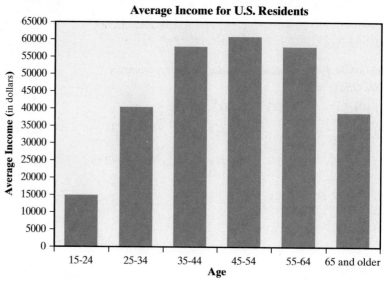

Average Income for U.S. Residents

SOURCE: *Statistical Abstract*, 2012

The polynomial function $I(a) = -54.42a^2 + 4853.38a - 46{,}106.66$ can be used to approximate the average income I, where a is the age of the individual.

(a) Use the function to estimate the average income of a 20-year-old in 2009.

(b) Use the function to estimate the average income of a 55-year-old in 2009.

---

105. Profit Function Suppose that the revenue R from selling x cell phones is $R(x) = -1.2x^2 + 220x$. The cost C of selling x cell phones is $C(x) = 0.05x^3 - 2x^2 + 65x + 500$.

(a) Find the profit function, $P(x)$.

(b) Find the profit if $x = 15$ cell phones are sold.

(c) Determine and interpret $P(100)$.

106. Profit Function Suppose that the revenue R from selling x clocks is $R(x) = -0.3x^2 + 30x$. The cost C of selling x clocks is $C(x) = 0.1x^2 + 7x + 400$.

(a) Find the profit function, $P(x)$.

(b) Find the profit if $x = 15$ clocks are sold.

(c) Determine and interpret $P(40)$.

Extending the Concepts

107. If $f(x) = 3x + 2$ and $g(x) = ax - 5$, find a such that $(f + g)(3) = 12$.

108. If $f(x) = -5x + 1$ and $g(x) = ax - 3$, find a such that $(f - g)(2) = 10$.

109. The graph of two functions, f and g, is shown to the right. Use the graph to answer parts (a)–(d).

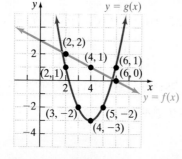

(a) $(f + g)(2)$

(b) $(f + g)(4)$

(c) $(f - g)(6)$

(d) $(g - f)(6)$

110. Work-Related Disability Suppose that $M(a)$ represents the number of American males with a work-related disability who are a years of age. Suppose that $F(a)$ represents the number of American females with a work-related disability who are a years of age. Determine a function T that represents the total number of Americans with work-related disabilities who are a years of age.

111. Taxes Let $T(x)$ represent the total taxes paid by everyone who worked in the United States in year x. Let $F(x)$ represent the taxes paid to the federal government in year x. Write a function S that represents the total taxes paid other than federal (state, property, sales, and so on) in year x.

Explaining the Concepts

112. Explain the difference between a term and a monomial.

113. What is the degree of a polynomial in one variable that is linear?

114. How many terms are in a binomial? How many terms are in a trinomial?

115. Give a definition of *polynomial* using your own words. Provide examples of polynomials that are monomials, binomials, and trinomials. In addition, give examples of polynomials that are linear.

116. What is the difference between a polynomial and a polynomial function?

117. Explain why the degree of the sum of two polynomials is at most the degree of the polynomial of highest degree.

The Graphing Calculator

Graphing calculators can be used to evaluate any function. Figure 2 shows the result obtained in Examples 9(a) and (b) on a TI-84 Plus graphing calculator with the function to be evaluated, $P(x) = 2x^3 - 5x^2 + x - 3$ in Y_1.

Figure 2

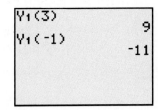

In Problems 118–121, use a graphing calculator to find the following values for each polynomial function.

     (a) $f(4)$ **(b)** $f(-2)$ **(c)** $f(6)$

118. $f(x) = -2x^2 + 5x + 3$

119. $f(x) = 4x^2 - 7x + 1$

120. $f(x) = -3x^3 + 5x^2 - 8x + 1$

121. $f(x) = 2x^3 - 5x^2 + x + 5$

4.2 Multiplying Polynomials

Objectives

1 Multiply a Monomial and a Polynomial

2 Multiply Two Binomials

3 Multiply Two Polynomials

4 Multiply Special Products

5 Multiply Polynomial Functions

Are You Ready for This Section?

Before getting started, take the following readiness quiz. If you get a problem wrong, go back to the section cited and review the material.

R1. Simplify: $4x^2 \cdot 3x^3$ [Getting Ready, p. 293]

R2. Simplify: $(-3x)^2$ [Getting Ready, pp. 297–298]

R3. Use the Distributive Property to remove the parentheses: $4(x - 5)$ [Section R.3, pp. 29–30]

Multiplying polynomials is based on the Product Rule for Exponents (Getting Ready, p. 293) and the Distributive Property (Section R.3, p. 29). The following example will help you review these concepts.

EXAMPLE 1 **Review the Product Rule for Exponents and the Distributive Property**

     (a) Simplify: $(2a^3b)(-6a^2b^4)$

     (b) Remove the parentheses: $2(z + 3)$

Solution

     (a)
$$\begin{aligned}(2a^3b)(-6a^2b^4) &= (2 \cdot (-6))(a^3 \cdot a^2)(b \cdot b^4)\\ \text{Product Rule: } a^m \cdot a^n = a^{m+n}: \quad &= -12a^{3+2}b^{1+4}\\ &= -12a^5b^5\end{aligned}$$

     (b) The Distributive Property is used to remove parentheses:

$$\begin{aligned}2(z + 3) &= 2 \cdot z + 2 \cdot 3\\ &= 2z + 6\end{aligned}$$

In Problems 1–4, simplify each expression completely. All exponents should be positive.

1. $(3x^5)(2x^2)$

2. $(-7a^3b^2)(3ab^4)$

3. $\left(\frac{2}{3}x^4\right)\left(\frac{15}{8}x\right)$

4. $-3(x + 2)$

▶ ❶ Multiply a Monomial and a Polynomial

Example 1(b) shows that when we multiply a binomial by a monomial, we use the Distributive Property to remove the parentheses. In general, when we multiply a polynomial by a monomial, we use the following property:

> **In Words**
> The Extended Form of the Distributive Property says to multiply each term in parentheses by a.

Extended Form of the Distributive Property

$$a(b_1 + b_2 + \cdots + b_n) = a \cdot b_1 + a \cdot b_2 + \cdots + a \cdot b_n$$

where $a, b_1, b_2, \ldots, b_n$ are real numbers.

EXAMPLE 2 **Using the Extended Form of the Distributive Property**

Multiply and simplify each of the following expressions:

(a) $3x^2(x^2 + 4x + 2)$

(b) $\frac{1}{2}xy^3\left(\frac{2}{3}xy^2 + \frac{6}{5}y + \frac{3}{4}\right)$

Solution

(a)
$$3x^2(x^2 + 4x + 2) = 3x^2 \cdot x^2 + 3x^2 \cdot 4x + 3x^2 \cdot 2$$
$$= 3x^4 + 12x^3 + 6x^2$$

(b)
$$\frac{1}{2}xy^3\left(\frac{2}{3}xy^2 + \frac{6}{5}y + \frac{3}{4}\right) = \frac{1}{2}xy^3 \cdot \frac{2}{3}xy^2 + \frac{1}{2}xy^3 \cdot \frac{6}{5}y + \frac{1}{2}xy^3 \cdot \frac{3}{4}$$
$$= \frac{1}{3}x^2y^5 + \frac{3}{5}xy^4 + \frac{3}{8}xy^3$$

In Problems 5–7, multiply and simplify each expression.

5. $5x(x^2 + 3x + 2)$

6. $2xy(3x^2 - 5xy + 2y^2)$

7. $\frac{3}{4}y^2\left(\frac{4}{3}y^2 + \frac{2}{9}y + \frac{16}{3}\right)$

▶ ❷ Multiply Two Binomials

To find the product of two binomials, we distribute the first binomial to each term in the second binomial. Just as with addition and subtraction of polynomials, we can use either the horizontal or the vertical format.

EXAMPLE 3 **Multiplying Two Binomials**

Find the product: $(3x + 4)(2x - 5)$

Solution

VERTICAL MULTIPLICATION

$$
\begin{array}{r}
3x + 4 \\
\times\ 2x - 5 \\
\hline
-15x - 20 \quad \leftarrow -5(3x + 4) \\
6x^2 + 8x \quad \longleftarrow 2x(3x + 4) \\
\hline
6x^2 - 7x - 20
\end{array}
$$

HORIZONTAL MULTIPLICATION

We distribute $3x + 4$ to each term in $2x - 5$.

$(3x + 4)(2x - 5) = (3x + 4)(2x) + (3x + 4)(-5)$

Distribute: $= 3x \cdot 2x + 4 \cdot 2x + 3x \cdot (-5) + 4 \cdot (-5)$

$= 6x^2 + 8x - 15x - 20$

Combine like terms: $= 6x^2 - 7x - 20$

In either case, $(3x + 4)(2x - 5) = 6x^2 - 7x - 20$.

The FOIL Method

To multiply two binomials, we can also use a method known as the **FOIL method.** The acronym FOIL stands for **F**irst, **O**uter, **I**nner, **L**ast and is illustrated below.

First   Last   F   O   I   L

$(ax + b)(cx + d) = ax \cdot cx + ax \cdot d + b \cdot cx + b \cdot d$

Inner

Outer

EXAMPLE 4 **Using the FOIL Method to Multiply Two Binomials**

Find the product:

(a) $(4y - 3)(2y + 5)$ **(b)** $(2m + n)(m - 3n)$

Solution

First   Last   F   O   I   L

(a) $(4y - 3)(2y + 5) = 4y \cdot 2y + 4y \cdot 5 - 3 \cdot 2y - 3 \cdot 5$

Inner

Outer $= 8y^2 + 20y - 6y - 15$

$= 8y^2 + 14y - 15$

F   O   I   L

(b) $(2m + n)(m - 3n) = 2m \cdot m + 2m \cdot (-3n) + n \cdot m + n \cdot (-3n)$

$= 2m^2 - 6mn + mn - 3n^2$

$= 2m^2 - 5mn - 3n^2$

Quick ✓

8. The acronym FOIL stands for ____, ____, ____, ____.

In Problems 9–11, find the product.

9. $(x + 4)(x + 1)$ **10.** $(3v + 5)(2v - 3)$

11. $(2a - b)(a + 5b)$

▶ ❸ **Multiply Two Polynomials**

Work Smart

When multiplying polynomials, it is a good idea to write the polynomials in descending order of degree.

When multiplying a trinomial by a binomial or a trinomial by a trinomial, we make repeated use of the Extended Form of the Distributive Property. The approach is similar to the one used in Example 3. It is a good idea to write each polynomial in standard form. Although you may use either the horizontal or the vertical format, we present only the horizontal format.

EXAMPLE 5 **Multiplying Two Polynomials**

Find the product:

(a) $(3x + 1)(x^2 + 4x - 3)$ **(b)** $(x^2 + 5x + 2)(2x^2 - x + 3)$

Solution

(a) We begin by distributing the binomial $3x + 1$ to each term in the trinomial.

$$(3x + 1)(x^2 + 4x - 3) = (3x + 1) \cdot x^2 + (3x + 1) \cdot 4x + (3x + 1) \cdot (-3)$$

Distributive Property: $= 3x^3 + x^2 + 12x^2 + 4x - 9x - 3$

Combine like terms: $= 3x^3 + 13x^2 - 5x - 3$

(b) Distribute $x^2 + 5x + 2$ to each term in the second trinomial.

$$(x^2 + 5x + 2)(2x^2 - x + 3) = (x^2 + 5x + 2) \cdot 2x^2 + (x^2 + 5x + 2) \cdot (-x)$$
$$+ (x^2 + 5x + 2) \cdot 3$$

Distributive Property: $= x^2 \cdot 2x^2 + 5x \cdot 2x^2 + 2 \cdot 2x^2 + x^2 \cdot (-x)$
$$+ 5x \cdot (-x) + 2 \cdot (-x) + x^2 \cdot 3 + 5x \cdot 3 + 2 \cdot 3$$
$$= 2x^4 + 10x^3 + 4x^2 - x^3 - 5x^2 - 2x + 3x^2$$
$$+ 15x + 6$$

Combine like terms: $= 2x^4 + 9x^3 + 2x^2 + 13x + 6$ ●

Quick ✓

In Problems 12 and 13, find the product.

12. $(2y - 3)(y^2 + 4y + 5)$ **13.** $(z^2 - 3z + 2)(2z^2 + z + 6)$

▶ ❹ **Multiply Special Products**

Certain binomials have products that result in patterns. We call these products *special products*.

EXAMPLE 6 **Products of the Form $(A - B)(A + B)$**

Find the product: $(x - 7)(x + 7)$

Solution

We use FOIL and obtain

$$\begin{matrix} & \text{F} & \text{O} & \text{I} & \text{L} \\ (x - 7)(x + 7) = & x \cdot x & + x \cdot 7 & - 7 \cdot x & - 7 \cdot 7 \end{matrix}$$
$$= x^2 + 7x - 7x - 49$$
$$= x^2 - 49$$ ●

Work Smart

When we multiply products of the form $(A - B)(A + B)$, the "middle terms" will always be "opposites" and therefore sum to 0. Also, don't forget multiplication is commutative. Therefore, $(A + B)(A - B)$
$= (A - B)(A + B) = A^2 - B^2$

We have the following rule based on the results of Example 6.

Difference of Two Squares

$$(A - B)(A + B) = A^2 - B^2$$

EXAMPLE 7 **Using the Difference of Two Squares Formula**

(a) $(A + B)(A - B) = A^2 - B^2$

$$(3x + 2)(3x - 2) = (3x)^2 - 2^2$$
$$= 9x^2 - 4$$

(b)
$$A = 2m \text{ and } B = 5n^2$$
$$\downarrow$$

$$(2m - 5n^2)(2m + 5n^2) = (2m)^2 - (5n^2)^2$$
$$= 4m^2 - 25n^4$$

Quick ✓

14. *True or False* The product of a binomial and a binomial is always a trinomial.

15. $(A - B)(A + B) =$ _____.

In Problems 16 and 17, find each product.

16. $(5y + 2)(5y - 2)$ **17.** $(7y + 2z^3)(7y - 2z^3)$

(EXAMPLE 8) **Products of the Form $(A + B)^2$ or $(A - B)^2$**

Find each product:

(a) $(4x + 3)^2$ (b) $(3z - 5)^2$

Solution

(a) $(4x + 3)^2 = (4x + 3)(4x + 3)$
$$= (4x)^2 + \underbrace{(4x)(3) + (4x)(3)}_{= 2(4x)(3)} + 3^2$$
$$= 16x^2 + 24x + 9$$

Work Smart

$(A + B)^2 \neq A^2 + B^2$
$(A - B)^2 \neq A^2 - B^2$

Whenever you feel the urge to perform an operation that you're not quite sure about, try it with actual numbers. For example, does
$$(3 + 2)^2 = 3^2 + 2^2?$$
NO! So
$$(A + B)^2 \neq A^2 + B^2$$

(b) $(3z - 5)^2 = (3z - 5)(3z - 5)$
$$= (3z)^2 \underbrace{- (3z)(5) - (3z)(5)}_{= -2(3z)(5)} + 5^2$$
$$= 9z^2 - 30z + 25$$

Example 8 leads to some general results.

> **Squares of Binomials, or Perfect Square Trinomials**
> $$(A + B)^2 = A^2 + 2AB + B^2$$
> $$(A - B)^2 = A^2 - 2AB + B^2$$

(EXAMPLE 9) **Using the Perfect Square Trinomial Formulas**

$$(A + B)^2 = A^2 + 2AB + B^2$$
$$\downarrow$$

(a) $(w + 5)^2 = w^2 + 2 \cdot w \cdot 5 + 5^2$
$$= w^2 + 10w + 25$$

$$(A - B)^2 = A^2 - 2AB + B^2$$
$$\downarrow$$

(b) $(6p - 5)^2 = (6p)^2 - 2 \cdot 6p \cdot 5 + 5^2$
$$= 36p^2 - 60p + 25 \quad ,$$

(c) $(3x + 5y^2)^2 = (3x)^2 + 2 \cdot 3x \cdot 5y^2 + (5y^2)^2$
$$= 9x^2 + 30xy^2 + 25y^4$$

Work Smart

If you can't remember the formulas for a perfect square, don't panic! Use the fact that

$$(A + B)^2 = (A + B)(A + B)$$

and then FOIL. Use the same reasoning with perfect squares of the form $(A - B)^2$.

Quick ✓

18. $(A - B)^2 = $ _____; $(A + B)^2 = $ _____

19. $x^2 + 2xy + y^2$ is referred to as a _____ _____ trinomial.

20. *True or False* $(x + a)^2 = x^2 + a^2$

In Problems 21–23, find each product.

21. $(z - 8)^2$ **22.** $(6p + 5)^2$ **23.** $(4a - 3b)^2$

▶ **❺ Multiply Polynomial Functions**

In Section 4.1, we added and subtracted functions. Now we multiply functions.

Definition

Let f and g be two functions. The **product** $f \cdot g$ is the function defined by

$$(f \cdot g)(x) = f(x) \cdot g(x)$$

EXAMPLE 10 **Finding the Product of Two Functions**

Suppose that $f(x) = 2x + 5$ and $g(x) = x^2 - 7x + 5$.

(a) Find $f(4) \cdot g(4)$.

(b) Find $(f \cdot g)(x)$.

(c) Use the result from part (b) to determine $(f \cdot g)(4)$.

Solution

(a) We will evaluate the functions at $x = 4$ separately.

$$
\begin{aligned}
f(4) &= 2(4) + 5 & g(4) &= (4)^2 - 7(4) + 5 \\
&= 13 & &= 16 - 28 + 5 \\
& & &= -7
\end{aligned}
$$

$$f(4) \cdot g(4) = (13)(-7) = -91$$

(b)
$$(f \cdot g)(x) = f(x) \cdot g(x)$$

$$= (2x + 5)(x^2 - 7x + 5)$$

Distribute:
$$= (2x + 5)x^2 + (2x + 5)(-7x) + (2x + 5) \cdot 5$$

Distribute:
$$= 2x \cdot x^2 + 5x^2 + (2x)(-7x) + 5(-7x) + 2x \cdot 5 + 5 \cdot 5$$

Simplify:
$$= 2x^3 + 5x^2 - 14x^2 - 35x + 10x + 25$$

Combine like terms:
$$= 2x^3 - 9x^2 - 25x + 25$$

(c) $(f \cdot g)(4) = 2(4)^3 - 9(4)^2 - 25(4) + 25$

$$= 2(64) - 9(16) - 100 + 25$$

$$= 128 - 144 - 100 + 25$$

$$= -91$$

Notice that $(f \cdot g)(4) = f(4) \cdot g(4)$, as we would expect.

Quick ✓

24. $(f \cdot g)(x) = $ ____ $\cdot$ ____.

25. Suppose that $f(x) = 5x - 3$ and $g(x) = x^2 + 3x + 1$.

(a) Find $f(2) \cdot g(2)$.

(b) Find $(f \cdot g)(x)$.

(c) Use the result from part (b) to determine $(f \cdot g)(2)$.

Back in Section 2.2, we learned to evaluate functions when the argument (the value at which we are evaluating the function) was an algebraic expression.

EXAMPLE 11 **Evaluating Functions**

For the function $f(x) = x^2 + 5x$, find

(a) $f(x + 3)$ **(b)** $f(x + h) - f(x)$

Solution

$$f(\text{input}) = (\text{input})^2 + 5(\text{input})$$

(a) $f(x + 3) = (x + 3)^2 + 5(x + 3)$
$$= x^2 + 6x + 9 + 5x + 15$$
$$= x^2 + 11x + 24$$

(b) To evaluate $f(x + h) - f(x)$, we first evaluate $f(x + h)$ by replacing x in $f(x)$ with $x + h$. We then subtract $f(x)$ from this result.

$$f(x + h) - f(x) = \overbrace{(x + h)^2 + 5(x + h)}^{f(x + h)} - \overbrace{(x^2 + 5x)}^{f(x)}$$
$$= x^2 + 2xh + h^2 + 5x + 5h - x^2 - 5x$$

Combine like terms: $= 2xh + h^2 + 5h$

Quick ✓

26. For the function $f(x) = x^2 - 2x$, find

(a) $f(x - 3)$

(b) $f(x + h) - f(x)$

4.2 Exercises   MyMathLab®   Math**XL**   Exercise numbers in green have complete video solutions in MyMathLab.
PRACTICE

Problems 1–26 are the Quick✓s that follow the EXAMPLES.

Building Skills

In Problems 27–38, find the product. See Objective 1.

27. $(5xy^2)(-3x^2y^3)$ **28.** $(9a^3b^2)(-3a^2b^5)$

29. $\left(\frac{3}{4}yz^3\right)\left(\frac{20}{9}y^3z^2\right)$ **30.** $\left(\frac{12}{5}x^2y\right)\left(\frac{15}{4}x^4y^3\right)$

31. $5x(x^2 + 4x + 2)$ **32.** $6y(y^2 - 4y + 3)$

33. $-4a^2b(3a^2 + 2ab - b^2)$

34. $-3mn^3(4m^2 - mn + 5n^2)$

35. $\frac{2}{3}ab\left(\frac{3}{4}a^2b - \frac{9}{8}ab^3 + 6ab\right)$

36. $\frac{5}{2}xy\left(\frac{4}{15}x^2y - \frac{6}{5}xy + \frac{3}{10}xy^2\right)$

37. $0.4x^2(1.2x^2 - 0.8x + 1.5)$

38. $0.8y(0.4y^2 + 1.1y - 2.5)$

In Problems 39–52, find the product of the two binomials.
See Objective 2.

39. $(x + 3)(x + 5)$

40. $(y - 2)(y - 6)$

41. $(a + 5)(a - 3)$

42. $(z - 8)(z + 3)$

43. $(4a + 3)(3a - 1)$

44. $(5x - 3)(x + 4)$

45. $(4 - 5x)(3 + 2x)$

46. $(2 - 7y)(5 + 2y)$

47. $\left(\frac{2}{3}x + 2\right)\left(\frac{1}{2}x - 4\right)$

48. $\left(\frac{3}{2}y + 4\right)\left(\frac{4}{3}y - 1\right)$

49. $(4a + 3b)(a - 5b)$

50. $(3m - 5n)(m + 2n)$

51. $(2x^2 + 1)(x^2 - 3)$

52. $(3x^2 - 5)(x^2 + 2)$

In Problems 53–68, find the product of the polynomials.
See Objective 3.

53. $(x + 1)(x^2 + 4x + 2)$

54. $(y - 2)(y^2 + 5y - 3)$

55. $(3a - 2)(2a^2 + a - 5)$

56. $(2b + 3)(3b^2 - 2b + 1)$

57. $(5z^2 + 3z + 2)(4z + 3)$

58. $(3p^2 - 5p + 3)(7p - 2)$

59. $-\frac{1}{2}x(2x + 6)(x - 3)$

60. $-\frac{4}{3}k(k + 7)(3k - 9)$

61. $(4 + y)(2y^2 - 3 + 5y)$

62. $(3 + 2z)(z^2 + 5 - 3z)$

63. $(w^2 + 2w + 1)(2w^2 - 3w + 1)$

64. $(a^2 + 4a + 4)(3a^2 - a - 2)$

65. $(b + 1)(b - 2)(b + 3)$

66. $(2a - 1)(a + 4)(a + 1)$

67. $(2ab + 5)(4a^2 - 2ab + b^2)$

68. $(xy - 2)(x^2 + 2xy + 4y^2)$

In Problems 69–82, find the special product. See Objective 4.

69. $(x - 6)(x + 6)$

70. $(y + 9)(y - 9)$

71. $(a + 8)^2$

72. $(b + 3)^2$

73. $(3y - 1)^2$

74. $(4z - 5)^2$

75. $(5a + 3b)(5a - 3b)$

76. $(8y + 3z)(8y - 3z)$

77. $(8z + y)^2$

78. $(4a + 7b)^2$

79. $(10x^2 - y)^2$

80. $(7p - 3q^2)^2$

81. $(a^3 + 2b)(a^3 - 2b)$

82. $(m^2 - 2n^3)(m^2 + 2n^3)$

In Problems 83–88, for the given functions find
    (a) $(f \cdot g)(x)$ **(b)** $(f \cdot g)(3)$
See Objective 5.

83. $f(x) = x + 4; g(x) = x - 1$

84. $f(x) = x + 5; g(x) = 2x + 1$

85. $f(x) = 4x - 3; g(x) = 2x + 5$

86. $f(x) = 5x - 1; g(x) = 4x + 5$

87. $f(x) = x - 2; g(x) = x^2 + 5x - 3$

88. $f(x) = x + 5; g(x) = x^2 - 2x + 3$

In Problems 89–94, for the given functions, find
    (a) $f(x + 2)$ **(b)** $f(x + h) - f(x)$
See Objective 5.

89. $f(x) = x^2 + 1$

90. $f(x) = x^2 - 4$

91. $f(x) = x^2 + 5x - 2$

92. $f(x) = x^2 - 2x + 3$

93. $f(x) = 3x^2 - x + 1$

94. $f(x) = -2x^2 + x - 5$

Mixed Practice
In Problems 95–116, simplify the expression.

95. $5ab(a - b)^2$

96. $-3x(x - 3)^2$

97. $(x^2 + 3)(x^2 - 3)$

98. $(b^3 - 10)(b^3 + 10)$

99. $(z^2 + 9)(z - 3)(z + 3)$

100. $(y^2 + 4)(y + 2)(y - 2)$

101. $(2x^3 + 3)^2$ $(2x^3 + 2)$

102. $(3a^4 - 2)^2$

103. $(2m - 3n)(4m + n) - (m - 2n)^2$

104. $(6p + q)(5p - 2q) - (p + 3q)^2$

105. $(x + 3)(x^2 - 3x + 9)$

106. $(2y + 3)(4y^2 - 6y + 9)$

107. $\left(2x - \frac{1}{2}\right)^2$

108. $\left(3x + \frac{1}{3}\right)^2$

109. $(p + 2)^3$

110. $(z - 3)^3$

111. $(7x - 5y + 2)(3x - 2y + 1)$

112. $(2a + b - 5)(4a - 2b + 1)$

113. $(2p - 1)(p + 3) + (p - 3)(p + 3)$

114. $(3z + 2)(z - 2) + (z + 2)(z - 2)$

115. $(x + 3)(x - 3)(x^2 - 9) - (x + 1)(x^2 - 3)$

116. $(a + 2)(a - 2)(a^2 - 4) - (a + 3)(a^2 - 3)$

Applying the Concepts

△ *We can visualize the product of polynomials by using the area of rectangles. In Problems 117–120, find a polynomial expression for the total area of each of the following figures.*

117.

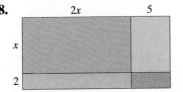

118.

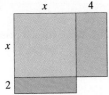

119.

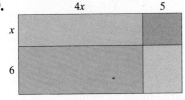

120.

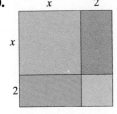

△ *In Problems 121 and 122, write a polynomial expression for the area of the shaded region of the figure.*

121.

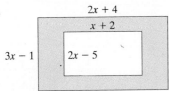

122.

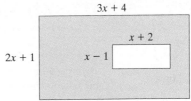

△ **123. Perfect Square** Why is the expression $(a + b)^2$ called a perfect square? Consider the figure below.

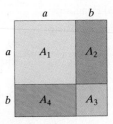

(a) Find the area of each of the four quadrilaterals.
(b) Use the result from part (a) to find the area of the entire region.
(c) Find the length and width of the entire region in terms of a and b. Use this result to find the area of the entire region. What do you notice?

△ **124. Area** Express as a polynomial in standard form the area of the shaded region in the figure shown.

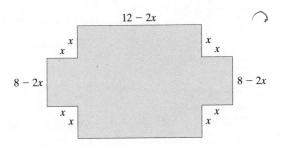

Extending the Concepts

In Problems 125–132, find the product.

125. $[3x - (y + 1)][3x + (y + 1)]$

126. $[5 - (a + b)][5 + (a + b)]$

127. $[2a + (b - 3)]^2$

128. $[(m + 4) - n]^2$

129. $(2^x + 3)(2^x - 4)$

130. $(3^x - 1)(3^x - 9)$

131. $(5^y - 1)^2$

132. $(2^z - 4)^2$

4.3 Dividing Polynomials; Synthetic Division

Objectives

1. Divide a Polynomial by a Monomial
2. Divide Polynomials Using Long Division
3. Divide Polynomials Using Synthetic Division
4. Divide Polynomial Functions
5. Use the Remainder and Factor Theorems

Are You Ready for This Section?

Before getting started, take the following readiness quiz. If you get a problem wrong, go back to the section cited and review the material.

R1. Simplify: $\dfrac{15x^5}{12x^3}$ [Getting Ready, p. 294]

R2. Add: $\dfrac{2}{7} + \dfrac{5}{7}$ [Section R.3, p. 27]

We have learned how to add, subtract and multiply polynomials. Now we can divide them! We begin with dividing a polynomial by a monomial.

When we added, subtracted, and multiplied polynomials, the result was also a polynomial. A polynomial divided by a polynomial, however, may not be a polynomial.

▶ ❶ Divide a Polynomial by a Monomial

Dividing a polynomial by a monomial is based on the Quotient Rule for Exponents (see Getting Ready, p. 294). For example,

$$\frac{24z^5}{18z^2} = \frac{6 \cdot 4}{6 \cdot 3}z^{5-2}$$

$$= \frac{4}{3}z^3$$

Recall that

$$\frac{A}{C} + \frac{B}{C} = \frac{A + B}{C}$$

We reverse this process when dividing a polynomial by a monomial.

For example, if A, B, and C are monomials, we can write $\dfrac{A + B}{C}$ as shown below.

$$\frac{A + B}{C} = \frac{A}{C} + \frac{B}{C}$$

We can extend this result to polynomials with three or more terms.

EXAMPLE 1 **Dividing a Polynomial by a Monomial**

Divide and simplify:

(a) $\dfrac{12p^4 + 15p^3 + 3p^2}{3p^2}$ **(b)** $\dfrac{6a^2b^2 - 4a^2b + 3ab^2}{2a^2b^2}$

Solution

Divide the monomial in the denominator into each term in the numerator.

(a) $\dfrac{12p^4 + 15p^3 + 3p^2}{3p^2} = \dfrac{12p^4}{3p^2} + \dfrac{15p^3}{3p^2} + \dfrac{3p^2}{3p^2}$

$\dfrac{a^m}{a^n} = a^{m-n}: \quad = \dfrac{12}{3}p^{4-2} + \dfrac{15}{3}p^{3-2} + \dfrac{3}{3} \cdot p^{2-2}$

Simplify: $\quad = 4p^2 + 5p + 1$

(b)

$$\frac{6a^2b^2 - 4a^2b + 3ab^2}{2a^2b^2} = \frac{6a^2b^2}{2a^2b^2} - \frac{4a^2b}{2a^2b^2} + \frac{3ab^2}{2a^2b^2}$$

$$\frac{a^m}{a^n} = a^{m-n}: \quad = \frac{6}{2}a^{2-2}b^{2-2} - \frac{4}{2}a^{2-2}b^{1-2} + \frac{3}{2}a^{1-2}b^{2-2}$$

$$\text{Simplify:} \quad = 3a^0b^0 - 2a^0b^{-1} + \frac{3}{2}a^{-1}b^0$$

$$a^0 = 1; b^0 = 1; a^{-1} = \frac{1}{a}; b^{-1} = \frac{1}{b}: \quad = 3 - \frac{2}{b} + \frac{3}{2a}$$

Quick ✓

1. The first step in simplifying $\dfrac{4x^4 + 8x^2}{2x}$ is to rewrite $\dfrac{4x^4 + 8x^2}{2x}$ as $\dfrac{\overline{}}{2x} + \dfrac{\overline{}}{2x}$.

In Problems 2–4, find the quotient.

2. $\dfrac{9p^4 - 12p^3 + 3p^2}{3p}$

3. $\dfrac{20a^5 - 10a^2 + 1}{5a^3}$

4. $\dfrac{x^4y^4 + 8x^2y^2 - 4xy}{4x^3y}$

▶ ❷ Divide Polynomials Using Long Division

Dividing two polynomials is similar to dividing two integers. Although this procedure should be familiar to you, we review it below.

EXAMPLE 2 **Dividing an Integer by an Integer**

Divide 645 by 14.

Solution

$$
\begin{array}{r}
46 \quad \leftarrow \text{Quotient} \\
\text{Divisor} \rightarrow 14\overline{)645} \quad \leftarrow \text{Dividend} \\
-56 \quad \leftarrow 4 \cdot 14 \ (\text{Subtract}) \\
\hline
85 \\
-84 \quad \leftarrow 6 \cdot 14 \ (\text{Subtract}) \\
\hline
1 \quad \leftarrow \text{Remainder}
\end{array}
$$

Thus 645 divided by 14 is 46 with a remainder of 1. We can write this as

$$\frac{645}{14} = 46 + \frac{1}{14}$$

In Example 2, the number 14 is the **divisor,** 645 is the **dividend,** 46 is the **quotient,** and 1 is the **remainder.**

We can always check a division problem by multiplying the quotient by the divisor and adding this product to the remainder. The result should be the dividend:

$$(\text{Quotient})\,(\text{Divisor}) + \text{Remainder} = \text{Dividend}$$

To check Example 2, we have:

$$(46)\,(14) + 1 = 644 + 1 = 645$$

To divide two polynomials, we must first write each polynomial in standard form (descending order of degree). Use the pattern in Example 2 to guide your work.

EXAMPLE 3 **How to Divide Two Polynomials**

Find the quotient and remainder when

$$4x^2 + 7x + 3 \text{ is divided by } x + 3$$

Step-by-Step Solution

Each polynomial is in standard form. The dividend is $4x^2 + 7x + 3$ and the divisor is $x + 3$.

Step 1: Divide the leading term of the dividend, $4x^2$, by the leading term of the divisor, x. Enter the result over the term $4x^2$.

$$\frac{4x^2}{x} = 4x$$

$$
\begin{array}{r}
4x \\
x + 3\overline{)4x^2 + 7x + 3}
\end{array}
$$

(continued)

Step 2: Multiply $4x$ by $x + 3$. Be sure to vertically align like terms.

$$
\begin{array}{r}
4x \\
x + 3\overline{)4x^2 + 7x + 3} \\
\underline{4x^2 + 12x} \quad \leftarrow 4x(x + 3) = 4x^2 + 12x
\end{array}
$$

Step 3: Subtract $4x^2 + 12x$ from $4x^2 + 7x + 3$.

$$
\begin{array}{r}
4x \\
x + 3\overline{)4x^2 + 7x + 3} \\
\underline{-(4x^2 + 12x)} \\
-5x + 3 \leftarrow (4x^2 + 7x - 3) - (4x^2 + 12x) \\
= -5x + 3
\end{array}
$$

Step 4: Repeat Steps 1–3, treating $-5x + 3$ as the dividend by dividing x into $-5x$ to obtain -5.

$$
\begin{array}{r}
4x - 5 \quad \leftarrow \dfrac{-5x}{x} = -5 \\
x + 3\overline{)4x^2 + 7x + 3} \\
\underline{-(4x^2 + 12x)} \\
-5x + 3 \\
\underline{-(-5x - 15)} \leftarrow -5(x + 3) = -5x - 15 \\
18 \leftarrow -5x + 3 - (-5x - 15) = 18
\end{array}
$$

Work Smart

When the degree of the remainder is less than the degree of the divisor, you are finished dividing.

Because the degree of 18 is less than the degree of the divisor, $x + 3$, the process ends. The quotient is $4x - 5$ and the remainder is 18.

Step 5: Check Verify that (Quotient)(Divisor) + Remainder = Dividend.

$$(4x - 5)(x + 3) + 18 = 4x^2 + 12x - 5x - 15 + 18$$
$$\text{Combine like terms: } = 4x^2 + 7x + 3$$

The answer checks, so

$$\frac{4x^2 + 7x + 3}{x + 3} = 4x - 5 + \frac{18}{x + 3}$$

Always write the results of polynomial division as follows:

$$\frac{\text{Dividend}}{\text{Divisor}} = \text{Quotient} + \frac{\text{Remainder}}{\text{Divisor}}$$

(**EXAMPLE 4**) **Dividing Two Polynomials**

Simplify by performing long division: $\dfrac{6x^3 - 11x^2 - 7x + 2}{3x + 2}$

Solution

$$
\begin{array}{r}
2x^2 - 5x + 1 \\
3x + 2\overline{)6x^3 - 11x^2 - 7x + 2} \\
\underline{-(6x^3 + 4x^2)} \quad \leftarrow 2x^2(3x + 2) \\
-15x^2 - 7x + 2 \\
\underline{-(-15x^2 - 10x)} \quad \leftarrow -5x(3x + 2) \\
3x + 2 \\
\underline{-(3x + 2)} \quad \leftarrow 1(3x + 2) \\
0 \quad \leftarrow \text{Remainder}
\end{array}
$$

$$\frac{6x^3}{3x} = 2x^2; \quad \frac{-15x^2}{3x} = -5x; \quad \frac{3x}{3x} = 1$$

The quotient is $2x^2 - 5x + 1$ and the remainder is 0.

Check (Quotient) (Divisor) + Remainder = Dividend

$$(2x^2 - 5x + 1)(3x + 2) + 0 = 6x^3 - 11x^2 - 7x + 2$$

Our answer checks, so

$$\frac{6x^3 - 11x^2 - 7x + 2}{3x + 2} = 2x^2 - 5x + 1$$

In Example 4, the remainder was 0. Therefore,

$$6x^3 - 11x^2 - 7x + 2 = (2x^2 - 5x + 1)(3x + 2)$$

which means that $2x^2 - 5x + 1$ and $3x + 2$ are *factors* of $6x^3 - 11x^2 - 7x + 2$. This result is true in general: If the remainder is zero, then the divisor and quotient are factors of the dividend.

EXAMPLE 5 **Dividing Two Polynomials with Missing Terms**

Simplify by performing long division: $\dfrac{8 - 15x + x^2 + 4x^3 + 3x^5}{3 + x^2}$

Solution

As usual, the dividend and divisor need to be in standard form. Write the missing x^4 term in the dividend as $0x^4$.

$$
\begin{array}{r}
3x^3 \qquad\quad -5x \ + 1 \\
x^2 + 3\overline{)\,3x^5 + 0x^4 + 4x^3 + x^2 - 15x + 8} \\
-(3x^5 \qquad\ + 9x^3) \qquad\qquad\leftarrow 3x^3(x^2 + 3)\\
\overline{\qquad -5x^3 + x^2 - 15x + 8} \\
-(-5x^3 \qquad - 15x) \qquad\leftarrow -5x(x^2 + 3)\\
\overline{\qquad\qquad x^2 \qquad + 8} \\
-(x^2 \qquad + 3) \ \leftarrow 1(x^2 + 3)\\
\overline{\qquad\qquad\qquad 5} \ \leftarrow \text{Remainder}
\end{array}
$$

The quotient is $3x^3 - 5x + 1$ and the remainder is 5.

Check (Quotient)(Divisor) + Remainder = Dividend

$$(3x^3 - 5x + 1)(x^2 + 3) + 5 = 3x^5 + 4x^3 + x^2 - 15x + 8$$

Our answer checks, so

$$\frac{8 - 15x + x^2 + 4x^3 + 3x^5}{3 + x^2} = 3x^3 - 5x + 1 + \frac{5}{3 + x^2}$$

Quick ✓

5. *True or False* To begin polynomial long division, write the divisor and dividend in standard form.

6. Because $\dfrac{3x^2 + 2x - 1}{x + 1} = 3x - 1$, the remainder when dividing $3x^2 + 2x - 1$ by $x + 1$ is ____, and $3x^2 + 2x - 1 = $ _____ · _____. We call $x + 1$ and $3x - 1$ _____ of $3x^2 + 2x - 1$.

7. Given that $\dfrac{6x^3 - x^2 - 9x + 8}{x + 1} = 6x^2 - 7x - 2 + \dfrac{10}{x + 1}$, we call $6x^3 - x^2 - 9x + 8$ the _____, $x + 1$ the _____, and 10 the _____.

8. To check a division problem, we verify _____ · _____ + _____ = _____.

In Problems 9–11, simplify by performing long division.

9. $\dfrac{x^3 + 3x^2 - 31x + 21}{x - 4}$ **10.** $\dfrac{2x^3 + 7x^2 - 7x - 12}{2x - 3}$ **11.** $\dfrac{2 + 12x^2 - 2x^3 - 5x^4 + x^5}{x^2 - 2}$

▶ ❸ Divide Polynomials Using Synthetic Division

To find the quotient and remainder of a polynomial of degree 1 or higher divided by $x - c$, we use an easier version of long division called **synthetic division.**

To see how synthetic division works, we will use long division to divide the polynomial $2x^3 - 5x^2 - 7x + 20$ by $x - 3$. Synthetic division is just long division written in a more compact form. For example, in the long division below, the terms in red ink are not really necessary because they are identical to the terms directly above them. The subtraction signs are not necessary, because subtraction is understood. With these items removed, we have the division shown on the right.

$$
\begin{array}{r}
2x^2 + x - 4 \quad \leftarrow \text{Quotient} \\
x - 3\overline{)2x^3 - 5x^2 - 7x + 20} \\
\underline{-(2x^3 - 6x^2)} \\
x^2 - 7x \\
\underline{-(x^2 - 3x)} \\
-4x + 20 \\
\underline{-(-4x + 12)} \\
8 \quad \leftarrow \text{Remainder}
\end{array}
$$

$$
\begin{array}{r}
2x^2 + x - 4 \\
x - 3\overline{)2x^3 - 5x^2 - 7x + 20} \\
\underline{- 6x^2} \\
x^2 \\
-3x \\
\underline{-4x} \\
12 \\
8
\end{array}
$$

The x's that appear in the division on the right are not necessary if we are careful about positioning each coefficient. As long as the rightmost number under the division symbol is the constant, the number to its left is the coefficient of x, and so on, we can remove the x's. Now we have

$$
\begin{array}{r}
2x^2 + x - 4 \\
x - 3\overline{)2 -5 - 7 20} \\
\underline{-6} \\
\boxed{1} \\
\underline{-3} \\
\boxed{-4} \\
\underline{12} \\
\boxed{8}
\end{array}
$$

We can make this display more compact by moving the rows up so that the boxed numbers align horizontally.

$$
\begin{array}{r}
2x^2 + x - 4 \\
x - 3\overline{)2 -5 -7 20} \\
\underline{-6 -3 12} \\
\square 1 -4 8
\end{array}
$$

Because the leading coefficient of the divisor is always 1, we know that the leading coefficient of the dividend will always be the leading coefficient of the quotient. Thus we place the leading coefficient of the quotient, 2, in the boxed position.

$$
\begin{array}{r}
2x^2 + x - 4 \\
x - 3\overline{)2 -5 -7 20} \\
\underline{-6 -3 12} \\
2 1 -4 8
\end{array}
$$

Notice that the first three numbers in Row 3 are the coefficients of the quotient. The last number in Row 3 is the remainder. Now, the top row above is not needed.

$$
\begin{array}{rl}
x - 3\overline{)2 -5 -7 20} & \text{Row 1} \\
\underline{-6 -3 12} & \text{Row 2 (subtract)} \\
2 1 -4 8 & \text{Row 3}
\end{array}
$$

Remember, the entries in Row 3 are obtained by subtracting the entries in Row 2 from the entries in Row 1. Rather than subtracting the entries in Row 2, we can change the sign of each entry and then add. With this modification, our display becomes

$$
\begin{array}{r|rrrr}
x - 3) & 2 & -5 & -7 & 20 \\
& & 6 & 3 & -12 \\
\hline
& 2 & 1 & -4 & 8 \\
\end{array}
$$
Row 1
Row 2 (add)
Row 3

Notice that each entry in Row 2 is 3 times the entry that is one column to the left in Row 3 (for example, the 6 in Row 2 is 3 times 2; the 3 in Row 2 is 3 times 1, and so on). We remove the $x - 3$ and replace it with 3. The entries in Row 3 give us the quotient and remainder.

$$
\begin{array}{r|rrrr}
3) & 2 & -5 & -7 & 20 \\
& & 6 & 3 & -12 \\
\hline
& 2^{3(2)\nearrow} & 1^{3(1)\nearrow} & -4^{3(-4)\nearrow} & 8 \\
& 2x^2 & + x & - 4 & 8
\end{array}
$$
Row 1
Row 2 (add)
Row 3
Remainder

$\underbrace{}_{\text{Quotient}}$

Work Smart

If there are any missing powers of x in the dividend, insert a coefficient of 0 for the missing term when doing synthetic division.

Let's go over an example step by step.

EXAMPLE 6 **How to Use Synthetic Division to Divide Polynomials**

Use synthetic division to find the quotient and remainder when $3x^3 + 11x^2 + 14$ is divided by $x + 4$.

Step-by-Step Solution

Step 1: Write the dividend in standard form. Then copy the coefficients of the dividend. Remember to insert a 0 for any missing power of x.

$$3x^3 + 11x^2 + 14 = 3x^3 + 11x^2 + 0x + 14$$
$$\quad\quad\quad\quad\quad 3 \quad\quad 11 \quad\quad 0 \quad\quad 14 \quad \text{Row 1}$$

Step 2: Insert the division symbol. Rewrite the divisor in the form $x - c$ and insert the value of c to the left of the division symbol.

$$x + 4 = x - (-4)$$

$$
\begin{array}{r|rrrr}
-4) & 3 & 11 & 0 & 14 \\
\end{array}
$$
Row 1

Step 3: Bring the 3 down two rows and enter it in Row 3.

$$
\begin{array}{r|rrrr}
-4) & 3 & 11 & 0 & 14 \\
& \downarrow & & & \\
\hline
& 3 & & &
\end{array}
$$
Row 1
Row 2
Row 3

Step 4: Multiply the latest entry in Row 3 by -4 and place the result in Row 2, one column over to the right.

$$
\begin{array}{r|rrrr}
-4) & 3 & 11 & 0 & 14 \\
& \downarrow & -12 & & \\
\hline
& 3^{-4(3)\nearrow} & & &
\end{array}
$$
Row 1
Row 2
Row 3

Step 5: Add the entry in Row 2 to the entry above it in Row 1. Enter the sum in Row 3.

$$
\begin{array}{r|rrrr}
-4) & 3 & 11 & 0 & 14 \\
& \downarrow & -12 & & \\
\hline
& 3^{-4(3)\nearrow} & -1 & &
\end{array}
$$
Row 1
Row 2
Row 3

Step 6: Repeat Steps 4 and 5 until no more entries are available in Row 1.

$$
\begin{array}{r|rrrr}
-4) & 3 & 11 & 0 & 14 \\
& \downarrow & -12 & 4 & -16 \\
\hline
& 3^{-4(3)\nearrow} & -1^{-4(-1)\nearrow} & 4^{-4(4)\nearrow} & -2
\end{array}
$$
Row 1
Row 2
Row 3

(continued)

Step 7: The final entry in Row 3, -2, is the remainder; the other entries in Row 3 ($3, -1$ and 4) are the coefficients of the quotient, in descending order of degree. The quotient is a polynomial whose degree is one less than the degree of the dividend.

Quotient: $3x^2 - x + 4$
Remainder: -2

Step 8: Check
(Quotient)(Divisor) + Remainder = Dividend

$$(3x^2 - x + 4)(x + 4) + (-2)$$
$$= 3x^3 + 12x^2 - x^2 - 4x + 4x + 16 - 2$$
$$= 3x^3 + 11x^2 + 14$$

Therefore, $\dfrac{3x^3 + 11x^2 + 14}{x + 4} = 3x^2 - x + 4 - \dfrac{2}{x + 4}$.

Let's do one more example, where we consolidate all the steps given in Example 6.

EXAMPLE 7 **Dividing Two Polynomials Using Synthetic Division**

Use synthetic division to find the quotient and remainder when $x^4 - 5x^3 - 6x^2 + 33x - 15$ is divided by $x - 5$.

Solution

The divisor is $x - 5$, so $c = 5$.

$$
\begin{array}{r|rrrrr}
5) & 1 & -5 & -6 & 33 & -15 \\
\downarrow & & 5 & 0 & -30 & 15 \\
\hline
& 1 & 0 & -6 & 3 & 0
\end{array}
$$

The dividend is a fourth-degree polynomial, so the quotient is a third-degree polynomial, $x^3 + 0x^2 - 6x + 3 = x^3 - 6x + 3$, and the remainder is 0. Thus

$$\dfrac{x^4 - 5x^3 - 6x^2 + 33x - 15}{x - 5} = x^3 - 6x + 3$$

In Example 7, because $\dfrac{x^4 - 5x^3 - 6x^2 + 33x - 15}{x - 5} = x^3 - 6x + 3$, we know that $x - 5$ and $x^3 - 6x + 3$ are factors of $x^4 - 5x^3 - 6x^2 + 33x - 15$. Therefore, we can write

$$x^4 - 5x^3 - 6x^2 + 33x - 15 = (x - 5)(x^3 - 6x + 3)$$

Work Smart: Study Skills

Knowing when a method **does not** apply is as essential as knowing when the method **does** apply. When can synthetic division be used to divide polynomials? When can't synthetic division be used?

Quick ✓

12. *True or False* We can divide $-4x^3 + 5x^2 + 10x - 3$ by $x^2 - 2$ using synthetic division.

13. *True or False* We can divide $-4x^3 + 5x^2 + 10x - 3$ by $2x + 1$ using synthetic division.

14. To divide a polynomial by $x + 7$, we use $c =$ ___

In Problems 15–17, use synthetic division to find the quotient.

15. $\dfrac{2x^3 + x^2 - 7x - 13}{x - 2}$

16. $\dfrac{x + 1 - 3x^2 + 4x^3}{x + 2}$

17. $\dfrac{x^4 + 8x^3 + 15x^2 - 2x - 6}{x + 3}$

▶ ❹ **Divide Polynomial Functions**

We have discussed adding, subtracting, and multiplying polynomial functions. All that is left is dividing polynomial functions.

Definition

If f and g are functions, then the **quotient** $\dfrac{f}{g}$ is the function defined by

$$\left(\frac{f}{g}\right)(x) = \frac{f(x)}{g(x)} \quad g(x) \neq 0$$

EXAMPLE 8 **Dividing Two Polynomial Functions**

If $f(x) = 2x^4 - x^3 - 3x^2 + 13x - 4$ and $g(x) = x^2 - 2$, find

(a) $\left(\dfrac{f}{g}\right)(x)$ **(b)** $\left(\dfrac{f}{g}\right)(2)$

Solution

(a) $\left(\dfrac{f}{g}\right)(x) = \dfrac{f(x)}{g(x)} = \dfrac{2x^4 - x^3 - 3x^2 + 13x - 4}{x^2 - 2}$

We use long division to find the quotient because the divisor is not of the form $x - c$.

$$
\begin{array}{r}
2x^2 - x + 1 \\
x^2 - 2 \overline{)\, 2x^4 - x^3 - 3x^2 + 13x - 4} \\
-\underline{(2x^4 \qquad\; - 4x^2)} \\
-x^3 + x^2 + 13x - 4 \\
-\underline{(-x^3 \qquad\; + 2x)} \\
x^2 + 11x - 4 \\
-\underline{(x^2 \qquad\; - 2)} \\
11x - 2
\end{array}
$$

Thus $\left(\dfrac{f}{g}\right)(x) = \dfrac{f(x)}{g(x)} = 2x^2 - x + 1 + \dfrac{11x - 2}{x^2 - 2}$.

(b) $\left(\dfrac{f}{g}\right)(2) = 2(2)^2 - 2 + 1 + \dfrac{11(2) - 2}{(2)^2 - 2}$

$\qquad\qquad = 8 - 2 + 1 + \dfrac{20}{2}$

$\qquad\qquad = 17$

Quick ✓

In Problem 18, find **(a)** $\left(\dfrac{f}{g}\right)(x)$ *and* **(b)** $\left(\dfrac{f}{g}\right)(3)$.

18. $f(x) = 3x^4 - 4x^3 - 3x^2 + 10x - 5; g(x) = x^2 - 2$

⑤ Use the Remainder and Factor Theorems

▶ In Example 6, we used synthetic division to find the quotient and remainder when $3x^3 + 11x^2 + 14$ is divided by $x + 4$. If we let $f(x) = 3x^3 + 11x^2 + 14$, we find that $f(-4) = -2$, which is equal to the remainder when $3x^3 + 11x^2 + 14$ is divided by $x + 4$. So the value of the function f at $x = -4$ is the same as the remainder when f is divided by $x + 4 = x - (-4)$. This result is true in general! It is called the *Remainder Theorem*.

In Words

A theorem is a big idea that can be shown to be true in general. The word comes from a Greek verb meaning "to view."

The Remainder Theorem

Let f be a polynomial function. When $f(x)$ is divided by $x - c$, the remainder is $f(c)$.

(EXAMPLE 9) **Using the Remainder Theorem**

Use the Remainder Theorem to find the remainder when $f(x) = 2x^3 - 3x + 8$ is divided by $x + 3$.

Solution

The divisor is $x + 3 = x - (-3)$, so the Remainder Theorem says that the remainder is $f(-3)$:

$$f(x) = 2x^3 - 3x + 8$$
$$f(-3) = 2(-3)^3 - 3(-3) + 8$$
$$= 2(-27) + 9 + 8$$
$$= -54 + 9 + 8$$
$$= -37$$

When $f(x) = 2x^3 - 3x + 8$ is divided by $x + 3$, the remainder is -37.

Check Using synthetic division, we find that the remainder is, in fact, -37.

$$
\begin{array}{r|rrrr}
-3) & 2 & 0 & -3 & 8 \\
   &   & -6 & 18 & -45 \\
\hline
   & 2 & -6 & 15 & -37 \leftarrow \text{Remainder}
\end{array}
$$

Quick ✓
19. Use the Remainder Theorem to find the remainder when
$f(x) = 3x^3 + 10x^2 - 9x - 4$ is divided by

(a) $x - 2$ **(b)** $x + 4$

   We saw from Example 7 that when the remainder is 0, the quotient and divisor are factors of the dividend. The Remainder Theorem can be used to determine whether an expression of the form $x - c$ is a factor of the dividend. This result is called the *Factor Theorem*.

Work Smart

"If and only if" statements are used to compress two statements into one. The Factor Theorem is two statements:

1. If $f(c) = 0$, then $x - c$ is a factor of f.
2. If $x - c$ is a factor of f, then $f(c) = 0$.

The Factor Theorem

Let f be a polynomial function. Then $x - c$ is a factor of $f(x)$ if and only if $f(c) = 0$.

We can use the Factor Theorem to determine whether a polynomial has a particular factor.

(EXAMPLE 10) **Using the Factor Theorem**

Use the Factor Theorem to determine whether the function
$f(x) = 2x^3 - 3x^2 - 18x - 8$ has the factor

(a) $x - 3$ **(b)** $x + 2$

Solution

The Factor Theorem states that if $f(c) = 0$, then $x - c$ is a factor of f.

(a) Because $x - 3$ is of the form $x - c$ with $c = 3$, we find the value of $f(3)$.

$$f(3) = 2(3)^3 - 3(3)^2 - 18(3) - 8 = -35 \neq 0$$

Since $f(3) \neq 0$, we know that $x - 3$ is not a factor of f.

(b) Because $x + 2 = x - (-2)$ is of the form $x - c$ with $c = -2$, we find the value of $f(-2)$. Rather than evaluating the function using substitution, we use synthetic division.

$$
\begin{array}{r|rrrr}
-2) & 2 & -3 & -18 & -8 \\
   &   & -4 & 14 & 8 \\
\hline
   & 2 & -7 & -4 & 0 \leftarrow \text{Remainder}
\end{array}
$$

The remainder is 0, so $f(-2) = 0$, and therefore $x + 2$ is a factor of f.

This means the dividend can be written as the product of the quotient and divisor. The quotient is $2x^2 - 7x - 4$ and the divisor is $x + 2$, so

$$2x^3 - 3x^2 - 18x - 8 = (x + 2)(2x^2 - 7x - 4)$$

> **In Words**
> If $f(c) = 0$, then $f(x)$ can be written in factored form as $f(x) = (x - c)$ (quotient)

Quick ✓

20. Use the Factor Theorem to determine whether $x - c$ is a factor of $f(x) = 2x^3 - 9x^2 - 6x + 5$ for the given values of c. If $x - c$ is a factor, then write f in factored form. That is, write $f(x) = (x - c)$ (quotient).

(a) $c = -2$ **(b)** $c = 5$

4.3 Exercises   MyMathLab®   PRACTICE

*Problems **1–20** are the Quick ✓s that follow the EXAMPLES.*

Building Skills

In Problems 21–28, divide and simplify. See Objective 1.

21. $\dfrac{8x^2 + 12x}{4x}$ **22.** $\dfrac{6z^3 + 9z^2}{3z^2}$

23. $\dfrac{2a^3 - 15a^2 + 10a}{5a}$ **24.** $\dfrac{4b^3 + 12b^2 + 24b}{6b}$

25. $\dfrac{2y^3 + 6y}{4y^2}$ **26.** $\dfrac{3z^4 + 12z^2}{6z^3}$

27. $\dfrac{4m^2n^2 + 6m^2n - 18mn^2}{4m^2n^2}$

28. $\dfrac{2x^2y^3 - 9xy^3 + 16x^2y}{2x^2y^2}$

In Problems 29–46, divide using long division. See Objective 2.

29. $\dfrac{x^2 + 5x + 6}{x + 2}$ **30.** $\dfrac{x^2 - 4x - 21}{x + 3}$

31. $\dfrac{2x^2 + x - 4}{x - 2}$ **32.** $\dfrac{3z^2 - 7z - 28}{z - 4}$

33. $\dfrac{2w^2 + 5w - 49}{2w - 7}$ **34.** $\dfrac{4x^2 - 17x - 33}{4x + 7}$

35. $\dfrac{x^3 + 8x^2 + x - 42}{x + 3}$ **36.** $\dfrac{x^3 + x^2 - 22x - 40}{x + 2}$

37. $\dfrac{w^3 - 21w - 20}{w + 4}$ **38.** $\dfrac{a^3 - 49a + 120}{a + 8}$

39. $\dfrac{6x^3 - 13x^2 - 80x - 25}{2x + 5}$ **40.** $\dfrac{4p^3 + 20p^2 + 19p - 15}{2p + 5}$

41. $\dfrac{x^3 - 7x^2 + 3x - 21}{x^2 + 3}$ **42.** $\dfrac{x^3 - 5x^2 - 2x + 10}{x^2 - 2}$

43. $\dfrac{3z^3 + 21z^2 + 5z + 2}{3z^2 + 1}$

44. $\dfrac{2k^3 + 10k^2 - 6k - 8}{2k^2 - 3}$

45. $\dfrac{2x^4 - 11x^3 + 8x^2 - 22x + 144}{x^2 + 2x + 5}$

46. $\dfrac{2x^4 - 7x^3 - 50x^2 - 10x + 96}{x^2 + x - 3}$

In Problems 47–60, divide using synthetic division. See Objective 3.

47. $\dfrac{x^2 - 3x - 10}{x - 5}$ **48.** $\dfrac{x^2 + 4x - 12}{x - 2}$

49. $\dfrac{2x^2 + 11x + 12}{x + 4}$ **50.** $\dfrac{3x^2 + 19x - 40}{x + 8}$

51. $\dfrac{x^2 - 3x - 14}{x - 6}$ **52.** $\dfrac{x^2 + 2x - 17}{x - 4}$

53. $\dfrac{x^3 - 19x - 15}{x - 5}$ **54.** $\dfrac{x^3 - 13x - 17}{x + 3}$

55. $\dfrac{3x^4 - 5x^3 - 21x^2 + 17x + 25}{x - 3}$

56. $\dfrac{2x^4 - x^3 - 38x^2 + 16x + 103}{x + 4}$

57. $\dfrac{x^4 - 40x^2 + 109}{x + 6}$ **58.** $\dfrac{a^4 - 65a^2 + 55}{a - 8}$

59. $\dfrac{2x^3 + 3x^2 - 14x - 15}{x - \dfrac{5}{2}}$ **60.** $\dfrac{3x^3 + 13x^2 + 8x - 12}{x - \dfrac{2}{3}}$

In Problems 61–70, find (a) $\left(\dfrac{f}{g}\right)(x)$, (b) $\left(\dfrac{f}{g}\right)(2)$. See Objective 4.

61. $f(x) = 4x^3 - 8x^2 + 12x$; $g(x) = 4x$

62. $f(x) = 3x^3 - 9x^2 + 12x$; $g(x) = 3x$

63. $f(x) = x^2 - x - 12; g(x) = x - 4$

64. $f(x) = x^2 + 3x - 4; g(x) = x + 4$

65. $f(x) = 2x^2 + 5x - 1; g(x) = x + 3$

66. $f(x) = 3x^2 - 6x + 5; g(x) = 2x + 1$

67. $f(x) = 2x^3 + 9x^2 + x - 12; g(x) = 2x + 3$

68. $f(x) = 3x^3 - 2x^2 - 19x - 6; g(x) = 3x + 1$

69. $f(x) = x^3 - 13x - 12; g(x) = x^2 - 9$

70. $f(x) = x^3 - 19x + 30; g(x) = x^2 - x - 6$

In Problems 71–78, use the Remainder Theorem to find the remainder. See Objective 5.

71. $f(x) = x^2 - 5x + 1$ is divided by $x - 2$

72. $f(x) = x^2 + 4x - 5$ is divided by $x + 2$

73. $f(x) = x^3 - 2x^2 + 5x - 3$ is divided by $x + 4$

74. $f(x) = x^3 + 3x^2 - x + 1$ is divided by $x - 3$

75. $f(x) = 2x^3 - 4x + 1$ is divided by $x - 5$

76. $f(x) = 3x^3 + 2x^2 - 5$ is divided by $x + 3$

77. $f(x) = x^4 + 1$ is divided by $x - 1$

78. $f(x) = x^4 - 1$ is divided by $x - 1$

In Problems 79–86, use the Factor Theorem to determine whether $x - c$ is a factor of the given function for the given values of c. If $x - c$ is a factor, then write f in factored form. That is, write $f(x) = (x - c)(quotient)$. See Objective 5.

79. $f(x) = x^2 - 3x + 2; c = 2$

80. $f(x) = x^2 + 5x + 6; c = 3$

81. $f(x) = 2x^2 + 5x + 2; c = -2$

82. $f(x) = 3x^2 + x - 2; c = -1$

83. $f(x) = 4x^3 - 9x^2 - 49x - 30; c = 3$

84. $f(x) = 2x^3 - 9x^2 - 2x + 24; c = 1$

85. $f(x) = 4x^3 - 7x^2 - 5x + 6; c = -1$

86. $f(x) = 5x^3 + 8x^2 - 7x - 6; c = -2$

Mixed Practice

In Problems 87–100, divide using any appropriate method.

87. $\dfrac{3a^3b^2 - 9a^2b + 18ab}{3ab}$

88. $\dfrac{5s^4t^3 - 15s^3t^2 + 50s^2t}{5s^2t}$

89. $\dfrac{3y^2 + 11y + 6}{3y + 2}$

90. $\dfrac{4a^2 + 23a + 15}{4a + 3}$

91. $\dfrac{x^3 + 6x^2 + 8x + 32}{x^2 + 5}$

92. $\dfrac{x^3 - 3x^2 + 5x - 12}{x^2 + 3}$

93. $\dfrac{8x^3 + 6x}{12x^2}$

94. $\dfrac{3x^4 + 6x^2}{9x^3}$

95. $\dfrac{x^3 + 7x^2 + 2x - 46}{x + 4}$

96. $\dfrac{x^3 + 5x^2 - 29x - 97}{x - 5}$

97. $\dfrac{3x^3 + 5x^2 - 24x - 40}{3x + 5}$

98. $\dfrac{4b^3 + 11b^2 - 28b - 17}{4b + 3}$

99. $\dfrac{8x^3 - 27}{2x - 3}$

100. $\dfrac{8a^3 + 125}{2a + 5}$

101. If $\dfrac{f(x)}{x - 5} = 3x + 5$, find $f(x)$.

102. If $\dfrac{f(x)}{x + 3} = 2x + 7$, find $f(x)$.

103. If $\dfrac{f(x)}{x - 3} = x + 8 + \dfrac{4}{x - 3}$, find $f(x)$.

104. If $\dfrac{f(x)}{x - 3} = x^2 + 2 + \dfrac{7}{x - 3}$, find $f(x)$.

Applying the Concepts

△ **105. Area** The area of a rectangle is $15x^2 + x - 2$ square feet. If the width of the rectangle is $3x - 1$ feet, find the length.

△ **106. Area** The area of a rectangle is $10x^2 + 9x - 9$ square centimeters. If the width of the rectangle is $2x + 3$ centimeters, find the length.

△ **107. Volume** The volume of the box shown is $2x^3 + 9x^2 - 20x - 75$ cubic centimeters. Find the length if the width is $x + 5$ centimeters and the height is $x - 3$ centimeters.

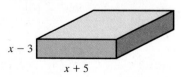

△ **108. Volume** The volume of the box shown is $4x^3 + 35x^2 + 52x + 21$ cubic feet. Find the height if the width is $4x + 3$ feet and the length is $x + 1$ feet.

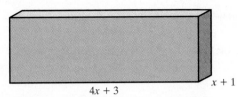

$x + 1$

$4x + 3$

109. Average Cost The average cost function is defined in economics as $\overline{C}(x) = \dfrac{C(x)}{x}$, where C is the cost of manufacturing x units of a good. Suppose that $C(x) = 0.01x^3 - 0.4x^2 + 13x + 400$, where x is the number of wristwatches manufactured in a day.

(a) What is $\overline{C}(x) = \dfrac{C(x)}{x}$?

(b) What is the average cost of manufacturing $x = 50$ wristwatches in a day?

110. Average Cost The average cost function is defined in economics as $\overline{C}(x) = \dfrac{C(x)}{x}$, where C is the cost of manufacturing x units of a good. Suppose that $C(x) = 0.01x^3 - 0.45x^2 + 16.5x + 600$, where x is the number of wagons manufactured in a day.

(a) What is $\overline{C}(x) = \dfrac{C(x)}{x}$?

(b) What is the average cost of manufacturing $x = 50$ wagons in a day?

Extending the Concepts

111. Find the sum of $a, b, c,$ and d if
$$\frac{2x^3 - 3x^2 - 26x - 37}{x + 2} = ax^2 + bx + c + \frac{d}{x + 2}.$$

112. What is the remainder when $f(x) = 2x^{30} - 3x^{20} + 4x^{10} - 2$ is divided by $x - 1$?

Explaining the Concepts

113. If f is a polynomial of degree n and it is divided by $x + 4$, the quotient will be a polynomial of degree $n - 1$. Explain why.

114. Explain the Remainder Theorem in your own words. Explain the Factor Theorem in your own words.

115. Given that $\dfrac{6x^3 - x^2 - 9x + 4}{3x + 4} = 2x^2 - 3x + 1$, is $3x + 4$ a factor of $6x^3 - x^2 - 9x + 4$? If so, write $6x^3 - x^2 - 9x + 4$ in factored form.

116. Suppose that you were asked to divide $8x^3 - 3x + 1$ by $x + 3$. Would you use long division or synthetic division? Why?

Putting the Concepts Together (Sections 4.1–4.3)

We designed these problems so that you can review Sections 4.1 through 4.3 and show your mastery of the concepts. Take time to work these problems before proceeding with the next section. The answers are located at the back of the text on page AN-28.

1. Write the polynomial in standard form and determine its degree:
$$3m + 5m^4 - 2m^3 + 8$$

2. Add: $(7a^2 - 4a^3 + 7a - 1) + (2a^2 - 6a - 7)$

3. Subtract: $\left(\dfrac{1}{5}y^2 + 2y - 6\right) - (4y^2 - y + 2)$

4. For $f(x) = 2x^3 - x^2 + 4x + 9$, find $f(2)$.

5. For $f(x) = 6x + 5$ and $g(x) = -x^2 + 2x + 3$, find $(f + g)(-3)$.

6. For $f(x) = 2x^2 + 7$ and $g(x) = x^2 - 4x - 3$, find $(f - g)(x)$.

7. Multiply: $2mn^3(m^2n - 4mn + 6)$

8. Multiply: $(3a - 5b)^2$

9. Multiply: $(7n^2 + 3)(7n^2 - 3)$

10. Multiply: $(3a + 2b)(6a^2 - 2ab + b^2)$

11. For $f(x) = x + 2$ and $g(x) = x^2 - 4x + 11$, find $(f \cdot g)(x)$.

12. Divide using long division: $\dfrac{10z^3 + 41z^2 + 7z - 49}{2z + 7}$

13. Divide using synthetic division:
$$\frac{2x^3 + 25x^2 + 62x - 6}{x + 9}$$

14. For $f(x) = x^3 + 2x^2 - 4x + 5$ and $g(x) = x - 1$, find $\left(\dfrac{f}{g}\right)(x)$.

15. Use the Factor Theorem to determine whether $x + 5$ is a factor of $f(x) = 3x^3 + 8x^2 - 23x + 60$. If so, write the function in factored form—that is, in the form $f(x) = (x + 5)(\text{quotient})$.

4.4 Greatest Common Factor; Factoring by Grouping

Objectives

1. Factor Out the Greatest Common Factor
2. Factor by Grouping

Are You Ready for This Section?

Before getting started, take the following readiness quiz. If you get a problem wrong, go back to the section cited and review the material.

R1. Write 24 as the product of prime factors. [Section R.3, p. 28]

R2. Distribute: $4(3x - 5)$ [Section R.3, pp. 29–30]

Consider the following products:

$$5y(y^2 - 2y + 5) = 5y^3 - 10y^2 + 25y$$
$$(3x + 1)(x - 5) = 3x^2 - 14x - 5$$

The polynomials on the left side are called **factors** of the polynomial on the right side. Expressing a polynomial with integer coefficients as the product of two or more other polynomials with integer coefficients is called **factoring over the integers.**

A polynomial with integer coefficients is **prime** if it cannot be written as the product of two other polynomials with integer coefficients (excluding 1 and −1). A polynomial that has been written as a product of prime factors is said to be **factored completely.** The word *prime* has the same meaning as it does for integers. For example, 3, 7, and 13 are prime numbers, while 4 $(= 2 \cdot 2)$, 12 $(=2 \cdot 6)$, and 35 $(= 5 \cdot 7)$ are not prime. If we write 4 as $2 \cdot 2$, we have factored 4 completely. If we write 12 as $2 \cdot 6$, we have not factored 12 completely, because 6 can be further factored as $2 \cdot 3$. Thus, 12 factored completely would be written $2 \cdot 2 \cdot 3$.

1 Factor Out the Greatest Common Factor

▶ The first step in factoring any polynomial is to look for its **greatest common factor (GCF),** the largest polynomial that is a factor of all the terms in the polynomial.

EXAMPLE 1   **Finding the Greatest Common Factor**

Find the greatest common factor (GCF) of the terms.

(a) $4x$, 12   **(b)** $6x^3$, $12x^2$, $15x$ **(c)** $4x^3y^4$, $8x^2y^3$, $12xy^2$

Solution

In each case, look for the largest polynomial that is a factor of each polynomial.

(a) Look at the coefficients first. The largest number that divides into 4 and 12 evenly is 4, so 4 is part of the GCF. Because 12 does not have a variable factor, the GCF is 4.

(b) The largest number that divides evenly into 6, 12, and 15 is 3, so 3 is part of the GCF. Now, look at the variable expressions, x^3, x^2, and x. Since $x^3 = x \cdot x \cdot x$, $x^2 = x \cdot x$, and $x = x$, we choose the variable expression with the smallest exponent, $x(= x^1)$, as part of the GCF. The GCF is $3x$.

(c) The largest number that divides into 4, 8, and 12 is 4, so 4 is part of the GCF. Look at the expression involving x. The smallest exponent involving x is 1, so x is part of the GCF. The smallest exponent involving y is y^2, so y^2 is part of the GCF. The GCF is $4xy^2$. ●

Quick ✓

1. In $(3x + 1)(x - 5) = 3x^2 - 14x - 5$, the polynomials on the left side are called _____ of the polynomial on the right side.

2. If a polynomial cannot be written as the product of two other polynomials (excluding 1 and −1), then the polynomial is said to be _____.

Ready?...Answers
R1. $24 = 2 \cdot 2 \cdot 2 \cdot 3$ **R2.** $12x - 20$

3. The _____ _____ _____ of a polynomial is the largest polynomial that is a factor of all the terms in the polynomial.

4. To _____ a polynomial means to write the polynomial as the product of two or more polynomials.

5. *True or False* If we write 8 as $2 \cdot 4$, it is factored completely.

In Problems 6–8, find the greatest common factor (GCF) of the terms.

6. $5y, 15$ **7.** $4z^3, 10z^2, 12z$ **8.** $6x^3y^5, 9x^2y^3, 12xy^4$

▶ Once the GCF is identified, we use the Distributive Property to factor out the GCF.

(EXAMPLE 2)   **How to Factor Out the Greatest Common Factor in a Polynomial**

Factor out the greatest common factor: $4a^2b^2 - 10ab^3 + 18a^3b^4$

Step-by-Step Solution

Step 1: Find the GCF. $\text{GCF} = 2ab^2$

Step 2: Rewrite each term as the product of the GCF and the remaining factor. $4a^2b^2 - 10ab^3 + 18a^3b^4 = 2ab^2 \cdot 2a - 2ab^2 \cdot 5b + 2ab^2 \cdot 9a^2b^2$

Step 3: Factor out the GCF. $= 2ab^2(2a - 5b + 9a^2b^2)$

Step 4: **Check** $2ab^2(2a - 5b + 9a^2b^2) = 2ab^2 \cdot 2a - 2ab^2 \cdot 5b + 2ab^2 \cdot 9a^2b^2$
$$= 4a^2b^2 - 10ab^3 + 18a^3b^4$$

Thus $4a^2b^2 - 10ab^3 + 18a^3b^4 = 2ab^2(2a - 5b + 9a^2b^2)$.

> **Factoring a Polynomial Using the Greatest Common Factor**
>
> **Step 1:** Identify the greatest common factor (GCF) of the terms.
> **Step 2:** Rewrite each term as the product of the GCF and the remaining factor.
> **Step 3:** Use the Distributive Property to factor out the GCF.
> **Step 4:** Use the Distributive Property to verify that the factorization is correct.

Quick ✓

In Problems 9 –11, factor out the greatest common factor.

9. $7z^2 - 14z$ **10.** $6y^3 - 14y^2 + 10y$ **11.** $2m^4n^2 + 8m^3n^4 - 6m^2n^5$

▶ When the coefficient of the term of highest degree is negative, we factor a negative number out of the polynomial as part of the GCF.

(EXAMPLE 3)   **Factoring Out a Negative GCF**

Factor out the greatest common factor.

     (a) $-8a + 16$ **(b)** $-2b^3 + 10b^2 + 8b$

Solution

$$\overset{\text{GCF} = -8}{}$$
(a) $-8a + 16 = -8 \cdot a + (-8) \cdot (-2)$
$$= -8(a - 2)$$

Check $-8(a - 2) = -8 \cdot a + (-8) \cdot (-2)$
$$= -8a + 16$$

$$\text{GCF} = -2b$$

(b) $-2b^3 + 10b^2 + 8b = -2b \cdot b^2 + (-2b) \cdot (-5b) + (-2b) \cdot (-4)$
$$= -2b(b^2 - 5b - 4)$$

Check $-2b(b^2 - 5b - 4) = -2b \cdot b^2 + (-2b) \cdot (-5b) + (-2b) \cdot (-4)$
$$= -2b^3 + 10b^2 + 8b$$

Quick ✓

In Problems 12 and 13, factor out the greatest common factor.

12. $-5y^2 + 10y$ **13.** $-3a^3 + 6a^2 - 12a$

Sometimes the greatest common factor is a binomial.

(EXAMPLE 4) **Factoring Out a Binomial as the Greatest Common Factor**

Factor out the greatest common factor.

(a) $4x(x - 3) + 5(x - 3)$

(b) $3y(2y + 1) - 5(2y + 1)^2$

(c) $(c + 4)(c - 1) + (5c - 2)(c - 1)$

Solution

$$\text{GCF} = (x - 3)$$

(a) $4x(x - 3) + 5(x - 3) = (x - 3) \cdot 4x + (x - 3) \cdot 5$
$$= (x - 3)(4x + 5)$$

(b) $3y(2y + 1) - 5(2y + 1)^2 = (2y + 1) \cdot 3y - (2y + 1) \cdot 5(2y + 1)$
$$= (2y + 1)(3y - 5(2y + 1))$$
$$= (2y + 1)(3y - 10y - 5)$$
$$= (2y + 1)(-7y - 5)$$

Factor out −1: $= -(2y + 1)(7y + 5)$

(c) $(c + 4)(c - 1) + (5c - 2)(c - 1) = (c - 1) \cdot (c + 4) + (c - 1) \cdot (5c - 2)$
$$= (c - 1)(c + 4 + 5c - 2)$$
$$= (c - 1)(6c + 2)$$

GCF = 2 in $(6c + 2)$: $= 2(c - 1)(3c + 1)$

Quick ✓

In Problems 14 and 15, factor out the greatest common factor.

14. $4a(a - 3) + 3(a - 3)$ **15.** $(w + 2)(w - 5) + (2w + 1)(w - 5)$

▶ ❷ **Factor by Grouping**

Sometimes a common factor does not occur in every term of a polynomial, but a common factor does occur in some of its terms, and a second common factor occurs in its remaining terms. When this happens, the common factor can be factored out of each group using the Distributive Property. This technique is called **factoring by grouping** and is used often when a polynomial contains four terms.

EXAMPLE 5 How to Factor by Grouping

Factor by grouping: $4x - 4y + ax - ay$

Step-by-Step Solution

Step 1: Group terms with common factors. In this problem the first two terms have a common factor of 4, and the last two terms have a common factor of a.

$$4x - 4y + ax - ay = (4x - 4y) + (ax - ay)$$

Step 2: In each grouping, factor out the common factor.

$$= 4(x - y) + a(x - y)$$

Step 3: Factor out the common factor that remains.

$$= (x - y)(4 + a)$$

Step 4: Check

$$
\begin{array}{c}
\text{F} \quad \text{O} \quad \text{I} \quad \text{L} \\
(x - y)(4 + a) = 4x + ax - 4y - ay
\end{array}
$$

$$\text{Rearrange terms:} \quad = 4x - 4y + ax - ay$$

Thus $4x - 4y + ax - ay = (x - y)(4 + a)$.

●

> **Factoring a Polynomial by Grouping**
>
> **Step 1:** Group the terms with common factors. Sometimes it will be necessary to rearrange the terms.
>
> **Step 2:** In each grouping, factor out the common factor.
>
> **Step 3:** Factor out the common factor that remains.
>
> **Step 4:** Check your work.

EXAMPLE 6 Factoring by Grouping

Factor by grouping:

(a) $x^3 + 3x^2 + 2x + 6$ **(b)** $6x^2 + 9x - 10x - 15$

Solution

(a) First, group terms with common factors. Note that the first two terms, x^3 and $3x^2$, have a common factor of x^2; the last two terms have a common factor of 2.

$$x^3 + 3x^2 + 2x + 6 = (x^3 + 3x^2) + (2x + 6)$$

$$\text{Factor out common factor:} \quad = x^2(x + 3) + 2(x + 3)$$

$$= (x + 3)(x^2 + 2)$$

Work Smart

In Example 6(a), grouping the first three terms would result in a common factor of x, but factoring out an x would not result in a common factor, such as $x + 3$.

Check $(x + 3)(x^2 + 2) = x^3 + 2x + 3x^2 + 6$

$$\text{Rearrange terms:} \quad = x^3 + 3x^2 + 2x + 6$$

So $x^3 + 3x^2 + 2x + 6 = (x + 3)(x^2 + 2)$.

Work Smart

In Example 6(b), notice how we changed $-10x - 15$ to $+ (-10x - 15)$ before we grouped so we would not change the original problem.

(b) $6x^2 + 9x - 10x - 15 = (6x^2 + 9x) + (-10x - 15)$

$$\text{Factor out } -5 \text{ in the second grouping:} \quad = 3x(2x + 3) + (-5)(2x + 3)$$

$$= (2x + 3)(3x - 5)$$

●

We leave the check to you.

Quick ✓

In Problems 16–18, factor by grouping.

16. $5x + 5y + bx + by$ **17.** $w^3 - 3w^2 + 4w - 12$ **18.** $2x^2 + x - 10x - 5$

4.4 Exercises MyMathLab®   PRACTICE

*Problems **1–18** are the Quick ✓s that follow the EXAMPLES.*

Building Skills

In Problems 19–36, factor out the greatest common factor. See Objective 1.

19. $5a + 35$　　　　　**20.** $8z + 48$

21. $-3y + 21$　　　　**22.** $-4b + 32$

23. $14x^2 - 21x$　　　**24.** $12a^2 + 45a$

25. $3z^3 - 6z^2 + 18z$

26. $2w^3 + 10w^2 - 14$

27. $-5p^4 + 10p^3 - 25p^2$

28. $-6q^3 + 36q^2 - 48q$

29. $49m^3n + 84mn^3 - 35m^4n^2$

30. $64x^4y^2 - 40x^3y^4 + 96xy^5$

31. $-18z^3 + 14z^2 + 4z$

32. $-18b^3 + 10b^2 + 6b$

33. $5c(3c - 2) - 3(3c - 2)$

34. $6z(5z + 3) + 5(5z + 3)$

35. $(4a + 3)(a - 3) + (2a - 7)(a - 3)$

36. $(5b + 3)(b + 4) + (3b + 1)(b + 4)$

In Problems 37–48, factor by grouping. See Objective 2.

37. $5x + 5y + ax + ay$

38. $8x - 8y + bx - by$

39. $2z^3 + 10z^2 - 5z - 25$

40. $3y^3 + 9y^2 - 5y - 15$

41. $w^2 - 5w + 3w - 15$

42. $p^2 - 3p + 8p - 24$

43. $2x^2 - 8x - 4x + 16$

44. $3a^2 - 15a - 9a + 45$

45. $3x^3 + 15x^2 - 12x^2 - 60x$

46. $2y^3 + 14y^2 - 4y^2 - 28y$

47. $2ax - 2ay - bx + by$

48. $15x^2 - 5xy + 18xy - 6y^2$

Mixed Practice

In Problems 49–58, factor each polynomial completely.

49. $(w + 3)(w - 3) - (w - 2)(w - 3)$

50. $(x + 5)(x - 3) - (x - 1)(x - 3)$

51. $2y^2 + 5y - 4y - 10$　　**52.** $3q^2 + 5q - 12q - 20$

53. $6x^3y^3 + 9x^2y - 21x^3y^2$　　**54.** $8a^4b^2 + 12a^3b^3 - 36ab^4$

55. $x^3 + x^2 + 3x + 3$　　**56.** $c^3 - c^2 + 5c - 5$

57. $x(x - 2) + 3(x - 2)^2$　　**58.** $2y(y + 4) + 3(y + 4)^2$

Math for the Future

In Problems 59–66, expressions that occur in calculus are given. Factor and simplify each expression.

59. $3x^2(4x + 1)^2 + 8x^3(4x + 1)$

60. $2x(3x + 5)^2 + 6x^2(3x + 5)$

61. $3(x + 9)^2(2x + 5) + 2(x + 9)^3$

62. $4(x - 3)^3(3x + 1) + 3(x - 3)^4$

63. $4(2x - 1)(x - 5)^3 + 3(x - 5)^2(2x - 1)^2$

64. $2(x + 3)(6x + 5)^3 + 18(6x + 5)^2(x + 3)^2$

65. $2(x^2 + 1) \cdot 2x(4x - 3)^3 + 3(4x - 3)^2 \cdot 4(x^2 + 1)^2$

66. $3(x^2 + 3)^2 \cdot 2x(2x + 1)^2 + 2(2x + 1) \cdot 2(x^2 + 3)^3$

Applying the Concepts

△ **67. Area** Write the area of the shaded region in factored form.

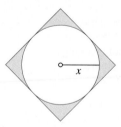

△ **68. Area** Write the area of the shaded region in factored form.

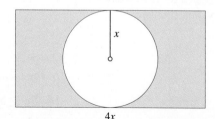

△ 69. **Surface Area** The surface area of a cylindrical can whose radius is r inches and height is 4 inches is given by $S = 2\pi r^2 + 8\pi r$. Express the surface area in factored form.

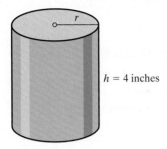

$h = 4$ inches

△ 70. **Volume** The volume of a right circular cylinder of height h and radius r inscribed in a sphere of fixed radius R is given by $V = \pi hR^2 - \dfrac{\pi h^3}{4}$. Express the volume of the cylinder in factored form.

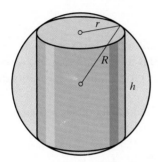

71. **Markups and Discounts** Suppose that a clothing store marks up its clothes 40% when it buys them from the supplier.

(a) Let x represent the cost of designer shirts purchased from the supplier. Write an algebraic expression representing the selling price of the shirt.

(b) After 1 month, the manager of the clothing store discounts the shirts by 40%. Write an algebraic expression representing the sale price of the shirt in terms of x.

(c) Write the algebraic expression in factored form.

(d) Based on your answer to part (c), is the store selling the shirt for the same price that it paid for it?

72. **Summer Clearance** Suppose that an electronics store decides to sell last year's model televisions for a 20% discount.

(a) Let x represent the original price of the television. Write an algebraic expression representing the selling price of the television.

(b) After 1 month, the manager of the store discounts the TVs by another 15%. Write an algebraic expression representing the sale price of the televisions in terms of x, the original selling price.

(c) Write the algebraic expression in factored form.

(d) If the original price of the television was $650, what is the sale price after the second discount?

73. **Stock Prices** Suppose that Christina purchased a stock for x dollars. During the first year, the stock's price rose 15%.

(a) Write an algebraic expression for the price of the stock after the first year in terms of x.

(b) During the second year, the stock's price rose 10%. Write an algebraic expression for the price of the stock after the second year in terms of x.

(c) Write the algebraic expression found in part (b) in factored form.

(d) If the stock was originally purchased for $x = 20, what is the value of the stock after 2 years?

Extending the Concepts

In this section, we factored polynomials over the integers. However, we could also factor polynomials with rational coefficients, such as $\dfrac{1}{3}x + \dfrac{2}{3}$. This polynomial can be factored over the rational numbers by factoring out the greatest common factor $\dfrac{1}{3}$:

$\dfrac{1}{3}x + \dfrac{2}{3} = \dfrac{1}{3}(x + 2)$. *Use this idea to factor out the greatest common factor in Problems 74–77.*

74. $\dfrac{1}{2}x + \dfrac{3}{2}$ 75. $\dfrac{1}{4}x - \dfrac{7}{4}$

76. $\dfrac{2}{3}x^2 + \dfrac{4}{9}x$ 77. $\dfrac{1}{5}b^3 + \dfrac{8}{25}b$

78. **The Better Deal** Which is the better deal: (a) receiving a 30% discount or (b) receiving a 15% discount and then another 15% discount after the first 15% discount was applied? Prove it!

In Problems 79–81, factor out the greater common factor of each expression. Assume n is a positive integer.

79. $x^n + 3x^{n+1} + 6x^{2n}$

80. $2x^{3n} - 8x^{4n} + 16x^{2n}$

81. $4y^{n+3} - 8y^{n+2} + 6y^{n+5}$

4.5 Factoring Trinomials

Objectives

1 Factor Trinomials of the Form $x^2 + bx + c$

2 Factor Trinomials of the Form $ax^2 + bx + c, a \neq 1$

3 Factor Trinomials Using Substitution

Are You Ready for This Section?

Before getting started, take the following readiness quiz. If you get a problem wrong, go back to the section cited and review the material.

R1. Determine the coefficients of $4x^2 - 9x + 2$. [Section R.5, pp. 41–42]

R2. Find two integers whose sum is 5 and product is 6.

R3. Find two integers whose sum is -4 and product is -32.

R4. Find two integers whose sum is 6 and product is -40.

R5. Find two integers whose sum is -12 and product is 32.

R6. List the factors of 18 whose sum is 11.

R7. List the factors of -24 whose sum is -2.

In this section we will factor trinomials of the form $ax^2 + bx + c$, where a, b, and c are integers. Because the word **quadratic** means "relating to a square," these trinomials are called *quadratic trinomials*.

> **Definition**
>
> A **quadratic trinomial** is a polynomial of the form $ax^2 + bx + c$, $a \neq 0$, where a represents the coefficient of the squared (second-degree) term, b represents the coefficient of the linear (first-degree) term, and c represents the constant.

In the trinomial $ax^2 + bx + c$, a is called the leading coefficient. We begin by looking at quadratic trinomials whose leading coefficient, a, is 1.

1 Factor Trinomials of the Form $x^2 + bx + c$

▶ The goal in factoring a second-degree polynomial of the form $x^2 + bx + c$ is to write it as the product of two first-degree polynomials.

For example,

$$\text{Multiplication} \atop \rightarrow$$

$$\text{Factored Form} \rightarrow (x - 5)(x + 2) = x^2 - 3x - 10 \leftarrow \text{Product}$$

$$\underset{\leftarrow}{\text{Factoring}}$$

The factors of $x^2 - 3x - 10$ are $x - 5$ and $x + 2$. Notice the following:

$$x^2 - 3x - 10 = (x - 5)(x + 2)$$

The sum of -5 and 2 is -3

The product of -5 and 2 is -10

Ready?...Answers **R1.** 4, -9, 2
R2. 2, 3   **R3.** -8, 4   **R4.** -4, 10
R5. -8, -4 **R6.** 9 and 2   **R7.** -6 and 4

In general, if $x^2 + bx + c = (x + m)(x + n)$, **then** $mn = c$ **and** $m + n = b$.

EXAMPLE 1 **How to Factor a Trinomial of the Form** $x^2 + bx + c$

Factor: $x^2 + 8x + 12$

Step-by-Step Solution

Step 1: Look for factors of $c = 12$ whose sum is $b = 8$. Begin by listing all factors of 12 and computing the sum of these factors.

Factors of 12	1, 12	2, 6	3, 4	$-1, -12$	$-2, -6$	$-3, -4$
Sum	13	8	7	-13	-8	-7

We can see that $2 \cdot 6 = 12$ and $2 + 6 = 8$, so $m = 2$ and $n = 6$.

Step 2: Write the trinomial in the form $(x + m)(x + n)$.

$$x^2 + 8x + 12 = (x + 2)(x + 6)$$

Step 3: Check Use FOIL to verify our solution.

$$(x + 2)(x + 6) = x^2 + 6x + 2x + 2(6)$$
$$= x^2 + 8x + 12$$

Thus $x^2 + 8x + 12 = (x + 2)(x + 6)$. Because multiplication is commutative, we can also write the factorization as $(x + 6)(x + 2)$.

Below, we summarize the steps used in Example 1.

> **Factoring a Trinomial of the Form $x^2 + bx + c$**
>
> **Step 1:** Find the pair of integers whose product is c and whose sum is b. That is, determine m and n such that $mn = c$ and $m + n = b$.
> **Step 2:** Write $x^2 + bx + c = (x + m)(x + n)$.
> **Step 3:** Check your work by multiplying the binomials.

Work Smart

In a trinomial $x^2 + bx + c$, if both b and c are positive, then m and n must both be positive.

Example 1 shows that if the coefficient of the middle term and the constant are both positive, then both m and n must be positive.

Quick ✓

1. A _____ _____ is a polynomial of the form $ax^2 + bx + c$, where a, b, and c are integers.

2. When factoring a trinomial of the form $x^2 + bx + c$, we need to find pairs of integers m and n such that ____ = c and _____ = b.

3. *True or False* $x^2 + 12x + 11 = (x + 11)(x + 1)$

In Problems 4 and 5, factor each trinomial.

4. $y^2 + 9y + 18$ 5. $p^2 + 14p + 24$

EXAMPLE 2 **Factoring a Trinomial of the Form $x^2 + bx + c$**

Factor: $p^2 - 10p + 24$

Solution

We want factors of $c = 24$ whose sum is $b = -10$. Begin by listing all factors of 24 and computing the sum of these factors.

Factors of 24	1, 24	2, 12	3, 8	4, 6	−1, −24	−2, −12	−3, −8	−4, −6
Sum	25	14	11	10	−25	−14	−11	−10

We can see that $-4 \cdot (-6) = 24$ and $-4 + (-6) = -10$, so $m = -4$ and $n = -6$. Write the trinomial in the form $(p + m)(p + n)$.

$$p^2 - 10p + 24 = (p + (-4))(p + (-6))$$
$$= (p - 4)(p - 6)$$

Check

$$(p - 4)(p - 6) = p^2 - 6p - 4p + (-4)(-6)$$
$$= p^2 - 10p + 24$$

Work Smart

In a trinomial $x^2 + bx + c$, if b is negative and c is positive, then both m and n must be negative.

As illustrated in Example 2, if the coefficient of the middle term is negative and the constant is positive, then m and n must both be negative.

Quick ✓

In Problems 6 and 7, factor each trinomial.

6. $q^2 - 6q + 8$ **7.** $x^2 - 8x + 12$

▶ **EXAMPLE 3**   **Factoring a Trinomial of the Form $x^2 + bx + c, c < 0$**

Factor: $z^2 - 3z - 28$

Solution

Find factors of $c = -28$ whose sum is $b = -3$. We proceed as usual.

Factors of -28	$-1, 28$	$-2, 14$	$-4, 7$	$1, -28$	$2, -14$	$4, -7$
Sum	27	12	3	-27	-12	-3

Since $4 \cdot (-7) = -28$ and $4 + (-7) = -3$, then $m = 4$ and $n = -7$. Write the trinomial in the form $(z + m)(z + n)$.

$$z^2 - 3z - 28 = (z + 4)(z + (-7))$$
$$= (z + 4)(z - 7)$$

Check

$$(z + 4)(z - 7) = z^2 - 7z + 4z + 4(-7) = z^2 - 3z - 28 \qquad \bullet$$

Work Smart

In a trinomial $x^2 + bx + c$, if c is negative, then m and n must have opposite signs.

Example 3 reveals that if the constant is negative, then m and n must be opposite in sign. In addition, if the coefficient of the middle term is also negative, then the factor with the larger absolute value must be negative.

Quick ✓

In Problems 8 and 9, factor each trinomial.

8. $w^2 - 4w - 21$ **9.** $q^2 - 9q - 36$

▶ Remember, a polynomial that cannot be written as the product of two other polynomials is said to be prime.

EXAMPLE 4   **Identifying a Prime Trinomial**

Factor: $y^2 + 4y + 12$

Solution

Look for factors of $c = 12$ whose sum is $b = 4$. Because both b and c are positive, both m and n must be positive, so we list only positive factors of 12.

Factors of 12	$1, 12$	$2, 6$	$3, 4$
Sum	13	8	7

There are no factors of 12 whose sum is 4. Therefore, $y^2 + 4y + 12$ is prime. $\qquad \bullet$

Quick ✓

10. A polynomial that cannot be written as the product of two other polynomials (other than 1 or -1) is a _____ polynomial.

In Problems 11 and 12, factor the trinomial. If the trinomial cannot be factored, say it is prime.

11. $z^2 - 5z + 8$ **12.** $q^2 + 4q - 45$

If a trinomial has more than one variable, we take the same approach used for trinomials with one variable. Thus trinomials of the form

$$x^2 + bxy + cy^2$$

factor as

$$(x + my)(x + ny)$$

where $mn = c$ and $m + n = b$.

▶ **EXAMPLE 5** **Factoring Trinomials with Two Variables**

Factor: $p^2 + 6pq - 16q^2$

Solution

The trinomial $p^2 + 6pq - 16q^2$ will factor as $(p + mq)(p + nq)$, where $mn = -16$ and $m + n = 6$. We want factors of $c = -16$ whose sum is $b = 6$. We list factors of -16, but because $c = -16$ and $b = 6$, the larger factor of 16 will be positive.

Factors of -16	$-1, 16$	$-2, 8$	$-4, 4$
Sum	15	6	0

We can see that $-2 \cdot 8 = -16$ and $-2 + 8 = 6$, so $m = -2$ and $n = 8$. Write the trinomial in the form $(p + mq)(p + nq)$.

$$p^2 + 6pq - 16q^2 = (p + (-2)q)(p + 8q)$$
$$= (p - 2q)(p + 8q)$$

Check $(p - 2q)(p + 8q) = p^2 + 8pq - 2pq - 16q^2$
$$= p^2 + 6pq - 16q^2$$ ●

Quick ✓

In Problems 13 and 14, factor each trinomial completely.

13. $x^2 + 8xy + 15y^2$ **14.** $m^2 + mn - 20n^2$

We say a polynomial is factored completely if each factor in the final factorization is prime. For example, $3x^2 - 6x - 45 = (x - 5)(3x + 9)$ is not factored completely because $(3x + 9)$ has a common factor of 3. However, $3x^2 - 6x - 45 = 3(x - 5)(x + 3)$ is factored completely. The first step in any factoring problem is to look for a greatest common factor.

Work Smart

What's the first thing to look for when factoring any polynomial? A greatest common factor!

▶ **EXAMPLE 6** **Factoring Trinomials with a Common Factor**

Factor: $3u^3 - 9u^2 - 120u$

Solution

We start by noticing the common factor of $3u$ in the trinomial. Factor the $3u$ out:

$$3u^3 - 9u^2 - 120u = 3u(u^2 - 3u - 40)$$

Now factor the trinomial in parentheses, $u^2 - 3u - 40$. We want factors of $c = -40$ whose sum is $b = -3$. We list all factors of -40, but since $c = -40$ and $b = -3$, the larger factor (in absolute value) will be negative.

Factors of -40	$1, -40$	$2, -20$	$4, -10$	$5, -8$
Sum		-39	-18	-6	-3

Since $5 \cdot (-8) = -40$ and $5 + (-8) = -3$, then $m = 5$ and $n = -8$. Write the trinomial in the form $(u + m)(u + n)$.

$$u^2 - 3u - 40 = (u + 5)(u + (-8))$$
$$= (u + 5)(u - 8)$$

Remember, we already factored out a common factor, so

$$3u^3 - 9u^2 - 120u = 3u(u^2 - 3u - 40)$$
$$= 3u(u + 5)(u - 8)$$

Check

$$3u(u + 5)(u - 8) = 3u(u^2 - 8u + 5u - 40)$$
$$= 3u(u^2 - 3u - 40)$$
$$= 3u^3 - 9u^2 - 120u$$

Quick ✓

15. *True or False* The polynomial $(x + 5)(3x - 15)$ is factored completely.

In Problems 16 and 17, factor each trinomial completely.

16. $2x^3 - 12x^2 - 54x$

17. $-3z^2 - 21z - 30$

❷ Factor Trinomials of the Form $ax^2 + bx + c, a \neq 1$

When it comes to factoring trinomials of the form $ax^2 + bx + c$, where a is not 1, we have a choice of two methods:

1. Factoring by grouping

2. Trial and error

Both methods have pros and cons, which we will point out.

▶ Factoring $ax^2 + bx + c, a \neq 1$, by Grouping

Example 7 illustrates how to factor $ax^2 + bx + c, a \neq 1$, by grouping.

EXAMPLE 7 **How to Factor $ax^2 + bx + c, a \neq 1$, by Grouping**

Factor: $3x^2 + 14x + 8$

Step-by-Step Solution

First, notice that $3x^2 + 14x + 8$ has no common factors and that $a = 3$, $b = 14$, and $c = 8$.

Step 1: Find the value of ac.   The value of $a \cdot c = 3 \cdot 8 = 24$.

Step 2: Find the integers m and n, whose product is 24 and whose sum is 14. Because both 24 and 14 are positive, list only the positive factors of 24.

Factors of 24	1, 24	2, 12	3, 8	4, 6
Sum	25	14	11	10

The integers whose product is 24 and whose sum is 14 are 2 and 12.

Step 3: Write

$$ax^2 + bx + c = ax^2 + mx + nx + c.$$

Write $3x^2 + 14x + 8$ as $3x^2 + 2x + 12x + 8$

$14x = 2x + 12x$

Step 4: Factor the expression in Step 3 by grouping.

$$3x^2 + 2x + 12x + 8 = (3x^2 + 2x) + (12x + 8)$$
$$= x(3x + 2) + 4(3x + 2)$$

Factor out $(3x + 2)$: $= (3x + 2)(x + 4)$

Step 5: Check

$$(3x + 2)(x + 4) = 3x^2 + 12x + 2x + 8$$
$$= 3x^2 + 14x + 8$$

Thus $3x^2 + 14x + 8 = (3x + 2)(x + 4)$.

> **Factoring** $ax^2 + bx + c, a \neq 1$**, by Grouping, Where** *a, b,* **and** *c*
> **Have No Common Factors**
>
> **Step 1:** Find the value of *ac*.
>
> **Step 2:** Find the pair of integers whose product is *ac* and whose sum is *b*. That is, find *m* and *n* so that $mn = ac$ and $m + n = b$.
>
> **Step 3:** Write $ax^2 + bx + c = ax^2 + mx + nx + c$.
>
> **Step 4:** Factor the expression in Step 3 by grouping.
>
> **Step 5:** Check by multiplying the factors.

(**EXAMPLE 8**)   **Factoring** $ax^2 + bx + c, a \neq 1$**, by Grouping**

Factor: $12x^2 - x - 6$

Solution

First, notice that $12x^2 - x - 6$ has no common factors and that $a = 12$, $b = -1$, and $c = -6$. The value of $a \cdot c = 12 \cdot (-6) = -72$.

We want to find the factors of -72 whose sum is -1. Because $-72 < 0$, one integer will be positive and the other negative. Because $-1 < 0$, the factor of -72 with the larger absolute value will be negative.

Factors of -72	$1, -72$	$2, -36$	$3, -24$	$4, -18$	$6, -12$	$8, -9$
Sum	-71	-34	-21	-14	-6	-1

The integers whose product is -72 and whose sum is -1 are 8 and -9.

Work Smart

Be careful with the negative sign on $-9x$.

$$-x = 8x - 9x$$

$$12x^2 - x - 6 = 12x^2 + 8x - 9x - 6$$
$$= (12x^2 + 8x) + (-9x - 6)$$
$$= 4x(3x + 2) - 3(3x + 2)$$

Factor out $(3x + 2)$: $= (3x + 2)(4x - 3)$

Check $(3x + 2)(4x - 3) = 12x^2 - 9x + 8x - 6$
$$= 12x^2 - x - 6$$

Thus $12x^2 - x - 6 = (3x + 2)(4x - 3)$.

●

Quick ✓

18. To factor $2x^2 - 13x + 6$ using grouping, begin by finding factors whose product is ___, and whose sum is ___.

In Problems 19 and 20, factor each trinomial completely.

19. $2b^2 + 7b - 15$ **20.** $10x^2 + 27x + 18$

The advantage of factoring by grouping when factoring trinomials of the form $ax^2 + bx + c$, $a \neq 1$, is that it is algorithmic (that is, step by step). However, if the product $a \cdot c$ is large, you must find the sum of many pairs of factors. This can be overwhelming. If this is the case, it may be better to try the second method, trial and error.

Using Trial and Error to Factor $ax^2 + bx + c, a \neq 1$

▶ In the trial and error method, you list various binomials and use FOIL to find their product until you find the pair of binomials that results in the original trinomial. This method may sound haphazard, but experience and logic minimize the number of possibilities you must try.

EXAMPLE 9 **How to Factor $ax^2 + bx + c, a \neq 1$, Using Trial and Error**

Factor: $10x^2 + 19x + 6$

Step-by-Step Solution

Always check for common factors first. There are no common factors in $10x^2 + 19x + 6$.

Step 1: List the possibilities for the first terms of each binomial whose product is ax^2.

List all possible ways of representing the first term, $10x^2$.

$$(10x + \_)(x + \_)$$
$$(5x + \_)(2x + \_)$$

Step 2: List the possibilities for the last terms of each binomial whose product is c.

The last term, 6, has the following factors:

$$1 \cdot 6, 2 \cdot 3, -1 \cdot (-6), \text{ or } -2 \cdot (-3).$$

Step 3: Write out all the combinations of factors from Steps 1 and 2. Multiply the pairs of binomials until we find a product that equals the trinomial.

Possible Factorization	Product	Possible Factorization	Product
$(10x + 1)(x + 6)$	$10x^2 + 61x + 6$	$(5x + 1)(2x + 6)$	$10x^2 + 32x + 6$
$(10x + 6)(x + 1)$	$10x^2 + 16x + 6$	$(5x + 6)(2x + 1)$	$10x^2 + 17x + 6$
$(10x + 2)(x + 3)$	$10x^2 + 32x + 6$	$(5x + 2)(2x + 3)$	$10x^2 + 19x + 6$
$(10x + 3)(x + 2)$	$10x^2 + 23x + 6$	$(5x + 3)(2x + 2)$	$10x^2 + 16x + 6$
$(10x - 1)(x - 6)$	$10x^2 - 61x + 6$	$(5x - 1)(2x - 6)$	$10x^2 - 32x + 6$
$(10x - 6)(x - 1)$	$10x^2 - 16x + 6$	$(5x - 6)(2x - 1)$	$10x^2 - 17x + 6$
$(10x - 2)(x - 3)$	$10x^2 - 32x + 6$	$(5x - 2)(2x - 3)$	$10x^2 - 19x + 6$
$(10x - 3)(x - 2)$	$10x^2 - 23x + 6$	$(5x - 3)(2x - 2)$	$10x^2 - 16x + 6$

The highlighted row is the factorization that works, so $10x^2 + 19x + 6 = (5x + 2)(2x + 3)$. ●

The solution to Example 9 may make you feel a little overwhelmed—so many possibilities! However, we could have eliminated many of the possibilities with a little thought. For example, both the middle and the last terms are positive, so the factors of c must be positive—this alone eliminates half the possibilities listed in Example 9. Further, because the original polynomial has no common factors, the binomials in the factored form cannot have common factors either. This eliminates four additional possibilities.

When we use these hints, the list in Example 9 becomes

Possible Factorization	Product
$(10x + 1)(x + 6)$	$10x^2 + 61x + 6$
$(10x + 3)(x + 2)$	$10x^2 + 23x + 6$
$(5x + 6)(2x + 1)$	$10x^2 + 17x + 6$
$(5x + 2)(2x + 3)$	$10x^2 + 19x + 6$

Not too bad! There are only four possibilities. We summarize the steps to factor using trial and error and some helpful hints.

Factoring $ax^2 + bx + c, a \neq 1$, **Using Trial and Error:** a, b, **and** c **Have No Common Factors**

Step 1: List the possibilities for the first terms of each binomial whose product is ax^2.

$$(\square x +)(\square x +) = ax^2 + bx + c$$

Step 2: List the possibilities for the last terms of each binomial whose product is c.

$$(\underline{} x + \square)(\underline{} x + \square) = ax^2 + bx + c$$

Step 3: Write all the combinations of factors found in Steps 1 and 2. Multiply the pairs of binomials until you find a product that equals the trinomial.

Hints for Using Trial and Error to Factor $ax^2 + bx + c, a \neq 1$

- Look first for a common factor. If there is a GCF, factor it out. If the leading coefficient is negative, factor it out as part of the GCF.
- If the constant c is positive, then the factors of c must be the same sign as b.
- If $ax^2 + bx + c$ has no common factor, then the binomials in the factored form cannot have common factors either.
- If the value of b is small, then choose factors of ac that are close to each other. If the value of b is large, then choose factors of ac that are far from each other.
- If the value of b is correct except for the sign, then interchange the signs in the binomial factors.

▶ (EXAMPLE 10)   **Factoring** $ax^2 + bx + c, a \neq 1$, **Using Trial and Error**

Factor: $18x^2 + 3x - 10$

Solution

There are no common factors in $18x^2 + 3x - 10$. List all possible ways of representing the first term, $18x^2$.

$$(18x + \underline{})(x + \underline{})$$
$$(9x + \underline{})(2x + \underline{})$$
$$(6x + \underline{})(3x + \underline{})$$

The last term, -10, has the factor pairs $1 \cdot (-10), 2 \cdot (-5), -1 \cdot 10$, and $-2 \cdot 5$.

Before we list some combinations of factors from Steps 1 and 2, we notice that the middle term is $3x$. Because this term is positive and small, the binomial factors we list should have outer and inner products that sum to a positive, small number. Therefore, we will start with $(6x + \underline{})(3x + \underline{})$ and the factors $2 \cdot (-5)$ and $-2 \cdot 5$. We do not use $6x + 2$ or $6x - 2$ as a possible factor, because there is a common factor in these binomials.

Let's try $(6x - 5)(3x + 2)$.

$$(6x - 5)(3x + 2) = 18x^2 + 12x - 15x - 10$$
$$= 18x^2 - 3x - 10$$

Close! The only problem is that the middle term has the opposite sign from what we want. Therefore, interchange the signs on -5 and 2 in the binomial factors:

$$(6x + 5)(3x - 2) = 18x^2 - 12x + 15x - 10$$
$$= 18x^2 + 3x - 10$$

Thus $18x^2 + 3x - 10 = (6x + 5)(3x - 2)$.

●

The moral of the story in Example 10 is that the name *trial and error* is a bit misleading. You won't have to choose binomial factors haphazardly "until the cows come home," provided that you use the helpful hints and some careful thought.

Quick ✓

In Problems 21 and 22, factor each trinomial completely.

21. $8x^2 + 14x + 5$ **22.** $12y^2 + 32y - 35$

EXAMPLE 11   **Factoring Trinomials with Two Variables**

Factor: $24x^2 + 13xy - 2y^2$

Solution

There are no common factors in $24x^2 + 13xy - 2y^2$. The trinomial will factor in the form $24x^2 + 13xy - 2y^2 = (\_x + \_y)(\_x + \_y)$.

     We list all possible ways of representing the first term, $24x^2$.

$$(24x + \_y)(x + \_y)$$
$$(12x + \_y)(2x + \_y)$$
$$(8x + \_y)(3x + \_y)$$
$$(6x + \_y)(4x + \_y)$$

The last term, -2, has the factor pairs $1 \cdot (-2)$ or $2 \cdot (-1)$.

     Notice that the middle term, $13xy$, is positive and neither large nor small, so the binomial factors we list should have outer and inner products that sum to a positive, midsize number. The only coefficients on y in the factored form are 1 and 2 (ignoring their sign for a moment). We cannot have 2 as a factor in the forms $(12x + \_y)(2x + \_y)$ or $(6x + \_y)(4x + \_y)$ because 2 would create a common factor. Therefore, we will start with $(8x + \_y)(3x + \_y)$ and the factors $1 \cdot (-2)$ and $2 \cdot (-1)$.

     Let's try $(8x + 1y)(3x - 2y)$.

$$(8x + 1y)(3x - 2y) = 24x^2 - 16xy + 3xy - 2y^2$$
$$= 24x^2 - 13xy - 2y^2$$

Close! The only problem is that the middle term has the opposite sign from what we want. Therefore, interchange the signs in the binomial factors:

$$(8x - 1y)(3x + 2y) = 24x^2 + 16xy - 3xy - 2y^2$$
$$= 24x^2 + 13xy - 2y^2$$

Thus $24x^2 + 13xy - 2y^2 = (8x - y)(3x + 2y)$.   ●

Quick ✓

In Problems 23 and 24, factor each trinomial completely.

23. $30x^2 + 7xy - 2y^2$ **24.** $8x^2 - 10xy - 42y^2$

EXAMPLE 12   **Factoring Trinomials with a Negative Leading Coefficient**

Factor: $-14x^2 + 29x + 15$

Solution

There are no common factors in $-14x^2 + 29x + 15$, but the coefficient of the square term is negative. We will factor -1 out of the trinomial to obtain

$$-14x^2 + 29x + 15 = -1(14x^2 - 29x - 15)$$

Now factor the expression in parentheses, using either the grouping technique or trial and error, and obtain

$$-14x^2 + 29x + 15 = -1(14x^2 - 29x - 15)$$
$$= -(7x + 3)(2x - 5)$$

Quick ✓

In Problems 25 and 26, factor each trinomial completely.

25. $-6y^2 + 23y + 4$ **26.** $-9x^2 - 21xy - 10y^2$

▶ ❸ Factor Trinomials Using Substitution

Sometimes we can factor a complicated-looking polynomial by substituting one variable for another. This approach is called **factoring by substitution.**

(EXAMPLE 13) **Factoring by Substitution**

Factor: $2n^6 - n^3 - 15$

Solution

Notice that $2n^6 - n^3 - 15$ can be written as $2(n^3)^2 - n^3 - 15$ so that the trinomial is in the form $au^2 + bu + c$, where $u = n^3$. Then we substitute u for n^3, as shown in the second line below.

$$2n^6 - n^3 - 15 = 2(n^3)^2 - n^3 - 15$$

Work Smart

To determine whether a trinomial can be factored using substitution, check to see whether the trinomial can be written in the form

$$a(☺)^2 + b(☺) + c$$

where ☺ is some algebraic expression.

Let $n^3 = u$: $= 2u^2 - u - 15$

Factor: $= (2u + 5)(u - 3)$

Let $u = n^3$: $= (2n^3 + 5)(n^3 - 3)$

Thus $2n^6 - n^3 - 15 = (2n^3 + 5)(n^3 - 3)$.

(EXAMPLE 14) **Factoring by Substitution**

Factor: $3(x - 3)^2 + 11(x - 3) - 4$

Solution

Notice that $3(x - 3)^2 + 11(x - 3) - 4$ can be written in the form $au^2 + bu + c$, where $u = x - 3$. If we substitute u for $x - 3$, we obtain

$$3(x - 3)^2 + 11(x - 3) - 4 = 3u^2 + 11u - 4$$

Factor: $= (3u - 1)(u + 4)$

We do not want to factor an expression in u; we want to factor an expression in x. Therefore, substitute $x - 3$ for u, as shown in the second line below.

$$3u^2 + 11u - 4 = (3u - 1)(u + 4)$$

Let $u = x - 3$: $= [3(x - 3) - 1][(x - 3) + 4]$

$$= (3x - 9 - 1)(x + 1)$$
$$= (3x - 10)(x + 1)$$

Thus $3(x - 3)^2 + 11(x - 3) - 4 = (3x - 10)(x + 1)$.

Quick ✓

27. When factoring $6x^4 - x^2 - 2$ by substitution, let $u =$ ___.

28. When factoring $3(2x - 3)^2 + 5(2x - 3) + 2$ by substitution, let $u =$ _____.

In Problems 29 and 30, factor each trinomial by substitution.

29. $y^4 - 2y^2 - 24$ **30.** $4(x - 3)^2 + 5(x - 3) - 6$

4.5 Exercises MyMathLab® Math XP PRACTICE

Problems 1–30 are the Quick ✓ s that follow the EXAMPLES.

Building Skills

In Problems 31–46, factor each trinomial completely. See Objective 1.

31. $x^2 + 8x + 15$ **32.** $x^2 + 8x + 12$

33. $p^2 + 3p - 18$ **34.** $z^2 + 3z - 28$

35. $r^2 + 10r + 25$ **36.** $y^2 - 12y + 36$

37. $s^2 + 7s - 60$ **38.** $q^2 + 2q - 80$

39. $2x^2 - 30x + 112$ **40.** $3z^2 - 48z + 144$

41. $-w^2 - 2w + 24$ **42.** $-p^2 + 3p + 54$

43. $x^2 + 7xy + 12y^2$ **44.** $m^2 + 7mn + 10n^2$

45. $p^3 + 2p^2q - 24pq^2$ **46.** $x^2y - 4xy^2 - 21y^3$

In Problems 47–62, factor each trinomial completely. See Objective 2.

47. $2p^2 - 15p - 8$ **48.** $3z^2 - 13z - 10$

49. $4y^2 - 11y + 6$ **50.** $6x^2 - 37x + 6$

51. $8s^2 + 2s - 3$ **52.** $12r^2 + 11r - 15$

53. $16z^2 + 8z - 15$ **54.** $18y^2 + 43y - 5$

55. $-18y^2 - 17y - 4$ **56.** $-20r^2 - 23r - 6$

57. $2x^2 + 11xy - 21y^2$ **58.** $3m^2 + 7mn - 6n^2$

59. $12r^2 - 69rs + 45s^2$ **60.** $12r^2 - 50rs + 8s^2$

61. $24r^2 + 23rs - 12s^2$ **62.** $18x^2 + 37xy - 20y^2$

In Problems 63–72, factor each trinomial completely. See Objective 3.

63. $x^4 + 3x^2 + 2$ **64.** $y^4 + 5y^2 + 6$

65. $m^2n^2 + 5mn - 14$ **66.** $r^2s^2 + 8rs - 48$

67. $(x + 1)^2 - 6(x + 1) - 16$

68. $(y - 3)^2 + 3(y - 3) + 2$

69. $(3r - 1)^2 - 9(3r - 1) + 20$

70. $(5z - 3)^2 - 12(5z - 3) + 32$

71. $2(y - 3)^2 + 13(y - 3) + 15$

72. $3(z + 3)^2 + 14(z + 3) + 8$

Mixed Practice

In Problems 73–104, factor each polynomial completely. If the polynomial cannot be factored, say it is prime. Be sure to look for a greatest common factor.

73. $10w^2 + 41w + 21$

74. $8q^2 + 26q + 11$

75. $4(2y + 1)^2 - 3(2y + 1) - 1$

76. $2(3z - 1)^2 + 3(3z - 1) + 1$

77. $12x^2 + 23xy + 10y^2$

78. $24m^2 + 58mn + 9n^2$

79. $y^2 + 2y - 27$

80. $t^2 - 5t + 8$

81. $x^4 + 6x^2 + 8$

82. $a^6 + 7a^3 + 12$

83. $z^6 + 9z^3 + 20$

84. $r^6 - 6r^3 + 8$

85. $r^2 - 12rs + 32s^2$

86. $p^2 - 14pq + 45q^2$

87. $8(z + 1)^2 + 2(z + 1) - 1$

88. $9(a + 2)^2 - 10(a + 2) + 1$

89. $3x^2 - 7x - 12$

90. $5w^2 - 10w + 12$

91. $2x^2 + 12x - 54$

92. $3y^2 - 6y - 189$

93. $-3r^2 + 39r - 120$

94. $-4s^2 - 32s - 48$

95. $-16m^2 + 12m + 70$

96. $-24y^2 - 39y + 18$

97. $48z^2 + 124z + 28$

98. $48w^2 + 20w - 42$

99. $3x^3 - 6x^2 - 240x$

100. $4x^3 - 52x^2 + 144x$

101. $8x^3y^2 - 76x^2y^2 + 140xy^2$

102. $-24m^3n - 18m^2n + 27mn$

103. $70r^4s - 36r^3s - 16r^2s$

104. $54x^3y + 33x^2y - 72xy$

Applying the Concepts

105. How to Make an Open Box An open box with a rectangular base is to be made from a piece of cardboard that is 24 by 30 inches by cutting a square piece of cardboard from each corner and turning up the sides. See the illustration. The volume V of the box, as a function of the length x of the side of the square cut from each corner, is

$$V(x) = 4x^3 - 108x^2 + 720x$$

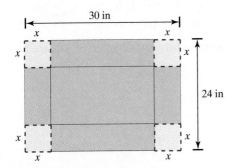

(a) Find and interpret $V(3)$.

(b) Completely factor the function $V(x) = 4x^3 - 108x^2 + 720x$.

(c) Find $V(3)$ using the factored form found in part (b).

(d) Did you find it easier to evaluate $V(3)$ in part (a) or in part (c)?

106. Projectile Motion A boy is standing on a cliff that is 240 feet high and throws a rock out toward the ocean. The height of the rock above the sea can be described by the function in which t is time in seconds and s is the height in feet:

$$s(t) = -16t^2 + 32t + 240$$

(a) Find and interpret $s(3)$.

(b) Completely factor the function $s(t) = -16t^2 + 32t + 240$.

(c) Find $s(3)$ using the factored form found in part (b).

(d) Did you find it easier to evaluate $s(3)$ in part (a) or in part (c)?

107. Suppose that we know one factor of $6x^2 - 11x - 10$ is $3x + 2$. What is the other factor?

108. Suppose that we know one factor of $8x^2 + 22x - 21$ is $2x + 7$. What is the other factor?

Extending the Concepts

In this section, we factored polynomials with integer coefficients. We can also factor polynomials with rational coefficients. One technique for doing this is to first factor out the coefficient of the highest-degree term as a greatest common factor as follows:

$$\frac{1}{3}x^2 + \frac{4}{3}x + 1 = \frac{1}{3}(x^2 + 4x + 3)$$
$$= \frac{1}{3}(x + 3)(x + 1)$$

Use this technique in Problems 109–114.

109. $\frac{1}{2}x^2 + 3x + 4$

110. $\frac{1}{4}x^2 + 2x + 3$

111. $\frac{1}{3}p^2 - \frac{2}{3}p - 1$

112. $\frac{1}{4}z^2 + \frac{1}{2}z - \frac{15}{4}$

113. $\frac{4}{3}a^2 - \frac{8}{3}a - 32$

114. $\frac{3}{8}b^2 - \frac{15}{8}b - 9$

Math for the Future

In College Algebra, you will be required to factor expressions such as those found in Problems 115–118. Completely factor the following expressions. Hint: Factor by substitution.

115. $2^{2n} - 4 \cdot 2^n - 5$

116. $3^{2x} - 11 \cdot 3^x + 10$

117. $4^{2x} - 12 \cdot 4^x + 32$

118. $4^x - 6 \cdot 2^x + 8$

Explaining the Concepts

119. State the circumstances for which trial and error is a better approach than grouping when factoring $ax^2 + bx + c, a \neq 1$.

120. When factoring any polynomial, what is always the first step?

121. A student factored the trinomial $3p^2 - 9p - 30$ as $(3p + 6)(p - 5)$ on an exam but received only partial credit. Can you explain why?

122. How can you tell whether a trinomial is prime? Make up an example to demonstrate your reasoning.

4.6 Factoring Special Products

Objectives

❶ Factor Perfect Square Trinomials

❷ Factor the Difference of Two Squares

❸ Factor the Sum or Difference of Two Cubes

Are You Ready for This Section?

Before getting started, take the following readiness quiz. If you get a problem wrong, go back to the section cited and review the material.

R1. What is 1^2? 2^2? 3^2? 4^2? 5^2? [Section R.4, pp. 33–34]

R2. What is $\left(-\dfrac{3}{2}\right)^2$? [Section R.4, pp. 33–34]

In this section, we look at polynomials that can be categorized as having special formulas for factoring. We begin with perfect square trinomials.

▶ ❶ **Factor Perfect Square Trinomials**

Recall the perfect square trinomials from Section 4.2:

$$(A + B)^2 = A^2 + 2AB + B^2$$
$$(A - B)^2 = A^2 - 2AB + B^2$$

Reversing these formulas, we obtain a method for factoring perfect square trinomials.

Perfect Square Trinomials

$$A^2 + 2AB + B^2 = (A + B)^2$$
$$A^2 - 2AB + B^2 = (A - B)^2$$

Work Smart

The perfect squares are $1^2 = 1$, $2^2 = 4$, $3^2 = 9$, and so on. Any variable raised to an even exponent is a perfect square. So $x^2, x^4 = (x^2)^2, x^6 = (x^3)^2$ are all perfect squares.

A polynomial must satisfy two conditions to be a perfect square trinomial:

1. The first and last terms must be perfect squares. Examples of perfect squares are

$$81 = 9^2, 4x^2 = (2x)^2, \text{ and } 25a^4 = (5a^2)^2$$

2. The middle term must equal 2 or -2 times the product of the expressions being squared in the first and last terms.

Perfect square trinomials can be factored using the methods introduced in the last section; however, they can be factored more quickly using the formulas above.

(**EXAMPLE 1**) **Factoring Perfect Square Trinomials**

Factor:

(a) $z^2 - 10z + 25$ **(b)** $9x^2 + 48xy + 64y^2$ **(c)** $32m^4 - 48m^2 + 18$

Solution

(a) The first term, z^2, and the third term, $25 = 5^2$, are perfect squares. Because the middle term, $-10z$, equals $-2(z \cdot 5)$, we have a perfect square trinomial, so

$$z^2 - 10z + 25 = z^2 - 2 \cdot z \cdot 5 + 5^2$$

$$A = z, B = 5; A^2 - 2AB + B^2 = (A - B)^2: = (z - 5)^2$$

(b) The first term, $9x^2 = (3x)^2$, and the third term, $64y^2 = (8y)^2$, are perfect squares. Because the middle term, $48xy$, equals $2(3x \cdot 8y)$, we have a perfect square trinomial, so

$$9x^2 + 48xy + 64y^2 = (3x)^2 + 2 \cdot (3x \cdot 8y) + (8y)^2$$

$A = 3x; B = 8y; A^2 + 2AB + B^2 = (A + B)^2: = (3x + 8y)^2$

(c) Remember, in any factoring problem first look for a common factor.

$$32m^4 - 48m^2 + 18 = 2(16m^4 - 24m^2 + 9)$$

The first term in parentheses, $16m^4 = (4m^2)^2$, and the third term, $9 = 3^2$, are perfect squares. Because the middle term equals $-2(4m^2 \cdot 3)$, we have a perfect square trinomial, so

$$16m^4 - 24m^2 + 9 = (4m^2)^2 - 2 \cdot (4m^2 \cdot 3) + 3^2$$
$$= (4m^2 - 3)^2$$

Therefore,

$$32m^4 - 48m^2 + 18 = 2(16m^4 - 24m^2 + 9)$$
$$= 2(4m^2 - 3)^2$$

Quick ✓

1. A trinomial of the form $A^2 + 2AB + B^2$ or $A^2 - 2AB + B^2$ is called a _____ _____ _____.

2. $A^2 - 2AB + B^2 = $ _____.

3. *True or False* $4x^2 + 6x + 9$ is a perfect square trinomial.

In Problems 4–6, factor each trinomial completely.

4. $x^2 - 18x + 81$ **5.** $4x^2 + 20xy + 25y^2$ **6.** $18p^4 - 84p^2 + 98$

▶ ❷ Factor the Difference of Two Squares

Another special product introduced in Section 4.2 was the difference of two squares:

$$(A - B)(A + B) = A^2 - B^2$$

> **Difference of Two Squares**
>
> $$A^2 - B^2 = (A - B)(A + B)$$

EXAMPLE 2 **Factoring the Difference of Two Squares**

Factor:

(a) $y^2 - 100$ **(b)** $9x^2 - 16y^4$

Solution

(a) Notice that $y^2 - 100$ is the difference of two squares, y^2 and $100 = 10^2$. So

$$y^2 - 100 = y^2 - 10^2$$
$$= (y - 10)(y + 10)$$

(b) Notice that $9x^2 - 16y^4$ is the difference of two squares, $9x^2 = (3x)^2$ and $16y^4 = (4y^2)^2$.

$$9x^2 - 16y^4 = (3x)^2 - (4y^2)^2$$
$$= (3x - 4y^2)(3x + 4y^2)$$

Work Smart

Don't forget that you can always check your answer by multiplying out the factored form.

Quick ✓

7. $16z^2 - 25$ is called a _____ of _____ and factors into two binomials.

8. $P^2 - Q^2 = $ _____.

In Problems 9–11, factor completely.

9. $z^2 - 16$ 10. $16m^2 - 81n^2$ 11. $4a^2 - 9b^4$

EXAMPLE 3   **Factoring the Difference of Two Squares**

Factor:

(a) $32x^4 - 2$ **(b)** $x^2 + 10x + 25 - y^2$

Solution

(a) Remember to look for a common factor first. This polynomial has a common factor of 2.

$$32x^4 - 2 = 2(16x^4 - 1)$$

Difference of Two Squares, $16x^4 = (4x^2)^2, 1 = 1^2$: $= 2((4x^2)^2 - 1^2)$

$A = 4x^2, B = 1; A^2 - B^2 = (A - B)(A + B)$: $= 2(4x^2 - 1)(4x^2 + 1)$

We are not quite finished. Notice that $4x^2 - 1$ is the difference of two squares, $4x^2 = (2x)^2$ and $1 = 1^2$: $= 2((2x)^2 - 1^2)(4x^2 + 1)$

$A = 2x, B = 1; A^2 - B^2 = (A - B)(A + B)$: $= 2(2x - 1)(2x + 1)(4x^2 + 1)$

Work Smart

Perfect square trinomials are trinomials of the form
$A^2 + 2AB + B^2 = (A + B)^2$
$A^2 - 2AB + B^2 = (A - B)^2$

Check $2(2x - 1)(2x + 1)(4x^2 + 1) = 2(4x^2 - 1)(4x^2 + 1)$

Use FOIL: $= 2(16x^4 + 4x^2 - 4x^2 - 1)$

$= 2(16x^4 - 1)$

$= 32x^4 - 2$

Thus $32x^4 - 2 = 2(2x - 1)(2x + 1)(4x^2 + 1)$.

(b) Remember, when we have four or more terms, we should try to factor by grouping. However, notice that the first three terms, $x^2 + 10x + 25$, form a perfect square trinomial (which factors into a perfect square) and the last term, y^2, is a perfect square.

$$x^2 + 10x + 25 - y^2 = (x^2 + 10x + 25) - y^2$$

$$= (x^2 + 2(5)(x) + 5^2) - y^2$$

$x^2 + 2(5)x + 5^2 = (x + 5)^2$: $= (x + 5)^2 - y^2$

$(x + 5)^2 - y^2$ is the difference of two squares, where $A = (x + 5)$ and $B = y$: $\overset{A}{\overbrace{(x + 5}} \overset{B}{- y)}(\overset{A}{\overbrace{x + 5}} \overset{B}{+ y})$

Check

$$(x + 5 - y)(x + 5 + y) = (x + 5 - y) \cdot x + (x + 5 - y) \cdot 5 + (x + 5 - y) \cdot y$$

$$= x^2 + 5x - xy + 5x + 25 - 5y + xy + 5y - y^2$$

$$= x^2 + 10x + 25 - y^2$$

Thus $x^2 + 10x + 25 - y^2 = (x + 5 - y)(x + 5 + y)$. ●

You may wonder whether the sum of two squares factors. **The sum of two squares, $a^2 + b^2$, is prime and does not factor over the integers.** Therefore, binomials such as $x^2 + 4$ or $4y^2 + 81$ are prime.

12. *True or False* The sum of two squares, such as $z^2 + 25$, is prime.

In Problems 13 and 14, factor completely.

13. $3b^4 - 48$ **14.** $p^2 - 8p + 16 - q^2$

▶ ❸ Factor the Sum or Difference of Two Cubes

Work Smart

The perfect cubes are $1^3 = 1$, $2^3 = 8$, $3^3 = 27$, and so on. Any variable raised to a multiple of 3 is a perfect cube. Thus $x^3, x^6 = \left(x^2\right)^3, x^9 = \left(x^3\right)^3$ are all perfect cubes.

Consider the following products:

$$(A + B)(A^2 - AB + B^2) = A^3 - A^2B + AB^2 + A^2B - AB^2 + B^3$$
$$= A^3 + B^3$$
$$(A - B)(A^2 + AB + B^2) = A^3 + A^2B + AB^2 - A^2B - AB^2 - B^3$$
$$= A^3 - B^3$$

These products show us that we can factor the sum or difference of two cubes as follows:

The Sum of Two Cubes

$$A^3 + B^3 = (A + B)(A^2 - AB + B^2)$$

The Difference of Two Cubes

$$A^3 - B^3 = (A - B)(A^2 + AB + B^2)$$

EXAMPLE 4   **Factoring the Sum or Difference of Two Cubes**

Factor:

(a) $x^3 - 27$ **(b)** $8m^3 + 125n^6$

Solution

(a) We have the difference of two cubes, x^3 and $27 = 3^3$. Let $A = x$ and $B = 3$ in the factoring formula for the difference of two cubes.

$$A^3 - B^3 = (A - B)(A^2 + AB + B^2)$$
$$x^3 - 27 = x^3 - 3^3 = (x - 3)(x^2 + x(3) + 3^2)$$
$$= (x - 3)(x^2 + 3x + 9)$$

(b) We have the sum of two cubes, $8m^3 = (2m)^3$ and $125n^6 = (5n^2)^3$. Let $A = 2m$ and $B = 5n^2$ in the factoring formula for the sum of two cubes.

$$8m^3 + 125n^6 = ((2m)^3 + (5n^2)^3)$$
$$= (2m + 5n^2)((2m)^2 - (2m)(5n^2) + (5n^2)^2)$$
$$= (2m + 5n^2)(4m^2 - 10mn^2 + 25n^4)$$

15. $A^3 + B^3 = ($____$)($_____$)$.

16. $A^3 - B^3 = ($____$)($_____$)$.

In Problems 17 and 18, factor completely.

17. $z^3 + 64$ **18.** $125p^3 - 216q^6$

EXAMPLE 5 **Factoring the Sum or Difference of Two Cubes**

Factor:

(a) $27x^3 - 216y^6$

(b) $(x - 4)^3 + 8x^3$

Solution

(a) We have the difference of two cubes, $27x^3 = (3x)^3$ and $216y^6 = (6y^2)^3$. However, there is a common factor of 27 in the binomial $27x^3 - 216y^6$. How did we recognize this common factor? Notice that our two cubes, $(3x)^3$ and $(6y^2)^3$, contain a 3 and a 6. Because the common factor, 3, is being cubed, we know that $3^3 = 27$ is a common factor. Therefore, we factor out the 27.

$$27x^3 - 216y^6 = 27(x^3 - 8y^6)$$
$$= 27(x^3 - (2y^2)^3)$$
$$A = x, B = 2y^2: \quad = 27(x - 2y^2)(x^2 + x \cdot 2y^2 + (2y^2)^2)$$
$$= 27(x - 2y^2)(x^2 + 2xy^2 + 4y^4)$$

(b) Notice that we have the sum of two cubes, $(x - 4)^3$ and $8x^3 = (2x)^3$. Let $A = x - 4$ and $B = 2x$ in the factoring formula for the sum of two cubes.

$$(x - 4)^3 + 8x^3 = (x - 4)^3 + (2x)^3$$
$$= (x - 4 + 2x)((x - 4)^2 - (x - 4) \cdot 2x + (2x)^2)$$
Combine like terms; multiply: $= (3x - 4)(x^2 - 8x + 16 - 2x^2 + 8x + 4x^2)$
Combine like terms: $= (3x - 4)(3x^2 + 16)$

Quick ✓

In Problems 19 and 20, factor completely.

19. $32m^3 + 500n^6$

20. $(x + 1)^3 - 27x^3$

4.6 Exercises MyMathLab® PRACTICE

Exercise numbers in green have complete video solutions in MyMathLab.

*Problems **1–20** are the Quick✓s that follow the EXAMPLES.*

35. $-5t^2 - 70t - 245$

36. $-2a^2 - 32a - 128$

37. $32a^2 - 80ab + 50b^2$

38. $12x^2 - 84xy + 147y^2$

Building Skills

39. $z^4 - 6z^2 + 9$

40. $b^4 + 8b^2 + 16$

In Problems 21–40, factor each perfect square trinomial completely. See Objective 1.

21. $x^2 + 4x + 4$

22. $y^2 + 6y + 9$

In Problems 41–56, factor the difference of two squares completely. See Objective 2.

23. $36 + 12w + w^2$

24. $49 - 14d + d^2$

41. $x^2 - 9$

42. $z^2 - 64$

25. $4x^2 + 4x + 1$

26. $9z^2 - 6z + 1$

43. $4 - y^2$

44. $81 - a^2$

27. $9p^2 - 30p + 25$

28. $16y^2 - 24y + 9$

45. $4z^2 - 9$

46. $16y^2 - 81$

29. $25a^2 + 90a + 81$

30. $36b^2 + 84b + 49$

47. $100m^2 - 81n^2$

48. $x^4 - 9y^2$

31. $9x^2 + 24xy + 16y^2$

32. $4a^2 + 20ab + 25b^2$

49. $m^4 - 36n^6$

50. $x^8 - 100y^4$

33. $3w^2 - 30w + 75$

34. $4c^2 - 24c + 36$

51. $8p^2 - 18q^2$

52. $12m^2 - 75n^2$

53. $80p^2r - 245b^2r$

54. $36x^2z - 64y^2z$

55. $(x + y)^2 - 9$

56. $16 - (x - y)^2$

In Problems 57–72, factor the sum or difference of two cubes completely. See Objective 3.

57. $x^3 - 8$

58. $z^3 + 64$

59. $125 + m^3$

60. $216 - n^3$

61. $x^6 - 64y^3$

62. $m^6 - 27n^3$

63. $24x^3 - 375y^3$

64. $16m^3 + 54n^3$

65. $(p + 1)^3 - 27$

66. $(y - 2)^3 - 8$

67. $(3y + 1)^3 + 8y^3$

68. $(2z + 3)^3 + 27z^3$

69. $y^6 + z^9$

70. $m^9 + n^{12}$

71. $y^9 - 1$

72. $x^{12} - 1$

Mixed Practice

In Problems 73–98, factor each polynomial completely.

73. $25x^2 - y^2$

74. $9a^2 - b^2$

75. $8x^3 + 27$

76. $64x^3 - 125$

77. $z^2 - 8z + 16$

78. $p^2 - 20p + 100$

79. $5x^4 - 40xy^3$

80. $3m^4 - 81mn^3$

81. $49m^2 - 42mn + 9n^2$

82. $81p^2 - 72pq + 16q^2$

83. $4x^2 + 16$

84. $-3y^2 - 27$

85. $y^4 - 8y^2 + 16$

86. $p^4 - 18p^2 + 81$

87. $4a^2b^2 + 12ab + 9$

88. $9m^2n^2 - 30mn + 25$

89. $p^2 + 2p + 4$

90. $y^2 - 3y + 9$

91. $n^2 + 13n + 36$

92. $t^2 + 10t + 16$

93. $2n^2 - 2m^2 + 40m - 200$

94. $2a^2 - 2b^2 - 24b - 72$

95. $3y^3 - 24$

96. $-5a^3 - 40$

97. $16x^2 + 24xy + 9y^2 - 100$

98. $36m^2 + 12mn + n^2 - 81$

Applying the Concepts

△ *In Problems 99–102, find an expression in factored form for the area of the shaded region.*

99.
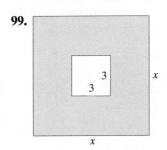

100.
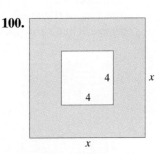

101.
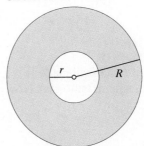

102.
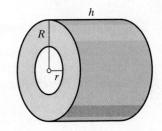

△ *In Problems 103–106, find an expression in factored form for the area or volume of the shaded region.*

103. Circle: *Area* $= \pi r^2$

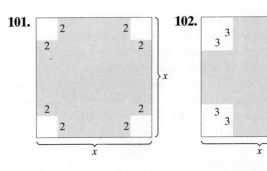

104. Cylinder: *Volume* $= \pi r^2 h$

105. Rectangular solid: *Volume* $= lwh$

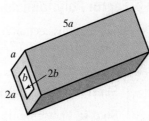

106. Sphere: *Volume* $= \dfrac{4}{3}\pi r^3$

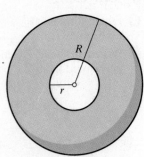

△ 107. **A Perfect Square** What is a perfect square? The figure below shows a square whose dimensions are $(a + b)$ by $(a + b)$. The area of the square is therefore $(a + b)^2$. Write the area of the square as the sum of the areas of the four quadrilaterals in the figure. Then show that this sum is equal to $(a + b)^2$.

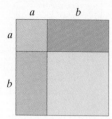

△ 108. **Difference of Two Squares** What is the difference of two squares? The figure below shows two squares. The length of the sides on the "outer" square is a and the length of the sides on the "inner" square (in orange) is b. The area of the shaded region (in blue) is $a^2 - b^2$. Express the area of the blue shaded region as the sum of the two shaded regions in terms of a and b. Conclude that $a^2 - b^2 = (a + b)(a - b)$.

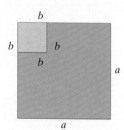

Extending the Concepts

109. Determine two values of b that will make $4x^2 + bx + 81$ a perfect square trinomial. How did you find these values?

110. Determine the value of c that will make $16y^2 + 24y + c$ a perfect square trinomial. How did you find this value?

111. What has to be added to $x^2 + 18x$ to make it a perfect square trinomial?

112. What has to be added to $4x^2 + 36x$ to make it a perfect square trinomial?

In this section, we factored polynomials with integer coefficients. We can also factor polynomials with rational number coefficients. For example,

$$\frac{x^2}{25} - \frac{1}{4} = \left(\frac{x}{5}\right)^2 - \left(\frac{1}{2}\right)^2$$
$$= \left(\frac{x}{5} - \frac{1}{2}\right)\left(\frac{x}{5} + \frac{1}{2}\right)$$

Use this technique to factor Problems 113–120.

113. $b^2 - 0.4b + 0.04$ 114. $x^2 + 0.6x + 0.09$

115. $9b^2 - \dfrac{1}{25}$ 116. $100x^2 - \dfrac{1}{81}$

117. $\dfrac{x^2}{9} - \dfrac{y^2}{25}$ 118. $\dfrac{a^2}{36} - \dfrac{b^2}{49}$

119. $\dfrac{x^3}{8} - \dfrac{y^3}{27}$ 120. $\dfrac{a^3}{27} + \dfrac{b^3}{64}$

4.7 Factoring: A General Strategy

Objectives

1. Factor Polynomials Completely
2. Write Polynomial Functions in Factored Form

1 Factor Polynomials Completely

▶ We begin this section by putting together all the factoring techniques we have discussed in Sections 4.4–4.6. The following steps should be followed for any factoring problem.

Steps for Factoring

Step 1: Factor out the greatest common factor (GCF), if any exists.

Step 2: Count the number of terms.

Step 3: (a) Two terms
- Is it the difference of two squares? If so,

$$A^2 - B^2 = (A - B)(A + B)$$

- Is it the sum of two squares? If so, stop! The expression is prime.

Work Smart

When factoring, check that the binomial factors do not factor further. See Example 6.

- Is it the difference of two cubes? If so,

$$A^3 - B^3 = (A - B)(A^2 + AB + B^2)$$

- Is it the sum of two cubes? If so,

$$A^3 + B^3 = (A + B)(A^2 - AB + B^2)$$

(b) Three terms

- Is it a perfect square trinomial? If so,

$$A^2 + 2AB + B^2 = (A + B)^2 \quad \text{or} \quad A^2 - 2AB + B^2 = (A - B)^2$$

- Is the coefficient of the square term 1? If so,

$$x^2 + bx + c = (x + m)(x + n), \text{ where } mn = c \text{ and } m + n = b$$

- Is the coefficient of the square term different from 1? If so,
  (a) Use factoring by grouping or
  (b) Use trial and error

(c) Four terms

- Use factoring by grouping

Step 4: Check your work by multiplying the factors.

▶ **EXAMPLE 1** **How to Factor Completely**

Factor: $8x^2 - 16x - 42$

Step-by-Step Solution

Step 1: Factor out the greatest common factor (GCF), if any exists.

The GCF is 2, so we factor out 2. $8x^2 - 16x - 42 = 2(4x^2 - 8x - 21)$

Step 2: Count the number of terms of the polynomial in parentheses.

The polynomial in parentheses has three terms.

Step 3: The trinomial in parentheses, $4x^2 - 8x - 21$, is not a perfect square trinomial. Because the leading coefficient, 4, and the constant, -21, aren't that big, we choose to factor by grouping.

$$ac = 4(-21) = -84$$

Because $6 \cdot (-14) = -84$ and $6 + (-14) = -8$,

$$4x^2 - 8x - 21 = 4x^2 + 6x - 14x - 21$$
$$= (4x^2 + 6x) + (-14x - 21)$$

GCF in 1st grouping: $2x$;
GCF in 2nd grouping: -7 $= 2x(2x + 3) - 7(2x + 3)$

Factor out $2x + 3$: $= (2x + 3)(2x - 7)$

Step 4: Check

$$2(2x + 3)(2x - 7) = 2(4x^2 - 14x + 6x - 21)$$
$$= 2(4x^2 - 8x - 21)$$

Distribute: $= 8x^2 - 16x - 42$

The answer checks, so $8x^2 - 16x - 42 = 2(2x + 3)(2x - 7)$.   ●

Quick ✓

In Problems 1 and 2, factor each polynomial completely.

1. $2p^2q - 8pq^2 - 90q^3$

2. $-45x^2y + 66xy + 27y$

▶ (**EXAMPLE 2**) **How to Factor Completely**

Factor: $9p^2 - 25q^2$

Step-by-Step Solution

Step 1: Factor out the greatest common factor (GCF), if any exists.

There is no GCF.

Step 2: Count the number of terms.

There are two terms.

Step 3: Because both the first term, $9p^2 = (3p)^2$, and the second term, $25q^2 = (5q)^2$, are perfect squares, we have a difference of two squares.

$$9p^2 - 25q^2 = \overset{A}{\overbrace{(3p)}}^2 - \overset{B}{\overbrace{(5q)}}^2$$
$$A^2 - B^2 = (A - B)(A + B): \quad = (3p - 5q)(3p + 5q)$$

Step 4: Check

$$(3p - 5q)(3p + 5q) = 9p^2 + 15pq - 15pq - 25q^2$$
$$\text{Combine like terms:} \quad = 9p^2 - 25q^2$$

Thus $9p^2 - 25q^2 = (3p - 5q)(3p + 5q)$.

Quick ✓

In Problems 3 and 4, factor each polynomial completely.

3. $81x^2 - 100y^2$

4. $-3m^2n + 147n$

(**EXAMPLE 3**) **Factoring Completely**

Factor: $16x^2 + 112xy + 196y^2$

Solution

We first look for, and find, the greatest common factor, 4.

$$16x^2 + 112xy + 196y^2 = 4(4x^2 + 28xy + 49y^2)$$

The polynomial in parentheses has three terms. The first term is a perfect square, $4x^2 = (2x)^2$. The third term is also a perfect square, $49y^2 = (7y)^2$. The middle term equals $2 \cdot 2x \cdot 7y$. This polynomial is a perfect square trinomial.

$$A = 2x; B = 7y: \quad 4(4x^2 + 28xy + 49y^2) = 4((2x)^2 + 2(2x)(7y) + (7y)^2)$$
$$A^2 + 2AB + B^2 = (A + B)^2: \quad = 4(2x + 7y)^2$$

Check $\quad 4(2x + 7y)^2 = 4(2x + 7y)(2x + 7y)$
$$\text{Use FOIL:} \quad = 4(4x^2 + 14xy + 14xy + 49y^2)$$
$$\text{Combine like terms:} \quad = 4(4x^2 + 28xy + 49y^2)$$
$$\text{Distribute:} \quad = 16x^2 + 112xy + 196y^2$$

Thus $16x^2 + 112xy + 196y^2 = 4(2x + 7y)^2$.

Quick ✓

In Problems 5 and 6, factor each polynomial completely.

5. $p^2 - 16pq + 64q^2$

6. $20x^2 + 60x + 45$

▶ **EXAMPLE 4** **Factoring Completely**

Factor: $8m^3 + 27n^6$

Solution

The polynomial does not have a common factor and it has two terms. Because the first term is a perfect cube, $8m^3 = (2m)^3$, and the second term is a perfect cube, $27n^6 = (3n^2)^3$, we have the sum of two cubes.

$$A = 2m; B = 3n^2: \qquad 8m^3 + 27n^6 = (2m)^3 + (3n^2)^3$$
$$A^3 + B^3 = (A + B)(A^2 - AB + B^2): \quad = (2m + 3n^2)((2m)^2 - (2m)(3n^2) + (3n^2)^2)$$
$$= (2m + 3n^2)(4m^2 - 6mn^2 + 9n^4)$$

Check $\quad (2m + 3n^2)(4m^2 - 6mn^2 + 9n^4) = 8m^3 - 12m^2n^2 + 18mn^4$
$$+ 12m^2n^2 - 18mn^4 + 27n^6$$

Combine like terms: $\qquad = 8m^3 + 27n^6$

Thus $8m^3 + 27n^6 = (2m + 3n^2)(4m^2 - 6mn^2 + 9n^4)$. ●

Quick ✓

In Problems 7 and 8, factor each polynomial completely.

7. $64y^3 - 125$ **8.** $-16m^3 - 2n^3$

EXAMPLE 5 **Factoring Completely**

Factor: $-4xy^2 + 12xy + 132x$

Solution

Factor out the GCF, $-4x$.

$$-4xy^2 + 12xy + 132x = -4x(y^2 - 3y - 33)$$

The polynomial in parentheses has three terms. Because 33 is not a perfect square, $y^2 - 3y - 33$ is not a perfect square trinomial. We need to find two factors of -33 whose sum is -3. There are no such factors. Therefore, $y^2 - 3y - 33$ is prime, so

$$-4xy^2 + 12xy + 132x = -4x(y^2 - 3y - 33)$$ ●

Quick ✓

In Problems 9 and 10, factor each polynomial completely.

9. $10z^2 - 15z + 35$ **10.** $6xy^2 + 81x^3$

EXAMPLE 6 **Factoring Completely**

Factor: $6x^3 - 4x^2 - 24x + 16$

Solution

Factor out the GCF, 2.

$$6x^3 - 4x^2 - 24x + 16 = 2(3x^3 - 2x^2 - 12x + 8)$$

Because there are four terms in the parentheses, try to factor by grouping. Group the first two terms and the last two terms. Watch out for the subtraction sign in front of the third term!

Factor by grouping: $\quad 2(3x^3 - 2x^2 - 12x + 8) = 2[(3x^3 - 2x^2) + (-12x + 8)]$

Factor out x^2; factor out -4: $\quad = 2[x^2(3x - 2) - 4(3x - 2)]$

Factor out $(3x - 2)$: $\quad = 2(3x - 2)(x^2 - 4)$

$x^2 - 4$ is the difference of two squares: $\quad = 2(3x - 2)(x - 2)(x + 2)$

$$\textbf{Check} \quad 2(3x - 2)(x - 2)(x + 2) = 2(3x - 2)(x^2 - 4)$$

$$\text{FOIL:} \quad = 2(3x^3 - 12x - 2x^2 + 8)$$

$$\text{Distribute; rearrange terms:} \quad = 6x^3 - 4x^2 - 24x + 16$$

Quick ✓

In Problems 11 and 12, factor each polynomial completely.

11. $2x^3 + 5x^2 + 4x + 10$

12. $9x^3 + 3x^2 - 9x - 3$

EXAMPLE 7 **Factoring Completely**

Factor: $x^2 - 6xy + 9y^2 - 25$

Solution

There are no common factors and there are four terms, so we try factoring by grouping. Attempts to form two groups of two terms fail. Before thinking that the polynomial is prime, notice that the first three terms form a perfect square trinomial, and the last term is a perfect square.

$$\text{Group the first three terms:} \quad x^2 - 6xy + 9y^2 - 25 = (x^2 - 6xy + 9y^2) - 25$$

$$= [x^2 - 2x(3y) + (3y)^2] - 25$$

$$A = x; B = 3y; A^2 - 2AB + B^2 = (A - B)^2: \quad = (x - 3y)^2 - 5^2$$

$$A = x - 3y; B = 5; A^2 - B^2 = (A - B)(A + B): \quad = (x - 3y - 5)(x - 3y + 5)$$

$$\textbf{Check} \quad (x - 3y - 5)(x - 3y + 5) = x^2 - 3xy + 5x - 3xy + 9y^2$$

$$-15y - 5x + 15y - 25$$

$$= x^2 - 6xy + 9y^2 - 25$$

Thus $x^2 - 6xy + 9y^2 - 25 = (x - 3y - 5)(x - 3y + 5)$.

Quick ✓

In Problems 13 and 14, factor each polynomial completely.

13. $4x^2 + 4xy + y^2 - 81$

14. $16 - m^2 - 8mn - 16n^2$

▶ ❷ **Write Polynomial Functions in Factored Form**

Later in this course and in other mathematics courses, you will be required to write polynomial functions in factored form.

EXAMPLE 8 **Writing a Polynomial Function in Factored Form**

Write each of the following polynomial functions in factored form.

(a) $f(x) = 8x^2 + 14x - 15$

(b) $H(z) = -3z^3 - 15z^2 - 27z - 135$

Solution

(a) To write $f(x) = 8x^2 + 14x - 15$ in factored form, we factor $8x^2 + 14x - 15$. First, we see that there are three terms in the polynomial, and the polynomial has no common factor. We will factor the polynomial using trial and error. We can represent the first term, $8x^2$, as

$$(x + \_\_)(8x + \_\_) \quad \text{or} \quad (2x + \_\_)(4x + \_\_)$$

The last term, -15, has the factor pairs $1 \cdot (-15), 3 \cdot (-5), -1 \cdot 15,$ and $-5 \cdot 3$. We want the middle term's coefficient to be 14, which is positive, but not too big. As we look for the correct combination of factors to obtain this result, we will start with $(2x + \_\_)(4x + \_\_)$. Since the middle term's coefficient is positive, let's try $+5$ and -3.

$$(2x + 5)(4x - 3) = 8x^2 - 6x + 20x - 15$$
$$= 8x^2 + 14x - 15$$

It works! Therefore,

$$f(x) = 8x^2 + 14x - 15$$
$$= (2x + 5)(4x - 3)$$

(b) To write $H(z) = -3z^3 - 15z^2 - 27z - 135$ in factored form, we factor $-3z^3 - 15z^2 - 27z - 135$.

$$H(z) = -3z^3 - 15z^2 - 27z - 135$$

Factor out common factor, -3: $\quad = -3(z^3 + 5z^2 + 9z + 45)$

Four terms; factor by grouping: $\quad = -3[(z^3 + 5z^2) + (9z + 45)]$

Factor out z^2; factor out 9: $\quad = -3[z^2(z + 5) + 9(z + 5)]$

Factor out $z + 5$ as a common factor: $\quad = -3(z + 5)(z^2 + 9)$

We leave the check to you.

Quick ✓

In Problems 15 and 16, write each polynomial function in factored form.

15. $G(x) = 2x^2 + 3x - 14$ **16.** $F(p) = 27p^2 - 12$

4.7 Exercises   MyMathLab®   MathXL PRACTICE

Problems 1–16 are the Quick ✓s that follow the EXAMPLES.

Building Skills

In Problems 17–64, factor each polynomial completely. See Objective 1.

17. $2x^2 - 12x - 144$ **18.** $3x^2 + 6x - 105$

19. $-3y^2 + 27$ **20.** $-5a^2 + 80$

21. $4b^2 + 20b + 25$ **22.** $8m^2 - 42m + 49$

23. $16w^3 + 2y^6$ **24.** $54p^6 - 2q^3$

25. $-3z^2 + 12z - 18$ **26.** $-4c^3 + 16c^2 - 28c$

27. $20y^2 - 9y - 18$ **28.** $18t^2 - 9t - 20$

29. $x^3 - 4x^2 + 5x - 20$ **30.** $p^3 + 7p^2 - 3p - 21$

31. $200x^2 + 18y^2$ **32.** $12p^2 + 50q^2$

33. $x^4 - 81$ **34.** $16w^4 - 1$

35. $3x^2 - 7x - 16$ **36.** $4w^2 - 3w - 6$

37. $36q^3 + 24q^2 + 4q$ **38.** $20k^3 - 60k^2 + 45k$

39. $24m^3n - 66m^2n - 63mn$

40. $20p^3q - 2p^2q - 4pq$

41. $3r^5 - 24r^2s^3$

42. $54p^5 + 16p^2q^3$

43. $2x^3 + 8x^2 - 18x - 72$

44. $3x^3 - 6x^2 - 48x + 96$

45. $9x^4 - 1$

46. $4z^4 - 25$

47. $3w^4 + 4w^2 - 15$

48. $4b^4 + 4b^2 - 15$

49. $(2y + 3)^2 - 5(2y + 3) + 6$

50. $(3x + 5)^2 + 4(3x + 5) - 21$

51. $p^2 - 10p + 25 - 36q^2$

52. $a^2 + 12a + 36 - 4b^2$

53. $y^6 + 6y^3 - 16$

54. $w^6 + 4w^3 - 5$

55. $p^6 - 1$

56. $q^6 + 1$

57. $-3x^3 - 15x^2 + 27x + 135$

58. $-2y^3 - 4y^2 + 32y + 64$

59. $3a - 27a^3$

60. $-5z - 20z^3$

61. $8t^5 + 14t^3 - 72t$

62. $18h^5 + 154h^3 - 72h$

63. $2x^4y + 10x^3y - 18x^2y - 90xy$

64. $2p^4q + 14p^3q - 32p^2q - 224pq$

In Problems 65–74, factor each polynomial function. See Objective 2.

65. $f(x) = x^2 - 2x - 63$

66. $F(x) = x^2 + 3x - 40$

67. $P(m) = 7m^2 + 31m + 12$

68. $H(p) = 5p^2 + 28p - 12$

69. $G(x) = -12x^2 + 147$

70. $g(x) = -100x^2 + 36$

71. $s(t) = -16t^2 + 96t + 256$

72. $S(y) = 2y^3 - 8y^2 - 42y$

73. $H(a) = 2a^3 + 5a^2 - 32a - 80$

74. $G(x) = 2x^3 - x^2 - 18x + 9$

Applying the Concepts

△ *In Problems 75–78, write an algebraic expression in completely factored form for the area that is shaded.*

75.

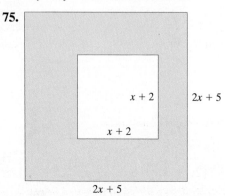

76.

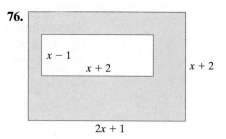

77.

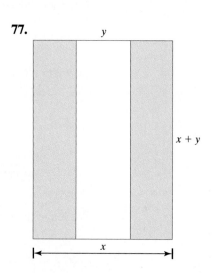

78.

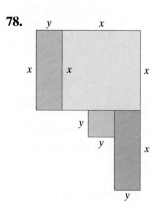

△ *In Problems 79 and 80, find an algebraic expression in factored form for the difference in the volumes of the two cubes shown.*

79.

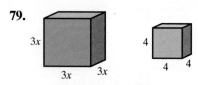

80.

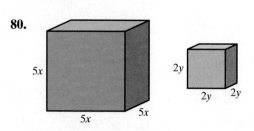

Extending the Concepts

81. Show that $x^2 + 9$ is prime.

82. Show that $x^2 - 4x - 8$ is prime.

In Problems 83–86, factor out the indicated common factor of the expression. Then completely factor the expression.

83. $x^{\frac{5}{2}} - 9x^{\frac{1}{2}}; x^{\frac{1}{2}}$

84. $x^{\frac{3}{2}} - 4x^{-\frac{1}{2}}; x^{-\frac{1}{2}}$

85. $1 + 6x^{-1} + 8x^{-2}; x^{-2}$

86. $3 - 2x^{-1} - x^{-2}; x^{-2}$

Explaining the Concepts

87. Write out the steps for factoring any polynomial.

88. What does factored completely mean?

4.8 Polynomial Equations

Objectives

1 Solve Polynomial Equations Using the Zero-Product Property

2 Solve Equations Involving Polynomial Functions

3 Model and Solve Problems Involving Polynomials

Are You Ready for This Section?

Before getting started, take the following readiness quiz. If you get a problem wrong, go back to the section cited and review the material.

R1. Solve: $x + 4 = 0$ [Section 1.1, pp. 50–52]

R2. Solve: $3(x - 2) - 12 = 0$ [Section 1.1, pp. 50–53]

R3. Evaluate $2x^2 + 3x + 1$
when (a) $x = 2$ (b) $x = -1$.   [Section R.5, pp. 40–41]

R4. If $f(x) = 4x + 3$, solve $f(x) = 11$.
What point is on the graph of f?   [Section 2.3, pp. 165–167]

R5. If $f(x) = -2x + 8$, find $f(4)$.
What point is on the graph of f?   [Section 2.3, pp. 165–167]

R6. Find the zero of $f(x) = \dfrac{2}{3}x - 6$.   [Section 2.4, pp. 175–176]

▶ In Sections 4.4–4.7, we learned how to factor polynomial expressions. Factoring is an important skill as you will see repeatedly in this course. We present one important use of factoring now. It turns out that factoring is really handy for solving *polynomial equations*. A **polynomial equation** is an equation that can be written in the form polynomial expression equals zero. The **degree of a polynomial equation** is the degree of the polynomial expression in the equation. Some examples of polynomial equations are

$$4x + 5 = 17 \qquad 2x^2 - 5x - 3 = 0 \qquad y^3 + 4y^2 = 3y + 18$$
Polynomial equation of degree 1   Polynomial equation of degree 2   Polynomial equation of degree 3

Work Smart

Remember, the degree of a polynomial is the value of the largest exponent on a variable. For example, the degree of $4x^3 - 9x^2 + 1$ is 3.

Notice that the polynomial equation $4x + 5 = 17$ is a linear, or first-degree, equation. We studied linear equations back in Section 1.1. In this section, we will learn methods for solving polynomial equations when the polynomial expression is of degree 2 or higher and factorable over the integers.

1 Solve Polynomial Equations Using the Zero-Product Property

In this section, we will use the following property to solve equations.

The Zero-Product Property

If the product of two factors is zero, then at least one of the factors is 0. That is, if $ab = 0$, then $a = 0$ or $b = 0$ or both a and b are 0.

▶ (EXAMPLE 1) **Using the Zero-Product Property**

Solve: $(x + 5)(2x - 3) = 0$

Solution

The product of two numbers, $x + 5$ and $2x - 3$, is equal to 0. By the Zero-Product Property, at least one of the numbers must equal 0. Therefore, we set each factor equal to 0 and solve each equation separately:

$$x + 5 = 0 \qquad \text{or} \qquad 2x - 3 = 0$$

Subtract 5 from both sides: $\quad x = -5 \qquad$ Add 3 to both sides: $\quad 2x = 3$

Divide both sides by 2: $\quad x = \dfrac{3}{2}$

Check

$x = -5$: $\quad (x + 5)(2x - 3) = 0$
$$(-5 + 5)(2(-5) - 3) \stackrel{?}{=} 0$$
$$0(-13) \stackrel{?}{=} 0$$
$$0 = 0 \quad \text{True}$$

$x = \dfrac{3}{2}$: $\quad (x + 5)(2x - 3) = 0$
$$\left(\frac{3}{2} + 5\right)\left(2 \cdot \left(\frac{3}{2}\right) - 3\right) \stackrel{?}{=} 0$$
$$\left(\frac{13}{2}\right) \cdot (3 - 3) \stackrel{?}{=} 0$$
$$\frac{13}{2} \cdot 0 \stackrel{?}{=} 0$$
$$0 = 0 \quad \text{True}$$

The solution set is $\left\{-5, \dfrac{3}{2}\right\}$.

Quick ✓

1. A _____ _____ is an equation that can be written in the form polynomial expression equals zero.

2. The Zero-Product Property states that if $a \cdot b = 0$, then _____ or _____.

In Problems 3 and 4, use the Zero-Product Property to solve the equation.

3. $x(x + 7) = 0$

4. $(x - 3)(4x + 3) = 0$

Using the Zero-Product Property to Solve Quadratic Equations

The Zero-Product Property can be used to solve *quadratic equations*.

Work Smart

Why can't a equal 0 in the definition? If a were equal to zero, the equation would be $bx + c = 0$, a linear equation.

> **Definition**
>
> A **quadratic equation** is polynomial equation that can be written in the form
> $$ax^2 + bx + c = 0$$
> where a, b, and c are real numbers and $a \neq 0$.

Some quadratic equations are

$$3x^2 + 5x + 2 = 0 \qquad -7z^2 + 14z = 0 \qquad y^2 - 16 = 0 \qquad p^2 + 8p = 16$$

A quadratic equation written in the form $ax^2 + bx + c = 0$ is said to be in **standard form.** The first three equations listed above are in standard form; the equation $p^2 + 8p = 16$ is not in standard form.

Sometimes, a quadratic equation is called a **second-degree equation** because the equation contains a polynomial of degree 2.

When a quadratic equation is written in standard form, $ax^2 + bx + c = 0$, it may be possible to factor the expression $ax^2 + bx + c$ as the product of two first-degree polynomials. In Sections 7.2 and 7.3, we will present methods for solving $ax^2 + bx + c = 0$ when we cannot factor the expression $ax^2 + bx + c$.

▶ **EXAMPLE 2** **How to Solve a Quadratic Equation by Factoring**

Solve: $2x^2 - 5x = 3$

Step-by-Step Solution

Step 1: Write the quadratic equation in standard form.

$$2x^2 - 5x = 3$$
Subtract 3 from both sides: $2x^2 - 5x - 3 = 0$

Step 2: Factor the expression on the left side of the equation.

$$(2x + 1)(x - 3) = 0$$

Step 3: Set each factor equal to 0.

Use the Zero-Product Property: $2x + 1 = 0$ or $x - 3 = 0$

Step 4: Solve each first-degree equation.

$$2x = -1 \quad \text{or} \quad x = 3$$
$$x = -\frac{1}{2}$$

Step 5: Check: Substitute $-\frac{1}{2}$ and 3 into the original equation.

$x = -\frac{1}{2}:\ 2x^2 - 5x = 3$

$$2\left(-\frac{1}{2}\right)^2 - 5\left(-\frac{1}{2}\right) \overset{?}{=} 3$$

$$2\left(\frac{1}{4}\right) + \frac{5}{2} \overset{?}{=} 3$$

$$\frac{1}{2} + \frac{5}{2} \overset{?}{=} 3$$

$$3 = 3 \quad \text{True}$$

$x = 3:\ 2x^2 - 5x = 3$

$$2(3)^2 - 5(3) \overset{?}{=} 3$$

$$18 - 15 \overset{?}{=} 3$$

$$3 = 3 \quad \text{True}$$

The solution set is $\left\{ -\frac{1}{2}, 3 \right\}$.

Below, we summarize the steps in solving a quadratic equation by factoring.

Solving a Quadratic Equation by Factoring

Step 1: Write the quadratic equation in standard form, $ax^2 + bx + c = 0$.

Step 2: Factor the expression on the left side of the equation.

Step 3: Set each factor found in Step 2 equal to zero using the Zero-Product Property.

Step 4: Solve each first-degree equation for the variable.

Step 5: Check your answers by substituting into the *original* equation.

Quick ✓

5. A _____ _____ is an equation equivalent to one of the form $ax^2 + bx + c = 0$, where $a, b,$ and c are real numbers and $a \neq 0$.

6. Quadratic equations are also known as _____ degree equations.

7. *True or False* The equation $7x - x^2 = 4$ is written in standard form.

In Problems 8–10, solve each quadratic equation by factoring.

8. $p^2 - 5p + 6 = 0$ **9.** $3t^2 - 14t = 5$

10. $4y^2 + 8y + 3 = y^2 - 1$

▶ EXAMPLE 3   Solving a Quadratic Equation by Factoring

Solve: $(m - 1)(3m + 5) = 16m$

Solution

First, write the equation in standard form.

$$(m - 1)(3m + 5) = 16m$$

FOIL: $3m^2 + 2m - 5 = 16m$

Subtract 16m from both sides: $3m^2 - 14m - 5 = 0$

Factor: $(3m + 1)(m - 5) = 0$

Set each factor equal to 0: $3m + 1 = 0$ or $m - 5 = 0$

$3m = -1$ or $m = 5$

$$m = -\frac{1}{3}$$

Work Smart

Do not attempt to solve the equation $(m - 1)(3m + 5) = 16m$ by setting each factor equal to 16m. The Zero-Product Property can be applied only when the product is equal to zero.

Check Substitute $m = -\dfrac{1}{3}$ and $m = 5$ into the original equation.

$m = -\dfrac{1}{3}$: $(m - 1)(3m + 5) = 16m$

$$\left(-\frac{1}{3} - 1\right)\left(3 \cdot \left(-\frac{1}{3}\right) + 5\right) \stackrel{?}{=} 16 \cdot \left(-\frac{1}{3}\right)$$

$$\left(-\frac{4}{3}\right)(-1 + 5) \stackrel{?}{=} -\frac{16}{3}$$

$$-\frac{16}{3} = -\frac{16}{3} \quad \text{True}$$

$m = 5$: $(m - 1)(3m + 5) = 16m$

$$(5 - 1)(3(5) + 5) \stackrel{?}{=} 16(5)$$

$$4(20) \stackrel{?}{=} 80$$

$$80 = 80 \quad \text{True}$$

The solution set is $\left\{-\dfrac{1}{3}, 5\right\}$.

Quick ✓

11. *True or False* The equation $(x - 2)(x + 3) = 12$ means that $x - 2 = 12$ or $x + 3 = 12$.

In Problems 12 and 13, solve each quadratic equation by factoring.

12. $x(x + 3) = -2$ 13. $(x - 3)(x + 5) = 9$

▶ **Using the Zero-Product Property to Solve Equations of Degree 3 or Higher**

We can use an extended form of the Zero-Product Property to solve polynomial equations of degree 3 or higher. The idea is to write the equation (in standard form), factor the expression that equals zero, set each factor equal to zero, and solve.

EXAMPLE 4   **How to Solve a Third-Degree Equation Using the Zero-Product Property**

Solve: $w^3 + 5w^2 - 4w = 20$

Step-by-Step Solution

Step 1: Write the equation in standard form by subtracting 20 from both sides.

$$w^3 + 5w^2 - 4w = 20$$

$$w^3 + 5w^2 - 4w - 20 = 0$$

Step 2: Factor the expression on the left side. Because there are four terms, we factor by grouping.

Group the 1st two terms; group the last two terms:

Factor out the common factor in each group:

Factor out $w + 5$:

Factor $w^2 - 4$:

$$(w^3 + 5w^2) + (-4w - 20) = 0$$

$$w^2(w + 5) - 4(w + 5) = 0$$

$$(w + 5)(w^2 - 4) = 0$$

$$(w + 5)(w + 2)(w - 2) = 0$$

Step 3: Set each factor equal to 0.

$w + 5 = 0$ or $w + 2 = 0$ or $w - 2 = 0$

Step 4: Solve each first-degree equation.

$w = -5$ or $w = -2$ or $w = 2$

Step 5: Check: Substitute $w = -5, w = -2$, and $w = 2$ into the original equation.

$w = -5$:

$w^3 + 5w^2 - 4w = 20$

$(-5)^3 + 5(-5)^2 - 4(-5) \stackrel{?}{=} 20$

$-125 + 125 + 20 \stackrel{?}{=} 20$

$20 = 20$

True

$w = -2$:

$w^3 + 5w^2 - 4w = 20$

$(-2)^3 + 5(-2)^2 - 4(-2) \stackrel{?}{=} 20$

$-8 + 20 + 8 \stackrel{?}{=} 20$

$20 = 20$

True

$w = 2$:

$w^3 + 5w^2 - 4w = 20$

$(2)^3 + 5(2)^2 - 4(2) \stackrel{?}{=} 20$

$8 + 20 - 8 \stackrel{?}{=} 20$

$20 = 20$

True

The solution set is $\{-5, -2, 2\}$. ●

Quick ✓

14. *True or False* The equation $x^3 - 2x^2 - 8x = 0$ can be solved using the Zero-Product Property.

In Problems 15 and 16, solve the polynomial equation.

15. $y^3 - y^2 + 9 = 9y$

16. $2n^3 - 3n^2 - 2n = 0$

▶ ❷ Solve Equations Involving Polynomial Functions

Suppose we were given the function $f(x) = x^2 + 6x - 3$ and wanted to know the values of x such that $f(x) = 4$. This requires solving the equation

$$\overbrace{x^2 + 6x - 3}^{f(x)} = 4$$

EXAMPLE 5 **Solving an Equation Involving a Polynomial Function**

If $f(x) = x^2 + 6x - 3$, find the values of x such that $f(x) = 4$. What points on the graph of f do these values correspond to?

Solution

We want to solve $f(x) = 4$. That is, we want to solve $x^2 + 6x - 3 = 4$. Start by putting the equation in standard form.

$$x^2 + 6x - 3 = 4$$

Subtract 4 from both sides:

$$x^2 + 6x - 7 = 0$$

Factor:

$$(x + 7)(x - 1) = 0$$

Set each factor equal to 0: $x + 7 = 0$ or $x - 1 = 0$

$x = -7$ or $x = 1$

Check

$$x = -7: \quad x^2 + 6x - 3 = 4$$
$$(-7)^2 + 6(-7) - 3 \overset{?}{=} 4$$
$$49 - 42 - 3 \overset{?}{=} 4$$
$$4 = 4 \quad \text{True}$$

$$x = 1: \quad x^2 + 6x - 3 = 4$$
$$(1)^2 + 6(1) - 3 \overset{?}{=} 4$$
$$1 + 6 - 3 \overset{?}{=} 4$$
$$4 = 4 \quad \text{True}$$

The values of x such that $f(x) = 4$ are -7 and 1. These values correspond to the points $(-7, 4)$ and $(1, 4)$ on the graph of f. ●

Figure 3

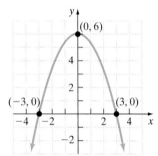

Quick ✓

17. If $g(x) = x^2 - 8x + 3$, find the values of x such that

(a) $g(x) = 12$

(b) $g(x) = -4$

What points do these values correspond to on the graph of g?

In Section 2.4 we discussed how to find the zero of a linear function. Recall that a zero of a function f is any number r such that $f(r) = 0$. Also, if r is a zero of a function, then $(r, 0)$ is also an x-intercept of the graph of a function; that is, the point $(r, 0)$ is on the graph of the function. Figure 3, which shows the graph of $y = f(x)$, shows that $(-3, 0)$ and $(3, 0)$ are x-intercepts. Therefore, the zeros of f are -3 and 3.

EXAMPLE 6 **Finding the Zeros of a Quadratic Function**

Find the zeros of $f(x) = 4x^2 - 5x - 6$. What are the x-intercepts of the graph of the function?

Solution

The zeros are found by solving the equation $f(x) = 0$, or $4x^2 - 5x - 6 = 0$. The solutions to this equation are $-\dfrac{3}{4}$ and 2, so the zeros of $f(x) = 4x^2 - 5x - 6$ are $-\dfrac{3}{4}$ and 2. Because the zeros are $-\dfrac{3}{4}$ and 2, the x-intercepts of the graph of the function are $\left(-\dfrac{3}{4}, 0\right)$ and $(2, 0)$. ●

Quick ✓

For Problems 18 and 19, find the zeros of the function, and state the x-intercepts of the graph of the function.

18. $h(x) = 2x^2 + 3x - 20$

19. $F(x) = 5x^2 - 3x - 2$

▶ ❸ Model and Solve Problems Involving Polynomials

Many applied problems require solving polynomial equations by factoring. For example, the height of a projectile over time can be described by a polynomial equation. We can use the equation to determine the time at which the projectile is a certain height. As always, we will use the problem-solving strategy from Section 1.2.

EXAMPLE 7 **Geometry: Area of a Rectangle**

The length of a rectangle is 8 feet more than its width. If the area of the rectangle is 84 square feet, what are the dimensions of the rectangle? See Figure 4.

Solution

Step 1: Identify This is a geometry problem involving the area of a rectangle.

Step 2: Name Let w represent the width of the rectangle and l represent its length.

Figure 4

ℓ

Area = 84 square feet

w

Step 3: Translate Because the length is 8 feet more than the width, we know that $l = w + 8$. We also know that the area of the rectangle is 84 square feet.

$$\text{Area} = (\text{length})(\text{width})$$
$$\text{Area} = lw$$
$$84 = (w + 8)w \quad \text{The Model}$$

Step 4: Solve Do you see that this equation is a quadratic equation? The first step is to write it in standard form, $ax^2 + bx + c = 0$.

$$w(w + 8) = 84$$

Distribute: $\qquad\qquad w^2 + 8w = 84$

Subtract 84 from both sides: $\qquad w^2 + 8w - 84 = 0$

Factor: $\qquad (w + 14)(w - 6) = 0$

Set each factor equal to 0: $\quad w + 14 = 0 \quad$ or $\quad w - 6 = 0$

Solve: $\qquad\qquad w = -14 \quad$ or $\qquad w = 6$

Step 5: Check Since w represents the width of the rectangle, we discard the solution $w = -14$. If the width is 6 feet, then the length would be $6 + 8 = 14$ feet. The area of a rectangle that is 6 feet by 14 feet would be $6(14) = 84$ square feet. We have the right answer!

Step 6: Answer The dimensions of the rectangle are 6 feet by 14 feet.

Quick ✓

20. The width of a rectangular plot of land is 6 miles less than its length. If the area of the land is 135 square miles, what are the dimensions of the land?

EXAMPLE 8 **Pricing a Charter**

Chicago Tours offers boat charters along the Chicago coastline on Lake Michigan. John Alfirivich wants to thank his employees by taking his company on a tour. Normally, a ticket costs $20 per person, but for each person John brings in excess of 30 people, Chicago Tours will lower the ticket price for everyone by $0.10. Assuming that more than 30 employees will go on the trip and that the capacity of the boat is 120 passengers, how many employees can attend if John is willing to spend $900 for the tour?

Solution

Step 1: Identify This is a direct translation problem involving revenue. Remember, revenue is price times quantity.

Step 2: Name We let x represent the number of employees in excess of 30 that attend.

Step 3: Translate Revenue is equal to price times quantity. If John brings 30 employees, the revenue to Chicago Tours will be $20(30)$. If John brings 31 employees, revenue will be $19.90(31)$. If John brings 32 employees, revenue will be $19.80(32)$. In general, if John brings x employees in excess of 30, revenue will be $(20 - 0.1x)(x + 30)$. Because John wants to spend $900, we have

$$(20 - 0.1x)(x + 30) = 900 \quad \text{The Model}$$

Step 4: Solve

$$(20 - 0.1x)(x + 30) = 900$$

Use FOIL:	$20x + 600 - 0.1x^2 - 3x = 900$
Combine like terms; rearrange terms:	$-0.1x^2 + 17x + 600 = 900$
Subtract 900 from both sides:	$-0.1x^2 + 17x - 300 = 0$
Multiply both sides by -10 to make the coefficient of x^2 equal to 1:	$x^2 - 170x + 3000 = 0$
Factor:	$(x - 150)(x - 20) = 0$
Set each factor equal to 0:	$x - 150 = 0$ or $x - 20 = 0$
Solve:	$x = 150$ or $x = 20$

Step 5: Check Remember that x represents the number of passengers in excess of 30. We discard the solution $x = 150$ because the boat holds only 120 passengers. Therefore, $30 + 20 = 50$ passengers can go on the trip. The cost per ticket is $\$20 - 0.1(20) = \$20 - \$2 = \$18$. The cost per ticket times the number of passengers is $\$18(50) = \$900$. The answer checks.

Step 6: Answer A total of 50 employees can attend. ●

Quick ✓

21. A compact disk manufacturer charges $100 for each box of CDs ordered. However, for orders in excess of 30 boxes, but less than 65 boxes, it reduces the price for all boxes by $1 per box. If a customer placed an order that qualified for the discount pricing and the bill was $4200, how many boxes of CDs were ordered?

EXAMPLE 9   **Projectile Motion**

Figure 5

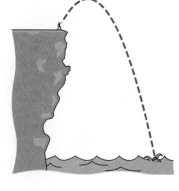

A ball is thrown off a cliff by a child from a height of 240 feet above sea level, as pictured in Figure 5. The height s of the ball above the water (in feet) as a function of time (in seconds) can be modeled by the function

$$s(t) = -16t^2 + 32t + 240$$

(a) When will the ball be 240 feet above sea level?

(b) When will the ball strike the water?

Solution

(a) To determine when the height of the ball will be 240 feet, solve the equation $s(t) = 240$.

	$s(t) = 240$
	$-16t^2 + 32t + 240 = 240$
Subtract 240 from both sides:	$-16t^2 + 32t = 0$
Factor out $-16t$:	$-16t(t - 2) = 0$
Set each factor equal to 0:	$-16t = 0$ or $t - 2 = 0$
Solve each equation:	$t = 0$ or $t = 2$

The ball will be at a height of 240 feet the instant it leaves the child's hand and, again, after 2 seconds of flight.

(b) The ball strikes the water when its height is 0, so we need to solve the equation $s(t) = 0$.

$$s(t) = 0$$
$$-16t^2 + 32t + 240 = 0$$

Factor out -16: $\quad -16(t^2 - 2t - 15) = 0$

Factor: $\quad -16(t - 5)(t + 3) = 0$

Set each factor equal to 0: $\quad -16 = 0 \quad \text{or} \quad t - 5 = 0 \quad \text{or} \quad t + 3 = 0$

Solve each equation: $\qquad\qquad\qquad\qquad t = 5 \quad \text{or} \qquad t = -3$

The equation $-16 = 0$ is false, and $t = -3$ makes no sense (since t is time). Therefore, the ball will strike the water 5 seconds after it is thrown. ●

Quick ✓

22. A model rocket is fired straight up from the ground. The height s of the rocket (in feet) as a function of time (in seconds) can be modeled by the function $s(t) = -16t^2 + 160t$.

(a) When will the rocket be 384 feet from the ground?

(b) When will the rocket strike the ground?

4.8 Exercises MyMathLab® Exercise numbers in green are complete video solutions in MyMathLab

*Problems **1–22** are the Quick ✓ s that follow the EXAMPLES.*

Building Skills

In Problems 23–62, solve each equation. See Objective 1.

23. $(x - 3)(x + 1) = 0$

24. $(x + 3)(x - 8) = 0$

25. $2x(3x + 4) = 0$

26. $4x(2x - 3) = 0$

27. $y(y - 5)(y + 3) = 0$

28. $3a(a - 9)(a + 11) = 0$

29. $3p^2 - 12p = 0$

30. $5c^2 + 15c = 0$

31. $2w^2 = 16w$

32. $4t^2 = -20t$

33. $m^2 + 2m - 15 = 0$

34. $x^2 + 3x - 40 = 0$

35. $w^2 - 13w = -36$

36. $y^2 + 13y = -42$

37. $p^2 - 6p + 9 = 0$

38. $a^2 + 12a + 36 = 0$

39. $5x^2 = 2x + 3$

40. $4c^2 + 6 = 25c$

41. $m^2 + 7m + 9 = m$

42. $n^2 - 8n = 2n - 25$

43. $3p^2 + 9p - 120 = 0$

44. $4y^2 - 20y - 56 = 0$

45. $-4b^2 - 14b + 60 = 0$

46. $-6n^2 - 9n + 60 = 0$

47. $\dfrac{1}{2}x^2 + 2x - 6 = 0$

48. $\dfrac{1}{2}t^2 - 3t - 8 = 0$

49. $\dfrac{2}{3}x^2 + x = \dfrac{14}{3}$

50. $\dfrac{2}{3}x^2 + \dfrac{7}{3}x = 5$

51. $x(x + 8) = 33$

52. $y(y + 4) = 45$

53. $(x + 6)(x - 1) = 9 + 3x$

54. $(x + 2)(x + 5) = 3x + 10$

55. $2z^3 - 5z^2 = 3z$

56. $7q^3 + 31q^2 = -12q$

57. $2p^3 + 5p^2 - 8p - 20 = 0$

58. $w^3 + 5w^2 - 16w - 80 = 0$

59. $-30b^3 - 38b^2 = 12b$

60. $-24b^3 + 27b = 18b^2$

61. $(x - 2)^3 = x^3 - 2x$

62. $(x + 2)^3 = x^3 - 2x$

In Problems 63–68, see Objective 2.

63. Suppose that $f(x) = x^2 + 7x + 12$. Find the values of x such that

    (a) $f(x) = 2$ **(b)** $f(x) = 20$

What points on the graph of f do these values correspond to?

64. Suppose that $f(x) = x^2 + 5x + 3$. Find the values of x such that

    (a) $f(x) = 3$ **(b)** $f(x) = 17$

What points on the graph of f do these values correspond to?

65. Suppose that $g(x) = 2x^2 - 6x - 5$. Find the values of x such that

    (a) $g(x) = 3$ **(b)** $g(x) = 15$

What points on the graph of g do these values correspond to?

66. Suppose that $h(x) = 3x^2 - 9x - 8$. Find the values of x such that

    (a) $h(x) = -8$ **(b)** $h(x) = 22$

What points on the graph of h do these values correspond to?

67. Suppose that $F(x) = -3x^2 + 12x + 5$. Find the values of x such that

    (a) $F(x) = 5$ **(b)** $F(x) = -10$

What points on the graph of F do these values correspond to?

68. Suppose that $G(x) = -x^2 + 4x + 6$. Find the values of x such that

    (a) $G(x) = 1$ **(b)** $G(x) = 9$

What points on the graph of G do these values correspond to?

In Problems 69–74, find the zeros of the function. What are the x-intercepts of the graph of the function? See Objective 2.

69. $f(x) = x^2 + 9x + 14$ **70.** $f(x) = x^2 - 13x + 42$

71. $g(x) = 6x^2 - 25x - 9$

72. $h(x) = 8x^2 - 18x - 35$

73. $s(x) = 2x^3 + 2x^2 - 40x$

74. $f(x) = 3x^3 - 15x^2 - 42x$

Mixed Practice

In Problems 75–92, solve each equation.

75. $(x + 3)(x - 5) = 9$ **76.** $(x + 7)(x - 3) = 11$

77. $2q^2 + 3q - 14 = 0$ **78.** $3t^2 + 7t - 20 = 0$

79. $-3b^2 + 21b = 0$ **80.** $-7z^2 + 42z = 0$

81. $(x + 2)(x + 3) = x(x - 2)$

82. $(x + 7)(x - 6) = x(x + 3)$

83. $x^3 + 5x^2 - 4x - 20 = 0$

84. $2c^3 + 3c^2 - 8c - 12 = 0$

85. $(2x + 1)(x - 3) - x^2 = (x - 2)(x - 3)$

86. $(3x - 2)(x + 4) - x(x + 1) = (2x + 1)(x + 4) - 12$

87. $x^3 + x^2 + x + 6 = 3x^2 + 6x$

88. $2x^3 + 16x^2 = 5x^2 - 15x$

89. $4x^4 - 17x^2 + 4 = 0$

90. $9z^4 - 13z^2 + 4 = 0$

91. $(a + 3)^2 - 5(a + 3) = -6$

92. $(2b + 1)^2 + 7(2b + 1) = -12$

In Problems 93–96, find the domain of each function.

93. $f(x) = \dfrac{5}{x^2 - 4}$

94. $f(x) = \dfrac{-9}{x^2 + 6x + 5}$

95. $g(x) = \dfrac{4x + 3}{2x^2 - 3x + 1}$

96. $h(x) = \dfrac{x + 4}{3x^2 - 7x - 6}$

Applying the Concepts

△ **97. Area** The length of a rectangle is 8 centimeters less than its width. What are the dimensions of the rectangle if its area is 128 square centimeters?

△ **98. Area** The length of a rectangle is twice the sum of its width and 3. What are the dimensions of the rectangle if its area is 216 square inches?

△ **99. Area** The height of a triangle is 12 feet more than its base. What are the height and base of the triangle if its area is 110 square feet?

△ **100. Area** The base of a triangle is 4 meters shorter than its height. What are the height and base of the triangle if its area is 48 square meters?

△ **101. Convex Polygons** A **convex polygon** is a polygon whose interior angles are between 0° and 180°. The number of diagonals D in a convex polygon with n sides is given by the formula $D = \dfrac{n(n-3)}{2}$. Determine the number of sides n in a convex polygon that has 20 diagonals.

102. Consecutive Integers The sum S of the consecutive integers 1, 2, 3, ... , n is given by the formula $S = \dfrac{n(n+1)}{2}$. That is, $1 + 2 + 3 + \cdots + n = \dfrac{n(n+1)}{2}$. How many consecutive integers must be added together to obtain a sum of 36?

103. Enclosing an Area with a Fence A farmer has 100 meters of fencing and wants to enclose a rectangular plot that borders a river. If the farmer does not fence the side along the river, what are the dimensions of the land enclosed if the area enclosed is 800 square meters?

104. Enclosing an Area with a Fence A farmer has 300 feet of fencing and wants to enclose a rectangular corral that borders his barn on one side and then divide it into two plots with a fence parallel to one of the sides (see the figure). Assuming that the farmer will not fence the side along the barn, what are the lengths of the parts of the fence if the total area enclosed is 4800 square feet?

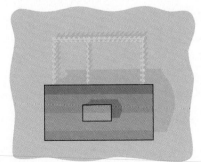

105. Landscape Design Robert Boehm just designed a cloister (a rectangular garden surrounded by a covered walkway on all four sides). The outside dimensions of the garden are 12 feet by 8 feet, and the area of the garden and the walkway together is 252 square feet. What is the width of the walkway?

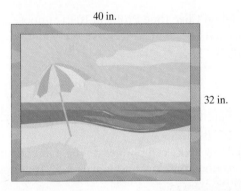

106. Picture Frame The outside dimensions of a picture frame are 40 inches by 32 inches. The area of the picture within the frame is 1008 square inches. Find the width of the frame.

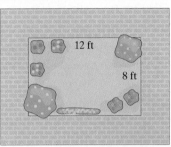

107. Making a Box A box is to be made from a rectangular piece of corrugated cardboard, where the length is 5 more inches than the width, by cutting a square piece 2 inches on each side from each corner. The volume of the box is to be 168 cubic inches. Find the dimensions of the rectangular piece of cardboard.

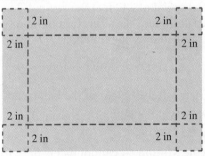

108. Making a Box A box is to be made from a rectangular piece of corrugated cardboard, where the length is 8 inches more than the width, by cutting a square piece 3 inches on each side from each corner. The volume of the box is to be 315 cubic inches. Find the dimensions of the rectangular piece of cardboard.

109. Marginal Cost Marginal cost can be thought of as the cost of producing one additional unit of output. For example, if the marginal cost of producing the 30th unit of output is $9.30, then it cost $9.30 to increase production from 29 to 30 units. The marginal cost C (in dollars) to produce x bicycles is given by $C(x) = x^2 - 40x + 600$.

(a) Find the marginal cost of producing 30 bicycles.

(b) How many bicycles can be manufactured so that the marginal cost equals $200? That is, solve $C(x) = 200$.

(c) Economic theory states that, to maximize profit, production should continue until marginal revenue equals marginal cost. Marginal revenue is the additional revenue received for each bicycle sold. In certain situations, the marginal revenue is the price of the product. Assuming that marginal revenue equals $225, how many bicycles should be manufactured?

110. Marginal Cost (See Problem 109.) Suppose that the marginal cost of manufacturing x cellular telephones is given by

$$C(x) = \frac{1}{2}x^2 - 30x + 475$$

(a) Find the marginal cost of producing 30 cell phones.

(b) How many cell phones can be manufactured so that marginal cost equals $75? That is, solve $C(x) = 75$.

(c) Economic theory states that, to maximize profit, production should continue until marginal revenue equals marginal cost. Assuming that marginal revenue equals $97, how many cell phones should be manufactured?

111. Projectile Motion Tiger Woods hits a golf ball with an initial speed of 240 feet per second. The height s of the ball (in feet) as a function of time (in seconds) can be modeled by the function

$$s(t) = -16t^2 + 120t$$

(a) When will the height of the ball be 200 feet?

(b) When will the ball hit the ground?

112. Projectile Motion A cannonball is fired from a cliff that is 260 feet high with an initial speed of 128 feet per second. The height s of the cannonball (in feet) as a function of time (in seconds) can be modeled by the function

$$s(t) = -16t^2 + 64t + 260$$

(a) When will the height of the cannonball be 320 feet?

(b) When will the cannonball hit the ground?

The Graphing Calculator

The graphing calculator can be used to find solutions to any equation. The ZERO (or ROOT) feature of a graphing calculator can be used to find solutions of an equation when one side of the equation is 0. Solving an equation for x when one side of the equation is 0 is equivalent to finding where the graph of the corresponding function crosses or touches the x-axis. For example, to solve the equation in Example 2, $2x^2 - 5x = 3$, we would graph $Y_1 = 2x^2 - 5x - 3$ and then use ZERO (or ROOT) to determine where $Y_1 = 0$. See Figures 6(a) and (b).

Figure 6

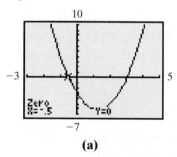

(a)

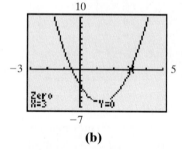

(b)

In Problems 113–118, solve each equation using a graphing calculator. Round your answers to two decimal places, if necessary.

113. $2x^2 - 7x - 5 = 0$

114. $2x^2 - x - 10 = 0$

115. $0.2x^2 - 5.1x + 3 = 0$

116. $0.4x^2 - 2.7x + 1 = 0$

117. $-x^2 + 0.6x = -2$

118. $-3.1x^2 - 0.4x = -3$

Chapter 4 Activity: What Is the Question?

Focus: Performing operations and solving equations with polynomials

Time: 15–20 minutes

Group size: 2 or 4

In this activity you will work as a team to solve eight multiple-choice questions. However, these questions are different from most multiple-choice questions. You are given the answer to a problem and must determine which of the multiple-choice options has the correct question for the given answer.

Before beginning the activity, decide how you will approach this task as a team. For example:

- If there are two members on your team… one member will always examine choices (a) and (b) and the other will always examine choices (c) and (d).

- If there are four members on your team… one member will always examine choice (a), another member will always examine choice (b), and so on.

1. The answer is $-3x^2 - 10x$. What is the question?

  (a) Simplify: $2x - 3x(x^2 + 4)$

  (b) Find the quotient: $(6x^3 - 20x^2) \div (-2x)$

  (c) Simplify: $-3x(x^2 + 3) + 1$

  (d) Find the quotient: $(-6x^4 - 20x^3) \div (2x^2)$

2. Find $(f + g)(-1)$. The answer is 5. What are f and g?

  (a) $f(x) = 2x + 2, g(x) = -3x - 1$

  (b) $f(x) = 2x + 3, g(x) = -3x + 1$

  (c) $f(x) = -2x + 5, g(x) = -3x + 4$

  (d) $f(x) = -2x - 2, g(x) = 3x - 4$

3. The answer is $(6x + 1)(2x - 3)$. What is the question?

  (a) Factor: $12x^2 - 20x + 3$

  (b) Factor: $12x^2 + 16x - 3$

  (c) Factor: $12x^2 - 16x - 3$

  (d) Factor: $12x^2 + 20x + 3$

4. The answer is $x^2 + 5x + 6$. What is the question?

  (a) Find the product: $(x + 6)(x - 1)$

  (b) Simplify: $2x^2 + 7x + 9 - (x^2 - 2x - 3)$

  (c) Find the product: $(x + 2)(x + 3)$

  (d) Simplify: $(x + 6)^2$

5. The answer is 3. What is the question?

  (a) What is the name of the variable in $16z^2 + 3z - 5$?

  (b) What is the degree of the polynomial $2mn + 6m - 3$?

  (c) How many terms are in the polynomial $2mn + 6m - 3$?

  (d) What is the coefficient of b in the polynomial $3a^2b - 9a + 5b$?

6. The answer is $x = -5$ or $x = 3$. What is the question?

  (a) Solve: $x(x + 2) = 15$

  (b) Find the values of x such that $f(x) = 8$ if $f(x) = x^2 + 4x + 3$.

  (c) Solve: $\frac{2}{3}x + 5 = \frac{1}{3}x^2$

  (d) Find the zeros of the function $f(x) = x^2 - 2x - 15$.

7. The answer is $8x^2$. What is the question?

  (a) Simplify: $9x^2(x + 1) - 3x^2(3x + 5)$

  (b) Find the greatest common factor: $16x^2y^2 - 8x^3y - 24x^2$

  (c) Find the quotient: $(-8x^3 - 8x^2) \div (x + 1)$

  (d) Factor by grouping: $8x^3 + 8x^2 - x - 1$

8. The answer is $x + 2$. What is the question?

  (a) Find the quotient: $(x^3 + x^2 - 7x - 2) \div (x^2 + 3x - 1)$

  (b) Find the binomial factor: $x^3 - 8$

  (c) Find the quotient: $(x^2 + 2x - 3) \div (x - 1)$

  (d) Find the binomial factor: $x^3 + 8$

Chapter 4 Review

Section 4.1   Adding and Subtracting Polynomials

KEY CONCEPTS

- **Monomial**

  A monomial in one variable is the product of a constant and a variable raised to a nonnegative integer power. A monomial in one variable is of the form ax^k, where a is a constant, x is a variable, and $k \geq 0$ is an integer. The constant a is called the **coefficient** of the monomial. If $a \neq 0$, then k is called the **degree** of the monomial.

- **Polynomial**

  A polynomial is a monomial or the sum of monomials.

KEY TERMS

Monomial
Coefficient
Degree
Polynomial
Standard form
Binomial

- **Polynomial Function**
  A polynomial function is a function whose rule is a polynomial. The domain of all polynomial functions is the set of all real numbers.

- **Sum or Difference of Two Functions**
  If f and g are two functions:
  The sum $f + g$ is the function defined by $(f + g)(x) = f(x) + g(x)$.
  The difference $f - g$ is the function defined by $(f - g)(x) = f(x) - g(x)$.

Trinomial
Polynomial function
Evaluate
Sum of two functions
Difference of two
   functions

You Should Be Able To...	EXAMPLE	Review Exercises
① Define monomial and determine the coefficient and degree of a monomial (p. 308)	Examples 1 through 3	1–2
② Define polynomial and determine the degree of a polynomial (p. 309)	Examples 4 and 5	3–4
③ Simplify polynomials by combining like terms (p. 310)	Examples 6 through 8	5–10
④ Evaluate polynomial functions (p. 312)	Examples 9 and 10	11–12, 13(b), 14(b), 15(b), 16
⑤ Add and subtract polynomial functions (p. 314)	Examples 11 and 12	13(a), 14(a), 15(a)

In Problems 1 and 2, determine the coefficient and degree of each monomial.

1. $-7x^4$

2. $\frac{1}{9}w^3$ $+ \frac{2}{3}x^4$

In Problems 3 and 4, write each polynomial in standard form. Then determine the degree of each polynomial.

3. $x + 7x^3 - 8 - 2x^2$ **4.** $3 + 2y - 3y^2 + y^4$

In Problems 5–10, add or subtract as indicated. Express your answer as a single polynomial in standard form.

5. $(x^2 + 2x - 7) + (3x^2 - x - 4)$

6. $(4x^3 - 3x^2 + x - 5) - (x^4 + 2x^2 - 7x + 1)$

7. $\left(\frac{1}{4}x^2 - \frac{1}{2}x\right) - \left(4x - \frac{1}{6}\right)$

8. $\left(\frac{1}{2}x^2 - x + \frac{1}{4}\right) + \left(\frac{1}{3}x^2 + \frac{2}{5}\right)$

9. $(x^3y^2 + 6x^2y^2 - xy) + (-x^3y^2 + 4x^2y^2 + xy)$

10. $(a^2b - 4ab^2 + 3) - (2a^2b + 2ab^2 + 7)$

In Problems 11–14, find the indicated function or function value.

11. $f(x) = -3x^2 + 2x - 8$
    (a) $f(-2)$ **(b)** $f(0)$ **(c)** $f(3)$

12. $f(x) = x^3 - 5x^2 + 3x - 1$
    (a) $f(-3)$ **(b)** $f(0)$ **(c)** $f(2)$

13. $f(x) = 4x - 3; g(x) = x^2 + 3x + 2$
    (a) $(f + g)(x)$ **(b)** $(f + g)(3)$

14. $f(x) = 2x^3 + x^2 - 7; g(x) = 3x^2 - x + 5$
    (a) $(f - g)(x)$ **(b)** $(f - g)(2)$

15. Profit Suppose that the revenue function R from selling x graphing calculators is $R(x) = -1.5x^2 + 180x$. The cost C of selling x graphing calculators is $C(x) = x^2 - 100x + 3290$.
    (a) Find the profit function.

    (b) Find the profit if $x = 25$ calculators are sold.

16. Area The area A of the region shown in quadrant I is given by $A(x) = -x^2 + 5x$, where (x, y) is a point in quadrant I on the graph of the line $y = -2x + 5$.

    (a) Find the area of the region when the given point is $(2, 1)$.

    (b) Find the area of the region when the given point is $(1, 3)$.

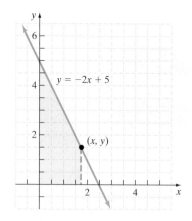

Section 4.2 Multiplying Polynomials

KEY CONCEPTS

- **Product Rule for Exponents**
  If a is a real number and m and n are integers, then $a^m \cdot a^n = a^{m+n}$.
  If m, n, or $m + n$ is 0 or negative, then a cannot be 0.

- **Distributive Property**
  If a, b, and c are real numbers, then $a \cdot (b + c) = a \cdot b + a \cdot c$ and
  $(a + b) \cdot c = a \cdot c + b \cdot c$.

- **Extended Form of the Distributive Property**
  $a(b_1 + b_2 + \ldots + b_n) = a \cdot b_1 + a \cdot b_2 + \ldots + a \cdot b_n$, where
  $a, b_1, b_2, \ldots, b_n$ are real numbers.

- **Difference of Two Squares**
  $(A - B)(A + B) = A^2 - B^2$

- **Squares of Binomials or Perfect Square Trinomials**
  $(A + B)^2 = A^2 + 2AB + B^2$
  $(A - B)^2 = A^2 - 2AB + B^2$

- **Product of Two Functions**
  Let f and g be two functions. The product $f \cdot g$ is the function defined
  by $(f \cdot g)(x) = f(x) \cdot g(x)$.

KEY TERMS

FOIL
Special products

You Should Be Able To...	EXAMPLE	Review Exercises
1 Multiply a monomial and a polynomial (p. 320)	Example 2	17–20
2 Multiply two binomials (p. 320)	Examples 3 and 4	21–24
3 Multiply two polynomials (p. 321)	Example 5	25–28
4 Multiply special products (p. 322)	Examples 6 through 9	29–34
5 Multiply polynomial functions (p. 324)	Example 10	35–36

In Problems 17–20, find the product.

17. $(-3x^3y)(4xy^2)$

18. $\left(\frac{1}{3}mn^4\right)(18m^3n^3)$

19. $5ab(-2a^2b + ab^2 - 3ab)$

20. $0.5c(1.7c^2 + 4.3c + 8.9)$

In Problems 21–24, find the product of the binomials.

21. $(x + 2)(x - 9)$

22. $(-3x + 1)(2x - 8)$

23. $(m - 4n)(2m + n)$

24. $(2a + 15)(-a + 3)$

In Problems 25–28, find the product of the polynomials.

25. $(x + 2)(3x^2 - 5x + 1)$

26. $(w - 4)(w^2 + w - 8)$

27. $(m^2 - 2m + 3)(2m^2 + 5m - 7)$

28. $(2p - 3q)(p^2 + 7pq - 4q^2)$

In Problems 29–34, find the special products.

29. $(3w + 1)(3w - 1)$

30. $(2x - 5y)(2x + 5y)$

31. $(6k - 5)^2$

32. $(3a + 2b)^2$

33. $(x + 2)(x^2 - 2x + 4)$

34. $(2x - 3)(4x^2 + 6x + 9)^2$

In Problems 35–38, find the indicated function or value.

35. $f(x) = 3x - 7; g(x) = 6x + 5$
   (a) $(f \cdot g)(x)$ **(b)** $(f \cdot g)(-2)$

36. $f(x) = x + 2; g(x) = 3x^2 - x + 1$
   (a) $(f \cdot g)(x)$ **(b)** $(f \cdot g)(4)$

37. $f(x - 3)$ when $f(x) = 5x^2 + 8$.

38. $f(x + h) - f(x)$ when $f(x) = -x^2 + 3x - 5$.

Section 4.3 Dividing Polynomials; Synthetic Division

KEY CONCEPTS

KEY TERMS

Divisor
Dividend
Quotient
Remainder
Synthetic division

- **Quotient Rule for Exponents**

  If a is a real number and m and n are integers, then $\dfrac{a^m}{a^n} = a^{m-n}, a \neq 0$.

- **Quotient of Two Functions**

  Let f and g be two functions. The quotient $\dfrac{f}{g}$ is the function defined by

  $$\left(\frac{f}{g}\right)(x) = \frac{f(x)}{g(x)}, g(x) \neq 0.$$

- **The Remainder Theorem**

  Let f be a polynomial function. If $f(x)$ is divided by $x - c$, then the remainder is $f(c)$.

- **The Factor Theorem**

  Let f be a polynomial function. Then $x - c$ is a factor of $f(x)$ if and only if $f(c) = 0$.

You Should Be Able To...	EXAMPLE	Review Exercises
① Divide a polynomial by a monomial (p. 328)	Example 1	39–42
② Divide polynomials using long division (p. 329)	Examples 3 through 5	43–48, 63
③ Divide polynomials using synthetic division (p. 332)	Examples 6 and 7	49–54, 64
④ Divide polynomial functions (p. 334)	Example 8	55–58
⑤ Use the Remainder and Factor Theorems (p. 335)	Examples 9 and 10	59–62

In Problems 39–42, divide and simplify.

39. $\dfrac{12x^3 - 6x^2}{3x}$

40. $\dfrac{15w^5 - 5w^3 + 25w^2 + 10w}{5w}$

41. $\dfrac{7y^3 + 12y^2 - 6y}{2y}$

42. $\dfrac{2m^3n^2 + 8m^2n^2 - 14mn^3}{4m^2n^3}$

In Problems 43–48, divide using long division.

43. $\dfrac{3x^2 - 2x - 8}{x - 2}$

44. $\dfrac{-2x^2 - 3x + 40}{x + 5}$

45. $\dfrac{6z^3 + 9z^2 + 4z - 6}{2z + 3}$

46. $\dfrac{12k^3 - 29k^2 - 14k + 16}{3k - 8}$

47. $\dfrac{16x^4 - 81}{2x - 3}$

48. $\dfrac{2x^4 - 11x^3 + 35x^2 - 54x + 55}{x^2 - 3x + 4}$

In Problems 49–54, divide using synthetic division.

49. $\dfrac{5x^2 + 11x + 8}{x + 2}$

50. $\dfrac{9a^2 - 14a - 8}{a - 2}$

51. $\dfrac{3m^3 + 11m^2 - 5m - 33}{m + 3}$

52. $\dfrac{n^3 + 2n^2 - 39n + 67}{n - 4}$

53. $\dfrac{x^4 + 6x^2 - 7}{x + 1}$

54. $\dfrac{2x^3 + 5x - 8}{x + 2}$

In Problems 55–58, find the indicated function or value.

55. $f(x) = 5x^3 + 25x^2 - 15x; g(x) = 5x$

  (a) $\left(\dfrac{f}{g}\right)(x)$ **(b)** $\left(\dfrac{f}{g}\right)(2)$

56. $f(x) = 9x^2 + 54x - 31; g(x) = 3x - 2$

  (a) $\left(\dfrac{f}{g}\right)(x)$ **(b)** $\left(\dfrac{f}{g}\right)(-3)$

57. $f(x) = 2x^3 + 12x^2 + 9x - 28;$
    $g(x) = x + 4$

  (a) $\left(\dfrac{f}{g}\right)(x)$ **(b)** $\left(\dfrac{f}{g}\right)(-2)$

58. $f(x) = 3x^4 - 14x^3 + 31x^2 - 58x + 22;$
    $g(x) = x^2 - x + 5$

  (a) $\left(\dfrac{f}{g}\right)(x)$ **(b)** $\left(\dfrac{f}{g}\right)(4)$

In Problems 59 and 60, use the Remainder Theorem to find the remainder.

59. $f(x) = 4x^2 - 7x + 23$ is divided by $x - 4$.

60. $f(x) = x^3 - 2x^2 + 12x - 5$ is divided by $x + 2$.

In Problems 61 and 62, use the Factor Theorem to determine whether $x - c$ is a factor of the given function for the given value of c. If $x - c$ is a factor, then write the function in factored form.

61. $f(x) = 3x^2 + x - 14; c = 2$

62. $f(x) = 2x^2 + 13x + 22; c = -4$

63. The area of a rectangle is $20x^2 - 11x - 3$ square meters. If the width of the rectangle is $4x - 3$ meters, find an expression for the length.

64. The volume of a rectangular box is $2x^3 + x^2 - 7x - 6$ cubic centimeters. If the height of the box is $x - 2$ centimeters, find an expression for the area of the top of the box.

Section 4.4 Greatest Common Factor; Factoring by Grouping

KEY CONCEPTS

- **Factoring out the greatest common factor**
  Identify the greatest common factor (GCF) of each term. Rewrite each term as the product of the GCF and the remaining factor. Use the Distributive Property to factor out the GCF. Check your work using the Distributive Property.

- **Factoring by grouping**
  Group the terms with common factors. Sometimes it will be necessary to rearrange the terms. In each grouping, factor out the common factor. Factor out the common factor that remains. Check your work.

KEY TERMS

Factors
Factoring a polynomial
Prime
Factored completely
Greatest common factor (GCF)
Factoring by grouping

You Should Be Able To...	EXAMPLE	Review Exercises
1 Factor out the greatest common factor (p. 340)	Examples 1 through 4	65–72, 79, 80
2 Factor by grouping (p. 342)	Examples 5 and 6	73–78

In Problems 65–72, factor out the greatest common factor.

65. $4z + 24$

66. $-7y^2 + 91y$

67. $14x^3y^2 + 2xy^2 - 8x^2y$

68. $30a^4b^3 + 15a^3b - 25a^2b^2$

69. $3x(x + 5) - 4(x + 5)$

70. $-4c(2c + 9) + 3(2c + 9)$

71. $(5x + 3)(x - 5y) + (x + 2)(x - 5y)$

72. $(3a - b)(a + 7) - (a + 1)(a + 7)$

In Problems 73–78, factor by grouping.

73. $x^2 + 6x - 3x - 18$

74. $c^2 + 2c - 5c - 10$

75. $14z^2 + 16z - 21z - 24$

76. $21w^2 - 28w + 6w - 8$

77. $2x^3 + 2x^2 - 18x^2 - 18x$

78. $10a^4 + 15a^3 + 70a^3 + 105a^2$

79. Integers The sum of the first n positive integers is given by $\frac{1}{2}n^2 + \frac{1}{2}n$.

(a) Write this expression in factored form.

(b) Use the factored form to determine the sum of the first 32 positive integers.

80. Revenue A computer manufacturer estimates that its revenue for selling x computer systems can be approximated by the function $R(x) = 5200x - 2x^3$. Express the revenue function in factored form.

Section 4.5 Factoring Trinomials

KEY CONCEPTS

- **Factoring $x^2 + bx + c$**
  $x^2 + bx + c = (x + m)(x + n)$, where $mn = c$ and $m + n = b$

- **Factoring $ax^2 + bx + c$ by grouping**
  See page 351.

- **Factoring $ax^2 + bx + c$ by trial and error**
  See page 353.

KEY TERMS

Factoring by substitution

You Should Be Able To...	EXAMPLE	Review Exercises
❶ Factor trinomials of the form $x^2 + bx + c$ (p. 346)	Examples 1 through 6	81–86
❷ Factor trinomials of the form $ax^2 + bx + c, a \neq 1$ (p. 350)	Examples 7 through 12	87–94
❸ Factor trinomials using substitution (p. 355)	Examples 13 and 14	95–98

In Problems 81–98, factor each trinomial completely. If the polynomial cannot be factored, say it is prime.

81. $w^2 - 11w - 26$

82. $x^2 - 9x + 15$

83. $-t^2 + 6t + 72$

84. $m^2 + 10m + 21$

85. $x^2 + 4xy - 320y^2$

86. $r^2 - 5rs + 6s^2$

87. $5x^2 + 13x - 6$

88. $6m^2 + 41m + 44$

89. $4y^2 - 5y + 7$

90. $8t^2 + 22t - 6$

91. $6x^2 - 13x + 5$

92. $21r^2 - rs - 2s^2$

93. $20x^2 - 57xy + 27y^2$

94. $-2s^2 + 12s + 14$

95. $x^4 - 10x^2 - 11$

96. $10x^2y^2 + 41xy + 4$

97. $(a + 4)^2 - 9(a + 4) - 36$

98. $2(w - 1)^2 + 11(w - 1) + 9$

Section 4.6 Factoring Special Products

KEY CONCEPTS

- **Perfect Square Trinomials**
  $A^2 + 2AB + B^2 = (A + B)^2$
  $A^2 - 2AB + B^2 = (A - B)^2$

- **Difference of Two Squares**
  $A^2 - B^2 = (A - B)(A + B)$

- **Sum or Difference of Two Cubes**
  $A^3 + B^3 = (A + B)(A^2 - AB + B^2)$
  $A^3 - B^3 = (A - B)(A^2 + AB + B^2)$

KEY TERMS

Perfect square trinomial
Difference of two squares
Sum of two squares
Sum of two cubes
Difference of two cubes

You Should Be Able To...	EXAMPLE	Review Exercises
❶ Factor perfect square trinomials (p. 358)	Example 1	99–104
❷ Factor the difference of two squares (p. 359)	Examples 2 and 3	105–110
❸ Factor the sum or difference of two cubes (p. 361)	Examples 4 and 5	111–116

In Problems 99–116, factor completely.

99. $x^2 + 22x + 121$

100. $w^2 - 34w + 289$

101. $144 - 24c + c^2$

102. $x^2 - 8x + 16$

103. $64y^2 + 80y + 25$

104. $12z^2 + 48z + 48$

105. $x^2 - 196$

106. $49 - y^2$

107. $t^2 - 225$

108. $4w^2 - 81$

109. $36x^4 - 25y^2$

110. $80mn^2 - 20m$

111. $x^3 - 343$

112. $729 - y^3$

113. $27x^3 - 125y^3$

114. $8m^6 + 27n^3$

115. $2a^6 - 2b^6$

116. $(y - 1)^3 + 64$

Section 4.7 Factoring: A General Strategy

KEY CONCEPT

Steps for Factoring

Step 1: Factor out the greatest common factor (GCF), if any exists.

Step 2: Count the number of terms.

Step 3: (a) Two terms

- Is it the difference of two squares? If so,
  $A^2 - B^2 = (A - B)(A + B)$
- Is it the sum of two squares? If so, stop!
  The expression is prime.

- Is it the difference of two cubes? If so,
  $A^3 - B^3 = (A - B)(A^2 + AB + B^2)$
- Is it the sum of two cubes? If so,
  $A^3 + B^3 = (A + B)(A^2 - AB + B^2)$

(b) Three terms

- Is it a perfect square trinomial? If so,
  $A^2 + 2AB + B^2 = (A + B)^2$ or $A^2 - 2AB + B^2 = (A - B)^2$
- Is the coefficient of the square term 1? If so,
  $x^2 + bx + c = (x + m)(x + n)$, where $mn = c$ and $m + n = b$
- Is the coefficient of the square term different from 1? If so,
  (a) Use factoring by grouping.
  (b) Use trial and error.

(c) Four terms

- Use factoring by grouping.

Step 4: Check your work by multiplying out the factors.

You Should Be Able To...	EXAMPLE	Review Exercises
❶ Factor polynomials completely (p. 364)	Examples 1 through 7	117–131
❷ Write polynomial functions in factored form (p. 368)	Example 8	132–136

In Problems 117–131, factor each polynomial completely.

117. $x^2 + 7x + 6$

118. $-8x^2y^3 + 12xy^3$

119. $7x^3 - 35x^2 + 28x$

120. $3x^2 - 3x - 18$

121. $4z^2 - 60z + 225$

122. $12x^2 + 7x - 49$

123. $10n^2 - 33n - 7$

124. $8 - 2y - y^2$

125. $2x^3 - 10x^2 + 6x - 30$

126. $(3h + 2)^3 + 64$

127. $5p^3q^2 - 80p$

128. $m^4 - 5m^2 + 4$

129. $686 - 16m^6$

130. $h^3 + 2h^2 - h - 2$

131. $108x^3 + 4y^3$

In Problems 132–136, factor each polynomial function.

132. $F(c) = c^2 - 24c + 144$

133. $f(x) = -3x^2 + 6x + 45$

134. $g(x) = 16x^2 - 100$

135. $G(y) = 16y^3 + 250$

136. $f(x) = -4x^2 - 16$

In Problems 137 and 138, write an expression for the shaded area in factored form.

137.

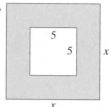

138.

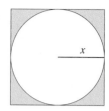

Section 4.8 Polynomial Equations

KEY CONCEPTS

- **The Zero-Product Property**

  If the product of two numbers is zero, then at least one of the numbers is 0. That is, if $ab = 0$, then $a = 0$ or $b = 0$ or both a and b are 0.

- **Zero of a Function**

  A zero of a function f is any number r such that $f(r) = 0$. If r is a zero of a function, then $(r, 0)$ is also an x-intercept of the graph of the function.

KEY TERMS

Polynomial equation
Degree of a polynomial
  equation
Zero-product property
Quadratic equation
Standard form
Second-degree equation
Zero

You Should Be Able To...	EXAMPLE	Review Exercises
① Solve polynomial equations using the Zero-Product Property (p. 371)	Examples 1 through 4	139–148
② Solve equations involving polynomial functions (p. 375)	Examples 5 and 6	149–152
③ Model and solve problems involving polynomials (p. 376)	Examples 7 through 9	153–154

In Problems 139–148, solve each equation.

139. $(w + 5)(w - 13) = 0$ **140.** $x^2 + 21x + 54 = 0$

141. $y^2 + 2y = 15$ **142.** $5a^2 = -20a$

143. $x(x + 1) = 110$ **144.** $15x^2 + 29x - 14 = 0$

145. $\frac{1}{2}x^2 + 5x + 12 = 0$ **146.** $(b + 1)(b - 3) = 5$

147. $2x^3 + 5x^2 - 8x = 20$ **148.** $5x^3 + x^2 - 45x - 9 = 0$

149. Suppose that $f(x) = x^2 + 5x - 18$. Find values of x such that

(a) $f(x) = 6$ (b) $f(x) = -4$

What points on the graph of f do these points correspond to?

150. Suppose that $f(x) = 5x^2 - 4x + 3$. Find values of x such that

(a) $f(x) = 3$ (b) $f(x) = 4$

What points on the graph of f do these values correspond to?

In Problems 151 and 152, find the zeros of the function. What are the x-intercepts of the graph of the function?

151. $f(x) = 3x^3 + 18x^2 + 24x$

152. $f(x) = -4x^2 + 22x + 42$

153. Falling Object At one point, a 4-foot flagpole on top of the KXJB-TV mast in Galesburg, North Dakota, made it the world's tallest structure, standing 2064 feet tall. If an object is dropped from the top of this mast, the height s of the object (in feet) as a function of time (in seconds) can be modeled by the function $s(t) = -16t^2 + 2064$. When will the object be 1280 feet above the ground?

154. Reliability A simple parallel system with two identical components has a reliability given by $R = 1 - (1 - r)^2$, where r is the reliability of the individual components.

(a) What is the reliability of the individual components if the system reliability is $R = 0.96$?

(b) What is the reliability of the individual components if the system reliability is $R = 0.99$?

1. Write the polynomial in standard form and determine its degree.

$$7x^2 + x^4 - 5x^7 + 1 - x$$

2. Add:

$$\left(-2a^3b^2 + 5a^2b + ab + 1\right) + \left(\frac{1}{3}a^3b^2 + 4a^2b - 6ab - 5\right)$$

3. For $f(x) = x^3 + 3x^2 - x + 1$, find $f(-2)$.

4. For $f(x) = 7x^3 - 1$ and $g(x) = 4x^2 + 3x - 2$, find $(f - g)(x)$.

In Problems 5–7, find each product.

5. $\frac{1}{2}a^2b(4ab^2 - 6ab + 8)$

6. $(3x - 1)(4x + 17)$

7. $(2m - n)^2$

8. Divide using long division: $\dfrac{6z^3 - 14z^2 + z + 4}{2z^2 + 1}$

9. Divide using synthetic division: $\dfrac{5x^2 - 27x - 18}{x - 6}$

10. For $f(x) = 6x^2 + x - 12$ and $g(x) = 2x + 3$, find $\left(\dfrac{f}{g}\right)(2)$.

11. Use the Remainder Theorem to determine the remainder when $f(x) = 2x^3 - 3x^2 - 4x + 7$ is divided by $x - 3$.

12. Factor out the greatest common factor: $12a^3b^2 + 8a^2b^2 - 16ab^3$

In Problems 13–18, factor completely.

13. $6c^2 + 21c - 4c - 14$

14. $x^2 - 13x - 48$

15. $-14p^2 - 17p + 6$

16. $5(z - 1)^2 + 17(z - 1) - 12$

17. $-98x^2 + 112x - 32$

18. $16x^2 - 196$

19. Solve: $3m^2 - 5m = 5m - 7$

20. One side of a rectangular patio is 3 meters longer than the other. If the area of the patio is 108 square meters, what are the dimensions of the patio?

Getting Ready for Chapter 5:
A Review of Operations on Rational Numbers

Objectives

① Write Rational Numbers in Lowest Terms

② Multiply and Divide Rational Numbers

③ Add and Subtract Rational Numbers

The purpose of this Getting Ready section is to review the operations on rational numbers. We will use these same methods in Chapter 5 when we discuss operations on rational expressions.

▶ **①** **Write Rational Numbers in Lowest Terms**

We prefer to write rational numbers in **lowest terms**—that is, without any common factors in the numerator and the denominator. We obtain rational numbers in lowest terms using the *Reduction Property*.

> **Reduction Property**
>
> If a, b, and c are real numbers, then
>
> $$\frac{ac}{bc} = \frac{a}{b} \qquad \text{if } b \neq 0, c \neq 0$$

(**EXAMPLE 1**) **Writing Rational Numbers in Lowest Terms**

$$\overset{\text{Divide out the common factor}}{\underset{\downarrow}{}}$$

$$\frac{45}{18} = \frac{9 \cdot 5}{9 \cdot 2} = \frac{5}{2}$$

Quick ✓

1. When a rational number is written so there are no common factors in the numerator and the denominator, we say that the rational number is in _____ _____.

In Problems 2 and 3, write each rational number in lowest terms.

2. $\dfrac{13 \cdot 5}{13 \cdot 6}$

3. $\dfrac{80}{12}$

▶ **②** **Multiply and Divide Rational Numbers**

We now review the methods for multiplying and dividing rational numbers.

> **Multiplying Rational Numbers**
>
> **Step 1:** Completely factor each integer in the numerator and denominator.
>
> **Step 2:** Use the fact that if $\dfrac{a}{b}$ and $\dfrac{c}{d}$, $b \neq 0$, $d \neq 0$, are two rational numbers, then $\dfrac{a}{b} \cdot \dfrac{c}{d} = \dfrac{ac}{bd}$.
>
> **Step 3:** Divide out common factors in the numerator and denominator.

In Words

To multiply two rational numbers, multiply the numerators and then multiply the denominators. To divide two rational numbers, multiply the rational number in the numerator by the reciprocal of the rational number in the denominator.

> **Dividing Rational Numbers**
>
> To divide rational numbers, use the fact that if $\dfrac{a}{b}$ and $\dfrac{c}{d}$, $b \neq 0$, $c \neq 0$, $d \neq 0$, are two rational numbers, then $\dfrac{\dfrac{a}{b}}{\dfrac{c}{d}} = \dfrac{a}{b} \div \dfrac{c}{d} = \dfrac{a}{b} \cdot \dfrac{d}{c} = \dfrac{ad}{bc}$.

EXAMPLE 2 **Multiplying and Dividing Rational Numbers**

Perform the indicated operation. Be sure to express the result in lowest terms.

(a) $\dfrac{10}{3} \cdot \dfrac{18}{25}$

(b) $\dfrac{\frac{14}{5}}{\frac{21}{10}}$

Solution

$$10 = 2 \cdot 5; \; 18 = 9 \cdot 2 = 3 \cdot 3 \cdot 2; \; 25 = 5 \cdot 5$$

Work Smart

In Example 2(a), rather than factoring each numerator and denominator individually and then dividing out like factors, we might proceed as follows:

$$\dfrac{10}{3} \cdot \dfrac{18}{25} = \dfrac{\overset{2}{\cancel{10}} \cdot \overset{6}{\cancel{18}}}{\underset{1}{\cancel{3}} \cdot \underset{5}{\cancel{25}}}$$

Multiply numerators; multiply denominators:
$$= \dfrac{2 \cdot 6}{1 \cdot 5}$$
$$= \dfrac{12}{5}$$

(a)

$$\dfrac{10}{3} \cdot \dfrac{18}{25} = \dfrac{2 \cdot 5}{3} \cdot \dfrac{3 \cdot 3 \cdot 2}{5 \cdot 5}$$

Multiply; use the Reduction Property to divide out common factors:
$$= \dfrac{2 \cdot \cancel{5} \cdot \cancel{3} \cdot 3 \cdot 2}{\cancel{3} \cdot \cancel{5} \cdot 5}$$

$$= \dfrac{2 \cdot 3 \cdot 2}{5}$$

Multiply:
$$= \dfrac{12}{5}$$

(b) Rewrite the division problem as a multiplication problem by multiplying the numerator, $\dfrac{14}{5}$, by the reciprocal of the denominator, $\dfrac{21}{10}$. The reciprocal of $\dfrac{21}{10}$ is $\dfrac{10}{21}$.

$$\dfrac{\frac{14}{5}}{\frac{21}{10}} = \dfrac{14}{5} \cdot \dfrac{10}{21}$$

Factor:
$$= \dfrac{7 \cdot 2}{5} \cdot \dfrac{5 \cdot 2}{7 \cdot 3}$$

Multiply; divide out common factors:
$$= \dfrac{\cancel{7} \cdot 2 \cdot \cancel{5} \cdot 2}{\cancel{5} \cdot \cancel{7} \cdot 3}$$

$$= \dfrac{2 \cdot 2}{3} = \dfrac{4}{3}$$

Quick ✓

In Problems 4–7, perform the indicated operation. Express your answer in lowest terms.

4. $\dfrac{5}{7} \cdot \left(-\dfrac{21}{10} \right)$

5. $\dfrac{35}{15} \cdot \dfrac{3}{14}$

6. $\dfrac{\frac{4}{5}}{\frac{12}{25}}$

7. $\dfrac{24}{35} \div \left(-\dfrac{8}{7} \right)$

❸ Add and Subtract Rational Numbers

▶ We now review addition and subtraction of rational numbers.

Adding or Subtracting Rational Numbers

In Words
To add two rational numbers with a common denominator, add the numerators and write the sum over the common denominator.

Step 1: If $\dfrac{a}{c}$ and $\dfrac{b}{c}, c \neq 0$, are two rational numbers, then $\dfrac{a}{c} + \dfrac{b}{c} = \dfrac{a+b}{c}$ and $\dfrac{a}{c} - \dfrac{b}{c} = \dfrac{a-b}{c}$.

Step 2: Write the result in lowest terms.

EXAMPLE 3 **Adding or Subtracting Rational Numbers with Common Denominators**

Perform the indicated operation. Be sure to express the result in lowest terms.

(a) $\dfrac{7}{24} + \dfrac{11}{24}$ (b) $\dfrac{2}{15} - \dfrac{7}{15}$

Solution

(a)

$$\dfrac{7}{24} + \dfrac{11}{24} = \dfrac{7 + 11}{24}$$

$$= \dfrac{18}{24}$$

Factor numerator and denominator: $\qquad = \dfrac{6 \cdot 3}{6 \cdot 4}$

Divide out common factors: $\qquad = \dfrac{3}{4}$

(b)

$$\dfrac{2}{15} - \dfrac{7}{15} = \dfrac{2 - 7}{15}$$

$$= \dfrac{-5}{15}$$

Factor numerator and denominator: $\qquad = \dfrac{-1 \cdot 5}{3 \cdot 5}$

Divide out common factors: $\qquad = \dfrac{-1}{3} = -\dfrac{1}{3}$ ●

Quick ✓

In Problems 8 and 9, perform the indicated operation. Express your answer in lowest terms.

8. $\dfrac{11}{12} + \dfrac{5}{12}$ **9.** $\dfrac{3}{18} - \dfrac{13}{18}$

▶ What if the denominators of the rational numbers to be added or subtracted are not the same? In this case, we must rewrite each rational number over a *least common denominator*. The **least common denominator (LCD)** is the smallest integer that is a multiple of each denominator.

EXAMPLE 4 **How to Find the Least Common Denominator**

Find the least common denominator of the rational numbers $\dfrac{7}{30}$ and $\dfrac{5}{12}$. Then rewrite each rational number with the least common denominator.

Step-by-Step Solution

Step 1: Factor each denominator as a product of prime factors, arranging like factors vertically.

$$30 = 5 \cdot 6 = 5 \cdot 3 \cdot 2 \cdot$$
$$12 = 3 \cdot 4 = \quad 3 \cdot 2 \cdot 2$$

Work Smart

Line up the factors vertically to find the LCD.

Step 2: Find the product of each of the prime factors the greatest number of times they appear in any factorization.

$$\text{LCD} = 5 \cdot 3 \cdot 2 \cdot 2$$
$$= 60$$

We use the Multiplicative Identity Property, $1 \cdot a = a$, to write $\dfrac{7}{30}$ and $\dfrac{5}{12}$ with the denominator 60. Multiply $\dfrac{7}{30}$ by $1 = \dfrac{2}{2}$ and multiply $\dfrac{5}{12}$ by $1 = \dfrac{5}{5}$:

$$\frac{7}{30} = \frac{7}{30} \cdot \frac{2}{2} = \frac{7 \cdot 2}{30 \cdot 2} = \frac{14}{60}$$

$$\frac{5}{12} = \frac{5}{12} \cdot \frac{5}{5} = \frac{5 \cdot 5}{12 \cdot 5} = \frac{25}{60}$$

Quick ✓

In Problems 10 and 11, find the least common denominator (LCD) of each pair of rational numbers. Then rewrite each rational number with the LCD.

10. $\dfrac{3}{25}$ and $\dfrac{2}{15}$

11. $\dfrac{5}{18}$ and $-\dfrac{1}{63}$

Now that we know how to find the least common denominator, we can add or subtract rational numbers that have unlike denominators.

EXAMPLE 5   **How to Add or Subtract Fractions Using the Least Common Denominator**

Perform the indicated operation:

(a) $\dfrac{5}{2} + \dfrac{4}{3}$ **(b)** $\dfrac{5}{28} - \dfrac{5}{12}$

Step-by-Step Solution

(a) $\dfrac{5}{2} + \dfrac{4}{3}$

Step 1: Find the least common denominator.   Each denominator is prime, so the LCD $= 2 \cdot 3 = 6$.

Step 2: Rewrite each rational number with the common denominator.

$$\frac{5}{2} + \frac{4}{3} = \frac{5}{2} \cdot \frac{3}{3} + \frac{4}{3} \cdot \frac{2}{2}$$

$$= \frac{15}{6} + \frac{8}{6}$$

Step 3: Add the numerators, and write the result over the common denominator.

$$= \frac{15 + 8}{6}$$

$$= \frac{23}{6}$$

Step 4: Write the result in lowest terms.   The rational number is already in lowest terms.

$$\text{So,} \ \frac{5}{2} + \frac{4}{3} = \frac{23}{6}$$

(continued)

(b) $\dfrac{5}{28} - \dfrac{5}{12}$

Step 1: Find the least common denominator.

$$28 = 7 \cdot 4$$
$$12 = 4 \cdot 3$$
$$\downarrow \quad \downarrow \quad \downarrow$$
$$\text{LCD} = 7 \cdot 4 \cdot 3 = 84$$

Step 2: Rewrite each rational number with the common denominator.

$$\dfrac{5}{28} - \dfrac{5}{12} = \dfrac{5}{28} \cdot \dfrac{3}{3} - \dfrac{5}{12} \cdot \dfrac{7}{7}$$
$$= \dfrac{15}{84} - \dfrac{35}{84}$$

Step 3: Subtract the numerators, and write the result over the common denominator.

$$= \dfrac{15 - 35}{84}$$
$$= \dfrac{-20}{84}$$

Step 4: Write the result in lowest terms.

$$= \dfrac{-5 \cdot 4}{21 \cdot 4}$$
$$= \dfrac{-5}{21}$$
$$= -\dfrac{5}{21}$$

We summarize the steps used in Example 5 below.

> **Adding or Subtracting Rational Numbers with Unlike Denominators**
>
> **Step 1:** Find the least common denominator.
> **Step 2:** Rewrite each rational number with the common denominator.
> **Step 3:** Add or subtract the numerators, and write the result over the common denominator.
> **Step 4:** Write the result in lowest terms.

Quick ✓

In Problems 12–14, perform the indicated operation. Express your answer in lowest terms.

12. $\dfrac{3}{4} + \dfrac{1}{5}$

13. $\dfrac{3}{20} + \dfrac{2}{15}$

14. $\dfrac{5}{14} - \dfrac{11}{21}$

Getting Ready for Chapter 5 Exercises

MyMathLab® PRACTICE

Exercise numbers in green
are complete video solutions
in MyMathLab

Problems **1–14** *are the* Quick ✓*s that follow the* EXAMPLES.

Building Skills

*In Problems 15–20, write each rational number in lowest terms.
See Objective 1.*

15. $\dfrac{4}{12}$ **16.** $\dfrac{6}{18}$

17. $-\dfrac{15}{35}$ **18.** $-\dfrac{12}{28}$

19. $\dfrac{-50}{-10}$ **20.** $\dfrac{81}{-27}$

*In Problems 21–32, multiply or divide the rational numbers.
Express each product or quotient as a rational number in lowest
terms. See Objective 2.*

21. $\dfrac{3}{4} \cdot \dfrac{20}{9}$ **22.** $\dfrac{2}{3} \cdot \dfrac{15}{6}$

23. $-\dfrac{5}{6} \cdot \dfrac{18}{5}$ **24.** $-\dfrac{9}{8} \cdot \dfrac{16}{3}$

25. $\dfrac{5}{8} \cdot \dfrac{2}{15}$ **26.** $\dfrac{3}{14} \cdot \dfrac{7}{12}$

27. $\dfrac{5}{2} \div \dfrac{25}{4}$ **28.** $\dfrac{2}{3} \div \dfrac{8}{9}$

29. $\dfrac{-\dfrac{6}{5}}{\dfrac{8}{15}}$ **30.** $\dfrac{\dfrac{12}{7}}{-\dfrac{18}{21}}$

31. $\dfrac{-\dfrac{9}{2}}{-\dfrac{3}{4}}$ **32.** $\dfrac{-\dfrac{10}{7}}{\dfrac{5}{14}}$

*In Problems 33–51, add or subtract the rational numbers. Express
each sum or difference as a rational number in lowest terms. See
Objective 3.*

33. $\dfrac{5}{3} + \dfrac{1}{3}$ **34.** $\dfrac{3}{4} + \dfrac{9}{4}$

35. $\dfrac{11}{6} - \dfrac{1}{6}$ **36.** $\dfrac{19}{6} - \dfrac{5}{6}$

37. $-\dfrac{3}{4} + \dfrac{1}{3}$ **38.** $\dfrac{-1}{7} + \dfrac{4}{9}$

39. $\dfrac{1}{4} + \dfrac{5}{6}$ **40.** $\dfrac{5}{8} + \dfrac{5}{12}$

41. $-\dfrac{3}{10} - \dfrac{7}{15}$ **42.** $\dfrac{7}{16} - \dfrac{9}{20}$

43. $\dfrac{5}{18} - \dfrac{11}{15}$ **44.** $\dfrac{5}{24} + \dfrac{7}{32}$

45. $-\dfrac{7}{8} + \dfrac{3}{10}$ **46.** $-\dfrac{7}{12} + \dfrac{2}{15}$

47. $\dfrac{7}{24} - \dfrac{3}{20}$ **48.** $\dfrac{3}{28} + \dfrac{5}{12}$

49. $-\dfrac{7}{9} - \dfrac{2}{15}$ **50.** $-\dfrac{5}{18} - \dfrac{1}{45}$

51. $\dfrac{-4}{25} - \dfrac{7}{30}$

Explaining the Concepts

52. Explain how to write a rational number in lowest
terms.

53. Explain how to find the least common denominator
of two rational numbers.

54. Explain how to add two rational numbers that do
not have a common denominator.

5 Rational Expressions and Rational Functions

Can a mathematical formula be used to predict the number of runs a major league baseball team will score in a season? Yes! In fact, his knowledge of this mathematical formula is one of the reasons why Billy Beane, general manager of the Oakland A's, seeks out baseball players who get a lot of walks. (Billy's baseball philosophy was the basis for the 2011 movie *Moneyball*, starring Brad Pitt.) See Problem 71 in Section 5.4.

The Big Picture: Putting It Together

In Chapter 4, we learned how to add, subtract, multiply, and divide polynomials. We then learned how to factor polynomial expressions and use the result (along with the Zero-Product Property) to solve a polynomial equation.

We will do the same thing with *rational expressions*. A rational expression is a polynomial divided by another polynomial. Factoring comes in handy when working with rational expressions. The methods for reducing, adding, subtracting, multiplying, and dividing rational expressions are identical to the methods used to do these operations on rational numbers. The point is this—the algebra of rational expressions can be thought of as generic arithmetic of rational numbers.

In this chapter, we introduce a new category of problem called Synthesis Review. These problems will help you see the "big picture" of algebra. For example, we might ask you to add various objects (polynomials, rational expressions, and so on) and then ask you to discuss similarities and differences in performing the operation on the objects.

Outline

5.1 Multiplying and Dividing Rational Expressions

Objectives

1 Determine the Domain of a Rational Expression

2 Simplify Rational Expressions

3 Multiply Rational Expressions

4 Divide Rational Expressions

5 Work with Rational Functions

Are You Ready for This Section?

Before getting started, take the following readiness quiz. If you get a problem wrong, go back to the section cited and review the material.

R1. Factor: $2x^2 - 11x - 21$ [Section 4.5, pp. 350–355]

R2. Solve: $q^2 - 16 = 0$ [Section 4.8, pp. 371–375]

R3. Determine the reciprocal of $\dfrac{5}{2}$.   [Section R.3, pp. 23–24]

R4. Explain what *domain* means.   [Section 2.1, pp. 147–149]

The quotient of two polynomials is a **rational expression.** Some examples of rational expressions are

    (a) $\dfrac{x - 5}{2x + 1}$ **(b)** $\dfrac{x^2 - 7x - 18}{x^2 - 4}$ **(c)** $\dfrac{2a^2 + 5ab + 2b^2}{a^2 - 6ab + 8b^2}$ **(d)** $\dfrac{1}{x - 3}$

Expressions (a), (b), and (d) are rational expressions in one variable, x. Expression (c) is a rational expression in two variables, a and b.

Rational expressions are described the same way as rational numbers. In the expression $\dfrac{x - 5}{2x + 1}$, we call $x - 5$ the **numerator** and $2x + 1$ the **denominator.** When the numerator and denominator have no common factors (except 1 and -1), we say that the rational expression is written in **lowest terms,** or **simplified.**

▶ **1** Determine the Domain of a Rational Expression

To find the domain of a rational expression, we find all values of the variable that cause the denominator to equal 0 and exclude these values from the domain because division by 0 is not defined. Knowing the domain of a rational expression will help you solve rational equations later in this chapter.

EXAMPLE 1   **Determining the Domain of a Rational Expression**

Determine the domain of each of the following rational expressions.

     (a) $\dfrac{2x}{x + 3}$ **(b)** $\dfrac{p^2 + 5p + 6}{p^2 - 4}$

Solution

(a) We want to find all values of x that cause $x + 3$ to equal 0, so we solve

$$x + 3 = 0$$

           Subtract 3 from both sides: $x = -3$

Work Smart

Because we are working with the set of real numbers, the domain of a variable is understood to be all real numbers except those listed. For example, the notation $\{x \mid x \neq -3\}$ means x is any real number except -3.

Since -3 causes the denominator, $x + 3$, to equal 0, the domain of $\dfrac{2x}{x + 3}$ is $\{x \mid x \neq -3\}$.

(b) Find all values of p that cause $p^2 - 4$ to equal 0.

$$p^2 - 4 = 0$$

    Factor the difference of two squares: $(p - 2)(p + 2) = 0$

            Zero-Product Property: $p - 2 = 0$ or $p + 2 = 0$

                                     $p = 2$ or $p = -2$

The domain of $\dfrac{p^2 + 5p + 6}{p^2 - 4}$ is $\{p \mid p \neq -2, p \neq 2\}$.   ●

Ready?...Answers

R1. $(2x + 3)(x - 7)$

R2. $\{-4, 4\}$ **R3.** $\dfrac{2}{5}$

R4. The set of all inputs for which an algebraic expression is defined.

Quick ✓

1. The quotient of two polynomials is called a _____ _____.

2. In the expression $\dfrac{x+3}{3x-5}$, we call $x+3$ the _____, and $3x-5$ is called the _____.

3. *True or False* The domain of all rational functions is the set of all real numbers.

In Problems 4 and 5, determine the domain of the rational expression.

4. $\dfrac{x-4}{x+6}$

5. $\dfrac{z^2-9}{z^2+3z-28}$

▶ ❷ **Simplify Rational Expressions**

A rational expression is simplified by completely factoring the numerator and the denominator and dividing out any common factors using the Reduction Property.

$$\frac{ac}{bc} = \frac{a\cancel{c}}{b\cancel{c}} = \frac{a}{b} \qquad \text{if } b \neq 0, c \neq 0$$

We simplify rational expressions in the same way that we write rational numbers in lowest terms. For example, we write $\dfrac{12}{20}$ in lowest terms as follows:

$$\frac{12}{20} = \frac{\cancel{4}\cdot 3}{\cancel{4}\cdot 5} = \frac{3}{5}$$

EXAMPLE 2 **Simplifying a Rational Expression**

Simplify each rational expression:

(a) $\dfrac{x^2+2x-15}{2x^2-3x-9}$ **(b)** $\dfrac{q^3-8}{3q^2-6q}$

Solution

(a) Factor the numerator and denominator, and divide out common factors using the Reduction Property.

$$\frac{x^2+2x-15}{2x^2-3x-9} = \frac{(x+5)\cancel{(x-3)}}{(2x+3)\cancel{(x-3)}}$$

$$= \frac{x+5}{2x+3} \qquad x \neq -\frac{3}{2}, x \neq 3$$

Work Smart

When we divide out like factors, the quotient is 1. For example,

$$\frac{\overset{1}{\cancel{(x+3)}}}{2x\underset{1}{\cancel{(x+3)}}} = \frac{1}{2x}$$

(b) Factor the numerator and denominator. Then divide out common factors.

$$\frac{q^3-8}{3q^2-6q} = \frac{\cancel{(q-2)}(q^2+2q+4)}{3q\cancel{(q-2)}}$$

$$= \frac{q^2+2q+4}{3q} \qquad q \neq 0, q \neq 2$$ ●

To keep the original rational expression and the simplified rational expression equivalent, we must restrict from the domain all values of the variable that are not in the domain of the *original* rational expression.

Consider Example 2(a). We include the restriction $x \neq -\dfrac{3}{2}, x \neq 3$ for two reasons:

1. To remind us of the restrictions on the variable x.

2. To keep the rational expressions equal. Without the restriction $x \neq -\dfrac{3}{2}, x \neq 3$, the expression $\dfrac{x^2+2x-15}{2x^2-3x-9}$ is not equal to $\dfrac{x+5}{2x+3}$ because in $\dfrac{x^2+2x-15}{2x^2-3x-9}$, x cannot take on the value 3, while in $\dfrac{x+5}{2x+3}$, x can take on the value 3. By not allowing x to equal 3 in both instances, we keep the expressions equal.

The same logic justifies the restrictions on q in Example 2(b). For the remainder of the text, we will not specify the restrictions on the variable, but you should be aware that the restrictions are necessary to maintain equality.

Sometimes we can find common factors in the numerator and denominator of a rational expression by factoring -1 out of one of the factors. Consider the rational expression $\dfrac{3 - 4x}{4x - 3}$. If we factor -1 out of $3 - 4x$, we have $-1(-3 + 4x)$ or $-1(4x - 3)$ so that we can now divide out the common factor, $4x - 3$.

EXAMPLE 3 **Simplifying a Rational Expression**

Simplify: $\dfrac{3x^2 + 11x - 4}{1 - 3x}$

Solution

$$\frac{3x^2 + 11x - 4}{1 - 3x} = \frac{(3x - 1)(x + 4)}{1 - 3x}$$

Factor -1 from $1 - 3x$; divide out common factors: $\quad = \dfrac{\cancel{(3x - 1)}(x + 4)}{-1\cancel{(3x - 1)}}$

$$= \frac{(x + 4)}{-1}$$

$\dfrac{a}{-1} = -a: \quad = -(x + 4)$

Work Smart

When simplifying, we can only divide out common factors, not common terms!

WRONG! $\dfrac{x + 1}{x} = \dfrac{\cancel{x} + 1}{\cancel{x}} = 1$ WRONG! $\dfrac{x^2 + x + 2}{x + 2} = \dfrac{x^2 + \cancel{x} + \cancel{2}}{\cancel{x} + \cancel{2}} = x^2$

If you aren't sure what you can divide out, try the computation with numbers to see whether it works. For example, does $\dfrac{4}{3} = \dfrac{3 + 1}{3} = \dfrac{\cancel{3} + 1}{\cancel{3}} = 1$? NO! So, $\dfrac{x + 1}{x} \neq 1$.

Quick ✓

6. *True or False* $\dfrac{x^2 + 3x + 5}{x^2 + 7} = \dfrac{3x + 5}{7}$

In Problems 7–9, simplify each rational expression.

7. $\dfrac{x^2 - 7x + 12}{x^2 + 4x - 21}$ **8.** $\dfrac{z^3 - 64}{2z^2 - 3z - 20}$ **9.** $\dfrac{3w^2 + 13w - 10}{2 - 3w}$

▶ ❸ **Multiply Rational Expressions**

We show how to multiply rational expressions in the next example.

EXAMPLE 4 **How to Multiply Rational Expressions**

Multiply $\dfrac{x^2 + 2x - 15}{x + 1} \cdot \dfrac{x^2 + 7x}{x^2 + 4x - 21}$. Simplify the product.

Step-by-Step Solution

Step 1: Completely factor each polynomial in the numerator and denominator.	$\dfrac{x^2 + 2x - 15}{x + 1} \cdot \dfrac{x^2 + 7x}{x^2 + 4x - 21} = \dfrac{(x + 5)(x - 3)}{x + 1} \cdot \dfrac{x(x + 7)}{(x + 7)(x - 3)}$
Step 2: Multiply.	$= \dfrac{(x + 5)(x - 3) \cdot x(x + 7)}{(x + 1) \cdot (x + 7)(x - 3)}$

(continued)

Step 3: Divide out common factors in the numerator and denominator.

$$= \frac{(x + 5)\,\cancel{(x - 3)}\,x\,\cancel{(x + 7)}}{(x + 1)\,\cancel{(x + 7)}\,\cancel{(x - 3)}}$$

$$= \frac{x(x + 5)}{x + 1}$$

The steps for multiplying rational expressions are the same as the steps for multiplying rational numbers.

> **Multiplying Rational Expressions**
>
> **Step 1:** Completely factor each polynomial in the numerator and denominator.
>
> **Step 2:** Use the fact that if $\dfrac{a}{b}$ and $\dfrac{c}{d}$, where $b \neq 0$, and $d \neq 0$, are two rational expressions, then $\dfrac{a}{b} \cdot \dfrac{c}{d} = \dfrac{ac}{bd}$.
>
> **Step 3:** Divide out common factors in the numerator and denominator.

Always leave your product in factored form. You will need the factored form to solve rational equations (Section 5.4) and rational inequalities (Section 5.5).

Quick ✓

In Problem 10, multiply and simplify the rational expressions.

10. $\dfrac{p^2 - 9}{p^2 + 5p + 6} \cdot \dfrac{3p^2 - p - 2}{2p - 6}$

EXAMPLE 5 **Multiplying Rational Expressions**

Multiply and simplify each of the following rational expressions.

(a) $\dfrac{y^2 + 2y + 1}{3y^2 + y - 2} \cdot \dfrac{2 - 3y}{y^2 + 5y + 4}$ **(b)** $\dfrac{p^2 + 5pq + 6q^2}{2p^2 + 7pq + 3q^2} \cdot \dfrac{2p + q}{3p + 6q}$

Solution

Factor the numerator and denominator

(a) $\dfrac{y^2 + 2y + 1}{3y^2 + y - 2} \cdot \dfrac{2 - 3y}{y^2 + 5y + 4} = \dfrac{(y + 1)(y + 1)}{(3y - 2)(y + 1)} \cdot \dfrac{-1(3y - 2)}{(y + 4)(y + 1)}$

Multiply: $= \dfrac{(y + 1)(y + 1)(-1)(3y - 2)}{(3y - 2)(y + 1)(y + 4)(y + 1)}$

Divide out common factors: $= \dfrac{\cancel{(y + 1)}\,\cancel{(y + 1)}\,(-1)\,\cancel{(3y - 2)}}{\cancel{(3y - 2)}\,\cancel{(y + 1)}\,(y + 4)\,\cancel{(y + 1)}}$

$= -\dfrac{1}{y + 4}$

Work Smart: Study Skills

It is a good idea to use a different slash mark for each pair of factors that divides out. Or use colored pencils to highlight the like factors that divide out.

Factor the numerator and denominator

(b) $\dfrac{p^2 + 5pq + 6q^2}{2p^2 + 7pq + 3q^2} \cdot \dfrac{2p + q}{3p + 6q} = \dfrac{(p + 2q)(p + 3q)}{(2p + q)(p + 3q)} \cdot \dfrac{2p + q}{3(p + 2q)}$

Multiply: $= \dfrac{(p + 2q)(p + 3q)(2p + q)}{(2p + q)(p + 3q)(3)(p + 2q)}$

Divide out common factors: $= \dfrac{\cancel{(p + 2q)}\,\cancel{(p + 3q)}\,\cancel{(2p + q)}}{\cancel{(2p + q)}\,\cancel{(p + 3q)}\,(3)\,\cancel{(p + 2q)}}$

$= \dfrac{1}{3}$

Work Smart

Did you notice in Example 5(b) that all the factors in the numerator divide out? This means that a factor of 1 remains in the numerator, not 0!

Quick ✓

In Problems 11 and 12, multiply and simplify the rational expressions.

11. $\dfrac{2x + 8}{2x^2 + 11x + 12} \cdot \dfrac{2x^2 - 3x - 9}{6 - 2x}$

12. $\dfrac{m^2 + 2mn + n^2}{2m^2 + 3mn + n^2} \cdot \dfrac{2m^2 - 5mn - 3n^2}{3n - m}$

▶ ④ Divide Rational Expressions

The rule for dividing rational expressions is the same as the rule for dividing rational numbers.

Dividing Rational Expressions

To divide rational expressions, use the fact that if $\dfrac{a}{b}$ and $\dfrac{c}{d}$, where $b \neq 0, c \neq 0$,

and $d \neq 0$, are two rational expressions, then $\dfrac{\frac{a}{b}}{\frac{c}{d}} = \dfrac{a}{b} \cdot \dfrac{d}{c}$. Then follow the steps for multiplying two rational expressions.

In Words
To divide two rational expressions, multiply the rational expression in the numerator by the reciprocal of the rational expression in the denominator.

EXAMPLE 6 **Dividing Rational Expressions**

Divide each of the following rational expressions. Simplify the quotient, if possible.

(a) $\dfrac{\frac{20x^5}{3y}}{\frac{4x^2}{15y^5}}$

(b) $\dfrac{\frac{x^2 - 4x - 12}{4x^3 - 6x^2}}{\frac{x^3 + 8}{2x^3 - 4x^2 + 8x}}$

Solution

Rewrite the division problem as a multiplication problem by multiplying the numerator by the reciprocal of the denominator.

Work Smart

When you see $\dfrac{\frac{20x^5}{3y}}{\frac{4x^2}{15y^5}}$, you may find it

helpful to think

$$\dfrac{20x^5}{3y} \div \dfrac{4x^2}{15y^5}$$

(a)

$$\dfrac{\frac{20x^5}{3y}}{\frac{4x^2}{15y^5}} = \dfrac{20x^5}{3y} \cdot \dfrac{15y^5}{4x^2}$$

Multiply: $= \dfrac{20 \cdot 15 \cdot x^5 \cdot y^5}{3 \cdot 4 \cdot y \cdot x^2}$

Divide out common factors: $= \dfrac{\overset{5}{\cancel{20}} \cdot \overset{5}{\cancel{15}}}{\underset{1}{\cancel{3}} \cdot \underset{1}{\cancel{4}}} x^{5-2} y^{5-1}$

Simplify: $= 25x^3 y^4$

(b)

$$\dfrac{\frac{x^2 - 4x - 12}{4x^3 - 6x^2}}{\frac{x^3 + 8}{2x^3 - 4x^2 + 8x}} = \dfrac{x^2 - 4x - 12}{4x^3 - 6x^2} \cdot \dfrac{2x^3 - 4x^2 + 8x}{x^3 + 8}$$

Factor the numerator and denominator: $= \dfrac{(x - 6)(x + 2)}{2x^2(2x - 3)} \cdot \dfrac{2x(x^2 - 2x + 4)}{(x + 2)(x^2 - 2x + 4)}$

Multiply; divide out common factors: $= \dfrac{(x - 6)\cancel{(x + 2)} \cdot \cancel{2x}\,\cancel{(x^2 - 2x + 4)}}{\cancel{2}x^2(2x - 3) \cdot \cancel{(x + 2)}\,\cancel{(x^2 - 2x + 4)}}$

Simplify: $= \dfrac{x - 6}{x(2x - 3)}$

Quick ✓

In Problems 13 and 14, divide the rational expressions. Simplify the quotient, if possible.

13. $\dfrac{\dfrac{12a^4}{5b^2}}{\dfrac{4a^2}{15b^5}}$

14. $\dfrac{\dfrac{m^2 - 5m}{m - 7}}{\dfrac{2m}{m^2 - 6m - 7}}$

⑤ Work with Rational Functions

▶ Ratios of integers, such as $\dfrac{3}{5}$, are called *rational numbers*. Ratios of polynomial expressions are called *rational expressions*. Similarly, ratios of polynomial functions are called *rational functions*.

> **Definition**
>
> A **rational function** is a function of the form
>
> $$R(x) = \frac{p(x)}{q(x)}$$
>
> where p and q are polynomial functions and q is not the zero polynomial. The domain consists of all real numbers except those for which the denominator q is 0.

We want to find the domain of a rational function because the behavior of the graph of a rational function is interesting near values excluded from its domain. See Problem 97.

(**EXAMPLE 7**)   **Finding the Domain of a Rational Function**

Find the domain of $R(x) = \dfrac{x + 2}{x^2 - 5x - 14}$.

Solution

The domain is the set of all real numbers such that the denominator, $x^2 - 5x - 14$, does not equal zero. Find the values of x such that $x^2 - 5x - 14 = 0$ and exclude these values from the domain.

$$x^2 - 5x - 14 = 0$$

Factor: $(x - 7)(x + 2) = 0$

Set each factor equal to 0: $x - 7 = 0$ or $x + 2 = 0$

Solve: $x = 7$ or $x = -2$

The values $x = 7$ and $x = -2$ cause the denominator to equal 0, so the domain of R is $\{x \mid x \neq 7, x \neq -2\}$. ●

Quick ✓

In Problem 15, find the domain of the rational function.

15. $R(x) = \dfrac{2x}{x^2 + x - 30}$

We can multiply and divide rational functions just as we multiplied and divided polynomial functions. The domain of the product or quotient of two rational functions is the set of all numbers that are in the domains of the functions being multiplied or divided. In addition, for $R(x) = \dfrac{f(x)}{g(x)}$, we exclude values of x such that $g(x) = 0$.

EXAMPLE 8 **Multiplying and Dividing Rational Functions**

Given that $f(x) = \dfrac{x^2 - 4}{3x^2 + 9x}$, $g(x) = \dfrac{x + 3}{x^2 - 2x - 8}$, and $h(x) = \dfrac{2x^2 + 7x + 6}{x^2 + 5x}$, find

▶ **(a)** $R(x) = f(x) \cdot g(x)$ ▶ **(b)** $P(x) = \dfrac{f(x)}{h(x)}$

and state the domain of each function.

Solution

(a) $R(x) = f(x) \cdot g(x) = \dfrac{x^2 - 4}{3x^2 + 9x} \cdot \dfrac{x + 3}{x^2 - 2x - 8}$

Factor the numerator
and denominator: $= \dfrac{(x + 2)(x - 2)}{3x(x + 3)} \cdot \dfrac{x + 3}{(x - 4)(x + 2)}$

Multiply; divide out common factors: $= \dfrac{\cancel{(x + 2)}(x - 2)\cancel{(x + 3)}}{3x\cancel{(x + 3)}(x - 4)\cancel{(x + 2)}}$

Simplify: $= \dfrac{x - 2}{3x(x - 4)}$

The domain of $f(x)$ is $\{x \mid x \neq -3, x \neq 0\}$. The domain of $g(x)$ is $\{x \mid x \neq -2, x \neq 4\}$. Therefore, the domain of $R(x)$ is $\{x \mid x \neq -3, x \neq -2, x \neq 0, x \neq 4\}$.

(b) $P(x) = \dfrac{f(x)}{h(x)} = \dfrac{\dfrac{x^2 - 4}{3x^2 + 9x}}{\dfrac{2x^2 + 7x + 6}{x^2 + 5x}}$

$= \dfrac{x^2 - 4}{3x^2 + 9x} \cdot \dfrac{x^2 + 5x}{2x^2 + 7x + 6}$

Factor the numerator
and denominator: $= \dfrac{(x - 2)(x + 2)}{3x(x + 3)} \cdot \dfrac{x(x + 5)}{(2x + 3)(x + 2)}$

Multiply; divide out common factors: $= \dfrac{(x - 2)\cancel{(x + 2)}\cancel{x}(x + 5)}{3\cancel{x}(x + 3)(2x + 3)\cancel{(x + 2)}}$

Simplify: $= \dfrac{(x - 2)(x + 5)}{3(x + 3)(2x + 3)}$

The domain of $f(x)$ is $\{x \mid x \neq -3, x \neq 0\}$. The domain of $h(x)$ is $\{x \mid x \neq -5, x \neq 0\}$. Because the denominator of $P(x)$ cannot equal 0, we also exclude those values of x such that $2x^2 + 7x + 6 = 0$. These values are $x = -2$ and $x = -\dfrac{3}{2}$. Therefore, the domain of $P(x)$ is $\left\{x \mid x \neq -5, x \neq -3, x \neq -2, x \neq -\dfrac{3}{2}, x \neq 0\right\}$.

Work Smart

In Example 8(b), $3x^2 + 9x$ cannot equal 0. Why? In addition, $2x^2 + 7x + 6$ cannot equal 0. Why? And $x^2 + 5x$ cannot equal 0. Why?

Quick ✓

16. Given that $f(x) = \dfrac{x^2 - 4x - 5}{3x - 5}$, $g(x) = \dfrac{3x^2 + 4x - 15}{x^2 - 2x - 15}$, and $h(x) = \dfrac{4x^2 + 7x + 3}{9x^2 - 15x}$, find

(a) $R(x) = f(x) \cdot g(x)$ **(b)** $H(x) = \dfrac{f(x)}{h(x)}$

and state the domain of each function.

5.1 Exercises MyMathLab® PRACTICE

Exercise numbers in green are complete video solutions in MyMathLab

Problems **1–16** are the Quick ✓s that follow the **EXAMPLES**.

Building Skills

In Problems 17–26, state the domain of each rational expression. See Objective 1.

17. $\dfrac{3}{x + 5}$

18. $\dfrac{4}{x - 7}$

19. $\dfrac{x - 1}{x^2 - 6x - 16}$

20. $\dfrac{2x + 1}{x^2 + 4x - 45}$

21. $\dfrac{p^2 - 4}{2p^2 + p - 10}$

22. $\dfrac{m^2 + 5m + 6}{3m^2 + 4m - 4}$

23. $\dfrac{x + 1}{x^2 + 1}$

24. $\dfrac{x - 2}{x^2 + 4}$

25. $\dfrac{3x - 2}{(x - 1)^2}$

26. $\dfrac{x + 5}{x^2 + 8x + 16}$

In Problems 27–46, simplify each rational expression. See Objective 2.

27. $\dfrac{2x + 8}{x^2 - 16}$

28. $\dfrac{x^2 - 3x}{x^2 - 9}$

29. $\dfrac{p^2 + 4p + 3}{p + 1}$

30. $\dfrac{a^2 - 2a - 24}{a + 4}$

31. $\dfrac{5x + 25}{x^3 + 5x^2}$

32. $\dfrac{6x - 42}{x^3 - 7x^2}$

33. $\dfrac{q^2 - 3q - 18}{q^2 - 8q + 12}$

34. $\dfrac{w^2 + 5w - 14}{w^2 + 6w - 16}$

35. $\dfrac{2y^2 - 3y - 20}{2y^2 + 15y + 25}$

36. $\dfrac{3n^2 + n - 2}{3n^2 - 20n + 12}$

37. $\dfrac{9 - x^2}{x^2 + 2x - 15}$

38. $\dfrac{25 - k^2}{k^2 + 2k - 35}$

39. $\dfrac{x^3 + 2x^2 - 8x}{2x^4 - 32x^2}$

40. $\dfrac{2z^2 - 10z - 28}{4z^3 - 32z^2 + 28z}$

41. $\dfrac{x^2 - xy - 6y^2}{x^2 - 4y^2}$

42. $\dfrac{a^2 + 5ab + 4b^2}{a^2 + 8ab + 16b^2}$

43. $\dfrac{x^3 - 5x^2 + 3x - 15}{x^2 - 10x + 25}$

44. $\dfrac{v^3 + 3v^2 - 5v - 15}{v^2 + 6v + 9}$

45. $\dfrac{x^3 + 8}{x^2 - 5x - 14}$

46. $\dfrac{27q^3 + 1}{6q^2 - 7q - 3}$

In Problems 47–58, multiply and simplify each rational expression. See Objective 3.

47. $\dfrac{3x}{x^2 - x - 12} \cdot \dfrac{x - 4}{12x^2}$

48. $\dfrac{5x^2}{x + 3} \cdot \dfrac{x^2 + 7x + 12}{20x}$

49. $\dfrac{2x^2 - x - 6}{x^2 + 3x - 4} \cdot \dfrac{x^2 - x - 20}{2x^2 - 7x - 15}$

50. $\dfrac{3x^2 + 14x - 5}{x^2 + x - 30} \cdot \dfrac{x^2 - 2x - 15}{3x^2 + 8x - 3}$

51. $\dfrac{x^2 - 9}{x^2 - 25} \cdot \dfrac{x^2 - 2x - 15}{x^2 + 4x - 21}$

52. $\dfrac{p^2 - 16}{p^2 - 25} \cdot \dfrac{p^2 + 2p - 24}{p^2 + 3p - 4}$

53. $\dfrac{2q^2 - 5q - 3}{3q^2 + 19q + 6} \cdot \dfrac{3q^2 + 7q + 2}{3 - q}$

54. $\dfrac{2y^2 - 5y - 12}{2y^2 - y - 6} \cdot \dfrac{4y^2 - 5y - 6}{4 - y}$

55. $\dfrac{x^2 - 5x + 6}{x^2 + 2x - 8} \cdot (x + 4)$

56. $\dfrac{p^2 - 4p - 5}{p^2 - 5p - 6} \cdot (p - 6)$

57. $\dfrac{m^2 - n^2}{5m - 5n} \cdot \dfrac{10m + 5n}{2m^2 + 3mn + n^2}$

58. $\dfrac{a^2 + 2ab + b^2}{3a + 3b} \cdot \dfrac{b - a}{a^2 - b^2}$

In Problems 59–66, divide each rational expression. Simplify the quotient, if possible. See Objective 4.

59. $\dfrac{\dfrac{x + 3}{2x - 8}}{\dfrac{4x}{9}}$

60. $\dfrac{\dfrac{x - 2}{3x}}{\dfrac{5x - 10}{x}}$

61. $\dfrac{\dfrac{4a}{b^2}}{\dfrac{2a^2}{b}}$

62. $\dfrac{\dfrac{9m^3}{2n^2}}{\dfrac{3m}{8n^4}}$

63. $\dfrac{\dfrac{p^2 - 4p - 5}{2p^2 - 3p - 2}}{\dfrac{p^2 + p}{p^2 + p - 6}}$

64. $\dfrac{\dfrac{y^2 - 9}{2y^2 - y - 15}}{\dfrac{3y^2 + 10y + 3}{2y^2 + y - 10}}$

65. $\dfrac{\dfrac{x^3 - 1}{x^2 - 1}}{\dfrac{3x^2 + 3x + 3}{x^2 + 3x + 1}}$

66. $\dfrac{\dfrac{8x^3 + 1}{2x}}{\dfrac{x^3 + 2x^2 - 15x}{2x^2 - 5x - 3}}$

In Problems 67–76, determine the domain of each rational function. See Objective 5.

67. $R(x) = \dfrac{2}{x - 1}$

68. $R(x) = \dfrac{5}{x + 3}$

69. $R(x) = \dfrac{x - 2}{(2x + 1)(x - 4)}$

70. $R(x) = \dfrac{3x + 2}{(4x - 1)(x + 5)}$

71. $R(x) = \dfrac{x + 9}{x^2 + 6x + 5}$

72. $R(x) = \dfrac{5x - 2}{x^2 - 6x - 16}$

73. $R(x) = \dfrac{x - 2}{2x^2 - 9x + 10}$

74. $R(x) = \dfrac{x + 3}{3x^2 + 7x - 6}$

75. $R(x) = \dfrac{x - 1}{x^2 + 1}$

76. $R(x) = \dfrac{4x}{4x^2 + 1}$

For Problems 77–80, see Objective 5.

77. If $f(x) = \dfrac{x^2 - 2x - 15}{x + 6}, g(x) = \dfrac{x^2 + 5x - 6}{2x^2 - 7x - 15}$, and

$h(x) = \dfrac{x + 3}{3x^2 + 17x - 6}$, find **(a)** $R(x) = f(x) \cdot g(x)$

and state its domain **(b)** $R(x) = \dfrac{f(x)}{h(x)}$ and state its domain.

78. If $f(x) = \dfrac{x^2 - 7x - 8}{2x - 5}, g(x) = \dfrac{2x^2 + 3x - 20}{x^2 - 10x + 16}$, and

$h(x) = \dfrac{x^2 - 3x - 40}{x + 9}$, find **(a)** $R(x) = f(x) \cdot g(x)$

and state its domain **(b)** $R(x) = \dfrac{f(x)}{h(x)}$ and state its domain.

79. If $f(x) = \dfrac{3x^2 - x - 10}{x^3 - 1}, g(x) = \dfrac{x^2 + 5x - 6}{2x^2 + 3x - 14}$, and

$h(x) = \dfrac{3x^2 + 8x + 5}{x^2 - 1}$, find **(a)** $R(x) = (f \cdot g)(x)$

and state its domain **(b)** $R(x) = \left(\dfrac{f}{h}\right)(x)$ and state its domain.

80. If $f(x) = \dfrac{4x^2 - 9x - 9}{x^3 - 8}, g(x) = \dfrac{x^2 + 7x - 18}{5x^2 - 14x - 3}$,

and $h(x) = \dfrac{x^2 - 6x + 9}{x^2 - 4}$, **(a)** find $R(x) = (f \cdot g)(x)$

and state its domain **(b)** find $R(x) = \left(\dfrac{f}{h}\right)(x)$ and state its domain.

Mixed Practice

In Problems 81–88, multiply or divide each rational expression, as indicated. Simplify the product or quotient, if possible.

81. $\dfrac{z^3 + 8}{z^2 - 3z - 10} \cdot \dfrac{z^2 - 2z - 15}{2z^2 - 4z + 8}$

82. $\dfrac{x^3 - 27}{2x^2 + 5x - 25} \cdot \dfrac{x^2 + 2x - 15}{x^3 + 3x^2 + 9x}$

83. $\dfrac{\dfrac{m^2 - 4n^2}{m^3 - n^3}}{\dfrac{2m + 4n}{m^2 + mn + n^2}}$

84. $\dfrac{\dfrac{x^2 + 2xy + y^2}{x^2 + 3xy + 2y^2}}{\dfrac{x^2 - y^2}{x + 2y}}$

85. $\dfrac{4w + 8}{w^2 - 4w} \cdot \dfrac{w^2 - 3w - 4}{w^2 + 3w + 2}$

86. $\dfrac{5m - 5}{m^2 + 6m} \cdot \dfrac{m^2 + 2m - 24}{m^2 + 3m - 4}$

87. $\dfrac{\dfrac{2x - 6}{x^2 + x}}{\dfrac{x^2 - 4x + 3}{x^2}} \cdot \dfrac{x^2 + 3x + 2}{x^2 + x}$

88. $\dfrac{\dfrac{3x + 15}{2x + 4}}{\dfrac{x + 5}{x^2 - 4}} \cdot \dfrac{4x + 8}{3x^2 - 12}$

Applying the Concepts

89. Make up a rational expression that is undefined at $x = 3$.

90. Make up a rational expression that is undefined at $x = -2$.

91. Make up a rational expression that is undefined at $x = -4$ and at $x = 5$.

92. Make up a rational expression that is undefined at $x = -6$ and at $x = 0$.

93. Develop a rational function R that is undefined at $x = -2$ and at $x = 1$ such that $R(-3) = 1$.

94. Develop a rational function R that is undefined at $x = -4$ and at $x = 3$ such that $R(4) = 1$.

95. Gravity In physics, it is established that the acceleration due to gravity g (in meters per second^2) at a height h meters above sea level is given by

$$g(h) = \frac{3.99 \times 10^{14}}{(6.374 \times 10^6 + h)^2}$$

where 6.374×10^6 is the radius of Earth in meters.

(a) What is the acceleration due to gravity at sea level?

(b) What is the acceleration due to gravity in Denver, Colorado, elevation 1600 meters?

(c) What is the acceleration due to gravity on the peak of Mount Everest, elevation 8848 meters?

96. Economics The gross domestic product (GDP) is the total value of all goods and services manufactured within the United States. A model for determining the change in GDP of a government spending plan is given by the function

$$G(s) = \frac{s}{1 - b}$$

where G is the change in gross domestic product, s is the amount spent by the government, and b is the marginal propensity to consume with $0 < b < 1$. The marginal propensity to consume can be thought of as the amount an individual would spend for each additional dollar of income earned. For example, if $b = 0.9$, then individuals will spend $0.90 for each additional dollar earned. If $b = 0.95$, then individuals will spend $0.95 for each additional dollar earned.

(a) Suppose the government spends $100 million on highway infrastructure and the marginal propensity to consume in the United States is $b = 0.9$. What will be the change in GDP?

(b) Suppose the government spends $100 million on highway infrastructure and the marginal propensity to consume in the United States is $b = 0.95$. What will be the change in GDP?

Extending the Concepts

97. Math for the Future Consider the function

$$f(x) = \frac{1}{x - 2}$$

(a) Determine the domain of f.

(b) Fill in the following table. What happens to the values of f as x approaches 2 but remains greater than 2?

x	3	2.5	2.1	2.01	2.001	2.0001
$f(x)$						

(c) Fill in the following table. What happens to the values of f as x approaches 2 but remains less than 2?

x	1	1.5	1.9	1.99	1.999	1.9999
$f(x)$						

(d) Below is the graph of $f(x) = \dfrac{1}{x - 2}$. What happens to the graph of the function as x approaches 2 for values of x larger than 2? What happens to the graph of the function as x approaches 2 for values of x smaller than 2? Compare your results to the results obtained in parts (b) and (c).

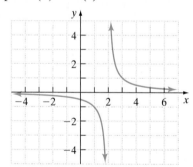

98. Math for the Future Consider the function

$$R(x) = \frac{2x + 1}{x - 2}$$

(a) Fill in the following table. What happens to the values of R as x gets larger in the positive direction?

x	5	10	50	100	1000
$R(x)$					

(b) Fill in the following table. What happens to the values of R as x gets larger in the negative direction?

x	-5	-10	-50	-100	-1000
$R(x)$					

(c) What is the term of highest degree in the numerator? What is the term of highest degree in the denominator? What is the ratio of the coefficients on the terms of highest degree in the numerator and denominator? Compare this result to your results in parts (a) and (b).

(d) Below is the graph of $R(x) = \dfrac{2x + 1}{x - 2}$. What happens to the graph of the function as x gets larger? That is, what happens to the graph of the function as x approaches ∞? What happens to the graph of the function as x gets smaller? That is, what happens to the graph of the function as x approaches $-\infty$? Compare your results to the results obtained in parts (b) and (c).

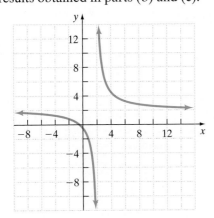

99. Write $\dfrac{x^{-1}}{x + 1}$ as a rational expression.

Explaining the Concepts

100. Define rational expression. Explain how a rational expression is related to a rational number.

101. What does it mean when we say that a rational expression is simplified?

102. Explain why we can divide out only common factors, not common terms.

103. Why is $\dfrac{\sqrt{x}}{x + 1}$ not a rational expression?

104. What is the difference between $f(x) = \dfrac{x^2 - 3x + 2}{x - 1}$ and $g(x) = x - 2$?

Synthesis Review

In Problems 105–110, graph each function.

105. $f(x) = 2x - 3$ **106.** $g(x) = 3x - 6$

107. $F(x) = -4x + 8$ **108.** $G(x) = -5x + 10$

109. $h(x) = x^2 - 4$ **110.** $H(x) = -x^2 + 4$

111. Discuss the methods that you used to graph each function. What methods were the same? What was different?

5.2 Adding and Subtracting Rational Expressions

Objectives

1 Add or Subtract Rational Expressions with a Common Denominator

2 Find the Least Common Denominator of Two or More Rational Expressions

3 Add or Subtract Rational Expressions with Different Denominators

Are You Ready for This Section?
Before getting started, take the following readiness quiz. If you get a problem wrong, go back to the section cited and review the material.

R1. Evaluate: $\dfrac{1}{6} - \dfrac{5}{8}$ [Getting Ready, pp. 395–396]

R2. Determine the additive inverse of **(a)** 5 **(b)** $x - 2$ [Section R.3, p. 21]

In Section 5.1, we learned how to multiply and divide rational expressions. We now learn how to add and subtract rational expressions.

Ready?... Answers **R1.** $-\dfrac{11}{24}$

R2. **(a)** -5 **(b)** $-(x - 2)$ or $2 - x$

▶ **①** **Add or Subtract Rational Expressions with a Common Denominator**

EXAMPLE 1 **How to Add or Subtract Rational Expressions with a Common Denominator**

Perform the indicated operation.

(a) $\dfrac{x^2 - 3x + 6}{x + 3} + \dfrac{7x - 3}{x + 3}, x \neq -3$

(b) $\dfrac{3x - 5}{x + 1} - \dfrac{x + 3}{x + 1}, x \neq -1$

Step-by-Step Solution

(a) The rational expressions in the sum have a common denominator, $x + 3$.

Step 1: Add the numerators, and write the result over the common denominator.

$$\dfrac{a}{c} + \dfrac{b}{c} = \dfrac{a + b}{c} : \dfrac{x^2 - 3x + 6}{x + 3} + \dfrac{7x - 3}{x + 3} = \dfrac{x^2 - 3x + 6 + (7x - 3)}{x + 3}$$

Combine like terms in the numerator: $= \dfrac{x^2 + 4x + 3}{x + 3}$

Step 2: Simplify the rational expression.

Factor the numerator: $= \dfrac{(x + 3)(x + 1)}{x + 3}$

Divide out like factors: $= x + 1$

(b) The rational expressions in the difference have a common denominator, $x + 1$.

Step 1: Subtract the numerators, and write the result over the common denominator.

$$\dfrac{a}{c} - \dfrac{b}{c} = \dfrac{a - b}{c} : \dfrac{3x - 5}{x + 1} - \dfrac{x + 3}{x + 1} = \dfrac{3x - 5 - (x + 3)}{x + 1}$$

Distribute the -1: $= \dfrac{3x - 5 - x - 3}{x + 1}$

Combine like terms in the numerator: $= \dfrac{2x - 8}{x + 1}$

Step 2: Simplify the rational expression.

Factor the numerator: $= \dfrac{2(x - 4)}{x + 1}$ ●

The rules for adding and subtracting rational expressions are the same as the rules for adding and subtracting rational numbers.

In Words
To add or subtract rational
expressions with the same
denominator, add or subtract
the numerators and write the
result over the common
denominator.

Adding or Subtracting Rational Expressions

Step 1: If $\dfrac{a}{c}$ and $\dfrac{b}{c}$, $c \neq 0$, are two rational expressions, then $\dfrac{a}{c} + \dfrac{b}{c} = \dfrac{a + b}{c}$ and $\dfrac{a}{c} - \dfrac{b}{c} = \dfrac{a - b}{c}$.

Step 2: Simplify the result.

Quick ✓

In Problems 1 and 2, perform the indicated operation. Be sure to simplify the result.

1. $\dfrac{x^2 - 3x - 1}{x - 2} + \dfrac{x^2 - 2x + 3}{x - 2}$

2. $\dfrac{4x + 3}{x + 5} - \dfrac{x - 6}{x + 5}$

EXAMPLE 2 **Adding Rational Expressions with Denominators That Are Additive Inverses**

Perform the indicated operation and simplify the result.

$$\frac{3x}{x-2} + \frac{2}{2-x}, x \neq 2$$

Solution

Notice that the denominators are additive inverses (opposites) of each other.

$$2 - x = -x + 2 = -1(x-2)$$

$$\frac{3x}{x-2} + \frac{2}{2-x} = \frac{3x}{x-2} + \frac{2}{-1(x-2)}$$

Use $\frac{a}{-b} = \frac{-a}{b}$: $= \frac{3x}{x-2} + \frac{-2}{x-2}$

Use $\frac{a}{c} + \frac{b}{c} = \frac{a+b}{c}$: $= \frac{3x-2}{x-2}$

Quick ✓

In Problem 3, perform the indicated operation. Be sure to simplify the result.

3. $\dfrac{4x}{x-5} + \dfrac{3}{5-x}$

▶ ❷ **Find the Least Common Denominator of Two or More Rational Expressions**

If the denominators of the rational expressions to be added or subtracted are not the same, then we must rewrite each rational expression with a *least common denominator*. The **least common denominator (LCD)** is the smallest polynomial that is a multiple of each denominator in the rational expressions to be added or subtracted. The idea is the same one we used to add rational numbers without common denominators. In fact, it would be a great idea to review Example 4 on pages 394–395 in Getting Ready for Chapter 5 before reading Example 3.

EXAMPLE 3 **How to Find the Least Common Denominator of Rational Expressions**

Find the least common denominator of each expression.

(a) $\dfrac{4}{3x^2y^2}$ and $\dfrac{5}{6xy^3}$ **(b)** $\dfrac{3x-5}{x^3+2x^2+x}$ and $\dfrac{x-1}{x^2+4x+3}$

Step-by-Step Solution

(a) $\dfrac{4}{3x^2y^2}$ and $\dfrac{5}{6xy^3}$

Step 1: Write each denominator as the product of prime factors.

$3x^2y^2 = 3 \cdot x^2 \cdot y^2$
$6xy^3 = 2 \cdot 3 \cdot x \cdot y^3$

Step 2: Find the product of each prime factor the greatest number of times it appears in any factorization.

$$\text{LCD} = 2 \cdot 3 \cdot x^2 \cdot y^3$$
$$= 6x^2y^3$$

(continued)

(b) $\dfrac{3x - 5}{x^3 + 2x^2 + x}$ and $\dfrac{x - 1}{x^2 + 4x + 3}$

Step 1: Write each denominator as the product of prime factors.

$$x^3 + 2x^2 + x = x\,(x^2 + 2x + 1) = x\,(x + 1)^2$$

$$x^2 + 4x + 3 = \qquad\qquad (x + 1)\,(x + 3)$$

Step 2: Find the product of each prime factor the greatest number of times it appears in any factorization.

$$\text{LCD} = x\,(x + 1)^2(x + 3)$$

Finding the Least Common Denominator

Step 1: Factor each denominator completely. When factoring, write the factored form using powers. For example, write $x^2 + 4x + 4$ as $(x + 2)^2$.

Step 2: The LCD is the product of each prime factor the greatest number of times it appears in any factorization.

Quick ✓

4. The ____ _____ _____ is the smallest polynomial that is a multiple of each denominator in the rational expressions to be added or subtracted.

In Problems 5 and 6, find the least common denominator of each expression.

5. $\dfrac{5}{8x^2y}$ and $\dfrac{1}{12xy^3}$ **6.** $\dfrac{4x - 3}{x^2 - 5x - 14}$ and $\dfrac{x + 1}{x^2 + 4x + 4}$

▶ ❸ Add or Subtract Rational Expressions with Different Denominators

Now that we know how to obtain the least common denominator, we can add or subtract rational expressions that have unlike denominators.

EXAMPLE 4 How to Add Rational Expressions with Unlike Denominators

Add $\dfrac{3}{8x^2} + \dfrac{1}{12x}$. Simplify the result.

Step-by-Step Solution

Step 1: Find the least common denominator.

$$8x^2 = 4 \cdot 2 \quad \cdot x^2$$
$$12x = 4 \quad \cdot 3 \cdot x$$

The LCD is $4 \cdot 2 \cdot 3 \cdot x^2 = 24x^2$.

Step 2: Rewrite each rational expression with the common denominator.

We need the denominator to be $24x^2$, so multiply $\dfrac{3}{8x^2}$ by $1 = \dfrac{3}{3}$.
Multiply $\dfrac{1}{12x}$ by $1 = \dfrac{2x}{2x}$. Do you see why?

$$\frac{3}{8x^2} = \frac{3}{8x^2} \cdot \frac{3}{3} = \frac{9}{24x^2}$$

$$\frac{1}{12x} = \frac{1}{12x} \cdot \frac{2x}{2x} = \frac{2x}{24x^2}$$

Step 3: Add the rational expressions found in Step 2.

$$\frac{3}{8x^2} + \frac{1}{12x} = \frac{9}{24x^2} + \frac{2x}{24x^2}$$

$$\frac{a}{c} + \frac{b}{c} = \frac{a + b}{c}: \quad = \frac{9 + 2x}{24x^2}$$

$$= \frac{2x + 9}{24x^2}$$

Step 4: Simplify the result.

The rational expression is simplified.

$$\text{Thus } \frac{3}{8x^2} + \frac{1}{12x} = \frac{2x + 9}{24x^2}.$$

Adding or Subtracting Rational Expressions with Unlike Denominators

Step 1: Find the least common denominator.

Step 2: Rewrite each rational expression as an equivalent rational expression with the common denominator. Multiply out the numerator, but leave the denominator in factored form.

Step 3: Add or subtract the rational expressions found in Step 2.

Step 4: Simplify the result.

Quick ✓

In Problems 7 and 8, perform the indicated operation and simplify the result.

7. $\dfrac{3}{10a} + \dfrac{4}{15a^2}$

8. $\dfrac{3}{8y} - \dfrac{13}{24y}$

EXAMPLE 5 **Adding Rational Expressions with Unlike Denominators**

Perform the indicated operation and simplify the result.

(a) $\dfrac{x - 1}{x + 3} + \dfrac{x}{x + 2}$

(b) $\dfrac{x - 1}{x^2 + 2x - 8} + \dfrac{x - 1}{x^2 - 16}$

Solution

(a) $\dfrac{x - 1}{x + 3} + \dfrac{x}{x + 2}$

The LCD is $(x + 3)(x + 2)$. To rewrite each rational expression as an equivalent rational expression with the LCD, multiply $\dfrac{x - 1}{x + 3}$ by $1 = \dfrac{x + 2}{x + 2}$ and multiply $\dfrac{x}{x + 2}$ by $\dfrac{x + 3}{x + 3}$.

$$\frac{x - 1}{x + 3} + \frac{x}{x + 2} = \frac{x - 1}{x + 3} \cdot \frac{x + 2}{x + 2} + \frac{x}{x + 2} \cdot \frac{x + 3}{x + 3}$$

Multiply the numerators: $= \dfrac{x^2 + x - 2}{(x + 3)(x + 2)} + \dfrac{x^2 + 3x}{(x + 3)(x + 2)}$

Use $\dfrac{a}{c} + \dfrac{b}{c} = \dfrac{a + b}{c}$: $= \dfrac{x^2 + x - 2 + (x^2 + 3x)}{(x + 3)(x + 2)}$

Combine like terms: $= \dfrac{2x^2 + 4x - 2}{(x + 3)(x + 2)}$

Factor the numerator: $= \dfrac{2(x^2 + 2x - 1)}{(x + 3)(x + 2)}$

(b) $\dfrac{x - 1}{x^2 + 2x - 8} + \dfrac{x - 1}{x^2 - 16}$

First, factor the denominators to find the LCD.

$$x^2 + 2x - 8 = (x + 4)(x - 2)$$
$$x^2 - 16 = (x + 4)(x - 4)$$
$$\text{LCD} = (x + 4)(x - 2)(x - 4)$$

$$\text{Multiply } \frac{x-1}{(x+4)(x-2)} \text{ by } \frac{x-4}{x-4}; \text{ Multiply } \frac{x-1}{(x+4)(x-4)} \text{ by } \frac{x-2}{x-2}$$

$$\frac{x-1}{x^2+2x-8} + \frac{x-1}{x^2-16} = \frac{x-1}{(x+4)(x-2)} \cdot \frac{x-4}{x-4} + \frac{x-1}{(x+4)(x-4)} \cdot \frac{x-2}{x-2}$$

Multiply the numerators:
$$= \frac{x^2-5x+4}{(x+4)(x-2)(x-4)} + \frac{x^2-3x+2}{(x+4)(x-2)(x-4)}$$

Use $\dfrac{a}{c} + \dfrac{b}{c} = \dfrac{a+b}{c}$:
$$= \frac{x^2-5x+4+(x^2-3x+2)}{(x+4)(x-2)(x-4)}$$

Combine like terms:
$$= \frac{2x^2-8x+6}{(x+4)(x-2)(x-4)}$$

Factor the numerator:
$$= \frac{2(x-3)(x-1)}{(x+4)(x-2)(x-4)}$$

Quick ✓

In Problems 9 and 10, perform the indicated operation and simplify the result.

9. $\dfrac{3x}{x-1} + \dfrac{x+5}{x+2}$

10. $\dfrac{x-1}{2x^2+7x+6} + \dfrac{x-1}{x^2+6x+8}$

EXAMPLE 6 **How to Subtract Rational Expressions with Unlike Denominators**

Perform the indicated operation and simplify the result.

$$\frac{2x-1}{2x^2-7x-4} - \frac{x-1}{2x^2+3x+1}$$

Step-by-Step Solution

Step 1: Find the least common denominator.

$$2x^2-7x-4 = (2x+1)(x-4)$$
$$2x^2+3x+1 = (2x+1)(x+1)$$
$$\text{LCD} = (2x+1)(x-4)(x+1)$$

Step 2: Rewrite each rational expression as an equivalent rational expression with the common denominator. Multiply the numerators, but leave the denominators in factored form.

$$\frac{2x-1}{2x^2-7x-4} - \frac{x-1}{2x^2+3x+1} = \frac{2x-1}{(2x+1)(x-4)} \cdot \frac{x+1}{x+1} - \frac{x-1}{(2x+1)(x+1)} \cdot \frac{x-4}{x-4}$$

Multiply the numerators:
$$= \frac{2x^2+x-1}{(2x+1)(x-4)(x+1)} - \frac{x^2-5x+4}{(2x+1)(x-4)(x+1)}$$

Step 3: Subtract the rational expressions found in Step 2.

Use $\dfrac{a}{c} - \dfrac{b}{c} = \dfrac{a-b}{c}$:
$$= \frac{2x^2+x-1-(x^2-5x+4)}{(2x+1)(x-4)(x+1)}$$

Distribute the -1:
$$= \frac{2x^2+x-1-x^2+5x-4}{(2x+1)(x-4)(x+1)}$$

Combine like terms:
$$= \frac{x^2+6x-5}{(2x+1)(x-4)(x+1)}$$

Step 4: Simplify the rational expression.

The rational expression in Step 3 above is simplified.

Quick ✓

In Problem 11, perform the indicated operation and simplify the result.

11. $\dfrac{3x + 4}{2x^2 + x - 6} - \dfrac{x - 1}{x^2 + 4x + 4}$

EXAMPLE 7 **Adding and Subtracting Three Rational Expressions**

Perform the indicated operations and simplify the result.

$$\frac{6}{x^2 - 9} + \frac{x + 1}{x + 3} - \frac{x - 2}{x - 3}$$

Solution

We factor each denominator.

$$x^2 - 9 = (x + 3)(x - 3)$$
$$x + 3 = x + 3$$
$$x - 3 = x - 3$$
$$\text{LCD} = (x + 3)(x - 3)$$

$$\frac{6}{x^2 - 9} + \frac{x + 1}{x + 3} - \frac{x - 2}{x - 3} = \frac{6}{(x + 3)(x - 3)} + \frac{x + 1}{x + 3} \cdot \frac{x - 3}{x - 3} - \frac{x - 2}{x - 3} \cdot \frac{x + 3}{x + 3}$$

Multiply the numerators:
$$= \frac{6}{(x + 3)(x - 3)} + \frac{x^2 - 2x - 3}{(x + 3)(x - 3)} - \frac{x^2 + x - 6}{(x + 3)(x - 3)}$$

Use $\dfrac{a}{c} + \dfrac{b}{c} = \dfrac{a + b}{c}; \dfrac{a}{c} - \dfrac{b}{c} = \dfrac{a - b}{c}$:
$$= \frac{6 + x^2 - 2x - 3 - (x^2 + x - 6)}{(x + 3)(x - 3)}$$

Distribute the -1:
$$= \frac{6 + x^2 - 2x - 3 - x^2 - x + 6}{(x + 3)(x - 3)}$$

Combine like terms:
$$= \frac{-3x + 9}{(x + 3)(x - 3)}$$

Factor:
$$= \frac{-3(x - 3)}{(x + 3)(x - 3)}$$

Divide out like factors:
$$= \frac{-3}{x + 3}$$

Quick ✓

In Problem 12, perform the indicated operations and simplify the result.

12. $\dfrac{4}{x^2 - 4} - \dfrac{x + 3}{x - 2} + \dfrac{x + 3}{x + 2}$

5.2 Exercises MyMathLab®
PRACTICE

Exercise numbers in green
are complete video solutions
in MyMathLab

Problems **1–12** *are the* Quick ✓ *s that follow the* EXAMPLES.

Building Skills

In Problems 13–24, perform the indicated operation and simplify the result. See Objective 1.

13. $\dfrac{3x}{x + 1} + \dfrac{5}{x + 1}$

14. $\dfrac{5x}{x - 3} + \dfrac{2}{x - 3}$

15. $\dfrac{2x}{2x + 5} - \dfrac{1}{2x + 5}$

16. $\dfrac{9x}{6x - 5} - \dfrac{2}{6x - 5}$

17. $\dfrac{2x}{2x^2 - 7x - 15} + \dfrac{3}{2x^2 - 7x - 15}$

18. $\dfrac{x}{3x^2 + 8x - 3} + \dfrac{3}{3x^2 + 8x - 3}$

19. $\dfrac{2x^2 - 5x + 7}{x^2 - 2x - 15} - \dfrac{x^2 + 3x - 8}{x^2 - 2x - 15}$

20. $\dfrac{3x^2 + 8x - 1}{x^2 - 3x - 28} - \dfrac{2x^2 + 2x - 9}{x^2 - 3x - 28}$

21. $\dfrac{3x}{x - 5} + \dfrac{1}{5 - x}$

22. $\dfrac{3x}{x - 6} + \dfrac{2}{6 - x}$

23. $\dfrac{2x^2 - 4x - 1}{x - 3} - \dfrac{x^2 - 6x + 4}{3 - x}$

24. $\dfrac{x^2 + 2x - 5}{x - 4} - \dfrac{x^2 - 5x - 15}{4 - x}$

In Problems 25–34, find the least common denominator. See Objective 2.

25. $\dfrac{3}{4x^3}$ and $\dfrac{9}{8x}$

26. $\dfrac{5}{3a^3}$ and $\dfrac{2}{9a^2}$

27. $\dfrac{1}{15xy^2}$ and $\dfrac{7}{18x^3y}$

28. $\dfrac{1}{8a^3b}$ and $\dfrac{5}{12ab^2}$

29. $\dfrac{5x}{x - 4}$ and $\dfrac{3}{x + 2}$

30. $\dfrac{x - 3}{x + 2}$ and $\dfrac{x + 7}{x - 5}$

31. $\dfrac{x - 4}{x^2 - x - 12}$ and $\dfrac{2x + 1}{x^2 - 9x + 20}$

32. $\dfrac{2m - 7}{m^2 + 3m - 18}$ and $\dfrac{5m + 1}{m^2 - 7m + 12}$

33. $\dfrac{p + 1}{2p^2 + 3p - 2}$ and $\dfrac{4p - 1}{p^3 + 2p^2}$

34. $\dfrac{x - 6}{x^2 - 9}$ and $\dfrac{3x}{x^3 - 3x^2}$

In Problems 35–60, add or subtract, as indicated, and simplify the result. See Objective 3.

35. $\dfrac{3}{4x^2} + \dfrac{5}{8x}$

36. $\dfrac{2}{9x} + \dfrac{5}{3x^2}$

37. $\dfrac{5}{12a^2 b} - \dfrac{4}{15ab^2}$

38. $\dfrac{3}{14mn^3} - \dfrac{2}{21m^2 n}$

39. $\dfrac{y + 2}{y - 5} - \dfrac{y - 4}{y + 3}$

40. $\dfrac{x + 2}{x - 3} - \dfrac{x + 2}{x + 1}$

41. $\dfrac{a + 5}{a - 2} - \dfrac{5a + 18}{a^2 - 4}$

42. $\dfrac{z + 1}{z + 3} - \dfrac{z + 17}{z^2 - z - 12}$

43. $\dfrac{3}{(x - 2)(x + 3)} - \dfrac{5}{(x + 3)(x + 4)}$

44. $\dfrac{1}{(x - 1)(x + 3)} + \dfrac{5}{(x + 1)(x - 1)}$

45. $\dfrac{x - 3}{x^2 + 3x + 2} + \dfrac{x - 1}{x^2 - 4}$

46. $\dfrac{x - 5}{x^2 + 4x + 3} + \dfrac{x - 2}{x^2 - 1}$

47. $\dfrac{w - 4}{2w^2 + 3w + 1} - \dfrac{w + 3}{2w^2 - 5w - 3}$

48. $\dfrac{y + 4}{3y^2 - y - 2} - \dfrac{1}{3y^2 + 14y + 8}$

49. $\dfrac{x + y}{x^2 - 6xy + 9y^2} + \dfrac{x + 2y}{x^2 - 2xy - 3y^2}$

50. $\dfrac{m - 2n}{m^2 + 4mn + 4n^2} + \dfrac{m - n}{m^2 - mn - 6n^2}$

51. $\dfrac{3}{x^2 - 4x - 5} - \dfrac{2}{x^2 - 6x + 5}$

52. $\dfrac{3}{x^2 + 7x + 10} - \dfrac{4}{x^2 + 6x + 5}$

53. $\dfrac{p^2 - 3p - 10}{p^2 - 16} + \dfrac{p^2 - 3p - 10}{16 - p^2}$

54. $\dfrac{y^2 + 4y + 4}{y^2 - 9} + \dfrac{y^2 + 4y + 4}{9 - y^2}$

55. $\dfrac{2}{w + 2} - \dfrac{3}{w} + \dfrac{w + 10}{w^2 - 4}$

56. $\dfrac{7}{m - 3} - \dfrac{5}{m} - \dfrac{2m + 6}{m^2 - 9}$

57. $\dfrac{p - 2}{p^2 + 6p + 9} + \dfrac{1}{p + 3} - \dfrac{2p + 1}{2p^2 + p - 15}$

58. $\dfrac{x - 1}{x^2 - 16} + \dfrac{1}{x + 4} - \dfrac{4x + 1}{3x^2 - 7x - 20}$

59. $\dfrac{2}{x} - \dfrac{2}{x - 1} + \dfrac{3}{(x - 1)^2}$

60. $\dfrac{2}{x} - \dfrac{2}{x + 2} + \dfrac{2}{(x + 2)^2}$

Mixed Practice

In Problems 61–72, perform the indicated operation and simplify the result.

61. $\dfrac{1}{x - 3} - \dfrac{x^2 + 18}{x^3 - 27}$

62. $\dfrac{1}{x + 2} + \dfrac{x - 10}{x^3 + 8}$

63. $\dfrac{2x^2 - x}{x + 3} + \dfrac{3x}{x + 3} - \dfrac{x^2 + 3}{x + 3}$

64. $\dfrac{2x^2}{x - 1} - \dfrac{x^2 - 2x}{x - 1} + \dfrac{x - 4}{x - 1}$

65. $6 + \dfrac{x - 3}{x + 3}$

66. $3 + \dfrac{x + 4}{x - 4}$

67. $\dfrac{b + 3}{b^2 + 2b - 8} - \dfrac{b + 2}{b^2 - 4}$

68. $\dfrac{a + 3}{a^2 - 8a + 15} + \dfrac{a + 3}{a^2 - 9}$

69. $\dfrac{y - 1}{y + 4} + \dfrac{y - 2}{y + 3} - \dfrac{y^2 + 3y + 1}{y^2 + 7y + 12}$

70. $\dfrac{z + 3}{z - 6} + \dfrac{z - 1}{z - 2} - \dfrac{6z}{z^2 - 8z + 12}$

71. $\dfrac{x + 4}{x^2 - 5x + 6} + \dfrac{x - 1}{x^2 - 2x - 3} - \dfrac{2x + 1}{x^2 - x - 2}$

72. $\dfrac{x - 1}{x^2 + 4x - 5} + \dfrac{3x - 1}{x^2 + 3x - 10} - \dfrac{4x + 1}{x^2 - 3x + 2}$

73. Given that $f(x) = \dfrac{3}{x - 2}$ and $g(x) = \dfrac{2}{x + 1}$,

(a) find $R(x) = f(x) + g(x)$,

(b) state the domain of $R(x)$.

74. Given that $f(x) = \dfrac{5}{x + 2}$ and $g(x) = \dfrac{3}{x - 1}$,

(a) find $R(x) = f(x) + g(x)$,

(b) state the domain of $R(x)$.

75. Given that $f(x) = \dfrac{x + 1}{x^2 - 3x - 4}$

and $g(x) = \dfrac{x + 4}{x^2 - x - 12}$,

(a) find $R(x) = f(x) + g(x)$,

(b) state the domain of $R(x)$,

(c) find $H(x) = f(x) - g(x)$,

(d) state the domain of $H(x)$.

76. Given that $f(x) = \dfrac{x + 5}{x^2 - 5x + 6}$ and

$g(x) = \dfrac{x + 1}{x^2 - 4x - 12}$,

(a) find $R(x) = f(x) + g(x)$,

(b) state the domain of $R(x)$,

(c) find $H(x) = f(x) - g(x)$,

(d) state the domain of $H(x)$.

Applying the Concepts

△ **77. Surface Area of a Box** The volume of a closed box with a square base is 2000 cubic inches. Its surface area S, as a function of the length of the base x, is given by

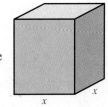

$$S(x) = 2x^2 + \dfrac{8000}{x}$$

(a) Write S over a common denominator. That is, write S so that the rule is a single rational expression.

(b) Find and interpret $S(10)$.

△ **78. Surface Area of a Can** The volume of a cylindrical can is 200 cubic centimeters. Its surface area S, as a function of the radius r of the can, is given by

$$S(r) = 2\pi r^2 + \dfrac{400}{r}$$

(a) Write S over a common denominator. That is, write S so that the rule is a single rational expression.

(b) Find and interpret $S(4)$. Round your answer to two decimal places.

79. Road Trip Suppose you and a group of your friends decide to go on a road trip to a neighboring university that is 200 miles away. Your average speed for the first 50 miles of the trip is 10 miles an hour slower than the average speed for the remaining 150 miles of the trip. If we let s represent your average speed for the first 50 miles of the trip, the function

$$T(s) = \dfrac{50}{s} + \dfrac{150}{s + 10}$$

represents the amount of time T it will take to get to the neighboring university.

(a) Write T over a common denominator. That is, write T so that the rule is a single rational expression.

(b) Find and interpret $T(50)$.

80. Vacation The distance from Chicago, Illinois, to Naples, Florida, is approximately 1200 miles. Atlanta, Georgia, is approximately the midpoint between Chicago and Naples. On a recent trip to Naples, the Sullivan family averaged s miles per hour between Chicago and Atlanta, and they averaged 5 miles per hour faster between Atlanta and Naples. The time T of their trip, as a function of their average speed s between Chicago and Atlanta, is given by the function

$$T(s) = \dfrac{600}{s} + \dfrac{600}{s + 5}$$

(a) Write T over a common denominator. That is, write T so that the rule is a single rational expression.

(b) Find and interpret $T(50)$. Round your answer to two decimal places.

(c) Using the result from part (b), compute the average speed of the Sullivans for the entire trip. Are you surprised by the result?

Extending the Concepts

81. Write $x^{-1} + y^{-1}$ as a single rational expression with no negative exponents.

82. Write $\left(\dfrac{a}{b}\right)^{-1} + \left(\dfrac{b}{a}\right)^{-1}$ as a single rational expression with no negative exponents.

Explaining the Concepts

83. Explain how to find the least common denominator when adding or subtracting two rational expressions with unlike denominators.

84. Explain how to add or subtract rational expressions with unlike denominators.

Synthesis Review

In Problems 85–91, multiply the expressions.

85. $4a(a - 3)$

86. $-5z(z + 4)$

87. $(p - 3)(p + 3)$

88. $(3q + 1)(3q - 1)$

89. $(w - 2)(w^2 + 2w + 4)$

90. $(2v + 1)(4v^2 - 2v + 1)$

91. $\dfrac{x^3 - 8}{2x^2 + 5x + 2} \cdot \dfrac{2x^2 - x - 1}{x^2 - 4} \cdot \dfrac{x^2 + 4x + 4}{x - 1}$

5.3 Complex Rational Expressions

Objectives

1 Simplify a Complex Rational Expression by Simplifying the Numerator and Denominator Separately (Method I)

2 Simplify a Complex Rational Expression Using the Least Common Denominator (Method II)

> **In Words**
>
> A complex rational expression is simplified when it is of the form polynomial over polynomial and the polynomials have no common factors.

Ready? ...Answers

R1. $(3y + 2)(2y - 3)$ **R2.** $\dfrac{4b^6}{9a^4}$

Are You Ready for This Section?

Before getting started, take the following readiness quiz. If you get a problem wrong, go back to the section cited and review the material.

R1. Factor: $6y^2 - 5y - 6$ [Section 4.5, pp. 350–355]

R2. Simplify: $\left(\dfrac{3ab^2}{2a^{-1}b^5}\right)^{-2}$ [Getting Ready, pp. 299–300]

A **complex rational expression** is a quotient whose numerator and/or denominator contains a rational expression. The following are examples of complex rational expressions.

$$\dfrac{3}{1 + \dfrac{x}{7}} \quad \text{and} \quad \dfrac{3 - \dfrac{1}{x}}{1 + \dfrac{1}{x}} \quad \text{and} \quad \dfrac{\dfrac{x + 1}{x - 2} - \dfrac{3}{x + 2}}{\dfrac{2x + 3}{x - 2} + 1}$$

To **simplify** a complex rational expression means to write the rational expression in the form $\dfrac{p}{q}$, where p and q are polynomials that have no common factors. This can be done using one of two methods. In Method I, simplify the numerator and the denominator separately. In Method II, simplify by using the least common denominator. We'll begin with Method I.

▶ **1** Simplify a Complex Rational Expression by Simplifying the Numerator and Denominator Separately (Method I)

EXAMPLE 1   **How to Simplify a Complex Rational Expression Using Method I**

Simplify: $\dfrac{\dfrac{1}{3} + \dfrac{1}{x}}{\dfrac{x + 3}{2}}, x \neq -3, 0$

Step-by-Step Solution

Notice that x cannot equal -3 because that would make the denominator, $\dfrac{x + 3}{2}$, equal zero; x cannot equal 0 because it would make the expression $\dfrac{1}{x}$ in the numerator undefined.

Step 1: Write the numerator of the complex rational expression as a single rational expression.

The least common denominator of $\dfrac{1}{3}$ and $\dfrac{1}{x}$ is $3x$.

$$\frac{1}{3} + \frac{1}{x} = \frac{1}{3} \cdot \frac{x}{x} + \frac{1}{x} \cdot \frac{3}{3}$$

$$= \frac{x}{3x} + \frac{3}{3x}$$

$$\frac{a}{c} + \frac{b}{c} = \frac{a+b}{c}: \qquad = \frac{x+3}{3x}$$

Step 2: Write the denominator of the complex rational expression as a single rational expression.

This is already done.

Step 3: Rewrite the complex rational expression using the rational expressions found in Steps 1 and 2.

$$\frac{\dfrac{1}{3} + \dfrac{1}{x}}{\dfrac{x+3}{2}} = \frac{\dfrac{x+3}{3x}}{\dfrac{x+3}{2}}$$

Step 4: Simplify the rational expression using the techniques for dividing rational expressions from Section 5.1.

Multiply the numerator by the reciprocal of the denominator:

$$= \frac{x+3}{3x} \cdot \frac{2}{x+3}$$

Multiply; divide out like factors:

$$= \frac{(x+3) \cdot 2}{3x(x+3)}$$

$$= \frac{2}{3x}$$

We summarize the steps to simplify a complex rational expression using Method I.

> **Simplifying a Complex Rational Expression by Simplifying the Numerator and Denominator Separately (Method I)**
>
> **Step 1:** Write the numerator of the complex rational expression as a single rational expression.
>
> **Step 2:** Write the denominator of the complex rational expression as a single rational expression.
>
> **Step 3:** Rewrite the complex rational expression using the rational expressions found in Steps 1 and 2.
>
> **Step 4:** Simplify the rational expression using the techniques for dividing rational expressions from Section 5.1.

Quick ✓

1. An expression such as $\dfrac{\dfrac{x}{5} - \dfrac{y}{7}}{\dfrac{xy}{35}}$ is called a _____ _____ _____.

2. *True or False* To simplify a complex rational expression means to write it in the form $\dfrac{p}{q}$ where p and q are polynomials with no common factors.

In Problems 3 and 4, simplify each expression using Method I.

3. $\dfrac{\dfrac{3}{2} - \dfrac{1}{3}}{\dfrac{5}{6}}$

4. $\dfrac{\dfrac{z}{4} - \dfrac{4}{z}}{\dfrac{z+4}{16}}, z \neq 0, -4$

We will not state the domain restrictions for the remaining examples, but you should be aware that the restrictions are needed to maintain equality.

EXAMPLE 2 **Simplifying a Complex Rational Expression Using Method I**

Simplify: $\dfrac{\dfrac{2x}{x+4} - \dfrac{x-7}{x^2-16}}{x - \dfrac{x^2+4}{x+4}}$

Solution

Write the numerator and denominator of the complex rational expression as a single rational expression.

NUMERATOR:

$$\text{LCD} = (x-4)(x+4)$$

$$\frac{2x}{x+4} - \frac{x-7}{x^2-16} \overset{\downarrow}{=} \frac{2x}{x+4} \cdot \frac{x-4}{x-4} - \frac{x-7}{x^2-16}$$

$$= \frac{2x^2-8x}{x^2-16} - \frac{x-7}{x^2-16}$$

$$\text{Use } \frac{a}{c} - \frac{b}{c} = \frac{a-b}{c}: \ = \frac{2x^2-9x+7}{x^2-16}$$

DENOMINATOR:

$$\text{LCD} = x+4$$

$$x - \frac{x^2+4}{x+4} \overset{\downarrow}{=} \frac{x}{1} \cdot \frac{x+4}{x+4} - \frac{x^2+4}{x+4}$$

$$= \frac{x^2+4x}{x+4} - \frac{x^2+4}{x+4}$$

$$\text{Use } \frac{a}{c} - \frac{b}{c} = \frac{a-b}{c}: \ = \frac{4x-4}{x+4}$$

Rewrite the complex rational expression using the numerator and denominator just found, and then simplify.

$$\frac{\dfrac{2x}{x+4} - \dfrac{x-7}{x^2-16}}{x - \dfrac{x^2+4}{x+4}} = \frac{\dfrac{2x^2-9x+7}{x^2-16}}{\dfrac{4x-4}{x+4}}$$

$$\text{Rewrite the division problem as a multiplication problem:} = \frac{2x^2-9x+7}{x^2-16} \cdot \frac{x+4}{4x-4}$$

$$\text{Factor; Multiply; Divide out like factors:} = \frac{(2x-7)\,\cancel{(x-1)}\,\cancel{(x+4)}}{\cancel{(x-4)}\,\cancel{(x+4)} \cdot 4\,\cancel{(x-1)}}$$

$$= \frac{2x-7}{4(x-4)}$$

Quick ✓

5. *True or False* $\dfrac{x-y}{\dfrac{1}{x} + \dfrac{1}{y}} = \dfrac{x-y}{1} \cdot \left(\dfrac{x}{1} + \dfrac{y}{1}\right)$

6. Simplify using Method I: $\dfrac{\dfrac{2x}{x+1} - \dfrac{x^2-3}{x^2+3x+2}}{4 + \dfrac{4}{x+2}}$

▶ ❷ **Simplify a Complex Rational Expression Using the Least Common Denominator (Method II)**

We now simplify complex rational expressions using the least common denominator. We will redo Example 1 using this second method so that you can compare the two methods.

EXAMPLE 3 **How to Simplify a Complex Rational Expression Using Method II**

Simplify: $\dfrac{\dfrac{1}{3} + \dfrac{1}{x}}{\dfrac{x+3}{2}}$

Step-by-Step Solution

Step 1: Find the least common denominator of all the denominators in the complex rational expression.

The denominators of the complex rational expression are 3, x, and 2. The least common denominator is $2 \cdot 3 \cdot x = 6x$.

Step 2: Multiply both the numerator and the denominator of the complex rational expression by the LCD found in Step 1.

$$\dfrac{\dfrac{1}{3} + \dfrac{1}{x}}{\dfrac{x+3}{2}} \cdot \dfrac{6x}{6x} = \dfrac{\left(\dfrac{1}{3} + \dfrac{1}{x}\right) \cdot 6x}{\left(\dfrac{x+3}{2}\right) \cdot 6x}$$

Distribute $6x$ to each term: $= \dfrac{\dfrac{1}{3} \cdot 6x + \dfrac{1}{x} \cdot 6x}{\dfrac{x+3}{2} \cdot 6x}$

Step 3: Simplify the rational expression.

Simplify: $= \dfrac{2x + 6}{3x(x+3)}$

Factor and divide out like factors: $= \dfrac{2\cancel{(x+3)}}{3x\cancel{(x+3)}} = \dfrac{2}{3x}$

Example 3 has the same result as Example 1!

> **Simplifying a Complex Rational Expression Using the Least Common Denominator (Method II)**
>
> **Step 1:** Find the least common denominator of all the denominators in the complex rational expression.
>
> **Step 2:** Multiply both the numerator and the denominator of the complex rational expression by the least common denominator found in Step 1.
>
> **Step 3:** Simplify the rational expression.

Quick ✓

In Problems 7 and 8, simplify each expression using Method II.

7. $\dfrac{\dfrac{3}{2} - \dfrac{1}{3}}{\dfrac{5}{6}}$ **8.** $\dfrac{\dfrac{z}{4} - \dfrac{4}{z}}{\dfrac{z+4}{16}}$

EXAMPLE 4 **Simplifying Complex Rational Expressions Using Method II**

Simplify: $\dfrac{\dfrac{1}{x} + \dfrac{1}{x-2}}{\dfrac{x}{x^2-4} + \dfrac{1}{x-2}}$

Solution

Since $x^2 - 4 = (x - 2)(x + 2)$, the least common denominator among all denominators is $x(x - 2)(x + 2)$. So multiply the numerator and denominator by $x(x - 2)(x + 2)$.

$$\dfrac{\dfrac{1}{x} + \dfrac{1}{x - 2}}{\dfrac{x}{x^2 - 4} + \dfrac{1}{x - 2}} = \dfrac{\dfrac{1}{x} + \dfrac{1}{x - 2}}{\dfrac{x}{x^2 - 4} + \dfrac{1}{x - 2}} \cdot \dfrac{x(x - 2)(x + 2)}{x(x - 2)(x + 2)}$$

Distribute the LCD
to each term:
$$= \dfrac{\dfrac{1}{x} \cdot x(x - 2)(x + 2) + \dfrac{1}{x - 2} \cdot x(x - 2)(x + 2)}{\dfrac{x}{x^2 - 4} \cdot x(x - 2)(x + 2) + \dfrac{1}{x - 2} \cdot x(x - 2)(x + 2)}$$

Factor and divide
out like factors:
$$= \dfrac{\dfrac{1}{\cancel{x}} \cdot \cancel{x} (x - 2)(x + 2) + \dfrac{1}{\cancel{x - 2}} \cdot x \cancel{(x - 2)}(x + 2)}{\dfrac{x}{\cancel{(x - 2)(x + 2)}} \cdot x \cancel{(x - 2)(x + 2)} + \dfrac{1}{\cancel{x - 2}} \cdot x \cancel{(x - 2)}(x + 2)}$$

$$= \dfrac{(x - 2)(x + 2) + x(x + 2)}{x^2 + x(x + 2)}$$

Factor out $(x + 2)$:
$$= \dfrac{(x + 2)(x - 2 + x)}{x^2 + x^2 + 2x}$$

$$= \dfrac{(x + 2)(2x - 2)}{2x^2 + 2x}$$

Factor and divide
out like factors:
$$= \dfrac{\cancel{2}(x + 2)(x - 1)}{\cancel{2}x(x + 1)}$$

$$= \dfrac{(x + 2)(x - 1)}{x(x + 1)}$$

Quick ✓

9. Simplify using Method II: $\dfrac{\dfrac{x + 2}{x + 5} - \dfrac{x + 2}{x + 1}}{\dfrac{2x + 1}{x + 1} - 1}$

▶ Comparing Methods

We will use both methods in the next example. As you work through the exercise set, be sure to start developing a sense about when you prefer Method I over Method II, and vice versa.

EXAMPLE 5 **Comparing Methods I and II**

Simplify $\dfrac{x^{-1} + y^{-1}}{x^{-3} + y^{-3}}$ as a rational expression that contains no negative exponents.

Work Smart

Remember, $x^{-1} = \dfrac{1}{x}$ and $y^{-1} = \dfrac{1}{y}$,

but $x^{-1} + y^{-1} \neq \dfrac{1}{x + y}$.

Rather, $x^{-1} + y^{-1} = \dfrac{1}{x} + \dfrac{1}{y}$.

Solution

First, rewrite the expression without negative exponents.

$$\dfrac{x^{-1} + y^{-1}}{x^{-3} + y^{-3}} = \dfrac{\dfrac{1}{x} + \dfrac{1}{y}}{\dfrac{1}{x^3} + \dfrac{1}{y^3}}$$

Method I

Write the numerator of the complex rational expression as a single quotient. The LCD is xy.

$$\frac{1}{x} + \frac{1}{y} = \frac{1}{x} \cdot \frac{y}{y} + \frac{1}{y} \cdot \frac{x}{x}$$

$$= \frac{y}{xy} + \frac{x}{xy}$$

$$= \frac{y + x}{xy}$$

$$= \frac{x + y}{xy}$$

Write the denominator of the complex rational expression as a single quotient. The LCD is $x^3 y^3$.

$$\frac{1}{x^3} + \frac{1}{y^3} = \frac{1}{x^3} \cdot \frac{y^3}{y^3} + \frac{1}{y^3} \cdot \frac{x^3}{x^3}$$

$$= \frac{y^3}{x^3 y^3} + \frac{x^3}{x^3 y^3}$$

$$= \frac{y^3 + x^3}{x^3 y^3}$$

$$= \frac{x^3 + y^3}{x^3 y^3}$$

Now rewrite the complex rational expression using the numerator and denominator just found and then simplify.

$$\frac{x^{-1} + y^{-1}}{x^{-3} + y^{-3}} = \frac{\dfrac{1}{x} + \dfrac{1}{y}}{\dfrac{1}{x^3} + \dfrac{1}{y^3}}$$

$$= \frac{\dfrac{x + y}{xy}}{\dfrac{x^3 + y^3}{x^3 y^3}}$$

Multiply the numerator by the reciprocal of the denominator:
$$= \frac{x + y}{xy} \cdot \frac{x^3 y^3}{x^3 + y^3}$$

Factor:
$$= \frac{x + y}{xy} \cdot \frac{x^3 y^3}{(x + y)(x^2 - xy + y^2)}$$

Divide out like factors:
$$= \frac{\cancel{x + y}}{\cancel{xy}} \cdot \frac{\overset{2}{\cancel{x^3}}\,\overset{2}{\cancel{y^3}}}{\cancel{(x + y)}(x^2 - xy + y^2)}$$

$$= \frac{x^2 y^2}{x^2 - xy + y^2}$$

Method II

The LCD of all the denominators is $x^3 y^3$. Multiply the numerator and denominator of the complex rational expression by $x^3 y^3$.

$$\frac{\dfrac{1}{x} + \dfrac{1}{y}}{\dfrac{1}{x^3} + \dfrac{1}{y^3}} = \frac{\dfrac{1}{x} + \dfrac{1}{y}}{\dfrac{1}{x^3} + \dfrac{1}{y^3}} \cdot \frac{x^3 y^3}{x^3 y^3}$$

Distribute the LCD to each term:
$$= \frac{\dfrac{1}{x} \cdot x^3 y^3 + \dfrac{1}{y} \cdot x^3 y^3}{\dfrac{1}{x^3} \cdot x^3 y^3 + \dfrac{1}{y^3} \cdot x^3 y^3}$$

$$= \frac{x^2 y^3 + x^3 y^2}{y^3 + x^3}$$

$$\text{Factor out the GCF in the numerator:}$$
$$\text{Factor the sum of two cubes in the denominator:} = \frac{x^2 y^2 (x + y)}{(x + y)(x^2 - xy + y^2)}$$

$$\text{Divide out like factors:} = \frac{x^2 y^2}{x^2 - xy + y^2}$$

In this particular problem, Method II seems more efficient.

Quick ✓

Simplify the expression so that it does not contain any negative exponents. Use both methods and decide which method you prefer for this problem.

10. $\dfrac{3a^{-1} + b^{-1}}{9a^{-2} - b^{-2}}$

5.3 Exercises MyMathLab® Math XP PRACTICE

Exercise numbers in green are complete video solutions in MyMathLab

*Problems **1–10** are the Quick ✓s that follow the EXAMPLES.*

Building Skills

In Problems 11–18, simplify the expression using Method I. See Objective 1.

11. $\dfrac{\dfrac{5}{6} + \dfrac{1}{9}}{\dfrac{5}{2} - \dfrac{3}{8}}$

12. $\dfrac{\dfrac{3}{10} - \dfrac{3}{4}}{\dfrac{2}{5} + \dfrac{7}{10}}$

13. $\dfrac{1 + \dfrac{1}{x}}{1 - \dfrac{1}{x}}$

14. $\dfrac{1 + \dfrac{1}{x^2}}{1 - \dfrac{1}{x^2}}$

15. $\dfrac{1 - \dfrac{a}{a - 2}}{2 - \dfrac{a + 2}{a}}$

16. $\dfrac{\dfrac{a}{a + 1} - 1}{\dfrac{a + 3}{a} - 2}$

17. $\dfrac{\dfrac{x + 2}{x - 1} - \dfrac{x + 5}{x + 3}}{x + 11}$

18. $\dfrac{\dfrac{x + 5}{x - 2} - \dfrac{x + 3}{x - 1}}{3x + 1}$

In Problems 19–26, simplify the expression using Method II. See Objective 2.

19. $\dfrac{\dfrac{5}{6} + \dfrac{7}{10}}{\dfrac{1}{5} - 1}$

20. $\dfrac{2 + \dfrac{7}{4}}{\dfrac{3}{8} - 2}$

21. $\dfrac{w - \dfrac{1}{w}}{w + \dfrac{1}{w}}$

22. $\dfrac{\dfrac{7}{w} + \dfrac{9}{x}}{\dfrac{9}{w} - \dfrac{7}{x}}$

23. $\dfrac{\dfrac{x - 1}{x - 4} - \dfrac{x}{x - 2}}{1 - \dfrac{3}{x - 4}}$

24. $\dfrac{\dfrac{x - 4}{x - 1} - \dfrac{x}{x - 3}}{3 + \dfrac{12}{x - 3}}$

25. $\dfrac{\dfrac{z + 2}{z - 2} + \dfrac{z - 2}{z + 2}}{\dfrac{z + 2}{z - 2} - \dfrac{z - 2}{z + 2}}$

26. $\dfrac{\dfrac{m + 3}{m - 3} - \dfrac{m - 3}{m + 3}}{\dfrac{m + 3}{m - 3} + \dfrac{m - 3}{m + 3}}$

Mixed Practice

In Problems 27–52, simplify the complex rational expression using either Method I or Method II.

27. $\dfrac{\dfrac{3}{y} - 1}{\dfrac{9}{y} - y}$

28. $\dfrac{1 - \dfrac{4}{z}}{z - \dfrac{16}{z}}$

29. $\dfrac{\dfrac{4n}{m} - \dfrac{4m}{n}}{\dfrac{2}{m} + \dfrac{2}{n}}$

30. $\dfrac{\dfrac{n^2}{m} - \dfrac{m^2}{n}}{\dfrac{1}{m} - \dfrac{1}{n}}$

31. $\dfrac{2 + \dfrac{3}{x}}{\dfrac{2x^2}{x + 3} - 3}$

32. $\dfrac{1 + \dfrac{5}{x}}{1 + \dfrac{1}{x + 4}}$

33. $\dfrac{\dfrac{x}{3} - \dfrac{3}{x}}{\dfrac{3}{x^2} - \dfrac{1}{3}}$

34. $\dfrac{\dfrac{5}{x} - \dfrac{x}{5}}{\dfrac{1}{5} - \dfrac{5}{x^2}}$

35. $\dfrac{\dfrac{2x + 1}{x - 1} - \dfrac{x - 1}{x - 3}}{x^2 - 3x - 4}$

36. $\dfrac{\dfrac{x + 5}{x - 3} - \dfrac{x}{x + 4}}{3x^2 - 4x - 15}$

37. $\dfrac{\dfrac{x-4}{x+4}+\dfrac{x-4}{x-2}}{1-\dfrac{2}{x-2}}$

38. $\dfrac{\dfrac{x-3}{x+3}+\dfrac{x-3}{x-4}}{1+\dfrac{x+3}{x-4}}$

39. $\dfrac{\dfrac{b^2}{b^2-25}-\dfrac{b}{b+5}}{\dfrac{b}{b^2-25}-\dfrac{1}{b-5}}$

40. $\dfrac{\dfrac{-6}{x^2+5x+6}}{\dfrac{2}{x+3}-\dfrac{3}{x+2}}$

41. $\dfrac{3}{\dfrac{2}{x}-1}$

42. $\dfrac{4}{\dfrac{5}{x}+3}$

43. $\dfrac{1}{\dfrac{4}{x-1}+2}$

44. $\dfrac{1}{\dfrac{1}{x+2}+2}$

45. $\dfrac{\dfrac{4}{x-1}-1}{\dfrac{4}{x-1}+3}$

46. $\dfrac{\dfrac{7}{x-2}+4}{\dfrac{3}{x-2}-1}$

47. $\dfrac{3x^{-1}+3y^{-1}}{x^{-2}-y^{-2}}$

48. $\dfrac{2x^{-1}+2y^{-1}}{xy^{-1}-x^{-1}y}$

49. $\dfrac{(m+n)^{-1}}{m^{-1}+n^{-1}}$

50. $\dfrac{(x-y)^{-1}}{x^{-1}-y^{-1}}$

51. $\dfrac{a^{-2}b^{-1}-a^{-1}b^{-2}}{4a^{-2}-4b^{-2}}$

52. $\dfrac{a^{-3}+8b^{-3}}{a^{-2}-4b^{-2}}$

Applying the Concepts

53. Electric Circuits An electric circuit contains two resistors connected in parallel, as shown in the figure. If the resistances of these resistors are R_1 and R_2 ohms, respectively, then their combined resistance R is given by the formula

$$R=\dfrac{1}{\dfrac{1}{R_1}+\dfrac{1}{R_2}}$$

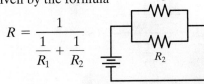

(a) Express R as a simplified rational expression.

(b) Evaluate the rational expression if $R_1 = 4$ ohms and $R_2 = 10$ ohms.

54. Future Value of Money The value of an account V in which P dollars is deposited every year for the next 5 years paying an interest rate i (expressed as a decimal) is given by the formula

$$V=P\cdot\dfrac{1-\dfrac{1}{(1+i)^5}}{\dfrac{i}{(1+i)^5}}$$

(a) Express V as a simplified rational expression.

(b) Determine the value of an account paying 5% when the annual deposit is $1000. Express your answer to the nearest penny.

55. The Lensmaker's Equation The focal length f of a convex lens with index of refraction n is given by

$$f=\dfrac{1}{(n-1)\left[\dfrac{1}{R_1}+\dfrac{1}{R_2}\right]}$$

where R_1 and R_2 are the radii of curvature of the front and back surfaces of the lens. See the figure.

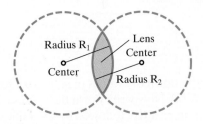

(a) Express f as a simplified rational expression.

(b) Determine the focal length of a lens for $n = 1.5$, $R_1 = 0.5$ meter, and $R_2 = 0.3$ meter.

56. Harmonic Mean The harmonic mean is used to determine an average value when data are measured as a rate of change, such as 50 miles per hour. The harmonic mean is found using the formula

$$H=\dfrac{n}{\dfrac{1}{x_1}+\dfrac{1}{x_2}+\cdots+\dfrac{1}{x_n}}$$

where $x_1, x_2, \ldots, x_n$ are the n rates of change.

(a) Suppose that a family drove from Chicago, Illinois, to Naples, Florida. The distance each way is about 1200 miles. The trip from Chicago to Naples resulted in an average speed of 48 miles per hour, while the return trip resulted in an average speed of 52 miles per hour. Compute the average speed of the entire trip.

(b) A 300-KB file is downloaded four separate times, resulting in download speeds of 4.3 KB/s, 4.1 KB/s, 3.8 KB/s, and 4.3 KB/s. Compute the average download time.

Extending the Concepts

57. Fibonacci Strikes Again! Write each of the following expressions in the form $\dfrac{ax+b}{bx+c}$. (*Hint:* Use Method I.)

(a) $1+\dfrac{1}{1+\dfrac{1}{x}}$

(b) $1 + \dfrac{1}{1 + \dfrac{1}{1 + \dfrac{1}{x}}}$

(c) $1 + \dfrac{1}{1 + \dfrac{1}{1 + \dfrac{1}{1 + \dfrac{1}{x}}}}$

(d) $1 + \dfrac{1}{1 + \dfrac{1}{1 + \dfrac{1}{1 + \dfrac{1}{1 + \dfrac{1}{x}}}}}$

(e) Write down the values of $a, b,$ and c from part (a). Now write down the values of $a, b,$ and c from part (b), followed by the values of $a, b,$ and c from part (c), followed by the values of $a, b,$ and c from part (d). What is the pattern? Write the sequence of numbers in increasing order. This sequence of numbers forms the first six numbers in the **Fibonacci sequence.**

Math for the Future *In Calculus, you will be asked to simplify the expression* $\dfrac{f(x + h) - f(x)}{h}$, *which is called* the **difference quotient.** *In Problems 58–61, simplify the difference quotient for the given functions.*

58. $f(x) = \dfrac{1}{x}$

59. $f(x) = \dfrac{1}{x^2}$

60. $f(x) = \dfrac{3}{x^2}$

61. $f(x) = \dfrac{1}{x - 1}$

Explaining the Concepts

62. In your own words, provide a definition for a complex rational expression.

63. Which of the two methods for simplifying complex rational expressions do you prefer? Write a paragraph supporting your opinion.

Synthesis Review

In Problems 64–69, solve each equation.

64. $4x + 3 = 15$

65. $-5a + 2 = 22$

66. $\dfrac{1}{2}w + 3 = 5$

67. $\dfrac{2}{3}x - \dfrac{5}{7}(x + 21) = \dfrac{11}{21}x + \dfrac{4}{7}$

68. $y^2 - 5y = 50$

69. $3p^2 + 19p = 14$

Putting the Concepts Together (Sections 5.1–5.3)

We designed these problems so that you can review Sections 5.1 through 5.3 and show your mastery of the concepts. Take time to work these problems before proceeding with the next section. The answers are located at the back of the text on page AN-31.

1. Determine the domain of the rational function
$$g(x) = \dfrac{3x + 1}{3x^2 - 17x - 6}$$

In Problems 2 and 3, write each rational expression in lowest terms.

2. $\dfrac{24n - 4n^2}{2n^2 - 9n - 18}$

3. $\dfrac{2p^2 - pq - 10q^2}{3p^2 + 2pq - 8q^2}$

In Problems 4–8, perform the indicated operations.

4. $\dfrac{a^2 - 16}{12a^2 + 48a} \cdot \dfrac{6a^3 - 30a^2}{a^2 + 2a - 24}$

5. $\dfrac{\dfrac{x^2 + x - 2}{3x^2 - 5x - 2}}{\dfrac{3x^2 - 2x - 1}{x^2 - 9x + 14}}$

6. $\dfrac{x^2 - 10}{x^2 - 4} - \dfrac{3x}{x^2 - 4}$

7. $\dfrac{3n}{n^2 - 7n + 10} - \dfrac{2n}{n^2 - 8n + 15}$

8. $\dfrac{3y + 2}{y^2 + 5y - 24} + \dfrac{7}{y^2 + 4y - 32}$

In Problems 9 and 10, use the functions $f(x) = \dfrac{2x + 1}{x^2 - 11x + 28}$, $g(x) = \dfrac{3x - 12}{4x^2 + 4x + 1}$, *and* $h(x) = \dfrac{3x}{x - 7}$ *to find each difference or product.*

9. $P(x) = f(x) \cdot g(x)$

10. $D(x) = h(x) - f(x)$

In Problems 11 and 12, simplify each complex rational expression using either method.

11. $\dfrac{\dfrac{1}{m^2} - \dfrac{1}{n^2}}{\dfrac{1}{m} - \dfrac{1}{n}}$

12. $\dfrac{\dfrac{z^2 - 2}{z^2 - 4} + \dfrac{7}{z - 2}}{\dfrac{z^2 + z - 24}{z^2 - 4} - \dfrac{2}{z + 2}}$

5.4 Rational Equations

Objectives

1 Solve Equations Containing Rational Expressions

2 Solve Equations Involving Rational Functions

Are You Ready for This Section?

Before getting started, take the following readiness quiz. If you get a problem wrong, go back to the section cited and review the material.

R1. Solve: $\dfrac{2}{3}x + \dfrac{1}{2} = \dfrac{3}{4}$ [Section 1.1, pp. 50–55]

R2. Factor: $3z^2 + 11z - 4$ [Section 4.5, pp. 350–355]

R3. Solve: $6y^2 - y - 12 = 0$ [Section 4.8, pp. 371–375]

R4. Determine which of the following is in the domain of $\dfrac{x + 4}{x^2 - 5x - 24}$.

    (a) $x = -4$ **(b)** $x = 8$ [Section R.5, p. 44]

R5. If $f(x) = x^2 - 3x - 15$, solve $f(x) = 3$. [Section 4.8, pp. 375–376]

R6. If $g(4) = 3$, what point is on the graph of g? [Section 2.3, pp. 163–164]

1 Solve Equations Containing Rational Expressions

So far, we have solved linear equations (Section 1.1), quadratic equations (Section 4.8), and equations that contain polynomial expressions that can be factored (Section 4.8). We now look at another type of equation, the *rational equation*. A **rational equation** is an equation that contains a rational expression. Examples of rational equations are

$$\frac{3}{x + 4} = \frac{5}{x - 1} + \frac{1}{x^2 + 3x - 4} \quad \text{and} \quad \frac{x - 5}{x^2 + 4x - 12} = 3$$

As usual, we exclude from the domain all values of the variable that result in division by zero.

Ready?...Answers R1. $\left\{\dfrac{3}{8}\right\}$

R2. $(3z - 1)(z + 4)$ **R3.** $\left\{-\dfrac{4}{3}, \dfrac{3}{2}\right\}$

R4. (a) Yes **(b)** No **R5.** $\{-3, 6\}$

R6. $(4, 3)$

EXAMPLE 1 **How to Solve a Rational Equation**

Solve: $\dfrac{2x - 1}{x - 3} = \dfrac{2(x + 1)}{x - 2}$

Step-by-Step Solution

Step 1: Determine the domain of the variable in the rational equation.

Because $x = 2$ and $x = 3$ result in division by zero, the domain of x is

$$\{x \mid x \neq 2, x \neq 3\}$$

Step 2: Determine the least common denominator (LCD) of all the denominators.

The LCD is $(x - 3)(x - 2)$.

Step 3: Multiply both sides of the equation by the LCD, and simplify the expressions on each side.

$$\frac{2x - 1}{x - 3} = \frac{2(x + 1)}{x - 2}$$

Multiply both sides by $(x - 3)(x - 2)$: $(x - 3)(x - 2) \cdot \dfrac{2x - 1}{x - 3} = (x - 3)(x - 2) \cdot \dfrac{2(x + 1)}{x - 2}$

(continued)

Divide out like factors: $(x - 2)(2x - 1) = 2(x - 3)(x + 1)$

Multiply: $2x^2 - 5x + 2 = 2(x^2 - 2x - 3)$

Distribute the 2: $2x^2 - 5x + 2 = 2x^2 - 4x - 6$

Step 4: Solve the resulting equation.

Subtract $2x^2$ from both sides: $-5x + 2 = -4x - 6$

Add $4x$ to both sides;
subtract 2 from both sides: $-x = -8$

Divide both sides by -1: $x = 8$

Step 5: Check Verify your solution using the original equation.

Let $x = 8$: $\dfrac{2(8) - 1}{8 - 3} \overset{?}{=} \dfrac{2(8 + 1)}{8 - 2}$

$\dfrac{16 - 1}{5} \overset{?}{=} \dfrac{2(9)}{6}$

$\dfrac{15}{5} \overset{?}{=} \dfrac{18}{6}$

$3 = 3$ True

The solution checks, so the solution set is $\{8\}$.

We summarize the steps that can be used to solve any rational equation.

In Words

The purpose of Step 3 is to "clear the fractions" so that we transform the equation into one that we already know how to solve, such as a linear equation.

Solving a Rational Equation

Step 1: Determine the domain of the variable in the rational equation.

Step 2: Determine the least common denominator (LCD) of all the denominators.

Step 3: Multiply both sides of the equation by the LCD, and simplify the expressions on each side of the equation.

Step 4: Solve the resulting equation.

Step 5: Verify your solution using the original equation.

Quick ✓

1. A _____ _____ is an equation that contains a rational expression.

In Problems 2 and 3, solve each equation. Be sure to verify your results.

2. $\dfrac{x - 4}{x^2 + 4} = \dfrac{3}{3x + 2}$

3. $\dfrac{4}{x + 2} = \dfrac{7}{x + 4}$

EXAMPLE 2   **Solving a Rational Equation**

Solve: $\dfrac{2}{x} - \dfrac{1}{6} = \dfrac{5}{2x} - \dfrac{1}{3}$

Solution

The domain of the variable is $\{x \mid x \neq 0\}$. The LCD of all denominators is $6x$.

$$\frac{2}{x} - \frac{1}{6} = \frac{5}{2x} - \frac{1}{3}$$

Multiply both sides by the LCD, $6x$: $6x \cdot \left(\dfrac{2}{x} - \dfrac{1}{6}\right) = 6x \cdot \left(\dfrac{5}{2x} - \dfrac{1}{3}\right)$

Distribute the 6x: $\quad 6x \cdot \dfrac{2}{x} - 6x \cdot \dfrac{1}{6} = 6x \cdot \dfrac{5}{2x} - 6x \cdot \dfrac{1}{3}$

Simplify: $\quad 12 - x = 15 - 2x$

Add 2x to both sides: $\quad 12 + x = 15$

Subtract 12 from both sides: $\qquad\qquad x = 3$

Check Let $x = 3$ in the original equation: $\dfrac{2}{3} - \dfrac{1}{6} \overset{?}{=} \dfrac{5}{2 \cdot 3} - \dfrac{1}{3}$

$$\dfrac{4}{6} - \dfrac{1}{6} \overset{?}{=} \dfrac{5}{6} - \dfrac{2}{6}$$

$$\dfrac{3}{6} = \dfrac{3}{6} \quad \text{True}$$

The solution checks, so the solution set is $\{3\}$.

Quick ✓

In Problems 4 and 5, solve each equation. Be sure to verify your results.

4. $\dfrac{5}{x} + \dfrac{1}{4} = \dfrac{3}{2x} - \dfrac{3}{2}$

5. $\dfrac{5}{x} + 2 = \dfrac{10}{3x} + 1$

EXAMPLE 3 **Solving a Rational Equation**

Solve: $\dfrac{3}{p^2 - 4p + 3} + \dfrac{6}{p^2 - 2p - 3} = \dfrac{5}{p^2 - 1}$

Solution

First, factor the denominator of each term to find the domain of the variable, p.

$$p^2 - 4p + 3 = (p - 3)(p - 1), \text{ so } p \neq 3, p \neq 1 \text{ in the first term.}$$
$$p^2 - 2p - 3 = (p - 3)(p + 1), \text{ so } p \neq 3, p \neq -1 \text{ in the second term.}$$
$$p^2 - 1 = (p - 1)(p + 1), \text{ so } p \neq 1, p \neq -1 \text{ in the third term.}$$

The domain of the variable p is $\{p \mid p \neq -1, p \neq 1, p \neq 3\}$. We know from the factored form of each denominator above that the LCD of all the denominators is $(p - 1)(p + 1)(p - 3)$, so multiply both sides of the equation by $(p - 1)(p + 1)(p - 3)$.

$$\dfrac{3}{p^2 - 4p + 3} + \dfrac{6}{p^2 - 2p - 3} = \dfrac{5}{p^2 - 1}$$

$$(p - 1)(p + 1)(p - 3) \cdot \left(\dfrac{3}{(p - 3)(p - 1)} + \dfrac{6}{(p - 3)(p + 1)} \right) = (p - 1)(p + 1)(p - 3) \cdot \dfrac{5}{(p - 1)(p + 1)}$$

Distribute the LCD and simplify: $\quad 3(p + 1) + 6(p - 1) = 5(p - 3)$

Distribute: $\qquad\qquad 3p + 3 + 6p - 6 = 5p - 15$

Combine like terms: $\qquad\qquad\qquad 9p - 3 = 5p - 15$

Subtract 5p from both sides; add 3 to both sides: $\qquad\qquad 4p = -12$

Divide both sides by 4: $\qquad\qquad\qquad p = -3$

We leave the check to you. The solution set is $\{-3\}$.

Quick ✓

6. Solve: $\dfrac{3}{x^2 + 5x + 4} + \dfrac{2}{x^2 - 3x - 4} = \dfrac{4}{x^2 - 16}$

▶ **EXAMPLE 4** **Solving a Rational Equation with No Solution**

Solve: $\dfrac{3}{y^2 - 5y + 4} + \dfrac{2}{y^2 - 10y + 24} = \dfrac{2}{y^2 - 7y + 6}$

Solution

First, factor the denominator of each term to find the domain of the variable, y.

$y^2 - 5y + 4 = (y - 4)(y - 1)$, so $y \neq 4$, $y \neq 1$ in the first term.
$y^2 - 10y + 24 = (y - 6)(y - 4)$, so $y \neq 6$, $y \neq 4$ in the second term.
$y^2 - 7y + 6 = (y - 6)(y - 1)$, so $y \neq 6$, $y \neq 1$ in the third term.

The domain of the variable y is $\{y \mid y \neq 1, y \neq 4, y \neq 6\}$. The LCD of all the denominators is $(y - 1)(y - 4)(y - 6)$. Multiply both sides of the equation by the LCD.

$$\frac{3}{y^2 - 5y + 4} + \frac{2}{y^2 - 10y + 24} = \frac{2}{y^2 - 7y + 6}$$

$$(y-1)(y-4)(y-6) \cdot \left(\frac{3}{(y-4)(y-1)} + \frac{2}{(y-6)(y-4)}\right) = (y-1)(y-4)(y-6) \cdot \frac{2}{(y-6)(y-1)}$$

Distribute the LCD and simplify: $3(y - 6) + 2(y - 1) = 2(y - 4)$

Distribute: $3y - 18 + 2y - 2 = 2y - 8$

Combine like terms: $5y - 20 = 2y - 8$

Subtract 2y from both sides; add 20 to both sides: $3y = 12$

Divide both sides by 3: $y = 4$

Notice that $y = 4$ is not in the domain of the variable y, so the equation has no solution. The solution set is $\varnothing$ or $\{\ \}$. ●

We call $y = 4$ an *extraneous solution*. **Extraneous solutions** are results that develop through the solution process but do not satisfy the original equation.

In Words
The word *extraneous* means "not constituting a vital part."

Quick ✓

7. _____ _____ are results that develop through the solution process but do not satisfy the original equation.

8. *True or False* Some rational equations have no solution.

In Problems 9 and 10, solve each equation. Be sure to verify your results.

9. $\dfrac{5}{z^2 + 2z - 3} - \dfrac{3}{z^2 + z - 2} = \dfrac{1}{z^2 + 5z + 6}$ 10. $\dfrac{5}{x - 4} + \dfrac{3}{x - 2} = \dfrac{11}{x - 4}$

▶ **EXAMPLE 5** **Solving a Rational Equation That Leads to a Quadratic Equation**

Solve: $\dfrac{w + 3}{w - 1} + \dfrac{w + 5}{w} = \dfrac{3w + 1}{w - 1}$

Solution

The domain of the variable w is $\{w \mid w \neq 0, w \neq 1\}$. The LCD of all the denominators is $w(w - 1)$, so multiply both sides of the equation by $w(w - 1)$.

$$\frac{w + 3}{w - 1} + \frac{w + 5}{w} = \frac{3w + 1}{w - 1}$$

$$w(w - 1) \cdot \left(\frac{w + 3}{w - 1} + \frac{w + 5}{w}\right) = w(w - 1) \cdot \frac{3w + 1}{w - 1}$$

Work Smart

Notice that this equation has a squared variable, so we know the equation is a quadratic equation. This is why we rewrite it with 0 on one side (standard form) rather than isolating the variable (as we do in solving linear equations).

Distribute the LCD: $\quad w(w + 3) + (w - 1)(w + 5) = w(3w + 1)$

$$w^2 + 3w + w^2 + 4w - 5 = 3w^2 + w$$

Combine like terms: $\qquad\qquad 2w^2 + 7w - 5 = 3w^2 + w$

Put equation in standard form: $\qquad\qquad 0 = w^2 - 6w + 5$

If $a = b$, then $b = a$: $\qquad\qquad w^2 - 6w + 5 = 0$

Factor: $\qquad\qquad (w - 5)(w - 1) = 0$

Use the Zero-Product Property: $\qquad\qquad w = 5 \quad \text{or} \quad w = 1$

Since $w = 1$ is not in the domain of the variable, it is an extraneous solution. The only potential solution is 5.

Check Let $w = 5$ in the original equation.

$$\frac{5 + 3}{5 - 1} + \frac{5 + 5}{5} \stackrel{?}{=} \frac{3(5) + 1}{5 - 1}$$

$$\frac{8}{4} + \frac{10}{5} \stackrel{?}{=} \frac{16}{4}$$

$$2 + 2 = 4 \ \text{True}$$

The solution set is $\{5\}$.

Quick ✓

In Problems 11 and 12, solve each equation. Be sure to verify your results.

11. $2 - \dfrac{3}{p + 2} = \dfrac{6}{p}$

12. $\dfrac{z + 1}{z + 4} + \dfrac{z + 1}{z - 3} = \dfrac{z^2 + z + 16}{z^2 + z - 12}$

▶ ② **Solve Equations Involving Rational Functions**

Now let's look at a problem involving a rational function that leads to a rational equation.

(**EXAMPLE 6**) **Working with Rational Functions**

For the function $f(x) = x + \dfrac{4}{x}$, solve $f(x) = 5$. What point(s) on the graph of f do these values correspond to?

Solution

We wish to solve the equation $x + \dfrac{4}{x} = 5$. The domain of the variable is $\{x \mid x \neq 0\}$.

The LCD of all denominators is x, so multiply both sides of the equation by x.

$$x \cdot \left(x + \frac{4}{x} \right) = 5 \cdot x$$

Distribute: $\qquad\qquad x^2 + 4 = 5x$

Subtract $5x$ from both sides: $\qquad x^2 - 5x + 4 = 0$

Factor: $\quad (x - 4)(x - 1) = 0$

Use the Zero-Product Property: $\qquad\qquad x - 4 = 0 \quad \text{or} \quad x - 1 = 0$

$$x = 4 \quad \text{or} \qquad x = 1$$

We leave the check to you. We have $f(1) = 5$, so the point $(1, 5)$ is on the graph of f. We have $f(4) = 5$, so the point $(4, 5)$ is on the graph of f.

Quick ✓

13. *True or False* To solve $f(x) = 4$ when $f(x) = x + \dfrac{3}{x}$, substitute 4 for x in the function and evaluate.

14. For the function $g(x) = x - \dfrac{5}{x}$, solve $g(x) = 4$. What point(s) on the graph of g do these values correspond to?

15. For the function $f(x) = 2x - \dfrac{3}{x}$, solve $f(x) = 1$. What point(s) on the graph of f do these values correspond to?

EXAMPLE 7 **An Application of Rational Functions: Drug Concentration**

The concentration C of a drug in a patient's bloodstream in milligrams per liter t hours after ingestion is modeled by

$$C(t) = \frac{40t}{t^2 + 9}$$

When will the concentration of the drug be 4 milligrams per liter?

Solution

To know when the concentration of the drug is 4, solve the equation $C(t) = 4$.

$$\frac{40t}{t^2 + 9} = 4$$

Multiply both sides by $t^2 + 9$:	$40t = 4(t^2 + 9)$
Divide both sides by 4:	$10t = t^2 + 9$
Subtract $10t$ from both sides:	$0 = t^2 - 10t + 9$
Factor:	$0 = (t - 1)(t - 9)$
Use the Zero-Product Property:	$t = 1$ or $t = 9$

The concentration of the drug will be 4 milligrams per liter 1 hour and 9 hours after ingestion. ●

Quick ✓

16. The concentration C of a drug in a patient's bloodstream in milligrams per liter t hours after ingestion is modeled by $C(t) = \dfrac{50t}{t^2 + 6}$. When will the concentration of the drug be 4 milligrams per liter?

5.4 Exercises MyMathLab® Math XL PRACTICE

Exercise numbers in green are complete video solutions in MyMathLab

*Problems **1–16** are the Quick ✓s that follow the EXAMPLES.*

Building Skills

In Problems 17–42, solve each equation. Be sure to verify your results. See Objective 1.

17. $\dfrac{3}{z} - \dfrac{1}{2z} = -\dfrac{5}{8}$

18. $\dfrac{8}{p} + \dfrac{1}{4p} = \dfrac{11}{8}$

19. $\dfrac{y + 2}{y - 5} = \dfrac{y + 6}{y + 1}$

20. $\dfrac{w - 4}{w + 1} = \dfrac{w - 3}{w + 3}$

21. $\dfrac{x + 8}{x + 4} = \dfrac{x + 2}{x - 2}$

22. $\dfrac{2x + 1}{x + 3} = \dfrac{4(x - 1)}{2x + 3}$

23. $a - \dfrac{5}{a} = 4$

24. $m + \dfrac{8}{m} = 6$

25. $6p - \dfrac{3}{p} = 7$

26. $8b - \dfrac{3}{b} = 2$

27. $\dfrac{5-p}{p-5} + 2 = \dfrac{1}{p}$

28. $\dfrac{3-y}{y-3} + 2 = \dfrac{2}{y}$

29. $\dfrac{3}{2} + \dfrac{5}{x-3} = \dfrac{x+9}{2x-6}$

30. $\dfrac{4}{3} + \dfrac{7}{x-4} = \dfrac{x-1}{3x-12}$

31. $1 + \dfrac{3}{x+3} = \dfrac{4}{x-3}$

32. $\dfrac{5}{x+2} = 1 - \dfrac{3}{x-2}$

33. $\dfrac{4}{x-5} + \dfrac{3}{x-2} = \dfrac{x+1}{x^2-7x+10}$

34. $\dfrac{4}{x+3} + \dfrac{5}{x-6} = \dfrac{4x+1}{x^2-3x-18}$

35. $\dfrac{1}{x+5} = \dfrac{2x}{x^2-25} - \dfrac{3}{x-5}$

36. $\dfrac{3}{x-4} = \dfrac{5x+4}{x^2-16} - \dfrac{4}{x+4}$

37. $\dfrac{3}{x-1} - \dfrac{2}{x+4} = \dfrac{x^2+8x+6}{x^2+3x-4}$

38. $\dfrac{5}{z-4} + \dfrac{3}{z-2} = \dfrac{z^2-z-2}{z^2-6z+8}$

39. $\dfrac{7}{y^2+y-12} - \dfrac{4y}{y^2+7y+12} = \dfrac{6}{y^2-9}$

40. $\dfrac{3}{x^2-5x-6} + \dfrac{3}{x^2-7x+6} = \dfrac{6}{x^2-1}$

41. $\dfrac{2x+3}{x-3} + \dfrac{x+6}{x-4} = \dfrac{x+6}{x-3}$

42. $\dfrac{x+5}{x+1} + 1 = \dfrac{x-5}{x-2}$

In Problems 43–50, answer each question regarding the rational function given. See Objective 2.

43. For the function $f(x) = x + \dfrac{9}{x}$, solve $f(x) = 10$. What point(s) on the graph of f do these values correspond to?

44. For the function $f(x) = x + \dfrac{7}{x}$, solve $f(x) = 8$. What point(s) on the graph of f do these values correspond to?

45. For the function $f(x) = 2x + \dfrac{4}{x}$, solve $f(x) = -9$. What point(s) on the graph of f do these values correspond to?

46. For the function $f(x) = 2x + \dfrac{8}{x}$, solve $f(x) = -10$. What point(s) on the graph of f do these values correspond to?

47. For the function $f(x) = \dfrac{x+3}{x-4}$, solve $f(x) = \dfrac{9}{2}$. What point(s) on the graph of f do these values correspond to?

48. For the function $f(x) = \dfrac{x+5}{x-3}$, solve $f(x) = \dfrac{1}{5}$. What point(s) on the graph of f do these values correspond to?

49. Let $f(x) = \dfrac{x+2}{2x+9}$ and $g(x) = \dfrac{x-1}{x+3}$. For what value(s) of x does $f(x) = g(x)$? What are the point(s) of intersection of the graphs of f and g?

50. Let $f(x) = \dfrac{4x+1}{8x+5}$ and $g(x) = \dfrac{x-4}{2x-7}$. For what value(s) of x does $f(x) = g(x)$? What are the point(s) of intersection of the graphs of f and g?

Mixed Practice

In Problems 51–60, solve each equation. Be sure to verify your results.

51. $\dfrac{4}{z+4} - \dfrac{3}{4} = \dfrac{5z+2}{4z+16}$

52. $\dfrac{2b-1}{b+5} - \dfrac{2}{3} = \dfrac{1}{3b+15}$

53. $x + \dfrac{9}{x} = 6$

54. $p + \dfrac{25}{p} = 10$

55. $\dfrac{2}{z^2+2z-3} + \dfrac{3}{z^2+4z+3} = \dfrac{6}{z^2-1}$

56. $\dfrac{3}{a^2+3a-10} + \dfrac{2}{a^2+7a+10} = \dfrac{4}{a^2-4}$

57. $\dfrac{4}{x} - \dfrac{5}{2x} = \dfrac{3}{4}$

58. $\dfrac{9}{b} + \dfrac{4}{5b} = \dfrac{7}{10}$

59. $\dfrac{3y+1}{y-1} + 3 = \dfrac{y+2}{y+1}$

60. $\dfrac{x+3}{x-2} + 4 = \dfrac{x+2}{x+1}$

Recall that to find the zero of a function f, we solve the equation $f(x) = 0$. The zeros of the function also represent the x-intercepts of the graph of the function. To find the zeros of a rational function R, the function must first be simplified. In Problems 61–66, find the zeros of each rational function and list the x-intercept(s) of the graph of the function.

61. $R(x) = \dfrac{3x+1}{x^2-4}$

62. $R(x) = \dfrac{2x+5}{x^2-9}$

63. $R(x) = \dfrac{2x^2 + 5x - 12}{3x^2 + 5x + 2}$

64. $R(x) = \dfrac{4x^2 + 7x - 2}{6x^2 + 5x - 4}$

65. $R(x) = \dfrac{x^3 + 3x^2 - 4x - 12}{4x^3 + 12x^2 + x + 3}$

66. $R(x) = \dfrac{x^3 - 2x^2 - 9x + 18}{x^3 - 2x^2 + 4x - 8}$

Applying the Concepts

67. Average Cost Suppose that the average daily cost $\overline{C}$ of manufacturing x bicycles is given by the function

$$\overline{C}(x) = \frac{x^2 + 75x + 5000}{x}$$

Determine the level of production for which the average daily cost will be $225.

68. Population When loggers began cutting in a region in the Amazon rain forest, a rare insect species was discovered. To protect the species, government scientists declared the insects endangered and moved them into a protected area. The population P of the insect t months after being transplanted is modeled by

$$P(t) = \frac{200(1 + 0.4t)}{2(1 + 0.01t)}$$

Predict when the population will be 1350 insects. Round your answer to two decimal places.

69. Cost-Benefit Model Environmental scientists often use cost-benefit models to estimate the cost of removing a pollutant from the environment as a function of the percentage of pollutant removed. Suppose a cost-benefit function for the cost C (in millions of dollars) of removing x percent of the pollutants from Maple Lake is given by

$$C(x) = \frac{25x}{100 - x}$$

(a) If the federal government budgets $100 million to clean up the lake, what percent of the pollutants can be removed?

(b) If the federal government budgets $225 million to clean up the lake, what percent of the pollutants can be removed?

70. The Learning Curve Suppose that a student is given 500 vocabulary words to learn. The function

$$P(x) = \frac{0.8x - 0.8}{0.8x + 0.1}$$

models the proportion P of words learned after x hours of studying.

(a) How long would a student need to study to learn 70% (0.7) of the words?

(b) How long would a student need to study to learn 400 words?

71. Runs in Baseball In his book *Moneyball*, author Michael Lewis cites a formula for predicting the number of runs a team will score in a season. According to the formula, the number of runs R a team will score is given by

$$R = \frac{(h + w)t}{b + w}$$

where h is the number of hits, w is the number of walks, t is the total number of bases, and b is the number of official at-bats. Suppose that the Oakland Athletics scored 750 runs in a season and had 1400 hits, 2250 total bases, and 5500 total at-bats. Use the formula to predict the number of walks that the Athletics had.

△ **72. Regular Polygons** A regular polygon is a polygon that is both equilateral and equiangular. The measure I of each interior angle of a regular polygon with n sides is $I = \dfrac{180°(n - 2)}{n}$. Find the number of sides of a regular polygon whose interior angles measure $135°$.

Extending the Concepts

73. Make up a rational equation that has one real solution.

74. Make up a rational equation that has no real solution.

75. Solve: $\left(\dfrac{4}{x + 3}\right)^2 - 5\left(\dfrac{4}{x + 3}\right) + 6 = 0$

76. Solve: $2 + 11a^{-1} = -12a^{-2}$

Explaining the Concepts

77. Explain the role that domain plays in solving a rational equation.

78. Is the solution set to the equation $\dfrac{x - 6}{x - 6} = 1$ the set of all real numbers? Explain.

Synthesis Review

In Problems 79–83, simplify or solve.

79. $\dfrac{2a^6}{(a^3)^2} - \dfrac{5a^2}{a^3} = \dfrac{3a}{a^3}$

80. $\left(\dfrac{z^{-2}}{2z^{-3}}\right)^{-1} + 3(z - 1)^{-1}$

81. $\dfrac{3}{x - 2} - \dfrac{2x + 1}{x + 1}$

82. $\dfrac{5}{x - 6} + \dfrac{2}{x + 2} = \dfrac{1}{x^2 - 4x - 12}$

83. $\dfrac{x + 1}{2x + 3} - \dfrac{3}{x - 4} = \dfrac{-3}{2x^2 - 5x - 12}$

84. Write a sentence or two explaining the difference between "simplify" and "solve."

The Graphing Calculator

We can approximate solutions to rational equations using the INTERSECT or ZERO (or ROOT) feature of the graphing calculator. We use the ZERO or ROOT feature of the graphing calculator when one side of the equation is 0; we use the INTERSECT feature when neither side of the equation is 0. For example, to solve $\dfrac{x-4}{x+1} = 4$, we would graph $Y_1 = \dfrac{x-4}{x+1}$ and $Y_2 = 4$. The x-coordinates of the point(s) of intersection represent the solution set, as shown in Figure 1.

Figure 1

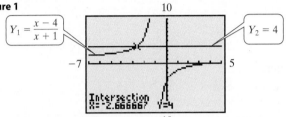

When using a graphing calculator to approximate solutions to equations, we will express the solution as a decimal rounded to two decimal places, if necessary. The solution to the equation $\dfrac{x-4}{x+1} = 4$ is $x = -2.67$ rounded to two decimal places.

In Problems 85–90, use a graphing utility to solve the equation by graphing each side of the equation and finding the point(s) of intersection.

85. $\dfrac{x-4}{x+4} = \dfrac{1}{2}$

86. $\dfrac{x-6}{x+1} = \dfrac{2}{3}$

87. $\dfrac{3}{5} + \dfrac{4}{x+6} = \dfrac{x+12}{5x+30}$

88. $\dfrac{4}{3} + \dfrac{7}{x-4} = \dfrac{-7}{3x-12}$

89. $\dfrac{2x^2 + 11x + 12}{x+4} = -5$

90. $\dfrac{3x^2 + 10x + 3}{x+3} = -8$

5.5 Rational Inequalities

Objective

❶ Solve a Rational Inequality

Are You Ready for This Section?

Before getting started, take the following readiness quiz. If you get a problem wrong, go back to the section cited and review the material.

R1. Write $-1 < x \le 8$ in interval notation. [Section 1.4, pp. 81–84]

R2. Solve: $2x + 3 > 4x - 9$ [Section 1.4, pp. 85–87]

In Section 1.4, we solved linear inequalities in one variable, such as $2x - 3 > 4x + 5$. We were able to solve these inequalities using methods that were similar to those used in solving linear equations. We used set-builder notation and interval notation to represent the solution sets for these inequalities.

Although the approach to solving inequalities involving rational expressions is not a simple extension of solving rational equations, we will use the skills developed in solving rational equations to solve rational inequalities.

❶ Solve a Rational Inequality

▶ A **rational inequality** is an inequality that contains a rational expression. Examples of rational inequalities include

$$\frac{1}{x} > 1 \qquad \frac{x-1}{x+5} \le 0 \qquad \frac{x^2 + 3x + 2}{x-5} > 0 \qquad \frac{3}{x-5} < \frac{4x}{2x-1} + \frac{1}{x}$$

To solve rational inequalities, remember the following:

1. The quotient of two positive numbers is positive; the quotient of a positive and a negative number is negative; and the quotient of two negative numbers is positive.
2. The value of a rational expression may be positive, negative, zero, or undefined, depending on the value of its variable.

For example, if $x = 4$, then the value of the rational expression $\dfrac{x-4}{x+5}$ is $\dfrac{4-4}{4+5} = \dfrac{0}{9} = 0$.

If $x = 3$, then the value of $\dfrac{x-4}{x+5}$ is $\dfrac{3-4}{3+5} = \dfrac{-1}{8} = -\dfrac{1}{8}$, a negative number. If $x = 5$,

Ready?...Answers **R1.** $(-1, 8]$
R2. $\{x \mid x < 6\}$ or $(-\infty, 6)$

then the value of $\dfrac{x-4}{x+5}$ is $\dfrac{5-4}{5+5} = \dfrac{1}{10}$, a positive number. Notice that the value of a rational expression may change signs on either side of a value of the variable that makes the rational expression equal to 0.

The value of a rational expression may also change signs on either side of a value of the variable that makes the rational expression undefined. The rational expression $\dfrac{x-4}{x+5}$ is undefined at $x = -5$. If $x = -6$, then the value of $\dfrac{x-4}{x+5}$ is $\dfrac{-6-4}{-6+5} = \dfrac{-10}{-1} = 10$, and if $x = -4$, then $\dfrac{-4-4}{-4+5} = \dfrac{-8}{1} = -8$.

EXAMPLE 1   **How to Solve a Rational Inequality**

Solve $\dfrac{x+3}{x-4} \geq 0$. Graph the solution set.

Step-by-Step Solution

Step 1: Write the inequality with 0 on one side of the inequality and with a rational expression, written as a single quotient, on the other side.

The inequality is in the form we need.

$$\frac{x+3}{x-4} \geq 0$$

Step 2: Determine the numbers for which the rational expression equals 0 or is undefined.

The rational expression equals 0 when $x = -3$. The rational expression is undefined when $x = 4$.

Step 3: Use the numbers found in Step 2 to separate the real number line into intervals.

We separate the real number line into the following intervals:

Because the rational expression is undefined at $x = 4$, we plot an open circle at 4.

Step 4: Choose a test point within each interval formed in Step 3 to determine the sign of the numerator, $x + 3$, and that of the denominator, $x - 4$. Then determine the sign of the quotient.

When we choose a test point in each interval, if one number in an interval satisfies the inequality, then all the numbers in that interval will satisfy the inequality.

- In the interval $(-\infty, -3)$, choose a test point of -4. When $x = -4$, the numerator, $x + 3$, equals -1, and the denominator, $x - 4$, equals -8. Since the quotient of two negatives is positive, the expression $\dfrac{x+3}{x-4} = \dfrac{-1}{-8} = \dfrac{1}{8}$ is positive when $x = -4$. Furthermore, the expression $\dfrac{x+3}{x-4}$ is positive for all x in the interval $(-\infty, -3)$.

- In the interval $(-3, 4)$, let 0 be the test point. When $x = 0$, $x + 3$ is positive, and $x - 4$ is negative. A positive divided by a negative is negative, so $\dfrac{x+3}{x-4} = \dfrac{0+3}{0-4} = \dfrac{3}{-4} = -\dfrac{3}{4}$ is negative when $x = 0$. The expression $\dfrac{x+3}{x-4}$ is negative for all x in the interval $(-3, 4)$.

- In the interval $(4, \infty)$, let 5 be the test point. When $x = 5$, both $x + 3$ and $x - 4$ are positive. A positive divided by a positive is positive, so $\dfrac{x+3}{x-4} = \dfrac{5+3}{5-4} = \dfrac{8}{1} = 8$ is positive. Furthermore, $\dfrac{x+3}{x-4}$ is positive for all x in $(4, \infty)$.

Table 1 shows these results and the sign of $\dfrac{x+3}{x-4}$ in each interval.

Table 1

Interval	$(-\infty, -3)$		$(-3, 4)$		$(4, \infty)$
	$x < -3$	$x = -3$	$-3 < x < 4$	$x = 4$	$x > 4$
Test Point	-4	-3	0	4	5
Sign of $x + 3$	Negative	0	Positive	Positive	Positive
Sign of $x - 4$	Negative	Negative	Negative	0	Positive
Sign of $\dfrac{x + 3}{x - 4}$	Positive	0	Negative	Undefined	Positve
Conclusion about $\dfrac{x + 3}{x - 4}$	$\dfrac{x + 3}{x - 4}$ is positive, so $(-\infty, -3)$ is part of the solution set.	Because the inequality is nonstrict, -3 is part of the solution set.	$\dfrac{x + 3}{x - 4}$ is negative, so $(-3, 4)$ is not part of the solution set.	4 cannot be part of the solution set because it causes division by 0.	$\dfrac{x + 3}{x - 4}$ is positive, so $(4, \infty)$ is part of the solution set.

We want to know where $\dfrac{x + 3}{x - 4}$ is greater than or equal to zero. The solution set is $\{x \mid x \le -3 \text{ or } x > 4\}$ using set-builder notation, and $(-\infty, -3] \cup (4, \infty)$ using interval notation. Notice that -3 is part of the solution set because $\dfrac{-3 + 3}{-3 - 4} = \dfrac{0}{-7} = 0$, but 4 is not part of the solution set because it is not in the domain of $\dfrac{x + 3}{x - 4}$. Figure 2 shows the graph of the solution set.

Figure 2

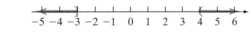

Solving Rational Inequalities

Step 1: Write the inequality so that a rational expression is on one side of the inequality and 0 is on the other. Be sure to write the rational expression as a single quotient in factored form.

Step 2: Determine the values for which the rational expression equals 0 or is undefined.

Step 3: Use the values found in Step 2 to separate the real number line into intervals.

Step 4: Choose a test point within each interval formed in Step 3 to determine the sign of each factor in the numerator and denominator. Then determine the sign of the quotient.

- If the quotient is positive, then the rational expression is positive for all values x in the interval.

- If the quotient is negative, then the rational expression is negative for all values x in the interval.

Also determine the value of the rational expression at each value found in Step 2. If the inequality is not strict ($\le$ or $\ge$), include the values of the variable for which the rational expression equals 0 in the solution set, but do not include the values for which the rational expression is undefined!

Quick ✔

1. The inequality $\dfrac{2x - 3}{x + 6} > 1$ is an example of a(n) _____ inequality.

2. Solve $\dfrac{x - 7}{x + 3} \ge 0$. Graph the solution set.

3. Solve $\dfrac{1 - x}{x + 5} > 0$. Graph the solution set.

▶ (EXAMPLE 2) **Solving a Rational Inequality**

Solve $\dfrac{x + 3}{x - 1} > 2$. Graph the solution set.

Solution

First, write the inequality so that a rational expression is on one side of the inequality and 0 is on the other.

$$\frac{x + 3}{x - 1} > 2$$

Subtract 2 from both sides: $\quad \dfrac{x + 3}{x - 1} - 2 > 0$

LCD $= x - 1$; multiply -2 by $\dfrac{x - 1}{x - 1}$: $\quad \dfrac{x + 3}{x - 1} - 2 \cdot \dfrac{x - 1}{x - 1} > 0$

Write the rational expression with the LCD: $\quad \dfrac{x + 3 - 2(x - 1)}{x - 1} > 0$

Distribute -2: $\quad \dfrac{x + 3 - 2x + 2}{x - 1} > 0$

Combine like terms in the numerator: $\quad \dfrac{-x + 5}{x - 1} > 0$

The rational expression equals 0 when $x = 5$. The rational expression is undefined when $x = 1$. Separate the real number line into the intervals as shown in Figure 3.

Figure 3

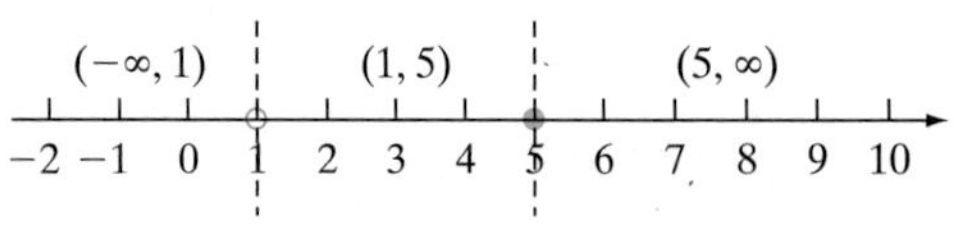

Table 2 shows the sign of $-x + 5$, $x - 1$, and $\dfrac{-x + 5}{x - 1}$ in each interval. It also shows the value of $\dfrac{-x + 5}{x - 1}$ at $x = 1$ and $x = 5$.

Table 2

Interval	$(-\infty, 1)$		$(1, 5)$		$(5, \infty)$
Test Point	0	1	3	5	6
Sign of $-x + 5$	Positive	Positive	Positive	0	Negative
Sign of $x - 1$	Negative	0	Positive	Positive	Positive
Sign of $\dfrac{-x + 5}{x - 1}$	Negative	Undefined	Positive	0	Negative
Conclusion about $\dfrac{-x + 5}{x - 1}$	$\dfrac{-x + 5}{x - 1}$ is negative, so $(-\infty, 1)$ is not part of the solution set.	Because $\dfrac{-x + 5}{x - 1}$ is undefined at $x = 1$, $x = 1$ is not part of the solution set.	$\dfrac{-x + 5}{x - 1}$ is positive, so $(1, 5)$ is part of the solution set.	Because the inequality is strict, 5 is not part of the solution set.	$\dfrac{-x + 5}{x - 1}$ is negative, so $(5, \infty)$ is not part of the solution set.

Figure 4

We want to know the values of x such that $\dfrac{x+3}{x-1} > 2$. This is equivalent to determining where $\dfrac{-x+5}{x-1} > 0$. Thus the solution set is $\{x \mid 1 < x < 5\}$, or $(1, 5)$ using interval notation. The endpoints of the interval are not in the solution set because the inequality in the original problem is strict. Figure 4 shows the graph of the solution set.

Quick ✓

4. Solve $\dfrac{4x+5}{x+2} < 3$. Graph the solution set.

5.5 Exercises MyMathLab®
PRACTICE

Exercise numbers in green
are complete video solutions
in MyMathLab

*Problems **1–4** are the* Quick ✓*s that follow the* EXAMPLES.

Building Skills

In Problems 5–20, solve each rational inequality. Graph the solution set. See Objective 1.

5. $\dfrac{x-4}{x+1} > 0$

6. $\dfrac{x+5}{x-2} > 0$

7. $\dfrac{x+9}{x-3} < 0$

8. $\dfrac{x+8}{x+2} < 0$

9. $\dfrac{x+10}{x-4} \geq 0$

10. $\dfrac{x+12}{x-2} \geq 0$

11. $\dfrac{(3x+5)(x+8)}{x-2} \leq 0$

12. $\dfrac{(3x-2)(x-6)}{x+1} \geq 0$

13. $\dfrac{x-5}{x+1} < 1$

14. $\dfrac{x+3}{x-4} > 1$

15. $\dfrac{2x-9}{x-3} > 4$

16. $\dfrac{3x+20}{x+6} < 5$

17. $\dfrac{3}{x-4} + \dfrac{1}{x} \geq 0$

18. $\dfrac{2}{x+3} + \dfrac{2}{x} \leq 0$

19. $\dfrac{3}{x-2} \leq \dfrac{4}{x+5}$

20. $\dfrac{1}{x-4} \geq \dfrac{3}{2x+1}$

Mixed Practice

In Problems 21–30, solve each inequality. Graph the solution set.

21. $\dfrac{(2x-1)(x+3)}{x-5} > 0$

22. $\dfrac{(5x-2)(x+4)}{x-5} < 0$

23. $3 - 4(x+1) < 11$

24. $2x + 3(x-2) \geq x + 2$

25. $\dfrac{x+7}{x-8} \leq 0$

26. $\dfrac{x-10}{x+5} \leq 0$

27. $(x-2)(2x+1) \geq 2(x-1)^2$

28. $(x+2)^2 < 3x^2 - 2(x+1)(x-2)$

29. $\dfrac{3x-1}{x+4} \geq 2$

30. $\dfrac{3x-7}{x+2} \leq 2$

In Problems 31–34, for each function, find the values of x that satisfy the given condition.

31. Solve $R(x) \leq 0$ if $R(x) = \dfrac{x-6}{x+1}$.

32. Solve $R(x) \geq 0$ if $R(x) = \dfrac{x+3}{x-8}$.

33. Solve $R(x) < 0$ if $R(x) = \dfrac{2x-5}{x+2}$.

34. Solve $R(x) < 0$ if $R(x) = \dfrac{3x+2}{x-4}$.

Applying the Concepts

35. Average Cost Suppose that the daily cost C of manufacturing x bicycles is given by $C(x) = 80x + 5000$. Then the average daily cost $\overline{C}$ is given by $\overline{C}(x) = \dfrac{80x + 5000}{x}$. How many bicycles must be produced each day in order for the average cost to be no more than \$130?

36. Average Cost See Problem 35. Suppose that the government imposes a \$10 tax on each bicycle manufactured so that the daily cost C of manufacturing x bicycles is now given by $C(x) = 90x + 5000$. Now the average daily cost $\overline{C}$ is given by $\overline{C}(x) = \dfrac{90x + 5000}{x}$. How many bicycles must be produced each day in order for the average cost to be no more than \$130?

Extending the Concepts

37. Write a rational inequality that has $(2, \infty)$ as the solution set.

38. Write a rational inequality that has $(-2, 5]$ as the solution set.

Explaining the Concepts

39. In solving the rational inequality $\dfrac{x - 4}{x + 1} \le 0$, a student determines that the only interval that makes the inequality true is $(-1, 4)$. He states that the solution set is $\{x \mid -1 \le x \le 4\}$. What is wrong with this solution?

40. In Step 2 of the steps for solving a rational inequality, we determine the values for which the rational expression equals 0 or is undefined. We then use these values to form intervals on the real number line. Explain why this guarantees that there is not a change in the sign of the rational expression within any given interval.

Synthesis Review

In Problems 41–46, find the x-intercept(s) of the graph of each function.

41. $F(x) = 6x - 12$

42. $G(x) = 5x + 30$

43. $f(x) = 2x^2 + 3x - 14$

44. $h(x) = -3x^2 - 7x + 20$

45. $R(x) = \dfrac{3x - 2}{x + 4}$

46. $R(x) = \dfrac{x^2 + 5x + 6}{x + 2}$

The Graphing Calculator

We can approximate solutions to rational inequalities using the INTERSECT or ZERO (or ROOT) feature of the graphing calculator. We use the ZERO or ROOT feature of the graphing calculator when one side of the inequality is 0; we use the INTERSECT feature when neither side of the inequality is 0. For example, to solve $\dfrac{x - 4}{x + 1} > \dfrac{7}{2}$, we would graph $Y_1 = \dfrac{x - 4}{x + 1}$ and $Y_2 = \dfrac{7}{2}$.

To determine the x-values such that the graph of Y_1 is above that of Y_2, we find x-coordinates of the point(s) of intersection. The graph of Y_1 is above that of Y_2 between $x = -3$ and $x = -1$. See Figure 5.

Figure 5

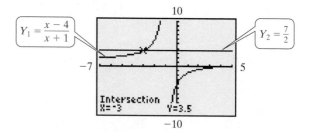

When using a graphing calculator to approximate solutions to inequalities, we typically express the solution as a decimal rounded to two decimal places, if exact answers cannot be found. The solution to the inequality $\dfrac{x - 4}{x + 1} > \dfrac{7}{2}$ is $\{x \mid -3 < x < -1\}$, or $(-3, -1)$ using interval notation.

In Problems 47–50, solve each inequality using a graphing calculator.

47. $\dfrac{x - 5}{x + 1} \le 3$

48. $\dfrac{x + 2}{x - 5} > -2$

49. $\dfrac{2x + 5}{x - 7} > 3$

50. $\dfrac{2x - 1}{x + 5} \le 4$

5.6 Models Involving Rational Expressions

Objectives

1 Solve for a Variable in a Rational Expression

2 Model and Solve Ratio and Proportion Problems

3 Model and Solve Work Problems

4 Model and Solve Uniform Motion Problems

Are You Ready for This Section?

Before getting started, take the following readiness quiz. If you get a problem wrong, go back to the section cited and review the material.

R1. Solve for y: $4x - 2y = 10$ [Section 1.3, pp. 73–76]

▶ **1** Solve for a Variable in a Rational Expression

To "solve for the variable" means to isolate the variable on one side of the equation, with all other variables and constants, if any, on the other side. We use the same steps to solve formulas for a certain variable that we use to solve rational equations.

EXAMPLE 1   **Solving for a Variable in a Lens Construction Formula**

The formula $\dfrac{1}{f} = \dfrac{1}{p} + \dfrac{1}{q}$ is used in telescope and camera construction, where f is the focal length of the lens. In general, the larger the focal length of the lens, f, the more power the telescope has. The variable p is the distance between the object we wish to see and the lens; the variable q is the distance from the lens to the point of focus (such as the film or your eye). See Figure 6.

Figure 6

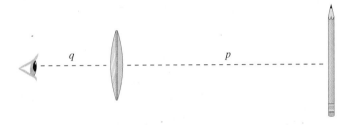

    (a) Solve the formula for q.

    (b) Suppose a camera has a focal length of 100 mm and the camera is focusing on an object 5000 mm away. What is q, the distance from the lens to the point of focus? Round your answer to the nearest millimeter.

Solution

    (a) Our goal is to isolate q. Follow the same steps that we used to solve a rational equation. First, note that none of the variables can equal 0. Multiply both sides of the equation by pqf, the LCD of all the denominators.

$$\frac{1}{f} = \frac{1}{p} + \frac{1}{q}$$

$$pqf \cdot \frac{1}{f} = pqf \cdot \left(\frac{1}{p} + \frac{1}{q}\right)$$

Distribute pqf: $pqf \cdot \dfrac{1}{f} = pqf \cdot \dfrac{1}{p} + pqf \cdot \dfrac{1}{q}$

Simplify: $pq = qf + pf$

To get all terms involving q on one side; subtract qf from both sides: $pq - qf = qf + pf - qf$

$$pq - qf = pf$$

Factor out q: $q(p - f) = pf$

Divide both sides by $p - f$: $q = \dfrac{pf}{p - f}$

(b) Substitute $f = 100$ mm and $p = 5000$ mm in $q = \dfrac{pf}{p-f}$.

$$q = \frac{5000 \cdot 100}{5000 - 100}$$

$$\approx 102 \text{ mm}$$

The distance from the lens to the point of focus is approximately 102 mm. ●

Quick ✓

1. The formula $Y = \dfrac{G}{1-b}$ is used in economics to determine the impact on gross domestic product (GDP) Y of increasing government spending by G dollars if the proportion of additional income that people spend is b.

(a) Solve the formula for b.

(b) Find b if the government increased spending by \$100 billion and GDP increased by \$1000 billion.

▶ ❷ Model and Solve Ratio and Proportion Problems

The problems in this objective focus on the idea of ratio and proportion. The **ratio** of two numbers a and b can be written as

$$a{:}b \qquad \text{or} \qquad \frac{a}{b}$$

When solving algebraic problems, we write ratios as $\dfrac{a}{b}$. A **proportion** is a statement (equation) that two ratios are equal. That is, proportions are equations of the form $\dfrac{a}{b} = \dfrac{c}{d}$. You may be familiar with using proportions to solve problems involving similar figures, such as triangles, from geometry. Two figures are **similar** if their angles have the same measure and their corresponding sides are proportional. Figure 7 shows examples of similar figures.

In Figure 7(a), $\triangle ABC$ is similar to $\triangle DEF$, and in Figure 7(b), quadrilateral $ABCD$ is similar to quadrilateral $EFGH$. Because $\triangle ABC$ is similar to $\triangle DEF$, we know that the ratio of AB to AC equals the ratio of DE to DF. That is,

$$\frac{AB}{AC} = \frac{DE}{DF}$$

We will use the principle that the ratios of corresponding sides are equal to find unknown lengths in similar figures.

Figure 7

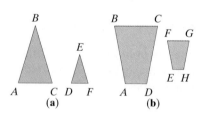

(a)   (b)

EXAMPLE 2 **Similar Figures**

Suppose you want to know the height of a building. A 20-foot-tall light post casts a shadow that is 12 feet long. The shadow cast by the building is 660 feet long. Find the height of the building.

Solution

The ratio of the length of the building's shadow to its height equals the ratio of the length of the light post's shadow to its height because the building and its shadow form a triangle that is similar to the triangle formed by the light post and its shadow. See Figure 8.

Figure 8

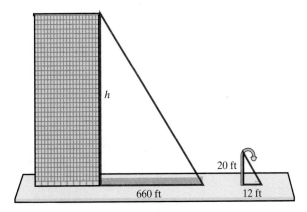

We set up the proportion problem as

                              building post

                                ↓   ↓

$$\frac{\text{shadow length}}{\text{object height}} : \frac{660}{h} = \frac{12}{20}$$

and solve the equation. The domain of h is $\{h \mid h > 0\}$ since h represents the height of the building. The LCD of all denominators is $20h$, so we multiply both sides of the equation by the LCD, $20h$.

$$20h \cdot \frac{660}{h} = 20h \cdot \frac{12}{20}$$

$$13{,}200 = 12h$$

Divide both sides by 12: $1100 = h$

The height of the building is 1100 feet.

Figure 9

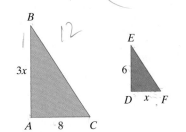

Quick ✓

2. A(n) _____ is a statement (equation) that two ratios are equal. Two figures are _____ if their angles have the same measure and their corresponding sides are proportional.

3. Suppose a man who is 6 feet tall casts a shadow that is 2 feet long. At the same time, a building casts a shadow that is 40 feet long. How tall is the building?

4. Suppose that $\triangle ABC$ is similar to $\triangle DEF$, as shown in Figure 9. Find the length of AB and DF.

Proportions come up in direct translation problems as well.

(**EXAMPLE 3**)   **A Proportion Problem**

According to the National Vital Statistics Report dated November 30, 2011, the birth rate for unmarried women (15–44 years of age) was 50.5 live births per 1000 population in the United States in 2009. In 2009, there were 1,694,000 births to unmarried women. Determine the population of unmarried women between 15 and 44 years of age in 2009.

Solution

Step 1: Identify This is a direct translation problem involving proportions. We are looking for the population of unmarried women in 2009.

Step 2: Name We let p represent the population of unmarried women in 2009.

Step 3: Translate Since we know that the rate of live births was 50.5 per 1000 and that there were 1,694,000 births, we have the model

$$\frac{\text{number of births}}{\text{population}}: \quad \frac{50.5}{1000} = \frac{1,694,000}{p}$$

Step 4: Solve We now proceed to solve the equation.

$$\frac{50.5}{1000} = \frac{1,694,000}{p}$$

Multiply both sides by the LCD, 1000p: $\quad 1000p \cdot \dfrac{50.5}{1000} = 1000p \cdot \dfrac{1,694,000}{p}$

Divide out common factors: $\qquad\qquad 50.5p = 1,694,000,000$

Divide both sides by 50.5: $\qquad\qquad\quad p \approx 33,544,554$

Step 5: Check It is always a good idea to make sure your answer is reasonable. According to the U.S. Census Bureau, there were over 60 million women 15–44 years of age in the United States in 2009, so the answer seems reasonable.

Step 6: Answer The population of unmarried women 15–44 years of age in the United States in 2009 was 33,544,554.

Quick ✓

5. According to a Gallup poll, approximately 170 of every 1000 Americans had no health insurance in 2011. There were approximately 53,300,000 uninsured Americans in 2011. Find the population of the United States in 2011.

▶ ❸ **Model and Solve Work Problems**

We are now going to solve work or "constant-rate jobs" problems. These problems assume that jobs, even those performed by people, are performed at a **constant rate.** Although this assumption is reasonable for machines, it is not likely to be true for people simply because of the old phrase "too many chefs spoil the broth." Think of it this way—if you continually add more people to paint a room, the time to complete the job may decrease initially, but eventually the painters get in each other's way and the time to completion actually increases. We could build models that take this into account, but instead we will make the "constant-rate" assumption for humans as well, in order to keep the mathematics manageable.

The constant-rate assumption states that if it takes t units of time to complete a job, then $\dfrac{1}{t}$ of the job is done in 1 unit of time. For example, if it takes 5 hours to paint a room, then $\dfrac{1}{5}$ of the room should be painted in 1 hour.

Work Smart

Remember, when we model we make simplifying assumptions to make the math easier to deal with.

EXAMPLE 4 **Working Together on a Job**

It's Saturday and Kevin needs to cut and edge the grass. At 9 A.M., Michael asks Kevin to go golfing at 11:00 A.M. Typically, it takes Kevin 3 hours to cut and edge the grass. When Michael cuts and edges the grass, it takes 4 hours. If they work together, will they be able to finish the lawn and still make the golf date?

Solution

Step 1: Identify We want to know how long it will take Michael and Kevin to finish the lawn, so that we can tell whether they can make the golf date 2 hours after they start the lawn work.

Step 2: Name We let t represent the time (in hours) that it takes to finish the lawn working together. Then, in 1 hour they will complete $\dfrac{1}{t}$ of the job.

Step 3: Translate Since we know that Kevin can finish the job in 3 hours, Kevin will finish $\frac{1}{3}$ of the job in 1 hour. Michael can finish the job in 4 hours, so Michael will finish $\frac{1}{4}$ of the job in 1 hour. We set up the model using the following logic:

$$\left(\begin{array}{c}\text{Part of job done by Kevin}\\\text{in 1 hour}\end{array}\right) + \left(\begin{array}{c}\text{Part of job done by Michael}\\\text{in 1 hour}\end{array}\right) = \left(\begin{array}{c}\text{Part of job done together}\\\text{in 1 hour}\end{array}\right)$$

$$\frac{1}{3} \quad + \quad \frac{1}{4} \quad = \quad \frac{1}{t} \quad \text{The Model}$$

Step 4: Solve

$$\frac{1}{3} + \frac{1}{4} = \frac{1}{t}$$

Multiply both sides by the LCD, 12t:
$$12t \cdot \left(\frac{1}{3} + \frac{1}{4}\right) = 12t \cdot \frac{1}{t}$$

Distribute:
$$12t \cdot \frac{1}{3} + 12t \cdot \frac{1}{4} = 12t \cdot \frac{1}{t}$$

Divide like factors:
$$4t + 3t = 12$$

Combine like terms:
$$7t = 12$$

Divide both sides by 7:
$$t = \frac{12}{7} \approx 1.714$$

Work Smart

We convert 0.714 hour to minutes by multiplying 0.714 by 60 minutes, and obtain 43 minutes.

Step 5: Check Is the answer reasonable? We expect the time to be greater than 0 but less than 3 (because it takes Kevin 3 hours working by himself). Our answer of 1.714 hours, or approximately 1 hour, 43 minutes, seems reasonable.

Step 6: Answer If they start right away, they should finish at 10:43 A.M. If they can get to the course in 17 minutes, they can make the tee time. ●

Quick ✓

6. Juan and Maria have a pool in their backyard. They can fill their pool in 30 hours using their hose alone. Their neighbor's pool is the same size and can be filled in 24 hours. Suppose their neighbor lets them borrow his hose. How long will it take Juan and Maria to fill their pool with both hoses?

EXAMPLE 5 **The Kitchen Sink**

Suppose the kitchen sink can be filled in 5 minutes. If the sink is full, it takes 8 minutes to drain the sink when the drain is left partially open. If the sink's drain is accidentally left partially open, how long will it take to fill the sink?

Solution

Step 1: Identify We want to know how long it will take to fill the sink with the drain partially open.

Step 2: Name We let t represent the time (in minutes) that it takes to fill the sink. Then, in 1 minute $\frac{1}{t}$ of the sink will be full.

Step 3: Translate We know that the sink can be filled in 5 minutes when the drain is closed, so after 1 minute, $\frac{1}{5}$ of the sink is full. It takes 8 minutes to drain, so after 1 minute, $\frac{1}{8}$ of the sink is drained. We set up the model using the following logic:

Work Smart

Note that we subtract the portion of sink drained after 1 minute, because it is "working against us."

$$\begin{pmatrix} \text{Portion of sink filled} \\ \text{in 1 minute with} \\ \text{closed drain} \end{pmatrix} - \begin{pmatrix} \text{Portion of sink drained} \\ \text{after 1 minute} \end{pmatrix} = \begin{pmatrix} \text{Portion of sink filled} \\ \text{after 1 minute with} \\ \text{open drain} \end{pmatrix}$$

$$\frac{1}{5} \qquad - \qquad \frac{1}{8} \qquad = \qquad \frac{1}{t} \quad \text{The Model}$$

Step 4: Solve

$$\frac{1}{5} - \frac{1}{8} = \frac{1}{t}$$

Multiply both sides by the LCD: 40t:
$$40t \cdot \left(\frac{1}{5} - \frac{1}{8}\right) = 40t \cdot \frac{1}{t}$$

Distribute:
$$40t \cdot \frac{1}{5} - 40t \cdot \frac{1}{8} = 40t \cdot \frac{1}{t}$$

Simplify:
$$8t - 5t = 40$$

Combine like terms:
$$3t = 40$$

Divide both sides by 3:
$$t = \frac{40}{3} \approx 13.3 \text{ minutes}$$

Step 5: Check Is this answer reasonable? We expect it to be greater than 5 since this is the time it takes to fill the sink with the drain closed. Our answer of 13.3 minutes or about 13 minutes, 20 seconds seems reasonable.

Step 6: Answer It will take 13.3 minutes or 13 minutes, 20 seconds to fill the sink. ●

Quick ✓

7. A children's inflatable pool takes 20 minutes to fill with an electric air pump. It takes 50 minutes to let the air out of the pool. If the pool's valve is accidentally left open, how long will it take to fill the pool?

▶ ❹ Model and Solve Uniform Motion Problems

We introduced uniform motion problems in Section 1.2. Recall that uniform motion problems use the fact that distance equals rate times time—that is, $d = rt$. In this section, we often use an alternative form of this model, $t = \dfrac{d}{r}$.

EXAMPLE 6   **A Round-Trip Flight**

A plane flies 990 miles west (into the wind) and makes the return trip following the same flight path. The effect of the wind on the plane is 20 miles per hour. The round trip takes 10 hours. What is the speed of the plane in still air?

Solution

Step 1: Identify In this uniform motion problem, we want to find the speed of the plane in still air.

Step 2: Name Let r represent the speed of the plane in still air.

Step 3: Translate Going west, the plane is flying into the wind. The speed of the plane is its rate in still air less the impact of the wind, or $r - 20$. Similar logic tells us that the speed of the plane going east is $r + 20$. We set up Table 3. Remember, since $d = r \cdot t$, then $t = \dfrac{d}{r}$ or $\dfrac{d}{r} = t$.

Table 3

	Distance (miles)	Rate (miles per hour)	Time (hours)
West	990	$r - 20$	$\dfrac{990}{r - 20}$
East	990	$r + 20$	$\dfrac{990}{r + 20}$

The round trip takes 10 hours, so we set up the equation:

$$\text{Time going west} + \text{Time going east} = 10$$

$$\frac{990}{r - 20} + \frac{990}{r + 20} = 10 \quad \text{The Model}$$

Step 4: Solve Solve for r:

$$\frac{990}{r - 20} + \frac{990}{r + 20} = 10$$

Multiply both sides by the LCD, $(r - 20)(r + 20)$:

$$(r - 20)(r + 20)\left(\frac{990}{r - 20} + \frac{990}{r + 20}\right) = 10(r - 20)(r + 20)$$

Distribute:

$$(r - 20)(r + 20)\frac{990}{r - 20} + (r - 20)(r + 20)\frac{990}{r + 20} = 10(r - 20)(r + 20)$$

Divide out common factors: $\quad 990(r + 20) + 990(r - 20) = 10(r - 20)(r + 20)$

Distribute: $\quad 990r + 19{,}800 + 990r - 19{,}800 = 10(r^2 - 400)$

Combine like terms: $\quad 1980r = 10(r^2 - 400)$

Divide both sides by 10: $\quad 198r = r^2 - 400$

Write the equation in standard form: $\quad 0 = r^2 - 198r - 400$

Factor: $\quad 0 = (r - 200)(r + 2)$

Use the Zero-Product Property: $\quad r = 200 \text{ or } r = -2$

Step 5: Check We reject -2 because the rate of the plane must be positive. Thus it appears that the plane will travel at a rate of 200 miles per hour in still air. Flying west, the plane travels at 180 miles per hour, so the trip takes $\dfrac{990 \text{ miles}}{180 \text{ miles per hour}} = 5.5$ hours.

Flying east, the plane travels at 220 miles per hour, so the trip takes $\dfrac{990 \text{ miles}}{220 \text{ miles per hour}} = 4.5$ hours. The total trip takes $5.5 + 4.5 = 10$ hours. It checks!

Step 6: Answer the Question The plane travels 200 miles per hour in still air. ●

Quick ✓

8. A canoe travels on a river whose current is running at 2 miles per hour. After traveling 12 miles upstream, the canoe turns around and makes the 12-mile trip back downstream. The trip up and back takes 8 hours. What is the speed of the canoe in still water?

5.6 Exercises MyMathLab® Math XP
PRACTICE

Exercise numbers in green are complete video solutions in MyMathLab

*Problems **1–8** are the* Quick ✓*s that follow the* EXAMPLES.

Building Skills

In Problems 9–18, solve each formula for the indicated variable. See Objective 1.

9. Chemistry (Gas Laws) Solve $\dfrac{V_1}{V_2} = \dfrac{P_2}{P_1}$ for P_1.

10. Chemistry (Gas Laws) Solve $\dfrac{V_1}{V_2} = \dfrac{P_2}{P_1}$ for V_2.

11. Finance Solve $R = \dfrac{r}{1-t}$ for t.

12. Finance Solve $P = \dfrac{A}{1+r}$ for r.

13. Slope Solve $m = \dfrac{y-y_1}{x-x_1}$ for x.

14. Slope Solve $m = \dfrac{y-y_1}{x-x_1}$ for x_1.

15. Physics Solve $\omega = \dfrac{rmv}{I+mr^2}$ for v.

16. Physics Solve $\omega = \dfrac{rmv}{I+mr^2}$ for I.

17. Physics Solve $V = \dfrac{mv}{M+m}$ for m.

18. Physics Solve $v_2 = \dfrac{2m_1}{m_1+m_2} v_1$ for m_1.

Applying the Concepts

In Problems 19–26, solve the proportion problem. See Objective 2.

△ **19.** Suppose that △ABC is similar to △DEF, as shown in the figure. Find the length of AB and DF.

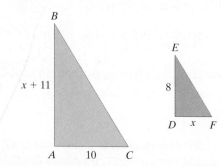

△ **20.** Suppose that △ABC is similar to △DEF, as shown in the figure. Find the length of AB and DF.

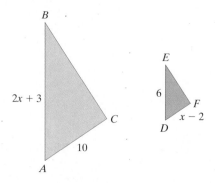

21. Motor Vehicle Death Rates According to the Centers for Disease Control, the death rate in the United States as a result of a motor vehicle accident is 11.9 per 100,000 population. In 2009, there were 36,216 fatalities in motor vehicle accidents. What was the population of the United States in 2009?

22. Flight Accidents According to the *Statistical Abstract of the United States,* in 2010, there were 1.27 fatal airplane accidents per 100,000 flight hours. Also in 2010, there were a total of 267 fatal accidents. How many flight hours were flown in 2010?

23. Road Trip At current prices, Roberta can drive her car 8.2 miles per dollar of gasoline that she buys. Roberta and three of her friends decide to go on a road trip to a neighboring university that is 105 miles away and agree to split the cost evenly. To the nearest cent, how much money will each have to contribute to get to the university and back?

24. Car Payments At current rates, a 60-month term car loan is being offered where the monthly payments are $0.0191 per dollar borrowed. Suppose that Eduardo's car payment is $340 per month. To the nearest dollar, how much did Eduardo borrow?

25. Pascal's Principle Pascal's Principle applied to a hydraulic lever states that the ratio of the force exerted on an input piston F_1 to the area displaced A_1 equals the ratio of the force on the output piston F_2 to the area displaced A_2. See the figure. If a force of 30 pounds is exerted with an area of 12 square feet of water displaced in the right pipe and an area of

5 square feet is displaced in the left pipe, determine the force exerted by the left pipe.

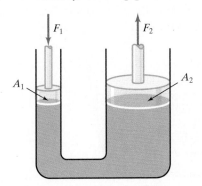

26. Pascal's Principle See Problem 25. If a force of 40 pounds is exerted with an area of 15 square feet of water displaced in the right pipe and an area of 8 square feet is displaced in the left pipe, determine the force exerted by the left pipe.

In Problems 27–32, solve the work problem. See Objective 3.

27. Sharing a Paper Route Amiri can deliver his newspapers in 80 minutes. It takes Horus 60 minutes to do the same route. How long would it take them to deliver the newspapers if they worked together?

28. Painting a Room Latoya can paint five 10-foot-by-14-foot rooms by herself in 14 hours. Lisa can paint five 10-foot-by-14-foot rooms by herself in 10 hours. Working together, how long would it take to paint five 10-foot-by-14-foot rooms?

29. Cutting the Grass Avery can cut the grass working by himself in 3 hours. When Avery cuts the grass with his younger brother Connor, it takes 2 hours. How long would it take Connor to cut the grass if he worked by himself?

30. Assembling a Swing Set Alexandra and Frank can assemble a King Kong swing set working together in 6 hours. One day, when Frank called in sick, Alexandra was able to assemble a King Kong swing set in 10 hours. How long would it take Frank to assemble a King Kong swing set if he worked by himself?

31. Emptying a Pool A swimming pool can be emptied in 6 hours using a 10-horsepower pump along with a 6-horsepower pump. The 6-horsepower pump requires 5 hours more than the 10-horsepower pump to empty the pool when working by itself. How long would it take to empty the pool using just the 10-horsepower pump?

32. Draining a Pond A pond can be emptied in $3.75 \left(= \dfrac{15}{4} \right)$ hours using a 10-horsepower pump along with a 4-horsepower pump. The 4-horsepower pump requires 4 hours more than the 10-horsepower pump to empty the pond when working by itself. How long would it take to empty the pond using just the 10-horsepower pump?

In Problems 33–42, solve the uniform motion problem. See Objective 4.

33. Tough Commute You have a 20-mile commute into work. Since you leave very early, the trip going to work is easier than the trip home. You can travel to work in the same time that it takes for you to make it 16 miles on the trip back home. Your average speed coming home is 7 miles per hour slower than your average speed going to work. What is your average speed going to work?

34. Riding Your Bicycle Every weekend, you ride your bicycle on a forest preserve path. The path is 20 miles long and ends at a waterfall, at which point you relax and then make the trip back to the starting point. One weekend, you find that in the same time it takes you to travel to the waterfall, you are only able to return 12 miles. Your average speed going to the waterfall is 4 miles per hour faster than your average speed on the return trip. What was your average speed going to the waterfall?

35. Moving Walkway In order to access the outer part of Terminal 1 at O'Hare International Airport, you must walk quite some distance in a tunnel that travels under part of the airport. To make the walk less difficult, there is a moving walkway that travels at 2 feet per second. Suppose that Hana can travel 152 feet while walking on the walkway in the same amount of time it takes her to travel 72 feet while walking on the pavement without the aid of the moving sidewalk. How fast does Hana walk?

36. Escalator When exiting Terminal 1 at O'Hare International Airport, you can either take an escalator up to the main level or take traditional stairs. Suppose that the escalator travels 1.5 feet per second. Karli can walk up the 50-foot escalator in the same amount of time it takes her to walk 30 feet up the stairs. How fast does Karli walk up stairs?

37. **Football** Suppose that Rob Gronkowski of the New England Patriots can run 100 yards in 12 seconds. Further suppose that Brian Urlacher of the Chicago Bears can run 100 yards in 9 seconds. Suppose that Gronkowski catches a pass at his own 20-yard line in stride and starts running away from Urlacher, who is at the 15-yard line directly behind Gronkowski. See the figure. At what yard line will Urlacher catch up to Gronkowski?

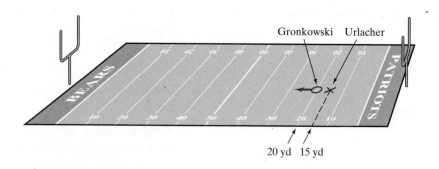

38. **Running a Race** Roger can run one mile in 8 minutes. Jeff can run one mile in 6 minutes. If Jeff gives Roger a 1-minute head start, how long will it take before Jeff catches up to Roger? How far will each have run?

39. **Uphill/Downhill** A bicyclist rides his bicycle 12 miles up a hill and then comes back down. His speed coming downhill is 8 miles per hour faster than his speed going uphill. The roundtrip takes 2 hours and 15 minutes $\left(= \frac{9}{4} \text{hours} \right)$. What was the speed of the bicyclist going uphill?

40. **Round Trip** A plane flies 600 miles west (into the wind) and makes the return trip following the same flight path. The effect of the jet stream on the plane is 15 miles per hour. The round trip takes 9 hours. What is the speed of the plane in still air?

41. **Scenic Drive** Joe and Nancy live in Morro Bay, California, right off of Highway 1. They decide to take a trip north to Monterey. The first 50 miles of the drive is pretty easy, while the last 68 miles of the drive is filled with curves. They drive at an average of 9 miles per hour faster for the first 50 miles of the trip. The entire trip takes 3 hours. How fast did Joe and Nancy drive for the first 50 miles of the trip?

42. **A Race** Dirk and Garret decide to have a 40-mile bicycle race. During the race, Dirk averages 2 miles per hour faster than Garret and beats Garret by $\frac{2}{3}$ of an hour. What was Dirk's average speed?

Extending the Concepts

43. **The Olympics** The current world record holder in the 100-meter dash is Usain Bolt, with a time of 9.58 seconds. In 1999, Maurice Green set the world record in the 100-meter dash with a time of 9.79 seconds. If these two athletes ran in the same race repeating their respective times, by how many meters would Bolt beat Greene?

Explaining the Concepts

44. When solving work problems, we assume that each individual works at a constant rate and that there is no gain or loss of efficiency when additional individuals are added to the job. Explain what "no gain or loss of efficiency" means. Do you think this assumption is reasonable? If not, then why do we make the assumption?

Synthesis Review

In Problems 45–50, use the Laws of Exponents to simplify each expression.

45. $(a^3)^5$

46. $a(a^2 b^{-3})^3$

47. $(ab)^{-2} \cdot \left(\frac{a^3}{b^2} \right)^2$

48. $\left(\frac{13a^5 b^2}{ab^{-7}} \right)^0$

49. $\left(\frac{3m^3 n^{-1}}{mn^5} \right)^{-2}$

50. $\left(\frac{12pq^{-3}}{3p^4 q^{-4}} \right)^2$

5.7 Variation

Objectives

1. Model and Solve Direct Variation Problems
2. Model and Solve Inverse Variation Problems
3. Model and Solve Joint Variation and Combined Variation Problems

Are You Ready for This Section?

Before getting started, take this readiness quiz. If you get a problem wrong, go back to the section cited and review the material.

R1. Solve: $30 = 5x$ [Section 1.1, pp. 50–53]

R2. Solve: $4 = \dfrac{k}{3}$ [Section 1.1, pp. 50–53]

R3. Graph: $y = 3x$ [Section 1.6, pp. 113–114]

▶ ❶ Model and Solve Direct Variation Problems

Often two variables are related in terms of proportions. For example, we say "Revenue is proportional to sales" or "Force is proportional to acceleration." When we say that one variable is proportional to another variable, we are talking about *variation*. **Variation** refers to how one quantity varies in relation to some other quantity. Quantities may vary *directly*, *inversely*, or *jointly*. We will discuss direct variation first.

In Words

If y is directly proportional to x, then y and x are related through a linear equation whose y-intercept is $(0, 0)$ and whose slope is k, the constant of proportionality.

> **Definition**
>
> If x and y represent two quantities, then we say that y **varies directly** with x, or y is **directly proportional to** x, if there is a nonzero number k such that
>
> $$y = kx$$
>
> The number k is called the **constant of proportionality** or the **constant of variation**.

Figure 10

$y = kx, k > 0, x \geq 0$

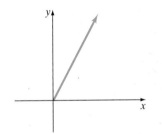

If y varies directly with x, then y is a linear function of x. The graph in Figure 10 shows the relationship between y and x if y varies directly with x and $k > 0, x \geq 0$. The constant of proportionality is the slope of the line, and the y-intercept is $(0, 0)$.

If two quantities vary directly, then knowing the value of each quantity in one instance enables us to write a formula that is true in all cases.

EXAMPLE 1 **Hooke's Law**

Hooke's Law states that the force (or weight) on a spring is directly proportional to the length that the spring stretches from its "at rest" position. That is, $F = kx$, where F is the force exerted, x is the extension of the spring, and k is the proportionality constant that varies from spring to spring.

(a) Suppose a 20-pound weight causes a spring to stretch 10 inches. See Figure 11. Find the constant of proportionality, k.

Figure 11

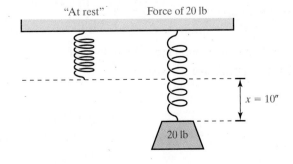

Ready?...Answers   **R1.** $\{6\}$

R2. $\{12\}$

R3.

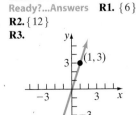

(b) Write the relation between force and stretch length using function notation.

(c) What amount of weight would make the spring stretch 8 inches?

(d) Graph the relation between force and the length the spring stretches.

Solution

(a) We know that $F = kx$ and that $F = 20$ pounds when $x = 10$ inches. We can use the given information to find k, the constant of proportionality.

$$F = kx$$
$$20 = k(10)$$
$$k = 2$$

(b) Because $k = 2$, we know that $F = 2x$. Using function notation, we write

$$F(x) = 2x$$

(c) If the spring stretches $x = 8$ inches, then the force exerted is

$$F(8) = 2(8) = 16 \text{ pounds}$$

(d) Figure 12 shows the relation between force and the length of the spring. ●

Figure 12

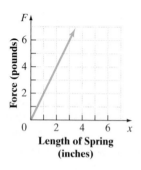

Length of Spring (inches)

(EXAMPLE 2) **Car Payments**

Suppose Dulce just purchased a used car for $10,000. She decides to put $1000 down on the car and borrows the remaining $9000. The bank lends Dulce $9000 at 4.9% interest for 48 months. Her monthly payment is $206.86. The monthly payment p on a car varies directly with the amount borrowed b.

(a) Find a function that relates the monthly payment p to the amount borrowed b for any car loan with the same terms.

(b) Suppose that Dulce put $2000 down on the car instead. What would her monthly payment be?

(c) Graph the relation between monthly payment and amount borrowed.

Solution

(a) Because p varies directly with b, we know that for some constant k,

$$p = kb$$

Because $p = \$206.86$ when $b = \$9000$, it follows that

$$206.86 = k(9000)$$

Divide by 9000: $k = 0.022984$

We rewrite the equation as

$$p = 0.022984b$$

We write this as a linear function:

$$p(b) = 0.022984b$$

(b) When $b = \$8000$,

$$p(8000) = 0.022984(8000)$$
$$= \$183.87$$

If Dulce put $2000 down, her monthly payment would be $183.87.

(c) Figure 13 shows the relation between the monthly payment p and the amount borrowed b. ●

Work Smart

To avoid round-off error, do not round the value of k to fewer than five decimal places.

Figure 13

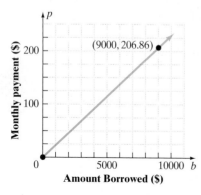

Amount Borrowed ($)

1. _____ refers to how one quantity varies in relation to some other quantity.

2. If x and y are two quantities, then y is directly proportional to x if there is a nonzero number k such that _____.

3. The cost of gas C varies directly with the number of gallons pumped, g. Suppose that the cost of pumping 8 gallons of gas is \$33.60.

   (a) Find a function that relates the cost of gas C to the number of gallons pumped g.

   (b) Suppose that 4.6 gallons are pumped into your car. What will the cost be?

   (c) Graph the relation between cost and number of gallons pumped.

▶ ❷ Model and Solve Inverse Variation Problems

We now discuss another kind of variation.

Figure 14

$$y = \frac{k}{x}, k > 0, x > 0$$

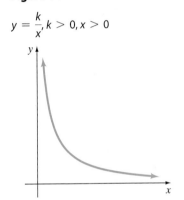

> **Definition**
>
> If x and y represent two quantities, then we say that y **varies inversely** with x, or y is **inversely proportional to** x, if there is a nonzero number k such that
>
> $$y = \frac{k}{x}$$
>
> where k is the constant of variation (or constant of proportionality).

The graph in Figure 14 illustrates the relationship between y and x if y varies inversely with x, where $k > 0$ and $x > 0$. Notice from the graph that as x increases, the value of y decreases.

EXAMPLE 3 | Weight That Can Be Supported by a Beam

The weight W that can be safely supported by a 2-inch by 4-inch (2-by-4) piece of lumber varies inversely with its length l. See Figure 15.

Figure 15

(a) Experiments indicate that the maximum weight a 12-foot pine 2-by-4 can support is 400 pounds. Find a function that relates the maximum weight to the length l for any pine 2-by-4.

(b) Determine the maximum weight a 15-foot pine 2-by-4 can sustain.

Solution

(a) Because W varies inversely with l, we know that for some constant k,

$$W = \frac{k}{l}$$

Because $W = 400$ when $l = 12$,

$$400 = \frac{k}{12}$$

Multiply both sides by 12: $\qquad k = 4800$

Substitute into $W = \dfrac{k}{l}$: $\qquad W = \dfrac{4800}{l}$

Write as a function: $\quad W(l) = \dfrac{4800}{l}$

(b) When $l = 15$ feet,

$$W(15) = \frac{4800}{15}$$

$$= 320 \text{ pounds}$$

The maximum weight that a 15-foot pine 2-by-4 can sustain is 320 pounds.   ●

Quick ✓

4. Suppose we let x and y represent two quantities. We say that y varies inversely with x, or y is inversely proportional to x, if there is a nonzero number k such

that

_____.

5. Suppose that y varies inversely with x for $x > 0$.

(a) Find an equation that relates x and y if $y = 2$ when $x = 3$.

(b) Use the equation in part (a) to find y when $x = 4$.

6. The rate of vibration (in oscillations per second) V of a string under constant tension varies inversely with the length l.

(a) If a string is 30 inches long and vibrates 500 times per second, find a function that relates the rate of vibration of a string to its length.

(b) What is the rate of vibration of a string that is 50 inches long?

▶ ❸ **Model and Solve Joint Variation and Combined Variation Problems**

When a variable quantity Q is proportional to the product of two or more other variables, we say that Q **varies jointly** with these quantities. For example, we can read $y = kxz$ as "y varies jointly with x and z." When direct and inverse variation occur at the same time, we have **combined variation**. For example, we can read $y = \dfrac{kx}{z}$ as "y varies directly with x and inversely with z." Similarly, $y = \dfrac{kmn}{p}$ can be read as "y varies jointly with m and n and inversely with p."

EXAMPLE 4   **Force of the Wind—Joint Variation**

The force F of the wind on a flat surface positioned at a right angle to the direction of the wind varies jointly with the area A of the surface and the square of the speed v of the wind. A wind of 30 miles per hour blowing on a window measuring 4 feet by 5 feet has a force of 150 pounds. What is the force on a window measuring 2 feet by 6 feet caused by a hurricane-force wind of 100 miles per hour?

Solution

Because F varies jointly with A and the square of v, we know that for some constant k,

$$F = kAv^2$$

The area A of the window is $(4 \text{ feet})(5 \text{ feet}) = 20$ square feet. Substitute $F = 150$, $A = 20$, and $v = 30$ to get

$$150 = k(20)(30^2)$$

We solve this equation for k.

$$150 = 18{,}000k$$

$$\text{Divide by 18,000:} \quad k = \frac{1}{120}$$

$$\text{Substitute into } F = kAv^2: \quad F = \frac{1}{120}Av^2$$

Thus, for a 100-mile-per-hour wind blowing on a window whose area is $A = (2\text{ feet})(6\text{ feet}) = 12$ square feet, the force F is

$$F = \frac{1}{120}(12)(100)^2$$

$$= 1000 \text{ pounds}$$

Quick ✓

7. The equation $t = ksp$ is an example of ____ variation.

8. The kinetic energy K of an object varies jointly with its mass and the square of its velocity. The kinetic energy of a 110-kg linebacker running at 9 meters per second is 4455 joules. Determine the kinetic energy of a 140-kg linebacker running at 5 meters per second.

EXAMPLE 5   Centripetal Force—Combined Variation

The force required to keep an object traveling in a circular motion is called centripetal force. The centripetal force F required to keep an object of a fixed mass in circular motion varies directly with the square of the velocity v of the object and inversely with the radius r of the circle. See Figure 16.

Figure 16

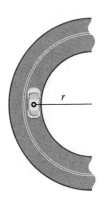

The force required to keep a car traveling 30 meters per second on a circular road with radius 50 meters is 21,600 newtons. Determine the force required to keep the same car on a circular road if it is traveling 40 meters per second on a road whose radius is 30 meters.

Solution

The force varies directly with the square of the velocity of the object and inversely with the radius of the circle. Then for some constant k, we have combined variation with

$$F = \frac{kv^2}{r}$$

Because $F = 21,600$ when $v = 30$ and $r = 50$,

$$21,600 = \frac{k \cdot 30^2}{50}$$

Solving for k, we find that $k = 1200$. Therefore,

$$F = \frac{1200v^2}{r}$$

For a car traveling 40 meters per second on a circular road with radius 30 meters, the force required to keep the car on the road is

$$F = \frac{1200 \cdot 40^2}{30}$$

$$= 64,000 \text{ newtons}$$

Quick ✓

9. When direct and inverse variation occur at the same time, we have _____ _____.

10. The electrical resistance R of a wire varies directly with the length l of the wire and inversely with the square of the diameter d of the wire. If a wire 432 feet long and 4 millimeters in diameter has a resistance of 1.24 ohms, find the resistance in a wire that is 282 feet long with a diameter of 3 millimeters. Round your answer to two decimal places.

5.7 Exercises MathLab®

PRACTICE

Exercise numbers in green
are complete video solutions
in MyMathLab

Problems **1–10** *are the* Quick ✓s *that follow the* EXAMPLES.

Building Skills

In Problems 11–16, (a) find the constant of proportionality k,
(b) write the linear function relating the two variables, and (c) find
the quantity indicated. See Objective 1

11. Suppose that y varies directly with x. When $x = 5$, $y = 30$. Find y when $x = 7$.

12. Suppose that y varies directly with x. When $x = 3$, $y = 15$. Find y when $x = 5$.

13. Suppose that y is directly proportional to x. When $x = 7$, $y = 3$. Find y when $x = 28$.

14. Suppose that y is directly proportional to x. When $x = 20$, $y = 4$. Find y when $x = 35$.

15. Suppose that y is directly proportional to x. When $x = 8$, $y = 4$. Find y when $x = 30$.

16. Suppose that y is directly proportional to x. When $x = 12$, $y = 8$. Find y when $x = 20$.

In Problems 17–20, (a) find the constant of proportionality k,
(b) write the function relating the two variables, and (c) find the
quantity indicated. See Objective 2.

17. Suppose that y varies inversely with x. When $x = 10$, $y = 2$. Find y if $x = 5$.

18. Suppose that y varies inversely with x. When $x = 3$, $y = 15$. Find y if $x = 5$.

19. Suppose that y is inversely proportional to x. When $x = 7$, $y = 3$. Find y if $x = 28$.

20. Suppose that y is inversely proportional to x. When $x = 20$, $y = 4$. Find y if $x = 35$.

In Problems 21–24 (a) find the constant of proportionality k,
(b) write the function relating the variables, and (c) find the
quantity indicated. See Objective 3.

21. Suppose that y varies jointly with x and z. When $y = 10$, $x = 8$ and $z = 5$. Find y if $x = 12$ and $z = 9$.

22. Suppose that y varies jointly with x and z. When $y = 20$, $x = 6$ and $z = 10$. Find y if $x = 8$ and $z = 15$.

23. Suppose that Q varies directly with x and inversely with y. When $Q = \dfrac{13}{12}$, $x = 5$ and $y = 6$. Find Q if $x = 9$ and $y = 4$.

24. Suppose that Q varies directly with x and inversely with y. When $Q = \dfrac{14}{5}$, $x = 4$ and $y = 3$. Find Q if $x = 8$ and $y = 3$.

Applying the Concepts

25. Mortgage Payments The monthly payment p on a mortgage varies directly with the amount borrowed b. Suppose that you decide to borrow \$120,000 using a 30-year mortgage at 5.75% interest. You are told that your payment is \$700.29.

   (a) Write a linear function that relates the monthly payment p to the amount borrowed b for a mortgage with the same terms.

   (b) Assume that you have decided to buy a more expensive home that requires you borrow \$140,000. What will your monthly payment be?

   (c) Graph the relation between monthly payment and amount borrowed.

26. Mortgage Payments The monthly payment p on a mortgage varies directly with the amount borrowed b. Suppose that you decide to borrow \$120,000 using a 15-year mortgage at 5.5% interest. You are told that your payment is \$980.50.

   (a) Write a linear function that relates the monthly payment p to the amount borrowed b for a mortgage with the same terms.

   (b) Assume that you have decided to buy a more expensive home that requires you borrow \$150,000. What will your monthly payment be?

   (c) Graph the relation between monthly payment and amount borrowed.

27. Cost Function The cost C of purchasing chocolate-covered almonds varies directly with the weight w in pounds. Suppose that the cost of purchasing 5 pounds of chocolate-covered almonds is \$28.

   (a) Write a linear function that relates the cost C to the number of pounds of chocolate-covered almonds purchased w.

   (b) What would it cost to purchase 3.5 pounds of chocolate-covered almonds?

   (c) Graph the relation between cost and weight.

28. Conversion Suppose that you are planning a trip to Europe, so you need to obtain some euros. The amount received in euros varies directly with the amount in U.S. dollars. Your friend just converted \$600 into 405 euros.

   (a) Write a linear function that relates the number of euros E to the number of U.S. dollars d.

(b) If you convert $700 into euros, how many euros will you receive?

(c) Graph the relation between euros and U.S. dollars.

29. Falling Objects The velocity of a falling object (ignoring air resistance) v is directly proportional to the time t of the fall. If, after 2 seconds, the velocity of the object is 64 feet per second, what will its velocity be after 3 seconds?

30. Circumference of a Circle The circumference of a circle C is directly proportional to its radius r. If the circumference of a circle whose radius is 5 inches is 10π inches, what is the circumference of a circle whose radius is 8 inches?

31. Demand Suppose that the demand D for candy at the movie theater is inversely related to the price p.

(a) When the price of candy is $2.50 per bag, the theater sells 150 bags of candy. Express the demand for candy as a function of its price.

(b) Determine the number of bags of candy that will be sold if the price is raised to $3 a bag.

32. Driving to School The time t that it takes you to get to school varies inversely with your average speed s.

(a) Suppose that it takes you 30 minutes to drive to school when your average speed is 35 miles per hour. Express your driving time to school as a function of your average speed.

(b) Suppose that your average speed driving to school is 30 miles per hour. How long will it take you to get to school?

33. Pressure The volume of a gas V held at a constant temperature in a closed container varies inversely with its pressure P. If the volume of a gas is 600 cubic centimeters (cc) when the pressure is 150 millimeters of mercury (mm Hg), find the volume when the pressure is 200 mm Hg.

34. Resistance The current i in a circuit is inversely proportional to its resistance R measured in ohms. Suppose that when the current in a circuit is 30 amperes, the resistance is 8 ohms. Find the current in the same circuit when the resistance is 10 ohms.

35. Weight The weight of an object above the surface of Earth varies inversely with the square of the distance from the center of Earth. Maria weighs 120 pounds when she is on the surface of Earth (3960 miles from the center). Determine Maria's weight if she is at the top of Mount McKinley (3.8 miles from the surface of Earth).

36. Intensity of Light The intensity I of light (measured in foot-candles) varies inversely with the square of the distance from the bulb. Suppose the intensity of a 100-watt light bulb at a distance of 2 meters is 0.075 foot-candle. Determine the intensity of the bulb at a distance of 3 meters.

37. Drag Force When an object moves through air, a frictionlike drag force tends to slow the object down. The drag force D on a free-falling parachutist varies jointly with the surface area of the parachutist and the square of his velocity. The drag force on a parachutist with surface area 2 square meters falling at 40 meters per second is 1152 newtons. Find the drag force on a parachutist whose surface area is 2.5 square meters falling at 50 meters per second.

38. Kinetic Energy The kinetic energy K (measured in joules) of a moving object varies jointly with the mass of the object and the square of its velocity v. The kinetic energy of a linebacker weighing 110 kilograms and running at a speed of 8 meters per second is 3520 joules. Find the kinetic energy of a wide receiver weighing 90 kilograms and running at a speed of 10 meters per second.

39. Newton's Law of Gravitation According to Newton's law of universal gravitation, the force F of gravity between any two objects varies jointly with the masses of the objects m_1 and m_2 and inversely with the square of the distance between the objects r. The force of gravity between a 105-kg man and his 80-kg wife when they are separated by a distance of 5 meters is 2.24112×10^{-8} newton. Find the force of gravity between the man and his wife when they are 2 meters apart.

40. Electrical Resistance The electrical resistance of a wire varies directly with the length of the wire and inversely with the square of the diameter of the wire. If a wire 50 feet long and 3 millimeters in diameter has a resistance of 0.255 ohm, find the length of a wire of the same material whose resistance is 0.147 ohm and whose diameter is 2.5 millimeters.

41. Stress of Material The stress in the material of a pipe subject to internal pressure varies jointly with the internal pressure and internal diameter of the pipe and inversely with the thickness of the pipe. The stress is 100 pounds per square inch when the diameter is 5 inches, the thickness is 0.75 inch, and the internal pressure is 25 pounds per square inch. Find the stress when the internal pressure is 50 pounds per square inch, the diameter is 6 inches, and the thickness is 0.5 inch.

42. Gas Laws The volume V of an ideal gas varies directly with the temperature T and inversely with the pressure P. If a cylinder contains oxygen at a temperature of 300 kelvin (K) and a pressure of 15 atmospheres in a volume of 100 liters, what is the constant of proportionality k? If a piston is lowered into the cylinder, decreasing the volume occupied by the gas to 70 liters and raising the temperature to 315 K, what is the pressure?

Extending the Concepts

43. David and Goliath The force F (in newtons) required to maintain an object in a circular path varies jointly with the mass m (in kilograms) of the object and the square of its speed (measured in meters per second) and inversely with the radius r (in meters) of the circular path. Suppose that David has a rope that is 3 meters long. On the end of the rope he has attached a pouch that holds a 0.5-kilogram stone. Suppose that David is able to spin the rope in a circular motion at the rate of 50 revolutions per minute.

(a) The spinning rate of 50 revolutions per minute can be converted into an *angular velocity* ω (lowercase Greek letter omega) using the formula $\omega = 2\pi \cdot$ (revolutions per minute.) Write the spinning rate as an angular velocity rounded to two decimal places.

(b) Use the fact that $v = \omega r$, where r is the radius of the circle, to find the linear velocity (in meters per minute) of the stone if it were released. Now convert the linear velocity to meters per second.

(c) Suppose that the force on the rope required to keep the rock in a circular motion is 2.3 newtons. Use the result from part (b) to find the constant of proportionality k.

(d) David is fairly certain that he will require more force than this to beat Goliath in a battle, so he increases the circular motion to 80 revolutions per minute and increases the length of the rope to 4 meters. What is the force required to keep the stone in a circular motion?

44. Suppose that y is directly proportional to x^2. If x is doubled, what happens to the value of y?

Chapter 5 Activity: Correct the Quiz

Focus: Performing operations and solving equations and inequalities with rational expressions
Time: 20 minutes
Group size: 2

In this activity you will work as a team to grade the student quiz shown below. One of you will grade the odd questions, and the

other will grade the even questions. If an answer is correct, mark it correct. If an answer is wrong, mark it wrong and show the correct answer.

Once all of the quiz questions are graded, explain your results to each other and compute the final score for the quiz. Be prepared to discuss your results with the rest of the class.

Student Quiz

Name: Ima Student   Quiz Score: _____

(1) Multiply: $\dfrac{2x^3 + 54}{5x^2 + 5x - 30} \cdot \dfrac{6x + 12}{3x^2 - 9x + 27}$	Answer: $\dfrac{4(x + 2)}{5(x - 2)}$
(2) Subtract: $\dfrac{xy}{x^2 - y^2} - \dfrac{y}{x + y}$	Answer: $\dfrac{y^2}{x^2 - y^2}$
(3) Solve: $\dfrac{3}{x - 3} + \dfrac{4}{x} = \dfrac{-12}{x^2 - 3x}$	Answer: $\varnothing$
(4) Divide: $\dfrac{x^2 + x - 6}{5x^2 - 7x - 6} \div \dfrac{3x^2 + 13x + 12}{6x^2 + 17x + 12}$	Answer: $\dfrac{2}{5}$
(5) Joe can mow his lawn in 2 hours. Mike can mow the same lawn in 3 hours. If they work together, how long will it take them to mow the lawn?	Answer: 1 hr, 12 min
(6) Simplify: $\dfrac{\dfrac{1}{x + 5} - \dfrac{2}{x - 7}}{\dfrac{4}{x - 7} + \dfrac{1}{x + 5}}$	Answer: $\dfrac{x + 17}{5x + 13}$
(7) Add: $\dfrac{x}{x^2 + 10x + 25} + \dfrac{4}{x^2 + 6x + 5}$	Answer: $\dfrac{x^2 + 5x + 20}{(x + 5)(x + 1)}$
(8) Solve: $\dfrac{4x}{x - 3} \geq 5$	Answer: $(3, 15]$

Chapter 5 Review

Section 5.1 Multiplying and Dividing Rational Expressions

KEY CONCEPTS

- **Multiplying Rational Expressions**

  If $\dfrac{a}{b}$ and $\dfrac{c}{d}$, $b \neq 0, d \neq 0$, are two rational expressions, then $\dfrac{a}{b} \cdot \dfrac{c}{d} = \dfrac{ac}{bd}$.

- **Dividing Rational Expressions**

  If $\dfrac{a}{b}$ and $\dfrac{c}{d}$, $b \neq 0, c \neq 0, d \neq 0$, are two rational expressions, then $\dfrac{\dfrac{a}{b}}{\dfrac{c}{d}} = \dfrac{a}{b} \cdot \dfrac{d}{c} = \dfrac{ad}{bc}$.

KEY TERMS

Rational expression
Numerator
Denominator
Lowest terms
Simplified
Rational function

You Should Be Able To...	EXAMPLE	Review Exercises
❶ Determine the domain of a rational expression (p. 399)	Examples 1	1–4
❷ Simplify rational expressions (p. 400)	Example 2 and 3	5–10
❸ Multiply rational expression (p. 401)	Examples 4 and 5	11–16
❹ Divide rational expressions (p. 403)	Examples 6	17–26
❺ Work with rational functions (p. 404)	Examples 7 through 8	23–26

In Problems 1–4, state the domain of each rational expression.

1. $\dfrac{x-5}{3x-2}$

2. $\dfrac{a^2-16}{a^2-3a-28}$

3. $\dfrac{m-3}{m^2+9}$

4. $\dfrac{n^2+7n+10}{n^2-2n-8}$

In Problems 5–10, simplify each rational expression.

5. $\dfrac{6x+30}{x^2-25}$

6. $\dfrac{4y^2-28y}{2y^5-14y^4}$

7. $\dfrac{w^2-4w-21}{w^2+7w+12}$

8. $\dfrac{6a^2-7ab-3b^2}{10a^2-11ab-6b^2}$

9. $\dfrac{7-m}{3m^2-20m-7}$

10. $\dfrac{n^3-4n^2+3n-12}{n^2-8n+16}$

In Problems 11–22, multiply or divide each rational expression, as indicated. Simplify the product or quotient.

11. $\dfrac{4p^2}{p^2-3p-18} \cdot \dfrac{p+3}{8p}$

12. $\dfrac{q^2+6q}{6q+12} \cdot \dfrac{4q+8}{q^2+q-30}$

13. $\dfrac{x^3-4x^2}{x^2-4} \cdot \dfrac{x^2+4x-12}{x^3+2x^2}$

14. $\dfrac{y^2-3y-28}{y^3+4y^2} \cdot \dfrac{2y^2+10y}{y^2-12y+35}$

15. $\dfrac{6a^2+ab-b^2}{3a^2+2ab-b^2} \cdot \dfrac{3a^2+4ab+b^2}{4a^2-b^2}$

16. $\dfrac{m^2+m-20}{m^3-64} \cdot \dfrac{3m^2+12m+48}{m^2+3m-10}$

17. $\dfrac{\dfrac{4c^2}{3d^4}}{\dfrac{8c}{27d}}$

18. $\dfrac{\dfrac{6z-24}{7z+21}}{\dfrac{z-4}{z^2-9}}$

19. $\dfrac{\dfrac{x^2 - 11x + 30}{x^2 - 8x + 15}}{\dfrac{x^2 - 5x - 6}{x^2 + 8x + 7}}$

20. $\dfrac{\dfrac{m^2 + mn - 12n^2}{m^3 - 27n^3}}{\dfrac{m + 5n}{m^2 + 3mn + 9n^2}}$

21. $\dfrac{\dfrac{4p^3 - 4pq^2}{p^2 - 5pq - 24q^2}}{\dfrac{2p^3 + 4p^2q + 2pq^2}{p^2 - 7pq - 8q^2}}$

22. $\dfrac{\dfrac{15a^2 + 11a - 14}{25a^2 - 49}}{\dfrac{27a^3 - 8}{10a^2 + 11a - 35}}$

In Problems 23–26, use the functions

$$f(x) = \frac{2x^2 + 3x - 2}{x - 5}, g(x) = \frac{x^2 - 3x - 10}{2x - 1}, \text{ and}$$

$$h(x) = \frac{2x - 1}{x^2 + 9x + 14} \text{ to find each product or quotient.}$$

State the domain of each product or quotient.

23. $P(x) = f(x) \cdot g(x)$

24. $R(x) = g(x) \cdot h(x)$

25. $Q(x) = \dfrac{g(x)}{f(x)}$ **26.** $T(x) = \dfrac{f(x)}{h(x)}$

Section 5.2 Adding and Subtracting Rational Expressions

KEY CONCEPTS

KEY TERM

- **Adding/Subtracting Rational Expressions**

  If $\dfrac{a}{c}$ and $\dfrac{b}{c}, c \neq 0$, are two rational expressions, then $\dfrac{a}{c} + \dfrac{b}{c} = \dfrac{a + b}{c}$ and $\dfrac{a}{c} - \dfrac{b}{c} = \dfrac{a - b}{c}$.

  Note: If the rational expressions do not have a common denominator, then the least common denominator can be found using the steps on page 412. Then follow the steps listed on page 413 to add or subtract rational expressions with unlike denominators.

Least common denominator

You Should Be Able To...	EXAMPLE	Review Exercises
❶ Add or subtract rational expressions with a common denominator (p. 410)	Examples 1 and 2	27–32
❷ Find the least common denominator of two or more rational expressions (p. 411)	Example 3	33–36
❸ Add or subtract rational expressions with different denominators (p. 412)	Examples 4 through 7	37–50

In Problems 27–32, perform the indicated operation and simplify the result.

27. $\dfrac{4x}{x - 5} + \dfrac{3}{x - 5}$

28. $\dfrac{4y}{y - 3} - \dfrac{12}{y - 3}$

29. $\dfrac{a^2 - 2a - 4}{a^2 - 6a + 8} + \dfrac{4a - 20}{a^2 - 6a + 8}$

30. $\dfrac{3b^2 + 8b - 5}{2b^2 - 5b - 12} - \dfrac{2b^2 + 7b + 15}{2b^2 - 5b - 12}$

31. $\dfrac{5c^2 - 8c}{c - 8} + \dfrac{2c^2 + 16c}{8 - c}$

32. $\dfrac{2d^2 + d}{d^2 - 1} - \dfrac{d^2 + 1}{d^2 - 1} + \dfrac{d - 2}{d^2 - 1}$

In Problems 33–36, find the least common denominator.

33. $\dfrac{4}{9x^4}$ and $\dfrac{5}{12x^2}$

34. $\dfrac{2y + 1}{y - 9}$ and $\dfrac{3y}{y + 2}$

35. $\dfrac{3p + 4}{2p^2 - 3p - 20}$ and $\dfrac{7p^2}{2p^3 + 5p^2}$

36. $\dfrac{q - 4}{q^2 + 4q - 5}$ and $\dfrac{q - 6}{q^2 + 2q - 15}$

In Problems 37–48, perform the indicated operation and simplify the result.

37. $\dfrac{1}{mn^4} + \dfrac{4}{m^3 n^2}$

38. $\dfrac{3}{2xy^3} - \dfrac{7}{6x^2 y}$

39. $\dfrac{p}{p - q} - \dfrac{q}{p + q}$

40. $\dfrac{x + 8}{x^2 - 10x + 21} - \dfrac{x - 5}{x^2 - 3x - 28}$

41. $\dfrac{3}{y^2 - 2y + 1} - \dfrac{2}{y^2 + y - 2}$

42. $\dfrac{3a - 5b}{4a^2 - 9b^2} + \dfrac{4}{2a - 3b}$

43. $\dfrac{4x^2 - 10x}{x^2 - 9} + \dfrac{8x - 2x^2}{9 - x^2}$

44. $\dfrac{1}{n + 5} - \dfrac{n^2 - 10n}{n^3 + 125}$

45. $\dfrac{m + n}{m + 3n} - \dfrac{m - 4n}{m - 7n} + \dfrac{7mn + n^2}{m^2 - 4mn - 21n^2}$

46. $\dfrac{z^2 + 10z + 3}{z^2 - 9} - \dfrac{2z}{z - 3} + \dfrac{z}{z + 3}$

47. $\dfrac{y - 1}{y - 2} - \dfrac{y + 1}{y + 2} + \dfrac{y - 6}{y^2 - 4}$

48. $\dfrac{2a}{a^2 - 16} - \dfrac{1}{a - 4} - \dfrac{1}{a^2 + 2a - 8}$

49. $f(x) = \dfrac{5}{x - 4}$ and $g(x) = \dfrac{x}{x + 2}$

    (a) Find $S(x) = f(x) + g(x)$.
    (b) State the domain of $S(x)$.

50. $f(x) = \dfrac{x + 3}{2x^2 + x - 15}$ and $g(x) = \dfrac{x - 7}{4x^2 - 8x - 5}$

    (a) Find $D(x) = f(x) - g(x)$.
    (b) State the domain of $D(x)$.

Section 5.3 Complex Rational Expressions

KEY CONCEPT

- There are two methods that can be used to simplify a complex rational expression. The steps for Method I are presented on page 419, and the steps for Method II are presented on page 421.

KEY TERMS

Complex rational expression
Complex fraction
Simplify

You Should Be Able To...	EXAMPLE	Review Exercises
1 Simplify a complex rational expression by simplifying the numerator and denominator separately (Method I) (p. 418)	Examples 1, 2, and 5	51–54, 59–66
2 Simplify a complex rational expression using the least common denominator (Method II) (p. 420)	Examples 3 through 5	55–58, 59–66

In Problems 51–54, simplify each complex rational expression by using Method I (that is, by simplifying the numerator and denominator separately).

51. $\dfrac{x - \dfrac{1}{x}}{1 - \dfrac{1}{x}}$

52. $\dfrac{\dfrac{1}{x} - \dfrac{1}{y}}{\dfrac{1}{x^2} - \dfrac{1}{y^2}}$

53. $\dfrac{\dfrac{a}{b} - \dfrac{a - b}{a + b}}{\dfrac{a}{b} + \dfrac{a + b}{a - b}}$

54. $\dfrac{\dfrac{2}{a + 2} - 1}{\dfrac{1}{a + 2} + 1}$

In Problems 55–58, simplify each complex rational expression using Method II (that is, by using the least common denominator).

55. $\dfrac{\dfrac{3}{t} + \dfrac{4}{t^2}}{5 + \dfrac{1}{t^2}}$

56. $\dfrac{\dfrac{1}{a} - \dfrac{1}{b}}{\dfrac{b}{a} - \dfrac{a}{b}}$

57. $\dfrac{\dfrac{1}{z - 1} - \dfrac{1}{z}}{\dfrac{1}{z} - \dfrac{1}{z + 1}}$

58. $\dfrac{1 + \dfrac{x}{x + 1}}{\dfrac{2x + 1}{x - 1}}$

In Problems 59–66, simplify the complex rational expression using either Method I or Method II.

59. $\dfrac{\dfrac{x}{y} + 1}{\dfrac{x}{y} - 1}$

60. $\dfrac{\dfrac{a}{a - b} - \dfrac{b}{a + b}}{\dfrac{b}{a - b} + \dfrac{a}{a + b}}$

61. $\dfrac{\dfrac{1}{x - 2} - \dfrac{x}{x^2 - 4}}{1 - \dfrac{2}{x + 2}}$

62. $\dfrac{z - \dfrac{5z}{z + 5}}{z + \dfrac{5z}{z - 5}}$

63. $\dfrac{\dfrac{m - n}{m + n} + \dfrac{n}{m}}{\dfrac{m}{n} - \dfrac{m - n}{m + n}}$

64. $\dfrac{\dfrac{x + 4}{x - 2} - \dfrac{x - 3}{x + 1}}{5x^2 + 4x - 1}$

65. $\dfrac{3x^{-1} - 3y^{-1}}{(x + y)^{-1}}$

66. $\dfrac{2c^{-1} - (3d)^{-1}}{(6d)^{-1}}$

Section 5.4 Rational Equations

KEY CONCEPT

- The steps for solving any rational equation are given on page 428.

KEY TERMS

Rational equation
Extraneous solution

You Should Be Able To...	EXAMPLE	Review Exercises
1 Solve equations containing rational expressions (p. 427)	Examples 1 through 5	67–78
2 Solve equations involving rational functions (p. 431)	Examples 6 and 7	79, 80

In Problems 67–78, solve each equation. Be sure to verify your results.

67. $\dfrac{2}{z} - \dfrac{1}{3z} = \dfrac{1}{6}$

68. $\dfrac{4}{m - 4} = \dfrac{-5}{m + 2}$

69. $m - \dfrac{14}{m} = 5$

70. $\dfrac{2}{n + 3} = \dfrac{1}{n - 3}$

71. $\dfrac{s}{s - 1} = 1 + \dfrac{2}{s}$

72. $\dfrac{3}{x^2 - 7x + 10} + 2 = \dfrac{x - 4}{x - 5}$

73. $\dfrac{1}{k - 1} + \dfrac{1}{k + 2} = \dfrac{3}{k^2 + k - 2}$

74. $x + \dfrac{3x}{x - 3} = \dfrac{9}{x - 3}$

75. $\dfrac{2}{a + 3} - \dfrac{4}{a^2 - 4} = \dfrac{a + 1}{a^2 + 5a + 6}$

76. $\dfrac{2}{z^2 + 2z - 8} = \dfrac{1}{z^2 + 9z + 20} + \dfrac{4}{z^2 + 3z - 10}$

77. $\dfrac{x - 3}{x + 4} = \dfrac{14}{x^2 + 6x + 8}$

78. $\dfrac{5}{y - 5} + 4 = \dfrac{3y - 10}{y - 5}$

79. For the function $f(x) = \dfrac{6}{x - 2}$, solve $f(x) = 2$. What point(s) are on the graph of f?

80. For the function $g(x) = x - \dfrac{21}{x}$, solve $g(x) = 4$. What point(s) are on the graph of g?

Section 5.5 Rational Inequalities

KEY CONCEPT

- The steps for solving any rational inequality are given on page 437.

KEY TERM

Rational inequality

You Should Be Able To...	EXAMPLE	Review Exercises
1 Solve a rational inequality (p. 435)	Examples 1 and 2	81–90

In Problems 81–88, solve each rational inequality. Graph the solution set

81. $\dfrac{x - 4}{x + 2} \geq 0$

82. $\dfrac{y - 5}{y + 4} < 0$

83. $\dfrac{4}{z^2 - 9} \leq 0$

84. $\dfrac{w^2 + 5w - 14}{w - 4} < 0$

85. $\dfrac{m - 5}{m^2 + 3m - 10} \geq 0$

86. $\dfrac{4}{n - 2} \leq -2$

87. $\dfrac{a + 1}{a - 2} > 3$

88. $\dfrac{4}{c - 2} - \dfrac{3}{c} < 0$

In Problems 89 and 90, for each function, find the values of x that satisfy the given condition.

89. Solve $Q(x) < 0$ if $Q(x) = \dfrac{2x + 3}{x - 4}$.

90. Solve $R(x) \geq 0$ if $R(x) = \dfrac{x + 5}{x + 1}$.

Section 5.6 Models Involving Rational Expressions

KEY TERMS

Ratio	Proportion	Similar	Constant rate

You Should Be Able To...	EXAMPLE	Review Exercises
❶ Solve for a variable in a rational expression (p. 441)	Example 1	91–94
❷ Model and solve ratio and proportion problems (p. 442)	Examples 2 and 3	95–98
❸ Model and solve work problems (p. 444)	Examples 4 and 5	99–102
❹ Model and solve uniform motion problems (p. 446)	Example 6	103–106

In Problems 91–94, solve each formula for the indicated variable.

91. Electronics (Capacitance) Solve $\dfrac{1}{C_1} + \dfrac{1}{C_2} = \dfrac{1}{C}$ for C.

92. Chemistry (Ideal Gas Law) Solve $\dfrac{P_1 V_1}{T_1} = \dfrac{P_2 V_2}{T_2}$ for T_2.

93. Physics (Kepler's Third Law) Solve $T = \dfrac{4\pi^2 a^2}{MG}$ for G.

94. Statistics (z-score) Solve $z = \dfrac{x - \mu}{\sigma}$ for x.

In Problems 95–98, solve each proportion problem.

95. Suppose that $\triangle ABC$ is similar to $\triangle DEF$, as shown in the figure. Find the lengths of AB and DF.

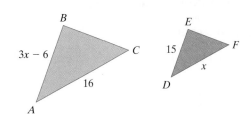

96. Casting Shadows At a particular time of day, a pine tree casts a 30-foot shadow. At the exact same time, a nearby 5-foot post casts an 8-foot shadow. Find the height of the tree.

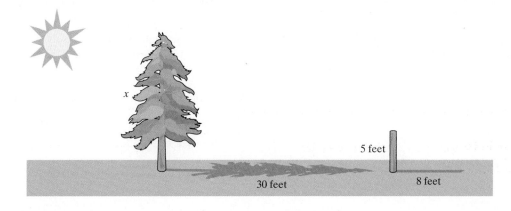

97. Nutrition According to the nutrition facts on a box of Honey Nut Chex® (source: *General Mills*), a $\frac{3}{4}$-cup serving contains 26 grams of total carbohydrates. How many grams of total carbohydrates are in a 3-cup bowl of the cereal?

98. Payroll One day, Jeri earns $48.75 for 5 hours of work. How much will she earn for 8 hours of work?

In Problems 99–102, solve each work problem.

99. Filling a Tank One pipe can fill a tank in 48 minutes. Another pipe can fill the tank in 1 hour and 12 minutes. If both pipes are used, how long will it take to fill the tank?

100. Mowing the Lawn Together, Diane and Craig can mow their lawn in 1 hour and 10 minutes. Working alone, Diane can mow the lawn in 2 hours. How long will it take Craig to mow the lawn when working alone?

101. Carpeting Together, Rick and John can carpet a large room in 12 hours. Alone, Rick can carpet the same size room in 7 hours less time than John. How long would it take Rick to carpet the same size room if working alone? How long would it take John if working alone?

102. Draining a Sink A faucet can fill a sink in 1 minute when the drain is plugged, but when the drain is unplugged, it takes 1 minute and 30 seconds to fill the sink. With the faucet off, how long would it take the drain to empty a full sink?

In Problems 103–106, solve each motion problem.

103. Pleasure Flight In his private plane, Nick can fly 180 miles per hour if the wind is not blowing. One day, Nick took a pleasure flight. He flew 100 miles directly against the wind and then returned (flying with the wind) to his point of origin. If the total time of the flight was 1 hour and 15 minutes, what was the speed of the wind?

104. Boating In his motorboat, Jesse can travel 20 miles downstream in the same amount of time it takes him to travel 10 miles upstream. If the speed of the current is 5 miles per hour, how fast can Jesse's motorboat travel in still water?

105. Running/Walking To stay in shape, Todd first runs 3 miles and then walks 1 mile every morning. Todd's average running speed is 4 times his average walking speed. If Todd spends a total of 35 minutes running and walking each morning, find the average speed at which he walks and the average speed at which he runs.

106. Road Trip Because of heavy traffic, Danielle averaged only 30 miles per hour for the first 20 miles of her trip. If she averaged 50 miles per hour for the entire 100-mile trip, what was her average speed for the last 80 miles?

Section 5.7 Variation

You Should Be Able To...	EXAMPLE	Review Exercises
❶ Model and solve direct variation problems (p. 451)	Examples 1 and 2	107, 108, 113, 114
❷ Model and solve inverse variation problems (p. 453)	Example 3	109, 111, 115, 116
❸ Model and solve joint variation and combined variation problems (p. 454)	Examples 4 and 5	110, 112, 117, 118

In Problems 107–112, (a) find the constant of proportionality k, (b) write the function relating the variables, and (c) find the quantity indicated.

107. Suppose that y varies directly with x. If $y = 30$ when $x = 6$, find y when $x = 10$.

108. Suppose that y varies directly with x. If $y = 18$ when $x = -3$, find y when $x = 8$.

109. Suppose that y varies inversely with x. If $y = 15$ when $x = 4$, find y when $x = 5$.

110. Suppose that y varies jointly with x and z. If $y = 45$ when $x = 6$ and $z = 10$, find y when $x = 8$ and $z = 7$.

111. Suppose that s varies inversely with the square of t. If $s = 18$ when $t = 2$, find s when $t = 3$.

112. Suppose that w varies directly with x and inversely with z. If $w = \frac{4}{3}$ when $x = 10$ and $z = 12$, find w when $x = 9$ and $z = 16$.

113. Snow-Water Equivalent The amount of water in snow is directly proportional to the depth of the snow. Suppose the amount of water in 40 inches of snow is 4.8 inches. How much water is contained in 50 inches of snow?

114. Car Payments Roberta is buying a car. The monthly payment p for the car varies directly with the amount borrowed b. Suppose the dealership tells her that if she borrows $15,000, her monthly payment will be $293.49. What will be Roberta's monthly payment if she borrows $18,000?

115. Radio Signals The frequency of a radio signal varies inversely with the wavelength. A signal of 800 kilohertz has a wavelength of 375 meters. What frequency has a signal of wavelength 250 meters?

116. Ohm's Law The electrical current flowing through a wire varies inversely with the resistance of the wire.

If the current is 8 amperes when the resistance is 15 ohms, for what resistance will the current be 10 amperes?

117. Volume of a Cylinder The volume V of a right circular cylinder varies jointly with the height h and the square of the diameter d. If the volume of a cylinder is 231 cubic centimeters when the diameter is 7 centimeters and the height is 6 centimeters, find the volume when the diameter is 8 centimeters and the height is 14 centimeters.

118. Volume of a Pyramid The volume V of a pyramid varies jointly with the base area B and the height h. If the volume of a pyramid is 270 cubic inches when the base area is 81 square inches and the height is 10 inches, find the volume of a pyramid with base area 125 square inches and height 9 inches.

Chapter 5 Test   *Step-by-step test solutions are found on the Chapter Test Prep Videos available in* MyMathLab® *or on* You Tube.

1. Determine the domain of $f(x) = \dfrac{2x + 1}{2x^2 - 13x - 7}$.

In Problems 2 and 3, simplify each rational expression.

2. $\dfrac{2m^2 + 5m - 12}{3m^2 + 11m - 4}$ **3.** $\dfrac{2b - 3a}{3a^2 + 10ab - 8b^2}$

In Problems 4–7, perform the indicated operations.

4. $\dfrac{4x^2 - 12x}{x^2 - 9} \cdot \dfrac{2x^2 + 11x + 15}{8x^3 - 32x^2}$

5. $\dfrac{\dfrac{y^2 + 2y - 8}{4y^2 - 5y - 6}}{\dfrac{3y^2 - 14y - 5}{4y^2 - 17y - 15}}$

6. $\dfrac{3p^2 + 3pq}{p^2 - q^2} - \dfrac{3p - 2q}{p - q}$

7. $\dfrac{9c + 2}{3c^2 - 2c - 8} + \dfrac{7}{3c^2 + c - 4}$

In Problems 8 and 9, use the functions $f(x) = \dfrac{3x}{x^2 - 4}$,

$g(x) = \dfrac{6}{x^2 + 2x}$, *and* $h(x) = \dfrac{9x^2 - 45x}{x^2 - 2x - 8}$ *to find each sum or quotient. State the domain of each.*

8. $Q(x) = \dfrac{f(x)}{h(x)}$ **9.** $S(x) = f(x) + g(x)$

In Problems 10 and 11, simplify each complex rational expression using either Method I or Method II.

10. $\dfrac{1 - \dfrac{1}{a}}{1 - \dfrac{1}{a^2}}$ **11.** $\dfrac{\dfrac{5}{d + 2} - \dfrac{1}{d - 2}}{\dfrac{3}{d + 2} - \dfrac{6}{d - 2}}$

In Problems 12 and 13, solve each rational equation. Be sure to verify the results.

12. $\dfrac{1}{6x} - \dfrac{1}{3} = \dfrac{5}{4x} + \dfrac{3}{4}$

13. $\dfrac{7n}{n + 3} + \dfrac{21}{n - 3} = \dfrac{126}{n^2 - 9}$

14. Solve $\dfrac{x + 5}{x - 2} \geq 3$. Graph the solution set.

15. Electronics (Coulomb's Law) Solve $\dfrac{1}{F} = \dfrac{D^2}{kq_1q_2}$ for k.

16. Printing Documents A particular laser printer can print out a 10-page document in 25 seconds. How long will it take to print out a 48-page document?

17. Cleaning House Linnette can clean the house in 4 hours. Her husband Darrell can do the same job in 6 hours. If the two work together, how long will it take them to clean the house?

18. Kayaking Chuck kayaked 4 miles upstream in the same time it took him to kayak 10 miles downstream. If Chuck can average 7 miles per hour in still water, what was the rate of the current?

19. Using a Lever Using a lever, the force F required to lift a weight is inversely proportional to the length l of the force arm of the lever (assuming all other factors are constant). If a force of 50 pounds is required to lift a granite boulder when the force arm length is 4 feet, how much force will be required to lift the boulder if the force arm length is 10 feet?

20. Lateral Surface Area of a Cylinder The lateral surface area L of a right circular cylinder varies jointly with its radius r and height h. If the lateral surface area of a cylinder with radius 7 centimeters and height 12 centimeters is 528 square centimeters, what would be the lateral surface area if the radius were 9 centimeters and the height were 14 centimeters?

Cumulative Review Chapters R–5

1. Evaluate: $\dfrac{4^3 - 6 \cdot 7 + 14}{4 - 1^2}$

2. Simplify: $2x(x - 3) + 4(x - 2) + 15$

3. Evaluate $\dfrac{3x^2 - 4x - 5}{x - 3}$ for $x = -2$.

4. Solve: $7x + 9 = 3x - 23$

5. Solve: $|3x + 7| \leq 8$

6. Determine the domain of $h(x) = \dfrac{x - 5}{2x^2 - 7x - 15}$.

7. For $f(x) = x^2 - 5x$, find each of the following.

   (a) $f(-3)$ **(b)** $f\left(\dfrac{1}{4}\right)$ **(c)** $f(x + 2)$

8. Graph the linear equation $4x + 3y = 15$.

9. Find the equation of the line that passes through the points $(-5, 1)$ and $(10, -8)$. Write your answer in either slope-intercept or standard form, whichever you prefer.

10. Find the equation of the line that passes through the point $(3, 5)$ and is perpendicular to the graph of $x + 4y = 20$. Write your answer in either slope-intercept or standard form, whichever you prefer.

11. The recommended dosage D of a prescription drug varies directly with a person's weight w. If the recommended dosage for a 125-pound person is 1500 milligrams, find the recommended dosage for a 180-pound person.

12. Write the system of linear equations that corresponds to the following augmented matrix.
$$\begin{bmatrix} 2 & -3 & | & 7 \\ 5 & 2 & | & 8 \end{bmatrix}$$

13. Perform each row operation on the given augmented matrix. $\begin{bmatrix} 1 & 1 & 0 & | & -3 \\ 0 & 2 & -1 & | & -1 \\ 5 & 0 & 1 & | & 1 \end{bmatrix}$

   (a) $R_3 = -5r_1 + r_3$ followed by

   (b) $R_2 = \dfrac{1}{2}r_2$

14. Solve the following system of equations
$$\begin{cases} x + 4y = -2 \\ 2x - 12y = -9 \end{cases}$$

15. Evaluate: $\begin{vmatrix} 2 & 0 & 4 \\ 1 & -1 & -2 \\ 2 & -2 & 3 \end{vmatrix}$

16. Graph the following system of linear inequalities:
$$\begin{cases} 3x + 2y < 8 \\ x - 4y \geq 12 \end{cases}$$

In Problems 17–20, perform the indicated operation.

17. $(3x^2 - 4xy + 7y^2) + (5x^2 - 9xy + 2y^2)$

18. $(5x^2 - 3x + 12) - (2x^2 - 4x - 15)$

19. $(2x - 3)(x^2 - 4x + 6)$

20. $\dfrac{4x^3 - 7x + 45}{2x + 5}$

In Problems 21–22, factor completely.

21. $x^4 + 5x^3 - 8x - 40$ **22.** $6x^2 + x - 15$

In Problems 23–24, perform the indicated operations. Be sure to simplify each result.

23. $\dfrac{3x^2 - 2x - 1}{3x^2 - 5x - 2} \cdot \dfrac{x^2 - 9x + 14}{x^2 + x - 2}$

24. $\dfrac{3}{x^2 + x - 6} - \dfrac{2}{x^2 + 2x - 3}$

25. Grading Tests Bill can grade a set of test papers in 80 minutes. Karl can grade the same number of test papers in 120 minutes. How long will it take Bill and Karl to grade the test papers if they work together?

Getting Ready for Chapter 6:
Square Roots

Objectives

❶ Evaluate Square Roots of Perfect Squares

❷ Determine Whether a Square Root Is Rational, Irrational, or Not a Real Number

❸ Find Square Roots of Variable Expressions

In Words

Taking the square root of a number "undoes" squaring a number.

In Words

The notation $b = \sqrt{a}$ means "give me the number b greater than or equal to 0 whose square is a."

In Section R.4, we introduced exponents. Exponents indicate repeated multiplication. For example, 4^2 means $4 \cdot 4$, so $4^2 = 16$; $(-6)^2$ means $(-6) \cdot (-6) = 36$. Now, we will "undo" the process of raising a number to the second power and ask questions such as, "What number, or numbers, when squared, give me 16?"

▶ ❶ **Evaluate Square Roots of Perfect Squares**

A real number is squared when it is raised to the power 2. The inverse of squaring a number is finding the **square root.** For example, since $5^2 = 25$ and $(-5)^2 = 25$, the square roots of 25 are -5 and 5. The square roots of $\dfrac{16}{49}$ are $-\dfrac{4}{7}$ and $\dfrac{4}{7}$.

If we want only the positive square root of a number, we use the symbol $\sqrt{}$, called a **radical sign,** to denote the **principal square root,** or nonnegative (zero or positive) square root.

> **Definition**
>
> If a is a nonnegative real number, the nonnegative real number b such that $b^2 = a$ is the **principal square root** of a and is denoted by $b = \sqrt{a}$.

For example the positive square root of 25 is written $\sqrt{25} = 5$. We read $\sqrt{25} = 5$ as "the principal (or positive) square root of 25 is 5." If we want the negative square root of 25, we use the expression $-\sqrt{25} = -5$.

> **Properties of Square Roots**
>
> - Every positive real number has two square roots, one positive and one negative.
> - The square root of 0 is 0. That is, $\sqrt{0} = 0$.
> - We use the symbol $\sqrt{}$, called a radical sign, to denote the nonnegative square root of a real number. The nonnegative square root is called the principal square root.
> - The number under the radical sign is called the **radicand**. For example, the radicand in $\sqrt{25}$ is 25.
> - For any real number c, such that $c \geq 0$, $(\sqrt{c})^2 = c$. For example, $(\sqrt{4})^2 = 4$ and $(\sqrt{8.3})^2 = 8.3$.

To **evaluate** a square root, we ask ourselves, "What is the nonnegative number whose square is equal to the radicand?"

EXAMPLE 1 **Evaluating Square Roots**

Evaluate each square root.

(a) $\sqrt{36}$ (b) $\sqrt{\dfrac{1}{9}}$ (c) $\sqrt{0.01}$ (d) $(\sqrt{2.3})^2$

Solution

(a) Is there a positive number whose square is 36? Because $6^2 = 36$, $\sqrt{36} = 6$.

(b) $\sqrt{\dfrac{1}{9}} = \dfrac{1}{3}$ because $\left(\dfrac{1}{3}\right)^2 = \dfrac{1}{9}$.

(c) $\sqrt{0.01} = 0.1$ because $0.1^2 = 0.01$.

(d) $(\sqrt{2.3})^2 = 2.3$ because $(\sqrt{c})^2 = c$ when $c \geq 0$.

Figure 1

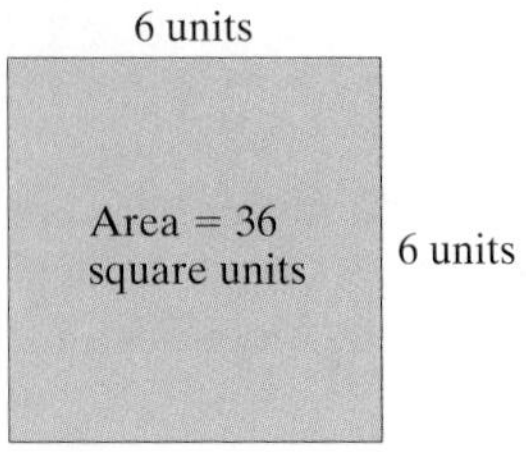

A rational number is a **perfect square** if it is the square of a rational number. Examples 1(a), (b), and (c) are square roots of perfect squares since $6^2 = 36$, $\left(\dfrac{1}{3}\right)^2 = \dfrac{1}{9}$, and $0.1^2 = 0.01$. We can think of perfect squares geometrically as shown in Figure 1, where we have a square whose area is 36 square units. The square root of the area, $\sqrt{36}$, gives us the length of each side of the square, 6 units.

Quick ✓

1. The symbol $\sqrt{}$ is called a _____ .

2. If a is a nonnegative real number, the nonnegative number b such that $b^2 = a$ is the _____ of a and is denoted by $b = \sqrt{a}$.

3. The square roots of 16 are ___ and _.

In Problems 4–8, evaluate each square root.

4. $\sqrt{81}$ 5. $\sqrt{900}$ 6. $\sqrt{\dfrac{9}{4}}$ 7. $\sqrt{0.16}$ 8. $(\sqrt{13})^2$

EXAMPLE 2 **Evaluating an Expression Containing Square Roots**

Evaluate each expression:

(a) $-4\sqrt{36}$ (b) $\sqrt{9} + \sqrt{16}$ (c) $\sqrt{9 + 16}$ (d) $\sqrt{64 - 4 \cdot 7 \cdot 1}$

Solution

(a) The expression $-4\sqrt{36}$ is asking us to find -4 times the positive square root of 36. We first find the positive square root of 36 and then multiply this result by -4.

$$-4\sqrt{36} = -4 \cdot 6$$
$$= -24$$

Work Smart

In Examples 2(b) and (c), notice that
$$\sqrt{9} + \sqrt{16} \neq \sqrt{9 + 16}$$
In general,
$$\sqrt{a} + \sqrt{b} \neq \sqrt{a + b}$$
The radical sign acts as a grouping symbol, so always simplify the radicand before taking the square root.

(b) $\sqrt{9} + \sqrt{16} = 3 + 4$
$$= 7$$

(c) $\sqrt{9 + 16} = \sqrt{25}$
$$= 5$$

(d) $\sqrt{64 - 4 \cdot 7 \cdot 1} = \sqrt{64 - 28}$
$$= \sqrt{36}$$
$$= 6$$

Quick ✓

In Problems 9–12, evaluate each expression.

9. $5\sqrt{9}$ 10. $\sqrt{36 + 64}$ 11. $\sqrt{36} + \sqrt{64}$ 12. $\sqrt{25 - 4 \cdot 3 \cdot (-2)}$

▷ **2 Determine Whether a Square Root Is Rational, Irrational, or Not a Real Number**

Not all radical expressions will simplify to a rational number. For example, because there is no rational number whose square is 5, $\sqrt{5}$ is not a rational number. In fact, $\sqrt{5}$ is an *irrational* number. Remember, an irrational number is a number that cannot be written as the quotient of two integers.

What if we wanted to evaluate $\sqrt{-16}$? Because any positive real number squared is positive, any negative real number squared is also positive, and 0 squared is 0, there is no real number whose square is -16. We conclude: **Negative real numbers do not have square roots that are real numbers!**

Work Smart

The square roots of negative real numbers are not real.

We summarize these points below.

> **More Properties of Square Roots**
>
> - The square root of a perfect square is a rational number.
> - The square root of a positive rational number that is not a perfect square is an irrational number. For example, $\sqrt{20}$ is an irrational number because 20 is not a perfect square.
> - The square root of a negative real number is not a real number. For example, $\sqrt{-2}$ is not a real number.

If a radical has a radicand that is not a perfect square, we can do one of two things:

1. Write a decimal approximation of the radical.

2. Simplify the radical using properties of radicals, if possible (Section 6.3).

Figure 2

```
√(5)
        2.236067977
```

(**EXAMPLE 3**) **Writing a Radical as a Decimal Using a Calculator**

Write $\sqrt{5}$ as a decimal rounded to two decimal places.

Solution

Figure 2 shows the result from a TI-84 Plus graphing calculator: $\sqrt{5} \approx 2.24$. ●

(**EXAMPLE 4**) **Determining Whether a Square Root of an Integer Is Rational, Irrational, or Not a Real Number**

Determine whether each square root is rational, irrational, or not a real number. Then evaluate each real square root. For each square root that is irrational, express the square root as a decimal rounded to two decimal places.

(a) $\sqrt{51}$ **(b)** $\sqrt{169}$ **(c)** $\sqrt{-81}$

Solution

Work Smart

$\sqrt{-81}$ is not a real number, but $-\sqrt{81}$ is a real number because $-\sqrt{81} = -9$. Note the placement of the negative sign!

(a) $\sqrt{51}$ is irrational because 51 is not a perfect square. That is, there is no rational number whose square is 51. Using a calculator, we find $\sqrt{51} \approx 7.14$.

(b) $\sqrt{169}$ is a rational number because $13^2 = 169$, so $\sqrt{169} = 13$.

(c) $\sqrt{-81}$ is not a real number. There is no real number whose square is -81. ●

Quick ✓

13. *True or False* Negative numbers do not have square roots that are real numbers.

In Problems 14–17, determine whether each square root is rational, irrational, or not a real number. Evaluate each square root that is rational. For each square root that is irrational, approximate the square root rounded to two decimal places.

14. $\sqrt{400}$ **15.** $\sqrt{40}$ **16.** $\sqrt{-25}$ **17.** $-\sqrt{196}$

❸ Find Square Roots of Variable Expressions

What is $\sqrt{4^2}$? Because $4^2 = 16$, we know that $\sqrt{4^2} = \sqrt{16} = 4$. This result suggests that $\sqrt{a^2} = a$ for any real number a. But wait. Does $\sqrt{(-4)^2} = -4$? No! $\sqrt{(-4)^2} = \sqrt{16} = 4$, so both $\sqrt{4^2}$ and $\sqrt{(-4)^2}$ equal 4. Regardless of whether the "a" in $\sqrt{a^2}$ is positive or negative, the result is positive. Therefore, $\sqrt{a^2} = a$ is incorrect. How can we fix our "formula"? In Section R.3, we learned that $|a|$ is a positive number if a is nonzero. From this, we have the following result:

In Words

The square root of a nonzero number squared will always be positive. The absolute value ensures this.

For any **real number** a,

$$\sqrt{a^2} = |a|$$

The bottom line is this—the square root of a variable expression raised to the second power is the absolute value of the variable expression.

EXAMPLE 5 **Evaluating Square Roots**

Evaluate each square root.

(a) $\sqrt{7^2}$ **(b)** $\sqrt{(-15)^2}$ **(c)** $\sqrt{x^2}$ **(d)** $\sqrt{(3x-1)^2}$
(e) $\sqrt{x^2 + 6x + 9}$

Solution

(a) $\sqrt{7^2} = 7$
(b) $\sqrt{(-15)^2} = |-15| = 15$
(c) We don't know whether the real number x is positive, negative, or zero. To ensure that the result is positive or zero, we write $\sqrt{x^2} = |x|$.
(d) $\sqrt{(3x-1)^2} = |3x - 1|$
(e) Notice that the radicand, $x^2 + 6x + 9$, factors to $(x + 3)^2$. Therefore,
$$\sqrt{x^2 + 6x + 9} = \sqrt{(x+3)^2}$$
$$= |x + 3|$$

Quick ✓

18. $\sqrt{a^2} = $ ___ .

In Problems 19–22, evaluate each square root.

19. $\sqrt{(-14)^2}$ **20.** $\sqrt{z^2}$ **21.** $\sqrt{(2x+3)^2}$ **22.** $\sqrt{p^2 - 12p + 36}$

Getting Ready for Chapter 6 Exercises

MyMathLab® PRACTICE

Exercise numbers in green are complete video solutions in MyMathLab

*Problems **1–22** are the Quick ✓s that follow the EXAMPLES.*

Building Skills

In Problems 23–32, evaluate each square root. See Objective 1.

23. $\sqrt{1}$

24. $\sqrt{9}$

25. $-\sqrt{100}$

26. $-\sqrt{144}$

27. $\sqrt{\dfrac{1}{4}}$

28. $\sqrt{\dfrac{4}{81}}$

29. $\sqrt{0.36}$

30. $\sqrt{0.25}$

31. $\left(\sqrt{1.6}\right)^2$

32. $\left(\sqrt{3.7}\right)^2$

In Problems 33–44, tell whether the square root is rational, irrational, or not a real number. If the square root is rational, find the exact value; if the square root is irrational, write the approximate value rounded to two decimal places. See Objective 2.

33. $\sqrt{-14}$

34. $\sqrt{-50}$

35. $\sqrt{64}$

36. $\sqrt{121}$

37. $\sqrt{\dfrac{1}{16}}$

38. $\sqrt{\dfrac{49}{100}}$

39. $\sqrt{44}$

40. $\sqrt{24}$

41. $\sqrt{50}$

42. $\sqrt{12}$

43. $\sqrt{-16}$

44. $\sqrt{-64}$

In Problems 45–56, simplify each square root. See Objective 3.

45. $\sqrt{8^2}$

46. $\sqrt{5^2}$

47. $\sqrt{(-19)^2}$

48. $\sqrt{(-13)^2}$

49. $\sqrt{r^2}$

50. $\sqrt{w^2}$

51. $\sqrt{(x+4)^2}$

52. $\sqrt{(x-8)^2}$

53. $\sqrt{(4x-3)^2}$

54. $\sqrt{(5x+2)^2}$

55. $\sqrt{4y^2+12y+9}$

56. $\sqrt{9z^2-24z+16}$

Mixed Practice

In Problems 57–74, simplify each expression.

57. $\sqrt{25+144}$

58. $\sqrt{9+16}$

59. $\sqrt{25}+\sqrt{144}$

60. $\sqrt{9}+\sqrt{16}$

61. $\sqrt{-144}$

62. $\sqrt{-36}$

63. $3\sqrt{25}$

64. $-10\sqrt{16}$

65. $5\sqrt{\dfrac{16}{25}}-\sqrt{144}$

66. $2\sqrt{\dfrac{9}{4}}-\sqrt{4}$

67. $\sqrt{8^2-4\cdot1\cdot7}$

68. $\sqrt{9^2-4\cdot1\cdot20}$

69. $\sqrt{(-5)^2-4\cdot2\cdot5}$

70. $\sqrt{(-3)^2-4\cdot3\cdot2}$

71. $\dfrac{-(-1)+\sqrt{(-1)^2-4\cdot6\cdot(-2)}}{2\cdot(-1)}$

72. $\dfrac{-7+\sqrt{7^2-4\cdot2\cdot6}}{2\cdot2}$

73. $\sqrt{(6-1)^2+(15-3)^2}$

74. $\sqrt{(2-(-1))^2+(6-2)^2}$

75. What are the square roots of 36? What is $\sqrt{36}$?

76. What are the square roots of 64? What is $\sqrt{64}$?

Math for the Future: Statistics *For Problems 77 and 78, use the formula* $Z = \dfrac{X-\mu}{\dfrac{\sigma}{\sqrt{n}}}$ *from statistics (a formula used to determine the value of one observation relative to that of another) to evaluate the expression for the given values. Write the exact value and then write your answer rounded to two decimal places.*

77. $X=120, \mu=100, \sigma=15, n=13$

78. $X=40, \mu=50, \sigma=10, n=5$

Explaining the Concepts

79. Explain why $\sqrt{a^2}=|a|$. Provide examples to support your explanation.

CHAPTER

6 Radicals and Rational Exponents

 Wind Chill Chart

Calm	40	35	30	25	20	15	10	5	0	-5	-10	-15	-20	-25	-30	-35	-40	-45
5	36	31	25	19	13	7	1	-5	-11	-16	-22	-28	-34	-40	-46	-52	-57	-63
10	34	27	21	15	9	3	-4	-10	-16	-22	-28	-35	-41	-47	-53	-59	-66	-72
15	32	25	19	13	6	0	-7	-13	-19	-26	-32	-39	-45	-51	-58	-64	-71	-77
20	30	24	17	11	4	-2	-9	-15	-22	-29	-35	-42	-48	-55	-61	-68	-74	-81
25	29	23	16	9	3	-4	-11	-17	-24	-31	-37	-44	-51	-58	-64	-71	-78	-84
30	28	22	15	8	1	-5	-12	-19	-26	-33	-39	-46	-53	-60	-67	-73	-80	-87
35	28	21	14	7	0	-7	-14	-21	-27	-34	-41	-48	-55	-62	-69	-76	-82	-89
40	27	20	13	6	-1	-8	-15	-22	-29	-36	-43	-50	-57	-64	-71	-78	-84	-91
45	26	19	12	5	-2	-9	-16	-23	-30	-37	-44	-51	-58	-65	-72	-79	-86	-93
50	26	19	12	4	-3	-10	-17	-24	-31	-38	-45	-52	-60	-67	-74	-81	-88	-95
55	25	18	11	4	-3	-11	-18	-25	-32	-39	-46	-54	-61	-68	-75	-82	-89	-97
60	25	17	10	3	-4	-11	-19	-26	-33	-40	-48	-55	-62	-69	-76	-84	-91	-98

Temperature (°F) — Wind (mph)

Frostbite Times ☐ 30 minutes ☐ 10 minutes ☐ 5 minutes

Wind Chill (°F) = 35.74 + 0.6215T - 35.75(V$^{0.16}$) + 0.4275T(V$^{0.16}$)
Where, T= Air Temperature (°F) V= Wind Speed (mph)   *Effective 11/01/01*

According to the National Weather Service, the wind chill temperature (WCT) index uses science, technology, and computer modeling to provide an accurate, understandable, and useful formula for calculating the dangers from winter winds and freezing temperatures. The wind chill temperature is defined only for temperatures at or below 50 degrees Fahrenheit (50°F) and wind speeds above 3 miles per hour.

What is the wind chill temperature if the temperature is 0°F and the wind speed is 10 miles per hour? See Problem 137 in Section 6.1.

The Big Picture: Putting It Together

In Chapter 4 we simplified polynomial expressions by adding, subtracting, multiplying, and dividing. We also factored polynomials. In Chapter 5 we used the skills learned in Chapter 4 to simplify rational expressions and perform operations on rational expressions.

We now present a similar discussion with radical expressions. We will learn how to add, subtract, multiply, and divide radical expressions. In addition, we will use factoring to simplify radical expressions. Throughout this discussion, keep in mind that radicals perform the "inverse" of raising a real number to a positive integer exponent. For example, a square root undoes the squaring operation.

Outline

6.1 *n*th Roots and Rational Exponents

6.2 Simplifying Expressions Using the Laws of Exponents

6.3 Simplifying Radical Expressions Using Properties of Radicals

6.4 Adding, Subtracting, and Multiplying Radical Expressions

6.5 Rationalizing Radical Expressions

Putting the Concepts Together (Sections 6.1–6.5)

6.6 Functions Involving Radicals

6.7 Radical Equations and Their Applications

6.8 The Complex Number System

Chapter 6 Activity: Which One Does Not Belong?

Chapter 6 Review

Chapter 6 Test

472

6.1 *nth* Roots and Rational Exponents

Objectives

1. Evaluate *nth* Roots
2. Simplify Expressions of the Form $\sqrt[n]{a^n}$
3. Evaluate Expressions of the Form $a^{\frac{1}{n}}$
4. Evaluate Expressions of the Form $a^{\frac{m}{n}}$

Are You Ready for This Section?

Before getting started, take the following readiness quiz. If you get a problem wrong, go to the section cited and review the material.

R1. Simplify: $\left(\dfrac{x^2 y}{xy^{-2}}\right)^{-3}$ [Getting Ready: Laws of Exponents, pp. 299–300]

R2. Simplify: $(\sqrt{7})^2$ [Getting Ready: Square Roots, pp. 467–468]

R3. Evaluate: $\sqrt{64}$ [Getting Ready: Square Roots, pp. 467–468]

R4. Evaluate: $\sqrt{(x+1)^2}$ [Getting Ready: Square Roots, pp. 469–470]

R5. Simplify: **(a)** 3^{-2} **(b)** x^{-4} [Getting Ready: Laws of Exponents, pp. 294–296]

In the Getting Ready: Square Roots section, we reviewed skills for evaluating square roots. We now extend this skill to other types of roots.

▶ 1 Evaluate *nth* Roots

A real number is cubed when it is raised to the power 3. The inverse of cubing a number is finding the *cube root*. For example, since $2^3 = 8$, the cube root of 8 is 2; since $(-2)^3 = -8$, the cube root of -8 is -2. In general, we can find *nth* roots of numbers.

In Words

The notation $\sqrt[n]{a} = b$ means a number *b* such that raising that number to the *n*th power gives *a."*

Thus $\sqrt[3]{125} = x$ means a number *x* such that raising *x* to the third power gives 125.

Definition

The **principal *nth* root of a number *a***, symbolized by $\sqrt[n]{a}$, where $n \geq 2$ is an integer, is defined as follows:

$$\sqrt[n]{a} = b \qquad \text{means} \qquad a = b^n$$

- If $n \geq 2$ and even, then *a* and *b* must be greater than or equal to 0.
- If $n \geq 3$ and odd, then *a* and *b* can be any real number.

Work Smart

If the index is even, then the radicand must be greater than or equal to zero in order for a radical to simplify to a real number. If the index is odd, the radicand can be any real number.

In the notation $\sqrt[n]{a}$, the integer $n, n \geq 2$, is called the **index.** A radical written without the index means the square root, so $\sqrt{a}$ represents the square root of *a*. If the index is 3, we call $\sqrt[3]{a}$ the **cube root** of *a*.

If the index is even, then the radicand must be greater than or equal to 0. If the index is odd, then the radicand can be any real number. Do you know why? Since $\sqrt[n]{a} = b$ means $b^n = a$, if the index *n* is even, then $b^n \geq 0$ so $a \geq 0$. If we have an odd index *n*, then b^n can be any real number, so *a* can be any real number.

Before we evaluate *nth* roots, we list some "perfect" powers of 2, 3, 4, and 5. This list will help you find roots in the following examples. Notice in the display that follows that perfect cubes and perfect fifths can be negative, but perfect squares and perfect fourths cannot. Do you know why?

Perfect Squares	Perfect Cubes		Perfect Fourths	Perfect Fifths	
$1^2 = 1$	$1^3 = 1$	$(-1)^3 = -1$	$1^4 = 1$	$1^5 = 1$	$(-1)^5 = -1$
$2^2 = 4$	$2^3 = 8$	$(-2)^3 = -8$	$2^4 = 16$	$2^5 = 32$	$(-2)^5 = -32$
$3^2 = 9$	$3^3 = 27$	$(-3)^3 = -27$	$3^4 = 81$	$3^5 = 243$	$(-3)^5 = -243$
$4^2 = 16$	$4^3 = 64$	$(-4)^3 = -64$	$4^4 = 256$	$4^5 = 1024$	$(-4)^5 = -1024$
$5^2 = 25$	$5^3 = 125$	$(-5)^3 = -125$	$5^4 = 625$	$5^5 = 3125$	$(-5)^5 = -3125$
and so on	and so on		and so on	and so on	

EXAMPLE 1   **Evaluating *n*th Roots of Real Numbers**

Evaluate:

(a) $\sqrt[3]{1000}$ (b) $\sqrt[4]{16}$ (c) $\sqrt[3]{-8}$ (d) $\sqrt[4]{-81}$

Solution

(a) We are looking for a number whose cube is 1000. Since $10^3 = 1000$, we have $\sqrt[3]{1000} = 10$.

(b) $\sqrt[4]{16} = 2$ since $2^4 = 16$.

(c) $\sqrt[3]{-8} = -2$ since $(-2)^3 = -8$.

(d) Because there is no real number b such that $b^4 = -81$, $\sqrt[4]{-81}$ is not a real number.

Quick ✓

1. In the notation $\sqrt[n]{a}$, the integer n, $n \geq 2$, is called the _____.

In Problems 2–6, evaluate each root.

2. $\sqrt[3]{64}$ 3. $\sqrt[4]{81}$ 4. $\sqrt[3]{-216}$

5. $\sqrt[4]{-32}$ 6. $\sqrt[5]{\dfrac{1}{32}}$

The *n*th roots in Examples 1(a)–(c) are all rational numbers, but some *n*th roots are not. Just as we can approximate square roots using a calculator, we can also approximate *n*th roots.

EXAMPLE 2   **Approximating an *n*th Root Using a Calculator**

(a) Write $\sqrt[3]{25}$ as a decimal rounded to two decimal places.

(b) Write $\sqrt[4]{18}$ as a decimal rounded to two decimal places.

Solution

We have used a TI-84 Plus graphing calculator to approximate both values.

(a) Because $\sqrt[3]{8} = 2$ and $\sqrt[3]{27} = 3$, we expect $\sqrt[3]{25}$ to be between 2 and 3 (closer to 3). Figure 1(a) shows that $\sqrt[3]{25} \approx 2.92$.

(b) Because $\sqrt[4]{16} = 2$ and $\sqrt[4]{81} = 3$, we expect $\sqrt[4]{18}$ to be between 2 and 3 (closer to 2). Figure 1(b) shows that $\sqrt[4]{18} \approx 2.06$.

Figure 1

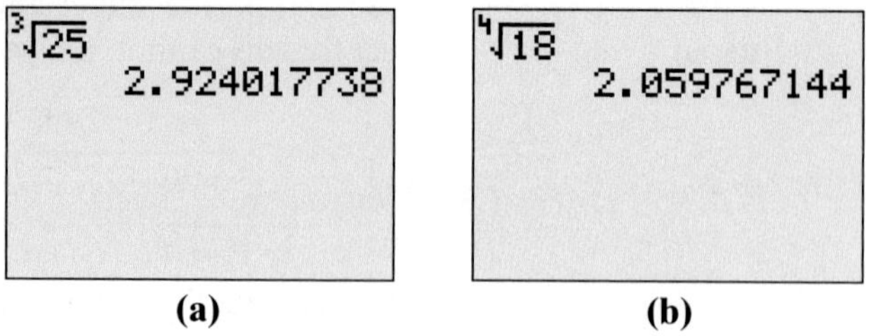

$\sqrt[3]{25}$	$\sqrt[4]{18}$
2.924017738	2.059767144

(a)   (b)

Quick ✓

In Problems 7 and 8, use a calculator to write the approximate value of each radical rounded to two decimal places.

7. $\sqrt[3]{50}$ 8. $\sqrt[4]{80}$

▶ ❷ Simplify Expressions of the Form $\sqrt[n]{a^n}$

We have already seen that $\sqrt{a^2} = |a|$. But what about $\sqrt[3]{a^3}$ or $\sqrt[4]{a^4}$? Recall that the definition of the principal *n*th root, $\sqrt[n]{a}$, requires that *a* be nonnegative (greater than or equal to zero) when *n* is even but *a* can be any real number when *n* is odd.

> **Simplifying $\sqrt[n]{a^n}$**
>
> If $n \geq 2$ is a positive integer and *a* is a real number, then
>
> $$\sqrt[n]{a^n} = |a| \quad \text{if } n \geq 2 \text{ is even}$$
> $$\sqrt[n]{a^n} = a \quad \text{if } n \geq 3 \text{ is odd}$$

EXAMPLE 3 — **Simplifying Radicals**

Simplify:

(a) $\sqrt[3]{x^3}$ (b) $\sqrt[4]{(x-7)^4}$ (c) $-\sqrt[6]{(-3)^6}$ (d) $\sqrt[3]{-\dfrac{8}{125}}$

Solution

(a) Because the index, 3, is odd, $\sqrt[3]{x^3} = x$.

(b) Because the index, 4, is even, $\sqrt[4]{(x-7)^4} = |x-7|$.

(c) $-\sqrt[6]{(-3)^6} = -|-3| = -3$

(d) $\sqrt[3]{-\dfrac{8}{125}} = \sqrt[3]{\dfrac{-8}{125}} = \sqrt[3]{\dfrac{(-2)^3}{5^3}} = \sqrt[3]{\left(\dfrac{-2}{5}\right)^3} = -\dfrac{2}{5}$

Quick ✓

In Problems 9–13, simplify each radical.

9. $\sqrt[4]{5^4}$ 10. $\sqrt[6]{z^6}$ 11. $\sqrt[7]{(3x-2)^7}$

12. $\sqrt[8]{(-2)^8}$ 13. $\sqrt[5]{-\dfrac{32}{243}}$

▶ ❸ Evaluate Expressions of the Form $a^{\frac{1}{n}}$

Work Smart

Remember, a rational number is a number of the form $\dfrac{p}{q}$, where *p* and *q* are integers and $q \neq 0$.

In the Getting Ready: Laws of Exponents section in Chapter 4, we developed methods for simplifying algebraic expressions that contain integer exponents. Now we will extend the rules that apply to integer exponents to rational exponents.

We start by defining "*a* raised to the power $\dfrac{1}{n}$," where *a* is a real number and *n* is a positive integer. This definition needs to obey all the laws of exponents presented earlier. For example, we know that $a^2 = a \cdot a$, so

$$\left(5^{\frac{1}{2}}\right)^2 = 5^{\frac{1}{2}} \cdot 5^{\frac{1}{2}}$$
$$a^m \cdot a^n = a^{m+n}: \quad = 5^{\frac{1}{2}+\frac{1}{2}}$$
$$= 5^1$$
$$= 5$$

We also know that $(\sqrt{5})^2 = 5$, so it is reasonable to conclude that

$$5^{\frac{1}{2}} = \sqrt{5}$$

This suggests the following definition:

> **Definition of $a^{\frac{1}{n}}$**
>
> If *a* is a real number and *n* is an integer with $n \geq 2$, then
>
> $$a^{\frac{1}{n}} = \sqrt[n]{a}$$
>
> provided that $\sqrt[n]{a}$ exists.

EXAMPLE 4

Evaluating Expressions Containing Exponents of the Form $a^{\frac{1}{n}}$

Write each of the following expressions as a radical and simplify, if possible.

(a) $9^{\frac{1}{2}}$ **(b)** $(-64)^{\frac{1}{3}}$ **(c)** $-100^{\frac{1}{2}}$ **(d)** $(-100)^{\frac{1}{2}}$ **(e)** $z^{\frac{1}{2}}$

Solution

(a) $9^{\frac{1}{2}} = \sqrt{9}$
      $= 3$

(b) $(-64)^{\frac{1}{3}} = \sqrt[3]{-64}$
      $= -4$

(c) $-100^{\frac{1}{2}} = -1 \cdot 100^{\frac{1}{2}}$
      $= -10$

(d) $(-100)^{\frac{1}{2}} = \sqrt{-100}$ is not a real number because there is no real number whose square is -100.

(e) $z^{\frac{1}{2}} = \sqrt{z}$

Work Smart

Notice how parentheses are used or not used in Examples 4(c) and (d). In Example 4(c), $-100^{\frac{1}{2}}$ means we evaluate $100^{\frac{1}{2}}$ first and then multiply the result by -1. Remember, evaluate exponents before multiplying.

Quick ✓

14. If a is a nonnegative real number and $n \geq 2$ is an integer, then $a^{\frac{1}{n}} = $ ___ .

In Problems 15–19, write each of the following expressions as a radical, and simplify, if possible.

15. $25^{\frac{1}{2}}$ **16.** $(-27)^{\frac{1}{3}}$ **17.** $-64^{\frac{1}{2}}$ **18.** $(-64)^{\frac{1}{2}}$ **19.** $b^{\frac{1}{2}}$

EXAMPLE 5

Writing Radicals with Rational Exponents

Rewrite each of the following radicals with a rational exponent.

(a) $\sqrt[4]{7a}$ **(b)** $\sqrt[5]{\dfrac{xy^3}{4}}$

Solution

(a) The index on the radical is 4, so 4 becomes the denominator of the rational exponent. Use parentheses because the radicand is $7a$, and the exponent $\frac{1}{4}$ is applied to both 7 and a.

$$\sqrt[4]{7a} = (7a)^{\frac{1}{4}}$$

(b) The index on the radical is 5, so the rational exponent is $\frac{1}{5}$.

$$\sqrt[5]{\frac{xy^3}{4}} = \left(\frac{xy^3}{4}\right)^{\frac{1}{5}}$$

Quick ✓

In Problems 20–22, rewrite each of the following radicals with a rational exponent.

20. $\sqrt[5]{8b}$ **21.** $\sqrt[8]{\dfrac{mn^5}{3}}$ **22.** $4\sqrt[3]{y}$

④ Evaluate Expressions of the Form $a^{\frac{m}{n}}$

We now define $a^{\frac{m}{n}}$, where m and n are integers.

> **Definition of $a^{\frac{m}{n}}$**
>
> If a is a real number and $\dfrac{m}{n}$ is a rational number in lowest terms with $n \geq 2$, then
>
> $$a^{\frac{m}{n}} = \sqrt[n]{a^m} = (\sqrt[n]{a})^m$$
>
> provided that $\sqrt[n]{a}$ exists.

In Words

The expression $a^{\frac{m}{n}} = \sqrt[n]{a^m}$ means that we will raise a to the mth power first, and then take the nth root of the result. The expression $a^{\frac{m}{n}} = (\sqrt[n]{a})^m$ means that we will take the nth root of a first, and then raise that result to the power of m.

In the expression $a^{\frac{m}{n}}$, $\frac{m}{n}$ is expressed in lowest terms, and $n \geq 2$. The definition obeys all the laws of exponents presented earlier. For example,

$$a^{\frac{m}{n}} = a^{m \cdot \frac{1}{n}} = (a^m)^{\frac{1}{n}} = \sqrt[n]{a^m}$$

and

$$a^{\frac{m}{n}} = a^{\frac{1}{n} \cdot m} = (a^{\frac{1}{n}})^m = (\sqrt[n]{a})^m$$

When evaluating or simplifying $a^{\frac{m}{n}}$, you may use either $\sqrt[n]{a^m}$ or $(\sqrt[n]{a})^m$, whichever makes simplifying the expression easier. Generally, taking the root first, as in $(\sqrt[n]{a})^m$, is easier.

EXAMPLE 6 **Evaluating Expressions of the Form $a^{\frac{m}{n}}$**

Evaluate each of the following expressions, if possible.

(a) $25^{\frac{3}{2}}$ **(b)** $64^{\frac{2}{3}}$ **(c)** $-9^{\frac{5}{2}}$ **(d)** $(-8)^{\frac{4}{3}}$ **(e)** $(-81)^{\frac{7}{2}}$

Solution

(a) $25^{\frac{3}{2}} = (\sqrt{25})^3 = 5^3 = 125$

(b) $64^{\frac{2}{3}} = (\sqrt[3]{64})^2 = 4^2 = 16$

(c) $-9^{\frac{5}{2}} = -1 \cdot 9^{\frac{5}{2}} = -1 \cdot (\sqrt{9})^5 = -1 \cdot 3^5 = -1 \cdot 243 = -243$

(d) $(-8)^{\frac{4}{3}} = (\sqrt[3]{-8})^4 = (-2)^4 = 16$

(e) $(-81)^{\frac{7}{2}}$ is not a real number because $(-81)^{\frac{7}{2}} = (\sqrt{-81})^7$, and $\sqrt{-81}$ is not a real number.

●

Quick ✓

23. If a is a real number and $\dfrac{m}{n}$ is a rational number in lowest terms with $n \geq 2$, then
$$a^{\frac{m}{n}} = \underline{\quad\quad} \text{ or } \underline{\quad\quad}, \text{ provided that } \sqrt[n]{a} \text{ exists.}$$

In Problems 24–28, evaluate each expression, if possible.

24. $16^{\frac{3}{2}}$ **25.** $27^{\frac{2}{3}}$ **26.** $-16^{\frac{3}{4}}$

27. $(-64)^{\frac{2}{3}}$ **28.** $(-25)^{\frac{5}{2}}$

The expressions in Examples 6(a)–(d) all simplified to rational numbers. Not all expressions involving rational exponents simplify to rational numbers.

EXAMPLE 7 **Approximating Expressions Involving Rational Exponents**

Write $34^{\frac{3}{4}}$ as a decimal rounded to two decimal places.

Solution

Figure 2 shows the results obtained from a TI-84 Plus graphing calculator. So $34^{\frac{3}{4}} \approx 14.08$.

Figure 2

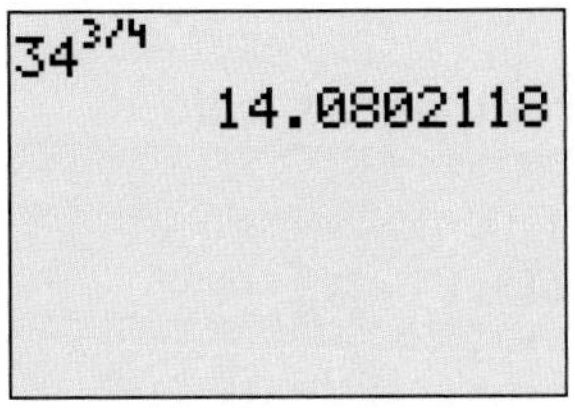

```
34^3/4
        14.0802118
```

Quick ✓

In Problems 29 and 30, approximate the expression rounded to two decimal places.

29. $50^{\frac{2}{3}}$ **30.** $40^{\frac{3}{20}}$

EXAMPLE 8   **Writing Radicals with Rational Exponents**

Rewrite each of the following radicals with a rational exponent.

(a) $\sqrt[3]{x^2}$ (b) $(\sqrt[5]{10a^2b})^4$

Solution

(a) The index, 3, is the denominator of the rational exponent, and the power on the radicand, 2, is the numerator of the rational exponent.
$$\sqrt[3]{x^2} = x^{\frac{2}{3}}$$

(b) The index, 5, is the denominator of the rational exponent, and the power, 4, is the numerator of the rational exponent.
$$(\sqrt[5]{10a^2b})^4 = (10a^2b)^{\frac{4}{5}}$$

Quick ✓

In Problems 31–33, write each radical with a rational exponent.

31. $\sqrt[8]{a^3}$ **32.** $\sqrt[4]{t^{12}}$ **33.** $(\sqrt[4]{12ab^3})^9$

If a rational exponent is negative, then we can use the rule for negative rational exponents given below.

> **NEGATIVE-EXPONENT RULE**
>
> If $\dfrac{m}{n}$ is a rational number, and if a is a nonzero real number (that is, if $a \neq 0$), then we define
>
> $$a^{-\frac{m}{n}} = \frac{1}{a^{\frac{m}{n}}} \quad \text{and} \quad \frac{1}{a^{-\frac{m}{n}}} = a^{\frac{m}{n}} \quad \text{if } a \neq 0, \quad \text{and } \sqrt[n]{a} \text{ is defined.}$$

EXAMPLE 9   **Evaluating Expressions with Negative Rational Exponents**

Rewrite each of the following with positive exponents, and completely simplify, if possible.

(a) $36^{-\frac{1}{2}}$ (b) $\dfrac{1}{27^{-\frac{2}{3}}}$ (c) $(6a)^{-\frac{5}{4}}$

Solution

(a) $36^{-\frac{1}{2}} = \dfrac{1}{36^{\frac{1}{2}}} = \dfrac{1}{\sqrt{36}} = \dfrac{1}{6}$

(b) Because the negative exponent is in the denominator, use $\dfrac{1}{a^{-\frac{m}{n}}} = a^{\frac{m}{n}}$ to simplify:
$$\frac{1}{27^{-\frac{2}{3}}} = 27^{\frac{2}{3}} = (\sqrt[3]{27})^2 = 3^2 = 9$$

(c) $(6a)^{-\frac{5}{4}} = \dfrac{1}{(6a)^{\frac{5}{4}}}$

Quick ✓

In Problems 34–36, rewrite each of the following with positive exponents, and completely simplify, if possible.

34. $81^{-\frac{1}{2}}$ **35.** $\dfrac{1}{8^{-\frac{2}{3}}}$ **36.** $(13x)^{-\frac{3}{2}}$

6.1 Exercises   MyMathLab®

Exercise numbers in green are complete video solutions in MyMathLab

*Problems **1–36** are the Quick ✓s that follow the* **EXAMPLES.**

Building Skills

In Problems 37–46, simplify each radical. See Objective 1.

37. $\sqrt[3]{125}$ **38.** $\sqrt[3]{216}$

39. $\sqrt[3]{-27}$ **40.** $\sqrt[3]{-64}$

41. $-\sqrt[4]{625}$ **42.** $-\sqrt[4]{256}$

43. $\sqrt[3]{-\dfrac{1}{8}}$ **44.** $\sqrt[3]{\dfrac{8}{125}}$

45. $-\sqrt[5]{-243}$ **46.** $-\sqrt[5]{-1024}$

In Problems 47–50, use a calculator to write each expression as a decimal rounded to two decimal places. See Objective 1.

47. $\sqrt[3]{25}$ **48.** $\sqrt[3]{85}$

49. $\sqrt[4]{12}$ **50.** $\sqrt[4]{2}$

In Problems 51–58, simplify each radical. See Objective 2.

51. $\sqrt[3]{5^3}$ **52.** $\sqrt[4]{6^4}$

53. $\sqrt[4]{m^4}$ **54.** $\sqrt[5]{n^5}$

55. $\sqrt[9]{(x-3)^9}$ **56.** $\sqrt[6]{(2x-3)^6}$

57. $-\sqrt[4]{(3p+1)^4}$ **58.** $-\sqrt[3]{(6z-5)^3}$

In Problems 59–72, evaluate each expression, if possible. See Objective 3.

59. $4^{\frac{1}{2}}$ **60.** $16^{\frac{1}{2}}$

61. $-36^{\frac{1}{2}}$ **62.** $-25^{\frac{1}{2}}$

63. $8^{\frac{1}{3}}$ **64.** $27^{\frac{1}{3}}$

65. $-16^{\frac{1}{4}}$ **66.** $-81^{\frac{1}{4}}$

67. $\left(\dfrac{4}{25}\right)^{\frac{1}{2}}$ **68.** $\left(\dfrac{8}{27}\right)^{\frac{1}{3}}$

69. $(-125)^{\frac{1}{3}}$ **70.** $(-216)^{\frac{1}{3}}$

71. $(-4)^{\frac{1}{2}}$ **72.** $(-81)^{\frac{1}{2}}$

In Problems 73–76, rewrite each of the following radicals with a rational exponent. See Objective 3.

73. $\sqrt[3]{3x}$ **74.** $\sqrt[5]{2y}$

75. $\sqrt[4]{\dfrac{x}{3}}$ **76.** $\sqrt{\dfrac{w}{2}}$

In Problems 77–92, evaluate each expression. See Objective 4.

77. $4^{\frac{5}{2}}$ **78.** $25^{\frac{3}{2}}$

79. $-16^{\frac{3}{2}}$ **80.** $-100^{\frac{5}{2}}$

81. $8^{\frac{4}{3}}$ **82.** $27^{\frac{4}{3}}$

83. $(-64)^{\frac{2}{3}}$ **84.** $(-125)^{\frac{2}{3}}$

85. $-(-32)^{\frac{3}{5}}$ **86.** $-(-216)^{\frac{2}{3}}$

87. $144^{-\frac{1}{2}}$ **88.** $121^{-\frac{1}{2}}$

89. $\dfrac{1}{25^{-\frac{3}{2}}}$ **90.** $\dfrac{1}{49^{-\frac{3}{2}}}$

91. $\dfrac{1}{8^{-\frac{5}{3}}}$ **92.** $27^{-\frac{4}{3}}$

In Problems 93–100, rewrite each of the following radicals with a rational exponent. See Objective 4.

93. $\sqrt[4]{x^3}$ **94.** $\sqrt[3]{p^5}$

95. $(\sqrt[5]{3x})^2$ **96.** $(\sqrt[4]{6z})^3$

97. $\sqrt{\left(\dfrac{5x}{y}\right)^3}$ **98.** $\sqrt[6]{\left(\dfrac{2a}{b}\right)^5}$

99. $\sqrt[3]{(9ab)^4}$ **100.** $\sqrt[4]{(3pq)^7}$

In Problems 101–106, use a calculator to write each expression as a decimal rounded to two decimal places. See Objective 4.

101. $20^{\frac{1}{2}}$ **102.** $5^{\frac{1}{2}}$

103. $4^{\frac{5}{3}}$ **104.** $100^{\frac{3}{4}}$

105. $10^{0.1}$ **106.** $100^{0.25}$

Mixed Practice

In Problems 107–130, evaluate each expression, if possible.

107. $\sqrt[3]{x^3} + 4\sqrt[5]{x^5}$ **108.** $\sqrt[3]{(x-1)^3} + \sqrt[5]{32}$

109. $-16^{\frac{3}{4}} - \sqrt[5]{32}$ **110.** $\dfrac{5\sqrt[7]{x^7}}{\sqrt[3]{x^3}}, x \neq 0$

111. $\sqrt[3]{512}$ **112.** $\sqrt[3]{-125}$

113. $9^{2.5}$ **114.** $100^{1.5}$

115. $\sqrt[4]{-16}$ **116.** $\sqrt[4]{-1}$

117. $144^{-\frac{1}{2}} \cdot 3^2$ **118.** $125^{-\frac{1}{3}}$

119. $\sqrt[3]{0.008}$ **120.** $\sqrt[4]{0.0081}$

121. $4^{0.5} + 25^{1.5}$

122. $100^{0.5} - 4^{1.5}$

123. $(-25)^{\frac{5}{2}}$

124. $(-125)^{-\frac{1}{3}}$

125. $\sqrt[3]{(3p - 5)^3}$

126. $\sqrt[5]{(6b - 1)^5}$

127. $-9^{\frac{3}{2}} + \dfrac{1}{27^{-\frac{2}{3}}}$

128. $\dfrac{1}{64^{\frac{1}{2}}} - 4^{-\frac{3}{2}}$

129. $\sqrt[6]{(-2)^6}$

130. $\sqrt[4]{(-10)^4}$

In Problems 131–134, evaluate each function.

131. $f(x) = x^{\frac{3}{2}}$; find $f(4)$

132. $g(x) = x^{-\frac{3}{2}}$; find $g(16)$

133. $F(z) = z^{\frac{4}{3}}$; find $F(-8)$

134. $G(a) = a^{\frac{5}{3}}$; find $G(-8)$

Applying the Concepts

135. What is the cube root of 1000?
What is $\sqrt[3]{1000}$?

136. What is the cube root of 729? What is $\sqrt[3]{729}$?

137. Wind Chill According to the National Weather Service, the wind chill temperature is how cold people and animals feel when outside. Wind chill is based on the rate of heat loss from exposed skin caused by wind and cold. The formula for computing wind chill W is

$$W = 35.74 + 0.6215T - 35.75v^{0.16} + 0.4275Tv^{0.16}$$

where T is the air temperature in degrees Fahrenheit and v is the wind speed in miles per hour.

(a) What is the wind chill if it is 30°F and the wind speed is 10 miles per hour?

(b) What is the wind chill if it is 30°F and the wind speed is 20 miles per hour?

(c) What is the wind chill if it is 0°F and the wind speed is 10 miles per hour?

138. Money The annual rate of interest r (expressed as a decimal) required to have A dollars after t years from an initial deposit of P dollars is given by

$$r = \left(\frac{A}{P}\right)^{\frac{1}{t}} - 1$$

(a) If you deposit $100 in a mutual fund today and have $144 in the account in 2 years, what was your annual rate of interest earned?

(b) If you deposit $100 in a mutual fund today and have $337.50 in 3 years, what was your annual rate of interest earned?

(c) The Rule of 72 states that your money will double in $\dfrac{72}{100r}$ years, where r is the rate of interest earned (expressed as a decimal). Suppose that you deposit $1000 in a mutual fund today and have $2000 in 8 years. What rate of interest did you earn? Compute $\dfrac{72}{100r}$ for this rate of interest. Is it close?

139. Terminal Velocity Terminal velocity is the maximum speed that a body falling through air can reach (because of air resistance). Terminal velocity is given by the formula $v_t = \sqrt{\dfrac{2mg}{C\rho A}}$, where m is the mass of the falling object, g is acceleration due to gravity (≈ 9.81 meters per second^2), C is a drag coefficient with $0.5 \le C \le 1.0$, ρ is the density of air (≈ 1.2 kg/m^3), and A is the cross-sectional area of the object. Suppose that a raindrop whose radius is 1.5 mm falls from the sky. The mass of the raindrop is given by $m = \dfrac{4}{3}\pi r^3 \rho_w$, where r is its radius and $\rho_w = 1000$ kg/m^3. The cross-sectional area of the raindrop is $A = \pi r^2$.

(a) Substitute the formulas for the mass and area of a raindrop into the formula for terminal speed, and simplify the expression.

(b) Determine the terminal velocity of a raindrop whose radius is 0.0015 m with $C = 0.6$.

140. Kepler's Law Early in the seventeenth century, Johannes Kepler (1571–1630) discovered that the square of the period T of a planet to rotate around the Sun varies directly with the cube of its mean distance r from the Sun. The period of a planet is the amount of time (in Earth years) for the planet to complete one orbit around the Sun. Kepler's Law can be expressed using rational exponents as $T = kr^{\frac{3}{2}}$, where k is the constant of proportionality.

(a) The period of Mercury is 0.241 year, and its mean distance from the Sun is 5.79×10^{10} meters. Use this information to state Kepler's Law (find the value of k).

(b) The mean distance of Mars from the Sun is 2.28×10^{11} m. Use this information, along with the result of part (a), to find the amount of time it takes Mars to complete one orbit around the Sun.

Explaining the Concepts

141. Explain why $(-9)^{\frac{1}{2}}$ is not a real number, but $-9^{\frac{1}{2}}$ is a real number.

142. In your own words, provide a justification for why $a^{\frac{1}{n}} = \sqrt[n]{a}$.

143. Under what conditions is $a^{\frac{m}{n}}$ a real number?

Synthesis Review

In Problems 144–147, simplify each expression completely.

144. $\left(\dfrac{x^2 y}{y^{-2}}\right)^3$

145. $\dfrac{(x + 2)^2 (x - 1)^4}{(x + 2)(x - 1)}$

146. $\dfrac{(3a^2 + 5a - 3) - (a^2 - 2a - 9)}{4a^2 + 12a + 9}$

147. $\dfrac{(4z^2 - 7z + 3) + (-3z^2 - z + 9)}{(4z^2 - 2z - 7) + (-3z^2 - z + 9)}$

6.2 Simplifying Expressions Using the Laws of Exponents

Objectives

1 Simplify Expressions Involving Rational Exponents

2 Simplify Radical Expressions

3 Factor Expressions Containing Rational Exponents

Are You Ready for This Section?

Before getting started, take the following readiness quiz. If you get a problem wrong, go to the section cited and review the material.

R1. Simplify: z^{-3} [Getting Ready: Laws of Exponents, pp. 294–296]

R2. Simplify: $x^{-2} \cdot x^5$ [Getting Ready: Laws of Exponents, pp. 296–297]

R3. Simplify: $\left(\dfrac{2a^2}{b^{-1}}\right)^3$ [Getting Ready: Laws of Exponents, pp. 299–300]

R4. Evaluate: $\sqrt{144}$ [Getting Ready: Square Roots, pp. 467–468]

▶ 1 Simplify Expressions Involving Rational Exponents

The Laws of Exponents presented in the Getting Ready: Laws of Exponents section on pages 299–300 applied to integer exponents. These laws apply to rational exponents as well.

The Laws of Exponents

If a and b are real numbers and if r and s are rational numbers, then, assuming the expression is defined, the following rules apply.

Zero-Exponent Rule:	$a^0 = 1$	if $a \neq 0$
Negative-Exponent Rule:	$a^{-r} = \dfrac{1}{a^r}$	if $a \neq 0$
Product Rule:	$a^r \cdot a^s = a^{r+s}$	
Quotient Rule:	$\dfrac{a^r}{a^s} = a^{r-s} = \dfrac{1}{a^{s-r}}$	if $a \neq 0$
Power Rule:	$(a^r)^s = a^{r \cdot s}$	
Product-to-a-Power Rule:	$(a \cdot b)^r = a^r \cdot b^r$	
Quotient-to-a-Power Rule:	$\left(\dfrac{a}{b}\right)^r = \dfrac{a^r}{b^r}$	if $b \neq 0$
Quotient-to-a-Negative-Power Rule:	$\left(\dfrac{a}{b}\right)^{-r} = \left(\dfrac{b}{a}\right)^r$	if $a \neq 0, b \neq 0$

Work Smart

We *simplify* expressions (no equal sign) and *solve* equations.

The direction **simplify** shall mean the following:

- All the exponents are positive.
- Each base occurs only once.
- There are no powers written to powers.

Ready?...Answers   **R1.** $\dfrac{1}{z^3}, z \neq 0$

R2. x^3 **R3.** $8a^6 b^3$ **R4.** 12

EXAMPLE 1 **Simplifying Expressions Involving Rational Exponents**

Simplify each of the following:

(a) $27^{\frac{1}{2}} \cdot 27^{\frac{5}{6}}$

(b) $\dfrac{8^{\frac{1}{3}}}{8^{\frac{5}{3}}}$

Solution

(a)

$$a^r \cdot a^s = a^{r+s}$$
$$\downarrow$$
$$27^{\frac{1}{2}} \cdot 27^{\frac{5}{6}} = 27^{\frac{1}{2}+\frac{5}{6}}$$
$$= 27^{\frac{3}{6}+\frac{5}{6}}$$
$$= 27^{\frac{8}{6}}$$
$$= 27^{\frac{4}{3}}$$
$$a^{\frac{m}{n}} = (\sqrt[n]{a})^m: = (\sqrt[3]{27})^4$$
$$= 3^4$$
$$= 81$$

(b)

$$\dfrac{a^r}{a^s} = a^{r-s}$$
$$\downarrow$$
$$\dfrac{8^{\frac{1}{3}}}{8^{\frac{5}{3}}} = 8^{\frac{1}{3}-\frac{5}{3}}$$
$$= 8^{-\frac{4}{3}}$$
$$= \dfrac{1}{8^{\frac{4}{3}}}$$
$$a^{\frac{m}{n}} = (\sqrt[n]{a})^m: = \dfrac{1}{(\sqrt[3]{8})^4}$$
$$= \dfrac{1}{2^4}$$
$$= \dfrac{1}{16}$$

Work Smart

In Example 1(b), we could also simplify as follows:

$$\dfrac{8^{\frac{1}{3}}}{8^{\frac{5}{3}}} = \dfrac{1}{8^{\frac{5}{3}-\frac{1}{3}}}$$
$$= \dfrac{1}{8^{\frac{4}{3}}}$$
$$= \dfrac{1}{(\sqrt[3]{8})^4}$$
$$= \dfrac{1}{16}$$

EXAMPLE 2 **Simplifying Expressions Involving Rational Exponents**

Simplify each of the following:

(a) $\left(36^{\frac{2}{5}}\right)^{\frac{5}{4}}$

(b) $\left(x^{\frac{1}{2}} \cdot y^{\frac{2}{3}}\right)^{\frac{3}{2}}$

Solution

(a)

$$(a^r)^s = a^{r \cdot s}$$
$$\downarrow$$
$$\left(36^{\frac{2}{5}}\right)^{\frac{5}{4}} = 36^{\frac{2}{5} \cdot \frac{5}{4}}$$
$$= 36^{\frac{10}{20}}$$
$$= 36^{\frac{1}{2}}$$
$$= 6$$

(b)

$$(ab)^r = a^r \cdot b^r$$
$$\downarrow$$
$$\left(x^{\frac{1}{2}} \cdot y^{\frac{2}{3}}\right)^{\frac{3}{2}} = \left(x^{\frac{1}{2}}\right)^{\frac{3}{2}} \cdot \left(y^{\frac{2}{3}}\right)^{\frac{3}{2}}$$
$$(a^r)^s = a^{r \cdot s}: = x^{\frac{1}{2} \cdot \frac{3}{2}} \cdot y^{\frac{2}{3} \cdot \frac{3}{2}}$$
$$= x^{\frac{3}{4}}y$$

Quick ✓

1. If a and b are real numbers and if r is a rational number, then assuming the expression is defined, $(ab)^r = $ ____ .

2. If a and b are real numbers and if r and s are rational numbers, then assuming the expression is defined, $a^r \cdot a^s = $ ____ .

In Problems 3–7, simplify each expression.

3. $5^{\frac{3}{4}} \cdot 5^{\frac{1}{6}}$

4. $\dfrac{32^{\frac{6}{5}}}{32^{\frac{3}{5}}}$

5. $\left(100^{\frac{3}{8}}\right)^{\frac{4}{3}}$

6. $\left(a^{\frac{3}{2}} \cdot b^{\frac{5}{4}}\right)^{\frac{2}{3}}$

7. $\dfrac{x^{\frac{1}{2}} \cdot x^{\frac{1}{3}}}{\left(x^{\frac{1}{12}}\right)^2}$

EXAMPLE 3 **Simplifying Expressions Involving Rational Exponents**

Simplify each of the following:

(a) $\left(x^{\frac{2}{3}}y^{-1}\right) \cdot \left(x^{-1}y^{\frac{1}{2}}\right)^{\frac{2}{3}}$

(b) $\left(\dfrac{9xy^{\frac{4}{3}}}{x^{\frac{5}{6}}y^{-\frac{2}{3}}}\right)^{\frac{1}{2}}$

Solution

(a)

Product-to-a-Power Rule: $(ab)^r = a^r b^r$

$$\left(x^{\frac{2}{3}}y^{-1}\right) \cdot \left(x^{-1}y^{\frac{1}{2}}\right)^{\frac{2}{3}} = x^{\frac{2}{3}}y^{-1}\left(x^{-1}\right)^{\frac{2}{3}}\left(y^{\frac{1}{2}}\right)^{\frac{2}{3}}$$

Power Rule: $(a^r)^s = a^{rs}$: $\quad = x^{\frac{2}{3}}y^{-1}x^{-\frac{2}{3}}y^{\frac{1}{3}}$

Product Rule: $a^r \cdot a^s = a^{r+s}$: $\quad = x^{\frac{2}{3}+\left(-\frac{2}{3}\right)}y^{-1+\frac{1}{3}}$

$$= x^0 y^{-\frac{2}{3}}$$

$a^0 = 1$; Negative-Exponent Rule: $a^{-r} = \dfrac{1}{a^r}$: $\quad = \dfrac{1}{y^{\frac{2}{3}}}$

(b)

Quotient Rule: $\dfrac{a^r}{a^s} = a^{r-s}$

$$\left(\frac{9xy^{\frac{4}{3}}}{x^{\frac{5}{6}}y^{-\frac{2}{3}}}\right)^{\frac{1}{2}} = \left(9x^{1-\frac{5}{6}}y^{\frac{4}{3}-\left(-\frac{2}{3}\right)}\right)^{\frac{1}{2}}$$

$x^{1-\frac{5}{6}} = x^{\frac{6}{6}-\frac{5}{6}} = x^{\frac{1}{6}}$;

$y^{\frac{4}{3}-\left(-\frac{2}{3}\right)} = y^{\frac{4}{3}+\frac{2}{3}} = y^{\frac{6}{3}} = y^2$: $\quad = \left(9x^{\frac{1}{6}}y^2\right)^{\frac{1}{2}}$

Product-to-a-Power Rule: $(a \cdot b)^r = a^r \cdot b^r$: $\quad = 9^{\frac{1}{2}} \cdot \left(x^{\frac{1}{6}}\right)^{\frac{1}{2}} \cdot \left(y^2\right)^{\frac{1}{2}}$

$9^{\frac{1}{2}} = \sqrt{9} = 3$; Power Rule: $(a^r)^s = a^{rs}$: $\quad = 3x^{\frac{1}{12}}y$

Quick ✓

In Problems 8–10, simplify each expression.

8. $\left(8x^{\frac{3}{4}}y^{-1}\right)^{\frac{2}{3}}$

9. $\left(\dfrac{25x^{\frac{1}{2}}y^{\frac{3}{4}}}{x^{-\frac{3}{4}}y}\right)^{\frac{1}{2}}$

10. $8\left(125a^{\frac{3}{4}}b^{-1}\right)^{\frac{2}{3}}$

▶ ❷ **Simplify Radical Expressions**

Rational exponents can be used to simplify radicals.

EXAMPLE 4　**Simplifying Radicals Using Rational Exponents**

Use rational exponents to simplify the radicals.

(a) $\sqrt[8]{16^4}$　　**(b)** $\sqrt[3]{64x^6y^3}$　　**(c)** $\dfrac{\sqrt{x}}{\sqrt[3]{x^2}}$　　**(d)** $\sqrt{\sqrt[3]{z}}$

Solution

The idea in all these problems is to rewrite the radical as an expression involving a rational exponent. Then use the Laws of Exponents to simplify the expression. Finally, write the simplified expression as a radical.

(a)

Write radical as a rational exponent using $\sqrt[n]{a^m} = a^{\frac{m}{n}}$.

$$\sqrt[8]{16^4} = 16^{\frac{4}{8}}$$

Simplify the exponent: $\quad = 16^{\frac{1}{2}}$

Write rational exponent as a radical: $\quad = \sqrt{16}$

$$= 4$$

(b)

$$\sqrt[n]{a} = a^{\frac{1}{n}}$$

$$\sqrt[3]{64x^6y^3} = \left(64x^6y^3\right)^{\frac{1}{3}}$$

$(ab)^r = a^r \cdot b^r$: $\quad = 64^{\frac{1}{3}} \cdot \left(x^6\right)^{\frac{1}{3}} \cdot \left(y^3\right)^{\frac{1}{3}}$

$(a^r)^s = a^{rs}$: $\quad = 64^{\frac{1}{3}} \cdot x^{6 \cdot \frac{1}{3}} \cdot y^{3 \cdot \frac{1}{3}}$

$64^{\frac{1}{3}} = \sqrt[3]{64} = 4$: $\quad = 4x^2y$

(c)

$$\sqrt{a} = a^{\frac{1}{2}};\ \sqrt[n]{a^m} = a^{\frac{m}{n}}$$
$$\downarrow$$
$$\frac{\sqrt{x}}{\sqrt[3]{x^2}} = \frac{x^{\frac{1}{2}}}{x^{\frac{2}{3}}}$$

$\dfrac{a^r}{a^s} = a^{r-s}: \qquad = x^{\frac{1}{2}-\frac{2}{3}}$

$\dfrac{1}{2} - \dfrac{2}{3} = \dfrac{3}{6} - \dfrac{4}{6} = -\dfrac{1}{6}: \qquad = x^{-\frac{1}{6}}$

$a^{-r} = \dfrac{1}{a^r}: \qquad = \dfrac{1}{x^{\frac{1}{6}}}$

Write rational exponent as a radical: $\quad = \dfrac{1}{\sqrt[6]{x}}$

(d)

Write radicand with a rational exponent.
$$\downarrow$$
$$\sqrt{\sqrt[3]{z}} = \sqrt{z^{\frac{1}{3}}}$$

$\sqrt{a} = a^{\frac{1}{2}}: \qquad = \left(z^{\frac{1}{3}}\right)^{\frac{1}{2}}$

$(a^r)^s = a^{rs}: \qquad = z^{\frac{1}{3}\cdot\frac{1}{2}}$

$\qquad = z^{\frac{1}{6}}$

Write rational exponent as a radical: $\quad = \sqrt[6]{z}$

Quick ✔

In Problems 11–14, use rational exponents to simplify each radical.

11. $\sqrt[10]{36^5}$

12. $\sqrt[4]{16a^8b^{12}}$

13. $\dfrac{\sqrt[3]{x^2}}{\sqrt[4]{x}}$

14. $\sqrt[4]{\sqrt[3]{a^2}}$

▶ ❸ Factor Expressions Containing Rational Exponents

Often, expressions involving rational exponents contain a common factor. When this occurs, factor out the common factor to write the expression in implified form. The goal of this type of problem is to write the expression as either a single product or a single quotient. The following examples illustrate the idea.

EXAMPLE 5 **Writing an Expression Containing Rational Exponents as a Single Product**

Simplify $9x^{\frac{4}{3}} + 4x^{\frac{1}{3}}(3x + 5)$ by factoring out $x^{\frac{1}{3}}$.

Solution

Clearly, $x^{\frac{1}{3}}$ is a factor of the second term, $4x^{\frac{1}{3}}(3x + 5)$. It is also a factor of the first term, $9x^{\frac{4}{3}}$. We can see this by rewriting $9x^{\frac{4}{3}}$ as $9x^{\frac{3}{3}+\frac{1}{3}} = 9x^{\frac{3}{3}} \cdot x^{\frac{1}{3}} = 9x \cdot x^{\frac{1}{3}}$. Now factor out $x^{\frac{1}{3}}$.

$$9x^{\frac{4}{3}} + 4x^{\frac{1}{3}}(3x + 5) = 9x \cdot x^{\frac{1}{3}} + 4x^{\frac{1}{3}}(3x + 5)$$

Factor out $x^{1/3}$: $\qquad = x^{\frac{1}{3}}\left(9x + 4(3x + 5)\right)$

Distribute the 4: $\qquad = x^{\frac{1}{3}}(9x + 12x + 20)$

Combine like terms: $\qquad = x^{\frac{1}{3}}(21x + 20)$

Work Smart

When factoring out the greatest common factor, factor out the variable expression raised to the largest exponent that the expressions have in common. For example,

$$3x^5 + 12x^2 = 3x^2(x^3 + 4)$$

Quick ✔

15. Simplify $8x^{\frac{3}{2}} + 3x^{\frac{1}{2}}(4x + 3)$ by factoring out $x^{\frac{1}{2}}$.

EXAMPLE 6 **Writing an Expression Containing Rational Exponents as a Single Quotient**

Simplify $4x^{\frac{1}{2}} + x^{-\frac{1}{2}}(2x + 1)$ by factoring out $x^{-\frac{1}{2}}$.

Solution

Clearly, $x^{-\frac{1}{2}}$ is a factor of the second term, $x^{-\frac{1}{2}}(2x + 1)$. It is also a factor of the first term, $4x^{\frac{1}{2}}$. We can see this by rewriting $4x^{\frac{1}{2}}$ as $4x^{\frac{2}{2}-\frac{1}{2}} = 4x^{\frac{2}{2}} \cdot x^{-\frac{1}{2}} = 4x \cdot x^{-\frac{1}{2}}$. Factor out $x^{-\frac{1}{2}}$ as the greatest common factor (GCF).

$$4x^{\frac{1}{2}} + x^{-\frac{1}{2}}(2x + 1) = 4x \cdot x^{-\frac{1}{2}} + x^{-\frac{1}{2}}(2x + 1)$$

Factor out $x^{-1/2}$: $= x^{-\frac{1}{2}}(4x + (2x + 1))$

Combine like terms: $= x^{-\frac{1}{2}}(6x + 1)$

Rewrite without negative exponents: $= \dfrac{6x + 1}{x^{\frac{1}{2}}}$

Quick ✓

16. Simplify $9x^{\frac{1}{3}} + x^{-\frac{2}{3}}(3x + 1)$ by factoring out $x^{-\frac{2}{3}}$.

6.2 Exercises MyMathLab® MathXL® PRACTICE

*Problems **1–16** are the Quick ✓s that follow the* **EXAMPLES**.

Building Skills

In Problems 17–38, simplify each of the following expressions. See Objective 1.

17. $5^{\frac{1}{2}} \cdot 5^{\frac{3}{2}}$

18. $3^{\frac{1}{3}} \cdot 3^{\frac{5}{3}}$

19. $\dfrac{8^{\frac{5}{4}}}{8^{\frac{1}{4}}}$

20. $\dfrac{10^{\frac{7}{5}}}{10^{\frac{2}{5}}}$

21. $2^{\frac{1}{3}} \cdot 2^{-\frac{3}{2}}$

22. $9^{-\frac{5}{4}} \cdot 9^{\frac{1}{3}}$

23. $\dfrac{x^{\frac{1}{4}}}{x^{\frac{5}{6}}}$

24. $\dfrac{y^{\frac{1}{5}}}{y^{\frac{9}{10}}}$

25. $(4^{\frac{1}{3}})^{\frac{3}{8}}$

26. $(9^{\frac{3}{5}})^{\frac{5}{6}}$

27. $(25^{\frac{3}{4}} \cdot 4^{-\frac{3}{4}})^2$

28. $(36^{-\frac{1}{4}} \cdot 9^{\frac{1}{4}})^{-2}$

29. $(x^{\frac{3}{4}} \cdot y^{\frac{1}{3}})^{\frac{2}{3}}$

30. $(a^{\frac{5}{4}} \cdot b^{\frac{3}{2}})^{\frac{2}{5}}$

31. $(x^{-\frac{1}{3}} \cdot y)(x^{\frac{1}{2}} \cdot y^{-\frac{4}{3}})$

32. $(a^{\frac{4}{3}} \cdot b^{-\frac{1}{2}})(a^{-2} \cdot b^{\frac{5}{2}})$

33. $(4a^2b^{-\frac{3}{2}})^{\frac{1}{2}}$

34. $(25p^{\frac{2}{5}}q^{-1})^{\frac{1}{2}}$

35. $\left(\dfrac{x^{\frac{2}{3}}y^{-\frac{1}{3}}}{8x^{\frac{1}{2}}y}\right)^{\frac{1}{3}}$

36. $\left(\dfrac{64m^{\frac{1}{2}}n}{m^{-2}n^{\frac{4}{3}}}\right)^{\frac{1}{2}}$

37. $\left(\dfrac{50x^{\frac{3}{4}}y}{2x^{\frac{1}{2}}}\right)^{\frac{1}{2}} + \left(\dfrac{x^{\frac{1}{2}}y^{\frac{1}{2}}}{9x^{\frac{3}{4}}y^{\frac{3}{2}}}\right)^{-\frac{1}{2}}$

38. $\left(\dfrac{27x^{\frac{1}{2}}y^{-1}}{y^{-\frac{2}{3}}x^{-\frac{1}{2}}}\right)^{\frac{1}{3}} - \left(\dfrac{4x^{\frac{4}{3}}y^{\frac{4}{9}}}{x^{-\frac{1}{3}}y^{\frac{2}{3}}}\right)^{\frac{1}{2}}$

In Problems 39–54, use rational exponents to simplify each radical. Assume all variables are positive. See Objective 2.

39. $\sqrt{x^8}$

40. $\sqrt[3]{x^6}$

41. $\sqrt[12]{8^4}$

42. $\sqrt[9]{125^6}$

43. $\sqrt[3]{8a^3b^{12}}$

44. $\sqrt{25x^4y^6}$

45. $\dfrac{\sqrt{x}}{\sqrt[4]{x}}$

46. $\dfrac{\sqrt[3]{y^2}}{\sqrt{y}}$

47. $\sqrt{x} \cdot \sqrt[3]{x}$

48. $\sqrt[4]{p^3} \cdot \sqrt[3]{p}$

49. $\sqrt{\sqrt[4]{x^3}}$

50. $\sqrt[3]{\sqrt{x^3}}$

51. $\sqrt{3} \cdot \sqrt[3]{9}$

52. $\sqrt{5} \cdot \sqrt[3]{25}$

53. $\dfrac{\sqrt{6}}{\sqrt[4]{36}}$

54. $\dfrac{\sqrt[4]{49}}{\sqrt{7}}$

For Problems 55–64, see Objective 3.

55. Simplify $2x^{\frac{3}{2}} + 3x^{\frac{1}{2}}(x + 5)$ by factoring out $x^{\frac{1}{2}}$.

56. Simplify $6x^{\frac{4}{3}} + 4x^{\frac{1}{3}}(2x - 3)$ by factoring out $x^{\frac{1}{3}}$.

57. Simplify $5(x + 2)^{\frac{2}{3}}(3x - 2) + 9(x + 2)^{\frac{5}{3}}$ by factoring out $(x + 2)^{\frac{2}{3}}$.

58. Simplify $3(x - 5)^{\frac{1}{2}}(3x + 1) + 6(x - 5)^{\frac{3}{2}}$ by factoring out $(x - 5)^{\frac{1}{2}}$.

59. Simplify $x^{-\frac{1}{2}}(2x + 5) + 4x^{\frac{1}{2}}$ by factoring out $x^{-\frac{1}{2}}$.

60. Simplify $x^{-\frac{2}{3}}(3x + 2) + 9x^{\frac{1}{3}}$ by factoring out $x^{-\frac{2}{3}}$.

61. Simplify $2(x - 4)^{-\frac{1}{3}}(4x - 3) + 12(x - 4)^{\frac{2}{3}}$ by factoring out $2(x - 4)^{-\frac{1}{3}}$.

62. Simplify $4(x + 3)^{\frac{1}{2}} + (x + 3)^{-\frac{1}{2}}(2x + 1)$ by factoring out $(x + 3)^{-\frac{1}{2}}$.

63. Simplify $15x(x^2 + 4)^{\frac{1}{2}} + 5(x^2 + 4)^{\frac{3}{2}}$.

64. Simplify $24x(x^2 - 1)^{\frac{1}{3}} + 9(x^2 - 1)^{\frac{4}{3}}$.

Mixed Practice

In Problems 65–76, simplify each expression.

65. $\sqrt[8]{4^4}$

66. $\sqrt[6]{27^2}$

67. $(-2)^{\frac{1}{2}} \cdot (-2)^{\frac{3}{2}}$

68. $25^{\frac{3}{4}} \cdot 25^{\frac{3}{4}}$

69. $(100^{\frac{1}{3}})^{\frac{3}{2}}$

70. $(8^4)^{\frac{5}{12}}$

71. $(\sqrt[4]{25})^2$

72. $(\sqrt[6]{27})^2$

73. $\sqrt[4]{x^2} - \dfrac{\sqrt[4]{x^6}}{x}$

74. $\sqrt[9]{a^6} - \dfrac{\sqrt[6]{a^5}}{\sqrt[6]{a}}$

75. $(4 \cdot 9^{\frac{1}{4}})^{-2}$

76. $(4^{-1} \cdot 81^{\frac{1}{2}})^{\frac{1}{2}}$

In Problems 77–82, distribute and simplify.

77. $x^{\frac{1}{2}}(x^{\frac{3}{2}} - 2)$

78. $x^{\frac{1}{3}}(x^{\frac{5}{3}} + 4)$

79. $2y^{-\frac{1}{3}}(1 + 3y)$

80. $3a^{-\frac{1}{2}}(2 - a)$

81. $4z^{\frac{3}{2}}(z^{\frac{3}{2}} - 8z^{-\frac{3}{2}})$

82. $8p^{\frac{2}{3}}(p^{\frac{4}{3}} - 4p^{-\frac{2}{3}})$

Applying the Concepts

83. If $3^x = 25$, what does $3^{\frac{x}{2}}$ equal?

84. If $5^x = 64$, what does $5^{\frac{x}{3}}$ equal?

85. If $7^x = 9$, what does $\sqrt{7^x}$ equal?

86. If $5^x = 27$, what does $\sqrt[3]{5^x}$ equal?

Extending the Concepts

In Problems 87 and 88, simplify the expression using rational exponents.

87. $\sqrt[4]{\sqrt[3]{\sqrt{x}}}$

88. $\sqrt[5]{\sqrt[3]{\sqrt{x^2}}}$

89. Without using a calculator, determine the value of $(6^{\sqrt{2}})^{\sqrt{2}}$.

90. Determine the domain of $g(x) = (x - 3)^{\frac{1}{2}}(x - 1)^{-\frac{1}{2}}$.

91. Determine the domain of $f(x) = (x + 3)^{\frac{1}{2}}(x + 1)^{-\frac{1}{2}}$.

Synthesis Review

In Problems 92–95, simplify each expression.

92. $(2x - 1)(x + 4) - (x + 1)(x - 1)$

93. $3a(a - 3) + (a + 3)(a - 2)$

94. $\dfrac{\sqrt{x^2 + 4x + 4}}{x + 2}, x + 2 > 0$

95. $\dfrac{x^2 - 4}{x + 2} \cdot (x + 5) - (x + 4)(x - 1)$

6.3 Simplifying Radical Expressions Using Properties of Radicals

Objectives

1. Use the Product Property to Multiply Radical Expressions
2. Use the Product Property to Simplify Radical Expressions
3. Use the Quotient Property to Simplify Radical Expressions
4. Multiply Radicals with Unlike Indices

Are You Ready for This Section?

Before getting started, take the following readiness quiz. If you get a problem wrong, go back to the section cited and review the material.

R1. List the perfect squares that are less than 200.

R2. List the perfect cubes that are less than 200.

R3. Simplify: **(a)** $\sqrt{16}$ **(b)** $\sqrt{p^2}$ [Getting Ready: Square Roots, pp. 467–470]

Ready?...Answers **R1.** 1, 4, 9, 16, 25, 36, 49, 64, 81, 100, 121, 144, 169, 196 **R2.** 1, 8, 27, 64, 125 **R3.** (a) 4 (b) $|p|$

▶ ① Use the Product Property to Multiply Radical Expressions

Perhaps you are noticing a trend: After we introduce a new algebraic expression, we learn how to multiply, divide, add, and subtract the algebraic expression. Well, here we go again! First, we are going to learn how to multiply radical expressions when they have the same index.

Consider that

$$\sqrt{4 \cdot 25} = \sqrt{100} = 10 \quad \text{and} \quad \sqrt{4} \cdot \sqrt{25} = 2 \cdot 5 = 10$$

This suggests the following result:

> **In Words**
> $\sqrt[n]{a} \cdot \sqrt[n]{b} = \sqrt[n]{ab}$ means the product of the roots equals the root of the product, provided the index is the same.

Product Property Of Radicals

If $\sqrt[n]{a}$ and $\sqrt[n]{b}$ are real numbers and $n \geq 2$ is an integer, then

$$\sqrt[n]{a} \cdot \sqrt[n]{b} = \sqrt[n]{ab}$$

We can justify this formula using rational exponents.

$$\sqrt[n]{a} \cdot \sqrt[n]{b} = a^{\frac{1}{n}} \cdot b^{\frac{1}{n}}$$

Product-to-a-Power Rule: $\qquad = (a \cdot b)^{\frac{1}{n}}$

$a^{\frac{1}{n}} = \sqrt[n]{a}: \qquad = \sqrt[n]{a \cdot b}$

EXAMPLE 1) **Using the Product Property to Multiply Radicals**

Multiply.

(a) $\sqrt{5} \cdot \sqrt{3}$ **(b)** $\sqrt[3]{2} \cdot \sqrt[3]{13}$ **(c)** $\sqrt{x-3} \cdot \sqrt{x+3}$ **(d)** $\sqrt[5]{6c} \cdot \sqrt[5]{7c^2}$

Solution

(a) $\sqrt{5} \cdot \sqrt{3} = \sqrt{5 \cdot 3} = \sqrt{15}$

(b) $\sqrt[3]{2} \cdot \sqrt[3]{13} = \sqrt[3]{2 \cdot 13} = \sqrt[3]{26}$

(c) $\sqrt{x-3} \cdot \sqrt{x+3} = \sqrt{(x-3)(x+3)} = \sqrt{x^2 - 9}$ if $x \geq 3$

(d) $\sqrt[5]{6c} \cdot \sqrt[5]{7c^2} = \sqrt[5]{6c \cdot 7c^2} = \sqrt[5]{42c^3}$

> **Work Smart**
> In Example 1(c), notice that $\sqrt{x^2 - 9}$ does not equal $\sqrt{x^2} - \sqrt{9}$.

Quick ✓

1. If $\sqrt[n]{a}$ and $\sqrt[n]{b}$ are real numbers and $n \geq 2$ is an integer, then $\sqrt[n]{a} \cdot \sqrt[n]{b} =$ ___.

In Problems 2–5, multiply each radical expression.

2. $\sqrt{11} \cdot \sqrt{7}$ **3.** $\sqrt[4]{6} \cdot \sqrt[4]{7}$ **4.** $\sqrt{x-5} \cdot \sqrt{x+5}, x \geq 5$ **5.** $\sqrt[7]{5p} \cdot \sqrt[7]{4p^3}$

▶ ❷ **Use the Product Property to Simplify Radical Expressions**

So far, we have simplified radicals only when the radicand was a perfect square, such as $\sqrt{81} = 9$, or a perfect cube, such as $\sqrt[3]{\dfrac{1}{8}} = \dfrac{1}{2}$. When a radical does not simplify to a rational number, we can do one of two things:

1. Write a decimal approximation of the radical.

2. Simplify the radical using properties of radicals, if possible.

We have already found the decimal approximation of radicals using a calculator in the Getting Ready: Square Roots and in Section 6.1. Now we will learn how to use properties of radicals to write the radical in simplified form.

The advantage of simplifying over approximating is that simplifying maintains an *exact* radical rather than an approximate value.

Recall that a number that is the square of a rational number is called a perfect square. Thus $1^2 = 1, 2^2 = 4, 3^2 = 9$, and so on are perfect squares. A number that is the cube of a rational number is called a perfect cube. Thus $1^3 = 1, 2^3 = 8, 3^3 = 27$, $(-1)^3 = -1, (-2)^3 = -8$, and so on are perfect cubes. In general, if n is the index of a radical, then a^n is a perfect power of a, where a is a rational number.

Work Smart

index $\rightarrow \sqrt[n]{a} \leftarrow$ radicand

A radical expression with index n is **simplified** if the radicand does not contain any factors that are perfect nth powers. For example, $\sqrt{50}$ is not simplified because 25 is a factor of 50, and 25 is a perfect square; $\sqrt[3]{16}$ is not simplified because 8 is a factor of 16, and 8 is a perfect cube; $\sqrt[3]{x^5}$ is not simplified because $x^5 = x^3 \cdot x^2$, and x^3 is a perfect cube.

To simplify radicals, we use the Product Property of Radicals "in reverse." That is, we use $\sqrt[n]{ab} = \sqrt[n]{a} \cdot \sqrt[n]{b}$.

EXAMPLE 2   **How to Use the Product Property to Simplify a Radical**

Simplify: $\sqrt{18}$

Step-by-Step Solution

Step 1: What is the index on the radical? Since the index is 2, write each factor of the radicand as the product of two factors, one of which is a perfect square.

The perfect squares are 1, 4, 9, 16, 25, ...
Because 9 is a factor of 18, and 9 is a perfect square, write 18 as $9 \cdot 2$. $\sqrt{18} = \sqrt{9 \cdot 2}$

Step 2: Write the radicand as the product of two radicals, one of which contains a perfect square.

$= \sqrt{9} \cdot \sqrt{2}$

Step 3: Take the square root of each perfect square.

$= 3\sqrt{2}$ ●

We summarize the steps used in Example 2 below.

Work Smart

When performing Step 1 with real numbers, look for the *largest* factor of the radicand that is a perfect power of the index.

> **Simplifying a Radical Expression of the Form $\sqrt[n]{a}$**
>
> **Step 1:** Write each factor of the radicand as the product of two factors, one of which is a perfect nth power.
>
> **Step 2:** Using the Product Property of Radicals, write the radicand as the product of two radicals, one of which contains perfect nth powers.
>
> **Step 3:** Take the nth root of each perfect nth power.

EXAMPLE 3   **Using the Product Property to Simplify a Radical**

Simplify each of the following:

(a) $5\sqrt[3]{24}$ **(b)** $\sqrt{128x^2}$ **(c)** $\sqrt[4]{20}$

Solution

(a) First, notice that the index is 3. Therefore, we want the largest factor of 24 that is a perfect cube. The positive perfect cubes are 1, 8, 27, ... Because 8 is a factor of 24 and is a perfect cube, write 24 as $8 \cdot 3$.

$$5\sqrt[3]{24} = 5 \cdot \sqrt[3]{8 \cdot 3}$$
$$\sqrt[n]{ab} = \sqrt[n]{a} \cdot \sqrt[n]{b}: \ = 5 \cdot \sqrt[3]{8} \cdot \sqrt[3]{3}$$
$$= 5 \cdot 2 \cdot \sqrt[3]{3}$$
$$= 10\sqrt[3]{3}$$

(b) The index is 2. Because 64 is a factor of 128, and 64 is a perfect square, write 128 as $64 \cdot 2$; x^2 is a perfect square.

$$\sqrt{128x^2} = \sqrt{64x^2 \cdot 2}$$
$$\sqrt[n]{ab} = \sqrt[n]{a} \cdot \sqrt[n]{b}: \ = \sqrt{64x^2} \cdot \sqrt{2}$$
$$\sqrt{64x^2} = \sqrt{64} \cdot \sqrt{x^2} = 8|x|: \ = 8|x|\sqrt{2}$$

(c) In $\sqrt[4]{20}$, the index is 4. The fourth powers (or perfect fourths) are 1, 16, 81,.... There are no factors of 20 that are fourth powers, so the radical $\sqrt[4]{20}$ cannot be simplified any further.

Quick ✓

6. List the first six positive integers that are perfect squares.

7. List the first six positive integers that are perfect cubes.

In Problems 8–11, simplify each of the radical expressions.

8. $\sqrt{48}$ **9.** $4\sqrt[3]{54}$ **10.** $\sqrt{200a^2}$ **11.** $\sqrt[4]{40}$

EXAMPLE 4   **Simplifying an Expression Involving a Square Root**

Simplify: $\dfrac{4 - \sqrt{20}}{2}$

Solution

We can simplify this expression using two different approaches.

Method 1:

4 is the largest perfect square factor of 20

$$\frac{4 - \sqrt{20}}{2} = \frac{4 - \sqrt{4 \cdot 5}}{2}$$

Use $\sqrt{a \cdot b} = \sqrt{a} \cdot \sqrt{b}$: $= \dfrac{4 - \sqrt{4} \cdot \sqrt{5}}{2}$

$$= \frac{4 - 2 \cdot \sqrt{5}}{2}$$

Factor out the 2 in the numerator: $= \dfrac{2(2 - \sqrt{5})}{2}$

Divide out common factor: $= 2 - \sqrt{5}$

Method 2:

4 is the largest perfect square factor of 20

$$\frac{4 - \sqrt{20}}{2} = \frac{4 - \sqrt{4 \cdot 5}}{2}$$

Use $\sqrt{a \cdot b} = \sqrt{a} \cdot \sqrt{b}$: $= \dfrac{4 - \sqrt{4} \cdot \sqrt{5}}{2}$

$$= \frac{4 - 2 \cdot \sqrt{5}}{2}$$

Use $\dfrac{A + B}{C} = \dfrac{A}{C} + \dfrac{B}{C}$: $= \dfrac{4}{2} - \dfrac{2 \cdot \sqrt{5}}{2}$

Divide out common factor: $= 2 - \sqrt{5}$

Quick ✓

In Problems 12 and 13, simplify the expression.

12. $\dfrac{6 + \sqrt{45}}{3}$ **13.** $\dfrac{-2 + \sqrt{32}}{4}$

Recall how to simplify $\sqrt[n]{a^n}$ from Section 6.1.

Simplifying $\sqrt[n]{a^n}$

If $n \geq 2$ is a positive integer and a is a real number, then

$$\sqrt[n]{a^n} = |a| \quad \text{if } n \geq 2 \text{ is even}$$
$$\sqrt[n]{a^n} = a \quad \text{if } n \geq 3 \text{ is odd}$$

This means that

$$\sqrt{a^2} = |a| \quad \sqrt[3]{a^3} = a \quad \sqrt[4]{a^4} = |a| \quad \sqrt[5]{a^5} = a \quad \text{and so on}$$

In order to make our mathematical lives a little easier, throughout this section we shall assume that all variables in the radicand are greater than or equal to zero (nonnegative). Thus

$$\sqrt{a^2} = a \quad \sqrt[3]{a^3} = a \quad \sqrt[4]{a^4} = a \quad \sqrt[5]{a^5} = a \quad \text{and so on}$$

What if the exponent on the radicand is greater than the index, as in $\sqrt{x^3}$ or $\sqrt[3]{x^6}$? We could use the Laws of Exponents along with the rule for simplifying $\sqrt[n]{a^n}$, or we could use rational exponents.

$$\sqrt{x^6} = \sqrt{(x^3)^2} = x^3 \quad \text{or} \quad \sqrt{x^6} = x^{\frac{6}{2}} = x^3$$
$$\sqrt[3]{x^{12}} = \sqrt[3]{(x^4)^3} = x^4 \quad \text{or} \quad \sqrt[3]{x^{12}} = x^{\frac{12}{3}} = x^4$$

EXAMPLE 5 | **Simplifying a Radical with a Variable Radicand**

Simplify $\sqrt{20x^{10}}$. Assume $x \geq 0$.

Solution

Because 4 is the largest perfect square factor of 20, write 20 as $4 \cdot 5$.

$$\sqrt{20x^{10}} = \sqrt{4 \cdot 5 \cdot x^{10}}$$
$$= \sqrt{4x^{10}} \cdot \sqrt{5}$$
$$\sqrt{4} = 2, \ \sqrt{x^{10}} = \sqrt{(x^5)^2} = x^5$$
$$\text{or } \sqrt{x^{10}} = x^{\frac{10}{2}} = x^5: \quad = 2x^5\sqrt{5}$$

Quick ✓

14. Simplify $\sqrt{75a^6}$. Assume $a \geq 0$.

What if the index does not divide evenly into the exponent on the variable in the radicand, as in $\sqrt[3]{x^8}$? In this case, we rewrite the variable expression as the product of two variable expressions where one of the factors has an exponent that is a multiple of the index:

$$\text{3 divides evenly into 6:} \quad \sqrt[3]{x^8} = \sqrt[3]{x^6 \cdot x^2}$$

So we have

$$\sqrt[3]{x^8} = \sqrt[3]{x^6 \cdot x^2} = \sqrt[3]{x^6} \cdot \sqrt[3]{x^2} = x^2\sqrt[3]{x^2}$$

EXAMPLE 6 | **Simplifying Radicals**

Simplify:

(a) $\sqrt{80a^3}$ (b) $\sqrt[3]{27m^4n^{14}}$

Solution

(a)
$$80 = 16 \cdot 5; a^3 = a^2 \cdot a$$
$$\sqrt{80a^3} = \sqrt{16 \cdot 5 \cdot a^2 \cdot a}$$
$$= \sqrt{16a^2 \cdot 5a}$$
$$\sqrt[n]{ab} = \sqrt[n]{a} \cdot \sqrt[n]{b}: \quad = \sqrt{16a^2} \cdot \sqrt{5a}$$
$$\sqrt{a^2} = a \text{ assuming } a \geq 0: \quad = 4a\sqrt{5a}$$

(b)

$$m^4 = m^3 \cdot m; \; n^{14} = n^{12} \cdot n^2$$

$$\sqrt[3]{27m^4n^{14}} = \sqrt[3]{27 \cdot m^3 \cdot m \cdot n^{12} \cdot n^2}$$
$$= \sqrt[3]{27 \cdot m^3 \cdot n^{12} \cdot m \cdot n^2}$$

$\sqrt[n]{ab} = \sqrt[n]{a} \cdot \sqrt[n]{b}:$ $\quad = \sqrt[3]{27m^3n^{12}} \cdot \sqrt[3]{mn^2}$

$\sqrt[3]{27} = 3; \; \sqrt[3]{m^3} = m; \; \sqrt[3]{n^{12}} = n^{\frac{12}{3}} = n^4:$ $\quad = 3mn^4\sqrt[3]{mn^2}$

Quick ✓

In Problems 15–17, simplify each radical. Assume all variables are greater than or equal to zero.

15. $\sqrt{18a^5}$ **16.** $\sqrt[3]{128x^6y^{10}}$ **17.** $\sqrt[4]{16a^5b^{11}}$

In this next example, we first multiply radical expressions and then simplify the product.

▶ **EXAMPLE 7** **Multiplying and Simplifying Radicals**

Multiply and simplify:

     (a) $\sqrt{3} \cdot \sqrt{15}$ **(b)** $3\sqrt[3]{4x} \cdot \sqrt[3]{2x^4}$ **(c)** $\sqrt[4]{27a^2b^5} \cdot \sqrt[4]{6a^3b^6}$

Assume all variables are greater than or equal to zero.

Solution

Remember, to multiply two radicals, the index must be the same. When we have the same index, multiply the radicands and then simplify the product.

     (a) Start by looking to see whether the index is the same. The index on both radicals is 2, so multiply the radicands.

$$\sqrt{3} \cdot \sqrt{15} = \sqrt{3 \cdot 15}$$
$$= \sqrt{45}$$

9 is the largest perfect square factor of 45: $\quad = \sqrt{9 \cdot 5}$

Product Property of Radicals: $\quad = \sqrt{9} \cdot \sqrt{5}$

$$= 3\sqrt{5}$$

     (b) The index on both radicals is 3, so multiply the radicands.

$$3\sqrt[3]{4x} \cdot \sqrt[3]{2x^4} = 3\sqrt[3]{4x \cdot 2x^4}$$
$$= 3\sqrt[3]{8x^5}$$

x^3 is a perfect cube; 8 is a perfect cube: $\quad = 3\sqrt[3]{8x^3 \cdot x^2}$

Product Property of Radicals: $\quad = 3\sqrt[3]{8x^3} \cdot \sqrt[3]{x^2}$

$\sqrt[3]{8} = 2; \; \sqrt[3]{x^3} = x:$ $\quad = 3 \cdot 2 \cdot x \cdot \sqrt[3]{x^2}$

$$= 6x\sqrt[3]{x^2}$$

     (c) The index on both radicals is 4, so multiply the radicands.

$$\sqrt[4]{27a^2b^5} \cdot \sqrt[4]{6a^3b^6} = \sqrt[4]{3^3 \cdot a^2b^5 \cdot 3 \cdot 2 \cdot a^3b^6}$$
$$= \sqrt[4]{3^4 \cdot 2 \cdot a^5b^{11}}$$

3^4, a^4, and b^8 are perfect fourths: $\quad = \sqrt[4]{3^4 \cdot 2 \cdot a^4 \cdot a \cdot b^8 \cdot b^3}$

Product Property of Radicals: $\quad = \sqrt[4]{3^4a^4b^8} \cdot \sqrt[4]{2ab^3}$

$$= 3ab^2\sqrt[4]{2ab^3}$$

Work Smart

Notice that $27 = 3^3$ and that $6 = 3 \cdot 2$, so that $27 \cdot 6 = 3^3 \cdot 3 \cdot 2 = 3^4 \cdot 2$. This makes finding the perfect fourth powers a lot easier!

Quick ✓

In Problems 18–20, multiply and simplify the radicals. Assume all variables are greater than or equal to zero.

18. $\sqrt{6} \cdot \sqrt{8}$ **19.** $\sqrt[3]{12a^2} \cdot \sqrt[3]{10a^4}$ **20.** $4\sqrt[3]{8a^2b^5} \cdot \sqrt[3]{6a^2b^4}$

▶ ❸ Use the Quotient Property to Simplify Radical Expressions

Now consider that

$$\sqrt{\frac{64}{4}} = \sqrt{\frac{4 \cdot 16}{4}} = \sqrt{16} = 4 \quad \text{and} \quad \frac{\sqrt{64}}{\sqrt{4}} = \frac{8}{2} = 4$$

This suggests the following result:

> **In Words**
>
> $$\sqrt[n]{\frac{a}{b}} = \frac{\sqrt[n]{a}}{\sqrt[n]{b}}$$
>
> means "the root of the quotient equals the quotient of the roots" provided that the radicals have the same index.

Quotient Property of Radicals

If $\sqrt[n]{a}$ and $\sqrt[n]{b}$ are real numbers, $b \neq 0$, and $n \geq 2$ is an integer, then

$$\frac{\sqrt[n]{a}}{\sqrt[n]{b}} = \sqrt[n]{\frac{a}{b}}$$

We can justify this formula using rational exponents.

$$\frac{\sqrt[n]{a}}{\sqrt[n]{b}} = \frac{a^{\frac{1}{n}}}{b^{\frac{1}{n}}} = \left(\frac{a}{b}\right)^{\frac{1}{n}} = \sqrt[n]{\frac{a}{b}}$$

EXAMPLE 8   **Using the Quotient Property to Simplify Radicals**

Simplify:

(a) $\sqrt{\dfrac{18}{25}}$ **(b)** $\sqrt[3]{\dfrac{6z^3}{125}}$ **(c)** $\sqrt[4]{\dfrac{10a^2}{81b^4}}, b \neq 0$

Assume all variables are greater than or equal to zero.

Solution

In each of these problems, the expression in the denominator is a perfect nth power, and the index is n. Therefore, simplify the expression using the Quotient Rule "in reverse": $\sqrt[n]{\dfrac{a}{b}} = \dfrac{\sqrt[n]{a}}{\sqrt[n]{b}}$.

(a) $\sqrt{\dfrac{18}{25}} = \dfrac{\sqrt{18}}{\sqrt{25}}$

$= \dfrac{3\sqrt{2}}{5}$

(b) $\sqrt[3]{\dfrac{6z^3}{125}} = \dfrac{\sqrt[3]{6z^3}}{\sqrt[3]{125}}$

$= \dfrac{z\sqrt[3]{6}}{5}$

(c) $\sqrt[4]{\dfrac{10a^2}{81b^4}} = \dfrac{\sqrt[4]{10a^2}}{\sqrt[4]{81b^4}}$

$= \dfrac{\sqrt[4]{10a^2}}{3b}$ ●

Quick ✓

In Problems 21–23, simplify the radicals. Assume all variables are greater than or equal to zero.

21. $\sqrt{\dfrac{13}{49}}$ **22.** $\sqrt[3]{\dfrac{27p^3}{8}}$ **23.** $\sqrt[4]{\dfrac{3q^4}{16}}$

▶ **EXAMPLE 9**   **Using the Quotient Property to Simplify Radicals**

Simplify;

(a) $\dfrac{\sqrt{24a^3}}{\sqrt{6a}}$ **(b)** $\dfrac{-2\sqrt[3]{54a}}{\sqrt[3]{2a^4}}$ **(c)** $\dfrac{\sqrt[3]{-375x^2y}}{\sqrt[3]{3x^{-1}y^7}}$

Assume all variables are greater than zero.

Solution

In these problems, the radical expression in the denominator cannot be simplified. However, the index on the numerator and denominator of each expression is the same, so we can write each expression as a single radical.

(a) $\dfrac{\sqrt{24a^3}}{\sqrt{6a}} = \sqrt{\dfrac{24a^3}{6a}}$

$= \sqrt{4a^2}$

$= 2a$

(b) $\dfrac{-2\sqrt[3]{54a}}{\sqrt[3]{2a^4}} = -2 \cdot \sqrt[3]{\dfrac{54a}{2a^4}}$

$= -2 \cdot \sqrt[3]{\dfrac{27}{a^3}}$

$= -2 \cdot \dfrac{3}{a}$

$= -\dfrac{6}{a}$

(c) $\dfrac{\sqrt[3]{-375x^2y}}{\sqrt[3]{3x^{-1}y^7}} = \sqrt[3]{\dfrac{-375x^2y}{3x^{-1}y^7}}$

$= \sqrt[3]{\dfrac{-125x^{2-(-1)}}{y^{7-1}}}$

$= \sqrt[3]{\dfrac{-125x^3}{y^6}}$

$= -\dfrac{5x}{y^2}$

Quick ✓

In Problems 24–26, simplify the radicals. Assume all variables are greater than zero.

24. $\dfrac{\sqrt{12a^5}}{\sqrt{3a}}$

25. $\dfrac{\sqrt[3]{-24x^2}}{\sqrt[3]{3x^{-1}}}$

26. $\dfrac{\sqrt[3]{250a^5b^{-2}}}{\sqrt[3]{2ab}}$

▶ ❹ Multiply Radicals with Unlike Indices

We already know the rule $\sqrt[n]{a} \cdot \sqrt[n]{b} = \sqrt[n]{ab}$. This rule works only when the index on each radical is the same. What if the index on each radical is different? Can we still simplify the product? The answer is yes! We use the fact that $\sqrt[n]{a} = a^{\frac{1}{n}}$.

EXAMPLE 10 **Multiplying Radicals with Unlike Indices**

Multiply and simplify: $\sqrt[4]{8} \cdot \sqrt[3]{5}$

Solution

Notice that the indices are not the same, so we cannot use $\sqrt[n]{a} \cdot \sqrt[n]{b} = \sqrt[n]{ab}$. We will use rational exponents along with $\sqrt[n]{a} = a^{\frac{1}{n}}$ instead.

$$\sqrt[4]{8} \cdot \sqrt[3]{5} = 8^{\frac{1}{4}} \cdot 5^{\frac{1}{3}}$$

$$\text{LCD} = 12: \qquad = 8^{\frac{3}{12}} \cdot 5^{\frac{4}{12}}$$

$$a^{\frac{r}{s}} = (a^r)^{\frac{1}{s}}: \qquad = \left[(8^3)^{\frac{1}{12}} \cdot (5^4)^{\frac{1}{12}} \right]$$

$$a^r \cdot b^r = (ab)^r: \qquad = \left[(8^3)(5^4) \right]^{\frac{1}{12}}$$

$$= (320,000)^{\frac{1}{12}}$$

$$a^{\frac{1}{n}} = \sqrt[n]{a}: \qquad = \sqrt[12]{320,000}$$

Quick ✓

In Problems 27 and 28, multiply and simplify.

27. $\sqrt[4]{5} \cdot \sqrt[3]{3}$

28. $\sqrt{10} \cdot \sqrt[3]{12}$

6.3 Exercises

MyMathLab® PRACTICE

Exercise numbers in green are complete video solutions in MyMathLab

Problems 1–28 are the Quick ✓s that follow the EXAMPLES.

Building Skills

In Problems 29–36, use the Product Property to multiply. Assume that all variables can be any real number. See Objective 1.

29. $\sqrt[3]{6} \cdot \sqrt[3]{10}$

30. $\sqrt[3]{-5} \cdot \sqrt[3]{7}$

31. $\sqrt{3a} \cdot \sqrt{5b}$

32. $\sqrt[4]{6a^2} \cdot \sqrt[4]{7b^2}$

33. $\sqrt{x-7} \cdot \sqrt{x+7}, x \geq 7$

34. $\sqrt{p-5} \cdot \sqrt{p+5}, p \geq 5$

35. $\sqrt{\dfrac{5x}{3}} \cdot \sqrt{\dfrac{3}{x}}, x > 0$

36. $\sqrt[3]{\dfrac{-9x^2}{4}} \cdot \sqrt[3]{\dfrac{4}{3x}}, x \neq 0$

In Problems 37–64, simplify each radical using the Product Property. Assume that all variables can be any real number. See Objective 2.

37. $\sqrt{50}$

38. $\sqrt{32}$

39. $\sqrt[3]{54}$

40. $\sqrt[4]{162}$

41. $\sqrt{48x^2}$

42. $\sqrt{20a^2}$

43. $\sqrt[3]{-27x^3}$

44. $\sqrt[3]{-64p^3}$

45. $\sqrt[4]{32m^4}$

46. $\sqrt[4]{48z^4}$

47. $\sqrt{12p^2q}$

48. $\sqrt{45m^2n}$

49. $\sqrt{162m^4}$

50. $\sqrt{98w^8}$

51. $\sqrt{y^{13}}$

52. $\sqrt{s^9}$

53. $\sqrt[3]{c^8}$

54. $\sqrt[5]{x^{12}}$

55. $\sqrt{125p^3q^4}$

56. $\sqrt{243ab^5}$

57. $\sqrt[3]{-16x^9}$

58. $\sqrt[3]{-54q^{12}}$

59. $\sqrt[5]{-16m^8n^2}$

60. $\sqrt{75x^6y}$

61. $\sqrt[4]{(x-y)^5}, x > y$

62. $\sqrt[3]{(a+b)^5}$

63. $\sqrt[3]{8x^3 - 8y^3}$

64. $\sqrt[3]{8a^3 + 8b^3}$

In Problems 65–70, simplify each expression. See Objective 2.

65. $\dfrac{4 + \sqrt{36}}{2}$

66. $\dfrac{5 - \sqrt{100}}{5}$

67. $\dfrac{9 + \sqrt{18}}{3}$

68. $\dfrac{10 - \sqrt{75}}{5}$

69. $\dfrac{7 - \sqrt{98}}{14}$

70. $\dfrac{-6 + \sqrt{108}}{6}$

In Problems 71–88, multiply and simplify. Assume that all variables are greater than or equal to zero. See Objective 2.

71. $\sqrt{5} \cdot \sqrt{5}$

72. $\sqrt{6} \cdot \sqrt{6}$

73. $\sqrt{2} \cdot \sqrt{8}$

74. $\sqrt{3} \cdot \sqrt{12}$

75. $\sqrt[3]{4} \cdot \sqrt[3]{2}$

76. $\sqrt[3]{9} \cdot \sqrt[3]{3}$

77. $\sqrt{5x} \cdot \sqrt{15x}$

78. $\sqrt{6x} \cdot \sqrt{30x}$

79. $\sqrt[3]{4b^2} \cdot \sqrt[3]{6b^2}$

80. $\sqrt[3]{9a} \cdot \sqrt[3]{6a^2}$

81. $2\sqrt{6ab} \cdot 3\sqrt{15ab^3}$

82. $3\sqrt{14pq^3} \cdot 2\sqrt{7pq}$

83. $\sqrt[4]{27p^3q^2} \cdot \sqrt[4]{12p^2q^2}$

84. $\sqrt[3]{16m^2n} \cdot \sqrt[3]{27m^2n}$

85. $\sqrt[5]{-8a^3b^4} \cdot \sqrt[5]{12a^3b}$

86. $\sqrt[5]{-27x^4y^2} \cdot \sqrt[5]{18x^3y^4}$

87. $\sqrt[4]{8(x-y)^2} \cdot \sqrt[4]{6(x-y)^3}, x > y$

88. $\sqrt[3]{9(a+b)^2} \cdot \sqrt[3]{6(a+b)^5}$

In Problems 89–96, simplify each expression. Assume that all variables are greater than zero. See Objective 3.

89. $\sqrt{\dfrac{3}{16}}$

90. $\sqrt{\dfrac{5}{36}}$

91. $\sqrt[4]{\dfrac{5x^4}{16}}$

92. $\sqrt[4]{\dfrac{2a^8}{81}}$

93. $\sqrt{\dfrac{9y^2}{25x^2}}$

94. $\sqrt{\dfrac{4a^4}{81b^2}}$

95. $\sqrt[3]{\dfrac{-27x^9}{64y^{12}}}$

96. $\sqrt[5]{\dfrac{-32a^{15}}{243b^{10}}}$

In Problems 97–110, divide and simplify. Assume that all variables are greater than zero. See Objective 3.

97. $\dfrac{\sqrt{8}}{\sqrt{2}}$

98. $\dfrac{\sqrt{27}}{\sqrt{3}}$

99. $\dfrac{\sqrt[3]{128}}{\sqrt[3]{2}}$

100. $\dfrac{\sqrt[4]{64}}{\sqrt[4]{4}}$

101. $\dfrac{\sqrt{48a^3}}{\sqrt{6a}}$

102. $\dfrac{\sqrt{54y^5}}{\sqrt{3y}}$

103. $\dfrac{\sqrt{24a^5b}}{\sqrt{3ab^3}}$

104. $\dfrac{\sqrt{360m^7n^3}}{\sqrt{5mn^5}}$

105. $\dfrac{\sqrt{512a^7b}}{3\sqrt{2ab^3}}$

106. $\dfrac{\sqrt{375x^2y^7}}{10\sqrt{3y}}$

107. $\dfrac{\sqrt[3]{104a^5}}{\sqrt[3]{4a^{-1}}}$

108. $\dfrac{\sqrt[3]{-128x^8}}{\sqrt[3]{2x^{-1}}}$

109. $\dfrac{\sqrt{90x^3y^{-1}}}{\sqrt{2x^{-3}y}}$

110. $\dfrac{\sqrt{96a^5b^{-3}}}{\sqrt{3a^{-5}b}}$

In Problems 111–118, multiply and simplify. See Objective 4.

111. $\sqrt{3} \cdot \sqrt[3]{4}$

112. $\sqrt{2} \cdot \sqrt[3]{7}$

113. $\sqrt[3]{2} \cdot \sqrt[6]{3}$

114. $\sqrt[4]{3} \cdot \sqrt[8]{5}$

115. $\sqrt{3} \cdot \sqrt[3]{18}$

116. $\sqrt{6} \cdot \sqrt[3]{9}$

117. $\sqrt[4]{9} \cdot \sqrt[6]{12}$

118. $\sqrt[8]{8} \cdot \sqrt[10]{16}$

Mixed Practice

In Problems 119–134, perform the indicated operation and simplify. Assume all variables are greater than zero.

119. $\sqrt[3]{\dfrac{5x}{8}}$

120. $\sqrt[3]{\dfrac{7a^2}{64}}$

121. $\sqrt[3]{5a} \cdot \sqrt[3]{9a}$

122. $\sqrt[5]{8b^2} \cdot \sqrt[5]{3b}$

123. $\sqrt{72a^4}$

124. $\sqrt{24b^6}$

125. $\sqrt[3]{6a^2b} \cdot \sqrt[3]{9ab}$

126. $\sqrt[4]{8x^3y^2} \cdot \sqrt[4]{4x^2y^3}$

127. $\dfrac{\sqrt[3]{-32a}}{\sqrt[3]{2a^4}}$

128. $\dfrac{\sqrt[3]{-250p^2}}{\sqrt[3]{2p^5}}$

129. $-5\sqrt[3]{32m^3}$

130. $-7\sqrt[3]{250p^3}$

131. $\sqrt[5]{81a^4b^7}$

132. $\sqrt[5]{32p^7q^{11}}$

133. $\sqrt[3]{12} \cdot \sqrt[3]{18}$

134. $\sqrt[4]{8} \cdot \sqrt[4]{18}$

Applying the Concepts

△ **135. Length of a Line Segment** The length of the line segment joining the points $(2, 5)$ and $(-1, -1)$ is given by

$$\sqrt{(5 - (-1))^2 + (2 - (-1))^2}$$

(a) Plot the points in the Cartesian plane, and draw a line segment connecting the points.

(b) Express the length of the line segment as a radical in simplified form.

△ **136. Length of a Line Segment** The length of the line segment joining the points $(4, 2)$ and $(-2, 4)$ is given by

$$\sqrt{(4 - 2)^2 + (-2 - 4)^2}$$

(a) Plot the points in the Cartesian plane, and draw a line segment connecting the points.

(b) Express the length of the line segment as a radical in simplified form.

137. Revenue Growth Suppose that the annual revenue R (in millions of dollars) of a company after t years of operating is modeled by the function

$$R(t) = \sqrt[3]{\dfrac{t}{2}}$$

(a) Predict the revenue of the company after 8 years of operation.

(b) Predict the revenue of the company after 27 years of operation.

△ **138. Sphere** The radius r of a sphere whose volume is V is given by

$$r = \sqrt[3]{\dfrac{3V}{4\pi}}$$

(a) Write the radius of a sphere whose volume is 9 cubic centimeters as a radical in simplified form.

(b) Write the radius of a sphere whose volume is 32π cubic meters as a radical in simplified form.

Extending the Concepts

139. Suppose that $f(x) = \sqrt{2x}$ and $g(x) = \sqrt{8x^3}$.

(a) Find $(f \cdot g)(x)$.

(b) Evaluate $(f \cdot g)(3)$.

In Problems 140–143, evaluate the formula

$$x = \dfrac{-b \pm \sqrt{b^2 - 4ac}}{2a}$$

for the given values of a, b, and c. Note that the symbol $\pm$ is shorthand notation to indicate that there are two solutions. One solution is obtained when you add the quantity after the $\pm$ symbol, and another is obtained when you subtract. This formula can be used to solve any equation of the form $ax^2 + bx + c = 0$.

140. $a = 1, b = 4, c = 1$

141. $a = 1, b = 6, c = 3$

142. $a = 2, b = 1, c = -1$

143. $a = 3, b = 4, c = -1$

Explaining the Concepts

144. Explain how you would simplify $\sqrt[3]{16a^5}$.

145. In order for us to use the Product Property to multiply radicals, what must be true about the index in each radical?

146. In Example 1(c) we included the restriction $x \geq 3$.

    (a) Find the domain of each factor in $(\sqrt{x + 3})(\sqrt{x - 3})$.

    (b) The domain of the product, $\sqrt{x^2 - 9}$, is $x \leq -3$ or $x \geq 3$. Show that the product is a real number if $x = -4$ or $x = 4$ but is not a real number if $x = 0$. What conclusion can you draw about the domain of the product of two square roots?

Synthesis Review

In Problems 147–150, solve the following.

147. $4x + 3 = 13$

148. $2|7x - 1| + 4 = 16$

149. $\dfrac{3}{5}x + 2 \leq 28$

150. $\dfrac{5}{2}|x + 1| + 1 \leq 11$

151. How are the solutions in Problems 147 and 149 similar? How are the solutions in Problems 148 and 150 similar?

The Graphing Calculator

152. Exploration To understand the circumstances under which absolute value symbols are required when simplifying radicals, do the following.

    (a) Graph $Y_1 = \sqrt{x^2}$ and $Y_2 = x$. Do you think that $\sqrt{x^2} = x$? Now graph $Y_1 = \sqrt{x^2}$ and $Y_2 = |x|$. Do you think that $\sqrt{x^2} = |x|$?

    (b) Graph $Y_1 = \sqrt[3]{x^3}$ and $Y_2 = x$. Do you think that $\sqrt[3]{x^3} = x$? Now graph $Y_1 = \sqrt[3]{x^3}$ and $Y_2 = |x|$. Do you think that $\sqrt[3]{x^3} = |x|$?

    (c) Graph $Y_1 = \sqrt[4]{x^4}$ and $Y_2 = x$. Do you think that $\sqrt[4]{x^4} = x$? Now graph $Y_1 = \sqrt[4]{x^4}$ and $Y_2 = |x|$. Do you think that $\sqrt[4]{x^4} = |x|$?

    (d) In your own words, make a generalization about $\sqrt[n]{x^n}$.

6.4 Adding, Subtracting, and Multiplying Radical Expressions

Objectives

❶ Add or Subtract Radical Expressions

❷ Multiply Radical Expressions

Are You Ready for This Section?

Before getting started, take the following readiness quiz. If you get a problem wrong, go back to the section cited and review the material.

R1. Add: $4y^3 - 2y^2 + 8y - 1 + (-2y^3 + 7y^2 - 3y + 9)$ [Section 4.1, pp. 310–311]

R2. Subtract: $5z^2 + 6 - (3z^2 - 8z - 3)$ [Section 4.1, pp. 311–312]

R3. Multiply: $(4x + 3)(x - 5)$ [Section 4.2, pp. 320–321]

R4. Multiply: $(2y - 3)(2y + 3)$ [Section 4.2, pp. 322–323]

❶ Add or Subtract Radical Expressions

Recall that a radical expression is an algebraic expression that contains a radical. **Like radicals** have the same index and the same radicand. For example, $3\sqrt{5}$ and $8\sqrt{5}$ are like radicals because each has the same index, 2, and the same radicand, 5; $4\sqrt[3]{x - 4}$ and $10\sqrt[3]{x - 4}$ are like radicals because each has the same index, 3, and the same radicand, $x - 4$.

Just as we added or subtracted like terms using the Distributive Property, we add or subtract radical expressions by combining like radicals using the Distributive Property. For example, $3\sqrt{5} + 8\sqrt{5} = (3 + 8)\sqrt{5} = 11\sqrt{5}$ and $4\sqrt[3]{x - 4} + 10\sqrt[3]{x - 4} = (4 + 10)\sqrt[3]{x - 4} = 14\sqrt[3]{x - 4}$

    (**EXAMPLE 1**)   **Adding and Subtracting Radical Expressions**

Add or subtract, as indicated. Assume all variables are greater than or equal to zero.

    (a) $5\sqrt{2x} + 9\sqrt{2x}$

    (b) $3\sqrt[3]{10} + 7\sqrt[3]{10} - 5\sqrt[3]{10}$

Work Smart

Remember that for us to add or subtract radicals, both the index and the radicand must be the same.

Solution

(a) Both radicals have the same index, 2, and the same radicand, $2x$.

$$5\sqrt{2x} + 9\sqrt{2x} = (5 + 9)\sqrt{2x}$$

$5 + 9 = 14$: $\quad = 14\sqrt{2x}$

(b) All three radicals have the same index, 3, and the same radicand, 10.

$$3\sqrt[3]{10} + 7\sqrt[3]{10} - 5\sqrt[3]{10} = (3 + 7 - 5)\sqrt[3]{10}$$

$3 + 7 - 5 = 5$: $\quad = 5\sqrt[3]{10}$

Quick ✓

1. Two radicals are ___ _____ if each radical has the same index and the same radicand.

2. *True or False* $\sqrt{7} + \sqrt{5} = \sqrt{7 + 5}$

In Problems 3 and 4, add or subtract, as indicated.

3. $9\sqrt{13y} + 4\sqrt{13y}$

4. $\sqrt[4]{5} + 9\sqrt[4]{5} - 3\sqrt[4]{5}$

Sometimes, before adding or subtracting, we have to simplify the radical so that the radicands are the same.

EXAMPLE 2 **Adding and Subtracting Radical Expressions**

Add or subtract, as indicated. Assume all variables are greater than or equal to zero.

(a) $3\sqrt{12} + 7\sqrt{3}$ (b) $3x\sqrt{20x} - 7\sqrt{5x^3}$ (c) $3\sqrt{5} + 7\sqrt{13}$

Solution

(a) The index on each radical is the same, but the radicands are different. However, we can simplify the radicals to make the radicands the same.

$$\sqrt{12} = \sqrt{4} \cdot \sqrt{3}$$

$3\sqrt{12} + 7\sqrt{3} = 3\sqrt{4} \cdot \sqrt{3} + 7\sqrt{3}$

$\sqrt{4} = 2$: $\quad = 3 \cdot 2\sqrt{3} + 7\sqrt{3}$

$\quad = 6\sqrt{3} + 7\sqrt{3}$

Factor out $\sqrt{3}$: $\quad = (6 + 7)\sqrt{3}$

$\quad = 13\sqrt{3}$

(b) The index on each radical is the same, but the radicands are different. However, we can simplify each square root and then combine like radicals.

$$\sqrt{20x} = \sqrt{4} \cdot \sqrt{5x}; \sqrt{5x^3} = \sqrt{x^2} \cdot \sqrt{5x}$$

$3x\sqrt{20x} - 7\sqrt{5x^3} = 3x \cdot \sqrt{4} \cdot \sqrt{5x} - 7 \cdot \sqrt{x^2} \cdot \sqrt{5x}$

Simplify radicals: $\quad = 3x \cdot 2 \cdot \sqrt{5x} - 7 \cdot x \cdot \sqrt{5x}$

Multiply: $\quad = 6x\sqrt{5x} - 7x\sqrt{5x}$

Factor out $\sqrt{5x}$: $\quad = (6x - 7x)\sqrt{5x}$

$\quad = -x\sqrt{5x}$

(c) For $3\sqrt{5} + 7\sqrt{13}$, the index on each radical is the same, but the radicands are different and cannot be simplified to make the radicands the same.

Quick ✓

In Problems 5–7, add or subtract, as indicated. Assume all variables are greater than or equal to zero.

5. $4\sqrt{18} - 3\sqrt{8}$ **6.** $-5x\sqrt[3]{54x} + 7\sqrt[3]{2x^4}$ **7.** $7\sqrt{10} - 6\sqrt{3}$

(EXAMPLE 3)   **Adding or Subtracting Radical Expressions**

Add or subtract, as indicated.

(a) $\sqrt[3]{16x^4} - 7x\sqrt[3]{-2x} + \sqrt[3]{54x}$ **(b)** $3\sqrt[4]{m^4 n} - 5m\sqrt[8]{n^2}$

Solution

(a) The index on each radical is the same. The radicands are different but can be simplified to make the radicands the same.

$$\sqrt[3]{16x^4} = \sqrt[3]{8x^3} \cdot \sqrt[3]{2x};\ \sqrt[3]{-2x} = \sqrt[3]{-1} \cdot \sqrt[3]{2x};\ \sqrt[3]{54x} = \sqrt[3]{27} \cdot \sqrt[3]{2x}$$

$$\sqrt[3]{16x^4} - 7x\sqrt[3]{-2x} + \sqrt[3]{54x} = \sqrt[3]{8x^3} \cdot \sqrt[3]{2x} - 7x\sqrt[3]{-1} \cdot \sqrt[3]{2x} + \sqrt[3]{27} \cdot \sqrt[3]{2x}$$

$\sqrt[3]{8x^3} = 2x;\ \sqrt[3]{-1} = -1;\ \sqrt[3]{27} = 3:$ $= 2x\sqrt[3]{2x} - 7x(-1)\sqrt[3]{2x} + 3\sqrt[3]{2x}$

$-7x(-1) = 7x:$ $= 2x\sqrt[3]{2x} + 7x\sqrt[3]{2x} + 3\sqrt[3]{2x}$

Factor out $\sqrt[3]{2x}:$ $= (2x + 7x + 3)\sqrt[3]{2x}$

Simplify: $= (9x + 3)\sqrt[3]{2x}$

Factor out 3: $= 3(3x + 1)\sqrt[3]{2x}$

(b) Here, the index on the radical and the radicand are different. We start by dealing with the index using rational exponents.

$$\sqrt[n]{a^m} = a^{\frac{m}{n}}$$

$$3\sqrt[4]{m^4 n} - 5m\sqrt[8]{n^2} = 3\sqrt[4]{m^4 n} - 5m \cdot n^{\frac{2}{8}}$$

Simplify the rational exponent: $= 3\sqrt[4]{m^4 n} - 5m \cdot n^{\frac{1}{4}}$

Rewrite as radical: $= 3\sqrt[4]{m^4 n} - 5m \cdot \sqrt[4]{n}$

Now the index is the same, but the radicands are different.

$\sqrt[4]{m^4 n} = \sqrt[4]{m^4} \cdot \sqrt[4]{n}:$ $= 3\sqrt[4]{m^4} \cdot \sqrt[4]{n} - 5m \cdot \sqrt[4]{n}$

Simplify: $= 3m \cdot \sqrt[4]{n} - 5m \cdot \sqrt[4]{n}$

Factor out $\sqrt[4]{n}:$ $= (3m - 5m) \cdot \sqrt[4]{n}$

Simplify: $= -2m\sqrt[4]{n}$

Work Smart: Study Skills

Contrast adding radicals with multiplying radicals:

Add: $3\sqrt{5} + 8\sqrt{5} = (3 + 8)\sqrt{5}$
$= 11\sqrt{5}$

Multiply:
$3\sqrt{5} \cdot 8\sqrt{5} = 3 \cdot 8 \cdot \sqrt{5} \cdot \sqrt{5}$
$= 24\sqrt{25}$
$= 24 \cdot 5$
$= 120$

Ask yourself these questions:
How must the radicals be "like" to be added?
How must the radicals be "like" to be multiplied?

Quick ✓

In Problems 8 and 9, add or subtract, as indicated. Assume all variables are greater than or equal to zero.

8. $\sqrt[3]{8z^4} - 2z\sqrt[3]{-27z} + \sqrt[3]{125z}$ **9.** $\sqrt{25m} - 3\sqrt[4]{m^2}$

❷ Multiply Radical Expressions

We have already multiplied a single radical expression by a second single radical expression. Now we will multiply radical expressions involving more than one radical. We multiply these expressions the same way we multiplied polynomials.

EXAMPLE 4 **Multiplying Radical Expressions**

Multiply and simplify:

(a) $\sqrt{5}(3 - 4\sqrt{5})$ **(b)** $\sqrt[3]{2}(3 + \sqrt[3]{4})$ **(c)** $(3 + 2\sqrt{7})(2 - 3\sqrt{7})$

Solution

(a) Use the Distributive Property and multiply each term in the parentheses by $\sqrt{5}$.

$$\sqrt{5}(3 - 4\sqrt{5}) = \sqrt{5} \cdot 3 - \sqrt{5} \cdot 4\sqrt{5}$$

Multiply radicals: $= 3\sqrt{5} - 4 \cdot \sqrt{25}$

$\sqrt{25} = 5$: $= 3\sqrt{5} - 4 \cdot 5$

Simplify: $= 3\sqrt{5} - 20$

(b) Distribute $\sqrt[3]{2}$ to each term in the parentheses.

$$\sqrt[3]{2}(3 + \sqrt[3]{4}) = \sqrt[3]{2} \cdot 3 + \sqrt[3]{2} \cdot \sqrt[3]{4}$$

Multiply radicals: $= 3\sqrt[3]{2} + \sqrt[3]{8}$

$\sqrt[3]{8} = 2$: $= 3\sqrt[3]{2} + 2$

(c) Treat this just like the product of two binomials and multiply using FOIL.

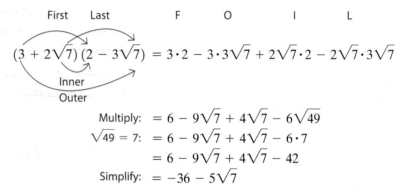

$$(3 + 2\sqrt{7})(2 - 3\sqrt{7}) = 3 \cdot 2 - 3 \cdot 3\sqrt{7} + 2\sqrt{7} \cdot 2 - 2\sqrt{7} \cdot 3\sqrt{7}$$

Multiply: $= 6 - 9\sqrt{7} + 4\sqrt{7} - 6\sqrt{49}$

$\sqrt{49} = 7$: $= 6 - 9\sqrt{7} + 4\sqrt{7} - 6 \cdot 7$

$= 6 - 9\sqrt{7} + 4\sqrt{7} - 42$

Simplify: $= -36 - 5\sqrt{7}$

Work Smart

$-36 - 5\sqrt{7} \neq -41\sqrt{7}$
Do you know why?

Quick ✓

In Problems 10–12, multiply and simplify.

10. $\sqrt{6}(3 - 5\sqrt{6})$ **11.** $\sqrt[3]{12}(3 - \sqrt[3]{2})$ **12.** $(2 - 7\sqrt{3})(5 + 4\sqrt{3})$

We can use some of our special products formulas (Section 4.2) to multiply radicals as well. We will use the formulas for perfect squares, $(A + B)^2 = A^2 + 2AB + B^2$ and $(A - B)^2 = A^2 - 2AB + B^2$, as well as the formula for the difference of two squares, $(A + B)(A - B) = A^2 - B^2$.

EXAMPLE 5 **Multiplying Radical Expressions Involving Special Products**

Multiply and simplify.

(a) $(2\sqrt{3} + \sqrt{5})^2$ **(b)** $(3 + \sqrt{7})(3 - \sqrt{7})$

Solution $(A + B)^2 = A^2 + 2 \cdot A \cdot B + B^2$

(a) $(2\sqrt{3} + \sqrt{5})^2 = (2\sqrt{3})^2 + 2 \cdot 2\sqrt{3} \cdot \sqrt{5} + (\sqrt{5})^2$

Multiply: $= 4\sqrt{9} + 4\sqrt{15} + \sqrt{25}$

Simplify: $= 4 \cdot 3 + 4\sqrt{15} + 5$

Work Smart

Notice, in Example 5(a), that
$(2\sqrt{3} + \sqrt{5})^2 \neq (2\sqrt{3})^2 + (\sqrt{5})^2$

Combine like terms: $= 17 + 4\sqrt{15}$

(b) Notice that $(3 + \sqrt{7})(3 - \sqrt{7})$ is in the form $(A + B)(A - B)$, so

$$\underset{(A\ +\quad B}{} \underset{)(A-\quad B)}{} = \underset{A^2\ -}{} \underset{B^2}{}$$

$$(3 + \sqrt{7})(3 - \sqrt{7}) = 3^2 - (\sqrt{7})^2$$
$$= 9 - 7$$
$$= 2 \qquad \bullet$$

Notice that the product found in Example 5(b) is an integer. There are no radicals in the product. Radical expressions such as $3 + \sqrt{7}$ and $3 - \sqrt{7}$ are called **conjugates** of each other. When we multiply square roots that are conjugates, the result never contains a radical. This result plays a huge role in the next section.

Quick ✓

13. *True or False* $(\sqrt{a} + \sqrt{b})^2 = (\sqrt{a})^2 + (\sqrt{b})^2$

14. The radical expressions $4 + \sqrt{5}$ and $4 - \sqrt{5}$ are examples of _____.

15. *True or False* The conjugate of $-5 + \sqrt{2}$ is $5 - \sqrt{2}$.

In Problems 16–18, multiply and simplify.

16. $(2 - \sqrt{5})^2$ **17.** $(\sqrt{7} - 3\sqrt{2})^2$ **18.** $(\sqrt{3} + \sqrt{2})(\sqrt{3} - \sqrt{2})$

6.4 Exercises

MyMathLab® PRACTICE

Exercise numbers in green are complete video solutions in MyMathLab

*Problems **1–18** are the* Quick ✓s *that follow the* EXAMPLES.

Building Skills

In Problems 19–26, add or subtract as indicated. Assume all variables are greater than or equal to zero. See Objective 1.

19. $3\sqrt{2} + 7\sqrt{2}$ **20.** $6\sqrt{3} + 8\sqrt{3}$

21. $5\sqrt[3]{x} - 3\sqrt[3]{x}$ **22.** $12\sqrt[4]{z} - 5\sqrt[4]{z}$

23. $8\sqrt{5x} - 3\sqrt{5x} + 9\sqrt{5x}$

24. $4\sqrt[3]{3y} + 8\sqrt[3]{3y} - 10\sqrt[3]{3y}$

25. $4\sqrt[3]{5} - 3\sqrt{5} + 7\sqrt[3]{5} - 8\sqrt{5}$

26. $12\sqrt{7} + 5\sqrt[4]{7} - 5\sqrt{7} + 6\sqrt[4]{7}$

In Problems 27–48, add or subtract as indicated. Assume all variables are greater than or equal to zero. See Objective 1.

27. $\sqrt{8} + 6\sqrt{2}$ **28.** $6\sqrt{3} + \sqrt{12}$

29. $\sqrt[3]{24} - 4\sqrt[3]{3}$ **30.** $\sqrt[3]{32} - 5\sqrt[3]{4}$

31. $\sqrt[3]{54} - 7\sqrt[3]{128}$ **32.** $7\sqrt[4]{48} - 4\sqrt[4]{243}$

33. $5\sqrt{54x} - 3\sqrt{24x}$ **34.** $2\sqrt{48z} - \sqrt{75z}$

35. $2\sqrt{8} + 3\sqrt{10}$ **36.** $4\sqrt{12} + 2\sqrt{20}$

37. $\sqrt{12x^3} + 5x\sqrt{108x}$ **38.** $3\sqrt{63z^3} + 2z\sqrt{28z}$

39. $\sqrt{12x^2} + 3x\sqrt{2} - 2\sqrt{98x^2}$

40. $\sqrt{48y^2} - 4y\sqrt{12} + \sqrt{108y^2}$

41. $\sqrt[3]{-54x^3} + 3x\sqrt[3]{16} - 2\sqrt[3]{128}$

42. $2\sqrt[3]{-5x^3} + 4x\sqrt[3]{40} - \sqrt[3]{135}$

43. $\sqrt{9x - 9} + \sqrt{4x - 4}$

44. $\sqrt{4x + 12} - \sqrt{9x + 27}$

45. $\sqrt{16x} - \sqrt[6]{x^3}$ **46.** $\sqrt{25x} - \sqrt[4]{x^2}$

47. $\sqrt[3]{27x} + 2\sqrt[9]{x^3}$ **48.** $\sqrt[4]{16y} + \sqrt[8]{y^2}$

In Problems 49–82, multiply and simplify. Assume all variables are greater than or equal to zero. See Objective 2.

49. $\sqrt{3}(2 - 3\sqrt{2})$ **50.** $\sqrt{5}(5 + 3\sqrt{3})$

51. $\sqrt{3}(\sqrt{2} + \sqrt{6})$ **52.** $\sqrt{2}(\sqrt{5} - 2\sqrt{10})$

53. $\sqrt[3]{4}(\sqrt[3]{3} - \sqrt[3]{6})$ **54.** $\sqrt[3]{6}(\sqrt[3]{2} + \sqrt[3]{12})$

55. $\sqrt{2x}(3 - \sqrt{10x})$ **56.** $\sqrt{5x}(6 + \sqrt{15x})$

57. $(3 + \sqrt{2})(4 + \sqrt{3})$ **58.** $(5 + \sqrt{5})(3 + \sqrt{6})$

59. $(6 + \sqrt{3})(2 - \sqrt{7})$ **60.** $(7 - \sqrt{3})(6 + \sqrt{5})$

61. $(4 - 2\sqrt{7})(3 + 3\sqrt{7})$

62. $(9 + 5\sqrt{10})(1 - 3\sqrt{10})$

63. $(\sqrt{2} + 3\sqrt{6})(\sqrt{3} - 2\sqrt{2})$

64. $(2\sqrt{3} + \sqrt{10})(\sqrt{5} - 2\sqrt{2})$

65. $(2\sqrt{5} + \sqrt{3})(4\sqrt{5} - 3\sqrt{3})$

66. $(\sqrt{6} - 2\sqrt{2})(2\sqrt{6} + 3\sqrt{2})$

67. $(1 + \sqrt{3})^2$ **68.** $(2 - \sqrt{3})^2$

69. $(\sqrt{2} - \sqrt{5})^2$ **70.** $(\sqrt{7} - \sqrt{3})^2$

71. $(\sqrt{x} - \sqrt{2})^2$ **72.** $(\sqrt{z} + \sqrt{5})^2$

73. $(\sqrt{2} - 1)(\sqrt{2} + 1)$

74. $(\sqrt{3} - 1)(\sqrt{3} + 1)$

75. $(3 - 2\sqrt{5})(3 + 2\sqrt{5})$

76. $(6 + 3\sqrt{2})(6 - 3\sqrt{2})$

77. $(\sqrt{2x} + \sqrt{3y})(\sqrt{2x} - \sqrt{3y})$

78. $(\sqrt{5a} + \sqrt{7b})(\sqrt{5a} - \sqrt{7b})$

79. $(\sqrt[3]{x} + 4)(\sqrt[3]{x} - 3)$

80. $(\sqrt[3]{y} - 6)(\sqrt[3]{y} + 3)$

81. $(\sqrt[3]{2a} - 5)(\sqrt[3]{2a} + 5)$

82. $(\sqrt[3]{4p} - 1)(\sqrt[3]{4p} + 3)$

Mixed Practice

In Problems 83–106, perform the indicated operation and simplify. Assume all variables are greater than or equal to zero.

83. $\sqrt{5}(\sqrt{3} + \sqrt{10})$

84. $\sqrt{7}(\sqrt{14} + \sqrt{3})$

85. $\sqrt{28x^5} - x\sqrt{7x^3} + 5\sqrt{175x^5}$

86. $\sqrt{180a^5} + a^2\sqrt{20} - a\sqrt{80a^3}$

87. $(2\sqrt{3} + 5)(2\sqrt{3} - 5)$

88. $(4\sqrt{2} - 2)(4\sqrt{2} + 2)$

89. $\sqrt[3]{7}(2 + \sqrt[3]{4})$

90. $\sqrt[3]{9}(5 + 2\sqrt[3]{2})$

91. $(2\sqrt{2} + 5)(4\sqrt{2} - 4)$

92. $(5\sqrt{5} - 3)(3\sqrt{5} - 4)$

93. $4\sqrt{18} + 2\sqrt{32}$

94. $5\sqrt{20} + 2\sqrt{80}$

95. $(\sqrt{5} - \sqrt{3})^2 - \sqrt{60}$

96. $(\sqrt{2} - \sqrt{7})^2 - \sqrt{56}$

97. $3\sqrt[3]{5x^3y} + \sqrt[3]{40y}$

98. $5\sqrt[3]{3m^3n} + \sqrt[3]{81n}$

99. $(\sqrt{2x} - \sqrt{7y})(\sqrt{2x} + \sqrt{7y}) - 2\sqrt{x^2}$

100. $(\sqrt{3a} - \sqrt{4b})(\sqrt{3a} + \sqrt{4b}) + 4\sqrt{b^2}$

101. $(2 + \sqrt{x + 1})^2$

102. $(4 + \sqrt{2x + 3})^2$

103. $(5 - \sqrt{x + 2})^2$

104. $(1 - \sqrt{3x - 4})^2$

105. $-\dfrac{3}{5} \cdot \left(-\dfrac{\sqrt{5}}{5}\right) - \dfrac{4}{5} \cdot \left(-\dfrac{2\sqrt{5}}{5}\right)$

106. $\dfrac{4}{5} \cdot \left(-\dfrac{\sqrt{5}}{5}\right) + \left(-\dfrac{3}{5}\right) \cdot \left(-\dfrac{2\sqrt{5}}{5}\right)$

107. Suppose that $f(x) = \sqrt{3x}$ and $g(x) = \sqrt{12x}$; find

   (a) $(f + g)(x)$ **(b)** $(f + g)(4)$

   (c) $(f \cdot g)(x)$

108. Suppose that $f(x) = \sqrt{4x - 4}$ and $g(x) = \sqrt{25x - 25}$; find

   (a) $(f + g)(x)$ **(b)** $(f + g)(10)$

   (c) $(f \cdot g)(x)$

109. Show that $-2 + \sqrt{5}$ is a solution to the equation $x^2 + 4x - 1 = 0$. Show that $-2 - \sqrt{5}$ is also a solution.

110. Show that $3 + \sqrt{7}$ is a solution to the equation $x^2 - 6x + 2 = 0$. Show that $3 - \sqrt{7}$ is also a solution.

Applying the Concepts

In Problems 111 and 112, find the perimeter and area of the figures shown. Express your answer as a radical in simplified form.

111.   **112.**

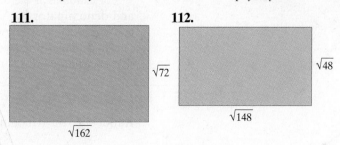

Problems 113 and 114, use **Heron's Formula** *for finding the area of a triangle whose sides are known. Heron's Formula states that the area A of a triangle with sides a, b, and c is*

$$A = \sqrt{s(s-a)(s-b)(s-c)}$$

where

$$s = \frac{1}{2}(a+b+c)$$

Find the area of the shaded region by computing the difference in the areas of the triangles. That is, compute "area of larger triangle minus area of smaller triangle." Write your answer as a radical in simplified form.

113.   **114.**

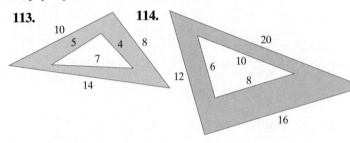

Explaining the Concepts

115. Explain how to add or subtract radicals.

116. Multiply $(\sqrt{a} - \sqrt{b})(\sqrt{a} + \sqrt{b})$ and provide a general result regarding the product of conjugates involving square roots.

Synthesis Review

In Problems 117–122, multiply each of the following.

117. $(3a^3b)(4a^2b^4)$

118. $(3p - 1)(2p + 3)$

119. $(3y + 2)(2y - 1)$

120. $(m - 4)(m + 4)$

121. $(5w + 2)(5w - 2)$

122. $(\sqrt{x} + 2)(\sqrt{x} - 2)$

6.5 Rationalizing Radical Expressions

Objectives

1 Rationalize a Denominator Containing One Term

2 Rationalize a Denominator Containing Two Terms

> **In Words**
> We call the process "rationalizing the denominator" because we are making the denominator a rational number (no radicals).

Are You Ready for This Section?

Before getting started, take the following readiness quiz. If you get a problem wrong, go back to the section cited and review the material.

R1. What would we need to multiply 12 by in order to make it the smallest perfect square that is a multiple of 12?

R2. Simplify: $\sqrt{25x^2}, x > 0$ [Section 6.3, pp. 487–489]

When radical expressions appear in the denominator of a quotient, it is customary to rewrite the quotient without radicals in the denominator. This process is called **rationalizing the denominator.** In this section, we will rationalize denominators that contain one or two terms.

▶ **1** Rationalize a Denominator Containing One Term

To rationalize a denominator containing a single square root, we multiply the numerator and denominator of the quotient by a square root so that the radicand in the denominator becomes a perfect square.

For example, if the denominator of a quotient contains $\sqrt{5}$, multiply the numerator and denominator by $\sqrt{5}$ because $5 \cdot 5 = 25$, a perfect square. If the denominator contains $\sqrt{8}$, multiply the numerator and denominator by $\sqrt{2}$ because $8 \cdot 2 = 16$, a perfect square. Since we are multiplying both the numerator and the denominator by the same value, we are multiplying the radical expression by 1.

(EXAMPLE 1)   **Rationalizing a Denominator Containing a Square Root**

Rationalize the denominator of each expression:

(a) $\dfrac{1}{\sqrt{7}}$ **(b)** $\dfrac{\sqrt{5}}{\sqrt{12}}$ **(c)** $\dfrac{2}{3\sqrt{2x}}, x > 0$

Solution

(a) The denominator contains $\sqrt{7}$. We ask "What can I multiply 7 by to obtain a perfect square?" Because $7 \cdot 7 = 49$, a perfect square, multiply the numerator and denominator by $\sqrt{7}$.

$$\frac{1}{\sqrt{7}} = \frac{1}{\sqrt{7}} \cdot \frac{\sqrt{7}}{\sqrt{7}}$$

Multiply numerators; multiply denominators: $\quad = \dfrac{\sqrt{7}}{\sqrt{49}}$

$\sqrt{49} = 7: \quad = \dfrac{\sqrt{7}}{7}$

(b) We have $\sqrt{12}$ in the denominator. Although multiplying 12 by itself would result in a perfect square, is this the best choice? No, we want to find the smallest integer that makes the radicand a perfect square. Thus multiply 12 by 3 to get 36, a perfect square.

$$\frac{\sqrt{5}}{\sqrt{12}} = \frac{\sqrt{5}}{\sqrt{12}} \cdot \frac{\sqrt{3}}{\sqrt{3}}$$

Multiply numerators; multiply denominators: $\quad = \dfrac{\sqrt{15}}{\sqrt{36}}$

$\sqrt{36} = 6: \quad = \dfrac{\sqrt{15}}{6}$

(c)

Multiply by $1 = \dfrac{\sqrt{2x}}{\sqrt{2x}}$

$$\frac{2}{3\sqrt{2x}} = \frac{2}{3\sqrt{2x}} \cdot \frac{\sqrt{2x}}{\sqrt{2x}}$$

Multiply numerators; multiply denominators: $\quad = \dfrac{2\sqrt{2x}}{3\sqrt{4x^2}}$

$\sqrt{4x^2} = 2x$ since $x > 0$: $\quad = \dfrac{2\sqrt{2x}}{3 \cdot 2x}$

Divide out common factor, 2: $\quad = \dfrac{\sqrt{2x}}{3x}$

Work Smart

Remember that $a = 1 \cdot a$ and that 1 can take many forms. In Example 1(a), $1 = \dfrac{\sqrt{7}}{\sqrt{7}}$.

Work Smart

An alternative approach to Example 1(b) would be to simplify $\sqrt{12}$ first, as follows:
$$\frac{\sqrt{5}}{\sqrt{12}} = \frac{\sqrt{5}}{\sqrt{4 \cdot 3}}$$
$$= \frac{\sqrt{5}}{2\sqrt{3}}$$
$$= \frac{\sqrt{5}}{2\sqrt{3}} \cdot \frac{\sqrt{3}}{\sqrt{3}}$$
$$= \frac{\sqrt{15}}{6}$$

Quick ✓

1. Rewriting a quotient to remove radicals from the denominator is called _____ ___ _____.

2. To rationalize the denominator of $\dfrac{\sqrt{5}}{\sqrt{11}}$, multiply the numerator and denominator by ____.

In Problems 3–5, rationalize each denominator.

3. $\dfrac{1}{\sqrt{3}}$ **4.** $\dfrac{\sqrt{5}}{\sqrt{8}}$ **5.** $\dfrac{5}{\sqrt{10x}}$

In general, to rationalize a denominator containing a single radical with index n, we multiply the numerator and denominator of the quotient by a radical so that the product in the denominator has a radicand that is a perfect nth power. So, if the denominator contains a radical whose index is 3, we multiply the numerator and denominator by a cube root so that the radicand in the denominator becomes a perfect cube. For example, if the denominator of a quotient contains $\sqrt[3]{4}$, multiply the numerator and denominator of the quotient by $\sqrt[3]{2}$ because $4 \cdot 2 = 8$, a perfect cube.

EXAMPLE 2

Rationalizing a Denominator Containing Cube Roots and Fourth Roots

Rationalize the denominator of each expression:

(a) $\dfrac{1}{\sqrt[3]{6}}$

(b) $\sqrt[3]{\dfrac{5}{18}}$

(c) $\dfrac{6}{\sqrt[4]{4z^3}}$

Assume all variables represent positive real numbers.

Solution

(a) We have $\sqrt[3]{6}$ in the denominator. We want the radicand in the denominator to be a perfect cube (since the index is 3), so we multiply the numerator and denominator by $\sqrt[3]{6^2} = \sqrt[3]{36}$ since $6 \cdot 36 = 6 \cdot 6^2 = 6^3 = 216$, a perfect cube.

$$\frac{1}{\sqrt[3]{6}} = \frac{1}{\sqrt[3]{6}} \cdot \frac{\sqrt[3]{6^2}}{\sqrt[3]{6^2}}$$

Multiply numerators;
multiply denominators: $\quad = \dfrac{\sqrt[3]{36}}{\sqrt[3]{6^3}}$

$\sqrt[3]{6^3} = 6:\quad = \dfrac{\sqrt[3]{36}}{6}$

Work Smart

The radicand in $\sqrt[3]{6}$ is 6, or 6^1. To make it a perfect cube, 6^3, multiply by 6^2, or 36. Then $6 \cdot 36 = 6^1 \cdot 6^2 = 6^3$. The cube root of 6^3, or 216, is 6.

(b) First, use the Quotient Property $\left(\sqrt[n]{\dfrac{a}{b}} = \dfrac{\sqrt[n]{a}}{\sqrt[n]{b}}\right)$ to rewrite the radical as the quotient of two radicals.

$$\sqrt[3]{\frac{5}{18}} = \frac{\sqrt[3]{5}}{\sqrt[3]{18}}$$

What do we need to multiply $\sqrt[3]{18}$ by to make it a perfect cube? First, write 18 as $9 \cdot 2 = 3^2 \cdot 2^1$. If we multiply $3^2 \cdot 2^1$ by $3^1 \cdot 2^2 = 12$, we will have a perfect cube as the radicand in the denominator.

$$\sqrt[3]{\frac{5}{18}} = \frac{\sqrt[3]{5}}{\sqrt[3]{18}} = \frac{\sqrt[3]{5}}{\sqrt[3]{3^2 \cdot 2}} \cdot \frac{\sqrt[3]{3 \cdot 2^2}}{\sqrt[3]{3 \cdot 2^2}}$$

Multiply numerators;
multiply denominators: $\quad = \dfrac{\sqrt[3]{60}}{\sqrt[3]{3^3 \cdot 2^3}}$

$\sqrt[3]{3^3 \cdot 2^3} = 3 \cdot 2 = 6:\quad = \dfrac{\sqrt[3]{60}}{6}$

(c) Rewrite the denominator as $\sqrt[4]{2^2 \cdot z^3}$. To make the radicand a perfect fourth power, multiply $\sqrt[4]{2^2 \cdot z^3}$ by $\sqrt[4]{2^2 \cdot z}$ to obtain $\sqrt[4]{2^4 z^4}$ in the denominator.

$$\frac{6}{\sqrt[4]{4z^3}} = \frac{6}{\sqrt[4]{2^2 \cdot z^3}} \cdot \frac{\sqrt[4]{2^2 \cdot z}}{\sqrt[4]{2^2 \cdot z}}$$

Multiply numerators;
multiply denominators: $\quad = \dfrac{6\sqrt[4]{4z}}{\sqrt[4]{2^4 \cdot z^4}}$

$\sqrt[4]{2^4 \cdot z^4} = 2z:\quad = \dfrac{6\sqrt[4]{4z}}{2z}$

Simplify: $\quad = \dfrac{3\sqrt[4]{4z}}{z}$

Quick ✓

In Problems 6–8, rationalize each denominator. Assume all variables are positive.

6. $\dfrac{4}{\sqrt[3]{3}}$

7. $\sqrt[3]{\dfrac{3}{20}}$

8. $\dfrac{3}{\sqrt[4]{p}}$

▶ ❷ Rationalize a Denominator Containing Two Terms

To rationalize a denominator containing two terms involving square roots, we use the fact that

$$(A + B)(A - B) = A^2 - B^2$$

and multiply both the numerator and the denominator of the quotient by the conjugate of the denominator. For example, if the quotient is $\dfrac{3}{\sqrt{3} + 2}$, we multiply both the numerator and the denominator by the conjugate of $\sqrt{3} + 2$, which is $\sqrt{3} - 2$. We know from the last section that the product $(\sqrt{3} + 2)(\sqrt{3} - 2)$ will not contain a radical.

EXAMPLE 3 **Rationalizing a Denominator Containing Two Terms**

Rationalize the denominator: $\dfrac{\sqrt{2}}{\sqrt{6} + 2}$

Solution

Since $\sqrt{6} + 2$ is in the denominator of the quotient, multiply the numerator and denominator by the conjugate of $\sqrt{6} + 2$, $\sqrt{6} - 2$.

$$\frac{\sqrt{2}}{\sqrt{6} + 2} = \frac{\sqrt{2}}{\sqrt{6} + 2} \cdot \frac{\sqrt{6} - 2}{\sqrt{6} - 2}$$

Multiply the numerators
and denominators:
$$= \frac{\sqrt{2}(\sqrt{6} - 2)}{(\sqrt{6} + 2)(\sqrt{6} - 2)}$$

Distribute $\sqrt{2}$ in the numerator;
$(A + B)(A - B) = A^2 - B^2$ in the denominator:
$$= \frac{\sqrt{12} - 2\sqrt{2}}{(\sqrt{6})^2 - 2^2}$$

$\sqrt{12} = 2\sqrt{3}$:
$$= \frac{2\sqrt{3} - 2\sqrt{2}}{6 - 4}$$

Factor out the GCF of 2 in numerator:
$$= \frac{2(\sqrt{3} - \sqrt{2})}{2}$$

Divide out the 2s:
$$= \sqrt{3} - \sqrt{2}$$

Quick ✓

9. To rationalize the denominator of $\dfrac{4 - \sqrt{3}}{-2 + \sqrt{7}}$, multiply the numerator and denominator by _____.

In Problems 10 and 11, rationalize the denominator.

10. $\dfrac{4}{\sqrt{3} + 1}$

11. $\dfrac{\sqrt{2}}{\sqrt{6} - \sqrt{2}}$

EXAMPLE 4 | **Rationalizing a Denominator Containing Two Terms**

Rationalize the denominator: $\dfrac{\sqrt{6} - 3}{\sqrt{10} - \sqrt{6}}$

Solution

Multiply the numerator and denominator by the conjugate of the denominator, $\sqrt{10} + \sqrt{6}$.

$$\frac{\sqrt{6} - 3}{\sqrt{10} - \sqrt{6}} = \frac{\sqrt{6} - 3}{\sqrt{10} - \sqrt{6}} \cdot \frac{\sqrt{10} + \sqrt{6}}{\sqrt{10} + \sqrt{6}}$$

Multiply the numerators and denominators: $= \dfrac{(\sqrt{6} - 3)(\sqrt{10} + \sqrt{6})}{(\sqrt{10} - \sqrt{6})(\sqrt{10} + \sqrt{6})}$

Use FOIL in the numerator;
$(A + B)(A - B) = A^2 - B^2$ in the denominator: $= \dfrac{\sqrt{60} + \sqrt{36} - 3\sqrt{10} - 3\sqrt{6}}{(\sqrt{10})^2 - (\sqrt{6})^2}$

Simplify radicals: $= \dfrac{2\sqrt{15} + 6 - 3\sqrt{10} - 3\sqrt{6}}{10 - 6}$

$$= \frac{2\sqrt{15} + 6 - 3\sqrt{10} - 3\sqrt{6}}{4}$$

Quick ✓

In Problem 12, rationalize the denominator.

12. $\dfrac{\sqrt{5} + 4}{\sqrt{5} - \sqrt{2}}$

6.5 Exercises   MyMathLab®   MathXL   PRACTICE

Exercise numbers in green
are complete video solutions
in MyMathLab

*Problems **1–12** are the Quick ✓s that follow the EXAMPLES.*

Building Skills

In Problems 13–36, rationalize each denominator. Assume all variables are positive. See Objective 1.

13. $\dfrac{1}{\sqrt{2}}$

14. $\dfrac{2}{\sqrt{3}}$

15. $-\dfrac{6}{5\sqrt{3}}$

16. $-\dfrac{3}{2\sqrt{3}}$

17. $\dfrac{3}{\sqrt{12}}$

18. $\dfrac{5}{\sqrt{20}}$

19. $\dfrac{\sqrt{2}}{\sqrt{6}}$

20. $\dfrac{\sqrt{3}}{\sqrt{11}}$

21. $\sqrt{\dfrac{2}{p}}$

22. $\sqrt{\dfrac{5}{z}}$

23. $\dfrac{\sqrt{8}}{\sqrt{y^3}}$

24. $\dfrac{\sqrt{32}}{\sqrt{a^5}}$

25. $\dfrac{2}{\sqrt[3]{2}}$

26. $\dfrac{5}{\sqrt[3]{3}}$

27. $\sqrt[3]{\dfrac{7}{q}}$

28. $\sqrt[3]{\dfrac{-4}{p}}$

29. $\sqrt[3]{\dfrac{-3}{50}}$

30. $\sqrt[3]{\dfrac{-5}{72}}$

31. $\dfrac{2}{\sqrt[3]{20y}}$

32. $\dfrac{8}{\sqrt[3]{36z^2}}$

33. $\dfrac{-4}{\sqrt[4]{3x^3}}$

34. $\dfrac{6}{\sqrt[4]{9b^2}}$

35. $\dfrac{12}{\sqrt[5]{m^3n^2}}$

36. $\dfrac{-3}{\sqrt[5]{ab^3}}$

In Problems 37–56, rationalize each denominator. Assume all variables are positive. See Objective 2.

37. $\dfrac{4}{\sqrt{6} - 2}$

38. $\dfrac{6}{\sqrt{7} - 2}$

39. $\dfrac{5}{\sqrt{5} + 2}$

40. $\dfrac{10}{\sqrt{10} + 3}$

41. $\dfrac{8}{\sqrt{7} - \sqrt{3}}$

42. $\dfrac{12}{\sqrt{11} - \sqrt{7}}$

43. $\dfrac{\sqrt{2}}{\sqrt{10} - \sqrt{6}}$

44. $\dfrac{\sqrt{3}}{\sqrt{15} - \sqrt{6}}$

45. $\dfrac{\sqrt{p}}{\sqrt{p} + \sqrt{q}}$

46. $\dfrac{\sqrt{a}}{\sqrt{a} + \sqrt{b}}$

47. $\dfrac{18}{2\sqrt{3} + 3\sqrt{2}}$

48. $\dfrac{15}{3\sqrt{5} + 4\sqrt{3}}$

49. $\dfrac{\sqrt{7}+3}{\sqrt{7}-3}$ 　　**50.** $\dfrac{\sqrt{5}+3}{\sqrt{5}-3}$ 　　**51.** $\dfrac{\sqrt{3}-4\sqrt{2}}{2\sqrt{3}+5\sqrt{2}}$

52. $\dfrac{3\sqrt{6}+5\sqrt{7}}{2\sqrt{6}-3\sqrt{7}}$ 　**53.** $\dfrac{\sqrt{p}+2}{\sqrt{p}-2}$ 　　**54.** $\dfrac{\sqrt{x}-4}{\sqrt{x}+4}$

55. $\dfrac{\sqrt{2}-3}{\sqrt{8}-\sqrt{2}}$ 　　**56.** $\dfrac{2\sqrt{3}+3}{\sqrt{12}-\sqrt{3}}$

Mixed Practice

In Problems 57–64, perform the indicated operation and simplify.

57. $\sqrt{3}+\dfrac{1}{\sqrt{3}}$ 　　　　**58.** $\sqrt{5}-\dfrac{1}{\sqrt{5}}$

59. $\dfrac{\sqrt{10}}{2}-\dfrac{1}{\sqrt{2}}$ 　　　**60.** $\dfrac{\sqrt{5}}{2}+\dfrac{3}{\sqrt{5}}$

61. $\sqrt{\dfrac{1}{3}}+\sqrt{12}+\sqrt{75}$ 　**62.** $\sqrt{\dfrac{2}{5}}+\sqrt{20}-\sqrt{45}$

63. $\dfrac{3}{\sqrt{18}}-\sqrt{\dfrac{1}{2}}$ 　　　**64.** $\sqrt{\dfrac{4}{3}}+\dfrac{4}{\sqrt{48}}$

In Problems 65–76, simplify each expression so that the denominator does not contain a radical. Work smart because in some of the problems, it will be easier if you divide the radicands before attempting to rationalize the denominator.

65. $\dfrac{\sqrt{3}}{\sqrt{12}}$ 　　**66.** $\dfrac{\sqrt{2}}{\sqrt{18}}$ 　　**67.** $\dfrac{3}{\sqrt{72}}$

68. $\dfrac{7}{\sqrt{98}}$ 　　**69.** $\sqrt{\dfrac{4}{3}}$ 　　**70.** $\sqrt{\dfrac{9}{5}}$

71. $\dfrac{\sqrt{3}-3}{\sqrt{3}+3}$ 　**72.** $\dfrac{\sqrt{2}-5}{\sqrt{2}+5}$ 　**73.** $\dfrac{2}{\sqrt{5}+2}$

74. $\dfrac{5}{\sqrt{6}+4}$ 　**75.** $\dfrac{\sqrt{8}}{\sqrt{2}}$ 　**76.** $\dfrac{\sqrt{75}}{\sqrt{3}}$

In Problems 77–82, find the reciprocal of the given number. Be sure to rationalize the denominator.

77. $\sqrt{3}$ 　　　**78.** $\sqrt{7}$ 　　　**79.** $\sqrt[3]{12}$

80. $\sqrt[3]{18}$ 　　**81.** $\sqrt{3}+5$ 　　**82.** $7-\sqrt{2}$

Applying the Concepts

Math for the Future: Trigonometry *In Problems 83 and 84, simplify each expression.*

83. $\dfrac{1}{\sqrt{2}}\cdot\dfrac{\sqrt{3}}{2}-\dfrac{1}{\sqrt{2}}\cdot\dfrac{1}{2}$

84. $-\sqrt{\dfrac{2}{3}}\cdot\left(-\dfrac{2}{\sqrt{5}}\right)+\dfrac{1}{\sqrt{3}}\cdot\dfrac{1}{\sqrt{5}}$

Sometimes we are asked to rationalize a numerator. In Problems 85–88, rationalize each expression by multiplying the numerator and denominator by the conjugate of the numerator.

85. $\dfrac{\sqrt{2}+1}{3}$ 　　　　**86.** $\dfrac{\sqrt{3}+2}{2}$

87. $\dfrac{\sqrt{x}-\sqrt{h}}{\sqrt{x}}$ 　　　**88.** $\dfrac{\sqrt{a}-\sqrt{b}}{\sqrt{2}}$

Extending the Concepts

89. When two quantities a and b are positive, we can verify that $a=b$ by showing that $a^2=b^2$. Verify that
$$\dfrac{\sqrt{6}+\sqrt{2}}{4}=\dfrac{\sqrt{2}+\sqrt{3}}{2}$$
by squaring each side.

90. Rationalize the denominator: $\dfrac{2}{\sqrt{2}+\sqrt{3}-\sqrt{9}}$

91. Math for the Future: Calculus The following problem comes up in calculus. Consider the rational expression:
$$\dfrac{\sqrt{x+h}-\sqrt{x}}{h}$$

(a) Rationalize the numerator by multiplying the numerator and denominator by $\sqrt{x+h}+\sqrt{x}$. Be sure to simplify the expression completely.

(b) Evaluate the expression found in part (a) at $h=0$.

(c) The expression found in part (b) represents the formula for the slope of the line tangent to the function $f(x)=\sqrt{x}$ at any value of $x\geq 0$. Find the slope of the line tangent to the function $f(x)=\sqrt{x}$ at $x=4$.

(d) If $f(x)=\sqrt{x}$, what is $f(4)$? What point is on the graph of $f(x)=\sqrt{x}$?

(e) Find the equation of the line tangent to $f(x)=\sqrt{x}$ at $x=4$, using the slope found in part (c) and the point found in part (d).

(f) Graph the function $f(x)=\sqrt{x}$ and the equation of the tangent line in the same Cartesian plane.

Explaining the Concepts

92. Explain why it is necessary to multiply the numerator and denominator by the conjugate of the denominator when rationalizing a denominator containing two terms.

93. Explain why removing irrational numbers from the denominator is called "rationalization."

Synthesis Review

In Problems 94–97, graph each of the following functions using point plotting.

94. $f(x) = 5x - 3$

95. $g(x) = -3x + 9$

96. $G(x) = x^2$

97. $F(x) = x^3$

Putting the Concepts Together (Sections 6.1–6.5)

We designed these problems so that you can review Sections 6.1 through 6.5 and show your mastery of the concepts. Take time to work these problems before proceeding with the next section. The answers are located at the back of the text on page AN-36.

1. Evaluate: $-25^{\frac{1}{2}}$

2. Evaluate: $(-64)^{-\frac{2}{3}}$

3. Write the expression $\sqrt[4]{3x^3}$ with a rational exponent.

4. Write the expression $7z^{\frac{4}{5}}$ as a radical expression.

5. Simplify using rational exponents: $\sqrt[3]{\sqrt{64x^3}}$

6. Distribute and simplify: $c^{\frac{1}{2}}\left(c^{\frac{3}{2}} + c^{\frac{5}{2}}\right)$

In Problems 7–9, use Laws of Exponents to simplify each expression. Assume all variables in the radicand are greater than or equal to zero. Express answers with positive exponents.

7. $\left(a^{\frac{2}{3}}b^{-\frac{1}{3}}\right)\left(a^{\frac{4}{3}}b^{-\frac{5}{3}}\right)$

8. $\dfrac{x^{\frac{3}{4}}}{x^{\frac{1}{8}}}$

9. $\left(x^{\frac{3}{4}}y^{-\frac{1}{8}}\right)^8$

In Problems 10–19, perform the indicated operation and simplify. Assume all variables in the radicand are greater than or equal to zero.

10. $\sqrt{15a} \cdot \sqrt{2b}$

11. $\sqrt{10m^3n^2} \cdot \sqrt{20mn}$

12. $\sqrt[3]{\dfrac{-32xy^4}{4x^{-2}y}}$

13. $2\sqrt{108} - 3\sqrt{75} + \sqrt{48}$

14. $-5b\sqrt{8b} + 7\sqrt{18b^3}$

15. $\sqrt[3]{16y^4} - y\sqrt[3]{2y}$

16. $(3\sqrt{x})(4\sqrt{x})$

17. $3\sqrt{x} + 4\sqrt{x}$

18. $(2 - 3\sqrt{2})(10 + \sqrt{2})$

19. $(4\sqrt{2} - 3)^2$

In Problems 20 and 21, rationalize the denominator.

20. $\dfrac{3}{2\sqrt{32}}$

21. $\dfrac{4}{\sqrt{3} - 8}$

6.6 Functions Involving Radicals

Objectives

1 Evaluate Functions Involving Radicals

2 Find the Domain of a Function Involving a Radical

3 Graph Functions Involving Square Roots

4 Graph Functions Involving Cube Roots

Are You Ready for This Section?

Before getting started, take the following readiness quiz. If you get a problem wrong, go back to the section cited and review the material.

R1. Simplify: $\sqrt{121}$ [Getting Ready: Square Roots, pp. 467–468]

R2. Simplify: $\sqrt{p^2}$ [Getting Ready: Square Roots, pp. 469–470]

R3. Given $f(x) = x^2 - 4$, find $f(3)$. [Section 2.2, pp. 156–158]

R4. Solve: $-2x + 3 \geq 0$ [Section 1.4, pp. 85–87]

R5. Graph $f(x) = x^2 + 1$ using point plotting. [Section 2.3, pp. 163–164]

▶ ① Evaluate Functions Involving Radicals

We can evaluate functions whose rule contains a radical by substituting the value of the independent variable into the rule, just as we did in Section 2.2.

EXAMPLE 1 **Evaluating Functions for Which the Rule Is a Radical Expression**

For the functions $f(x) = \sqrt{x + 2}$ and $g(x) = \sqrt[3]{3x + 1}$, find

(a) $f(7)$ **(b)** $f(10)$ **(c)** $g(-3)$

Solution

(a) $f(x) = \sqrt{x + 2}$
$f(7) = \sqrt{7 + 2}$
$= \sqrt{9}$
$= 3$

(b) $f(x) = \sqrt{x + 2}$
$f(10) = \sqrt{10 + 2}$
$= \sqrt{12}$
$= 2\sqrt{3}$

(c) $g(x) = \sqrt[3]{3x + 1}$
$g(-3) = \sqrt[3]{3(-3) + 1}$
$= \sqrt[3]{-8}$
$= -2$ ●

Quick ✓

In Problems 1 and 2, find the values for each function.

1. $f(x) = \sqrt{3x + 7}$

   (a) $f(3)$ **(b)** $f(7)$

2. $g(x) = \sqrt[3]{2x + 7}$

   (a) $g(-4)$ **(b)** $g(10)$

▶ ② Find the Domain of a Function Involving a Radical

Recall $\sqrt[n]{a} = b$ means $a = b^n$. From this definition we learned that for $n \geq 2$ and even, the radicand, a, must be greater than or equal to 0. For $n \geq 3$ and odd, the radicand, a, can be any real number. This leads to a procedure for finding the domain of a function whose rule contains a radical.

> **Finding the Domain of a Function Involving a Radical**
>
> - If the index on a radical is even, then the radicand must be greater than or equal to zero.
> - If the index on a radical is odd, then the radicand can be any real number.

EXAMPLE 2 **Finding the Domain of a Radical Function**

Find the domain of each of the following functions:

   (a) $f(x) = \sqrt{x - 5}$ **(b)** $G(x) = \sqrt[3]{2x + 1}$ **(c)** $h(t) = \sqrt[4]{5 - 2t}$

Solution

(a) The function $f(x) = \sqrt{x - 5}$ tells us to take the square root of $x - 5$. We can only take square roots of numbers greater than or equal to zero, so the radicand, $x - 5$, must be greater than or equal to zero. This requires that

$$x - 5 \geq 0$$

Add 5 to both sides: $x \geq 5$

The domain of f is $\{x | x \geq 5\}$ or the interval $[5, \infty)$.

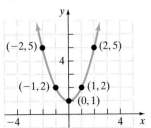

(b) The function $G(x) = \sqrt[3]{2x + 1}$ tells us to take the cube root of $2x + 1$. We can take the cube root of any real number, so the domain of G is $\{x | x \text{ is any real number}\}$, or the interval $(-\infty, \infty)$.

(c) The function $h(t) = \sqrt[4]{5 - 2t}$ tells us to take the fourth root of $5 - 2t$. We can only take fourth roots of numbers greater than or equal to zero, so the radicand, $5 - 2t$, must be greater than or equal to zero:

$$5 - 2t \geq 0$$

Subtract 5 from both sides: $\quad -2t \geq -5$

Divide both sides by -2 (Don't forget to change the direction of the inequality!): $\quad t \leq \dfrac{5}{2}$

The domain of h is $\left\{ t \mid t \leq \dfrac{5}{2} \right\}$ or the interval $\left(-\infty, \dfrac{5}{2} \right]$.

Quick ✓

3. If the index on a radical expression is ____, then the radicand must be greater than or equal to zero. If the index on a radical expression is ____, then the radicand can be any real number.

In Problems 4–6, find the domain of each function.

4. $H(x) = \sqrt{x + 6}$ **5.** $g(t) = \sqrt[5]{3t - 1}$ **6.** $F(m) = \sqrt[4]{6 - 3m}$

▶ ❸ Graph Functions Involving Square Roots

The **square root function** is given by $f(x) = \sqrt{x}$. The domain of the square root function is $\{x | x \geq 0\}$ or, in interval notation, $[0, \infty)$. We can graph $f(x) = \sqrt{x}$ by finding some ordered pairs (x, y) such that $y = \sqrt{x}$. We then plot the ordered pairs in the xy-plane and connect the points. To make life easy, we choose values of x that are perfect squares (0, 1, 4, 9, and so on). Table 1 shows some points on the graph of $f(x) = \sqrt{x}$. Figure 3 shows the graph of $f(x) = \sqrt{x}$ and illustrates that the range of $f(x) = \sqrt{x}$ is $[0, \infty)$.

Table 1

x	$f(x) = \sqrt{x}$	(x, y) or $(x, f(x))$
0	$f(0) = 0$	$(0, 0)$
1	$f(1) = 1$	$(1, 1)$
4	$f(4) = 2$	$(4, 2)$
9	$f(9) = 3$	$(9, 3)$
16	$f(16) = 4$	$(16, 4)$

Figure 3
$f(x) = \sqrt{x}$

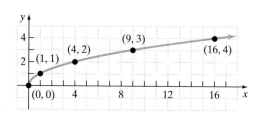

The point-plotting method can be used to graph a variety of functions involving square roots.

EXAMPLE 3   **Graphing a Function Involving a Square Root**

Consider the function $f(x) = \sqrt{x - 2}$.

(a) Find the domain.

(b) Graph the function using point plotting.

(c) Based on the graph, determine the range.

Solution

(a) Because $f(x) = \sqrt{x - 2}$ is a square root function, the radicand, $x - 2$, must be greater than or equal to zero. This requires that

$$x - 2 \geq 0$$

Add 2 to both sides: $x \geq 2$

The domain of f is $\{x \mid x \geq 2\}$, or the interval $[2, \infty)$.

(b) Choose values of x that are greater than or equal to 2 and that will make the radicand a perfect square. See Table 2. Figure 4 shows the graph of $f(x) = \sqrt{x - 2}$.

Table 2

x	$f(x) = \sqrt{x - 2}$	(x, y) or $(x, f(x))$
2	$f(2) = 0$	$(2, 0)$
3	$f(3) = 1$	$(3, 1)$
6	$f(6) = 2$	$(6, 2)$
11	$f(11) = 3$	$(11, 3)$
18	$f(18) = 4$	$(18, 4)$

Figure 4

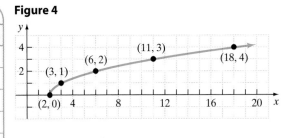

(c) From the graph of $f(x) = \sqrt{x - 2}$ given in Figure 4, we can see that the range of $f(x) = \sqrt{x - 2}$ is $[0, \infty)$.

Quick ✓

7. Consider the function $f(x) = \sqrt{x + 3}$.
   (a) Find the domain.
   (b) Graph the function using point plotting.
   (c) Based on the graph, determine the range.

▶ ❹ Graph Functions Involving Cube Roots

The **cube root function** is given by $f(x) = \sqrt[3]{x}$. The domain of the cube root function is $\{x \mid x \text{ is any real number}\}$ or, in interval notation, $(-\infty, \infty)$. We can graph $f(x) = \sqrt[3]{x}$ by finding some ordered pairs (x, y) such that $y = \sqrt[3]{x}$. Then plot the ordered pairs in the xy-plane and connect the points. To make life easy, choose values of x that are perfect cubes $(-8, -1, 0, 1, 8,$ and so on$)$. See Table 3. Figure 5 shows the graph of $f(x) = \sqrt[3]{x}$.

Table 3

x	$f(x) = \sqrt[3]{x}$	(x, y) or $(x, f(x))$
-8	$f(-8) = -2$	$(-8, -2)$
-1	$f(-1) = -1$	$(-1, -1)$
0	$f(0) = 0$	$(0, 0)$
1	$f(1) = 1$	$(1, 1)$
8	$f(8) = 2$	$(8, 2)$

Figure 5
$f(x) = \sqrt[3]{x}$

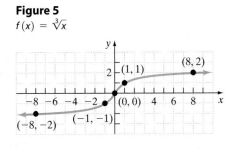

From the graph of $f(x) = \sqrt[3]{x}$ in Figure 5, we can see that the range of the function is the set of all real numbers or $(-\infty, \infty)$.

The point-plotting method can be used to graph a variety of functions involving cube roots.

EXAMPLE 4 Graphing a Function Involving a Cube Root

Consider the function $g(x) = \sqrt[3]{x} + 2$.

(a) Find the domain.

(b) Graph the function using point plotting.

(c) Based on the graph, determine the range.

Solution

(a) The function $g(x) = \sqrt[3]{x} + 2$ tells us to take the cube root of x and then add 2. We can take the cube root of any real number, so the domain of g is $\{x|x \text{ is any real number}\}$, or the interval $(-\infty, \infty)$.

(b) Choose values of x that make the radicand a perfect cube. See Table 4. Figure 6 shows the graph of $g(x) = \sqrt[3]{x} + 2$.

Table 4

x	$g(x) = \sqrt[3]{x} + 2$	(x, y) or $(x, g(x))$
-8	$g(-8) = \sqrt[3]{-8} + 2 = 0$	$(-8, 0)$
-1	$g(-1) = 1$	$(-1, 1)$
0	$g(0) = 2$	$(0, 2)$
1	$g(1) = 3$	$(1, 3)$
8	$g(8) = 4$	$(8, 4)$

Figure 6

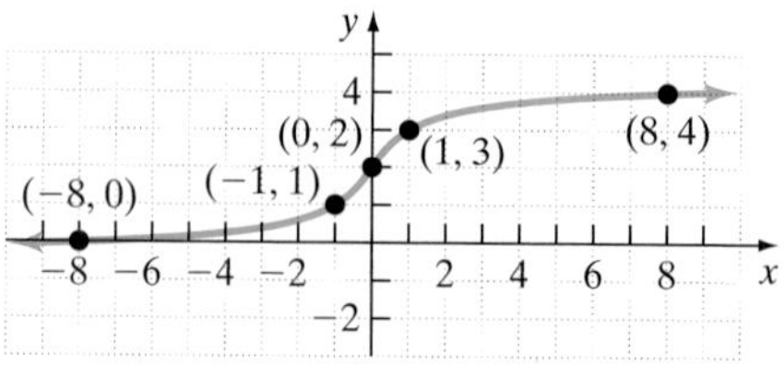

(c) The graph of $g(x) = \sqrt[3]{x} + 2$ shows that the range of $g(x) = \sqrt[3]{x} + 2$ is the set of all real numbers or $(-\infty, \infty)$.

Quick ✓

8. Consider the function $G(x) = \sqrt[3]{x} - 1$.

(a) Find the domain.

(b) Graph the function using point plotting.

(c) Based on the graph, determine the range.

6.6 Exercises

MyMathLab® MathXL PRACTICE

Exercise numbers in green are complete video solutions in MyMathLab

*Problems **1–8** are the Quick ✓s that follow the EXAMPLES.*

Building Skills

In Problems 9–20, evaluate each radical function at the indicated values. See Objective 1.

9. $f(x) = \sqrt{x + 6}$
(a) $f(3)$
(b) $f(8)$
(c) $f(-2)$

10. $f(x) = \sqrt{x + 10}$
(a) $f(6)$
(b) $f(2)$
(c) $f(-6)$

11. $g(x) = -\sqrt{2x + 3}$
(a) $g(11)$
(b) $g(-1)$
(c) $g\left(\frac{1}{8}\right)$

12. $g(x) = -\sqrt{4x + 5}$
(a) $g(1)$
(b) $g(10)$
(c) $g\left(\frac{1}{8}\right)$

13. $G(m) = 2\sqrt{5m - 1}$
(a) $G(1)$
(b) $G(5)$
(c) $G\left(\frac{1}{2}\right)$

14. $G(p) = 3\sqrt{4p + 1}$
(a) $G(2)$
(b) $G(11)$
(c) $G\left(\frac{1}{8}\right)$

15. $H(z) - \sqrt[3]{z+4}$

  (a) $H(4)$

  (b) $H(-12)$

  (c) $H(-20)$

16. $G(t) = \sqrt[3]{t-6}$

  (a) $G(7)$

  (b) $G(-21)$

  (c) $G(22)$

17. $f(x) = \sqrt{\dfrac{x-2}{x+2}}$

  (a) $f(7)$

  (b) $f(6)$

  (c) $f(10)$

18. $f(x) = \sqrt{\dfrac{x-4}{x+4}}$

  (a) $f(5)$

  (b) $f(8)$

  (c) $f(12)$

19. $g(z) = \sqrt[3]{\dfrac{2z}{z-4}}$

  (a) $g(-4)$

  (b) $g(8)$

  (c) $g(12)$

20. $H(z) = \sqrt[3]{\dfrac{3z}{z+5}}$

  (a) $H(3)$

  (b) $H(4)$

  (c) $H(-1)$

In Problems 21–36, find the domain of the radical function. See Objective 2.

21. $f(x) = \sqrt{x-7}$

22. $f(x) = \sqrt{x+4}$

23. $g(x) = \sqrt{2x+7}$

24. $g(x) = \sqrt{3x+7}$

25. $F(x) = \sqrt{4-3x}$

26. $G(x) = \sqrt{5-2x}$

27. $H(z) = \sqrt[3]{2z+1}$

28. $G(z) = \sqrt[3]{5z-3}$

29. $W(p) = \sqrt[4]{7p-2}$

30. $C(y) = \sqrt[4]{3y-2}$

31. $g(x) = \sqrt[5]{x-3}$

32. $g(x) = \sqrt[5]{x+9}$

33. $f(x) = \sqrt{\dfrac{3}{x+5}}$

34. $f(x) = \sqrt{\dfrac{3}{x-3}}$

35. $H(x) = \sqrt{\dfrac{x+3}{x-3}}$

36. $H(x) = \sqrt{\dfrac{x-5}{x}}$

In Problems 37–52, (a) determine the domain of the function; (b) graph the function using point plotting; and (c) based on the graph, determine the range of the function. See Objective 3.

37. $f(x) = \sqrt{x-4}$

38. $f(x) = \sqrt{x-1}$

39. $g(x) = \sqrt{x+2}$

40. $g(x) = \sqrt{x+5}$

41. $G(x) = \sqrt{2-x}$

42. $F(x) = \sqrt{4-x}$

43. $f(x) = \sqrt{x}+3$

44. $f(x) = \sqrt{x}+1$

45. $g(x) = \sqrt{x}-4$

46. $g(x) = \sqrt{x}-2$

47. $H(x) = 2\sqrt{x}$

48. $h(x) = 3\sqrt{x}$

49. $f(x) = \dfrac{1}{2}\sqrt{x}$

50. $g(x) = \dfrac{1}{4}\sqrt{x}$

51. $G(x) = -\sqrt{x}$

52. $F(x) = \sqrt{-x}$

In Problems 53–58, (a) determine the domain of the function; (b) graph the function using point plotting; and (c) based on the graph, determine the range of the function. See Objective 4.

53. $h(x) = \sqrt[3]{x+2}$

54. $g(x) = \sqrt[3]{x-4}$

55. $f(x) = \sqrt[3]{x}-3$

56. $H(x) = \sqrt[3]{x}+3$

57. $G(x) = 2\sqrt[3]{x}$

58. $F(x) = 3\sqrt[3]{x}$

Applying the Concepts

59. Distance to a Point on a Graph Suppose that $P = (x, y)$ is a point on the graph of $y = x^2 - 4$. The distance from P to $(0, 1)$ is given by the function

$$d(x) = \sqrt{x^4 - 9x^2 + 25}$$

See the figure.

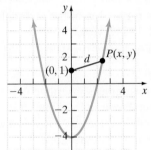

  (a) What is the distance from $P = (0, -4)$ to $(0, 1)$? That is, what is $d(0)$?

  (b) What is the distance from $P = (1, -3)$ to $(0, 1)$? That is, what is $d(1)$?

  (c) What is the distance from $P = (5, 21)$ to $(0, 1)$? That is, what is $d(5)$?

60. Distance to a Point on a Graph Suppose that $P = (x, y)$ is a point on the graph of $y = x^2 - 2$. The distance from P to $(0, 2)$ is given by the function

$$d(x) = \sqrt{x^4 - 7x^2 + 16}$$

See the figure.

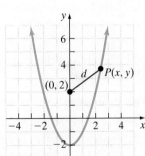

  (a) What is the distance from $P = (0, -2)$ to $(0, 2)$? That is, what is $d(0)$?

  (b) What is the distance from $P = (1, -1)$ to $(0, 2)$? That is, what is $d(1)$?

(c) What is the distance from $P = (4, 14)$ to $(0, 2)$? That is, what is $d(4)$?

△ **61. Area** A rectangle is inscribed in a semicircle of radius 3, as shown in the figure. Let $P = (x, y)$ be the point in quadrant I that is a vertex of the rectangle and is on the circle.

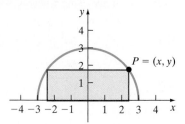

The area A of the rectangle as a function of x is given by

$$A(x) = 2x\sqrt{9 - x^2}$$

(a) What is the area of the rectangle whose vertex is at $(1, 2\sqrt{2})$?

(b) What is the area of the rectangle whose vertex is at $(2, \sqrt{5})$?

(c) What is the area of the rectangle whose vertex is at $(\sqrt{2}, \sqrt{7})$?

△ **62. Area** A rectangle is inscribed in a semicircle of radius 4 as shown in the figure. Let $P = (x, y)$ be the point in quadrant I that is a vertex of the rectangle and is on the circle.

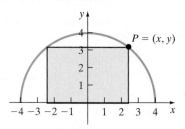

The area A of the rectangle as a function of x is given by

$$A(x) = 2x\sqrt{16 - x^2}$$

(a) What is the area of the rectangle whose vertex is at $(1, \sqrt{15})$?

(b) What is the area of the rectangle whose vertex is at $(2, 2\sqrt{3})$?

(c) What is the area of the rectangle whose vertex is at $(2\sqrt{2}, 2\sqrt{2})$?

Extending the Concepts

63. Use the results of Problems 37–40 to make a generalization about how to obtain the graph of $g(x) = \sqrt{x} + c$ from the graph of $f(x) = \sqrt{x}$.

64. Use the results of Problems 43–46 to make a generalization about how to obtain the graph of $g(x) = \sqrt{x} + c$ from the graph of $f(x) = \sqrt{x}$.

Synthesis Review

In Problems 65-69, add each of the following.

65. $\dfrac{1}{3} + \dfrac{1}{2}$

66. $\dfrac{1}{5} + \dfrac{3}{4}$

67. $\dfrac{1}{x} + \dfrac{3}{x + 1}$

68. $\dfrac{5}{x - 3} + \dfrac{2}{x + 1}$

69. $\dfrac{4}{x - 1} + \dfrac{3}{x + 1}$

70. Explain how adding rational numbers that do not have denominators with any common factors is similar to adding rational expressions that do not have denominators with any common factors.

The Graphing Calculator

The graphing calculator can graph square root and cube root functions. The figure below shows the graph of $f(x) = \sqrt{x - 2}$ using a TI-84 Plus graphing calculator. Note how the graph shown exists only for $x \geq 2$. Graphing functions is useful in verifying the domain that we found algebraically.

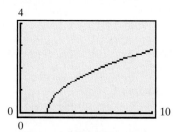

In Problems 71–92, graph the function using a graphing calculator. Compare the graphs obtained on the calculator to the hand-drawn graphs in Problems 37–58.

71. $f(x) = \sqrt{x - 4}$

72. $f(x) = \sqrt{x - 1}$

73. $g(x) = \sqrt{x + 2}$

74. $g(x) = \sqrt{x + 5}$

75. $G(x) = \sqrt{2 - x}$

76. $F(x) = \sqrt{4 - x}$

77. $f(x) = \sqrt{x} + 3$

78. $f(x) = \sqrt{x} + 1$

79. $g(x) = \sqrt{x} - 4$

80. $g(x) = \sqrt{x} - 2$

81. $H(x) = 2\sqrt{x}$

82. $h(x) = 3\sqrt{x}$

83. $f(x) = \dfrac{1}{2}\sqrt{x}$

84. $g(x) = \dfrac{1}{4}\sqrt{x}$

85. $G(x) = -\sqrt{x}$

86. $F(x) = \sqrt{-x}$

87. $h(x) = \sqrt[3]{x + 2}$

88. $g(x) = \sqrt[3]{x - 4}$

89. $f(x) = \sqrt[3]{x} - 3$

90. $H(x) = \sqrt[3]{x} + 3$

91. $G(x) = 2\sqrt[3]{x}$

92. $F(x) = 3\sqrt[3]{x}$

6.7 Radical Equations and Their Applications

Objectives

1. Solve Radical Equations Containing One Radical
2. Solve Radical Equations Containing Two Radicals
3. Solve for a Variable in a Radical Equation

Are You Ready for This Section?

Before getting started, take the following readiness quiz. If you get a problem wrong, go back to the section cited and review the material.

R1. Solve: $3x - 5 = 0$ [Section 1.1, pp. 50–52]

R2. Solve: $2p^2 + 4p - 6 = 0$ [Section 4.8, pp. 372–374]

R3. Simplify: $\sqrt[3]{(x - 5)^3}$ [Section 6.1, p. 475]

When the variable in an equation occurs in a radicand, the equation is called a **radical equation.** Examples of radical equations are

$$\sqrt{3x + 1} = 5 \qquad \sqrt[3]{x - 5} - 5 = 12 \qquad \sqrt{x - 2} - \sqrt{2x + 5} = 2$$

In this section, we are going to solve radical equations involving one or two radicals.

▶ 1 Solve Radical Equations Containing One Radical

To solve an equation containing a single radical, isolate the radical and then use the fact that $(\sqrt[n]{a})^n = a$.

EXAMPLE 1   **How to Solve a Radical Equation Containing One Radical**

Solve: $\sqrt{2x - 3} - 5 = 0$

Step-by-Step Solution

Step 1: Isolate the radical.

$$\sqrt{2x - 3} - 5 = 0$$

Add 5 to both sides: $\sqrt{2x - 3} = 5$

Step 2: Raise both sides to the power of the index.

The index is 2, so we square both sides: $(\sqrt{2x - 3})^2 = 5^2$

$$2x - 3 = 25$$

Step 3: Solve the equation that results.

Add 3 to both sides: $2x - 3 + 3 = 25 + 3$

$$2x = 28$$

Divide both sides by 2: $\dfrac{2x}{2} = \dfrac{28}{2}$

$$x = 14$$

Step 4: Check

Let $x = 14$ in the original equation:

$$\sqrt{2x - 3} - 5 = 0$$
$$\sqrt{2 \cdot 14 - 3} - 5 \stackrel{?}{=} 0$$
$$\sqrt{28 - 3} - 5 \stackrel{?}{=} 0$$
$$5 - 5 = 0 \quad \text{True}$$

The solution set is $\{14\}$.

We have always emphasized that you should check your answers whenever solving an equation. This is particularly important when you are solving equations involving radicals. Why? When we solve radical equations containing an even index, apparent solutions that are not solutions to the original equation can creep in. As we saw when we solved rational equations, these solutions are extraneous.

Ready?...Answers   **R1.** $\left\{\dfrac{5}{3}\right\}$

R2. $\{-3, 1\}$ **R3.** $x - 5$

> **Solving a Radical Equation Containing One Radical**
>
> **Step 1:** Isolate the radical. That is, get the radical by itself on one side of the equation.
>
> **Step 2:** Raise both sides of the equation to the power of the index. This will eliminate the radical from the equation.
>
> **Step 3:** Solve the equation that results.
>
> **Step 4:** Check your answer in the original equation.

Quick ✓

1. When the variable in an equation occurs in a radical, the equation is called a _____ _____.

2. When an apparent solution is not a solution of the original equation, we say the apparent solution is a(n) _____ solution.

3. *True or False* The first step in solving $x + \sqrt{x - 3} = 5$ is to square both sides of the equation.

4. Solve: $\sqrt{3x + 1} - 4 = 0$

EXAMPLE 2   **Solving a Radical Equation Containing One Radical**

Solve:

(a) $\sqrt{3x + 10} + 2 = 4$ **(b)** $\sqrt{5x - 1} + 7 = 5$

Solution

(a)

$$\sqrt{3x + 10} + 2 = 4$$

Subtract 2 from both sides: $\sqrt{3x + 10} + 2 - 2 = 4 - 2$

$$\sqrt{3x + 10} = 2$$

The index is 2, so we square both sides: $(\sqrt{3x + 10})^2 = 2^2$

$$3x + 10 = 4$$

Subtract 10 from both sides: $3x + 10 - 10 = 4 - 10$

$$3x = -6$$

Divide both sides by 3: $\dfrac{3x}{3} = \dfrac{-6}{3}$

$$x = -2$$

Work Smart

Just because the apparent solution is negative does not automatically make it extraneous. Determine whether the apparent solution makes the radicand negative.

Check

$$\sqrt{3x + 10} + 2 = 4$$

Let $x = -2$ in the original equation: $\sqrt{3 \cdot (-2) + 10} + 2 \overset{?}{=} 4$

$$\sqrt{-6 + 10} + 2 \overset{?}{=} 4$$

$$\sqrt{4} + 2 \overset{?}{=} 4$$

$$2 + 2 \overset{?}{=} 4$$

$$4 = 4 \quad \text{True}$$

The solution set is $\{-2\}$.

(b)

$$\sqrt{5x - 1} + 7 = 5$$

Subtract 7 from both sides: $\sqrt{5x - 1} + 7 - 7 = 5 - 7$

$$\sqrt{5x - 1} = -2$$

The equation has no real solution because the principal square root of a number cannot be less than 0. Put another way, there is no real number whose square root is -2, so the equation has no real solution. The solution set is $\varnothing$ or $\{\ \}$. ●

Quick ✓

5. Solve: $\sqrt{2x + 35} - 2 = 3$ **6.** Solve: $\sqrt{2x + 3} + 8 = 6$

▶ **EXAMPLE 3**   **Solving a Radical Equation Containing One Radical**

Solve: $\sqrt{x + 5} = x - 1$

Solution

$$\sqrt{x + 5} = x - 1$$

The index is 2, so we square both sides: $(\sqrt{x + 5})^2 = (x - 1)^2$

Work Smart

$(x - 1)^2 = x^2 - 2x + 1$

Do not write

$(x - 1)^2 = x^2 + 1$

$$x + 5 = x^2 - 2x + 1$$

Subtract x and 5 from both sides: $x + 5 - x - 5 = x^2 - 2x + 1 - x - 5$

$$0 = x^2 - 3x - 4$$

Factor: $0 = (x - 4)(x + 1)$

Zero-Product Property: $x - 4 = 0$ or $x + 1 = 0$

$$x = 4 \text{ or } \qquad x = -1$$

Check $\sqrt{x + 5} = x - 1$

$x = 4:$ $\sqrt{4 + 5} \overset{?}{=} 4 - 1$ $x = -1:$ $\sqrt{-1 + 5} \overset{?}{=} -1 - 1$

$\sqrt{9} \overset{?}{=} 3$ $\sqrt{4} \overset{?}{=} -2$

$3 = 3$ True $2 = -2$ False

The apparent solution $x = -1$ does not check, so it is an extraneous solution. The solution set is $\{4\}$.

EXAMPLE 4   **Solving a Radical Equation Containing One Radical**

Solve: $\sqrt[3]{3x - 12} + 4 = 1$

Solution

$$\sqrt[3]{3x - 12} + 4 = 1$$

Subtract 4 from both sides: $\sqrt[3]{3x - 12} + 4 - 4 = 1 - 4$

$$\sqrt[3]{3x - 12} = -3$$

The index is 3, so we cube both sides: $(\sqrt[3]{3x - 12})^3 = (-3)^3$

$$3x - 12 = -27$$

Add 12 to both sides: $3x - 12 + 12 = -27 + 12$

$$3x = -15$$

Divide both sides by 3: $x = -5$

Check

$$\sqrt[3]{3x - 12} + 4 = 1$$

Let $x = -5$ in the original equation: $\sqrt[3]{3 \cdot (-5) - 12} + 4 \overset{?}{=} 1$

$$\sqrt[3]{-15 - 12} + 4 \overset{?}{=} 1$$

$$\sqrt[3]{-27} + 4 \overset{?}{=} 1$$

$$-3 + 4 \overset{?}{=} 1$$

$$1 = 1 \quad \text{True}$$

The solution set is $\{-5\}$.

Quick ✓

7. Solve: $\sqrt{2x + 1} = x - 1$ **8.** Solve: $\sqrt[3]{3x + 1} - 4 = -6$

Sometimes, an equation will contain rational exponents rather than radicals. When solving these problems, we can rewrite the equation with a radical or use the fact that $(a^r)^s = a^{r \cdot s}$.

EXAMPLE 5 **Solving an Equation Containing a Rational Exponent**

Solve: $(5x - 1)^{\frac{1}{2}} + 3 = 10$

Solution

$$(5x - 1)^{\frac{1}{2}} + 3 = 10$$

Subtract 3 from both sides: $(5x - 1)^{\frac{1}{2}} = 7$

Square both sides: $((5x - 1)^{\frac{1}{2}})^2 = 7^2$

Use $(a^r)^s = a^{r \cdot s}$: $(5x - 1)^{\frac{1}{2} \cdot 2} = 49$

$$5x - 1 = 49$$

Add 1 to both sides: $5x = 50$

Divide both sides by 5: $x = 10$

Check

$$(5x - 1)^{\frac{1}{2}} + 3 = 10$$

Let $x = 10$ in the original equation: $(5 \cdot 10 - 1)^{\frac{1}{2}} + 3 \stackrel{?}{=} 10$

$$(49)^{\frac{1}{2}} + 3 \stackrel{?}{=} 10$$

$$7 + 3 = 10$$

$$10 = 10 \quad \text{True}$$

The solution set is $\{10\}$.

Quick ✓

9. Solve: $(2x - 3)^{\frac{1}{3}} - 7 = -4$

▶ ❷ **Solve Radical Equations Containing Two Radicals**

EXAMPLE 6 **How to Solve a Radical Equation Containing Two Radicals**

Solve: $\sqrt[3]{p^2 - 4p - 4} = \sqrt[3]{-3p + 2}$

Step-by-Step-Solution

Step 1: Isolate one of the radicals.

The radical on the left side of the equation is isolated: $\sqrt[3]{p^2 - 4p - 4} = \sqrt[3]{-3p + 2}$

Step 2: Raise both sides to the power of the index.

The index is 3, so we cube both sides:
$$(\sqrt[3]{p^2 - 4p - 4})^3 = (\sqrt[3]{-3p + 2})^3$$
$$p^2 - 4p - 4 = -3p + 2$$

Step 3: Because there is no radical, solve the equation that results.

Add $3p$ to both sides; subtract 2 from both sides: $p^2 - 4p - 4 + 3p - 2 = -3p + 2 + 3p - 2$

Combine like terms: $p^2 - p - 6 = 0$

Factor: $(p - 3)(p + 2) = 0$

Use the Zero-Product Property: $p - 3 = 0 \quad \text{or} \quad p + 2 = 0$

$$p = 3 \quad \text{or} \quad p = -2$$

Step 4: Check

$$\sqrt[3]{p^2 - 4p - 4} = \sqrt[3]{-3p + 2}$$

$p = -2$:

$$\sqrt[3]{(-2)^2 - 4(-2) - 4} \stackrel{?}{=} \sqrt[3]{-3(-2) + 2}$$
$$\sqrt[3]{4 + 8 - 4} \stackrel{?}{=} \sqrt[3]{6 + 2}$$
$$\sqrt[3]{8} = \sqrt[3]{8} \quad \text{True}$$

$p = 3$:

$$\sqrt[3]{(3)^2 - 4(3) - 4} \stackrel{?}{=} \sqrt[3]{-3(3) + 2}$$
$$\sqrt[3]{9 - 12 - 4} \stackrel{?}{=} \sqrt[3]{-9 + 2}$$
$$\sqrt[3]{-7} = \sqrt[3]{-7} \quad \text{True}$$

Both apparent solutions check. The solution set is $\{-2, 3\}$.

When a radical equation contains two radicals, use the following steps to solve the equation.

> **Solving a Radical Equation Containing Two Radicals**
>
> **Step 1:** Isolate one of the radicals. That is, get one of the radicals by itself on one side of the equation.
>
> **Step 2:** Raise both sides of the equation to the power of the index. This will eliminate one or both radicals from the equation.
>
> **Step 3:** If a radical remains in the equation, then follow the steps for solving a radical equation containing one radical. Otherwise, solve the equation that results.
>
> **Step 4:** Check your answer. Discard any extraneous solutions.

Quick ✓

10. Solve: $\sqrt[3]{m^2 + 4m + 4} = \sqrt[3]{2m + 7}$

EXAMPLE 7 **Solving a Radical Equation—Squaring Twice**

Solve: $\sqrt{3x + 6} - \sqrt{x + 6} = 2$

Solution

$$\sqrt{3x + 6} - \sqrt{x + 6} = 2$$

Add $\sqrt{x+6}$ to both sides: $\qquad \sqrt{3x + 6} = 2 + \sqrt{x + 6}$

Square both sides: $\qquad (\sqrt{3x + 6})^2 = (2 + \sqrt{x + 6})^2$

Use $(A + B)^2 = A^2 + 2AB + B^2$: $\qquad 3x + 6 = 4 + 4\sqrt{x + 6} + (\sqrt{x + 6})^2$

$(\sqrt{x + 6})^2 = x + 6$: $\qquad 3x + 6 = 4 + 4\sqrt{x + 6} + x + 6$

Work Smart

$(2 + \sqrt{x + 6})^2 \neq 2^2 + (\sqrt{x + 6})^2$

Remember,

$(A + B)^2 = (A + B)(A + B)$
$\qquad\quad = A^2 + 2AB + B^2$

Isolate the radical: $\qquad 2x - 4 = 4\sqrt{x + 6}$

Factor out 2: $\qquad 2(x - 2) = 4\sqrt{x + 6}$

Divide both sides by 2: $\qquad x - 2 = 2\sqrt{x + 6}$

Square both sides: $\qquad (x - 2)^2 = (2\sqrt{x + 6})^2$

$$x^2 - 4x + 4 = 4(x + 6)$$

Work Smart

When there is more than one radical, it is best to isolate the radical with the more complicated radicand.

Distribute: $\qquad x^2 - 4x + 4 = 4x + 24$

Subtract $4x$ and 24 from both sides: $\qquad x^2 - 8x - 20 = 0$

Factor: $\qquad (x - 10)(x + 2) = 0$

Use the Zero-Product Property: $\qquad x - 10 = 0 \quad \text{or} \quad x + 2 = 0$

$$x = 10 \quad \text{or} \qquad x = -2$$

Check $x = -2$: $\quad \sqrt{3 \cdot (-2) + 6} - \sqrt{-2 + 6} \overset{?}{=} 2 \qquad x = 10$: $\quad \sqrt{3 \cdot 10 + 6} - \sqrt{10 + 6} \overset{?}{=} 2$

$$\sqrt{0} - \sqrt{4} \overset{?}{=} 2 \qquad\qquad\qquad \sqrt{36} - \sqrt{16} \overset{?}{=} 2$$

$$0 - 2 \overset{?}{=} 2 \qquad\qquad\qquad\qquad 6 - 4 \overset{?}{=} 2$$

$$-2 = 2 \ \text{False} \qquad\qquad\qquad\qquad 2 = 2 \ \text{True}$$

The apparent solution $x = -2$ does not check, so it is an extraneous solution. The solution set is $\{10\}$.

Quick ✓

11. Solve: $\sqrt{2x + 1} - \sqrt{x + 4} = 1$

▶ ❸ Solve for a Variable in a Radical Equation

In many situations, you will need to solve for a variable in a formula. For instance, in Example 8 we are assessing how much error there is in an estimate based on a statistical study. This commonly used formula from statistics contains a radical.

EXAMPLE 8 **Solving for a Variable**

A formula from statistics for finding the margin of error in estimating a population mean is given by

$$E = z \cdot \frac{\sigma}{\sqrt{n}}$$

where E represents the margin of error, σ (sigma) represents a measure of variability, and z is a measure of relative position.

 (a) Solve this equation for n, the sample size.

 (b) Find n when $\sigma = 12$, $z = 2$, and $E = 3$.

Solution

 (a)
$$E = z \cdot \frac{\sigma}{\sqrt{n}}$$

Multiply both sides by $\sqrt{n}$: $\quad \sqrt{n} \cdot E = z\sigma$

Divide both sides by E: $\quad \sqrt{n} = \dfrac{z\sigma}{E}$

Square both sides: $\quad n = \left(\dfrac{z\sigma}{E}\right)^2$

 (b) $n = \left(\dfrac{z\sigma}{E}\right)^2 = \left(\dfrac{2 \cdot 12}{3}\right)^2$

$\qquad = 64$

We need the sample size to be $n = 64$ for E to equal 3. ●

Quick ✓

12. The period of a pendulum is the time it takes to complete one trip back and forth. The period T, in seconds, of a pendulum of length L, in feet, may be approximated using the formula $T = 2\pi\sqrt{\dfrac{L}{32}}$.

 (a) Solve the equation for L.

 (b) Determine the length of a pendulum whose period is 2π seconds.

6.7 Exercises MyMathLab® PRACTICE

Exercise numbers in green are complete video solutions in MyMathLab

*Problems **1–12** are the Quick ✓s that follow the* **EXAMPLES.**

Building Skills

In Problems 13–44, solve each equation. See Objective 1.

13. $\sqrt{x} = 4$

14. $\sqrt{p} = 6$

15. $\sqrt{x - 3} = 2$

16. $\sqrt{y - 5} = 3$

17. $\sqrt{2t + 3} = 5$

18. $\sqrt{3w - 2} = 4$

19. $\sqrt{4x + 3} = -2$

20. $\sqrt{6p - 5} = -5$

21. $\sqrt[3]{4t} = 2$

22. $\sqrt[3]{9w} = 3$

23. $\sqrt[3]{5q + 4} = 4$

24. $\sqrt[3]{7m + 20} = 5$

25. $\sqrt{y} + 3 = 8$

26. $\sqrt{q} - 5 = 2$

27. $\sqrt{x + 5} - 3 = 1$

28. $\sqrt{x - 4} + 4 = 7$

29. $\sqrt{2x + 9} + 5 = 6$

30. $\sqrt{4x + 21} + 2 = 5$

31. $3\sqrt{x} + 5 = 8$

32. $4\sqrt{t} - 2 = 10$

33. $\sqrt{4 - x} - 3 = 0$

34. $\sqrt{6 - w} - 3 = 1$

35. $\sqrt{p} = 2p$

36. $\sqrt{q} = 3q$

37. $\sqrt{x + 6} = x$

38. $\sqrt{2p + 8} = p$

39. $\sqrt{w} = 6 - w$

40. $\sqrt{m} = 12 - m$

41. $\sqrt{17 - 2x} + 1 = x$

42. $\sqrt{1 - 4x} - 5 = x$

43. $\sqrt{w^2 - 11} + 5 = w + 4$

44. $\sqrt{z^2 - z - 7} + 3 = z + 2$

In Problems 45–58, solve each equation. See Objective 2.

45. $\sqrt{x + 9} = \sqrt{2x + 5}$

46. $\sqrt{3x + 1} = \sqrt{2x + 7}$

47. $\sqrt[3]{4x - 3} = \sqrt[3]{2x - 9}$

48. $\sqrt[3]{3y - 2} = \sqrt[3]{5y + 8}$

49. $\sqrt{2w^2 - 3w - 4} = \sqrt{w^2 + 6w + 6}$

50. $\sqrt{2x^2 + 7x - 10} = \sqrt{x^2 + 4x + 8}$

51. $\sqrt{3w + 4} = 2 + \sqrt{w}$

52. $\sqrt{3y - 2} = 2 + \sqrt{y}$

53. $\sqrt{x + 1} - \sqrt{x - 2} = 1$

54. $\sqrt{2x - 1} - \sqrt{x - 1} = 1$

55. $\sqrt{2x + 6} - \sqrt{x - 1} = 2$

56. $\sqrt{2x + 6} - \sqrt{x - 6} = 3$

57. $\sqrt{2x + 5} - \sqrt{x - 1} = 2$

58. $\sqrt{4x + 1} - \sqrt{2x + 1} = 2$

In Problems 59–72, solve each equation.

59. $(2x + 3)^{\frac{1}{2}} = 3$

60. $(4x + 1)^{\frac{1}{2}} = 5$

61. $(6x - 1)^{\frac{1}{4}} = (2x + 15)^{\frac{1}{4}}$

62. $(6p + 3)^{\frac{1}{5}} = (4p - 9)^{\frac{1}{5}}$

63. $(x + 3)^{\frac{1}{2}} - (x - 5)^{\frac{1}{2}} = 2$

64. $(3x + 1)^{\frac{1}{2}} - (x - 1)^{\frac{1}{2}} = 2$

65. $(x - 1)^{2/3} = 4$

66. $(2x + 3)^{3/5} = 8$

67. $(x - 2)^{3/2} = 16$

68. $(3x - 2)^{3/4} = 8$

69. $(2x^4 - 81)^{1/4} + 3x = 4x$

70. $(17x^4 - 256)^{1/4} + 5x = 7x$

71. $\dfrac{1}{(x - 4)^{-2/3}} - 3 = 6$

72. $\dfrac{1}{(x + 3)^{-2/3}} + 5 = 9$

For Problems 73–78, see Objective 3.

73. Finance Solve $A = P\sqrt{1 + r}$ for r.

74. Centripetal Acceleration Solve $v = \sqrt{ar}$ for a.

75. Volume of a Sphere Solve $r = \sqrt[3]{\dfrac{3V}{4\pi}}$ for V.

76. Surface Area of a Sphere Solve $r = \sqrt{\dfrac{S}{4\pi}}$ for S.

77. Coulomb's Law Solve $r = \sqrt{\dfrac{4F\pi\epsilon_0}{q_1 q_2}}$ for F.

78. Potential Energy Solve $V = \sqrt{\dfrac{2U}{C}}$ for U.

Mixed Practice

In Problems 79–94, solve each equation.

79. $\sqrt{5p - 3} + 7 = 3$

80. $\sqrt{3b - 2} + 8 = 5$

81. $\sqrt{x + 12} = x$

82. $\sqrt{x + 20} = x$

83. $\sqrt{2p + 12} = 4$

84. $\sqrt{3a - 5} = 2$

85. $\sqrt[4]{x + 7} = 2$

86. $\sqrt[5]{x + 23} = 2$

87. $(3x + 1)^{\frac{1}{3}} + 2 = 0$

88. $(5x - 2)^{\frac{1}{3}} + 3 = 0$

89. $\sqrt{10 - x} - x = 10$

90. $\sqrt{9 - x} - x = 11$

91. $\sqrt{2x + 5} = \sqrt{3x - 4}$

92. $\sqrt{4c - 5} = \sqrt{3c + 1}$

93. $\sqrt{x - 1} + \sqrt{x + 4} = 5$ **94.** $\sqrt{x - 3} + \sqrt{x + 4} = 7$

95. Solve $\sqrt{x^2} = x + 4$ using the techniques of this section. Now solve $\sqrt{x^2} = x + 4$ using the fact that $\sqrt{x^2} = |x|$. Which approach do you like better?

96. Solve $\sqrt{p^2} = 3p + 4$ using the techniques of this section. Now solve $\sqrt{p^2} = 3p + 4$ using the fact that $\sqrt{p^2} = |p|$. Which approach do you like better?

97. Suppose that $f(x) = \sqrt{x - 2}$.

   (a) Solve $f(x) = 0$. What point is on the graph of f?

   (b) Solve $f(x) = 1$. What point is on the graph of f?

   (c) Solve $f(x) = 2$. What point is on the graph of f?

   (d) Use the information obtained in parts (a)–(c) to graph $f(x) = \sqrt{x - 2}$.

   (e) Use the graph and the concept of the range of a function to explain why the equation $f(x) = -1$ has no solution.

98. Suppose that $g(x) = \sqrt{x + 3}$.

   (a) Solve $g(x) = 0$. What point is on the graph of g?

   (b) Solve $g(x) = 1$. What point is on the graph of g?

   (c) Solve $g(x) = 2$. What point is on the graph of g?

   (d) Use the information obtained in parts (a)–(c) to graph $g(x) = \sqrt{x + 3}$.

   (e) Use the graph and the concept of the range of a function to explain why the equation $g(x) = -1$ has no solution.

Applying the Concepts

△ **99. Finding a y-Coordinate** The solutions to the equation

$$\sqrt{4^2 + (y - 2)^2} = 5$$

represent the y-coordinates such that the distance from the point $(3, 2)$ to $(-1, y)$ in the Cartesian plane is 5 units.

(a) Solve the equation for y.

(b) Plot the points in the Cartesian plane, and label the lengths of the sides of the figure formed.

△ **100. Finding an x-Coordinate** The solutions to the equation

$$\sqrt{x^2 + 4^2} = 5$$

represent the x-coordinates such that the distance from the point $(0, 3)$ to $(x, -1)$ in the Cartesian plane is 5 units.

(a) Solve the equation for x.

(b) Plot the points in the Cartesian plane, and label the lengths of the sides of the figure formed.

101. Revenue Growth Suppose that the annual revenue R (in millions of dollars) of a company after t years of operating is modeled by the function

$$R(t) = \sqrt[3]{\frac{t}{2}}$$

(a) After how many years can the company expect to have annual revenue of $1 million?

(b) After how many years can the company expect to have annual revenue of $2 million?

△ **102. Sphere** The radius r of a sphere whose volume is V is given by

$$r = \sqrt[3]{\frac{3V}{4\pi}}$$

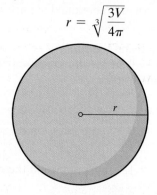

(a) Find the volume of a sphere whose radius is 3 meters.

(b) Find the volume of a sphere whose radius is 2 meters.

103. Birth Rates A plural birth is a live birth to twins, triplets, and so forth. The function $R(t) = 26 \cdot \sqrt[10]{t}$ models the plural birth rate R (live births per 1000 live births), where t is the number of years since 1995.

(a) Use the model to predict the year in which the plural birth rate will be 39. Round to the nearest year.

(b) Use the model to predict the year in which the plural birth rate will be 36. Round to the nearest year

104. Money The annual rate of interest r (expressed as a decimal) required to have A dollars after t years from an initial deposit of P dollars is given by

$$r = \sqrt[t]{\frac{A}{P}} - 1$$

(a) Suppose that you deposit $1000 in an account that pays 5% annual interest so that $r = 0.05$. How much will you have after $t = 2$ years?

(b) Suppose that you deposit $1000 in an account that pays 5% annual interest so that $r = 0.05$. How much will you have after $t = 3$ years?

Extending the Concepts

105. Solve: $\sqrt{3\sqrt{x + 1}} = \sqrt{2x + 3}$

106. Solve: $\sqrt[3]{2\sqrt{x - 2}} = \sqrt[3]{x - 1}$

Explaining the Concepts

107. Why is it always necessary to check solutions when solving radical equations?

108. How can you tell by inspection that the equation $\sqrt{x - 2} + 5 = 0$ will have no real solution?

109. Using the concept of domain, explain why radical equations with an even index may have extraneous solutions, but radical equations with an odd index do not have extraneous solutions.

110. Which step in the process of solving a radical equation leads to the possibility of extraneous solutions?

Synthesis Review

In Problems 111–114, consider the set

$$\left\{ 0, -4, 12, \frac{2}{3}, 1.\overline{56}, \sqrt{2^3}, \pi, \sqrt{-5}, \sqrt[3]{-4} \right\}$$

Indicate which of the numbers in the set are . . .

111. Integers

112. Rational numbers

113. Irrational numbers

114. Real numbers

115. State the difference between a rational number and an irrational number. Why is $\sqrt{-1}$ not real?

The Graphing Calculator

A graphing calculator can be used to verify solutions obtained algebraically. To solve $\sqrt{2x - 3} = 5$ presented in Example 1, we graph $Y_1 = \sqrt{2x - 3}$ and $Y_2 = 5$ and determine the x-coordinate of the point of intersection.

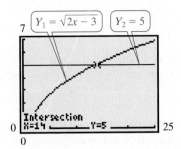

The x-coordinate of the point of intersection is 14, so the solution set is $\{14\}$.

116. Verify your solution to Problem 17 by graphing $Y_1 = \sqrt{2x + 3}$ and $Y_2 = 5$ and then finding the x-coordinate of their point of intersection.

117. Verify your solution to Problem 18 by graphing $Y_1 = \sqrt{3x - 2}$ and $Y_2 = 4$ and then finding the x-coordinate of their point of intersection.

118. When you solved Problem 19 algebraically, you should have determined that the equation has no real solution. Verify this result by graphing $Y_1 = \sqrt{4x + 3}$ and $Y_2 = -2$. Explain what the algebraic solution means graphically.

119. When you solved Problem 20 algebraically, you should have determined that the equation has no real solution. Verify this result by graphing $Y_1 = \sqrt{6x - 5}$ and $Y_2 = -5$. Explain what the algebraic solution means graphically.

6.8 The Complex Number System

Objectives

1. Evaluate the Square Root of Negative Real Numbers
2. Add or Subtract Complex Numbers
3. Multiply Complex Numbers
4. Divide Complex Numbers
5. Evaluate the Powers of i

Are You Ready for This Section?

Before getting started, take the following readiness quiz. If you get a problem wrong, go back to the section cited and review the material.

R1. List the numbers in the set $\left\{ 8, -\dfrac{1}{3}, -23, 0, \sqrt{2}, 1.\overline{26}, -\dfrac{12}{3}, \sqrt{-5} \right\}$ that are

    (a) Natural numbers   **(b)** Whole numbers

    (c) Integers   **(d)** Rational numbers

    (e) Irrational numbers   **(f)** Real numbers   [Section R.2, pp. 11–14]

R2. Distribute: $3x(4x - 3)$ [Section 4.2, p. 320]

R3. Multiply: $(z + 4)(3z - 2)$ [Section 4.2, pp. 320–321]

R4. Multiply: $(2y + 5)(2y - 5)$ [Section 4.2, pp. 322–323]

▶ If you look back at Section R.2, where we introduced the various number systems, you should notice that each time we encounter a situation where a number system can't handle a problem, we expand the number system. For example, if we considered only the whole numbers, we could not describe a negative balance in a checking account, so we introduced integers. If the world could be described only by integers, then we could not talk about parts of a whole, as in $\frac{1}{2}$ a pizza or $\frac{3}{4}$ of a dollar, so we introduced rational numbers. If we considered only rational numbers, then we wouldn't be able to find a number whose square is 2. So we introduced the irrational numbers, so that $(\sqrt{2})^2 = 2$. By combining the rational numbers with the irrational numbers, we created the real number system.

Now, suppose we wanted to determine a number whose square is -1. We know that the square of any real number is never negative. This is the *Nonnegativity Property*.

Nonnegativity Property of Real Numbers

For any real number a, $a^2 \geq 0$.

Because of the Nonnegativity Property, there is no real number solution to the equation

$$x^2 = -1$$

To remedy this situation, we introduce a new number.

Definition

The **imaginary unit,** denoted by ***i***, is the number whose square is -1. That is,

$$i^2 = -1$$

By introducing the number i, we now establish a new number system called the **complex number system.**

Definition

Complex numbers are numbers of the form ***a + bi***, where a and b are real numbers. The real number a is called the **real part** of the number $a + bi$; the real number b is called the **imaginary part** of $a + bi$.

In Words

The real number system is a subset of the complex number system. This means that all real numbers are, more generally, complex numbers.

For example, the complex number $6 + 2i$ has the real part 6 and the imaginary part 2. The complex number $4 - 3i = 4 + (-3)i$ has the real part 4 and the imaginary part -3.

When a complex number is written in the form $a + bi$, where a and b are real numbers, we say that it is in **standard form.** The complex number $a + 0i$ is typically written as a. This fact reminds us that the real number system is a subset of the complex number system. The complex number $0 + bi$ is usually written as bi. Any number of the form bi is called a **pure imaginary number.** Figure 7 shows the relation between the number systems.

Figure 7
The Complex Number System

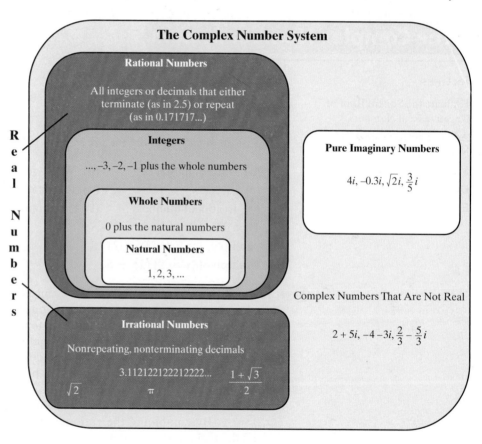

ⓘ ❶ Evaluate the Square Root of Negative Real Numbers

First, we define square roots of negative numbers.

Definition

If N is a positive real number, we define the **principal square root of** $-N$, denoted by $\sqrt{-N}$, as

$$\sqrt{-N} = \sqrt{N}i$$

where i is the number whose square is -1.

EXAMPLE 1   **Evaluating the Square Root of a Negative Number**

Write each of the following as a pure imaginary number.

(a) $\sqrt{-16}$ (b) $\sqrt{-3}$ (c) $\sqrt{-18}$

Solution

(a) $\sqrt{-16} = \sqrt{16}i$ (b) $\sqrt{-3} = \sqrt{3}i$ (c) $\sqrt{-18} = \sqrt{18}i$

$\quad\quad\quad = 4i$ $\quad\quad\quad\quad\quad\quad\quad = 3\sqrt{2}i$ ●

Work Smart

When writing complex numbers whose imaginary part is a radical, as in $\sqrt{3}i$, be sure that the "i" is not written under the radical.

Quick ✓

1. The _____, denoted by i, is the number whose square is -1.

2. Any number of the form bi is called a ____ _____ _____.

3. If N is a positive real number, we define the principal square root of $-N$, denoted by $\sqrt{-N}$, as $\sqrt{-N} = $ ___.

4. *True or False* All real numbers are complex numbers.

In Problems 5–7, write each radical as a pure imaginary number.

5. $\sqrt{-36}$ 6. $\sqrt{-5}$ 7. $\sqrt{-12}$

EXAMPLE 2   **Writing Complex Numbers in Standard Form**

Write each of the following in standard form.

(a) $2 - \sqrt{-25}$ (b) $3 + \sqrt{-50}$ (c) $\dfrac{4 - \sqrt{-12}}{2}$

Solution

The standard form of a complex number is $a + bi$.

(a) $2 - \sqrt{-25} = 2 - \sqrt{25}i$

$\quad\quad\quad\quad\quad = 2 - 5i$

(b) $3 + \sqrt{-50} = 3 + \sqrt{50}i$

$\sqrt{50} = \sqrt{25} \cdot \sqrt{2} = 5\sqrt{2}: \quad = 3 + 5\sqrt{2}i$

(c) $\dfrac{4 - \sqrt{-12}}{2} = \dfrac{4 - \sqrt{12}i}{2}$

$\sqrt{12} = \sqrt{4} \cdot \sqrt{3} = 2\sqrt{3}: \quad = \dfrac{4 - 2\sqrt{3}i}{2}$

Factor out 2: $\quad = \dfrac{2(2 - \sqrt{3}i)}{2}$

Divide out the 2's: $\quad = 2 - \sqrt{3}i$ ●

Quick ✓

In Problems 8–10, write each expression in the standard form of a complex number, $a + bi$.

8. $4 + \sqrt{-100}$ 9. $-2 - \sqrt{-8}$

10. $\dfrac{6 - \sqrt{-72}}{3}$

▶ ❷ Add or Subtract Complex Numbers

You have learned to add, subtract, multiply, and divide using various number systems. We will now see how to perform these operations on complex numbers.

We add complex numbers by adding the real parts and then adding the imaginary parts.

In Words

To add two complex numbers, add the real parts, and then add the imaginary parts. To subtract two complex numbers, subtract the real parts, and then subtract the imaginary parts.

Sum of Complex Numbers

$$(a + bi) + (c + di) = (a + c) + (b + d)i$$

To subtract two complex numbers, we use this rule:

Difference of Complex Numbers

$$(a + bi) - (c + di) = (a - c) + (b - d)i$$

EXAMPLE 3 | **Adding Complex Numbers**

Add:

(a) $(4 - 3i) + (-2 + 5i)$　　　　**(b)** $(4 + \sqrt{-25}) + (6 - \sqrt{-16})$

Solution

(a) $(4 - 3i) + (-2 + 5i) = [4 + (-2)] + (-3 + 5)i$
$$= 2 + 2i$$

$$\sqrt{-25} = 5i; \; \sqrt{-16} = 4i$$
$$\downarrow \qquad\qquad \downarrow$$

(b) $(4 + \sqrt{-25}) + (6 - \sqrt{-16}) = (4 + 5i) + (6 - 4i)$
$$= (4 + 6) + (5 - 4)i$$
$$= 10 + 1i$$
$$= 10 + i$$　　●

EXAMPLE 4 | **Subtracting Complex Numbers**

Subtract:

(a) $(-3 + 7i) - (5 - 4i)$　　　　**(b)** $(3 + \sqrt{-12}) - (-2 - \sqrt{-27})$

Solution

(a) $(-3 + 7i) - (5 - 4i) = (-3 - 5) + (7 - (-4))i$
$$= -8 + 11i$$

$$\sqrt{-12} = 2\sqrt{3}i; \quad \sqrt{-27} = 3\sqrt{3}i$$
$$\downarrow \qquad\qquad\qquad \downarrow$$

(b) $(3 + \sqrt{-12}) - (-2 - \sqrt{-27}) = (3 + 2\sqrt{3}i) - (-2 - 3\sqrt{3}i)$
$$= (3 + 2) + \left[2\sqrt{3} - (-3\sqrt{3})\right]i$$
$$= 5 + 5\sqrt{3}i$$　　●

Work Smart

Adding or subtracting complex numbers is just like combining like terms. For example,

$(4 - 3x) + (-2 + 5x)$
$= 4 + (-2) - 3x + 5x$
$= 2 + 2x$

so

$(4 - 3i) + (-2 + 5i)$
$= 4 + (-2) - 3i + 5i$
$= 2 + 2i$

Quick ✓

In Problems 11–13, add or subtract, as indicated.

11. $(4 + 6i) + (-3 + 5i)$　　　　**12.** $(4 - 2i) - (-2 + 7i)$

13. $(4 - \sqrt{-4}) + (-7 + \sqrt{-9})$

▷ **③ Multiply Complex Numbers**

We multiply complex numbers using the same ideas we used to multiply polynomials.

(EXAMPLE 5)   **Multiplying Complex Numbers**

Multiply:

    (a) $4i(3 - 6i)$ **(b)** $(-2 + 4i)(3 - i)$

Solution

    (a) Distribute the $4i$ to each term in the parentheses.

$$4i(3 - 6i) = 4i \cdot 3 - 4i \cdot 6i$$
$$= 12i - 24i^2$$
$$i^2 = -1: \quad = 12i - 24 \cdot (-1)$$
$$= 24 + 12i$$

    (b)

$$\text{F}\text{O}\text{I}\text{L}$$
$$(-2 + 4i)(3 - i) = -2 \cdot 3 - 2 \cdot (-i) + 4i \cdot 3 + 4i \cdot (-i)$$
$$= -6 + 2i + 12i - 4i^2$$
$$\text{Combine like terms; } i^2 = -1: \quad = -6 + 14i - 4(-1)$$
$$= -6 + 14i + 4$$
$$= -2 + 14i$$

Quick ✓

In Problems 14 and 15, multiply.

14. $3i(5 - 4i)$ **15.** $(-2 + 5i)(4 - 2i)$

Look back at the Product Property of Radicals on page 487. Notice that the property applies only when $\sqrt[n]{a}$ and $\sqrt[n]{b}$ are real numbers. This means that

$$\sqrt{a} \cdot \sqrt{b} \neq \sqrt{ab} \quad \text{if } a < 0 \text{ or } b < 0$$

How, then, do we perform this multiplication? First write the radical as a complex number, using the fact that $\sqrt{-N} = \sqrt{N}i$, and then multiply.

(EXAMPLE 6)   **Multiplying Square Roots of Negative Numbers**

Multiply:

    (a) $\sqrt{-25} \cdot \sqrt{-4}$ **(b)** $(2 + \sqrt{-16})(1 - \sqrt{-4})$

Solution

    (a) We cannot use the Product Property of Radicals on this product because $\sqrt{-25}$ and $\sqrt{-4}$ are not real numbers. Therefore, we express the radicals as pure imaginary numbers and then multiply.

$$\sqrt{-25} \cdot \sqrt{-4} = 5i \cdot 2i$$
$$= 10i^2$$
$$i^2 = -1: \quad = 10(-1)$$
$$= -10$$

Work Smart

$\sqrt{-25} \cdot \sqrt{-4} \neq \sqrt{(-25)(-4)}$ because the Product Property of Radicals applies only when the radical is a real number.

(b) First, rewrite each expression as a complex number in standard form.

$$(2 + \sqrt{-16})(1 - \sqrt{-9}) = (2 + 4i)(1 - 3i)$$

$$\text{FOIL:} \quad = 2 \cdot 1 + 2 \cdot (-3i) + 4i \cdot 1 + 4i \cdot (-3i)$$

$$= 2 - 6i + 4i - 12i^2$$

$$\text{Combine like terms; } i^2 = -1: \quad = 2 - 2i - 12(-1)$$

$$= 2 - 2i + 12$$

$$= 14 - 2i$$

Quick ✓

In Problems 16 and 17, multiply.

16. $\sqrt{-9} \cdot \sqrt{-36}$

17. $(2 + \sqrt{-36})(4 - \sqrt{-25})$

▶ Complex Conjugates

We now introduce a special product that involves the *conjugate* of a complex number.

Complex Conjugate

If $a + bi$ is a complex number, then its **conjugate** is defined as $a - bi$.

For example,

Complex Number	Conjugate
$4 + 7i$	$4 - 7i$
$-10 - 3i$	$-10 + 3i$

In Words

The complex conjugate of $a + bi$ is $a - bi$. The complex conjugate of $a - bi$ is $a + bi$.

EXAMPLE 7 **Multiplying a Complex Number by Its Conjugate**

Find the product of $4 + 3i$ and its conjugate, $4 - 3i$.

Solution

$$(4 + 3i)(4 - 3i) = 4 \cdot 4 + 4 \cdot (-3i) + 3i \cdot 4 + 3i \cdot (-3i)$$

$$= 16 - 12i + 12i - 9i^2$$

$$= 16 - 9(-1)$$

$$= 16 + 9$$

$$= 25$$

Wow! The product of $4 + 3i$ and its conjugate $4 - 3i$ is 25—a real number! In fact, the result of Example 7 is true in general:

Product of a Complex Number and its Conjugate

The product of a complex number and its conjugate is a nonnegative real number. That is,

$$(a + bi)(a - bi) = a^2 + b^2$$

Notice how multiplying a complex number by its conjugate is different from multiplying $(a + b)(a - b)$:

$$(a + b)(a - b) = a^2 - b^2, \quad \text{whereas} \quad (a + bi)(a - bi) = a^2 + b^2$$

Quick ✓

18. The complex conjugate of $-3 + 5i$ is _____.

In Problems 19 and 20, multiply.

19. $(3 - 8i)(3 + 8i)$ **20.** $(-2 + 5i)(-2 - 5i)$

▶ ④ Divide Complex Numbers

Now that we know that the product of a complex number and its conjugate is a nonnegative real number, we can divide complex numbers. In Example 8, notice how the approach is similar to rationalizing a denominator. (See Examples 3 and 4 in Section 6.5.)

EXAMPLE 8 **How to Divide Complex Numbers**

Divide: $\dfrac{-3 + i}{5 + 3i}$

Step-by-Step Solution

Step 1: Write the numerator and denominator in standard form, $a + bi$.

The numerator and denominator are already in standard form.

Step 2: Multiply the numerator and denominator by the complex conjugate of the denominator.

$$\text{Multiply by } 1 = \frac{5 - 3i}{5 - 3i}$$

$$\frac{-3 + i}{5 + 3i} = \frac{-3 + i}{5 + 3i} \cdot \frac{5 - 3i}{5 - 3i}$$

$$= \frac{(-3 + i)(5 - 3i)}{(5 + 3i)(5 - 3i)}$$

Step 3: Simplify by writing the quotient in standard form, $a + bi$.

Multiply numerator and denominator; $(a + bi)(a - bi) = a^2 + b^2$:

$$= \frac{-3 \cdot 5 - 3 \cdot (-3i) + i \cdot 5 + i \cdot (-3i)}{5^2 + 3^2}$$

$$= \frac{-15 + 9i + 5i - 3i^2}{25 + 9}$$

Combine like terms; $i^2 = -1$:

$$= \frac{-15 + 14i + 3}{34}$$

$$= \frac{-12 + 14i}{34}$$

Divide 34 into each term in the numerator to write in standard form:

$$= \frac{-12}{34} + \frac{14}{34}i$$

Write each fraction in lowest terms:

$$= -\frac{6}{17} + \frac{7}{17}i$$

●

Below, we summarize the steps used in Example 8.

Dividing Complex Numbers

Step 1: Write the numerator and denominator in standard form, $a + bi$.

Step 2: Multiply the numerator and denominator by the complex conjugate of the denominator.

Step 3: Simplify by writing the quotient in standard form, $a + bi$.

EXAMPLE 9 **Dividing Complex Numbers**

Divide: $\dfrac{3 + 4i}{2i}$

Solution

The denominator, $2i$, can be written as $0 + 2i$, and the conjugate of $0 + 2i$ is $0 - 2i$. Since $0 - 2i = -2i$, multiply the numerator and denominator by $-2i$.

Work Smart

We could also have multiplied the numerator and denominator in Example 9 by i. Do you know why? Try it yourself! Which approach do you prefer?

$$\frac{3 + 4i}{2i} = \frac{3 + 4i}{0 + 2i} = \frac{3 + 4i}{2i} \cdot \frac{-2i}{-2i}$$

Multiply numerator; multiply denominator: $\quad = \dfrac{(3 + 4i)(-2i)}{(2i)(-2i)}$

Distribute $-2i$: $\quad = \dfrac{-6i - 8i^2}{-4i^2}$

$i^2 = -1$: $\quad = \dfrac{-6i + 8}{4}$

Divide 4 into each term in the numerator to write in standard form: $\quad = \dfrac{8}{4} - \dfrac{6}{4}i$

Write each fraction in lowest terms: $\quad = 2 - \dfrac{3}{2}i$

In Example 5(b), we multiplied to find $(-2 + 4i)(3 - i) = -2 + 14i$. Now divide to find $-2 + 14i$ by $3 - i$. What result do you expect? Verify that $\dfrac{-2 + 14i}{3 - i} = -2 + 4i$. Remember, we can always check that $\dfrac{A}{B} = C$ by verifying that $A = B \cdot C$.

Quick ✓

In Problems 21 and 22, divide.

21. $\dfrac{-4 + i}{3i}$

22. $\dfrac{4 + 3i}{1 - 3i}$

▶ ⑤ **Evaluate the Powers of i**

The **powers of i** follow a pattern.

$i^1 = i$ $\qquad$ $i^2 = -1$ $\qquad$ $i^3 = i^2 \cdot i^1 = -1 \cdot i = -i$ $\qquad$ $i^4 = i^2 \cdot i^2 = (-1)(-1) = 1$

$i^5 = i^4 \cdot i = 1 \cdot i = i$ $\qquad$ $i^6 = i^4 \cdot i^2 = 1 \cdot (-1) = -1$ $\qquad$ $i^7 = i^4 \cdot i^3 = 1 \cdot (-i) = -i$ $\qquad$ $i^8 = i^4 \cdot i^4 = 1 \cdot 1 = 1$

$i^9 = i^8 \cdot i = 1 \cdot i = i$ $\qquad$ $i^{10} = i^8 \cdot i^2 = 1 \cdot (-1) = -1$ $\qquad$ $i^{11} = i^8 \cdot i^3 = 1 \cdot (-i) = -i$ $\qquad$ $i^{12} = (i^4)^3 = (1)^3 = 1$

Do you see the pattern? In the first column, all the expressions simplify to i; in the second column, all the expressions simplify to -1; in the third column, all the expressions simplify to $-i$; in the fourth column, all the expressions simplify to 1. That is, the powers of i repeat with every fourth power. Figure 8 shows the pattern.

Figure 8

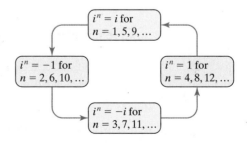

Thus any power of i can be expressed as $1, -1, i,$ or $-i$. We use the following steps to simplify any power of i.

> **Simplifying the Powers of _i_**
>
> **Step 1:** Divide the exponent of _i_ by 4. Rewrite i^n as $(i^4)^q \cdot i^r$, where q is the quotient and r is the remainder of the division.
>
> **Step 2:** Simplify the product in Step 1 to i^r since $i^4 = 1$.

(EXAMPLE 10) **Simplifying Powers of _i_**

Simplify:

(a) i^{34}　　　　　　　　　　　　　　　　　**(b)** i^{101}

Solution

(a) Divide 34 by 4 and obtain a quotient of $q = 8$ and a remainder of $r = 2$, so

$$i^{34} = (i^4)^8 \cdot i^2$$
$$= (1)^8 \cdot (-1)$$
$$= -1$$

Work Smart

An alternative approach to Example 10:

(a) $i^{34} = (i^2)^{17} = (-1)^{17} = -1$

(b) $i^{101} = i^{100} \cdot i^1 = (i^2)^{50} \cdot i$
$$= (-1)^{50} \cdot i$$
$$= 1 \cdot i$$
$$= i$$

(b) Divide 101 by 4 and obtain a quotient of $q = 25$ and a remainder of $r = 1$, so

$$i^{101} = (i^4)^{25} \cdot i^1$$
$$= (1)^{25} \cdot i$$
$$= i$$

Quick ✓

In Problems 23 and 24, simplify the power of i.

23. i^{43}　　　　　　　　　　　　　　　　　**24.** i^{98}

6.8 Exercises   MyMathLab®   MathXL PRACTICE

Exercise numbers in green are complete video solutions in MyMathLab

*Problems **1–24** are the Quick ✓ s that follow the* EXAMPLES.

Building Skills

In Problems 25–34, write each expression as a pure imaginary number. See Objective 1.

25. $\sqrt{-4}$　　　**26.** $\sqrt{-25}$　　　**27.** $-\sqrt{-81}$

28. $-\sqrt{-100}$　　**29.** $\sqrt{-45}$　　　**30.** $\sqrt{-48}$

31. $\sqrt{-300}$　　**32.** $\sqrt{-162}$　　　**33.** $\sqrt{-7}$

34. $\sqrt{-13}$

In Problems 35–42, write each expression as a complex number in standard form. See Objective 1.

35. $5 + \sqrt{-49}$　　　　　　　**36.** $4 - \sqrt{-36}$

37. $-2 - \sqrt{-28}$　　　　　　**38.** $10 + \sqrt{-32}$

39. $\dfrac{4 + \sqrt{-4}}{2}$　　　　　　　**40.** $\dfrac{10 - \sqrt{-25}}{5}$

41. $\dfrac{4 + \sqrt{-8}}{12}$ 　　　　　**42.** $\dfrac{15 - \sqrt{-50}}{5}$

In Problems 43–50, add or subtract, as indicated. See Objective 2.

43. $(4 + 5i) + (2 - 7i)$　　**44.** $(-6 + 2i) + (3 + 12i)$

45. $(4 + i) - (8 - 5i)$　　**46.** $(-7 + 3i) - (-3 + 2i)$

47. $(4 - \sqrt{-4}) - (2 + \sqrt{-9})$

48. $(-4 + \sqrt{-25}) + (1 - \sqrt{-16})$

49. $(-2 + \sqrt{-18}) + (5 - \sqrt{-50})$

50. $(-10 + \sqrt{-20}) - (-6 + \sqrt{-45})$

In Problems 51–74, multiply. See Objective 3.

51. $6i(2 - 4i)$　　　　　　**52.** $3i(-2 - 6i)$

53. $-\dfrac{1}{2}i(4 - 10i)$　　　　**54.** $\dfrac{1}{3}i(12 + 15i)$

55. $(2 + i)(4 + 3i)$　　　　**56.** $(3 - i)(1 + 2i)$

57. $(-3 - 5i)(2 + 4i)$　　　**58.** $(5 - 2i)(-1 + 2i)$

59. $(2 - 3i)(4 + 6i)$　　　　**60.** $(6 + 8i)(-3 + 4i)$

61. $(3 - \sqrt{2}i)(-2 + \sqrt{2}i)$ **62.** $(1 + \sqrt{3}i)(-4 - \sqrt{3}i)$

63. $\left(\dfrac{1}{2} - \dfrac{1}{4}i\right)\left(\dfrac{2}{3} + \dfrac{3}{4}i\right)$ **64.** $\left(-\dfrac{2}{3} + \dfrac{4}{3}i\right)\left(\dfrac{1}{2} - \dfrac{3}{2}i\right)$

65. $(3 + 2i)^2$ **66.** $(2 + 5i)^2$

67. $(-4 - 5i)^2$ **68.** $(2 - 7i)^2$

69. $\sqrt{-9} \cdot \sqrt{-4}$ **70.** $\sqrt{-36} \cdot \sqrt{-4}$

71. $\sqrt{-8} \cdot \sqrt{-10}$ **72.** $\sqrt{-12} \cdot \sqrt{-15}$

73. $(2 + \sqrt{-81})(-3 - \sqrt{-100})$

74. $(1 - \sqrt{-64})(-2 + \sqrt{-49})$

In Problems 75–80, (a) find the conjugate of the complex number, and (b) multiply the complex number by its conjugate. See Objective 3.

75. $3 + 5i$ **76.** $5 + 2i$

77. $2 - 7i$ **78.** $9 - i$

79. $-7 + 2i$ **80.** $-1 - 4i$

In Problems 81–94, divide. See Objective 4.

81. $\dfrac{1 + i}{3i}$ **82.** $\dfrac{2 - i}{2i}$

83. $\dfrac{-5 + 2i}{5i}$ **84.** $\dfrac{-4 + 5i}{6i}$

85. $\dfrac{3}{2 + i}$ **86.** $\dfrac{2}{4 + i}$

87. $\dfrac{-2}{-3 - 7i}$ **88.** $\dfrac{-4}{-5 - 3i}$

89. $\dfrac{2 + 3i}{3 - 2i}$ **90.** $\dfrac{2 + 5i}{5 - 2i}$

91. $\dfrac{4 + 2i}{1 - i}$ **92.** $\dfrac{-6 + 2i}{1 + i}$

93. $\dfrac{4 - 2i}{1 + 3i}$ **94.** $\dfrac{5 - 3i}{2 + 4i}$

In Problems 95–102, simplify. See Objective 5.

95. i^{53} **96.** i^{72}

97. i^{43} **98.** i^{110}

99. i^{153} **100.** i^{131}

101. i^{-45} **102.** i^{-26}

Mixed Practice

In Problems 103–116, perform the indicated operation.

103. $(-4 - i)(4 + i)$ **104.** $(-5 + 2i)(5 - 2i)$

105. $(3 + 2i)^2$ **106.** $(-3 + 2i)^2$

107. $\dfrac{-3 + 2i}{3i}$ **108.** $\dfrac{5 - 3i}{4i}$

109. $\dfrac{-4 + i}{-5 - 3i}$ **110.** $\dfrac{-4 + 6i}{-5 - i}$

111. $(10 - 3i) + (2 + 3i)$ **112.** $(-4 + 5i) + (4 - 2i)$

113. $5i^{37}(-4 + 3i)$ **114.** $2i^{57}(3 - 4i)$

115. $\sqrt{-10} \cdot \sqrt{-15}$ **116.** $\sqrt{-8} \cdot \sqrt{-12}$

In Problems 117–122, find the reciprocal of the complex number. Write each number in standard form.

117. $5i$ **118.** $7i$

119. $2 - i$ **120.** $3 - 5i$

121. $-4 + 5i$ **122.** $-6 + 2i$

123. Suppose that $f(x) = x^2$; find **(a)** $f(i)$ **(b)** $f(1 + i)$.

124. Suppose that $f(x) = x^2 + x$; find **(a)** $f(i)$ **(b)** $f(1 + i)$.

125. Suppose that $f(x) = x^2 + 2x + 2$; find **(a)** $f(3i)$
(b) $f(1 - i)$.

126. Suppose that $f(x) = x^2 + x - 1$; find **(a)** $f(2i)$
(b) $f(2 + i)$.

Applying the Concepts

127. Impedance (Series Circuit)   The total impedance, Z, of an ac circuit containing components in series is equivalent to the sum of the individual impedances. Impedance is measured in ohms (Ω) and is expressed as an imaginary number of the form $Z = R + i \cdot X$. Here, R represents resistance and X represents reactance.
   (a) If the impedance in one part of a series circuit is $7 + 3i$ ohms and the impedance of the remainder of the circuit is $3 - 4i$ ohms, find the total impedance of the circuit.
   (b) What is the total resistance of the circuit?
   (c) What is the total reactance of the circuit?

128. Impedance (Parallel Circuit)   The total impedance, Z, of an ac circuit consisting of two parallel pathways is given by the formula $\dfrac{1}{Z} = \dfrac{1}{Z_1} + \dfrac{1}{Z_2}$, where Z_1 and Z_2 are the impedances of the pathways. If the impedances of the individual pathways are $Z_1 = 5$ ohms and $Z_2 = 1 - 2i$ ohms, find the total impedance of the circuit.

Extending the Concepts

129. For the function $f(x) = x^2 + 4x + 5$, find
   (a) $f(-2 + i)$ **(b)** $f(-2 - i)$.

130. For the function $f(x) = x^2 - 2x + 2$, find
   (a) $f(1 + i)$ **(b)** $f(1 - i)$.

131. For the function $f(x) = x^3 + 1$, find **(a)** $f(-1)$
   (b) $f\left(\dfrac{1}{2} + \dfrac{\sqrt{3}}{2}i\right)$ **(c)** $f\left(\dfrac{1}{2} - \dfrac{\sqrt{3}}{2}i\right)$.

132. For the function $f(x) = x^3 - 1$, find **(a)** $f(1)$
(b) $f\left(-\dfrac{1}{2} + \dfrac{\sqrt{3}}{2}i\right)$ **(c)** $f\left(-\dfrac{1}{2} - \dfrac{\sqrt{3}}{2}i\right)$.

133. Any complex number z such that $f(z) = 0$ is called a **complex zero** of f. Look at the complex zeros in Problems 129–132. Conjecture a general result regarding the complex zeros of polynomials that have real coefficients.

134. The Complex Plane We are able to plot real numbers on a real number line. We are able to plot ordered pairs (x, y) in a Cartesian plane. Can we plot complex numbers? Yes! A complex number $x + yi$ can be interpreted geometrically as the point (x, y) in the xy-plane. Each point in the plane corresponds to a complex number, and each complex number corresponds to a point in the plane (just as in ordered pairs). In the **complex plane,** the x-axis is called the **real axis** and the y-axis is called the **imaginary axis.**

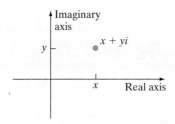

(a) Plot the complex number $4 + 2i$ in the complex plane.
(b) Plot the complex number $-2 + 6i$ in the complex plane.
(c) Plot the complex number $-5i$ in the complex plane.
(d) Plot the complex number -4 in the complex plane.

Explaining the Concepts

135. Explain the relations among the natural numbers, whole numbers, integers, rational numbers, real numbers, and complex number system.

136. Explain why the product of a complex number and its conjugate is a nonnegative real number. That is, explain why $(a + bi)(a - bi) = a^2 + b^2$.

137. How is multiplying two complex numbers related to multiplying two binomials?

138. How is the method used to rationalize denominators related to the method used to write the quotient of two complex numbers in standard form?

Synthesis Review

139. Expand: $(x + 2)^3$ **140.** Expand: $(y - 4)^2$

141. Evaluate: $(3 + i)^3$ **142.** Evaluate: $(4 - 3i)^2$

143. How is raising a complex number to a positive integer power related to raising a binomial to a positive integer power?

The Graphing Calculator

Graphing calculators have the ability to add, subtract, multiply, and divide complex numbers. First, put the calculator into complex mode, as shown in Figure 9(a). Figure 9(b) shows the results of Examples 3(a) and 3(b). Figure 9(c) shows the results of Examples 5(b) and 8.

Figure 9

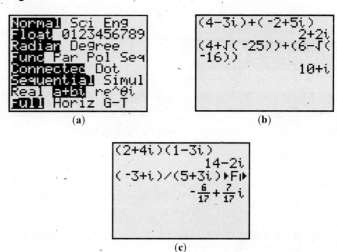

(a)   (b)

(c)

In Problems 144–151, use a graphing calculator to simplify each expression. Write your answers in standard form, $a + bi$.

144. $(1.3 - 4.3i) + (-5.3 + 0.7i)$

145. $(-3.4 + 1.9i) - (6.5 - 5.3i)$

146. $(0.3 - 5.2i)(1.2 + 3.9i)$

147. $(-4.3 + 0.2i)(7.2 - 0.5i)$

148. $\dfrac{4}{3 - 8i}$ **149.** $\dfrac{1 - 7i}{-4 + i}$

150. $3i^6 + (5 - 0.3i)(3 + 2i)$

151. $-6i^{14} + 5i^6 - (2 + i)(3 + 11i)$

Chapter 6 Activity: Which One Does Not Belong?

Focus: Simplifying Radicals and Rational Exponents

Time: 15–25 minutes

Group Size: 2–4

5–10 Minutes: Individually, evaluate each row below. Decide which item does not belong and why. Be creative with your reasons!

10–15 Minutes: Discuss each row, and make a list of all members' responses. Is there any example that everyone agrees with?

	A	B	C	D
1	$16^{\frac{3}{2}}$	$-16^{\frac{3}{2}}$	$-16^{\frac{2}{3}}$	$-16^{-\frac{2}{3}}$
2	$\sqrt{(a-b)^3}, a > b$	$\sqrt{a^3 - b^3}$	$\sqrt{(a-b)^3}, a < b$	$\sqrt{a-b}, a > b$
3	The range of $f(x) = x^2$	The range of $f(x) = \sqrt[3]{x}$	The range of $f(x) = |x|$	The range of $f(x) = x^3$
4	i^{212}	i^0	i^{-108}	i

Chapter 6 Review

Section 6.1   nth Roots and Rational Exponents

KEY CONCEPTS

- **Simplifying $\sqrt[n]{a^n}$**

  If $n \geq 2$ is a positive integer and a is a real number, then
  $$\sqrt[n]{a^n} = |a| \quad \text{if } n \geq 2 \text{ is even}$$
  $$\sqrt[n]{a^n} = a \quad \text{if } n \geq 3 \text{ is odd}$$

- If a is a real number and $n \geq 2$ is an integer, then $a^{\frac{1}{n}} = \sqrt[n]{a}$ provided that $\sqrt[n]{a}$ exists.

- If a is a real number, $\dfrac{m}{n}$ is a rational number in lowest terms with $n \geq 2$, then $a^{\frac{m}{n}} = \sqrt[n]{a^m} = (\sqrt[n]{a})^m$ provided that $\sqrt[n]{a}$ exists.

- If $\dfrac{m}{n}$ is a rational number and if a is a nonzero real number, then $a^{-\frac{m}{n}} = \dfrac{1}{a^{\frac{m}{n}}}$ or $\dfrac{1}{a^{-\frac{m}{n}}} = a^{\frac{m}{n}}$, if $a \neq 0$ and the nth root of a is defined.

KEY TERMS

Principal nth root of a number a

Index

Cube root

You Should Be Able To...	EXAMPLE	Review Exercises
❶ Evaluate nth roots (p. 473)	Examples 1 and 2	1–5, 19, 20
❷ Simplify expressions of the form $\sqrt[n]{a^n}$ (p. 475)	Example 3	6–8
❸ Evaluate expressions of the form $a^{\frac{1}{n}}$ (p. 475)	Examples 4 and 5	9–12, 17, 21
❹ Evaluate expressions of the form $a^{\frac{m}{n}}$ (p. 476)	Examples 6 through 9	13–16, 18, 22–24

In Problems 1–8, simplify each radical.

1. $\sqrt[3]{343}$

2. $\sqrt[3]{-125}$

3. $\sqrt[3]{\dfrac{8}{27}}$

4. $\sqrt[4]{81}$

5. $-\sqrt[5]{-243}$

6. $\sqrt[3]{10^3}$

7. $\sqrt[5]{z^5}$

8. $\sqrt[4]{(5p-3)^4}$

In Problems 9–16, evaluate each of the following expressions.

9. $81^{\frac{1}{2}}$

10. $(-256)^{\frac{1}{4}}$

11. $-4^{\frac{1}{2}}$

12. $729^{\frac{1}{3}}$

13. $16^{\frac{7}{4}}$

14. $-(-27)^{\frac{2}{3}}$

15. $-121^{\frac{3}{2}}$

16. $\dfrac{1}{36^{-\frac{1}{2}}}$

In Problems 17–20, use a calculator to write each expression rounded to two decimal places.

17. $(-65)^{\frac{1}{3}}$

18. $4^{\frac{3}{5}}$

19. $\sqrt[3]{100}$

20. $\sqrt[4]{10}$

In Problems 21–24, rewrite each of the following radicals with a rational exponent.

21. $\sqrt[3]{5a}$

22. $\sqrt[5]{p^7}$

23. $(\sqrt[4]{10z})^3$

24. $\sqrt[6]{(2ab)^5}$

Section 6.2 Simplifying Expressions Using the Laws of Exponents

KEY CONCEPTS

- **If a and b are real numbers and if r and s are rational numbers, then, assuming the expression is defined, the following rules apply.**

Zero-Exponent Rule:	$a^0 = 1$	if $a \neq 0$
Negative-Exponent Rule:	$a^{-r} = \dfrac{1}{a^r}$	if $a \neq 0$
Product Rule:	$a^r \cdot a^s = a^{r+s}$	
Quotient Rule:	$\dfrac{a^r}{a^s} = a^{r-s} = \dfrac{1}{a^{s-r}}$	if $a \neq 0$
Power Rule:	$(a^r)^s = a^{r \cdot s}$	
Product-to-a-Power Rule:	$(a \cdot b)^r = a^r \cdot b^r$	
Quotient-to-a-Power Rule:	$\left(\dfrac{a}{b}\right)^r = \dfrac{a^r}{b^r}$	if $b \neq 0$
Quotient-to-a-Negative-Power Rule:	$\left(\dfrac{a}{b}\right)^{-r} = \left(\dfrac{b}{a}\right)^r$	if $a \neq 0, b \neq 0$

KEY TERM

Simplify

You Should Be Able To...	EXAMPLE	Review Exercises
1 Simplify expressions involving rational exponents (p. 481)	Examples 1 through 3	25–30
2 Simplify radical expressions (p. 483)	Example 4	31–34
3 Factor expressions containing rational exponents (p. 484)	Examples 5 and 6	35, 36

In Problems 25–34, simplify the expression. Assume all variables are positive.

25. $4^{\frac{2}{3}} \cdot 4^{\frac{7}{3}}$

26. $\dfrac{k^{\frac{1}{2}}}{k^{\frac{3}{4}}}$

27. $(p^{\frac{4}{3}} \cdot q^4)^{\frac{3}{2}}$

28. $(32a^{-\frac{3}{2}} \cdot b^{\frac{1}{4}})^{\frac{1}{5}}$

29. $5m^{-\frac{2}{3}}(2m + m^{-\frac{1}{3}})$

30. $\left(\dfrac{16x^{\frac{1}{3}}}{x^{-\frac{1}{3}}}\right)^{-\frac{1}{2}} + \left(\dfrac{x^{-\frac{3}{2}}}{64x^{-\frac{1}{2}}}\right)^{\frac{1}{3}}$

31. $\sqrt[8]{x^6}$

32. $\sqrt{121x^4y^{10}}$

33. $\sqrt[3]{m^2} \cdot \sqrt{m^3}$

34. $\dfrac{\sqrt[3]{c}}{\sqrt[6]{c^4}}$

In Problems 35 and 36, factor the expression.

35. $2(3m - 1)^{\frac{1}{4}} + (m - 7)(3m - 1)^{\frac{5}{4}}$

36. $3(x^2 - 5)^{\frac{1}{3}} - 4x(x^2 - 5)^{-\frac{2}{3}}$

Section 6.3 Simplifying Radical Expressions Using Properties of Radicals

KEY CONCEPTS

KEY TERM

- **Product Property of Radicals**
  If $\sqrt[n]{a}$ and $\sqrt[n]{b}$ are real numbers and $n \geq 2$ is an integer, then $\sqrt[n]{a} \cdot \sqrt[n]{b} = \sqrt[n]{ab}$.

- **Quotient Property of Radicals**
  If $\sqrt[n]{a}$ and $\sqrt[n]{b}$ are real numbers, $b \neq 0$, and $n \geq 2$ is an integer, then $\dfrac{\sqrt[n]{a}}{\sqrt[n]{b}} = \sqrt[n]{\dfrac{a}{b}}$.

Simplify

You Should Be Able To...	EXAMPLE	Review Exercises
❶ Use the Product Property to multiply radical expressions (p. 486)	Example 1	37, 38
❷ Use the Product Property to simplify radical expressions (p. 487)	Examples 2 through 7	39–52
❸ Use the Quotient Property to simplify radical expressions (p. 492)	Examples 8 and 9	53–60
❹ Multiply radicals with unlike indices (p. 493)	Example 10	61, 62

In Problems 37 and 38, use the Product Property to multiply. Assume that all variables can be any real number.

37. $\sqrt{15} \cdot \sqrt{7}$

38. $\sqrt[4]{2ab^2} \cdot \sqrt[4]{6a^2b}$

In Problems 39–44, simplify each radical using the Product Property. Assume that all variables can be any real number.

39. $\sqrt{80}$

40. $\sqrt[3]{-500}$

41. $\sqrt[3]{162m^6n^4}$

42. $\sqrt[4]{50p^8q^4}$

43. $2\sqrt{16x^6y}$

44. $\sqrt{(2x+1)^3}$

In Problems 45–48, simplify each radical using the Product Property. Assume that all variables are greater than or equal to zero.

45. $\sqrt{w^3z^2}$

46. $\sqrt{45x^4yz^3}$

47. $\sqrt[3]{16a^{12}b^5}$

48. $\sqrt{4x^2+8x+4}$

In Problems 49–52, multiply and simplify. Assume that all variables are greater than or equal to zero.

49. $\sqrt{15} \cdot \sqrt{18}$

50. $\sqrt[3]{20} \cdot \sqrt[3]{30}$

51. $\sqrt[3]{-3x^4y^7} \cdot \sqrt[3]{24x^3y^2}$

52. $3\sqrt{4xy^2} \cdot 5\sqrt{3x^2y}$

In Problems 53–60, simplify. Assume that all variables are greater than zero.

53. $\sqrt{\dfrac{121}{25}}$

54. $\sqrt{\dfrac{5a^4}{64b^2}}$

55. $\sqrt[3]{\dfrac{54k^2}{9k^5}}$

56. $\sqrt[3]{\dfrac{-160w^{11}}{343w^{-4}}}$

57. $\dfrac{\sqrt{12h^3}}{\sqrt{3h}}$

58. $\dfrac{\sqrt{50a^3b^3}}{\sqrt{8a^5b^{-3}}}$

59. $\dfrac{\sqrt[3]{-8x^7y}}{\sqrt[3]{27xy^4}}$

60. $\dfrac{\sqrt[4]{48m^2n^7}}{\sqrt[4]{3m^6n}}$

In Problems 61 and 62, multiply and simplify.

61. $\sqrt{5} \cdot \sqrt[3]{2}$

62. $\sqrt[4]{8} \cdot \sqrt[6]{4}$

Section 6.4 Adding, Subtracting, and Multiplying Radical Expressions

KEY CONCEPTS

KEY TERM

- To add or subtract radicals, the radicals must have the same index and the same radicand.

- To multiply radicals, the index on each radical must be the same. Then multiply the radicands.

Like radicals

You Should Be Able To...	EXAMPLE	Review Exercises
❶ Add or subtract radical expressions (p. 496)	Examples 1 through 3	63–72
❷ Multiply radical expressions (p. 498)	Examples 4 and 5	73–82

In Problems 63–72, add or subtract as indicated. Assume all variables are positive or zero.

63. $2\sqrt[4]{x} + 6\sqrt[4]{x}$

64. $7\sqrt[3]{4y} + 2\sqrt[3]{4y} - 3\sqrt[3]{4y}$

65. $5\sqrt{2} - 2\sqrt{12}$

66. $\sqrt{18} + 2\sqrt{50}$

67. $\sqrt[3]{-16z} + \sqrt[3]{54z}$

68. $7\sqrt[3]{8x^2} - \sqrt[3]{-27x^2}$

69. $\sqrt{16a} + \sqrt[6]{729a^3}$

70. $\sqrt{27x^2} - x\sqrt{48} + 2\sqrt{75x^2}$

71. $5\sqrt[3]{4m^5y^2} - \sqrt[6]{16m^{10}y^4}$

72. $\sqrt{y^3 - 4y^2} - 2\sqrt{y - 4} + \sqrt[4]{y^2 - 8y + 16}$

In Problems 73–82, multiply and simplify.

73. $\sqrt{3}(\sqrt{5} - \sqrt{15})$

74. $\sqrt[3]{5}(3 + \sqrt[3]{4})$

75. $(3 + \sqrt{5})(4 - \sqrt{5})$

76. $(7 + \sqrt{3})(6 + \sqrt{2})$

77. $(1 - 3\sqrt{5})(1 + 3\sqrt{5})$

78. $(\sqrt[3]{x} + 1)(9\sqrt[3]{x} - 4)$

79. $(\sqrt{x} - \sqrt{5})^2$

80. $(11\sqrt{2} + \sqrt{5})^2$

81. $(\sqrt{2a} - b)(\sqrt{2a} + b)$

82. $(\sqrt[3]{6s} + 2)(\sqrt[3]{6s} - 7)$

Section 6.5 Rationalizing Radical Expressions

KEY CONCEPTS

- To rationalize the denominator containing a single radical with index n, multiply the numerator and denominator by a radical so that the radicand in the denominator is a perfect nth power.

- To rationalize a denominator containing two terms, use the fact that $(A + B)(A - B) = A^2 - B^2$.

KEY TERM

Rationalizing the denominator

You Should Be Able To...	EXAMPLE	Review Exercises
❶ Rationalize a denominator containing one term (p. 502)	Examples 1 and 2	83–90, 99
❷ Rationalize a denominator containing two terms (p. 505)	Examples 3 and 4	91–98, 100

In Problems 83–98, rationalize the denominator.

83. $\dfrac{2}{\sqrt{6}}$

84. $\dfrac{6}{\sqrt{3}}$

85. $\dfrac{\sqrt{48}}{\sqrt{p^3}}$

86. $\dfrac{5}{\sqrt{2a}}$

87. $\dfrac{-2}{\sqrt{6y^3}}$

88. $\dfrac{3}{\sqrt[3]{5}}$

89. $\sqrt[3]{\dfrac{-4}{45}}$

90. $\dfrac{27}{\sqrt[5]{8p^3q^4}}$

91. $\dfrac{6}{7 - \sqrt{6}}$

92. $\dfrac{3}{\sqrt{3} - 9}$

93. $\dfrac{\sqrt{3}}{3 + \sqrt{2}}$

94. $\dfrac{\sqrt{k}}{\sqrt{k} - \sqrt{m}}$

95. $\dfrac{\sqrt{10} + 2}{\sqrt{10} - 2}$

96. $\dfrac{3 - \sqrt{y}}{3 + \sqrt{y}}$

97. $\dfrac{4}{2\sqrt{3} + 5\sqrt{2}}$

98. $\dfrac{\sqrt{5} - \sqrt{6}}{\sqrt{10} + \sqrt{3}}$

99. Simplify: $\dfrac{\sqrt{7}}{3} + \dfrac{6}{\sqrt{7}}$

100. Find the reciprocal: $4 - \sqrt{7}$

Section 6.6 Functions Involving Radicals

KEY CONCEPT

KEY TERMS

- **Finding the Domain of a Function Involving Radicals**
  1. If the index on a radical expression is even, then the radicand must be greater than or equal to zero.
  2. If the index on a radical expression is odd, then the radicand can be any real number.

Square root function
Cube root function

You Should Be Able To...	EXAMPLE	Review Exercises
❶ Evaluate functions involving radicals (p. 509)	Example 1	101–104
❷ Find the domain of a function involving a radical (p. 509)	Example 2	105–110, 111(a), 112(a), 113(a), 114(a)
❸ Graph functions involving square roots (p. 510)	Example 3	111(b)–113(b)
❹ Graph functions involving cube roots (p. 511)	Example 4	114(b)

In Problems 101–104, evaluate each radical function at the indicated values.

101. $f(x) = \sqrt{x + 4}$
   (a) $f(-3)$
   (b) $f(0)$
   (c) $f(5)$

102. $g(x) = \sqrt{3x - 2}$
   (a) $g\left(\dfrac{2}{3}\right)$
   (b) $g(2)$
   (c) $g(6)$

103. $H(t) = \sqrt[3]{t + 3}$
   (a) $H(-2)$
   (b) $H(-4)$
   (c) $H(5)$

104. $G(z) = \sqrt{\dfrac{z - 1}{z + 2}}$
   (a) $G(1)$
   (b) $G(-3)$
   (c) $G(2)$

In Problems 105–110, find the domain of the radical function.

105. $f(x) = \sqrt{3x - 5}$ **106.** $g(x) = \sqrt[3]{2x - 7}$

107. $h(x) = \sqrt[4]{6x + 1}$ **108.** $F(x) = \sqrt[5]{2x - 9}$

109. $G(x) = \sqrt{\dfrac{4}{x - 2}}$ **110.** $H(x) = \sqrt{\dfrac{x - 3}{x}}$

In Problems 111–114, (a) determine the domain of the function; (b) graph the function using point plotting; and (c) based on the graph, determine the range of the function.

111. $f(x) = \dfrac{1}{2}\sqrt{1 - x}$ **112.** $g(x) = \sqrt{x + 1} - 2$

113. $h(x) = -\sqrt{x + 3}$ **114.** $F(x) = \sqrt[3]{x + 1}$

Section 6.7 Radical Equations and Their Applications

KEY TERMS

Radical equation   Extraneous solution

You Should Be Able To...	EXAMPLE	Review Exercises
❶ Solve radical equations containing one radical (p. 515)	Examples 1 through 5	115–124, 129, 130
❷ Solve radical equations containing two radicals (p. 518)	Examples 6 and 7	125–128
❸ Solve for a variable in a radical equation (p. 520)	Example 8	131, 132

In Problems 115–130, solve each equation.

115. $\sqrt{m} = 13$

116. $\sqrt[3]{3t + 1} = -2$

117. $\sqrt[4]{3x - 8} = 3$

118. $\sqrt{2x + 5} + 4 = 2$

119. $\sqrt{4 - k} - 3 = 0$

120. $3\sqrt{t} - 4 = 11$

121. $2\sqrt[3]{m} + 5 = -11$

122. $\sqrt{q + 2} = q$

123. $\sqrt{w + 11} + 3 = w + 2$

124. $\sqrt{p^2 - 2p + 9} = p + 1$

125. $\sqrt{a + 10} = \sqrt{2a - 1}$

126. $\sqrt{5x + 9} = \sqrt{7x - 3}$

127. $\sqrt{c - 8} + \sqrt{c} = 4$

128. $\sqrt{x + 2} - \sqrt{x + 9} = 7$

129. $(4x - 3)^{1/3} - 3 = 0$

130. $(x^2 - 9)^{1/4} = 2$

131. Height of a Cone Solve $r = \sqrt{\dfrac{3V}{\pi h}}$ for h.

132. Ball Slide Speed Factor Solve $f_s = \sqrt[3]{\dfrac{30}{v}}$ for v.

Section 6.8 The Complex Number System

KEY CONCEPTS	KEY TERMS

KEY CONCEPTS

- **Nonnegativity Property of Real Numbers**
  For any real number a, $a^2 \geq 0$.
- **Imaginary Unit**
  The imaginary unit, denoted by i, is the number whose square is -1.
  That is, $i^2 = -1$.
- **Complex numbers**
  Complex numbers are numbers of the form $a + bi$, where a and b are real numbers. The real number a is called the real part of the number $a + bi$; the real number b is called the imaginary part of $a + bi$.
- **Square Roots of Negative Numbers**
  If N is a positive real number, then the principal square root of $-N$, denoted by $\sqrt{-N}$, is $\sqrt{-N} = \sqrt{N}i$, where i is the number whose square is -1.
- **Sum of Complex Numbers**
  $(a + bi) + (c + di) = (a + c) + (b + d)i$
- **Difference of Complex Numbers**
  $(a + bi) - (c + di) = (a - c) + (b - d)i$
- **Complex Conjugate**
  If $a + bi$ is a complex number, then its conjugate is $a - bi$.
- **Product of a Complex Number and Its Conjugate**
  $(a + bi)(a - bi) = a^2 + b^2$

KEY TERMS

Imaginary unit
Complex number system
Complex number
Real part
Imaginary part
Standard form
Pure imaginary number
Principal square root
Conjugate
Powers of i

You Should Be Able To...	EXAMPLE	Review Exercises
❶ Evaluate the square root of negative real numbers (p. 524)	Examples 1 and 2	133–136
❷ Add or subtract complex numbers (p. 526)	Examples 3 and 4	137–140
❸ Multiply complex numbers (p. 527)	Examples 5 through 7	141–146
❹ Divide complex numbers (p. 529)	Examples 8 and 9	147–150
❺ Evaluate the powers of i (p. 530)	Example 10	151–152

In Problems 133 and 134, write each expression as a pure imaginary number.

133. $\sqrt{-29}$

134. $\sqrt{-54}$

In Problems 135 and 136, write each expression as a complex number in standard form.

135. $14 - \sqrt{-162}$

136. $\dfrac{6 + \sqrt{-45}}{3}$

In Problems 137–150, perform the indicated operation.

137. $(3 - 7i) + (-2 + 5i)$

138. $(4 + 2i) - (9 - 8i)$

139. $(8 - \sqrt{-45}) - (3 + \sqrt{-80})$

140. $(1 + \sqrt{-9}) + (-6 + \sqrt{-16})$

141. $(4 - 5i)(3 + 7i)$

142. $\left(\dfrac{1}{2} + \dfrac{2}{3}i\right)(4 - 9i)$

143. $\sqrt{-3} \cdot \sqrt{-27}$

144. $(1 + \sqrt{-36})(-5 - \sqrt{-144})$

145. $(1 + 12i)(1 - 12i)$

146. $(7 + 2i)(5 + 4i)$

147. $\dfrac{4}{3 + 5i}$

148. $\dfrac{-3}{7 - 2i}$

149. $\dfrac{2 - 3i}{5 + 2i}$

150. $\dfrac{4 + 3i}{1 - i}$

In Problems 151 and 152, simplify.

151. i^{59}

152. i^{173}

Chapter 6 Test

Step-by-step test solutions are found on the Chapter Test Prep Videos available in MyMathLab® or on YouTube.

1. Evaluate: $49^{-\frac{1}{2}}$

In Problems 2 and 3, simplify using rational exponents.

2. $\sqrt[3]{8x^{\frac{1}{2}}y^3} \cdot \sqrt{9xy^{\frac{1}{2}}}$

3. $\sqrt[5]{(2a^4b^3)^7}$

In Problems 4–9, perform the indicated operation and simplify. Assume all variables in the radicand are greater than or equal to zero.

4. $\sqrt{3m} \cdot \sqrt{13n}$

5. $\sqrt{32x^7y^4}$

6. $\dfrac{\sqrt{9a^3b^{-3}}}{\sqrt{4ab}}$

7. $\sqrt{5x^3} + 2\sqrt{45x}$

8. $\sqrt{9a^2b} - \sqrt[4]{16a^4b^2}$

9. $(11 + 2\sqrt{x})(3 - \sqrt{x})$

In Problems 10 and 11, rationalize the denominator.

10. $\dfrac{-2}{3\sqrt{72}}$

11. $\dfrac{\sqrt{5}}{\sqrt{5} + 2}$

12. For $f(x) = \sqrt{-2x + 3}$, find the following:
(a) $f(1)$
(b) $f(-3)$

13. Determine the domain of the function $g(x) = \sqrt{-3x + 5}$.

14. For $f(x) = \sqrt{x} - 3$, do the following:
(a) Determine the domain of the function.
(b) Graph the function using point plotting.
(c) From the graph, determine the range of the function.

In Problems 15–17, solve the given equations.

15. $\sqrt{x + 3} = 4$

16. $\sqrt{x + 13} - 4 = x - 3$

17. $\sqrt{x - 1} + \sqrt{x + 2} = 3$

In Problems 18–20, perform the indicated operation.

18. $(13 + 2i) + (4 - 15i)$

19. $(4 - 7i)(2 + 3i)$

20. $\dfrac{7 - i}{12 + 11i}$

7 Quadratic Equations and Functions

One of the more unusual sports—Punkin Chunkin—is played in Millsboro, Delaware. Participants catapult or fire pumpkins to see who can toss them the farthest (and most accurately). Interestingly, this bizarre ritual is an application of a quadratic function at work. See Problems 79–82 in Section 7.5.

The Big Picture: Putting It Together

In Part I of Chapter 1, we reviewed solving linear equations and inequalities in one variable. In Part II of Chapter 1, we completed our discussion of "everything linear" by covering linear equations and inequalities in two variables.

This chapter is dedicated to extending our discussion of quadratic equations, $ax^2 + bx + c = 0$, that began in Section 4.8. There, we solved quadratic equations $ax^2 + bx + c = 0$, where the expression $ax^2 + bx + c$ was factorable. Remember, if the expression $ax^2 + bx + c$ was not factorable, we did not have a method for solving the equation. The missing piece for solving this type of quadratic equation was the idea of a radical. Having learned how to work with radicals in Chapter 6, we now have the tools necessary to solve all quadratic equations $ax^2 + bx + c = 0$, regardless of whether the expression $ax^2 + bx + c$ is factorable. In addition to solving quadratic equations, we will graph quadratic functions and solve quadratic and polynomial inequalities.

Outline

7.1 Solving Quadratic Equations by Completing the Square

Objectives

1 Solve Quadratic Equations Using the Square Root Property

2 Complete the Square in One Variable

3 Solve Quadratic Equations by Completing the Square

4 Solve Problems Using the Pythagorean Theorem

Are You Ready for This Section?

Before getting started, take the following readiness quiz. If you get a problem wrong, go back to the section cited and review the material.

R1. Multiply: $(2p + 3)^2$ [Section 4.2, pp. 323–324]

R2. Factor: $y^2 - 8y + 16$ [Section 4.6, pp. 358–359]

R3. Solve: $x^2 + 5x - 14 = 0$ [Section 4.8, pp. 372–374]

R4. Solve: $x^2 - 16 = 0$ [Section 4.8, pp. 372–374]

R5. Simplify: **(a)** $\sqrt{36}$ **(b)** $\sqrt{45}$ **(c)** $\sqrt{-12}$ [Getting Ready, pp. 467–468; Section 6.3, pp. 487–489; Section 6.8, pp. 524–525]

R6. Find the complex conjugate of $-3 + 2i$. [Section 6.8, p. 528]

R7. Simplify: $\sqrt{x^2}$ [Getting Ready, pp. 469–470]

R8. Define: $|x|$ [Section R.3, p. 19]

▶ **1** Solve Quadratic Equations Using the Square Root Property

Suppose we want to solve the quadratic equation

$$x^2 = p$$

where p is any real number. This equation is saying, "Give me all numbers whose square is p." For example, the equation $x^2 = 16$ means we want all numbers whose square is 16. Since those numbers are -4 and 4, the solution set to the equation $x^2 = 16$ is $\{-4, 4\}$.

In general, to solve an equation of the form $x^2 = p$, we can take the square root of both sides of the equation.

$$x^2 = p$$

Take the square root of both sides: $\sqrt{x^2} = \sqrt{p}$

$\sqrt{x^2} = |x|$: $\quad |x| = \sqrt{p}$

$|x| = -x$ if $x < 0$; $|x| = x$ if $x \geq 0$: $\quad -x = \sqrt{p} \quad$ or $\quad x = \sqrt{p}$

Solve for x: $\quad x = -\sqrt{p} \quad$ or $\quad x = \sqrt{p}$

This gives us the following result.

Work Smart

The Square Root Property is useful for solving equations of the form "some unknown squared equals a real number." To solve this equation, take the square root of both sides of the equation, but don't forget the $\pm$ symbol to obtain the positive and negative square root.

The Square Root Property

If $x^2 = p$, then $x = \sqrt{p}$ or $x = -\sqrt{p}$.

When using the Square Root Property to solve an equation such as $x^2 = p$, we usually write the solutions as $x = \pm\sqrt{p}$, which is read "x equals plus or minus the square root of p." For example, the two solutions of the equation

$$x^2 = 16$$

are

$$x = \pm\sqrt{16}$$
$$= \pm 4$$

Let's use the Square Root Property in an example now.

Ready?...Answers

R1. $4p^2 + 12p + 9$

R2. $(y - 4)^2$ **R3.** $\{-7, 2\}$ **R4.** $\{-4, 4\}$

R5. (a) 6 **(b)** $3\sqrt{5}$ **(c)** $2\sqrt{3}i$ **R6.** $-3 - 2i$

R7. $|x|$ **R8.** $|x| = -x$ if $x < 0$
$|x| = x$ if $x \geq 0$

EXAMPLE 1   **How to Solve a Quadratic Equation Using the Square Root Property**

Solve: $x^2 - 9 = 0$

Step-by-Step Solution

Step 1: Isolate the expression containing the square term.

$$x^2 - 9 = 0$$

Add 9 to both sides: $x^2 = 9$

Step 2: Use the Square Root Property. Don't forget the $\pm$ symbol.

$$x = \pm\sqrt{9}$$

Simplify the radical: $= \pm 3$

Step 3: Isolate the variable, if necessary.

The variable is already isolated.

Step 4: Verify your solution(s).

$x = -3:$ $(-3)^2 - 9 \overset{?}{=} 0$ $x = 3:$ $3^2 - 9 \overset{?}{=} 0$

$9 - 9 = 0$ True $9 - 9 = 0$ True

The solution set is $\{-3, 3\}$.

We summarize the steps that are used to solve a quadratic equation using the Square Root Property.

> **Solving a Quadratic Equation Using the Square Root Property**
>
> **Step 1:** Isolate the expression containing the squared term.
> **Step 2:** Use the Square Root Property: If $x^2 = p$, then $x = \pm\sqrt{p}$.
> **Step 3:** Isolate the variable, if necessary.
> **Step 4:** Verify your solution(s).

You could also solve the equation in Example 1 by factoring the difference of two squares:

$$p^2 - 9 = 0$$
$$(p - 3)(p + 3) = 0$$

Zero-Product Property: $p - 3 = 0$ or $p + 3 = 0$

$p = 3$ or $p = -3$

Work Smart

As our mathematical knowledge develops, we often find there is more than one way to solve a problem.

So there is more than one way to find the solution! However, in solving equations of the form $x^2 = p$, factoring works nicely only when p is a perfect square.

EXAMPLE 2   **Solving a Quadratic Equation Using the Square Root Property**

Solve: $3x^2 - 60 = 0$

Solution

$$3x^2 - 60 = 0$$

Add 60 to both sides of the equation: $3x^2 = 60$

Divide both sides by 3: $x^2 = 20$

Use the Square Root Property: $x = \pm\sqrt{20}$

Simplify the radical: $= \pm 2\sqrt{5}$

Check

$$x = -2\sqrt{5}: \quad 3\left(-2\sqrt{5}\right)^2 - 60 \overset{?}{=} 0 \qquad x = 2\sqrt{5}: \quad 3\left(2\sqrt{5}\right)^2 - 60 \overset{?}{=} 0$$

$$(ab)^2 = a^2b^2: \quad 3(-2)^2\left(\sqrt{5}\right)^2 - 60 \overset{?}{=} 0 \qquad\qquad 3 \cdot (2)^2\left(\sqrt{5}\right)^2 - 60 \overset{?}{=} 0$$

$$3 \cdot 4 \cdot 5 - 60 \overset{?}{=} 0 \qquad\qquad\qquad 3 \cdot 4 \cdot 5 - 60 \overset{?}{=} 0$$

$$60 - 60 = 0 \quad \text{True} \qquad\qquad\qquad 60 - 60 = 0 \quad \text{True}$$

The solution set is $\left\{-2\sqrt{5}, 2\sqrt{5}\right\}$.

Quick ✓

1. If $x^2 = p$, then $x = $ ___ or $x = $ ____.

In Problems 2–4, solve the quadratic equation using the Square Root Property.

2. $p^2 = 48$ **3.** $3b^2 = 75$ **4.** $s^2 - 81 = 0$

As the next example illustrates, the solution to a quadratic equation does not have to be real.

EXAMPLE 3 **Solving a Quadratic Equation Using the Square Root Property**

Solve: $y^2 + 14 = 2$

Solution

$$y^2 + 14 = 2$$

Subtract 14 from both sides: $\qquad y^2 = -12$

Use the Square Root Property: $\qquad y = \pm\sqrt{-12}$

Simplify the radical: $\qquad = \pm 2\sqrt{3}i$

Check

$$y = -2\sqrt{3}i: \quad \left(-2\sqrt{3}i\right)^2 + 14 \overset{?}{=} 2 \qquad y = 2\sqrt{3}i: \quad \left(2\sqrt{3}i\right)^2 + 14 \overset{?}{=} 2$$

$$\left(-2\sqrt{3}\right)^2 i^2 + 14 \overset{?}{=} 2 \qquad\qquad \left(2\sqrt{3}\right)^2 i^2 + 14 \overset{?}{=} 2$$

$$i^2 = -1: \quad 4 \cdot 3 \cdot (-1) + 14 \overset{?}{=} 2 \qquad\qquad 4 \cdot 3 \cdot (-1) + 14 \overset{?}{=} 2$$

$$-12 + 14 \overset{?}{=} 2 \qquad\qquad\qquad -12 + 14 \overset{?}{=} 2$$

$$2 = 2 \quad \text{True} \qquad\qquad\qquad 2 = 2 \quad \text{True}$$

The solution set is $\left\{-2\sqrt{3}i, 2\sqrt{3}i\right\}$.

Quick ✓

In Problems 5 and 6, solve the quadratic equation using the Square Root Property.

5. $d^2 = -72$ **6.** $3q^2 + 27 = 0$

EXAMPLE 4 **Solving Quadratic Equations Using the Square Root Property**

Solve:

(a) $(x - 2)^2 = 25$ **(b)** $(y + 5)^2 + 24 = 0$

Solution

(a)

$$(x - 2)^2 = 25$$

Use the Square Root Property: $\qquad x - 2 = \pm\sqrt{25}$

Simplify the radical: $\qquad x - 2 = \pm 5$

Add 2 to each side: $\qquad x = 2 \pm 5$

2 ± 5 means $2 - 5$ or $2 + 5$: $\quad x = 2 - 5 \quad$ or $\quad x = 2 + 5$

$$= -3 \qquad\qquad = 7$$

Check

$$x = -3: \quad (-3 - 2)^2 \overset{?}{=} 25 \qquad\qquad x = 7: \quad (7 - 2)^2 \overset{?}{=} 25$$

$$(-5)^2 \overset{?}{=} 25 \qquad\qquad\qquad\qquad 5^2 \overset{?}{=} 25$$

$$25 = 25 \quad \text{True} \qquad\qquad\qquad 25 = 25 \quad \text{True}$$

The solution set is $\{-3, 7\}$.

(b)
$$(y + 5)^2 + 24 = 0$$

Subtract 24 from both sides: $\qquad\qquad (y + 5)^2 = -24$

Use the Square Root Property: $\qquad\qquad y + 5 = \pm\sqrt{-24}$

$\sqrt{-24} = \sqrt{24} \cdot i = \sqrt{4 \cdot 6}i = 2\sqrt{6}i: \qquad y + 5 = \pm 2\sqrt{6}i$

Subtract 5 from each side: $\qquad\qquad\qquad y = -5 \pm 2\sqrt{6}i$

$$y = -5 - 2\sqrt{6}i \quad \text{or} \quad y = -5 + 2\sqrt{6}i$$

Check

$$y = -5 - 2\sqrt{6}i: \qquad\qquad\qquad y = -5 + 2\sqrt{6}i:$$

$$\left(-5 - 2\sqrt{6}i + 5\right)^2 + 24 \overset{?}{=} 0 \qquad \left(-5 + 2\sqrt{6}i + 5\right)^2 + 24 \overset{?}{=} 0$$

$$\left(-2\sqrt{6}i\right)^2 + 24 \overset{?}{=} 0 \qquad\qquad \left(2\sqrt{6}i\right)^2 + 24 \overset{?}{=} 0$$

$$4 \cdot 6 \cdot i^2 + 24 \overset{?}{=} 0 \qquad\qquad\quad 4 \cdot 6 \cdot i^2 + 24 \overset{?}{=} 0$$

$$-24 + 24 = 0 \quad \text{True} \qquad\qquad -24 + 24 = 0 \quad \text{True}$$

The solution set is $\left\{-5 - 2\sqrt{6}i, -5 + 2\sqrt{6}i\right\}$. ●

Quick ✓

In Problems 7 and 8, solve the quadratic equation using the Square Root Property.

7. $(y + 3)^2 = 100$ **8.** $(q - 5)^2 + 20 = 4$

▶ ❷ **Complete the Square in One Variable**

The idea behind **completing the square** in one variable is to "adjust" the left side of a quadratic equation of the form $x^2 + bx + c = 0$ to make it a perfect square trinomial. Recall that perfect square trinomials have the form

$$A^2 + 2AB + B^2 = (A + B)^2 \quad \text{or} \quad A^2 - 2AB + B^2 = (A - B)^2$$

For example, $x^2 + 6x + 9$ is a perfect square trinomial because $x^2 + 6x + 9 = x^2 + 2 \cdot x \cdot 3 + 3^2 = (x + 3)^2$.

We "adjust" $x^2 + bx + c$ by adding a number to make it a perfect square trinomial. For example, to make $x^2 + 6x$ a perfect square, we would add 9. Why did we choose 9? If we divide the coefficient of the first-degree term, 6, by 2 and then square the result, we have 9. This approach works in general.

Work Smart

To complete the square, the coefficient of x^2 must be 1.

Obtaining a Perfect Square Trinomial

Multiply the coefficient of the first-degree term by $\frac{1}{2}$. Square the result. That is, for $x^2 + bx$, compute $\left(\frac{1}{2}b\right)^2$, and add the result to the expression $x^2 + bx$ to obtain

$$x^2 + bx + \left(\frac{1}{2}b\right)^2.$$

(**EXAMPLE 5**) **Obtaining a Perfect Square Trinomial**

Determine the number that must be added to each expression to make it a perfect square trinomial. Then factor the expression.

Start	Add	Result	Factored Form
$y^2 + 8y$	$\left(\dfrac{1}{2} \cdot 8\right)^2 = 16$	$y^2 + 8y + 16$	$(y + 4)^2$
$x^2 + 12x$	$\left(\dfrac{1}{2} \cdot 12\right)^2 = 36$	$x^2 + 12x + 36$	$(x + 6)^2$
$a^2 - 20a$	$\left(\dfrac{1}{2} \cdot (-20)\right)^2 = 100$	$a^2 - 20a + 100$	$(a - 10)^2$
$p^2 - 5p$	$\left(\dfrac{1}{2} \cdot (-5)\right)^2 = \dfrac{25}{4}$	$p^2 - 5p + \dfrac{25}{4}$	$\left(p - \dfrac{5}{2}\right)^2$

●

Work Smart

It is common to write the value $\left(\dfrac{1}{2}b\right)^2$ as a fraction, not a decimal.

Did you notice in the factored form that the perfect square trinomial always factors so that

$$x^2 + bx + \left(\frac{b}{2}\right)^2 = \left(x + \frac{b}{2}\right)^2 \quad \text{or} \quad x^2 - bx + \left(\frac{b}{2}\right)^2 = \left(x - \frac{b}{2}\right)^2$$

That is, the perfect square trinomial always factors as $\left(x \pm \dfrac{b}{2}\right)^2$, where we use the $+$ if the coefficient of the first-degree term is positive and we use the $-$ if the coefficient of the first-degree term is negative. The $\dfrac{b}{2}$ represents $\dfrac{1}{2}$ the value of the coefficient of the first-degree term.

Quick ✓

In Problems 9 and 10, determine the number that must be added to the expression to make it a perfect square trinomial. Then factor the expression.

9. $p^2 + 14p$
10. $w^2 - 3w$

Figure 1

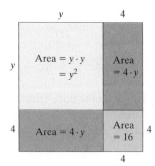

Are you wondering why we call this process "completing the square"? Consider the expression $y^2 + 8y$ given in Example 5, which we have geometrically represented in Figure 1. The yellow area is y^2 and each orange area is $4y$ (for a total area of $8y$). But what is the area of the green region to make the square complete? The dimensions of the green region are 4 by 4, so its area is 16. The area of the entire square region, $(y + 4)^2$, equals the sum of the area of the four regions that make up the square: $y^2 + 4y + 4y + 16 = y^2 + 8y + 16$.

❸ **Solve Quadratic Equations by Completing the Square**

Until now, we have solved quadratic equations of the form $ax^2 + bx + c = 0$ only when $ax^2 + bx + c$ was factorable. Now we can solve $ax^2 + bx + c = 0$ when $ax^2 + bx + c$ is not factorable. We begin with equations of the form $x^2 + bx + c = 0$. That is, the coefficient of the square term is 1.

⊙ **EXAMPLE 6**   **How to Solve a Quadratic Equation by Completing the Square**

Solve: $x^2 + 6x + 1 = 0$

Step-by-Step Solution

Step 1: Rewrite $x^2 + bx + c = 0$ as $x^2 + bx = -c$ by subtracting the constant from both sides of the equation.

Subtract 1 from both sides:

$$x^2 + 6x + 1 = 0$$
$$x^2 + 6x = -1$$

Step 2: Complete the square in the expression $x^2 + bx$ by making it a perfect square trinomial.

$\left(\dfrac{1}{2} \cdot 6\right)^2 = 9$; add 9 to both sides:

$$x^2 + 6x + 9 = -1 + 9$$
$$x^2 + 6x + 9 = 8$$

Step 3: Factor the perfect square trinomial on the left side of the equation.

$x^2 + 6x + 9 = (x + 3)^2$:

$$(x + 3)^2 = 8$$

Step 4: Solve the equation using the Square Root Property.

$$x + 3 = \pm\sqrt{8}$$

$\sqrt{8} = 2\sqrt{2}$:

$$x + 3 = \pm 2\sqrt{2}$$

Subtract 3 from both sides:

$$x = -3 \pm 2\sqrt{2}$$

$a \pm b$ means $a - b$ or $a + b$: $x = -3 - 2\sqrt{2}$ or $x = -3 + 2\sqrt{2}$

Step 5: Verify your solution(s).

$$x^2 + 6x + 1 = 0$$

$x = -3 - 2\sqrt{2}$:

$$\left(-3 - 2\sqrt{2}\right)^2 + 6\left(-3 - 2\sqrt{2}\right) + 1 \overset{?}{=} 0$$
$$9 + 12\sqrt{2} + 8 - 18 - 12\sqrt{2} + 1 \overset{?}{=} 0$$
$$0 = 0 \quad \text{True}$$

$x = -3 + 2\sqrt{2}$:

$$\left(-3 + 2\sqrt{2}\right)^2 + 6\left(-3 + 2\sqrt{2}\right) + 1 \overset{?}{=} 0$$
$$9 - 12\sqrt{2} + 8 - 18 + 12\sqrt{2} + 1 \overset{?}{=} 0$$
$$0 = 0 \quad \text{True}$$

The solution set is $\left\{-3 - 2\sqrt{2}, -3 + 2\sqrt{2}\right\}$.

Solving a Quadratic Equation By Completing The Square

Step 1: Rewrite $x^2 + bx + c = 0$ as $x^2 + bx = -c$ by subtracting the constant from both sides of the equation.

Step 2: Complete the square in the expression $x^2 + bx$ by making it a perfect square trinomial. Add the same number to each side of the equation.

Step 3: Factor the perfect square trinomial on the left side of the equation.

Step 4: Solve the equation using the Square Root Property.

Step 5: Verify your solutions.

Quick ✓

In Problems 11 and 12, solve the equation by completing the square.

11. $b^2 + 2b - 8 = 0$ **12.** $z^2 - 8z + 9 = 0$

Work Smart

If the coefficient of the square term is not 1, then before using the method of completing the square, divide each side of the equation by the coefficient of the square term so that it becomes 1.

When the coefficient of the square term is not 1, we multiply or divide both sides of the equation by a nonzero constant so this coefficient becomes 1. The next example demonstrates this method.

EXAMPLE 7 **Solving a Quadratic Equation by Completing the Square When the Coefficient of the Square Term Is Not 1**

Solve: $2x^2 + 4x + 3 = 0$

Solution

To make the coefficient of the square term be 1, divide both sides of the equation by 2.

$$2x^2 + 4x + 3 = 0$$

Divide both sides of the equation by 2: $\dfrac{2x^2 + 4x + 3}{2} = \dfrac{0}{2}$

Simplify: $x^2 + 2x + \dfrac{3}{2} = 0$

Now solve the equation by completing the square.

Subtract $\dfrac{3}{2}$ from both sides: $x^2 + 2x = -\dfrac{3}{2}$

$\left(\dfrac{1}{2} \cdot 2\right)^2 = 1$; add 1 to both sides: $x^2 + 2x + 1 = -\dfrac{3}{2} + 1$

Simplify: $x^2 + 2x + 1 = -\dfrac{1}{2}$

Factor: $(x + 1)^2 = -\dfrac{1}{2}$

Use the Square Root Property: $x + 1 = \pm\sqrt{-\dfrac{1}{2}}$

$\sqrt{-N} = \sqrt{N}\,i$ $x + 1 = \pm\sqrt{\dfrac{1}{2}}\,i$

$\sqrt{\dfrac{1}{2}} = \dfrac{\sqrt{1}}{\sqrt{2}} = \dfrac{\sqrt{1}}{\sqrt{2}} \cdot \dfrac{\sqrt{2}}{\sqrt{2}} = \dfrac{\sqrt{2}}{2}$: $x + 1 = \pm\dfrac{\sqrt{2}}{2}\,i$

Subtract 1 from both sides: $x = -1 \pm \dfrac{\sqrt{2}}{2}\,i$

$a \pm b$ means $a - b$ or $a + b$: $x = -1 - \dfrac{\sqrt{2}}{2}\,i$ or $x = -1 + \dfrac{\sqrt{2}}{2}\,i$

Work Smart

Notice the solutions in Example 7 are complex conjugates of each other.

The solution set is $\left\{-1 - \dfrac{\sqrt{2}}{2}\,i, -1 + \dfrac{\sqrt{2}}{2}\,i\right\}$. We leave the check to you.

Quick ✓

In Problems 13 and 14, solve the quadratic equation by completing the square.

13. $2q^2 + 6q - 1 = 0$ **14.** $3m^2 + 2m + 7 = 0$

Figure 2

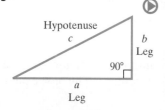

4 Solve Problems Using the Pythagorean Theorem

The Pythagorean Theorem is a statement about *right triangles*. A **right triangle** is one that contains a **right angle**—that is, an angle of 90°. The side of the triangle opposite the 90° angle is the **hypotenuse;** the remaining two sides are the **legs.** In Figure 2, c represents the length of the hypotenuse; a and b represent the lengths of the legs. Notice the use of the symbol $\llcorner$ to show the 90° angle.

The Pythagorean Theorem

In a right triangle, the square of the length of the hypotenuse equals the sum of the squares of the lengths of the legs. That is, in the right triangle shown in Figure 2,

$$c^2 = a^2 + b^2$$

EXAMPLE 8 | **Finding the Hypotenuse of a Right Triangle**

In a right triangle, one leg is of length 5 inches and the other is of length 12 inches. What is the length of the hypotenuse?

Solution

Since this is a right triangle, use the Pythagorean Theorem with $a = 5$ and $b = 12$ to find the length c of the hypotenuse.

$$c^2 = a^2 + b^2$$
$$c^2 = 5^2 + 12^2$$
$$= 25 + 144$$
$$= 169$$

Now use the Square Root Property to find c, the length of the hypotenuse.

$$c = \sqrt{169} = 13$$

The length of the hypotenuse is 13 inches. Notice that we find only the positive square root of 169 since c represents a length, which must be positive. ●

Quick ✓

15. The side of a right triangle opposite the 90° angle is called the _____; the remaining two sides are called ___.

16. *True or False* The Pythagorean Theorem states that for any triangle, the length of the hypotenuse is equal to the sum of the squares of the lengths of the legs.

In Problem 17, the lengths of the legs of a right triangle are given. Find the length of the hypotenuse.

17. $a = 3, b = 4$

EXAMPLE 9 | **How Far Can You See?**

The Currituck Lighthouse is located in Corolla, in the Outer Banks of North Carolina. The lighthouse, completed in 1875, stands 162 feet tall, however, a person standing in the observation deck is located 158 feet above the ground. See Figure 3.

Figure 3

The website for the Currituck Lighthouse states that a person standing on the observation deck could see approximately 18 miles. See Figure 4. Assuming that the radius of the Earth is 3960 miles, verify this claim.

Figure 4

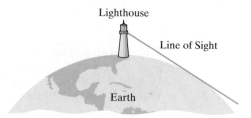

Lighthouse

Line of Sight

Earth

Solution

Step 1: Identify We want to know how far a person can see from the lighthouse.

Step 2: Name We will call this unknown distance d.

Step 3: Translate To help with the translation, draw a picture. From the center of Earth, draw two lines: one through the lighthouse and the other to the farthest point a person can see from the lighthouse. See Figure 5.

Figure 5

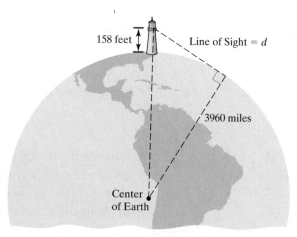

158 feet

Line of Sight = d

3960 miles

Center of Earth

The line of sight and the two lines drawn from the center of Earth form a right triangle, so the angle where the line of sight touches the horizon measures 90°. From the Pythagorean Theorem, we know that

$$\text{Hypotenuse}^2 = \text{Leg}^2 + \text{Leg}^2$$

The length of the hypotenuse is 3960 miles plus 158 feet. We add 3960 miles to 158 feet by converting the height of the tower to miles. Since 158 feet =

$158 \text{ feet} \cdot \dfrac{1 \text{ mile}}{5280 \text{ feet}} = \dfrac{158}{5280}$ mile, the hypotenuse is $\left(3960 + \dfrac{158}{5280}\right)$ miles. One of the

legs is 3960 miles. The length of the other leg is our unknown, d. Thus we have

$$3960^2 + d^2 = \left(3960 + \frac{158}{5280}\right)^2 \quad \text{The Model}$$

Step 4: Solve

$$3960^2 + d^2 = \left(3960 + \frac{158}{5280}\right)^2$$

Subtract 3960^2 from both sides: $d^2 = \left(3960 + \dfrac{158}{5280}\right)^2 - 3960^2$

Use a calculator: $d^2 \approx 237.000895$

Square Root Property: $d \approx \sqrt{237.000895}$

$$\approx 15.39 \text{ miles}$$

Step 5: Check Our answer is less than the distance given on the website.

Step 6: Answer The distance given on the Currituck Lighthouse website appears to overstate the actual distance a person could see. Someone standing on the observation deck of the lighthouse could see about 15.39 miles. ●

Quick ✓

18. The USS *Constitution* (aka *Old Ironsides*) is perhaps the most famous ship from United States Naval history. The mainmast of the Constitution is 220 feet high. Suppose that a sailor climbs the mainmast to a height of 200 feet in order to look for enemy vessels. How far could the sailor see? Assume the radius of the Earth is 3960 miles.

7.1 Exercises   MyMathLab®   Math XP PRACTICE

Exercise numbers in green have complete video solutions in MyMathLab.

*Problems **1–18** are the* Quick ✓s *that follow the* EXAMPLES.

Building Skills

In Problems 19–44, solve each equation using the Square Root Property. See Objective 1.

19. $y^2 = 100$

20. $x^2 = 81$

21. $p^2 = 50$

22. $z^2 = 48$

23. $m^2 = -25$

24. $n^2 = -49$

25. $w^2 = \dfrac{5}{4}$

26. $z^2 = \dfrac{8}{9}$

27. $x^2 + 5 = 13$

28. $w^2 - 6 = 14$

29. $3z^2 = 48$

30. $4y^2 = 100$

31. $3x^2 = 8$

32. $5y^2 = 32$

33. $2p^2 + 23 = 15$

34. $-3x^2 - 5 = 22$

35. $(d - 1)^2 = -18$

36. $(z + 4)^2 = -24$

37. $3(q + 5)^2 - 1 = 8$

38. $5(x - 3)^2 + 2 = 27$

39. $(3q + 1)^2 = 9$

40. $(2p + 3)^2 = 16$

41. $\left(x - \dfrac{2}{3}\right)^2 = \dfrac{5}{9}$

42. $\left(y + \dfrac{3}{2}\right)^2 = \dfrac{3}{4}$

43. $x^2 + 8x + 16 = 81$

44. $q^2 - 6q + 9 = 16$

In Problems 45–52, complete the square in each expression. Then factor the perfect square trinomial. See Objective 2.

45. $x^2 + 10x$

46. $y^2 + 16y$

47. $z^2 - 18z$

48. $p^2 - 4p$

49. $y^2 + 7y$

50. $x^2 + x$

51. $w^2 + \dfrac{1}{2}w$

52. $z^2 - \dfrac{1}{3}z$

In Problems 53–72, solve each quadratic equation by completing the square. See Objective 3.

53. $x^2 + 4x = 12$

54. $y^2 + 3y = 18$

55. $x^2 - 4x + 1 = 0$

56. $p^2 - 6p + 4 = 0$

57. $a^2 - 4a + 5 = 0$

58. $m^2 - 2m + 5 = 0$

59. $b^2 + 5b - 2 = 0$

60. $q^2 + 7q + 7 = 0$

61. $m^2 = 8m + 3$

62. $n^2 = 10n + 5$

63. $p^2 - p + 3 = 0$

64. $z^2 - 3z + 5 = 0$

65. $2y^2 - 5y - 12 = 0$

66. $3a^2 - 4a - 4 = 0$

67. $3y^2 - 6y + 2 = 0$

68. $2y^2 - 2y - 1 = 0$

69. $2z^2 - 5z + 1 = 0$

70. $2x^2 - 7x + 2 = 0$

71. $2x^2 + 4x + 5 = 0$

72. $2z^2 + 6z + 5 = 0$

In Problems 73–82, the lengths of the legs of a right triangle are given. Find the hypotenuse. Give exact answers and decimal approximations rounded to two decimal places. See Objective 4.

73. $a = 6, b = 8$

74. $a = 7, b = 24$

75. $a = 12, b = 16$

76. $a = 15, b = 8$

77. $a = 5, b = 5$

78. $a = 3, b = 3$

79. $a = 1, b = \sqrt{3}$

80. $a = 2, b = \sqrt{5}$

81. $a = 6, b = 10$

82. $a = 8, b = 10$

In Problems 83–86, use the right triangle shown below and find the missing length. Give exact answers and decimal approximations rounded to two decimal places. See Objective 4.

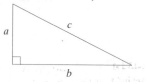

83. $a = 4, c = 8$

84. $a = 4, c = 10$

85. $b = 8, c = 12$

86. $b = 2, c = 10$

Mixed Practice

87. Given that $f(x) = (x - 3)^2$, find all x such that $f(x) = 36$. What points are on the graph of f?

88. Given that $f(x) = (x - 5)^2$, find all x such that $f(x) = 49$. What points are on the graph of f?

89. Given that $g(x) = (x + 2)^2$, find all x such that $g(x) = 18$. What points are on the graph of g?

90. Given that $h(x) = (x + 1)^2$, find all x such that $h(x) = 32$. What points are on the graph of h?

Applying the Concepts

In Problems 91 and 92, find the exact length of the diagonal in each rectangle.

△ **91.**

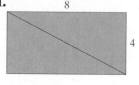

△ **92.**

In Problems 93–100, express your answer as a decimal rounded to three decimal places.

93. Golf A golfer hits an errant tee shot that lands in the rough. The golfer finds that the ball is exactly 30 yards to the right of the 100-yard marker, which indicates the distance to the center of the green as shown in the figure. How far is the ball from the center of the green?

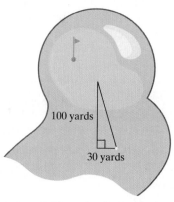

94. Baseball Alex Rios plays right field for the Chicago White Sox. He catches a fly ball 40 feet from the right field foul line, as indicated in the figure. How far is it to home plate?

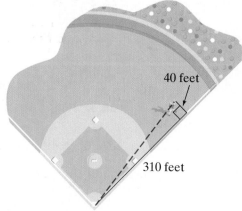

95. Guy Wire A guy wire is a wire used to support telephone poles. Suppose that a guy wire is located 30 feet up a telephone pole and is anchored to the ground 10 feet from the base of the pole. How long is the guy wire?

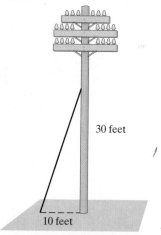

96. Guy Wire A guy wire is used to support an antenna on a rooftop. The wire is located 40 feet up on the antenna and anchored to the roof 8 feet from the base of the antenna. What is the length of the guy wire?

97. Ladder Bob needs to wash the windows on his house. He has a 25-foot ladder and places the base of the ladder 10 feet from the wall on the house.

(a) How far up the wall will the ladder reach?
(b) If his windows are 20 feet above the ground, what is the farthest distance the base of the ladder can be from the wall?

98. Fire Truck Ladder A fire truck has a 75-foot ladder. If the truck can safely park 20 feet from a building, how far up the building can the ladder reach, assuming that the base of the ladder is resting on top of the truck and the truck is 10 feet tall?

99. Gravity The distance s that an object falls (in feet) after t seconds, ignoring air resistance, is given by the equation $s = 16t^2$.

(a) How long does it take an object to fall 16 feet?
(b) How long does it take an object to fall 48 feet?
(c) How long does it take an object to fall 64 feet?

△ **100. Equilateral Triangles** An equilateral triangle is one whose sides are all the same length. The area A of an equilateral triangle whose sides are each length x is given by $A = \dfrac{\sqrt{3}}{4}x^2$.

(a) What is the length of each side of an equilateral triangle whose area is $\dfrac{8\sqrt{3}}{9}$ square feet?

(b) What is the length of each side of an equilateral triangle whose area is $\dfrac{25\sqrt{3}}{4}$ square meters?

Problems 101 and 102 are based on the following information. If P dollars are invested today at an annual interest rate r compounded once a year, then the value of the account A after 2 years is given by the formula $A = P(1 + r)^2$.

101. Value of Money Find the rate of interest required to turn an investment of $1000 into $1200 after 2 years. Express your answer as a percent rounded to two decimal places.

102. Value of Money Find the rate of interest required to turn an investment of $5000 into $6200 after 2 years. Express your answer as a percent rounded to two decimal places.

Extending the Concepts

If you look carefully at the Pythagorean Theorem, it states, "If we have a right triangle, then $c^2 = a^2 + b^2$, where c is the length of the hypotenuse." In this theorem "a right triangle" represents the hypothesis, and "$c^2 = a^2 + b^2$" represents the conclusion. The **converse** *of a theorem interchanges the hypothesis and conclusion. The converse of the Pythagorean Theorem is true.*

Converse of The Pythagorean Theorem

In a triangle, if the square of the length of one side equals the sum of the squares of the lengths of the other two sides, then the triangle is a right triangle. The 90° angle is opposite the longest side.

In Problems 103–106, the lengths of the sides of a triangle are given. Determine whether the triangle is a right triangle. If it is, identify the hypotenuse.

103. 8, 15, 17   **104.** 4, 6, 8

105. 14, 18, 20   **106.** 20, 48, 52

107. Pythagorean Triples Suppose that m and n are positive integers with $m > n$. If $a = m^2 - n^2$, $b = 2mn$, and $c = m^2 + n^2$, show that a, b, and c are the lengths of the sides of a right triangle using the Converse of the Pythagorean Theorem. We call any numbers a, b, and c found from the above formulas **Pythagorean Triples.**

108. Solve $ax^2 + bx + c = 0$ for x by completing the square.

Synthesis Review

In Problems 109–112, solve each equation.

109. $a^2 - 5a - 36 = 0$ **110.** $p^2 + 4p = 32$

111. $|4q + 1| = 3$ **112.** $\left|\dfrac{3}{4}w - \dfrac{2}{3}\right| = \dfrac{5}{2}$

113. In Problems 109–112, you are asked to solve quadratic and absolute value equations. In solving both types of equations, we reduce the equation to a simpler equation. What is the simpler equation?

7.2 Solving Quadratic Equations by the Quadratic Formula

Objectives

1 Solve Quadratic Equations Using the Quadratic Formula

2 Use the Discriminant to Determine the Nature of Solutions of a Quadratic Equation

3 Model and Solve Problems Involving Quadratic Equations

Are You Ready for This Section?

Before getting started, take the following readiness quiz. If you get a problem wrong, go back to the section cited and review the material.

R1. Simplify: **(a)** $\sqrt{121}$ **(b)** $\sqrt{54}$ [Getting Ready, pp. 467–468; Section 6.3, pp. 487–489]

R2. Simplify: **(a)** $\sqrt{-9}$ **(b)** $\sqrt{-72}$ [Section 6.8, pp. 524–525]

R3. Simplify: $\dfrac{3 + \sqrt{18}}{6}$ [Section 6.3, pp. 487–489]

So far, we have used three methods for solving quadratic equations: (1) factoring, (2) the Square Root Property, and (3) completing the square. Why do we need three methods? Because each method provides the quickest route to the solution when used appropriately. For example, if the quadratic expression is easy to factor, the method of factoring will get you to the solution most easily. If the equation is in the form $x^2 = p$ or $(ax \pm b)^2 = p$, using the Square Root Property is fastest. If the quadratic expression is not factorable, we must complete the square. But the method of completing the square is tedious. Is there an alternative to this method? Yes!

1 Solve Quadratic Equations Using the Quadratic Formula

▶ The method of completing the square leads to a general formula for solving the quadratic equation

$$ax^2 + bx + c = 0 \quad a \neq 0$$

To complete the square, first get the constant c on the right-hand side of the equation.

$$ax^2 + bx = -c \quad a \neq 0$$

Divide both sides of the equation by a: $\quad x^2 + \dfrac{b}{a}x = -\dfrac{c}{a}$

Now complete the square on the left side by adding the square of $\dfrac{1}{2}$ of the coefficient of x to both sides of the equation. That is, add

$$\left(\frac{1}{2} \cdot \frac{b}{a}\right)^2 = \frac{b^2}{4a^2}$$

to both sides of the equation. We now have

$$x^2 + \frac{b}{a}x + \frac{b^2}{4a^2} = -\frac{c}{a} + \frac{b^2}{4a^2}$$

$$x^2 + \frac{b}{a}x + \frac{b^2}{4a^2} = \frac{b^2}{4a^2} - \frac{c}{a}$$

Combine like terms on the right-hand side. The least common denominator on the right-hand side is $4a^2$, so multiply $-\dfrac{c}{a}$ by $\dfrac{4a}{4a}$:

$$x^2 + \frac{b}{a}x + \frac{b^2}{4a^2} = \frac{b^2}{4a^2} - \frac{c}{a} \cdot \frac{4a}{4a}$$

$$x^2 + \frac{b}{a}x + \frac{b^2}{4a^2} = \frac{b^2}{4a^2} - \frac{4ac}{4a^2}$$

$$x^2 + \frac{b}{a}x + \frac{b^2}{4a^2} = \frac{b^2 - 4ac}{4a^2}$$

Work Smart

To factor any perfect square trinomial of the form $x^2 + bx + c$, we write
$$\left(x + \frac{b}{2}\right)^2.$$

Factor the left-hand side and obtain

$$\left(x + \frac{b}{2a}\right)^2 = \frac{b^2 - 4ac}{4a^2}$$

Assume that $a > 0$ (you'll see why in a little while). This assumption does not affect the results because if $a < 0$, we could multiply both sides of the equation $ax^2 + bx + c = 0$ by -1 to make it positive. With this assumption, use the Square Root Property and get

$$x + \frac{b}{2a} = \pm\sqrt{\frac{b^2 - 4ac}{4a^2}}$$

$$\sqrt{\frac{a}{b}} = \frac{\sqrt{a}}{\sqrt{b}}: \quad x + \frac{b}{2a} = \pm\frac{\sqrt{b^2 - 4ac}}{\sqrt{4a^2}}$$

$$\sqrt{4a^2} = 2a \text{ since } a > 0: \quad x + \frac{b}{2a} = \pm\frac{\sqrt{b^2 - 4ac}}{2a}$$

Work Smart: Study Skill

When solving homework problems, always write the quadratic formula as part of the solution. This will help you memorize the formula.

Subtract $\dfrac{b}{2a}$ from both sides: $\qquad x = -\dfrac{b}{2a} \pm \dfrac{\sqrt{b^2 - 4ac}}{2a}$

Write over a common denominator: $\qquad x = \dfrac{-b \pm \sqrt{b^2 - 4ac}}{2a}$

This gives us the *quadratic formula*.

In Words

The quadratic formula says that the solution(s) to the equation $ax^2 + bx + c = 0$ is (are) "the opposite of b plus or minus the square root of b squared minus $4ac$, all over $2a$."

The Quadratic Formula

The solution(s) to the quadratic equation $ax^2 + bx + c = 0$, $a \neq 0$, is (are) given by the **quadratic formula**

$$x = \frac{-b \pm \sqrt{b^2 - 4ac}}{2a}$$

▶ **EXAMPLE 1** **How to Solve a Quadratic Equation Using the Quadratic Formula**

Solve: $12x^2 + 5x - 3 = 0$

Step-by-Step Solution

$$\boxed{a = 12} \qquad \boxed{b = 5}\,\boxed{c = -3}$$

Step 1: Write the equation in standard form $ax^2 + bx + c = 0$ and identify the values of a, b, and c.

$$12x^2 + 5x - 3 = 0$$

Step 2: Substitute the values of a, b, and c into the quadratic formula.

$$x = \frac{-b \pm \sqrt{b^2 - 4ac}}{2a}$$

$$x = \frac{-5 \pm \sqrt{5^2 - 4(12)(-3)}}{2(12)}$$

Step 3: Simplify the expression found in Step 2.

$$= \frac{-5 \pm \sqrt{25 + 144}}{24}$$

$$= \frac{-5 \pm \sqrt{169}}{24}$$

$\sqrt{169} = 13: \quad = \dfrac{-5 \pm 13}{24}$

(continued)

$$a \pm b \text{ means } a - b \text{ or } a + b: \quad x = \frac{-5 - 13}{24} \quad \text{or} \quad x = \frac{-5 + 13}{24}$$

$$= \frac{-18}{24} \quad \text{or} \quad = \frac{8}{24}$$

$$\text{Simplify:} \quad = -\frac{3}{4} \quad \text{or} \quad = \frac{1}{3}$$

Step 4: *Check*

$$12x^2 + 5x - 3 = 0$$

$$x = -\frac{3}{4}: \quad 12\left(-\frac{3}{4}\right)^2 + 5\left(-\frac{3}{4}\right) - 3 \overset{?}{=} 0 \qquad x = \frac{1}{3}: \quad 12\left(\frac{1}{3}\right)^2 + 5\left(\frac{1}{3}\right) - 3 \overset{?}{=} 0$$

$$12 \cdot \frac{9}{16} - \frac{15}{4} - 3 \overset{?}{=} 0 \qquad\qquad 12 \cdot \frac{1}{9} + \frac{5}{3} - 3 \overset{?}{=} 0$$

$$\frac{27}{4} - \frac{15}{4} - 3 \overset{?}{=} 0 \qquad\qquad \frac{4}{3} + \frac{5}{3} - 3 \overset{?}{=} 0$$

$$\frac{12}{4} - 3 \overset{?}{=} 0 \qquad\qquad \frac{9}{3} - 3 \overset{?}{=} 0$$

$$0 = 0 \quad \text{True} \qquad\qquad 0 = 0 \quad \text{True}$$

The solution set is $\left\{-\frac{3}{4}, \frac{1}{3}\right\}$.

Work Smart

If $b^2 - 4ac$ is a perfect square, then the quadratic equation can be solved by factoring. In Example 1, $b^2 - 4ac = 169$, a perfect square.

$$12x^2 + 5x - 3 = 0$$
$$(4x + 3)(3x - 1) = 0$$
$$4x + 3 = 0 \text{ or } 3x - 1 = 0$$
$$x = -\frac{3}{4} \text{ or } x = \frac{1}{3}$$

Notice that the solutions to the equation in Example 1 are rational numbers and that the expression $b^2 - 4ac$ under the radical in the quadratic formula, 169, is a perfect square. This leads to a generalization. Whenever the expression $b^2 - 4ac$ is a perfect square, the quadratic equation will have rational solutions, and the quadratic equation can be solved by factoring (provided the coefficients of the quadratic equation are rational numbers).

With that said, we can solve any quadratic equation using the quadratic formula.

Solving A Quadratic Equation Using The Quadratic Formula

Step 1: Write the equation in standard form $ax^2 + bx + c = 0$ and identify the values of a, b, and c.

Step 2: Substitute the values of a, b, and c into the quadratic formula.

Step 3: Simplify the expression found in Step 2.

Step 4: Verify your solution(s).

Quick ✓

1. The solution(s) to the quadratic equation $ax^2 + bx + c = 0, a \neq 0$, are given

by the quadratic formula $x = $ _____ .

In Problems 2 and 3, solve each equation using the quadratic formula.

2. $2x^2 - 3x - 9 = 0$ **3.** $2x^2 + 7x = 4$

▶ (**EXAMPLE 2**)   **Solving a Quadratic Equation Using the Quadratic Formula**

Solve: $3p^2 = 6p - 1$

Solution

First, write the equation in standard form to identify a, b, and c.

$$\text{Subtract } 6p \text{ from both sides;} \qquad 3p^2 = 6p - 1$$
$$\text{Add 1 to both sides:} \quad 3p^2 - 6p + 1 = 0$$

Write the quadratic equation using "$p =$": $\qquad p = \dfrac{-b \pm \sqrt{b^2 - 4ac}}{2a}$

$a = 3, b = -6, c = 1: \qquad p = \dfrac{-(-6) \pm \sqrt{(-6)^2 - 4(3)(1)}}{2(3)}$

$\qquad\qquad\qquad\qquad = \dfrac{6 \pm \sqrt{36 - 12}}{6}$

$\qquad\qquad\qquad\qquad = \dfrac{6 \pm \sqrt{24}}{6}$

$\sqrt{24} = \sqrt{4 \cdot 6} = 2\sqrt{6}: \qquad = \dfrac{6 \pm 2\sqrt{6}}{6}$

$\dfrac{a + b}{c} = \dfrac{a}{c} + \dfrac{b}{c}: \qquad = \dfrac{6}{6} \pm \dfrac{2\sqrt{6}}{6}$

Simplify: $\qquad = 1 \pm \dfrac{\sqrt{6}}{3}$

$a \pm b$ means $a - b$ or $a + b$: $\qquad p = 1 - \dfrac{\sqrt{6}}{3}$ or $p = 1 + \dfrac{\sqrt{6}}{3}$

Work Smart

We could also simplify $\dfrac{6 \pm 2\sqrt{6}}{6}$ by factoring:

$\dfrac{6 \pm 2\sqrt{6}}{6} = \dfrac{2(3 \pm \sqrt{6})}{6}$

$\qquad\qquad = \dfrac{3 \pm \sqrt{6}}{3}$

This is equivalent to $1 \pm \dfrac{\sqrt{6}}{3}$. Ask your instructor which form of the solution is preferred, if any.

We leave it to you to verify the solutions. The solution set is $\left\{ 1 - \dfrac{\sqrt{6}}{3}, 1 + \dfrac{\sqrt{6}}{3} \right\}$. ●

Notice in Example 2 that the value of $b^2 - 4ac$, 24, is positive, but not a perfect square. There are two solutions to the quadratic equation and they are irrational.

Quick ✓

4. *True or False* For $3x^2 - 2x = 9$, $a = 3$, $b = -2$, and $c = 9$.

5. Solve: $4z^2 + 1 = 8z$

EXAMPLE 3 **Solving a Rational Equation That Leads to a Quadratic Equation**

Solve: $9m + \dfrac{4}{m} = 12$

Solution

We first note that m cannot equal 0. To clear the equation of rational expressions, multiply both sides of the equation by the LCD, m.

$$m\left(9m + \frac{4}{m} \right) = 12 \cdot m$$

Distribute m: $\qquad 9m^2 + 4 = 12m$

Subtract $12m$ from both sides: $\quad 9m^2 - 12m + 4 = 0$

$$m = \frac{-b \pm \sqrt{b^2 - 4ac}}{2a}$$

$a = 9, b = -12, c = 4: \qquad m = \dfrac{-(-12) \pm \sqrt{(-12)^2 - 4(9)(4)}}{2(9)}$

Work Smart

Notice that the entire expression $-b \pm \sqrt{b^2 - 4ac}$ is in the numerator of the quadratic formula.

$\qquad\qquad\qquad = \dfrac{12 \pm \sqrt{144 - 144}}{18}$

$$= \frac{12 \pm \sqrt{0}}{18}$$

$$= \frac{12}{18} = \frac{2}{3}$$

We leave it to you to verify the solution. The solution set is $\left\{\dfrac{2}{3}\right\}$.

In Example 3, we had one solution rather than two (as in Examples 1 and 2). The solution $x = \dfrac{2}{3}$ is called a **repeated root** because it actually occurs twice, once for $\dfrac{12 + 0}{18} = \dfrac{12}{18} = \dfrac{2}{3}$ and once for $\dfrac{12 - 0}{18} = \dfrac{12}{18} = \dfrac{2}{3}$. Notice that the value of $b^2 - 4ac$ equals 0, which is the reason for the repeated root. We will have more to say about this soon.

Quick ✓

In Problems 6 and 7, solve each equation.

6. $4w + \dfrac{25}{w} = 20$

7. $2x = 8 - \dfrac{3}{x}$

▶ (EXAMPLE 4)   **Solving a Quadratic Equation Using the Quadratic Formula**

Solve: $y^2 - 4y + 13 = 0$

Solution

$$y^2 - 4y + 13 = 0$$

$$1y^2 - 4y + 13 = 0$$

$$y = \frac{-b \pm \sqrt{b^2 - 4ac}}{2a}$$

$a = 1, b = -4, c = 13$: $y = \dfrac{-(-4) \pm \sqrt{(-4)^2 - 4(1)(13)}}{2(1)}$

$$= \frac{4 \pm \sqrt{16 - 52}}{2}$$

$\sqrt{-36} = 6i$: $= \dfrac{4 \pm \sqrt{-36}}{2} = \dfrac{4 \pm 6i}{2}$

$\dfrac{a + b}{c} = \dfrac{a}{c} + \dfrac{b}{c}$: $= \dfrac{4}{2} \pm \dfrac{6}{2}i = 2 \pm 3i$

$a \pm b$ means $a - b$ or $a + b$: $y = 2 - 3i$ or $y = 2 + 3i$

We leave it to you to verify the solution. The solution set is $\{2 - 3i, 2 + 3i\}$.

Notice in Example 4 that the value of $b^2 - 4ac$ is negative and that the equation has two complex solutions that are not real.

Quick ✓

8. Solve: $z^2 + 2z + 26 = 0$

---

❷ Use the Discriminant to Determine the Nature of Solutions of a Quadratic Equation

In the quadratic formula $x = \dfrac{-b \pm \sqrt{b^2 - 4ac}}{2a}$, the quantity $b^2 - 4ac$ is called the **discriminant** because its value tells us the number and the type of solutions to expect from the quadratic equation when a, b, and c are rational numbers.

The Discriminant and the Nature of the Solutions of a Quadratic Equation $ax^2 + bx + c = 0$, where a, b, and c are Rational Numbers

Example	Value of Discriminant	Description of Discriminant	Number of Solutions	Type of Solution
1	169	Positive and a perfect square	Two	Rational
2	24	Positive and not a perfect square	Two	Irrational
3	0	Zero	One (repeated root)	Real
4	−36	Negative	Two	Complex, nonreal

Work Smart

The rules in the box to the right apply only if the coefficients of the quadratic equation are rational numbers.

In Example 4, the solutions are complex conjugates. In general, for any quadratic equation of the form $ax^2 + bx + c = 0$, where a, b, and c are real numbers and $b^2 - 4ac < 0$, the equation has two complex solutions that are not real and are complex conjugates.

This result is a consequence of the quadratic formula. Suppose that $b^2 - 4ac = -N < 0$. Then, by the quadratic formula, the solutions are

$$x = \frac{-b \pm \sqrt{b^2 - 4ac}}{2a} = \frac{-b \pm \sqrt{-N}}{2a}$$

$$= \frac{-b \pm \sqrt{N}i}{2a} = \frac{-b}{2a} \pm \frac{\sqrt{N}}{2a}i$$

which are complex conjugates.

▶ **EXAMPLE 5** **Determining the Nature of the Solutions of a Quadratic Equation**

For each quadratic equation, determine the discriminant. Use the value of the discriminant to determine whether the quadratic equation has two rational solutions, two irrational solutions, one repeated real solution, or two complex solutions that are not real.

(a) $x^2 - 5x + 2 = 0$ **(b)** $9y^2 + 6y + 1 = 0$ **(c)** $3p^2 - p = -5$

Solution

(a) We compare $x^2 - 5x + 2 = 0$ to the standard form $ax^2 + bx + c = 0$.

$$x^2 - 5x + 2 = 0$$

$$\boxed{a = 1} \quad \boxed{b = -5} \quad \boxed{c = 2}$$

We have $a = 1$, $b = -5$, and $c = 2$. Substituting the values for a, b, and c shown above into the discriminant formula $b^2 - 4ac$, we obtain

$$b^2 - 4ac = (-5)^2 - 4(1)(2) = 25 - 8 = 17$$

Because $b^2 - 4ac = 17$ is positive but not a perfect square, the quadratic equation has two irrational solutions.

(b) For the quadratic equation $9y^2 + 6y + 1 = 0$, we have $a = 9$, $b = 6$, and $c = 1$. Substituting these values into $b^2 - 4ac$ gives us

$$b^2 - 4ac = 6^2 - 4(9)(1) = 36 - 36 = 0$$

Because $b^2 - 4ac = 0$, the quadratic equation has one repeated real solution.

(c) Is the quadratic equation $3p^2 - p = -5$ in standard form? No! Add 5 to both sides of the equation and write the equation as $3p^2 - p + 5 = 0$. Thus we have $a = 3$, $b = -1$, and $c = 5$. The value of the discriminant, $b^2 - 4ac$, is

$$b^2 - 4ac = (-1)^2 - 4(3)(5) = 1 - 60 = -59$$

Because $-59 < 0$, the quadratic equation has two complex solutions that are not real and are conjugates.

Quick ✓

9. In the quadratic formula, the quantity $b^2 - 4ac$ is called the _____ of the quadratic equation.

10. *True or False* The discriminant of the quadratic equation $4x^2 - 5x + 1 = 0$ is $\sqrt{25 - 4(4)(1)} = \sqrt{9} = 3$.

11. If the discriminant of a quadratic equation is _____, then the quadratic equation has two complex solutions that are not real.

12. *True or False* If the discriminant of a quadratic equation is zero, then the equation has no solution.

13. *True or False* The solutions to a quadratic equation in which the solutions are complex numbers that are not real are complex conjugates.

In Problems 14–16, use the value of the discriminant to determine whether the quadratic equation has two rational solutions, two irrational solutions, one repeated real solution, or two complex solutions that are not real.

14. $2z^2 + 5z + 4 = 0$ 15. $4y^2 + 12y = -9$ 16. $2x^2 - 4x + 1 = 0$

▶ Which Method Should I Use?

We now have four methods for solving quadratic equations:

1. Factoring

2. Square Root Property

3. Completing the Square

4. The Quadratic Formula

You are probably asking yourself, "Which method should I use?" and "Does it matter which method I use?" The answer to the second question is that it does not matter which method you use, but one method may be more efficient than the others. Table 1 contains guidelines to help you solve any quadratic equation. The method that is most efficient for solving a quadratic equation depends on the equation. The value of the discriminant can be used to guide you in choosing the most efficient method.

Notice in Table 1 that we did not recommend completing the square as one of the methods to use in solving a quadratic equation. This is because the quadratic formula is based on completing the square of $ax^2 + bx + c = 0$. Besides, completing the square may be a cumbersome task, whereas the quadratic formula is fairly straightforward. Learning how to complete the square was worth your time, however, because it is used to derive the quadratic formula. Also, completing the square is a skill that you will need later in this course and in future math courses.

Table 1

Form of the Quadratic Equation	Most Efficient Method	Example
$x^2 = p$, where p is any real number	Square Root Property	$$x^2 = 45$$ $$\text{Square Root Property:} \quad x = \pm\sqrt{45}$$ $$= \pm 3\sqrt{5}$$
$ax^2 + c = 0$	Square Root Property	$$3p^2 + 12 = 0$$ Subtract 12 from both sides: $3p^2 = -12$ Divide both sides by 3: $p^2 = -4$ Square Root Property: $p = \pm\sqrt{-4}$ $= \pm 2i$
$(ax + b)^2 = p$, where p is any real number	Square Root Property	$$(2x + 7)^2 = 18$$ Square Root Property: $2x + 7 = \pm 3\sqrt{2}$ Subtract 7 from both sides: $2x = -7 \pm 3\sqrt{2}$ Divide by 2: $x = \dfrac{-7 \pm 3\sqrt{2}}{2}$
$ax^2 + bx + c = 0$, where $b^2 - 4ac$ is a perfect square	Factoring or the quadratic formula. Factor if the quadratic expression is easy to factor. Otherwise, use the quadratic formula.	$a = 2, b = 1, c = -10$: $\quad 2m^2 + m - 10 = 0$ $b^2 - 4ac = 1^2 - 4(2)(-10) = 1 + 80 = 81$ 81 is a perfect square, so we can use factoring: $$2m^2 + m - 10 = 0$$ $$(2m + 5)(m - 2) = 0$$ $$2m + 5 = 0 \quad \text{or} \quad m - 2 = 0$$ $$m = -\frac{5}{2} \quad \text{or} \quad m = 2$$
$ax^2 + bx + c = 0$, where $b^2 - 4ac$ is not a perfect square	Quadratic formula	$a = 2, b = 4, c = -1$: $\quad 2x^2 + 4x - 1 = 0$ $b^2 - 4ac = 4^2 - 4(2)(-1) = 16 + 8 = 24$ 24 is not a perfect square, so we use the quadratic formula (since it's easier than completing the square): $$x = \frac{-b \pm \sqrt{b^2 - 4ac}}{2a}$$ $$= \frac{-4 \pm \sqrt{24}}{2(2)}$$ $$= \frac{-4 \pm 2\sqrt{6}}{4}$$ $$= -1 \pm \frac{\sqrt{6}}{2}$$ $$x = -1 - \frac{\sqrt{6}}{2} \text{ or } x = -1 + \frac{\sqrt{6}}{2}$$
$ax^2 + bx + c = 0$, where $b^2 - 4ac$ is negative	Quadratic formula	$a = 3, b = -4, c = 7$: $\quad 3x^2 - 4x + 7 = 0$ $b^2 - 4ac = (-4)^2 - 4(3)(7) = 16 - 84 = -68$ The discriminant is negative, so we use the quadratic formula: $$x = \frac{-b \pm \sqrt{b^2 - 4ac}}{2a}$$ $$x = \frac{-(-4) \pm \sqrt{-68}}{2(3)}$$ $$= \frac{4 \pm \sqrt{-1 \cdot 4 \cdot 17}}{6}$$ $$= \frac{4 \pm 2\sqrt{17}i}{6}$$ $$= \frac{2}{3} \pm \frac{1}{3}\sqrt{17}i$$ $$x = \frac{2}{3} - \frac{1}{3}\sqrt{17}i \text{ or } x = \frac{2}{3} + \frac{1}{3}\sqrt{17}i$$

Quick ✓

17. *True or False* If the discriminant of a quadratic equation is a perfect square, then the equation can be solved by factoring.

In Problems 18–20, solve each quadratic equation using any appropriate method.

18. $5n^2 - 45 = 0$ **19.** $-2y^2 + 5y - 6 = 0$ **20.** $3w^2 + 2w = 5$

▶ ❸ Model and Solve Problems Involving Quadratic Equations

Many applied problems, such as the one below, require solving quadratic equations. As always, we will use the problem-solving strategy from Section 1.2.

EXAMPLE 6   **Revenue**

The revenue R received by a company selling x specialty T-shirts per week is given by the function $R(x) = -0.005x^2 + 30x$.

(a) How many T-shirts must be sold in order for revenue to be $25,000 per week?

(b) How many T-shirts must be sold in order for revenue to be $45,000 per week?

Solution

(a) Step 1: Identify We want to determine the number x of T-shirts required so that $R = \$25,000$.

Step 2: Name We know that x represents the number of T-shirts sold.

Step 3: Translate We need to solve the equation $R(x) = 25,000$.

$$R(x) = 25,000$$
$$-0.005x^2 + 30x = 25,000$$
$$-0.005x^2 + 30x - 25,000 = 0$$

Step 4: Solve Letting $a = -0.005$, $b = 30$, and $c = -25,000$, we have

$$b^2 - 4ac = 30^2 - 4(-0.005)(-25,000)$$
$$= 400$$

Because 400 is a perfect square, we can solve the equation by factoring or using the quadratic formula. It is not obvious how to factor $-0.005x^2 + 30x - 25000$, so we will use the quadratic formula:

$$\overset{b^2 - 4ac = 400}{\underset{\downarrow}{}}$$
$$x = \frac{-30 \pm \sqrt{400}}{2(-0.005)}$$
$$= \frac{-30 \pm 20}{-0.01}$$

$$x = \frac{-30 - 20}{-0.01} = 5000 \quad \text{or} \quad x = \frac{-30 + 20}{-0.01} = 1000$$

Step 5: Check If 1000 T-shirts are sold, then revenue is $R(1000) = -0.005(1000)^2 + 30(1000) = \$25,000$. If 5000 T-shirts are sold, then revenue is $R(5000) = -0.005(5000)^2 + 30(5000) = \$25,000$.

Step 6: Answer The company needs to sell either 1000 or 5000 T-shirts each week to earn $25,000 in revenue.

(b) Step 1: Identify We want to determine the number x of T-shirts required so that $R = \$45,000$. That is, we wish to solve the equation $R(x) = 45{,}000$.

Step 2: Name We know that x represents the number of T-shirts sold.

Step 3: Translate We need to solve the equation $R(x) = 45{,}000$.

$$R(x) = 45{,}000$$
$$-0.005x^2 + 30x = 45{,}000$$
$$-0.005x^2 + 30x - 45{,}000 = 0$$

Step 4: Solve Letting $a = -0.005$, $b = 30$, and $c = -45{,}000$, we have

$$b^2 - 4ac = 30^2 - 4(-0.005)(-45{,}000) = 0$$

Because the discriminant is 0, the quadratic equation has a single real solution. We can solve the equation by factoring or using the quadratic formula. It is not obvious how to factor $-0.005x^2 + 30x - 45{,}000$, so we will use the quadratic formula to solve the equation.

$$x = \frac{-30 \pm \sqrt{0}}{2(-0.005)} \quad b^2 - 4ac = 0$$
$$= \frac{-30 \pm 0}{-0.01}$$
$$= \frac{-30}{-0.01} = 3000$$

Step 5: Check If 3000 T-shirts are sold, then revenue is $R(3000) = -0.005(3000)^2 + 30(3000) = \$45{,}000$.

Step 6: Answer The company needs to sell 3000 T-shirts each week to earn $45,000 in revenue.

Quick ✓

21. The revenue R received by a video store renting x DVDs per day is given by the function $R(x) = -0.005x^2 + 4x$.
   (a) How many DVDs must be rented in order for revenue to be $600 per day?
   (b) How many DVDs must be rented in order for revenue to be $800 per day?

EXAMPLE 7   **Designing a Window**

A window designer wishes to design a window so that the diagonal is 20 feet. In addition, the width of the window needs to be 4 feet more than the height. What are the dimensions of the window?

Solution

Step 1: Identify We want to know the dimensions of the window. That is, we want to know the width and height of the window.

Step 2: Name Let h represent the height of the window such that $h + 4$ is the width (since the width is 4 feet more than the height).

Step 3: Translate Figure 6 illustrates the situation. Notice that the three sides form a right triangle. We can express the relation among the three sides using the Pythagorean Theorem.

$$\text{leg}^2 + \text{leg}^2 = \text{hypotenuse}^2: \qquad h^2 + (h+4)^2 = 20^2$$
$$(A + B)^2 = A^2 + 2AB + B^2: \quad h^2 + h^2 + 8h + 16 = 400$$

Figure 6

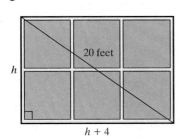

h

20 feet

$h + 4$

Combine like terms: $2h^2 + 8h + 16 = 400$

Subtract 400 from both sides: $2h^2 + 8h - 384 = 0$

Divide both sides by 2: $h^2 + 4h - 192 = 0$

Step 4: Solve In the model, we have $a = 1, b = 4, c = -192$. The discriminant is $b^2 - 4ac = 4^2 - 4(1)(-192) = 784 = 28^2$. Thus we can solve the equation by factoring.

$$h^2 + 4h - 192 = 0$$
$$(h + 16)(h - 12) = 0$$
$$h + 16 = 0 \quad \text{or} \quad h - 12 = 0$$
$$h = -16 \quad \text{or} \quad h = 12$$

Step 5: Check We disregard the solution $h = -16$ because h represents the height of the window. We see whether a window whose dimensions are 12 feet by $12 + 4 = 16$ feet has a diagonal that is 20 feet by verifying that $12^2 + 16^2 = 20^2$.

$$12^2 + 16^2 \overset{?}{=} 20^2$$
$$144 + 256 \overset{?}{=} 400$$
$$400 = 400$$

Step 6: Answer The dimensions of the window are 12 feet by 16 feet.   ●

Quick ✓

22. A rectangular plot of land is designed so that its length is 14 meters more than its width. The diagonal of the land is known to be 34 meters. What are the dimensions of the land?

7.2 Exercises MyMathLab®   PRACTICE

Exercise numbers in green have complete video solutions in MyMathLab.

Problems 1–22 are the Quick ✓s that follow the EXAMPLES.

Building Skills

In Problems 23–40, solve each equation using the quadratic formula. See Objective 1.

23. $x^2 - 4x - 12 = 0$

24. $p^2 - 4p - 32 = 0$

25. $6y^2 - y - 15 = 0$

26. $10x^2 + x - 2 = 0$

27. $4m^2 - 8m + 1 = 0$

28. $2q^2 - 4q + 1 = 0$

29. $3w - 6 = \dfrac{1}{w}$

30. $x + \dfrac{1}{x} = 3$

31. $3p^2 = -2p + 4$

32. $5w^2 = -3w + 1$

33. $x^2 - 2x + 7 = 0$

34. $y^2 - 4y + 5 = 0$

35. $2z^2 + 7 = 2z$

36. $2z^2 + 7 = 4z$

37. $4x^2 = 2x + 1$

38. $6p^2 = 4p + 1$

39. $1 = 3q^2 + 4q$

40. $1 = 5w^2 + 6w$

In Problems 41–50, determine the discriminant of each quadratic equation. Use the value of the discriminant to determine whether the quadratic equation has two rational solutions, two irrational solutions, one repeated real solution, or two complex solutions that are not real. See Objective 2.

41. $x^2 - 5x + 1 = 0$

42. $p^2 + 4p - 2 = 0$

43. $3z^2 + 2z + 5 = 0$

44. $2y^2 - 3y + 5 = 0$

45. $9q^2 - 6q + 1 = 0$

46. $16x^2 + 24x + 9 = 0$

47. $3w^2 = 4w - 2$

48. $6x^2 - x = -4$

49. $6x = 2x^2 - 1$

50. $10w^2 = 3$

Mixed Practice

In Problems 51–76, solve each equation.

51. $w^2 - 5w + 5 = 0$

52. $q^2 - 7q + 7 = 0$

53. $3x^2 + 5x = 8$

54. $4p^2 + 5p = 9$

55. $2x^2 = 3x + 35$

56. $3x^2 + 5x = 2$

57. $q^2 + 2q + 8 = 0$

58. $w^2 + 4w + 9 = 0$

59. $2z^2 = 2(z + 3)^2$

60. $3z^2 = 3(z + 1)(z - 2)$

61. $7q - 2 = \dfrac{4}{q}$

62. $5m - 4 = \dfrac{5}{m}$

63. $5a^2 - 80 = 0$

64. $4p^2 - 100 = 0$

65. $8n^2 + 1 = 4n$

66. $4q^2 + 1 = 2q$

67. $27x^2 + 36x + 12 = 0$

68. $8p^2 - 40p + 50 = 0$

69. $\dfrac{1}{3}x^2 + \dfrac{2}{9}x - 1 = 0$

70. $\dfrac{1}{2}x^2 + \dfrac{3}{4}x - 1 = 0$

71. $(x - 5)(x + 1) = 4$

72. $(a - 3)(a + 1) = 2$

73. $\dfrac{x - 2}{x + 2} = x - 3$

74. $\dfrac{x - 5}{x + 3} = x - 3$

75. $\dfrac{x - 4}{x^2 + 2} = 2$

76. $\dfrac{x - 1}{x^2 + 4} = 1$

77. Suppose that $f(x) = x^2 + 4x - 21$.

(a) Solve $f(x) = 0$ for x.

(b) Solve $f(x) = -21$ for x. What points are on the graph of f?

78. Suppose that $f(x) = x^2 + 2x - 8$.
(a) Solve $f(x) = 0$ for x.
(b) Solve $f(x) = -8$ for x. What points are on the graph of f?

79. Suppose that $H(x) = -2x^2 - 4x + 1$.
(a) Solve $H(x) = 0$ for x.
(b) Solve $H(x) = 2$ for x.

80. Suppose that $g(x) = 3x^2 + x - 1$.
(a) Solve $g(x) = 0$ for x.
(b) Solve $g(x) = 4$ for x.

81. What are the zeros of $G(x) = 3x^2 + 2x - 2$?

82. What are the zeros of $F(x) = x^2 + 3x - 3$?

Applying the Concepts

In Problems 83–86, use the Pythagorean Theorem to determine the value of x and the measurements of all sides of the right triangle.

△ **83.**

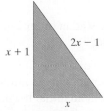

△ **84.**

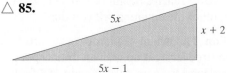

△ **85.**

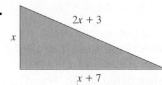

△ **86.**

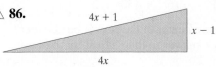

△ **87. Area** The area of a rectangle is 40 square inches. The width of the rectangle is 4 inches more than the length. What are the dimensions of the rectangle?

△ **88. Area** The area of a rectangle is 60 square inches. The width of the rectangle is 6 inches more than the length. What are the dimensions of the rectangle?

△ 89. **Area** The area of a triangle is 25 square inches. The height of the triangle is 3 inches less than the base. What are the base and height of the triangle?

△ 90. **Area** The area of a triangle is 35 square inches. The height of the triangle is 2 inches less than the base. What are the base and height of the triangle?

91. **Revenue** The revenue R received by a company selling x pairs of sunglasses per week is given by the function $R(x) = -0.1x^2 + 70x$.
   (a) Find and interpret the values of $R(17)$ and $R(25)$.
   (b) How many pairs of sunglasses must be sold in order for revenue to be $10,000 per week?
   (c) How many pairs of sunglasses must be sold in order for revenue to be $12,250 per week?

92. **Revenue** The revenue R received by a company selling x "all-day passes" to a small amusement park per day is given by the function $R(x) = -0.02x^2 + 24x$.
   (a) Find and interpret the values of $R(300)$ and $R(800)$.
   (b) How many tickets must be sold in order for revenue to be $4000 per day?
   (c) How many tickets must be sold in order for revenue to be $7200 per day?

93. **Projectile Motion** The height s of a ball after t seconds, when thrown straight up with an initial speed of 70 feet per second from an initial height of 5 feet, can be modeled by the function

$$s(t) = -16t^2 + 70t + 5$$

   (a) When will the height of the ball be 40 feet? Round your answer to the nearest tenth of a second.
   (b) When will the height of the ball be 70 feet? Round your answer to the nearest tenth of a second.
   (c) Will the ball ever reach a height of 150 feet? How does the result of the equation tell you this?

94. **Projectile Motion** The height s of a toy rocket after t seconds, when fired straight up with an initial speed of 150 feet per second from an initial height of 2 feet, can be modeled by the function

$$s(t) = -16t^2 + 150t + 2$$

   (a) When will the height of the rocket be 200 feet? Round your answer to the nearest tenth of a second.
   (b) When will the height of the rocket be 300 feet? Round your answer to the nearest tenth of a second.
   (c) Will the rocket ever reach a height of 500 feet?

△ 95. **Similar Triangles** Consult the figure. Suppose that $\triangle ABC$ is similar to $\triangle DEC$. The length of $\overline{BC}$ is 24 inches and the length of $\overline{DE}$ is 6 inches. If the length of $\overline{AB}$ equals the length of $\overline{CE}$, which we call x, find x.

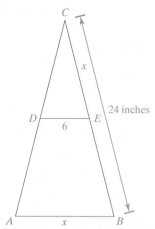

96. **Number Sense** Three times the square of a number equals the sum of two times the number and 5. Find the number(s).

97. **Life Cycle Hypothesis** The Life Cycle Hypothesis from economics was presented by Franco Modigliani in 1954. One of its components states that income is a function of age. The function $I(a) = -55a^2 + 5119a - 54,448$ represents the relation between average annual income I and age a.
   (a) For what age does average income I equal $40,000? Round your answer to the nearest year.
   (b) For what age does average income I equal $50,000? Round your answer to the nearest year.

98. **Population** The function $P(a) = 0.006a^2 - 4.694a + 317.167$ represents the population (in millions) of Americans in 2010, P, that are a years of age or older.
   (SOURCE: *United States Census Bureau*)
   (a) For what age range was the population 200 million in 2010? Round your answer to the nearest year.
   (b) For what age range was the population 50 million in 2010? Round your answer to the nearest year.

99. **Upstream and Back** Zene decides to canoe 4 miles upstream on a river to a waterfall and then canoe back. The total trip (excluding the time spent at the waterfall) takes 6 hours. Zene knows she can canoe at an average speed of 5 miles per hour in still water. What is the speed of the current?

100. Round Trip A Cessna aircraft flies 200 miles due west into the jet stream and flies back home on the same route. The total time of the trip (excluding the time on the ground) is 4 hours. The Cessna aircraft can fly 120 miles per hour in still air. What is the net effect of the jet stream on the aircraft?

101. Work Robert and Susan have a newspaper route. When they work the route together, it takes 2 hours to deliver all the newspapers. One morning Robert tells Susan he is too sick to deliver the papers. Susan doesn't remember how long it takes her to deliver the newspapers working alone, but she does remember that Robert can finish the route one hour sooner than she can when working alone. How long will it take Susan to finish the route?

102. Work Demitrius needs to fill up his pool. When he rents a water tanker to fill the pool with the help of the hose from his house, it takes 5 hours to fill the pool. This year, money is tight and he can't afford to rent the water tanker to fill the pool. He doesn't remember how long it takes for his house hose to fill the pool, but he does remember that the tanker hose filling the pool alone can finish the job in 8 fewer hours than using his house hose alone. How long will it take Demitrius to fill his pool using only his house hose?

Extending the Concepts

103. Show that the sum of the solutions to a quadratic equation is $-\dfrac{b}{a}$.

104. Show that the product of the solutions to a quadratic equation is $\dfrac{c}{a}$.

105. Show that the real solutions to the equation $ax^2 + bx + c = 0$ are the negatives of the real solutions to the equation $ax^2 - bx + c = 0$. Assume that $b^2 - 4ac \geq 0$.

106. Show that the real solutions to the equation $ax^2 + bx + c = 0$ are the reciprocals of the real solutions to the equation $cx^2 + bx + a = 0$. Assume that $b^2 - 4ac \geq 0$.

Explaining the Concepts

107. Explain the circumstances for which you would use factoring to solve a quadratic equation.

108. Explain the circumstances for which you would use the Square Root Property to solve a quadratic equation.

Synthesis Review

109. (a) Graph $f(x) = x^2 + 3x + 2$ by plotting points.
   (b) Solve the equation $x^2 + 3x + 2 = 0$.

(c) Compare the solutions to the equation in part (b) to the x-intercepts of the graph drawn in part (a). What do you notice?

110. (a) Graph $f(x) = x^2 - x - 6$ by plotting points.
   (b) Solve the equation $x^2 - x - 6 = 0$.
   (c) Compare the solutions to the equation in part (b) to the x-intercepts of the graph drawn in part (a). What do you notice?

111. (a) Graph $g(x) = x^2 - 2x + 1$ by plotting points.
   (b) Solve the equation $x^2 - 2x + 1 = 0$.
   (c) Compare the solutions to the equation in part (b) to the x-intercepts of the graph drawn in part (a). What do you notice?

112. (a) Graph $g(x) = x^2 + 4x + 4$ by plotting points.
   (b) Solve the equation $x^2 + 4x + 4 = 0$.
   (c) Compare the solutions to the equation in part (b) to the x-intercepts of the graph drawn in part (a). What do you notice?

The Graphing Calculator

In Problems 113–116, the graph of the quadratic function f is given. For each function, determine the discriminant of the equation $f(x) = 0$ in order to determine the nature of the solutions of the equation $f(x) = 0$. Compare the nature of solutions based on the discriminant to the graph of the function.

113. $f(x) = x^2 - 7x + 3$ **114.** $f(x) = -x^2 - 3x + 1$

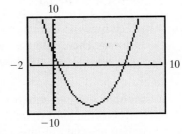   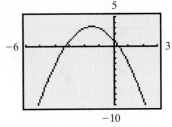

115. $f(x) = -x^2 - 3x - 4$ **116.** $f(x) = x^2 - 6x + 9$

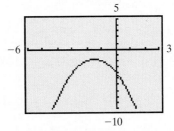   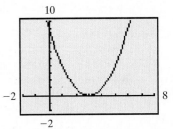

117. (a) Solve the equation $x^2 - 5x - 24 = 0$ algebraically.
   (b) Graph $Y_1 = x^2 - 5x - 24$. Compare the x-intercepts of the graph to the solutions found in part (a).

118. (a) Solve the equation $x^2 - 4x - 45 = 0$ algebraically.
   (b) Graph $Y_1 = x^2 - 4x - 45$. Compare the x-intercepts of the graph to the solutions found in part (a).

119. (a) Solve the equation $x^2 - 6x + 9 = 0$ algebraically.

(b) Graph $Y_1 = x^2 - 6x + 9$. Compare the x-intercepts of the graph to the solutions found in part (a).

120. (a) Solve the equation $x^2 + 10x + 25 = 0$ algebraically.

(b) Graph $Y_1 = x^2 + 10x + 25$. Compare the x-intercepts of the graph to the solutions found in part (a).

121. (a) Solve the equation $x^2 + 5x + 8 = 0$ algebraically.

(b) Graph $Y_1 = x^2 + 5x + 8$. How is the result of part (a) related to the graph?

122. (a) Solve the equation $x^2 + 2x + 5 = 0$ algebraically.

(b) Graph $Y_1 = x^2 + 2x + 5$. How is the result of part (a) related to the graph?

7.3 Solving Equations Quadratic in Form

Objective

1 Solve Equations That Are Quadratic in Form

Are You Ready for This Section?

Before getting started, take the following readiness quiz. If you get a problem wrong, go back to the section cited and review the material.

R1. Factor: $x^4 - 5x^2 - 6$ [Section 4.5, p. 355]

R2. Factor: $2(p+3)^2 + 3(p+3) - 5$ [Section 4.5, p. 355]

R3. Simplify: **(a)** $(x^2)^2$ **(b)** $(p^{-1})^2$ [Getting Ready, pp. 297–298]

▶ **1** Solve Equations That Are Quadratic in Form

Consider the equation $x^4 - 4x^2 - 12 = 0$. Even though this equation is not in the form of a quadratic equation, $ax^2 + bx + c = 0$, we can write the equation as $(x^2)^2 - 4x^2 - 12 = 0$. Then, if we let $u = x^2$ in the equation, we get $u^2 - 4u - 12 = 0$, which is of the form $ax^2 + bx + c = 0$. Now we can solve $u^2 - 4u - 12 = 0$ for u by factoring. Then, using the fact that $u = x^2$, we can find x, which was our goal in the first place.

In general, if a substitution u transforms an equation into one of the form

$$au^2 + bu + c = 0$$

then the original equation is an **equation quadratic in form.**

It is often hard to tell whether an equation is quadratic in form. Table 2 shows some equations that are quadratic in form and the appropriate substitution.

Table 2

Original Equation	Substitution	Equation with Substitution
$2x^4 - 3x^2 + 5 = 0$ $2(x^2)^2 - 3x^2 + 5 = 0$	$u = x^2$	$2u^2 - 3u + 5 = 0$
$3(z - 5)^2 + 4(z - 5) + 1 = 0$	$u = z - 5$	$3u^2 + 4u + 1 = 0$
$-2y + 5\sqrt{y} - 2 = 0$ $-2(\sqrt{y})^2 + 5\sqrt{y} - 2 = 0$	$u = \sqrt{y}$	$-2u^2 + 5u - 2 = 0$

EXAMPLE 1   How to Solve Equations That Are Quadratic in Form

Solve: $x^4 + x^2 - 12 = 0$

Step-by-Step Solution

Step 1: Determine the appropriate substitution and write the equation in the form $au^2 + bu + c = 0$.

$$x^4 + x^2 - 12 = 0$$
$$(x^2)^2 + x^2 - 12 = 0$$

Let $u = x^2$: $u^2 + u - 12 = 0$

Step 2: Solve the equation $au^2 + bu + c = 0$.

$$(u + 4)(u - 3) = 0$$
$$u + 4 = 0 \quad \text{or} \quad u - 3 = 0$$
$$u = -4 \quad \text{or} \quad u = 3$$

Step 3: Rewrite the solution in terms of the original variable.

We want to know x, so replace u with x^2: $\quad x^2 = -4 \quad$ or $\quad x^2 = 3$

Square Root Property: $\quad x = \pm\sqrt{-4} \quad$ or $\quad x = \pm\sqrt{3}$

$$= \pm 2i$$

Step 4: Verify your solutions.

$x = 2i$:

$$(2i)^4 + (2i)^2 - 12 \overset{?}{=} 0$$
$$2^4 i^4 + 2^2 i^2 - 12 \overset{?}{=} 0$$
$$16(1) + 4(-1) - 12 \overset{?}{=} 0$$
$$16 - 4 - 12 \overset{?}{=} 0$$
$$0 = 0 \quad \text{True}$$

$x = -2i$:

$$(-2i)^4 + (-2i)^2 - 12 \overset{?}{=} 0$$
$$(-2)^4 i^4 + (-2)^2 i^2 - 12 \overset{?}{=} 0$$
$$16(1) + 4(-1) - 12 \overset{?}{=} 0$$
$$16 - 4 - 12 \overset{?}{=} 0$$
$$0 = 0 \quad \text{True}$$

$x = \sqrt{3}$:

$$(\sqrt{3})^4 + (\sqrt{3})^2 - 12 \overset{?}{=} 0$$
$$\sqrt{3^4} + \sqrt{3^2} - 12 \overset{?}{=} 0$$
$$\sqrt{81} + 3 - 12 \overset{?}{=} 0$$
$$9 + 3 - 12 \overset{?}{=} 0$$
$$0 = 0 \quad \text{True}$$

$x = -\sqrt{3}$:

$$(-\sqrt{3})^4 + (-\sqrt{3})^2 - 12 \overset{?}{=} 0$$
$$(-1)^4\sqrt{3^4} + (-1)^2\sqrt{3^2} - 12 \overset{?}{=} 0$$
$$\sqrt{81} + 3 - 12 \overset{?}{=} 0$$
$$9 + 3 - 12 \overset{?}{=} 0$$
$$0 = 0 \quad \text{True}$$

The solution set is $\left\{ 2i, -2i, \sqrt{3}, -\sqrt{3} \right\}$.

We summarize the steps used to solve an equation that is quadratic in form.

> **Solving Equations Quadratic in Form**
>
> **Step 1:** Determine the appropriate substitution and write the equation in the form $au^2 + bu + c = 0$.
>
> **Step 2:** Solve the equation $au^2 + bu + c = 0$.
>
> **Step 3:** Rewrite the solution in terms of the original variable.
>
> **Step 4:** Verify your solutions.

Quick ✓

1. If a substitution u transforms an equation into one of the form $au^2 + bu + c = 0$, then the original equation is called an equation _____ — ____.

2. For the equation $2(3x + 1)^2 - 5(3x + 1) + 2 = 0$, an appropriate substitution would be $u =$ ____.

3. *True or False* The equation $3\left(\dfrac{x}{x-2}\right)^2 - \dfrac{5x}{x-2} + 3 = 0$ is quadratic in form.

4. What is the appropriate choice for u when solving the equation $2 \cdot \dfrac{1}{x^2} - 6 \cdot \dfrac{1}{x} + 3 = 0$?

In Problems 5 and 6, solve each equation.

5. $x^4 - 13x^2 + 36 = 0$

6. $p^4 - 7p^2 = 18$

Ready?...Answers

R1. $(x^2 - 6)(x^2 + 1)$

R2. $(2p + 11)(p + 2)$

R3. (a) x^4 (b) $p^{-2} = \dfrac{1}{p^2}$

EXAMPLE 2   **Solving Equations That Are Quadratic in Form**

Solve: $(z^2 - 5)^2 - 3(z^2 - 5) - 4 = 0$

Solution

$$(z^2 - 5)^2 - 3(z^2 - 5) - 4 = 0$$

Let $u = z^2 - 5$: $u^2 - 3u - 4 = 0$

$$(u - 4)(u + 1) = 0$$

$u - 4 = 0$ or $u + 1 = 0$

$u = 4$ or $u = -1$

Replace u with $z^2 - 5$ and solve for z: $z^2 - 5 = 4$ or $z^2 - 5 = -1$

$z^2 = 9$ or $z^2 = 4$

Square Root Property: $z = \pm 3$ or $z = \pm 2$

Check

$z = -3$:

$((-3)^2 - 5)^2 - 3((-3)^2 - 5) - 4 \overset{?}{=} 0$

$(9 - 5)^2 - 3(9 - 5) - 4 \overset{?}{=} 0$

$4^2 - 3(4) - 4 \overset{?}{=} 0$

$16 - 12 - 4 \overset{?}{=} 0$

$0 = 0$ True

$z = 3$:

$((3)^2 - 5)^2 - 3((3)^2 - 5) - 4 \overset{?}{=} 0$

$(9 - 5)^2 - 3(9 - 5) - 4 \overset{?}{=} 0$

$4^2 - 3(4) - 4 \overset{?}{=} 0$

$16 - 12 - 4 \overset{?}{=} 0$

$0 = 0$ True

$z = -2$:

$((-2)^2 - 5)^2 - 3((-2)^2 - 5) - 4 \overset{?}{=} 0$

$(4 - 5)^2 - 3(4 - 5) - 4 \overset{?}{=} 0$

$(-1)^2 - 3(-1) - 4 \overset{?}{=} 0$

$1 + 3 - 4 \overset{?}{=} 0$

$0 = 0$ True

$z = 2$:

$((2)^2 - 5)^2 - 3((2)^2 - 5) - 4 \overset{?}{=} 0$

$(4 - 5)^2 - 3(4 - 5) - 4 \overset{?}{=} 0$

$(-1)^2 - 3(-1) - 4 \overset{?}{=} 0$

$1 + 3 - 4 \overset{?}{=} 0$

$0 = 0$ True

The solution set is $\{-3, -2, 2, 3\}$.   ●

Quick ✓

In Problems 7 and 8, solve each equation.

7. $(p^2 - 2)^2 - 9(p^2 - 2) + 14 = 0$ **8.** $2(2z^2 - 1)^2 + 5(2z^2 - 1) - 3 = 0$

When we raise both sides of an equation to an even power (such as when squaring both sides of the equation), we might introduce extraneous solutions to the equation. Therefore, it is important that we verify our solutions.

EXAMPLE 3   **Solving Equations That Are Quadratic in Form**

Solve: $3x - 5\sqrt{x} - 2 = 0$

Solution

$$3x - 5\sqrt{x} - 2 = 0$$

$$3(\sqrt{x})^2 - 5\sqrt{x} - 2 = 0$$

Let $u = \sqrt{x}$: $3u^2 - 5u - 2 = 0$

$$(3u + 1)(u - 2) = 0$$

$$3u + 1 = 0 \quad \text{or} \quad u - 2 = 0$$
$$3u = -1 \quad \text{or} \quad u = 2$$
$$u = -\frac{1}{3}$$

Replace u with $\sqrt{x}$ and solve for x: $\quad \sqrt{x} = -\frac{1}{3} \quad \text{or} \quad \sqrt{x} = 2$

Square both sides: $\quad x = \frac{1}{9} \quad \text{or} \quad x = 4$

Check

$x = \frac{1}{9}$: $\quad 3 \cdot \frac{1}{9} - 5\sqrt{\frac{1}{9}} - 2 = 0$

$$\frac{1}{3} - 5 \cdot \frac{1}{3} - 2 = 0$$

$$\frac{1}{3} - \frac{5}{3} - \frac{6}{3} = 0$$

$$-\frac{10}{3} = 0 \quad \text{False}$$

$x = 4$: $\quad 3 \cdot 4 - 5\sqrt{4} - 2 = 0$

$$12 - 5 \cdot 2 - 2 = 0$$

$$12 - 10 - 2 = 0$$

$$0 = 0 \quad \text{True}$$

The apparent solution $x = \frac{1}{9}$ is extraneous. The solution set is $\{4\}$. ●

Work Smart

The equation $\sqrt{x} = -\frac{1}{3}$ has no solution because the square root of a positive number is positive, the square root of 0 is 0, and the square root of a negative number is a non-real, complex number.

We could also have solved the equation in Example 3 using the methods introduced in Section 6.7 by isolating the radical and squaring both sides.

Quick ✓

In Problems 9 and 10, solve the equation.

9. $3w - 14\sqrt{w} + 8 = 0$

10. $2q - 9\sqrt{q} - 5 = 0$

EXAMPLE 4 **Solving Equations That Are Quadratic in Form**

Solve: $4x^{-2} + 13x^{-1} - 12 = 0$

Solution

$$4x^{-2} + 13x^{-1} - 12 = 0$$

$$4(x^{-1})^2 + 13x^{-1} - 12 = 0$$

Let $u = x^{-1}$: $\quad 4u^2 + 13u - 12 = 0$

Factor: $(4u - 3)(u + 4) = 0$

$$4u - 3 = 0 \quad \text{or} \quad u + 4 = 0$$
$$4u = 3 \quad \text{or} \quad u = -4$$
$$u = \frac{3}{4}$$

Replace u with x^{-1} and solve for x: $\quad x^{-1} = \frac{3}{4} \quad \text{or} \quad x^{-1} = -4$

$x^{-1} = \frac{1}{x}$: $\quad \frac{1}{x} = \frac{3}{4} \quad \text{or} \quad \frac{1}{x} = -4$

Take the reciprocal of both sides of the equation: $\quad x = \frac{4}{3} \quad \text{or} \quad x = \frac{1}{-4} = -\frac{1}{4}$

Check

$$x = \frac{4}{3}:$$

$$4\left(\frac{4}{3}\right)^{-2} + 13 \cdot \left(\frac{4}{3}\right)^{-1} - 12 \stackrel{?}{=} 0$$

$$4\left(\frac{3}{4}\right)^2 + 13 \cdot \frac{3}{4} - 12 \stackrel{?}{=} 0$$

$$4 \cdot \frac{9}{16} + \frac{39}{4} - 12 \stackrel{?}{=} 0$$

$$\frac{9}{4} + \frac{39}{4} - \frac{48}{4} \stackrel{?}{=} 0$$

$$0 = 0 \quad \text{True}$$

$$x = -\frac{1}{4}:$$

$$4\left(-\frac{1}{4}\right)^{-2} + 13 \cdot \left(-\frac{1}{4}\right)^{-1} - 12 \stackrel{?}{=} 0$$

$$4(-4)^2 + 13 \cdot (-4) - 12 \stackrel{?}{=} 0$$

$$4 \cdot 16 - 52 - 12 \stackrel{?}{=} 0$$

$$64 - 52 - 12 \stackrel{?}{=} 0$$

$$0 = 0 \quad \text{True}$$

The solution set is $\left\{-\frac{1}{4}, \frac{4}{3}\right\}$.

Quick ✓

In Problem 11, solve the equation.

11. $5x^{-2} + 12x^{-1} + 4 = 0$

EXAMPLE 5 **Solving Equations That Are Quadratic in Form**

Solve: $a^{\frac{2}{3}} + 3a^{\frac{1}{3}} - 28 = 0$

Solution

$$a^{\frac{2}{3}} + 3a^{\frac{1}{3}} - 28 = 0$$

$$\left(a^{\frac{1}{3}}\right)^2 + 3a^{\frac{1}{3}} - 28 = 0$$

Let $u = a^{\frac{1}{3}}$:

$$u^2 + 3u - 28 = 0$$

$$(u + 7)(u - 4) = 0$$

$$u + 7 = 0 \quad \text{or} \quad u - 4 = 0$$

$$u = -7 \quad \text{or} \quad u = 4$$

Replace u with $a^{\frac{1}{3}}$ and solve for a:

$$a^{\frac{1}{3}} = -7 \quad \text{or} \quad a^{\frac{1}{3}} = 4$$

Cube both sides of the equation:

$$\left(a^{\frac{1}{3}}\right)^3 = (-7)^3 \quad \text{or} \quad \left(a^{\frac{1}{3}}\right)^3 = 4^3$$

$$a = -343 \quad \text{or} \quad a = 64$$

Check $a = -343:$

$$(-343)^{\frac{2}{3}} + 3(-343)^{\frac{1}{3}} - 28 = 0$$

$$\left(\sqrt[3]{-343}\right)^2 + 3 \cdot \sqrt[3]{-343} - 28 \stackrel{?}{=} 0$$

$$(-7)^2 + 3 \cdot (-7) - 28 \stackrel{?}{=} 0$$

$$49 - 21 - 28 \stackrel{?}{=} 0$$

$$0 = 0 \quad \text{True}$$

$a = 64:$

$$(64)^{\frac{2}{3}} + 3(64)^{\frac{1}{3}} - 28 = 0$$

$$\left(\sqrt[3]{64}\right)^2 + 3 \cdot \sqrt[3]{64} - 28 \stackrel{?}{=} 0$$

$$4^2 + 3 \cdot 4 - 28 \stackrel{?}{=} 0$$

$$16 + 12 - 28 \stackrel{?}{=} 0$$

$$0 = 0 \quad \text{True}$$

The solution set is $\{-343, 64\}$.

Quick ✓

In Problem 12, solve the equation.

12. $p^{\frac{2}{3}} - 4p^{\frac{1}{3}} - 5 = 0$

7.3 Exercises   MyMathLab PRACTICE

Exercise numbers in green have complete video solutions in MyMathLab.

*Problems **1–12** are the* Quick ✓ *s that follow the* EXAMPLES.

Building Skills

In Problems 13–48, solve each equation. See Objective 1.

13. $x^4 - 5x^2 + 4 = 0$

14. $x^4 - 10x^2 + 9 = 0$

15. $q^4 + 13q^2 + 36 = 0$

16. $z^4 + 10z^2 + 9 = 0$

17. $4a^4 - 17a^2 + 4 = 0$

18. $4b^4 - 5b^2 + 1 = 0$

19. $p^4 + 6 = 5p^2$

20. $q^4 + 15 = 8q^2$

21. $(x - 3)^2 - 6(x - 3) - 7 = 0$

22. $(x + 2)^2 - 3(x + 2) - 10 = 0$

23. $(x^2 - 1)^2 - 11(x^2 - 1) + 24 = 0$

24. $(p^2 - 2)^2 - 8(p^2 - 2) + 12 = 0$

25. $(y^2 + 2)^2 + 7(y^2 + 2) + 10 = 0$

26. $(q^2 + 4)^2 + 3(q^2 + 4) - 4 = 0$

27. $x - 3\sqrt{x} - 4 = 0$

28. $x - 5\sqrt{x} - 6 = 0$

29. $w + 5\sqrt{w} + 6 = 0$

30. $z + 7\sqrt{z} + 6 = 0$

31. $2x + 5\sqrt{x} = 3$

32. $3x = 11\sqrt{x} + 4$

33. $x^{-2} + 3x^{-1} = 28$

34. $q^{-2} + 2q^{-1} = 15$

35. $10z^{-2} + 11z^{-1} = 6$

36. $10a^{-2} + 23a^{-1} = 5$

37. $x^{\frac{2}{3}} + 3x^{\frac{1}{3}} - 4 = 0$

38. $y^{\frac{2}{3}} - 2y^{\frac{1}{3}} - 3 = 0$

39. $z^{\frac{2}{3}} - z^{\frac{1}{3}} = 2$

40. $w^{\frac{2}{3}} + 2w^{\frac{1}{3}} = 3$

41. $a + a^{\frac{1}{2}} = 30$

42. $b + 3b^{\frac{1}{2}} = 28$

43. $\dfrac{1}{x^2} - \dfrac{5}{x} + 6 = 0$

44. $\dfrac{1}{x^2} - \dfrac{7}{x} + 12 = 0$

45. $\left(\dfrac{1}{x + 2}\right)^2 + \dfrac{4}{x + 2} = 5$

46. $\left(\dfrac{1}{x + 2}\right)^2 + \dfrac{6}{x + 2} = 7$

47. $p^6 - 28p^3 + 27 = 0$

48. $y^6 - 7y^3 - 8 = 0$

Mixed Practice

In Problems 49–62, solve each equation.

49. $8a^{-2} + 2a^{-1} = 1$

50. $6b^{-2} - b^{-1} = 1$

51. $z^4 = 4z^2 + 32$

52. $x^4 + 3x^2 = 4$

53. $x^{\frac{1}{2}} + x^{\frac{1}{4}} - 6 = 0$

54. $c^{\frac{1}{2}} + c^{\frac{1}{4}} - 12 = 0$

55. $w^4 - 5w^2 - 36 = 0$

56. $p^4 - 15p^2 - 16 = 0$

57. $\left(\dfrac{1}{x + 3}\right)^2 + \dfrac{2}{x + 3} = 3$

58. $\left(\dfrac{1}{x - 1}\right)^2 + \dfrac{7}{x - 1} = 8$

59. $x - 7\sqrt{x} + 12 = 0$

60. $x - 8\sqrt{x} + 12 = 0$

61. $2(x - 1)^2 - 7(x - 1) = 4$

62. $3(y - 2)^2 - 4(y - 2) = 4$

63. Suppose that $f(x) = x^4 + 7x^2 + 12$. Find the values of x such that
   (a) $f(x) = 12$ **(b)** $f(x) = 6$

64. Suppose that $f(x) = x^4 + 5x^2 + 3$. Find the values of x such that
   (a) $f(x) = 3$ **(b)** $f(x) = 17$

65. Suppose that $g(x) = 2x^4 - 6x^2 - 5$. Find the values of x such that
    (a) $g(x) = -5$ **(b)** $g(x) = 15$

66. Suppose that $h(x) = 3x^4 - 9x^2 - 8$. Find the values of x such that
    (a) $h(x) = -8$ **(b)** $h(x) = 22$

67. Suppose that $F(x) = x^{-2} - 5x^{-1}$. Find the values of x such that
    (a) $F(x) = 6$ **(b)** $F(x) = 14$

68. Suppose that $f(x) = x^{-2} - 3x^{-1}$. Find the values of x such that
    (a) $f(x) = 4$ **(b)** $f(x) = 18$

In Problems 69–74, find the zeros of the function. (Hint: Remember, r is a zero if f(r) = 0.)

69. $f(x) = x^4 + 9x^2 + 14$

70. $f(x) = x^4 - 13x^2 + 42$

71. $g(t) = 6t - 25\sqrt{t} - 9$

72. $h(p) = 8p - 18\sqrt{p} - 35$

73. $s(d) = \dfrac{1}{(d+3)^2} - \dfrac{4}{d+3} + 3$

74. $f(a) = \dfrac{1}{(a-2)^2} + \dfrac{3}{a-2} - 4$

Applying the Concepts

75. (a) Solve $x^2 - 5x + 6 = 0$.

    (b) Solve $(x-3)^2 - 5(x-3) + 6 = 0$. Compare the solutions to part (a).

    (c) Solve $(x+2)^2 - 5(x+2) + 6 = 0$. Compare the solutions to part (a).

    (d) Solve $(x-5)^2 - 5(x-5) + 6 = 0$. Compare the solutions to part (a).

    (e) Conjecture a generalization for the solution of $(x-a)^2 - 5(x-a) + 6 = 0$.

76. (a) Solve $x^2 + 3x - 18 = 0$.

    (b) Solve $(x-1)^2 + 3(x-1) - 18 = 0$. Compare the solutions to part (a).

    (c) Solve $(x+5)^2 + 3(x+5) - 18 = 0$. Compare the solutions to part (a).

    (d) Solve $(x-3)^2 + 3(x-3) - 18 = 0$. Compare the solutions to part (a).

    (e) Conjecture a generalization for the solution of $(x-a)^2 + 3(x-a) - 18 = 0$.

77. Consider the function $f(x) = 2x^2 - 3x + 1$.

    (a) Solve $f(x) = 0$.

    (b) Solve $f(x-2) = 0$. Compare the solutions to part (a).

    (c) Solve $f(x-5) = 0$. Compare the solutions to part (a).

    (d) Conjecture a generalization for the zeros of $f(x-a)$.

78. Consider the function $f(x) = 3x^2 - 5x - 2$.

    (a) Solve $f(x) = 0$.

    (b) Solve $f(x-1) = 0$. Compare the solutions to part (a).

    (c) Solve $f(x-4) = 0$. Compare the solutions to part (a).

    (d) Conjecture a generalization for the zeros of $f(x-a)$.

79. Revenue The function
$$R(x) = \frac{(x-1990)^2}{2} + \frac{3(x-1990)}{2} + 3000$$
models the revenue R (in thousands of dollars) of a start-up computer consulting firm in year x, where $x \geq 1990$.

    (a) Determine and interpret $R(1990)$.

    (b) Solve and interpret $R(x) = 3065$.

    (c) According to the model, in what year can the firm expect to receive $3350 thousand in revenue?

80. Revenue The function
$$R(x) = \frac{(x-2000)^2}{3} + \frac{5(x-2000)}{3} + 2000$$
models the revenue R (in thousands of dollars) of a start-up computer software firm in year x, where $x \geq 2000$.

    (a) Determine and interpret $R(2000)$.

    (b) Solve and interpret $R(x) = 2250$.

    (c) According to the model, in what year can the firm expect to receive $2350 thousand in revenue?

Extending the Concepts

All of the problems given in this section resulted in equations quadratic in form that could be factored after the appropriate substitution. However, this is not a necessary requirement to solving equations quadratic in form. In Problems 81–84, determine the appropriate substitution, and then use the quadratic formula to find the value of u. Finally, determine the value of the variable in the equation.

81. $x^4 + 5x^2 + 2 = 0$

82. $x^4 + 7x^2 + 4 = 0$

83. $2(x - 2)^2 + 8(x - 2) - 1 = 0$

84. $3(x + 1)^2 + 6(x + 1) - 1 = 0$

Explaining the Concepts

85. The equation $x - 5\sqrt{x} - 6 = 0$ can be solved either by using the methods of this section or by isolating the radical and squaring both sides. Solve it both ways and explain which approach you prefer.

86. Explain the steps required to solve an equation quadratic in form. Be sure to include an explanation of how to identify the appropriate substitution.

87. Under what circumstances might extraneous solutions occur when one is solving equations quadratic in form?

Synthesis Review

In Problems 88–91, add or subtract the expressions.

88. $(4x^2 - 3x - 1) + (-3x^2 + x + 5)$

89. $(3p^{-2} - 4p^{-1} + 8) - (2p^{-2} - 8p^{-1} - 1)$

90. $3\sqrt{2x} - \sqrt{8x} + \sqrt{50x}$

91. $\sqrt[3]{16a} + \sqrt[3]{54a} - \sqrt[3]{128a^4}$

92. Write a sentence or two that discusses how to add or subtract algebraic expressions, in general.

The Graphing Calculator

In Problems 93–98, use a graphing calculator to find the real solutions to the equations using either the ZERO or the INTERSECT feature. Round your answers to two decimal places, if necessary.

93. $x^4 + 5x^2 - 14 = 0$ **94.** $x^4 - 4x^2 - 12 = 0$

95. $2(x - 2)^2 = 5(x - 2) + 1$

96. $3(x + 3)^2 = 2(x + 3) + 6$

97. $x - 5\sqrt{x} = -3$ **98.** $x + 4\sqrt{x} = 5$

99. (a) Graph $Y_1 = x^2 - 5x - 6$. Find the x-intercepts of the graph.
  (b) Graph $Y_1 = (x + 2)^2 - 5(x + 2) - 6$. Find the x-intercepts of the graph.
  (c) Graph $Y_1 = (x + 5)^2 - 5(x + 5) - 6$. Find the x-intercepts of the graph.
  (d) Make a generalization based upon the results of parts (a), (b), and (c).

100. (a) Graph $Y_1 = x^2 + 4x + 3$. Find the x-intercepts of the graph.
  (b) Graph $Y_1 = (x - 3)^2 + 4(x - 3) + 3$. Find the x-intercepts of the graph.
  (c) Graph $Y_1 = (x - 6)^2 + 4(x - 6) + 3$. Find the x-intercepts of the graph.
  (d) Make a generalization based upon the results of parts (a), (b), and (c).

Putting the Concepts Together (Sections 7.1–7.3)

We designed these problems so that you can review Sections 7.1–7.3 and show your mastery of the concepts. Take time to work these problems before proceeding with the next section. The answers are located at the back of the text on page AN-41.

In Problems 1–3, complete the square in the given expression. Then factor the perfect square trinomial.

1. $z^2 + 10z$

2. $x^2 + 7x$

3. $n^2 - \frac{1}{4}n$

In Problems 4–6, solve each quadratic equation using the stated method.

4. $(2x - 3)^2 - 5 = -1$; Square Root Property

5. $x^2 + 8x + 4 = 0$; completing the square

6. $x(x - 6) = -7$; quadratic formula

In Problems 7–10, solve each equation using the method you prefer.

7. $49x^2 - 80 = 0$

8. $p^2 - 8p + 6 = 0$

9. $3y^2 + 6y + 4 = 0$

10. $\frac{1}{4}n^2 + n = \frac{1}{6}$

In Problems 11–13, determine the discriminant of each quadratic equation. Use the value of the discriminant to determine whether the equation has two rational solutions, two irrational solutions, one repeated real solution, or two complex solutions that are not real.

11. $9x^2 + 12x + 4 = 0$ **12.** $3x^2 + 6x - 2 = 0$

13. $2x^2 + 6x + 5 = 0$

14. Find the missing length in the right triangle shown below.

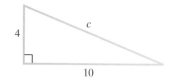

In Problems 15 and 16, solve each equation.

15. $2m + 7\sqrt{m} - 15 = 0$

16. $p^{-2} - 3p^{-1} - 18 = 0$

17. Revenue The revenue R received by a company selling x microwave ovens per day is given by the function $R(x) = -0.4x^2 + 140x$. How many microwave ovens must be sold for revenue to be $12,000 per day?

18. Airplane Ride An airplane flies 300 miles into the wind and then flies home against the wind. The total time of the trip (excluding time on the ground) is 5 hours. If the plane can fly 140 miles per hour in still air, what was the speed of the wind? Round your answer to the nearest tenth.

7.4 Graphing Quadratic Functions Using Transformations

Objectives

❶ Graph Quadratic Functions of the Form $f(x) = x^2 + k$

❷ Graph Quadratic Functions of the Form $f(x) = (x - h)^2$

❸ Graph Quadratic Functions of the Form $f(x) = ax^2$

❹ Graph Quadratic Functions of the Form $f(x) = ax^2 + bx + c$

❺ Find a Quadratic Function from Its Graph

Are You Ready for This Section?

Before getting started, take the following readiness quiz. If you get a problem wrong, go back to the section cited and review the material.

R1. Graph $y = x^2$ using point plotting.   [Section 1.5, pp. 96–97]

R2. Use the point-plotting method to graph $y = x^2 - 3$.   [Section 1.5, pp. 96–98]

R3. What is the domain of $f(x) = 2x^2 + 5x + 1$?   [Section 2.2, pp. 158–159]

We begin with a definition.

Definition

A **quadratic function** is a function of the form

$$f(x) = ax^2 + bx + c$$

where a, b, and c are real numbers and $a \neq 0$. The domain of a quadratic function consists of all real numbers.

Many situations can be modeled using quadratic functions. For example, we saw in Example 10 of Section 2.2 that Franco Modigliani used the quadratic function $I(a) = -55a^2 + 5119a - 54,448$ to model the relation between average annual income, I, and age, a.

Also, if we ignore the effect of air resistance on a projectile, then the height of the projectile as a function of horizontal distance traveled can be modeled using a quadratic function. See Figure 7.

In this section and the next section, we will learn new techniques for graphing quadratic functions. In Section 1.5 we learned how to graph equations using

Ready?...Answers

R1.

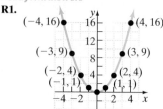

R2.

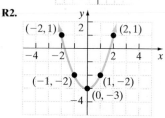

R3. The set of all real numbers

Figure 7

point plotting, but we also discovered that this method is inefficient and can lead to incomplete graphs. Remember, a complete graph shows all of its "interesting features," such as its intercepts and high and low points. We will present two methods for graphing quadratic functions that are superior to the point-plotting method. The first method uses *transformations* and is the subject of this section. The second method, discussed in the next section, uses properties of quadratic functions.

▶ ① Graph Quadratic Functions of the Form $f(x) = x^2 + k$

Work Smart

Consider the quadratic function $f(x) = ax^2 + bx + c$, where $a = 1$, $b = 0$, and c is any real number. This is a function of the form $f(x) = x^2 + k$.

We begin by learning how to graph any quadratic function of the form $f(x) = x^2 + k$, such as $g(x) = x^2 + 3$ or $h(x) = x^2 - 4$.

First, let's review the graph of $y = g(x) = x^2$ in Figure 8, which we first saw in Section 1.5. What effect does adding a real number k to the function $g(x) = x^2$ have on its graph? Let's see!

Figure 8
$y = g(x) = x^2$

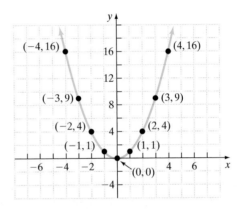

EXAMPLE 1 | **Graphing a Quadratic Function of the Form $f(x) = x^2 + k$**

In the same Cartesian plane, graph $g(x) = x^2$ and $f(x) = x^2 + 3$.

Solution

Begin by obtaining some points on the graphs of g and f. For example, when $x = 0$, then $y = g(0) = 0$ and $y = f(0) = 0^2 + 3 = 3$. When $x = 1$, then $y = g(1) = 1$ and $y = f(1) = 1^2 + 3 = 4$. Table 3 lists these points along with a few others. Notice that the y-coordinates on the graph of $f(x) = x^2 + 3$ are exactly 3 units greater than the corresponding y-coordinates on the graph of $g(x) = x^2$. Figure 9 shows the graphs of f and g.

Figure 9

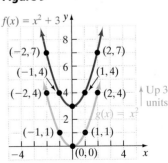

Table 3

x	$g(x) = x^2$	$(x, g(x))$	$f(x) = x^2 + 3$	$(x, f(x))$
-2	$(-2)^2 = 4$	$(-2, 4)$	$(-2)^2 + 3 = 7$	$(-2, 7)$
-1	$(-1)^2 = 1$	$(-1, 1)$	$(-1)^2 + 3 = 4$	$(-1, 4)$
0	0	$(0, 0)$	3	$(0, 3)$
1	1	$(1, 1)$	4	$(1, 4)$
2	4	$(2, 4)$	7	$(2, 7)$

The graph of f is identical to the graph of g, except that it is shifted up 3 units.

Let's look at another example.

EXAMPLE 2 **Graphing a Quadratic Function of the Form** $f(x) = x^2 - k$

In the same Cartesian plane, graph $g(x) = x^2$ and $f(x) = x^2 - 4$.

Solution

Table 4 lists some points on the graphs of g and f. Notice that the y-coordinates on the graph of $f(x) = x^2 - 4$ are exactly 4 units less than the corresponding y-coordinates on the graph of $g(x) = x^2$. Figure 10 shows the graphs of f and g.

Figure 10

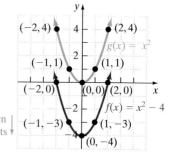

Table 4

x	$g(x) = x^2$	$(x, g(x))$	$f(x) = x^2 - 4$	$(x, f(x))$
−2	4	(−2, 4)	$(-2)^2 - 4 = 0$	(−2, 0)
−1	1	(−1, 1)	$(-1)^2 - 4 = -3$	(−1, −3)
0	0	(0, 0)	−4	(0, −4)
1	1	(1, 1)	−3	(1, −3)
2	4	(2, 4)	0	(2, 0)

The graph of f is identical to the graph of g except that f is shifted down 4 units.

The results of Examples 1 and 2 lead us to the following conclusion.

Graphing a Function of the Form $f(x) = x^2 + k$ **or** $f(x) = x^2 - k$

To obtain the graph of $f(x) = x^2 + k$, $k > 0$, from the graph of $y = x^2$, shift the graph of $y = x^2$ up k units. To obtain the graph of $f(x) = x^2 - k$, $k > 0$, from the graph of $y = x^2$, shift the graph of $y = x^2$ down k units.

Quick ✓

1. A _____ _____ is a function of the form $f(x) = ax^2 + bx + c$, where a, b, and c are real numbers and $a \neq 0$.

2. To graph $f(x) = x^2 + k$, $k > 0$, from the graph of $y = x^2$, shift the graph of $y = x^2$ ___ k units. To graph $f(x) = x^2 - k$, $k > 0$, from the graph of $y = x^2$ shift the graph of $y = x^2$ _____ k units.

In Problems 3 and 4, use the graph of $y = x^2$ to graph the quadratic function. Show at least three points on the graph.

3. $f(x) = x^2 + 5$ 　　　　　　**4.** $f(x) = x^2 - 2$

▶ ❷ **Graph Quadratic Functions of the Form** $f(x) = (x - h)^2$

We now look at the graph of any quadratic function of the form $f(x) = (x - h)^2$ such as $f(x) = (x + 3)^2$ or $f(x) = (x - 2)^2$. How does subtracting a real positive number h from x affect the graph of the function $g(x) = x^2$?

EXAMPLE 3   **Graphing a Quadratic Function of the Form** $f(x) = (x - h)^2$

In the same Cartesian plane, graph $g(x) = x^2$ and $f(x) = (x - 2)^2$.

Solution

Again, use the point-plotting method. Table 5 lists some points on the graphs of g and f. Note that when $g(x) = 0$, $x = 0$, and when $f(x) = 0$, $x = 2$. Also, when $g(x) = 4$, $x = -2$ or 2, and when $f(x) = 4$, $x = 0$ or 4. The graph of f is identical to that of g, except that f is shifted 2 units to the right of g. See Figure 11.

Figure 11

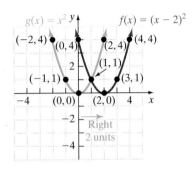

Table 5

x	$g(x) = x^2$	$(x, g(x))$	$f(x) = (x - 2)^2$	$(x, f(x))$
-2	4	$(-2, 4)$	$(-2 - 2)^2 = 16$	$(-2, 16)$
-1	1	$(-1, 1)$	$(-1 - 2)^2 = 9$	$(-1, 9)$
0	0	$(0, 0)$	4	$(0, 4)$
1	1	$(1, 1)$	1	$(1, 1)$
2	4	$(2, 4)$	0	$(2, 0)$
3	9	$(3, 9)$	1	$(3, 1)$
4	16	$(4, 16)$	4	$(4, 4)$

Why does this happen? Because we are subtracting 2 from each x-value in the function $f(x) = (x - 2)^2$, the x-values must be greater by 2 in order to obtain the same y-value that was obtained in the graph of $g(x) = x^2$.

What if we add a positive number h to x?

EXAMPLE 4   **Graphing a Quadratic Function of the Form** $f(x) = (x + h)^2$

In the same Cartesian plane, graph $g(x) = x^2$ and $f(x) = (x + 3)^2$.

Solution

Table 6 lists some points on the graphs of g and f. Notice that when $g(x) = 0$, then $x = 0$, and when $f(x) = 0$, then $x = -3$. Also, when $g(x) = 4$, then $x = -2$ or 2, and when $f(x) = 4$, then $x = -5$ or -1. The graph of f is identical to that of g, except that f is shifted 3 units to the left of g. See Figure 12.

Figure 12

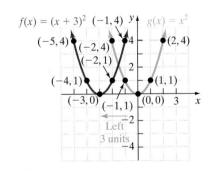

Table 6

x	$g(x) = x^2$	$(x, g(x))$	$f(x) = (x + 3)^2$	$(x, f(x))$
-5	25	$(-5, 25)$	$(-5 + 3)^2 = 4$	$(-5, 4)$
-4	16	$(-4, 16)$	$(-4 + 3)^2 = 1$	$(-4, 1)$
-3	9	$(-3, 9)$	$(-3 + 3)^2 = 0$	$(-3, 0)$
-2	4	$(-2, 4)$	1	$(-2, 1)$
-1	1	$(-1, 1)$	4	$(-1, 4)$
0	0	$(0, 0)$	9	$(0, 9)$
1	1	$(1, 1)$	16	$(1, 16)$
2	4	$(2, 4)$	25	$(2, 25)$

Why does this happen? Because we are adding 3 to each x-value in the function $f(x) = (x + 3)^2$, the x-values must be smaller by 3 in order to obtain the same y-value that was obtained in the graph of $g(x) = x^2$.

Examples 3 and 4 lead us to the following conclusion.

> **Graphing a Function of the Form $f(x) = (x - h)^2$ or $f(x) = (x + h)^2$**
>
> To obtain the graph of $f(x) = (x - h)^2, h > 0,$ from the graph of $y = x^2$, shift the graph of $y = x^2$ to the right h units. To obtain the graph of $f(x) = (x + h)^2, h > 0,$ from the graph of $y = x^2$, shift the graph of $y = x^2$ to the left h units.

Quick ✓

5. *True or False* To obtain the graph of $f(x) = (x + 12)^2$ from the graph of $y = x^2$, shift the graph of $y = x^2$ to the right 12 units.

In Problems 6 and 7, use the graph of $y = x^2$ to graph the quadratic function. Show at least three points on the graph.

6. $f(x) = (x + 5)^2$

7. $f(x) = (x - 1)^2$

Let's do an example where we combine a horizontal shift and a vertical shift.

EXAMPLE 5 **Combining Horizontal and Vertical Shifts**

Graph the function $f(x) = (x + 2)^2 - 3$.

Solution

We will graph f in steps. Begin with the graph of $y = x^2$ as shown in Figure 13(a). Shift that graph 2 units to the left to get the graph of $y = (x + 2)^2$ shown in Figure 13(b). Then shift the graph of $y = (x + 2)^2$ down 3 units to get the graph of $y = (x + 2)^2 - 3$ shown in Figure 13(c). Notice that we keep track of key points plotted on each graph.

Figure 13

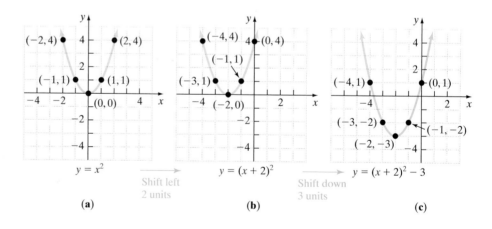

Note: The order in which we perform the steps to obtain the graph does not matter. In Example 5, we could just as easily have shifted down 3 units first and then shifted left 2 units.

Quick ✓

In Problems 8 and 9, graph each quadratic function using horizontal and vertical shifts. Show at least three points on the graph.

8. $f(x) = (x - 3)^2 + 2$

9. $f(x) = (x + 1)^2 - 4$

▶ ❸ Graph Quadratic Functions of the Form $f(x) = ax^2$

So far, the coefficient of the square term has been equal to 1. What impact does the value of a have on the graph of $f(x) = ax^2 + bx + c$? To answer this question, we will consider quadratic functions of the form $f(x) = ax^2, a \neq 0$.

First, let's consider situations in which the value of a is positive. Table 7 shows points on the graphs of $f(x) = x^2, g(x) = \frac{1}{2}x^2$, and $h(x) = 2x^2$. Figure 14 shows the graphs of f, g, and h. First, notice that all three graphs open "up." Next, notice that the values of the y-coordinates on the graph of g are exactly $\frac{1}{2}$ of the values of the y-coordinates on the graph of f. The values of the y-coordinates on the graph of h are exactly 2 times the values of the y-coordinates on the graph of f. Put another way, the larger the value of a, the "taller" the graph is, and the smaller the value of a, the "shorter" the graph is.

Work Smart

By "taller," we mean that points with the same x-values have greater y-values. By "shorter," we mean that points with the same x-values have smaller y-values.

Figure 14
$f(x) = ax^2$. Since $a > 0$, the graphs open up.

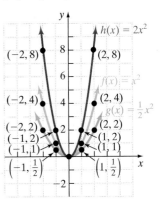

Table 7

x	$f(x) = x^2$	$g(x) = \frac{1}{2}x^2$	$h(x) = 2x^2$
-2	$(-2)^2 = 4$	$\frac{1}{2}(-2)^2 = 2$	$2(-2)^2 = 8$
-1	$(-1)^2 = 1$	$\frac{1}{2}(-1)^2 = \frac{1}{2}$	$2(-1)^2 = 2$
0	0	0	0
1	1	$\frac{1}{2}$	2
2	4	2	8

Now let's consider what happens when a is negative. Table 8 shows points on the graphs of $f(x) = -x^2, g(x) = -\frac{1}{2}x^2$, and $h(x) = -2x^2$. Figure 15 shows the graphs of f, g, and h. First, notice that all three graphs open "down." Next, notice that the values of the y-coordinates on the graph of g are exactly $\frac{1}{2}$ times the values of the y-coordinates on the graph of f. The values of the y-coordinates on the graph of h are exactly 2 times the values of the y-coordinates on the graph of f. Put another way, the larger the value of $|a|$, the "taller" the graph is, and the smaller the value of $|a|$, the "shorter" the graph is.

Figure 15
$f(x) = ax^2$. Since $a < 0$, the graphs open down.

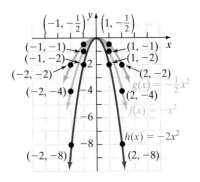

Table 8

x	$f(x) = -x^2$	$g(x) = -\frac{1}{2}x^2$	$h(x) = -2x^2$
-2	$-(-2)^2 = -4$	$-\frac{1}{2}(-2)^2 = -2$	$-2(-2)^2 = -8$
-1	$-(-1)^2 = -1$	$-\frac{1}{2}(-1)^2 = -\frac{1}{2}$	$-2(-1)^2 = -2$
0	0	0	0
1	-1	$-\frac{1}{2}$	-2
2	-4	-2	-8

We summarize these conclusions on the following page.

Properties of the Graph of $f(x) = ax^2$

- If $a > 0$, the graph of $f(x) = ax^2$ opens up. In addition, if $0 < a < 1$ (a is between 0 and 1), the graph will be "shorter" than that of $y = x^2$. If $a > 1$, the graph will be "taller" than that of $y = x^2$.
- If $a < 0$, the graph of $f(x) = ax^2$ will open down. In addition, if $0 < |a| < 1$, the graph will be "shorter" than that of $y = x^2$. If $|a| > 1$, the graph will be "taller" than that of $y = x^2$.
- When $|a| > 1$, we say that the graph is **vertically stretched** by a factor of $|a|$. When $0 < |a| < 1$, we say that the graph is **vertically compressed** by a factor of $|a|$.

Graphing a Function of the Form $f(x) = ax^2$

To obtain the graph of $f(x) = ax^2$ from the graph of $y = x^2$, multiply each y-coordinate on the graph of $y = x^2$ by a.

EXAMPLE 6   **Graphing a Quadratic Function of the Form $f(x) = ax^2$**

Use the graph of $y = x^2$ to obtain the graph of $f(x) = -2x^2$.

Solution

To obtain the graph of $f(x) = -2x^2$ from the graph of $y = x^2$, we multiply each y-coordinate on the graph of $y = x^2$ by -2 (the value of a). See Figure 16.

Figure 16

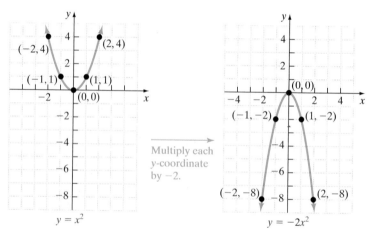

Quick ✓

10. When obtaining the graph of $f(x) = ax^2$ from the graph of $y = x^2$, we multiply each _-coordinate on the graph of $y = x^2$ by _. If $|a| > 1$, we say that the graph is _____ _____ by a factor of $|a|$. If $0 < |a| < 1$, we say that the graph is _____ _____ by a factor of $|a|$.

In Problems 11 and 12, use the graph of $y = x^2$ to graph each quadratic function. Label at least 3 points on the graph.

11. $f(x) = 3x^2$ 12. $f(x) = -\dfrac{1}{4}x^2$

❹ Graph Quadratic Functions of the Form $f(x) = ax^2 + bx + c$

▶ The graphs in Examples 1–6 are typical of graphs of all quadratic functions. We call the graph of a quadratic function a **parabola** (pronounced puh-răb-uh-luh). Figure 17 shows two parabolas.

Figure 17

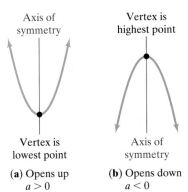

Axis of symmetry

Vertex is highest point

Vertex is lowest point

Axis of symmetry

(a) Opens up
$a > 0$

(b) Opens down
$a < 0$

The parabola in Figure 17(a) **opens up** (since $a > 0$) and has a lowest point; the parabola in Figure 17(b) **opens down** (since $a < 0$) and has a highest point. The lowest or highest point of a parabola is the **vertex**. The vertical line passing through the vertex of a parabola is its **axis of symmetry**. If we were to take the portion of the parabola to the right of the vertex and fold it over the axis of symmetry, it would lie directly on top of the portion of the parabola to the left of the vertex. Therefore, we say the parabola is symmetric about its axis of symmetry. Note that the axis of symmetry is not part of the graph of the quadratic function, but it will be useful when we graph quadratic functions using the methods we present in the next section.

Our goal right now is to combine the techniques from Examples 1–6 to graph any quadratic function. The techniques of shifting horizontally, shifting vertically, stretching, and compressing are collectively referred to as **transformations.** To graph any quadratic function of the form $f(x) = ax^2 + bx + c$ using transformations, we take the following steps.

Work Smart

Although order does not matter, we recommend that, when graphing parabolas using transformations, you obtain the graph that results from the vertical compression or stretch first, followed by the horizontal shift, followed by the vertical shift.

> **Graphing Quadratic Functions Using Transformations**
>
> **Step 1:** Write the function $f(x) = ax^2 + bx + c$ as $f(x) = a(x - h)^2 + k$ by completing the square in x.
>
> **Step 2:** Graph the function $f(x) = a(x - h)^2 + k$ using transformations.

Notice that we have to write the quadratic function $f(x) = ax^2 + bx + c$ as $f(x) = a(x - h)^2 + k$ so that we can determine the horizontal and vertical shifts.

EXAMPLE 7 **How to Graph a Quadratic Function of the Form $f(x) = ax^2 + bx + c$ Using Transformations**

Graph $f(x) = x^2 + 4x + 3$ using transformations. Identify the vertex and axis of symmetry of the parabola. Based on the graph, determine the domain and range of the quadratic function.

Step-by-Step Solution

Step 1: Write the function $f(x) = ax^2 + bx + c$ as $f(x) = a(x - h)^2 + k$ by completing the square in x.

$$f(x) = x^2 + 4x + 3$$
$$= (x^2 + 4x) + 3$$

Group the terms involving x:

Complete the square in x by taking $\dfrac{1}{2}$ the coefficient of x and squaring the result: $\left(\dfrac{1}{2} \cdot 4\right)^2 = 4$.

Because we added 4, we must also subtract 4 so that we don't change the function:

$$= (x^2 + 4x + 4) + 3 - 4$$
$$= (x^2 + 4x + 4) - 1$$

Factor the perfect square trinomial in parentheses:

$$= (x + 2)^2 - 1$$

Step 2: Graph the function $f(x) = a(x - h)^2 + k$ using transformations.

The graph of $f(x) = (x + 2)^2 - 1$ is the graph of $y = x^2$ shifted 2 units left and 1 unit down. See Figure 18.

Figure 18

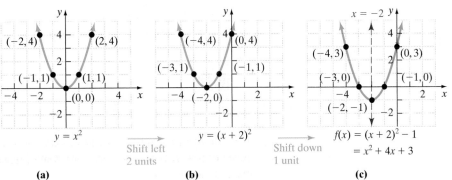

$y = x^2$

$(-2, 4)$ $(2, 4)$ $(-1, 1)$ $(1, 1)$ $(0, 0)$

(a)

Shift left 2 units

$y = (x + 2)^2$

$(-4, 4)$ $(0, 4)$ $(-3, 1)$ $(-1, 1)$ $(-2, 0)$

(b)

Shift down 1 unit

$x = -2$

$f(x) = (x + 2)^2 - 1$
$= x^2 + 4x + 3$

$(-4, 3)$ $(0, 3)$ $(-3, 0)$ $(-1, 0)$ $(-2, -1)$

(c)

From the graph in Figure 18(c), we can see that the vertex of the parabola is $(-2, -1)$. The axis of symmetry is the line $x = -2$. The domain is the set of all real numbers, or $(-\infty, \infty)$. The range is $\{y | y \geq -1\}$, or $[-1, \infty)$.

▶ **EXAMPLE 8** **Graphing a Quadratic Function of the Form $f(x) = ax^2 + bx + c$ Using Transformations**

Graph $f(x) = -2x^2 + 4x + 1$ using transformations. Identify the vertex and axis of symmetry of the parabola. Based on the graph, determine the domain and range of the quadratic function.

Solution

Write the function $f(x) = ax^2 + bx + c$ as $f(x) = a(x - h)^2 + k$ by completing the square in x.

$$f(x) = -2x^2 + 4x + 1$$

Group the terms involving x: $\quad = (-2x^2 + 4x) + 1$

Factor out the coefficient of x^2, -2, from the parentheses: $\quad = -2(x^2 - 2x) + 1$

Complete the square in x by taking $\dfrac{1}{2}$ the coefficient of x and squaring the result: $\left(\dfrac{1}{2} \cdot -2\right)^2 = 1$. Add 1 inside the parentheses. Because 1 is multiplied by -2, we really subtracted 2, so we must add 2 to offset this: $\quad = -2(x^2 - 2x + 1) + 1 + 2$

Factor the perfect square trinomial in parentheses: $\quad = -2(x - 1)^2 + 3$

Now, graph the function $f(x) = a(x - h)^2 + k$ using transformations. Since $a = -2$, the parabola opens down and is stretched by a factor of 2. Since we are subtracting 1 from x, the parabola shifts 1 unit to the right; since $k = 3$, the parabola shifts 3 units up. See Figure 19.

Figure 19

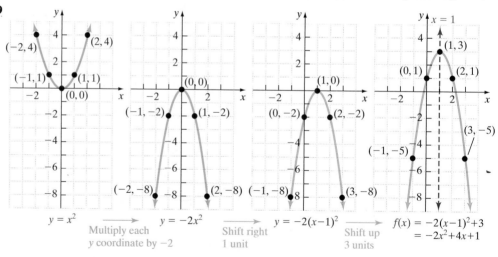

$$y = x^2 \xrightarrow[\substack{\text{Multiply each} \\ y \text{ coordinate by } -2}]{} y = -2x^2 \xrightarrow[\substack{\text{Shift right} \\ 1 \text{ unit}}]{} y = -2(x-1)^2 \xrightarrow[\substack{\text{Shift up} \\ 3 \text{ units}}]{} \begin{aligned} f(x) &= -2(x-1)^2 + 3 \\ &= -2x^2 + 4x + 1 \end{aligned}$$

The vertex of the parabola is $(1, 3)$. The axis of symmetry is the line $x = 1$. The domain is the set of all real numbers, or $(-\infty, \infty)$. The range is $\{y | y \leq 3\}$, or $(-\infty, 3]$.

Work Smart: Study Skills

Be sure you know what $y = a(x - h)^2 + k$ represents in reference to the graph of $y = x^2$:

- $|a|$ represents the vertical stretch or compression factor: how tall or short the graph appears.
- The sign of a determines whether the parabola opens up or down.
- h represents the number of units the graph is shifted horizontally.
- k represents the number of units the graph is shifted vertically.
- The vertex of the parabola has coordinates (h, k).

For example, $y = 4(x + 2)^2 - 3$ means that the graph of $y = x^2$ is stretched vertically by a factor of 4, is shifted 2 units to the left, and is shifted down 3 units. The vertex of $y = 4(x + 2)^2 - 3$ is at $(-2, -3)$ and represents the low point of the graph since $a > 0$.

Quick ✓

13. *True or False* The graph of $f(x) = -3x^2 + x + 6$ opens down.

In Problems 14 and 15, graph each quadratic function using transformations. Be sure to label at least three points on the graph. Based on the graph, determine the domain and range of each function.

14. $f(x) = -3(x + 2)^2 + 1$

15. $f(x) = 2x^2 - 8x + 5$

▶ ⑤ Find a Quadratic Function from Its Graph

In Example 7, we graphed the quadratic function $f(x) = x^2 + 4x + 3 = (x + 2)^2 - 1$. Notice that the vertex of the parabola is $(-2, -1)$. If we write f as $f(x) = (x - (-2))^2 - 1$ and compare it to $f(x) = a(x - h)^2 + k$, we see that $h = -2$ and $k = -1$. Thus, the vertex of a quadratic function in the form $f(x) = a(x - h)^2 + k$ is (h, k). Now look at Example 8, where $f(x) = -2x^2 + 4x + 1 = -2(x - 1)^2 + 3$. The vertex is $(h, k) = (1, 3)$. If we are given the vertex, (h, k), and one additional point on the graph of a quadratic function, we can find the quadratic function $f(x) = a(x - h)^2 + k$ that has this graph.

EXAMPLE 9   **Finding the Quadratic Function Given Its Vertex and One Other Point**

Figure 20

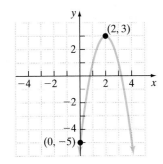

Determine the quadratic function whose graph is given in Figure 20. Write the function in the form $f(x) = a(x - h)^2 + k$.

Solution

The vertex is $(2, 3)$, so $h = 2$ and $k = 3$. Substitute these values into $f(x) = a(x - h)^2 + k$.

$$f(x) = a(x - h)^2 + k$$

$h = 2, k = 3:$ $= a(x - 2)^2 + 3$

To determine the value of a, use the fact that $f(0) = -5$ (the y-intercept).

$$f(x) = a(x - 2)^2 + 3$$

$x = 0, y = f(0) = -5:$ $-5 = a(0 - 2)^2 + 3$

$$-5 = a(4) + 3$$

$$-5 = 4a + 3$$

Subtract 3 from both sides: $-8 = 4a$

$$a = -2$$

The quadratic function whose graph is shown in Figure 20 is $f(x) = -2(x - 2)^2 + 3$.

●

Quick ✓

In Problem 16, find the quadratic function whose graph is given. Write the function in the form $f(x) = a(x - h)^2 + k$.

16.

7.4 Exercises MyMathLab® PRACTICE

Exercise numbers in green have complete video solutions in MyMathLab.

*Problems **1–16** are the* Quick ✓s *that follow the* EXAMPLES.

Building Skills

17. Match each quadratic function to its graph.

(I) $f(x) = x^2 + 3$ **(II)** $f(x) = (x + 3)^2$

(III) $f(x) = x^2 - 3$ **(IV)** $f(x) = (x - 3)^2$

(A)   **(B)**

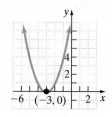

(C)   **(D)**

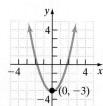   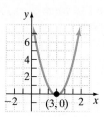

18. Match each quadratic function to its graph.

(I) $f(x) = (x - 2)^2 - 4$ **(II)** $f(x) = -(x - 2)^2 + 4$

(III) $f(x) = -(x + 2)^2 + 4$ **(IV)** $f(x) = 2(x - 2)^2 - 4$

(A)   **(B)**

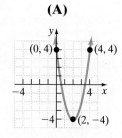   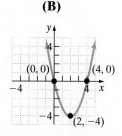

(C)   **(D)**

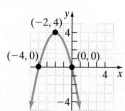   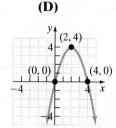

*In Problems **19–26**, verbally explain how to obtain the graph of the given quadratic function from the graph of $y = x^2$. For example, to obtain the graph of $f(x) = (x - 3)^2 - 6$ from the graph of $y = x^2$,*

we take the graph of $y = x^2$ and shift it 3 units to the right and 6 units down.

19. $f(x) = (x + 10)^2$ **20.** $G(x) = (x - 9)^2$

21. $F(x) = x^2 + 12$ **22.** $g(x) = x^2 - 8$

23. $H(x) = 2(x - 5)^2$ **24.** $h(x) = 4(x + 7)^2$

25. $f(x) = -3(x + 5)^2 + 8$ **26.** $F(x) = -\dfrac{1}{2}(x - 3)^2 - 5$

*In Problems **27–30**, use the graph of $y = x^2$ to graph the quadratic function. See Objective 1.*

27. $f(x) = x^2 + 1$ **28.** $h(x) = x^2 + 6$

29. $f(x) = x^2 - 1$ **30.** $g(x) = x^2 - 7$

*In Problems **31–34**, use the graph of $y = x^2$ to graph each quadratic function. See Objective 2.*

31. $F(x) = (x - 3)^2$ **32.** $F(x) = (x - 2)^2$

33. $h(x) = (x + 2)^2$ **34.** $f(x) = (x + 4)^2$

*In Problems **35–40**, use the graph of $y = x^2$ to graph each quadratic function. See Objective 3.*

35. $g(x) = 4x^2$ **36.** $G(x) = 5x^2$

37. $H(x) = \dfrac{1}{3}x^2$ **38.** $h(x) = \dfrac{3}{2}x^2$

39. $p(x) = -x^2$ **40.** $P(x) = -3x^2$

*In Problems **41–54**, use the graph of $y = x^2$ to graph each quadratic function. See Objective 4.*

41. $f(x) = (x - 1)^2 - 3$ **42.** $g(x) = (x + 2)^2 - 1$

43. $F(x) = (x + 3)^2 + 1$ **44.** $G(x) = (x - 4)^2 + 2$

45. $h(x) = -(x + 3)^2 + 2$ **46.** $H(x) = -(x - 3)^2 + 5$

47. $G(x) = 2(x + 1)^2 - 2$ **48.** $F(x) = 3(x - 2)^2 - 1$

49. $H(x) = -\dfrac{1}{2}(x + 5)^2 + 3$ **50.** $f(x) = -\dfrac{1}{2}(x + 6)^2 + 2$

51. $f(x) = x^2 + 2x - 4$ **52.** $f(x) = x^2 + 4x - 1$

53. $g(x) = x^2 - 4x + 8$ **54.** $G(x) = x^2 - 2x + 7$

*In Problems **55–60**, determine the quadratic function whose graph is given. See Objective 5.*

55.   **56.**

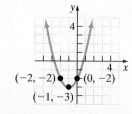   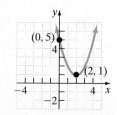

57.

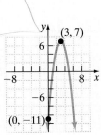

58.

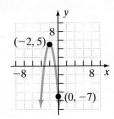

59.

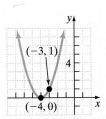

60.

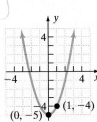

82. Opens up; vertex at $(4, -2)$

83. Opens down; vertex at $(5, -1)$

84. Opens down; vertex at $(-4, -7)$

85. Opens up; vertically stretched by a factor of 4; vertex at $(9, -6)$.

86. Opens up; vertically compressed by a factor of $\frac{1}{2}$; vertex at $(-5, 0)$.

87. Opens down; vertically compressed by a factor of $\frac{1}{3}$; vertex at $(0, 6)$.

88. Opens down; vertically stretched by a factor of 5; vertex at $(5, 8)$.

Explaining the Concepts

89. What is the lowest or highest point on a parabola called? How do we know whether this point is a high point or a low point?

90. Why does the graph of a quadratic function open up if $a > 0$ and down if $a < 0$?

91. Can a quadratic function have a range of $(-\infty, \infty)$? Justify your answer.

92. Can the graph of a quadratic function have more than one y-intercept? Justify your answer.

Mixed Practice

In Problems 61–78, write each function in the form
$f(x) = a(x - h)^2 + k$. *Then graph each quadratic function using transformations. Determine the vertex and axis of symmetry. Based on the graph, determine the domain and range of the quadratic function.*

61. $f(x) = x^2 + 6x - 16$

62. $f(x) = x^2 + 4x + 5$

63. $F(x) = x^2 + x - 12$

64. $h(x) = x^2 - 7x + 10$

65. $H(x) = 2x^2 - 4x - 1$

66. $g(x) = 2x^2 + 4x - 3$

67. $P(x) = 3x^2 + 12x + 13$

68. $f(x) = 3x^2 + 18x + 25$

69. $F(x) = -x^2 - 10x - 21$

70. $g(x) = -x^2 - 8x - 14$

71. $g(x) = -x^2 + 6x - 1$

72. $f(x) = -x^2 + 10x - 17$

73. $H(x) = -2x^2 + 8x - 4$

74. $h(x) = -2x^2 + 12x - 17$

75. $f(x) = \frac{1}{3}x^2 - 2x + 4$

76. $f(x) = \frac{1}{2}x^2 + 2x - 1$

77. $G(x) = -12x^2 - 12x + 1$

78. $h(x) = -4x^2 + 4x$

Synthesis Review

In Problems 93–95, divide.

93. $\dfrac{349}{12}$

94. $\dfrac{4x^2 + 19x - 1}{x + 5}$

95. $\dfrac{2x^4 - 11x^3 + 13x^2 - 8x}{2x - 1}$

96. Explain how division of real numbers is related to division of polynomials.

The Graphing Calculator

In Problems 97–104, graph each quadratic function. Determine the vertex and axis of symmetry. Based on the graph, determine the range of the function.

97. $f(x) = x^2 + 1.3$

98. $f(x) = x^2 - 3.5$

99. $g(x) = (x - 2.5)^2$

100. $G(x) = (x + 4.5)^2$

101. $h(x) = 2.3(x - 1.4)^2 + 0.5$

102. $H(x) = 1.2(x + 0.4)^2 - 1.3$

103. $F(x) = -3.4(x - 2.8)^2 + 5.9$

104. $f(x) = 0.3(x + 3.8)^2 - 8.9$

Applying the Concepts

In Problems 79–88, write a quadratic function in the form
$f(x) = a(x - h)^2 + k$ *with the properties given.*

79. Opens up; vertex at $(3, 0)$.

80. Opens up; vertex at $(0, 2)$.

81. Opens up; vertex at $(-3, 1)$

7.5 Graphing Quadratic Functions Using Properties

Objectives

1. Graph Quadratic Functions of the Form $f(x) = ax^2 + bx + c$
2. Find the Maximum or Minimum Value of a Quadratic Function
3. Model and Solve Optimization Problems Involving Quadratic Functions

Are You Ready for This Section?

Before getting started, take the following readiness quiz. If you get a problem wrong, go back to the section cited and review the material.

R1. Find the intercepts of the graph of $2x + 5y = 20$.   [Section 1.6, pp. 105–106]

R2. Solve: $2x^2 - 3x - 20 = 0$ [Section 4.8, pp. 372–374]

R3. Find the zeros of $f(x) = x^2 - 3x - 4$.   [Section 4.8, p. 376]

In Section 7.4, we graphed quadratic functions using transformations. We now graph a quadratic function by using its intercepts, axis of symmetry, and vertex.

1 Graph Quadratic Functions of the Form $f(x) = ax^2 + bx + c$

We saw in Section 7.4 that a quadratic function $f(x) = ax^2 + bx + c$ can be written in the form $f(x) = a(x - h)^2 + k$ by completing the square in x. We also learned that the value of a determines whether the graph of the quadratic function (the parabola) opens up or down. In addition, we know that the point with coordinates (h, k) is the vertex of the quadratic function.

We can find another formula for the vertex of a parabola by completing the square in $f(x) = ax^2 + bx + c, a \neq 0$, as follows:

$$f(x) = ax^2 + bx + c$$

Group terms involving x: $\quad = (ax^2 + bx) + c$

Factor out a: $\quad = a\left(x^2 + \dfrac{b}{a}x\right) + c$

Complete the square in x by taking $\frac{1}{2}$ the coefficient of x and squaring the result: $\left(\dfrac{1}{2} \cdot \dfrac{b}{a}\right)^2 = \dfrac{b^2}{4a^2}$. Because we add $\dfrac{b^2}{4a^2}$ inside the parentheses, we subtract $a \cdot \dfrac{b^2}{4a^2} = \dfrac{b^2}{4a}$ outside the parentheses:

$$= a\left(x^2 + \dfrac{b}{a}x + \dfrac{b^2}{4a^2}\right) + c - \dfrac{b^2}{4a}$$

Factor the perfect square trinomial; multiply c by $\dfrac{4a}{4a}$ to get a common denominator:

$$= a\left(x + \dfrac{b}{2a}\right)^2 + c \cdot \dfrac{4a}{4a} - \dfrac{b^2}{4a}$$

Write expression in the form $f(x) = a(x - h)^2 + k$:

$$= a\left(x - \left(-\dfrac{b}{2a}\right)\right)^2 + \dfrac{4ac - b^2}{4a}$$

If we compare $f(x) = a\left(x - \left(-\dfrac{b}{2a}\right)\right)^2 + \dfrac{4ac - b^2}{4a}$ to $f(x) = a(x - h)^2 + k$, we come to the following conclusion:

Work Smart

Another formula for the vertex is

$$\left(-\dfrac{b}{2a}, \dfrac{-D}{4a}\right)$$

where $D = b^2 - 4ac$, the discriminant.

The Vertex of a Parabola

Any quadratic function $f(x) = ax^2 + bx + c, a \neq 0$, has vertex

$$\left(-\dfrac{b}{2a}, \dfrac{4ac - b^2}{4a}\right)$$

Because the y-coordinate on the graph of any function can be found by evaluating the function at the corresponding x-coordinate, we can restate the coordinates of the vertex as

$$\left(-\dfrac{b}{2a}, f\left(-\dfrac{b}{2a}\right)\right)$$

Ready?...Answers **R1.** $(0, 4)$, $(10, 0)$
R2. $\left\{-\dfrac{5}{2}, 4\right\}$ **R3.** -1 and 4

For example, in Example 7 on page 583 from Section 7.4, we learned that $f(x) = x^2 + 4x + 3 = (x + 2)^2 - 1$ has a vertex of $(-2, -1)$. In $f(x) = x^2 + 4x + 3$, $a = 1$, $b = 4$, and $c = 3$, so the x-coordinate of the vertex is given by $x = -\dfrac{b}{2a} = -\dfrac{4}{2(1)} = -2$. The y-coordinate of the vertex is $f(-2) = (-2)^2 + 4(-2) + 3 = -1$.

Because the axis of symmetry intersects the vertex, the axis of symmetry is $x = -\dfrac{b}{2a}$. In addition, the parabola will open up if $a > 0$ and down if $a < 0$. By using this information along with the intercepts of the graph, we can obtain a complete graph.

The y-intercept is the value of the quadratic function $f(x) = ax^2 + bx + c$ at $x = 0$—that is, $f(0) = c$.

Any x-intercepts are found by solving the quadratic equation

$$f(x) = ax^2 + bx + c = 0$$

This equation has two, one, or no real solutions, depending on the value of the discriminant $b^2 - 4ac$. Use the value of the discriminant to determine how many x-intercepts the graph of the quadratic function has.

The x-Intercepts of the Graph of a Quadratic Function

1. If the discriminant $b^2 - 4ac > 0$, the graph of $f(x) = ax^2 + bx + c$ has two x-intercepts. The graph crosses the x-axis at the solutions to the equation $ax^2 + bx + c = 0$.
2. If the discriminant $b^2 - 4ac = 0$, the graph of $f(x) = ax^2 + bx + c$ has one x-intercept. The graph touches the x-axis at the solution to the equation $ax^2 + bx + c = 0$.
3. If the discriminant $b^2 - 4ac < 0$, the graph of $f(x) = ax^2 + bx + c$ has no x-intercepts. The graph does not cross or touch the x-axis.

Figure 21 illustrates these possibilities for parabolas that open up.

Figure 21
$f(x) = ax^2 + bx + c, a > 0$

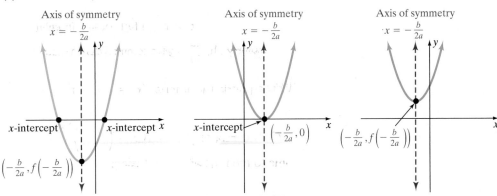

$\boxed{\text{EXAMPLE 1}}$ **How to Graph a Quadratic Function Using Its Properties**

Graph $f(x) = x^2 + 2x - 15$ using its properties.

Step-by-Step Solution

We compare $f(x) = x^2 + 2x - 15$ to $f(x) = ax^2 + bx + c$ and see that $a = 1$, $b = 2$, and $c = -15$.

Step 1: Determine whether the parabola opens up or down.

The parabola opens up because $a = 1 > 0$.

(continued)

Step 2: Determine the vertex and axis of symmetry.

The x-coordinate of the vertex is

$$x = -\frac{b}{2a} = -\frac{2}{2(1)} = -1$$

The y-coordinate of the vertex is

$$f\left(-\frac{b}{2a}\right) = f(-1)$$
$$= (-1)^2 + 2(-1) - 15$$
$$= 1 - 2 - 15$$
$$= -16$$

The vertex is $(-1, -16)$. The axis of symmetry is the line $x = -\frac{b}{2a} = -1$.

Step 3: Determine the y-intercept, $f(0)$.

$$f(0) = 0^2 + 2(0) - 15$$
$$= -15$$

The y-intercept is $(0, -15)$.

Step 4: Find the discriminant, $b^2 - 4ac$, to determine the number of the x-intercepts. Then determine the x-intercepts, if any.

We have $a = 1$, $b = 2$, and $c = -15$, so $b^2 - 4ac = (2)^2 - 4(1)(-15) = 64 > 0$. Because the value of the discriminant is positive, the parabola has two x-intercepts. We find the x-intercepts by solving

$$f(x) = 0$$
$$x^2 + 2x - 15 = 0$$

Factor: $\qquad (x + 5)(x - 3) = 0$

$\qquad\qquad\qquad x + 5 = 0 \quad\text{or}\quad x - 3 = 0$

Zero-Product Property: $\qquad x = -5 \quad\text{or}\quad x = 3$

The x-intercepts are $(-5, 0)$ and $(3, 0)$.

Step 5: Plot the vertex, y-intercept, and x-intercepts. Use the axis of symmetry to find an additional point. Draw the graph of the quadratic function.

Use the axis of symmetry to find the additional point $(-2, -15)$ by recognizing that the y-intercept, $(0, -15)$, is 1 unit to the right of the axis of symmetry, so there must be a point 1 unit to the left of the axis of symmetry. Plot the points and draw the graph. See Figure 22.

Figure 22

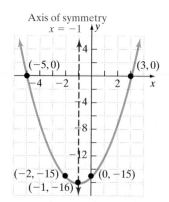

Graphing A Quadratic Function Using Its Properties

To graph any quadratic function of the form $f(x) = ax^2 + bx + c, a \neq 0$, use the following steps:

Step 1: Determine whether the parabola opens up or down.

Step 2: Determine the vertex and axis of symmetry.

Step 3: Determine the y-intercept, $f(0) = c$.

Step 4: Determine the discriminant, $b^2 - 4ac$.
- If $b^2 - 4ac > 0$, then the parabola has two x-intercepts, which are found by solving $f(x) = 0$ $(ax^2 + bx + c = 0)$.
- If $b^2 - 4ac = 0$, the vertex is the x-intercept.
- If $b^2 - 4ac < 0$, there are no x-intercepts.

Step 5: Plot the vertex, y-intercept, and any x-intercept(s). Use the axis of symmetry to find an additional point. Draw the graph of the quadratic function.

Quick ✓

1. Any quadratic function $f(x) = ax^2 + bx + c, a \neq 0$, will have a vertex whose x-coordinate is $x = $ ____.

2. The graph of $f(x) = ax^2 + bx + c$ will have two x-intercepts if $b^2 - 4ac$ __ 0.

3. How many x-intercepts does the graph of $f(x) = -2x^2 - 3x + 6$ have?

4. What is the vertex of $f(x) = x^2 + 4x - 3$?

5. Graph $f(x) = x^2 - 4x - 12$ using its properties.

In Example 1, the function was factorable, so the x-intercepts were rational numbers. When the quadratic function cannot be factored, we can use the quadratic formula to find the x-intercepts. For the purpose of graphing the quadratic function, we will approximate the x-intercepts rounded to two decimal places.

(EXAMPLE 2) **Graphing a Quadratic Function Using Its Properties**

Graph $f(x) = -2x^2 + 12x - 5$ using its properties.

Solution

In $f(x) = -2x^2 + 12x - 5$, we see that $a = -2, b = 12$, and $c = -5$. The parabola opens down because $a = -2 < 0$. The x-coordinate of the vertex is

$$x = -\frac{b}{2a} = -\frac{12}{2(-2)} = 3$$

The y-coordinate of the vertex is

$$y = f\left(-\frac{b}{2a}\right)$$

$$= f(3)$$

$$= -2(3)^2 + 12(3) - 5$$

$$= -18 + 36 - 5$$

$$= 13$$

The vertex is (3, 13). The axis of symmetry is the line $x = -\dfrac{b}{2a} = 3$.

The y-intercept is $f(0) = -2(0)^2 + 12(0) - 5 = -5$. Find the discriminant, $b^2 - 4ac$, to determine the number of the x-intercepts. We see that $a = -2, b = 12$, and $c = -5$, so $b^2 - 4ac = (12)^2 - 4(-2)(-5) = 104 > 0$. The parabola will have two x-intercepts. We find the x-intercepts by solving

$$f(x) = 0$$
$$-2x^2 + 12x - 5 = 0$$

The equation cannot be solved by factoring (since $b^2 - 4ac$ is not a perfect square), so we use the quadratic formula:

$$x = \frac{-b \pm \sqrt{b^2 - 4ac}}{2a}$$

$a = -2, b = 12, b^2 - 4ac = 104$: $\qquad = \dfrac{-12 \pm \sqrt{104}}{2(-2)}$

$\sqrt{104} = \sqrt{4 \cdot 26} = 2\sqrt{26}$: $\qquad = \dfrac{-12 \pm 2\sqrt{26}}{-4}$

Divide -4 into each term in the numerator and simplify: $\qquad x = 3 \pm \dfrac{\sqrt{26}}{2}$ Exact solution

Figure 23

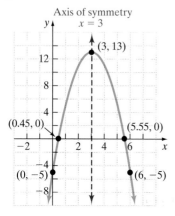

Evaluate $3 \pm \dfrac{\sqrt{26}}{2}$ and find the x-intercepts are approximately 0.45 and 5.55. Use the axis of symmetry to find the additional point $(6, -5)$. (The y-intercept, $(0, -5)$, is 3 units to the left of the axis of symmetry, therefore, there must be a point 3 units to the right of the axis of symmetry.) Next, plot the vertex, the intercepts, and an additional point. Draw the graph of the quadratic function. See Figure 23.

Work Smart

Notice that the vertex in Example 2 lies in quadrant I (above the x-axis) and the graph opens down. This tells us the graph must have two x-intercepts.

Quick ✓

6. Graph $f(x) = -3x^2 + 12x - 7$ using its properties.

EXAMPLE 3 **Graphing a Quadratic Function Using Its Properties**

Graph $g(x) = x^2 - 8x + 16$ using its properties.

Solution

In $g(x) = x^2 - 8x + 16$, $a = 1, b = -8$, and $c = 16$. The parabola opens up because $a = 1 > 0$. The x-coordinate of the vertex is

$$x = -\frac{b}{2a} = -\frac{-8}{2(1)} = 4$$

The y-coordinate of the vertex is

$$y = f\left(-\frac{b}{2a}\right)$$
$$= f(4)$$
$$= 4^2 - 8(4) + 16$$
$$= 16 - 32 + 16$$
$$= 0$$

The vertex is (4, 0). The axis of symmetry is the line $x = -\dfrac{b}{2a} = 4$.

Figure 24

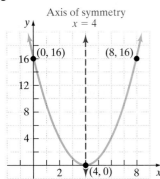

The y-intercept is $f(0) = 0^2 - 8(0) + 16 = 16$. We have $a = 1, b = -8$, and $c = 16$, so $b^2 - 4ac = (-8)^2 - 4(1)(16) = 0$. Since the vertex $(4, 0)$ is on the x-axis, the parabola has one x-intercept. Verify this by solving

$$f(x) = 0$$

$$x^2 - 8x + 16 = 0$$

Factor: $\qquad (x - 4)^2 = 0$

$$x = 4$$

Use the point $(0, 16)$ and the axis of symmetry to find the additional point $(8, 16)$. The parabola is shown in Figure 24.

Quick ✓

7. Graph $f(x) = x^2 + 6x + 9$ using its properties.

▶ **EXAMPLE 4** **Graphing a Quadratic Function Using Its Properties**

Graph $F(x) = 2x^2 + 6x + 5$ using its properties.

Solution

In $F(x) = 2x^2 + 6x + 5$, $a = 2, b = 6$, and $c = 5$. The parabola opens up because $a = 2 > 0$. The x-coordinate of the vertex is

$$x = -\frac{b}{2a} = -\frac{6}{2(2)} = -\frac{3}{2}$$

The y-coordinate of the vertex is

$$y = f\left(-\frac{b}{2a}\right) = f\left(-\frac{3}{2}\right)$$

$$= 2\left(-\frac{3}{2}\right)^2 + 6\left(-\frac{3}{2}\right) + 5$$

$$= 2\left(\frac{9}{4}\right) - 9 + 5$$

$$= \frac{1}{2}$$

Work Smart

Notice that the vertex lies in quadrant II (above the x-axis) and the graph opens up. This tells us the graph has no x-intercepts.

Figure 25

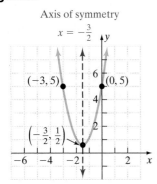

The vertex is $\left(-\frac{3}{2}, \frac{1}{2}\right)$. The axis of symmetry is the line $x = -\frac{b}{2a} = -\frac{3}{2}$.

The y-intercept is $f(0) = 2(0)^2 + 6(0) + 5 = 5$. Because the parabola opens up and the vertex is in the second quadrant, we expect that the parabola has no x-intercepts. We verify this by computing the discriminant with $a = 2, b = 6$, and $c = 5$:

$$b^2 - 4ac = 6^2 - 4(2)(5) = -4 < 0$$

The parabola has no x-intercepts. Use the point $(0, 5)$ and the axis of symmetry to find the additional point $(-3, 5)$. The parabola is shown in Figure 25.

Quick ✓

8. Graph $G(x) = -3x^2 + 9x - 8$ using its properties.

Figure 26

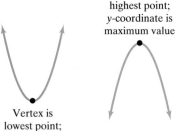

Vertex is highest point; y-coordinate is maximum value

Vertex is lowest point; y-coordinate is minimum value

(a) Opens up
$a > 0$

(b) Opens down
$a < 0$

▶ ❷ Find the Maximum or Minimum Value of a Quadratic Function

Recall, the graph of a quadratic function $f(x) = ax^2 + bx + c$ is a parabola with vertex at $\left(-\dfrac{b}{2a}, f\left(-\dfrac{b}{2a}\right)\right)$. The vertex is the highest point on the graph if $a < 0$ and the lowest point on the graph if $a > 0$. If the vertex is the highest point then $f\left(-\dfrac{b}{2a}\right)$, the y-coordinate of the vertex, is the **maximum value** of f. If the vertex is the lowest point, then $f\left(-\dfrac{b}{2a}\right)$ is the **minimum value** of f. See Figure 26.

We can now answer questions involving **optimization,** the process of finding the maximum or minimum value(s) of a function. In the case of quadratic functions, the maximum $(a < 0)$ or minimum $(a > 0)$ is found at the vertex.

EXAMPLE 5

Finding the Maximum or Minimum Value of a Quadratic Function

Determine whether the quadratic function

$$f(x) = 3x^2 + 12x - 7$$

has a maximum or a minimum value. Then find the maximum or minimum value of the function, and indicate when it occurs.

Solution

In $f(x) = 3x^2 + 12x - 7$, we see that $a = 3$, $b = 12$, and $c = -7$. Because $a = 3 > 0$, the graph of the quadratic function opens up, so the function has a minimum value. The minimum value of the function occurs at

$$x = -\frac{b}{2a} = -\frac{12}{2(3)} = -2$$

Work Smart

The vertex is not the maximum or minimum value—the y-coordinate of the vertex is the maximum or minimum value of the function.

The minimum value of the function is

$$
\begin{aligned}
f\left(-\frac{b}{2a}\right) &= f(-2) \\
&= 3(-2)^2 + 12(-2) - 7 \\
&= 3(4) - 24 - 7 \\
&= -19
\end{aligned}
$$

The minimum value of the function is -19 and occurs at $x = -2$.

Quick ✓

9. *True or False* For the quadratic function $f(x) = ax^2 + bx + c$, if $a < 0$ then $f\left(-\dfrac{b}{2a}\right)$ is the maximum value of f.

In Problems 10 and 11, determine whether the quadratic function has a maximum or a minimum value. Then find the maximum or minimum value of the function and where it occurs.

10. $f(x) = 2x^2 - 8x + 1$

11. $G(x) = -x^2 + 10x + 8$

We can use properties of quadratic functions to answer questions regarding applications modeled by quadratic functions.

EXAMPLE 6 **Maximizing Revenue**

Suppose the marketing department of Dell Computer has found that, when a certain model of computer is sold at a price of p dollars, the daily revenue R (in dollars) as a function of the price p is the function

$$R(p) = -\frac{1}{4}p^2 + 400p$$

(a) For what price will the revenue be maximized?

(b) What is the maximum daily revenue?

Solution

(a) Notice that the revenue function is a quadratic function whose graph opens down because $a = -\frac{1}{4} < 0$. Therefore, the function has a maximum at

$$p = -\frac{b}{2a} = -\frac{400}{2 \cdot \left(-\dfrac{1}{4}\right)} = -\frac{400}{-\dfrac{1}{2}} = 800$$

Revenue will be maximized when the price is $p = \$800$.

(b) The maximum daily revenue is found by letting $p = \$800$ in the revenue function.

$$R(800) = -\frac{1}{4} \cdot (800)^2 + 400 \cdot 800$$

$$= \$160{,}000$$

The maximum daily revenue is $\$160{,}000$. See Figure 27 for an illustration.

Figure 27

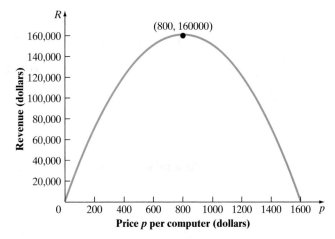

Quick ✓

12. Suppose that the marketing department of Texas Instruments has found that, when a certain model of calculator is sold at a price of p dollars, the daily revenue R (in dollars) as a function of the price p is the function $R(p) = -0.5p^2 + 75p$.

(a) For what price will the daily revenue be maximized?

(b) What is the maximum daily revenue?

▶ ❸ **Model and Solve Optimization Problems Involving Quadratic Functions**

We now discuss models that result in quadratic functions. As always, we shall use the problem-solving strategy introduced in Section 1.2.

(**EXAMPLE 7**) **Maximizing the Area Enclosed by a Fence**

A farmer has 3000 feet of fence with which to enclose a rectangular field. What is the maximum area that can be enclosed by the fence? What are the dimensions of the rectangle that encloses the most area?

Solution

Step 1: Identify We need the dimensions of a rectangle that maximize the area.

Step 2: Name We let w represent the width of the rectangle and let l represent its length.

Step 3: Translate See Figure 28.

Figure 28

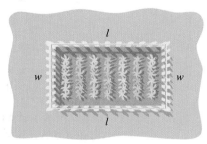

Since 3000 feet of fence is available, the perimeter of the rectangle will be 3000 feet. Using the perimeter formula of a rectangle we have

$$2l + 2w = 3000$$

The area of the rectangle, $A = lw$, uses two variables, l and w. To express A in terms of only one variable, solve the equation $2l + 2w = 3000$ for l and then substitute for l in the area formula $A = lw$.

$$2l + 2w = 3000$$

Subtract 2w from both sides: $2l = 3000 - 2w$

Divide both sides by 2: $l = \dfrac{3000 - 2w}{2}$

Simplify: $l = 1500 - w$

Next let $l = 1500 - w$ in the formula $A = lw$.

$$A = (1500 - w)w$$
$$= -w^2 + 1500w$$

Now, A is a quadratic function of w.

$$A(w) = -w^2 + 1500w \quad \text{The Model}$$

Step 4: Solve We wish to find the dimensions that result in a maximum area enclosed by the fence. The model developed in Step 3 is a quadratic function that opens down (because $a = -1 < 0$), so the vertex is a maximum point on the graph of A. The maximum value occurs at

$$w = -\frac{b}{2a} = -\frac{1500}{2(-1)} = 750$$

Work Smart

We could just as easily have solved $2l + 2w = 3000$ for w. We ultimately would obtain the same solution.

The maximum value of A is

$$A\left(-\frac{b}{2a}\right) = A(750) = -750^2 + 1500(750)$$

$$= 562,500 \text{ square feet}$$

We know that $l = 1500 - w$, so if the width is 750 feet, then the length will be $l = 1500 - 750 = 750$ feet.

Step 5: Check With a length and width of 750 feet, the perimeter is $2(750) + 2(750) = 3000$ feet. The area is $(750 \text{ feet})(750 \text{ feet}) = 562,500$ square feet. Everything checks!

Step 6: Answer The largest area that can be enclosed by 3000 feet of fence has an area of 562,500 square feet. Its dimensions are 750 feet by 750 feet.

Quick ✓

13. Roberta has 1000 yards of fence to enclose a rectangular field. What is the maximum area that can be enclosed by the fence? What are the dimensions of the rectangle that encloses the most area?

(EXAMPLE 8) **Pricing a Charter**

Chicago Tours offers boat charters along the Chicago coastline on Lake Michigan. Normally, a ticket costs $20 per person, but for any group, Chicago Tours will lower the price of a ticket by $0.10 per person for each person in excess of 30. Determine the group size that will maximize revenue. What is the maximum revenue that can be earned from a group sale?

Solution

Step 1: Identify This is a direct translation problem involving revenue. Remember, revenue is price times quantity.

Step 2: Name Let x represent the number in the group in excess of 30.

Step 3: Translate Revenue is price times quantity. If 30 individuals make up a group, the revenue to Chicago Tours will be $20(30)$. If 31 individuals make up a group, revenue will be $19.90(31)$. The revenue for 32 individuals will be $19.80(32)$. In general, for $30 + x$ individuals, revenue will be $(20 - 0.1x)(30 + x)$. Thus the revenue R for a group that has $30 + x$ people in it is given by

$$R(x) = (20 - 0.1x)(30 + x)$$

$$\text{FOIL:} \quad = 600 + 20x - 3x - 0.1x^2$$

$$= -0.1x^2 + 17x + 600 \qquad \text{The Model}$$

Step 4: Solve We want to find the revenue-maximizing number of individuals in the group. The function R is a quadratic function with a graph that opens down (because $a = -0.1 < 0$), so the vertex is a maximum point. The value of x that results in a maximum is given by

$$x = -\frac{b}{2a} = -\frac{17}{2(-0.1)}$$

$$= 85$$

The maximum revenue is

$$R(85) = -0.1(85)^2 + 17(85) + 600$$

$$= \$1322.50$$

Step 5: Check Since x represents the number of passengers in excess of 30, $30 + 85 = 115$ tickets should be sold to maximize revenue. The cost per ticket is $\$20 - 0.1(85) = \$20 - \$8.50 = \$11.50$. The cost per ticket times the number of passengers is $\$11.50(115) = \$1322.50$. The answer checks!

Step 6: Answer A group sale of 115 passengers will maximize revenue. The maximum revenue is $\$1322.50$.

Quick ✓

14. A compact disk manufacturer charges \$100 for each box of CDs ordered. However, it reduces the price by \$1 per box for each box in excess of 30 boxes but less than 90 boxes. Determine the number of boxes of CDs that should be sold to maximize revenue. What is the maximum revenue?

Exercises 7.5   MyMathLab®   MathXP   PRACTICE

*Problems **1–14** are the Quick ✓s that follow the EXAMPLES.*

Building Skills

In Problems 15–22, (a) find the vertex of each parabola, (b) use the discriminant to determine the number of x-intercepts the graph will have. Then determine the x-intercepts. See Objective 1.

15. $f(x) = x^2 - 6x - 16$

16. $g(x) = x^2 + 4x - 12$

17. $G(x) = -2x^2 + 4x - 5$

18. $H(x) = x^2 - 4x + 5$

19. $h(x) = 4x^2 + 4x + 1$

20. $f(x) = x^2 - 6x + 9$

21. $F(x) = 4x^2 - x - 1$

22. $P(x) = -2x^2 + 3x + 1$

In Problems 23–62, graph each quadratic function using its properties by following Steps 1–5 on page 591. Based on the graph, determine the domain and range of the quadratic function. See Objective 1.

23. $f(x) = x^2 - 4x - 5$

24. $f(x) = x^2 - 2x - 8$

25. $G(x) = x^2 + 12x + 32$

26. $g(x) = x^2 - 12x + 27$

27. $F(x) = -x^2 + 2x + 8$

28. $g(x) = -x^2 + 2x + 15$

29. $H(x) = x^2 - 4x + 4$

30. $h(x) = x^2 + 6x + 9$

31. $g(x) = x^2 + 2x + 5$

32. $f(x) = x^2 - 4x + 7$

33. $h(x) = -x^2 - 10x - 25$

34. $P(x) = -x^2 - 12x - 36$

35. $p(x) = -x^2 + 2x - 5$

36. $f(x) = -x^2 + 4x - 6$

37. $F(x) = 4x^2 - 4x - 3$

38. $f(x) = 4x^2 - 8x - 21$

39. $G(x) = -9x^2 + 18x + 7$

40. $g(x) = -9x^2 - 36x - 20$

41. $H(x) = 4x^2 - 4x + 1$

42. $h(x) = 9x^2 + 12x + 4$

43. $f(x) = -16x^2 - 24x - 9$

44. $F(x) = -4x^2 - 20x - 25$

45. $f(x) = 2x^2 + 8x + 11$

46. $F(x) = 3x^2 + 6x + 7$

47. $P(x) = -4x^2 + 6x - 3$

48. $p(x) = -2x^2 + 6x + 5$

49. $h(x) = x^2 + 5x + 3$

50. $H(x) = x^2 + 3x + 1$

51. $G(x) = -3x^2 + 8x + 2$

52. $F(x) = -2x^2 + 6x + 1$

53. $f(x) = 5x^2 - 5x + 2$

54. $F(x) = 4x^2 + 4x - 1$

55. $H(x) = -3x^2 + 6x$

56. $h(x) = -4x^2 + 8x$

57. $f(x) = x^2 - \dfrac{5}{2}x - \dfrac{3}{2}$

58. $g(x) = x^2 + \dfrac{5}{2}x - 6$

59. $G(x) = \dfrac{1}{2}x^2 + 2x - 6$

60. $H(x) = \dfrac{1}{4}x^2 + x - 8$

61. $F(x) = -\dfrac{1}{4}x^2 + x + 15$

62. $G(x) = -\dfrac{1}{2}x^2 - 8x - 24$

In Problems 63–74, determine whether the quadratic function has a maximum or a minimum value. Then find the maximum or minimum value and where it occurs. See Objective 2.

63. $f(x) = x^2 + 8x + 13$

64. $f(x) = x^2 - 6x + 3$

65. $G(x) = -x^2 - 10x + 3$

66. $g(x) = -x^2 + 4x + 12$

67. $F(x) = -2x^2 + 12x + 5$ **68.** $H(x) = -3x^2 + 12x - 1$

69. $h(x) = 4x^2 + 16x - 3$ **70.** $G(x) = 5x^2 + 10x - 1$

71. $f(x) = 2x^2 - 5x + 1$ **72.** $F(x) = 3x^2 + 4x - 3$

73. $H(x) = -3x^2 + 4x + 1$ **74.** $h(x) = -4x^2 - 6x + 1$

Applying the Concepts

75. Revenue Function Suppose that the marketing department of Samsung has found that, when a certain model of DVD player is sold at a price of p dollars, the daily revenue R (in dollars) as a function of the price p is $R(p) = -2.5p^2 + 600p$.
  (a) For what price will the daily revenue be maximized?
  (b) What is the maximum daily revenue?

76. Revenue Function Suppose that the marketing department of Samsung has found that, when a certain model of cellular telephone is sold at a price of p dollars, the daily revenue R (in dollars) as a function of the price p is $R(p) = -5p^2 + 600p$.
  (a) For what price will the daily revenue be maximized?
  (b) What is the maximum daily revenue?

77. Marginal Cost The marginal cost of a product can be thought of as the cost of producing one additional unit of output. For example, if the marginal cost of producing the fiftieth unit of a product is $6.30, then it costs $6.30 to increase production from 49 to 50 units of output. Suppose that the marginal cost C (in dollars) to produce x digital cameras is given by $C(x) = 0.05x^2 - 6x + 215$. How many digital cameras should be produced to minimize marginal cost? What is the minimum marginal cost?

78. Marginal Cost (See Problem 77.) The marginal cost C (in dollars) of manufacturing x portable CD players is given by $C(x) = 0.05x^2 - 9x + 435$. How many portable CD players should be manufactured to minimize marginal cost? What is the minimum marginal cost?

79. Punkin Chunkin Suppose that an air cannon in the Punkin Chunkin contest whose muzzle is 10 feet above the ground fires a pumpkin at an angle of 45° to the horizontal with a muzzle velocity of 335 feet per second. The model

$s(t) = -16t^2 + 240t + 10$ can be used to estimate the height s of a pumpkin after t seconds.
  (a) Determine the time at which the pumpkin is at its maximum height.
  (b) Determine the maximum height of the pumpkin.
  (c) After how long will the pumpkin strike the ground?

80. Punkin Chunkin Suppose that a catapult in the Punkin Chunkin contest releases a pumpkin 8 feet above the ground at an angle of 45° to the horizontal with an initial speed of 220 feet per second. The model $s(t) = -16t^2 + 155t + 8$ can be used to estimate the height s of a pumpkin after t seconds.
  (a) Determine the time at which the pumpkin is at its maximum height.
  (b) Determine the maximum height of the pumpkin.
  (c) After how long will the pumpkin strike the ground?

81. Punkin Chunkin Suppose that an air cannon in the Punkin Chunkin contest whose muzzle is 10 feet above the ground fires a pumpkin at an angle of 45° to the horizontal with a muzzle velocity of 335 feet per second. The model $h(x) = \dfrac{-32}{335^2}x^2 + x + 10$ can be used to estimate the height h of the pumpkin after it has traveled x feet.
  (a) How far from the cannon will the pumpkin reach its maximum height?
  (b) What is the maximum height of the pumpkin?
  (c) How far will the pumpkin travel before it strikes the ground?
  (d) Compare your answer in part (b) of this problem with the answer found in part (b) of Problem 79. Why might the answers differ?

82. Punkin Chunkin Suppose that a catapult in the Punkin Chunkin contest releases a pumpkin 8 feet above the ground at an angle of 45° to the horizontal with an initial speed of 220 feet per second. The model $h(x) = \dfrac{-32}{220^2}x^2 + x + 8$ can be used to estimate the height h of the pumpkin after it has traveled x feet.
  (a) How far from the cannon will the pumpkin reach its maximum height?
  (b) What is the maximum height of the pumpkin?
  (c) How far will the pumpkin travel before it strikes the ground?
  (d) Compare your answer in part (b) of this problem with the answer found in part (b) of Problem 80. Why might the answers differ?

83. **Life Cycle Hypothesis** The Life Cycle Hypothesis from economics was presented by Franco Modigliani in 1954. One of its components states that income is a function of age. The function $I(a) = -55a^2 + 5119a - 54{,}448$ represents the relation between average annual income I and age a.
    (a) According to the model, at what age will average income be a maximum?
    (b) According to the model, what is the maximum average income?

84. **Advanced Degrees** The function $P(x) = -0.007x^2 + 0.603x - 2.012$ models the percentage of the United States population whose age is x who have earned an advanced degree (more than a bachelor's degree) as of March 2009. (SOURCE: *Based on data obtained from the U.S. Census Bureau*)
    (a) To the nearest year, what is the age for which the highest percentage of Americans have earned an advanced degree?
    (b) According to the model, what is the percentage of Americans who have earned an advanced degree at the age found in part (a)?

85. **Fun with Numbers** The sum of two numbers is 36. Find the numbers such that their product is a maximum.

86. **Fun with Numbers** The sum of two numbers is 50. Find the numbers such that their product is a maximum.

87. **Fun with Numbers** The difference of two numbers is 18. Find the numbers such that their product is a minimum.

88. **Fun with Numbers** The difference of two numbers is 10. Find the numbers such that their product is a minimum.

89. **Enclosing a Rectangular Field** Maurice has 500 yards of fencing and wishes to enclose a rectangular area. What is the maximum area that can be enclosed by the fence? What are the dimensions of the area enclosed?

90. **Enclosing a Rectangular Field** Maude has 800 yards of fencing and wishes to enclose a rectangular area. What is the maximum area that can be enclosed by the fence? What are the dimensions of the area enclosed?

91. **Maximizing an Enclosed Area** A farmer with 2000 meters of fencing wants to enclose a rectangular plot that borders a river. If the farmer does not fence the side along the river, what is the largest area that can be enclosed? What are the dimensions of the enclosed area? See the figure.

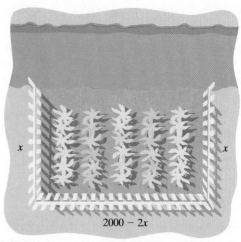

2000 − 2x

92. **Maximizing an Enclosed Area** A farmer with 8000 meters of fencing wants to enclose a rectangular plot and then divide it into two plots with a fence parallel to one of the sides. See the figure. What is the largest area that can be enclosed? What are the lengths of the sides of each part of the enclosed area?

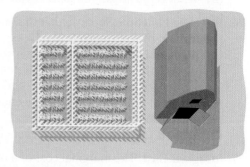

93. **Constructing Rain Gutters** A rain gutter is to be made of aluminum sheets that are 20 inches wide by turning up the edges 90°. What depth will provide maximum cross-sectional area and hence allow the most water to flow?

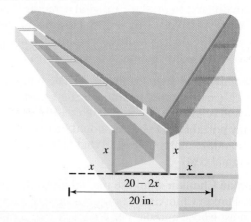

20 − 2x
20 in.

94. **Maximizing the Volume of a Box** A box with a rectangular base is to be constructed such that the perimeter of the base of the box is to be 40 inches. The height of the box must be 15 inches. Find the dimensions of the box such that the volume is maximized. What is the maximum volume?

95. Revenue Function Weekly demand for jeans at a department store obeys the demand equation

$$x = -p + 110$$

where x is the quantity demanded and p is the price (in dollars).
(a) Express the revenue R as a function of p. (*Hint: $R = xp$*)
(b) What price p maximizes revenue? What is the maximum revenue?
(c) How many pairs of jeans will be sold at the revenue-maximizing price?

96. Revenue Function Demand for hot dogs at a baseball game obeys the demand equation

$$x = -800p + 8000$$

where x is the quantity demanded and p is the price (in dollars).
(a) Express the revenue R as a function of p. (*Hint: $R = xp$*)
(b) What price p maximizes revenue? What is the maximum revenue?
(c) How many hot dogs will be sold at the revenue-maximizing price?

Extending the Concepts

Answer Problems 97 and 98 using the following information:
A quadratic function of the form $f(x) = ax^2 + bx + c$
with $b^2 - 4ac > 0$ may also be written in the form $f(x) = a(x - r_1)(x - r_2)$, where r_1 and r_2 are the x-intercepts of the graph of the quadratic function.

97. (a) Find a quadratic function whose x-intercepts are 2 and 6 with $a = 1$, $a = 2$, and $a = -2$.
(b) How does the value of a affect the intercepts?
(c) How does the value of a affect the axis of symmetry?
(d) How does the value of a affect the vertex?

98. (a) Find a quadratic function whose x-intercepts are -1 and 5 with $a = 1$, $a = 2$, and $a = -2$.
(b) How does the value of a affect the intercepts?
(c) How does the value of a affect the axis of symmetry?
(d) How does the value of a affect the vertex?

Explaining the Concepts

99. Explain how the discriminant is used to determine the number of x-intercepts the graph of a quadratic function will have.

100. Provide two methods for finding the vertex of any quadratic function $f(x) = ax^2 + bx + c$.

101. Refer to Example 6 on page 595. Notice that if the price charged for the computer is $0 or $1600, the revenue is $0. It is easy to explain why revenue

would be $0 if the price charged were $0, but how can revenue be $0 if the price charged is $1600?

Synthesis Review

In Problems 102–105, graph each function using point plotting.

102. $f(x) = -2x + 12$

103. $G(x) = \dfrac{1}{4}x - 2$

104. $f(x) = x^2 - 5$

105. $f(x) = (x + 2)^2 + 4$

106. For each function in Problems 102–105, explain an alternative method for graphing the function. Which method do you prefer? Why?

The Graphing Calculator

Graphing calculators have a MAXIMUM feature and a MINIMUM feature that enable us to determine the coordinates of the vertex of a parabola. For example, to find the vertex of $f(x) = x^2 + 2x - 15$ using a TI-84 graphing calculator, we use the MINIMUM feature. See Figure 29. The vertex is $(-1, -16)$.

Figure 29

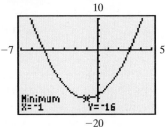

In Problems 107–114, use a graphing calculator to graph each quadratic function. Using the MAXIMUM or MINIMUM feature on the calculator, determine the vertex. If necessary, round your answers to two decimal places.

107. $f(x) = x^2 - 7x + 3$

108. $f(x) = x^2 + 3x + 8$

109. $G(x) = -2x^2 + 14x + 13$

110. $g(x) = -4x^2 - x + 11$

111. $F(x) = 5x^2 + 3x - 20$

112. $F(x) = 3x^2 + 2x - 21$

113. $H(x) = \dfrac{1}{2}x^2 - \dfrac{2}{3}x + 5$

114. $h(x) = \dfrac{3}{4}x^2 + \dfrac{4}{3}x - 1$

115. On the same screen, graph the family of parabolas $f(x) = x^2 + 2x + c$ for $c = -3$, $c = 0$, and $c = 1$. Describe the role that c plays in the graph for this family of functions.

116. On the same screen, graph the family of parabolas $f(x) = x^2 + bx + 1$ for $b = -4$, $b = 0$, and $b = 4$. Describe the role that b plays in the graph for this family of functions.

7.6 Polynomial Inequalities

Objectives

1. Solve Quadratic Inequalities
2. Solve Polynomial Inequalities

Are You Ready for This Section?

Before getting started, take the following readiness quiz. If you get a problem wrong, go back to the section cited and review the material.

R1. Write $-4 \leq x < 5$ in interval notation.   [Section 1.4, pp. 81–84]

R2. Solve: $3x + 5 > 5x - 3$ [Section 1.4, pp. 85–87]

In Section 1.4, we solved linear inequalities in one variable, such as $2x - 3 > 4x + 5$, by using methods that were similar to solving linear equations. We represented the solution sets of inequalities using either set-builder notation or interval notation.

Solving *polynomial inequalities* is not a simple extension of solving polynomial equations. However, we do use the skills developed in solving polynomial equations to solve polynomial inequalities.

> **Definition**
>
> A **polynomial inequality** is an inequality of the form
>
> $$f(x) > 0 \quad f(x) < 0 \quad f(x) \geq 0 \quad f(x) \leq 0$$
>
> where f is a polynomial function.

Some examples of polynomial inequalities are

$$x^2 - 3x - 4 > 0 \quad 2x^2 - 5x + 1 < 0 \quad x^3 - 6x^2 + 7x + 1 \geq 0$$

We begin by looking at quadratic inequalities—a special case of polynomial inequalities.

1 Solve Quadratic Inequalities

> **Definition**
>
> A **quadratic inequality** is an inequality of the form
>
> $$ax^2 + bx + c > 0 \quad \text{or} \quad ax^2 + bx + c < 0 \quad \text{or}$$
> $$ax^2 + bx + c \geq 0 \quad \text{or} \quad ax^2 + bx + c \leq 0$$
>
> where $a \neq 0$.

We will first present a graphical approach to the solution of a quadratic inequality and then present an algebraic method.

To understand the logic behind the graphical approach, consider the following. Suppose we need to solve the inequality $ax^2 + bx + c > 0$. If we let $f(x) = ax^2 + bx + c$, then we want all x-values such that $f(x) > 0$. Since f represents the y-values of the graph of the function $f(x) = ax^2 + bx + c$, we are looking for all x-values such that the graph of f is *above* the x-axis—that is, where y-values are positive. This occurs when the graph lies in either quadrant I or quadrant II of the Cartesian plane. If we were asked to solve $f(x) < 0$, we would look for the x-values such that the graph is *below* the x-axis—that is, where the y-values are negative. This occurs when the graph lies in either quadrant III or quadrant IV of the Cartesian plane. See Figure 30 for an illustration of this idea.

Ready?...Answers   **R1.** $[-4, 5)$
R2. $\{x \mid x < 4\}$ or $(-\infty, 4)$

Figure 30

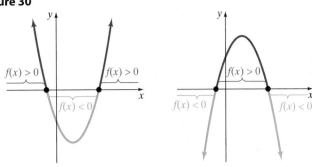

EXAMPLE 1   How to Solve a Quadratic Inequality Using the Graphical Method

Solve $x^2 - 4x - 5 \geq 0$ using the graphical method.

Step-by-Step Solution

Step 1: Write the inequality so that $ax^2 + bx + c$ is on one side of the inequality and 0 is on the other.

The inequality is already in the form we need.

$$x^2 - 4x - 5 \geq 0$$

Step 2: Graph the function $f(x) = ax^2 + bx + c$. Label the x-intercepts on the graph.

Graph $f(x) = x^2 - 4x - 5$. Because $a = 1 > 0$, the graph opens up.

x-intercepts: $f(x) = 0$ **y-intercept:** $f(0) = -5$

$$x^2 - 4x - 5 = 0$$ The y-intercept is $(0, -5)$.
$$(x - 5)(x + 1) = 0$$
$$x = 5 \quad \text{or} \quad x = -1$$

The x-intercepts are $(-1, 0)$ and $(5, 0)$.

$$\textbf{Vertex:} \quad x = -\frac{b}{2a} = -\frac{-4}{2(1)} = 2$$

$$y = f\left(-\frac{b}{2a}\right) = f(2) = -9$$

The vertex is at $(2, -9)$. Figure 31 shows the graph.

Figure 31

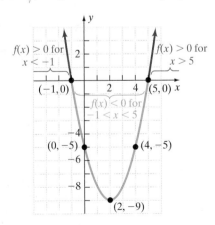

Step 3: From the graph, determine where the function is positive and where it is negative. This information will give you the solution set to the inequality.

Figure 31 shows that the graph of $f(x) = x^2 - 4x - 5$ is greater than 0 for $x < -1$ or $x > 5$. Because the inequality is nonstrict $(\geq)$, we include the x-intercepts in the solution. The solution is $\{x | x \leq -1 \text{ or } x \geq 5\}$ using set-builder notation, or $(-\infty, -1] \cup [5, \infty)$ using interval notation. See Figure 32 for a graph of the solution set.

Figure 32

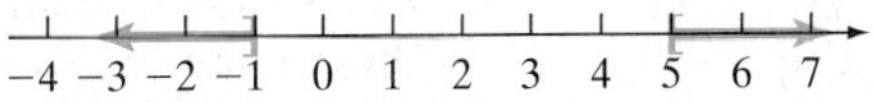

Quick ✓

In Problem 1, solve the quadratic inequality using the graphical method. Graph the solution set.

1. $x^2 + 3x - 10 \geq 0$

The second method for solving inequalities is algebraic. This method uses the Rules of Signs for multiplying or dividing real numbers. Recall that the product of two positive real numbers is positive, the product of a positive real number and a negative real number is negative, and the product of two negative real numbers is positive.

 Let's solve the inequality from Example 1 using the algebraic method.

EXAMPLE 2 **How to Solve a Quadratic Inequality Using the Algebraic Method**

Solve the inequality from Example 1, $x^2 - 4x - 5 \geq 0$, using the algebraic method.

Step-by-Step Solution

Step 1: Write the inequality so that $ax^2 + bx + c$ is on one side of the inequality and 0 is on the other.

The inequality is already in the form we need.

$$x^2 - 4x - 5 \geq 0$$

Step 2: Determine the solutions to the equation $ax^2 + bx + c = 0$.

$$x^2 - 4x - 5 = 0$$

Factor: $(x - 5)(x + 1) = 0$

Use the Zero-Product Property: $x - 5 = 0$ or $x + 1 = 0$

$x = 5$ or $x = -1$

Step 3: Use the solutions to the equation from Step 2 to separate the real number line into intervals.

We separate the real number line into intervals:

$$(-\infty, -1) \quad (-1, 5) \quad (5, \infty)$$

Step 4: Write $x^2 - 4x - 5$ in factored form as $(x - 5)(x + 1)$. Choose a test point within each interval formed in Step 3 to determine the sign of each factor. Then determine the sign of the product. Also write the value of $x^2 - 4x - 5$ at each solution found in Step 2.

The sign of a binomial factor will change only on either side of its corresponding zero. For example, the sign of $x + 1$ changes from negative to positive on either side of $x = -1$. Therefore, when we choose a test point in each interval, we know that if one number in an interval satisfies the inequality, then all the numbers in that interval will satisfy the inequality.

- In the interval $(-\infty, -1)$, we choose a test point of -2. When $x = -2$, the factor $x - 5$ equals $-2 - 5 = -7$, and the factor $x + 1$ equals $-2 + 1 = -1$. Since the product of two negatives is positive, the expression $(x - 5)(x + 1)$ is positive for all x in the interval $(-\infty, -1)$.

- In the interval $(-1, 5)$, let 0 be the test point. When $x = 0$, $x - 5$ is negative and $x + 1$ is positive. Therefore, $(x - 5)(x + 1)$ is negative for all x in the interval $(-1, 5)$.

- In the interval $(5, \infty)$, let 6 be the test point. For $x = 6$, both $x - 5$ and $x + 1$ are positive, so $(x - 5)(x + 1)$ is positive in the interval $(5, \infty)$.

- For $x = -1$ and $x = 5$, the value of $x^2 - 4x - 5 = (x - 5)(x + 1)$ is zero.

Table 9 shows the sign of $(x - 5)(x + 1)$ in each interval, as well as the value of $x^2 - 4x - 5$ at $x = -1$ and at $x = 5$.

Table 9

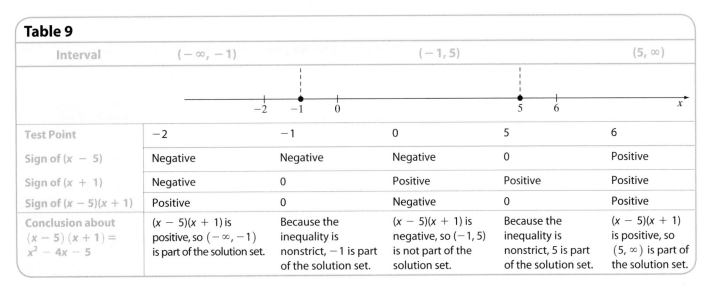

Interval	$(-\infty, -1)$		$(-1, 5)$		$(5, \infty)$
Test Point	-2	-1	0	5	6
Sign of $(x - 5)$	Negative	Negative	Negative	0	Positive
Sign of $(x + 1)$	Negative	0	Positive	Positive	Positive
Sign of $(x - 5)(x + 1)$	Positive	0	Negative	0	Positive
Conclusion about $(x - 5)(x + 1) = x^2 - 4x - 5$	$(x - 5)(x + 1)$ is positive, so $(-\infty, -1)$ is part of the solution set.	Because the inequality is nonstrict, -1 is part of the solution set.	$(x - 5)(x + 1)$ is negative, so $(-1, 5)$ is not part of the solution set.	Because the inequality is nonstrict, 5 is part of the solution set.	$(x - 5)(x + 1)$ is positive, so $(5, \infty)$ is part of the solution set.

Figure 33

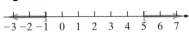

We want to know where $x^2 - 4x - 5$ is greater than or equal to zero, so we include -1 and 5 in the solution. The solution set is $\{x \mid x \le -1 \text{ or } x \ge 5\}$ in set-builder notation, or $(-\infty, 1] \cup [5, \infty)$ in interval notation. See Figure 33 for a graph of the solution set. ●

Quick ✓

In Problem 2, solve the quadratic inequality using the algebraic method.

2. $x^2 + 3x - 10 \ge 0$

Solving Quadratic Inequalities

Graphical Method	Algebraic Method
Step 1: Write the inequality so that $ax^2 + bx + c$ is on one side of the inequality and 0 is on the other.	**Step 1:** Write the inequality so that $ax^2 + bx + c$ is on one side of the inequality and 0 is on the other.
Step 2: Graph the function $f(x) = ax^2 + bx + c$. Label the x-intercepts of the graph.	**Step 2:** Determine the solutions to the equation $ax^2 + bx + c = 0$.
Step 3: From the graph, determine where the function is positive and where it is negative. This information will give you the solution set to the inequality.	**Step 3:** Use the solutions to the equation from Step 2 to separate the real number line into intervals.
	Step 4: Write $ax^2 + bx + c$ in factored form. Within each interval formed in Step 3, determine the sign of each factor. Then determine the sign of the product. Also write the value of $ax^2 + bx + c$ at each solution found in Step 2.
	(a) If the product of the factors is positive, then $ax^2 + bx + c > 0$ for all numbers x in the interval.
	(b) If the product of the factors is negative, then $ax^2 + bx + c < 0$ for all numbers x in the interval.

If the inequality is not strict ($\le$ or $\ge$), include the solutions of $ax^2 + bx + c = 0$ in the solution set.

Now let's do an example where we use both methods.

⊳ (EXAMPLE 3) **Solving a Quadratic Inequality**

Solve $-x^2 + 10 > 3x$ using both the graphical and the algebraic methods.

Solution

Graphical Method:
To make our lives easier, rearrange the inequality so that the coefficient of x^2 is positive.

$$-x^2 + 10 > 3x$$

Add x^2 to both sides; subtract 10 from both sides: $\qquad 0 > x^2 + 3x - 10$

$0 > b$ is equivalent to $b < 0$: $\quad x^2 + 3x - 10 < 0$

Graph $f(x) = x^2 + 3x - 10$. Because $a = 1 > 0$, the graph opens up. Find the intercepts and vertex of the function.

Figure 34

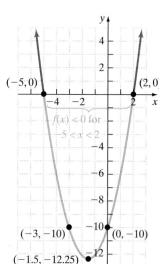

x-intercepts: $\qquad\qquad f(x) = 0 \qquad\qquad$ **y-intercept:** $\quad f(0) = -10$

$$x^2 + 3x - 10 = 0$$
$$(x + 5)(x - 2) = 0$$
$$x + 5 = 0 \quad \text{or} \quad x - 2 = 0$$
$$x = -5 \quad \text{or} \quad x = 2$$

Vertex:

$$x = -\frac{b}{2a} = -\frac{3}{2(1)} = -\frac{3}{2} = -1.5$$

$$y = f\left(-\frac{b}{2a}\right) = f\left(-\frac{3}{2}\right) = -\frac{49}{4} = -12.25$$

The vertex is at $(-1.5, -12.25)$.

Figure 34 shows the graph. The graph of $f(x) = x^2 + 3x - 10$ is below the x-axis (and therefore $x^2 + 3x - 10 < 0$) for $-5 < x < 2$. Because the inequality is strict ($<$) we do not include the x-intercepts in the solution. Thus the solution is $\{x | -5 < x < 2\}$ in set-builder notation, or $(-5, 2)$ in interval notation.

Algebraic Method:
Rearrange the inequality so that the coefficient of x^2 is positive.

$$-x^2 + 10 > 3x$$

Add x^2 to both sides; subtract 10 from both sides: $\qquad 0 > x^2 + 3x - 10$

$0 > b$ Is equivalent to $b < 0$: $\quad x^2 + 3x - 10 < 0$

Now solve the equation

$$x^2 + 3x - 10 = 0$$
$$(x + 5)(x - 2) = 0$$
$$x + 5 = 0 \quad \text{or} \quad x - 2 = 0$$
$$x = -5 \quad \text{or} \quad x = 2$$

Use the solutions to separate the real number line into the following intervals:

$$(-\infty, -5) \qquad (-5, 2) \qquad (2, \infty)$$

The factored form of $x^2 + 3x - 10$ is $(x + 5)(x - 2)$. Table 10 shows the sign of each factor and of $(x + 5)(x - 2)$ in each interval. We also list the value of $(x + 5)(x - 2)$ at $x = -5$ and at $x = 2$.

Table 10

Interval	$(-\infty, -5)$		$(-5, 2)$		$(2, \infty)$

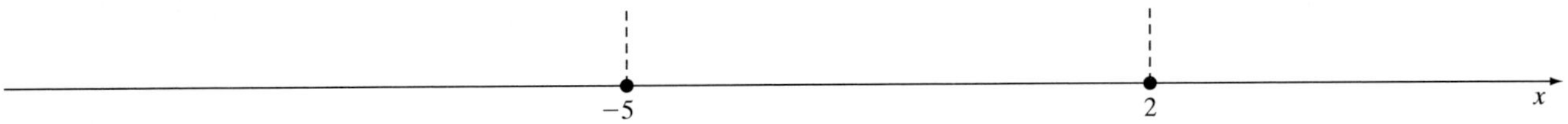

Test Point	-6	-5	0	2	3
Sign of $(x + 5)$	Negative	0	Positive	Positive	Positive
Sign of $(x - 2)$	Negative	Negative	Negative	0	Positive
Sign of $(x + 5)(x - 2)$	Positive	0	Negative	0	Positive
Conclusion about $(x + 5)(x - 2) < 0$	$(x + 5)(x - 2)$ is positive in the interval $(-\infty, -5)$, so it is not part of the solution set.	The inequality is strict, so -5 is not part of the solution set.	$(x + 5)(x - 2)$ is negative in the interval $(-5, 2)$, so it is part of the solution set.	The inequality is strict, so 2 is not part of the solution set.	$(x + 5)(x - 2)$ is positive in the interval $(2, \infty)$, so it is not part of the solution set.

The inequality is strict, so we do not include -5 or 2 in the solution. The solution set is $\{x | -5 < x < 2\}$ in set-builder notation, or $(-5, 2)$ in interval notation. Figure 35 shows the graph of the solution set.

Figure 35

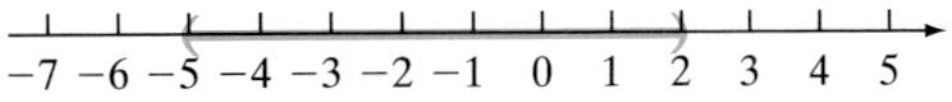

Quick ✓

In Problem 3, solve the quadratic inequality. Graph the solution set.

3. $-x^2 > 2x - 24$

EXAMPLE 4   **Solving a Quadratic Inequality**

Solve $2x^2 > 4x - 1$ using both the graphical and the algebraic methods. Graph the solution set.

Solution

Graphical Method:

$$2x^2 > 4x - 1$$

Subtract 4x from both sides; add 1 to both sides: $2x^2 - 4x + 1 > 0$

Graph $f(x) = 2x^2 - 4x + 1$ to determine where $f(x) > 0$. Because $a = 2 > 0$, the graph opens up.

x-intercepts: $f(x) = 0$ 　　　　　　　　　 **y-intercept:** $f(0) = 1$

$$2x^2 - 4x + 1 = 0$$

$a = 2; b = -4; c = 1$ 　　$x = \dfrac{-(-4) \pm \sqrt{(-4)^2 - 4(2)(1)}}{2(2)}$

$$= \frac{4 \pm \sqrt{8}}{4}$$

$$= 1 \pm \frac{\sqrt{2}}{2}$$

$$\approx 0.29 \text{ or } 1.71$$

Vertex: $\qquad x = -\dfrac{b}{2a} = -\dfrac{-4}{2(2)} = 1$

$$f\left(-\frac{b}{2a}\right) = f(1) = -1$$

The vertex is at $(1, -1)$. Figure 36 shows the graph.

Figure 36

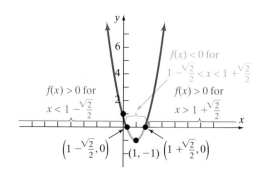

The graph shown in Figure 36 shows that the graph of $f(x) = 2x^2 - 4x + 1$ is greater than 0 for $x < 1 - \dfrac{\sqrt{2}}{2}$ or $x > 1 + \dfrac{\sqrt{2}}{2}$. Because the inequality is strict, we do not include the x-intercepts in the solution. Thus the solution is $\left\{x \,\middle|\, x < 1 - \dfrac{\sqrt{2}}{2} \text{ or } x > 1 + \dfrac{\sqrt{2}}{2}\right\}$ in set-builder notation, or $\left(-\infty, 1 - \dfrac{\sqrt{2}}{2}\right) \cup \left(1 + \dfrac{\sqrt{2}}{2}, \infty\right)$ in interval notation.

Algebraic Method:

First, write the inequality in the form $ax^2 + bx + c > 0$.

$$2x^2 > 4x - 1$$

Subtract 4x from both sides; add 1 to both sides: $\qquad 2x^2 - 4x + 1 > 0$

Now determine the solutions to the equation $2x^2 - 4x + 1 = 0$.

$$x = \frac{-(-4) \pm \sqrt{(-4)^2 - 4(2)(1)}}{2(2)}$$

$$= \frac{4 \pm \sqrt{8}}{4}$$

$$= 1 \pm \frac{\sqrt{2}}{2}$$

$$\approx 0.29 \text{ or } 1.71$$

Separate the real number line into the following intervals:

$$\left(-\infty, 1 - \frac{\sqrt{2}}{2}\right) \quad \left(1 - \frac{\sqrt{2}}{2}, 1 + \frac{\sqrt{2}}{2}\right) \quad \left(1 + \frac{\sqrt{2}}{2}, \infty\right)$$

Work Smart

The factor $\left(x - \left(1 - \dfrac{\sqrt{2}}{2}\right)\right)$ is close to the factor $x - 0.29$. Similarly, use $x - 1.71$ as an approximation of $\left(x - \left(1 + \dfrac{\sqrt{2}}{2}\right)\right)$.

The solutions to the equation $2x^2 - 4x + 1 = 0$ enable us to factor $2x^2 - 4x + 1$ as $\left(x - \left(1 - \dfrac{\sqrt{2}}{2}\right)\right)\left(x - \left(1 + \dfrac{\sqrt{2}}{2}\right)\right)$. Set up Table 11, using 0.29 as an approximation of $1 - \dfrac{\sqrt{2}}{2}$ and 1.71 as an approximation for $1 + \dfrac{\sqrt{2}}{2}$ to help determine the sign of each factor.

Table 11

Interval	$\left(-\infty, 1 - \dfrac{\sqrt{2}}{2}\right)$		$\left(1 - \dfrac{\sqrt{2}}{2}, 1 + \dfrac{\sqrt{2}}{2}\right)$		$\left(1 + \dfrac{\sqrt{2}}{2}, \infty\right)$

$1 - \dfrac{\sqrt{2}}{2} \approx 0.29$ $1 + \dfrac{\sqrt{2}}{2} \approx 1.71$

Test Point	0	$1 - \dfrac{\sqrt{2}}{2}$	1	$1 + \dfrac{\sqrt{2}}{2}$	2
Sign of $\left(x - \left(1 - \dfrac{\sqrt{2}}{2}\right)\right)$ $\approx (x - 0.29)$	Negative	0	Positive	Positive	Positive
Sign of $\left(x - \left(1 + \dfrac{\sqrt{2}}{2}\right)\right)$ $\approx (x - 1.71)$	Negative	Negative	Negative	0	Positive
Sign of $\left(x - \left(1 - \dfrac{\sqrt{2}}{2}\right)\right)$ $\left(x - \left(1 + \dfrac{\sqrt{2}}{2}\right)\right)$	Positive	0	Negative	0	Positive
Conclusion about $2x^2 - 4x + 1 > 0$	$2x^2 - 4x + 1$ is positive in the interval $\left(-\infty, 1 - \dfrac{\sqrt{2}}{2}\right)$, so it is part of the solution set.	Since the inequality is strict, $1 - \dfrac{\sqrt{2}}{2}$ is not part of the solution set.	$2x^2 - 4x + 1$ is negative in the interval $\left(1 - \dfrac{\sqrt{2}}{2}, 1 + \dfrac{\sqrt{2}}{2}\right)$, so it is not part of the solution set.	Since the inequality is strict, $1 + \dfrac{\sqrt{2}}{2}$ is not part of the solution set.	$2x^2 - 4x + 1$ is positive in the interval $\left(1 + \dfrac{\sqrt{2}}{2}, \infty\right)$, so it is part of the solution set.

Work Smart

Instead of using the signs of the factors, we could simply evaluate $2x^2 - 4x + 1$ at each test point. For example, at $x = 0$, $2x^2 - 4x + 1 = 2(0)^2 - 4(0) + 1 = 1 > 0$. Therefore, $2x^2 - 4x + 1 > 0$ for all $x < 1 - \dfrac{\sqrt{2}}{2}$.

We want to know where $2x^2 - 4x + 1$ is positive, so we do not include $1 - \dfrac{\sqrt{2}}{2}$ or $1 + \dfrac{\sqrt{2}}{2}$ in the solution. Thus the solution set is $\left\{x \,\middle|\, x < 1 - \dfrac{\sqrt{2}}{2} \text{ or } x > 1 + \dfrac{\sqrt{2}}{2}\right\}$ in set-builder notation, or $\left(-\infty, 1 - \dfrac{\sqrt{2}}{2}\right) \cup \left(1 + \dfrac{\sqrt{2}}{2}, \infty\right)$ in interval notation.

Figure 37 shows the graph of the solution set.

Figure 37

Quick ✓

4. Solve $3x^2 > -x + 5$ using both the graphical and the algebraic methods. Graph the solution set.

② Solve Polynomial Inequalities

We now discuss solving polynomial inequalities. Because we do not have experience graphing polynomial functions, we focus only on the algebraic method. To solve polynomial inequalities algebraically, we use the same steps used to solve quadratic inequalities algebraically.

EXAMPLE 5 **Solving a Polynomial Inequality of Degree 3**

Solve $x^3 + x^2 - 9x - 9 > 0$ algebraically and graph the solution set.

Solution

The inequality is already written so that the polynomial is on one side of the inequality and 0 is on the other side. Therefore, we solve the equation $x^3 + x^2 - 9x - 9 = 0$.

$$x^3 + x^2 - 9x - 9 = 0$$

Factor by grouping: $\quad x^2(x + 1) - 9(x + 1) = 0$

$$(x^2 - 9)(x + 1) = 0$$

$$(x + 3)(x - 3)(x + 1) = 0$$

Set each factor equal to 0: $\quad x + 3 = 0 \quad$ or $\quad x - 3 = 0 \quad$ or $\quad x + 1 = 0$

$$x = -3 \qquad\qquad x = 3 \qquad\qquad x = -1$$

Notice that the solutions of the equation $x^3 + x^2 - 9x - 9 = 0$ separate the number line into *four* intervals: $(-\infty, -3)$, $(-3, -1)$, $(-1, 3)$ and $(3, \infty)$. Determine the sign of each of the factors $(x + 3)$, $(x - 3)$, and $(x + 1)$, and then find the sign of the product. Table 12 shows the sign of each factor and the sign of $(x + 3)(x - 3)(x + 1)$ in each interval. We also list the value of $(x + 3)(x - 3)(x + 1)$ at $x = -3, x = -1$, and $x = 3$.

The inequality is strict, so we do not include –3, –1, or 3 in the solution set. The solution of the inequality $x^3 + x^2 - 9x - 9 > 0$ is $\{x | -3 < x < -1 \text{ or } x > 3\}$ in set-builder notation, or $(-3, -1) \cup (3, \infty)$ in interval notation. Figure 38 shows the graph of the solution set.

Figure 38

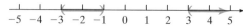

Quick ✓

Solve Problems 5 and 6 algebraically. Graph the solution set.

5. $(2x - 3)(x + 4)(x - 5) \leq 0$

6. $x^3 + 7x^2 - 4x - 28 > 0$

Table 12

Interval	$(-\infty, -3)$		$(-3, -1)$		$(-1, 3)$		$(3, \infty)$
Test Point	-4	-3	-2	-1	1	3	4
Sign of $(x + 3)$	Negative	0	Positive	Positive	Positive	Positive	Positive
Sign of $(x - 3)$	Negative	Negative	Negative	Negative	Negative	0	Positive
Sign of $(x + 1)$	Negative	Negative	Negative	0	Positive	Positive	Positive
Sign of $(x + 3)(x - 3)(x + 1)$	Negative	0	Positive	0	Negative	0	Positive
Conclusion about $(x + 3)(x - 3)(x + 1) > 0$	The product of the factors is negative, so $(-\infty, -3)$ is not part of the solution set.	Since the inequality is strict, $x = -3$, is not in the solution set.	The product of the factors is positive, so $(-3, -1)$ is part of the solution set.	Since the inequality is strict, $x = -1$, is not in the solution set.	The product is negative, so $(-1, 3)$ is not part of the solution set.	Since the inequality is strict, $x = 3$ is not in the solution set.	The product of factors is positive, so $(3, \infty)$ is part of the solution set.

7.6 Exercises MyMathLab® PRACTICE | Exercise numbers in green have complete video solutions in MyMathLab.

*Problems **1–6** are the* Quick ✓s *that follow the* EXAMPLES.

Building Skills

In Problems 7–10, use the graphs of the quadratic function f to determine the solution.

7. (a) Solve $f(x) > 0$. **(b)** Solve $f(x) \le 0$.

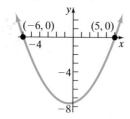

8. (a) Solve $f(x) > 0$. **(b)** Solve $f(x) \le 0$.

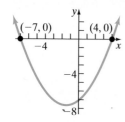

9. (a) $f(x) \ge 0$ **(b)** $f(x) < 0$

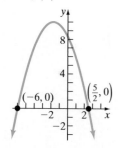

10. (a) $f(x) > 0$ **(b)** $f(x) \le 0$

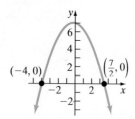

In Problems 11–38, solve each quadratic inequality. Graph the solution set. See Objective 1.

11. $(x - 5)(x + 2) \ge 0$ **12.** $(x - 8)(x + 1) \le 0$

13. $(x + 3)(x + 7) < 0$ **14.** $(x - 4)(x - 10) > 0$

15. $x^2 - 2x - 35 > 0$ **16.** $x^2 + 3x - 18 \ge 0$

17. $n^2 - 6n - 8 \le 0$ **18.** $p^2 + 5p + 4 < 0$

19. $m^2 + 5m \ge 14$ **20.** $z^2 > 7z + 8$

21. $2q^2 \ge q + 15$ **22.** $2b^2 + 5b < 7$

23. $3x + 4 \ge x^2$ **24.** $x + 6 < x^2$

25. $-x^2 + 3x < -10$ **26.** $-x^2 > 4x - 21$

27. $-3x^2 \le -10x - 8$ **28.** $-3m^2 \ge 16m + 5$

29. $x^2 + 4x + 1 < 0$ **30.** $x^2 - 3x - 5 \ge 0$

31. $-2a^2 + 7a \ge -4$ **32.** $-3p^2 < 3p - 5$

33. $z^2 + 2z + 3 > 0$ **34.** $y^2 + 3y + 5 \ge 0$

35. $2b^2 + 5b \le -6$ **36.** $3w^2 + w < -2$

37. $x^2 - 6x + 9 > 0$ **38.** $p^2 - 8p + 16 \le 0$

In Problems 39–48, solve each polynomial inequality. Graph the solution set. See Objective 2.

39. $(x + 1)(x - 2)(x - 5) > 0$

40. $(x + 3)(x - 1)(x - 3) < 0$

41. $(2x + 1)(x - 4)(x - 9) \le 0$

42. $(3x + 4)(x - 2)(x - 6) \ge 0$

43. $(x + 3)(2x^2 - x - 1) \ge 0$

44. $(x - 2)(3x^2 + x - 2) \le 0$

45. $x^3 + 3x^2 - 4x - 12 \le 0$

46. $x^3 + 4x^2 - 9x - 36 \geq 0$

47. $4x^3 + 16x^2 - 9x - 36 > 0$

48. $3x^3 + 5x^2 - 12x - 20 < 0$

Mixed Practice

In Problems 49–56, for each function find the values of x that satisfy the given condition.

49. Solve $f(x) < 0$ if $f(x) = x^2 - 5x$.

50. Solve $f(x) > 0$ if $f(x) = x^2 + 4x$.

51. Solve $f(x) \geq 0$ if $f(x) = x^2 - 3x - 28$.

52. Solve $f(x) \leq 0$ if $f(x) = x^2 + 2x - 48$.

53. Solve $g(x) > 0$ if $g(x) = 2x^2 + x - 10$.

54. Solve $F(x) < 0$ if $F(x) = 2x^2 + 7x - 15$.

55. Solve $f(x) < 0$ if $f(x) = 4x^3 - x^2 - 14x$.

56. Solve $g(x) > 0$ if $g(x) = 2x^3 - 7x^2 - 9x$.

In Problems 57–60, find the domain of the given function.

57. $f(x) = \sqrt{x^2 + 8x}$ **58.** $f(x) = \sqrt{x^2 - 5x}$

59. $g(x) = \sqrt{x^2 - x - 30}$ **60.** $G(x) = \sqrt{x^2 + 2x - 63}$

Applying the Concepts

61. Physics A ball is thrown vertically upward with an initial speed of 80 feet per second from a cliff 500 feet above sea level. The height s (in feet) of the ball from the ground after t seconds is $s(t) = -16t^2 + 80t + 500$. For what time t is the ball more than 596 feet above sea level?

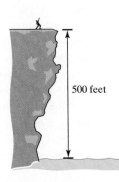

500 feet

62. Physics A water balloon is thrown vertically upward with an initial speed of 64 feet per second from the top of a building 200 feet above the ground. The height s (in feet) of the balloon from the ground after t seconds is $s(t) = -16t^2 + 64t + 200$. For what time t is the balloon more than 248 feet above the ground?

63. Revenue Function Suppose that the marketing department of Panasonic has found that, when a certain model of DVD player is sold at a price of p dollars, the daily revenue R (in dollars) as a function of the price p is $R(p) = -2.5p^2 + 600p$. Determine the prices for which revenue will exceed \$35,750. That is, solve $R(p) > 35,750$.

64. Revenue Function Suppose that the marketing department of Samsung has found that, when a certain model of cellular telephone is sold at a price of p dollars, the daily revenue R (in dollars) as a function of the price p is $R(p) = -5p^2 + 600p$. Determine the prices for which revenue will exceed \$17,500. That is, solve $R(p) > 17,500$.

Extending the Concepts

In Problems 65–68, solve each inequality algebraically by inspection. Then provide a verbal explanation of the solution.

65. $(x + 3)^2 \leq 0$

66. $(x - 4)^2 > 0$

67. $(x + 8)^2 > -2$

68. $(3x + 1)^2 < -2$

69. Write a quadratic inequality that has $[-3, 2]$ as the solution set.

70. Write a quadratic inequality that has $(0, 5)$ as the solution set.

In Problems 71–76, use the techniques presented in this section to solve each inequality algebraically.

71. $(x + 1)(x - 2)(x + 3)(x - 4) > 0$

72. $(x - 4)(x - 1)(x + 5)(x + 6) < 0$

73. $x^4 - 29x^2 + 100 \leq 0$

74. $x^4 - 10x^2 + 9 \geq 0$

75. $\dfrac{x^2 + 5x + 6}{x - 2} > 0$

76. $\dfrac{x^2 - 3x - 10}{x + 1} < 0$

Explaining the Concepts

77. The inequalities $(3x + 2)^2 < 2$ and $(3x + 2)^{-2} > \dfrac{1}{2}$ have the same solution set. Why?

78. The inequality $x^2 + 3 < -2$ has no solution. Explain why.

79. The inequality $x^2 - 1 \geq -1$ has the set of all real numbers as the solution. Explain why.

80. Explain when the endpoints of an interval are included in the solution set of a quadratic inequality.

81. Is the inequality $x^2 + 1 > 1$ true for all real numbers? Explain.

82. Explain how to solve the quadratic inequality $f(x) < 0$ from the graph of $y = f(x)$, where f is a quadratic function.

83. Explain how to solve the quadratic inequality $f(x) > 0$ from the graph of $y = f(x)$, where f is a quadratic function.

Synthesis Review

In Problems 84–87, simplify each expression.

84. $\dfrac{3a^4 b}{12a^{-3} b^5}$

85. $(4mn^{-3})(-2m^4 n)$

86. $\left(\dfrac{3x^4 y}{6x^{-2} y^5}\right)^{1/2}$

87. $\left(\dfrac{9a^{2/3} b^{1/2}}{a^{-1/9} b^{3/4}}\right)^{-1}$

88. Do the Laws of Exponents presented in the "Getting Ready: Laws of Exponents" section also apply to the Laws of Exponents for rational exponents presented in Section 6.1?

The Graphing Calculator

A graphing calculator may be used to solve the quadratic inequality in Example 1 by graphing $Y_1 = x^2 - 4x - 5$. We use the ZERO feature of the calculator to find the x-intercepts of the graph. See Figures 37(a) and (b).

Figure 37

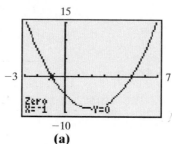
(a)

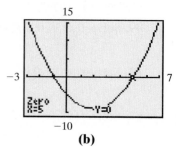
(b)

From Figures 37(a) and (b), we can see that $x^2 - 4x - 5 \geq 0$ for $x \leq -1$ or $x \geq 5$.

In Problems 89–92, solve each quadratic inequality using a graphing calculator.

89. $2x^2 + 7x - 49 > 0$

90. $2x^2 + 3x - 27 < 0$

91. $6x^2 + x \leq 40$

92. $8x^2 + 18x \geq 81$

Chapter 7 Activity: Presidential Decision Making

Focus: Developing quadratic equations.

Time: 30–35 minutes

Group size: 2–4

1. Your boss, Huntington Corporation's president, Gerald Cain, is very concerned about his approval rating with his employees. Last year, on January 1, he made some policy changes and saw his approval rating drop. In fact, his approval rating was 48% just before he made the policy changes. One month later, his rating was at 41%, two months later it was at 40%, and at three months it began to climb and was 45%. He discovered that the following function described his approval rating for that year, where x is the month:

$$R(x) = 3x^2 - 10x + 48$$

(a) As a group, use the above function to find when President Cain's approval rating will return to the original rating.

(b) If his approval rating continues to climb, when will he reach a 68% approval rating?

2. On January 1 of this year, President Cain surveyed his employees again and found that 68% of them approved of his leadership skills. At this time, President Cain decided to become very strict with his employees and began a series of new policies. He noticed that his approval rating began steadily to slip and reached an all-time low of 38% on March 30 (3 months later). President Cain is not worried because he knows from last year that this drop in popularity will bottom out and eventually rise. He believes his popularity can be modeled by a quadratic function and needs your help. He has more bad news to deliver but does not want to begin the next round of policy changes until his approval rating is back to approximately 50%.

(a) As a group, write a quadratic function that would model President Cain's approval rating.

(b) As a group, develop different ways to advise President Cain on what date to begin his policy changes. Use graphs and computations to prove your point.

Chapter 7 Review

Section 7.1   Solving Quadratic Equations by Completing the Square

KEY CONCEPTS

- **Square Root Property**
  If $x^2 = p$, then $x = \sqrt{p}$ or $x = -\sqrt{p}$.

- **Pythagorean Theorem**
  In a right triangle, the square of the length of the hypotenuse is equal to the sum of the squares of the lengths of the legs. That is, $\text{leg}^2 + \text{leg}^2 = \text{hypotenuse}^2$.

KEY TERMS

Completing the square
Right triangle
Right angle
Hypotenuse
Legs

You Should Be Able To...	EXAMPLE	Review Exercises
① Solve quadratic equations using the Square Root Property (p. 542)	Examples 1 through 4	1–10
② Complete the square in one variable (p. 545)	Example 5	11–16
③ Solve quadratic equations by completing the square (p. 546)	Examples 6 and 7	17–26
④ Solve problems using the Pythagorean Theorem (p. 548)	Examples 8 and 9	27–36

In Problems 1–10, solve each equation using the Square Root Property.

1. $m^2 = 169$

2. $n^2 = 75$

3. $a^2 = -16$

4. $b^2 = \dfrac{8}{9}$

5. $(x - 8)^2 = 81$

6. $(y - 2)^2 - 62 = 88$

7. $(3z + 5)^2 = 100$

8. $7p^2 = 18$

9. $3q^2 + 251 = 11$

10. $\left(x + \dfrac{3}{4}\right)^2 = \dfrac{13}{16}$

In Problems 11–16, complete the square in each expression. Then factor the perfect square trinomial.

11. $a^2 + 30a$

12. $b^2 - 14b$

13. $c^2 - 11c$

14. $d^2 + 9d$

15. $m^2 - \dfrac{1}{4}m$

16. $n^2 + \dfrac{6}{7}n$

In Problems 17–26, solve each quadratic equation by completing the square.

17. $x^2 - 10x + 16 = 0$

18. $y^2 - 3y - 28 = 0$

19. $z^2 - 6z - 3 = 0$

20. $a^2 - 5a - 7 = 0$

21. $b^2 + b + 7 = 0$

22. $c^2 - 6c + 17 = 0$

23. $2d^2 - 7d + 3 = 0$

24. $2w^2 + 2w + 5 = 0$

25. $3x^2 - 9x + 8 = 0$

26. $3x^2 + 4x - 2 = 0$

In Problems 27–32, the lengths of the legs of a right triangle are given. Find the hypotenuse.

27. $a = 9, b = 12$

28. $a = 8, b = 8$

29. $a = 3, b = 6$

30. $a = 10, b = 24$

31. $a = 5, b = \sqrt{11}$

32. $a = 6, b = \sqrt{13}$

In Problems 33–35, use the right triangle shown below to find the missing length.

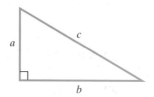

33. $a = 9, c = 12$ **34.** $b = 5, c = 10$

35. $b = 6, c = 17$

36. Baseball Diamond A baseball diamond is really a square that is 90 feet long on each side. (See the figure.) What is the distance between home plate and second base? Round your answer to the nearest tenth of a foot.

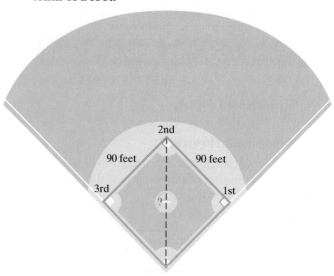

Section 7.2 Solving Quadratic Equations by the Quadratic Formula

KEY CONCEPTS

- **The Quadratic Formula**

  The solutions to the equation $ax^2 + bx + c = 0, a \neq 0$, are given by $x = \dfrac{-b \pm \sqrt{b^2 - 4ac}}{2a}$

- **Discriminant**

  For the quadratic equation $ax^2 + bx + c = 0, a \neq 0$:

  - If $b^2 - 4ac > 0$, the equation has two unequal real solutions.
    - If $b^2 - 4ac$ is a perfect square, the equation has two rational solutions provided a, b, and c are rational numbers.
    - If $b^2 - 4ac$ is not a perfect square, the equation has two irrational solutions provided a, b, and c are rational numbers.
  - If $b^2 - 4ac = 0$, the equation has a repeated real solution.
  - If $b^2 - 4ac < 0$, the equation has two complex solutions that are not real.

KEY TERM

Discriminant

You Should Be Able To...	EXAMPLE	Review Exercises
❶ Solve quadratic equations using the quadratic formula (p. 554)	Examples 1 through 4	37–46
❷ Use the discriminant to determine the nature of solutions of a quadratic equation (p. 558)	Example 5	47–52
❸ Model and solve problems involving quadratic equations (p. 562)	Examples 6 and 7	63–68

In Problems 37–46, solve each equation using the quadratic formula.

37. $x^2 - x - 20 = 0$

38. $4y^2 = 8y + 21$

39. $3p^2 + 8p = -3$

40. $2q^2 - 3 = 4q$

41. $3w^2 + w = -3$

42. $9z^2 + 16 = 24z$

43. $m^2 - 4m + 2 = 0$

44. $5n^2 + 4n + 1 = 0$

45. $5x + 13 = -x^2$

46. $-2y^2 = 6y + 7$

In Problems 47–52, determine the discriminant of each quadratic equation. Use the value of the discriminant to determine whether the quadratic equation has two rational solutions, two irrational solutions, one repeated real solution, or two complex solutions that are not real.

47. $p^2 - 5p - 8 = 0$

48. $m^2 + 8m + 16 = 0$

49. $3n^2 + n = -4$

50. $7w^2 + 3 = 8w$

51. $4x^2 + 49 = 28x$

52. $11z - 12 = 2z^2$

In Problems 53–62, solve each equation using any appropriate method.

53. $x^2 + 8x - 9 = 0$

54. $6p^2 + 13p = 5$

55. $n^2 + 13 = -4n$

56. $5y^2 - 60 = 0$

57. $\frac{1}{4}q^2 - \frac{1}{2}q - \frac{3}{8} = 0$

58. $\frac{1}{8}m^2 + m + \frac{5}{2} = 0$

59. $(w - 8)(w + 6) = -33$

60. $(x - 3)(x + 1) = -2$

61. $9z^2 = 16$

62. $\dfrac{1 - 2x}{x^2 + 5} = 1$

63. Pythagorean Theorem Use the Pythagorean Theorem to determine the value of x and the measurements of the right triangle shown below.

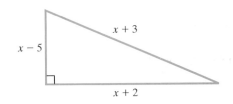

64. Area The area of a rectangle is 108 square centimeters. The width of the rectangle is 3 centimeters less than the length. What are the dimensions of the rectangle?

65. Revenue The revenue R received by a company selling x cellular phones per week is given by the function $R(x) = -0.2x^2 + 180x$.

  (a) How many cellular phones must be sold in order for revenue to be \$36,000 per week?

  (b) How many cellular phones must be sold in order for revenue to be \$40,500 per week?

66. Projectile Motion The height s of a ball after t seconds, when it is thrown straight up with an initial speed of 50 feet per second from an initial height of 180 feet, can be modeled by the function $s(t) = -16t^2 + 50t + 180$.

  (a) When will the height of the ball be 200 feet? Round your answer to the nearest tenth of a second.

  (b) When will the height of the ball be 100 feet? Round your answer to the nearest tenth of a second.

  (c) Will the ball ever reach a height of 300 feet? How does the result of the equation tell you this?

67. Pleasure Boat Ride A pleasure boat carries passengers 10 miles upstream and then returns to the starting point. The total time of the trip (excluding the time on land) is 2 hours. If the speed of the current is 3 miles per hour, find the speed of the boat in still water. Round your answer to the nearest tenth of a mile per hour.

68. Work Together, Tom and Beth can wash their car in 30 minutes. By himself, Tom can wash the car in 14 minutes less time than Beth can by herself. How long will it take Beth to wash the car by herself? Round your answer to the nearest tenth of a minute.

Section 7.3 Solving Equations Quadratic in Form

KEY TERM

Equation quadratic in form

You Should Be Able To...	EXAMPLE	Review Exercises
❶ Solve equations that are quadratic in form (p. 568)	Examples 1 through 5	69–80

In Problems 69–78, solve each equation.

69. $x^4 + 7x^2 - 144 = 0$

70. $4w^4 + 5w^2 - 6 = 0$

71. $3(a + 4)^2 - 11(a + 4) + 6 = 0$

72. $(q^2 - 11)^2 - 2(q^2 - 11) - 15 = 0$

73. $y - 13\sqrt{y} + 36 = 0$

74. $5z + 2\sqrt{z} - 3 = 0$

75. $p^{-2} - 4p^{-1} - 21 = 0$

76. $2b^{\frac{2}{3}} + 13b^{\frac{1}{3}} - 7 = 0$

77. $m^{\frac{1}{2}} + 2m^{\frac{1}{4}} - 8 = 0$

78. $\left(\dfrac{1}{x + 5}\right)^2 + \dfrac{3}{x + 5} = 28$

In Problems 79 and 80, find the zeros of the function.

79. $f(x) = 4x - 20\sqrt{x} + 21$

80. $g(x) = x^4 - 17x^2 + 60$

Section 7.4 Graphing Quadratic Functions Using Transformations

KEY CONCEPTS

- **Graphing a Function of the Form $f(x) = x^2 + k$ or $f(x) = x^2 - k$**
  To obtain the graph of $f(x) = x^2 + k, k > 0$, from the graph of $y = x^2$, shift the graph of $y = x^2$ vertically up k units. To obtain the graph of $f(x) = x^2 - k, k > 0$, from the graph of $y = x^2$ shift the graph of $y = x^2$ vertically down k units.

- **Graphing a Function of the Form $f(x) = (x - h)^2$ or $f(x) = (x + h)^2$**
  To obtain the graph of $f(x) = (x - h)^2, h > 0$, from the graph of $y = x^2$, shift the graph of $y = x^2$ horizontally to the right h units. To obtain the graph of $f(x) = (x + h)^2, h > 0$, from the graph of $y = x^2$ shift the graph of $y = x^2$ horizontally left h units.

- **Graphing a Function of the Form $f(x) = ax^2$**
  To obtain the graph of $f(x) = ax^2$ from the graph of $y = x^2$, multiply each y-coordinate on the graph of $y = x^2$ by a.

KEY TERMS

Quadratic function
Vertically stretched
Vertically compressed
Parabola
Opens up
Opens down
Vertex
Axis of symmetry
Transformations

You Should Be Able To...	EXAMPLE	Review Exercises
① Graph quadratic functions of the form $f(x) = x^2 + k$ (p. 577)	Examples 1 and 2	81–82; 87–90
② Graph quadratic functions of the form $f(x) = (x - h)^2$ (p. 578)	Examples 3 through 5	83–84; 87–90
③ Graph quadratic functions of the form $f(x) = ax^2$ (p. 581)	Example 6	85–86; 89–90
④ Graph quadratic functions of the form $f(x) = ax^2 + bx + c$ (p. 582)	Examples 7 and 8	91–96
⑤ Find a quadratic function from its graph (p. 585)	Example 9	97–100

In Problems 81–90, use the graph of $y - x^2$ to graph the quadratic function.

81. $f(x) = x^2 + 4$

82. $g(x) = x^2 - 5$

83. $h(x) = (x + 1)^2$

84. $F(x) = (x - 4)^2$

85. $G(x) = -4x^2$

86. $H(x) = \dfrac{1}{5}x^2$

87. $p(x) = (x - 4)^2 - 3$

88. $P(x) = (x + 4)^2 + 2$

89. $f(x) = -(x - 1)^2 + 4$

90. $F(x) = \dfrac{1}{2}(x + 2)^2 - 1$

In Problems 91–96, graph each quadratic function using transformations. Determine the vertex and axis of symmetry. Based on the graph, determine the domain and range of each function.

91. $g(x) = x^2 - 6x + 10$

92. $G(x) = x^2 + 8x + 11$

93. $h(x) = 2x^2 - 4x - 3$

94. $H(x) = -x^2 - 6x - 10$

95. $p(x) = -3x^2 + 12x - 8$

96. $P(x) = \dfrac{1}{2}x^2 - 2x + 5$

In Problems 97–100, determine the quadratic function whose graph is given.

97.

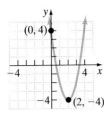

98.

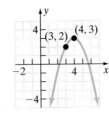

99.

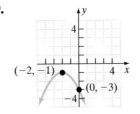

100.

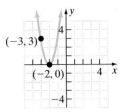

Section 7.5 Graphing Quadratic Functions Using Properties

KEY CONCEPTS

- **Vertex of a Parabola**

  Any quadratic function of the form $f(x) = ax^2 + bx + c$, $a \neq 0$, will have vertex

  $$\left(-\frac{b}{2a}, f\left(-\frac{b}{2a}\right)\right)$$

- **The *x*-Intercepts of the Graph of a Quadratic Function**

  1. If $b^2 - 4ac > 0$, the graph of $f(x) = ax^2 + bx + c$ has two *x*-intercepts.

  2. If $b^2 - 4ac = 0$, the graph of $f(x) = ax^2 + bx + c$ has one *x*-intercept.

  3. If $b^2 - 4ac < 0$, the graph of $f(x) = ax^2 + bx + c$ has no *x*-intercepts.

KEY TERMS

Maximum value
Minimum value
Optimization

You Should Be Able To...	EXAMPLE	Review Exercises
① Graph quadratic functions of the form $f(x) = ax^2 + bx + c$ (p. 588)	Examples 1 through 4	101—108
② Find the maximum or minimum value of a quadratic function (p. 594)	Examples 5 and 6	109–112
③ Model and solve optimization problems involving quadratic functions (p. 596)	Examples 7 and 8	113–118

In Problems 101–108, graph each quadratic function using its properties by following Steps 1–5 on page 591. Based on the graph, determine the domain and range of each function.

101. $f(x) = x^2 + 2x - 8$

102. $F(x) = 2x^2 - 5x + 3$

103. $g(x) = -x^2 + 6x - 7$

104. $G(x) = -2x^2 + 4x + 3$

105. $h(x) = 4x^2 - 12x + 9$

106. $H(x) = \frac{1}{3}x^2 + 2x + 3$

107. $p(x) = \frac{1}{4}x^2 + 3x + 10$

108. $P(x) = -x^2 + 4x - 9$

In Problems 109–112, determine whether the quadratic function has a maximum or a minimum value. Then find the maximum or minimum value and where it occurs.

109. $f(x) = -2x^2 + 16x - 10$

110. $g(x) = 6x^2 - 3x - 1$

111. $h(x) = -4x^2 + 8x + 3$

112. $F(x) = -\frac{1}{3}x^2 + 4x - 7$

113. Revenue Suppose that the marketing department of Zenith has found that, when a certain model of television is sold for a price of p dollars, the daily revenue R (in dollars) as a function of the price p is

$$R(p) = -\frac{1}{3}p^2 + 150p.$$

(a) For what price will the daily revenue be maximized?

(b) What is this maximum daily revenue?

114. Electrical Power In a 120-volt electrical circuit having a resistance of 16 ohms, the available power P (in watts) is given by the function $P(I) = -16I^2 + 120I$, where I represents the current (in amperes).

(a) What current will produce the maximum power in the circuit?

(b) What is this maximum power?

115. Fun with Numbers The sum of two numbers is 24. Find the numbers such that their product is a maximum.

116. Maximizing an Enclosed Area Becky has 15 yards of fencing to make a rectangular kennel for her dog. She will build the kennel next to her garage, so she only needs to enclose three sides. (See the figure.)

(a) What dimensions maximize the area of the kennel?

(b) What is this maximum area?

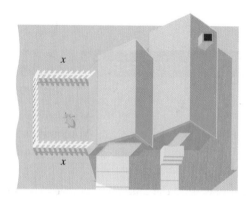

117. Kicking a Football Ted kicks a football at a 45° angle to the horizontal with an initial velocity of 80 feet per second. The model $h(x) = -0.005x^2 + x$ can be used to estimate the height h of the ball after it has traveled x feet.

(a) How far from Ted will the ball reach a maximum height?

(b) What is the maximum height of the ball?

(c) How far will the ball travel before it strikes the ground?

118. Revenue Monthly demand for automobiles at a certain dealership obeys the demand equation $x = -0.002p + 60$, where x is the quantity and p is the price (in dollars).

(a) Express the revenue R as a function of p. (*Hint*: $R = xp$)

(b) What price p maximizes revenue? What is the maximum revenue?

(c) How many automobiles will be sold at the revenue-maximizing price?

Section 7.6 Polynomial Inequalities

KEY CONCEPT

- The steps for solving a quadratic inequality are given on page 605.

KEY TERMS

Quadratic inequality Polynomial inequality

You Should Be Able To...	EXAMPLE	Review Exercises
❶ Solve quadratic inequalities (p. 602)	Examples 1 through 4	119–126
❷ Solve polynomial inequalities (p. 610)	Example 5	127, 128

In Problems 119 and 120, use the graphs of the quadratic function f to determine the solution.

119. (a) $f(x) > 0$ **(b)** $f(x) < 0$

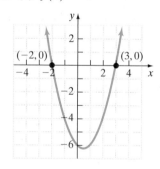

120. (a) $f(x) \geq 0$ **(b)** $f(x) \leq 0$

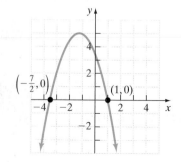

In Problems 121–128, solve each inequality. Graph the solution set.

121. $x^2 - 2x - 24 \leq 0$ **122.** $y^2 + 7y - 8 \geq 0$

123. $3z^2 - 19z + 20 > 0$ **124.** $p^2 + 4p - 2 < 0$

125. $4m^2 - 20m + 25 \geq 0$ **126.** $6w^2 - 19w - 7 \leq 0$

127. $(2x - 3)(x + 1)(x - 2) < 0$

128. $x^3 + 5x^2 - 9x - 45 \geq 0$

In Problems 1 and 2, complete the square in the given expression. Then factor the perfect square trinomial.

1. $x^2 - 3x$

2. $m^2 + \dfrac{2}{5}m$

In Problems 3–6, solve each equation using any appropriate method you prefer.

3. $9\left(x + \dfrac{4}{3}\right)^2 = 1$

4. $m^2 - 6m + 4 = 0$

5. $2w^2 - 4w + 3 = 0$

6. $\dfrac{1}{2}z^2 - \dfrac{3}{2}z = -\dfrac{7}{6}$

7. Determine the discriminant of $2x^2 + 5x = 4$. Use the value of the discriminant to determine whether the quadratic equation has two rational solutions, two irrational solutions, one repeated real solution, or two complex solutions that are not real.

8. Find the missing length in the right triangle shown below.

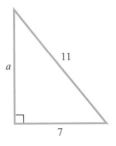

In Problems 9 and 10, solve each equation.

9. $x^4 - 5x^2 - 36 = 0$

10. $6y^{\frac{1}{2}} + 13y^{\frac{1}{4}} - 5 = 0$

In Problems 11 and 12, graph each quadratic function by determining the vertex, intercepts, and axis of symmetry. Based on the graph, determine the domain and range of the quadratic function.

11. $f(x) = (x + 2)^2 - 5$

12. $g(x) = -2x^2 - 8x - 3$

13. Determine the quadratic function whose graph is given.

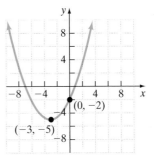

14. Determine whether the quadratic function
$$h(x) = -\dfrac{1}{4}x^2 + x + 5$$
has a maximum or a minimum value. Then find the maximum or minimum value.

In Problems 15 and 16, solve each inequality. Graph the solution set.

15. $2m^2 + m - 15 > 0$

16. $x^3 + 5x^2 - 4x - 20 \le 0$

17. Projectile Motion The height s of a rock after t seconds, when propelled straight up with an initial speed of 80 feet per second from an initial height of 20 feet, can be modeled by the function $s(t) = -16t^2 + 80t + 20$. When will the height of the rock be 50 feet? Round your answer to the nearest tenth of a second.

18. Work Together, Lex and Rupert can roof a house in 16 hours. By himself, Rex can roof the house in 4 hours less time than Rupert can by himself. How long will it take Rupert to roof the house by himself? Round your answer to the nearest tenth of an hour.

19. Revenue A small company has found that, when one of its products is sold for a price of p dollars, the weekly revenue (in dollars) as a function of price p is $R(p) = -0.25p^2 + 170p$.

(a) For what price will the weekly revenue from this product be maximized?

(b) What is this maximum weekly revenue?

20. Maximizing Volume A box with a rectangular base is to be constructed such that the perimeter of the base of the box is 50 inches. The height of the box must be 12 inches.

(a) Find the dimensions that maximize the volume of the box.

(b) What is this maximum volume?

Cumulative Review Chapters R–7

1. Evaluate: $\dfrac{2^4 - 5 \cdot 7 + 3}{1 - 3^2}$

2. Simplify: $2(3c + 1) - (c - 9) - 2c$

3. Solve: $p - 5 = 3(p - 2) - 9$

4. Solve: $5 + 2|x + 1| > 9$

5. Determine the domain of $h(x) = \dfrac{x^2 - 16}{x^2 + x - 12}$.

6. Graph the linear equation: $3x - 4y = 8$

7. Find the equation of the line that passes through the point $(5, -1)$ and is parallel to the graph of $2x + 5y = -15$. Write your answer in either slope-intercept or standard form, whichever you prefer.

8. Solve the following system of equations:
$$\begin{cases} 5x + 2y = 0 \\ 2x - y = -9 \end{cases}$$

9. Evaluate: $\begin{vmatrix} 4 & 1 & 0 \\ 1 & 2 & -1 \\ 3 & -2 & 1 \end{vmatrix}$

10. Graph the following system of linear inequalities:
$$\begin{cases} 2x + 3y < -3 \\ 2x + y > -5 \end{cases}$$

In Problems 11 and 12, add, subtract, multiply, or divide as indicated.

11. $(a^3 - 9a^2 + 11) + (7a^2 - 5a - 8)$

12. $\dfrac{8x^4 + 12x^2 + 17x - 18}{2x^2 - x + 5}$

In Problems 13–14, factor completely.

13. $8m^3 + 27n^3$ **14.** $7y^2 + 23y - 20$

In Problems 15 and 25, perform the indicated operations. Be sure to express the final answer in lowest terms.

15. $\dfrac{\dfrac{2w^2 - 11w + 12}{w^2 - 16}}{\dfrac{2w^2 - 7w + 6}{w^2 + 9w + 20}}$

16. $\dfrac{3}{k^2 - 5k + 4} - \dfrac{2}{k^2 + 4k - 5}$

17. Solve: $\dfrac{1}{2n} - \dfrac{1}{6} = \dfrac{5}{4n} + \dfrac{1}{3}$

18. Solve the inequality and graph the solution set:
$$\dfrac{x - 4}{x + 5} \ge 0$$

19. Plane Speeds Two private planes take off from the same airport at the same time and travel in opposite directions. One plane travels 16 miles per hour faster than the other. After 2 hours, the planes are 536 miles apart. Find the speed of each plane.

20. Add: $\sqrt{48} + \sqrt{75} - 2\sqrt{3}$

21. Rationalize the denominator: $\dfrac{5}{\sqrt[3]{16}}$

22. Solve: $\sqrt{a - 12} + \sqrt{a} = 6$

23. Pythagorean Theorem Use the Pythagorean Theorem to determine the value of x for the given measurements of the right triangle shown below.

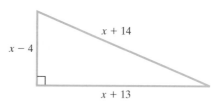

24. Solve: $x^2 - 20 = 8x$

25. Graph: $f(x) = -2x^2 + 4x + 3$

8 Exponential and Logarithmic Functions

We all dream of a long and happy retirement. To make this dream a reality, we must save money during our working years so that our retirement savings can grow through the power of compounding interest. To see how this phenomenon works, see Example 9 in Section 8.2 and Problems 89–92 in Section 8.2.

The Big Picture: Putting It Together

We have studied polynomial, rational, and radical expressions and functions. We learned how to evaluate, simplify, and solve equations involving these types of algebraic expressions. We also learned how to graph linear functions, quadratic functions, and certain types of radical functions.

We now introduce two more types of functions, the *exponential* and *logarithmic functions*. As usual, we will evaluate these functions, graph them, and learn about their properties. We also will solve equations involving exponential or logarithmic expressions.

Outline

8.1 Composite Functions and Inverse Functions

Objectives

1. Form the Composite Function
2. Determine Whether a Function Is One-to-One
3. Find the Inverse of a Function Defined by a Map or Set of Ordered Pairs
4. Obtain the Graph of the Inverse Function from the Graph of a Function
5. Find the Inverse of a Function Defined by an Equation

Are You Ready for This Section?

Before getting started, take the following readiness quiz. If you get a problem wrong, go back to the section cited and review the material.

R1. Determine the domain: $R(x) = \dfrac{x^2 - 9}{x^2 + 3x - 28}$ [Section 5.1, p. 404]

R2. If $f(x) = 2x^2 - x + 1$, find **(a)** $f(-2)$ **(b)** $f(a + 1)$. [Section 2.2, pp. 156–158]

R3. The graph of a relation is given. Does the relation represent a function? Why?   [Section 2.2, pp. 155–156]

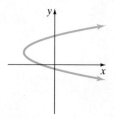

1 Form the Composite Function

Suppose you decide to buy four shirts at $20 per shirt, and the sales tax on your purchase is 5%. First, you determine the before-tax cost of the shirts to be $20(4) = $80. The after-tax cost of the shirts is $80 + $80(0.05) = 1.05($80) = $84. Thus $C(x) = 20x$ represents the cost C of buying x shirts at $20 per shirt. The after-tax cost S of the x shirts is $1.05(20x) = S(C(x))$.

In general, we can find the total cost as a function of the number of shirts by evaluating $S(C(x))$ and obtaining $S(C(x)) = S(20x) = 1.05(20x) = 21x$. The function $S(C(x))$ is a special type of function called a *composite function*.

Here's another example. Consider the function $y = (x - 3)^4$. If we write $y = f(u) = u^4$ and $u = g(x) = x - 3$, then, by a substitution process, we can obtain the original function: $y = f(u) = f(g(x)) = (x - 3)^4$. This process is called **composition.** To evaluate this function, we first evaluate $x - 3$ and then raise the result to the fourth power. Technically, then, two different functions form the function $y = (x - 3)^4$. Figure 1 illustrates the idea for $x = 5$.

In general, suppose that f and g are two functions and that x is a number in the domain of g. By evaluating g at x, we get $g(x)$. If $g(x)$ is in the domain of f, then we may evaluate f at $g(x)$ and obtain the expression $f(g(x))$. The correspondence from x to $f(g(x))$ is called a *composite function $f \circ g$.*

Figure 1

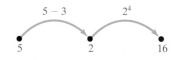

> **In Words**
>
> To find $f(g(x))$, first determine $g(x)$. This is the output of g. Then evaluate f at the output of g. The result is $f(g(x))$.

Definition

Given two functions f and g, the **composite function,** denoted by $f \circ g$ (read as "f composed with g"), is defined by

$$(f \circ g)(x) = f(g(x))$$

The notation $f(g(x))$ is read "f of g of x".

Figure 2 illustrates the definition. Notice that the "inside" function g in $f(g(x))$ is always evaluated first.

Figure 2

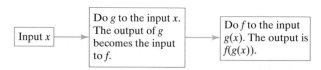

EXAMPLE 1   **Evaluating a Composite Function**

Suppose that $f(x) = x^2 - 3$ and $g(x) = 2x + 1$. Find

(a) $(f \circ g)(3)$ **(b)** $(g \circ f)(3)$ **(c)** $(f \circ f)(-2)$

Solution

(a) Using the flowchart in Figure 2, we evaluate $(f \circ g)(3)$ as follows:

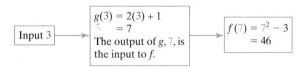

More directly,

$$(f \circ g)(3) = f(g(3)) = f(7) = 7^2 - 3 = 49 - 3 = 46$$

$f \circ g$

$$g(x) = 2x + 1 \quad f(x) = x^2 - 3$$
$$g(3) = 2(3) + 1$$
$$= 7$$

(b) $(g \circ f)(3) = g(f(3)) = g(6) = 2(6) + 1 = 13$

$$f(x) = x^2 - 3 \quad g(x) = 2x + 1$$
$$f(3) = 3^2 - 3$$
$$= 6$$

(c) $(f \circ f)(-2) = f(f(-2)) = f(1) = 1^2 - 3 = -2$

$$f(x) = x^2 - 3 \quad f(x) = x^2 - 3$$
$$f(-2) = (-2)^2 - 3$$
$$= 1$$

Quick ✓

1. Given two functions f and g, the _____ _____, denoted by $f \circ g$, is defined by $(f \circ g)(x) = f(g(x))$.

2. Suppose that $f(x) = 4x - 3$ and $g(x) = x^2 + 1$. Find

(a) $(f \circ g)(2)$ **(b)** $(g \circ f)(2)$ **(c)** $(f \circ f)(-3)$

Rather than evaluating a composite function at a specific value, we can also form a composite function that is written in terms of the independent variable, x.

▶ EXAMPLE 2   **Finding a Composite Function**

Suppose that $f(x) = x^2 + 2x$ and $g(x) = 2x - 1$. Find

(a) $(f \circ g)(x)$ **(b)** $(g \circ f)(x)$ **(c)** $(f \circ g)(2)$

Solution

(a) $$(f \circ g)(x) = f(g(x))$$

$g(x) = 2x - 1$: $= f(2x - 1)$

$f(x) = x^2 + 2x$: $= (2x - 1)^2 + 2(2x - 1)$

FOIL; distribute: $= 4x^2 - 4x + 1 + 4x - 2$

Combine like terms: $= 4x^2 - 1$

Work Smart

$(f \circ g)(x)$ does not mean $(f \cdot g)(x)$.

(b)
$$(g \circ f)(x) = g(f(x))$$

$f(x) = x^2 + 2x: \quad = g(x^2 + 2x)$

$g(x) = 2x - 1: \quad = 2(x^2 + 2x) - 1$

Distribute: $\quad = 2x^2 + 4x - 1$

(c) Instead of finding $(f \circ g)(2)$ using the approach presented in Example 1, we will find $(f \circ g)(2)$ using the results from part (a).

$$(f \circ g)(2) = f(g(2)) = 4(2)^2 - 1 = 4(4) - 1 = 15$$

Notice that $(f \circ g)(x) \neq (g \circ f)(x)$ in Examples 2(a) and 2(b).

Quick ✓

3. *True or False* $(f \circ g)(x) = f(x) \cdot g(x)$

4. If $f(g(x)) = (4x - 3)^5$ and $f(x) = x^5$, what is $g(x)$?

5. Suppose that $f(x) = x^2 - 3x + 1$ and $g(x) = 3x + 2$. Find

    (a) $(f \circ g)(x)$ **(b)** $(g \circ f)(x)$ **(c)** $(f \circ g)(-2)$

▶ ❷ Determine Whether a Function Is One-to-One

Figures 3 and 4 illustrate two different functions represented as mappings. The function in Figure 3 maps states to their population in millions. The function in Figure 4 maps animals to their life expectancy.

Figure 3

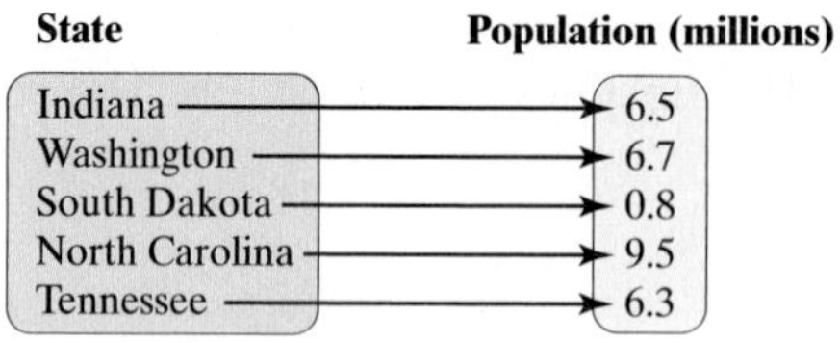

State	Population (millions)
Indiana	6.5
Washington	6.7
South Dakota	0.8
North Carolina	9.5
Tennessee	6.3

Figure 4

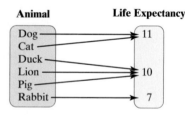

If we asked you to name the state with a population of 0.8 million based on the function in Figure 3, you would say "South Dakota." If we asked you to name the animal whose life expectancy is 11 years based on the function in Figure 4, you might say "dog" or "cat." What is the difference between the functions in Figures 3 and 4? In Figure 3, one (and only one) element in the domain corresponds to each element in the range. In Figure 4, this is not the case—there is more than one element in the domain that corresponds to an element in the range. For example, "dog" corresponds to "11," but "cat" also corresponds to "11." Functions such as the one in Figure 3 have a special name.

In Words

A function is NOT one-to-one if two different inputs correspond to the same output.

Definition

A function is **one-to-one** if any two different inputs in the domain correspond to two different outputs in the range. That is, if x_1 and x_2 are two different inputs to a function f, then $f(x_1) \neq f(x_2)$.

Thus a function is not one-to-one if two different elements in the domain correspond to the same element in the range. The function in Figure 4, then, is not one-to-one because two different elements in the domain, dog and cat, both correspond to 11.

EXAMPLE 3 Determining Whether a Function Is One-to-One

Determine which of the following functions is one-to-one.

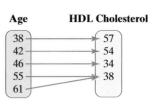

(a) For the function to the right, the domain represents the age of five males, and the range represents their HDL (good) cholesterol (mg/dL).

(b) {(−2, 6), (−1, 3), (0, 2), (1, 4), (2, 8)}

Solution

(a) The function is not one-to-one because two different inputs, 55 and 61, correspond to the same output, 38.

(b) The function is one-to-one because no two distinct inputs correspond to the same output.

Quick ✓

6. A function is ___ ___ ___ if any two different inputs in the domain correspond to two different outputs in the range. That is, if x_1 and x_2 are two different inputs to a function f, then $f(x_1) \neq f(x_2)$.

In Problems 7 and 8, determine whether the function is one-to-one.

7.
Friend	Birthday
Max	January 20
Alesia	March 3
Trent	July 6
Wanda	November 8
Yolanda	January 8
Elvis	

8. {(−3, 3), (−2, 2), (−1, 1), (0, 0), (1, −1)}

Consider the functions $y = 2x - 5$ and $y = x^2 - 4$ shown in Figure 5.

Figure 5

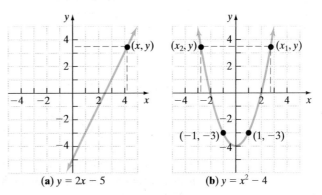

(a) $y = 2x - 5$ (b) $y = x^2 - 4$

Note, for the function $y = 2x - 5$ shown in Figure 5(a), that any output is the result of only one input. However, for the function $y = x^2 - 4$ shown in Figure 5(b), there are instances where a given output is the result of more than one input. For example, the output −3 is the result of input −1 and of input 1.

If a horizontal line intersects the graph of a function at more than one point, the function cannot be one-to-one because this would mean that two different inputs give the same output. We state this result formally as the *horizontal line test*.

Horizontal Line Test

If every horizontal line intersects the graph of a function f in at most one point, then f is one-to-one.

(**EXAMPLE 4**) **Using the Horizontal Line Test**

For each function, use the graph to determine whether the function is one-to-one.

(a) $f(x) = -2x^3 + 1$ **(b)** $g(x) = x^3 - 4x$

Solution

(a) Figure 6(a) illustrates the horizontal line test for $f(x) = -2x^3 + 1$. Because every horizontal line intersects the graph of f exactly once, f must be one-to-one.

(b) Figure 6(b) illustrates the horizontal line test for $g(x) = x^3 - 4x$. The horizontal line $y = 1$ intersects the graph three times, so g is not one-to-one.

Figure 6

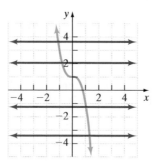

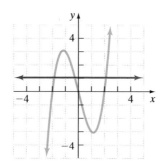

(a) Every horizontal line intersects the graph once; f is one-to-one.

(b) A horizontal line intersects the graph three times; g is not one-to-one.

Quick ✓

In Problem 9, use the graph to determine whether the given function is one-to-one.

9. (a) $f(x) = x^4 - 4x^2$ **(b)** $f(x) = x^5 + 4x$

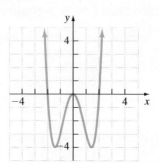

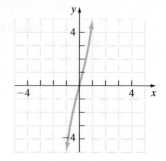

▶ ❸ **Find the Inverse of a Function Defined by a Map or Set of Ordered Pairs**

Now that we understand the concept of a one-to-one function, we can talk about *inverse functions.*

> **Definition**
>
> If f is a one-to-one function with ordered pairs of the form (a, b), then the **inverse function**, denoted f^{-1}, is the set of ordered pairs of the form (b, a).

Work Smart

The symbol f^{-1} represents the inverse function of f. The -1 used in f^{-1} is not an exponent. That is,

$$f^{-1}(x) \neq \frac{1}{f(x)}.$$

  In order for the inverse of a function also to be a function, the function must be one-to-one. We begin by finding inverses of functions represented by maps and sets of ordered pairs. To do this, we interchange the inputs and outputs, so for each ordered pair (a, b) that is defined in a function f, the ordered pair (b, a) is defined in the inverse function f^{-1}. The inverse undoes what the function does. Suppose a function takes the input 11 and gives the output 3. This can be represented as $(11, 3)$. The inverse would take as input 3 and give the output 11, which can be represented as $(3, 11)$.

EXAMPLE 5 **Finding the Inverse of a Function Defined by a Map**

Find the inverse of the following function. The domain of the function represents the states, and the range represents the states' population in millions. Give the domain and the range of the inverse function.

State	Population (millions)
Indiana	6.5
Washington	6.7
South Dakota	0.8
North Carolina	9.5
Tennessee	6.3

Solution

The elements in the domain represent the inputs to the function. The elements in the range represent the outputs of the function. The function is one-to-one, so the inverse is a function. To find the inverse function, we interchange the elements in the domain with the elements in the range. For example, the function receives as input Indiana and outputs 6.5, so the inverse receives as input 6.5 and outputs Indiana. The inverse function is shown below.

Population (millions)	State
6.5	Indiana
6.7	Washington
0.8	South Dakota
9.5	North Carolina
6.3	Tennessee

The domain of the inverse function is $\{6.5, 6.7, 0.8, 9.5, 6.3\}$. The range of the inverse function is $\{$ Indiana, Washington, South Dakota, North Carolina, Tennessee $\}$. ●

Quick ✓

10. Find the inverse of the one-to-one function shown below. The domain of the function represents the lengths of the right humerus (in mm), and the range represents the lengths of the right tibia (in mm) of rats sent to space. Give the domain and the range of the inverse function.

Right Humerus	Right Tibia
24.80	36.05
24.59	35.57
24.29	34.58
23.81	34.20
24.87	34.73

SOURCE: *NASA Life Sciences Data Archive*

EXAMPLE 6 **Finding the Inverse of a Function Defined by a Set of Ordered Pairs**

Find the inverse of the following function. State the domain and the range of the inverse function.

$$\{(-2, 6), (-1, 3), (0, 2), (1, 5), (2, 8)\}$$

Solution

The function is one-to-one, so the inverse is a function. We find the inverse by interchanging the entries in each ordered pair. Thus the inverse function is given by

$$\{(6, -2), (3, -1), (2, 0), (5, 1), (8, 2)\}$$

The domain of the inverse function is $\{6, 3, 2, 5, 8\}$, and the range of the inverse function is $\{-2, -1, 0, 1, 2\}$. ●

Quick ✓

11. Find the inverse of the following function. State the domain and the range of the inverse function.

$$\{(-3, 3), (-2, 2), (-1, 1), (0, 0), (1, -1)\}$$

In Examples 5 and 6, notice that the elements that are in the domain of f are the same elements that are in the range of its inverse. In addition, the elements that are in the range of f are the same elements that are in the domain of its inverse.

> **Relation Between the Domain and Range of a Function and Its Inverse**
>
> All elements in the domain of a function are elements in the range of its inverse.
>
> All elements in the range of a function are elements in the domain of its inverse.

Figure 7

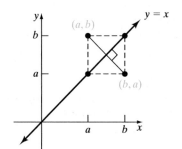

▶ ④ Obtain the Graph of the Inverse Function from the Graph of a Function

We know that if a function is defined by a set of ordered pairs, we can find the inverse by interchanging the entries. Thus if (a, b) is a point on the graph of a function, then (b, a) is a point on the graph of the inverse function. See Figure 7. From the graph, it follows that the point (b, a) on the graph of the inverse function is the reflection about the line $y = x$ of the point (a, b).

The graph of a function f and the graph of its inverse are symmetric with respect to the line $y = x$.

EXAMPLE 7   **Graphing the Inverse Function**

Figure 8(a) shows the graph of a one-to-one function $y = f(x)$. Draw the graph of its inverse.

Solution

We begin by adding the graph of $y = x$ to Figure 8(a). Since the points $(-4, -2)$, $(-3, -1)$, $(-1, 0)$, and $(4, 2)$ are on the graph of f, we know that the points $(-2, -4)$, $(-1, -3)$, $(0, -1)$, and $(2, 4)$ are on the graph of the inverse of f, f^{-1}. Using these points, along with the fact that the graph of f^{-1} is a reflection about the line $y = x$ of the graph of f, we draw the graph of f^{-1}. See Figure 8(b).

Figure 8

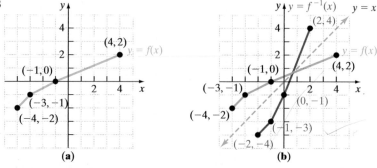

(a)   (b)

Quick ✓

12. Below is the graph of a one-to-one function $y = f(x)$. Draw the graph of its inverse.

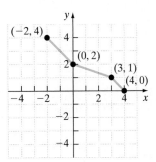

▶ ⑤ Find the Inverse of a Function Defined by an Equation

Work Smart

If $f(x) = 2x$, then each input is doubled. The inverse function is $f^{-1}(x) = x/2$, so each input is cut in half. If $x = 10$, then $f(10) = 20$. Now substitute the output of f, 20, into f^{-1} to get $f^{-1}(20) = 10$. Notice we are right back where we started!

We have learned how to find the inverse of a one-to-one function defined by a map, a set of ordered pairs, or a graph. We now discuss how to find the inverse of a function defined by an equation.

We know that when a function is defined by an equation, we use the notation $y = f(x)$. We can use the notation $y = f^{-1}(x)$ to denote the equation whose rule is the inverse function of f.

Because the inverse of a function "undoes" what the original function does, we have the following relation between a function f and its inverse, f^{-1}.

$$f^{-1}(f(x)) = x \text{ for every } x \text{ in the domain of } f$$

$$f(f^{-1}(x)) = x \text{ for every } x \text{ in the domain of } f^{-1}$$

Let's now find the inverse of a one-to-one function defined by an equation.

EXAMPLE 8 **How to Find the Inverse of a One-to-One Function**

Find the inverse of $f(x) = 2x - 3$.

Step-by-Step Solution

First, verify that the function is one-to-one. Because the graph of the function f is a line with y-intercept $(0, -3)$ and slope 2, we know by the horizontal line test that the function is one-to-one and therefore has an inverse function.

Step 1: Replace $f(x)$ with y in the equation for $f(x)$.	$f(x) = 2x - 3$ $y = 2x - 3$
Step 2: In $y = f(x)$, interchange the variables x and y to obtain $x = f(y)$.	$x = 2y - 3$
Step 3: Solve the equation found in Step 2 for y in terms of x.	Add 3 to both sides: $x + 3 = 2y$ Divide both sides by 2: $\dfrac{x + 3}{2} = y$
Step 4: Replace y with $f^{-1}(x)$.	$f^{-1}(x) = \dfrac{x + 3}{2}$

(continued)

Step 5: Check Verify your result by showing that

$$f^{-1}(f(x)) = x \text{ and}$$
$$f(f^{-1}(x)) = x.$$

$$f^{-1}(f(x)) = f^{-1}(2x - 3)$$

$$= \frac{2x - 3 + 3}{2}$$

$$= \frac{2x}{2}$$

$$= x$$

$$f(f^{-1}(x)) = f\left(\frac{x + 3}{2}\right)$$

$$= 2\left(\frac{x + 3}{2}\right) - 3$$

$$= x + 3 - 3$$

$$= x$$

Everything checks, so $f^{-1}(x) = \dfrac{x + 3}{2}$.

Now let's summarize the steps used in Example 8.

Finding The Inverse of a One-to-One Function Defined by an Equation

Step 1: Replace $f(x)$ with y in the equation for $f(x)$.
Step 2: In $y = f(x)$, interchange the variables x and y to obtain $x = f(y)$.
Step 3: Solve the equation found in Step 2 for y in terms of x.
Step 4: Replace y with $f^{-1}(x)$.
Step 5: Verify your result by showing that $f^{-1}(f(x)) = x$ and $f(f^{-1}(x)) = x$.

Consider the function $f(x) = 2x - 3$ and its inverse $f^{-1}(x) = \dfrac{x + 3}{2}$ from Example 8. Notice that $f(5) = 2(5) - 3 = 7$. What do you think $f^{-1}(7)$ will equal? Notice that $f^{-1}(7) = \dfrac{7 + 3}{2} = \dfrac{10}{2} = 5$. Thus f^{-1} "undoes" what f did!

Quick ✓

13. *True or False* If f is a one-to-one function so that its inverse is f^{-1}, then $f^{-1}(f(x)) = x$ for every x in the domain of f, and $f(f^{-1}(x)) = x$ for every x in the domain of f^{-1}.

14. *True or False* The notation $f^{-1}(x)$ is equivalent to $\dfrac{1}{f(x)}$.

15. Find the inverse of $g(x) = 5x - 1$.

EXAMPLE 9 **Finding the Inverse of a One-to-One Function**

Find the inverse of $h(x) = x^3 + 4$.

Solution

$$h(x) = x^3 + 4$$

Replace $h(x)$ with y in the equation for $h(x)$: $\quad y = x^3 + 4$

Interchange the variables x and y: $\quad x = y^3 + 4$

Solve the equation for y in terms of x: $\quad x - 4 = y^3$

Take the cube root of both sides: $\quad \sqrt[3]{x - 4} = y$

Replace y with $h^{-1}(x)$: $\quad h^{-1}(x) = \sqrt[3]{x - 4}$

Check Verify your result by showing that $h^{-1}(h(x)) = x$ and $h(h^{-1}(x)) = x.$

$$h^{-1}(h(x)) = h^{-1}(x^3 + 4) \qquad\qquad h(h^{-1}(x)) = h\left(\sqrt[3]{x-4}\,\right)$$
$$= \sqrt[3]{x^3 + 4 - 4} \qquad\qquad = \left(\sqrt[3]{x-4}\,\right)^3 + 4$$
$$= \sqrt[3]{x^3} \qquad\qquad\qquad = x - 4 + 4$$
$$= x \qquad\qquad\qquad\qquad = x$$

So $h^{-1}(x) = \sqrt[3]{x-4}.$

Quick ✓

16. Find the inverse of $f(x) = x^5 + 3.$

8.1 Exercises   MyMathLab® MathXP
PRACTICE

Exercise numbers in green
have complete video solutions
in MyMathLab.

Problems **1–16** *are the* Quick ✓ *s that follow the* EXAMPLES.

Building Skills

In Problems 17–24, for the given functions f and g, find

(a) $(f \circ g)(3)$ (b) $(g \circ f)(-2)$
(c) $(f \circ f)(1)$ (d) $(g \circ g)(-4)$

See Objective 1.

17. $f(x) = 2x + 5; g(x) = x - 4$

18. $f(x) = 4x - 3; g(x) = x + 2$

19. $f(x) = x^2 + 4; g(x) = 2x + 3$

20. $f(x) = x^2 - 3; g(x) = 5x + 1$

21. $f(x) = 2x^3; g(x) = -2x^2 + 5$

22. $f(x) = -2x^3; g(x) = x^2 + 1$

23. $f(x) = |x - 10|; g(x) = \dfrac{12}{x + 3}$

24. $f(x) = \sqrt{x + 8}; g(x) = x^2 - 4$

In Problems 25–36, for the given functions f and g, find

(a) $(f \circ g)(x)$ (b) $(g \circ f)(x)$
(c) $(f \circ f)(x)$ (d) $(g \circ g)(x)$

See Objective 1.

25. $f(x) = x + 1; g(x) = 2x$

26. $f(x) = x - 3; g(x) = 4x$

27. $f(x) = 2x + 7; g(x) = -4x + 5$

28. $f(x) = 3x - 1; g(x) = -2x + 5$

29. $f(x) = x^2; g(x) = x - 3$

30. $f(x) = x^2 + 1; g(x) = x + 1$

31. $f(x) = \sqrt{x}; g(x) = x + 4$

32. $f(x) = \sqrt{x + 2}; g(x) = x - 2$

33. $f(x) = |x + 4|; g(x) = x^2 - 4$

34. $f(x) = |x - 3|; g(x) = x^3 + 3$

35. $f(x) = \dfrac{2}{x + 1}; g(x) = \dfrac{1}{x}$

36. $f(x) = \dfrac{2}{x - 1}; g(x) = \dfrac{4}{x}$

In Problems 37–46, determine which of the following functions is one-to-one. See Objective 2.

37.

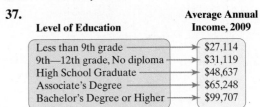

Level of Education	Average Annual Income, 2009
Less than 9th grade	$27,114
9th—12th grade, No diploma	$31,119
High School Graduate	$48,637
Associate's Degree	$65,248
Bachelor's Degree or Higher	$99,707

SOURCE: *United States Census Bureau*

38.

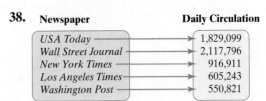

Newspaper	Daily Circulation
USA Today	1,829,099
Wall Street Journal	2,117,796
New York Times	916,911
Los Angeles Times	605,243
Washington Post	550,821

SOURCE: *Information Please Almanac*

39.

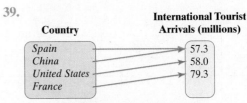

Country	International Tourist Arrivals (millions)
Spain	57.3
China	58.0
United States	79.3
France	

SOURCE: *Information Please Almanac*

40.

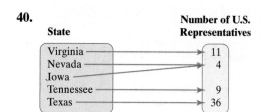

| State | Number of U.S. Representatives |

41. $\{(-3, 4), (-2, 6), (-1, 8), (0, 10), (1, 12)\}$

42. $\{(-2, 6), (-1, 3), (0, 0), (1, -3), (2, 6)\}$

43. $\{(-2, 4), (-1, 2), (0, 0), (1, 2), (2, 4)\}$

44. $\{(-2, -8), (-1, -1), (0, 0), (1, 1), (2, 8)\}$

45. $\{(0, -4), (-1, -1), (-2, 0), (1, 1), (2, 4)\}$

46. $\{(-3, 0), (-2, 3), (-1, 0), (0, -3)\}$

In Problems 47–52, use the horizontal line test to determine whether the function whose graph is given is one-to-one. See Objective 2.

47.

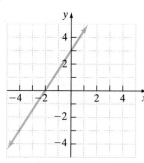

48.

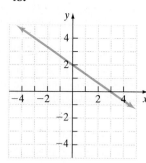

49.

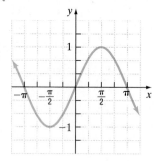

50.

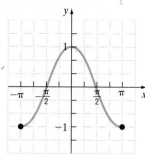

51.

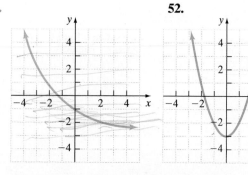

52.

In Problems 53–58, find the inverse of the following one-to-one functions. See Objective 3.

53.

U.S. Coin	Weight (g)
Cent	2.500
Nickel	5.000
Dime	2.268
Quarter	5.670
Half Dollar	11.340
Dollar	8.100

SOURCE: *U.S. Mint website*

54.

Price ($)	Quantity Demanded
2300	152
2000	159
1700	164
1500	171
1300	176

55. $\{(0, 3), (1, 4), (2, 5), (3, 6)\}$

56. $\{(-1, 4), (0, 1), (1, -2), (2, -5)\}$

57. $\{(-2, 3), (-1, 1), (0, -3), (1, 9)\}$

58. $\{(-10, 1), (-5, 4), (0, 3), (5, 2)\}$

In Problems 59–64, the graph of a one-to-one function f is given. Draw the graph of the inverse function f^{-1}. See Objective 4.

59.

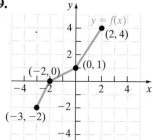

60.

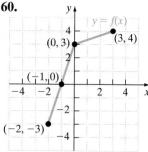

61.

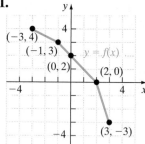

62.

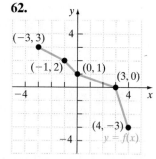

63.

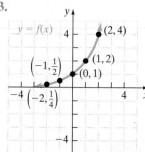

64.

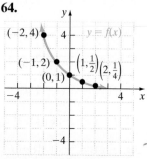

In Problems 65–72, verify that the functions f and g are inverses of each other. See Objective 5.

65. $f(x) = x + 5; g(x) = x - 5$

66. $f(x) = 10x; g(x) = \dfrac{x}{10}$

67. $f(x) = 5x + 7; g(x) = \dfrac{x - 7}{5}$

68. $f(x) = 3x - 5; g(x) = \dfrac{x + 5}{3}$

69. $f(x) = \dfrac{3}{x - 1}; g(x) = \dfrac{3}{x} + 1$

70. $f(x) = \dfrac{2}{x + 4}; g(x) = \dfrac{2}{x} - 4$

71. $f(x) = \sqrt[3]{x + 4}; g(x) = x^3 - 4$

72. $f(x) = \sqrt[3]{2x + 1}; g(x) = \dfrac{x^3 - 1}{2}$

In Problems 73–92, find the inverse function of the given one-to-one function. See Objective 5.

73. $f(x) = 6x$

74. $f(x) = 12x$

75. $f(x) = x + 4$

76. $g(x) = x + 6$

77. $h(x) = 2x - 7$

78. $H(x) = 3x + 8$

79. $G(x) = 2 - 5x$

80. $F(x) = 1 - 6x$

81. $g(x) = x^3 + 3$

82. $f(x) = x^3 - 2$

83. $p(x) = \dfrac{1}{x + 3}$

84. $P(x) = \dfrac{1}{x + 1}$

85. $F(x) = \dfrac{5}{2 - x}$

86. $G(x) = \dfrac{2}{3 - x}$

87. $f(x) = \sqrt[3]{x - 2}$

88. $f(x) = \sqrt[5]{x + 5}$

89. $R(x) = \dfrac{x}{x + 2}$

90. $R(x) = \dfrac{2x}{x + 4}$

91. $f(x) = \sqrt[3]{x - 1} + 4$

92. $g(x) = \sqrt[3]{x + 2} - 3$

Applying the Concepts

93. Environmental Disaster An oil tanker hits a rock that rips a hole in the hull of the ship. Oil leaking from the ship forms a circular region around the ship. If the radius r of the circle (in feet) as a function of time t (in hours) is $r(t) = 20t$, express the area A of the circular region contaminated with oil as a function of time. What will be the area of the circular region at 3 hours? (*Hint:* $A(r) = \pi r^2$)

94. Volume of a Balloon The volume V of a hot-air balloon (in cubic meters) as a function of its radius r is given by $V(r) = \dfrac{4}{3}\pi r^3$. If the radius r of the balloon is increasing as a function of time t (in minutes) according to $r(t) = 3\sqrt[3]{t}$, for $t \geq 0$, find the volume of the balloon as a function of time t. What will be the volume of the balloon after 30 minutes?

95. Buying Carpet You want to purchase new carpet for your family room. Carpet is sold by the square yard, but you have measured your family room in square feet. The function $A(x) = \dfrac{x}{9}$ converts the area of a room in square feet to an area in square yards. Suppose that the carpet you have selected is \$18 per square yard installed. Then the function $C(A) = 18A$ represents the cost C of installing carpet in a room that measures A square yards.
   (a) Find the cost C as a function of the square footage of the room x.
   (b) If your family room is 15 feet by 21 feet, what will it cost to install the carpet?

96. Tax Time You have a job that pays \$20 per hour. Your gross salary G as a function of hours worked h is given by $G(h) = 20h$. Federal tax withholding T on your paycheck is equal to 18% of gross earnings G, so federal tax withholding as a function of gross pay is given by the function $T(G) = 0.18G$.
   (a) Find federal tax withholding T as a function of hours worked h.
   (b) Suppose that you worked 28 hours last week. What will be the federal tax withholding on your paycheck?

97. If $f(4) = 12$, what is $f^{-1}(12)$?

98. If $g(-2) = 7$, what is $g^{-1}(7)$?

99. The domain of a function f is $[0, \infty)$, and its range is $[-5, \infty)$. State the domain and the range of f^{-1}.

100. The domain of a one-to-one function f is $[5, \infty)$, and its range is $[0, \infty)$. State the domain and the range of f^{-1}.

101. The domain of a one-to-one function g is $[-4, 10]$, and its range is $(-6, 12)$. State the domain and the range of g^{-1}.

102. The domain of a function g is $[0, 15]$, and its range is $(0, 8)$. State the domain and the range of g^{-1}.

103. Taxes The function $T(x) = 0.15(x - 8700) + 870$ represents the tax bill T of a single person whose adjusted gross income is x dollars for income between $8700 and $35,350, inclusive. (SOURCE: *Internal Revenue Service*) Find the inverse function that expresses adjusted gross income x as a function of taxes T. That is, find $x(T)$.

104. Health Costs The annual cost of health insurance H as a function of age a is given by the function $H(a) = 22.8a - 117.5$ for $15 \le a \le 90$. (SOURCE: *Statistical Abstract,*) Find the inverse function that expresses age a as a function of health insurance cost H. That is, find $a(H)$.

Extending the Concepts

105. If $f(x) = 2x^2 - x + 5$ and $g(x) = x + a$, find a so that the y-intercept of the graph of $(f \circ g)(x)$ is $(0, 20)$.

106. If $f(x) = x^2 - 3x + 1$ and $g(x) = x - a$, find a so that the y-intercept of the graph of $(f \circ g)(x)$ is $(0, -1)$.

Explaining the Concepts

107. Explain what it means for a function to be one-to-one. Why must a function be one-to-one in order for its inverse to be a function?

108. Are all linear functions one-to-one? Explain.

109. Explain why domain of f = range of f^{-1} and range of f = domain of f^{-1}.

110. State the horizontal line test. Why does it work?

The Graphing Calculator

Graphing calculators have the ability to evaluate composite functions. To obtain the results of Example 1, we would let $Y_1 = f(x) = x^2 - 3$ and $Y_2 = g(x) = 2x + 1$.

Figure 9 shows the results of Example 1 using a TI-84 Plus graphing calculator.

Figure 9

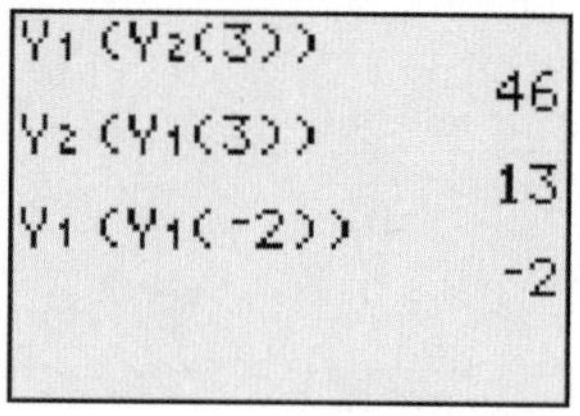

```
Y₁ (Y₂(3))
                    46
Y₂ (Y₁(3))
                    13
Y₁ (Y₁(-2))
                    -2
```

In Problems 111–118, use a graphing calculator to evaluate the composite functions. Compare your answers with those found in Problems 17–24.

   (a) $(f \circ g)(3)$ **(b)** $(g \circ f)(-2)$
   (c) $(f \circ f)(1)$ **(d)** $(g \circ g)(-4)$

111. $f(x) = 2x + 5; g(x) = x - 4$

112. $f(x) = 4x - 3; g(x) = x + 2$

113. $f(x) = x^2 + 4; g(x) = 2x + 3$

114. $f(x) = x^2 - 3; g(x) = 5x + 1$

115. $f(x) = 2x^3; g(x) = -2x^2 + 5$

116. $f(x) = -2x^3; g(x) = x^2 + 1$

117. $f(x) = |x - 10|; g(x) = \dfrac{12}{x + 3}$

118. $f(x) = \sqrt{x + 8}; g(x) = x^2 - 4$

In Problems 119–122, the functions f and g are inverses. Graph both functions on the same screen, along with the line $y = x$, to see the symmetry of the functions about the line $y = x$.

119. $f(x) = x + 5; g(x) = x - 5$

120. $f(x) = 10x; g(x) = \dfrac{x}{10}$

121. $f(x) = 5x + 7; g(x) = \dfrac{x - 7}{5}$

122. $f(x) = 3x - 5; g(x) = \dfrac{x + 5}{3}$

8.2 Exponential Functions

Objectives

1. Evaluate Exponential Expressions
2. Graph Exponential Functions
3. Define the Number e
4. Solve Exponential Equations
5. Use Exponential Models That Describe Our World

Are You Ready for This Section?

Before getting started, take the following readiness quiz. If you get a problem wrong, go back to the section cited and review the material.

R1. Evaluate: **(a)** 2^3 **(b)** 2^{-1} **(c)** 3^4 [Getting Ready, pp. 293–296]

R2. Graph: $f(x) = x^2$ [Section 7.4, p. 577]

R3. State the definition of a rational number.   [Section R.2, p. 12]

R4. State the definition of an irrational number.   [Section R.2, p. 13]

R5. Write 3.20349193 as a decimal **(a)** rounded to four decimal places **(b)** truncated to four decimal places.   [Section R.2, pp. 14–15]

R6. Simplify: **(a)** $m^3 \cdot m^5$ **(b)** $\dfrac{a^7}{a^2}$ **(c)** $(z^3)^4$ [Getting Ready, pp. 293–298]

R7. Solve: $x^2 - 5x = 14$ [Section 4.8, pp. 372–374]

Suppose editor Mary Beckwith has just hired you as a proofreader for Pearson Publishing. Mary offers you two options: Option A states that you will be paid $100 for each error you find in the final page proofs of a text. Option B states that you will get $2 for the first error you find in the final page proofs and your payment will double for each additional error you find. You know from experience that the final page proofs of a text typically have about 15–20 errors. Which option will you choose?

If there is one error, Option A pays $100, while Option B pays $2. For two errors, Option A pays $2($100$) = $200, while Option B pays $\$2^2 = \$4$. For three errors, Option A pays $3($100$) = $300, while Option B pays $\$2^3 = \$8$. Option A seems to be the way to go. To complete the analysis, we set up Table 1, which lists the payment amount as a function of the number of errors in the page proofs. Remember, in Option B, the payment amount doubles each time you find an error.

Ready?...Answers

R1. (a) 8 **(b)** $\dfrac{1}{2}$ **(c)** 81

R2.

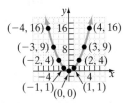

R3. A rational number is a number that can be expressed as a quotient $\dfrac{p}{q}$ of two integers. The integer p is called the numerator, and the integer q, which cannot be 0, is called the denominator. The set of rational numbers is the numbers

$$\mathbb{Q} = \left\{ x \,\middle|\, x = \frac{p}{q}, \text{ where } p, q \right.$$

are integers and $\left. q \neq 0 \right\}$.

R4. An irrational number has a decimal representation that neither repeats nor terminates.

R5. (a) 3.2035 **(b)** 3.2034

R6. (a) m^8 **(b)** a^5 **(c)** z^{12}

R7. $\{-2, 7\}$

Table 1

Number of Errors	Option A Payment	Option B Payment	Number of Errors	Option A Payment	Option B Payment
0	$0	$1	11	$1100	$2048
1	$100	$2	12	$1200	$4096
2	$200	$4	13	$1300	$8192
3	$300	$8	14	$1400	$16,384
4	$400	$16	15	$1500	$32,768
5	$500	$32	16	$1600	$65,536
6	$600	$64	17	$1700	$131,072
7	$700	$128	18	$1800	$262,144
8	$800	$256	19	$1900	$524,288
9	$900	$512	20	$2000	$1,048,576
10	$1000	$1024			

Holy cow! If you find 20 errors, you'll get paid over a million dollars! Mary Beckwith better reconsider her offer! If we let x represent the number of errors, we can express Option A as a linear function, $f(x) = 100x$; we can express Option B as an *exponential function*, $g(x) = 2^x$.

Definition

An **exponential function** is a function of the form

$$f(x) = a^x$$

where $a \neq 1$ is a positive real number $(a > 0)$. The domain of the exponential function is the set of all real numbers.

We will address the restrictions on the base a shortly. The key point is that *the independent variable is in the exponent of the exponential expression.* Contrast this idea with polynomial functions (such as $f(x) = x^2 - 4x$ or $g(x) = 2x^3 + x^2 - 5$), where the independent variable is the base of each exponential expression.

▶ ❶ Evaluate Exponential Expressions

In Section 6.1 we defined $a^{\frac{m}{n}}$, where the base a is a positive real number and the exponent $\frac{m}{n}$ is a rational number.

But what if you want to raise the base a to an irrational number? Although the answer to this question requires advanced mathematics, we can give you an intuitive explanation.

Suppose we want to find the value of $3^{\sqrt{2}}$. Our calculator tells us that $\sqrt{2} \approx 1.414213562$, so it should seem reasonable that we can approximate $3^{\sqrt{2}}$ as $3^{1.4}$, where the 1.4 comes from truncating the decimals to the right of the 4 in the tenths position. A better approximation of $3^{\sqrt{2}}$ would be $3^{1.4142}$, where the digits to the right of the ten-thousandths position have been truncated. The more decimals used in approximating $\sqrt{2}$, the better the approximation of $3^{\sqrt{2}}$.

To evaluate expressions of the form a^x using a scientific calculator, enter the base a, press the $\boxed{x^y}$ key, enter the exponent x, and press $\boxed{=}$. To evaluate expressions of the form a^x using a graphing calculator, enter the base a, press the caret $\boxed{\wedge}$ key, enter the exponent x, and press $\boxed{\text{ENTER}}$.

(**EXAMPLE 1**)　**Evaluating Exponential Expressions**

Using a calculator, evaluate each of the following expressions. Write as many decimals as your calculator allows.

(a) $3^{1.4}$ 　　(b) $3^{1.41}$ 　　(c) $3^{1.414}$ 　　(d) $3^{1.4142}$ 　　(e) $3^{\sqrt{2}}$

Solution

(a) $3^{1.4} \approx 4.655536722$ 　　　　(b) $3^{1.41} \approx 4.706965002$

(c) $3^{1.414} \approx 4.727695035$ 　　　(d) $3^{1.4142} \approx 4.72873393$

(e) $3^{\sqrt{2}} \approx 4.728804388$

Quick ✓

1. An exponential function is a function of the form $f(x) = a^x$ where $a \underline{\geq} 0$ and $a \underline{\neq} 1$.

In Problem 2, use a calculator to evaluate each of the following expressions. Write as many decimals as your calculator allows.

2. (a) $2^{1.7}$ 　　**(b)** $2^{1.73}$ 　　**(c)** $2^{1.732}$ 　　**(d)** $2^{1.7321}$ 　　**(e)** $2^{\sqrt{3}}$

Example 1 illustrates that we can approximate the value of an exponential expression at any real number. This is why the domain of any exponential function is the set of all real numbers.

In our definition of an exponential function, we ruled out $a = 1$ and required that a be positive. We exclude the base $a = 1$ because this function is the constant function $f(x) = 1^x = 1$. We also exclude bases that are negative because we would run into problems for exponents such as $\frac{1}{2}$ or $\frac{3}{4}$. For example, suppose $f(x) = (-2)^x$. In the real number system, we could not evaluate $f\left(\frac{1}{2}\right)$ because $f\left(\frac{1}{2}\right) = (-2)^{\frac{1}{2}} = \sqrt{-2}$, which is not a real number. Finally, we exclude a base of 0, because this function is $f(x) = 0^x$, which equals zero for $x > 0$ and is undefined when $x < 0$. When $x = 0, f(x) = 0^x$ is *indeterminate* because its value is not precisely determined.

❷ Graph Exponential Functions

▶ Let's learn about properties of exponential functions from their graphs.

EXAMPLE 2 **Graphing an Exponential Function**

Graph the exponential function $f(x) = 2^x$ using point plotting. From the graph, state the domain and the range of the function.

Solution

We begin by locating some points on the graph of $f(x) = 2^x$ as shown in Table 2. We plot the points in Table 2 and connect them in a smooth curve. Figure 10 shows the graph of $f(x) = 2^x$.

Figure 10

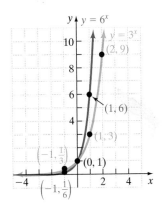

Table 2

x	$f(x) = 2^x$	$(x, f(x))$
-3	$f(-3) = 2^{-3} = \dfrac{1}{2^3} = \dfrac{1}{8}$	$\left(-3, \dfrac{1}{8}\right)$
-2	$f(-2) = 2^{-2} = \dfrac{1}{2^2} = \dfrac{1}{4}$	$\left(-2, \dfrac{1}{4}\right)$
-1	$f(-1) = 2^{-1} = \dfrac{1}{2^1} = \dfrac{1}{2}$	$\left(-1, \dfrac{1}{2}\right)$
0	$f(0) = 2^0 = 1$	$(0, 1)$
1	$f(1) = 2^1 = 2$	$(1, 2)$
2	$f(2) = 2^2 = 4$	$(2, 4)$
3	$f(3) = 2^3 = 8$	$(3, 8)$

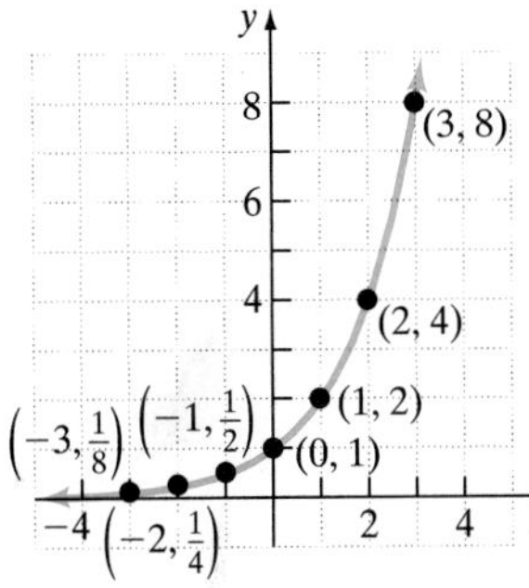

The domain of any exponential function is the set of all real numbers. Notice that no x caused 2^x to be less than or equal to 0. Based on this and the graph, we conclude that the range of $f(x) = 2^x$ is the set of all positive real numbers $\{y \mid y > 0\}$, or $(0, \infty)$ in interval notation.

Figure 11

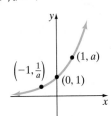

The graph of $f(x) = 2^x$ in Figure 10 is typical of all exponential functions that have a base larger than 1. Figure 11 shows the graphs of two other exponential functions whose bases are larger than 1: $y = 3^x$ and $y = 6^x$. Notice that the larger the base, the steeper the graph is for $x > 0$ and the closer the graph is to the x-axis for $x < 0$.

We summarize below the information we have about $f(x) = a^x$, where $a > 1$.

Figure 12
$f(x) = a^x, a > 1$

> **Properties of the Graph of an Exponential Function $f(x) = a^x, a > 1$**
>
> **1.** The domain is the set of all real numbers. The range is the set of all positive real numbers.
>
> **2.** There are no x-intercepts; the y-intercept is $(0, 1)$.
>
> **3.** The graph of f contains the points $\left(-1, \dfrac{1}{a}\right)$, $(0, 1)$, and $(1, a)$.
>
> See Figure 12.

Quick ✓

3. Graph the exponential function $f(x) = 4^x$ using point plotting. From the graph, state the domain and the range of the function.

Now we consider $f(x) = a^x, 0 < a < 1$.

EXAMPLE 3 **Graphing an Exponential Function**

Graph the exponential function $f(x) = \left(\dfrac{1}{2}\right)^x$ using point plotting. From the graph, state the domain and the range of the function.

Solution

We begin by locating some points on the graph of $f(x) = \left(\dfrac{1}{2}\right)^x$ as shown in Table 3. We plot the points in Table 3 and connect them in a smooth curve. Figure 13 shows the graph of $f(x) = \left(\dfrac{1}{2}\right)^x$.

Table 3

x	$f(x) = \left(\dfrac{1}{2}\right)^x$	$(x, f(x))$
-3	$f(-3) = \left(\dfrac{1}{2}\right)^{-3} = 2^3 = 8$	$(-3, 8)$
-2	$f(-2) = \left(\dfrac{1}{2}\right)^{-2} = 2^2 = 4$	$(-2, 4)$
-1	$f(-1) = \left(\dfrac{1}{2}\right)^{-1} = 2^1 = 2$	$(-1, 2)$
0	$f(0) = \left(\dfrac{1}{2}\right)^{0} = 1$	$(0, 1)$
1	$f(1) = \left(\dfrac{1}{2}\right)^{1} = \dfrac{1}{2}$	$\left(1, \dfrac{1}{2}\right)$
2	$f(2) = \left(\dfrac{1}{2}\right)^{2} = \dfrac{1}{4}$	$\left(2, \dfrac{1}{4}\right)$
3	$f(3) = \left(\dfrac{1}{2}\right)^{3} = \dfrac{1}{8}$	$\left(3, \dfrac{1}{8}\right)$

Figure 13

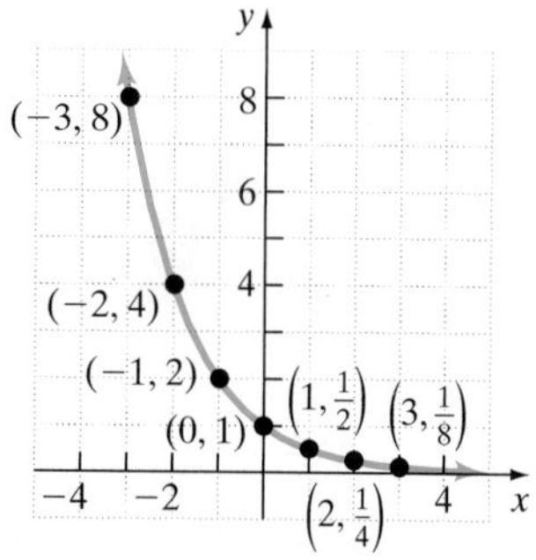

The domain of any exponential function is the set of all real numbers. From the graph, we conclude that the range of $f(x) = \left(\dfrac{1}{2}\right)^x$ is the set of all positive real numbers $\{y \mid y > 0\}$, or $(0, \infty)$ in interval notation.

Figure 14

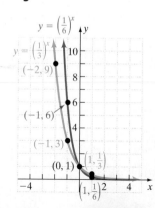

The graph of $f(x) = \left(\dfrac{1}{2}\right)^x$ in Figure 13 is typical of all exponential functions that have a base between 0 and 1. Figure 14 shows the graphs of two additional exponential functions whose bases are between 0 and 1: $y = \left(\dfrac{1}{3}\right)^x$ and $y = \left(\dfrac{1}{6}\right)^x$. Notice that the smaller the base, the closer the graph is to the x-axis for $x > 0$ and the steeper the graph is for $x < 0$.

We summarize the information we have about $f(x) = a^x$, where $0 < a < 1$.

Figure 15
$f(x) = a^x, 0 < a < 1$

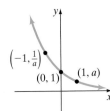

Properties of the Graph of an Exponential Function $f(x) = a^x$, $0 < a < 1$

1. The domain is the set of all real numbers. The range is the set of all positive real numbers.

2. There are no x-intercepts; the y-intercept is $(0, 1)$.

3. The graph of f contains the points $\left(-1, \dfrac{1}{a}\right)$, $(0, 1)$, and, $(1, a)$. See Figure 15.

Quick ✓

4. The graph of every exponential function $f(x) = a^x$ passes through three points:

_____ , _____, and _____.

5. *True or False* The domain of the exponential function $f(x) = a^x, a > 0, a \neq 1$, is the set of all real numbers.

6. *True or False* The range of the exponential function $f(x) = a^x, a > 0, a \neq 1$, is the set of all real numbers.

7. Graph the exponential function $f(x) = \left(\dfrac{1}{4}\right)^x$ using point plotting. From the graph, state the domain and the range of the function.

EXAMPLE 4 **Graphing an Exponential Function**

Use point plotting to graph $f(x) = 3^{x+1}$. From the graph, state the domain and the range of the function.

Solution

Choose values of x and find the corresponding values of the function. See Table 4. Then plot the ordered pairs and connect them in a smooth curve. See Figure 16.

Table 4

x	$f(x)$	$(x, f(x))$
-3	$f(-3) = 3^{-3+1} = 3^{-2} = \dfrac{1}{3^2} = \dfrac{1}{9}$	$\left(-3, \dfrac{1}{9}\right)$
-2	$f(-2) = 3^{-2+1} = 3^{-1} = \dfrac{1}{3^1} = \dfrac{1}{3}$	$\left(-2, \dfrac{1}{3}\right)$
-1	$f(-1) = 3^{-1+1} = 3^0 = 1$	$(-1, 1)$
0	$f(0) = 3^{0+1} = 3^1 = 3$	$(0, 3)$
1	$f(1) = 3^{1+1} = 3^2 = 9$	$(1, 9)$

Figure 16

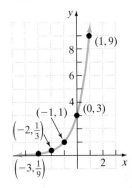

The domain is the set of all real numbers. The range is $\{y \mid y > 0\}$, or $(0, \infty)$ in interval notation.

Quick ✓

In Problems 8 and 9, graph each function using point plotting. From the graph, state the domain and the range of each function.

8. $f(x) = 2^{x-1}$

9. $f(x) = 3^x + 1$

▶ ❸ Define the Number *e*

Many models use an exponential function whose base is an irrational number symbolized by the letter *e*. The number *e* can be used to model the growth of a stock's price or to estimate the time of death of a carbon-based life form.

> **Definition**
>
> The **number *e*** is defined as the number that the expression
>
> $$\left(1 + \frac{1}{n}\right)^n$$
>
> approaches as *n* increases.

Table 5 shows some values of $\left(1 + \dfrac{1}{n}\right)^n$ as *n* increases. The last number in column 4 is the number *e* correct to nine decimal places.

Table 5

n	$\dfrac{1}{n}$	$1 + \dfrac{1}{n}$	$\left(1 + \dfrac{1}{n}\right)^n$
1	1	2	2
2	0.5	1.5	2.25
5	0.2	1.2	2.48832
10	0.1	1.1	2.59374246
100	0.01	1.01	2.704813829
1,000	0.001	1.001	2.716923932
10,000	0.0001	1.0001	2.718145927
100,000	0.00001	1.00001	2.718268237
1,000,000	0.000001	1.000001	2.718280469
1,000,000,000	10^{-9}	$1 + 10^{-9}$	2.718281827

The exponential function $f(x) = e^x$, whose base is the number *e*, occurs so often in applications that it is sometimes called *the* exponential function. Most calculators have the key $\boxed{e^x}$ or $\boxed{\exp{(x)}}$, which you can use to evaluate the exponential function $f(x) = e^x$ for a given value of *x*.

Use your calculator to approximate the values of $f(x) = e^x$ for $x = -1, 0$, and 1 as we have done to create Table 6. The graph of the exponential function is shown in Figure 17(a). Since $2 < e < 3$, the graph of $f(x) = e^x$ lies between the graph of $y = 2^x$ and that of $y = 3^x$. See Figure 17(b).

Figure 17

Table 6

x	$f(x) = e^x$
-2	$e^{-2} \approx 0.135$
-1	$e^{-1} \approx 0.368$
0	$e^0 = 1$
1	$e^1 \approx 2.718$
2	$e^2 \approx 7.389$

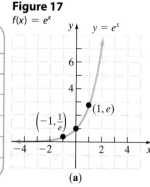

(a)

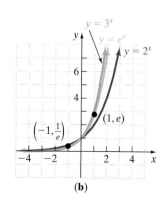

(b)

10. What is the value of e rounded to five decimal places?

11. Evaluate each of the following rounded to three decimal places:

    (a) e^4 **(b)** e^{-4}

▶ ❹ Solve Exponential Equations

Equations that involve terms of the form a^x, $a > 0$, $a \neq 1$, are called **exponential equations.** We can solve some exponential equations using the Laws of Exponents and the following property.

Work Smart

In order to use the Property for Solving Exponential Equations, both sides of the equation must have the same base.

Property For Solving Exponential Equations

$$\text{If} \quad a^u = a^v, \quad \text{then} \quad u = v.$$

This property results from the fact that exponential functions are one-to-one. It basically states that any output is the result of one (and only one) input. That is to say, two different inputs cannot yield the same output. For example, $y = x^2$ is not one-to-one because two different inputs, -2 and 2, correspond to the same output, 4.

EXAMPLE 5   **How to Solve an Exponential Equation**

Solve: $2^{x-3} = 32$

Step-by-Step Solution

Step 1: Use the Laws of Exponents to write both sides of the equation with the same base.

       $32 = 2^5$:
$$2^{x-3} = 32$$
$$2^{x-3} = 2^5$$

Step 2: Set the exponents on each side of the equation equal to each other.

       If $a^u = a^v$, then $u = v$:
$$x - 3 = 5$$

Step 3: Solve the equation resulting from Step 2.

       Add 3 to both sides:
$$x - 3 + 3 = 5 + 3$$
$$x = 8$$

Step 4: Verify your solution(s).

       Let $x = 8$:
$$2^{x-3} = 32$$
$$2^{8-3} \overset{?}{=} 32$$
$$2^5 \overset{?}{=} 32$$
$$32 = 32 \quad \text{True}$$

Work Smart

Did you notice that we used properties of algebra to reduce the exponential equation to a linear equation?

The solution set is $\{8\}$.

Solving Exponential Equations of the Form $a^u = a^v$

Step 1: Use the Laws of Exponents to write both sides of the equation with the same base.

Step 2: Set the exponents on each side of the equation equal to each other.

Step 3: Solve the equation resulting from Step 2.

Step 4: Verify your solution(s).

In Problems 12 and 13, solve each equation.

12. $5^{x-4} = 5^{-1}$ **13.** $3^{x+2} = 81$

EXAMPLE 6   **Solving Exponential Equations**

Solve the following equations:

(a) $4^{x^2} = 32$

(b) $\dfrac{e^{x^2}}{e^{2x}} = e^8$

Solution

(a)

	$4^{x^2} = 32$
Rewrite as exponential expressions with a common base:	$(2^2)^{x^2} = 2^5$
Use $(a^m)^n = a^{m \cdot n}$:	$2^{2x^2} = 2^5$
If $a^u = a^v$, then $u = v$:	$2x^2 = 5$
Divide both sides by 2:	$x^2 = \dfrac{5}{2}$
Take the square root of both sides:	$x = \pm\sqrt{\dfrac{5}{2}}$
Rationalize the denominator:	$x = \pm\dfrac{\sqrt{10}}{2}$

Check

$x = \dfrac{-\sqrt{10}}{2}$: $\quad 4^{\left(-\frac{\sqrt{10}}{2}\right)^2} \overset{?}{=} 32$

$\qquad\qquad 4^{\frac{10}{4}} \overset{?}{=} 32$

$\qquad\qquad 4^{\frac{5}{2}} \overset{?}{=} 32$

$a^{m/n} = \sqrt[n]{a^m}$: $\quad (\sqrt{4})^5 \overset{?}{=} 32$

$\qquad\qquad 2^5 \overset{?}{=} 32$

$\qquad\qquad 32 = 32$ True

$x = \dfrac{\sqrt{10}}{2}$: $\quad 4^{\left(\frac{\sqrt{10}}{2}\right)^2} \overset{?}{=} 32$

$\qquad\qquad 4^{\frac{10}{4}} \overset{?}{=} 32$

$\qquad\qquad 4^{\frac{5}{2}} \overset{?}{=} 32$

$\qquad\qquad (\sqrt{4})^5 \overset{?}{=} 32$

$\qquad\qquad 2^5 \overset{?}{=} 32$

$\qquad\qquad 32 = 32$ True

The solution set is $\left\{-\dfrac{\sqrt{10}}{2}, \dfrac{\sqrt{10}}{2}\right\}$.

(b)

	$\dfrac{e^{x^2}}{e^{2x}} = e^8$
Use $\dfrac{a^m}{a^n} = a^{m-n}$:	$e^{x^2 - 2x} = e^8$
If $a^u = a^v$, then $u = v$:	$x^2 - 2x = 8$
Subtract 8 from both sides:	$x^2 - 2x - 8 = 0$
Factor:	$(x - 4)(x + 2) = 0$
Zero-Product Property:	$x = 4$ or $x = -2$

Check

$x = 4$:

$\dfrac{e^{4^2}}{e^{2(4)}} \overset{?}{=} e^8$

$\dfrac{e^{16}}{e^8} \overset{?}{=} e^8$

$e^8 = e^8$

$x = -2$:

$\dfrac{e^{(-2)^2}}{e^{2(-2)}} \overset{?}{=} e^8$

$\dfrac{e^4}{e^{-4}} \overset{?}{=} e^8$

$e^8 = e^8$

The solution set is $\{-2, 4\}$.

In Problems 14 and 15, solve each equation.

14. $e^{x^2} = e^x \cdot e^{4x}$

15. $\dfrac{2^{x^2}}{8} = 2^{2x}$

▶ ❺ Use Exponential Models That Describe Our World

Exponential functions are used in many different disciplines, such as biology (half-life), chemistry (carbon dating), economics (time value of money), and psychology (learning curves). Exponential functions are also used in statistics, as shown in the next example.

(EXAMPLE 7) **Exponential Probability**

The manager of a crisis helpline knows that between 3:00 A.M. and 5:00 A.M., 3 calls per hour occur (that's 0.05 call per minute). The following formula from statistics gives the likelihood that a call will occur within t minutes of 3 A.M.

$$F(t) = 1 - e^{-0.05t}$$

Determine the likelihood that a person will call

(a) within 5 minutes of 3:00 A.M.

(b) within 20 minutes of 3:00 A.M.

Solution

(a) The likelihood that a call will occur within 5 minutes of 3:00 A.M. is found by evaluating the function $F(t) = 1 - e^{-0.05t}$ at $t = 5$.

$$F(5) = 1 - e^{-0.05(5)}$$
$$\approx 0.221$$

The likelihood that a call will occur in this time span is $0.221 = 22.1\%$.

(b) The likelihood that a call will occur within 20 minutes of 3:00 A.M. is found by evaluating the function $F(t) = 1 - e^{-0.05t}$ at $t = 20$, or $F(20)$.

$$F(20) = 1 - e^{-0.05(20)}$$
$$\approx 0.632$$

The likelihood that a call will occur in this time span is $0.632 = 63.2\%$. ●

16. A bank manager knows that between 3:00 P.M. and 5:00 P.M., 15 people arrive per hour (that's 0.25 people per minute). The following formula from statistics gives the likelihood that a person will arrive within t minutes of 3:00 P.M.

$$F(t) = 1 - e^{-0.25t}$$

Find the likelihood that a person will arrive
(a) within 10 minutes of 3 P.M.   (b) within 25 minutes of 3 P.M.

(EXAMPLE 8) **Radioactive Decay**

The radioactive **half-life** for a given radioisotope of an element is the time it takes for half the radioactive nuclei in any sample to decay to some other substance. For example, the half-life of plutonium-239 is 24,360 years. Plutonium-239 is particularly dangerous because it emits alpha particles that are absorbed into bone marrow. The maximum amount of plutonium-239 an adult can handle without significant injury is

0.13 microgram $(= 0.000000013$ gram$)$. Suppose a researcher has a 1-gram sample of plutonium-239. The amount A (in grams) of plutonium-239 after t years is given by

$$A(t) = 1 \cdot \left(\frac{1}{2}\right)^{\frac{t}{24,360}}$$

How much plutonium-239 is left in the sample after

(a) 500 years?   **(b)** 24,360 years?   **(c)** 73,080 years?

Solution

(a) The amount of plutonium-239 left in the sample after 500 years is found by evaluating A at $t = 500$. That is, we determine $A(500)$.

$$A(500) = 1 \cdot \left(\frac{1}{2}\right)^{\frac{500}{24,360}}$$

Use a calculator: ≈ 0.986 gram

After 500 years, approximately 0.986 gram of plutonium-239 will be left in the sample.

(b) The amount of plutonium-239 remaining after 24,360 years is found by evaluating A at $t = 24,360$. That is, we determine $A(24,360)$.

$$A(24,360) = 1 \cdot \left(\frac{1}{2}\right)^{\frac{24,360}{24,360}}$$

$$= 1 \cdot \left(\frac{1}{2}\right)^{1}$$

$$= 0.5 \text{ gram}$$

After 24,360 years, there will be 0.5 gram of plutonium-239 left.

(c) To find the amount left after 73,080 years, we evaluate $A(73,080)$.

$$A(73,080) = 1 \cdot \left(\frac{1}{2}\right)^{\frac{73,080}{24,360}}$$

$$= 1 \cdot \left(\frac{1}{2}\right)^{3}$$

$$= \frac{1}{8} \text{ gram}$$

After 73,080 years, there will be $\frac{1}{8} = 0.125$ gram of plutonium-239 left in the sample.

Quick ✓

17. The half-life of thorium-227 is 18.72 days. Suppose that a researcher possesses a 10-gram sample of thorium-227. The amount A (in grams) of thorium-227 left after t days is given by

$$A'(t) = 10 \cdot \left(\frac{1}{2}\right)^{\frac{t}{18.72}}$$

How much thorium-227 is left in the sample after
(a) 10 days?   **(b)** 18.72 days?
(c) 74.88 days?   **(d)** 100 days?

Table 7	
Payment Period	**Number of Times Interest Is Paid**
Annually	Once per year
Semiannually	Twice per year
Quarterly	4 times per year
Monthly	12 times per year
Daily	360 times per year

When we deposit money in a bank, the bank pays us interest on the balance in the account. When solving interest problems, we use the term **payment period** as shown in Table 7 to indicate how often the bank pays interest.

When the interest due at the end of a payment period is added to the principal so that the interest computed at the end of the *next* payment period is based on this new principal amount (old principal + interest), we say the interest has been **compounded.** **Compound interest** is interest paid on previously earned interest.

The following formula can be used to determine the value of an account after a certain period of time.

Work Smart

When using the compound interest formula, be sure to express the interest rate as a decimal.

Compound Interest Formula

The amount A after t years resulting from a principal P invested at an annual interest rate r compounded n times per year is

$$A = P\left(1 + \frac{r}{n}\right)^{nt}$$

For example, if you deposit $500 into an account paying 3% annual interest compounded monthly, then $P = \$500$, $r = 0.03$, and $n = 12$ (twelve compounding periods per year).

EXAMPLE 9 **Future Value of Money**

Suppose you deposit $3000 into a Roth IRA today. If the deposit earns 8% interest compounded quarterly, determine its future value A after

(a) 1 year **(b)** 10 years **(c)** 35 years, when you plan on retiring

Solution

We use the compound interest formula with $P = \$3000$, $r = 0.08$, and $n = 4$ (for quarterly compounding), so

$$A = \$3000\left(1 + \frac{0.08}{4}\right)^{4t}$$
$$= \$3000(1 + 0.02)^{4t}$$
$$= \$3000(1.02)^{4t}$$

Figure 18

```
3000(1.02)^4
        3247.29648
```

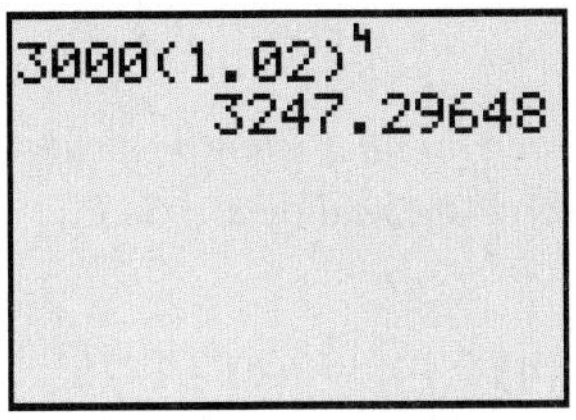

(a) The value of the account after $t = 1$ year is
$$A = \$3000(1.02)^{4(1)}$$
$$= \$3000(1.02)^4$$
Use a calculator; see Figure 18: $= \$3000(1.08243216)$
$$= \$3247.30$$

(b) The value after $t = 10$ years is
$$A = \$3000(1.02)^{4(10)}$$
$$= \$3000(1.02)^{40}$$
Use a calculator: $= \$3000(2.208039664)$
$$= \$6624.12$$

(c) The value after $t = 35$ years is
$$A = \$3000(1.02)^{4(35)}$$
$$= \$3000(1.02)^{140}$$
Use a calculator: $= \$3000(15.99646598)$
$$= \$47,989.40$$

Quick ✓

18. Suppose that you deposit $2000 into a Roth IRA today. Determine the future value A of the deposit if it earns 5% interest compounded monthly (twelve times per year) after

(a) 1 year **(b)** 15 years **(c)** 30 years, when you plan on retiring

8.2 Exercises MyMathLab® MathXL PRACTICE

*Problems **1–18** are the Quick ✓s that follow the EXAMPLES.*

Building Skills

In Problems 19–22, approximate each number using a calculator. Express your answer rounded to three decimal places. See Objective 1.

19. (a) $3^{2.2}$ **(b)** $3^{2.23}$ **(c)** $3^{2.236}$
  (d) $3^{2.2361}$ **(e)** $3^{\sqrt{5}}$

20. (a) $5^{1.4}$ **(b)** $5^{1.41}$ **(c)** $5^{1.414}$
  (d) $5^{1.4142}$ **(e)** $5^{\sqrt{2}}$

21. (a) $4^{3.1}$ **(b)** $4^{3.14}$ **(c)** $4^{3.142}$
  (d) $4^{3.1416}$ **(e)** 4^{π}

22. (a) $10^{2.7}$ **(b)** $10^{2.72}$ **(c)** $10^{2.718}$
  (d) $10^{2.7183}$ **(e)** 10^{e}

In Problems 23–30, the graph of an exponential function is given. Match each graph to one of the following functions. It may prove useful to create a table of values for each function to help you identify the correct graph. See Objective 2.

(a) $f(x) = 2^x$ **(b)** $f(x) = 2^{-x}$ **(c)** $f(x) = 2^{x+1}$
(d) $f(x) = 2^{x-1}$ **(e)** $f(x) = -2^x$ **(f)** $f(x) = 2^x + 1$
(g) $f(x) = 2^x - 1$ **(h)** $f(x) = -2^{-x}$

23.
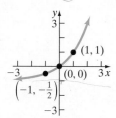

24.
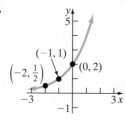

25.
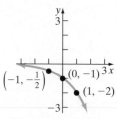

26.
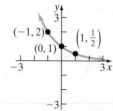

27.
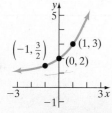

28.
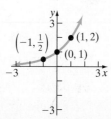

29.
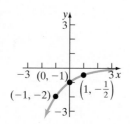

30.
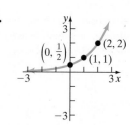

In Problems 31–42, graph each function. State the domain and the range of the function. See Objective 2.

31. $f(x) = 5^x$ **32.** $f(x) = 7^x$

33. $F(x) = \left(\dfrac{1}{5}\right)^x$ **34.** $F(x) = \left(\dfrac{1}{7}\right)^x$

35. $h(x) = 2^{x+2}$ **36.** $H(x) = 2^{x-2}$

37. $f(x) = 2^x + 3$ **38.** $F(x) = 2^x - 3$

39. $F(x) = \left(\dfrac{1}{2}\right)^x - 1$ **40.** $G(x) = \left(\dfrac{1}{2}\right)^x + 2$

41. $P(x) = \left(\dfrac{1}{3}\right)^{x-2}$ **42.** $p(x) = \left(\dfrac{1}{3}\right)^{x+2}$

In Problems 43–50, approximate each number using a calculator. Express your answer rounded to three decimal places. See Objective 3.

43. (a) $3.1^{2.7}$ **(b)** $3.14^{2.72}$ **(c)** $3.142^{2.718}$
  (d) $3.1416^{2.7183}$ **(e)** π^e

44. (a) $2.7^{3.1}$ **(b)** $2.72^{3.14}$ **(c)** $2.718^{3.142}$
  (d) $2.7183^{3.1416}$ **(e)** e^π

45. e^2 **46.** e^3 **47.** e^{-2}
48. e^{-3} **49.** $e^{2.3}$ **50.** $e^{1.5}$

In Problems 51–54, graph each function. State the domain and the range of the function. See Objective 3.

51. $g(x) = e^{x-1}$ **52.** $f(x) = e^x - 1$

53. $f(x) = -2e^x$ **54.** $F(x) = \dfrac{1}{2}e^x$

In Problems 55–80, solve each equation. See Objective 4.

55. $2^x = 2^5$

56. $3^x = 3^{-2}$

57. $3^{-x} = 81$

58. $4^{-x} = 64$

59. $\left(\dfrac{1}{2}\right)^x = \dfrac{1}{32}$

60. $\left(\dfrac{1}{3}\right)^x = \dfrac{1}{243}$

61. $5^{x-2} = 125$

62. $2^{x+3} = 128$

63. $4^x = 8$

64. $9^x = 27$

65. $2^{-x+5} = 16^x$

66. $3^{-x+4} = 27^x$

67. $3^{x^2-4} = 27^x$

68. $5^{x^2-10} = 125^x$

69. $4^x \cdot 2^{x^2} = 16^2$

70. $9^{2x} \cdot 27^{x^2} = 3^{-1}$

71. $2^x \cdot 8 = 4^{x-3}$

72. $3^x \cdot 9 = 27^x$

73. $\left(\dfrac{1}{5}\right)^x - 25 = 0$

74. $\left(\dfrac{1}{6}\right)^x - 36 = 0$

75. $(2^x)^x = 16$

76. $(3^x)^x = 81$

77. $e^x = e^{3x+4}$

78. $e^{3x} = e^2$

79. $(e^x)^2 = e^{3x-2}$

80. $(e^3)^x = e^2 \cdot e^x$

Mixed Practice

81. Suppose that $f(x) = 2^x$.

  (a) What is $f(3)$? What point is on the graph of f?

  (b) If $f(x) = \dfrac{1}{8}$, what is x? What point is on the graph of f?

82. Suppose that $f(x) = 3^x$.

  (a) What is $f(2)$? What point is on the graph of f?

  (b) If $f(x) = \dfrac{1}{81}$, what is x? What point is on the graph of f?

83. Suppose that $g(x) = 4^x - 1$.

  (a) What is $g(-1)$? What point is on the graph of g?

  (b) If $g(x) = 15$, what is x? What point is on the graph of g?

84. Suppose that $g(x) = 5^x + 1$.

  (a) What is $g(-1)$? What point is on the graph of g?

  (b) If $g(x) = 126$, what is x? What point is on the graph of g?

85. Suppose that $H(x) = 3 \cdot \left(\dfrac{1}{2}\right)^x$.

  (a) What is $H(-3)$? What point is on the graph of H?

  (b) If $H(x) = \dfrac{3}{4}$, what is x? What point is on the graph of H?

86. Suppose that $F(x) = -2 \cdot \left(\dfrac{1}{3}\right)^x$.

  (a) What is $F(-1)$? What point is on the graph of F?

  (b) If $F(x) = -18$, what is x? What point is on the graph of F?

Applying the Concepts

87. A Population Model According to the U.S. Census Bureau, the population of the United States in 2012 was 313 million people. In addition, the population of the United States was growing at a rate of 1.1% per year. Assuming that this growth rate continues, the model $P(t) = 313(1.011)^{t-2012}$ represents the population P (in millions of people) in year t.

  (a) According to this model, what will be the population of the United States in 2015?

  (b) According to this model, what will be the population of the United States in 2050?

  (c) The United States Census Bureau predicts that the United States population will be 439 million in 2050. Compare this estimate to the one obtained in part (b). What might account for any differences?

88. A Population Model According to the U.S. Census Bureau, the population of the world in 2012 was 7018 million people. In addition, the population of the world was growing at a rate of 1.26% per year. Assuming that this growth rate continues, the model $P(t) = 7018(1.0126)^{t-2012}$ represents the population P (in millions of people) in year t.

  (a) According to this model, what will be the population of the world in 2015?

  (b) According to this model, what will be the population of the world in 2025?

  (c) The United States Census Bureau predicts that the world population will be 8000 million (8 billion) in 2025. Compare this estimate to the one obtained in part (b). What might account for any differences?

89. Time Is Money Suppose that you deposit $5000 into a certificate of deposit (CD) today. Determine the future value A of the deposit if it earns 6% interest compounded monthly after

  (a) 1 year.

  (b) 3 years.

  (c) 5 years, when the CD comes due.

90. Time Is Money Suppose that you deposit $8000 into a certificate of deposit (CD) today. Determine the future value A of the deposit if it earns 4% interest compounded quarterly (4 times per year) after
(a) 1 year.　　　　(b) 3 years.
(c) 5 years, when the CD comes due.

91. Do the Compounding Periods Matter? Suppose that you deposit $2000 into an account that pays 3% annual interest. How much will you have after 5 years if interest is compounded
(a) annually?　　　　(b) quarterly?
(c) monthly?　　　　(d) daily?
(e) Based on the results of parts (a)–(d), what impact does increasing the number of compounding periods have on the future value, all other things equal?

92. Do the Compounding Periods Matter? Suppose that you deposit $1000 into an account that pays 6% annual interest. How much will you have after 3 years if interest is compounded
(a) annually?　　　　(b) quarterly?
(c) monthly?　　　　(d) daily?
(e) Based on the results of parts (a)–(d), what impact does increasing the number of compounding periods have on the future value, all other things equal?

93. Depreciation Based on data obtained from the *Kelley Blue Book*, the value V of a Ford Focus that is t years old can be modeled by $V(t) = 19{,}841(0.88)^t$.
(a) According to the model, what is the value of a brand-new Focus?
(b) According to the model, what is the value of a 2-year-old Focus?
(c) According to the model, what is the value of a 5-year-old Focus?

94. Depreciation Based on data obtained from the *Kelley Blue Book*, the value V of a Chevy Malibu that is t years old can be modeled by $V(t) = 25{,}258(0.84)^t$.
(a) According to the model, what is the value of a brand-new Chevy Malibu?
(b) According to the model, what is the value of a 2-year-old Chevy Malibu?
(c) According to the model, what is the value of a 5-year-old Chevy Malibu?

95. Radioactive Decay The half-life of beryllium-11 is 13.81 seconds. Suppose that a researcher possesses a 100-gram sample of beryllium-11. The amount A (in grams) of beryllium-11 after t seconds is given by

$$A(t) = 100 \cdot \left(\frac{1}{2}\right)^{\frac{t}{13.81}}$$

(a) How much beryllium-11 is left in the sample after 1 second?
(b) How much beryllium-11 is left in the sample after 13.81 seconds?

(c) How much beryllium-11 is left in the sample after 27.62 seconds?
(d) How much beryllium-11 is left in the sample after 100 seconds?

96. Radioactive Decay The half-life of carbon-10 is 19.255 seconds. Suppose that a researcher possesses a 100-gram sample of carbon-10. The amount A (in grams) of carbon-10 after t seconds is given by

$$A(t) = 100 \cdot \left(\frac{1}{2}\right)^{\frac{t}{19.255}}$$

(a) How much carbon-10 is left in the sample after 1 second?
(b) How much carbon-10 is left in the sample after 19.255 seconds?
(c) How much carbon-10 is left in the sample after 38.51 seconds?
(d) How much carbon-10 is left in the sample after 100 seconds?

For Exercises 97 and 98, use Newton's Law of Cooling, which states that the temperature of a heated object decreases exponentially over time toward the temperature of the surrounding medium.

97. Newton's Law of Cooling Suppose a pizza is removed from a 400°F oven and placed in a room whose temperature is 70°F. The temperature u (in °F) of the pizza at time t (in minutes) can be modeled by $u(t) = 70 + 330e^{-0.072t}$.
(a) According to the model, what will be the temperature of the pizza in 5 minutes?

(b) According to the model, what will be the temperature of the pizza in 10 minutes?

(c) If the pizza can be safely consumed when its temperature is 200°F, will it be ready to eat after cooling for 13 minutes?

98. Newton's Law of Cooling Suppose coffee at a temperature of 170°F is poured into a coffee mug and allowed to cool in a room whose temperature is 70°F. The temperature u (in °F) of the coffee at time t (in minutes) can be modeled by $u(t) = 70 + 100e^{-0.045t}$.
(a) According to the model, what will be the temperature of the coffee in 5 minutes?

(b) According to the model, what will be the temperature of the coffee in 10 minutes?

(c) If the coffee doesn't taste good once its temperature reaches 120°F, will it still be tasty after cooling for 20 minutes?

99. Learning Curve Suppose that a student has 200 vocabulary words to learn. If a student learns 20 words in 30 minutes, the function

$$L(t) = 200(1 - e^{-0.0035t})$$

models the number of words L that the student will learn in t minutes.

(a) How many words will the student learn in 45 minutes?

(b) How many words will the student learn in 60 minutes?

100. Learning Curve Suppose that a student has 50 biology terms to learn. If a student learns 10 terms in 30 minutes, the function

$$L(t) = 50(1 - e^{-0.0223t})$$

models the number of terms L that the student will learn in t minutes.

(a) How many words will the student learn in 45 minutes?

(b) How many words will the student learn in 60 minutes?

101. Current in an *RL* Circuit The equation governing the amount of current I (in amperes) after time t (in seconds) in a single *RL* circuit consisting of a resistance R (in ohms), an inductance L (in henrys), and an electromotive force E (in volts) is

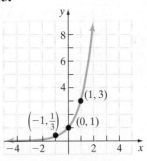

$$I = \frac{E}{R}[1 - e^{-(R/L)t}]$$

(a) If $E = 120$ volts, $R = 10$ ohms, and $L = 25$ henrys, how much current I is flowing in 0.05 second?

(b) If $E = 240$ volts, $R = 10$ ohms, and $L = 25$ henrys, how much current I is flowing in 0.05 second?

102. Current in an *RC* Circuit The equation governing the amount of current I (in amperes) after time t (in microseconds) in a single *RC* circuit consisting of a resistance R (in ohms), a capacitance C (in microfarads), and an electromotive force E (in volts) is

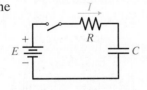

$$I = \frac{E}{R} e^{-t/(RC)}$$

(a) If $E = 120$ volts, $R = 2500$ ohms, and $C = 100$ microfarads, how much current I is flowing initially $(t = 0)$? In 50 microseconds?

(b) If $E = 240$ volts, $R = 2500$ ohms, and $C = 100$ microfarads, how much current I is flowing initially $(t = 0)$? In 50 microseconds?

Extending the Concepts

In Problems 103 and 104, find the exponential function whose graph is given.

103.

104.

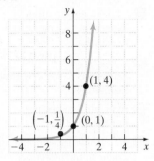

Explaining the Concepts

105. As the base a of an exponential function $f(x) = a^x$ increases (for $a > 1$), what happens to the graph of the exponential function for $x > 0$? What happens to the behavior of the graph for $x < 0$?

106. The graphs of $f(x) = 2^{-x}$ and $g(x) = \left(\dfrac{1}{2}\right)^x$ are identical. Why?

107. Explain the difference between exponential functions and polynomial functions.

108. Can we solve the equation $2^x = 12$ using the fact that if $a^u = a^v$, then $u = v$? Why or why not?

Synthesis Review

In Problems 109–114, evaluate each expression.

109. Evaluate $x^2 - 5x + 1$ at
   (a) $x = -2$ **(b)** $x = 3$

110. Evaluate $x^3 - 5x + 2$ at
   (a) $x = -4$ **(b)** $x = 2$

111. Evaluate $\dfrac{4}{x + 2}$ at
   (a) $x = -2$ **(b)** $x = 3$

112. Evaluate $\dfrac{2x}{x + 1}$ at
   (a) $x = -4$ **(b)** $x = 5$

113. Evaluate $\sqrt{2x + 5}$ at
   (a) $x = 2$ **(b)** $x = 11$

114. Evaluate $\sqrt[3]{3x - 1}$ at
   (a) $x = 3$ **(b)** $x = 0$

The Graphing Calculator

In Problems 115–122, graph each function using a graphing calculator. State the domain and the range of each function.

115. $f(x) = 1.5^x$ **116.** $G(x) = 3.1^x$

117. $H(x) = 0.9^x$ **118.** $F(x) = 0.3^x$

119. $g(x) = 2.5^x + 3$ **120.** $f(x) = 1.7^x - 2$

121. $F(x) = 1.6^{x-3}$ **122.** $g(x) = 0.3^{x+2}$

8.3 Logarithmic Functions

Objectives

1. Change Exponential Equations to Logarithmic Equations
2. Change Logarithmic Equations to Exponential Equations
3. Evaluate Logarithmic Functions
4. Determine the Domain of a Logarithmic Function
5. Graph Logarithmic Functions
6. Work with Natural and Common Logarithms
7. Solve Logarithmic Equations
8. Use Logarithmic Models That Describe Our World

Work Smart

We require that a be positive and not equal to 1 for the same reasons that we had these restrictions for the exponential function.

In Words

The logarithm to the base a of x is the number y that we must raise a to in order to obtain x.

Work Smart: Study Skills

In doing problems similar to Examples 1 and 2, it is a good idea to say, "y equals the logarithm to the base a of x is equivalent to a to the y equals x" so that you memorize the definition of a logarithm.

Are You Ready for This Section?

Before getting started, take the following readiness quiz. If you get a problem wrong, go back to the section cited and review the material.

R1. Solve: $3x + 2 > 0$ [Section 1.4, pp. 85–87]

R2. Solve: $\sqrt{x+2} = x$ [Section 6.7, pp. 515–518]

R3. Solve: $x^2 = 6x + 7$ [Section 4.8, pp. 372–374]

We know about the "squaring function," $f(x) = x^2$, and we also know about the square root function, $g(x) = \sqrt{x}$. How are these two functions related? Well, we know that $f(5) = 25$ and $g(25) = 5$. That is, the input to f is the output of g and the output of f is the input to g, so g "undoes" what f does.

In general, whenever we introduce a function in mathematics, we would also like to find a second function that "undoes" it. For example, a square root function undoes the squaring function. A cube root function undoes the cubing function. Do we have a function that undoes an exponential function? Yes! This function is the *logarithmic function*.

> **Definition**
>
> The **logarithmic function to the base a,** where $a > 0$ and $a \neq 1$, is denoted by $y = \log_a x$ (read as "y is the logarithm to the base a of x"), and
> $$y = \log_a x \quad \text{is equivalent to} \quad x = a^y$$

To evaluate logarithmic functions, we convert them into their equivalent exponential form. Therefore, we must be able to go from logarithmic form to exponential form, and back. For example,

$$0 = \log_3 1 \quad \text{is equivalent to} \quad 3^0 = 1$$
$$2 = \log_5 25 \quad \text{is equivalent to} \quad 5^2 = 25$$
$$-2 = \log_4 \frac{1}{16} \quad \text{is equivalent to} \quad 4^{-2} = \frac{1}{16}$$

Notice that the base of the logarithm is the base of the exponential; the argument of the logarithm is what the exponential equals; and the value of the logarithm is the exponent of the exponential expression. A logarithm is just a fancy way of writing an exponential expression.

To see how the logarithmic function undoes the exponential function, consider the function $y = 2^x$. If we input $x = 3$, then the output is $y = 2^3 = 8$. To undo this function would require that an input of 8 give an output of 3. If $2^3 = 8$, then $3 = \log_2 8$ using the definition of a logarithm. The input of $\log_2 8$ is 8, and its output is 3.

▶ 1 Change Exponential Equations to Logarithmic Equations

We can use the definition of a logarithm to rewrite exponential equations as logarithmic equations.

(EXAMPLE 1)   **Changing Exponential Equations to Logarithmic Equations**

Rewrite each exponential equation as an equivalent logarithmic equation.

     (a) $6^2 = 36$ **(b)** $x^2 = 9$ **(c)** $4^3 = y$

Solution

Use the fact that $y = \log_a x$ is equivalent to $a^y = x$ provided that $a > 0$ and $a \neq 1$.

   (a) If $6^2 = 36$, then $2 = \log_6 36$.   **(b)** If $x^2 = 9$ then $2 = \log_x 9$.

   (c) If $4^3 = y$ then $3 = \log_4 y$

Ready?...Answers

R1. $\left\{x \mid x > -\dfrac{2}{3}\right\}$ or $\left(-\dfrac{2}{3}, \infty\right)$

R2. $\{2\}$ **R3.** $\{-1, 7\}$

Quick ✓

1. The logarithm to the base a of x, denoted $y = \log_a x$, can be expressed in exponential form as _____, where a _ 1 and a _ 1.

In Problems 2 and 3, rewrite each exponential equation as an equivalent equation involving a logarithm.

2. $4^3 = 64$ **3.** $p^{-2} = 8$

▶ ❷ Change Logarithmic Equations to Exponential Equations

We can also use the definition of a logarithm to rewrite logarithmic equations as exponential equations.

EXAMPLE 2   **Changing Logarithmic Equations to Exponential Equations**

Change each logarithmic equation to an equivalent exponential equation.

(a) $4 = \log_3 81$ (b) $-3 = \log_a \dfrac{1}{27}$ (c) $2 = \log_4 x$

Solution

In each of these problems, use the fact that $y = \log_a x$ is equivalent to $a^y = x$ provided that $a > 0$ and $a \neq 1$.

(a) If $4 = \log_3 81$, then $3^4 = 81$.

(b) If $-3 = \log_a \dfrac{1}{27}$, then $a^{-3} = \dfrac{1}{27}$.

(c) If $2 = \log_4 x$, then $4^2 = x$.

Quick ✓

In Problems 4–6, rewrite each logarithmic equation as an equivalent equation involving an exponent.

4. $4 = \log_2 16$ **5.** $5 = \log_a 20$ **6.** $-3 = \log_5 z$

▶ ❸ Evaluate Logarithmic Functions

To find the exact value of a logarithm, write the logarithm in exponential notation and use the fact that if $a^u = a^v$, then $u = v$.

EXAMPLE 3   **Finding the Exact Value of a Logarithmic Expression**

Find the exact value of

(a) $\log_2 32$ (b) $\log_4 \dfrac{1}{16}$

Solution

(a) Let $y = \log_2 32$ and convert this equation into an exponential equation.

$$y = \log_2 32$$

Write the logarithm as an exponent: $\quad 2^y = 32$

$32 = 2^5$: $\quad 2^y = 2^5$

If $a^u = a^v$, then $u = v$: $\quad y = 5$

Therefore, $\log_2 32 = 5$.

(b) Let $y = \log_4 \dfrac{1}{16}$ and convert this equation into an exponential equation.

$$y = \log_4 \dfrac{1}{16}$$

Write the logarithm as an exponent: $\quad 4^y = \dfrac{1}{16}$

$$\dfrac{1}{16} = 4^{-2}: \quad 4^y = 4^{-2}$$

If $a^u = a^v$, then $u = v$: $\quad y = -2$

Therefore, $\log_4 \dfrac{1}{16} = -2$.

Quick ✓

In Problems 7 and 8, find the exact value of each logarithmic expression.

7. $\log_5 25$

8. $\log_2 \dfrac{1}{8}$

We could also write $y = \log_a x$ using function notation as $f(x) = \log_a x$. We use this notation in the next example to evaluate a logarithmic function.

EXAMPLE 4   **Evaluating a Logarithmic Function**

Find the value of each of the following, given that $f(x) = \log_2 x$.

(a) $f(2)$

(b) $f\left(\dfrac{1}{4}\right)$

Solution

(a) Finding $f(2)$ means evaluating $\log_2 x$ at $x = 2$, or evaluating $\log_2 2$. To determine this value, we follow the approach of Example 3 by letting $y = \log_2 2$ and converting the equation into an exponential equation.

$$y = \log_2 2$$

Write the logarithm as an exponent: $\quad 2^y = 2$

$$2 = 2^1: \quad 2^y = 2^1$$

If $a^u = a^v$, then $u = v$: $\quad y = 1$

Therefore, $f(2) = 1$.

(b) $f\left(\dfrac{1}{4}\right)$ means evaluating $\log_2 x$ at $x = \dfrac{1}{4}$ or $\log_2\left(\dfrac{1}{4}\right)$. Let $y = \log_2\left(\dfrac{1}{4}\right)$.

$$y = \log_2\left(\dfrac{1}{4}\right)$$

Write the logarithm as an exponent: $\quad 2^y = \dfrac{1}{4}$

$$\dfrac{1}{4} = 2^{-2}: \quad 2^y = 2^{-2}$$

If $a^u = a^v$, then $u = v$: $\quad y = -2$

Therefore, $f\left(\dfrac{1}{4}\right) = -2$.

Quick ✓

In Problems 9 and 10, evaluate the function, given that $g(x) = \log_5 x$.

9. $g(25)$

10. $g\left(\dfrac{1}{5}\right)$

▶ ④ **Determine the Domain of a Logarithmic Function**

The domain of a function $y = f(x)$ is the set of all x such that the function makes sense, and the range is the set of all outputs of the function. To find the range of the logarithmic function, we recognize that $y = f(x) = \log_a x$ is equivalent to $x = a^y$. Because we can raise a to any real number (since $a > 0$ and $a \neq 1$), we conclude that y can be any real number in the function $y = f(x) = \log_a x$. In addition, because a^y is positive for any real number, we conclude that x must be positive. Since x represents the input of the logarithmic function, the domain of the logarithmic function is the set of all positive real numbers.

Work Smart

Notice that the domain of the logarithmic function is the same as the range of the exponential function. The range of the logarithmic function is the same as the domain of the exponential function.

Domain and Range of The Logarithmic Function

Domain of the logarithmic function $= (0, \infty)$

Range of the logarithmic function $= (-\infty, \infty)$

Because the domain of the logarithmic function is the set of all positive real numbers, the argument of the logarithmic function must be greater than zero. For example, $f(x) = \log_{10} x$ is defined for $x = 2$, but not for $x = -1$, $x = -8$, or any other $x \leq 0$.

(EXAMPLE 5) Finding the Domain of a Logarithmic Function

Find the domain of each logarithmic function.

(a) $f(x) = \log_6(x - 5)$
(b) $G(x) = \log_3(3x + 1)$

Solution

(a) The argument of the function $f(x) = \log_6(x - 5)$ is $x - 5$. The domain of f is the set of all real numbers x such that $x - 5 > 0$. We solve this inequality:

$$x - 5 > 0$$
Add 5 to both sides: $\quad x > 5$

The domain of f is $\{x \mid x > 5\}$, or $(5, \infty)$ in interval notation.

(b) The argument of the function $G(x) = \log_3(3x + 1)$ is $3x + 1$. The domain of G is the set of all real numbers x such that $3x + 1 > 0$.

$$3x + 1 > 0$$
Subtract 1 from both sides: $\quad 3x > -1$

Divide both sides by 3: $\quad x > -\dfrac{1}{3}$

The domain of G is $\left\{ x \mid x > -\dfrac{1}{3} \right\}$, or $\left(-\dfrac{1}{3}, \infty \right)$ in interval notation.

Quick ✓

In Problems 11 and 12, find the domain of each logarithmic function.

11. $g(x) = \log_8(x + 3)$
12. $F(x) = \log_2(5 - 2x)$

▶ ⑤ **Graph Logarithmic Functions**

To graph a logarithmic function $y = \log_a x$, it helps to rewrite the function in exponential form: $x = a^y$. Then we choose "nice" values of y and use the expression $x = a^y$ to find the corresponding values of x, as shown in Example 6.

EXAMPLE 6 **Graphing a Logarithmic Function**

Graph $f(x) = \log_2 x$ using point plotting. From the graph, state the domain and the range of the function.

Solution

We rewrite $y = f(x) = \log_2 x$ as $x = 2^y$. Table 8 shows various values of y, the corresponding values of x, and points on the graph of $y = f(x) = \log_2 x$. We plot these ordered pairs, connect them in a smooth curve, and obtain the graph of $f(x) = \log_2 x$ in Figure 19. The domain of f is $\{x \mid x > 0\}$, or $(0, \infty)$ in interval notation. The range of f is the set of all real numbers, or $(-\infty, \infty)$ in interval notation.

Table 8

y	$x = 2^y$	(x, y)
-2	$\dfrac{1}{4}$	$\left(\dfrac{1}{4}, -2\right)$
-1	$\dfrac{1}{2}$	$\left(\dfrac{1}{2}, -1\right)$
0	1	$(1, 0)$
1	2	$(2, 1)$
2	4	$(4, 2)$

Figure 19

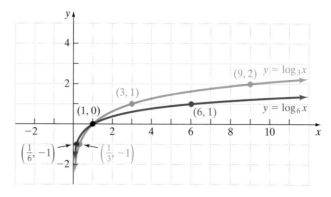

The graph of $f(x) = \log_2 x$ in Figure 19 is typical of all logarithmic functions that have a base larger than 1. Figure 20 shows the graphs of two other logarithmic functions with bases larger than 1, $y = \log_3 x$ and $y = \log_6 x$.

Figure 20

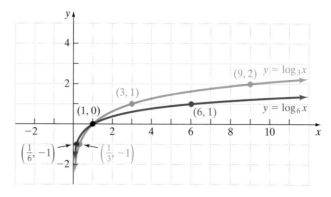

We summarize the information we have about $f(x) = \log_a x$, where $a > 1$.

Figure 21

$f(x) = \log_a x, a > 1$

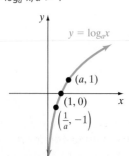

Properties of the Graph of a Logarithmic Function $f(x) = \log_a x, a > 1$

1. The domain is the set of all positive real numbers. The range is the set of all real numbers.
2. There is no y-intercept; the x-intercept is $(1, 0)$.
3. The graph of f contains the points $\left(\dfrac{1}{a}, -1\right)$, $(1, 0)$, and $(a, 1)$.

See Figure 21.

Quick ✓

13. Graph the logarithmic function $f(x) = \log_4 x$ using point plotting. From the graph, state the domain and the range of the function.

Now we consider $f(x) = \log_a x, 0 < a < 1$.

EXAMPLE 7 **Graphing a Logarithmic Function**

Graph $f(x) = \log_{1/2} x$ using point plotting. From the graph, state the domain and the range of the function.

Solution

Rewrite $y = f(x) = \log_{1/2} x$ as $x = \left(\frac{1}{2}\right)^y$. Table 9 shows various values of y, the corresponding values of x, and points on the graph of $y = f(x) = \log_{1/2} x$. Plot these points, connect them in a smooth curve, and obtain the graph of $f(x) = \log_{1/2} x$ in Figure 22. The domain of f is $\{x \mid x > 0\}$, or $(0, \infty)$ in interval notation. The range of f is the set of all real numbers, or $(-\infty, \infty)$ in interval notation.

Table 9

y	$x = \left(\frac{1}{2}\right)^y$	(x, y)
-2	4	$(4, -2)$
-1	2	$(2, -1)$
0	1	$(1, 0)$
1	$\frac{1}{2}$	$\left(\frac{1}{2}, 1\right)$
2	$\frac{1}{4}$	$\left(\frac{1}{4}, 2\right)$

Figure 22

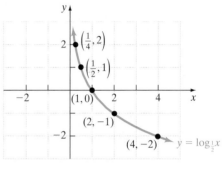

The graph of $f(x) = \log_{1/2} x$ in Figure 22 is typical of all logarithmic functions that have a base between 0 and 1. Figure 23 shows the graph of two more logarithmic functions whose bases are less than 1, $y = \log_{1/3} x$ and $y = \log_{1/6} x$.

Figure 23

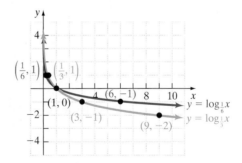

We summarize the information that we have about $f(x) = \log_a x$, where the base a is between 0 and 1 $(0 < a < 1)$.

Figure 24

$f(x) = \log_a x, 0 < a < 1$

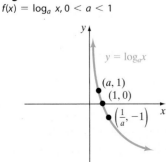

Properties of the Graph of a Logarithmic Function $f(x) = \log_a x, 0 < a < 1$

1. The domain is the set of all positive real numbers. The range is the set of all real numbers.

2. There is no y-intercept; the x-intercept is $(0, 1)$.

3. The graph of f contains the points $\left(\frac{1}{a}, -1\right)$, $(1, 0)$, and $(a, 1)$. See Figure 24.

Quick ✓

14. Graph the logarithmic function $f(x) = \log_{\frac{1}{4}} x$ using point plotting. From the graph, state the domain and the range of the function.

⑥ Work with Natural and Common Logarithms

If the base of a logarithmic function is the number e, then we have the **natural logarithm function.** This function occurs so frequently in applications that it is given a special symbol, **ln** (from the Latin *logarithmus naturalis*).

> ### Definition
>
> The natural logarithm: $y = \ln x$ if and only if $x = e^y$

Figure 25

$y = \ln x$

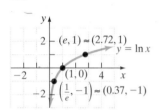

Figure 25 shows the graph of $y = \ln x$.

If the base of a logarithmic function is 10, then we have the **common logarithm function.** If the base a of the logarithmic function is not indicated, it is understood to be 10.

> ### Definition
>
> The common logarithm: $y = \log x$ if and only if $x = 10^y$

Figure 26 shows the graph of $y = \log x$.

Scientific and graphing calculators have both a natural logarithm button, $\boxed{\ln}$, and a common logarithm button, $\boxed{\log}$. This enables us to approximate the values of logarithms to the base e and the base 10, when the results are not exact.

To evaluate logarithmic expressions using a scientific calculator, enter the argument of the logarithm and then press the $\boxed{\ln}$ or $\boxed{\log}$ button, depending on the base of the logarithm. For example, to evaluate log 80, we would type in 80 and then press the $\boxed{\log}$ button. The display should show 1.90308999. Try it!

Figure 26

$y = \log x$

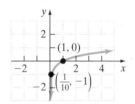

To evaluate logarithmic expressions using a graphing calculator, press the $\boxed{\ln}$ or $\boxed{\log}$ button, depending on the base of the logarithm, and then enter the argument of the logarithm. Finally, press $\boxed{\text{ENTER}}$. For example, to evaluate log 80, we would press the $\boxed{\log}$ button, type in 80, and then press $\boxed{\text{ENTER}}$. The display should show 1.903089987. Try it!

By the way, it shouldn't be surprising that log 80 is between 1 and 2 since log 10 = 1 (because $10^1 = 10$) and log 100 = 2 (because $10^2 = 100$). It is a good idea to develop number sense for logarithms prior to using your calculator to approximate the value.

EXAMPLE 8   **Evaluating Natural and Common Logarithms on a Calculator**

Using a calculator, evaluate each of the following. Round your answers to three decimal places.

(a) ln 20   (b) log 30   (c) ln 0.5

Solution

(a) $\ln 20 \approx 2.996$ (b) $\log 30 \approx 1.477$ (c) $\ln 0.5 \approx -0.693$ ●

Quick ✓

In Problems 15–17, evaluate each logarithm using a calculator. Round your answers to three decimal places.

    15. log 1400   **16.** ln 4.8   **17.** log 0.3

▶ ⑦ Solve Logarithmic Equations

Equations that contain logarithms are called **logarithmic equations.** Take care when solving logarithmic equations, because extraneous solutions (apparent solutions that are not solutions to the original equation) might creep in. To help locate extraneous solutions, remember that in the expression $\log_a M$, both a and M must be positive with $a \neq 1$.

To solve logarithmic equations, rewrite the logarithmic equation as an equivalent exponential equation, using the fact that $y = \log_a x$ is equivalent to $a^y = x$.

EXAMPLE 9 Solving Logarithmic Equations

Solve:

(a) $\log_2(3x + 4) = 4$ (b) $\log_x 25 = 2$

Solution

(a) Find the solution by writing the logarithmic equation as an exponential equation, using the fact that if $y = \log_a x$, then $a^y = x$.

$$\log_2(3x + 4) = 4$$

If $y = \log_a x$, then $a^y = x$: $\quad 2^4 = 3x + 4$

$2^4 = 16$: $\quad 16 = 3x + 4$

Subtract 4 from both sides: $\quad 12 = 3x$

Divide both sides by 3: $\quad 4 = x$

Verify this solution by letting $x = 4$ in the original equation:

$$\log_2(3 \cdot 4 + 4) \overset{?}{=} 4$$

$$\log_2(16) \overset{?}{=} 4$$

Since $2^4 = 16$, $\log_2 16 = 4$. Therefore, the solution set is $\{4\}$.

(b) Change the logarithmic equation to an exponential equation.

$$\log_x 25 = 2$$

If $y = \log_a x$, then $a^y = x$: $\quad x^2 = 25$

Use the Square Root Property: $\quad x = \pm\sqrt{25} = \pm 5$

Since the base of a logarithm must be positive, $x = -5$ is extraneous. We leave it to you to verify that the solution set is $\{5\}$.

Quick ✓

In Problems 18 and 19, solve each equation. Be sure to verify your solution.

18. $\log_3(5x + 1) = 4$ **19.** $\log_x 16 = 2$

EXAMPLE 10 Solving Logarithmic Equations

Solve each equation and state the exact solution.

(a) $\ln x = 3$ (b) $\log(x + 1) = -2$

Solution

Write each logarithmic equation as an exponential equation.

(a) $\qquad\qquad\qquad\qquad \ln x = 3$

Write as an exponent; if $y = \ln x$, then $e^y = x$: $\quad e^3 = x$

Verify this solution by letting $x = e^3$ in the original equation:

$$\ln e^3 \overset{?}{=} 3$$

We know that $\ln e^3 = 3$ can be written as $\log_e e^3 = 3$, which is equivalent to $e^3 = e^3$, so we have a true statement. The solution set is $\{e^3\}$.

(b) $\qquad\qquad\qquad\qquad \log(x + 1) = -2$

Write as an exponent; if $y = \log x$, then $10^y = x$: $\quad 10^{-2} = x + 1$

$10^{-2} = 0.01$: $\quad x + 1 = 0.01$

Subtract 1 from both sides: $\quad x = -0.99$ *(continued)*

Verify this solution by letting $x = -0.99$ in the original equation:

$$\log(-0.99 + 1) \stackrel{?}{=} -2$$

$$\log(0.01) \stackrel{?}{=} -2$$

$$10^{-2} = 0.01 \quad \text{True}$$

The solution set is $\{-0.99\}$.

Quick ✓

In Problems 20 and 21, solve each equation. Be sure to verify your solution.

20. $\ln x = -2$ **21.** $\log(x - 20) = 4$

❽ Use Logarithmic Models That Describe Our World

Common logarithms often are used when quantities vary from very large to very small. This is because the common logarithm can "scale down" the measurement. For example, if a certain quantity varied from $0.00000001 = 10^{-8}$ to $100{,}000{,}000 = 10^{8}$, the common logarithm of the same quantity would vary from $\log 10^{-8} = -8$ to $\log 10^{8} = 8$.

Physicists define the **intensity of a sound wave** as the amount of energy the sound wave transmits through a given area. The *loudness* L (measured in **decibels** in honor of Alexander Graham Bell) of a sound of intensity x (measured in watts per square meter) is defined as follows.

> **Definition**
>
> The **loudness** L, measured in decibels, of a sound of intensity x, measured in watts per square meter, is
>
> $$L(x) = 10 \log \frac{x}{10^{-12}}$$

The quantity 10^{-12} watt per square meter in the definition is the least intense sound that a human ear can detect. If we let $x = 10^{-12}$ watt per square meter, we obtain

$$L(10^{-12}) = 10 \log \frac{10^{-12}}{10^{-12}}$$

$$= 10 \log 1$$

$$= 10(0)$$

$$= 0$$

Thus the loudness of the least intense sound a human ear can detect is 0 decibels.

EXAMPLE 11   **Measuring the Loudness of a Sound**

What is the loudness, in decibels, of normal conversation, which has an intensity level of 10^{-6} watt per square meter?

Solution

We evaluate L at $x = 10^{-6}$.

$$L(10^{-6}) = 10 \log \frac{10^{-6}}{10^{-12}}$$

Laws of exponents; $\dfrac{a^m}{a^n} = a^{m-n}$: $\quad = 10 \log 10^{-6-(-12)}$

Simplify: $\quad = 10 \log 10^{6}$

If $y = \log 10^{6}$, then $10^{y} = 10^{6}$, so $y = 6$: $\quad = 10(6)$

$$= 60 \text{ decibels}$$

The loudness of normal conversation is 60 decibels.

Quick ✓

22. An MP3 player has an intensity level of 10^{-2} watt per square meter when set at its maximum level. What is the loudness, in decibels, of the MP3 player on "full blast"?

8.3 Exercises MyMathLab® PRACTICE

*Problems **1–22** are the* Quick ✓s *that follow the* EXAMPLES.

Building Skills

In Problems 23–30, change each exponential equation to an equivalent equation involving a logarithm. See Objective 1.

23. $64 = 4^3$

24. $16 = 2^4$

25. $\dfrac{1}{8} = 2^{-3}$

26. $\dfrac{1}{9} = 3^{-2}$

27. $a^3 = 19$

28. $b^4 = 23$

29. $5^{-6} = c$

30. $10^{-3} = z$

In Problems 31–40, change each logarithmic equation to an equivalent equation involving an exponent. See Objective 2.

31. $\log_2 16 = 4$

32. $\log_3 81 = 4$

33. $\log_3 \dfrac{1}{9} = -2$

34. $\log_2 \dfrac{1}{32} = -5$

35. $\log_5 a = -3$

36. $\log_6 x = -4$

37. $\log_a 4 = 2$

38. $\log_a 16 = 2$

39. $\log_{1/2} 12 = y$

40. $\log_{1/2} 18 = z$

In Problems 41–48, find the exact value of each logarithm without using a calculator. See Objective 3.

41. $\log_3 1$

42. $\log_5 5$

43. $\log_2 8$

44. $\log_4 16$

45. $\log_4 \left(\dfrac{1}{16} \right)$

46. $\log_5 \left(\dfrac{1}{125} \right)$

47. $\log_{\sqrt{2}} 4$

48. $\log_{\sqrt{3}} 3$

In Problems 49–52, evaluate each function, given that $f(x) = \log_3 x$ and $g(x) = \log_5 x$. See Objective 3.

49. $f(81)$

50. $f(9)$

51. $g\left(\sqrt{5}\right)$

52. $g\left(\sqrt[3]{5}\right)$

In Problems 53–62, find the domain of each function. See Objective 4.

53. $f(x) = \log_2(x - 4)$

54. $f(x) = \log_3(x - 2)$

55. $F(x) = \log_3(2x)$

56. $h(x) = \log_4(5x)$

57. $f(x) = \log_8(3x - 2)$

58. $F(x) = \log_2(4x - 3)$

59. $H(x) = \log_7(2x + 1)$

60. $f(x) = \log_3(5x + 3)$

61. $H(x) = \log_2(1 - 4x)$

62. $G(x) = \log_4(3 - 5x)$

In Problems 63–68, graph each function. From the graph, state the domain and the range of each function. See Objective 5.

63. $f(x) = \log_5 x$

64. $f(x) = \log_7 x$

65. $g(x) = \log_6 x$

66. $G(x) = \log_8 x$

67. $F(x) = \log_{1/5} x$

68. $F(x) = \log_{1/7} x$

In Problems 69–72, change each exponential equation to an equivalent equation involving a logarithm, and change each logarithmic equation to an equivalent equation involving an exponent. See Objective 6.

69. $e^x = 12$

70. $e^4 = M$

71. $\ln x = 4$

72. $\ln(x - 1) = 3$

In Problems 73–76, evaluate each function, given that $H(x) = \log x$ and $P(x) = \ln x$. See Objective 6.

73. $H(0.1)$

74. $H(100,000)$

75. $P(e^3)$

76. $P(e^{-3})$

In Problems 77–88, use a calculator to evaluate each expression. Round your answers to three decimal places. See Objective 6.

77. $\log 67$

78. $\log 106$

79. $\ln 5.4$

80. $\ln 10.4$

81. $\log 0.35$

82. $\log 0.78$

83. $\ln 0.2$

84. $\ln 0.4$

85. $\log \dfrac{5}{4}$

86. $\log \dfrac{10}{7}$

87. $\ln \dfrac{3}{8}$

88. $\ln \dfrac{1}{2}$

In Problems 89–108, solve each logarithmic equation. All answers should be exact. See Objective 7.

89. $\log_3(2x + 1) = 2$

90. $\log_3(5x - 3) = 3$

91. $\log_5(20x - 5) = 3$

92. $\log_4(8x + 10) = 3$

93. $\log_a 36 = 2$

94. $\log_a 81 = 2$

95. $\log_a 18 = 2$

96. $\log_a 28 = 2$

97. $\log_a 1000 = 3$

98. $\log_a 243 = 5$

99. $\ln x = 5$

100. $\ln x = 10$

101. $\log(2x - 1) = -1$

102. $\log(2x + 3) = 1$

103. $\ln e^x = -3$

104. $\ln e^{2x} = 8$

105. $\log_3 81 = x$

106. $\log_4 16 = x + 1$

107. $\log_2 (x^2 - 1) = 3$

108. $\log_3 (x^2 + 1) = 2$

Mixed Practice

109. Suppose that $f(x) = \log_2 x$.
  (a) What is $f(16)$? What point is on the graph of f?
  (b) If $f(x) = -3$, what is x? What point is on the graph of f?

110. Suppose that $f(x) = \log_5 x$.
  (a) What is $f(5)$? What point is on the graph of f?
  (b) If $f(x) = -2$, what is x? What point is on the graph of f?

111. Suppose that $G(x) = \log_4(x + 1)$.
  (a) What is $G(7)$? What point is on the graph of G?
  (b) If $G(x) = 2$, what is x? What point is on the graph of G?

112. Suppose that $F(x) = \log_2 x - 3$.
  (a) What is $F(8)$? What point is on the graph of f?
  (b) If $F(x) = -1$, what is x? What point is on the graph of F?

113. Find a so that the graph of $f(x) = \log_a x$ contains the point $(16, 2)$.

114. Find a so that the graph of $f(x) = \log_a x$ contains the point $\left(\dfrac{1}{4}, -2\right)$.

In Problems 115–120, state the domain of each logarithmic function.

115. $f(x) = \log_3(x^2 - 4x - 5)$

116. $f(x) = \log_2(x^2 - 2x - 8)$

117. $f(x) = \log\left(\dfrac{x - 4}{x + 1}\right)$

118. $f(x) = \log\left(\dfrac{x + 3}{x - 2}\right)$

119. $f(x) = \ln|x - 3|$

120. $f(x) = \ln|x + 1|$

Applying the Concepts

121. Loudness of a Whisper A whisper has an intensity level of 10^{-10} watt per square meter. How many decibels is a whisper?

122. Loudness of a Concert If you sit in the front row of a rock concert, you will experience an intensity level of 10^{-1} watt per square meter. How many decibels is a rock concert in the front row? If you move back to the 15th row, you will experience an intensity level of 10^{-2} watt per square meter. How many decibels is a rock concert in the 15th row?

123. Threshold of Pain The threshold of pain has an intensity level of 10^1 watt per square meter. How many decibels is the threshold of pain?

124. Exploding Eardrum Instant perforation of the eardrum occurs at an intensity level of 10^4 watts per square meter. At how many decibels does instant perforation of the eardrum occur?

*Problems 125–128 use the following discussion: The **Richter scale** is one way of converting seismographic readings into numbers that provide an easy reference for measuring the **magnitude M** of an earthquake. All earthquakes are compared to a **zero–level earthquake** whose seismographic reading measures 0.001 millimeter at a distance of 100 kilometers from the epicenter. An earthquake whose seismographic reading measures x millimeters has magnitude M given by*

$$M(x) = \log\left(\frac{x}{10^{-3}}\right)$$

where 10^{-3} is the reading of a zero–level earthquake 100 kilometers from its epicenter.

125. San Francisco, 1906 According to the United States Geological Survey, the San Francisco earthquake of 1906 resulted in a seismographic reading of 63,096 millimeters 100 kilometers from its epicenter. What was the magnitude of this earthquake?

126. Alaska, 1964 According to the United States Geological Survey, an earthquake on March 28, 1964, in Prince William Sound, Alaska, resulted in a seismographic reading of 1,584,893 millimeters 100 kilometers from its epicenter. What was the magnitude of this earthquake? This earthquake was the second largest ever recorded. The largest was the Great Chilean Earthquake of 1960, whose magnitude was 9.5 on the Richter scale.

127. Japan, 2011 According to the United States Geological Survey, an earthquake on March 11, 2011, off the coast of Japan had a magnitude of 8.9. What was the seismographic reading 100 kilometers from its epicenter?

128. South Carolina, 1886 According to the United States Geological Survey, an earthquake on September 1, 1886, in Charleston, South Carolina, had a magnitude of 7.3. What was the seismographic reading 100 kilometers from its epicenter?

129. pH The pH of a chemical solution is given by the formula

$$pH = -\log[H^+]$$

where $[H^+]$ is the concentration of hydrogen ions in moles per liter. Values of pH range from 0 to 14. A solution whose pH is 7 is considered neutral. The pH of pure water at 25 degrees Celsius is 7. A solution whose pH is less than 7 is considered acidic, while a solution whose pH is greater than 7 is considered basic.

(a) What is the pH of household ammonia for which $[H^+]$ is 10^{-12}? Is ammonia basic or acidic?

(b) What is the pH of black coffee for which $[H^+]$ is 10^{-5}? Is black coffee basic or acidic?

(c) What is the pH of lemon juice for which $[H^+]$ is 10^{-2}? Is lemon juice basic or acidic?

(d) What is the concentration of hydrogen ions in human blood (pH = 7.4)?

130. Energy of an Earthquake The magnitude and the seismic moment are related to the amount of energy that is given off by an earthquake. The relationship between magnitude and energy is

$$\log E_S = 11.8 + 1.5M$$

where E_S is the energy (in ergs) for an earthquake whose magnitude is M. Note that E_S is the amount of energy given off from the earthquake as seismic waves.

(a) How much energy is given off by an earthquake that measures 5.8 on the Richter scale?

(b) The earthquake on the Rat Islands in Alaska on February 4, 1965, measured 8.7 on the Richter scale. How much energy was given off by this earthquake?

Explaining the Concepts

131. Explain why the base in the logarithmic function $f(x) = \log_a x$ is not allowed to equal 1.

132. Explain why the base in the logarithmic function $f(x) = \log_a x$ must be greater than 0.

133. Explain why the domain of the function $f(x) = \log_a(x^2 + 1)$ is the set of all real numbers.

134. What are the domain and the range of the exponential function $f(x) = a^x$? What are the domain and the range of the logarithmic function $f(x) = \log_a x$? What is the relation between the domain and the range of each function?

Synthesis Review

In Problems 135–140, add or subtract each expression.

135. $(2x^2 - 6x + 1) - (5x^2 + x - 9)$

136. $(x^3 - 2x^2 + 10x + 2) + (3x^3 - 4x - 5)$

137. $\dfrac{3x}{x^2 - 1} + \dfrac{x - 3}{x^2 + 3x + 2}$

138. $\dfrac{x + 5}{x^2 + 5x + 6} - \dfrac{2}{x^2 + 2x - 3}$

139. $\sqrt{8x^3} + x\sqrt{18x}$

140. $\sqrt[3]{27a} + 2\sqrt[3]{a} - \sqrt[3]{8a}$

The Graphing Calculator

In Problems 141–146, graph each logarithmic function using a graphing calculator. State the domain and the range of each function.

141. $f(x) = \log(x + 1)$

142. $g(x) = \log(x - 2)$

143. $G(x) = \ln(x) + 1$

144. $F(x) = \ln(x) - 4$

145. $f(x) = 2\log(x - 3) + 1$

146. $G(x) = -\log(x + 1) - 3$

Putting the Concepts Together (Sections 8.1–8.3)

We designed these problems so that you can review Sections 8.1–8.3 and show your mastery of the concepts. Take time to work these problems before proceeding with the next section. The answers are located at the back of the text on page AN-50.

1. Given the functions $f(x) = 2x + 3$ and $g(x) = 2x^2 - 4x$, find
   (a) $(f \circ g)(x)$ **(b)** $(g \circ f)(x)$
   (c) $(f \circ g)(3)$ **(d)** $(g \circ f)(-2)$
   (e) $(f \circ f)(1)$

2. Find the inverse of the given one-to-one functions.
   (a) $f(x) = 3x + 4$ **(b)** $g(x) = x^3 - 4$

3. Sketch the graph of the inverse of the given one-to-one function.

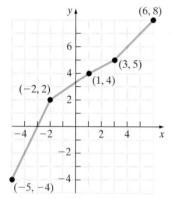

4. Approximate each number using a calculator. Express your answer rounded to three decimal places.
   (a) $2.7^{2.7}$ **(b)** $2.72^{2.72}$ **(c)** $2.718^{2.718}$
   (d) $2.7183^{2.7183}$ **(e)** e^e

5. Change each exponential equation to an equivalent equation involving a logarithm.
   (a) $a^4 = 6.4$ **(b)** $10^x = 278$

6. Change each logarithmic equation to an equivalent equation involving an exponent.
   (a) $\log_2 x = 7$ **(b)** $\ln 16 = M$

7. Find the exact value of each expression without using a calculator.
   (a) $\log_5 625$ **(b)** $\log_{\frac{2}{3}}\left(\dfrac{9}{4}\right)$

8. Determine the domain of the logarithmic function $f(x) = \log_{13}(2x + 12)$.

In Problems 9 and 10, graph each function. State the domain and range of the function.

9. $f(x) = \left(\dfrac{1}{6}\right)^x$ 10. $g(x) = \log_{\frac{3}{2}} x$

In Problems 11–14, solve each equation. State the exact solution.

11. $3^{-x+2} = 27$ 12. $e^x = e^{2x+5}$

13. $\log_2(2x + 5) = 4$

14. $\ln x = 7$

15. Suppose that a student has 150 anatomy and physiology terms to learn. If the student learns 40 terms in 60 minutes, the function $L(t) = 150(1 - e^{-0.0052t})$ models the number of terms L that the student will learn after t minutes. According to the model, how many terms will the student have learned after 90 minutes?

8.4 Properties of Logarithms

Objectives

1. Understand the Properties of Logarithms
2. Write a Logarithmic Expression as a Sum or Difference of Logarithms
3. Write a Logarithmic Expression as a Single Logarithm
4. Evaluate a Logarithm Whose Base Is Neither 10 Nor e

Are You Ready for This Section?

Before getting started, take the following readiness quiz. If you get a problem wrong, go back to the section cited and review the material.

R1. Round 3.03468 to three decimal points.   [Section R.2, pp. 14–15]
R2. Evaluate $a^0, a \neq 0$.   [Getting Ready, p. 295]

▶ 1 Understand the Properties of Logarithms

Logarithms have some very useful properties. We will derive these properties directly from the definition of a logarithm and the laws of exponents.

EXAMPLE 1 **Deriving Properties of Logarithms**

Determine the value of the following logarithmic expressions: **(a)** $\log_a 1$ **(b)** $\log_a a$.

Solution

(a) Remember, $y = \log_a x$ is equivalent to $x = a^y$, where $a > 0, a \neq 1$.

$$y = \log_a 1$$

Change to an exponent: $\quad a^y = 1$

Since $a \neq 0, a^0 = 1$: $\quad a^y = a^0$

If $a^u = a^v$, then $u = v$: $\quad y = 0$

Thus $\log_a 1 = 0$.

(b) Again, we need to write the logarithm as an exponent in order to evaluate the logarithm. Let $y = \log_a a$, so that

$$y = \log_a a$$

Change to an exponent: $\quad a^y = a$

$a = a^1$: $\quad a^y = a^1$

If $a^u = a^v$, then $u = v$: $\quad y = 1$

Thus $\log_a a = 1$.

We summarize the results of Example 1 below.

$$\log_a 1 = 0 \qquad \log_a a = 1$$

Quick ✓

In Problems 1–4, evaluate each logarithm.

1. $\log_5 1$ **2.** $\ln 1$ **3.** $\log_4 4$ **4.** $\log 10$

How can we evaluate $3^{\log_3 81}$? The exponent, $\log_3 81$, is 4 (because $y = \log_3 81$ means $3^y = 81$, or $3^y = 3^4$, so $y = 4$). Now $3^{\log_3 81} = 3^4 = 81$, so $3^{\log_3 81} = 81$. This result is true in general.

> **In Words**
> If we raise the number a to $\log_a M$, we obtain M.

An Inverse Property of Logarithms

If a and M are positive real numbers, with $a \neq 1$, then

$$a^{\log_a M} = M$$

This is an Inverse Property of Logarithms because if we compute the logarithm to the base a of a positive number M and then compute a raised to this power, we end up right back where we started, with M.

EXAMPLE 2 **Using an Inverse Property of Logarithms**

(a) $5^{\log_5 20} = 20$ **(b)** $0.8^{\log_{0.8} \sqrt{23}} = \sqrt{23}$

Quick ✓

In Problems 5 and 6, evaluate each logarithm.

5. $12^{\log_{12} \sqrt{2}}$ **6.** $10^{\log 0.2}$

How can we evaluate $\log_5 5^6$? If we let $y = \log_5 5^6$, then $5^y = 5^6$ (since $y = \log_a x$ means $a^y = x$), so $y = 6$. Thus $\log_5 5^6 = 6$. This result is true in general.

An Inverse Property of Logarithms

If a is a positive real number, with $a \neq 1$, and r is any real number, then

$$\log_a a^r = r$$

This is another Inverse Property of Logarithms because if we compute a raised to some power r and then compute the logarithm to the base a of a^r, we end up right back where we started, with r.

EXAMPLE 3 **Using an Inverse Property of Logarithms**

(a) $\log_4 4^3 = 3$ **(b)** $\ln e^{-0.5} = -0.5$

Quick ✓

In Problems 7 and 8, evaluate each logarithm.

7. $\log_8 8^{1.2}$ **8.** $\log 10^{-4}$

▶ ❷ **Write a Logarithmic Expression as a Sum or Difference of Logarithms**

The next two properties deal with logarithms whose arguments are products or quotients.
Notice that $\log_2 8 = 3$ and that $\log_2 2 + \log_2 4 = 1 + 2 = 3$. Therefore, $\log_2 8 = \log_2 2 + \log_2 4$ and $8 = 4 \cdot 2$. This suggests the following result.

The Product Rule of Logarithms

If M, N, and a are positive real numbers, with $a \neq 1$, then
$$\log_a (MN) = \log_a M + \log_a N$$

EXAMPLE 4 **Using the Product Rule of Logarithms**

Simplify each of the following logarithms by writing it as the sum of logarithms.

(a) $\log_2 (5 \cdot 3)$ **(b)** $\ln(6z)$

Solution

Do you notice that each argument contains a product? To simplify, use the Product Rule of Logarithms.

(a) $\log_2 (5 \cdot 3) = \log_2 5 + \log_2 3$ **(b)** $\ln (6z) = \ln 6 + \ln z$

Quick ✓

9. *True or False* $\log(x + 4) = \log x + \log 4$

In Problems 10 and 11, write each logarithm as the sum of logarithms.

10. $\log_4 (9 \cdot 5)$ **11.** $\log (5w)$

Notice that $\log_2 8 = 3$ and that $\log_2 16 - \log_2 2 = 4 - 1 = 3$. Therefore, $\log_2 8 = \log_2 16 - \log_2 2$ and $8 = \dfrac{16}{2}$. This suggests the following result.

The Quotient Rule of Logarithms

If M, N, and a are positive real numbers, with $a \neq 1$, then

$$\log_a\left(\frac{M}{N}\right) = \log_a M - \log_a N$$

The relationships in the product and quotient rules for logarithms are not coincidental. After all, when we multiply exponential expressions with the same base, we add the exponents; when we divide exponential expressions with the same base, we subtract the exponents. The same thing is going on here!

EXAMPLE 5 **Using the Quotient Rule of Logarithms**

Simplify each of the following logarithms by writing it as the difference of logarithms.

(a) $\log_2\left(\frac{5}{3}\right)$ **(b)** $\log\left(\frac{y}{5}\right)$

Solution

Do you notice that each argument contains a quotient? To simplify, use the Quotient Rule of Logarithms.

(a) $\log_2\left(\frac{5}{3}\right) = \log_2 5 - \log_2 3$ **(b)** $\log\left(\frac{y}{5}\right) = \log y - \log 5$ ●

Work Smart

$\log_a (M - N)$ does not equal $\log_a M - \log_a N$.

Quick ✓

In Problems 12 and 13, write each logarithm as the difference of logarithms.

12. $\log_7\left(\frac{9}{5}\right)$ **13.** $\ln\left(\frac{p}{3}\right)$

We can write a single logarithm as the sum or difference of logs when the argument of the logarithm contains both products and quotients.

EXAMPLE 6 **Writing a Single Logarithm as the Sum or Difference of Logs**

Write $\log_3\left(\frac{4x}{y}\right)$ as the sum or difference of logarithms.

Solution

Notice that the argument of the logarithm contains a product and a quotient. Thus we will use both the Product Rule and the Quotient Rule of Logarithms. When using both rules, it is typically easier to use the Quotient Rule first.

Work Smart

When you need to use both the Product Rule and the Quotient Rule, use the Quotient Rule first.

The Quotient Rule of Logarithms
↓

$$\log_3\left(\frac{4x}{y}\right) = \log_3 (4x) - \log_3 y$$

The Product Rule of Logarithms: $= \log_3 4 + \log_3 x - \log_3 y$ ●

Quick ✓

In Problems 14 and 15, write each logarithm as the sum or difference of logarithms.

14. $\log_2\left(\frac{3m}{n}\right)$ **15.** $\ln\left(\frac{q}{3p}\right)$

Another useful property of logarithms allows us to express powers on the argument of a logarithm as factors.

The Power Rule of Logarithms

If M and a are positive real numbers, with $a \neq 1$, and r is any real number, then

$$\log_a M^r = r \log_a M$$

EXAMPLE 7 **Using the Power Rule of Logarithms**

Use the Power Rule of Logarithms to express all powers as factors.

    (a) $\log_8 3^5$ **(b)** $\ln x^{\sqrt{3}}$

$5 \log_8 3$

Solution

    (a) $\log_8 3^5 = 5 \log_8 3$ **(b)** $\ln x^{\sqrt{3}} = \sqrt{3} \ln x$ ●

Quick ✓

In Problems 16 and 17, write each logarithm so that all powers are factors.

16. $\log_2 5^{1.6}$ **17.** $\log b^5$

We will use the direction *expand the logarithm* to mean "Write a logarithm as a sum or difference with all exponents written as factors."

EXAMPLE 8 **Expanding a Logarithm**

Expand the logarithm.

    (a) $\log_2 (x^2 y^3)$ **(b)** $\log\left(\dfrac{100x}{\sqrt{y}}\right)$

Solution

    (a) The argument of the logarithm contains a product, so we use the Product Rule of Logarithms to write the single log as the sum of two logs.

$$\overset{\text{Product Rule}}{\downarrow}$$
$$\log_2 (x^2 y^3) = \log_2 x^2 + \log_2 y^3$$
$$\text{Write exponents as factors:} \quad = 2 \log_2 x + 3 \log_2 y$$

    (b) The argument of the logarithm contains a quotient and a product.

$$\overset{\text{Quotient Rule}}{\downarrow}$$
$$\log\left(\frac{100x}{\sqrt{y}}\right) = \log(100x) - \log\sqrt{y}$$
$$\sqrt{y} = y^{\frac{1}{2}}: \quad = \log(100x) - \log y^{\frac{1}{2}}$$
$$\text{Product Rule:} \quad = \log 100 + \log x - \log y^{\frac{1}{2}}$$
$$\text{Write exponents as factors:} \quad = \log 100 + \log x - \frac{1}{2} \log y$$
$$\log 100 = 2: \quad = 2 + \log x - \frac{1}{2} \log y$$

 ●

Quick ✓

In Problems 18 and 19, expand each logarithm.

18. $\log_4 (a^2 b)$ **19.** $\log_3 \left(\dfrac{9m^4}{\sqrt[3]{n}}\right)$

▶ ❸ **Write a Logarithmic Expression as a Single Logarithm**

We can also use the Product Rule, Quotient Rule, and Power Rule of Logarithms to write the sums and/or differences of logarithms that have the same base as a single logarithm. This skill will be useful when solving certain logarithmic equations in Section 8.5.

EXAMPLE 9 **Writing Expressions as Single Logarithms**

Write each of the following as a single logarithm.

(a) $\log_6 3 + \log_6 12$ (b) $\log(x - 2) - \log x$

Solution

(a) The base of each logarithm is the same, 6. Use the Product Rule to write the sum of the logs as a single log.

$$\overset{\text{Product Rule}}{\downarrow}$$
$$\log_6 3 + \log_6 12 = \log_6 (3 \cdot 12)$$
$$3 \cdot 12 = 36: \qquad = \log_6 36$$
$$6^2 = 36, \text{ so } \log_6 36 = 2: \qquad = 2$$

(b) The base of each logarithm is the same, 10. Use the Quotient Rule to write the difference of the logs as a single log.

$$\log(x - 2) - \log x = \log\left(\frac{x - 2}{x}\right)$$

Quick ✓

In Problems 20 and 21, write each expression as a single logarithm.

20. $\log_8 4 + \log_8 16$ **21.** $\log_3 (x + 4) - \log_3 (x - 1)$

In order for us to use the Product Rule or the Quotient Rule to write the sum or difference of logs as a single log, the coefficients of the logarithms must be 1. Therefore if logarithms have coefficients, first use the Power Rule to write the coefficient as a power. For example, write $2 \log x$ as $\log x^2$.

EXAMPLE 10 **Writing Expressions as Single Logarithms**

Write each of the following as a single logarithm.

(a) $2 \log_2 (x - 1) + \dfrac{1}{2} \log_2 x$ (b) $\log(x - 1) + \log(x + 1) - 3 \log x$

Solution

(a) Each logarithm has the same base. Because the logarithms have coefficients, use the Power Rule to write the coefficients as exponents.

$$2 \log_2 (x - 1) + \frac{1}{2} \log_2 x = \log_2 (x - 1)^2 + \log_2 x^{\frac{1}{2}}$$
$$x^{\frac{1}{2}} = \sqrt{x}: \qquad = \log_2 (x - 1)^2 + \log_2 \sqrt{x}$$
$$\text{Product Rule:} \qquad = \log_2\left[(x - 1)^2 \sqrt{x}\right]$$

(b) We will work from left to right. Since the first two logs are being added, use the Product Rule to write these logs as a single log.

$$\underset{\underset{\downarrow}{\text{Product Rule}}}{\log(x-1) + \log(x+1) - 3\log x} = \log[(x-1)(x+1)] - 3\log x$$

$$(x-1)(x+1) = x^2 - 1: \qquad = \log(x^2 - 1) - 3\log x$$

Write the coefficient as an exponent: $\qquad = \log(x^2 - 1) - \log x^3$

Quotient Rule: $\qquad = \log\left(\dfrac{x^2 - 1}{x^3}\right)$

Quick ✓

In Problems 22 and 23, write each expression as a single logarithm.

22. $\log_5 x - 3\log_5 2$ **23.** $\log_2(x+1) + \log_2(x+2) - 2\log_2 x$

▷ ④ Evaluate a Logarithm Whose Base Is Neither 10 Nor *e*

In Section 8.3, we learned how to use a calculator to approximate logarithms whose base was either 10 (the common logarithm) or *e* (the natural logarithm). But what if the base of the logarithm is neither 10 nor *e*? To determine how to evaluate these types of logarithms, we need the following property. In this property, M, N, and a are positive numbers and $a \neq 1$. The property is a result of the fact that the logarithmic function is a one-to-one function.

One-to-One Property of Logarithms

If $M = N,$ then $\log_a M = \log_a N.$

EXAMPLE 11   **Approximating a Logarithm Whose Base Is Neither 10 Nor *e***

Approximate $\log_2 5$. Round the answer to three decimal places.

Solution

We let $y = \log_2 5$ and then convert the logarithmic equation to an equivalent exponential equation using the fact that if $y = \log_a x$, then $x = a^y$.

$$y = \log_2 5$$
$$2^y = 5$$

Now, use the One-to-One Property and "take the logarithm of both sides." Because we can use a calculator, we will take either the common log or the natural log of both sides (it doesn't matter which). Let's take the common log of both sides.

$$\log 2^y = \log 5$$

Use the Power Rule, $\log a^r = r\log a$: $y\log 2 = \log 5$

Divide both sides by log 2: $y = \dfrac{\log 5}{\log 2}$

Using a calculator, we find $\dfrac{\log 5}{\log 2} \approx 2.322$. Thus $\log_2 5 \approx 2.322$.

> **In Words**
>
> "Taking the logarithm" of both sides of an equation is the same type of approach as squaring both sides of an equation.

There is an easier way to approximate a logarithm whose base is neither 10 nor *e*. It is called the **Change-of-Base Formula.**

Change-of-Base Formula

If $a \neq 1$, $b \neq 1$, and M are positive real numbers, then

$$\log_a M = \frac{\log_b M}{\log_b a}$$

Because calculators have keys for only the common logarithm, $\boxed{\log}$, and the natural logarithm, $\boxed{\ln}$, we use the Change-of-Base Formula with either $b = 10$ or $b = e$:

$$\log_a M = \frac{\log M}{\log a} \quad \text{or} \quad \log_a M = \frac{\ln M}{\ln a}$$

EXAMPLE 12   **Using the Change-of-Base Formula**

Approximate $\log_4 45$. Round your answer to three decimal places.

Solution

Use the Change-of-Base Formula.

Using common logarithms: $\log_4 45 = \dfrac{\log 45}{\log 4}$

≈ 2.746

Using natural logarithms: $\log_4 45 = \dfrac{\ln 45}{\ln 4}$

≈ 2.746

Thus $\log_4 45 \approx 2.746$.

Quick ✓

24. $\log_3 10 = \dfrac{\log\underline{}}{\log\underline{}} = \dfrac{\ln\underline{}}{\ln\underline{}}$

In Problems 25 and 26, approximate each logarithm. Round your answers to three decimal places.

25. $\log_3 32$ **26.** $\log_{\sqrt{2}} \sqrt{7}$

Let's review all the properties of logarithms.

Properties of Logarithms

In the following properties, M, N, a, and b are positive real numbers, $a \neq 1$, $b \neq 1$, and r is any real number.

- **Inverse Properties of Logarithms**
  $a^{\log_a M} = M$ and $\log_a a^r = r$
- **The Product Rule of Logarithms**
  $\log_a (MN) = \log_a M + \log_a N$
- **The Quotient Rule of Logarithms**
  $\log_a \left(\dfrac{M}{N} \right) = \log_a M - \log_a N$
- $\log_a 1 = 0$

- **The Power Rule of Logarithms**
  $\log_a M^r = r \log_a M$
- **Change-of-Base Formula**
  $\log_a M = \dfrac{\log_b M}{\log_b a} = \dfrac{\log M}{\log a} = \dfrac{\ln M}{\ln a}$
- **One-to-One Property**
  If $M = N$, then $\log_a M = \log_a N$
- $\log_a a = 1$

8.4 Exercises   MyMathLab®   Math XL PRACTICE

Exercise numbers in green have complete video solutions in MyMathLab.

*Problems **1–26** are the Quick ✓s that follow the EXAMPLES.*

Building Skills

In Problems 27–38, use properties of logarithms to find the exact value of each expression. Do not use a calculator. See Objective 1.

27. $\log_2 2^3$ **28.** $\log_5 5^{-3}$

29. $\ln e^{-7}$ **30.** $\ln e^9$

31. $3^{\log_3 5}$ **32.** $5^{\log_5 \sqrt{2}}$

33. $e^{\ln 2}$ **34.** $e^{\ln 10}$

35. $\log_7 7$ **36.** $\log_5 5$

37. $\log 1$ **38.** $\ln 1$

In Problems 39–46, suppose that $\ln 2 = a$ *and* $\ln 3 = b$. *Use properties of logarithms to write each logarithm in terms of a and b. See Objective 2.*

39. $\ln 6$

40. $\ln \dfrac{3}{2}$

41. $\ln 9$

42. $\ln 4$

43. $\ln 12$

44. $\ln 18$

45. $\ln \sqrt{2}$

46. $\ln \sqrt[4]{3}$

In Problems 47–68, write each expression as a sum or difference of logarithms. Express exponents as factors. See Objective 2.

47. $\log(ab)$

48. $\log_4 \left(\dfrac{a}{b} \right)$

49. $\log_5 x^4$

50. $\log_3 z^{-2}$

51. $\log_2 (xy^2)$

52. $\log_3 (a^3 b)$

53. $\log_5 (25x)$

54. $\log_2 (8z)$

55. $\log_7 \left(\dfrac{49}{y} \right)$

56. $\log_2 \left(\dfrac{16}{p} \right)$

57. $\ln(e^2 x)$

58. $\ln \left(\dfrac{x}{e^3} \right)$

59. $\log_3 (27 \sqrt{x})$

60. $\log_2 \left(32 \sqrt[4]{z} \right)$

61. $\log_5 \left(x^2 \sqrt{x^2 + 1} \right)$

62. $\log_3 \left(x^3 \sqrt{x^2 - 1} \right)$

63. $\log \left(\dfrac{x^4}{\sqrt[3]{x - 1}} \right)$

64. $\ln \left(\dfrac{\sqrt[5]{x}}{(x + 2)^2} \right)$

65. $\log_7 \sqrt{\dfrac{x + 1}{x}}$

66. $\log_6 \sqrt[3]{\dfrac{x - 2}{x + 1}}$

67. $\log_2 \left[\dfrac{x(x - 1)^2}{\sqrt{x + 1}} \right]$

68. $\log_4 \left[\dfrac{x^3(x - 3)}{\sqrt[3]{x + 1}} \right]$

In Problems 69–92, write each expression as a single logarithm. See Objective 3.

69. $\log 25 + \log 4$

70. $\log_4 32 + \log_4 2$

71. $\log x + \log 3$

72. $\log_2 6 + \log_2 z$

73. $\log_3 36 - \log_3 4$

74. $\log_2 48 - \log_2 3$

75. $10^{\log 8 - \log 2}$

76. $e^{\ln 24 - \ln 3}$

77. $3 \log_3 x$

78. $8 \log_2 z$

79. $\log_4(x + 1) - \log_4 x$

80. $\log_5(2y - 1) - \log_5 y$

81. $2 \ln x + 3 \ln y$

82. $4 \log_2 a + 2 \log_2 b$

83. $\dfrac{1}{2} \log_3 x + 3 \log_3 (x - 1)$

84. $\dfrac{1}{3} \log_4 z + 2 \log_4 (2z + 1)$

85. $\log x^5 - 3 \log x$

86. $\log_7 x^4 - 2 \log_7 x$

87. $\dfrac{1}{2} [3 \log x + \log y]$

88. $\dfrac{1}{3} [\ln (x - 1) + \ln (x + 1)]$

89. $\log_8 (x^2 - 1) - \log_8 (x + 1)$

90. $\log_5 (x^2 + 3x + 2) - \log_5 (x + 2)$

91. $18 \log \sqrt{x} + 9 \log \sqrt[3]{x} - \log 10$

92. $10 \log_4 \sqrt[5]{x} + 4 \log_4 \sqrt{x} - \log_4 16$

In Problems 93–100, use the Change-of-Base Formula and a calculator to evaluate each logarithm. Round your answer to three decimal places. See Objective 4.

93. $\log_2 10$

94. $\log_3 18$

95. $\log_8 3$

96. $\log_7 5$

97. $\log_{1/3} 19$

98. $\log_{1/4} 3$

99. $\log_{\sqrt{2}} 5$

100. $\log_{\sqrt{3}} \sqrt{6}$

Applying the Concepts

101. Find the value of
$\log_2 3 \cdot \log_3 4 \cdot \log_4 5 \cdot \log_5 6 \cdot \log_6 7 \cdot \log_7 8$.

102. Find the value of $\log_2 4 \cdot \log_4 6 \cdot \log_6 8$.

103. Find the value of
$\log_2 3 \cdot \log_3 4 \cdot \ \cdots \ \cdot \log_n (n + 1) \cdot \log_{n+1} 2$.

104. Find the value of
$\log_3 3 \cdot \log_3 9 \cdot \log_3 27 \cdot \ \cdots \ \cdot \log_3 3^n$.

Extending the Concepts

105. Show that
$$\log_a\left(x + \sqrt{x^2 - 1}\right) + \log_a\left(x - \sqrt{x^2 - 1}\right) = 0.$$

106. Show that
$$\log_a(\sqrt{x} + \sqrt{x - 1}) + \log_a(\sqrt{x} - \sqrt{x - 1}) = 0.$$

107. If $f(x) = \log_a x$, show that $f(AB) = f(A) + f(B)$.

108. Find the domain of $f(x) = \log_a x^2$ and the domain of $g(x) = 2\log_a x$. Since $\log_a x^2 = 2\log_a x$, how can it be that the domains are not equal? Write a brief explanation.

Explaining the Concepts

109. State the Product Rule for Logarithms in your own words.

110. State the Quotient Rule for Logarithms in your own words.

111. Write an example to illustrate
$$\log_2(x + y) \neq \log_2 x + \log_2 y.$$

112. Write an example to illustrate
$$(\log_a x)^r \neq r\log_a x.$$

Synthesis Review

In Problems 113–118, solve each equation.

113. $4x + 3 = 13$

114. $-3x + 10 = 4$

115. $x^2 + 4x + 2 = 0$

116. $3x^2 = 2x + 1$

117. $\sqrt{x + 2} - 3 = 4$

118. $\sqrt[3]{2x} - 2 = -5$

The Graphing Calculator

We can use the Change-of-Base Formula to graph any logarithmic function on a graphing calculator. For example, to graph
$$f(x) = \log_2 x \text{ we would graph } Y_1 = \frac{\log x}{\log 2} \text{ or } Y_1 = \frac{\ln x}{\ln 2}.$$
In Problems 119–122, graph each logarithmic function using a graphing calculator. State the domain and the range of each function.

119. $f(x) = \log_3 x$

120. $f(x) = \log_5 x$

121. $F(x) = \log_{1/2} x$

122. $G(x) = \log_{1/3} x$

8.5 Exponential and Logarithmic Equations

Objectives

1 Solve Logarithmic Equations Using the Properties of Logarithms

2 Solve Exponential Equations

3 Solve Equations Involving Exponential Models

Are You Ready for This Section?

Before getting started, take the following readiness quiz. If you get a problem wrong, go back to the section cited and review the material.

R1. Solve: $2x + 5 = 13$ [Section 1.1, pp. 50–53]

R2. Solve: $x^2 - 4x = -3$ [Section 4.8, pp. 372–374]

R3. Solve: $3a^2 = a + 5$ [Section 7.2, pp. 554–558]

R4. Solve: $(x + 3)^2 + 2(x + 3) - 8 = 0$ [Section 7.3, pp. 568–572]

▶ **1 Solve Logarithmic Equations Using the Properties of Logarithms**

In Section 8.3, we solved logarithmic equations of the form $\log_a x = y$ by changing the logarithmic equation to an equivalent exponential equation. If a logarithmic equation has more than one logarithm in it, however, we need to use properties of logarithms to solve the equation.

In the last section we learned the One-to-One Property that if $M = N$, then $\log_a M = \log_a N$. It turns out that the converse of this property is true as well.

One-to-One Property of Logarithms

In the following property, M, N, and a are positive real numbers, with $a \neq 1$.

$$\text{If } \log_a M = \log_a N, \text{ then } M = N.$$

This property is useful for solving equations that contain logarithms with the same base by setting the arguments equal to each other.

EXAMPLE 1 **Solving a Logarithmic Equation**

Solve: $2 \log_3 x = \log_3 25$

Solution

Both logarithms have the same base, 3, so if we can write the equation in the form $\log_a M = \log_a N$, we can use the One-to-One Property.

$$2 \log_3 x = \log_3 25$$

$r \log_a M = \log_a M^r$: $\quad \log_3 x^2 = \log_3 25$

Set the arguments equal to each other: $\quad x^2 = 25$

Square Root Property: $\quad x = -5 \quad \text{or} \quad x = 5$

The apparent solution $x = -5$ is extraneous because the argument of a logarithm must be greater than zero, and -5 causes the argument to be negative. We now check the other apparent solution.

Check $x = 5$: $\qquad\qquad\qquad 2 \log_3 5 \overset{?}{=} \log_3 25$

$$2 \log_3 5 \overset{?}{=} \log_3 5^2$$

$\log_a m^r = r \log_a m$: $\quad 2 \log_3 5 = 2 \log_3 5 \quad$ True

The solution set is $\{5\}$.

Quick ✓

1. If $\log_a M = \log_a N$, then _____. **2.** Solve: $2 \log_4 x = \log_4 9$

If a logarithmic equation contains more than one logarithm on one side of the equation, then we can use properties of logarithms to rewrite the equation as a single logarithm. Once again, we use properties to reduce an equation into a form that is familiar. In this case, we use properties of logarithms to express the sum or difference of logarithms as a single logarithm. We then rewrite the equation in exponential form and solve for the unknown.

EXAMPLE 2 **Solving a Logarithmic Equation**

Solve: $\log_2 (x - 2) + \log_2 x = 3$

Solution

Begin by rewriting the equation as a single logarithm.

$$\log_2 (x - 2) + \log_2 x = 3$$

$\log_a M + \log_a N = \log_a (MN)$: $\quad \log_2 [x(x - 2)] = 3$

If $y = \log_a M$, then $a^y = M$: $\quad 2^3 = x(x - 2)$

Distribute: $\quad 8 = x^2 - 2x$

Write in standard form: $\quad x^2 - 2x - 8 = 0$

Factor: $\quad (x - 4)(x + 2) = 0$

Zero-Product Property: $\quad x - 4 = 0 \quad \text{or} \quad x + 2 = 0$

$$x = 4 \quad \text{or} \quad x = -2$$

Work Smart

The apparent solution $x = -2$ in Example 2 is extraneous because it results in our attempting to find the log of a negative number, not because -2 is negative.

Check $x = 4$:

$$\log_2 (4 - 2) + \log_2 4 \overset{?}{=} 3$$
$$\log_2 2 + \log_2 4 \overset{?}{=} 3$$
$$1 + 2 = 3 \quad \text{True}$$

$x = -2$:

$$\log_2 (-2 - 2) + \log_2 (-2) \overset{?}{=} 3$$

$x = -2$ is extraneous because it causes the argument to be negative

The solution set is $\{4\}$.

Quick ✓
3. Solve: $\log_4(x - 6) + \log_4 x = 2$ **4.** Solve: $\log_3(x + 3) + \log_3(x + 5) = 1$

▶ ❷ Solve Exponential Equations

In Section 8.2, we solved exponential equations using the fact that if $a^u = a^v$, then $u = v$. However, in many situations we cannot write each side of the equation with the same base.

EXAMPLE 3 **Using Logarithms to Solve Exponential Equations**

Solve: $3^x = 5$

Solution

We cannot write 5 so that it is 3 raised to some integer power. Therefore, we write the equation $3^x = 5$ as a logarithm.

$$3^x = 5$$

If $a^y = x$, then $y = \log_a x$: $x = \log_3 5$ Exact solution

To find a decimal approximation to the solution, use the Change-of-Base Formula.

$$x = \log_3 5 = \frac{\log 5}{\log 3}$$

$$\approx 1.465 \quad \text{Approximate solution}$$

An alternative approach to solving the equation would be to take either the natural logarithm or the common logarithm of both sides of the equation. If we take the natural logarithm of both sides of the equation, we get the following:

$$3^x = 5$$

If $M = N$, then $\ln M = \ln N$: $\ln 3^x = \ln 5$

$\log_a M^r = r \log_a M$: $x \ln 3 = \ln 5$

Divide both sides by ln 3: $x = \dfrac{\ln 5}{\ln 3}$ Exact solution

$$\approx 1.465 \quad \text{Approximate solution}$$

The solution set is $\left\{ \dfrac{\ln 5}{\ln 3} \right\}$. If we had taken the common logarithm of both sides, the solution set would have been $\left\{ \dfrac{\log 5}{\log 3} \right\}$.

●

Quick ✓
In Problems 5 and 6, solve each equation. Express answers in exact form and as a decimal rounded to three decimal places.

5. $2^x = 11$ **6.** $5^{2x} = 3$

EXAMPLE 4 **Using Logarithms to Solve Exponential Equations**

Solve: $4e^{3x} = 10$

Solution

We first need to isolate the exponential expression by dividing both sides of the equation by 4.

$$4e^{3x} = 10$$

$$e^{3x} = \frac{5}{2}$$

We cannot express $\frac{5}{2}$ as e raised to an integer power. However, we can solve the equation by writing the exponential equation as an equivalent logarithmic equation.

$$e^{3x} = \frac{5}{2}$$

If $e^y = x$, then $y = \ln x$: $\ln\left(\frac{5}{2}\right) = 3x$

Divide both sides by 3: $x = \dfrac{\ln\left(\frac{5}{2}\right)}{3}$ Exact solution

$x \approx 0.305$ Approximate solution

We leave it to you to verify the solution. The solution set is $\left\{\dfrac{\ln\left(\frac{5}{2}\right)}{3}\right\}$.

Quick ✓

In Problems 7 and 8, solve each equation. Express answers in exact form and as a decimal rounded to three decimal places.

7. $e^{2x} = 5$ **8.** $3e^{-4x} = 20$

▶ ❸ Solve Equations Involving Exponential Models

In Section 8.2, we looked at a variety of models from areas such as statistics, biology, and finance. Now, rather than evaluating the models at certain values of the independent variable, we will solve equations involving the models.

EXAMPLE 5 **Radioactive Decay**

The half-life of plutonium-239 is 24,360 years. The maximum amount of plutonium-239 that an adult can handle without significant injury is 0.13 microgram (= 0.000000013 gram). Suppose a researcher has a 1-gram sample of plutonium-239. The amount A (in grams) of plutonium-239 after t years is given by

$$A(t) = 1 \cdot \left(\frac{1}{2}\right)^{\frac{t}{24,360}}$$

(a) How long will it take until 0.9 gram of plutonium-239 is left in the sample?

(b) How long will it take until the 1-gram sample is safe—that is, until 0.000000013 gram is left?

Solution

(a) To find the time until $A = 0.9$ gram, solve the equation

$$0.9 = 1 \cdot \left(\frac{1}{2}\right)^{\frac{t}{24,360}}$$

for t. How can we get the t out of the exponent? Use the fact that $\log_a M^r = r \log_a M$ to move the variable.

$$\log 0.9 = \log\left(\frac{1}{2}\right)^{\frac{t}{24,360}}$$

$$\log 0.9 = \frac{t}{24,360} \log\left(\frac{1}{2}\right)$$

Multiply both sides by 24,360: $\quad 24{,}360 \log 0.9 = t \log\left(\dfrac{1}{2}\right)$

Divide both sides by $\log\left(\dfrac{1}{2}\right)$: $\quad \dfrac{24{,}360 \log 0.9}{\log\left(\dfrac{1}{2}\right)} = t$

Thus $t = \dfrac{24{,}360 \log 0.9}{\log\left(\dfrac{1}{2}\right)} \approx 3702.8.$ After approximately 3703 years, there will be 0.9 gram of plutonium-239 left.

(b) To determine the time until $A = 0.000000013$ gram, solve the equation

$$0.000000013 = 1 \cdot \left(\dfrac{1}{2}\right)^{\frac{t}{24{,}360}}$$

Take the logarithm of both sides: $\quad \log\left(0.000000013\right) = \log\left(\dfrac{1}{2}\right)^{\frac{t}{24{,}360}}$

$\log_a M^r = r \log_a M$: $\quad \log\left(0.000000013\right) = \dfrac{t}{24{,}360} \log\left(\dfrac{1}{2}\right)$

Multiply both sides by 24,360: $\quad 24{,}360 \log\left(0.000000013\right) = t \log\left(\dfrac{1}{2}\right)$

Divide both sides by $\log\left(\dfrac{1}{2}\right)$: $\quad \dfrac{24{,}360 \log\left(0.000000013\right)}{\log\left(\dfrac{1}{2}\right)} = t$

Thus $t = \dfrac{24{,}360 \log\left(0.000000013\right)}{\log\left(\dfrac{1}{2}\right)} \approx 638{,}156.8.$ After approximately 638,157 years, the 1-gram sample will be safe!

Quick ✓

9. The half-life of thorium-227 is 18.72 days. Suppose a researcher has a 10-gram sample of thorium-227. The amount A (in grams) of thorium-227 after t days is given by

$$A(t) = 10 \cdot \left(\dfrac{1}{2}\right)^{\frac{t}{18.72}}$$

(a) How long will it take until 9 grams of thorium-227 is left in the sample?
(b) How long will it take until 3 grams of thorium-227 is left in the sample?

Now let's look at an example involving compound interest. Remember, the compound interest formula states that the future value of P dollars invested in an account paying an annual interest rate r, compounded n times per year for t years, is given by $A = P\left(1 + \dfrac{r}{n}\right)^{nt}$.

EXAMPLE 6   **Future Value of Money**

Suppose you deposit $3000 into a Roth IRA today. If the deposit earns 8% interest compounded quarterly, when will it be worth

(a) $4500?
(b) $6000? That is, when will your money have doubled?

Solution

With $P = 3000$, $r = 0.08$, and $n = 4$ (compounded quarterly),

$$A = 3000\left(1 + \frac{0.08}{4}\right)^{4t} \quad \text{or} \quad A = 3000(1.02)^{4t}$$

(a) We want to know the time t when $A = 4500$. That is, we want to solve

$$4500 = 3000(1.02)^{4t}$$

Divide both sides by 3000:	$1.5 = (1.02)^{4t}$
Take the logarithm of both sides:	$\log 1.5 = \log(1.02)^{4t}$
$\log_a M^r = r \log_a M$:	$\log 1.5 = 4t \log(1.02)$
Divide both sides by $4\log(1.02)$:	$\dfrac{\log 1.5}{4 \log(1.02)} = t$

So $t = \dfrac{\log 1.5}{4 \log(1.02)} \approx 5.12$. After approximately 5.12 years (5 years, 1.4 months), the account will be worth $4500.

(b) We want to know the time t when $A = 6000$. That is, we want to solve

$$6000 = 3000(1.02)^{4t}$$

Divide both sides by 3000:	$2 = (1.02)^{4t}$
Take the logarithm of both sides:	$\log 2 = \log(1.02)^{4t}$
$\log_a M^r = r \log_a M$:	$\log 2 = 4t \log(1.02)$
Divide both sides by $4\log(1.02)$:	$\dfrac{\log 2}{4 \log(1.02)} = t$

Thus $t = \dfrac{\log 2}{4 \log(1.02)} \approx 8.75$. After approximately 8.75 years (8 years, 9 months), the account will be worth $6000.

Quick ✓

10. Suppose that you deposit $2000 into a Roth IRA today. If the deposit earns 6% interest compounded monthly, how long will it be before the account is worth
   (a) $3000?
   (b) $4000? That is, when will your money have doubled?

8.5 Exercises MyMathLab® *Math XP* PRACTICE

Exercise numbers in green have complete video solutions in MyMathLab.

*Problems **1–10** are the* Quick ✓*s that follow the* EXAMPLES.

Building Skills

In Problems 11–28, solve each equation. See Objective 1.

11. $\log_2 x = \log_2 7$

12. $\log_5 x = \log_5 13$

13. $2 \log_3 x = \log_3 81$

14. $2 \log_3 x = \log_3 4$

15. $\log_6(3x + 1) = \log_6 10$

16. $\log(2x - 3) = \log 11$

17. $\dfrac{1}{2} \ln x = 2 \ln 3$

18. $\dfrac{1}{2} \log_2 x = 2 \log_2 2$

19. $\log_2(x + 3) + \log_2 x = 2$

20. $\log_2(x - 7) + \log_2 x = 3$

21. $\log_2(x + 2) + \log_2(x + 5) = \log_2 4$

22. $\log_2(x + 5) + \log_2(x + 4) = \log_2 2$

23. $\log(x + 3) - \log x = 1$

24. $\log_3(x + 5) - \log_3 x = 2$

25. $\log_4(x + 5) - \log_4(x - 1) = 2$

26. $\log_3(x + 2) - \log_3(x - 2) = 4$

27. $\log_4(x + 8) + \log_4(x + 6) = \log_4 3$

28. $\log_3(x + 8) + \log_3(x + 4) = \log_3 5$

In Problems 29–48, solve each equation. Express irrational solutions in exact form and as a decimal rounded to three decimal places. See Objective 2.

29. $2^x = 10$

30. $3^x = 8$

31. $5^x = 20$

32. $4^x = 20$

33. $\left(\dfrac{1}{2}\right)^x = 7$

34. $\left(\dfrac{1}{2}\right)^x = 10$

35. $e^x = 5$

36. $e^x = 3$

37. $10^x = 5$

38. $10^x = 0.2$

39. $3^{2x} = 13$

40. $2^{2x} = 5$

41. $\left(\dfrac{1}{2}\right)^{4x} = 3$

42. $\left(\dfrac{1}{3}\right)^{2x} = 4$

43. $4 \cdot 2^x + 3 = 8$

44. $3 \cdot 4^x - 5 = 10$

45. $-3e^x = -18$

46. $\dfrac{1}{2} e^x = 4$

47. $0.2^{x+1} = 3^x$

48. $0.4^x = 2^{x-3}$

Mixed Practice

In Problems 49–64, solve each equation. Express irrational solutions in exact form and as a decimal rounded to three decimal places.

49. $\log_4 x + \log_4(x - 6) = 2$

50. $\log_6 x + \log_6(x + 5) = 2$

51. $5^{3x} = 7$

52. $3^{2x} = 4$

53. $3 \log_2 x = \log_2 8$

54. $5 \log_4 x = \log_4 32$

55. $\dfrac{1}{3} e^x = 5$

56. $-4e^x = -16$

57. $\left(\dfrac{1}{4}\right)^{x+1} = 8^x$

58. $9^x = 27^{x-4}$

59. $\log_3 x^2 = \log_3 16$

60. $\log_7 x^2 = \log_7 8$

61. $\log_2(x + 4) + \log_2(x + 6) = \log_2 8$

62. $\log_2(x + 5) + \log_2(x + 7) = \log_2 3$

63. $\log_4 x + \log_4(x - 4) = \log_4 3$

64. $\log_6(x + 1) + \log_6(x - 5) = \log_6 2$

Applying the Concepts

65. A Population Model According to the U.S. Census Bureau, the population of the United States in 2012 was 313 million people. In addition, the population of the United States was growing at a rate of 1.1% per year. Assuming that this growth rate continues, the model $P(t) = 313(1.011)^{t-2012}$ represents the population P (in millions of people) in year t.

(a) According to this model, when will the population of the United States be 321 million people?

(b) According to this model, when will the population of the United States be 471 million people?

66. A Population Model According to the *United States Census Bureau*, the population of the world in 2012 was 7018 million people. In addition, the population of the world was growing at a rate of 1.26% per year. Assuming that this growth rate continues, the model $P(t) = 7018(1.0126)^{t-2012}$ represents the population P (in millions of people) in year t.

(a) According to this model, when will the population of the world be 9.84 billion people?

(b) According to this model, when will the population of the world be 11.58 billion people?

67. Time Is Money Suppose that you deposit $5000 in a certificate of deposit (CD) today. If the deposit earns 6% interest compounded monthly, when will the account be worth

(a) $7000?

(b) $10,000? That is, when will your money have doubled?

68. Time Is Money Suppose that you deposit $8000 in a certificate of deposit (CD) today. If the deposit earns 4% interest compounded quarterly, when will it be worth

(a) $10,000?

(b) $24,000? That is, when will your money have tripled?

69. Depreciation Based on data obtained from the *Kelley Blue Book*, the value V of a Ford Focus that is t years old can be modeled by $V(t) = 19,841(0.88)^t$.

(a) According to the model, when will the car be worth $15,000?

(b) According to the model, when will the car be worth $5000?

(c) According to the model, when will the car be worth $1000?

70. Depreciation Based on data obtained from the *Kelley Blue Book*, the value V of a Chevy Malibu that is t years old can be modeled by $V(t) = 25,258(0.84)^t$.

(a) According to the model, when will the car be worth $15,000?

(b) According to the model, when will the car be worth $5000?

(c) According to the model, when will the car be worth $1000?

71. Radioactive Decay The half-life of beryllium-11 is 13.81 seconds. Suppose that a researcher possesses a 100-gram sample of beryllium-11. The amount A (in grams) of beryllium-11 after t seconds is given by

$$A(t) = 100 \cdot \left(\frac{1}{2}\right)^{\frac{t}{13.81}}$$

(a) When will there be 90 grams of beryllium-11 left in the sample?

(b) When will 25 grams be left?

(c) When will 10 grams of beryllium-11 be left?

72. Radioactive Decay The half-life of carbon-10 is 19.255 seconds. Suppose that a researcher possesses a 100-gram sample of carbon-10. The amount A (in grams) of carbon-10 after t seconds is given by

$$A(t) = 100 \cdot \left(\frac{1}{2}\right)^{\frac{t}{19.255}}$$

(a) When will there be 90 grams of carbon-10 left in the sample?

(b) When will 25 grams be left?

(c) When will 10 grams be left?

For Exercises 73 and 74, use Newton's Law of Cooling, which states that the temperature of a heated object decreases exponentially over time toward the temperature of the surrounding medium.

73. Newton's Law of Cooling Suppose that a pizza is removed from a 400°F oven and placed in a room where the temperature is 70°F. The temperature u (in °F) of the pizza at time t (in minutes) can be modeled by $u(t) = 70 + 330e^{-0.072t}$.

(a) According to the model, when will the temperature of the pizza be 300°F?

(b) According to the model, when will the temperature of the pizza be 220°F?

74. Newton's Law of Cooling Suppose that coffee that is 170°F is poured into a coffee mug and allowed to cool in a room where the temperature is 70°F. The temperature u (in °F) of the coffee at time t (in minutes) can be modeled by $u(t) = 70 + 100e^{-0.045t}$.

(a) According to the model, when will the temperature of the coffee be 120°F?

(b) According to the model, when will the temperature of the coffee be 100°F?

75. Learning Curve Suppose that a student has 200 vocabulary words to learn. If a student learns 20 words in 30 minutes, the function

$$L(t) = 200(1 - e^{-0.0035t})$$

models the number of words L that the student will learn in t minutes.

(a) How long will it take the student to learn 50 words?

(b) How long will it take the student to learn 150 words?

76. Learning Curve Suppose that a student has 50 biology terms to learn. If a student learns 10 terms in 30 minutes, the function

$$L(t) = 50(1 - e^{-0.0223t})$$

models the number of terms L that the student will learn in t minutes.

(a) How long will it take the student to learn 10 words?

(b) How long will it take the student to learn 40 words?

Extending the Concepts

77. The Rule of 72 The Rule of 72 states that the time for an investment to double in value is approximately given by 72 divided by the annual interest rate. For example, an investment earning 10% annual interest will double in approximately $\dfrac{72}{10} = 7.2$ years.

(a) According to the Rule of 72, approximately how long will it take an investment to double if it earns 8% annual interest?

(b) Derive a formula that can be used to find the number of years required for an investment to double. (*Hint*: Let $A = 2P$ in the formula

$$A = P\left(1 + \frac{r}{n}\right)^{nt} \text{ and solve for } t.)$$

(c) Use the formula derived in part (b) to determine the exact amount of time it takes an investment to double that earns 8% interest compounded monthly. Compare the result to the results given by the Rule of 72.

78. Critical Thinking Suppose you need to open a savings account. Bank A offers 4% interest compounded daily, and Bank B offers 4.1% interest compounded quarterly. Which bank offers the better deal? Why?

79. Critical Thinking The bacteria in a 2-liter container double every minute. After 30 minutes the container is full. How long did it take to fill half the container?

Synthesis Review

In Problems 80–85, find the following values for each function.

   (a) $f(3)$ **(b)** $f(-2)$ **(c)** $f(0)$

80. $f(x) = 5x + 2$ **81.** $f(x) = -2x + 7$

82. $f(x) = \dfrac{x + 3}{x - 2}$ **83.** $f(x) = \dfrac{x}{x - 5}$

84. $f(x) = \sqrt{x + 5}$ **85.** $f(x) = 2^x$

The Graphing Calculator

The techniques for solving equations introduced in this chapter apply only to certain types of exponential or logarithmic equations. Solutions for other types of equations are usually studied in calculus using numerical methods. However, a graphing calculator can be used to approximate solutions using the INTERSECT feature. For example, to solve $e^x = 3x + 2$, graph $Y_1 = e^x$ and $Y_2 = 3x + 2$ on the same screen. Then use the INTERSECT feature to find the x-coordinate of each point of intersection. Each x-coordinate represents an approximate solution, as shown in Figure 27.

Figure 27

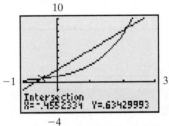
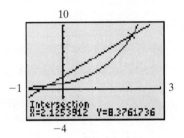

The solution set is $\{-0.46, 2.13\}$ rounded to two decimal places.

In Problems 86–93, solve each equation using a graphing calculator. Express your answer rounded to two decimal places.

86. $e^x = -3x + 2$ **87.** $e^x = -2x + 5$

88. $e^x = x + 2$ **89.** $e^x = x^2$

90. $e^x + \ln(x) = 2$ **91.** $e^x - \ln(x) = 4$

92. $\ln x = x^2 + 1$ **93.** $\ln x = x^2 - 1$

CHAPTER 8 Activity: Correct the Quiz

Focus: Solving exponential and logarithmic equations

Time: 10–15 minutes

Group size: 2

In this activity you will work as a team to grade the student quiz shown on the next page. One of you will grade the odd questions, and the other will grade the even questions. If an answer is correct, mark it correct. If an answer is wrong, mark it wrong and show the correct answer.

Once all of the quiz questions are graded, explain your results to each other and compute the final score for the quiz. Be prepared to discuss your results with the rest of the class.

Student Quiz	
Name: *Ima Student*	Quiz Score: _____
Solve the following equations. Express any irrational answers in exact form.	
(1) $\log_2 x = 3$	Answer: $\{9\}$
(2) $\log_{16} x = \dfrac{3}{4}$	Answer: $\{8\}$
(3) $\log(2x) - \log 6 = \log(x - 8)$	Answer: $\{12\}$
(4) $3^x = 27$	Answer: $\{9\}$
(5) $6^{2x} = 18$	Answer: $\left\{ \dfrac{\log 18}{2\log 6} \right\}$
(6) $4^{x+9} = 7$	Answer: $\left\{ \dfrac{\log 7}{\log 4} - 9 \right\}$
(7) $9^{3x-1} = 27^{4x}$	Answer: $\left\{ -\dfrac{1}{3} \right\}$
(8) $\log_3(2x - 3) = 2$	Answer: $\left\{ \dfrac{11}{2} \right\}$

CHAPTER 8 Review

Section 8.1 Composite Functions and Inverse Functions

KEY CONCEPTS

- $(f \circ g)(x) = f(g(x))$
- **Horizontal Line Test**

  If every horizontal line intersects the graph of a function f in at most one point, then f is one-to-one.
- For the inverse of a function also to be a function, the function must be one-to-one.
- **Relation Between the Domain and Range of a Function and Its Inverse**

  All elements in the domain of a function are also elements in the range of its inverse.

  All elements in the range of a function are also elements in the domain of its inverse.
- The graph of a function f and the graph of its inverse are symmetric with respect to the line $y = x$.
- $f^{-1}(f(x)) = x$ for every x in the domain of f and $f(f^{-1}(x)) = x$ for every x in the domain of f^{-1}

KEY TERMS

Composition
Composite function
One-to-one
Inverse function

You Should be Able To...	EXAMPLE	Review Exercises
❶ Form the composite function (p. 624)	Examples 1 and 2	1–8
❷ Determine whether a function is one-to-one (p. 626)	Examples 3 and 4	9–12
❸ Find the inverse of a function defined by a map or set of ordered pairs (p. 628)	Examples 5 and 6	13–16
❹ Obtain the graph of the inverse function from the graph of a function (p. 630)	Example 7	17, 18
❺ Find the inverse of a function defined by an equation (p. 631)	Examples 8 and 9	19–22

In Problems 1–4, for the given functions f and g, find:

   (a) $(f \circ g)(5)$ **(b)** $(g \circ f)(-3)$

   (c) $(f \circ f)(-2)$ **(d)** $(g \circ g)(4)$

1. $f(x) = 3x + 5; g(x) = 2x - 1$

2. $f(x) = x - 3; g(x) = 5x + 2$

3. $f(x) = 2x^2 + 1; g(x) = x + 5$

4. $f(x) = x - 3; g(x) = x^2 + 1$

In Problems 5–8, for the given functions f and g, find:

   (a) $(f \circ g)(x)$ **(b)** $(g \circ f)(x)$

   (c) $(f \circ f)(x)$ **(d)** $(g \circ g)(x)$

5. $f(x) = x + 1; g(x) = 5x$

6. $f(x) = 2x - 3; g(x) = x + 6$

7. $f(x) = x^2 + 1; g(x) = 2x + 1$

8. $f(x) = \dfrac{2}{x + 1}; g(x) = \dfrac{1}{x}$

In Problems 9–12, determine which of the functions are one-to-one.

9. $\{(-5, 8), (-3, 2), (-1, 8), (0, 12), (1, 15)\}$

10. $\{(-4, 2), (-2, 1), (0, 0), (1, -1), (2, 8)\}$

11.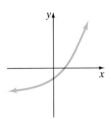

12.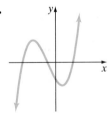

In Problems 13–16, find the inverse of each one-to-one function.

13.   **14.**

Age	Height (inches)
24	69
59	71
29	72
81	73
37	74

Price ($)	Quantity Demanded
300	112
200	129
170	144
150	161
130	176

15. $\{(-5, 3), (-3, 1), (1, -3), (2, 9)\}$

16. $\{(-20, 1), (-15, 4), (5, 3), (25, 2)\}$

In Problems 17 and 18, the graph of a one-to-one function f is given. Draw the graph of the inverse function f^{-1}.

17. 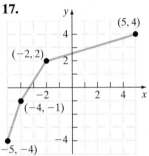   **18.**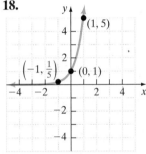

In Problems 19–22, find the inverse function of the given one-to-one function.

19. $f(x) = 5x$ **20.** $H(x) = 2x + 7$

21. $P(x) = \dfrac{4}{x + 2}$ **22.** $g(x) = 2x^3 - 1$

Section 8.2 Exponential Functions

KEY CONCEPTS

- **Properties of Exponential Functions of the Form $f(x) = a^x, a > 0, a \neq 1$**
  1. The domain is the set of all real numbers. The range is the set of all positive real numbers.
  2. There are no x-intercepts. The y-intercept is $(0, 1)$.
  3. The graph of an exponential function contains the points $\left(-1, \dfrac{1}{a}\right)$, $(0, 1)$, and $(1, a)$.
- **Property for Solving Exponential Equations of the Form $a^u = a^v$**
  If $a^u = a^v$, then $u = v$.

KEY TERMS

Exponential function

The number e

Exponential equation

Half-life

Payment period

Compound interest

You Should Be Able To...	EXAMPLE	Review Exercises
❶ Evaluate exponential expressions (p. 638)	Example 1	23–25
❷ Graph exponential functions (p. 639)	Examples 2 through 4	26–29
❸ Define the number *e* (p. 642)		30
❹ Solve exponential equations (p. 643)	Examples 5 and 6	31–36
❺ Use exponential models that describe our world (p. 645)	Examples 7 through 9	37–40

In Problems 23–25, approximate each number using a calculator. Express your answer rounded to three decimal places.

23. (a) $7^{1.7}$ **(d)** $7^{1.7321}$
  (b) $7^{1.73}$ **(e)** $7^{\sqrt{3}}$
  (c) $7^{1.732}$

24. (a) $10^{3.1}$ **(d)** $10^{3.1416}$
  (b) $10^{3.14}$ **(e)** 10^{π}
  (c) $10^{3.142}$

25. (a) $e^{0.5}$ **(d)** $e^{-0.8}$
  (b) e^{-1} **(e)** $e^{\sqrt{\pi}}$
  (c) $e^{1.5}$

In Problems 26–29, graph each function. State the domain and the range of the function.

26. $f(x) = 9^x$ **27.** $g(x) = \left(\dfrac{1}{9}\right)^x$

28. $H(x) = 4^{x-2}$ **29.** $h(x) = 4^x - 2$

30. State the definition of the number *e*.

In Problems 31–36, solve each equation.

31. $2^x = 64$ **32.** $25^{x-2} = 125$

33. $27^x \cdot 3^{x^2} = 9^2$

34. $\left(\dfrac{1}{4}\right)^x = 16$

35. $(e^2)^{x-1} = e^x \cdot e^7$ **36.** $(2^x)^x = 512$

37. Future Value of Money Suppose that you deposit $2500 into a traditional IRA that pays 4.5% annual interest. How much money will you have after 25 years if interest is compounded

(a) annually? **(b)** quarterly?
(c) monthly? **(d)** daily?

38. Radioactive Decay The half-life of the radioactive gas radon is 3.5 days. Suppose a researcher possesses a 100-gram sample of radon gas. The amount *A* (in grams) of radon after *t* days is given by
$A(t) = 100\left(\dfrac{1}{2}\right)^{\frac{t}{3.5}}$. How much radon gas is left in the sample after
(a) 1 day?
(b) 3.5 days?
(c) 7 days?
(d) 30 days?

39. A Population Model According to the U.S. Census Bureau, the population of Nevada in 2011 was 2.723 million people. In addition, the population of Nevada was growing at a rate of 5.2% per year. Assuming that this growth rate continues, the model $P(t) = 2.723(1.052)^{t-2011}$ represents the population (in millions of people) in year *t*. According to this model, what will be the population of Nevada
(a) in 2018?
(b) in 2025?

40. Newton's Law of Cooling A baker removes a cake from a 350°F oven and places it in a room whose temperature is 72°F. According to Newton's Law of Cooling, the temperature *u* (in °F) of the cake at time *t* (in minutes) can be modeled by $u(t) = 72 + 278e^{-0.0835t}$. According to this model, what will be the temperature of the cake
(a) after 15 minutes?
(b) after 30 minutes?

Section 8.3 Logarithmic Functions

KEY CONCEPTS

- **The Logarithmic Function to the Base *a***
  $y = \log_a x$ is equivalent to $x = a^y$.
- **Properties of the Logarithmic Function of the Form**
  $f(x) = \log_a x, a > 0, a \neq 1$
  1. The domain is the set of all positive real numbers. The range is the set of all real numbers.
  2. There is no *y*-intercept. The *x*-intercept is (0, 1).
  3. The graph of the logarithmic function contains the points $\left(\dfrac{1}{a}, -1\right)$, (1, 0), and (a, 1).

KEY TERMS

Logarithmic function
Natural logarithm function
Common logarithm function
Logarithmic equation
Intensity of a sound wave
Decibels
Loudness

You Should Be Able To...	EXAMPLE	Review Exercises
1 Change exponential equations to logarithmic equations (p. 652)	Example 1	41–44
2 Change logarithmic equations to exponential equations (p. 653)	Example 2	45–48
3 Evaluate logarithmic functions (p. 653)	Examples 3 and 4	49–52
4 Determine the domain of a logarithmic function (p. 655)	Example 5	53–56
5 Graph logarithmic functions (p. 655)	Examples 6 and 7	57, 58
6 Work with natural and common logarithms (p. 658)	Example 8	59–62
7 Solve logarithmic equations (p. 658)	Examples 9 and 10	63–68
8 Use logarithmic models that describe our world (p. 660)	Example 11	69, 70

In Problems 41–44, change each exponential equation to an equivalent equation involving a logarithm.

41. $3^4 = 81$ **42.** $4^{-3} = \dfrac{1}{64}$

43. $b^3 = 5$ **44.** $10^{3.74} = x$

In Problems 45–48, change each logarithmic equation to an equivalent equation involving an exponent.

45. $\log_8 2 = \dfrac{1}{3}$ **46.** $\log_5 18 = r$

47. $\ln(x + 3) = 2$ **48.** $\log x = -4$

In Problems 49–52, find the exact value of each logarithm without using a calculator.

49. $\log_8 128$ **50.** $\log_6 1$

51. $\log \dfrac{1}{100}$ **52.** $\log_9 27$

In Problems 53–56, find the domain of each function.

53. $f(x) = \log_2(x + 5)$ **54.** $g(x) = \log_8(7 - 3x)$

55. $h(x) = \ln(3x)$ **56.** $F(x) = \log_{\frac{1}{3}}(4x + 10)$

In Problems 57 and 58, graph each function.

57. $f(x) = \log_{\frac{5}{2}} x$ **58.** $g(x) = \log_{\frac{2}{5}} x$

In Problems 59–62, use a calculator to evaluate each expression. Round your answers to three decimal places.

59. $\ln 24$ **60.** $\ln \dfrac{5}{6}$

61. $\log 257$ **62.** $\log 0.124$

In Problems 63–68, solve each logarithmic equation.

63. $\log_7(4x - 19) = 2$ **64.** $\log_{\frac{1}{3}}(x^2 + 8x) = -2$

65. $\log_a \dfrac{4}{9} = -2$ **66.** $\ln e^{5x} = 30$

67. $\log(6 - 7x) = 3$ **68.** $\log_b 75 = 2$

69. Loudness of a Vacuum Cleaner A vacuum cleaner has an intensity level of 10^{-4} watt per square meter. What is the loudness, in decibels, of the vacuum cleaner?

70. The Great New Madrid Earthquake According to the United States Geological Survey, an earthquake on December 16, 1811, in New Madrid, Missouri, had a magnitude of approximately 8.0. What would have been the seismographic reading 100 kilometers from its epicenter?

Section 8.4 Properties of Logarithms

KEY CONCEPTS

- **Properties of Logarithms**

  For the following properties, a, M, and N are positive real numbers, with $a \neq 1$, and r is any real number.

  $\log_a a = 1$ $\log_a(MN) = \log_a M + \log_a N$

  $\log_a 1 = 0$

  $a^{\log_0 M} = M$ $\log_a M^r = r \log_a M$

  $\log_a a^r = r$ $\log_a\left(\dfrac{M}{N}\right) = \log_a M - \log_a N$

             If $M = N$, then $\log_a M = \log_a N$.

- **Change-of-Base Formula**

  If $a \neq 1$, $b \neq 1$, and M are positive real numbers, then

  $$\log_a M = \frac{\log_b M}{\log_b a} = \frac{\log M}{\log a} = \frac{\ln M}{\ln a}$$

You Should Be Able To...	EXAMPLE	Review Exercises
❶ Understand the properties of logarithms (p. 664)	Examples 1 through 3	71–76
❷ Write a logarithmic expression as a sum or difference of logarithms (p. 666)	Examples 4 through 8	77–80
❸ Write a logarithmic expression as a single logarithm (p. 669)	Examples 9 and 10	81–84
❹ Evaluate a logarithm whose base is neither 10 nor e (p. 670)	Examples 11 and 12	85–88

In Problems 71–76, use properties of logarithms to find the exact value of each expression. Do not use a calculator.

71. $\log_4 4^{21}$

72. $7^{\log_7 9.34}$

73. $\log_5 5$

74. $\log_9 1$

75. $\log_4 12 - \log_4 3$

76. $12^{\log_{12} 2 + \log_{12} 8}$

In Problems 77–80, write each expression as a sum and/or difference of logarithms. Write exponents as factors.

77. $\log_7\left(\dfrac{xy}{z}\right)$

78. $\log_3\left(\dfrac{81}{x^2}\right)$

79. $\log(1000r^4)$

80. $\ln\sqrt{\dfrac{x-1}{x}}$

In Problems 81–84, write each expression as a single logarithm.

81. $4\log_3 x + 2\log_3 y$

82. $\dfrac{1}{4}\ln x + \ln 7 - 2\ln 3$

83. $\log_2 3 - \log_2 6$

84. $\log_6(x^2 - 7x + 12) - \log_6(x - 3)$

In Problems 85–88, use the Change-of-Base Formula and a calculator to evaluate each logarithm. Round your answer to three decimal places.

85. $\log_6 50$

86. $\log_\pi 2$

87. $\log_{\frac{2}{3}} 6$

88. $\log_{\sqrt{5}} 20$

Section 8.5 Exponential and Logarithmic Equations

KEY CONCEPT

• In the following property, M, N, and a are positive real numbers, with $a \neq 1$.

If $\log_a M = \log_a N$ then $M = N$

You Should Be Able To...	EXAMPLE	Review Exercises
❶ Solve logarithmic equations using the properties of logarithms (p. 673)	Examples 1 and 2	89–92
❷ Solve exponential equations (p. 675)	Examples 3 and 4	93–96
❸ Solve equations involving exponential models (p. 676)	Examples 5 and 6	97, 98

In Problems 89–96, solve each equation. Express irrational solutions in exact form and as a decimal rounded to three decimal places.

89. $3\log_4 x = \log_4 1000$

90. $\log_3(x + 7) + \log_3(x + 6) = \log_3 2$

91. $\ln(x + 2) - \ln x = \ln(x + 1)$

92. $\dfrac{1}{3}\log_{12} x = 2\log_{12} 2$

93. $2^x = 15$

94. $10^{3x} = 27$

95. $\dfrac{1}{3}e^{7x} = 13$

96. $3^x = 2^{x+1}$

97. Radioactive Decay The half-life of the radioactive gas radon is 3.5 days. Suppose a researcher possesses a 100-gram sample of radon gas. The amount A (in grams) of radon after t days is given by

$$A(t) = 100\left(\frac{1}{2}\right)^{\frac{t}{3.5}}.$$

(a) When will 75 grams of radon gas be left in the sample?

(b) When will 1 gram of radon gas be left in the sample?

98. A Population Model According to the U.S. Census Bureau, the population of Nevada in 2011 was 2.723 million people. In addition, the population of Nevada was growing at a rate of 5.2% per year. Assuming that this growth rate continues, the model

$P(t) = 2.723(1.052)^{t-2011}$ represents the population (in millions of people) in year t. According to this model, when will the population of Nevada be

(a) 3.939 million people?

(b) 8.426 million people?

Chapter 8 Test

Step-by-step test solutions are found on the Chapter Test Prep Videos available in MyMathLab® *or on* You Tube .

1. Determine whether the following function is one-to-one:

   $\{(1, 4), (3, 2), (5, 8), (-1, 4)\}$

2. Find the inverse of $f(x) = 4x - 3$.

3. Approximate each number using a calculator. Express your answer rounded to three decimal places.

   (a) $3.1^{3.1}$ **(b)** $3.14^{3.14}$

   (c) $3.142^{3.142}$ **(d)** $3.1416^{3.1416}$

   (e) π^π

4. Change $4^x = 19$ to an equivalent equation involving a logarithm.

5. Change $\log_b x = y$ to an equivalent equation involving an exponent.

6. Find the exact value of each expression without using a calculator.

   (a) $\log_3\left(\dfrac{1}{27}\right)$ **(b)** $\log 10,000$

7. Determine the domain of $f(x) = \log_5(7 - 4x)$.

In Problems 8 and 9, graph each function. State the domain and the range of the function.

8. $f(x) = 6^x$ 9. $g(x) = \log_{\frac{1}{9}} x$

10. Use the properties of logarithms to find the exact value of each expression. Do not use a calculator.

    (a) $\log_7 7^{10}$ **(b)** $3^{\log_3 15}$

11. Write the expression $\log_4 \dfrac{\sqrt{x}}{y^3}$ as a sum or difference of logarithms. Express exponents as factors.

12. Write the expression $4 \log M + 3 \log N$ as a single logarithm.

13. Use the Change-of-Base Formula and a calculator to evaluate $\log_{\frac{3}{4}} 10$. Round to three decimal places.

In Problems 14–20, solve each equation. Express irrational solutions in exact form and as a decimal rounded to three places.

14. $4^{x+1} = 2^{3x+1}$ 15. $5^{x^2} \cdot 125 = 25^{2x}$

16. $\log_a 64 = 3$ 17. $\log_2(x^2 - 33) = 8$

18. $2 \log_7(x - 3) = \log_7 3 + \log_7 12$

19. $3^{x-1} = 17$

20. $\log(x - 2) + \log(x + 2) = 2$

21. According to the U.S. Census Bureau, International Data Base, the population of Canada in 2010 was 34.1 million people. In addition, the population of Canada was growing at a rate of 0.8% annually. Assuming that this growth rate continues, the model $P(t) = 34.1(1.008)^{t-2010}$ represents the population (in millions of people) in year t.

    (a) According to the model, what will be the population of Canada in 2015?

    (b) According to the model, in what year will the population of Canada be 50 million?

22. Rustling leaves have an intensity of 10^{-11} watt per square meter. How many decibels are rustling leaves?

9 Conics

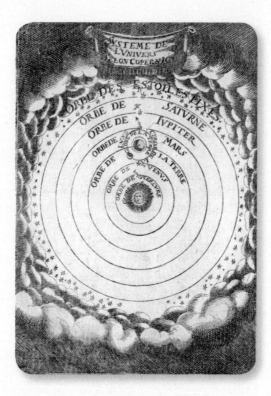

The Big Picture: Putting It Together

In Chapter 1, we introduced the Cartesian plane or rectangular coordinate system. We said this system lets us make connections between algebra and geometry. In this chapter, we develop this connection by showing how geometric definitions of certain figures lead to algebraic equations.

We start the chapter by showing how to use algebra to find the distance between any two points in the Cartesian plane. This method is a direct consequence of the Pythagorean Theorem we studied in Section 7.1. Knowing the distance formula enables us to present a complete discussion of the so-called *conic sections*.

How would you describe a circle to someone? You would probably draw a picture to illustrate your verbal description. The Cartesian plane and the distance formula enable us to take the geometric definition of a circle and develop an algebraic equation whose graph represents a circle. This powerful connection between geometry and algebra lets us answer all types of interesting questions. The methods we are about to present form the foundation of an area of mathematics called analytic geometry.

Earth is the center of the universe. Although this statement seems ludicrous to us now, it was commonly believed until the 1500s. A Polish astronomer named Nicolaus Copernicus published a book entitled *De Revolutionibus Orbium Coelestium* (On the Revolutions of the Celestial Spheres), which stated that Earth and the other planets orbited the Sun in a circular motion. Copernicus's ideas were not readily accepted by the geocentrists, who held on to the belief that Earth is at the center of the universe. Copernicus's model of planetary motion was later improved upon by the German astronomer Johannes Kepler. In 1609, Kepler published *Astronomia nova* (New Astronomy) in which he proved that the orbit of Mars is an ellipse, with the Sun occupying one of its two foci. See Problems 45–48 on page 718.

Outline

9.1 Distance and Midpoint Formulas

Objectives

1 Use the Distance Formula
2 Use the Midpoint Formula

Are You Ready for This Section?

Before getting started, take the following readiness quiz. If you get a problem wrong, go back to the section cited and review the material.

R1. Simplify: **(a)** $\sqrt{64}$ **(b)** $\sqrt{24}$ **(c)** $\sqrt{(x-2)^2}$ [Section 6.3, pp. 487–491]

R2. Find the length of the hypotenuse in a right triangle whose legs are 6 and 8.   [Section 7.1, pp. 548–551]

▶ **1** Use the Distance Formula

In Section 1.5, we learned how to plot points in the Cartesian plane. If we could algebraically compute the distance between any two points plotted in the Cartesian plane, we would have a connection between geometry (literally, "measuring the distance") and algebra. We *can* find this distance using the Pythagorean Theorem.

EXAMPLE 1   **Finding the Distance between Two Points**

Find the distance d between the points $(2, 4)$ and $(5, 8)$.

Solution

First plot the points in the Cartesian plane and connect them with a straight line, as shown in Figure 1(a). To find the length d, draw a horizontal line through the point $(2, 4)$ and a vertical line through the point $(5, 8)$ and form a right triangle. The right angle of this triangle is at the point $(5, 4)$. Do you see why? If we travel horizontally from the point $(2, 4)$, then there is no "up or down" movement. For this reason, the y-coordinate of the point at the right angle must be 4. Similarly, if we travel vertically straight down from the point $(5, 8)$, we find that the x-coordinate of the point at the right angle must be 5. See Figure 1(b).

Figure 1

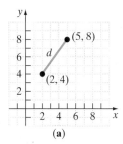
(a)

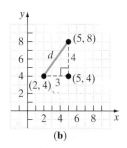
(b)

One leg of the triangle has length 3 (since $5 - 2 = 3$). The other leg has length 4 (since $8 - 4 = 4$). The Pythagorean Theorem says that

$$d^2 = 3^2 + 4^2$$
$$= 9 + 16$$
$$= 25$$

Square Root Property: $d = \pm 5$

Because d is the length of the hypotenuse, we discard the solution $d = -5$ and find that the hypotenuse is 5. Therefore, the distance between the points $(2, 4)$ and $(5, 8)$ is 5 units. ●

That was quite a bit of work to find the distance between $(2, 4)$ and $(5, 8)$. The *distance formula* lets us easily find the distance between any two points in the Cartesian plane.

The Distance Formula

The distance between two points $P_1 = (x_1, y_1)$ and $P_2 = (x_2, y_2)$, denoted by $d(P_1, P_2)$, is

$$d(P_1, P_2) = \sqrt{(x_2 - x_1)^2 + (y_2 - y_1)^2}$$

Figure 2

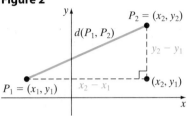

See Figure 2.

Let's justify the formula. Let point P_1 have coordinates (x_1, y_1), and let point P_2 have coordinates (x_2, y_2). If the line joining the points P_1 and P_2 is neither vertical nor horizontal, form a right triangle so that the vertex of the right angle is at the point $P_3 = (x_2, y_1)$, as shown in Figure 3(a).

The vertical distance from P_3 to P_2 is the absolute value of the difference of the y-coordinates, $|y_2 - y_1|$. The horizontal distance from P_1 to P_3 is the absolute value of the difference of the x-coordinates, $|x_2 - x_1|$. See Figure 3(b).

Figure 3

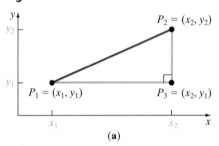 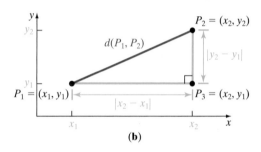

(a) (b)

The distance $d(P_1, P_2)$ that we seek is the length of the hypotenuse of the right triangle, so by the Pythagorean Theorem, it follows that

$$[d(P_1, P_2)]^2 = |x_2 - x_1|^2 + |y_2 - y_1|^2$$
$$= (x_2 - x_1)^2 + (y_2 - y_1)^2$$

Work Smart

When using the distance formula in applications, be sure that both the x-axis and the y-axis have the same unit of measure, such as inches.

Take the square root of both sides: $d(P_1, P_2) = \sqrt{(x_2 - x_1)^2 + (y_2 - y_1)^2}$

If the line joining P_1 and P_2 is horizontal, then the y-coordinate of P_1 equals the y-coordinate of P_2; that is, $y_1 = y_2$. See Figure 4(a). In this case, the distance formula still works, because, for $y_1 = y_2$, it becomes

$$d(P_1, P_2) = \sqrt{(x_2 - x_1)^2 + 0^2}$$
$$= \sqrt{(x_2 - x_1)^2} = |x_2 - x_1|$$

A similar argument holds if the line joining P_1 and P_2 is vertical. See Figure 4(b). The distance formula works in all cases.

Figure 4

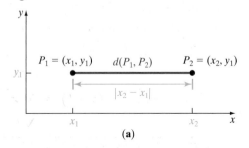 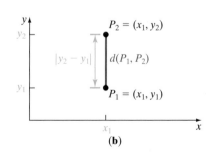

(a) (b)

The distance between two points $P_1 = (x_1, y_1)$ and $P_2 = (x_2, y_2)$ cannot be negative; it can be 0 only when the two points are identical—that is, when $x_1 = x_2$ and $y_1 = y_2$.

Also it does not matter whether we compute the distance from P_1 to P_2 or from P_2 to P_1 because $(x_2 - x_1)^2 = (x_1 - x_2)^2$. This should seem reasonable since the distance from P_1 to P_2 equals the distance from P_2 to P_1.

(EXAMPLE 2)   **Finding the Length of a Line Segment**

Find the length of the line segment shown in Figure 5.

Solution

The length of the line segment is the distance between the points $(-2, -1)$ and $(4, 3)$. We use the distance formula with $P_1 = (x_1, y_1) = (-2, -1)$ and $P_2 = (x_2, y_2) = (4, 3)$.

$$d = \sqrt{(x_2 - x_1)^2 + (y_2 - y_1)^2}$$

$x_1 = -2, y_1 = -1, x_2 = 4, y_2 = 3:\quad = \sqrt{(4 - (-2))^2 + (3 - (-1))^2}$

$$= \sqrt{6^2 + 4^2}$$
$$= \sqrt{36 + 16}$$
$$= \sqrt{52}$$
$$= 2\sqrt{13} \quad \text{Exact}$$
$$\approx 7.21 \quad \text{Approximate}$$

Figure 5

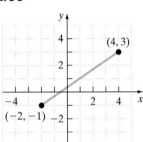

Quick ✓

1. The distance between two points $P_1 = (x_1, y_1)$ and $P_2 = (x_2, y_2)$, denoted by $d(P_1, P_2)$, is _____.

2. *True or False* The distance between two points can be a negative number.

In Problems 3 and 4, find the distance between the points.

3. $(3, 8)$ and $(0, 4)$ **4.** $(-2, -5)$ and $(4, 7)$

The next example shows how algebra (the distance formula) can be used to solve geometry problems.

(EXAMPLE 3)   **Using Algebra to Solve Geometry Problems**

Consider the three points $A = (-1, 1)$, $B = (2, -2)$, and $C = (3, 5)$.

  (a) Plot each point in the Cartesian plane and form the triangle ABC.
  (b) Find the length of each side of the triangle.
  (c) Verify that the triangle is a right triangle.
  (d) Find the area of the triangle.

Solution

  (a) We plot points A, B, and C in Figure 6 on the next page.
  (b) We use the distance formula to find the length of each side of the triangle.

$$d(A, B) = \sqrt{(2 - (-1))^2 + (-2 - 1)^2} = \sqrt{3^2 + (-3)^2}$$
$$= \sqrt{9 + 9} = \sqrt{18} = 3\sqrt{2}$$

$$d(A, C) = \sqrt{(3 - (-1))^2 + (5 - 1)^2} = \sqrt{4^2 + 4^2}$$
$$= \sqrt{16 + 16} = \sqrt{32} = 4\sqrt{2}$$

$$d(B, C) = \sqrt{(3 - 2)^2 + (5 - (-2))^2} = \sqrt{1^2 + 7^2} = \sqrt{1 + 49} = \sqrt{50} = 5\sqrt{2}$$

Figure 6

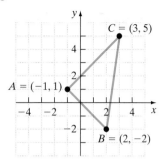

(c) The Pythagorean Theorem says that if you have a right triangle, the sum of squares of the two legs will equal the square of the hypotenuse. The converse of this statement is true as well. That is, if the sum of the squares of two sides of a triangle equals the square of the third side, then the triangle is a right triangle. Thus we need to show that the sum of the squares of two sides of the triangle in Figure 6 equals the square of the third side. In Figure 6, the right angle appears to be at vertex A, which means the side opposite vertex A should be the hypotenuse of the triangle. We want to show that

$$[d(B, C)]^2 = [d(A, B)]^2 + [d(A, C)]^2$$

We know each of these distances from part (b), so

$$(5\sqrt{2})^2 \stackrel{?}{=} (3\sqrt{2})^2 + (4\sqrt{2})^2$$

$(ab)^n = a^n \cdot b^n$: $\quad 25 \cdot 2 \stackrel{?}{=} 9 \cdot 2 + 16 \cdot 2$

$$50 \stackrel{?}{=} 18 + 32$$

$$50 = 50 \quad \text{True}$$

Since $[d(B, C)]^2 = [d(A, B)]^2 + [d(A, C)]^2$, we know that triangle ABC is a right triangle.

(d) The area of a triangle is $\frac{1}{2}$ times the product of the base and the height. From Figure 6, we will say that side AB forms the base and side AC forms the height. The length of side AB is $3\sqrt{2}$ and the length of side AC is $4\sqrt{2}$, so the area of triangle ABC is

$$\text{Area} = \frac{1}{2}(\text{Base})(\text{Height}) = \frac{1}{2}(3\sqrt{2})(4\sqrt{2}) = 12 \text{ square units}$$

Quick ✓

5. Consider the three points $A = (-2, -1)$, $B = (4, 2)$, and $C = (0, 10)$.

   (a) Plot each point in the Cartesian plane and form the triangle ABC.

   (b) Find the length of each side of the triangle.

   (c) Verify that the triangle is a right triangle.

   (d) Find the area of the triangle.

▶ ❷ Use the Midpoint Formula

Suppose we have two points $P_1 = (x_1, y_1)$ and $P_2 = (x_2, y_2)$ in the Cartesian plane. Further suppose that we want to find the point $M = (x, y)$ that is on the line segment joining P_1 and P_2 and is the same distance to each of these two points so that $d(P_1, M) = d(M, P_2)$. We can find this point M using the **midpoint formula.**

In Words

To find the midpoint of a line segment, average the x-coordinates of the endpoints and average the y-coordinates of the endpoints.

Midpoint Formula

The midpoint $M = (x, y)$ of the line segment from $P_1 = (x_1, y_1)$ to $P_2 = (x_2, y_2)$ is

$$M = \left(\frac{x_1 + x_2}{2}, \frac{y_1 + y_2}{2}\right)$$

EXAMPLE 4 Finding the Midpoint of a Line Segment

Find the midpoint of the line segment joining $P_1 = (-2, 3)$ and $P_2 = (4, 7)$. Plot the points P_1, P_2, and their midpoint. Check your answer.

Solution

Substitute $x_1 = -2$, $y_1 = 3$, $x_2 = 4$, and $y_2 = 7$ into the midpoint formula. The coordinates (x, y) of the midpoint M are

$$x = \frac{x_1 + x_2}{2} = \frac{-2 + 4}{2} = \frac{2}{2} = 1$$

and

$$y = \frac{y_1 + y_2}{2} = \frac{3 + 7}{2} = \frac{10}{2} = 5$$

Thus the midpoint of the line segment joining $P_1 = (-2, 3)$ and $P_2 = (4, 7)$ is $M = (1, 5)$. See Figure 7.

Figure 7

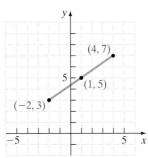

Quick ✓

6. The midpoint $M = (x, y)$ of the line segment from $P_1 = (x_1, y_1)$ to $P_2 = (x_2, y_2)$ is _____ .

In Problems 7 and 8, find the midpoint of the line segment joining the points.

7. $(3, 8)$ and $(0, 4)$

8. $(-2, -5)$ and $(4, 10)$

9.1 Exercises MyMathLab® Math XP PRACTICE

*Problems **1–8** are the Quick ✓s that follow the EXAMPLES.*

Building Skills

In Problems 9–24, find the distance $d(P_1, P_2)$ between the points P_1 and P_2. See Objective 1.

9.

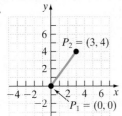

10.

11.

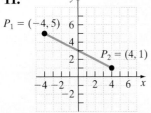

12.

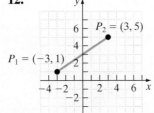

13. $P_1 = (2, 1)$; $P_2 = (6, 4)$

14. $P_1 = (1, 3)$; $P_2 = (4, 7)$

15. $P_1 = (-3, 2)$; $P_2 = (9, -3)$

16. $P_1 = (-10, -3)$; $P_2 = (14, 4)$

17. $P_1 = (-4, 2)$; $P_2 = (2, 2)$

18. $P_1 = (-1, 2)$; $P_2 = (-1, 0)$

19. $P_1 = (0, -3)$; $P_2 = (-3, 3)$

20. $P_1 = (5, 0)$; $P_2 = (-1, -4)$

21. $P_1 = \left(2\sqrt{2}, \sqrt{5}\right)$; $P_2 = \left(5\sqrt{2}, 4\sqrt{5}\right)$

22. $P_1 = \left(\sqrt{6}, -2\sqrt{2}\right)$; $P_2 = \left(3\sqrt{6}, 10\sqrt{2}\right)$

23. $P_1 = (0.3, -3.3)$; $P_2 = (1.3, 0.1)$

24. $P_1 = (-1.7, 1.3)$; $P_2 = (0.3, 2.6)$

In Problems 25–36, find the midpoint of the line segment formed by joining the points P_1 and P_2. See Objective 2.

25. $P_1 = (2, 2)$; $P_2 = (6, 4)$

26. $P_1 = (1, 3)$; $P_2 = (5, 7)$

27. $P_1 = (-3, 2)$; $P_2 = (9, -4)$

28. $P_1 = (-10, -3)$; $P_2 = (14, 7)$

29. $P_1 = (-4, 3)$; $P_2 = (2, 4)$

30. $P_1 = (-1, 2)$; $P_2 = (3, 9)$

31. $P_1 = (0, -3)$; $P_2 = (-3, 3)$

32. $P_1 = (5, 0)$; $P_2 = (-1, -4)$

33. $P_1 = \left(2\sqrt{2}, \sqrt{5}\right); P_2 = \left(5\sqrt{2}, 4\sqrt{5}\right)$

34. $P_1 = \left(\sqrt{6}, -2\sqrt{2}\right); P_2 = \left(3\sqrt{6}, 10\sqrt{2}\right)$

35. $P_1 = (0.3, -3.3); P_2 = (1.3, 0.1)$

36. $P_1 = (-1.7, 1.3); P_2 = (0.3, 2.6)$

Applying the Concepts

37. Consider the three points $A = (0, 3)$, $B = (2, 1)$, and $C = (6, 5)$.

   (a) Plot each point in the Cartesian plane and form the triangle ABC.

   (b) Find the length of each side of the triangle.

   (c) Verify that the triangle is a right triangle.

   (d) Find the area of the triangle.

38. Consider the three points $A = (0, 2)$, $B = (1, 4)$, and $C = (4, 0)$.

   (a) Plot each point in the Cartesian plane and form the triangle ABC.

   (b) Find the length of each side of the triangle.

   (c) Verify that the triangle is a right triangle.

   (d) Find the area of the triangle.

39. Consider the three points $A = (-2, -4)$, $B = (3, 1)$, and $C = (15, -11)$.

   (a) Plot each point in the Cartesian plane and form the triangle ABC.

   (b) Find the length of each side of the triangle.

   (c) Verify that the triangle is a right triangle.

   (d) Find the area of the triangle.

40. Consider the three points $A = (-2, 3)$, $B = (2, 0)$, and $C = (5, 4)$.

   (a) Plot each point in the Cartesian plane and form the triangle ABC.

   (b) Find the length of each side of the triangle.

   (c) Verify that the triangle is a right triangle.

   (d) Find the area of the triangle.

41. Find all points having an x-coordinate of 2 whose distance from the point $(5, 1)$ is 5.

42. Find all points having an x-coordinate of 4 whose distance from the point $(0, 3)$ is 5.

43. Find all points having a y-coordinate of -3 whose distance from the point $(2, 3)$ is 10.

44. Find all points having a y-coordinate of -3 whose distance from the point $(-4, 2)$ is 13.

45. The City of Chicago The city of Chicago's road system is set up like a Cartesian plane, where streets are indicated by the number of blocks they are from Madison Street and State Street. For example, Wrigley Field in Chicago is located at 1060 West Addison, which is 10 blocks west of State Street and 36 blocks north of Madison Street.

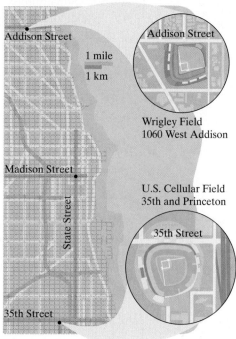

City of Chicago, Illinois

   (a) Find the distance "as the crow flies" from Madison and State Street to Wrigley Field. Use city blocks as the unit of measurement.

   (b) U.S. Cellular Field, home of the White Sox, is located at 35th and Princeton, which is 3 blocks west of State Street and 35 blocks south of Madison. Find the distance "as the crow flies" from Madison and State Street to U.S. Cellular Field.

   (c) Find the distance "as the crow flies" from Wrigley Field to U.S. Cellular Field.

46. Baseball A major league baseball "diamond" is actually a square, 90 feet on a side (see the figure). Overlay a Cartesian plane on a major league baseball diamond so that the origin is at home plate, the positive x-axis lies in the direction from home plate to first base, and the positive y-axis lies in the direction from home plate to third base.

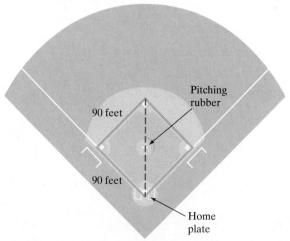

(a) What are the coordinates of home plate, first base, second base, and third base? Use feet as the unit of measurement.

(b) Suppose the center fielder is located at $(310, 260)$. How far is he from second base?

(c) Suppose the shortstop is located at $(60, 100)$. How far is he from second base?

Extending the Concepts

47. Baseball Refer to Problem 46.

(a) Suppose the right fielder catches a fly ball at $(320, 20)$. How many seconds will it take him to throw the ball to second base if he can throw 130 feet per second? (*Hint:* Time $=$ distance divided by speed.)

(b) Suppose a runner "tagging up" from first base can run 27 feet per second. Would you "send the runner" as the first base coach if the right fielder requires 0.8 second to catch and throw? Why?

48. Let $P = (x, y)$ be a point on the graph of $y = x^2 - 4$.

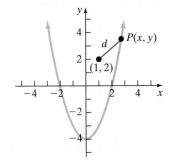

(a) Express the distance d from P to the point $(1, 2)$ as a function of x using the distance formula. See the figure for an illustration of the problem.

(b) What is d if $x = 0$?

(c) What is d if $x = 3$?

Explaining the Concepts

49. How is the distance formula related to the Pythagorean Theorem?

50. How can the distance formula be used to verify that a point is the midpoint of a line segment? What must be done in addition to using the distance formula to verify a midpoint? Why?

Synthesis Review

51. Evaluate 3^2. What is $\sqrt{9}$?

52. Evaluate 8^2. What is $\sqrt{64}$?

53. Evaluate $(-3)^4$. What is $\sqrt[4]{81}$?

54. Evaluate $(-3)^3$. What is $\sqrt[3]{-27}$?

55. Describe the relationship between raising a number to a positive integer power, n, and the nth root of a number.

9.2 Circles

Objectives

1 Write the Standard Form of the Equation of a Circle

2 Graph a Circle

3 Find the Center and Radius of a Circle Given an Equation in General Form

Are You Ready for This Section?

Before getting started, take the following readiness quiz. If you get the problem wrong, go back to the section cited and review the material.

R1. Complete the square in x: $x^2 - 8x$ [Section 7.1, pp. 545–546]

Conics, an abbreviation for **conic sections,** are curves that result from the intersection of a right circular cone and a plane. The four conics that we study are shown in Figure 8. These conics are *circles* (Figure 8(a)), *ellipses* (Figure 8(b)), *parabolas* (Figure 8(c)), and *hyperbolas* (Figure 8(d)).

Figure 8

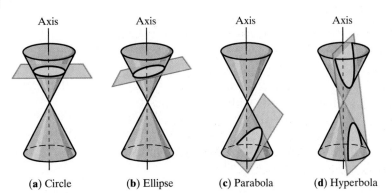

(a) Circle   (b) Ellipse   (c) Parabola   (d) Hyperbola

Ready?...Answer   **R1.** $x^2 - 8x + 16$

We study circles in this section, parabolas in Section 9.3, ellipses in Section 9.4, and hyperbolas in Section 9.5.

▶ ❶ Write the Standard Form of the Equation of a Circle

Because of the Cartesian plane, we can translate a geometric statement into an algebraic statement, and vice versa. Consider the following geometric statement that defines a circle.

Figure 9

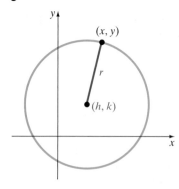

> **Definition**
>
> A **circle** is the set of all points in the Cartesian plane that are a fixed distance r from a fixed point (h, k). The fixed distance r is the **radius,** and the fixed point (h, k) is the **center** of the circle. See Figure 9.

To find an equation that has this graph, we let (x, y) represent the coordinates of any point on a circle with radius r and center (h, k). Then the distance between the points (x, y) and (h, k) must always equal r. That is, by the distance formula,

$$\sqrt{(x - h)^2 + (y - k)^2} = r$$

If we square both sides of the equation, then

$$(x - h)^2 + (y - k)^2 = r^2$$

and we have the equation of a circle.

> **Definition**
>
> The **standard form of an equation of a circle** with radius r and center (h, k) is
>
> $$(x - h)^2 + (y - k)^2 = r^2$$

The standard form of a circle of radius r with center at the origin $(0,0)$ is $x^2 + y^2 = r^2$.

EXAMPLE 1   **Writing the Standard Form of the Equation of a Circle**

Write the standard form of the equation of the circle with radius 4 and center $(2, -3)$.

Solution

We use the equation $(x - h)^2 + (y - k)^2 = r^2$ with $r = 4$, $h = 2$, and $k = -3$.

$$(x - 2)^2 + (y - (-3))^2 = 4^2$$
$$(x - 2)^2 + (y + 3)^2 = 16$$

●

Quick ✓

1. A _____ is the set of all points in the Cartesian plane that are a fixed distance r from a fixed point (h, k).

2. For a circle, the _____ is the distance from the center to any point on the circle.

In Problems 3 and 4, write the standard form of the equation of each circle whose radius is r and center is (h, k).

3. $r = 5$; $(h, k) = (2, 4)$

4. $r = \sqrt{2}$; $(h, k) = (-2, 0)$

▶ ❷ Graph a Circle

The graph of any equation of the form $(x - h)^2 + (y - k)^2 = r^2$ is a circle with radius r and center (h, k).

EXAMPLE 2 **Graphing a Circle**

Graph the equation: $(x + 2)^2 + (y - 3)^2 = 9$

Solution

The equation is of the form $(x - h)^2 + (y - k)^2 = r^2$, so its graph is a circle. To graph the equation, we first identify the center and radius of the circle by comparing the equation to the standard form of the equation of a circle.

$$(x + 2)^2 + (y - 3)^2 = 9$$
$$(x - (-2))^2 + (y - 3)^2 = 3^2$$
$$(x - h)^2 + (y - k)^2 = r^2$$

We see that $h = -2, k = 3$, and $r = 3$. The circle has center $(-2, 3)$ and radius 3. To graph this circle, we first plot the center $(-2, 3)$. Since the radius is 3 units, we can go 3 units in any direction from the center and find a point on the circle. It is easiest to find the four points left, right, up, and down from the center. These four points are $(-5, 3)$, $(1, 3), (-2, 6)$, and $(-2, 0)$, respectively. We plot these points in Figure 10(a) and use them to help us draw the graph of the circle in Figure 10(b).

Figure 10

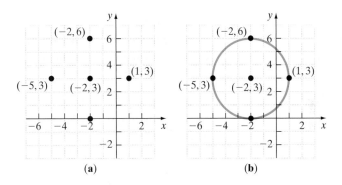

(a)   (b)

Notice that the graph of $(x + 2)^2 + (y - 3)^2 = (x - (-2))^2 + (y - 3)^2 = 9$ is the graph of $x^2 + y^2 = 9$ shifted 2 units left and 3 units down.

Quick ✓

5. *True or False* The center of the circle $(x + 1)^2 + (y - 3)^2 = 25$ is $(1, -3)$.

6. *True or False* The center of the circle $x^2 + y^2 = 9$ is $(0, 0)$; its radius is 3.

In Problems 7 and 8, graph the circle.

7. $(x - 3)^2 + (y - 1)^2 = 4$ **8.** $(x + 5)^2 + y^2 = 16$

▶ ❸ Find the Center and Radius of a Circle Given an Equation in General Form

If we eliminate the parentheses from the standard form of the equation of the circle given in Example 2, we get

$$(x + 2)^2 + (y - 3)^2 = 9$$
$$\text{FOIL:} \quad x^2 + 4x + 4 + y^2 - 6y + 9 = 9$$
$$\text{Subtract 9 from both sides:} \quad x^2 + y^2 + 4x - 6y + 4 = 0$$

Any equation of the form

$$x^2 + y^2 + ax + by + c = 0$$

has a graph that is a circle, a graph that is a point, or no graph at all. For example, the graph of the equation $x^2 + y^2 = 0$ is the single point $(0, 0)$. The equation $x^2 + y^2 + 4 = 0$, or $x^2 + y^2 = -4$, has no graph, because the sum of squares of real numbers is never negative.

> **In Words**
>
> The standard form of the equation of a circle is $(x - h)^2 + (y - k)^2 = r^2$. The general form is $x^2 + y^2 + ax + by + c = 0$.

Definition

The **general form of the equation of a circle** is given by the equation

$$x^2 + y^2 + ax + by + c = 0$$

when the graph exists.

If an equation of a circle is in general form, we use the method of completing the square to put the equation in standard form, $(x - h)^2 + (y - k)^2 = r^2$, in order to find the center and radius of the circle.

(**EXAMPLE 3**) **Graphing a Circle Whose Equation Is in General Form**

Graph the equation: $x^2 + y^2 + 8x - 2y - 8 = 0$

Solution

To find the center and radius of the circle, put the equation in standard form by completing the square in both x and y (covered in Section 7.1). To do this, group the terms involving x, group the terms involving y, and put the constant on the right side of the equation by adding 8 to both sides.

$$(x^2 + 8x) + (y^2 - 2y) = 8$$

Now complete the square of each expression in parentheses. Remember, any number added to the left side of the equation must also be added to the right.

$$(x^2 + 8x + 16) + (y^2 - 2y + 1) = 8 + 16 + 1$$

$$\left(\frac{1}{2} \cdot 8\right)^2 = 16 \qquad \left(\frac{1}{2} \cdot (-2)\right)^2 = 1$$

Factor: $(x + 4)^2 + (y - 1)^2 = 25$

The equation is now in the standard form of the equation of a circle. The center is $(-4, 1)$ and the radius is 5. To graph the equation, plot the center and the four points to the left, to the right, above, and below the center. Then draw in the circle. See Figure 11.

Figure 11

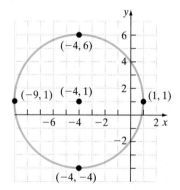

Notice that the graph of $(x + 4)^2 + (y - 1)^2 = 25$ is the graph of $x^2 + y^2 = 25$ shifted 4 units left and 1 unit up.

Quick ✓

In Problems 9 and 10, graph each circle.

9. $x^2 + y^2 - 6x - 4y + 4 = 0$

10. $2x^2 + 2y^2 - 16x + 4y - 38 = 0$

9.2 Exercises · MyMathLab® · MathXL PRACTICE

Exercise numbers in green are complete video solutions in MyMathLab

*Problems **1–10** are the Quick ✓s that follow the EXAMPLES.*

Building Skills

In Problems 11–14, find the center and radius of each circle. Write the standard form of the equation.

11.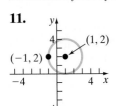

12.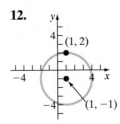

13.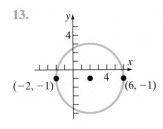

14.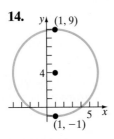

In Problems 15–26, write the standard form of the equation of each circle whose radius is r and center is (h, k). See Objective 1.

15. $r = 3; (h, k) = (0, 0)$ **16.** $r = 5; (h, k) = (0, 0)$

17. $r = 2; (h, k) = (1, 4)$ **18.** $r = 4; (h, k) = (3, 1)$

19. $r = 6; (h, k) = (-2, 4)$ **20.** $r = 3; (h, k) = (1, -4)$

21. $r = 4; (h, k) = (0, 3)$ **22.** $r = 2; (h, k) = (1, 0)$

23. $r = 5; (h, k) = (5, -5)$ **24.** $r = 4; (h, k) = (-4, 4)$

25. $r = \sqrt{5}; (h, k) = (1, 2)$ **26.** $r = \sqrt{7}; (h, k) = (5, 2)$

In Problems 27–36, find the center (h, k) and radius r of each circle. Graph each circle. See Objective 2.

27. $x^2 + y^2 = 36$

28. $x^2 + y^2 = 144$

29. $(x - 4)^2 + (y - 1)^2 = 25$

30. $(x - 2)^2 + (y - 3)^2 = 9$

31. $(x + 3)^2 + (y - 2)^2 = 81$

32. $(x - 5)^2 + (y + 2)^2 = 49$

33. $x^2 + (y - 3)^2 = 64$

34. $(x - 6)^2 + y^2 = 36$

35. $(x - 1)^2 + (y + 1)^2 = \dfrac{1}{4}$

36. $(x - 2)^2 + (y + 2)^2 = \dfrac{1}{4}$

In Problems 37–42, find the center (h, k) and radius r of each circle. Graph each circle. See Objective 3.

37. $x^2 + y^2 - 6x + 2y + 1 = 0$

38. $x^2 + y^2 + 2x - 8y + 8 = 0$

39. $x^2 + y^2 + 10x + 4y + 4 = 0$

40. $x^2 + y^2 + 4x - 12y + 36 = 0$

41. $2x^2 + 2y^2 - 12x + 24y - 72 = 0$

42. $2x^2 + 2y^2 - 28x + 20y + 20 = 0$

Mixed Practice

In Problems 43–48, find the standard form of the equation of each circle.

43. Center at the origin and containing the point $(4, -2)$

44. Center at $(0, 3)$ and containing the point $(3, 7)$

45. Center at $(-3, 2)$ and tangent to the y-axis

46. Center at $(2, -3)$ and tangent to the x-axis

47. With endpoints of a diameter at $(2, 3)$ and $(-4, -5)$

48. With endpoints of a diameter at $(-5, -3)$ and $(7, 2)$

Applying the Concepts

△ **49.** Find the area and circumference of the circle $(x - 3)^2 + (y - 8)^2 = 64$.

△ **50.** Find the area and circumference of the circle $(x - 1)^2 + (y - 4)^2 = 49$.

△ **51.** Find the area of the square in the figure.

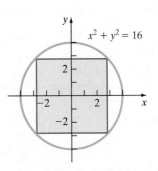

△ **52.** Find the area of the shaded region in the figure.

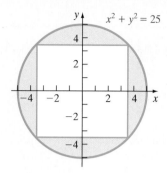

Extending the Concepts

53. Which of the following equations might have the graph shown? (More than one answer is possible.)

(a) $(x - 2)^2 + y^2 = 1$

(b) $x^2 + (y - 2)^2 = 1$

(c) $(x + 4)^2 + y^2 = 9$

(d) $(x - 5)^2 + y^2 = 25$

(e) $x^2 + y^2 - 8x + 7 = 0$

(f) $x^2 + y^2 + 10x + 18 = 0$

54. Which of the following equations might have the graph shown? (More than one answer is possible.)

(a) $(x - 2)^2 + (y + 3)^2 = 4$

(b) $(x - 3)^2 + (y - 4)^2 = 4$

(c) $(x + 3)^2 + (y + 4)^2 = 9$

(d) $(x - 5)^2 + (y - 5)^2 = 25$

(e) $x^2 + y^2 + 8x + 10y + 32 = 0$

(f) $x^2 + y^2 - 4x - 6y - 3 = 0$

Explaining the Concepts

55. How is the distance formula related to the definition of a circle?

56. Are circles functions? Why or why not?

57. Is $x^2 = 36 - y^2$ the equation of a circle? Why or why not? If so, what are the center and the radius?

58. Is $3x^2 - 12x + 3y^2 - 15 = 0$ the equation of a circle? Why or why not? If so, what is the center and radius?

Synthesis Review

In Problems 59–63, graph each function using either point plotting or properties of the function. For example, to graph a line, use the slope and y-intercept or the intercepts.

59. $f(x) = 4x - 3$

60. $2x + 5y = 20$

61. $g(x) = x^2 - 4x - 5$

62. $F(x) = -3x^2 + 12x - 12$

63. $G(x) = -2(x + 3)^2 - 5$

64. Present an argument that supports the approach you took to graphing each function in Problems 59–63. For example, in Problem 59, did you use point plotting? Intercepts? Slope? Which method did you use and why?

The Graphing Calculator

To graph a circle using a graphing calculator, we must first solve the equation for y. For example, to graph $(x + 2)^2 + (y - 3)^2 = 9$, solve for y as follows:

$$(x + 2)^2 + (y - 3)^2 = 9$$

Subtract $(x + 2)^2$ from both sides: $(y - 3)^2 = 9 - (x + 2)^2$

Take the square root of both sides: $y - 3 = \pm\sqrt{9 - (x + 2)^2}$

Add 3 to both sides: $y = 3 \pm \sqrt{9 - (x + 2)^2}$

Graph the top half $Y_1 = 3 + \sqrt{9 - (x + 2)^2}$ and the bottom half $Y_2 = 3 - \sqrt{9 - (x + 2)^2}$. To get an undistorted view of the graph, be sure to use the ZOOM SQUARE feature on your graphing calculator. The figure below shows the graph of $(x + 2)^2 + (y - 3)^2 = 9$.

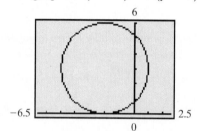

In Problems 65–74, graph each circle using a graphing calculator. Compare your graphs to the graphs drawn by hand in Problems 27–36.

65. $x^2 + y^2 = 36$

66. $x^2 + y^2 = 144$

67. $(x - 4)^2 + (y - 1)^2 = 25$

68. $(x - 2)^2 + (y - 3)^2 = 9$

69. $(x + 3)^2 + (y - 2)^2 = 81$

70. $(x - 5)^2 + (y + 2)^2 = 49$

71. $x^2 + (y - 3)^2 = 64$

72. $(x - 6)^2 + y^2 = 36$

73. $(x - 1)^2 + (y + 1)^2 = \dfrac{1}{4}$

74. $(x - 2)^2 + (y + 2)^2 = \dfrac{1}{4}$

9.3 Parabolas

Objectives

1. Graph Parabolas Whose Vertex Is the Origin
2. Find the Equation of a Parabola
3. Graph a Parabola Whose Vertex Is Not the Origin
4. Solve Applied Problems Involving Parabolas

Are You Ready for This Section?

Before getting started, take the following readiness quiz. If you get a problem wrong, go back to the section cited and review the material.

R1. Identify the vertex and axis of symmetry of $f(x) = -3(x + 4)^2 - 5$. Does the parabola open up or down? Why? [Section 7.4, pp. 582–585]

R2. Identify the vertex and axis of symmetry of the quadratic function $f(x) = 2x^2 - 8x + 1$. Does the parabola open up or down? Why? [Section 7.5, pp. 588–594]

R3. Complete the square of $x^2 - 12x$. [Section 7.1, pp. 545–546]

R4. Solve: $(x - 3)^2 = 25$ [Section 7.1, pp. 542–545]

We began a discussion of parabolas in Section 7.4 when we studied quadratic functions. You can refresh your memory with the summary below.

Summary Parabolas That Open Up or Down

The graph of $y = a(x - h)^2 + k$ or $y = ax^2 + bx + c$ is a parabola that

1. opens up if $a > 0$ and opens down if $a < 0$.
2. has vertex (h, k).
3. has a vertex whose x-coordinate is $x = -\dfrac{b}{2a}$. The y-coordinate is found by evaluating the equation at the x-coordinate of the vertex.

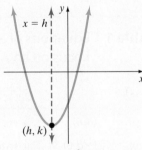 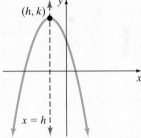

$$y = a(x - h)^2 + k, a > 0 \qquad y = a(x - h)^2 + k, a < 0$$

Our ideas about parabolas in Sections 7.4 and 7.5 relied on algebra. In this chapter, we will look at parabolas (and other conic sections) from a geometric point of view.

Definition

A **parabola** is the collection of all points P in the plane that are the same distance from a fixed point F as they are from a fixed line D. The point F is called the **focus** of the parabola, and the line D is its **directrix.** In other words, a parabola is the set of points P for which

$$d(F, P) = d(P, D)$$

Figure 12 shows a parabola. The line through the focus F and perpendicular to the directrix D is the **axis of symmetry** of the parabola. The point of intersection of the parabola with its axis of symmetry is the **vertex** V.

Figure 12

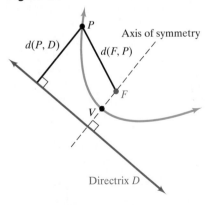

Ready?...Answers **R1.** Vertex: $(-4, -5)$; axis of symmetry: $x = -4$; opens down since $a = -3 < 0$.
R2. Vertex: $(2, -7)$; axis of symmetry: $x = 2$; opens up since $a = 2 > 0$.
R3. $x^2 - 12x + 36$ **R4.** $\{-2, 8\}$

▶ ① Graph Parabolas Whose Vertex Is the Origin

We want to develop an equation for the parabola based on the definition just given. For example, the distance from the focus to the vertex (a point on the parabola) must equal the distance from the vertex to the directrix. Let a represent the distance from the focus to the vertex. To develop an equation for a parabola, we start by looking at parabolas whose vertex is at the origin. We consider four possibilities—parabolas that open left, parabolas that open right, parabolas that open up, and parabolas that open down.

First we will obtain the equation of a parabola whose vertex is at the origin and opens right. To do this, position a parabola in a Cartesian plane so that its vertex is at the origin, $(0, 0)$, the focus is on the positive x-axis, and the directrix is a vertical line in quadrants II and III. Because a represents the distance from the vertex to the focus, the focus is located at $(a, 0)$. Also, because the distance from the vertex to the directrix is a, the directrix is the line $x = -a$. See Figure 13. If $P = (x, y)$ is any point on the parabola, then the distance from P to the focus F, $(a, 0)$, must equal the distance from P to the directrix, $x = -a$. We use the point $(-a, y)$ on the directrix in the distance formula. That is,

Figure 13

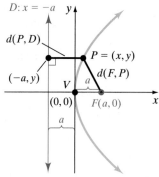

$$d(F, P) = d(P, D)$$

Use the distance formula: $\quad \sqrt{(x - a)^2 + (y - 0)^2} = \sqrt{(x - (-a))^2 + (y - y)^2}$

Square both sides; simplify: $\quad (x - a)^2 + y^2 = (x + a)^2$

Multiply out binomials: $\quad x^2 - 2ax + a^2 + y^2 = x^2 + 2ax + a^2$

Combine like terms: $\quad y^2 = 4ax$

Thus the equation of the parabola whose vertex is at the origin and opens to the right is $y^2 = 4ax$. We obtain the equations of the three other possibilities (opens left, opens up, or opens down) using similar reasoning. We summarize these possibilities in Table 1.

Table 1 Equations of a Parabola: Vertex at $(0, 0)$; Focus on an Axis; $a > 0$

Vertex	Focus	Directrix	Equation	Axis of Symmetry	Opens
$(0, 0)$	$(a, 0)$	$x = -a$	$y^2 = 4ax$	x-axis	Right
$(0, 0)$	$(-a, 0)$	$x = a$	$y^2 = -4ax$	x-axis	Left
$(0, 0)$	$(0, a)$	$y = -a$	$x^2 = 4ay$	y-axis	Up
$(0, 0)$	$(0, -a)$	$y = a$	$x^2 = -4ay$	y-axis	Down

The graphs of the four parabolas are given in Figure 14.

Figure 14

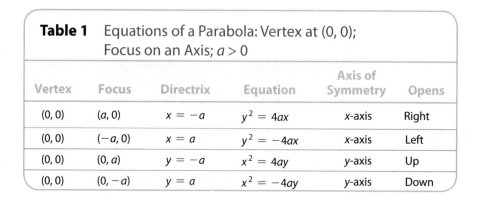

(a) $y^2 = 4ax$ (b) $y^2 = -4ax$ (c) $x^2 = 4ay$ (d) $x^2 = -4ay$

EXAMPLE 1   **Graphing a Parabola That Opens Left**

Graph the equation $y^2 = -12x$.

Solution

The equation $y^2 = -12x$ is of the form $y^2 = -4ax$, where $-4a = -12$, so $a = 3$. The graph of the equation is a parabola with vertex at $(0, 0)$ and focus at $(-a, 0) = (-3, 0)$, so the parabola opens to the left. The directrix is the line $x = 3$. To graph the parabola, it helps to plot the two points on the graph above and below the focus. Because the points are directly above and below the focus, let $x = -3$ in the equation $y^2 = -12x$ and solve for y.

$$y^2 = -12x$$
$$\text{Let } x = -3: \quad = -12(-3)$$
$$= 36$$
$$\text{Take the square root of both sides:} \quad y = \pm 6$$

Figure 15

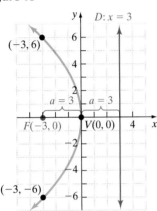

The points on the parabola above and below the focus are $(-3, 6)$ and $(-3, -6)$. These points help in graphing the parabola because they determine the "opening." See Figure 15.

Quick ✓

1. A _____ is the collection of all points P in the plane that are the same distance from a fixed point F as they are from a fixed line D.
2. The point of intersection of the parabola with its axis of symmetry is called the _____.
3. The line through the focus and perpendicular to the directrix is called the ___ __ _____.

In Problems 4 and 5, graph the equation.

4. $y^2 = 8x$ 5. $y^2 = -20x$

EXAMPLE 2   **Graphing a Parabola That Opens Up**

Graph the equation $x^2 = 8y$.

Solution

The equation $x^2 = 8y$ is of the form $x^2 = 4ay$, where $4a = 8$, so $a = 2$. The graph of the equation is a parabola with vertex at $(0, 0)$ and focus at $(0, a) = (0, 2)$, so the parabola opens up. The directrix is the line $y = -2$. To graph the parabola, we will plot the two points on the graph to the left and right of the focus. We let $y = 2$ in the equation $x^2 = 8y$ and solve for x.

$$x^2 = 8y = 8(2) = 16$$
$$\text{Take the square root of both sides:} \quad x = \pm 4$$

The points on the parabola to the left and right of the focus are $(-4, 2)$ and $(4, 2)$. See Figure 16.

Figure 16

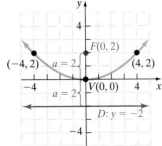

Quick ✓

In Problems 6 and 7, graph the equation.

6. $x^2 = 4y$ 7. $x^2 = -12y$

▶ ❷ **Find the Equation of a Parabola**

We are now going to change gears and use information about a parabola to obtain its equation.

EXAMPLE 3 Finding an Equation of a Parabola

Figure 17

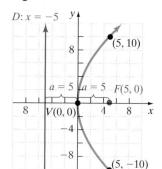

Find an equation of the parabola with vertex at $(0, 0)$ and focus at $(5, 0)$. Graph the equation.

Solution

The distance from the vertex $(0, 0)$ to the focus $(5, 0)$ is $a = 5$. Because the focus lies on the positive x-axis, we know that the parabola opens to the right. This means the equation of the parabola is of the form $y^2 = 4ax$ with $a = 5$:

$$y^2 = 4(5)x$$
$$= 20x$$

Figure 17 shows the graph of $y^2 = 20x$.

EXAMPLE 4 Finding the Equation of a Parabola

Find the equation of a parabola with vertex at $(0, 0)$ if its axis of symmetry is the y-axis and its graph contains the point $(4, 3)$. Graph the equation.

Solution

Which of the four equations of parabolas listed in Table 1 should we use? The vertex is at the origin and the axis of symmetry is the y-axis, so the parabola opens either up or down. Because the graph contains the point $(4, 3)$, which is in quadrant I, the parabola must open up. Therefore, the equation of the parabola is of the form $x^2 = 4ay$. Because the point $(4, 3)$ is on the parabola, we let $x = 4$ and $y = 3$ in the equation $x^2 = 4ay$ to determine a.

$$x^2 = 4ay$$
$$x = 4, y = 3: \quad 4^2 = 4a(3)$$
$$16 = 12a$$

Divide both sides by 12: $\quad a = \dfrac{16}{12} = \dfrac{4}{3}$

The equation of the parabola is

$$x^2 = 4\left(\frac{4}{3}\right)y, \quad \text{or} \quad x^2 = \frac{16}{3}y$$

Figure 18

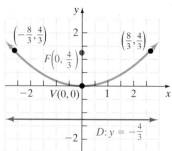

With $a = \dfrac{4}{3}$, we know that the focus is $\left(0, \dfrac{4}{3}\right)$ and the directrix is the line $y = -\dfrac{4}{3}$. We let $y = \dfrac{4}{3}$ to find points left and right of the focus to determine the "opening," so the points $\left(-\dfrac{8}{3}, \dfrac{4}{3}\right)$ and $\left(\dfrac{8}{3}, \dfrac{4}{3}\right)$ are on the graph. Figure 18 shows the graph of the parabola.

Quick ✓

In Problems 8 and 9, find the equation of the parabola described. Graph the equation.

8. Vertex at $(0, 0)$; focus at $(0, -8)$

9. Vertex at $(0, 0)$; axis of symmetry the x-axis; contains the point $(3, 2)$

▶ ❸ **Graph a Parabola Whose Vertex Is Not the Origin**

If a parabola with vertex at the origin and axis of symmetry along a coordinate axis is shifted horizontally h units and then vertically k units, the result is a parabola with vertex at (h, k) and axis of symmetry parallel to either the x- or the y-axis. The equations of these parabolas have the same form as those whose vertex is at the origin, except that x is replaced with $x - h$ (the horizontal shift) and y is replaced with $y - k$ (the vertical shift). Table 2 gives the equations of the four parabolas. Figure 19(a)–(d) illustrates the graphs for $h > 0$ and $k > 0$.

Table 2 Parabolas with Vertex at (h, k); Axis of Symmetry Parallel to a Coordinate Axis; $a > 0$

Vertex	Focus	Directrix	Equation	Axis of Symmetry	Opens
(h, k)	$(h + a, k)$	$x = h - a$	$(y - k)^2 = 4a(x - h)$	Parallel to x-axis,	Right
(h, k)	$(h - a, k)$	$x = h + a$	$(y - k)^2 = -4a(x - h)$	Parallel to x-axis	Left
(h, k)	$(h, k + a)$	$y = k - a$	$(x - h)^2 = 4a(y - k)$	Parallel to y-axis	Up
(h, k)	$(h, k - a)$	$y = k + a$	$(x - h)^2 = -4a(y - k)$	Parallel to y-axis	Down

Figure 19

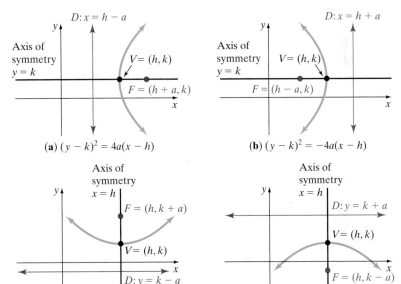

(a) $(y - k)^2 = 4a(x - h)$

(b) $(y - k)^2 = -4a(x - h)$

(c) $(x - h)^2 = 4a(y - k)$

(d) $(x - h)^2 = -4a(y - k)$

EXAMPLE 5 **Graphing a Parabola Whose Vertex Is Not the Origin**

Graph the parabola $x^2 - 2x + 8y + 25 = 0$.

Solution

Notice the equation is not in any of the forms given in Table 2. We need to complete the square in x to write the equation in standard form.

$$x^2 - 2x + 8y + 25 = 0$$

Isolate the terms involving x: $\qquad x^2 - 2x = -8y - 25$

Complete the square: $\qquad x^2 - 2x + 1 = -8y - 25 + 1$

Simplify: $\qquad x^2 - 2x + 1 = -8y - 24$

Factor: $\qquad (x - 1)^2 = -8(y + 3)$

The equation is of the form $(x - h)^2 = -4a(y - k)$. This is a parabola that opens down with vertex $(h, k) = (1, -3)$. Since $-4a = -8$, we know that $a = 2$. Because the parabola opens down, the focus is $a = 2$ units below the vertex at $(1, -5)$. To

Work Smart

Notice that the graph of
$(x - 1)^2 = -8(y + 3)$ is the graph
of $x^2 = -8y$ shifted right 1 unit and
down 3 units.

Figure 20

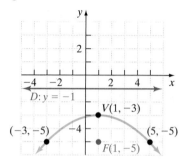

find two points on the graph to the left and right of the focus, let $y = -5$ in the equation of the parabola.

$$\text{Let } y = -5: \quad (x - 1)^2 = -8(-5 + 3)$$
$$(x - 1)^2 = -8(-2)$$
$$(x - 1)^2 = 16$$
$$\text{Take the square root of both sides:} \quad x - 1 = \pm 4$$
$$\text{Add 1 to both sides:} \quad x = 1 \pm 4$$
$$x = 1 - 4 \quad \text{or} \quad x = 1 + 4$$
$$x = -3 \quad \text{or} \quad x = 5$$

The points $(-3, -5)$ and $(5, -5)$ are on the graph of the parabola. The directrix is $a = 2$ units above the vertex, so $y = -1$ is the directrix. Figure 20 shows the graph. ●

Quick ✓

10. The vertex of the parabola $(x + 3)^2 = -14(y - 2)$ is _____.

11. Graph the parabola $y^2 - 4y - 12x - 32 = 0$.

▶ ❹ Solve Applied Problems Involving Parabolas

Work Smart: Study Skills

Which is the graph of a parabola
and which is the graph of a circle?
$x^2 + y^2 - 2y + 3 = 0$
$2y^2 - 12y - x + 22 = 0$
How can you tell the difference
between the equation of a circle and
the equation of a parabola?
The equation of a parabola has either
an x^2-term or a y^2-term but not both.
The equation of a circle has the
sum of an x^2-term and a y^2-term.

The variety of applications of parabolas is astounding. Suppose that a mirror is shaped like a parabola. If a light bulb is placed at the focus of the parabola, then all the rays from the bulb reflect off the mirror in lines parallel to the axis of symmetry. This concept is used in the design of car headlights, flashlights, and searchlights. See Figure 21.

Suppose that rays of light are received by a parabola. When the rays strike the surface of a parabolic mirror whose axis of symmetry is parallel to these rays, they are all reflected to a single point—the focus. This idea is used in the design of some telescopes and satellite dishes. See Figure 22.

Figure 21

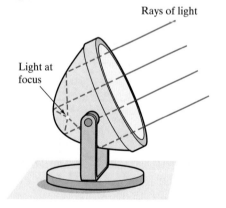

Figure 22

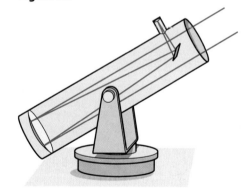

EXAMPLE 6   **A Satellite Dish**

A satellite dish is shaped like a parabola. See Figure 23(a). Signals strike the surface of the dish and are reflected to a single point, where the receiver of the dish is located. If a satellite dish is 2 feet across at its opening and 6 inches deep (0.5 foot) at its center, at what position should the receiver be placed?

Solution

We want to know where to locate the receiver of the satellite dish. This means we want to locate its focus. Draw a parabola in a Cartesian plane so that the vertex is at the origin and the focus is on the positive y-axis. The width of the dish is 2 feet across and its height is 0.5 feet, so we know two points on the graph, indicated in Figure 23(b).

Figure 23

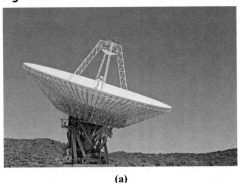

(a)

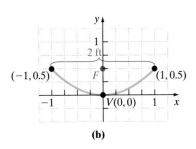

(b)

The parabola is an equation of the form $x^2 = 4ay$. Since $(1, 0.5)$ is a point on the graph, we have

$$x = 1, y = 0.5: \quad 1^2 = 4a\,(0.5)$$
$$1 = 2a$$

Divide both sides by 2: $\quad a = \dfrac{1}{2}$

The receiver should be located $\dfrac{1}{2}$ foot from the base of the dish along its axis of symmetry.

Quick ✓

12. A satellite dish is shaped like a parabola. Signals strike the surface of the dish and are reflected to a single point, where the receiver of the dish is located. If the dish is 4 feet across at its opening and 6 inches deep (0.5 foot) at its center, at what position should the receiver be placed?

9.3 Exercises   MyMathLab® Math·XP PRACTICE   Exercise numbers in green have complete video solutions in MyMathLab.

*Problems **1–12** are the Quick ✓s that follow the EXAMPLES.*

Building Skills

In Problems 13–20, the graph of a parabola is given. Match each graph to its equation.

(a) $y^2 = 8x$ **(b)** $y^2 = -8x$

(c) $x^2 = 8y$ **(d)** $x^2 = -8y$

(e) $(y - 2)^2 = 8(x + 1)$ **(f)** $(y - 2)^2 = -8(x + 1)$

(g) $(x + 1)^2 = 8(y - 2)$ **(h)** $(x + 1)^2 = -8(y - 2)$

13.

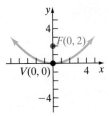

14.

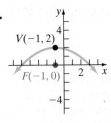

15.

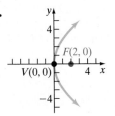

16.

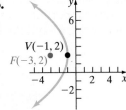

17.

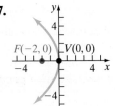

18.

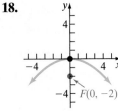

19.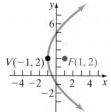

20.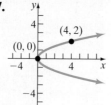

In Problems 37 and 38, write an equation for each parabola. See Objective 2.

37.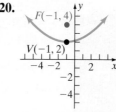

38.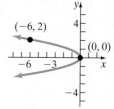

In Problems 21–26, find the vertex, focus, and directrix of each parabola. Graph the parabola. See Objective 1.

21. $x^2 = 24y$ **22.** $x^2 = 28y$

23. $y^2 = -6x$ **24.** $y^2 = 10x$

25. $x^2 = -8y$ **26.** $x^2 = -16y$

In Problems 27–36, find the equation of the parabola described. Graph the parabola. See Objective 2.

27. Vertex at $(0, 0)$; focus at $(5, 0)$

28. Vertex at $(0, 0)$; focus at $(0, 5)$

29. Vertex at $(0, 0)$; focus at $(0, -6)$

30. Vertex at $(0, 0)$; focus at $(-8, 0)$

31. Vertex at $(0, 0)$; contains the point $(6, 6)$; axis of symmetry the y-axis

32. Vertex at $(0, 0)$; contains the point $(2, 2)$; axis of symmetry the x-axis

33. Vertex at $(0, 0)$; directrix the line $y = 3$

34. Vertex at $(0, 0)$; directrix the line $x = -4$

35. Focus at $(-3, 0)$; directrix the line $x = 3$

36. Focus at $(0, -2)$; directrix the line $y = 2$

In Problems 39–50, find the vertex, focus, and directrix of each parabola. Graph the parabola. See Objective 3.

39. $(x - 2)^2 = 4(y - 4)$ **40.** $(x + 4)^2 = -4(y - 1)$

41. $(y + 3)^2 = -8(x + 2)$ **42.** $(y - 2)^2 = 12(x + 5)$

43. $(x + 5)^2 = -20(y - 1)$ **44.** $(x - 6)^2 = 2(y - 2)$

45. $x^2 + 4x + 12y + 16 = 0$ **46.** $x^2 + 2x - 8y + 25 = 0$

47. $y^2 - 8y - 4x + 20 = 0$ **48.** $y^2 - 8y + 16x - 16 = 0$

49. $x^2 + 10x + 6y + 13 = 0$ **50.** $x^2 - 4x + 10y + 4 = 0$

Applying the Concepts

51. A Headlight The headlight of a car is in the shape of a parabola. Its diameter is 4 inches and its depth is 1 inch. How far from the vertex should the light bulb be placed so that the rays will be reflected parallel to the axis?

52. A Headlight The headlight of a car is in the shape of a parabola. Suppose the engineers have designed the headlight to be 5 inches wide and want the bulb to be placed at the focus 1 inch from the vertex. What is the depth of the headlight?

53. Suspension Bridge The cables of a suspension bridge are in the shape of a parabola, as shown in the figure. The towers supporting the cable are 500 feet apart and 60 feet high. If the cables touch the road surface midway between the towers, what is the height of the cable at a point 150 feet from the center of the bridge?

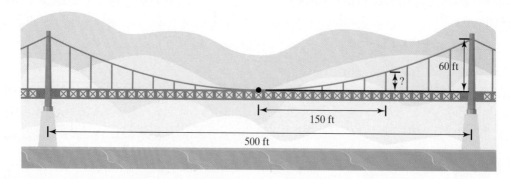

54. Suspension Bridge The cables of a suspension bridge are in the shape of a parabola. The towers supporting the cable are 400 feet apart and 80 feet high. If the cables touch the road surface midway between the towers, what is the height of the cable at a point 100 feet from the center of the bridge?

55. Parabolic Arch Bridge A bridge is built in the shape of a parabolic arch. The arch has a span of 100 feet and a maximum height of 30 feet. See the illustration. Choose a suitable rectangular coordinate system and find the height of the arch at distances of 10, 30, and 50 feet from the center.

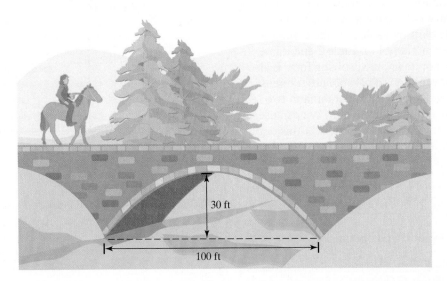

30 ft

100 ft

56. Parabolic Arch Bridge A bridge is to be built in the shape of a parabolic arch and is to have a span of 120 feet. The height of the arch at a distance of 30 feet from the center is to be 15 feet. Find the height of the arch at its center.

Extending the Concepts

In Problems 57–60, write an equation for each parabola.

57.

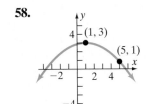

58.

(1, 3)

(5, 1)

59.

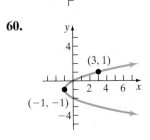

(2, 3)

(−2, −1)

60.

(3, 1)

(−1, −1)

61. For the parabola $x^2 = 8y$, **(a)** verify that $(4, 2)$ is a point on the parabola, and **(b)** show that the distance from the focus to $(4, 2)$ equals the distance from $(4, 2)$ to the directrix.

62. For the parabola $(y - 3)^2 = 12(x + 1)$, **(a)** verify that $(2, 9)$ is a point on the parabola, and **(b)** show that the distance from the focus to $(2, 9)$ equals the distance from $(2, 9)$ to the directrix.

Explaining the Concepts

63. The distance from a point on a parabola to its focus is 8 units. What is the distance from the same point on the parabola to the directrix? Why?

64. Write down the four equations that are parabolas with vertex at (h, k).

65. Draw a parabola and label the vertex, axis of symmetry, focus, and directrix.

66. Explain the difference between the discussion of parabolas presented in this section and the discussion presented in Sections 7.4 and 7.5.

Synthesis Review

67. Graph $y = (x + 3)^2$ using the methods of Section 7.4.

68. Graph $y = (x + 3)^2 = x^2 + 6x + 9$ using the methods of Section 7.5.

69. Graph $y = (x + 3)^2$ using the methods of this section.

70. Graph $4(y + 2) = (x - 2)^2$ using the methods of Section 7.4. (*Hint:* Write the equation in the form $y = a(x - h)^2 + k$.)

71. Graph $4(y + 2) = (x - 2)^2$ using the methods of this section.

72. Compare and contrast the methods of graphing a parabola using the approach in Section 7.4 and the approach in this section. Which do you prefer? Why?

The Graphing Calculator

To graph a parabola using a graphing calculator, we must solve the equation for y, just as we did for circles. This is fairly straightforward if the equation is in the form given in either Table 1 or Table 2. If the equation is not in the form given in Table 1 or Table 2 and it is a parabola that opens up or down, then solving for y is also straightforward. However, if it is a parabola that opens left or right, we will need to graph the parabola in two "pieces"—the top half and the bottom half. We can use the quadratic formula to accomplish this task. Consider the equation $y^2 + 8y + x - 4 = 0$. This equation is quadratic in y as shown:

$$y^2 + 8y + x - 4 = 0$$
$$\boxed{a = 1} \quad \boxed{b = 8} \quad \boxed{c = x - 4}$$

We use the quadratic formula and obtain

$$y = \frac{-8 \pm \sqrt{8^2 - 4(1)(x - 4)}}{2(1)}$$

so we graph

$$Y_1 = \frac{-8 - \sqrt{64 - 4(x - 4)}}{2} \quad \text{and}$$

$$Y_2 = \frac{-8 + \sqrt{64 - 4(x - 4)}}{2}$$

as shown in Figure 24.

Figure 24

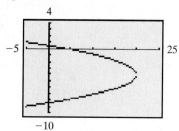

In Problems 73–84, graph each parabola using a graphing calculator.

73. $x^2 = 24y$ **74.** $x^2 = -8y$

75. $y^2 = -6x$ **76.** $y^2 = 10x$

77. $(x - 2)^2 = 4(y - 4)$ **78.** $(x + 4)^2 = -4(y - 1)$

79. $(y + 3)^2 = -8(x + 2)$ **80.** $(y - 2)^2 = 12(x + 5)$

81. $x^2 + 4x + 12y + 16 = 0$ **82.** $x^2 + 2x - 8y + 25 = 0$

83. $y^2 - 8y - 4x + 20 = 0$ **84.** $y^2 - 8y + 16x - 16 = 0$

9.4 Ellipses

Objectives

1. Graph an Ellipse Whose Center Is the Origin
2. Find the Equation of an Ellipse Whose Center Is the Origin
3. Graph an Ellipse Whose Center Is Not the Origin
4. Solve Applied Problems Involving Ellipses

Are you Ready for This Section?

Before getting started, take the following readiness quiz. If you get a problem wrong, go back to the section cited and review the material.

R1. Complete the square of $x^2 + 10x$.   [Section 7.1, pp. 545–546]

R2. Graph $f(x) = (x + 2)^2 - 1$ using transformations.   [Section 7.4, p. 580]

▶ ① Graph an Ellipse Whose Center Is the Origin

An ellipse is a conic section that is obtained through the intersection of a plane and a cone. See Figure 8(b) on page 695.

Ready?...Answers   **R1.** $x^2 + 10x + 25$
R2.

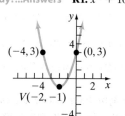

Definition

An **ellipse** is the collection of points in the plane such that the sum of the distances from two fixed points, called the **foci,** is a constant.

This definition enables us to physically draw an ellipse. To do this, find a piece of string (the length of the string is the constant in the definition). Now stick two thumbtacks on a

Figure 25

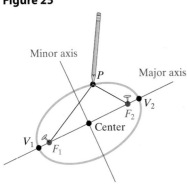

piece of cardboard so that the distance between them is less than the length of the string. The two thumbtacks represent the foci of the ellipse. Now attach the ends of the string to the thumbtacks and, using the point of a pencil, pull the string taut. Keeping the string taut, rotate the pencil around the two thumbtacks. The pencil traces out an ellipse, as shown in Figure 25.

In Figure 25, the foci are labeled F_1 and F_2. The line containing the foci is called the **major axis.** The midpoint of the line segment joining the foci is called the **center** of the ellipse. The line through the center and perpendicular to the major axis is called the **minor axis.**

The two points of intersection of the ellipse and the major axis are the **vertices,** V_1 and V_2, of the ellipse. The distance from one vertex to the other is called the **length of the major axis.**

In this text, we consider only ellipses whose major axis is parallel to (or coincides with) either the x-axis or the y-axis. The equations of an ellipse with center at the origin are given in Table 3.

Table 3	Ellipses with Center at the Origin			
Center	**Major Axis**	**Foci**	**Vertices**	**Equation**
$(0, 0)$	x-axis	$(-c, 0)$ and $(c, 0)$	$(-a, 0)$ and $(a, 0)$	$\dfrac{x^2}{a^2} + \dfrac{y^2}{b^2} = 1$
$(0, 0)$	y-axis	$(0, -c)$ and $(0, c)$	$(0, -a)$ and $(0, a)$	$\dfrac{x^2}{b^2} + \dfrac{y^2}{a^2} = 1$

The graphs of the two ellipses are given in Figure 26.

Figure 26

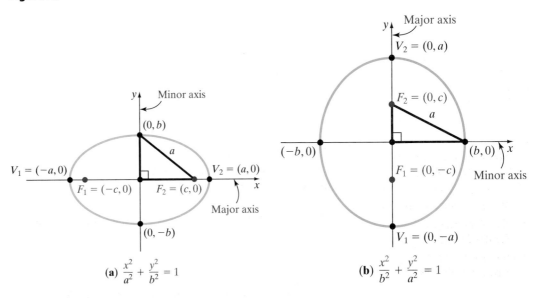

(a) $\dfrac{x^2}{a^2} + \dfrac{y^2}{b^2} = 1$

(b) $\dfrac{x^2}{b^2} + \dfrac{y^2}{a^2} = 1$

Did you notice that in the ellipse $\dfrac{x^2}{a^2} + \dfrac{y^2}{b^2} = 1$ in Figure 26(a) and in the ellipse $\dfrac{x^2}{b^2} + \dfrac{y^2}{a^2} = 1$ in Figure 26(b), the focus, the origin, and the endpoint of the minor axis form a right triangle? In both ellipses, $a > b$ and, from the Pythagorean Theorem, $b^2 + c^2 = a^2$, or $b^2 = a^2 - c^2$.

We use the fact that $a > b$ to determine whether the major axis is the x-axis or y-axis. If the larger denominator is associated with the x^2-term, then the major axis is

Work Smart

The term with the larger denominator tells us which axis is the major axis.

the x-axis; if the larger denominator is associated with the y^2-term, then the major axis is the y-axis.

> **Quick ✓**
>
> 1. An _____ is the collection of points in the plane such that the sum of the distances from two fixed points, called the ___, is a constant.
>
> 2. For an ellipse, the line containing the foci is called the _____ ___.
>
> 3. The two points of intersection of the ellipse and the major axis are the _____, V_1 and V_2, of the ellipse.
>
> 4. *True or False* If an ellipse has vertex $(a, 0)$ and focus $(c, 0)$, then $c^2 = a^2 + b^2$.

Graphing an ellipse whose center is at the origin is fairly straightforward because all we have to do is find the intercepts by letting $y = 0$ (x-intercepts) and letting $x = 0$ (y-intercepts). When you are asked to graph an ellipse, be sure to label the foci.

EXAMPLE 1   **Graphing an Ellipse**

Graph the ellipse: $\dfrac{x^2}{25} + \dfrac{y^2}{9} = 1$

Solution

First, notice that the x^2-term has the larger denominator 25. This means that the major axis is the x-axis and the equation of the ellipse has the form $\dfrac{x^2}{a^2} + \dfrac{y^2}{b^2} = 1$,

so $a^2 = 25$ and $b^2 = 9$. The center of the ellipse is the origin, $(0, 0)$. Because $b^2 = a^2 - c^2$, we know that $c^2 = a^2 - b^2 = 25 - 9 = 16$, so $c = 4$. Since the major axis is the x-axis, the foci are $(-4, 0)$ and $(4, 0)$. We now find the intercepts:

x-intercepts: Let $y = 0$: $\dfrac{x^2}{25} + \dfrac{0^2}{9} = 1$ **y-intercepts:**   Let $x = 0$: $\dfrac{0^2}{25} + \dfrac{y^2}{9} = 1$

$$\dfrac{x^2}{25} = 1$$ $$\dfrac{y^2}{9} = 1$$

$$x^2 = 25$$ $$y^2 = 9$$

$$x = \pm 5$$ $$y = \pm 3$$

The intercepts are $(-5, 0)$, $(5, 0)$, $(0, -3)$, and $(0, 3)$. Figure 27 shows the graph of the ellipse.

Figure 27

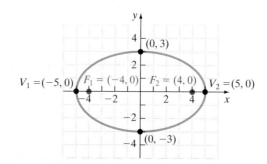

EXAMPLE 2   **Graphing an Ellipse**

Graph the ellipse: $\dfrac{x^2}{4} + \dfrac{y^2}{16} = 1$

Solution

First, notice that the y^2-term has the larger denominator, 16. This means that the major axis is the y-axis and the equation has the form $\dfrac{x^2}{b^2} + \dfrac{y^2}{a^2} = 1$, so $a^2 = 16$ and $b^2 = 4$. The center of the ellipse is the origin, $(0, 0)$. Because $b^2 = a^2 - c^2$, we know that $c^2 = a^2 - b^2 = 16 - 4 = 12$, so $c = \sqrt{12} = 2\sqrt{3}$. Since the major axis is the y-axis, the foci are $\left(0, -2\sqrt{3}\right)$ and $\left(0, 2\sqrt{3}\right)$. Now find the intercepts:

Figure 28

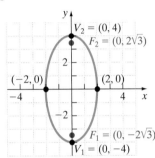

$V_2 = (0, 4)$
$F_2 = (0, 2\sqrt{3})$
$(-2, 0)$
$(2, 0)$
$F_1 = (0, -2\sqrt{3})$
$V_1 = (0, -4)$

x-intercepts: Let $y = 0$: $\dfrac{x^2}{4} + \dfrac{0^2}{16} = 1$

$\dfrac{x^2}{4} = 1$

$x^2 = 4$

$x = \pm 2$

y-intercepts: Let $x = 0$: $\dfrac{0^2}{4} + \dfrac{y^2}{16} = 1$

$\dfrac{y^2}{16} = 1$

$y^2 = 16$

$y = \pm 4$

The intercepts are $(-2, 0)$, $(2, 0)$, $(0, -4)$, and $(0, 4)$. Figure 28 shows the graph of the ellipse. ●

Quick ✓

In Problems 5 and 6, graph each ellipse.

5. $\dfrac{x^2}{9} + \dfrac{y^2}{4} = 1$ **6.** $\dfrac{x^2}{16} + \dfrac{y^2}{36} = 1$

▶ ❷ **Find the Equation of an Ellipse Whose Center Is the Origin**

Just as we did with parabolas, we are now going to use information about an ellipse to find its equation.

EXAMPLE 3   **Finding the Equation of an Ellipse**

Find the equation of the ellipse whose center is the origin with a focus at $(-2, 0)$ and vertex at $(-5, 0)$. Graph the ellipse.

Solution

The given focus and vertex lie on the x-axis, so, the major axis is the x-axis. The equation of the ellipse must be of the form $\dfrac{x^2}{a^2} + \dfrac{y^2}{b^2} = 1$. The distance from the center of the ellipse to the vertex is $a = 5$ units.

The distance from the center of the ellipse to the focus is $c = 2$ units. We know that $b^2 = a^2 - c^2 = 5^2 - 2^2 = 25 - 4 = 21$. Thus the equation of the ellipse is

$$\dfrac{x^2}{25} + \dfrac{y^2}{21} = 1$$

Figure 29 shows the graph. We use $\sqrt{21} \approx 4.6$ to graph the y-intercepts.

Figure 29

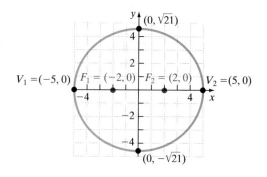

Quick ✓

7. Find the equation of the ellipse whose center is the origin with a focus at $(0, 3)$ and vertex at $(0, 7)$. Graph the ellipse.

▶ ❸ Graph an Ellipse Whose Center Is Not the Origin

If an ellipse with center at the origin and major axis coinciding with a coordinate axis is shifted horizontally h units and vertically k units, the result is an ellipse with center at (h, k) and major axis parallel to a coordinate axis. The equations of these ellipses have the same forms as those given for ellipses whose center is the origin, except that x is replaced by $x - h$ (the horizontal shift) and y is replaced by $y - k$ (the vertical shift). Table 4 gives the forms of the equations for these ellipses. Figure 30 shows their graphs.

Work Smart

Do not attempt to memorize Table 4. Instead, understand the roles of a and c in an ellipse—a is the distance from the center to each vertex, and c is the distance from the center to each focus.

Table 4 Ellipses with Center at (h, k) and Major Axis Parallel to a Coordinate Axis

Center	Major Axis	Foci	Vertices	Equation
(h, k)	Parallel to x-axis	$(h + c, k)$	$(h + a, k)$	$\dfrac{(x - h)^2}{a^2} + \dfrac{(y - k)^2}{b^2} = 1$
		$(h - c, k)$	$(h - a, k)$	$a > b$ and $b^2 = a^2 - c^2$
(h, k)	Parallel to y-axis	$(h, k + c)$	$(h, k + a)$	$\dfrac{(x - h)^2}{b^2} + \dfrac{(y - k)^2}{a^2} = 1$
		$(h, k - c)$	$(h, k - a)$	$a > b$ and $b^2 = a^2 - c^2$

Figure 30

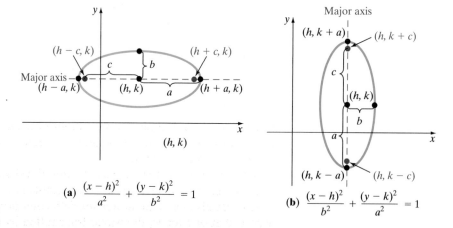

(a) $\dfrac{(x - h)^2}{a^2} + \dfrac{(y - k)^2}{b^2} = 1$

(b) $\dfrac{(x - h)^2}{b^2} + \dfrac{(y - k)^2}{a^2} = 1$

EXAMPLE 4   Graphing an Ellipse

Graph the equation: $x^2 + 16y^2 - 4x + 32y + 4 = 0$

Solution

We need to write the equation in one of the forms given in Table 4. To do this, complete the squares in x and in y.

$$x^2 + 16y^2 - 4x + 32y + 4 = 0$$

Group like variables; place the constant on the right side:

$$(x^2 - 4x) + (16y^2 + 32y) = -4$$

Factor out 16 from $16y^2 + 32y$:

$$(x^2 - 4x) + 16(y^2 + 2y) = -4$$

Complete each square:

$$(x^2 - 4x + 4) + 16(y^2 + 2y + 1) = -4 + 4 + 16$$

Factor:

$$(x - 2)^2 + 16(y + 1)^2 = 16$$

Divide both sides by 16:

$$\frac{(x - 2)^2}{16} + \frac{(y + 1)^2}{1} = 1$$

Work Smart

When we completed the square in y, we added 1, but because it was inside the parentheses with a factor of 16 in front, we must add $16 \cdot 1 = 16$ to the other side.

This is the equation of an ellipse with center at $(2, -1)$. Because the larger value, 16 is the denominator of the x^2-term, the major axis is parallel to the x-axis. Because $a^2 = 16$, and $b^2 = 1$, we have $c^2 = a^2 - b^2 = 16 - 1 = 15$. The vertices are $a = 4$ units to the left and right of center at $V_1 = (-2, -1)$ and $V_2 = (6, -1)$. The foci are $c = \sqrt{15}$ units to the left and right of center at $F_1 = (2 - \sqrt{15}, -1)$ and $F_2 = (2 + \sqrt{15}, -1)$. We then plot points $b = 1$ unit above and below the center at $(2, 0)$ and $(2, -2)$. We use $2 \pm \sqrt{15} \approx -1.9$ and -5.9 when plotting the foci of the ellipse. See Figure 31.

Figure 31

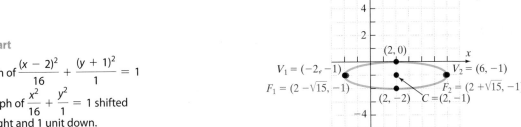

Work Smart

The graph of $\dfrac{(x - 2)^2}{16} + \dfrac{(y + 1)^2}{1} = 1$ is the graph of $\dfrac{x^2}{16} + \dfrac{y^2}{1} = 1$ shifted 2 units right and 1 unit down.

Quick ✓

8. The center of $\dfrac{(x - 3)^2}{25} + \dfrac{(y + 1)^2}{16} = 1$ is _____.

9. Graph the equation: $9x^2 + y^2 + 54x - 2y + 73 = 0$

❹ **Solve Applied Problems Involving Ellipses**

Ellipses are found in many applications in science and engineering. For example, the orbits of the planets around the Sun are elliptical, with the Sun's position at a focus. See Figure 32 on the next page.

Stone and concrete bridges are often the shape of semielliptical arches. Elliptical gears are used in machinery when a variable rate of motion is required.

Ellipses also have an interesting reflection property. If a source of sound (or light) is placed at one focus, the waves transmitted by the source reflect off the ellipse and

Figure 32

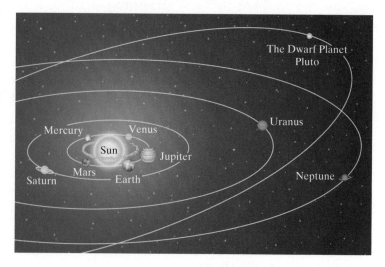

concentrate at the other focus. This is the principle behind whispering galleries, which are rooms with elliptical ceilings. A person standing at one focus of the ellipse can whisper and be heard by a person standing at the other focus, because all the sound waves that reach the ceiling are reflected to the other person. National Statuary Hall in the United States Capitol is a whispering gallery.

EXAMPLE 5 **A Whispering Gallery**

The whispering gallery in the Museum of Science and Industry in Chicago is 47.3 feet long. The distance from the center of the room to the foci is 20.3 feet. Find an equation that describes the shape of the room. How high is the room at its center?

Solution

We set up a Cartesian plane so that the center of the ellipse is at the origin and the major axis lies on the x-axis. The equation of the ellipse is

$$\frac{x^2}{a^2} + \frac{y^2}{b^2} = 1$$

Figure 33

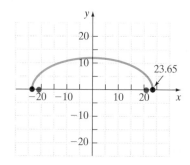

Since the length of the room is 47.3 feet, the distance from the center of the room to each vertex (the end of the room) will be $47.3/2 = 23.65$ feet, so $a = 23.65$ feet. The distance from the center of the room to each focus is $c = 20.3$ feet. See Figure 33. Since $b^2 = a^2 - c^2$, we have $b^2 = 23.65^2 - 20.3^2 = 147.2325$. An equation that describes the shape of the room is given by

$$\frac{x^2}{23.65^2} + \frac{y^2}{147.2325} = 1$$

The height of the room at its center is $b = \sqrt{147.2325} \approx 12.1$ feet.

Quick ✓

10. A hall 100 feet in length is to be designed as a whispering gallery. If the foci are to be located 30 feet from the center, determine an equation that describes the room. What is the height of the room at its center?

9.4 Exercises   MyMathLab® Math XL PRACTICE   Exercise numbers in green are complete video solutions in MyMathLab

Problems 1–10 are the Quick ✓s that follow the EXAMPLES.

Building Skills

In Problems 11–14, the graph of an ellipse is given. Match each graph to its equation. See Objective 1.

(a) $x^2 + \dfrac{y^2}{9} = 1$ **(b)** $\dfrac{x^2}{9} + y^2 = 1$

(c) $\dfrac{x^2}{16} + \dfrac{y^2}{9} = 1$ **(d)** $\dfrac{x^2}{9} + \dfrac{y^2}{16} = 1$

11.
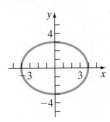

12.
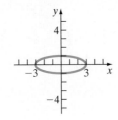

13.
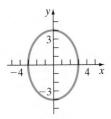

14.
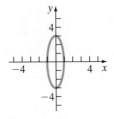

In Problems 15–24, find the vertices and foci of each ellipse. Graph each ellipse. See Objective 1.

15. $\dfrac{x^2}{25} + \dfrac{y^2}{16} = 1$ **16.** $\dfrac{x^2}{25} + \dfrac{y^2}{4} = 1$

17. $\dfrac{x^2}{36} + \dfrac{y^2}{100} = 1$ **18.** $\dfrac{x^2}{16} + \dfrac{y^2}{36} = 1$

19. $\dfrac{x^2}{49} + \dfrac{y^2}{4} = 1$ **20.** $\dfrac{x^2}{121} + \dfrac{y^2}{100} = 1$

21. $x^2 + \dfrac{y^2}{49} = 1$ **22.** $\dfrac{x^2}{64} + y^2 = 1$

23. $4x^2 + y^2 = 16$ **24.** $9x^2 + y^2 = 81$

In Problems 25–32, find an equation for each ellipse. Graph each ellipse. See Objective 2.

25. Center at $(0, 0)$; focus at $(4, 0)$; vertex at $(6, 0)$

26. Center at $(0, 0)$; focus at $(2, 0)$; vertex at $(5, 0)$

27. Center at $(0, 0)$; focus at $(0, -4)$; vertex at $(0, 7)$

28. Center at $(0, 0)$; focus at $(0, -1)$; vertex at $(0, 5)$

29. Foci at $(\pm 6, 0)$; vertices at $(\pm 10, 0)$

30. Foci at $(0, \pm 2)$; vertices at $(0, \pm 7)$

31. Foci at $(0, \pm 5)$; length of the major axis is 16

32. Foci at $(\pm 6, 0)$; length of the major axis is 20

In Problems 33–42, graph each ellipse. See Objective 3.

33. $\dfrac{(x - 3)^2}{9} + \dfrac{(y + 2)^2}{25} = 1$

34. $\dfrac{(x - 1)^2}{36} + \dfrac{(y + 4)^2}{100} = 1$

35. $\dfrac{(x + 2)^2}{16} + \dfrac{(y - 5)^2}{4} = 1$

36. $\dfrac{(x + 5)^2}{64} + \dfrac{(y + 1)^2}{16} = 1$

37. $(x - 5)^2 + \dfrac{(y + 1)^2}{49} = 1$

38. $\dfrac{(x + 8)^2}{81} + (y - 3)^2 = 1$

39. $4(x + 2)^2 + 16(y - 1)^2 = 64$

40. $9(x - 3)^2 + (y - 4)^2 = 81$

41. $4x^2 + y^2 - 24x + 2y - 63 = 0$

42. $16x^2 + 9y^2 - 128x + 54y - 239 = 0$

Applying the Concepts

43. Semielliptical Arch Bridge An arch in the shape of the upper half of an ellipse is used to support a bridge that is to span a river 30 meters wide. The center of the arch is 10 meters above the center of the river. See the figure.

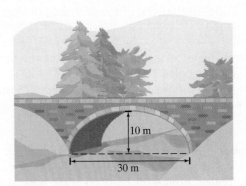

(a) Write the equation for the ellipse in which the x-axis coincides with the water and the y-axis passes through the center of the arch.

(b) Can a rectangular barge that is 18 meters wide and sits 7 meters above the surface of the water fit through the opening of the bridge?

(c) If heavy rains cause the river's level to increase 1.1 meters, will the barge make it through the opening?

44. London Bridge An arch in the shape of the upper half of an ellipse is used to support London Bridge. The main span is 45.6 meters wide. Suppose that the center of the arch is 15 meters above the center of the river.

(a) Write the equation for the ellipse in which the x-axis coincides with the water and the y-axis passes through the center of the arch.

(b) Can a rectangular barge that is 20 meters wide and sits 12 meters above the surface of the water fit through the opening of the bridge?

(c) If heavy rains cause the river's level to increase 1.5 meters, will the barge make it through the opening?

*In Problems 45–48, use the fact that the orbit of a planet about the Sun is an ellipse, with the Sun at one focus. The **aphelion** of a planet is its greatest distance from the Sun, and the **perihelion** is its shortest distance. The **mean distance** of a planet from the Sun is the distance from the center to a vertex of the elliptical orbit. See the illustration.*

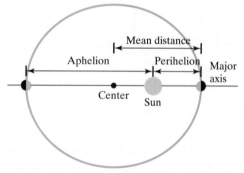

45. Earth The mean distance of Earth from the Sun is 93 million miles. If the aphelion of Earth is 94.5 million miles, what is the perihelion? Write an equation for the orbit of Earth around the Sun.

46. Mars The mean distance of Mars from the Sun is 142 million miles. If the perihelion of Mars is 128.5 million miles, what is the aphelion? Write an equation for the orbit of Mars about the Sun.

47. Jupiter The aphelion of Jupiter is 507 million miles. If the distance from the Sun to the center of its elliptical orbit is 23.2 million miles, what is the perihelion? What is the mean distance? Write an equation for the orbit of Jupiter around the Sun.

48. Dwarf Planet Pluto The perihelion of Pluto is 4551 million miles, and the distance of the Sun from the center of its elliptical orbit is 897.5 million miles. Find the aphelion of Pluto. What is the mean distance of Pluto from the Sun? Write an equation for the orbit of Pluto about the Sun.

Extending the Concepts

In Problems 49–52, write an equation for each ellipse.

49.

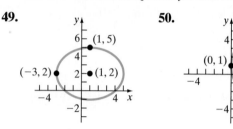

50.

51.

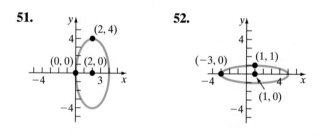

52.

53. Show that a circle is a special kind of ellipse by letting $a = b$ in the equation of an ellipse centered at the origin. What is the value of c in a circle? What does this mean regarding the location of the foci?

54. The **eccentricity** e of an ellipse is defined as the number $\frac{c}{a}$, where a is the distance from the center of an ellipse to a vertex and c is the distance from the center of an ellipse to a focus. Because $a > c$, it follows that $e < 1$ for an ellipse. Write a paragraph about the general shape of each of the following ellipses. Be sure to justify your conclusion.

(a) Eccentricity close to 0

(b) Eccentricity = 0.5

(c) Eccentricity close to 1

Explaining the Concepts

55. In the ellipse drawn below, the center is at the origin and the major axis is the x-axis. The point F is a focus. In the right triangle (drawn in red), label the lengths of the legs and the hypotenuse using a, b, and c.

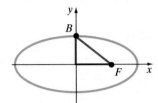

56. How does the ellipse given by the equation $\dfrac{x^2}{25} + \dfrac{y^2}{9} = 1$ differ from the ellipse given by the equation $\dfrac{x^2}{9} + \dfrac{y^2}{25} = 1$? How are they the same?

Synthesis Review

In Problems 57–60, fill in the following table for each function.

x	5	10	100	1000
$f(x)$				

57. $f(x) = \dfrac{5}{x + 2}$

58. $f(x) = \dfrac{x - 1}{x^2 + 4}$

59. $f(x) = \dfrac{2x + 1}{x - 3}$

60. $f(x) = \dfrac{3x^2 - x + 1}{x^2 + 1}$

In Problems 61 and 62, fill in the following table for each function.

x	5	10	100	1000
$f(x)$				
$g(x)$				

61. $f(x) = \dfrac{x^2 + 3x + 1}{x + 1}; g(x) = x + 2$

62. $f(x) = \dfrac{x^2 - 3x + 5}{x + 2}; g(x) = x - 5$

63. For the functions in Problems 57–60, compare the degree of the polynomial in the numerator to the degree of the polynomial in the denominator. Conjecture what happens to a rational function as x increases when the degree of the numerator is less than the degree of the denominator. Conjecture what happens to a rational function as x increases when the degree of the numerator equals the degree of the denominator.

64. For the functions in Problems 61 and 62, write f in the form quotient $+ \dfrac{\text{remainder}}{\text{dividend}}$. That is, perform the division indicated by the rational function. Now compare the values of f to those of g in the table. What does the function g represent?

The Graphing Calculator

A graphing calculator can be used to graph an ellipse by solving the equation for y. Because ellipses are not functions, we need to graph the upper half and the lower half of an ellipse in two pieces. For example, to graph $\dfrac{x^2}{4} + \dfrac{y^2}{9} = 1$, we would graph $Y_1 = 3\sqrt{1 - \dfrac{x^2}{4}}$ and $Y_2 = -3\sqrt{1 - \dfrac{x^2}{4}}$. We obtain the graph shown in Figure 34.

Figure 34

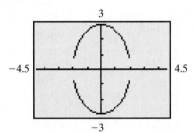

In Problems 65–72, graph each ellipse using a graphing calculator.

65. $\dfrac{x^2}{25} + \dfrac{y^2}{16} = 1$

66. $\dfrac{x^2}{25} + \dfrac{y^2}{4} = 1$

67. $4x^2 + y^2 = 16$

68. $9x^2 + y^2 = 81$

69. $\dfrac{(x - 3)^2}{9} + \dfrac{(y + 2)^2}{25} = 1$

70. $\dfrac{(x - 1)^2}{36} + \dfrac{(y + 4)^2}{100} = 1$

71. $\dfrac{(x + 2)^2}{16} + \dfrac{(y - 5)^2}{4} = 1$

72. $\dfrac{(x + 5)^2}{64} + \dfrac{(y + 1)^2}{16} = 1$

9.5 Hyperbolas

Objectives

1. Graph a Hyperbola Whose Center Is the Origin
2. Find the Equation of a Hyperbola Whose Center Is the Origin
3. Find the Asymptotes of a Hyperbola Whose Center Is the Origin

In Words

The distance from F_1 to P minus the distance from F_2 to P is a constant value for any point P on a hyperbola.

Figure 35

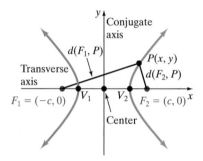

Work Smart

Notice for hyperbolas that $c^2 = a^2 + b^2$, but for ellipses $c^2 = a^2 - b^2$.

Are You Ready for This Section?

Before getting started, take the following readiness quiz. If you get a problem wrong, go back to the section cited and review the material.

R1. Complete the square: $x^2 - 5x$ [Section 7.1, pp. 545–546]

R2. Solve: $y^2 = 64$ [Section 7.1, pp. 542–545]

▶ 1 Graph a Hyperbola Whose Center Is the Origin

Recall from Section 9.2 that a hyperbola is a conic section that is obtained through the intersection of a plane and two cones. See Figure 8(d) on page 695.

> A **hyperbola** is the collection of all points in the plane the difference of whose distances from two fixed points, called the **foci**, is a positive constant.

Figure 35 illustrates a hyperbola with foci F_1 and F_2. The line containing the foci is the **transverse axis.** The midpoint of the line segment joining the foci is the **center** of the hyperbola. The line through the center and perpendicular to the transverse axis is the **conjugate axis.** The hyperbola consists of two separate curves called **branches.** The two points of intersection of the hyperbola and the transverse axis are the **vertices,** V_1 and V_2, of the hyperbola.

We can find the equation of a hyperbola using the distance formula. However, we won't present this information here. Instead, we will give you the equations of the hyperbolas whose branches open left and right and the equations of hyperbolas whose branches open up and down, with both hyperbolas centered at the origin. See Table 5.

Table 5   Hyperbolas with Center at the Origin

Center	Transverse Axis	Branches Open	Foci	Vertices	Equation
$(0, 0)$	x-axis	Left and right	$(-c, 0)$ and $(c, 0)$	$(-a, 0)$ and $(a, 0)$	$\dfrac{x^2}{a^2} - \dfrac{y^2}{b^2} = 1$, where $b^2 = c^2 - a^2$ or $c^2 = a^2 + b^2$
$(0, 0)$	y-axis	Up and down	$(0, -c)$ and $(0, c)$	$(0, -a)$ and $(0, a)$	$\dfrac{y^2}{a^2} - \dfrac{x^2}{b^2} = 1$, where $b^2 = c^2 - a^2$ or $c^2 = a^2 + b^2$

The graphs of the two hyperbolas are given in Figure 36.

Figure 36

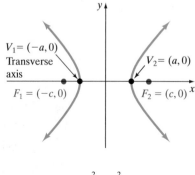

(a) $\dfrac{x^2}{a^2} - \dfrac{y^2}{b^2} = 1$

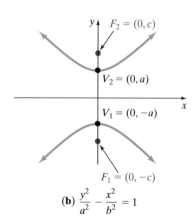

(b) $\dfrac{y^2}{a^2} - \dfrac{x^2}{b^2} = 1$

Ready?...Answers   **R1.** $x^2 - 5x + \dfrac{25}{4}$

R2. $\{-8, 8\}$

Notice the difference between the two equations given in Table 5. When the x^2-term is first, as in Figure 36(a), the transverse axis is the x-axis and the hyperbola opens left and right. When the y^2-term is first, as in Figure 36(b), the transverse axis is the y-axis and the hyperbola opens up and down. In both cases, the value of a^2 is in the denominator of the first term.

Quick ✓

1. A(n) _____ is the collection of points in the plane the difference of whose distances from two fixed points is a positive constant.

2. For a hyperbola, the foci lie on a line called the _____ ___.

3. The line through the center of a hyperbola that is perpendicular to the transverse axis is called the _____ ___.

EXAMPLE 1 **Graphing a Hyperbola**

Graph the equation: $\dfrac{x^2}{16} - \dfrac{y^2}{4} = 1$

Solution

The equation is of the form $\dfrac{x^2}{a^2} - \dfrac{y^2}{b^2} = 1$. The x^2-term is first, so the hyperbola opens left and right, and the transverse axis is the x-axis. The center of the hyperbola is the origin, $(0, 0)$. Because $a^2 = 16$ and $b^2 = 4$, then $c^2 = a^2 + b^2 = 16 + 4 = 20$, so $c = 2\sqrt{5}$. The vertices are at $(\pm a, 0) = (\pm 4, 0)$, and the foci are at $(\pm c, 0) = (\pm 2\sqrt{5}, 0)$.

To obtain the graph, we plot the vertices and foci. Then we locate points above and below the foci (as we did when graphing parabolas). Thus we let $x = \pm 2\sqrt{5}$ in the equation $\dfrac{x^2}{16} - \dfrac{y^2}{4} = 1$.

$$\frac{\left(\pm 2\sqrt{5}\right)^2}{16} - \frac{y^2}{4} = 1$$

$$\frac{20}{16} - \frac{y^2}{4} = 1$$

Reduce the fraction: $\quad \dfrac{5}{4} - \dfrac{y^2}{4} = 1$

Subtract $\dfrac{5}{4}$ from both sides: $\quad -\dfrac{y^2}{4} = -\dfrac{1}{4}$

Multiply both sides by -4: $\quad y^2 = 1$

Take the square root of both sides: $\quad y = \pm 1$

The points above and below the foci are $\left(\pm 2\sqrt{5}, -1\right)$ and $\left(\pm 2\sqrt{5}, 1\right)$. We use $\pm 2\sqrt{5} \approx \pm 4.5$ to graph the foci. See Figure 37 for the graph of the hyperbola.

Figure 37

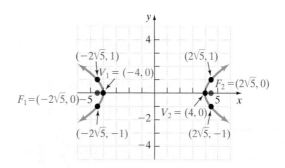

EXAMPLE 2 **Graph a Hyperbola**

Graph the equation: $4y^2 - 16x^2 = 16$

Solution

We need the equation to be in one of the forms given in Table 5, so we divide both sides of the equation by 16:

$$\frac{y^2}{4} - \frac{x^2}{1} = 1$$

Do you see that the y^2-term is first? This means that the hyperbola opens up and down and that the transverse axis is the y-axis. The center of this hyperbola is the origin, $(0, 0)$. Comparing the equation to $\frac{y^2}{a^2} - \frac{x^2}{b^2} = 1$, we find that $a^2 = 4$ and $b^2 = 1$; therefore, $c^2 = a^2 + b^2 = 4 + 1 = 5$. The vertices are at $(0, \pm a) = (0, \pm 2)$. The foci are at $(0, \pm c) = (0, \pm\sqrt{5})$. We locate four more points on the hyperbola to the left and right of each focus by letting $y = \pm\sqrt{5}$ in the equation $\frac{y^2}{4} - \frac{x^2}{1} = 1$ and solving for x.

$$\frac{(\pm\sqrt{5})^2}{4} - \frac{x^2}{1} = 1$$

$$\frac{5}{4} - \frac{x^2}{1} = 1$$

Subtract $\frac{5}{4}$ from both sides: $-x^2 = -\frac{1}{4}$

Multiply both sides by -1: $x^2 = \frac{1}{4}$

Take the square root of both sides: $x = \pm\frac{1}{2}$

Figure 38

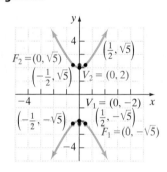

Four additional points on the graph are $\left(-\frac{1}{2}, \pm\sqrt{5}\right)$ and $\left(\frac{1}{2}, \pm\sqrt{5}\right)$. See Figure 38 for the graph of the hyperbola.

Quick ✓

In Problems 4 and 5, graph each hyperbola.

4. $\dfrac{x^2}{36} - \dfrac{y^2}{64} = 1$ **5.** $\dfrac{y^2}{9} - \dfrac{x^2}{16} = 1$

▶ ❷ Find the Equation of a Hyperbola Whose Center Is the Origin

Just as we did with parabolas and ellipses, we are now going to use information about a hyperbola to find its equation.

EXAMPLE 3 **Finding and Graphing the Equation of a Hyperbola**

Find an equation of the hyperbola with one focus at $(0, -3)$ and vertices at $(0, -2)$ and $(0, 2)$. Graph the equation.

Solution

The center of a hyperbola is located at the midpoint of the vertices (or foci). Because the vertices are at $(0, -2)$ and $(0, 2)$, the center must be at the origin, $(0, 0)$. Notice that the given focus and vertices must lie on the y-axis. Therefore, the transverse axis is the y-axis, and the hyperbola opens up and down. A vertex is at $(0, 2)$, so $a = 2$. A focus is at $(0, -3)$, so $c = 3$. With $a = 2$ and $c = 3$, $b^2 = c^2 - a^2 = 3^2 - 2^2 = 9 - 4 = 5$.

The equation must be of the form $\dfrac{y^2}{a^2} - \dfrac{x^2}{b^2} = 1$, so letting $a^2 = 4$ and $b^2 = 5$ we have

$$\frac{y^2}{4} - \frac{x^2}{5} = 1$$

To find points to the left and right of each focus, let $y = \pm 3$ in the equation $\dfrac{y^2}{4} - \dfrac{x^2}{5} = 1$.

$$\frac{(\pm 3)^2}{4} - \frac{x^2}{5} = 1$$

$$\frac{9}{4} - \frac{x^2}{5} = 1$$

Subtract $\dfrac{9}{4}$ from both sides: $-\dfrac{x^2}{5} = -\dfrac{5}{4}$

Multiply both sides by -5: $x^2 = \dfrac{25}{4}$

Take the square root of both sides: $x = \pm \dfrac{5}{2}$

Figure 39

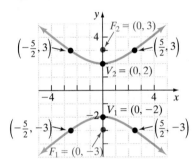

The points to the left and right of the foci are $\left(-\dfrac{5}{2}, \pm 3\right)$ and $\left(\dfrac{5}{2}, \pm 3\right)$. See Figure 39 for the graph of the hyperbola.

Look at the equations of the hyperbolas in Examples 1 and 3. For the hyperbola in Example 1, $a^2 = 16$ and $b^2 = 4$, so $a > b$; for the hyperbola in Example 3, $a^2 = 4$ and $b^2 = 5$, so $a < b$. We conclude that for hyperbolas, there are no requirements involving the relative sizes for a and b. For ellipses, however, the relative sizes of a and b determine the major axis.

Quick ✓

6. *True or False* In any hyperbola, it must be the case that $a > b$.

7. Find the equation of a hyperbola whose vertices are $(-4, 0)$ and $(4, 0)$ and has a focus at $(6, 0)$. Graph the hyperbola.

▶ ❸ **Find the Asymptotes of a Hyperbola Whose Center Is the Origin**

As x and y get larger in both the positive and the negative directions, the branches of the hyperbola approach two lines called **asymptotes** of the hyperbola. The asymptotes help us graph hyperbolas. Table 6 gives the asymptotes of the two hyperbolas discussed in this section.

Work Smart

Asymptotes provide an alternative method for determining the opening of each branch of the hyperbola, rather than finding and plotting four additional points.

Table 6 Asymptotes of a Hyperbola	
Hyperbola	Asymptotes
$\dfrac{x^2}{a^2} - \dfrac{y^2}{b^2} = 1$	$y = -\dfrac{b}{a}x$ and $y = \dfrac{b}{a}x$
$\dfrac{y^2}{a^2} - \dfrac{x^2}{b^2} = 1$	$y = -\dfrac{a}{b}x$ and $y = \dfrac{a}{b}x$

Figure 40 illustrates how the asymptotes can be used to help graph a hyperbola. It is important to remember that the asymptotes are not part of the hyperbola—they only serve as guides in graphing the hyperbola.

Figure 40

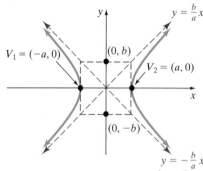

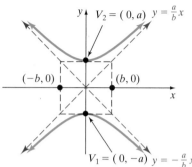

(a) Asymptotes on the graph of $\dfrac{x^2}{a^2} - \dfrac{y^2}{b^2} = 1$ (b) Asymptotes on the graph of $\dfrac{y^2}{a^2} - \dfrac{x^2}{b^2} = 1$

For example, to graph the equation $\dfrac{x^2}{a^2} - \dfrac{y^2}{b^2} = 1$, first plot the vertices $(-a, 0)$ and $(a, 0)$. Then plot the points $(0, -b)$ and $(0, b)$. Use these four points to construct a rectangle as shown in Figure 40(a).

The lines through the diagonals of this rectangle have slopes $\dfrac{b}{a}$ and $-\dfrac{b}{a}$. These lines are the asymptotes. The equations of these asymptotes are $y = -\dfrac{b}{a}x$ and $y = \dfrac{b}{a}x$. Using this technique enables us to avoid plotting the four additional points we plotted earlier.

EXAMPLE 4

Graphing a Hyperbola and Finding Its Asymptotes

Graph the equation $9x^2 - 4y^2 = 36$ using the asymptotes as a guide.

Solution

Divide both sides of the equation by 36 to put the equation in the form $\dfrac{x^2}{a^2} - \dfrac{y^2}{b^2} = 1$.

$$\frac{x^2}{4} - \frac{y^2}{9} = 1$$

Figure 41

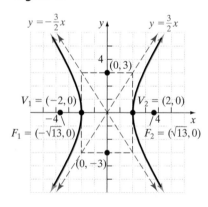

The center of the hyperbola is the origin, $(0, 0)$. Because the x^2-term is first, the hyperbola opens to the left and right. The transverse axis is along the x-axis. We can see that $a^2 = 4$ and $b^2 = 9$. Because $c^2 = a^2 + b^2$, then $c^2 = 4 + 9 = 13$, so $c = \pm\sqrt{13}$. The vertices are at $(\pm a, 0) = (\pm 2, 0)$; the foci are at $(\pm c, 0) = (\pm\sqrt{13}, 0)$. Since $b^2 = 9$, we know that $b = \pm 3$. The equations of the asymptotes are

$$y = \frac{b}{a}x = \frac{3}{2}x \quad \text{and} \quad y = -\frac{b}{a}x = -\frac{3}{2}x$$

To graph the hyperbola, form a rectangle using the points $(\pm a, 0) = (\pm 2, 0)$ and $(0, \pm b) = (0, \pm 3)$. Draw the asymptotes through the vertices of the rectangle. See Figure 41.   ●

Quick ✓

8. The asymptotes of the hyperbola $\dfrac{x^2}{a^2} - \dfrac{y^2}{b^2} = 1$ are _____ and _____.

9. Graph the equation $x^2 - 9y^2 = 9$ using the asymptotes as a guide.

10. Graph the equation $\dfrac{y^2}{16} - \dfrac{x^2}{9} = 1$ using the asymptotes as a guide.

1–10 *Problems are the Quick ✓s that follow the* **EXAMPLES**.

Building Skills

In Problems 11–14, the graph of a hyperbola is given. Match each graph to its equation. See Objective 1.

(a) $\dfrac{x^2}{4} - y^2 = 1$

(b) $x^2 - \dfrac{y^2}{4} = 1$

(c) $\dfrac{y^2}{4} - x^2 = 1$

(d) $y^2 - \dfrac{x^2}{4} = 1$

11.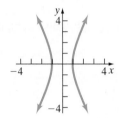

12.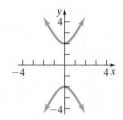

13.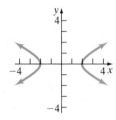

14.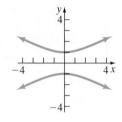

In Problems 15–22, graph each equation. See Objective 1.

15. $\dfrac{x^2}{4} - \dfrac{y^2}{16} = 1$

16. $\dfrac{x^2}{9} - \dfrac{y^2}{16} = 1$

17. $\dfrac{y^2}{25} - \dfrac{x^2}{36} = 1$

18. $\dfrac{y^2}{81} - \dfrac{x^2}{9} = 1$

19. $4x^2 - y^2 = 36$

20. $x^2 - 9y^2 = 36$

21. $25y^2 - x^2 = 100$

22. $4y^2 - 9x^2 = 36$

In Problems 23–28, find the equation for the hyperbola described. Graph the equation. See Objective 2.

23. Center at $(0, 0)$; focus at $(3, 0)$; vertex at $(2, 0)$

24. Center at $(0, 0)$; focus at $(-4, 0)$; vertex at $(-1, 0)$

25. Vertices at $(0, 5)$ and $(0, -5)$; focus at $(0, 7)$

26. Vertices at $(0, 6)$ and $(0, -6)$; focus at $(0, 8)$

27. Foci at $(-10, 0)$ and $(10, 0)$; vertex at $(-7, 0)$

28. Foci at $(-5, 0)$ and $(5, 0)$; vertex at $(-3, 0)$

In Problems 29–34, graph each equation using its asymptotes. See Objective 3.

29. $\dfrac{x^2}{25} - \dfrac{y^2}{9} = 1$

30. $\dfrac{x^2}{16} - \dfrac{y^2}{64} = 1$

31. $\dfrac{y^2}{4} - \dfrac{x^2}{100} = 1$

32. $\dfrac{y^2}{25} - \dfrac{x^2}{100} = 1$

33. $x^2 - y^2 = 4$

34. $y^2 - x^2 = 25$

Mixed Practice

In Problems 35–38, find the equation for the hyperbola described. Graph the equation.

35. Vertices at $(0, -8)$ and $(0, 8)$; asymptote the line $y = 2x$

36. Vertices at $(0, -4)$ and $(0, 4)$; asymptote the line $y = 2x$

37. Foci at $(-3, 0)$ and $(3, 0)$; asymptote the line $y = x$

38. Foci at $(-9, 0)$ and $(9, 0)$; asymptote the line $y = -3x$

Applying the Concepts

In Problems 39–42, write the equation of the hyperbola.

39.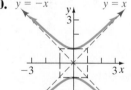

40.

41.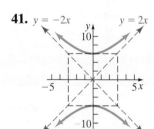

42.

Extending the Concepts

43. Two hyperbolas that have the same set of asymptotes are called **conjugate.** Show that the hyperbolas

$$\frac{x^2}{4} - y^2 = 1 \quad \text{and} \quad y^2 - \frac{x^2}{4} = 1$$

are conjugate. Graph each hyperbola in the same Cartesian plane.

44. The **eccentricity** e of a hyperbola is defined as the number $\frac{c}{a}$. Because $c > a$, it follows that $e > 1$. Describe the general shape of a hyperbola whose eccentricity is close to 1. What is the shape if e is very large?

Explaining the Concepts

45. Explain how the asymptotes of a hyperbola are helpful in obtaining its graph.

46. How can you tell the difference between the equation of a hyperbola and the equation of an ellipse just by looking at the equations?

Synthesis Review

In Problems 47–51, solve each system of equations using either substitution or elimination.

47. $\begin{cases} 2x - 3y = -9 \\ -x + 5y = 8 \end{cases}$

48. $\begin{cases} 3x + 4y = 3 \\ -6x + 2y = -\dfrac{7}{2} \end{cases}$

49. $\begin{cases} 2x - 3y = 6 \\ -6x + 9y = -18 \end{cases}$

50. $\begin{cases} -2x + y = 8 \\ x - \dfrac{1}{2}y = -4 \end{cases}$

51. $\begin{cases} 6x + 3y = 4 \\ -2x - y = -\dfrac{4}{3} \end{cases}$

52. In Problems 47–51, which method did you use more often? Why? Do you think there are situations where substitution is superior to elimination? Are there situations where elimination is superior to substitution? Describe these circumstances.

The Graphing Calculator

A graphing calculator can be used to graph hyperbolas by solving the equation for y. Because hyperbolas are not functions, we need to graph the upper half and lower half of the hyperbola in two pieces. For example, to graph $\frac{x^2}{4} - \frac{y^2}{9} = 1$, we would graph $Y_1 = 3\sqrt{\frac{x^2}{4} - 1}$ and $Y_2 = -3\sqrt{\frac{x^2}{4} - 1}$. We obtain the graph shown in Figure 42.

Figure 42

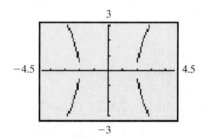

In Problems 53–60, graph each hyperbola using a graphing calculator.

53. $\dfrac{x^2}{4} - \dfrac{y^2}{16} = 1$

54. $\dfrac{x^2}{9} - \dfrac{y^2}{16} = 1$

55. $\dfrac{y^2}{25} - \dfrac{x^2}{36} = 1$

56. $\dfrac{y^2}{81} - \dfrac{x^2}{9} = 1$

57. $4x^2 - y^2 = 36$

58. $x^2 - 9y^2 = 36$

59. $25y^2 - x^2 = 100$

60. $4y^2 - 9x^2 = 36$

Putting The Concepts Together (Sections 9.1–9.5)

We designed these problems so that you can review Sections 9.1–9.5 and show your mastery of the concepts. Take time to work these problems before proceeding with the next section. The answers are located at the back of the text on page AN-56.

1. Find the exact distance $d(P_1, P_2)$ between points $P_1 = (-6, 4)$ and $P_2 = (3, -2)$.

2. Find the midpoint of the line segment formed by joining the points $P_1 = (-3, 1)$ and $P_2 = (5, -7)$.

In Problems 3 and 4, find the center (h, k) and the radius r of each circle. Graph each circle.

3. $(x + 2)^2 + (y - 8)^2 = 36$

4. $x^2 + y^2 + 6x - 4y - 3 = 0$

In Problems 5 and 6, find the standard form of the equation of each circle.

5. Center $(0, 0)$; contains the point $(-5, 12)$

6. With endpoints of a diameter at $(-1, 5)$ and $(5, -3)$.

In Problems 7 and 8, find the vertex, focus, and directrix of each parabola. Graph each parabola.

7. $(x + 2)^2 = -4(y - 4)$

8. $y^2 + 2y - 8x + 25 = 0$

In Problems 9 and 10, find an equation for each parabola described.

9. Vertex at $(-1, -2)$; focus at $(-1, -5)$

10. Vertex at $(-3, 3)$; contains the point $(-1, 7)$; axis of symmetry parallel to the x-axis

In Problems 11 and 12, find the center, vertices, and foci of each ellipse. Graph each ellipse.

11. $x^2 + 9y^2 = 81$

12. $\dfrac{(x + 1)^2}{36} + \dfrac{(y - 2)^2}{49} = 1$

In Problems 13 and 14, find an equation for each ellipse described.

13. Foci at $(0, \pm 6)$; vertices at $(0, \pm 9)$

14. Center at $(3, -4)$; vertex at $(7, -4)$; focus at $(6, -4)$

In Problems 15 and 16, find the vertices, foci, and asymptotes for each hyperbola. Graph each hyperbola using the asymptotes as a guide.

15. $\dfrac{y^2}{81} - \dfrac{x^2}{9} = 1$ **16.** $25x^2 - y^2 = 25$

17. Find an equation for a hyperbola with center at $(0, 0)$, focus at $(0, -5)$, and vertex at $(0, -2)$.

18. A large floodlight is in the shape of a parabola with diameter 36 inches and depth 12 inches. How far from the vertex should the light bulb be placed so that the rays will be reflected parallel to the axis?

9.6 Systems of Nonlinear Equations

Objectives

1 Solve a System of Nonlinear Equations Using Substitution

2 Solve a System of Nonlinear Equations Using Elimination

Are You Ready for This Section?

Before getting started, take the following readiness quiz. If you get a problem wrong, go back to the section cited and review the material.

R1. Solve the system using substitution: $\begin{cases} y = 2x - 5 \\ 2x - 3y = 7 \end{cases}$ [Section 3.1, pp. 221–223]

R2. Solve the system using elimination: $\begin{cases} 2x - 4y = -11 \\ -x + 5y = 13 \end{cases}$ [Section 3.1, pp. 223–226]

R3. Solve the system: $\begin{cases} 3x - 5y = 4 \\ -6x + 10y = -8 \end{cases}$ [Section 3.1, pp. 227–228]

Recall from Section 3.1 that a system of equations is a grouping of two or more equations, each containing one or more variables. We learned how to solve systems of linear equations in Chapter 3. We now learn how to solve systems of equations where the equations are not linear. We deal with systems containing only two equations with two unknowns. In a **system of nonlinear equations** in two variables, at least one of the equations is not linear. That is, at least one of the equations cannot be written in the form $Ax + By = C$. Some examples of nonlinear systems of equations containing two unknowns follow.

$\begin{cases} x + y^2 = 5 & \text{(1) A parabola} \\ 2x + y = 4 & \text{(2) A line} \end{cases}$ $\begin{cases} x^2 + y^2 = 9 & \text{(1) A circle} \\ -x^2 + y = 9 & \text{(2) A parabola} \end{cases}$

In Section 3.1, we solved some systems of linear equations geometrically by finding the point of intersection of the equations in the system. The same idea holds for nonlinear systems—the point(s) of intersection represent the solution(s) to the system.

As you worked through Chapter 3, you probably developed a sense of when substitution or when elimination was the best approach for solving a system. The same is true of nonlinear systems—sometimes substitution is best, sometimes elimination is best. Experience and a degree of imagination are your friends when solving these problems.

Ready?...Answers   **R1.** $(2, -1)$
R2. $(-1/2, 5/2)$
R3. $\{(x, y) \mid 3x - 5y = 4\}$

▶ ① Solve a System of Nonlinear Equations Using Substitution

The steps for solving a system of nonlinear equations using substitution are identical to the steps for solving a system of linear equations using substitution.

EXAMPLE 1 **How to Solve a System of Nonlinear Equations Using Substitution**

Solve the following system of equations using substitution: $\begin{cases} 3x - y = -2 & \text{(1)A line} \\ 2x^2 - y = 0 & \text{(2)A parabola} \end{cases}$

Step-by-Step Solution

Step 1: Solve equation (1) for y.

Equation (1): $3x - y = -2$

Add y to both sides; add 2 to both sides: $y = 3x + 2$

Step 2: Substitute $3x + 2$ for y in equation (2).

Equation (2): $2x^2 - y = 0$

$2x^2 - (3x + 2) = 0$

Distribute: $2x^2 - 3x - 2 = 0$

Step 3: Solve for x.

Factor: $(2x + 1)(x - 2) = 0$

Zero-Product Property: $2x + 1 = 0 \quad \text{or} \quad x - 2 = 0$

$x = -\dfrac{1}{2} \quad \text{or} \quad x = 2$

Step 4: Let $x = -\dfrac{1}{2}$ and $x = 2$ in equation (1) to determine y.

Equation (1): $3x - y = -2$

$x = -\dfrac{1}{2}$: $3\left(-\dfrac{1}{2}\right) - y = -2$

$-\dfrac{3}{2} - y = -2$

Add $\dfrac{3}{2}$ to both sides: $-y = -\dfrac{1}{2}$

Multiply both sides by -1: $y = \dfrac{1}{2}$

$x = 2$: $3(2) - y = -2$

$6 - y = -2$

Subtract 6 from both sides: $-y = -8$

Multiply both sides by -1: $y = 8$

The apparent solutions are $\left(-\dfrac{1}{2}, \dfrac{1}{2}\right)$ and $(2, 8)$.

Step 5: Check

Equation (1)

$3x - y = -2$

$x = -\dfrac{1}{2}; y = \dfrac{1}{2}:\quad 3\left(-\dfrac{1}{2}\right) - \dfrac{1}{2} \overset{?}{=} -2$

$-\dfrac{3}{2} - \dfrac{1}{2} \overset{?}{=} -2$

$-\dfrac{4}{2} = -2 \quad$ True

$x = 2; y = 8:\quad 3x - y = -2$

$3(2) - 8 \overset{?}{=} -2$

$6 - 8 = -2 \quad$ True

Equation (2)

$2x^2 - y = 0$

$2\left(-\dfrac{1}{2}\right)^2 - \dfrac{1}{2} \overset{?}{=} 0$

$2\left(\dfrac{1}{4}\right) - \dfrac{1}{2} \overset{?}{=} 0$

$\dfrac{1}{2} - \dfrac{1}{2} = 0 \quad$ True

$2x^2 - y = 0$

$2(2)^2 - 8 \overset{?}{=} 0$

$2(4) - 8 = 0 \quad$ True

Figure 43

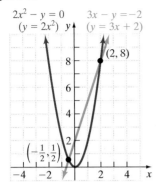

$2x^2 - y = 0$ $3x - y = -2$
$(y = 2x^2)$ y $(y = 3x + 2)$

Each solution checks. Figure 43 shows the graphs of each equation in the system with the points of intersection (the solutions of the system) labeled at $\left(-\dfrac{1}{2}, \dfrac{1}{2}\right)$ and $(2, 8)$.   ●

Quick ✓

1. Solve the following system of equations using substitution:
$$\begin{cases} 2x + y = -1 \\ x^2 - y = 4 \end{cases}$$

EXAMPLE 2 **Solving a System of Nonlinear Equations Using Substitution**

Solve the following system of equations using substitution:
$$\begin{cases} x + y = 2 & \text{(1) A line} \\ (x + 2)^2 + (y - 1)^2 = 9 & \text{(2) A circle} \end{cases}$$

Solution

Solve equation (1) for y.

$$x + y = 2$$
Subtract x from both sides: $y = -x + 2$

Substitute this expression for y in equation (2), $(x + 2)^2 + (y - 1)^2 = 9$, and then solve for x.

Substitute $-x + 2$ for y in equation (2):	$(x + 2)^2 + (-x + 2 - 1)^2 = 9$
Combine like terms:	$(x + 2)^2 + (1 - x)^2 = 9$
Use FOIL:	$x^2 + 4x + 4 + 1 - 2x + x^2 = 9$
Combine like terms:	$2x^2 + 2x + 5 = 9$
Put in standard form:	$2x^2 + 2x - 4 = 0$
Factor:	$2(x^2 + x - 2) = 0$
Divide both sides by 2:	$x^2 + x - 2 = 0$
Factor:	$(x + 2)(x - 1) = 0$
Zero-Product Property:	$x + 2 = 0$ or $x - 1 = 0$
Solve for x:	$x = -2$ or $x = 1$

Let $x = -2$ and $x = 1$ in equation (1) to determine y.

$$x + y = 2$$

$x = -2$: $-2 + y = 2$ $x = 1$: $1 + y = 2$
Add 2 to both sides: $y = 4$ Subtract 1 from both sides: $y = 1$

The apparent solutions are $(-2, 4)$ and $(1, 1)$.

Check

$x = -2; y = 4$: $x + y = 2$ $(x + 2)^2 + (y - 1)^2 = 9$
                   $-2 + 4 = 2$ True $(-2 + 2)^2 + (4 - 1)^2 \overset{?}{=} 9$
                                                    $3^2 = 9$ True

$x = 1; y = 1$: $x + y = 2$ $(x + 2)^2 + (y - 1)^2 = 9$
                  $1 + 1 = 2$ True $(1 + 2)^2 + (1 - 1)^2 \overset{?}{=} 9$
                                                    $3^2 = 9$ True

Figure 44

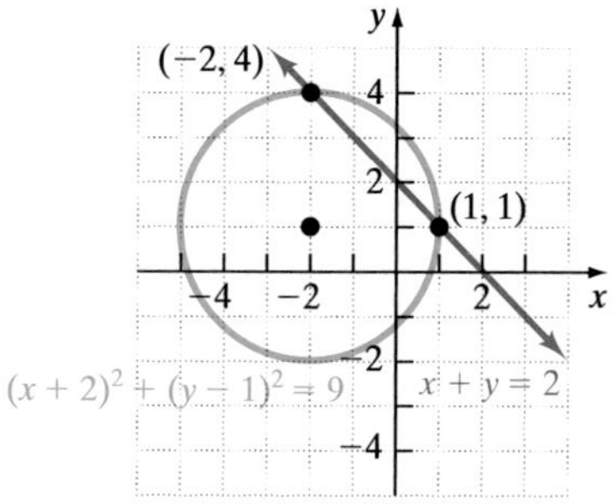

$(-2, 4)$

$(1, 1)$

$(x + 2)^2 + (y - 1)^2 = 9$ $x + y = 2$

Each solution checks. We have also drawn the graphs in Figure 44 with the points of intersection (the solutions of the system) labeled at $(-2, 4)$ and $(1, 1)$.

Quick ✓

2. Solve the following system of equations using substitution:

$$\begin{cases} 2x + y = 0 \\ (x - 4)^2 + (y + 2)^2 = 9 \end{cases}$$

▶ ❷ **Solve a System of Nonlinear Equations Using Elimination**

Now we discuss the method of elimination. We use elimination to solve a system of nonlinear equations in the same way we used it to solve systems of linear equations.

Remember that in using elimination, we want to get the coefficients of one variable to be additive inverses.

EXAMPLE 3 **How to Solve a System of Nonlinear Equations by Elimination**

Solve the following system of equations using elimination: $\begin{cases} x^2 + y^2 = 13 & \text{(1) A circle} \\ x^2 - y = 7 & \text{(2) A parabola} \end{cases}$

Step-by-Step Solution

Step 1: Multiply equation (2) by -1 so the coefficients of x^2 become additive inverses.

$$\begin{cases} x^2 + y^2 = 13 & \text{(1)} \\ -x^2 + y = -7 & \text{(2)} \end{cases}$$

Step 2: Add equations (1) and (2) to eliminate x^2. Solve the resulting equation for y.

Add:
$$\begin{cases} x^2 + y^2 = 13 & \text{(1)} \\ \underline{-x^2 + y = -7} & \text{(2)} \end{cases}$$
$$y^2 + y = 6$$

Put in standard form: $y^2 + y - 6 = 0$

Factor: $(y + 3)(y - 2) = 0$

Zero-Product Property: $y + 3 = 0 \quad \text{or} \quad y - 2 = 0$

$y = -3 \text{ or} \qquad y = 2$

Step 3: Solve for x using equation (2).

Work Smart

We use equation (2) in Step 3 because it's easier to work with.

$x^2 - y = 7 \text{ (2)}$

Using $y = -3$: $x^2 - (-3) = 7$

$x^2 + 3 = 7$

$x^2 = 4$

$x = \pm 2$

Using $y = 2$: $x^2 - 2 = 7$

$x^2 = 9$

$x = \pm 3$

The apparent solutions are $(-2, -3)$, $(2, -3)$, $(-3, 2)$, and $(3, 2)$.

Step 4: Check

We leave it to you to verify that all four of the apparent solutions are solutions to the system. The four points $(-2, -3)$, $(2, -3)$, $(-3, 2)$, and $(3, 2)$ are the points of intersection of the graphs. See Figure 45.

Figure 45

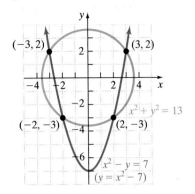

EXAMPLE 4 Solving a System of Nonlinear Equations by Elimination

Solve the following system of equations using elimination:

$$\begin{cases} x^2 - y^2 = 4 & \text{(1) A hyperbola} \\ x^2 - y = 0 & \text{(2) A parabola} \end{cases}$$

Solution

Multiply equation (2) by -1 to get the coefficients of x^2 to be additive inverses.

$$\begin{cases} x^2 - y^2 = 4 & \text{(1)} \\ -x^2 + y = 0 & \text{(2)} \end{cases}$$

Add equations (1) and (2) to eliminate x^2. Solve the resulting equation for y.

Figure 46

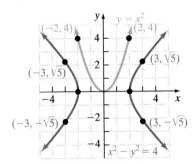

$$\begin{cases} x^2 - y^2 = 4 & \text{(1)} \\ \underline{-x^2 + y = 0} & \text{(2)} \\ -y^2 + y = 4 \end{cases}$$

Add y^2 to both sides; subtract y from both sides: $y^2 - y + 4 = 0$

This is a quadratic equation whose discriminant is $b^2 - 4ac = (-1)^2 - 4(1)(4) = 1 - 16 = -15$. The equation has no real solution. Therefore, the system of equations is inconsistent. The solution set is $\varnothing$ or $\{\ \}$. Figure 46 shows the graphs of $x^2 - y^2 = 4$ and $x^2 - y = 0$ $(y = x^2)$. Figure 46 confirms the algebraic result.

Quick ✓

Solve each system of nonlinear equations using elimination.

3. $\begin{cases} x^2 + y^2 = 16 \\ x^2 - 2y = 8 \end{cases}$

4. $\begin{cases} x^2 - y = -4 \\ x^2 + y^2 = 9 \end{cases}$

9.6 Exercises MyMathLab®   PRACTICE

Exercise numbers in green have complete video solutions in MyMathLab.

*Problems **1–4** are the* Quick ✓*s that follow the* EXAMPLES.

Building Skills

In Problems 5–12, solve the system of nonlinear equations using the method of substitution. See Objective 1.

5. $\begin{cases} y = x^2 + 4 \\ y = x + 4 \end{cases}$

6. $\begin{cases} y = x^3 + 2 \\ y = x + 2 \end{cases}$

7. $\begin{cases} y = \sqrt{25 - x^2} \\ x + y = 7 \end{cases}$

8. $\begin{cases} y = \sqrt{100 - x^2} \\ x + y = 14 \end{cases}$

9. $\begin{cases} x^2 + y^2 = 4 \\ y = x^2 - 2 \end{cases}$

10. $\begin{cases} x^2 + y^2 = 16 \\ y = x^2 - 4 \end{cases}$

11. $\begin{cases} xy = 4 \\ x^2 + y^2 = 8 \end{cases}$

12. $\begin{cases} xy = 1 \\ x^2 - y = 0 \end{cases}$

In Problems 13–20, solve the system of nonlinear equations using the method of elimination. See Objective 2.

13. $\begin{cases} x^2 + y^2 = 4 \\ y^2 - x = 4 \end{cases}$

14. $\begin{cases} x^2 + y^2 = 8 \\ x^2 + y^2 + 4y = 0 \end{cases}$

15. $\begin{cases} x^2 + y^2 = 7 \\ x^2 - y^2 = 25 \end{cases}$

16. $\begin{cases} 4x^2 + 16y^2 = 16 \\ 2x^2 - 2y^2 = 8 \end{cases}$

17. $\begin{cases} x^2 + y^2 = 6y \\ x^2 = 3y \end{cases}$

18. $\begin{cases} 2x^2 + y^2 = 18 \\ x^2 - y^2 = 9 \end{cases}$

19. $\begin{cases} x^2 - 2x - y = 8 \\ 6x + 2y = -4 \end{cases}$

20. $\begin{cases} 2x^2 - 5x + y = 12 \\ 14x - 2y = -16 \end{cases}$

Mixed Practice

In Problems 21–36, solve the system of nonlinear equations using any method you wish.

21. $\begin{cases} y = x^2 - 6x + 4 \\ 5x + y = 6 \end{cases}$

22. $\begin{cases} y = x^2 + 4x + 5 \\ x - y = 9 \end{cases}$

23. $\begin{cases} x^2 + y^2 = 16 \\ x^2 - y^2 = 16 \end{cases}$

24. $\begin{cases} x^2 + y^2 = 25 \\ x^2 - y^2 = 25 \end{cases}$

25. $\begin{cases} (x - 4)^2 + y^2 = 25 \\ x - y = -3 \end{cases}$

26. $\begin{cases} (x + 5)^2 + (y - 2)^2 = 100 \\ 8x + y = 18 \end{cases}$

27. $\begin{cases} (x - 1)^2 + (y + 2)^2 = 4 \\ y^2 + 4y - x = -1 \end{cases}$

28. $\begin{cases} (x + 2)^2 + (y - 1)^2 = 4 \\ y^2 - 2y - x = 5 \end{cases}$

29. $\begin{cases} (x + 3)^2 + 4y^2 = 4 \\ x^2 + 6x - y = 13 \end{cases}$

30. $\begin{cases} 9x^2 + 4y^2 = 36 \\ x^2 + (y - 7)^2 = 4 \end{cases}$

31. $\begin{cases} x^2 - y^2 = 21 \\ x + y = 7 \end{cases}$

32. $\begin{cases} y - 2x = 1 \\ 2x^2 + y^2 = 1 \end{cases}$

33. $\begin{cases} x^2 + 2y^2 = 16 \\ 4x^2 - y^2 = 24 \end{cases}$

34. $\begin{cases} 4x^2 + 3y^2 = 4 \\ 6y^2 - 2x^2 = 3 \end{cases}$

35. $\begin{cases} x^2 + y^2 = 25 \\ y = -x^2 + 6x - 5 \end{cases}$

36. $\begin{cases} x^2 + y^2 = 65 \\ y = -x^2 + 9 \end{cases}$

Applying the Concepts

37. Fun with Numbers The difference of two numbers is 2. The sum of their squares is 34. Find the numbers.

38. Fun with Numbers The sum of two numbers is 8. The sum of their squares is 160. Find the numbers.

△ **39. Perimeter and Area of a Rectangle** The perimeter of a rectangle is 48 feet. The area of the rectangle is 140 square feet. Find the dimensions of the rectangle.

△ **40. Perimeter and Area of a Rectangle** The perimeter of a rectangle is 64 meters. The area of the rectangle is 240 square meters. Find the dimensions of the rectangle.

41. Constructing a Box A rectangular piece of cardboard, whose area is 190 square centimeters, is made into an open box by cutting a 2-centimeter square from each corner and turning up the sides. See the figure. If the box is to have a volume of 180 cubic centimeters, what size cardboard should you start with?

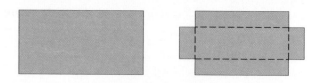

42. Fencing A farmer has 132 yards of fencing available to enclose a 900-square-yard region in the shape of adjoining squares with sides of length x and y. See the figure. Find x and y.

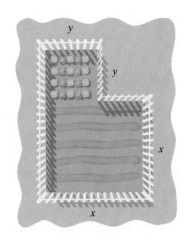

Extending the Concepts

In Problems 43–46, solve the system of nonlinear equations. Do not attempt to graph the equations in the system.

43. $\begin{cases} y^2 + y + x^2 - x - 2 = 0 \\ y + 1 + \dfrac{x-2}{y} = 0 \end{cases}$

44. $\begin{cases} x^3 - 2x^2 + y^2 + 3y - 4 = 0 \\ x - 2 + \dfrac{y^2 - y}{x^2} = 0 \end{cases}$

45. $\begin{cases} \ln x = 4 \ln y \\ \log_3 x = 2 + 2\log_3 y \end{cases}$

46. $\begin{cases} \ln x = 5 \ln y \\ \log_2 x = 3 + 2\log_2 y \end{cases}$

47. If r_1 and r_2 are two solutions of a quadratic equation $ax^2 + bx + c = 0$, then it can be shown that

$$r_1 + r_2 = -\frac{b}{a} \quad \text{and} \quad r_1 r_2 = \frac{c}{a}$$

Solve this system of equations for r_1 and r_2.

48. A circle and a line intersect at most twice. A circle and a parabola intersect at most four times. How many times do you think a circle and the graph of a polynomial of degree 3 can intersect? What about a circle and the graph of a polynomial of degree 4? What about a circle and the graph of a polynomial of degree n? Explain your conclusions using an algebraic argument.

Synthesis Review

In Problems 49–53, evaluate the functions given that $f(x) = 3x + 4$ and $g(x) = 2^x$.

49. (a) $f(1)$ **(b)** $g(1)$ **50. (a)** $f(2)$ **(b)** $g(2)$

51. (a) $f(3)$ **(b)** $g(3)$ **52. (a)** $f(4)$ **(b)** $g(4)$

53. (a) $f(5)$ **(b)** $g(5)$

54. Use the results of Problems 49–53 to compute
  (a) $f(2) - f(1)$ **(b)** $f(3) - f(2)$
  (c) $f(4) - f(3)$ **(d)** $f(5) - f(4)$
  (e) $\dfrac{g(2)}{g(1)}$ **(f)** $\dfrac{g(3)}{g(2)}$
  (g) $\dfrac{g(4)}{g(3)}$ **(h)** $\dfrac{g(5)}{g(4)}$
  (i) Make a generalization about $f(n+1) - f(n)$ for $n \geq 1$ an integer. Make a generalization about $\dfrac{g(n+1)}{g(n)}$ for $n \geq 1$ an integer.

The Graphing Calculator

In Problems 55–64, use a graphing calculator and the INTERSECT feature to solve the following systems of nonlinear equations. If necessary, round answers to three decimal places.

55. $\begin{cases} y = x^2 - 6x + 4 \\ 5x + y = 6 \end{cases}$

56. $\begin{cases} y = x^2 + 4x + 5 \\ x - y = 9 \end{cases}$

57. $\begin{cases} x^2 + y^2 = 16 \\ x^2 - y^2 = 16 \end{cases}$

58. $\begin{cases} x^2 + y^2 = 25 \\ x^2 - y^2 = 25 \end{cases}$

59. $\begin{cases} (x-4)^2 + y^2 = 25 \\ x - y = -3 \end{cases}$

60. $\begin{cases} (x+5)^2 + (y-2)^2 = 100 \\ 8x + y = 18 \end{cases}$

61. $\begin{cases} (x-1)^2 + (y+2)^2 = 4 \\ x^2 + 4y - x = -1 \end{cases}$

62. $\begin{cases} (x+2)^2 + (y-1)^2 = 4 \\ y^2 - 2y - x = 5 \end{cases}$

63. $\begin{cases} x^2 + 4y^2 = 4 \\ x^2 + 6x - y = -13 \end{cases}$

64. $\begin{cases} 9x^2 + 4y^2 = 36 \\ x^2 + (y-7)^2 = 4 \end{cases}$

Chapter 9 Activity: How Do You Know That ...?

Focus: A sharing of ideas to identify topics contained in the study of conics

Time: 30–35 minutes

Group size: 2–4

For each question, every member of the group should spend 1–2 minutes individually listing "How they know ... ." At the end of the allotted time, the group should convene and conduct a 2–3-minute discussion of the different responses.

How Do You Know ...

...that a triangle with vertices at $(2, 6)$, $(0, -2)$ and $(5, 1)$ is an isosceles triangle?

...the coordinates for the midpoint between two given ordered pairs?

...that $x^2 + y^2 + 4x - 8y = 16$ is an equation of a circle and not a parabola?

...which way a parabola opens?

...whether an ellipse's center is at the origin or another point?

...that a hyperbola will not intersect its asymptotes?

...that a circle and a parabola can have 0, 1, 2, 3, or 4 points of intersection? (If necessary, use a sketch to support your response.)

Chapter 9 Review

Section 9.1 Distance and Midpoint Formulas

KEY CONCEPTS

- **The Distance Formula**
  The distance between two points $P_1 = (x_1, y_1)$ and $P_2 = (x_2, y_2)$, denoted by $d(P_1, P_2)$, is
  $d(P_1, P_2) = \sqrt{(x_2 - x_1)^2 + (y_2 - y_1)^2}$.

- **The Midpoint Formula**
  The midpoint $M = (x, y)$ of the line segment from $P_1 = (x_1, y_1)$ to $P_2 = (x_2, y_2)$ is $M = \left(\dfrac{x_1 + x_2}{2}, \dfrac{y_1 + y_2}{2} \right)$.

You Should Be Able To...	EXAMPLE	Review Exercises
1 Use the Distance Formula (p. 689)	Examples 1 through 3	1–6, 12
2 Use the Midpoint Formula (p. 692)	Example 4	7–11

In Problems 1–6, find the distance $d(P_1, P_2)$ between points P_1 and P_2.

1. $P_1 = (0, 0)$ and $P_2 = (-4, -3)$

2. $P_1 = (-3, 2)$ and $P_2 = (5, -4)$

3. $P_1 = (-1, 1)$ and $P_2 = (5, 3)$

4. $P_1 = (6, -7)$ and $P_2 = (6, -1)$

5. $P_1 = \left(\sqrt{7}, -\sqrt{3} \right)$ and $P_2 = \left(4\sqrt{7}, 5\sqrt{3} \right)$

6. $P_1 = (-0.2, 1.7)$ and $P_2 = (1.3, 3.7)$

In Problems 7–11, find the midpoint of the line segment formed by joining the points P_1 and P_2.

7. $P_1 = (-1, 6)$ and $P_2 = (-3, 4)$

8. $P_1 = (7, 0)$ and $P_2 = (5, -4)$

9. $P_1 = \left(-\sqrt{3}, 2\sqrt{6} \right)$ and $P_2 = \left(-7\sqrt{3}, -8\sqrt{6} \right)$

10. $P_1 = (5, -2)$ and $P_2 = (0, 3)$

11. $P_1 = \left(\dfrac{1}{4}, \dfrac{2}{3} \right)$ and $P_2 = \left(\dfrac{5}{4}, \dfrac{1}{3} \right)$

12. Consider the three points
$A = (-2, 2)$, $B = (1, -1)$, $C = (-1, -3)$.

(a) Plot each point in the Cartesian plane and form the triangle ABC.

(b) Find the length of each side of the triangle.

(c) Verify that the triangle is a right triangle.

(d) Find the area of the triangle.

Section 9.2 Circles

KEY CONCEPTS

- **Standard Form of an Equation of a Circle**
  The standard form of the equation of a circle with radius r and center (h, k) is $(x - h)^2 + (y - k)^2 = r^2$.

- **General Form of the Equation of a Circle**
  The general form of the equation of a circle is $x^2 + y^2 + ax + by + c = 0$ when the graph exists.

KEY TERMS

Conic sections
Circle
Radius
Center

You Should Be Able To...	EXAMPLE	Review Exercises
1 Write the standard form of the equation of a circle (p. 696)	Example 1	13–20
2 Graph a circle (p. 697)	Example 2	15–18; 21–30
3 Find the center and radius of a circle given an equation in general form (p. 697)	Example 3	27–30

In Problems 13 and 14, find the center and radius of each circle. Write the standard form of the equation.

13.
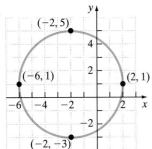

14.
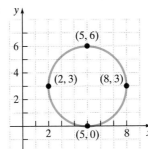

In Problems 15–18, write the standard form of the equation of each circle whose radius is r and center is (h, k). Graph each circle.

15. $r = 4$; $(h, k) = (0, 0)$ **16.** $r = 3$; $(h, k) = (-3, 1)$

17. $r = 1$; $(h, k) = (5, -2)$ **18.** $r = \sqrt{7}$; $(h, k) = (4, 0)$

In Problems 19 and 20, find the standard form of the equation of each circle.

19. Center at $(2, -1)$; contains the point $(5, 3)$.

20. Endpoints of a diameter at $(1, 7)$ and $(-3, -1)$

In Problems 21–26, find the center (h, k) and the radius r of each circle. Graph each circle.

21. $x^2 + y^2 = 25$

22. $(x - 1)^2 + (y - 2)^2 = 4$

23. $x^2 + (y - 4)^2 = 16$

24. $(x + 1)^2 + (y + 6)^2 = 49$

25. $(x + 2)^2 + \left(y - \dfrac{3}{2}\right)^2 = \dfrac{1}{4}$

26. $(x + 3)^2 + (y + 3)^2 = 4$

In Problems 27–30, find the center (h, k) and the radius r of each circle. Graph each circle.

27. $x^2 + y^2 + 6x + 10y - 2 = 0$

28. $x^2 + y^2 - 8x + 4y + 16 = 0$

29. $x^2 + y^2 + 2x - 4y - 4 = 0$

30. $x^2 + y^2 - 10x - 2y + 17 = 0$

Section 9.3 Parabolas

KEY CONCEPTS

- **Equations of a Parabola: Vertex at (0, 0); $a > 0$**
  See Table 1 on page 702.

- **Equations of a Parabola: Vertex at (h, k); $a > 0$**
  See Table 2 on page 705.

KEY TERMS

Parabola
Focus
Directrix
Axis of symmetry
Vertex

You Should Be Able To...	EXAMPLE	Review Exercises
❶ Graph parabolas whose vertex is the origin (p. 702)	Examples 1 and 2	31–34; 35, 36
❷ Find the equation of a parabola (p. 703)	Examples 3 and 4	31–34
❸ Graph a parabola whose vertex is not the origin (p. 704)	Example 5	37–39
❹ Solve applied problems involving parabolas (p. 706)	Example 6	40

In Problems 31–34, find the equation of the parabola described. Graph each parabola.

31. Vertex at $(0, 0)$; focus at $(0, -3)$

32. Focus at $(-4, 0)$; directrix the line $x = 4$

33. Vertex at $(0, 0)$; contains the point $(8, -2)$; axis of symmetry the x-axis

34. Vertex at $(0, 0)$; directrix the line $y = -2$; axis of symmetry the y-axis

In Problems 35–39, find the vertex, focus, and directrix of each parabola. Graph each parabola.

35. $x^2 = 2y$ **36.** $y^2 = 16x$

37. $(x + 1)^2 = 8(y - 3)$

38. $(y - 4)^2 = -2(x + 3)$

39. $x^2 - 10x + 3y + 19 = 0$

40. Radio Telescope The U.S. Naval Research Laboratory has a giant radio telescope with a dish that is shaped like a parabola. The signals that are received by the dish strike the surface of the dish and are reflected to a single point, where the receiver is located. If the giant dish is 300 feet across and 44 feet deep at its center, at what position should the receiver be placed?

Section 9.4 Ellipses

KEY CONCEPTS

- **Ellipses with Center at (0, 0)**
  See Table 3 on page 711.

- **Ellipses with Center at (h, k)**
  See Table 4 on page 714.

KEY TERMS

Ellipse
Foci
Major axis
Center
Minor axis
Vertices

You Should Be Able To...	EXAMPLE	Review Exercises
❶ Graph an ellipse whose center is the origin (p. 710)	Examples 1 and 2	41, 42
❷ Find the equation of an ellipse whose center is the origin (p. 713)	Example 3	43–45
❸ Graph an ellipse whose center is not the origin (p. 714)	Example 4	46–47
❹ Solve applied problems involving ellipses (p. 715)	Example 5	48

In Problems 41 and 42, find the vertices and foci of each ellipse. Graph each ellipse.

41. $\dfrac{x^2}{9} + y^2 = 1$ **42.** $9x^2 + 4y^2 = 36$

In Problems 43–45, find an equation for each ellipse. Graph each ellipse.

43. Center at $(0, 0)$; focus at $(0, 3)$; vertex at $(0, 5)$

44. Center at $(0, 0)$; focus at $(-2, 0)$; vertex at $(-6, 0)$

45. Foci at $(\pm 8, 0)$; vertices at $(\pm 10, 0)$

In Problems 46 and 47, find the center, vertices, and foci of each ellipse. Graph each ellipse.

46. $\dfrac{(x - 1)^2}{49} + \dfrac{(y + 2)^2}{25} = 1$

47. $25(x + 3)^2 + 9(y - 4)^2 = 225$

48. Semielliptical Arch Bridge An arch in the shape of the upper half of an ellipse is used to support a bridge that spans a river 60 feet wide. The center of the arch is 16 feet above the center of the river.

(a) Write the equation for the ellipse in which the x-axis coincides with the water and the y-axis passes through the center of the arch.

(b) Can a rectangular barge that is 25 feet wide and sits 12 feet above the surface of the water fit through the opening of the bridge?

Section 9.5 Hyperbolas

KEY CONCEPTS

- **Hyperbolas with Center at the Origin**
  See Table 5 on page 720.
- **Asymptotes of a Hyperbola**
  See Table 6 on page 723.

KEY TERMS

Hyperbola
Foci
Transverse axis
Center
Conjugate axis
Branches
Vertices

You Should Be Able To...	EXAMPLE	Review Exercises
❶ Graph a hyperbola whose center is the origin (p. 720)	Examples 1 and 2	49–53
❷ Find the equation of a hyperbola whose center is the origin (p. 722)	Example 3	54–56
❸ Find the asymptotes of a hyperbola whose center is the origin (p. 723)	Example 4	52, 53

In Problems 49–51, find the vertices and foci of each hyperbola. Graph each hyperbola.

49. $\dfrac{x^2}{4} - \dfrac{y^2}{9} = 1$ **50.** $\dfrac{y^2}{25} - \dfrac{x^2}{49} = 1$

51. $16y^2 - 25x^2 = 400$

In Problems 52 and 53, graph each hyperbola using the asymptotes as a guide.

52. $\dfrac{x^2}{36} - \dfrac{y^2}{36} = 1$ **53.** $\dfrac{y^2}{25} - \dfrac{x^2}{4} = 1$

In Problems 54–56, find an equation for each hyperbola described. Graph each hyperbola.

54. Center at $(0, 0)$; focus at $(-4, 0)$; vertex at $(-3, 0)$

55. Vertices at $(0, -3)$ and $(0, 3)$; focus at $(0, 5)$

56. Vertices at $(0, \pm 4)$; asymptote the line $y = \dfrac{4}{3}x$

Section 9.6 Systems of Nonlinear Equations

KEY TERM

System of nonlinear equations

You Should Be Able To...	EXAMPLE	Review Exercises
❶ Solve a system of nonlinear equations using substitution (p. 728)	Examples 1 and 2	57–60; 65–76
❷ Solve a system of nonlinear equations using elimination (p. 730)	Examples 3 and 4	61–64; 65–76

In Problems 57–60, solve the system of nonlinear equations using the method of substitution.

57. $\begin{cases} 4x^2 + y^2 = 10 \\ y = x \end{cases}$ **58.** $\begin{cases} y = 2x^2 + 1 \\ y = x + 2 \end{cases}$ **59.** $\begin{cases} 6x - y = 5 \\ xy = 1 \end{cases}$ **60.** $\begin{cases} x^2 + y^2 = 26 \\ x^2 - 2y^2 = 23 \end{cases}$

In Problems 61–64, solve the system of nonlinear equations using the method of elimination.

61. $\begin{cases} 4x - y^2 = 0 \\ 2x^2 + y^2 = 16 \end{cases}$

62. $\begin{cases} x^2 - y = -2 \\ x^2 + y = 4 \end{cases}$

63. $\begin{cases} 4x^2 - 2y^2 = 2 \\ -x^2 + y^2 = 2 \end{cases}$

64. $\begin{cases} x^2 + y^2 = 8x \\ y^2 = 3x \end{cases}$

In Problems 65–72, solve the system of nonlinear equations using the method you prefer.

65. $\begin{cases} y = x + 2 \\ y = x^2 \end{cases}$

66. $\begin{cases} x^2 + 2y = 9 \\ 5x - 2y = 5 \end{cases}$

67. $\begin{cases} x^2 + y^2 = 36 \\ x - y = -6 \end{cases}$

68. $\begin{cases} y = 2x - 4 \\ y^2 = 4x \end{cases}$

69. $\begin{cases} x^2 + y^2 = 9 \\ x + y = 7 \end{cases}$

70. $\begin{cases} 2x^2 + 3y^2 = 14 \\ x^2 - y^2 = -3 \end{cases}$

71. $\begin{cases} x^2 + y^2 = 16 \\ x^2 + 4y = 16 \end{cases}$

72. $\begin{cases} x = 4 - y^2 \\ x = 2y + 4 \end{cases}$

73. Fun with Numbers The sum of two numbers is 12. The difference of their squares is 24. Find the two numbers.

△ **74. Perimeter and Area of a Rectangle** The perimeter of a rectangle is 34 centimeters. The area of the rectangle is 60 square centimeters. Find the dimensions of the rectangle.

△ **75. Dimensions of a Rectangle** The area of a rectangle is 2160 square inches. The diagonal of the rectangle is 78 inches. Find the dimensions of the rectangle.

△ **76. Dimensions of a Triangle** A right triangle has a perimeter of 36 feet and a hypotenuse of 15 feet. Find the lengths of the legs of the right triangle.

Chapter 9 Test

Step-by-step test solutions are found on the Chapter Test Prep Videos available in MyMathLab® *or on* YouTube*.*

1. Find the distance $d(P_1, P_2)$ between points $P_1 = (-1, 3)$ and $P_2 = (3, -5)$.

2. Find the midpoint of the line segment formed by joining the points $P_1 = (-7, 6)$ and $P_2 = (5, -2)$.

In Problems 3 and 4, find the center (h, k) and the radius r of each circle. Graph each circle.

3. $(x - 4)^2 + (y + 1)^2 = 9$

4. $x^2 + y^2 + 10x - 4y + 13 = 0$

In Problems 5 and 6, find the standard form of the equation of each circle.

5. Radius $r = 6$ and center $(h, k) = (-3, 7)$

6. Center at $(-5, 8)$; contains the point $(3, 2)$.

In Problems 7 and 8, find the vertex, focus, and directrix of each parabola. Graph each parabola.

7. $(y + 2)^2 = 4(x - 1)$

8. $x^2 - 4x + 3y - 8 = 0$

In Problems 9 and 10, find an equation for each parabola described.

9. Vertex at $(0, 0)$; focus at $(0, -4)$

10. Focus at $(3, 4)$; directrix the line $x = -1$

In Problems 11 and 12, find the vertices and foci of each ellipse. Graph each ellipse.

11. $9x^2 + 25y^2 = 225$

12. $\dfrac{(x - 2)^2}{9} + \dfrac{(y + 4)^2}{16} = 1$

In Problems 13 and 14, find an equation for each ellipse described.

13. Center at $(0, 0)$; focus at $(0, -4)$; vertex at $(0, -5)$

14. Vertices at $(-1, 7)$ and $(-1, -3)$; focus at $(-1, -1)$

In Problems 15 and 16, find the vertices, foci, and asymptotes for each hyperbola. Graph each hyperbola using the asymptotes as a guide.

15. $x^2 - \dfrac{y^2}{4} = 1$

16. $16y^2 - 25x^2 = 1600$

17. Find an equation for a hyperbola with foci at $(\pm 8, 0)$ and vertex at $(-3, 0)$.

In Problems 18 and 19, solve the system of nonlinear equations using the method you prefer.

18. $\begin{cases} x^2 + y^2 = 17 \\ x + y = -3 \end{cases}$

19. $\begin{cases} x^2 + y^2 = 9 \\ 4x^2 - y^2 = 16 \end{cases}$

20. An arch in the shape of the upper half of an ellipse is used to support a bridge spanning a creek 30 feet wide. The center of the arch is 10 feet above the center of the creek.

   (a) Write the equation for the ellipse in which the x-axis coincides with the creek and the y-axis passes through the center of the arch.

   (b) What is the height of the arch at a distance 12 feet from the center of the creek?

Cumulative Review Chapters R–9

1. Determine the domain of $h(x) = \dfrac{4 - x}{7 - x}$.

2. Graph the linear equation: $5x - 2y = 6$.

3. Find the equation of the line that contains the point $(-3, -2)$ and is parallel to the line $4x - 3y = 15$. Write your answer in either slope-intercept or standard form, whichever you prefer.

4. Solve the following system of linear equations using the method you prefer:

$$\begin{cases} 2x + y = -2 \\ 5x + 2y = -1 \end{cases}$$

5. Evaluate: $\begin{vmatrix} 1 & 2 & 0 \\ 7 & -1 & -2 \\ 4 & -2 & 1 \end{vmatrix}$

In Problems 6 and 7, factor completely.

6. $8m^3 + n^3$

7. $9p^2 - 18p + 8$

In Problems 8 and 9, perform the indicated operations. Be sure to express the final answer in lowest terms.

8. $\dfrac{2a^2 - 7a - 4}{3a^2 - 13a + 4} \div \dfrac{a^2 - 2a - 35}{3a^2 + 14a - 5}$

9. $\dfrac{5}{3t^2 - 7t + 2} - \dfrac{2}{3t^2 - 4t + 1}$

10. Solve: $1 + \dfrac{1}{x - 1} = \dfrac{x + 2}{x}$

11. Solve: $\dfrac{x - 5}{x + 4} \geq 0$

12. Simplify: $\sqrt{75} + 7\sqrt{3} - \sqrt{27}$

13. Rationalize the denominator: $\dfrac{9}{2 + \sqrt{7}}$

14. Solve: $5 - \sqrt{x + 4} = 2$

15. Simplify: $(4 - i)(7 + 3i)$

16. Solve: $2x^2 - 4x - 3 = 0$

17. Graph: $f(x) = -x^2 + 2x + 8$

18. Solve: $x^2 - 3x - 4 \leq 0$

In Problems 19 and 20, graph each function. State the domain and the range of the function.

19. $f(x) = 2^{x+3}$

20. $g(x) = \log_4 x$

21. Evaluate: $\log_4 32$

In Problems 22 and 23, solve each equation.

22. $6^{x+1} = 216$

23. $\log_3(x^2 - 16) = 2$

24. Write the standard form of the equation of a circle with center at $(-3, 1)$ and containing the point $(0, 0)$.

25. Find the vertex, focus, and directrix, and then graph the following parabola: $(y - 3)^2 = 8(x + 3)$.

10 Sequences, Series, and the Binomial Theorem

Population growth has been a topic of debate among scientists for a long time. In fact, over 200 years ago, the English economist and mathematician Thomas Robert Malthus anonymously published a paper predicting that the world's population would overwhelm the Earth's capacity to sustain it—he claimed that food supplies increase arithmetically while population grows geometrically. See Problems 55 and 56 in Section 10.1 and Problem 99 in Section 10.3.

The Big Picture: Putting It Together

In Chapter 2, we defined the domain of a function (inputs) as the set of real numbers so that the outputs of the function make sense. In Sections 10.1–10.3, we introduce a special type of function called a *sequence*. The domain of a sequence is the set of natural numbers (that is, the positive integers). Functions whose domain is the set of all real numbers enable us to model situations in which the value of an independent variable, such as a population of bacteria, changes continuously. In contrast, sequences make it possible to model situations in which the value of a *dependent* variable changes at discrete intervals of time, such as weekly changes in the value of a deposit at a bank.

In Chapter 4, we gave a formula for expanding $(x + a)^2$. Does any formula exist for expanding $(x + a)^n$, where n is an integer greater than 2? Yes! We discuss this formula in Section 10.4.

10.1 Sequences

Objectives

1. Write the First Few Terms of a Sequence
2. Find a Formula for the nth Term of a Sequence
3. Use Summation Notation

Are You Ready for This Section?

Before getting started, take the following readiness quiz. If you get a problem wrong, go back to the section cited and review the material.

R1. Evaluate $f(x) = x^2 - 4$ at **(a)** $x = 3$ **(b)** $x = -7$. [Section 2.2, pp. 156–158]

R2. If $g(x) = 2x - 3$, find $g(1) + g(2) + g(3)$. [Section 2.2, pp. 156–158]

R3. In the function $f(n) = n^2 - 4$, what is the independent variable? [Section 2.2, pp. 156–158]

We begin with a definition.

> ### Definition
>
> A **sequence** is a function whose domain is the set of positive integers.

Because a sequence is a function, it has a graph. In Figure 1(a), we graph the function $f(x) = \dfrac{1}{x}$ for $x > 0$. If all the points on this graph were removed except those whose x-coordinates are positive integers—that is, if all points were removed except $(1, 1)$, $\left(2, \dfrac{1}{2}\right), \left(3, \dfrac{1}{3}\right)$, and so on, the remaining points would be the graph of the sequence $f(n) = \dfrac{1}{n}$, as shown in Figure 1(b). Notice that we use n to represent the independent variable in a sequence. This serves to remind us that n is a positive integer, or natural number.

Figure 1

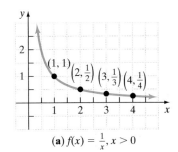

(a) $f(x) = \frac{1}{x}, x > 0$ **(b)** $f(n) = \frac{1}{n}, n$ a positive integer

A sequence may be represented by listing its values in order. For example, the sequence whose graph is given in Figure 1(b) might be represented as

$$f(1), f(2), f(3), f(4), \ldots \quad \text{or} \quad 1, \frac{1}{2}, \frac{1}{3}, \frac{1}{4}, \ldots$$

The numbers in the ordered list are the **terms** of the sequence. Notice that the list never ends, as the **ellipsis** (the three dots) indicates. A sequence that does not end is an **infinite sequence.** In contrast, **finite sequences** have a domain that is the first n positive integers. For example,

$$40, 44, 48, 52, 56, 60, 64$$

is a finite sequence because it contains $n = 7$ terms.

▶ 1 Write the First Few Terms of a Sequence

In a sequence, we do not use the traditional notation $f(n)$. This is done to distinguish sequences from functions whose domain is the set of all real numbers. Instead, we use the notation a_n to indicate that the function is a sequence. The value of the subscript represents the value of the independent variable, so a_4 means to evaluate the function defined by the sequence a_n at $n = 4$. For the sequence $f(n) = \dfrac{1}{n}$, we would write the function as $a_n = \dfrac{1}{n}$ and evaluate the function as follows:

$$a_1 = 1, \quad a_2 = \frac{1}{2}, \quad a_3 = \frac{1}{3}, \quad a_4 = \frac{1}{4}, \quad \ldots, \quad a_n = \frac{1}{n}$$

In addition, just as we can name functions f, g, F, G, and so on, we can name sequences. Usually, we use lowercase letters from the beginning of the alphabet to name a sequence, as in a_n, b_n, c_n, and so on although this naming scheme is not necessary.

When a formula for the nth term (sometimes called the **general term**) of a sequence is known, rather than write out the terms of the sequence, we can represent the entire sequence by placing braces around the formula for the nth term. For example, the sequence whose nth term is $b_n = \left(\dfrac{1}{3}\right)^n$ can be written as

$$\{b_n\} = \left\{\left(\frac{1}{3}\right)^n\right\}$$

or as

$$b_1 = \frac{1}{3}, \quad b_2 = \frac{1}{9}, \quad b_3 = \frac{1}{27}, \quad \ldots, \quad b_n = \left(\frac{1}{3}\right)^n$$

EXAMPLE 1 **Writing the First Five Terms of a Sequence**

Write the first five terms of the sequence $\{a_n\} = \left\{\dfrac{n}{n+1}\right\}$.

Solution

To find the first five terms of the sequence, evaluate the function at $n = 1, 2, 3, 4$, and 5.

$$a_1 = \frac{1}{1+1} = \frac{1}{2}, \quad a_2 = \frac{2}{2+1} = \frac{2}{3}, \quad a_3 = \frac{3}{3+1} = \frac{3}{4}, \quad a_4 = \frac{4}{4+1} = \frac{4}{5}, \quad a_5 = \frac{5}{5+1} = \frac{5}{6}$$

The first five terms of the sequence are $\dfrac{1}{2}, \dfrac{2}{3}, \dfrac{3}{4}, \dfrac{4}{5}$, and $\dfrac{5}{6}$. ●

EXAMPLE 2 **Writing the First Five Terms of a Sequence**

Write the first five terms of the sequence $\{b_n\} = \{(-1)^n \cdot n^2\}$.

Solution

The first five terms of the sequence are

$$b_1 = (-1)^1 \cdot 1^2 = -1, \quad b_2 = (-1)^2 \cdot 2^2 = 4, \quad b_3 = (-1)^3 \cdot 3^2 = -9,$$
$$b_4 = (-1)^4 \cdot 4^2 = 16, \quad b_5 = (-1)^5 \cdot 5^2 = -25$$

The first five terms of the sequence are $-1, 4, -9, 16$, and -25. ●

Notice in Example 2 that the signs of the terms in the sequence **alternate**. This occurs when we have factors such as $(-1)^{n+1}$, which equals 1 if n is odd and -1 if n is even, or $(-1)^n$, which equals -1 if n is odd and 1 if n is even.

Quick ✓

1. A(n) _____ is a function whose domain is the set of positive integers.

2. A sequence that does not end is said to be a(n) _____ sequence. A(n) _____ sequence has a domain that is the first n positive integers.

3. *True or False* A sequence is a function.

In Problems 4 and 5, write the first five terms of the given sequence.

4. $\{a_n\} = \{2n - 3\}$ 5. $\{b_n\} = \{(-1)^n \cdot 4n\}$

⊙ ❷ Find a Formula for the *n*th Term of a Sequence

Sometimes a sequence is indicated by an observed pattern in the first few terms that makes it possible to determine the formula for the *n*th term. In the example that follows, a sufficient number of terms of the sequence are given so that a natural choice for the *n*th term is suggested.

(EXAMPLE 3) **Determining a Sequence from a Pattern**

Find a formula for the *n*th term of each sequence.

    (a) $3, 7, 11, 15, \ldots$

    (b) $2, 4, 8, 16, \ldots$

    (c) $1, -8, 27, -64, 125, \ldots$

Solution

Our goal in all these problems is to find a formula in terms of *n* such that when $n = 1$ we get the first term, when $n = 2$ we get the second term, and so on.

    (a) When $n = 1$, then $a_1 = 3$; when $n = 2$, then $a_2 = 7$. Notice that each subsequent term increases by 4. A formula for the *n*th term is given by $a_n = 4n - 1$.

    (b) The terms of the sequence are all powers of 2 with the first term equaling 2^1, the second term equaling 2^2, and so on. A formula for the *n*th term is given by $b_n = 2^n$.

    (c) Notice that the sign of the terms alternate with the first term being positive. Thus $(-1)^{n+1}$ must be part of the formula. Ignoring the sign, notice that the terms are all perfect cubes. A formula for the *n*th term is given by $c_n = (-1)^{n+1} \cdot n^3$.   ●

Quick ✓

In Problems 6 and 7, find a formula for the nth term of each sequence.

    6. $5, 7, 9, 11, \ldots$ **7.** $\dfrac{1}{2}, -\dfrac{1}{3}, \dfrac{1}{4}, -\dfrac{1}{5}, \ldots$

⊙ ❸ Use Summation Notation

In other mathematics courses, such as statistics and calculus, it is important to find the sum of the first *n* terms of a sequence $\{a_n\}$ — that is,

$$a_1 + a_2 + a_3 + \cdots + a_n$$

Mathematicians express the sum using **summation notation.** With this notation, we write $a_1 + a_2 + a_3 + \cdots + a_n$ as

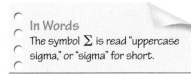

In Words
The symbol Σ is read "uppercase sigma," or "sigma" for short.

$$\sum_{i=1}^{n} a_i = a_1 + a_2 + a_3 + \cdots + a_n$$

The symbol Σ is an instruction to sum, or add up, the terms. The integer *i* is called the **index** of the sum; it tells you where to start the sum and where to end it. The expression

$$\sum_{i=1}^{n} a_i$$

means to add the terms of the sequence $\{a_i\}$ from $i = 1$ through $i = n$. We read the expression $\displaystyle\sum_{i=1}^{n} a_i$ as "the sum of *a* sub *i* from $i = 1$ to $i = n$." When a finite number of terms are to be added, the sum is called a **partial sum.**

EXAMPLE 4 **Finding a Partial Sum**

Write out the sum and determine its value.

(a) $\sum_{i=1}^{4} (2i + 5)$

(b) $\sum_{i=1}^{5} (i^2 - 5)$

Solution

(a) $\sum_{i=1}^{4} (2i + 5) = \underbrace{(2 \cdot 1 + 5)}_{i = 1} + \underbrace{(2 \cdot 2 + 5)}_{i = 2} + \underbrace{(2 \cdot 3 + 5)}_{i = 3} + \underbrace{(2 \cdot 4 + 5)}_{i = 4}$

$= 7 + 9 + 11 + 13$

$= 40$

(b) $\sum_{i=1}^{5} (i^2 - 5) = \underbrace{(1^2 - 5)}_{i = 1} + \underbrace{(2^2 - 5)}_{i = 2} + \underbrace{(3^2 - 5)}_{i = 3} + \underbrace{(4^2 - 5)}_{i = 4} + \underbrace{(5^2 - 5)}_{i = 5}$

$= -4 + (-1) + 4 + 11 + 20$

$- 30$

Quick ✓

8. When there is a finite number of terms to be added, the sum is called a
_____ ___.

In Problems 9 and 10, write out the sum and determine its value.

9. $\sum_{i=1}^{3} (4i - 1)$

10. $\sum_{i=1}^{5} (i^3 + 1)$

The index of summation does not have to begin with 1. In addition, the index of summation does not have to be *i*. For example, we might have

$$\sum_{k=3}^{6} (2k + 1) = \underbrace{(2 \cdot 3 + 1)}_{k = 3} + \underbrace{(2 \cdot 4 + 1)}_{k = 4} + \underbrace{(2 \cdot 5 + 1)}_{k = 5} + \underbrace{(2 \cdot 6 + 1)}_{k = 6}$$

$$= 7 + 9 + 11 + 13$$

$$= 40$$

Notice that the terms of this partial sum are identical to those in Example 4(a). What is the moral of the story? We can make the same sum look entirely different by changing the starting point of the index of summation and changing the variable that represents the index.

Now let's reverse the process. Rather than writing out a sum, we will express a sum using summation notation.

EXAMPLE 5 **Writing a Sum in Summation Notation**

Express each sum using summation notation.

(a) $1^2 + 2^2 + 3^2 + \cdots + 10^2$

(b) $2 + 1 + \dfrac{2}{3} + \dfrac{1}{2} + \dfrac{2}{5} + \cdots + \dfrac{1}{6}$

Solution

(a) The sum $1^2 + 2^2 + 3^2 + \cdots + 10^2$ has 10 terms, each of the form i^2, and starts at $i = 1$ and ends at $i = 10$:

$$1^2 + 2^2 + 3^2 + \cdots + 10^2 = \sum_{i=1}^{10} i^2$$

(b) First, figure out the pattern. With a little investigation, we discover that $2 + 1 + \frac{2}{3} + \frac{1}{2} + \frac{2}{5} + \cdots + \frac{1}{6}$ can be written as

$\frac{2}{1} + \frac{2}{2} + \frac{2}{3} + \frac{2}{4} + \frac{2}{5} + \cdots + \frac{2}{12}$, so that the nth term of the sum is $\frac{2}{n}$. There are $n = 12$ terms, so

$$2 + 1 + \frac{2}{3} + \frac{1}{2} + \frac{2}{5} + \cdots + \frac{1}{6} = \sum_{n=1}^{12}\left(\frac{2}{n}\right)$$

Work Smart

The index of summation can be any variable we desire and can start at any value we desire. Keep this in mind when checking your answers.

Quick ✓

In Problems 11 and 12, write each sum using summation notation.

11. $1 + 4 + 9 + \cdots + 144$

12. $1 + \frac{1}{2} + \frac{1}{4} + \cdots + \frac{1}{32}$

10.1 Exercises   MyMathLab®   PRACTICE

Exercise numbers in green are complete video solutions in MyMathLab

Problems 1–12 are the Quick ✓s that follow the EXAMPLES.

Building Skills

In Problems 13–24, write the first five terms of each sequence. See Objective 1.

13. $\{3n + 5\}$

14. $\{n - 4\}$

15. $\left\{\dfrac{n}{n + 2}\right\}$

16. $\left\{\dfrac{n + 4}{n}\right\}$

17. $\{(-1)^n\, n\}$

18. $\{(-1)^{n+1}\, n\}$

19. $\{2^n + 1\}$

20. $\{3^n - 1\}$

21. $\left\{\dfrac{2n}{2^n}\right\}$

22. $\left\{\dfrac{3n}{3^n}\right\}$

23. $\left\{\dfrac{n}{e^n}\right\}$

24. $\left\{\dfrac{n^2}{2}\right\}$

In Problems 25–32, the given pattern continues. Write the nth term of each sequence suggested by the pattern. See Objective 2.

25. $2, 4, 6, 8, \ldots$

26. $5, 10, 15, 20, \ldots$

27. $\dfrac{1}{2}, \dfrac{2}{3}, \dfrac{3}{4}, \dfrac{4}{5}, \ldots$

28. $\dfrac{1}{2}, 1, \dfrac{3}{2}, 2, \dfrac{5}{2}, \ldots$

29. $3, 6, 11, 18, \ldots$

30. $0, 7, 26, 63, \ldots$

31. $-1, 4, -9, 16, \ldots$

32. $1, -\dfrac{1}{2}, \dfrac{1}{4}, -\dfrac{1}{8}, \ldots$

In Problems 33–44, write out each sum and determine its value. See Objective 3.

33. $\displaystyle\sum_{i=1}^{4} (5i + 1)$

34. $\displaystyle\sum_{i=1}^{5} (3i + 2)$

35. $\displaystyle\sum_{i=1}^{5} \frac{i^2}{2}$

36. $\displaystyle\sum_{i=1}^{4} \frac{i^3}{2}$

37. $\displaystyle\sum_{k=1}^{3} 2^k$

38. $\displaystyle\sum_{k=1}^{4} 3^k$

39. $\displaystyle\sum_{k=1}^{5} [(-1)^{k+1} \cdot 2k]$

40. $\displaystyle\sum_{k=1}^{8} [(-1)^k \cdot k]$

41. $\displaystyle\sum_{j=1}^{10} 5$

42. $\displaystyle\sum_{j=1}^{8} 2$

43. $\displaystyle\sum_{k=3}^{7} (2k - 1)$

44. $\displaystyle\sum_{j=5}^{10} (k + 4)$

In Problems 45–52, express each sum using summation notation. See Objective 3.

45. $1 + 2 + 3 + \cdots + 15$

46. $1 + 3 + 5 + \cdots + 17$

47. $1 + \dfrac{1}{2} + \dfrac{1}{3} + \cdots + \dfrac{1}{12}$

48. $1 + \dfrac{1}{2} + \dfrac{1}{4} + \cdots + \dfrac{1}{2^{15}}$

49. $1 - \dfrac{1}{3} + \dfrac{1}{9} - \dfrac{1}{27} + \cdots + (-1)^{9+1}\left(\dfrac{1}{3^{9-1}}\right)$

50. $\dfrac{2}{3} - \dfrac{4}{9} + \dfrac{8}{27} + \cdots + (-1)^{15+1}\left(\dfrac{2}{3}\right)^{15}$

51. $5 + (5 + 2\cdot 1) + (5 + 2\cdot 2) + (5 + 2\cdot 3) + \cdots + (5 + 2\cdot 10)$

52. $3 + 3\cdot\dfrac{1}{2} + 3\cdot\dfrac{1}{4} + \cdots + 3\cdot\left(\dfrac{1}{2}\right)^{11}$

Applying the Concepts

53. The Future Value of Money Suppose that you place $12,000 in your company 401(k) plan, which pays 6% interest compounded quarterly. The balance in the account after n quarters is given by

$$a_n = 12,000\left(1 + \frac{0.06}{4}\right)^n$$

(a) Find the value in the account after 1 quarter.
(b) Find the value in the account after 1 year.
(c) Find the value in the account after 10 years.

54. The Future Value of Money Suppose that you place $5000 into a company 401(k) plan that pays 8% interest compounded monthly. The balance in the account after n months is given by

$$a_n = 5000\left(1 + \frac{0.08}{12}\right)^n$$

(a) Find the value in the account after 1 month.
(b) Find the value in the account after 1 year.
(c) Find the value in the account after 10 years.

55. Population Growth According to the U.S. Census Bureau, the population of the United States in 2012 was 313 million people. In addition, the population of the United States was growing at a rate of 1.1% per year. A model for the population of the United States is given by

$$p_n = 313(1.011)^n$$

where n is the number of years after 2012.
(a) Use this model to predict the United States population in 2015 to the nearest million.
(b) Use this model to predict the United States population in 2050 to the nearest million.

56. Population Growth According to the United States Census Bureau, the population of the world in 2012 was 7018 million people. In addition, the population of the world was growing at a rate of 1.26% per year. A model for the population of the world is given by

$$p_n = 7018(1.0126)^n$$

where n is the number of years after 2012.
(a) Use this model to predict the world population in 2015 to the nearest million.
(b) Use this model to predict the world population in 2050 to the nearest million.

57. Fibonacci Sequence Let

$$u_n = \frac{\left(1 + \sqrt{5}\right)^n - \left(1 - \sqrt{5}\right)^n}{2^n \cdot \sqrt{5}}$$

define the nth term of a sequence. Find the first 10 terms of the sequence. This sequence is called the

Fibonacci sequence. The terms of the sequence are called **Fibonacci numbers.**

58. Pascal's Triangle The triangular array of numbers shown below is called Pascal's Triangle. Each number in the triangle is found by adding the entries directly above to the left and right of the number. For example, the 4 in the fourth row is found by adding the 1 and 3 above the 4.

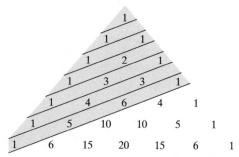

Divide the triangular array using diagonal lines (as shown). Now find the sum of the numbers in each of the highlighted diagonal rows. Do you recognize the sequence? (*Hint:* See Problem 57.)

Extending the Concepts

A second way of defining a sequence is to assign a value to the first (or first few) term(s) and specify the nth term by a formula or equation that involves one or more of the terms preceding it. Sequences defined this way are said to be defined **recursively**, *and the rule or formula is called a* **recursive formula**. *For example, $s_1 = 3$ and $s_n = 2s_{n-1}$ is a recursively defined sequence where $s_1 = 3, s_2 = 2s_1 = 2(3) = 6, s_3 = 2s_2 = 2(6) = 12, s_4 = 2s_3 = 2(12) = 24$, and so on.*

In Problems 59–62, a sequence is defined recursively. Write the first five terms of the sequence.

59. $a_1 = 10, a_n = 1.05a_{n-1}$ **60.** $b_1 = 20, b_n = 3b_{n-1}$

61. $b_1 = 8, b_n = n + b_{n-1}$ **62.** $c_1 = 1000,$
$\qquad\qquad\qquad\qquad\qquad\quad c_n = 1.01c_{n-1} + 100$

63. Fibonacci Sequence Use the result of Problem 57 to do the following problems:

(a) Compute the ratio $\dfrac{u_{n+1}}{u_n}$ for the first 10 terms.

(b) As n gets large, what number does this ratio approach? This number is referred to as the **golden ratio.** Rectangles whose sides are in this ratio were considered pleasing to the eye by the Greeks. For example, the façade of the Parthenon was constructed using the golden ratio.

(c) Compute the ratio $\dfrac{u_n}{u_{n+1}}$ for the first 10 terms.

(d) As n gets large, what number does this ratio approach? This number is also referred to as the **golden ratio.** This ratio is believed to have been used in the construction of the Great Pyramid in Egypt. The ratio equals the sum of the areas of the four face triangles divided by the total surface area of the Great Pyramid.

64. Investigate various applications that lead to a Fibonacci sequence, such as art, architecture, or financial markets. Write an essay on these applications.

Explaining the Concepts

65. Explain how a sequence and a function differ.

66. What does the graph of a sequence look like when compared to the graph of a function? Use the function $f(x) = 3x + 1$ and the sequence $\{a_n\} = \{3n + 1\}$ when doing the comparison.

67. Write a sentence that explains the meaning of the symbol Σ.

68. What does it mean when a sequence alternates?

Synthesis Review

In Problems 69–71, (a) determine the slope of the linear function, and (b) compute $f(1)$, $f(2)$, $f(3)$, and $f(4)$.

69. $f(x) = 4x - 6$

70. $f(x) = 2x - 10$

71. $f(x) = -5x + 8$

72. Compute $f(2) - f(1), f(3) - f(2)$, and $f(4) - f(3)$ for each of the functions in Problems 69–71. How is the difference in the value of the function for consecutive values of the independent variable related to the slope?

The Graphing Calculator

Graphing calculators can be used to write the terms of a sequence. For example, Figure 2 shows the first five terms of the sequence $\{b_n\} = \{(-1)^n \cdot n^2\}$ that we studied in Example 2. Figure 3 shows how the terms of the sequence can be listed in a table using SEQuence mode.

Figure 2

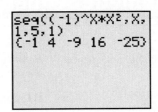

Figure 3

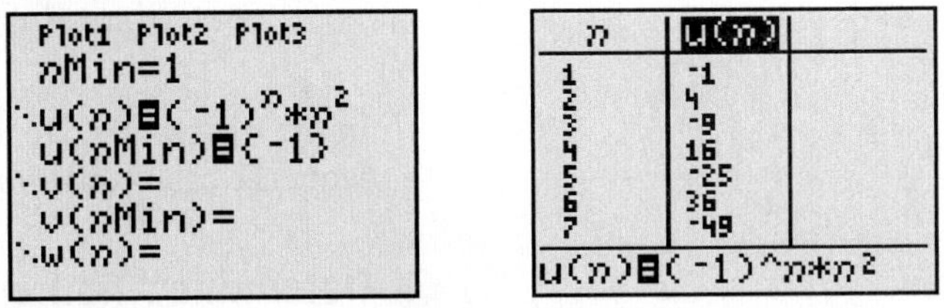

In Problems 73–80, use a graphing calculator to find the first five terms of the sequence.

73. $\{3n + 5\}$

74. $\{n - 4\}$

75. $\left\{\dfrac{n}{n + 2}\right\}$

76. $\left\{\dfrac{n + 4}{n}\right\}$

77. $\{(-1)^n \, n\}$

78. $\{(-1)^{n+1} \, n\}$

79. $\{2^n + 1\}$

80. $\{3^n - 1\}$

Graphing calculators can also be used to find the sum of a sequence. For example, Figure 4 shows the result of the sum $\displaystyle\sum_{i=1}^{5} (i^2 - 5)$ that we studied in Example 4(b).

Figure 4

In Problems 81–88, use a graphing calculator to find the sum.

81. $\displaystyle\sum_{i=1}^{4} (5i + 1)$

82. $\displaystyle\sum_{i=1}^{5} (3i + 2)$

83. $\displaystyle\sum_{i=1}^{5} \dfrac{i^2}{2}$

84. $\displaystyle\sum_{i=1}^{4} \dfrac{i^3}{2}$

85. $\displaystyle\sum_{k=1}^{3} 2^k$

86. $\displaystyle\sum_{k=1}^{4} 3^k$

87. $\displaystyle\sum_{k=1}^{5} [(-1)^{k+1} \cdot 2k]$

88. $\displaystyle\sum_{k=1}^{8} [(-1)^k \cdot k]$

10.2 Arithmetic Sequences

Objectives

① Determine Whether a Sequence Is Arithmetic

② Find a Formula for the nth Term of an Arithmetic Sequence

③ Find the Sum of an Arithmetic Sequence

Are You Ready for This Section?

Before getting started, take the following readiness quiz. If you get a problem wrong, go back to the section cited and review the material.

R1. Determine the slope of $y = -3x + 1$. [Section 1.6 pp. 113–114]

R2. If $g(x) = 5x + 2$, find $g(3)$. [Section 2.2, pp. 156–158]

R3. Solve: $\begin{cases} x - 3y = -17 \\ 2x + y = 1 \end{cases}$ [Section 3.1, pp. 221–226]

In the last section, we looked at sequences in general. Now we look at a specific type of sequence called an *arithmetic sequence.*

▶ ① Determine Whether a Sequence Is Arithmetic

When the difference between successive terms of a sequence is always the same number, the sequence is called an **arithmetic sequence** (sometimes called an **arithmetic progression**). For example, the sequence

$$2, 6, 10, 14, \ldots$$

is arithmetic because the constant difference between consecutive terms is 4. If we call the first term a_1 and the **common difference** between consecutive terms d, then the terms of an arithmetic sequence follow the pattern

$$a_1, a_1 + d, a_1 + 2d, a_1 + 3d, \ldots$$

EXAMPLE 1 **Determining Whether a Sequence Is Arithmetic**

Determine whether the sequence $3, 9, 15, 21, \ldots$ is arithmetic. If it is, find the first term a_1 and the common difference d.

Solution

If the difference between consecutive terms is constant, then the sequence is arithmetic. The sequence $3, 9, 15, 21, \ldots$ is arithmetic because this difference is 6 ($= 9 - 3$ or $15 - 9$ or $21 - 15$). The first term is $a_1 = 3$, and the common difference is $d = 6$. ●

Quick ✓

1. In a(n) _____ sequence, the difference between consecutive terms is constant.

In Problems 2 and 3, determine which of the following sequences is arithmetic. If the sequence is arithmetic, determine the first term a_1 and common difference d.

2. $-3, -1, 1, 3, 5, \ldots$ 3. $3, 9, 27, 81, \ldots$

EXAMPLE 2 **Determining Whether a Sequence Is Arithmetic**

Show that the sequence $\{s_n\} = \{2n + 7\}$ is arithmetic. Find the first term and the common difference.

Solution

We could list the first few terms and show that the difference between consecutive terms is the same, but this would be more of a demonstration than a proof. To prove the sequence is arithmetic, we must show that for *any* consecutive terms, the difference is the same number. We do this by evaluating the sequence at $(n - 1)$ and at n.

Ready? ...Answers
R1. -3 **R2.** 17 **R3.** $(-2, 5)$

Then compute the difference between these values. If it is constant, we've proved the sequence is arithmetic.

$$s_{n-1} = 2(n-1) + 7 = 2n - 2 + 7 = 2n + 5 \quad \text{and} \quad s_n = 2n + 7$$

Now we compute $s_n - s_{n-1}$.

$$s_n - s_{n-1} = (2n + 7) - (2n + 5)$$

$$\text{Distribute:} \quad = 2n + 7 - 2n - 5$$

$$\text{Combine like terms:} \quad = 2$$

The difference between *any* consecutive terms is 2, so the sequence is arithmetic with common difference $d = 2$. To find the first term, we evaluate s_1 and find $s_1 = 2(1) + 7 = 9$, so $a_1 = 9$. ●

EXAMPLE 3 **Determining Whether a Sequence Defined by a Function Is Arithmetic**

Show that the sequence $\{b_n\} = \{n^2\}$ is not arithmetic.

Solution

As in Example 2, evaluate the sequence at $(n - 1)$ and n. Then show that the difference between consecutive terms is not constant.

$$b_{n-1} = (n-1)^2 = n^2 - 2n + 1 \quad \text{and} \quad b_n = n^2$$

Now compute $b_n - b_{n-1}$.

$$b_n - b_{n-1} = n^2 - (n^2 - 2n + 1)$$

$$\text{Distribute:} \quad = n^2 - n^2 + 2n - 1$$

$$\text{Combine like terms:} \quad = 2n - 1$$

The difference between consecutive terms is not constant—its value depends on n. Therefore, the sequence is not arithmetic. ●

Quick ✓

In Problems 4–6, determine whether the sequence is arithmetic. If it is, state the first term a_1 and the common difference d.

4. $\{a_n\} = \{3n - 8\}$ **5.** $\{b_n\} = \{n^2 - 1\}$ **6.** $\{c_n\} = \{5 - 2n\}$

▶ ❷ **Find a Formula for the *n*th Term of an Arithmetic Sequence**

Suppose a_1 is the first term of an arithmetic sequence whose common difference is d. We want to find a formula for a_n, the nth term of the sequence. To do this, we list the first few terms of the sequence.

$$a_1 = a$$
$$a_2 = a_1 + d = a_1 + 1 \cdot d$$
$$a_3 = a_2 + d = (a_1 + d) + d = a_1 + 2d$$
$$a_4 = a_3 + d = (a_1 + 2d) + d = a_1 + 3d$$
$$a_5 = a_4 + d = (a_1 + 3d) + d = a_1 + 4d$$
$$\vdots$$
$$a_n = a_{n-1} + d = [a_1 + (n-2)d] + d = a_1 + (n-1)d$$

The *n*th Term of an Arithmetic Sequence

For an arithmetic sequence $\{a_n\}$ whose first term is a_1 and whose common difference is d, the nth term is determined by the formula

$$a_n = a_1 + (n-1)d$$

EXAMPLE 4 Finding a Formula for the *n*th Term of an Arithmetic Sequence

(a) Write a formula for the *n*th term of an arithmetic sequence whose 4th term is 8 and whose common difference is -3.

(b) Find the 14th term of the sequence.

Solution

(a) We wish to find a formula for the *n*th term of an arithmetic sequence. We know that $d = -3$ and that $a_4 = 8$.

$$n\text{th term:} \quad a_n = a_1 + (n-1)d$$

$$d = -3; a_4 = 8: \quad a_4 = 8 = a_1 + (4-1)(-3)$$

Solve this equation for a_1, the first term of the sequence.

$$8 = a_1 - 9$$

Add 9 to both sides: $17 = a_1$

The formula for the *n*th term is

$$a_n = a_1 + (n-1)d: \quad a_n = 17 + (n-1)(-3)$$

$$\text{Distribute:} \qquad = 17 - 3n + 3$$

$$\text{Combine like terms:} \qquad = -3n + 20$$

(b) To find the 14th term, let $n = 14$ in $a_n = -3n + 20$.

$$a_{14} = -3(14) + 20$$

$$= -42 + 20$$

$$= -22$$

The 14th term in the sequence is -22.

Quick ✓

7. For an arithmetic sequence $\{a_n\}$ whose first term is a_1 and whose common difference is d, the *n*th term is determined by the formula _____.

8. (a) Write a formula for the *n*th term of an arithmetic sequence whose 5th term is 25 and whose common difference is 6.

(b) Find the 14th term of the sequence.

EXAMPLE 5 Finding a Formula for the *n*th Term of an Arithmetic Sequence

The 4th term of an arithmetic sequence is 7, and the 10th term is 31.

(a) Find the first term and the common difference.

(b) Give a formula for the *n*th term of the sequence.

Solution

(a) The *n*th term of an arithmetic sequence is $a_n = a_1 + (n-1)d$, where a_1 is the first term and d is the common difference. Since $a_4 = 7$ and $a_{10} = 31$:

$$\begin{cases} a_4 = a_1 + (4-1)d \\ a_{10} = a_1 + (10-1)d \end{cases} \quad \text{or} \quad \begin{cases} 7 = a_1 + 3d & (1) \\ 31 = a_1 + 9d & (2) \end{cases}$$

This is a system of two linear equations with two variables, a and d. We can solve this system by elimination. If we subtract equation (2) from equation (1), we obtain

$$-24 = -6d$$

Divide both sides by -6: $4 = d$

Let $d = 4$ in equation (1) to find a_1.

$$7 = a_1 + 3(4)$$
$$7 = a_1 + 12$$

Subtract 12 from both sides: $\quad -5 = a_1$

The first term is $a_1 = -5$ and the common difference is $d = 4$.

(b) A formula for the nth term is

$$a_n = a_1 + (n - 1)d$$
$$a_1 = -5; d = 4: \quad = -5 + (n - 1)(4)$$
$$\text{Distribute:} \quad = -5 + 4n - 4$$
$$\text{Combine like terms:} \quad = 4n - 9$$

Quick ✓

9. The 5th term of an arithmetic sequence is 7, and the 13th term is 31.

(a) Find the first term and the common difference.

(b) Give a formula for the nth term of the sequence.

▶ ❸ Find the Sum of an Arithmetic Sequence

The next result gives a formula for finding the sum of the first n terms of an arithmetic sequence.

Work Smart

There are two formulas for finding the sum of the first n terms of an arithmetic sequence. Use Formula (1) if you know n, the first term a_1, and the common difference d; use Formula (2) if you know n, the first term a_1, and the last term a_n.

Sum of the First n Terms of an Arithmetic Sequence

Let $\{a_n\}$ be an arithmetic sequence with first term a_1 and common difference d. The sum S_n of the first n terms of $\{a_n\}$ is

$$S_n = \frac{n}{2}[2a_1 + (n - 1)d] \quad \text{or} \quad S_n = \frac{n}{2}(a_1 + a_n)$$

We next show where these results come from.

$$S_n = a_1 + a_2 + a_3 + \cdots + a_n$$
$$= \underbrace{a_1}_{a_1} + \underbrace{(a_1 + d)}_{a_2} + \underbrace{(a_1 + 2d)}_{a_3} + \cdots + \underbrace{(a_1 + (n - 1)d)}_{a_n}$$

We can also represent S_n by reversing the order in which we add the terms, so that

$$S_n = a_n + a_{n-1} + \cdots + a_1$$
$$= \underbrace{(a_1 + (n - 1)d)}_{a_n} + \underbrace{(a_1 + (n - 2)d)}_{a_{n-1}} + \cdots + \underbrace{a_1}_{a_1}$$

Add these two different representations of S_n as follows:

$$\begin{aligned}
S_n &= \quad a_1 \quad + \quad (a_1 + d) \quad + (a_1 + 2d) \quad + \cdots + (a_1 + (n - 2)d) + (a_1 + (n - 1)d) \\
S_n &= (a_1 + (n - 1)d) + (a_1 + (n - 2)d) + (a_1 + (n - 3)d) + \cdots + \quad (a_1 + d) \quad + \quad a_1 \\
\hline
2S_n &= 2a_1 + (n - 1)d \ + 2a_1 + (n - 1)d \ + 2a_1 + (n - 1)d \ + \cdots + 2a_1 + (n - 1)d \ + 2a_1 + (n - 1)d
\end{aligned}$$

Thus $2S_n$ is $2a_1 + (n - 1)d$ added to itself n times, or $n[2a_1 + (n - 1)d]$. Therefore,

$$2S_n = n[2a_1 + (n - 1)d]$$

Divide both sides by 2: $\quad S_n = \frac{n}{2}[2a_1 + (n - 1)d] \quad$ Formula (1)

This is the first formula in the box. If we rewrite the expression $2a_1 + (n - 1)d$ as $a_1 + a_1 + (n - 1)d$, we notice that $a_1 + (n - 1)d$ is a_n.

$$S_n = \frac{n}{2}[a_1 + a_n] \quad \text{Formula (2)}$$

We have two ways to find the sum of the first n terms of an arithmetic sequence. Notice that Formula (1) involves the first term a_1 and the common difference d, whereas Formula (2) involves the first term a_1 and the last term a_n. Use whichever is easier.

EXAMPLE 6 Finding the Sum of the First n Terms of an Arithmetic Sequence

Find the sum S_n of the first 50 terms of the arithmetic sequence $2, 5, 8, 11, \ldots$.

Solution

Because we know the first term is $a_1 = 2$ and the common difference is $d = 5 - 2 = 3$, we use the formula $S_n = \frac{n}{2}[2a_1 + (n - 1)d]$ to find the sum.

$$S_n = \frac{n}{2}[2a_1 + (n - 1)d]$$

$$n = 50, a_1 = 2, d = 3: \quad S_{50} = \frac{50}{2}[2(2) + (50 - 1)(3)]$$

$$= 25[4 + 49(3)]$$

$$= 25[151]$$

$$= 3775$$

The sum of the first 50 terms, S_{50}, of the arithmetic sequence $2, 5, 8, 11, \ldots$ is 3775. ●

Quick ✓

10. Find the sum S_n of the first 100 terms of the arithmetic sequence whose first term is 5 and whose common difference is 2.

11. Find the sum S_n of the first 70 terms of the arithmetic sequence $1, 5, 9, 13, \ldots$.

EXAMPLE 7 Finding the Sum of the First n Terms of an Arithmetic Sequence

Find the sum S_n of the first 40 terms of the arithmetic sequence $\{-2n + 50\}$.

Solution

We find that the first term is $a_1 = -2(1) + 50 = 48$ and the 40th term is $a_{40} = -2(40) + 50 = -30$. Since we know the first term and the last term of the sequence, we use the formula $S_n = \frac{n}{2}[a_1 + a_n]$ to find the sum.

$$S_n = \frac{n}{2}[a_1 + a_n]$$

$$n = 40, a_1 = 48, a_{40} = -30: \quad S_{40} = \frac{40}{2}[48 + (-30)]$$

$$= 360$$

The sum of the first 40 terms, S_{40}, of the arithmetic sequence $\{-2n + 50\}$ is 360. ●

Quick ✓

12. Find the sum S_n of the first 50 terms of the arithmetic sequence whose first term is 4 and 50th term 298.

13. Find the sum S_n of the first 75 terms of the arithmetic sequence $\{-3n + 100\}$.

EXAMPLE 8   **Creating a Floor Design**

A ceramic tile floor is designed in the shape of a trapezoid 10 feet wide at the base and 5 feet wide at the top. See Figure 5. The tiles, which measure 6 inches by 6 inches, are to be placed so that each successive row contains one fewer tile than the preceding row. How many tiles are required?

Figure 5

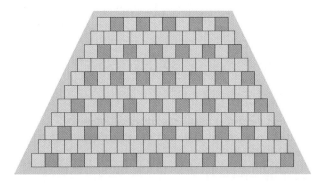

Solution

Each tile is 6 inches (0.5 foot) wide, and the bottom row is 10 feet wide, so the bottom row requires 20 tiles. Similar reasoning tells us the top row requires 10 tiles. Since each successive row has one fewer tile, the total number of tiles required is

$$S = 20 + 19 + 18 + \cdots + 11 + 10$$

This is the sum of an arithmetic sequence with common difference -1. The number of terms to be added is $n = 11$, the first term is $a_1 = 20$, and the last term is $a_{11} = 10$. The sum S is

$$S_n = \frac{n}{2}[a_1 + a_n] \qquad S_{11} = \frac{11}{2}(20 + 10) = 165$$

In all, 165 tiles will be required.

Quick ✓

14. In the corner section of a theater, the first row has 20 seats. Each subsequent row has 2 more seats, and there are a total of 30 rows. How many seats are in this section?

10.2 Exercises   MyMathLab® MathXL PRACTICE   Exercise numbers in green are complete video solutions in MyMathLab

*Problems **1–14** are the* Quick ✓s *that follow the* EXAMPLES.

Building Skills

In Problems 15–22, verify that the given sequence is arithmetic. Then find the first term and common difference. See Objective 1.

15. $\{n + 5\}$ **16.** $\{n - 1\}$ **17.** $\{7n + 2\}$

18. $\{10n + 1\}$ **19.** $\{7 - 3n\}$ **20.** $\{5 - 2n\}$

21. $\left\{\dfrac{1}{2}n + 5\right\}$ **22.** $\left\{\dfrac{1}{4}n + \dfrac{3}{4}\right\}$

In Problems 23–30, find a formula for the nth term of the arithmetic sequence whose first term a_1 and common difference d are given. What is the 5th term? See Objective 2.

23. $a_1 = 4; d = 3$ **24.** $a_1 = 8; d = 3$

25. $a_1 = 10; d = -5$ **26.** $a_1 = 12; d = -3$

27. $a_1 = 2; d = \dfrac{1}{3}$ **28.** $a_1 = -3; d = \dfrac{1}{2}$

29. $a_1 = 5; d = -\dfrac{1}{5}$ **30.** $a_1 = -\dfrac{4}{3}; d = -\dfrac{2}{3}$

In Problems 31–36, write a formula for the nth term of each arithmetic sequence. Use the formula to find the 20th term in each arithmetic sequence. See Objective 2.

31. 2, 7, 12, 17, . . .

32. −5, −1, 3, 7, . . .

33. 12, 9, 6, 3, . . .

34. 20, 14, 8, 2, . . .

35. $1, \dfrac{5}{4}, \dfrac{3}{2}, \dfrac{7}{4}, \ldots$

36. $10, \dfrac{19}{2}, 9, \dfrac{17}{2}, \ldots$

In Problems 37–44, find the first term and the common difference of the arithmetic sequence described. Give a formula for the nth term of the sequence. See Objective 2.

37. 3rd term is 17; 7th term is 37

38. 5th term is 7; 9th term is 19

39. 4th term is −2; 8th term is 26

40. 2nd term is −9; 8th term is 15

41. 5th term is −1; 12th term is −22

42. 6th term is −8; 12th term is −38

43. 3rd term is 3; 9th term is 0

44. 5th term is 5; 13th term is 7

For Problems 45–56, see Objective 3.

45. Find the sum of the first 30 terms of the sequence 2, 8, 14, 20, ...

46. Find the sum of the first 40 terms of the sequence 1, 8, 15, 22, ...

47. Find the sum of the first 25 terms of the sequence −8, −5, −2, 1, ...

48. Find the sum of the first 75 terms of the sequence −9, −5, −1, 3, ...

49. Find the sum of the first 40 terms of the sequence 10, 3, −4, −11, ...

50. Find the sum of the first 50 terms of the sequence 12, 4, −4, −12, ...

51. Find the sum of the first 40 terms of the arithmetic sequence $\{4n - 3\}$.

52. Find the sum of the first 80 terms of the arithmetic sequence $\{2n - 13\}$.

53. Find the sum of the first 75 terms of the arithmetic sequence $\{-5n + 70\}$.

54. Find the sum of the first 35 terms of the arithmetic sequence $\{-6n + 25\}$.

55. Find the sum of the first 30 terms of the arithmetic sequence $\left\{5 + \dfrac{2}{3}n\right\}$.

56. Find the sum of the first 28 terms of the arithmetic sequence $\left\{7 - \dfrac{3}{2}n\right\}$.

Applying the Concepts

57. Find x so that $x + 3, 2x + 1,$ and $5x + 2$ are consecutive terms of an arithmetic sequence.

58. Find x so that $2x, 3x + 2,$ and $5x + 3$ are consecutive terms of an arithmetic sequence.

59. A Stack of Cans Suppose that the bottom row in a stack of cans contains 35 cans. Each layer contains one can fewer than the layer below it. The top row has 1 can. How many cans are in the stack?

60. A Pile of Bricks Suppose that the bottom row in a pile of bricks contains 46 bricks. Each layer contains 2 bricks fewer than the layer below it. The top row has 2 bricks. How many bricks are in the stack?

61. The Theater An auditorium has 40 seats in the first row and 25 rows in all. Each successive row contains 2 additional seats. How many seats are in the auditorium?

62. Mosaic A mosaic is designed in the shape of an equilateral triangle, 20 feet on each side. Each tile in the mosaic is in the shape of an equilateral triangle, 12 inches to a side. The tiles are to alternate in color as shown in the illustration. How many tiles of each color will be required?

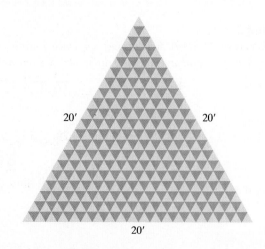

Extending the Concepts

In Problems 63–66, determine the number of terms that are in each arithmetic sequence.

63. $-5, -2, 1, \ldots, 244$

64. $-9, -5, -1, \ldots, 219$

65. $108, 101, 94, \ldots, -326$

66. $99, 93, 87, \ldots, -339$

67. Salary Suppose you have just been hired at the starting salary of $32,000 per year. Your contract guarantees you a $2500 raise each year. How many years will it take before your aggregate salary is $757,500? (*Hint:* Your aggregate salary is $32,000 + ($32,000 + $2500) + \cdots$.)

68. Stadium Construction How many rows are in the corner section of a stadium containing 2030 seats if the first row has 14 seats and each successive row has 4 additional seats?

Explaining the Concepts

69. Explain how you can determine whether a sequence is arithmetic.

70. Provide an explanation that justifies the formula for the sum of the first n terms of an arithmetic sequence.

Synthesis Review

In Problems 71–73, (a) determine the base a of the exponential function, and (b) compute f(1), f(2), f(3), and f(4).

71. $f(x) = 3^x$

72. $f(x) = 4^x$

73. $f(x) = 10\left(\dfrac{1}{2}\right)^x$

74. Compute $\dfrac{f(2)}{f(1)}, \dfrac{f(3)}{f(2)},$ and $\dfrac{f(4)}{f(3)}$ for each of the functions in Problems 71–73. How is the ratio in the value of the function for consecutive values of the independent variable related to the base?

The Graphing Calculator

In Problems 75–78, use a graphing calculator to find the sum of each sequence.

75. $\{3.45n + 4.12\}; n = 20$

76. $\{2.67n - 1.23\}; n = 25$

77. $85.9 + 83.5 + 81.1 + \cdots; n = 25$

78. $-11.8 + (-8.2) + (-4.6) + \cdots; n = 30$

10.3 Geometric Sequences and Series

Objectives

1 Determine Whether a Sequence Is Geometric

2 Find a Formula for the nth Term of a Geometric Sequence

3 Find the Sum of a Geometric Sequence

4 Find the Sum of a Geometric Series

5 Solve Annuity Problems

Are You Ready for This Section?

Before getting started, take the following readiness quiz. If you get a problem wrong, go back to the section cited and review the material.

R1. If $g(x) = 4^x$, evaluate $g(1), g(2),$ and $g(3)$.   [Section 8.2, pp. 638–641]

R2. Simplify: $\dfrac{x^4}{x^3}$ [Getting Ready, pp. 296–297]

▶ **1 Determine Whether a Sequence Is Geometric**

We have seen that the *difference* between consecutive terms in an arithmetic sequence is constant. If the *ratio* of consecutive terms in a sequence is constant, then the sequence is a **geometric sequence.** For example, the sequence

$$2, 4, 8, 16, \ldots$$

is geometric because the ratio of consecutive terms is $2\left(= \dfrac{4}{2} = \dfrac{8}{4} = \dfrac{16}{8}\right)$. If we call the first term a_1 and the **common ratio** of consecutive terms r, then the terms of a geometric sequence follow the pattern

$$a_1, a_1r, a_1r^2, a_1r^3, \ldots$$

Ready?...Answers

R1. $g(1) = 4, g(2) = 16, g(3) = 64$

R2. x

EXAMPLE 1   **Determining Whether a Sequence Is Geometric**

Determine whether the sequence 2, 6, 18, 54, 162, . . . is geometric. If it is, state the first term and the common ratio.

Solution

The sequence is geometric if the ratio of consecutive terms is a constant. Because the ratio of consecutive terms is $3 \left(= \dfrac{6}{2} = \dfrac{18}{6} = \dfrac{54}{18} = \dfrac{162}{54} \right)$, the sequence is geometric. The first term is $a_1 = 2$ and the common ratio is $r = 3$.   ●

Quick ✓

1. In a(n) _____ sequence the ratio of consecutive terms is constant.

In Problems 2–4, determine whether the sequence is geometric. If it is, state the first term and common ratio.

2. 4, 8, 16, 32, 64, . . .   **3.** 5, 10, 16, 23, 31, . . .   **4.** $9, 3, 1, \dfrac{1}{3}, \dfrac{1}{9}, \ldots$

EXAMPLE 2   **Determining Whether a Sequence Is Geometric**

Determine whether the sequence $\{b_n\} = \{3^{-n}\}$ is geometric. If it is, find the common ratio.

Solution

We must show that for *any* consecutive terms, the ratio is the same number. We do this by evaluating the sequence at $(n - 1)$ and n. Then compute the ratio of these two values. If the ratio is constant, we've shown the sequence is geometric.

$$b_{n-1} = 3^{-(n-1)} = 3^{-n+1} \quad \text{and} \quad b_n = 3^{-n}$$

Work Smart

Notice that the technique used in Example 2 is similar to the approach used in Section 10.2 to show a sequence is arithmetic.

Now compute $\dfrac{b_n}{b_{n-1}}$.

$$\frac{b_n}{b_{n-1}} = \frac{3^{-n}}{3^{-n+1}}$$

$$a^{m+n} = a^m \cdot a^n: \quad = \frac{3^{-n}}{3^{-n} \cdot 3^1}$$

$$\text{Divide out like factors:} \quad = \frac{1}{3}$$

The ratio of *any* two consecutive terms is the constant $\dfrac{1}{3}$, so the sequence is geometric with $r = \dfrac{1}{3}$.   ●

EXAMPLE 3   **Determining Whether a Sequence Is Geometric**

Determine whether the sequence $\{a_n\} = \{n^2 + 1\}$ is geometric. If it is, find the common ratio.

Solution

If the ratio $\dfrac{a_n}{a_{n-1}}$ is constant, then the sequence is geometric.

$$\frac{a_n}{a_{n-1}} = \frac{n^2 + 1}{(n-1)^2 + 1}$$

Use FOIL: $\quad = \dfrac{n^2 + 1}{n^2 - 2n + 1 + 1}$

Combine like terms: $\quad = \dfrac{n^2 + 1}{n^2 - 2n + 2}$

We cannot simplify the rational expression any further. Because the ratio $\dfrac{a_n}{a_{n-1}}$ depends on the value of n, it is not constant. Therefore, the sequence is not geometric. ●

> **Quick ✓**
>
> In Problems 5–7, determine whether the sequence is geometric. If it is geometric, find the common ratio.
>
> **5.** $\{a_n\} = \{5^n\}$ $\qquad\qquad$ **6.** $\{b_n\} = \{n^2\}$
>
> **7.** $\{c_n\} = \left\{ 5\left(\dfrac{2}{3}\right)^n \right\}$

▶ ❷ Find a Formula for the *n*th Term of a Geometric Sequence

If a_1 is the first term of a geometric sequence with common ratio $r \neq 0$, what is a formula for the *n*th term of a_n? Let's list the first few terms to find out.

$$a_1 = 1a_1 = a_1 r^0$$
$$a_2 = ra_1 = a_1 r^1$$
$$a_3 = ra_2 = a_1 r^2$$
$$a_4 = ra_3 = a_1 r^3$$
$$a_5 = ra_4 = a_1 r^4$$
$$\vdots$$
$$a_n = ra_{n-1} = r(a_1 r^{n-2}) = a_1 r^{n-1}$$

We are led to the following result.

> **The *n*th Term of a Geometric Sequence**
>
> If a geometric sequence $\{a_n\}$ has first term a_1 and common ratio r, then the *n*th term is determined by the formula
>
> $$a_n = a_1 r^{n-1} \qquad r \neq 0$$

EXAMPLE 4 **Finding a Particular Term of a Geometric Sequence**

Consider the geometric sequence $8, 6, \dfrac{9}{2}, \dfrac{27}{8}, \ldots$.

(a) Find a formula for the *n*th term.

(b) Find the 11th term of the sequence.

Solution

(a) The sequence is geometric. The first term is $a_1 = 8$. The common ratio, r, is the ratio of any two consecutive terms. So

$$r = \frac{6}{8} = \frac{9/2}{6} = \frac{27/8}{9/2} = \frac{3}{4}.$$ Now substitute these values in the formula for the nth term of a geometric sequence.

$$a_n = a_1 r^{n-1}$$

$$a_1 = 8; r = \frac{3}{4}: \qquad = 8\left(\frac{3}{4}\right)^{n-1}$$

(b) Find the eleventh term by letting $n = 11$ in the formula found in part (a).

$$a_n = 8\left(\frac{3}{4}\right)^{n-1}$$

$$a_{11} = 8\left(\frac{3}{4}\right)^{11-1}$$

$$= 8\left(\frac{3}{4}\right)^{10}$$

$$\approx 0.45051$$

Quick ✓

In Problems 8 and 9, find a formula for the nth term of each geometric sequence. Use this result to find the 9th term of the sequence.

8. $a_1 = 5, r = 2$ **9.** $\{50, 25, 12.5, 6.25, \ldots\}$

▶ ❸ Find the Sum of a Geometric Sequence

The next result gives us a formula for finding the sum of the first n terms of a geometric sequence.

Sum of the First *n* Terms of a Geometric Sequence

Let $\{a_n\}$ be a geometric sequence with first term a_1 and common ratio r, where $r \neq 0, r \neq 1$. The sum S_n of the first n terms of $\{a_n\}$ is

$$S_n = a_1 \cdot \frac{1 - r^n}{1 - r} \qquad r \neq 0, r \neq 1$$

Where does this formula come from? The sum S_n of the first n terms of $\{a_n\} = \{a_1 r^{n-1}\}$ is

$$S_n = a_1 + a_1 r + a_1 r^2 + \cdots + a_1 r^{n-1}$$

Multiply both sides by r to obtain

$$rS_n = a_1 r + a_1 r^2 + a_1 r^3 + \cdots + a_1 r^n$$

Now subtract rS_n from S_n:

$$\begin{aligned} S_n &= a_1 + a_1 r + a_1 r^2 + \cdots + a_1 r^{n-1} \\ rS_n &= a_1 r + a_1 r^2 + a_1 r^3 + \cdots + a_1 r^n \\ \hline S_n - rS_n &= a_1 - a_1 r^n \end{aligned}$$

Factor S_n from the expression on the left and factor a_1 from the expression on the right.

$$S_n(1 - r) = a_1(1 - r^n)$$

Divide both sides by $1 - r$ (since $r \neq 1$) and solve for S_n.

$$S_n = a_1 \cdot \frac{1 - r^n}{1 - r}$$

EXAMPLE 5 **Finding the Sum of the First n Terms of a Geometric Sequence**

Find the sum of the first 10 terms of the sequence $2, 6, 18, 54, \ldots$.

Solution

We want the sum of the first 10 terms of a geometric sequence with $a_1 = 2$ and common ratio $r = 3$.

$$S_n = a_1 \cdot \frac{1 - r^n}{1 - r}$$

$a_1 = 2, r = 3, n = 10:$ $\quad S_{10} = 2 \cdot \dfrac{1 - 3^{10}}{1 - 3}$

$$= 2 \cdot \frac{-59{,}048}{-2}$$

$$= 59{,}048 \qquad \bullet$$

EXAMPLE 6 **Finding the Sum of the First n Terms of a Geometric Sequence in Summation Notation**

Find the sum: $\displaystyle\sum_{n=1}^{8}\left[5 \cdot \left(\frac{1}{2}\right)^n\right]$

Express your answer to as many decimal places as your calculator allows.

Solution

We can expand this sum as follows:

$$\sum_{n=1}^{8}\left[5 \cdot \left(\frac{1}{2}\right)^n\right] = \frac{5}{2} + \frac{5}{4} + \cdots + \frac{5}{256}$$

Work Smart

$$\sum_{n=1}^{8}\left[5 \cdot \frac{1}{2}\right]^n =$$

$$5\left(\frac{1}{2}\right)^1 + 5\left(\frac{1}{2}\right)^2 + 5\left(\frac{1}{2}\right)^3 + 5\left(\frac{1}{2}\right)^4$$

$$+ \cdots + 5\left(\frac{1}{2}\right)^8$$

We want to find the sum of the first $n = 8$ terms of a geometric sequence with $a_1 = \dfrac{5}{2}$ and common ratio $r = \dfrac{1}{2}$.

$$S_n = a_1 \cdot \frac{1 - r^n}{1 - r}$$

$a_1 = \dfrac{5}{2}, r = \dfrac{1}{2}, n = 8:$ $\quad S_8 = \dfrac{5}{2} \cdot \dfrac{1 - \left(\dfrac{1}{2}\right)^8}{1 - \dfrac{1}{2}}$

$$= \frac{5}{2} \cdot \frac{0.99609375}{\dfrac{1}{2}}$$

$$= 4.98046875 \qquad \bullet$$

Quick ✓

10. For a geometric sequence with first term a_1 and common ratio r, where

$r \neq 0, r \neq 1$, the sum of the first n terms is _____.

In Problems 11 and 12, find the sum.

11. $3 + 6 + 12 + 24 + \cdots + 3 \cdot 2^{12}$

12. $\displaystyle\sum_{n=1}^{10} \left[8 \cdot \left(\frac{1}{2} \right)^n \right]$

▶ ❹ Find the Sum of a Geometric Series

An infinite sum of the form

$$a_1 + a_1 r + a_1 r^2 + \cdots + a_1 r^{n-1} + \cdots$$

whose first term is a_1 and whose common ratio is r, is called an **infinite geometric series** and is denoted by

$$\sum_{n=1}^{\infty} a_1 r^{n-1}$$

In Words

A sequence is a list of terms. A series is the sum of an infinite number of terms.

The sum of the first n terms of a geometric sequence is given by the formula

$$S_n = a_1 \cdot \frac{1 - r^n}{1 - r}$$

Distribute the a_1: $= \dfrac{a_1 - a_1 r^n}{1 - r}$

Divide each term by $1 - r$: $= \dfrac{a_1}{1 - r} - \dfrac{a_1 r^n}{1 - r}$

As n gets larger and larger, the expression S_n approaches the value $\dfrac{a_1}{1 - r}$ because r^n approaches 0 provided that $-1 < r < 1$.

Sum of a Geometric Series

If $-1 < r < 1$, the sum of the terms of an infinite geometric series with first term a_1 and common ratio r is

$$\sum_{n=1}^{\infty} a_1 r^{n-1} = \frac{a_1}{1 - r}$$

EXAMPLE 7 **Finding the Sum of a Geometric Series**

Find the sum of the geometric series: $4 + 2 + 1 + \dfrac{1}{2} + \cdots$

Solution

This is a geometric series with $a_1 = 4$ and common ratio $r = \dfrac{1}{2} \left(= \dfrac{2}{4} = \dfrac{1}{2} = \dfrac{\frac{1}{2}}{1} \right)$.

Since the common ratio r is between -1 and 1, we can use the formula for the sum of a geometric series to find that

$$4 + 2 + 1 + \frac{1}{2} + \cdots = \frac{a_1}{1 - r} = \frac{4}{1 - \dfrac{1}{2}} = 8$$

●

Quick ✓

13. If $-1 < r < 1$, then the sum of the geometric series $\displaystyle\sum_{n=1}^{\infty} a_1 r^{n-1} = $ _____.

In Problems 14 and 15, find the sum of the geometric series.

14. $10 + \dfrac{5}{2} + \dfrac{5}{8} + \dfrac{5}{32} + \cdots$ **15.** $\displaystyle\sum_{n=1}^{\infty} \left(\dfrac{1}{3}\right)^n$

Work Smart: Study Skills

Let's summarize the formulas for arithmetic and geometric sequences where a_1 is the first term of the sequence, d is the common difference, and r is the common ratio.

	Arithmetic	**Geometric**
*n*th term	$a_n = a_1 + (n-1)d$	$a_n = a_1 r^{n-1}, r \neq 0$
Sum of the first *n* terms	$S_n = \dfrac{n}{2}[2a_1 + (n-1)d]$ $= \dfrac{n}{2}(a_1 + a_n)$	$S_n = a_1 \cdot \dfrac{1-r^n}{1-r}, r \neq 0, r \neq 1$

If you were asked to find the 5th term of the sequence $5, 20, 80, 320, \ldots$, which formula would you use?

If you were asked to find the sum of the first 5 terms of the sequence $5, 9, 13, 17, \ldots$, which formula would you use?

EXAMPLE 8   **Writing a Repeating Decimal as a Fraction**

Express $0.\overline{1}$ as a fraction in lowest terms.

Solution

The line over the 1 indicates that the 1 repeats indefinitely. That is,

$$0.\overline{1} = 0.11111\cdots$$
$$= 0.1 + 0.01 + 0.001 + \cdots$$

This is a geometric series with $a_1 = 0.1$ and common ratio $r = 0.1 \left(=\dfrac{0.01}{0.1} = \dfrac{0.001}{0.01}\right)$.

Since $-1 < r < 1$, we can use the formula for the sum of a geometric series to find that

$$0.\overline{1} = 0.1 + 0.01 + 0.001 + \cdots$$

$a_1 = 0.1, r = 0.1$ in $\dfrac{a_1}{1-r}$: $= \dfrac{0.1}{1 - 0.1}$

$$= \dfrac{0.1}{0.9}$$

Multiply numerator and denominator by 10: $= \dfrac{1}{9}$

Thus $0.\overline{1} = \dfrac{1}{9}$.

Quick ✓

16. Express $0.\overline{2}$ as a fraction in lowest terms.

EXAMPLE 9 **The Multiplier**

Suppose that Americans spend 90% of every additional dollar they earn. Economists would say that an individual's **marginal propensity to consume** is 0.90. For example, if Roberta earns a dollar, she will spend $0.9(\$1) = \$0.90$ of it and save $0.10. Whoever earns $0.90 (from Roberta) will spend 90% of it, or $0.9(\$0.90) = \$0.81$. This process of spending continues and results in a geometric series as follows:

$$\$1 + \$0.90 + \$0.81 + \$0.729 + \cdots$$

The sum of this geometric series is called the **multiplier.** The multiplier is used to determine the total impact of spending on the U.S. economy. Suppose the government gives a child-tax rebate of $500 to Roberta. Determine the impact of the rebate on the U.S. economy if Americans spend 90% of every dollar they earn.

Solution

The total impact of the $500 tax rebate on the U.S. economy is

$$\$500 + \$500(0.9) + \$500(0.9)^2 + \$500(0.9)^3 + \cdots$$

This is a geometric series with first term $a_1 = 500$ and common ratio $r = 0.9$. The sum of this series is

$$\$500 + \$500(0.9) + \$500(0.9)^2 + \$500(0.9)^3 + \ldots = \frac{\$500}{1 - 0.9}$$
$$= \$5000$$

The United States economy will grow by $5000 because of the child-tax credit to Roberta.

Quick ✓

17. Redo Example 9 if the marginal propensity to consume is 95%.

▶ ❺ **Solve Annuity Problems**

In Section 8.2, we looked at the compound interest formula, which lets us calculate the future value of a lump sum of money that is deposited in an account that pays interest compounded periodically. Often, though, money is invested at periodic intervals of time. An **annuity** is a sequence of equal periodic deposits that may be made annually, quarterly, monthly, or daily.

When deposits are made at the same time the interest is credited, the annuity is called **ordinary.** We will discuss only ordinary annuities. The **amount of an annuity** is the sum of all deposits made plus all interest paid.

Suppose the interest an account earns is i percent per payment period (expressed as a decimal). For example, if an account pays 6% compounded monthly (12 times a year), then $i = \dfrac{0.06}{12} = 0.005$. Let's develop a formula for the amount of an annuity. Suppose $\$P$ is deposited each payment period for n payment periods in an account that earns $i\%$ per payment period. When the last deposit is made at the nth payment period, the first deposit has earned interest compounded for $n - 1$ payment periods, the second deposit of $\$P$ has earned interest compounded for $n - 2$ payment periods, and so on. Table 1 shows the value of each deposit after n deposits have been made.

Table 1

Deposit	1	2	3	$n-1$	n
Future Value of Deposit of $P	$P(1 + i)^{n-1}$	$P(1 + i)^{n-2}$	$P(1 + i)^{n-3}$	$P(1 + i)$	P

Work Smart

The common ratio is $1 + i$ because
$$\frac{(1 + i)^{n-2}}{(1 + i)^{n-3}} = (1 + i)^{n-2-(n-3)}$$
$$= (1 + i)^{n-2-n+3}$$
$$= (1 + i)^1$$
$$= 1 + i$$

The amount A of the annuity is the sum of the amounts shown in Table 1, namely,

$$A = P(1 + i)^{n-1} + P(1 + i)^{n-2} + P(1 + i)^{n-3} + \cdots + P(1 + i) + P$$
$$= P[(1 + i)^{n-1} + (1 + i)^{n-2} + (1 + i)^{n-3} + \cdots + (1 + i) + 1]$$
$$= P[1 + (1 + i) + \cdots + (1 + i)^{n-3} + (1 + i)^{n-2} + (1 + i)^{n-1}]$$

The expression is the sum of a geometric sequence with n terms, first term $a_1 = P$, and a common ratio $r = (1 + i)$. Using $S_n = a_1 \dfrac{1 - r^n}{1 - r}$, we have

$$A = P[1 + (1 + i) + \cdots + (1 + i)^{n-3} + (1 + i)^{n-2} + (1 + i)^{n-1}]$$
$$= P \cdot \frac{1 - (1 + i)^n}{1 - (1 + i)} = P \cdot \frac{1 - (1 + i)^n}{-i} = P \cdot \frac{(1 + i)^n - 1}{i}$$

We have the following result.

Amount of an Annuity

If P represents the deposit in dollars made at each payment period for an annuity at i percent interest per payment period, then the amount A of the annuity after n payment periods is

$$A = P \cdot \frac{(1 + i)^n - 1}{i}$$

EXAMPLE 10 **Determining the Amount of an Annuity**

To save for retirement, Alejandro decides to put $100 into a Roth Individual Retirement Account (IRA) every month for the next 30 years. What will be the value of the IRA after 30 years if his account earns 6% interest compounded monthly?

Solution

This is an ordinary annuity with $n = 12 \cdot 30 = 360$ payments and deposits of $P = \$100$. The rate of interest per payment period is $i = \dfrac{0.06}{12} = 0.005$. The amount after 30 years (360 deposits) is

$$A = \$100 \left[\frac{(1 + 0.005)^{360} - 1}{0.005} \right]$$
$$= \$100[1004.515042]$$
$$= \$100,451.50$$

●

Quick ✓

18. To save for retirement, Magglio decides to place $500 into a Roth Individual Retirement Account (IRA) every quarter (every three months) for the next 30 years. What will be the value of the IRA after 30 years if his account earns 8% interest compounded quarterly?

10.3 Exercises MyMathLab® PRACTICE

Exercise numbers in green have complete video solutions in MyMathLab.

*Problems **1–18** are the Quick ✓s that follow the EXAMPLES.*

Building Skills

In Problems 19–26, show that the given sequence is a geometric sequence. Then find the first term and common ratio. See Objective 1.

19. $\{4^n\}$

20. $\{(-2)^n\}$

21. $\left\{\left(\dfrac{2}{3}\right)^n\right\}$

22. $\left\{\dfrac{2^n}{3}\right\}$

23. $\{3 \cdot 2^{-n}\}$

24. $\left\{-10 \cdot \left(\dfrac{1}{2}\right)^n\right\}$

25. $\left\{\dfrac{5^{n-1}}{2^n}\right\}$

26. $\left\{\dfrac{3^{-n}}{2^{n-1}}\right\}$

In Problems 27–34, (a) find a formula for the nth term of the geometric sequence whose first term and common ratio are given, and (b) use the formula to find the 8th term. See Objective 2.

27. $a_1 = 10, r = 2$

28. $a_1 = 2, r = 3$

29. $a_1 = 100, r = \dfrac{1}{2}$

30. $a_1 = 30, r = \dfrac{1}{3}$

31. $a_1 = 1, r = -3$

32. $a_1 = 1, r = -4$

33. $a_1 = 100, r = 1.05$

34. $a_1 = 500, r = 1.04$

In Problems 35–40, find the indicated term of each geometric sequence. See Objective 2.

35. 10th term of $3, 6, 12, 24, \ldots$

36. 12th term of $1, 3, 9, 27, \ldots$

37. 15th term of $4, -2, 1, -\dfrac{1}{2}, \ldots$

38. 8th term of $10, -20, 40, -80, \ldots$

39. 9th term of $0.5, 0.05, 0.005, 0.0005, \ldots$

40. 10th term of $0.4, 0.04, 0.004, 0.0004, \ldots$

In Problems 41–48, find the sum. If necessary, express your answer to as many decimal places as your calculator allows. See Objective 3.

41. $2 + 4 + 8 + \cdots + 2^{12}$

42. $3 + 9 + 27 + \cdots + 3^{10}$

43. $50 + 20 + 8 + \dfrac{16}{5} + \cdots + 50\left(\dfrac{2}{5}\right)^{10-1}$

44. $10 + 5 + \dfrac{5}{2} + \cdots + 10\left(\dfrac{1}{2}\right)^{12-1}$

45. $\displaystyle\sum_{n=1}^{10} [3 \cdot 2^n]$

46. $\displaystyle\sum_{n=1}^{12} [5 \cdot 2^n]$

47. $\displaystyle\sum_{n=1}^{8} \left[\dfrac{4}{2^{n-1}}\right]$

48. $\displaystyle\sum_{n=1}^{14} \left[10 \cdot \left(\dfrac{1}{2}\right)^{n-1}\right]$

In Problems 49–58, find the sum of each geometric series. If necessary, express your answer to as many decimal places as your calculator allows. See Objective 4.

49. $1 + \dfrac{1}{2} + \dfrac{1}{4} + \cdots$

50. $1 + \dfrac{1}{3} + \dfrac{1}{9} + \cdots$

51. $10 + \dfrac{10}{3} + \dfrac{10}{9} + \cdots$

52. $20 + 5 + \dfrac{5}{4} + \cdots$

53. $6 - 2 + \dfrac{2}{3} - \dfrac{2}{9} + \cdots$

54. $12 - 3 + \dfrac{3}{4} - \dfrac{3}{16} + \cdots$

55. $\displaystyle\sum_{n=1}^{\infty} \left(5 \cdot \left(\dfrac{1}{5}\right)^n\right)$

56. $\displaystyle\sum_{n=1}^{\infty} \left(10 \cdot \left(\dfrac{1}{3}\right)^n\right)$

57. $\displaystyle\sum_{n=1}^{\infty} \left(12 \cdot \left(-\dfrac{1}{3}\right)^{n-1}\right)$

58. $\displaystyle\sum_{n=1}^{\infty} \left(100 \cdot \left(-\dfrac{1}{2}\right)^{n-1}\right)$

In Problems 59–62, express each repeating decimal as a fraction in lowest terms. See Objective 4.

59. $0.\overline{5}$

60. $0.\overline{3}$

61. $0.\overline{89}$

62. $0.\overline{45}$

Mixed Practice

In Problems 63–74, determine whether the given sequence is arithmetic, geometric, or neither. If the sequence is arithmetic, find the common difference; if it is geometric, find the common ratio.

63. $\{5n + 1\}$

64. $\{8 - 3n\}$

65. $\{2n^2\}$

66. $\{n^2 - 2\}$

67. $\left\{\dfrac{2^{-n}}{5}\right\}$

68. $\left\{\dfrac{2}{3^n}\right\}$

69. $54, 36, 24, 16, \ldots$

70. $100, 20, 4, \dfrac{4}{5}, \ldots$

71. $2, 6, 10, 14, \ldots$

72. $15, 12, 9, 6, \ldots$

73. $1, 2, 3, 5, 8, \ldots$

74. $5, -2, 3, -1, 2, \ldots$

Applying the Concepts

75. Find x so that $x, x + 2,$ and $x + 3$ are consecutive terms of a geometric sequence.

76. Find x so that $x - 1, x,$ and $x + 2$ are consecutive terms of a geometric sequence.

77. Salary Increases Suppose that you have been hired at an annual salary of $40,000 per year. You have been promised a raise of 5% for each of the next 10 years.
   (a) What will be your salary at the beginning of your 2nd year?
   (b) What will be your salary at the beginning of your 10th year?
   (c) How much will you have earned cumulatively once you have finished your 10th year?

78. Salary Increases Suppose that you have been hired at an annual salary of $45,000 per year. Historically, the typical raise is 4% each year. You expect to be at the company for the next 10 years.
   (a) What will be your salary at the beginning of your 2nd year?
   (b) What will be your salary at the beginning of your 10th year?
   (c) How much will you have earned cumulatively once you have finished your 10th year?

79. Depreciation of a Car Suppose that you have just purchased a Honda Accord for $20,000. Historically, the car depreciates by 8% each year, so that next year the car is worth $20,000(0.92). What will the value of the car be after you have owned it for 5 years?

80. Depreciation of a Car Suppose that you have just purchased a Chevy Impala for $16,000. Historically, the car depreciates by 10% each year, so that next year the car is worth $16,000(0.9). What will the value of the car be after you have owned it for 4 years?

81. Pendulum Swings Initially, a pendulum swings through an arc of 3 feet. On each successive swing, the length of the arc is 0.95 of the previous length.
   (a) What is the length of the arc after 10 swings?
   (b) On which swing is the length of the arc less than 1 foot for the first time?
   (c) After 10 swings, what total length will the pendulum have swung?
   (d) When it stops, what total length will the pendulum have swung?

82. Bouncing Balls A ball is dropped from a height of 30 feet. Each time it strikes the ground, it bounces up to 0.8 of the previous height.

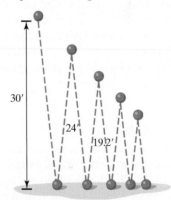

   (a) What height will the ball bounce up to after it strikes the ground for the 4th time?
   (b) How high will it bounce after it strikes the ground for the 5th time?
   (c) How many times does the ball need to strike the ground before its bounce is less than 6 inches?
   (d) What total distance does the ball travel before it stops bouncing?

83. A Job Offer You are interviewing for a job and receive two offers:

   A: $30,000 to start, with guaranteed annual increases of 5% for the first 5 years

   B: $31,000 to start with guaranteed annual increases of 4% for the first 5 years.

   Which offer is best if your goal is to be making as much money as possible after the 5th year? Which is best if your goal is to make as much money as possible over the entire 5th years of the contract?

84. Be an Agent Suppose that you are an agent for a professional baseball player. Management has just offered your client a 5-year contract with a first-year salary of $1,500,000. Beyond that, your client has three choices:

A: A bonus of $75,000 each year (including the first year)

B: An annual increase of 4% per year beginning after the first year

C: An annual increase of $80,000 per year beginning after the first year

Which option provides the most money over the 5-year period? Which the least? Which would you choose? Why?

85. **The Multiplier** Suppose the marginal propensity to consume throughout the U.S. economy is 0.98. What is the multiplier for the U.S. economy?

86. **The Multiplier** Suppose the marginal propensity to consume throughout the U.S. economy is 0.96. What is the multiplier for the U.S. economy?

87. **Stock Price** One method of pricing a stock is based on the stream of future dividends of the stock. Suppose that a stock pays $*P* per year in dividends and, historically, the dividend has been increased by *i*% per year. If you desire an annual rate of return of *r*%, this method of pricing a stock states that the price you should pay is the present value of an infinite stream of payments:

$$\text{Price} = P + P \cdot \frac{1+i}{1+r} + P \cdot \left(\frac{1+i}{1+r}\right)^2 + P \cdot \left(\frac{1+i}{1+r}\right)^3 + \cdots$$

The price of the stock is the sum of a geometric series. Suppose that a stock pays an annual dividend of $2.00 and, historically, the dividend has been increased by 2% per year. You desire an annual rate of return of 9%. What is the most you should pay for the stock?

88. **Stock Price** Refer to Problem 87. Suppose that a stock pays an annual dividend of $3.00 and, historically, the dividend has been increased by 3% per year. You desire an annual rate of return of 10%. What is the most you should pay for the stock?

89. **401(k)** Christine contributes $100 each month into her 401(k) retirement plan. What will be the value of Christine's 401(k) in 30 years if the per annum rate of return is assumed to be 8% compounded monthly?

90. **Saving for a Home** Jolene wants to purchase a new home. Suppose she invests $400 a month into a money market fund. If the per annum interest rate of return on the money market fund is 4% compounded monthly, how much will Jolene have for a down payment in 4 years?

91. **Roth IRA** Jackson contributes $500 each quarter into his Roth IRA. What will be the value of Jackson's IRA

in 25 years if the per annum rate of return is assumed to be 6% compounded quarterly?

92. **Retirement** Raymont is planning on retiring in 15 years, so he contributes $1500 into his IRA every 6 months (semiannually). What will be the value of the IRA when Raymont retires if the per annum interest rate is 10% compounded semiannually?

93. **What's My Payment?** Suppose that Aaliyah wants to have $1,500,000 in her 401(k) retirement account in 35 years. How much does she need to contribute each month if the account earns 10% interest compounded monthly?

94. **What's My Payment?** Suppose that Sophia wants to have $2,000,000 in her 401(k) retirement account in 25 years. How much does she need to contribute each quarter if the account earns 12% interest compounded quarterly?

Extending the Concepts

95. Express $0.4\overline{9}$ as a fraction in lowest terms.

96. Express $0.85\overline{9}$ as a fraction in lowest terms.

97. Find the sum: $2 + 4 + 8 + \cdots + 1{,}073{,}741{,}824$

98. Can a sequence be both arithmetic and geometric? Give reasons for your answer.

99. Which yields faster growth in the terms of a sequence—an arithmetic sequence or a geometric sequence with $r > 1$? Why? Explain why Thomas Robert Malthus's conjecture that food supplies grow arithmetically while population grows geometrically provides a recipe for disaster unless population growth is curbed.

Explaining the Concepts

100. How do you determine whether a sequence is geometric?

101. How do you determine whether a geometric series has a sum?

Synthesis Review

102. Express $\frac{1}{3}$ as a repeating decimal. Express $\frac{2}{3}$ as a repeating decimal.

103. What is $\frac{1}{3} + \frac{2}{3}$? Now add the repeating decimals found in Problem 102. Conjecture the value of $0.9999999\ldots$.

104. Prove that $0.9999\ldots = 0.\overline{9}$ equals 1.

The Graphing Calculator

In Problems 105–108, use a graphing calculator to find the sum of each sequence.

105. $4 + 4.8 + 5.76 + \cdots + 4(1.2)^{15-1}$

106. $3 + 4.8 + 7.68 + \cdots + 3(1.6)^{20-1}$

107. $\sum_{n=1}^{20} [1.2(1.05)^n]$

108. $\sum_{n=1}^{25} [1.3(0.55)^n]$

Putting the Concepts Together (Sections 10.1–10.3)

We designed these problems from Sections 10.1 through 10.3 so that you can review the chapter so far and show your mastery of the concepts. Take time to work these problems before proceeding with the next section. The answers are located at the back of the text on page AN-61.

In Problems 1–6, determine if the sequence is arithmetic, geometric, or neither. If arithmetic or geometric, determine the first term and the common difference or common ratio.

1. $\dfrac{3}{4}, \dfrac{3}{16}, \dfrac{3}{64}, \dfrac{3}{256}, \cdots$

2. $\{2(n+3)\}$

3. $\left\{\dfrac{7n+2}{9}\right\}$

4. $1, -4, 9, -16, \ldots$

5. $\{3 \cdot 2^{n+1}\}$

6. $\{n^2 - 5\}$

7. Write out the sum and evaluate: $\sum_{k=1}^{6} [3k + 4]$

8. Express the sum using summation notation:

$$\frac{1}{2(6+1)} + \frac{1}{2(6+2)} + \frac{1}{2(6+3)} + \cdots + \frac{1}{2(6+12)}$$

In Problems 9–12, write a formula for the nth term of the indicated sequence. Write the first five terms of each sequence.

9. arithmetic: $a_1 = 25, d = -2$

10. arithmetic: $a_4 = 9, d = 11$

11. geometric: $a_4 = \dfrac{9}{25}, r = \dfrac{1}{5}$

12. geometric: $a_1 = 150, r = 1.04$

In Problems 13–15, find the indicated sum.

13. $2 + 6 + 18 + \cdots + 2 \cdot (3)^{11-1}$

14. $2 + 7 + 12 + 17 + \cdots + [2 + (20 - 1) \cdot 5]$

15. $1000 + 100 + 10 + \cdots$

16. Table Seating A restaurant uses square tables in the main dining area that seat four people. For larger parties, tables can be placed together. Two tables will seat 6 people, three tables will seat 8 people, and so on (see diagram). How many tables are needed to seat a party of 24 people?

10.4 The Binomial Theorem

Objectives

1 Compute Factorials

2 Evaluate a Binomial Coefficient

3 Expand a Binomial

Are You Ready for This Section?

Before getting started, take the following readiness quiz. If you get a problem wrong, go back to the section cited and review the material.

R1. Multiply: $(x - 5)^2$ [Section 4.2, pp. 323–324]

R2. Multiply: $(2x + 3)^2$ [Section 4.2, pp. 323–324]

▶ **1 Compute Factorials**

Suppose we want to find the product of the first 12 integers, or

$$12 \cdot 11 \cdot 10 \cdot 9 \cdot 8 \cdot 7 \cdot 6 \cdot 5 \cdot 4 \cdot 3 \cdot 2 \cdot 1$$

This product equals 479,001,600. Not only is this a big number, but writing all the factors is time-consuming. A shorthand method for writing this product is *factorial notation*.

> **Definition**
>
> If $n \geq 0$ is an integer, the **factorial symbol $n!$** (read "n factorial") is defined as
>
> $$0! = 1 \qquad 1! = 1$$
> $$n! = n(n-1)(n-2) \cdot \cdots \cdot 3 \cdot 2 \cdot 1 \quad \text{if } n \geq 2$$

For example, $2! = 2 \cdot 1 = 2$, $3! = 3 \cdot 2 \cdot 1 = 6$, $4! = 4 \cdot 3 \cdot 2 \cdot 1 = 24$, and so on. Table 2 lists the values of $n!$ for $0 \leq n \leq 7$.

Table 2

n	0	1	2	3	4	5	6	7
$n!$	1	1	2	6	24	120	720	5040

Because

$$\underbrace{n(n-1)(n-2) \cdot \cdots \cdot 3 \cdot 2 \cdot 1}_{(n-1)!}$$

we can use the formula

$$n! = n(n-1)!$$

to find successive factorials. For example, because $7! = 5040$, we have

$$8! = 8 \cdot 7! = 8(5040) = 40{,}320$$

Use your calculator's factorial key to see how fast factorials increase in value. Find the value of $69!$. What happens when you try to find $70!$? In fact, $70!$ is larger than 10^{100} (a **googol**), the largest number that most calculators can display.

EXAMPLE 1 **Computing Factorials**

Compute the value of $\dfrac{12!}{9!}$.

Solution

We could directly compute $12!$ and then $9!$, but this would be inefficient. Let's use the properties of factorials instead.

$$\frac{12!}{9!} = \frac{12 \cdot 11 \cdot 10 \cdot 9!}{9!}$$

Divide out $9!$: $\qquad = 12 \cdot 11 \cdot 10$

$$= 1320$$

See Figure 6 for this calculation on a TI-84 Plus graphing calculator. ●

Figure 6

Quick ✓

1. If $n \geq 2$ is an integer, then $n! = $ _____.

2. $0! = $ ___; $1! = $ ___.

In Problems 3 and 4, find the value of each factorial.

3. $5!$

4. $\dfrac{7!}{3!}$

▶ ❷ Evaluate a Binomial Coefficient

We know the formula for expanding $(x + a)^n$ for $n = 2$. The *Binomial Theorem* is a formula for expanding $(x + a)^n$ for any positive integer n. If $n = 1, 2, 3$, or 4, the expansion of $(x + a)^n$ is straightforward:

$(x + a)^1 = x + a$		Two terms, beginning with x^1 and ending with a^1
$(x + a)^2 = x^2 + 2ax + a^2$		Three terms, beginning with x^2 and ending with a^2
$(x + a)^3 = x^3 + 3ax^2 + 3a^2x + a^3$		Four terms, beginning with x^3 and ending with a^3
$(x + a)^4 = x^4 + 4ax^3 + 6a^2x^2 + 4a^3x + a^4$		Five terms, beginning with x^4 and ending with a^4

Each expansion of $(x + a)^n$ begins with x^n and ends with a^n. As you read from left to right, the powers of x decrease by 1, while the powers of a increase by 1. The number of terms is $n + 1$. Notice, that the degree of each monomial in the expansion equals n. For example, in the expansion of $(x + a)^3$, each monomial ($x^3, 3ax^2, 3a^2x, a^3$) is of degree 3. As a result, we might conjecture that the expansion of $(x + a)^n$ would look like this:

$$(x + a)^n = x^n + \underline{\quad} ax^{n-1} + \underline{\quad} a^2 x^{n-2} + \cdots + \underline{\quad} a^{n-1} x + a^n$$

where the blanks are numbers to be found. This is correct, as we shall see shortly.

Before we go any further, we need to introduce the symbol $\binom{n}{j}$, which is read "n taken j at a time" or "n choose j":

Work Smart

Do not write $\binom{n}{j}$ as $\left(\dfrac{n}{j}\right)$.

> **Definition**
>
> If j and n are integers with $0 \le j \le n$, the symbol $\binom{n}{j}$ is defined as
>
> $$\binom{n}{j} = \frac{n!}{j!(n - j)!}$$

(**EXAMPLE 2**)

Evaluating $\binom{n}{j}$

Find:

(a) $\binom{4}{1}$ **(b)** $\binom{6}{2}$ **(c)** $\binom{5}{4}$

Solution

(a) Here, we have $n = 4$ and $j = 1$, so

$$\binom{4}{1} = \frac{4!}{1!(4 - 1)!} = \frac{4!}{1! \cdot 3!} = \frac{4 \cdot 3!}{1! \cdot 3!} = \frac{4 \cdot 3!}{1 \cdot 3!} = 4$$

$$6! = 6 \cdot 5 \cdot 4!$$
$$\downarrow$$

(b) $\binom{6}{2} = \frac{6!}{2!(6 - 2)!} = \frac{6!}{2! \cdot 4!} = \frac{6 \cdot 5 \cdot 4!}{2 \cdot 1 \cdot 4!} = \frac{6 \cdot 5 \cdot 4!}{2 \cdot 1 \cdot 4!} = \frac{30}{2} = 15$

(c) $\binom{5}{4} = \frac{5!}{4!(5 - 4)!} = \frac{5!}{4! \cdot 1!} = \frac{5 \cdot 4!}{4! \cdot 1} = \frac{5 \cdot 4!}{4! \cdot 1} = 5$

Quick ✓

5. *True or False* $\binom{n}{j} = \dfrac{j!}{n!(n - j)!}$ **6.** *True or False* $\binom{5}{1} = 5$

In Problems 7 and 8, evaluate each expression.

7. $\binom{7}{1}$ **8.** $\binom{6}{3}$

Four useful formulas involving the symbol $\binom{n}{j}$ are

$$\binom{n}{0} = 1 \qquad \binom{n}{1} = n \qquad \binom{n}{n-1} = n \qquad \binom{n}{n} = 1$$

Suppose that we arrange the various values of the symbol $\binom{n}{j}$ in a triangular display, as shown next and in Figure 7.

$$\binom{0}{0}$$

$$\binom{1}{0} \quad \binom{1}{1}$$

$$\binom{2}{0} \quad \binom{2}{1} \quad \binom{2}{2}$$

$$\binom{3}{0} \quad \binom{3}{1} \quad \binom{3}{2} \quad \binom{3}{3}$$

$$\binom{4}{0} \quad \binom{4}{1} \quad \binom{4}{2} \quad \binom{4}{3} \quad \binom{4}{4}$$

$$\binom{5}{0} \quad \binom{5}{1} \quad \binom{5}{2} \quad \binom{5}{3} \quad \binom{5}{4} \quad \binom{5}{5}$$

Figure 7
Pascal's Triangle

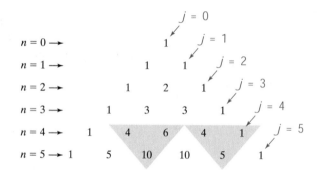

This display is called **Pascal's triangle,** named after Blaise Pascal (1623–1662), a French mathematician.

Pascal's triangle has 1s down the sides. To get any other entry, add the two nearest entries in the row above it. The shaded triangles in Figure 7 illustrate this feature. Using this feature, we can find the row corresponding to $n = 6$.

$$
\begin{array}{ccccccccccccc}
n = 5 \rightarrow & & 1 & & 5 & & 10 & & 10 & & 5 & & 1 \\
n = 6 \rightarrow & 1 & & 6 & & 15 & & 20 & & 15 & & 6 & & 1
\end{array}
$$

Although Pascal's triangle is an interesting and organized display of the symbol $\binom{n}{j}$, in practice it is not all that helpful. For example, to find the value of $\binom{12}{8}$, you would need to produce 12 rows of the triangle before seeing the answer. It is much faster to use the definition of $\binom{n}{j}$.

▶ ❸ Expand a Binomial

Now we are ready to state the **Binomial Theorem.**

The Binomial Theorem

Let x and a be real numbers. For any positive integer n,

$$(x + a)^n = \binom{n}{0}x^n + \binom{n}{1}ax^{n-1} + \binom{n}{2}a^2x^{n-2} + \cdots + \binom{n}{j}a^jx^{n-j} + \cdots + \binom{n}{n}a^n$$

You should now see why we needed to discuss the symbol $\binom{n}{j}$; these symbols are the numerical coefficients that appear in the expansion of $(x + a)^n$. Because of this, the symbol $\binom{n}{j}$ is called a **binomial coefficient.**

EXAMPLE 3 **Expanding a Binomial**

Use the Binomial Theorem to expand $(x + 3)^4$.

Solution

In the Binomial Theorem, let $a = 3$ and $n = 4$. Then

Work Smart

In the expansion, notice how the exponent on x **decreases** by 1 for each term as we move to the right, while the exponent on 3 **increases** by 1.

$$(x + 3)^4 = \binom{4}{0}x^4 + \binom{4}{1}3^1 \cdot x^{4-1} + \binom{4}{2}3^2 \cdot x^{4-2} + \binom{4}{3}3^3 \cdot x^{4-3} + \binom{4}{4}3^4$$

Use Pascal's triangle in Figure 7 or use $\binom{n}{j} = \dfrac{n!}{j!(n-j)!}$ to evaluate the binomial coefficients.

$$= 1 \cdot x^4 + 4 \cdot 3 \cdot x^3 + 6 \cdot 9 \cdot x^2 + 4 \cdot 27 \cdot x + 1 \cdot 81$$
$$= x^4 + 12x^3 + 54x^2 + 108x + 81$$

●

EXAMPLE 4 **Expanding a Binomial**

Expand $(2y - 3)^5$ using the Binomial Theorem.

Solution

First, rewrite the expression $(2y - 3)^5$ as $[2y + (-3)]^5$. Then use the Binomial Theorem with $n = 5$, $x = 2y$, and $a = -3$.

$$[2y + (-3)]^5 = \binom{5}{0}(2y)^5 + \binom{5}{1}(-3)^1(2y)^{5-1} + \binom{5}{2}(-3)^2(2y)^{5-2} + \binom{5}{3}(-3)^3(2y)^{5-3} + \binom{5}{4}(-3)^4(2y)^{5-4} + \binom{5}{5}(-3)^5$$

Use Pascal's triangle or use the formula $\binom{n}{j} = \dfrac{n!}{j!(n-j)!}$ to evaluate binomial coefficients.

$$= 1(2y)^5 + 5(-3)^1(2y)^4 + 10(-3)^2(2y)^3 + 10(-3)^3(2y)^2 + 5(-3)^4(2y)^1 + 1(-3)^5$$
$$= 32y^5 + 5 \cdot (-3) \cdot 16y^4 + 10 \cdot 9 \cdot 8y^3 + 10 \cdot (-27) \cdot 4y^2 + 5 \cdot 81 \cdot 2y - 243$$
$$= 32y^5 - 240y^4 + 720y^3 - 1080y^2 + 810y - 243$$

●

Quick ✓

In Problems 9 and 10, expand each binomial using the Binomial Theorem.

9. $(x + 2)^4$ **10.** $(2p - 1)^5$

10.4 Exercises MyMathLab® Math·XL PRACTICE

Exercise numbers in green have complete video solutions in MyMathLab.

*Problems **1–10** are the Quick ✓s that follow the EXAMPLES.*

Building Skills

In Problems 11–18, evaluate each factorial expression. See Objective 1.

11. $3!$ **12.** $4!$ **13.** $8!$ **14.** $9!$

15. $\dfrac{10!}{8!}$ **16.** $\dfrac{12!}{10!}$ **17.** $\dfrac{8!}{5!}$ **18.** $\dfrac{9!}{6!}$

In Problems 19–22, evaluate each expression. See Objective 2.

19. $\binom{7}{2}$ **20.** $\binom{8}{5}$ **21.** $\binom{10}{4}$ **22.** $\binom{12}{9}$

In Problems 23–38, expand each expression using the Binomial Theorem. See Objective 3.

23. $(x + 1)^5$

24. $(x - 1)^4$

25. $(x - 4)^4$

26. $(x + 5)^5$

27. $(3p + 2)^4$

28. $(2q + 3)^4$

29. $(2z - 3)^5$

30. $(3w - 4)^4$

31. $(x^2 + 2)^4$

32. $(y^2 - 3)^4$

33. $(2p^3 + 1)^5$

34. $(3b^2 + 2)^5$

35. $(x + 2)^6$

36. $(p - 3)^6$

37. $(2p^2 - q^2)^4$

38. $(3x^2 + y^3)^4$

Applying the Concepts

39. Use the Binomial Theorem to find the numerical value of $(1.001)^4$ correct to five decimal places. (*Hint*: $(1.001)^4 = (1 + 10^{-3})^4$.)

40. Use the Binomial Theorem to find the numerical value of $(1.001)^5$ correct to five decimal places. (*Hint*: $(1.001)^5 = (1 + 10^{-3})^5$.)

41. Use the Binomial Theorem to find the numerical value of $(0.998)^5$ correct to five decimal places.

42. Use the Binomial Theorem to find the numerical value of $(0.997)^5$ correct to five decimal places.

Extending the Concepts

Notice that in the formula for expanding a binomial $(x + a)^n$, the first term is $\binom{n}{0}a^0 \cdot x^n$, the second term is $\binom{n}{1}a^1 \cdot x^{n-1}$, the third term is $\binom{n}{2}a^2 \cdot x^{n-2}$, and so on. In general, the jth term in a binomial expansion of $(x + a)^n$ is $\binom{n}{j-1}a^{j-1}x^{n-j+1}$. Use this result to find the indicated term in Problems 43–46.

43. The 3rd term in the expansion of $(x + 2)^7$

44. The 4th term in the expansion of $(x - 1)^{10}$

45. The 6th term in the expansion of $(2p - 3)^8$

46. The 7th term in the expansion of $(3p + 1)^9$

47. Show that $\binom{n}{n-1} = n$ and $\binom{n}{n} = 1$.

48. Show that if n and j are integers with $0 \le j \le n$, then $\binom{n}{j} = \binom{n}{n-j}$.

Explaining the Concepts

49. Write the first four rows of Pascal's triangle.

50. Describe the pattern of exponents of x in the expansion of $(x + a)^n$. Describe the pattern of exponents of a in the expansion of $(x + a)^n$.

51. What is true about the degree of each monomial in the expansion of $(x + y)^n$?

52. Explain how you might find a particular term in a binomial expansion. For example, how might you find the 5th term in the expansion of $(x + 3)^8$?

Synthesis Review

53. If $f(x) = x^4$, find $f(a - 2)$ with the aid of the Binomial Formula.

54. If $g(x) = x^5 + 3$, find $g(z + 1)$ with the aid of the Binomial Formula.

55. If $H(x) = x^5 - 4x^4$, find $H(p + 1)$ with the aid of the Binomial Formula.

56. If $h(x) = 2x^5 + 5x^4$, find $h(a + 3)$ with the aid of the Binomial Formula.

Chapter 10 Activity: Discover the Relation

Focus: Review of objectives for arithmetic and geometric sequences and the Binomial Theorem
Time: 30–35 minutes
Group size: 2–4

As a group, discover the relationship between Column A and Column B (that is, $<$, $>$, or $=$). Be sure to discuss any differences in outcomes.

	Column A	Column B
1	The sum of the first five terms of $$a_n = \frac{3}{5}n + 1 \quad A < B$$	The sum of the first five terms of $$a_n = -\frac{1}{4}(n-1) + 4$$
2	$$\sum_{i=1}^{4} (i^2 + 2i) \quad A = B$$	$$\sum_{i=1}^{4} i^2 + \sum_{i=1}^{4} 2i$$
3	The 8th term of an arithmetic sequence when $a_1 = 3$ and $d = \dfrac{3}{2}$ $\quad A > B$	The 27th term of an arithmetic sequence when $a_1 = \dfrac{5}{3}$ and $d = \dfrac{1}{3}$
4	The 10th term of a geometric sequence with $a_1 = 3$ and $r = \sqrt{2}$ $\quad A < B$	The 9th term of a geometric sequence with $a_1 = 5$ and $r = \sqrt{3}$
5	The coefficient of the 4th term of $(x+y)^{10}$ $\quad A < B$	The coefficient of the 3rd term of $(8x - y)^4$

Chapter 10 Review

Section 10.1 Sequences

KEY TERMS

Sequence
Terms
Ellipsis
Infinite sequence

Finite sequence
General term
Alternate

Summation notation
Index
Partial sum

You Should Be Able To...	EXAMPLE	Review Exercises
1 Write the first few terms of a sequence (p. 741)	Examples 1 and 2	1–6
2 Find a formula for the nth term of a sequence (p. 743)	Example 3	7–12
3 Use summation notation (p. 743)	Examples 4 and 5	13–20

In Problems 1–6, write the first five terms of each sequence.

1. $\{-3n + 2\}$

2. $\left\{\dfrac{n-2}{n+4}\right\}$

3. $\{5^n + 1\}$

4. $\{(-1)^{n-1} \cdot 3n\}$

5. $\left\{\dfrac{n^2}{n+1}\right\}$

6. $\left\{\dfrac{\pi^n}{n}\right\}$

In Problems 7–12, the given pattern continues. Write the nth term of each sequence suggested by the pattern.

7. $-3, -6, -9, -12, -15, \ldots$

8. $\dfrac{1}{3}, \dfrac{2}{3}, 1, \dfrac{4}{3}, \dfrac{5}{3}, \ldots$

9. $5, 10, 20, 40, 80, \ldots$

10. $-\dfrac{1}{2}, 1, -\dfrac{3}{2}, 2, \ldots$

11. $6, 9, 14, 21, 30, \ldots$

12. $0, \dfrac{1}{3}, \dfrac{1}{2}, \dfrac{3}{5}, \ldots$

In Problems 13–16, write out each sum and determine its value.

13. $\displaystyle\sum_{k=1}^{5} (5k - 2)$

14. $\displaystyle\sum_{k=1}^{6} \left(\dfrac{k + 2}{2}\right)$

15. $\displaystyle\sum_{i=1}^{5} (-2i)$

16. $\displaystyle\sum_{i=1}^{4} \dfrac{i^2 - 1}{3}$

In Problems 17–20, express each sum using summation notation.

17. $(4 + 3 \cdot 1) + (4 + 3 \cdot 2) + (4 + 3 \cdot 3) + \cdots + (4 + 3 \cdot 15)$

18. $\dfrac{1}{3^1} + \dfrac{1}{3^2} + \dfrac{1}{3^3} + \cdots + \dfrac{1}{3^8}$

19. $\dfrac{1^3 + 1}{1 + 1} + \dfrac{2^3 + 1}{2 + 1} + \dfrac{3^3 + 1}{3 + 1} + \cdots + \dfrac{10^3 + 1}{10 + 1}$

20. $(-1)^{1-1} \cdot 1^2 + (-1)^{2-1} \cdot 2^2 + (-1)^{3-1} \cdot 3^2 + \cdots + (-1)^{7-1} \cdot 7^2$

Section 10.2 Arithmetic Sequences

KEY CONCEPTS

- **The nth Term of an Arithmetic Sequence**
  For an arithmetic sequence whose first term is a_1 and whose common difference is d, the nth term is $a_n = a_1 + (n - 1)d$
- **Sum of the First n Terms of an Arithmetic Sequence**
  For an arithmetic sequence whose first term is a_1 and whose common difference is d, the sum of the first n terms is
  $$S_n = \frac{n}{2}[2a_1 + (n - 1)d] \quad \text{or} \quad S_n = \frac{n}{2}(a_1 + a_n)$$

KEY TERMS

Arithmetic sequence
Common difference

You Should Be Able To...	EXAMPLE	Review Exercises
❶ Determine whether a sequence is arithmetic (p. 748)	Examples 1 through 3	21–26
❷ Find a formula for the nth term of an arithmetic sequence (p. 749)	Examples 4 and 5	27–32, 37
❸ Find the sum of an arithmetic sequence (p. 751)	Examples 6 through 8	33–36, 38

In Problems 21–26, determine whether the sequence is arithmetic. If so, find the common difference.

21. $4, 10, 16, 22, \ldots$

22. $-1, \dfrac{1}{2}, 2, \dfrac{7}{2}, \ldots$

23. $-2, -5, -9, -14, \ldots$

24. $-1, 3, -5, 7, \ldots$

25. $\{4n + 7\}$

26. $\left\{\dfrac{n + 1}{2n}\right\}$

In Problems 27–32, find a formula for the nth term of the arithmetic sequence. Use the formula to find the 25th term of the sequence.

27. $a_1 = 3; d = 8$

28. $a_1 = -4; d = -3$

29. $7, \dfrac{20}{3}, \dfrac{19}{3}, 6, \ldots$

30. $11, 17, 23, 29, \ldots$

31. The 3rd term is 7 and the 8th term is 25.

32. The 4th term is -20 and the 7th term is -32.

33. Find the sum of the first 30 terms of the sequence $-1, 9, 19, 29, \ldots$.

34. Find the sum of the first 40 terms of the sequence $5, 2, -1, -4, \ldots$.

35. Find the sum of the first 60 terms of the sequence $\{-2n - 7\}$.

36. Find the sum of the first 50 terms of the sequence $\left\{\dfrac{1}{4}n + 3\right\}$.

37. Cicadas Seventeen-year cicadas emerge every 17 years to mate, lay eggs, and start the next 17-year cycle. In 2004, the Brood X cicada (the largest brood of the 17-year cicada) emerged in Maryland, Kentucky, Tennessee, and parts of surrounding states. Determine when the Brood X cicada will first appear in the twenty-second century.

38. Wind Sprints At a certain football practice, players would run wind sprints for exercise. Starting at the goal line, players would sprint to the 10-yard line and back to the goal line. The players would then sprint to the 20-yard line and back to the goal line. This would continue for the 30-yard line, 40-yard line, and 50-yard line. Determine the total distance a player would run during the wind sprints.

Section 10.3 Geometric Sequences and Series

KEY CONCEPTS

- **The nth Term of a Geometric Sequence**
  For a geometric sequence whose first term is a_1 and whose common ratio is r, the nth term is $a_n = a_1 r^{n-1}$.

- **Sum of the First n Terms of a Geometric Sequence**
  For a geometric sequence whose first term is a_1 and whose common ratio is r, the sum of the first n terms is $S_n = a_1 \cdot \dfrac{1 - r^n}{1 - r}, r \neq 0, r \neq 1$.

- **Sum of a Geometric Series**
  If $-1 < r < 1$, then the sum of the geometric series $\displaystyle\sum_{n=1}^{\infty} a_1 r^{n-1} = \dfrac{a_1}{1 - r}$.

KEY TERMS

Geometric sequence
Common ratio
Geometric series
Marginal propensity to consume
Multiplier

You Should Be Able To...	EXAMPLE	Review Exercises
❶ Determine whether a sequence is geometric (p. 755)	Examples 1 through 3	39–44
❷ Find a formula for the nth term of a geometric sequence (p. 757)	Example 4	45–48, 57
❸ Find the sum of a geometric sequence (p. 758)	Examples 5 and 6	49–52, 58
❹ Find the sum of a geometric series (p. 760)	Examples 7 through 9	53–56
❺ Solve annuity problems (p. 762)	Example 10	59–62

In Problems 39–44, determine whether the given sequence is geometric. If so, determine the common ratio.

39. $\dfrac{1}{3}, 2, 12, 72, \ldots$

40. $-1, 3, -9, 27, \ldots$

41. $1, 1, 2, 6, \ldots$

42. $6, 4, \dfrac{8}{3}, \dfrac{16}{9}, \ldots$

43. $\{5 \cdot (-2)^n\}$

44. $\{3n - 14\}$

In Problems 45–48, find a formula for the nth term of the geometric sequence. Use the formula to find the 10th term of the sequence.

45. $a_1 = 4, r = 3$

46. $a_1 = 8, r = \dfrac{1}{4}$

47. $a_1 = 5, r = -2$

48. $a_1 = 1000, r = 1.08$

In Problems 49–52, find the sum. If necessary, express your answer to as many decimal places as your calculator allows.

49. $2 + 4 + 8 + \cdots + 2^{15}$

50. $40 + 5 + \dfrac{5}{8} + \cdots + 40\left(\dfrac{1}{8}\right)^{13-1}$

51. $\displaystyle\sum_{n=1}^{12} \left[\dfrac{3}{4} \cdot (2)^{n-1} \right]$

52. $\displaystyle\sum_{n=1}^{16} \left[-4 \cdot (3^n) \right]$

In Problems 53–56, find the sum of each geometric series. If necessary, express your answer to as many decimal places as your calculator allows.

53. $\displaystyle\sum_{n=1}^{\infty} \left[20 \cdot \left(\dfrac{1}{4}\right)^n \right]$

54. $\displaystyle\sum_{n=1}^{\infty} \left[50 \cdot \left(-\dfrac{1}{2}\right)^{n-1} \right]$

55. $1 + \dfrac{1}{5} + \dfrac{1}{25} + \cdots$

56. $0.8 + 0.08 + 0.008 + 0.0008 + \cdots$

57. Radioactive Decay The radioactive isotope tritium has a half-life of about 12 years. If there were 200 grams of the isotope initially, use a geometric sequence to determine how much would remain after 72 years.

58. Computer Virus In January 2004, the Mydoom e-mail worm was declared the worst e-mail worm incident in virus history, accounting for roughly 20–30% of worldwide e-mail traffic. Suppose the virus was initially sent to 5 e-mail addresses and that, upon receipt, it sends itself out to 5 e-mail addresses from the address book of the infected computer. If each cycle of e-mails, including the initial sending, takes 1 minute to complete, how many total e-mails will have been sent after 15 minutes?

59. 403(b) Each quarter, Scott contributes $900 to a 403(b) plan at work. His employer agrees to match half of employee contributions up to $600 per quarter. What will be the value of Scott's 403(b) in 25 years if the per annum rate of return is assumed to be 7% compounded quarterly?

60. Lottery Payment The winner of a state lottery has the option of receiving about $2 million per year for 26 years (after taxes) or a lump sum payment of about $28 million (after taxes). Assuming all winnings will be invested at a per annum rate of return of 6.5% compounded annually, which option yields the most money after 26 years?

61. What's My Payment? Sheri starts her career when she is 22 years old and wants to have $2,500,000 in her 401(k) retirement account when she retires in 40 years. How much does she need to contribute each month if the account earns 9% interest compounded monthly?

62. College Savings Plan On Samantha's 8th birthday, her parents open a 529 college savings plan for her. They plan to contribute $400 per month until she turns 18. The per annum rate of return is assumed to be 5.25% compounded monthly, and the cost per credit hour at a private 4-year university is locked in at a rate of $340 per hour. What will be the value of the plan when Samantha turns 18, and how many credit hours will the plan cover?

Section 10.4 The Binomial Theorem

KEY CONCEPTS

- **Factorial symbol $n!$**
  If $n \geq 0$ is an integer, the factorial symbol $n!$ is defined as
  $0! = 1 \qquad 1! = 1 \qquad n! = n(n-1)(n-2) \cdots \cdot 3 \cdot 2 \cdot 1 \quad$ if $n \geq 2$

- **The symbol $\dbinom{n}{j}$**
  If j and n are integers with $0 \leq j \leq n$, then $\dbinom{n}{j} = \dfrac{n!}{j!(n-j)!}$.

- **The Binomial Theorem**
  Let x and a be real numbers. For any positive integer n,
  $$(x + a)^n = \binom{n}{0}x^n + \binom{n}{1}ax^{n-1} + \binom{n}{2}a^2 x^{n-2} + \cdots + \binom{n}{j}a^j x^{n-j} + \cdots + \binom{n}{n}a^n$$

KEY TERMS

Factorial symbol
Googol
Pascal's triangle
Binomial coefficient

You Should Be Able To...	EXAMPLE	Review Exercises
❶ Compute factorials (p. 767)	Example 1	63–66
❷ Evaluate a binomial coefficient (p. 769)	Example 2	67–70
❸ Expand a binomial (p. 770)	Examples 3 and 4	71–78

In Problems 63–66, evaluate the expression.

63. $5!$

64. $\dfrac{11!}{7!}$

65. $\dfrac{10!}{6!}$

66. $\dfrac{13!}{6!7!}$

In Problems 67–70, evaluate each binomial coefficient.

67. $\dbinom{7}{3}$

68. $\dbinom{10}{5}$

69. $\dbinom{8}{8}$

70. $\dbinom{6}{0}$

In Problems 71–76, expand each expression using the Binomial Theorem.

71. $(z + 1)^4$

72. $(y - 3)^5$

73. $(3y + 4)^6$

74. $(2x^2 - 3)^4$

75. $(3p - 2q)^4$

76. $(a^3 + 3b)^5$

77. Find the 4th term in the expansion of $(x - 2)^8$.

78. Find the 7th term in the expansion of $(2x + 1)^{11}$.

Chapter 10 Test *Step-by-step test solutions are found on the Chapter Test Prep Videos available in* MyMathLab® *or on* YouTube.

In Problems 1–6, determine whether the sequence is arithmetic, geometric, or neither. If arithmetic or geometric, determine the first term and the common difference or common ratio.

1. $-15, -7, 1, 9, \ldots$

2. $\{ (-4)^n \}$

3. $\left\{ \dfrac{4}{n!} \right\}$

4. $\left\{ \dfrac{2n - 3}{5} \right\}$

5. $-3, 2, 0, 5, 3, \ldots$

6. $\{ 7 \cdot 3^n \}$

7. Write out the sum and evaluate: $\displaystyle\sum_{i=1}^{5} \left[\dfrac{3}{i^2} + 2 \right]$

8. Express the sum using summation notation:
$$\dfrac{3}{5} + \dfrac{2}{3} + \dfrac{5}{7} + \dfrac{3}{4} + \cdots + \dfrac{5}{6}$$

In Problems 9–12, write a formula for the nth term of the indicated sequence. Write the first five terms of each sequence.

9. arithmetic: $a_1 = 6, \ d = 10$

10. arithmetic: $a_1 = 0, \ d = -4$

11. geometric: $a_1 = 10, \ r = 2$

12. geometric: $a_3 = 9, \ r = -3$

In Problems 13–15, find the indicated sum.

13. $-2 + 2 + 6 + \cdots + [4 \cdot (20 - 1) - 2]$

14. $\dfrac{1}{9} - \dfrac{1}{3} + 1 - 3 + \cdots + \dfrac{1}{9} \cdot (-3)^{12-1}$

15. $216 + 72 + 24 + 8 + \cdots$

16. Evaluate $\dfrac{15!}{8!7!}$.

17. Evaluate $\dbinom{12}{5}$.

18. Expand $(5m - 2)^4$ using the Binomial Theorem.

19. A new car sold for \$31,000. If the vehicle loses 15% of its value each year, how much will it be worth after 10 years?

20. A weightlifter begins his routine by benching 100 pounds and increases the weight by 30 pounds for each set. If he does 10 repetitions in each set, what is the total weight lifted after 5 sets?

Answers to Selected Exercises

Chapter R Real Numbers and Algebraic Expressions

Section R.1 Success in Mathematics Answers will vary.

Section R.2 Sets and Classification of Numbers **1.** set **2.** elements **3.** set-builder: $\{x \mid x \text{ is a digit less than } 5\}$; roster: $\{0, 1, 2, 3, 4\}$
4. set-builder: $\{x \mid x \text{ is a digit greater than or equal to } 6\}$; roster: $\{6, 7, 8, 9\}$ **5.** True **6.** subset **7.** False **8.** True **9.** False **10.** False
11. True **12.** True **13.** False **14.** True **15.** whole **16.** rational **17.** irrational numbers **18.** False **19.** False **20.** True **21.** True
22. $10, \dfrac{12}{4}$ **23.** $10, \dfrac{0}{3}, \dfrac{12}{4}$ **24.** $-9, 10, \dfrac{0}{3}, \dfrac{12}{4}$ **25.** $\dfrac{7}{3}, -9, 10, 4.\overline{56}, \dfrac{0}{3}, -\dfrac{4}{7}, \dfrac{12}{4}$ **26.** π and $5.7377377737777\ldots, \sqrt{11}$ **27.** All **28. (a)** 5.694
(b) 5.694 **29. (a)** -4.93 **(b)** -4.94 **30.** **31.** $<$ **32.** $<$ **33.** $>$ **34.** $>$ **35.** $=$ **36.** $>$ **37.** $\{0, 1, 2, 3, 4, 5\}$

39. $\{-2, -1, 0, 1, 2, 3, 4\}$ **41.** $\varnothing$ or $\{ \ \}$ **43.** True **45.** False **47.** True **49.** True **51.** $\notin$ **53.** $\in$ **55. (a)** 4 **(b)** $-5, 4$ **(c)** $-5, 4, \dfrac{4}{3}, -\dfrac{7}{5}, 5.\overline{1}$
(d) π **(e)** $-5, 4, \dfrac{4}{3}, -\dfrac{7}{5}, 5.\overline{1}, \pi$ **57. (a)** 100 **(b)** $100, -64$ **(c)** $100, -5.423, \dfrac{8}{7}, -64$ **(d)** $\sqrt{2} + 4$ **(e)** $100, -5.423, \dfrac{8}{7}, \sqrt{2} + 4, -64$ **59. (a)** 19.9348
(b) 19.9348 **61. (a)** 0.0 **(b)** 0.1 **63.** **65.** $<$ **67.** $=$ **69.** $>$ **71.** -282 feet **73.** $-\$4.06$ **75.** -6

77. Answers will vary. **79.** Answers will vary. **81.** No, if a number is rational, it cannot be irrational, and vice versa, because a number cannot have a terminating or nonterminating/repeating decimal while simultaneously having a nonterminating/nonrepeating decimal. No; every real number is either a rational or an irrational number. **83.** A set is a well-defined collection of objects. The collection must be well defined so that it is easy to determine whether an object is in the set. **85.** If $A \subseteq B$, then all elements in A are also in B. A could also equal B. If $A \subset B$, then all elements in A are also in B, but $A \neq B$. **87.** If the digit *after* the specified final digit is 4 or less, then truncating and rounding will yield the same decimal approximation.
89. $1.143; 1.142$

Section R.3 Operations on Signed Numbers; Properties of Real Numbers **1.** factors **2.** False **3.** 6 **4.** 10 **5.** 12 **6.** -11 **7.** -6.6 **8.** -2.2
9. 0 **10.** additive inverse; opposite; 0 **11.** a **12.** -5 **13.** $-\dfrac{4}{5}$ **14.** 12 **15.** $\dfrac{5}{3}$ **16.** 0 **17.** 4 **18.** -9 **19.** -11 **20.** 9.1 **21.** -5.1 **22.** -16.1

23. True **24.** True **25.** -48 **26.** -60 **27.** 56 **28.** 105 **29.** 5.13 **30.** opposite; reciprocal **31.** $\dfrac{1}{10}$ **32.** $-\dfrac{1}{8}$ **33.** $\dfrac{5}{2}$ **34.** -5 **35.** lowest terms

36. $\dfrac{6}{5}$ **37.** $\dfrac{11}{8}$ **38.** $-\dfrac{6}{5}$ **39.** in lowest terms **40.** $\dfrac{4}{5}$ **41.** $-\dfrac{5}{6}$ **42.** $\dfrac{1}{2}$ **43.** 8 **44.** $\dfrac{5}{11}$ **45.** $-\dfrac{1}{3}$ **46.** 2 **47.** 1 **48.** least common denominator

49. LCD $= 60$; $\dfrac{3}{20} = \dfrac{9}{60}, \dfrac{2}{15} = \dfrac{8}{60}$ **50.** LCD $= 90$; $\dfrac{5}{18} = \dfrac{25}{90}, -\dfrac{1}{45} = -\dfrac{2}{90}$ **51.** $\dfrac{17}{60}$ **52.** $-\dfrac{1}{6}$ **53.** $-\dfrac{59}{150}$ **54.** $-\dfrac{3}{10}$ **55.** $ab + ac$ **56.** $-a - b$

57. $5x + 15$ **58.** $-6x - 6$ **59.** $-4z + 32$ **60.** $2x + 3$ **61.** $-11p - 8$ **62.** $9 + 4t$ **63.** $\dfrac{2}{3}$ **65.** $\dfrac{8}{3}$ **67.** -9 **69.** 7 **71.** -2.5 **73.** -32 **75.** 84

77. -36.55 **79.** $\dfrac{7}{3}$ **81.** $\dfrac{6}{5}$ **83.** $\dfrac{5}{3}$ **85.** $-\dfrac{3}{7}$ **87.** $\dfrac{2}{3}$ **89.** $\dfrac{5}{4}$ **91.** $\dfrac{1}{6}$ **93.** $\dfrac{11}{12}$ **95.** $2x + 8$ **97.** $-3z + 6$ **99.** $3x - 30$ **101.** $6x - 9$ **103.** $-5z - 17$

105. 3 **107.** $\dfrac{101}{70}$ **109.** $\dfrac{4}{3}$ **111.** 3.3 **113.** -32 **115.** $\dfrac{13}{10}$ **117.** $\dfrac{17}{15}$ **119.** $-\dfrac{139}{32}$ **121.** $\dfrac{1}{9}$ **123.** undefined **125.** 0 **127.** Commutative Property of
Multiplication **129.** Multiplicative Inverse Property **131.** Reduction Property **133.** Associative Property of Addition **135.** The difference in age of the oldest and youngest presidents at the time of inauguration is 27 years. **137.** The Bears gained a total of 9 yards in the first 3 plays. Since 9 is less than 10, they did not obtain a first down. **139.** The difference between the highest and lowest elevations is 20,602 feet. **141.** Consider $3 \cdot 6$. This means to add 6 to itself three times. Since the sum of three positive numbers is positive, the product will be positive. Consider $3 \cdot (-6)$. This means to add -6 to itself three times. Since the sum of three negative numbers is negative, the product will be negative.

143. $d(P, Q) = 14$ **145.** $d(P, Q) = 10.4$

147. $d(P, Q) = \dfrac{68}{15}$ **149. (a)** $a \cdot 0 = 0$ **(b)** $a \cdot (b + (-b)) = 0$ **(c)** $ab + a(-b) = 0$ **(d)** $ab < 0$ since the product of a negative and a positive is negative; $a(-b)$ must be positive so that $ab + a(-b) = 0$. **151.** Zero does not have a multiplicative inverse because division by zero is not defined. **153.** The Reduction Property only applies to dividing out factors and does not apply to sums.

155. No. For example, $4 - (5 - 3) \neq (4 - 5) - 3$ **157.** No. For example, $16 \div (8 \div 2) \neq (16 \div 8) \div 2$ **159.** -3.8 **161.** -19.2 **163.** $\dfrac{1}{30}$

165. 14.4 **167.** $\dfrac{9}{2}$

Section R.4 Order of Operations **1.** base; exponent; power **2.** False **3.** 64 **4.** 49 **5.** -1000 **6.** $\dfrac{8}{27}$ **7.** -64 **8.** 125 **9.** 16 **10.** 36 **11.** 32
12. 48 **13.** 40 **14.** 26 **15.** $\dfrac{10}{13}$ **16.** 18 **17.** 50 **18.** $\dfrac{4}{5}$ **19.** parentheses; exponents; multiplication; division; addition; subtraction **20.** 16 **21.** 48

22. $\frac{6}{7}$ **23.** 14 **24.** 7 **25.** 42 **26.** -9 **27.** 20 **28.** -18 **29.** 5 **30.** -15 **31.** 81 **33.** -625 **35.** 8 **37.** $-\frac{4}{9}$ **39.** 7 **41.** 15 **43.** 12 **45.** -28 **47.** 3

49. 14 **51.** -2 **53.** 60 **55.** 180 **57.** 9 **59.** 8 **61.** $\frac{5}{2}$ **63.** 2 **65.** $\frac{6}{5}$ **67.** $\frac{2}{3}$ **69.** $-\frac{7}{4}$ **71.** $\frac{3}{2}$ **73.** $\frac{53}{14}$ **75.** $3 \cdot (7-2) = 15$ **77.** $3 + 5 \cdot (6-3) = 18$

79. The surface area of the cylinder is about 534.07 square inches. **81.** After 3 seconds, the height of the ball is 6 feet. **83.** 100 **85.** We cannot use the Reduction Property across addition. **87.** Answers will vary. One suggestion: $5 \cdot 2^3$ is equivalent to $5 \cdot 2 \cdot 2 \cdot 2$, which equals 40. Since $5 \cdot 2^3 = 5 \cdot 8 = 40$, we can see that we evaluate exponents before multiplication. **89.** $\frac{16}{45}$ **91.** $\frac{17}{13}$ **93.** -212.96 **95.** -534.53

Section R.5 Algebraic Expressions **1.** variable **2.** constant **3.** $3 + 11$ **4.** $6 \cdot 7$ **5.** $\frac{y}{4}$ **6.** $3 - z$ **7.** $2(x - 3)$ **8.** $2x - 3$ **9.** evaluate **10.** -7

11. 41 **12.** 1 **13.** 3 **14.** 8000 yen; 80,000 yen; 800,000 yen **15.** 0°C; 30°C; 100°C **16.** like terms **17.** -1 **18.** $-5x$ **19.** $11x^2$ **20.** $-8x + 3$

21. $-4x + 8y$ **22.** $15y - 1$ **23.** $2.3x^2 + 0.9$ **24.** $-6z + 3$ **25.** $4x - 6$ **26.** $-5y + 11$ **27.** $-3z + 10$ **28.** $-6x + 4$ **29.** 1

30. $\frac{25x + 25}{6} = \frac{25(x + 1)}{6}$ **31.** domain **32.** (a) Yes (b) Yes (c) No (d) Yes **33.** (a) Yes (b) Yes (c) Yes (d) No **34.** (a) No (b) Yes

(c) Yes (c) No **35.** $5 + x$ **37.** $4z$ **39.** $y - 7$ **41.** $2(t + 4)$ **43.** $5x - 3$ **45.** $\frac{y}{3} + 6x$ **47.** 11 **49.** -9 **51.** -21 **53.** $\frac{3}{8}$ **55.** $\frac{6}{7}$ **57.** 29 **59.** 25

61. $\frac{3}{2}$ **63.** x **65.** $-6z + 3$ **67.** $-z - 5$ **69.** $\frac{11}{12}x$ **71.** $4x^2 - 3x$ **73.** $1.6x$ **75.** $-3x + 1$ **77.** $-14x + 7$ **79.** $-z + 10$ **81.** $4x - 3$

83. $12v - 11$ **85.** $2x + \frac{9}{10}$ **87.** $\frac{13}{36}x$ **89.** $14.46x - 15.49$ **91.** $-11.08x - 5.44$ **93.** (a) No (b) Yes (c) Yes (d) Yes **95.** (a) Yes (b) Yes

(c) Yes (d) Yes **97.** (a) No (b) Yes (c) Yes (d) No **99.** 1 in.^3; 8 in.^3; 27 in.^3; 64 in.^3 **101.** (a) 0 ft; 59 ft; 86 ft; 81 ft; 44 ft (b) The ball begins on the ground. When it is hit, it rises in the air (for somewhere around 2 seconds) and then begins to fall back toward the ground.
103. Let $x =$ Bob's age in years; Tony's age $= x + 5$; when Bob is 13 years old, Tony is 18 years old. **105.** Let $p =$ the original price in dollars; discounted price $= \frac{1}{2}p$; when the original price is $900, the discount price is $450. **107.** $\frac{4}{3}$ **109.** Answers will vary. One possible answer is *"Twice a number z decreased by 5."* **111.** Answers will vary. One possible answer is *"Twice the difference of a number z and 5."* **113.** Answers will vary. One possible answer is *"One-half the sum of a number z and 3."* **115.** A variable is a letter used to represent any number from a set of numbers; a constant is either a fixed number or a letter that represents a fixed number. **117.** Like terms are terms that have the same variable or variables, along with the same exponents on the variables. We use the Distributive Property "in reverse" to combine like terms, as in $3x + 4x = (3 + 4)x = 7x$. **119.** (a) 3
(b) 15 **121.** (a) 63 (b) 35 **123.** (a) $-\frac{7}{5}$ (b) $\frac{23}{65}$ **125.** (a) 67 (b) 32 **127.** The calculator displays an error message because $x = 5$ makes the denominator equal to 0.

Chapter 1 Linear Equations and Inequalities

Section 1.1 Linear Equations in One Variable **1.** linear; sides **2.** solution **3.** $x = 1$ **4.** $x = -7$ **5.** $z = -1$ **6.** True **7.** isolate **8.** $\{3\}$ **9.** $\{-2\}$ **10.** $\left\{\frac{1}{5}\right\}$

11. $\{2\}$ **12.** $\{-1\}$ **13.** $\left\{\frac{3}{2}\right\}$ **14.** $\{4\}$ **15.** $\{-4\}$ **16.** $\left\{\frac{3}{4}\right\}$ **17.** $\left\{\frac{1}{2}\right\}$ **18.** least common denominator **19.** $\{2\}$ **20.** $\{-5\}$ **21.** $\{-4\}$ **22.** $\left\{-\frac{3}{5}\right\}$
23. $\{10\}$ **24.** $\{32\}$ **25.** conditional equation **26.** contradiction; identity **27.** $\varnothing$ or $\{\ \}$; contradiction **28.** $\{x \mid x$ is any real number$\}$; identity
29. $\{0\}$; conditional **30.** $\{z \mid z$ is any real number$\}$; identity **31.** $x = 2$ **33.** $m = 1$ **35.** $x = 5$ **37.** $\{2\}$ **39.** $\left\{-\frac{1}{4}\right\}$ **41.** $\{9\}$ **43.** $\left\{-\frac{2}{3}\right\}$ **45.** $\{-4\}$
47. $\{2\}$ **49.** $\left\{-\frac{3}{5}\right\}$ **51.** $\{-40\}$ **53.** $\{3\}$ **55.** $\{-5\}$ **57.** $\{\ \}$ or $\varnothing$; contradiction **59.** $\{\ \}$ or $\varnothing$; contradiction **61.** $\{y \mid y$ is any real number$\}$ or $\mathbb{R}$; identity **63.** $\{-3\}$; conditional **65.** $\{\ \}$ or $\varnothing$; contradiction **67.** $\left\{\frac{5}{2}\right\}$; conditional **69.** $\{z \mid z$ is any real number$\}$ or $\mathbb{R}$; identity **71.** $\left\{-\frac{7}{2}\right\}$; conditional
73. $\left\{\frac{1}{7}\right\}$; conditional **75.** $\left\{\frac{7}{2}\right\}$; conditional **77.** $\{p \mid p$ is any real number$\}$ or $\mathbb{R}$; identity **79.** $\{\ \}$ or $\varnothing$; contradiction **81.** $\left\{-\frac{4}{3}\right\}$; conditional
83. $\{-14\}$; conditional **85.** $\{-3\}$; conditional **87.** $\{-1.6\}$; conditional **89.** $\{2\}$; conditional **91.** $a = -4$ **93.** $a = 3$ **95.** $x = -\frac{1}{2}$ **97.** $x = \frac{3}{4}$
99. $x = 1$ **101.** The card's annual interest rate is 0.15 or 15% **103.** Your adjusted income for 2011 was $32,250. **105.** $4(x + 1) - 2$ is an algebraic expression and $4(x + 1) = 2$ is an equation. An algebraic expression is any combination of variables, grouping symbols, and mathematical operations but does not contain an equal sign. An equation is a statement made up of two algebraic expressions that are equal. **107.** Answers will vary.

Section 1.2 An Introduction to Problem Solving **1.** equations **2.** $x + 7 = 12$ **3.** $3y = 21$ **4.** $2(n + 3) = 5$ **5.** $x - 10 = \frac{x}{2}$ **6.** $2n + 3 = 5$ **7.** False
8. 18, 20, 22 **9.** 25, 26, 27 **10.** $15 per hour **11.** $12 per hour **12.** 150 miles **13.** 240 minutes **14.** 100 **15.** 40 **16.** 160 **17.** 75% **18.** $30 **19.** $1.20
20. Interest; principal **21.** $32.50 **22.** $10.50; $1410.50 **23.** $67,500 in Aaa-rated bonds; $22,500 in B-rated bonds **24.** $5000 in CD; $20,000 in corporate bond **25.** 6 pounds of Tea A and 4 pounds of Tea B **26.** 10 pounds of cashews; 20 pounds of peanuts **27.** uniform motion **28.** rate **29.** After 4 hours; 240 miles **30.** After 5 hours; 450 miles **31.** 10 **33.** 40 **35.** 37.5% **37.** $x + 12 = 20$; 8 **39.** $2(y + 3) = 16$; 5 **41.** $w - 22 = 3w$; -11 **43.** $4x = 2x + 14$; 7
45. $0.8x = x + 5$; -25 **47.** 13 and 26 **49.** 24, 25, and 26 **51.** Kendra needs an 83 on her final exam to have an average of 80. **53.** Jacob would need to print 2500 pages for the cost to be the same for the two printers. **55.** Connor: $400,000; Olivia: $300,000; Avery: $100,000. **57.** The final bill will be $637.04. **59.** The dealer's cost is about $20,826.09. **61.** The flash drives originally cost $68.88. **63.** The Nissan Altima weighs 3193 pounds, the Mazda 6s weighs 2?29 pounds, and the Honda Accord EX weighs 3312 pounds. **65.** Adam will get $8500 and Krissy will get $11,500. **67.** You should invest $15,000 in stocks and $9000 in bonds. **69.** The interest charge after one month will be $29.17. **71.** The bank loaned $225,000 at 6% interest. **73.** Pedro

should invest $9375 in the 5% bond fund and $15,625 in the 9% stock fund. **75.** $50 - x$ pounds of coffee B **77.** Bobby has 15 dimes and 32 quarters saved. **79.** 36 grams of pure gold should be mixed with 36 grams of 12-karat gold. **81.** The race consists of running for 12 miles and biking for 50 miles. **83.** The slow car is traveling at 60 mph, while the faster car travels at 70 mph. **85.** The boats will be 155 miles apart after 2.5 hours. **87.** One person is walking at a rate of 4 miles per hour, and the other is walking at a rate of 6 miles per hour. **89.** Written answers will vary. The average speed of the trip to Florida and back is approximately 54.55 miles per hour. **91.** Answers will vary. **93.** The train is 0.3 mile or 1584 feet long. **95.** Mathematical modeling is the process of developing an equation or inequality to find a solution to a problem. Just as there is more than one way to solve a problem, there is typically more than one way to develop a mathematical model. **97.** Direct translation, mixture, geometry, uniform motion, and work problems. Two types of mixture problems are finance/investment and dry mixtures problems.

Section 1.3 Using Formulas to Solve Problems
1. formula **2.** $A = \pi r^2$ **3.** $V = \pi r^2 h$ **4.** $C = 175x + 7000$ **5.** $s = \dfrac{1}{2} gt^2$ **6.** area; volume **7.** False

8. (a) $h = \dfrac{2A}{b}$ **(b)** 5 inches **9. (a)** $b = \dfrac{P - 2a}{2}$ **(b)** 10 cm **10.** $P = \dfrac{I}{rt}$ **11.** $y = \dfrac{C - Ax}{B}$ **12.** $h = \dfrac{4x - 3}{2x - 3}$ **13.** $n = \dfrac{S + d}{a + d}$ **14.** Width: 40 feet; Length: 50 feet **15.** Height: 72 inches; Width: 40 inches **16.** 4.00 inches **17.** $F = ma$ **19.** $V = \dfrac{4}{3}\pi r^3$ **21.** $r = \dfrac{d}{t}$ **23.** $m = \dfrac{y - y_1}{x - x_1}$ **25.** $x = \mu + \sigma Z$

27. $m_1 = \dfrac{r^2 F}{Gm_2}$ **29.** $P = \dfrac{A}{1 + rt}$ **31.** $F = \dfrac{9}{5}C + 32$ **33.** $y = -2x + 13$ **35.** $y = 3x - 5$ **37.** $y = -\dfrac{4}{3}x + \dfrac{13}{3}$ **39.** $y = -3x + 12$ **41. (a)** $h = \dfrac{V}{\pi r^2}$

(b) The height of the cylinder is 8 inches. **43. (a)** $A = \dfrac{206.3 - M}{0.711}$ **(b)** An individual whose maximum heart rate is 160 should be about 65 years old.

45. (a) $P = \dfrac{A}{(1 + r)^t}$ **(b)** Approximately $4109.64 should be deposited today to have $5000 in 5 years in an account that pays 4% annual interest.

47. The smaller angle measures 75° and its supplement measures 105°. **49.** The smaller angle measures 20° and its complement measures 70°. **51.** The window is 5 feet long and 8 feet wide. **53.** The area of the circle is 25π square inches (approximately 78.54 square inches). **55.** The first angle measures 40°, the second measures 55°, and the third measures 85°. **57. (a)** The patio is 17.5 ft wide and 22.5 ft long. **(b)** You would need to purchase 131.25 cubic feet of cement. **59. (a)** The deck has an area of 84π square feet (approximately 264 square feet). **(b)** It would require approximately 97.39 feet of fencing to encircle the pool and deck. **(c)** The fence would cost about $2434.73. **61.** No, the area would increase by a factor of 4. If the length of a side of a cube is doubled, the volume increases by a factor of 8.

Section 1.4 Linear Inequalities in One Variable
1. closed interval **2.** left endpoint; right endpoint **3.** False

4. $[-3, 2]$ **5.** $[3, 6)$ **6.** $(-\infty, 3]$

7. $\left(\dfrac{1}{2}, \dfrac{7}{2}\right)$ **8.** $0 < x \le 5$

9. $-6 < x < 0$ **10.** $x > 5$ **11.** $x \le \dfrac{8}{3}$

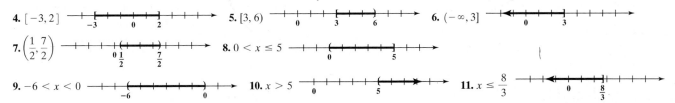

12. $9 < 12$; Addition Property of Inequalities **13.** $x > -9$; Addition Property of Inequalities **14.** $1 < 4$; Multiplication Property of Inequalities **15.** $2 > -3$; Multiplication Property of Inequalities **16.** $x < 6$; Multiplication Property of Inequalities

17. $\{x \mid x > 2\}$; $(2, \infty)$ **18.** $\{x \mid x \le 6\}$; $(-\infty, 6]$

19. $\{x \mid x \ge -6\}$; $[-6, \infty)$ **20.** $\{x \mid x > -3\}$; $(-3, \infty)$

21. $\{x \mid x \ge -2\}$; $[-2, \infty)$ **22.** $\left\{x \mid x > \dfrac{5}{2}\right\}$; $\left(\dfrac{5}{2}, \infty\right)$

23. $\{x \mid x < 4\}$; $(-\infty, 4)$ **24.** $\left\{x \mid x \le -\dfrac{7}{3}\right\}$; $\left(-\infty, -\dfrac{7}{3}\right]$

25. $\{x \mid x \ge 4\}$; $[4, \infty)$ **26.** $\{x \mid x \ge 3\}$; $[3, \infty)$

27. $\left\{x \mid x < \dfrac{1}{2}\right\}$; $\left(-\infty, \dfrac{1}{2}\right)$ **28.** $\{x \mid x > -17\}$; $(-17, \infty)$

29. Any balance over $480 **30.** For any more than 24 boxes, revenue exceeds cost.

31. $[2, 10]$;   **33.** $[-4, 0)$;   **35.** $[6, \infty)$;

37. $\left(-\infty, \dfrac{3}{2}\right)$;   **39.** $1 < x < 8$;

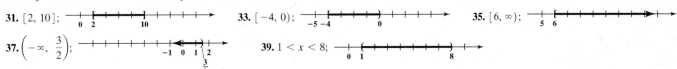

41. $-5 < x \le 1$;   **43.** $x < 5$;   **45.** $x \ge 3$;

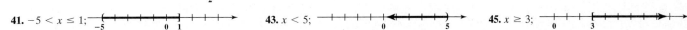

47. $<$; Addition Property of Inequalities **49.** $>$; Multiplication Property of Inequalities **51.** $\le$; Addition Property of Inequalities

53. $\le$; Multiplication Property of Inequalities **55.** $\{x \mid x \le 6\}$; $(-\infty, 6]$

57. $\{x|x < 4\}$; $(-\infty, 4)$

59. $\{x|x > -3\}$; $(-3, \infty)$

61. $\{x|x > 6\}$; $(6, \infty)$

63. $\{x|x > 3\}$; $(3, \infty)$

65. $\{x|x < -4\}$; $(-\infty, -4)$

67. $\{x|x \le -1\}$; $(-\infty, -1]$

69. $\{x|x > -2\}$; $(-2, \infty)$

71. $\{x|x < 17\}$; $(-\infty, 17)$

73. $\{x|x \le 4\}$; $(-\infty, 4]$

75. $\{x|x \le -30\}$; $(-\infty, -30]$

77. $\left\{x \middle| x < \frac{1}{3}\right\}$; $\left(-\infty, \frac{1}{3}\right)$

79. $\left\{x \middle| x < -\frac{11}{4}\right\}$; $\left(-\infty, -\frac{11}{4}\right)$

81. $\left\{x \middle| x < -\frac{16}{15}\right\}$; $\left(-\infty, -\frac{16}{15}\right)$

83. $\left\{x \middle| x \le \frac{15}{2}\right\}$; $\left(-\infty, \frac{15}{2}\right]$

85. $\left\{x \middle| x \ge \frac{21}{16}\right\}$; $\left[\frac{21}{16}, \infty\right)$

87. $\left\{x \middle| x > \frac{1}{10}\right\}$; $\left(\frac{1}{10}, \infty\right)$

89. $\{x|x < 3\}$; $(-\infty, 3)$

91. $\left\{x \middle| x < \frac{9}{2}\right\}$; $\left(-\infty, \frac{9}{2}\right)$

93. $\{y|y > -15\}$; $(-15, \infty)$

95. $\{a|a \ge 7\}$; $[7, \infty)$

97. $\left\{x \middle| x > -\frac{1}{3}\right\}$; $\left(-\frac{1}{3}, \infty\right)$

99. $\{x|x \le -14\}$; $(-\infty, -14]$

101. $\left\{x \middle| x < \frac{3}{2}\right\}$; $\left(-\infty, \frac{3}{2}\right)$

103. $\{x|x \ge -3\}$; $[-3, \infty)$

105. $\left\{x \middle| x \ge -\frac{3}{4}\right\}$; $\left[-\frac{3}{4}, \infty\right)$

107. $\{x|x \ge 4\}$; $[4, \infty)$ **109.** $\{z|z \le 3\}$; $(-\infty, 3]$

111. Jackie must earn at least 182 points on the final exam to earn an A in Mr. Ruffatto's class. **113.** You can order no more than 3 hamburgers to keep the fat content to no more than 69 grams. **115.** The plane can carry up to 18,836 pounds of luggage and cargo. **117.** The monthly benefit will exceed $1000 in 2012. **119.** Susan will need to sell at least $5,500,000 in computer systems to earn $100,000. **121.** Supply will exceed demand when the price is greater than $40. **123.** All real numbers are solutions; $\mathbb{R}$ **125.** When we multiply both sides of an inequality by a negative number. Or, if the sides of the inequality are interchanged. **127.** The inequality $4 < x > 7$ means $x > 4$ and $x > 7$, which is equivalent to $x > 7$.

Putting the Concepts Together (Sections 1.1–1.4) 1. (a) $x = -3$ is not a solution. **(b)** $x = 1$ is a solution. **2.** $\{-5\}$ **3.** $\{0\}$ **4.** identity **5.** $x - 3 = \frac{1}{2}x + 2$
6. $\frac{x}{2} < x + 5$ **7.** The chemist needs to mix 4 liters of the 20% solution with 12 liters of the 40% solution. **8.** After 3.4 hours, the two cars will be 255 miles apart. **9.** $y = \frac{3}{2}x - 2$ **10.** $r = \frac{A - P}{Pt}$ **11. (a)** $h = \frac{V}{\pi r^2}$ **(b)** 6 in. **12. (a)** Interval: $(-3, \infty)$ Graph:

(b) Interval: $(2, 5]$ Graph: **13. (a)** Inequality: $x \le -1.5$ Graph:
(b) Inequality: $-3 < x \le 1$ Graph: **14.** $[6, \infty)$
15. $(-\infty, 1)$ **16.** $(-\infty, 5]$ **17.** Logan can invite at most 9 children to the party.

Section 1.5 Rectangular Coordinates and Graphs of Equations **1.** origin **2.** True

3. A: Quadrant I;
B: Quadrant IV;
C: y-axis;
D: Quadrant III

4. A: Quadrant II;
B: x-axis;
C: Quadrant IV;
D: Quadrant I

5. True **6. (a)** No **(b)** Yes **(c)** Yes

7. (a) Yes **(b)** No **(c)** Yes

8. **9.** **10.** **11.** **12.**

13. intercepts **14.** False
15. Intercepts: $(-5, 0)$, $(0, -0.9)$, $(1, 0)$, $(6.7, 0)$; x-intercepts: $(-5, 0)$, $(1, 0)$, $(6.7, 0)$
y-intercept: $(0, -0.9)$

16. (a) $200 thousand **(b)** $350 thousand **(c)** The capacity of the refinery is 700 thousand gallons of gasoline per hour. **(d)** The intercept is $(0, 100)$. The cost of $100 thousand for producing 0 gallons of gasoline can be thought of as fixed costs. **17.** A: $(2, 3)$ I; B: $(-5, 2)$ II; C: $(0, -2)$ y-axis; D: $(-4, -3)$ III; E: $(3, -4)$ IV; F: $(4, 0)$ x-axis

19. *A*: quadrant I;
   B: quadrant III;
   C: *x*-axis;
   D: quadrant IV;
   E: *y*-axis;
   F: quadrant II

21. (a) yes **(b)** no **(c)** yes **(d)** yes **23. (a)** yes **(b)** no **(c)** no **(d)** yes

25. (a) no **(b)** yes **(c)** yes **(d)** yes

27. $y = 4x$ **29.** $y = -\dfrac{1}{2}x$ **31.** $y = x + 3$ **33.** $y = -3x + 1$ **35.** $y = \dfrac{1}{2}x - 4$

    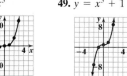

37. $2x + y = 7$ **39.** $y = -x^2$ **41.** $y = 2x^2 - 8$ **43.** $y = |x|$ **45.** $y = |x - 1|$ **47.** $y = x^3$ **49.** $y = x^3 + 1$

51. $x^2 - y = 4$ **53.** $x = y^2 - 1$ **55.** $(-2, 0)$ and $(0, 3)$

57. $(-2, 0)$, $(1, 0)$, and $(0, -4)$

59. $a = \dfrac{7}{4}$ **61.** $b = 4$ **63. (a)** 400 ft² **(b)** 25 feet; 625 ft² **(c)** The *x*-intercepts are $x = 0$ and $x = 50$. These values form the bounds for the width of the opening. The *y*-intercept is $y = 0$. The area of the opening will be 0 ft² when the width is 0 feet. **65. (a)** $100; $100 **(b)** $1600 **(c)** $(0, 100)$; The monthly cost will be $100 if no minutes are used.

67. Vertical line with an *x*-intercept of 4.

69. Answers will vary. One possible graph is shown.

71. Answers will vary. One possibility: $y = 0$

73. A complete graph is one that shows enough of the graph so that anyone who is looking at it will "see" the rest of it as an obvious continuation of what is shown. A complete graph should show all the interesting features of the graph, such as intercepts and high/low points.

75. The point-plotting method of graphing an equation requires one to choose certain values of one variable and use the equation to find the corresponding values of the other variable. These points are then plotted and connected in a smooth curve.

77. $y = 3x - 9$ **79.** $y = -x^2 + 8$ **81.** $y + 2x^2 = 13$ **83.** $y = x^3 - 6x + 1$

 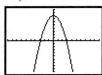 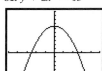 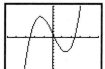

Section 1.6 Linear Equations in Two Variables   **1.** linear equation **2.** line

3.   **4.**   **5.**   **6.** True   **7.**   **8.**   **9.**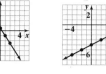

10. x; y **11.**   **12.**   **13.**   **14.** undefined   **15.** $\dfrac{2}{5}$ **16.** False   **17.** True   **18.** 3; For every 1-unit increase in *x*, *y* will increase by 3 units. **19.** $-\dfrac{7}{4}$; For every 4-unit increase in *x*, *y* will decrease by 7 units. **20.** 0; For every 1-unit increase in *x*, there is no change in *y*; Horizontal line **21.** Undefined; Vertical line **22.** True **23.** 0

24. L_1: $m = \dfrac{1}{5}$; L_2: *m* is undefined; L_3: $m = -1$; L_4: $m = 0$

25. (a)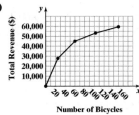

(b) $1120 per bicycle. For each bicycle sold, total revenue increased by $1120 when between 0 and 25 bicycles were sold.
(c) $120 per bicycle. For each bicycle sold, total revenue increased by $120 per bicycle when between 102 and 150 bicycles were sold.
(d) No, because the average rate of change (slope) is not constant.

26. (a) **(b)** **(c)**

27. point-slope

28. $y - 5 = 2(x - 3)$

29. $y - 3 = -4(x + 2)$

30. $y + 4 = \dfrac{1}{3}(x - 3)$

31. $y = -2$

32. slope-intercept

33. $m = 3$; y-intercept: $(0, -2)$

34. $m = -3$; y-intercept: $(0, 4)$

35. $m = \dfrac{3}{2}$; y-intercept: $\left(0, -\dfrac{7}{2}\right)$

36. $m = -\dfrac{7}{3}$; y-intercept: $(0, 0)$

37. $y = 2x + 1$

38. $y = -\dfrac{1}{2}x + 3$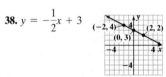

39. $y = 6$

40. $x = 3$

41. **43.** **45.** **47.** **49.** **51.** 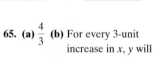 **53.**

55. **57.** **59.** **61.** **63.**

65. (a) $\dfrac{4}{3}$ **(b)** For every 3-unit increase in x, y will increase 4 units.

67. (a) $-\dfrac{8}{3}$ **(b)** For every 3-unit increase in x, y will decrease by 8 units. For every 3-unit decrease in x, y will increase by 8 units.

69. $m = 5$

71. $m = -3$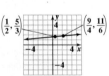

73. $m = \dfrac{4}{5}$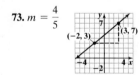

75. $m = 0$

77. m is undefined

79. $m = \dfrac{2}{21}$ $\left(\dfrac{1}{2}, \dfrac{5}{3}\right)$ $\left(\dfrac{9}{4}, \dfrac{11}{6}\right)$

81.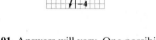

83.

85. **87.** **89.**

91. Answers will vary. One possibility: $(0, 8)$, $(2, 13)$, and $(4, 18)$

93. $y = -\dfrac{4}{9}x - \dfrac{7}{9}$ **95.** $y = 3$ **97.** $y = 2x$ **99.** $y = -3x - 2$

101. $y = \dfrac{4}{3}x - 2$ **103.** $y = -\dfrac{5}{4}x + \dfrac{3}{2}$ **105.** $x = 6$ **107.** $y = \dfrac{7}{5}x$

109. $y = 5x - 13$ **111.** $y = -\dfrac{3}{7}x + \dfrac{1}{7}$ **113.** $x = -1$ **115.** $y = \dfrac{5}{2}x + \dfrac{1}{2}$ **117.** $y = 4$

119. The slope is 2 and the y-intercept is $(0, -1)$.

121. The slope is -4 and the y-intercept is $(0, 0)$.

123. The slope is -2 and the y-intercept is $(0, 3)$.

125. The slope is -2 and the y-intercept is $(0, 4)$.

127. The slope is $\frac{1}{4}$ and the y-intercept is $\left(0, -\frac{1}{2}\right)$.

129. The slope is undefined and there is no y-intercept.

131. $y = 0$

133. (a)

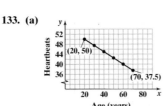

(b) -0.25 heartbeat per year; Between ages 20 and 30, the maximum number of heartbeats decreases at a rate of 0.25 heartbeat per year. (c) -0.25 heartbeat per year; Between ages 50 and 60, the maximum number of heartbeats decreases at a rate of 0.25 heartbeat per year. (d) Yes. The maximum number of heartbeats appears to be linearly related to age. The average rate of change (slope) is constant for the data provided.

135. (a)

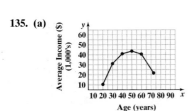

(b) $2187.80 per year; Between ages 20 and 30, the average individual's income increases at a rate of $2187.80 per year. (c) $-$343.50 per year; Between ages 50 and 60, the average individual's income decreases at a rate of $343.50 per year. (d) No. The average income is not linearly related to age. The average rate of change (slope) is not constant.

137. $C = \frac{5}{9}(F - 32)$; 60°F is equivalent to 15.6°C. **139.** (c)

141. Horizontal line: $y = b$; Vertical line: $x = a$; Point-slope: $y - y_1 = m(x - x_1)$; Slope-intercept: $y = mx + b$; Standard: $Ax + By = C$

143. Horizontal line of the form $y = b$, where b does not equal 0. **145.** No. All lines must travel through either the x-axis or the y-axis or both.

147. For positive slopes, the larger the slope, the steeper the line. **149.** Answers may vary. Answers may vary.

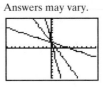

Section 1.7 Parallel and Perpendicular Lines   **1.** slope; y-intercepts   **2.** Not parallel   **3.** Parallel   **4.** Not parallel

5. $y = 3x - 7$

6. $y = -\frac{3}{2}x + 1$

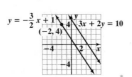

7. -1 **8.** $\frac{1}{3}$

9. Perpendicular
10. Not perpendicular
11. Perpendicular

12. $y = -\frac{1}{2}x$

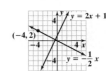

13. $y = -\frac{4}{3}x - 8$

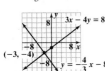

14. $y = -2$

15. (a) $m = 5$ (b) $m = -\frac{1}{5}$ **17.** (a) $m = -\frac{5}{6}$ (b) $m = \frac{6}{5}$ **19.** Parallel

21. Neither **23.** Perpendicular **25.** Parallel **27.** $y = \frac{3}{2}x - 3$ **29.** $y = \frac{1}{2}x + 2$ **31.** $y = 2$

33. $y = 2x - 5$

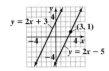

35. $y = \frac{1}{2}x + 2$

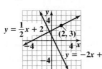

37. $y = -3$

39. $y = 3$

41. $y = 3x + 2$

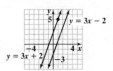

43. $y = \frac{3}{4}x + 4$

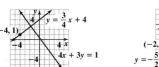

45. $y = -\frac{5}{2}x - 8$

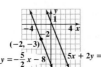

47. $m_1 = \frac{3}{5}; m_2 = -\frac{5}{3};$
Perpendicular

49. $m_1 = -\frac{7}{3}; m_2 = -\frac{7}{3};$
Parallel

51. $m_1 = -2; m_2 = 2;$
Neither

53. (a)

(b) Slope of $\overline{AB} = \dfrac{2}{3}$;

Slope of $\overline{BC} = -\dfrac{3}{2}$

Because the slopes are negative reciprocals, segments $\overline{AB}$ and $\overline{BC}$ are perpendicular. Thus, triangle ABC is a right triangle.

55. (a)

(b) Slope of $\overline{AB} = \dfrac{1}{5}$;

Slope of $\overline{BC} = 3$;

Slope of $\overline{CD} = \dfrac{1}{5}$;

Slope of $\overline{DA} = 3$

Because the slopes of $\overline{AB}$ and $\overline{CD}$ are equal, $\overline{AB}$ and $\overline{CD}$ are parallel. Because the slopes of $\overline{BC}$ and $\overline{DA}$ are equal, $\overline{BC}$ and $\overline{DA}$ are parallel. Thus, quadrilateral $ABCD$ is a parallelogram.

57. $A = -1$ **59.** (c) **61.** No. If the two nonvertical lines have the same x-intercept but different y-intercepts, then their slopes cannot be equal. Therefore, they cannot be parallel lines.

Section 1.8 Linear Inequalities in Two Variables   1. half-planes   2. (a) No   (b) Yes   (c) Yes   (d) Yes   3. False   4. True

5.

6.

7.

8. (a) $560x + 300y \le 900$ **(b)** Yes   **(c)** No
9. (a) yes   **(b)** no   **(c)** yes
11. (a) yes   **(b)** no   **(c)** yes

13.

15.

17.

19.

21.

23.

25.

27.

29.

31.

33. (a) Let $x = $ the number of Filet-o-Fish orders. Let $y = $ the number of orders of fries. $39x + 29y \le 125$ **(b)** yes **(c)** no **35. (a)** Let $x = $ the number of Switch A assemblies. Let $y = $ the number of Switch B assemblies. $2x + 1.5y \le 80$ **(b)** no **(c)** no **37.** $y > 2x - 2$ **39.** $y \le -x + 1$
41. Consider the inequality $Ax + By > C$. The equation $Ax + By = C$ separates the Cartesian plane into two half-planes. Any ordered pair (x, y) that makes $Ax + By$ equal to C lies on this line. Any other point will lie in one of the two half-planes. If a point on one side of the line does not satisfy the inequality, then none of the points satisfy the inequality. For this reason, if a point does not satisfy the inequality, then we shade the opposite side.

43. $y > 3$

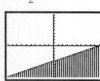

45. $y < 5x$

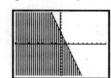

47. $y > 2x + 3$

49. $y \le \dfrac{1}{2}x - 5$

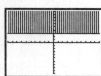

51. $3x + y \le 4$

$(y \le -3x + 4)$

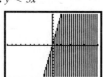

53. $2x + 5y \le -10$

$\left(y \le -\dfrac{2}{5}x - 2\right)$

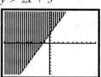

Chapter 1 Review **1.** $x = 5$ is a solution to the equation. $x = 6$ is not a solution to the equation. **2.** $x = -2$ is a solution to the equation. $x = -1$ is a solution to the equation. **3.** $y = -2$ is not a solution to the equation. $y = 0$ is a solution to the equation. **4.** $w = -14$ is not a solution to the equation. $w = 7$ is not a solution to the equation. **5.** conditional; $\{3\}$ **6.** conditional; $\{4\}$ **7.** conditional; $\{3\}$
8. conditional; $\{-8\}$ **9.** identity; $\{x \mid x$ is any real number$\}$ **10.** contradiction; $\{\ \}$ or $\varnothing$ **11.** contradiction; $\{\ \}$ or $\varnothing$ **12.** conditional; $\{-2\}$
13. conditional; $\{-47\}$ **14.** identity; $\{w \mid w$ is any real number$\}$ **15.** $x = -\dfrac{3}{2}$ must be excluded from the domain. **16.** $x = \dfrac{1}{2}$ must be excluded from the domain. **17.** Her Missouri taxable income was $43,250. **18.** The regular club price for a DVD is $21.95.
19. $3x + 7 = 22$ **20.** $x - 3 = \dfrac{x}{2}$ **21.** $0.2x = x - 12$ **22.** $6x = 2x - 4$ **23.** Payton is 5 years old and Shawn is 13 years old.
24. The five odd integers are 21, 23, 25, 27, and 29. **25.** Logan needs to get a score of 76.5 on the final exam to have an average of 80. **26.** After 1 month, Cherie will accrue about $68.63 in interest. **27.** The original price of the sleeping bag was $135.00. **28.** The federal minimum wage was

$6.55 (per hour). **29.** The store should mix 4 pounds of chocolate-covered blueberries with 8 pounds of chocolate-covered strawberries. **30.** The store should mix 7.5 pounds of baseball gumballs with 2.5 pounds of soccer gumballs. **31.** Angie should invest $4800 at 8% and $3200 at 18%.
32. About 2.14 quarts would need to be drained and replaced with pure antifreeze. **33.** Josh drove 90 miles at 60 miles per hour
34. The F14 is traveling at a speed of 1220 miles per hour, and the F15 is traveling at a speed of 1420 miles per hour.

35. $x = \dfrac{k}{y}$ **36.** $C = \dfrac{5}{9}(F - 32)$ **37.** $W = \dfrac{P - 2L}{2}$ **38.** $m_2 = \dfrac{\rho - m_1 v_1}{v_2}$ **39.** $T = \dfrac{PV}{nR}$ **40.** $W = \dfrac{S - 2LH}{2L + 2H}$ **41.** $y = -\dfrac{3}{4}x + \dfrac{1}{2}$

42. $y = \dfrac{5}{4}x + \dfrac{5}{2}$ **43.** $y = 4x - 5$ **44.** $y = -\dfrac{6}{5}x + 24$ **45.** The melting point of platinum is 1772°C.   **46.** The angles measure 70°, 70°, and 40°.

47. The window measures 15 feet by 23 feet.   **48.** (a) $x = 25C - 73.75$ (b) Debbie can talk for 426 minutes in one month and not spend more

than $20 on long distance.   **49.** The patio will be $\dfrac{10}{27}$ of a foot thick (about 4.44 inches).   **50.** (a) $r = \dfrac{A - \pi Rs}{\pi s}$ (b) The radius of the top of

the frustum is 2 feet.   **51.** (a) $x = \dfrac{C - 7.48}{0.08674}$ (b) Approximately 2289 kwh were used.   **52.** The angle measures 60°.

53. (2,7] 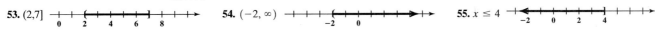   **54.** $(-2, \infty)$ **55.** $x \le 4$

56. $-1 \le x < 3$ **57.** $a = 7$ and $b = 15$ **58.** $a = -1$ and $b = 5$ **59.** Solution set: $\{x \mid x \le -4\}$

Interval: $(-\infty, -4]$ Graph: 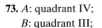   **60.** Solution set: $\left\{x \mid x < -\dfrac{1}{3}\right\}$ Interval: $\left(-\infty, -\dfrac{1}{3}\right)$

Graph: **61.** Solution set: $\left\{h \mid h \ge -\dfrac{2}{3}\right\}$ Interval: $\left[-\dfrac{2}{3}, \infty\right)$ Graph:

62. Solution set: $\{x \mid x > 2\}$ Interval: $(2, \infty)$ Graph: **63.** Solution set: $\{p \mid p > 2\}$ Interval: $(2, \infty)$

Graph: **64.** Solution set: $\{x \mid x \text{ is any real number}\}$ Interval: $(-\infty, \infty)$

Graph: **65.** $\{\ \}$ or $\varnothing$ **66.** Solution set: $\{x \mid x > 4.2\}$ Interval: $(4.2, \infty)$

Graph: **67.** Solution set: $\{w \mid w > 1\}$ Interval: $(1, \infty)$ Graph:
68. Solution set: $\{y \mid y < -120\}$ Interval: $(-\infty, -120)$ Graph: **69.** To stay within budget, no more

than 60 people can attend the banquet.   **70.** To stay within budget, you can drive an average of 216 miles per day.   **71.** The band must sell more
than 125 candy bars to make a profit.   **72.** You can purchase up to 5 DVDs and still be within budget.

73. A: quadrant IV;
B: quadrant III;
C: y-axis; D: quadrant II;
E: x-axis; F: quadrant I

74. A: x-axis;
B: quadrant I;
C: quadrant III; D: quadrant II;
E: quadrant IV;
F: y-axis

75. (a) yes (b) no (c) no (d) yes
76. (a) no (b) yes (c) yes (d) no
77. $y = x + 2$

78. $2x + y = 3$

79. $y = -x^2 + 4$

80. $y = |x + 2| - 1$

81. $y = x^3 + 2$

82. $x = y^2 + 1$

83. $(-3, 0), (0, -1), (0, 3)$
x-intercept: $(-3, 0)$;
y-intercepts: $(0, -1), (0, 3)$
84. (a) $40
(b) About $500

85.

86.

87.

88.

89.

90.

91.

92.

93.

94.

95.

96. (a) $m = \dfrac{5}{4}$ (b) For every 4-unit increase in x, y will increase by 5 units.

97. (a) $m = -\dfrac{1}{4}$ (b) For every 4-unit increase in x, y will decrease by
1 unit. For every 4-unit decrease in x, y will increase by 1 unit.

98. $m = -2$

99. $m = \dfrac{3}{2}$

100. (a)

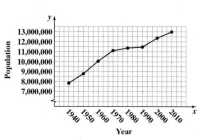

(b) 81,493.5 people per year; Between 1940 and 1950, the population of Illinois increased at an average rate of 81,493.5 people per year.   **(c)** 319.3 people per year; Between 1980 and 1990, the population of Illinois increased at an average rate of 319.3 people per year.   **(d)** 41,133.9 people per year; Between 2000 and 2010, the population of Illinois increased at an average rate of 41,133.9 people per year.   **(e)** No. The average rate of change (slope) is not constant.

101.   **102.**   **103.**   **104.**   **105.** $y = -\dfrac{1}{2}x + 2$ or $x + 2y = 4$

106. $y = 3x$ or $3x - y = 0$

107. $y = -x + 5$ or $x + y = 5$

108. $y = \dfrac{3}{5}x + 2$ or $3x - 5y = -10$ **109.** $y = -\dfrac{1}{3}x + 4$ or $x + 3y = 12$ **110.** $y = 3$ **111.** $y = 2x - 9$ or $2x - y = 9$

112. $y = -\dfrac{1}{3}x + \dfrac{5}{3}$ or $x + 3y = 5$

113. The slope is 4 and the y-intercept is $(0, -6)$.   **114.** The slope is $-\dfrac{2}{3}$ and the y-intercept is $(0,4)$.   **115.** $-\dfrac{3}{8}$ **116.** $\dfrac{8}{3}$ **117.** Perpendicular

118. Neither   **119.** Parallel

120. Perpendicular

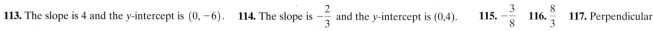

121. $y = -2x + 4$ **122.** $y = \dfrac{5}{2}x - 7$ **123.** $x = 1$ **124.** $y = -\dfrac{1}{3}x + 4$ **125.** $y = \dfrac{4}{3}x + 2$

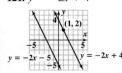   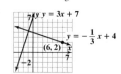   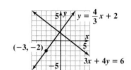

126. $y = -4$ **127. (a)** yes   **129.**   **130.**   **131.**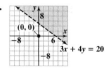

(b) yes

(c) no

128. (a) no

(b) yes

(c) no

132.   **133.** 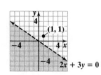   **134.**   **135. (a)** Let $x =$ the number of movie tickets. Let $y =$ the number of music downloads $7.50x + 2y \le 60$ **(b)** no   **(c)** yes

136. (a) Let $x =$ the number of candy bars. Let $y =$ the number of candles. $0.50x + 2y \ge 1000$ **(b)** no   **(c)** yes

Chapter 1 Test 1. (a) $x = 6$ is a solution to the equation.   **(b)** $x = -2$ is not a solution to the equation.

2. (a) Interval: $(-4, \infty)$    **(b)** Interval: $(3, 7]$ 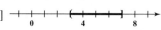   **3.** $3x - 8 = x + 4$

4. $\dfrac{2}{3}x + 2(x - 5) > 7$ **5.** $\{2\}$; conditional   **6.** $\{\ \}$ or $\varnothing$ contradiction;

7. $\{x \mid x \ge 3\}$; $[3, \infty)$    **8.** $\left\{x \mid x > -\dfrac{1}{5}\right\}$; $\left(-\dfrac{1}{5}, \infty\right)$

9. $\left\{x \mid x \ge \dfrac{1}{2}\right\}$; $\left[\dfrac{1}{2}, \infty\right)$ **10.** $y = -\dfrac{7}{4}x + \dfrac{3}{4}$ **11.** Glen's weekly sales must be at least \$4375 for him to earn

at least \$750.   **12.** There were 14 children at Payton's party.   **13.** The sandbox has a width of 4 feet and a length of 6 feet.   **14.** The chemist needs to mix 8 liters of the 10% solution with 4 liters of the 40% solution.   **15.** It will take Contestant B two hours to catch up to Contestant A.

16. *A*: quadrant IV;
  B: *y*-axis;
  C: *x*-axis;
  D: quadrant I;
  E: quadrant III;
  F: quadrant II

17. **(a)** no **(b)** yes **(c)** yes
18. $y = 4x - 1$

19. $y = 4x^2$

20. $(-3, 0)$, $(0, 1)$, $(0, 3)$
  x-intercept: $(-3, 0)$;
  y-intercepts: $(0, 1)$, $(0, 3)$

21. **(a)** At 6 seconds the car is traveling 30 miles per hour. **(b)** $(0, 0)$: At the start of the trip, the car is not moving. $(32, 0)$: After 32 seconds, the speed of the car is 0 miles per hour.

22. **23.** **24.** **25.** **26.**

27. $m = -\dfrac{4}{3}$

For every 3-unit increase in *x*, *y* will decrease by 4 units. For every 3-unit decrease in *x*, *y* will increase by 4 units.

28.

29. Perpendicular **30.** $y = 4x + 13$ or $4x - y = -13$ **31.** $y = -\dfrac{2}{3}x + 5$ or $2x + 3y = 15$
32. $y = \dfrac{1}{5}x - 3$ or $x - 5y = 15$ **33.** $y = -\dfrac{1}{3}x + 4$ or $x + 3y = 12$ **34.** **(a)** no **(b)** no **(c)** yes

35. **36.** **37.** **(a)**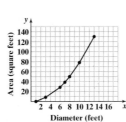

(b) 3.14 square feet per foot; Between diameters of 1 foot and 3 feet, the area of the circle increases at a rate of 3.14 square feet per foot. **(c)** Approximately 18.06 square feet per foot; Between diameters of 10 feet and 13 feet, the area of the circle increases at a rate of approximately 18.06 square feet per foot. **(d)** No. The average rate of change (slope) is not constant.

Cumulative Review Chapters R–1 **1.** **(a)** (i) 27.235 (ii) 27.236 **(b)** (i) 1.0 (ii) 1.1 **2.** **3.** -14

4. -6 **5.** 6 **6.** 81 **7.** 74 **8.** $\dfrac{11}{12}$ **9.** 9 **10.** $5a^2 - 4a - 13$ **11.** **(a)** no **(b)** yes **12.** $-5x + 18$

13. $x = 3$ is not a solution to the equation. **14.** $\{4\}$ **15.** $\left\{\dfrac{13}{2}\right\}$ **16.** $y = \dfrac{2}{5}x - \dfrac{6}{5}$

17. $\{x \mid x \ge 7\}$; $[7, \infty)$ **18.** $\{x \mid x \le 5\}$ or $(-\infty, 5]$

19. **20.** **21.** **22.** $y = -\dfrac{4}{3}x + 2$ or $4x + 3y = 6$ **24.**

23. $y = -3x - 8$ or $3x + y = -8$

25. Shawn needs to score at least 91 on the final exam to earn an A (assuming the maximum score on the exam is 100). **26.** A person 62 inches tall would be considered obese if she or he weighed 160 pounds or more. **27.** The angles measure $55°$ and $125°$. **28.** The cylinder should be about 5.96 inches tall. **29.** The three consecutive even integers are 24, 26, and 28.

Chapter 2 Relations, Functions, and More Inequalities

Section 2.1 Relations **1.** corresponds; depends **2.** {(Max, November 8), (Alesia, January 20), (Trent, March 3), (Yolanda, November 8), (Wanda, July 6), (Elvis, January 8)} **3.** **4.** domain; range **5.** Domain: {Max, Alesia, Trent, Yolanda, Wanda, Elvis}; Range: {January 20, March 3, July 6, November 8, January 8} **6.** Domain: {1, 5, 8, 10}; Range: {3, 4, 13} **7.** Domain: $\{-2, -1, 2, 3, 4\}$; Range: $\{-3, -2, 0, 2, 3\}$ **8.** True **9.** False **10.** Domain: $\{x \mid -2 \le x \le 4\}$ or $[-2, 4]$; Range: $\{y \mid -2 \le y \le 2\}$ or $[-2, 2]$ **11.** Domain: $\{x \mid x$ is a real number$\}$ or $(-\infty, \infty)$; Range: $\{y \mid y$ is a real number$\}$ or $(-\infty, \infty)$

12.

Domain: $\{x \mid x$ is a real number$\}$ or $(-\infty, \infty)$
Range: $\{y \mid y$ is a real number$\}$ or $(-\infty, \infty)$

13.

Domain: $\{x \mid x$ is a real number$\}$ or $(-\infty, \infty)$
Range: $\{y \mid y \ge -8\}$ or $[-8, \infty)$

14.

Domain: $\{x \mid x \ge 1\}$ or $[1, \infty)$
Range: $\{y \mid y$ is a real number$\}$ or $(-\infty, \infty)$

15. {(*USA Today*, 1.83.), (*Wall Street Journal*, 2.12), (*New York Times*, 0.92), (*Los Angeles Times*, 0.61), (*Washington Post*, 0.55)}; Domain: {*USA Today*, *Wall Street Journal*, *New York Times*, *Los Angeles Times*, *Washington Post*}; Range: {1.83, 2.12, 0.92, 0.61, 0.55}   **17.** {(Less than 9th Grade, $13,992), (9th–12th Grade – no diploma, $14,460), (High School Graduate, $23,520), (Associate's Degree, $36,012), (Bachelor's Degree or Higher, $51,108)}; Domain: {Less than 9th Grade, 9th–12th Grade – no diploma, High School Graduate, Associate;s Degree, Bachelor's Degree or Higher}; Range: {$13,992, $14,460, $23,520, $36,012, $51,108}

19.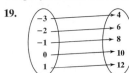
Domain: $\{-3, -2, -1, 0, 1\}$
Range: $\{4, 6, 8, 10, 12\}$

21.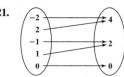
Domain: $\{-2, -1, 0, 1, 2\}$
Range: $\{0, 2, 4\}$

23.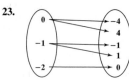
Domain: $\{-2, -1, 0\}$
Range: $\{-4, -1, 0, 1, 4\}$

25. Domain: $\{-3, -2, 0, 2, 3\}$; Range: $\{-3, -1, 2, 3\}$
27. Domain: $\{x | -4 \le x \le 4\}$ or $[-4, 4]$;
Range: $\{y | -2 \le y \le 2\}$ or $[-2, 2]$
29. Domain: $\{x | -1 \le x \le 3\}$ or $[-1, 3]$;
Range: $\{y | 0 \le y \le 4\}$ or $[0, 4]$
31. Domain: $\{x | x \text{ is a real number}\}$ or $(-\infty, \infty)$;
Range: $\{y | y \ge -3\}$ or $[-3, \infty)$

33. Domain: $\{x | x \text{ is a real number}\}$ or $(-\infty, \infty)$; Range: $\{y | y \text{ is a real number}\}$ or $(-\infty, \infty)$ **35.** Domain: $\{x | x \text{ is a real number}\}$ or $(-\infty, \infty)$; Range: $\{y | y \text{ is a real number}\}$ or $(-\infty, \infty)$ **37.** Domain: $\{x | x \text{ is a real number}\}$ or $(-\infty, \infty)$; Range: $\{y | y \text{ is a real number}\}$ or $(-\infty, \infty)$

39. Domain: $\{x | x \text{ is a real number}\}$ or $(-\infty, \infty)$; Range: $\{y | y \le 0\}$ or $(-\infty, 0]$ **41.** Domain: $\{x | x \text{ is a real number}\}$ or $(-\infty, \infty)$;
Range: $\{y | y \ge -8\}$ or $[-8, \infty)$ **43.** Domain: $\{x | x \text{ is a real number}\}$ or $(-\infty, \infty)$; Range: $\{y | y \ge 0\}$ or $[0, \infty)$

45. Domain: $\{x | x \text{ is a real number}\}$ or $(-\infty, \infty)$; Range: $\{y | y \ge 0\}$ or $[0, \infty)$ **47.** Domain: $\{x | x \text{ is a real number}\}$ or $(-\infty, \infty)$;
Range: $\{y | y \text{ is a real number}\}$ or $(-\infty, \infty)$ **49.** Domain: $\{x | x \text{ is a real number}\}$ or $(-\infty, \infty)$; Range: $\{y | y \text{ is a real number}\}$ or $(-\infty, \infty)$

51. Domain: $\{x | x \text{ is a real number}\}$ or $(-\infty, \infty)$; Range: $\{y | y \ge -4\}$ or $[-4, \infty)$ **53.** Domain: $\{x | x \ge -1\}$ or $[-1, \infty)$;
Range: $\{y | y \text{ is a real number}\}$ or $(-\infty, \infty)$ **55. (a)** Domain: $\{x | 0 < x < 50\}$ or $(0, 50)$; Range: $\{y | 0 < y \le 625\}$ or $(0, 625]$ **(b)** Width must be less than $\frac{1}{2}$ the perimeter.   **57. (a)** Domain: $\{m | 0 \le m \le 15{,}120\}$ or $[0, 15120]$; Range: $\{C | 100 \le C \le 3380\}$ or $[100, 3380]$

(b) $21 \cdot 12 \cdot 60 = 15{,}120$ minutes   **59.** Actual graphs will vary but all should be horizontal lines.   **61.** A relation is a correspondence between two sets called the domain and range. The domain is the set of all inputs, and the range is the set of all outputs.

Section 2.2 An Introduction to Functions   **1.** function   **2.** false   **3.** Function; Domain: {Max, Alesia, Trent, Yolanda, Wanda, Elvis}; Range: {January 20, March 3, July 6, November 8, January 8}   **4.** Not a function   **5.** Function; Domain: $\{-3, -2, -1, 0, 1\}$; Range: $\{0, 1, 2, 3\}$ **6.** Not a function
7. Function   **8.** Not a function   **9.** Function   **10.** True   **11.** Function   **12.** Not a function   **13.** 14   **14.** -4 **15.** -18 **16.** 2
17. dependent; independent; argument   **18.** $2x - 9$ **19.** $2x - 4$ **20.** domain   **21.** $(-\infty, \infty)$ **22.** $\{x | x \ne 3\}$ **23.** $\{r | r > 0\}$ or $(0, \infty)$
24. (a) Independent variable: t; dependent variable: A **(b)** $A(30) \approx 706.86$ square miles. After 30 days, the area contaminated with oil will be a circle covering about 706.86 square miles.   **25.** Function. Domain: {Virginia, Nevada, Arkansas, Tennessee, Texas}; Range: $\{4, 9, 11, 36\}$
27. Not a function. Domain: $\{150, 174, 180\}$; Range: $\{118, 130, 140\}$ **29.** Function; Domain: $\{0, 1, 2, 3\}$; Range: $\{3, 4, 5, 6\}$
31. Function; Domain: $\{-3, 1, 4, 7\}$; Range: $\{5\}$ **33.** Not a function. Domain: $\{-10, -5, 0\}$; Range: $\{1, 2, 3, 4\}$ **35.** Function   **37.** Function
39. Not a function   **41.** Function   **43.** Not a function   **45.** Function   **47.** Not a function   **49.** Function   **51.** Function   **53. (a)** $f(0) = 3$;
(b) $f(3) = 9$ **(c)** $f(-2) = -1$ **55. (a)** $f(0) = 2$ **(b)** $f(3) = -13$ **(c)** $f(-2) = 12$ **57. (a)** $f(0) = 0$ **(b)** $f(3) = 0$ **(c)** $f(-2) = 10$
59. (a) $f(0) = 3$ **(b)** $f(3) = -3$ **(c)** $f(-2) = -3$ **61. (a)** $f(-x) = -2x - 5$ **(b)** $f(x + 2) = 2x - 1$ **(c)** $f(2x) = 4x - 5$
(d) $-f(x) = -2x + 5$ **(e)** $f(x + h) = 2x + 2h - 5$ **63. (a)** $f(-x) = 7 + 5x$ **(b)** $f(x + 2) = -3 - 5x$ **(c)** $f(2x) = 7 - 10x$
(d) $-f(x) = -7 + 5x$ **(e)** $f(x + h) = 7 - 5x - 5h$ **65.** $f(2) = 7$ **67.** $s(-2) = 16$ **69.** $F(-3) = 5$ **71.** $F(4) = -6$
73. $\{x | x \text{ is any real number}\}$ or $(-\infty, \infty)$ **75.** $\{z | z \ne 5\}$ **77.** $\{x | x \text{ is any real number}\}$ or $(-\infty, \infty)$ **79.** $\left\{x | x \ne -\dfrac{1}{3}\right\}$ **81.** $C = -6$ **83.** $A = 5$
85. $A(r) = \pi r^2$; 50.27 in.^{2} **87.** $G(h) = 15h$; \$375   **89. (a)** The dependent variable is the population, P, and the independent variable is the age, a.
(b) $P(20) = 223{,}091$ thousand; The population of U.S. residents who were 20 years of age or older in 2010 was roughly 223 million.
(c) $P(0) = 321{,}783$ thousand; $P(0)$ represents the entire population of the U.S. since every member of the population is at least 0 years of age. The population of the U.S. in 2010 was roughly 322 million.   **91. (a)** The dependent variable is revenue, R, and the independent variable is price, p.
(b) $R(50) = 7500$; Selling MP3 players for \$50 will yield a daily revenue of \$7500 for the company.   **(c)** $R(120) = 9600$; Selling MP3 players for \$120 will yield a daily revenue of \$9600 for the company.   **93.** $\{r | r > 0\}$ or $(0, \infty)$ **95.** $\{h | 0 \le h \le 60\}$ or $[0, 60]$ **97.** $\{p | 0 \le p \le 120\}$ or $[0, 120]$
99. (a)(i) -5 **(ii)** 1 **(iii)** 1 **(b)(i)** 13 **(ii)** 4 **(iii)** 4   **101.** Answers will vary.   **103.** A function is a relation between two sets, the domain and the range. The domain is the set of all inputs to the function, and the range is the set of all outputs. In a function, each input in the domain corresponds to exactly one output in the range.   **105.** The four forms of a function are map, ordered pairs, equation, and graph.   **107.** $f(2) = 7$ **109.** $F(-3) = 5$
111. $H(7) = 5$ **113.** $F(4) = -6$

Section 2.3 Functions and Their Graphs   **1.** graph; function   **2.** 4; -7

3.   **4.**   **5.**   **6. (a)** Domain: $\{x | x \text{ is a real number}\}$; $(-\infty, \infty)$; Range: $\{y | y \le 1\}$; $(-\infty, 1]$ **(b)** x-intercepts: $(-2, 0)$ and $(2, 0)$; y-intercept: $(0, 1)$ **7.** $f(3) = 8$; $(-2, 4)$ **8. (a)** $f(-3) = -15$; $f(1) = -3$
(b) Domain: $\{x | x \text{ is a real number}\}$ or $(-\infty, \infty)$ **(c)** Range: $\{y | y \text{ is a real number}\}$ or $(-\infty, \infty)$ **(d)** x-intercepts: $(-2, 0)$, $(0, 0)$ and $(2, 0)$; y-intercept: $(0, 0)$ **(e)** $\{3\}$

9. (a) No **(b)** $f(3) = -2$; $(3, -2)$ is on the graph **(c)** $x = 5$; $(5, -8)$ is on the graph **10.** yes **11.** no **12.** yes **13.** -2 and 2

14.

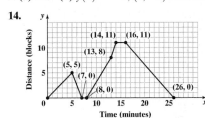

15. $f(x) = 4x - 6$ **17.** $h(x) = x^2 - 2$ **19.** $G(x) = |x - 1|$ **21.** $g(x) = x^3$

23. (a) Domain: $\{x \mid x$ is a real number$\}$ or $(-\infty, \infty)$; Range: $\{y \mid y$ is a real number$\}$ or $(-\infty, \infty)$ **(b)** $(0, 2)$ and $(1, 0)$ **(c)** 1

25. (a) Domain: $\{x \mid x$ is a real number$\}$ or $(-\infty, \infty)$; Range: $\{y \mid y \geq -2.25\}$ or $[-2.25, \infty)$ **(b)** $(-2, 0)$, $(4, 0)$, and $(0, -2)$ **(c)** $-2, 4$

27. (a) Domain: $\{x \mid x$ is a real number$\}$ or $(-\infty, \infty)$; Range: $\{y \mid y$ is a real number$\}$ or $(-\infty, \infty)$ **(b)** $(-3, 0)$, $(-1, 0)$, $(2, 0)$, and $(0, -3)$

(c) $-3, -1, 2$ **29. (a)** Domain: $\{x \mid x$ is a real number$\}$ or $(-\infty, \infty)$; Range: $\{y \mid y \geq 0\}$ or $[0, \infty)$ **(b)** $(-3, 0)$, $(3, 0)$, and $(0, 9)$ **(c)** $-3, 3$

31. (a) Domain: $\{x \mid x \leq 4\}$ or $(-\infty, 4]$; Range: $\{y \mid y \leq 3\}$ or $(-\infty, 3]$ **(b)** $(-2, 0)$ and $(0, 2)$ **(c)** -2 **33. (a)** $f(-7) = -2$ **(b)** $f(-3) = 3$

(c) $f(6) = 2$ **(d)** negative **(e)** $\{-6, -1, 4\}$ **(f)** $\{x \mid -7 \leq x \leq 6\}$ or $[-7, 6]$ **(g)** $\{y \mid -2 \leq y \leq 3\}$ or $[-2, 3]$ **(h)** $(-6, 0)$, $(-1, 0)$ and $(4, 0)$

(i) $(0, -1)$ **(j)** $\{-7, 2\}$ **(k)** $x = -3$ **(l)** $-6, -1, 4$ **35. (a)** $F(-2) = 3$ **(b)** $F(3) = -6$ **(c)** $x = -1$ **(d)** $(-4, 0)$ **(e)** $(0, 2)$ **37. (a)** no

(b) $f(3) = 3$; $(3, 3)$ **(c)** 4; $(4, 7)$ **(d)** no **39. (a)** yes **(b)** $g(6) = 1$; $(6, 1)$ **(c)** -12; $(-12, 10)$ **(d)** yes **41.** (c) **43.** (e) **45.** (d)

47. **49.** **51. (a)** III **(b)** I **(c)** IV **(d)** V **(e)** II

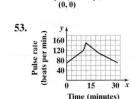 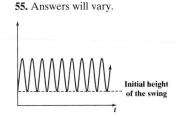

53. **55.** Answers will vary.

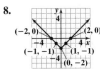

57. The person's weight increases until age 30, then oscillates back and forth between 158 pounds and 178 pounds, then slowly levels off at about 150 pounds. **59.** One possibility: Answers will vary.

61. A function cannot have more than one output for a given input. So, there cannot be two outputs for the input 0. **63.** The range of a function is the set of all outputs of the elements in the domain.

Putting the Concepts Together (Sections 2.1–2.3) **1.** The relation is a function because each element in the domain corresponds to exactly one element in the range. $\{(-2, -1), (-1, 0), (0, 1), (1, 2), (2, 3)\}$ **2. (a)** Function **(b)** Not a function **3.** Yes; Domain: $\{-4, -1, 0, 3, 6\}$;

Range: $\{-3, -2, 2, 6\}$ **4.** The graph passes the vertical line test. $f(5) = -6$ **5.** 4 **6. (a)** -17 **(b)** -34 **(c)** $-5x + 20$ **(d)** $-5x + 23$

7. (a) $\{h \mid h$ is any real number$\}$ or $(-\infty, \infty)$ **(b)** $\left\{w \mid w \neq -\dfrac{1}{3}\right\}$

8.

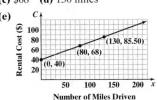

Domain: $\{x \mid x$ is any real number$\}$ or $(-\infty, \infty)$. Range: $\{y \mid y \geq -2\}$ or $[-2, \infty)$

9. (a) $h(2.5) = 80$; After 2.5 seconds, the height of the ball is 80 feet. **(b)** $\{t \mid 0 \leq t \leq 3.8\}$ or $[0, 3.8]$

(c) $\{h \mid 0 \leq h \leq 105\}$ or $[0, 105]$ **(d)** 1.25 seconds **10. (a)** no **(b)** -12; $(-2, -12)$ **(c)** -4; $(-4, -22)$ **(d)** yes

---

Section 2.4 Linear Functions and Models **1.** slope; y-intercept **2.** line **3.** False **4.** -2; $(0, 3)$

5. **6.** **7.** **8.** **9.** 5 **10.** -8 **11.** 12

12. (a) $\{x \mid x \geq 0\}$; $[0, \infty)$ **(b)** \$40 **13. (a)** $C(x) = 81x + 2000$ **14. (a)** $C(x) = 0.18x + 250$ **15. (a)**

(c) \$68 **(d)** 130 miles **(b)** \$2405 **(c)** 10 bicycles **(b)** $[0, \infty)$ **(c)** \$307.60

(e) **(d)** **(d)** 180 miles

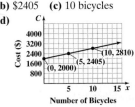 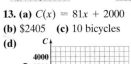 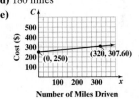 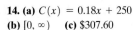 **(e)**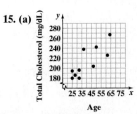

(f) You may drive between 0 and 250 miles.

(b) As age increases, total cholesterol also increases.

16. nonlinear   **17.** linear, positive slope   **18. (a)** Answers will vary. Using (25, 180) and (65, 269): $y = f(x) = 2.225x + 124.375$

(b)
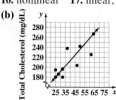

(c) 211 mg/dL   **(d)** Each year, a male's total cholesterol increases by 2.225 mg/dL; No

19. $F(x) = 5x - 2$ **21.** $G(x) = -3x + 7$ **23.** $H(x) = -2$

25. $f(x) = \dfrac{1}{2}x - 4$

27. $F(x) = -\dfrac{5}{2}x + 5$ **29.** $G(x) = -\dfrac{3}{2}x$ **31.** -5 **33.** 8 **35.** 6 **37.** 9 **39.** nonlinear   **41.** linear; positive slope

43. (a)
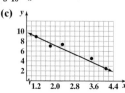

(b) Answers will vary. Using the points (4, 1.8) and (9, 2.6), the equation is $y = 0.16x + 1.16$.

(c)

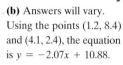
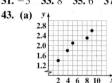

45. (a)

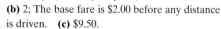

(b) Answers will vary. Using the points (1.2, 8.4) and (4.1, 2.4), the equation is $y = -2.07x + 10.88$.

(c)
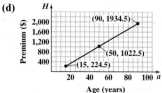

47. (a) 3   **(b)** (0, 2)   **(c)** $\dfrac{-2}{3}$ **(d)** 1; (1, 5)   **(e)** $\{x \mid x \le -1\}$; $(-\infty, -1]$

(f)
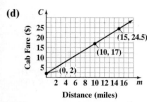

49. (a) $\{3\}$; -2; (3, -2); (3, -2)   **(b)** $\{x \mid x > 3\}$; (3, ∞)

(c)
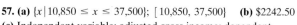

51. $f(x) = 2x + 2$; $f(-2) = -2$

53. $h(x) = -\dfrac{7}{4}x + \dfrac{49}{4}$; $h\left(\dfrac{1}{2}\right) = \dfrac{91}{8}$

55. (a) 3   **(b)** -1 **(c)** 2   **(d)** x-intercept: (2,0); y-intercept: (0, -2)   **(e)** $f(x) = x - 2$

57. (a) $\{x \mid 10{,}850 \le x \le 37{,}500\}$; $[\,10{,}850, 37{,}500\,]$ **(b)** $2242.50   **(c)** Independent variable: adjusted gross income; dependent variable: tax bill

(d)
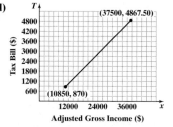
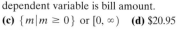

(e) $25,025

59. (a) $\{m \mid m \ge 0\}$ or $[0, \infty)$ **(b)** 2; The base fare is $2.00 before any distance is driven.   **(c)** $9.50.

(d)
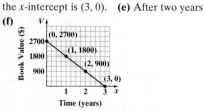

(e) A person can travel 7.5 miles in a cab for $13.25.   **(f)** A person can ride between 0 miles and 25 miles.

61. (a) The independent variable is age; the dependent variable is insurance cost.   **(b)** $\{a \mid 15 \le a \le 90\}$ or $[15, 90]$ **(c)** $566.50

(d)

(e) 48 years

63. (a) $B(m) = 0.05m + 5.95$ **(b)** The independent variable is minutes; the dependent variable is bill amount.   **(c)** $\{m \mid m \ge 0\}$ or $[0, \infty)$ **(d)** $20.95   **(e)** 240 minutes

(f)

(g) You can talk between 0 minutes and 250 minutes.

65. (a) $V(x) = -900x + 2700$ **(b)** $\{x \mid 0 \le x \le 3\}$ or $[0, 3]$ **(c)** $1800   **(d)** The V-intercept is (0, 2700), and the x-intercept is (3, 0).   **(e)** After two years

(f)

67. (a) $C(x) = 8350x - 2302$ **(b)** $4127.50   **(c)** The cost of diamonds increases at a rate of $8350 per carat.   **(d)** 0.91 carat

69. (a) $C(x) = 0.701x + 2371.884$ **(b)** $9201.0 billion   **(c)** If personal disposable income increases by $1, personal consumption increase by $0.70.   **(d)** $10,197 billion

71. (a)
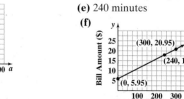

(b) Linear   **(c)** Answers will vary. Using the points (2300, 4070) and (3390, 5220), the equation is $y = 1.06x + 1632$.

(d)
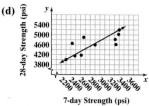

(e) 4812 psi   **(f)** If the 7-day strength is increased by 1 psi, then the 28-day strength will increase by 1.06 psi.

73. (a) No **(b)**

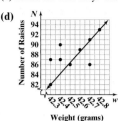

(c) Answers will vary. Using the points (42.3, 82) and (42.8, 93), the equation is $N = 22w - 848.6$.

(d)

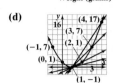

(e) $N(w) = 22w - 848.6$ **(f)** approximately 86 raisins
(g) If the weight increases by 1 gram, then the number of raisins increases by 22 raisins.

75. (a)

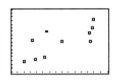

(d)

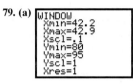

$x = 3$: slope $= 4$; $y = 4x - 5$; $x = 2$: slope $= 2$; $y = 2x - 3$; $x = 1.5$:
slope $= 1$; $y = x - 2$; $x = 1.1$: slope $= 0.2$; $y = 0.2x - 1.2$
(e) As x approaches 1, the slope gets closer to 0.

(b) 6 **(c)** $y = 6x - 7$

77. (a)

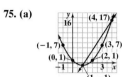

(b) $y = 0.676x + 2675.562$

79. (a)

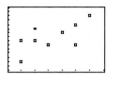

(b) $y = 11.449x - 399.123$

Section 2.5 Compound Inequalities

1. intersection **2.** and; or **3.** True **4.** False **5.** $\{1, 3, 5\}$ **6.** $\{2, 4, 6\}$ **7.** $\{1, 2, 3, 4, 5, 6, 7\}$
8. $\{1, 2, 3, 4, 5, 6, 8\}$ **9.** $\varnothing$ or $\{\ \}$ **10.** $\{1, 2, 3, 4, 5, 6, 7, 8\}$ **11.** $\{x \mid 2 < x < 7\}$; $(2, 7)$

12. $\{x \mid x \le -3 \text{ or } x > 2\}$; $(-\infty, -3] \cup (2, \infty)$ **13.** $\{x \mid x \ge 2\}$; $[2, \infty)$

14. $\{x \mid -3 < x < 3\}$; $(-3, 3)$ **15.** $\{x \mid 1 < x < 3\}$; $(1, 3)$ **16.** $\{\ \}$ or $\varnothing$

17. $\{1\}$ **18.** $\{x \mid -1 < x < 3\}$; $(-1, 3)$

19. $\left\{x \mid \dfrac{5}{4} < x \le 2\right\}$; $\left(\dfrac{5}{4}, 2\right]$ **20.** $\{x \mid -6 \le x \le -2\}$; $[-6, -2]$

21. $\{x \mid x < -2 \text{ or } x > 5\}$; $(-\infty, -2) \cup (5, \infty)$

22. $\{x \mid x \le 2 \text{ or } x > 6\}$; $(-\infty, 2] \cup (6, \infty)$ **23.** $\{x \mid x \ge 1\}$; $[1, \infty)$

24. $\{x \mid x < 4 \text{ or } x > 9\}$; $(-\infty, 4) \cup (9, \infty)$

25. $\{x \mid x \text{ is any real number}\}$; $(-\infty, \infty)$ **26.** $\{x \mid x \ge -1\}$; $[-1, \infty)$

27. Income between \$37,500 and \$87,000.

28. Total usage was between 45 and 245 minutes. **29.** $\{1, 4, 5, 6, 7, 8, 9\}$ **31.** $\{5, 7, 9\}$ **33.** $\varnothing$ or $\{\ \}$

35. (a) $A \cap B = \{x \mid -2 < x \le 5\}$; $(-2, 5]$

(b) $A \cup B = \{x \mid x \text{ is any real number}\}$; $(-\infty, \infty)$

37. (a) $E \cap F = \varnothing$ or $\{\ \}$

(b) $E \cup F = \{x \mid x < -1 \text{ or } x > 3\}$; $(-\infty, -1) \cup (3, \infty)$

39. (a) $\{x \mid -2 \le x \le 2\}$; $[-2, 2]$ **(b)** $(-\infty, -2) \cup (2, \infty)$

41. (a) $\{x \mid -3 < x < 3\}$; $(-3, 3)$ **(b)** $(-\infty, -3] \cup [3, \infty)$

43. $\{x \mid -2 \le x < 3\}$; $[-2, 3)$ **45.** $\varnothing$ or $\{\ \}$ **47.** $\{x \mid x < -2\}$; $(-\infty, -2)$

49. $\{1\}$ **51.** $\{x \mid -1 \le x < 3\}$; $[-1, 3)$

53. $\left\{x \mid -\dfrac{2}{3} \le x \le \dfrac{3}{2}\right\}$; $\left[-\dfrac{2}{3}, \dfrac{3}{2}\right]$ **55.** $\left\{x \mid -1 < x \le \dfrac{4}{5}\right\}$; $\left(-1, \dfrac{4}{5}\right]$

57. $\{x \mid 0 \le x \le 8\}$; $[0, 8]$ **59.** $\{x \mid -6 \le x \le -2\}$; $[-6, -2]$

61. $\varnothing$ or $\{\ \}$ **63.** $\left\{x \mid -\dfrac{5}{3} < x \le 5\right\}$; $\left(-\dfrac{5}{3}, 5\right]$ **65.** $\{x \mid -4 < x \le 3\}$; $(-4, 3]$

67. $\{x \mid x < -2 \text{ or } x > 3\}$; $(-\infty, -2) \cup (3, \infty)$

69. $\{x \mid x < -2 \text{ or } x > 5\}; (-\infty, -2) \cup (5, \infty)$

71. $\{x \mid x \text{ is any real number}\}; (-\infty, \infty)$

73. $\{x \mid x < -1 \text{ or } x > 4\}; (-\infty, -1) \cup (4, \infty)$

75. $\{x \mid x \leq -3 \text{ or } x > 6\}; (-\infty, -3] \cup (6, \infty)$

77. $\{x \mid x \text{ is any real number}\}; (-\infty, \infty)$

79. $\left\{x \mid x < 0 \text{ or } x > \frac{5}{2}\right\}; (-\infty, 0) \cup \left(\frac{5}{2}, \infty\right)$ **81.** $\{a \mid -3 \leq a < 0\}; [-3, 0)$

83. $\{x \mid x \text{ is any real number}\}; (-\infty, \infty)$ **85.** $\left\{x \mid -2 \leq x \leq \frac{8}{3}\right\}; \left[-2, \frac{8}{3}\right]$

87. $\{x \mid x < -10 \text{ or } x > 2\}; (-\infty, -10) \cup (2, \infty)$ **89.** $\{x \mid 2 < x < 5\}; (2, 5)$

91. $\left\{x \mid x \leq -3 \text{ or } x > \frac{15}{4}\right\}; \left(-\infty, -3\right] \cup \left(\frac{15}{4}, \infty\right)$

93. $\left\{x \mid -5 < x \leq \frac{1}{2}\right\}; \left(-5, \frac{1}{2}\right]$ **95.** $a = 1$ and $b = 8$ **97.** $a = 12$ and $b = 30$ **99.** $a = -1$ and $b = 23$

101. $90 < x < 140$ **103.** Joanna needs to score at least a 77 on the final. That is, $77 \leq x \leq 100$ (assuming 100 is the max score, otherwise $77 \leq x \leq 104$).
105. The amount withheld ranges between $113.35 and $138.35, inclusive.   **107.** Total sales between $50,000 and $250,000
109. The electrical usage ranged from 850 kwh and 1370 kwh.

111. Step 1:

$$a < b$$
$$a + a < a + b$$
$$2a < a + b$$
$$\frac{2a}{2} < \frac{a+b}{2}$$
$$a < \frac{a+b}{2}$$

Step 2:

$$a < b$$
$$a + b < b + b$$
$$a + b < 2b$$
$$\frac{a+b}{2} < \frac{2b}{2}$$
$$\frac{a+b}{2} < b$$

Step 3:

Since $a < \dfrac{a+b}{2}$ and $\dfrac{a+b}{2} < b$, it follows that $a < \dfrac{a+b}{2} < b$.

113. $\{\ \}$ or $\varnothing$ **115.** This is a contradiction. There is no solution. If, during simplification, the variable terms all cancel out and a contradiction results, then there is no solution to the inequality.   **117.** If $x < 2$ then $x - 2 < 2 - 2 \Rightarrow x - 2 < 0$. When multiplying both sides of the inequality by $x - 2$ in the second step, the direction of the inequality must switch.

Section 2.6 Absolute Value Equations and Inequalities   **1.** $\{-7, 7\}$ **2.** $\{-1, 1\}$ **3.** $a; -a$ **4.** $2x + 3 = -5$ **5.** $\{-2, 5\}$ **6.** $\left\{-\frac{5}{3}, 3\right\}$

7. $\left\{-1, \frac{9}{5}\right\}$ **8.** $\{-5, 1\}$ **9.** True   **10.** $\{\ \}$ or $\varnothing$ **11.** $\{\ \}$ or $\varnothing$ **12.** $\{-1\}$ **13.** $u; v; u; -v$ **14.** $\left\{-8, -\frac{2}{3}\right\}$ **15.** $\{-2, 3\}$ **16.** $\{-3, 0\}$

17. $\{2\}$ **18.** $-a < u < a$ **19.** $-10; 10$ **20.** $\{x \mid -5 \leq x \leq 5\}; [-5, 5]$

21. $\left\{x \mid -\frac{3}{2} < x < \frac{3}{2}\right\}; \left(-\frac{3}{2}, \frac{3}{2}\right)$ **22.** $\{x \mid -8 < x < 2\}; (-8, 2)$

23. $\{x \mid -2 \leq x \leq 5\}; [-2, 5]$ **24.** $\{\ \}$ or $\varnothing$

25. False   **26.** $\{x \mid -2 < x < 2\}; (-2, 2)$ **27.** $\{x \mid -1 \leq x \leq 7\}; [-1, 7]$

28. $\{x \mid -2 \leq x \leq 1\}; [-2, 1]$ **29.** $\left\{x \mid -\frac{7}{3} < x < 3\right\}; \left(-\frac{7}{3}, 3\right)$

30. $u < -a; u > a$ **31.** $-7; 7$ **32.** $\{x \mid x \leq -6 \text{ or } x \geq 6\}; (-\infty, -6] \cup [6, \infty)$

33. $\left\{x \mid x < -\frac{5}{2} \text{ or } x > \frac{5}{2}\right\}; \left(-\infty, -\frac{5}{2}\right) \cup \left(\frac{5}{2}, \infty\right)$ **34.** $<$

35. $\{x \mid x < -7 \text{ or } x > 1\}; (-\infty, -7) \cup (1, \infty)$

36. $\left\{x \mid x \leq -\frac{1}{2} \text{ or } x \geq 2\right\}; \left(-\infty, -\frac{1}{2}\right] \cup [2, \infty)$

37. $\left\{x \mid x < -\frac{5}{3} \text{ or } x > 3\right\}; \left(-\infty, -\frac{5}{3}\right) \cup (3, \infty)$

38. $\left\{x \mid x \neq -\frac{5}{2}\right\}; \left(-\infty, -\frac{5}{2}\right) \cup \left(-\frac{5}{2}, \infty\right)$

39. $\{x \mid x \text{ is any real number}\}; (-\infty, \infty)$

40. $\{x \mid x \text{ is any real number}\}; (-\infty, \infty)$ **41.** The acceptable belt width is between $\dfrac{127}{32}$ inches and $\dfrac{129}{32}$ inches.

42. The percentage of Americans that have been shot at is between 7.3% and 10.7%, inclusive.   **43.** $\{-10, 10\}$ **45.** $\{-1, 7\}$

47. $\left\{-1, \frac{13}{3}\right\}$ **49.** $\{-5, 5\}$ **51.** $\left\{-\frac{11}{2}, \frac{5}{2}\right\}$ **53.** $\{-4, 10\}$ **55.** $\{0\}$ **57.** $\left\{-\frac{7}{3}, 3\right\}$ **59.** $\left\{-7, \frac{3}{5}\right\}$ **61.** $\{1, 3\}$ **63.** $\{2\}$

65. $\{x\,|\,-9 < x < 9\}$; $(-9, 9)$

67. $\{x\,|\,-3 \le x \le 11\}$; $[-3, 11]$

69. $\left\{x\,\middle|\,-3 < x < \dfrac{7}{3}\right\}$; $\left(-3, \dfrac{7}{3}\right)$

71. $\varnothing$ or $\{\ \}$

73. $\{x\,|\,0 < x < 6\}$; $(0, 6)$

75. $\left\{x\,\middle|\,-1 < x < \dfrac{9}{5}\right\}$; $\left(-1, \dfrac{9}{5}\right)$

77. $\{x\,|\,1.995 < x < 2.005\}$; $(1.995, 2.005)$

79. $\{y\,|\,y < 3 \text{ or } y > 7\}$; $(-\infty, 3) \cup (7, \infty)$

81. $\left\{x\,\middle|\,x \le -2 \text{ or } x \ge \dfrac{1}{2}\right\}$; $(-\infty, -2] \cup \left[\dfrac{1}{2}, \infty\right)$

83. $\{y\,|\,y \text{ is any real number}\}$; $(-\infty, \infty)$

85. $\left\{x\,\middle|\,x < -2 \text{ or } x > \dfrac{4}{5}\right\}$; $(-\infty, -2) \cup \left(\dfrac{4}{5}, \infty\right)$

87. $\{x\,|\,x < 0 \text{ or } x > 1\}$; $(-\infty, 0) \cup (1, \infty)$

89. $\{x\,|\,x \le -2 \text{ or } x \ge 3\}$; $(-\infty, -2] \cup [3, \infty)$

91. (a) $\{-5, 5\}$ **(b)** $\{x\,|\,-5 \le x \le 5\}$; $[-5, 5]$ **(c)** $\{x\,|\,x < -5 \text{ or } x > 5\}$; $(-\infty, -5) \cup (5, \infty)$

93. (a) $\{-5, 1\}$ **(b)** $\{x\,|\,-5 < x < 1\}$; $(-5, 1)$ **(c)** $\{x\,|\,x \le -5 \text{ or } x \ge 1\}$; $(-\infty, -5] \cup [1, \infty)$

95. $\{x\,|\,x < -5 \text{ or } x > 5\}$; $(-\infty, -5) \cup (5, \infty)$

97. $\{-4, -1\}$ **99.** $\{-5, 5\}$ **101.** $\left\{x\,\middle|\,-2 \le x \le \dfrac{6}{5}\right\}$; $\left[-2, \dfrac{6}{5}\right]$ **103.** $\varnothing$ or $\{\ \}$

105. $\left\{x\,\middle|\,x \le -\dfrac{7}{3} \text{ or } x \ge 1\right\}$; $\left(-\infty, -\dfrac{7}{3}\right] \cup [1, \infty)$

107. $\left\{x\,\middle|\,x < 0 \text{ or } x > \dfrac{4}{3}\right\}$; $(-\infty, 0) \cup \left(\dfrac{4}{3}, \infty\right)$ **109.** $\{-1, 1\}$ **111.** $\varnothing$ or $\{\ \}$ **113.** $\left\{-8, \dfrac{4}{7}\right\}$

115. $|x - 5| < 3$ $\{x\,|\,2 < x < 8\}$; $(2, 8)$ **117.** $|2x - (-6)| > 3$ $\left\{x\,\middle|\,x < -\dfrac{9}{2} \text{ or } x > -\dfrac{3}{2}\right\}$; $\left(-\infty, -\dfrac{9}{2}\right) \cup \left(-\dfrac{3}{2}, \infty\right)$

119. The acceptable rod lengths are between 5.6995 inches and 5.7005 inches, inclusive. **121.** An unusual IQ score would be less than 70.6 or greater than 129.4. **123.** $\left\{-\dfrac{5}{2}\right\}$ **125.** $\{2\}$ **127.** $\varnothing$ or $\{\ \}$ **129.** $\{x\,|\,x \le -5\}$; $(-\infty, -5]$ **131.** The absolute value, when isolated, is equal to a negative number, which is not possible. **133.** The absolute value, when isolated, is less than -3. Since absolute values are always nonnegative, this is not possible.

Chapter 2 Review

1. {(Penny, 2.500), (Nickel, 5.000), (Dime, 2. 268), (Quarter, 5.670), (Half Dollar, 11.340), (Dollar, 8.100)}
Domain: {Penny, Nickel, Dime, Quarter, Half Dollar, Dollar}; Range: {2.268, 2.500, 5.000, 5.670, 8.100, 11.340}
2. { (16, \$12.99), (28, \$14.99) (30, \$14.99) (59, \$24.99) (85, \$29.99)}. Domain: {16, 28, 30, 59, 85}; Range: {\$12.99, \$14.99, \$24.99, \$29.99}

3. Domain: $\{-4, -2, 2, 3, 6\}$
Range: $\{-9, -1, 5, 7, 8\}$

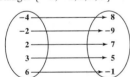

4. Domain: $\{-2, 1, 3, 5\}$
Range: $\{1, 4, 7, 8\}$

5. Domain: $\{x\,|\,x \text{ is a real number}\}$ or $(-\infty, \infty)$
Range: $\{y\,|\,y \text{ is a real number}\}$ or $(-\infty, \infty)$

6. Domain: $\{x\,|\,-6 \le x \le 4\}$ or $[-6, 4]$
Range: $\{y\,|\,-4 \le y \le 6\}$ or $[-4, 6]$

7. Domain: $\{2\}$; Range: $\{y\,|\,y \text{ is a real number}\}$ or $(-\infty, \infty)$

8. Domain: $\{x\,|\,x \ge -1\}$ or $[-1, \infty)$; Range: $\{y\,|\,y \ge -2\}$ or $[-2, \infty)$

9. $y = x + 2$
Domain: $\{x\,|\,x \text{ is a real number}\}$ or $(-\infty, \infty)$
Range: $\{y\,|\,y \text{ is a real number}\}$ or $(-\infty, \infty)$

10. $2x + y = 3$
Domain: $\{x\,|\,x \text{ is a real number}\}$ or $(-\infty, \infty)$
Range: $\{y\,|\,y \text{ is a real number}\}$ or $(-\infty, \infty)$

11. $y = -x^2 + 4$
Domain: $\{x\,|\,x \text{ is a real number}\}$ or $(-\infty, \infty)$; Range: $\{y\,|\,y \le 4\}$ or $(-\infty, 4]$

12. $y = |x + 2| - 1$
Domain: $\{x\,|\,x \text{ is a real number}\}$ or $(-\infty, \infty)$
Range: $\{y\,|\,y \ge -1\}$ or $[-1, \infty)$

13. $y = x^3 + 2$
Domain: $\{x\,|\,x \text{ is a real number}\}$ or $(-\infty, \infty)$
Range: $\{y\,|\,y \text{ is a real number}\}$ or $(-\infty, \infty)$

14. $x = y^2 + 1$
Domain: $\{x\,|\,x \ge 1\}$ or $[1, \infty)$
Range: $\{y\,|\,y \text{ is a real number}\}$ or $(-\infty, \infty)$

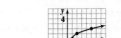

15. (a) Domain: $\{x \mid 0 \leq x \leq 44640\}$ or $[0, 44640]$; Range: $\{y \mid 40 \leq y \leq 2122\}$ or $[40, 2122]$ **(b)** Answers may vary.   **16.** Domain: $\{t \mid 0 \leq t \leq 4\}$ or $[0, 4]$; Range: $\{y \mid 0 \leq y \leq 121\}$ or $[0, 121]$ **17. (a)** Not a function. Domain: $\{-1, 5, 7, 9\}$; Range: $\{-2, 0, 2, 3, 4\}$ **(b)** Function. Domain: $\{$Camel, Macaw, Deer, Fox, Tiger, Crocodile$\}$; Range: $\{14, 22, 35, 45, 50\}$ **18. (a)** Function; Domain: $\{-3, -2, 2, 4, 5\}$; Range: $\{-1, 3, 4, 7\}$ **(b)** Not a function; Domain: $\{$Red, Blue, Green, Black$\}$; Range: $\{$Camry, Taurus, Windstar, Durango$\}$ **19.** Function   **20.** Not a function **21.** Not a function   **22.** Function   **23.** Not a function   **24.** Function   **25.** Function   **26.** Not a function   **27. (a)** $f(-2) = -5$ **(b)** $f(3) = 10$ **28. (a)** $g(0) = -\dfrac{1}{3}$ **(b)** $g(2) = -5$ **29. (a)** $F(5) = -3$ **(b)** $F(-x) = 2x + 7$ **30. (a)** $G(7) = 15$ **(b)** $G(x + h) = 2x + 2h + 1$

31. $\{x \mid x$ is a real number$\}$ or $(-\infty, \infty)$ **32.** $\left\{w \mid w \neq -\dfrac{5}{2}\right\}$ **33.** $\{t \mid t \neq 5\}$ **34.** $\{t \mid t$ is a real number$\}$ or $(-\infty, \infty)$ **35. (a)** The dependent variable is the population, P, and the independent variable is the number of years after 1900, t.   **(b)** $P(120) = 1370.907$; The population of Orange County will be roughly 1,370,907 in 2020.   **(c)** $P(-70) = 1025.487$; The population of Orange County was roughly 1,025,487 in 1830. This is not reasonable. (The population of the entire Florida territory was roughly 35,000 in 1830.)   **36. (a)** The dependent variable is percent of the population with an advanced degree, P, and the independent variable is age, a.   **(b)** $P(30) = 7.9$; According to the model, 7.9% of 30-year-olds have an advanced degree.

37. $f(x) = 2x - 5$ **38.** $g(x) = x^2 - 3x + 2$ **39.** $h(x) = (x - 1)^3 - 3$ **40.** $f(x) = |x + 1| - 4$ **41. (a)** Domain: $\{x \mid x$ is a real number$\}$ or $(-\infty, \infty)$; Range: $\{y \mid y$ is a real number$\}$ or $(-\infty, \infty)$ **(b)** $(0, 2)$ and $(4, 0)$

42. (a) Domain: $\{x \mid x$ is a real number$\}$ or $(-\infty, \infty)$; Range: $\{y \mid y \geq -3\}$ or $[-3, \infty)$ **(b)** $(-2, 0), (2, 0), (0, -3)$

43. (a) Domain: $\{x \mid x$ is a real number$\}$ or $(-\infty, \infty)$; Range: $\{y \mid y$ is a real number$\}$ or $(-\infty, \infty)$ **(b)** $(0, 0)$ and $(2, 0)$
44. (a) Domain: $\{x \mid x \geq -3\}$ or $[-3, \infty)$; Range: $\{y \mid y \geq 1\}$ or $[1, \infty)$ **(b)** $(0, 3)$ **45. (a)** 4   **(b)** 1   **(c)** -1 and 3

46. (a)   **(b)**   **47. (a)** yes   **48. (a)** no   **49.**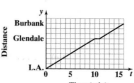
(b) $h(-2) = -11$; $(-2, -11)$ **(b)** $g(3) = \dfrac{29}{5}$; $\left(3, \dfrac{29}{5}\right)$
(c) $x = \dfrac{11}{2}$; $\left(\dfrac{11}{2}, 4\right)$ **(c)** $x = -10$; $(-10, -2)$

50. 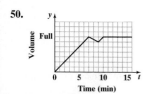   **51.**   **52.** 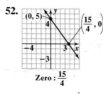   **53.**   **54.**

55. (a) $[0, 44640)$ **56. (a)** The independent variable is years; the dependent variable is value.   **57. (a)** $L(x) = -\dfrac{4}{75}x + 45$ **(b)** 7%   **(c)** The auto loan rate
(b) \$21.45   decreases approximately by 0.05% for every 1-unit increase in
(c) 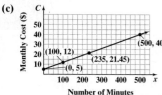   **(b)** $\{x \mid 0 \leq x \leq 5\}$ or $[0, 5]$ FICO score.   **(d)** 722
   (c) \$1800   **(d)** \$1080

   (e) 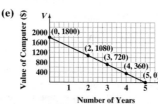   **58. (a)** $H(x) = -x + 220$ **(b)** 175 beats per minute
(d) 1000 minutes   **(c)** The maximum recommended heart rate for men under stress decreases at a rate of 1 beat per minute per year.
   (d) 52 years

   (f) After 5 years

59. (a) $C(m) = 0.12m + 35$ **60. (a)** $B(x) = 3.50x + 33.99$ **61. (a)**   **62. (a)**
(b) The independent variable is m;   **(b)** The independent variable is x;
the dependent variable is C.   the dependent variable is B.
(c) $\{m \mid m \geq 0\}$ or $[0, \infty)$ **(c)** $\{x \mid x \geq 0\}$ or $[0, \infty)$
(d) \$49.88   **(d)** \$51.49
(e) 268 miles were driven   **(e)** 7 pay-per-view movies
(f) 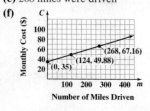   **(f)** 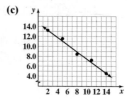   **(b)** Answers will vary. Using the points $(2, 13.3)$ and $(14, 4.6)$, the equation is $y = -0.725x + 14.75$.   **(b)** Answers will vary. Using the points $(0, 0.6)$ and $(4.2, 3.0)$, the equation is $y = \dfrac{4}{7}x + 0.6$.

   (c) 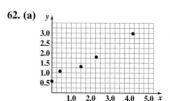   **(c)**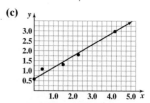

63. (a)

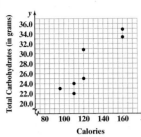

(b) Approximately linear.
(c) Answers will vary. Using the points (96, 23.2) and (160, 33.3), the equation is $y = 0.158x + 8.032$.

(d)

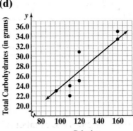

(e) 30.2 grams
(f) In a one-cup serving of cereal, total carbohydrates will increase by 0.158 gram for each 1-calorie increase.

64. (a)

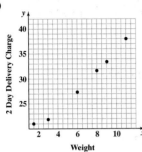

(b) Linear with positive slope
(c) Answers will vary. Using (3, 22.46) and (9, 33.92), $y = 1.91x + 16.73$

(d)

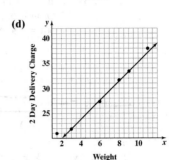

(e) $26.28
(f) If the weight increases by 1 pound, the shipping charge increases $1.91.

65. $\{-1, 0, 1, 2, 3, 4, 6, 8\}$ **66.** $\{2, 4\}$ **67.** $\{1, 2, 3, 4\}$ **68.** $\{1, 2, 3, 4, 6, 8\}$ **69. (a)** $\{x \mid 2 < x \le 4\}; (2, 4]$

(b) $\{x \mid x$ is any real number$\}; (-\infty, \infty)$ **70. (a)** $\{\ \}$ or $\varnothing$

(b) $\{x \mid x < -2$ or $x \ge 3\}; (-\infty, -2) \cup [3, \infty)$

71. $\{x \mid -1 < x < 4\}; (-1, 4)$ Graph:

72. $\{x \mid -5 < x < -1\}; (-5, -1)$ Graph:

73. $\{x \mid x < -2$ or $x > 2\}; (-\infty, -2) \cup (2, \infty)$ Graph:

74. $\{x \mid x \le 0$ or $x \ge 4\}; (-\infty, 0] \cup [4, \infty)$ Graph:

75. $\{\ \}$ or $\varnothing$ **76.** $\{x \mid -2 \le x < 4\}; [-2, 4)$ Graph:

77. $\{x \mid x \le -2$ or $x > 3\}; (-\infty, -2] \cup (3, \infty)$ Graph:

78. $\{x \mid x$ is any real number$\}; (-\infty, \infty)$ Graph:

79. $\{x \mid x < -10$ or $x > 6\}; (-\infty, -10) \cup (6, \infty)$ Graph:

80. $\left\{x \mid -\dfrac{3}{2} \le x < \dfrac{5}{8}\right\}; \left[-\dfrac{3}{2}, \dfrac{5}{8}\right)$ Graph: **81.** $70 \le x \le 75$

82. The electric usage varied from roughly 1033.8 kilowatt hours up to roughly 1952.5 kilowatt hours. (recall, x is the number *above* 800).

83. $\{-4, 4\}$ **84.** $\left\{\dfrac{1}{3}, 3\right\}$ **85.** $\{-5, 13\}$ **86.** $\{-3, -1\}$ **87.** $\{\ \}$ or $\varnothing$ **88.** $\left\{-\dfrac{1}{2}, 2\right\}$

89. $\{x \mid -2 < x < 2\}; (-2, 2)$

90. $\left\{x \mid x \le -\dfrac{7}{2}$ or $x \ge \dfrac{7}{2}\right\}; \left(-\infty, -\dfrac{7}{2}\right] \cup \left[\dfrac{7}{2}, \infty\right)$

91. $\{x \mid -5 \le x \le 1\}; [-5, 1]$ **92.** $\left\{x \mid x \le \dfrac{1}{2}$ or $x \ge 1\right\}; \left(-\infty, \dfrac{1}{2}\right] \cup [1, \infty)$

93. $\{x \mid x$ is a real number$\}; (-\infty, \infty)$ **94.** $\{\ \}$ or $\varnothing$

95. $\{x \mid 4.99 \le x \le 5.01\}; [4.99, 5.01]$

96. $\left\{x \mid x < -\dfrac{1}{2}$ or $x > \dfrac{7}{2}\right\}; \left(-\infty, -\dfrac{1}{2}\right) \cup \left(\dfrac{7}{2}, \infty\right)$

97. The acceptable diameters of the bearing are between 0.502 inch and 0.504 inch, inclusive. **98.** Tensile strengths below 36.08 lb/in.^{2} or above 43.92 lb/in.^{2} would be considered unusual.

Chapter 2 Test

1. Domain: $\{-4, 2, 5, 7\}$
Range: $\{-7, -2, -1, 3, 8, 12\}$

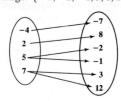

2. Domain: $\left\{x \,\middle|\, -\dfrac{5\pi}{2} \le x \le \dfrac{5\pi}{2}\right\}$ or $\left[-\dfrac{5\pi}{2}, \dfrac{5\pi}{2}\right]$
Range: $\{y \mid 1 \le y \le 5\}$ or $[1, 5]$

3. $y = x^2 - 3$

Domain: $\{x \mid x \text{ is a real number}\}$ or $(-\infty, \infty)$; Range: $\{y \mid y \ge -3\}$ or $[-3, \infty)$

4. Function. Domain: $\{-5, -3, 0, 2\}$; Range: $\{3, 7\}$ **5.** Not a function. Domain: $\{x \mid x \le 3\}$ or $(-\infty, 3]$; Range: $\{y \mid y \text{ is a real number}\}$ or $(-\infty, \infty)$
6. No   **7.** $f(x + h) = -3x - 3h + 11$ **8. (a)** $g(-2) = 5$ **(b)** $g(0) = -1$ **(c)** $g(3) = 20$

9. $f(x) = x^2 + 3$

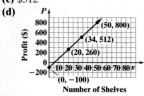

10. (a) The dependent variable is the ticket price, P, and the independent variable is the number of years after 1996, x.
(b) $P(14) = 7.7$; According to the model, the average ticket price in 2010 $(x = 14)$ was \$7.70.   **(c)** 2020
11. $\{x \mid x \ne -2\}$ **12. (a)** yes   **(b)** $h(3) = -3$; $(3, -3)$ **(c)** $x = -3$; $(-3, 27)$ **(d)** $\dfrac{12}{5}$
13. (a) The car stops accelerating when the speed stops increasing. Thus, the car stops accelerating after 6 seconds.
(b) The car has a constant speed when the graph is horizontal. Thus, the car maintains a constant speed for 18 seconds.

14. (a) $P(x) = 18x - 100$
(b) $\{x \mid x \ge 0\}$ or $[0, \infty)$
(c) \$512
(d)

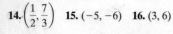

(e) 48 shelves

15. (a)

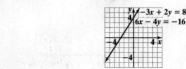

(b) Approximately linear.
(c) Answers will vary. Using the points $(6, 95)$ and $(18, 170)$, the equation is $y = 6.25x + 57.5$.

(d)

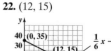

(e) 113.75 kilograms
(f) A Shetland pony's weight will increase by 6.25 kilograms for each one-month increase in age.

16. $\{-4, -1\}$ **17.** $\{x \mid -2 \le x < 6\}$; $[-2, 6)$

18. $\left\{x \,\middle|\, x < -\dfrac{3}{2} \text{ or } x > 4\right\}$; $\left(-\infty, -\dfrac{3}{2}\right) \cup (4, \infty)$

19. $\{x \mid 2 < x < 8\}$; $(2, 8)$

20. $\{x \mid x \le -2 \text{ or } x \ge 3\}$; $(-\infty, -2] \cup [3, \infty)$

Chapter 3 Systems of Linear Equations and Inequalities

Section 3.1 Systems of Linear Equations in Two Variables   **1.** system of linear equations   **2.** Solution   **3. (a)** no   **(b)** yes   **(c)** no

4. inconsistent   **5.** consistent; dependent   **6.** False   **7.** True   **8.** True   **9.** $(3, 1)$ **10.** $(-2, 3)$ **11.** $(-4, 7)$ **12.** $(-6, 10)$ **13.** additive inverses

14. $\left(\dfrac{1}{2}, \dfrac{7}{3}\right)$ **15.** $(-5, -6)$ **16.** $(3, 6)$

17. $\varnothing$ or $\{\}$ **18.** $\varnothing$ or $\{\}$ **19.** dependent   **20.** $\{(x, y) \mid -3x + 2y = 8\}$ **21.** $(-5, 2)$ **22.** $(12, 15)$

23. (a) no   **(b)** yes   **25. (a)** yes   **(b)** yes   **27.** consistent; independent   **29.** inconsistent
31. $(1, 3)$ **33.** $(3, -4)$ **35.** $(6, -2)$ **37.** $(-2, -3)$ **39.** $\left(\dfrac{1}{2}, -\dfrac{1}{4}\right)$ **41.** $(2500, 7500)$ **51.** $\varnothing$ or $\{\}$

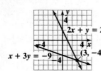

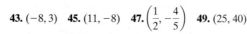

43. $(-8, 3)$ **45.** $(11, -8)$ **47.** $\left(\dfrac{1}{2}, -\dfrac{4}{5}\right)$ **49.** $(25, 40)$

53. ∅ or { }   **55.** $\{(x, y) \mid 2x - 4y = -4\}$ **57.** $\{(x, y) \mid x + 3y = 6\}$ **59.** $\left\{(x, y) \mid \dfrac{1}{3}x - 2y = 6\right\}$ **61.** $(-9, 3)$ **63.** $\{(x, y) \mid x = 5y - 3\}$

65. $\left(\dfrac{39}{11}, -\dfrac{30}{11}\right)$ **67.** ∅ or { }

69. $y = -2x - 5$; $y = -\dfrac{5}{3}x + \dfrac{1}{3}$; exactly one solution

71. $y = \dfrac{3}{2}x + 1$; $y = \dfrac{3}{2}x + 1$; infinite number of solutions

73. (a) $y = -\dfrac{1}{2}x + \dfrac{5}{2}$; $y = x + 1$ **(b)** $(1, 2)$ **75. (c)** and **(f)**   **77.** $A = \dfrac{7}{6}$ and $B = -\dfrac{1}{2}$ **79.** Answers will vary. One possibility follows: $\begin{cases} x + y = 3 \\ x - y = -5 \end{cases}$

81. $(1, 2)$ **83.** $(6, -1)$ **85.** Yes, Typically, we use substitution when one of the equations is solved for one of the variables or when the coefficient on one of the variables is 1 (which makes it easy to solve for that variable). Otherwise, we use elimination.   **89.** $(1.2, 2.6)$

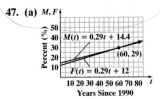   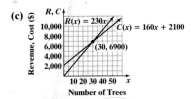

87. $(3, -2)$

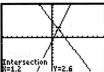

91. $(4, 13)$ **93.** $\{(x, y) \mid 4x - 3y = 1\}$ **95.** approximately $(0.35, -3.76)$

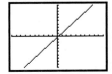   

Section 3.2 Problem Solving: Systems of Two Linear Equations Containing Two Unknowns   **1.** 15 and -7

2. Cheeseburger: $1.60: shake: $1.85   **3.** Length: 120 yards; width: 60 yards   **4.** $x = 15$, $y = 30$

5. $96,000 in Aaa-rated bonds: $24,000 in B-rated bonds   **6.** 10 pounds of cashews; 20 pounds of peanuts

7. Airspeed: 350 mph; wind resistance: 50 mph

8. (a) $R(x) = 230x$ **(b)** $C(x) = 160x + 2100$

9. 8 and 10   **11.** The first number is 12 and the second number is 23.

13. $60; $35   **15.** 40 meters by 20 meters   **17.** $x = 4$ and $y = 14$

19. Invest $22,500 in stocks and $13,500 in bonds.

21. 60 pounds of Arabica beans and 40 pounds of African Robusta beans

23. Jonathon and Samantha can paddle 8 miles per hour in still water.

25. The Lincoln traveled 400 miles and the Infiniti traveled 500 miles. The time for both trips was 10 hours.

(c)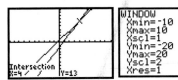

(d) 30 trees; $6900

27. (a)

(b) $(1520, 18240)$

29. 15 cm, 15 cm, and 5 cm   **31.** $250,000

33. 12 and 19

35. Drenell's normal walking speed is 4.5 feet per second.

37. $x = 4$ and $y = 8$

39. Mix 50 mg of Liquid A with 60 mg of Liquid B.

41. A hamburger has 280 calories and a medium Coke has 210 calories

43. Mix 60 g of the 10% alloy with 40 g of the 25% alloy

45. (a)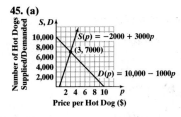

(b) $3; 7000 hot dogs

47. (a)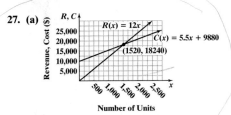

(b) 2050; 29.4%

49. (a) Let x represent the amount of sales, let A represent the pay from option A, and let B represent the pay from option B.

$\begin{cases} A(x) = 15,000 + 0.01x \\ B(x) = 25,000 + 0.0075x \end{cases}$

(b)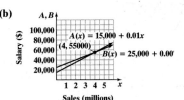

(c) Annual sales of $4,000,000 result in a salary of $55,000 for both options.

51. (a) $R(x) = 60x$ **(b)** $C(x) = 35x + 3500$

(c)

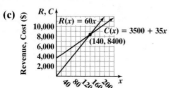

Number of Desks

(d) 140 desks: The cost and revenue will both be $8400.

53. (a) and **(b)** Let x represent the number of years since 1968.

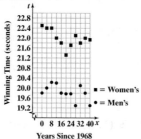

Years Since 1968

(c) Answers may vary. Using the points (4, 20.00) and (36, 19.79), the equation is $y = -0.0065625x + 20.02625$. See graph in part (d).

(d) Answers may vary. Using the points (0, 22.50) and (32, 21.84), the equation is $y = -0.020625x + 22.50$.

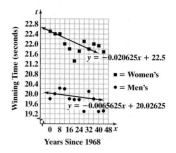

Years Since 1968

(e) In the year 2144: approximately 18.87 seconds
(f) Answers will vary.

Section 3.3 Systems of Linear Equations in Three Variables

1. inconsistent; consistent; dependent **2.** solution **3.** False **4.** True

5. (a) no **(b)** yes **6.** $(-3, 1, -1)$ **7.** $\left(-\frac{5}{2}, \frac{5}{4}, \frac{1}{2}\right)$ **8.** $\varnothing$ or { }; the system is inconsistent **9.** $\{(x, y, z)\mid x = -z + 1, y = 2z - 1, z \text{ is any real number}\}$ **10.** The company can manufacture fourteen 21-inch, eleven 24-inch, and 5 40-inch **11. (a)** no **(b)** yes **13.** $(3, -2, 4)$

15. $(3, -4, -2)$ **17.** $\left(\frac{1}{3}, -\frac{2}{3}, 2\right)$ **19.** $(3, -3, 1)$ **21.** $\varnothing$ or { } **23.** $\{(x, y, z)\mid x = 3z - 10, y = -4z + 14, z \text{ is any real number}\}$

25. $\{(x, y, z)\mid x = -3z + 5, y = -6z + 11, z \text{ is any real number}\}$ **27.** $\left(0, 2, \frac{3}{2}\right)$ **29.** $\varnothing$ or { } **31.** $(-5, 0, 3)$ **33.** $\left(-\frac{7}{2}, -\frac{7}{2}, \frac{3}{2}\right)$

35. $\{(x, y, z)\mid x = 5z - 5, y = -2z + 3, z \text{ is any real number}\}$ **37.** $\left(\frac{11}{2}, 0, -\frac{1}{2}\right)$ **39.** $\left\{(x, y, z)\mid x = \frac{1}{3}z + \frac{2}{3}, y = -\frac{4}{3}z + \frac{7}{3}, z \text{ is any real number}\right\}$

41. Answers will vary. One possibility follows. $\begin{cases} x + y + z = 4 \\ x - y + z = 6 \\ x + y - z = -2 \end{cases}$ **43. (a)** $a - b + c = -6$; $4a + 2b + c = 3$ **(b)** $a = -2, b = 5, c = 1$; $f(x) = -2x^2 + 5x + 1$ **45.** $i_1 = 1, i_2 = 4,$ and $i_3 = 3$ **47.** There are 1490 box seats, 970 reserved seats, and 1640 lawn seats in the stadium.

49. Nancy needs 1 serving of Chex® cereal, 2 servings of 2% milk, and 1.5 servings of orange juice. **51.** $12,000 in Treasury bills, $8000 in municipal bonds, and $5000 in corporate bonds. **53.** $\overline{AM} = 4, \overline{BN} = 2,$ and $\overline{OC} = 10$ **55.** $(10, -4, 6)$ **57.** $(-2, 1, 0, 4)$
59. To create a system of two equations and two unknowns, something we already know how to solve.

Putting the Concepts Together (Sections 3.1–3.3)

1. $(-3, 5)$

2. $(3, -2)$

3. $(5, -1)$ **4.** $\left(\frac{4}{5}, -\frac{7}{5}\right)$ **5.** $(-10, 6)$ **6.** $\{(x, y)\mid 8x - 4y = 12\}$ **7.** $(-1, -3, 5)$ **8.** $\varnothing$ or { }

9. 450 adult tickets and 375 youth tickets **10.** 290 orchestra seats, 380 mezzanine seats, and 530 balcony seats

---

Section 3.4 Using Matrices to Solve Systems

1. matrix **2.** augmented **3.** 4; 3 **4.** False **5.** $\begin{bmatrix} 3 & -1 & | & -10 \\ -5 & 2 & | & 0 \end{bmatrix}$

6. $\begin{bmatrix} 1 & 2 & -2 & | & 11 \\ -1 & 0 & -2 & | & 4 \\ 4 & -1 & 1 & | & 3 \end{bmatrix}$ **7.** $\begin{cases} x - 3y = 7 \\ -2x + 5y = -3 \end{cases}$ **8.** $\begin{cases} x - 3y + 2z = 4 \\ 3x \quad - z = -1 \\ -x + 4y \quad = 0 \end{cases}$ **9.** $\begin{bmatrix} 1 & -2 & | & 5 \\ 0 & -3 & | & 9 \end{bmatrix}$ **10.** $R_1 = -5r_2 + r_1$; $\begin{bmatrix} 1 & 0 & | & 3 \\ 0 & 1 & | & 2 \end{bmatrix}$

11. True **12.** $(6, -2)$ **13.** $(3, -2, 1)$ **14.** $\left(\frac{7}{2}, \frac{2}{3}, 4\right)$ **15.** $\{(x, y)\mid 2x + 5y = -6\}$.

16. $\{(x, y, z)\mid x = -z + 5, y = 4z + 3, z \text{ is any real number}\}$ **17.** $\varnothing$ or { } **18.** $\varnothing$ or { } **19.** $\begin{bmatrix} 1 & -3 & | & 2 \\ 2 & 5 & | & 1 \end{bmatrix}$ **21.** $\begin{bmatrix} 1 & 1 & 1 & | & 3 \\ 2 & -1 & 3 & | & 1 \\ -4 & 2 & -5 & | & -3 \end{bmatrix}$

23. $\begin{bmatrix} -1 & 1 & | & 2 \\ 5 & 1 & | & -5 \end{bmatrix}$ **25.** $\begin{bmatrix} 1 & 0 & 1 & | & 2 \\ 2 & 1 & 0 & | & 13 \\ 1 & -1 & 4 & | & -4 \end{bmatrix}$ **27.** $\begin{cases} 2x + 5y = 3 \\ -4x + y = 10 \end{cases}$ **29.** $\begin{cases} x + 5y - 3z = 2 \\ 3y - z = -5 \\ 4x \quad + 8z = 6 \end{cases}$ **31.** $\begin{cases} x - 2y + 9z = 2 \\ y - 5z = 8 \\ z = \frac{4}{3} \end{cases}$

33. (a) $\begin{bmatrix} 1 & -3 & | & 2 \\ 0 & -1 & | & 5 \end{bmatrix}$ **(b)** $\begin{bmatrix} 1 & -3 & | & 2 \\ 0 & 1 & | & -5 \end{bmatrix}$ **35. (a)** $\begin{bmatrix} 1 & 1 & -1 & | & 4 \\ 0 & 3 & 5 & | & -11 \\ -1 & -3 & 2 & | & 1 \end{bmatrix}$ **(b)** $\begin{bmatrix} 1 & 1 & -1 & | & 4 \\ 0 & 3 & 5 & | & -11 \\ 0 & -2 & 1 & | & 5 \end{bmatrix}$

37. (a) $\begin{bmatrix} 1 & 1 & 1 & | & 4 \\ 0 & 1 & 5 & | & 5 \\ 0 & -4 & 2 & | & 8 \end{bmatrix}$ **(b)** $\begin{bmatrix} 1 & 1 & 1 & | & 4 \\ 0 & 1 & 5 & | & 5 \\ 0 & -2 & 1 & | & 4 \end{bmatrix}$ **39.** $(8, -5)$ **41.** $(-2, 5, 0)$ **43.** $\left(\dfrac{3}{4}, -\dfrac{3}{5}, \dfrac{1}{2}\right)$ **45.** $\{(x, y)\,|\, x - 3y = 3\}$ **47.** $\varnothing$ or $\{\}$

49. $\left\{(x, y, z)\,|\, x = \dfrac{3}{4}z + 1, y = \dfrac{5}{4}z + 3, z \text{ is any real number}\right\}$ **51.** $\begin{cases} x + 4y = -5 & (1) \\ \quad\quad\; y = -2 & (2) \end{cases}$ consistent and independent; $(3, -2)$

53. $\begin{cases} x + 3y - 2z = 6 & (1) \\ \quad\quad y + 5z = -2 & (2) \\ \quad\quad\quad\quad 0 = 4 & (3) \end{cases}$ inconsistent; $\varnothing$ or $\{\}$ **55.** $\begin{cases} x - 2y - z = 3 & (1) \\ \quad\quad y - 2z = -8 & (2) \\ \quad\quad\quad\quad z = 5 & (3) \end{cases}$ consistent and independent; $(12, 2, 5)$

57. $(3, -5)$ **59.** $\varnothing$ or $\{\}$ **61.** $\left(\dfrac{1}{2}, -\dfrac{5}{4}\right)$ **63.** $\{(x, y)\,|\, 4x - y = 8\}$ **65.** $(3, -4, 1)$ **67.** $(4, 0, -5)$

69. $\{(x, y, z)\,|\, x = -2z - 0.2, y = -z - 1.4, z \text{ is any real number}\}$ **71.** $\varnothing$ or $\{\}$ **73.** $\left(\dfrac{3}{10}, \dfrac{1}{10}, -\dfrac{1}{2}\right)$ **75.** $\left(\dfrac{5}{3}, \dfrac{2}{5}, -\dfrac{1}{2}\right)$

77. $(-2, 1, -5)$ **79. (a)** $a + b + c = 0; 4a + 2b + c = 3$ **(b)** $a = 2, b = -3, c = 1; f(x) = 2x^2 - 3x + 1$
81. \$8000 in Treasury bills, \$7000 in municipal bonds, and \$5000 in corporate bonds **83.** $(-3, 7)$ **85.** $(2, 5, -4)$ **87.** Answers will vary.

89. Multiply each entry in row 2 by $\dfrac{1}{5}$: $R_2 = \dfrac{1}{5}r_2$. **91.** $(5, -3)$ **93.** $(4, -2, 7)$

Section 3.5 Determinants and Cramer's Rule **1.** $ad - bc$ **2.** square **3.** 18 **4.** -9 **5.** $(-3, 5)$ **6.** Cramer's Rule does not apply

7. -91 **8.** 20 **9.** $\left(\dfrac{1}{2}, -2, -1\right)$ **10.** $(-1, 4, 1)$ **11.** 10 **13.** -2 **15.** $(-8, 4)$ **17.** $(3, -1)$ **19.** $\left(-\dfrac{1}{6}, \dfrac{3}{8}\right)$ **21.** $\left(\dfrac{3}{4}, -\dfrac{7}{4}\right)$ **23.** -9 **25.** -163 **27.** 0

29. $(-2, 1, -1)$ **31.** $(3, -1, 2)$ **33.** Cramer's Rule does not apply. **35.** Cramer's Rule does not apply. **37.** $(12, -6, 3)$

39. $\left(\dfrac{7}{5}, -\dfrac{5}{3}, \dfrac{11}{3}\right)$ **41.** $(1,1,1)$ **43.** $x = 5$ **45.** $x = -2$

47. (a)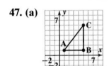

(b) The area of triangle ABC is 10 square units.

49. (a)

(b)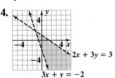

The area of triangle ABC is 4.5 square units.

(c) The area of triangle ADC is 4.5 square units.
(d) 9 square units

51. (a) $x + 2y = 7$ **(b)** $x + 2y = 7$
53. 14: -14: answers may vary.
55. $(-8, 4)$
57. $(3, -1)$
59. $(-2, 1, -1)$

Section 3.6 Systems of Linear Inequalities **1.** satisfies **2. (a)** no **(b)** yes **(c)** no

3. $2x + y = 5$ 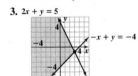 **4.** 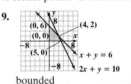 **5.** False **7.** corner points **8.** unbounded

6.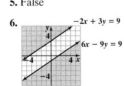

9.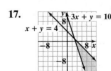
bounded

10. $\begin{cases} x + y \le 25{,}000 \\ x \quad\quad\; \ge 10{,}000 \\ \quad\; y \le 15{,}000 \\ \quad\; y \ge 0 \end{cases}$ where x is amount in Treasury notes and y is amount in corporate bonds

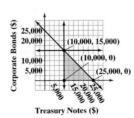

11. (a) no **(b)** yes
13. (a) no **(b)** no
15. (a) no **(b)** yes

17.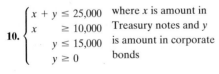

19.  **21.** 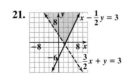 **23.**  **25.**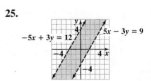

27.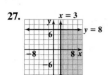

29.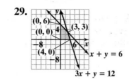

31.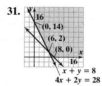

33.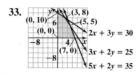

The graph is bounded.   The graph is unbounded.   The graph is bounded.

35.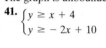

37. (a) $\begin{cases} 450x + 50y \geq 500 \\ 2x + 6y \geq 14 \\ x \geq 0 \\ y \geq 0 \end{cases}$ **(b)**

39. (a) $\begin{cases} x + y \leq 25{,}000 \\ 0.05x + 0.08y \geq 1{,}400 \\ x \geq 5{,}000 \\ y \leq 15{,}000 \\ y \geq 0 \end{cases}$ **(b)**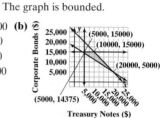

The graph is unbounded.

41. $\begin{cases} y \geq x + 4 \\ y \geq -2x + 10 \end{cases}$ **43.** $\begin{cases} y \leq -x + 12 \\ y \leq -2x + 16 \\ x \geq 0 \\ y \geq 0 \end{cases}$ **45. (a), (b)**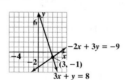

(c) $z = 18; x = 0, y = 8$ **47.** No, because the corner point is not a solution to the strict inequality.   **49.** Yes. Examples may vary. One possibility follows: $\begin{cases} x + y \leq 1 \\ x + y \geq 1 \end{cases}$

51.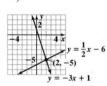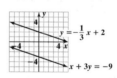

53.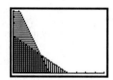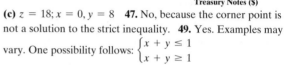

55. The inequalities $x \geq 0$ and $y \geq 0$ restrict the graph to the first quadrant.

Chapter 3 Review   **1. (a)** no   **(b)** yes   **2. (a)** yes   **(b)** no   **3. (a)** yes   **(b)** no   **4. (a)** no   **(b)** yes   **5.** $(4, 2)$ **6.** $(2, -1)$

7. $(2, -5)$ **8.** $(3, -1)$ **9.** $\varnothing$ or $\{\,\}$ **10.** $(-3, -2)$

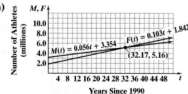

11. $(8, 0)$ **12.** $\left\{(x, y) \mid y = -\dfrac{3}{4}x + 2\right\}$ **13.** $(2, -3)$ **14.** $(-3, -5)$ **15.** $(4, -1)$ **16.** $(2, 2)$ **17.** $\{(x, y) \mid 2x - 4y = 8\}$ **18.** $(-5, -1)$

19. $(0, -4)$ **20.** $(-2, -4)$ **21.** $(1, -1)$ **22.** $\left(\dfrac{1}{2}, \dfrac{3}{4}\right)$ **23.** $\varnothing$ or $\{\,\}$ **24.** $(-7, 4)$ **25.** 35 and 21   **26.** 31 males and 42 females

27. 300 calories in a slice of pepperoni pizza and 340 calories in a slice of Italian sausage pizza   **28.** 21 inches by 13 inches

29. angle $x = 30°$ and angle $y = 60°$ **30.** $x = 15$ and $y = 5$ **31.** 28 nickels and 12 quarters   **32.** 8 liters of the 25%-hydrochloric-acid solution and 4 liters of the 40%-hydrochloric-acid solution   **33.** \$7000 in stocks and \$3000 in bonds   **34.** The plane's speed is 136 miles per hour, and the effect of the wind's resistance is 24 miles per hour.   **35.** $\dfrac{5}{6}$ hour, or 50 minutes; $66\dfrac{2}{3}$ miles   **36.** The speed of the boat in still water is 25 miles per hour, and the speed of the current is 5 miles per hour.

37. (a)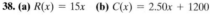

(b) approximately 2022

38. (a) $R(x) = 15x$ **(b)** $C(x) = 2.50x + 1200$

(c)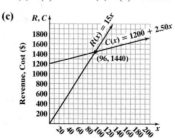

(d) 96 pies; \$1440

39. $(4, -2, 1)$
40. $(-1, 3, -2)$
41. $(5, -3, 4)$
42. $\{\,\}$ or $\varnothing$

43. $\left\{(x, y, z)\,|\, x = \dfrac{1}{2}z + 4,\ y = \dfrac{1}{2}z - 6,\ z \text{ is any real number}\right\}$ **44.** $\left(-\dfrac{2}{3}, \dfrac{1}{2}, \dfrac{3}{4}\right)$ **45.** $(1, 0, -6)$ **46.** $(-3, 4, 2)$ **47.** angle $x = 21°$, angle $y = 53°$, and angle $z = 106°$ **48.** A cheeseburger contains 300 calories, a medium order of fries contains 410 calories, and a medium Coke contains 290 calories.

49. $\begin{bmatrix} 3 & 1 & | & 7 \\ 2 & 5 & | & 9 \end{bmatrix}$ **50.** $\begin{bmatrix} 1 & -5 & | & 14 \\ -1 & 1 & | & -3 \end{bmatrix}$ **51.** $\begin{bmatrix} 5 & -1 & 4 & | & 6 \\ -3 & 0 & -3 & | & -1 \\ 1 & -2 & 0 & | & 0 \end{bmatrix}$ **52.** $\begin{bmatrix} 8 & -1 & 3 & | & 14 \\ -3 & 5 & -6 & | & -18 \\ 7 & -4 & 5 & | & 21 \end{bmatrix}$ **53.** $\begin{cases} x + 2y = 12 \\ 3y = 15 \end{cases}$

54. $\begin{cases} 3x - 4y = -5 \\ -x + 2y = 7 \end{cases}$ **55.** $\begin{cases} x + 3y + 4z = 20 \\ y - 2z = -16 \\ z = 7 \end{cases}$ **56.** $\begin{cases} -3x + 7y + 9z = 1 \\ 4x + 10y + 7z = 5 \\ 2x - 5y - 6z = -8 \end{cases}$ **57. (a)** $\begin{bmatrix} 1 & -5 & | & 22 \\ 0 & -1 & | & 4 \end{bmatrix}$ **(b)** $\begin{bmatrix} 1 & -5 & | & 22 \\ 0 & 1 & | & -4 \end{bmatrix}$

58. (a) $\begin{bmatrix} 1 & -4 & | & 7 \\ 0 & 5 & | & -15 \end{bmatrix}$ **(b)** $\begin{bmatrix} 1 & -4 & | & 7 \\ 0 & 1 & | & -3 \end{bmatrix}$ **59. (a)** $\begin{bmatrix} -1 & 2 & 1 & | & 1 \\ 0 & 3 & 5 & | & -1 \\ -1 & 5 & 6 & | & 2 \end{bmatrix}$ **(b)** $\begin{bmatrix} -1 & 2 & 1 & | & 1 \\ 0 & 3 & 5 & | & -1 \\ 0 & 3 & 5 & | & 1 \end{bmatrix}$ **60. (a)** $\begin{bmatrix} 1 & 3 & 4 & | & 4 \\ 0 & 1 & 2 & | & -3 \\ 0 & -4 & -7 & | & 7 \end{bmatrix}$

(b) $\begin{bmatrix} 1 & 3 & 4 & | & 4 \\ 0 & 1 & 2 & | & -3 \\ 0 & 0 & 1 & | & -5 \end{bmatrix}$ **61.** $(7, -3)$ **62.** $\left(\dfrac{1}{2}, 5\right)$ **63.** $(-4, 1)$ **64.** $\varnothing$ or $\{\,\}$ **65.** $\{(x, y)\,|\,4x - 2y = 6\}$ **66.** $\left(\dfrac{5}{2}, \dfrac{3}{8}\right)$ **67.** $(-5, 1, 4)$

68. $\varnothing$ or $\{\,\}$ **69.** $\left\{(x, y, z)\,|\, x = -\dfrac{5}{6}z + 1,\ y = \dfrac{4}{3}z + 1,\ z \text{ is any real number}\right\}$ **70.** $(-3, 1, 3)$ **71.** 10 **72.** 0 **73.** -2 **74.** 9

75. 111 **76.** 0 **77.** -31 **78.** 78 **79.** $(5, -3)$ **80.** $\left(-4, \dfrac{1}{3}\right)$ **81.** $(-1, 2)$ **82.** $\left(\dfrac{7}{3}, \dfrac{1}{3}\right)$ **83.** $\varnothing$ or $\{\,\}$ **84.** $\left(\dfrac{3}{10}, \dfrac{12}{5}\right)$

85. $(5, -2, 1)$ **86.** $(1, 7, -5)$ **87.** $(1, 2, -2)$ **88.** $\left(-\dfrac{1}{8}, 2, \dfrac{1}{2}\right)$ **89.** $\{(x, y, z)\,|\, x = -9z - 17,\ y = -13z - 24,\ z \text{ is any real number}\}$

90. $(2, -5, -9)$ **91. (a)** yes **(b)** no **92. (a)** no **(b)** yes **93. (a)** yes **(b)** no **94. (a)** no **(b)** yes

95. 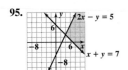 **96.** 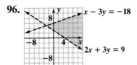 **97.** 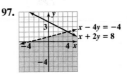 **98.**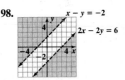

99. 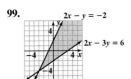 **100.** 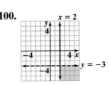 **101.**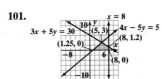

The graph is bounded. **102.**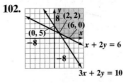

The graph is unbounded.

103. (a) $\begin{cases} x + y \leq 4000 \\ 0.06x + 0.08y \geq 275 \\ x \geq 500 \\ y \leq 2500 \\ y \geq 0 \end{cases}$ **(b)** 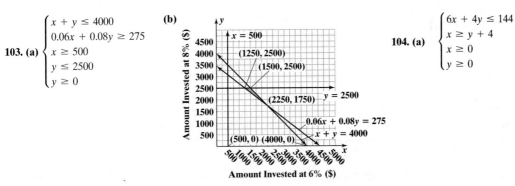 **104. (a)** $\begin{cases} 6x + 4y \leq 144 \\ x \geq y + 4 \\ x \geq 0 \\ y \geq 0 \end{cases}$ **(b)**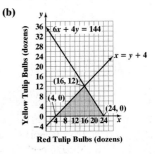

Chapter 3 Test **1.** $(-2, -4)$ **2.** $(1, -4)$ **3.** $\left(\dfrac{2}{3}, -\dfrac{5}{3}\right)$ **4.** $\varnothing$ or $\{\,\}$ **5.** $(-6, 8)$ **6.** $\{(x, y, z)\,|\, x = z + 5,\ y = 2z + 2,\ z \text{ is any real number}\}$

7. $(5, 1, -3)$ **8. (a)** $\begin{bmatrix} 1 & -3 & | & -2 \\ 0 & 2 & | & 12 \end{bmatrix}$ **(b)** $\begin{bmatrix} 1 & -3 & | & -2 \\ 0 & 1 & | & 6 \end{bmatrix}$ **9. (a)** $\begin{bmatrix} 1 & -2 & 1 & | & -2 \\ 0 & 1 & -1 & | & 7 \\ 0 & -4 & 5 & | & -32 \end{bmatrix}$ **(b)** $\begin{bmatrix} 1 & -2 & 1 & | & -2 \\ 0 & 1 & -1 & | & 7 \\ 0 & 0 & 1 & | & -4 \end{bmatrix}$

10. $\begin{bmatrix} 1 & -5 & | & 2 \\ 2 & 1 & | & 4 \end{bmatrix}; (2, 0)$ **11.** $\begin{bmatrix} 1 & 2 & 1 & | & 3 \\ 0 & 4 & 3 & | & 5 \\ 2 & 3 & 0 & | & 1 \end{bmatrix}; (2, -1, 3)$ **12.** -4 **13.** 14 **14.** $\left(-\dfrac{7}{4}, \dfrac{1}{4}\right)$ **15.** $(-5, 4, -1)$

16. 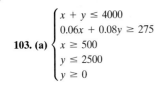 **17.** 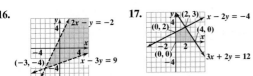 **18.** 32 twenty-five-ton bins and 18 twenty-ton bins

19. angle $x = 30°$, angle $y = 70°$, and angle $z = 80°$

20. (a) $\begin{cases} 12x + 18y \le 180 \\ \quad x + \quad y \le 13 \\ \quad x \qquad\quad \ge 0 \\ \qquad\quad y \ge 0 \end{cases}$ **(b)**

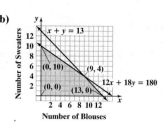

Cumulative Review Chapter R–3

1. (a) 11   **(b)** $-13, 0, 11$ **(c)** $-13, -\dfrac{7}{8}, 0, 2.7, 11$ **(d)** $\dfrac{\pi}{2}, 4\sqrt{2}$ **(e)** $-13, -\dfrac{7}{8}, 0, \dfrac{\pi}{2}, 2.7, 4\sqrt{2}, 11$ **2.** $2(x + 4) = 3x - 6$ **3.** -8

4. 3   **5.** $3x^2 - 11x - 27$ **6.** $\{-15, 1\}$ **7.** $\{x \mid x \le -9\}$ or $(-\infty, -9\,]$ **8.** $\{x \mid x < 2 \text{ or } x > 5\}$; $(-\infty, 2) \cup (5, \infty)$ **9.**   **(a)** $\{0, 6, 12\}$

(b) $\{0, 2, 3, 4, 6, 8, 9, 10, 12, 15\}$ **10.**   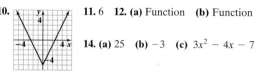   **11.** 6   **12. (a)** Function   **(b)** Function   **(c)** Not a function   **13.** $\left\{x \mid x \ne \dfrac{1}{2}\right\}$

14. (a) 25   **(b)** -3 **(c)** $3x^2 - 4x - 7$

15.   **16.** Parallel   **17.** $y = -\dfrac{5}{3}x + 4$; $5x + 3y = 12$ **18.** 196.7 g   **19. (a)** Independent: year; Dependent: population

(b) $\{x \mid 0 \le x \le 80\}$ or $[\,0, 80\,]$ **(c)** The population in the year 1990 was 517,653   **(d)** 344,113

20. (a) $C(x) = 0.08x + 4.95$ **(b)** $\{x \mid x \ge 0\}$ **(c)** $56.55   **(d)** 1575 minutes   **21.** $(-8, 5)$ **22.** $\left(\dfrac{2}{3}, -\dfrac{1}{2}, -1\right)$ **23.** -2

24. -13

25. [graph]

Getting Ready for Chapter 4: Laws of Exponents and Scientific Notation

1. base; power; exponent   **2.** a^{m+n} **3.** 125   **4.** -243 **5.** y^7 **6.** $-10x^7$ **7.** $-6y^5$ **8.** True   **9.** 25   **10.** y^2 **11.** $\dfrac{8}{5}a$ **12.** $-\dfrac{3}{2}b^2$ **13.** $1; 0$ **14.** $\dfrac{1}{a^n}; 0$

15. $\dfrac{1}{125}$ **16.** $\dfrac{5}{z^7}$ **17.** x^4 **18.** $5y^3$ **19.** -1 **20.** 1   **21.** $\dfrac{9}{16}$ **22.** -64 **23.** $\dfrac{x^2}{9}$ **24.** 20   **25.** $\dfrac{1}{36}$ **26.** 100   **27.** $\dfrac{20x^3}{y}$ **28.** $\dfrac{2}{3}ab^4$ **29.** $-\dfrac{3}{2}b^8$ **30.** $\dfrac{10t^5}{3s^3}$

31. 64   **32.** 1   **33.** 4096   **34.** a^{15} **35.** $\dfrac{1}{z^{18}}$ **36.** s^{21} **37.** $125y^3$ **38.** 1   **39.** $81x^8$ **40.** $\dfrac{1}{16a^6}$ **41.** $\dfrac{z^4}{81}$ **42.** $\dfrac{32}{x^5}$ **43.** $\dfrac{x^8}{y^{12}}$ **44.** $\dfrac{27}{a^6b^{12}}$ **45.** $\dfrac{3x^3y^4}{4}$ **46.** $\dfrac{3b^5}{4a}$

47. $\dfrac{x^6}{y^{10}}$ **48.** 5.32×10^2 **49.** -1.23×10^6 **50.** 3.4×10^{-2} **51.** -8.45×10^{-5} **52.** 500   **53.** 910,000   **54.** 0.00018   **55.** 0.000001   **56.** 6×10^8

57. 8×10^{-11} **58.** 2.4×10^4 **59.** 2×10^2 **60.** 2×10^{-3} **61.** 5×10^3 **62.** 3.75×10^{-13} **63.** 240,000,000,000   **64.** 0.003   **65.** 750   **66.** 0.000000000003

67. -25 **69.** $-\dfrac{1}{25}$ **71.** -1 **73.** $\dfrac{81}{16}$ **75.** $-\dfrac{1}{27}$ **77.** -64 **79.** 288   **81.** 1 (assuming $x \ne 0$)   **83.** $-10t^5$ **85.** $\dfrac{5x^2}{y}$ **87.** $4x^2y$ **89.** $\dfrac{3b^3}{2a}$ **91.** $\dfrac{1}{x^8}$ **93.** $27x^6y^3$

95. $\dfrac{64}{z^3}$ **97.** $\dfrac{a^6}{9}$ **99.** $\dfrac{1}{16a^8b^{12}}$ **101.** $\dfrac{2x^5}{3y^3}$ **103.** $\dfrac{b^4}{25a^{12}}$ **105.** 1   **107.** $-x^2y^4z$ **109.** $\dfrac{9y^8z}{x^5}$ **111.** $\dfrac{3b^{16}}{2a^9}$ **113.** -1.484×10^{-6} **115.** 6.4×10^{19} **117.** 6.25×10^{-3}

119. 8×10^6 **121.** 1   **123.** 12,000,000,000,000   **125.** 0.04   **127.** 4,000,000   **129.** x^6 cubic units   **131.** 1.276×10^{-5} meter   **133.** 1.276×10^7 meters

135. 0.000000001 meter   **137.** 40,000,000,000,000,000,000 possible states   **139.** $49,355   **141.** $5.25 \times 10^7 = 52,500,000$ visitors   **143. (a)** 79.4 people per square mile; On average, each square mile of land contained about 79 people.   **(b)** 87.9 people per square mile; on average, each square mile of land contained about 88 people.   **(c)** about 9 people per square mile; on average, the number of people living on each square mile of land increased by about 9, thereby making living space more crowded.   **145.** 2^{x+1} **147.** 3^{10x+3} **149.** 625   **151.** $\dfrac{1}{2401}$ **153.** No, your friend is incorrect. He added exponents instead of multiplying.   **155.** Explanations will vary. Essentially, if a were to equal zero and m, n, or $m + n$ is negative, we would end up with division by 0, which is not defined. For example, $0^{-2} = \dfrac{1}{0^2} = \dfrac{1}{0}$, which is undefined. If m, n or $m + n$ were to equal 0, we would end up with 0^0, which is indeterminate.

157. Explanations will vary. If a or b were to equal 0 when n was negative, we would end up with division by 0, which is not defined. If a or b were to equal 0 when n was zero, we end up with 0^0, which is indeterminate.   **159.** Explanations will vary. One possibility: $\dfrac{x^5}{x^2} = \dfrac{x \cdot x \cdot x \cdot x \cdot x}{x \cdot x} = x^3 = x^{5-2}$

161. Explanations will vary. One possibility: $(xy)^3 = (xy)(xy)(xy) = x \cdot x \cdot x \cdot y \cdot y \cdot y = x^3y^3$ **163.** Explanations will vary. Basically, using scientific notation enables us to use the Laws of Exponents to perform multiplication or division.

Chapter 4 Polynomials and Polynomial Functions

Section 4.1 Adding and Subtracting Polynomials **1.** monomial **2.** Monomial; 8; 5 **3.** Not a monomial **4.** Monomial; 12; 0 **5.** Not a monomial
6. $m + n$ **7.** Monomial; coefficient: 3, degree: 7 **8.** Monomial; coefficient: -2, degree: 4 **9.** Not a monomial **10.** Monomial; coefficient: -1, degree: 2
11. True **12.** Polynomial; 3 **13.** Not a polynomial **14.** Not a polynomial **15.** Polynomial; 2 **16.** Polynomial; 4 **17.** False **18.** $6x^2 + 2x - 2$

19. $3w^4 - 2w^3 - 7w^2 + w - 5$ **20.** $\frac{3}{4}x^2 - \frac{2}{3}x - 7$ **21.** $5x^2y + 7x^2y^2 - 4xy^2$ **22.** $x^3 - 16x^2 + 7x + 2$ **23.** $11y^3 - 5y^2 - 3y - 7$

24. $11x^2y - 3x^2y^2 - 10xy^2$ **25. (a)** 1 **(b)** -1 **(c)** 34 **26. (a)** In 2006, there were 394,248 first-time births to women 25 years of age. **(b)** In 2006,
there were 88,818 first-time births to women 40 years of age. **27. (a)** $2x^2 + 4x - 5$ **(b)** $4x^2 - 6x + 7$ **(c)** 1 **(d)** 35 **28. (a)** $4x - 1250$ **(b)** $1950;
If the company manufactures and sells 800 calculators, its profit will be $1950. **29.** Coefficient: 3; degree: 2 **31.** Coefficient: -8; degree: 5
33. Coefficient: $\frac{4}{3}$; degree: 6 **35.** Coefficient: 2; degree: 0 **37.** The exponent in the first term is not an integer that is greater than or equal to 0.
39. There is a variable in the denominator. The expression cannot be written in the standard form for a polynomial. **41.** Yes; $5x^2 - 9x + 1$;
2; trinomial **43.** No **45.** No **47.** Yes; $\frac{5}{8}$; 0; monomial **49.** Yes; $2y^2 - 8y + 5$; 2; trinomial **51.** No **53.** Yes; $2xy^4 + 3x^2y^2 + 4$; 5; trinomial
55. No **57.** $13z^3$ **59.** $4x^2 + 3x - 2$ **61.** $8p^3 - p^2 - 4p - 1$ **63.** $7x^3 + 2x^2 - 6x + 8$ **65.** $2x^2 + 4x + 3$ **67.** $3s^2t^3 - 4st^2 - 5t - 1$
69. $-4x^2 - 2x + 1$ **71.** $3y^3 - y^2 - 2y + 8$ **73.** $\frac{3}{4}x^2 + \frac{5}{4}x + 1$ **75.** $8x^2y^2 - 7x^2y - 3xy^2$ **77.** $x^2y + 11xy^2 + 2xy$ **79.** $4a^3 + 4a^2 - 6a - 2$
81. $x^3 + x^2 - 7x + 4$ **83.** $-2b^3 + 6b^2 - 4b + 4$ **85.** $-x^2 - x + 15$ **87.** $n^3 - 2n^2 + 11n - 15$ **89. (a)** 1 **(b)** -3 **(c)** 22 **91. (a)** 3 **(b)** 5
(c) -30 **93. (a)** 3 **(b)** 3 **(c)** 63 **95. (a)** $-3x + 6$ **(b)** $7x + 4$ **(c)** 0 **(d)** 11 **97. (a)** $3x^2 - 5x + 6$ **(b)** $-x^2 - 5x$ **(c)** 8 **(d)** -6
99. (a) $2x^3 + 6x^2 + 12x - 6$ **(b)** $6x^2 + 12x + 10$ **(c)** 58 **(d)** 28 **101. (a)** 4 square units **(b)** 4 square units **(c)** Since the rectangular region is in
quadrant I and has one corner at the origin, the coordinates of the vertex give us the length and width of the rectangle. Since the area of a rectangle
can be found by multiplying the length and width, we can just multiply the coordinates together to get the area of the region. **103. (a)** The average
price per square foot of a home in 1995 was $65.10. **(b)** We predict that the average price per square foot of a home in 2015 will be $29.02. This
result is not reasonable. Home prices will not likely continue to decline in value. **105. (a)** $P(x) = -0.05x^3 + 0.8x^2 + 155x - 500$ **(b)** $1836.25
(c) $P(100) = -27,000$; If 100 cell phones are sold, there would be a loss of $27,000. **107.** $a = 2$ **109. (a)** 3 **(b)** -2 **(c)** -1 **(d)** 1
111. $S(x) = T(x) - F(x)$ **113.** 1 **115.** A polynomial is the sum or difference of monomials. Monomial: $4x^2$; binomial: $-5y^3 + 8y$; trinomial:
$2z^5 - 8z^3 + 10z$. Linear polynomials are of the form $ax + b$, where a and b are constants. **117.** When we add two polynomials, we add the
coefficients of the like monomials. Therefore, the highest-degree term of the polynomials being added will be the degree of the polynomial.

119. (a)–(c)

```
Plot1 Plot2 Plot3
\Y1■4X²-7X+1
\Y2=
\Y3=
\Y4=
\Y5=
\Y6=
\Y7=
```
```
Y1(4)
            37
Y1(-2)
            31
Y1(6)
           103
```

$f(4) = 37$ $f(-2) = 31$ $f(6) = 103$

121. (a)–(c)

```
Plot1 Plot2 Plot3
\Y1■2X^3-5X²+X+5
\Y2=
\Y3=
\Y4=
\Y5=
\Y6=
```
```
Y1(4)
            57
Y1(-2)
           -33
Y1(6)
           263
```

$f(4) = 57$ $f(-2) = -33$ $f(6) = 263$

Section 4.2 Multiplying Polynomials **1.** $6x^7$ **2.** $-21a^4b^6$ **3.** $\frac{5}{4}x^5$ **4.** $-3x - 6$ **5.** $5x^3 + 15x^2 + 10x$ **6.** $6x^3y - 10x^2y^2 + 4xy^3$
7. $y^4 + \frac{1}{6}y^3 + 4y^2$ **8.** First, Outer, Inner, Last **9.** $x^2 + 5x + 4$ **10.** $6v^2 + v - 15$ **11.** $2a^2 + 9ab - 5b^2$ **12.** $2y^3 + 5y^2 - 2y - 15$
13. $2z^4 - 5z^3 + 7z^2 - 16z + 12$ **14.** False **15.** $A^2 - B^2$ **16.** $25y^2 - 4$ **17.** $49y^2 - 4z^6$ **18.** $A^2 - 2AB + B^2, A^2 + 2AB + B^2$ **19.** perfect square
20. False **21.** $z^2 - 16z + 64$ **22.** $36p^2 + 60p + 25$ **23.** $16a^2 - 24ab + 9b^2$ **24.** $f(x) \cdot g(x)$ **25. (a)** 77 **(b)** $5x^3 + 12x^2 - 4x - 3$ **(c)** 77
26. (a) $x^2 - 8x + 15$ **(b)** $2xh + h^2 - 2h$ **27.** $-15x^3y^5$ **29.** $\frac{5}{3}y^4z^5$ **31.** $5x^3 + 20x^2 + 10x$ **33.** $-12a^4b - 8a^3b^2 + 4a^2b^3$
35. $\frac{1}{2}a^3b^2 - \frac{3}{4}a^2b^4 + 4a^2b^2$ **37.** $0.48x^4 - 0.32x^3 + 0.6x^2$ **39.** $x^2 + 8x + 15$ **41.** $a^2 + 2a - 15$ **43.** $12a^2 + 5a - 3$ **45.** $-10x^2 - 7x + 12$
47. $\frac{1}{3}x^2 - \frac{5}{3}x - 8$ **49.** $4a^2 - 17ab - 15b^2$ **51.** $2x^4 - 5x^2 - 3$ **53.** $x^3 + 5x^2 + 6x + 2$ **55.** $6a^3 - a^2 - 17a + 10$ **57.** $20z^3 + 27z^2 + 17z + 6$
59. $-x^3 + 9x$ **61.** $2y^3 + 13y^2 + 17y - 12$ **63.** $2w^4 + w^3 - 3w^2 - w + 1$ **65.** $b^3 + 2b^2 - 5b - 6$ **67.** $8a^3b - 4a^2b^2 + 2ab^3 + 20a^2 - 10ab + 5b^2$
69. $x^2 - 36$ **71.** $a^2 + 16a + 64$ **73.** $9y^2 - 6y + 1$ **75.** $25a^2 - 9b^2$ **77.** $64z^2 + 16yz + y^2$ **79.** $100x^4 - 20x^2y + y^2$ **81.** $a^6 - 4b^2$
83. (a) $x^2 + 3x - 4$ **(b)** 14 **85. (a)** $8x^2 + 14x - 15$ **(b)** 99 **87. (a)** $x^3 + 3x^2 - 13x + 6$ **(b)** 21 **89. (a)** $x^2 + 4x + 5$ **(b)** $2xh + h^2$
91. (a) $x^2 + 9x + 12$ **(b)** $2xh + 5h + h^2$ **93. (a)** $3x^2 + 11x + 11$ **(b)** $6xh - h + 3h^2$ **95.** $5a^3b - 10a^2b^2 + 5ab^3$ **97.** $x^4 - 9$ **99.** $z^4 - 81$
101. $4x^6 + 12x^3 + 9$ **103.** $7m^2 - 6mn - 7n^2$ **105.** $x^3 + 27$ **107.** $4x^2 - 2x + \frac{1}{4}$ **109.** $p^3 + 6p^2 + 12p + 8$
111. $21x^2 - 29xy + 13x + 10y^2 - 9y + 2$ **113.** $3p^2 + 5p - 12$ **115.** $x^4 - x^3 - 19x^2 + 3x + 84$ **117.** $x^2 + 6x$ **119.** $4x^2 + 29x + 30$
121. $4x^2 + 11x + 6$ **123. (a)** $A_1 = a^2; A_2 = ab; A_3 = ab; A_4 = b^2$ **(b)** $a^2 + 2ab + b^2$ **(c)** The length of the region is $a + b$ and the width is also
$a + b$. The area of the region would be $A = (a + b)(a + b) = (a + b)^2$. The result from part (b) is obtained by multiplying this expression out.
125. $9x^2 - y^2 - 2y - 1$ **127.** $4a^2 + 4ab - 12a + b^2 - 6b + 9$ **129.** $2^{2x} - 2^x - 12$ **131.** $5^{2y} - 2(5^y) + 1$

Section 4.3 Dividing Polynomials; Synthetic Division **1.** $4x^4; 8x^2$ **2.** $3p^3 - 4p^2 + p$ **3.** $4a^2 - \frac{2}{a} + \frac{1}{5a^3}$ **4.** $\frac{xy^3}{4} + \frac{2y}{x} - \frac{1}{x^2}$ **5.** True

6. 0; $(3x - 1)(x + 1)$; factors **7.** dividend; divisor; remainder **8.** Quotient; Divisor; Remainder; Dividend **9.** $x^2 + 7x - 3 + \dfrac{9}{x - 4}$

10. $x^2 + 5x + 4$ **11.** $x^3 - 5x^2 + 2 + \dfrac{6}{x^2 - 2}$ **12.** False **13.** False **14.** -7 **15.** $2x^2 + 5x + 3 - \dfrac{7}{x - 2}$ **16.** $4x^2 - 11x + 23 - \dfrac{45}{x + 2}$

17. $x^3 + 5x^2 - 2$ **18. (a)** $3x^2 - 4x + 3 + \dfrac{2x + 1}{x^2 - 2}$ **(b)** 19 **19. (a)** 42 **(b)** 0 **20.(a)** $f(-2) = -35; x + 2$ is not a factor

(b) $f(5) = 0; f(x) = (x - 5)(2x^2 + x - 1)$ **21.** $2x + 3$ **23.** $\dfrac{2}{5}a^2 - 3a + 2$ **25.** $\dfrac{1}{2}y + \dfrac{3}{2y}$ **27.** $1 + \dfrac{3}{2n} - \dfrac{9}{2m}$ **29.** $x + 3$ **31.** $2x + 5 + \dfrac{6}{x - 2}$

33. $w + 6 - \dfrac{7}{2w - 7}$ **35.** $x^2 + 5x - 14$ **37.** $w^2 - 4w - 5$ **39.** $3x^2 - 14x - 5$ **41.** $x - 7$ **43.** $z + 7 + \dfrac{4z - 5}{3z^2 + 1}$

45. $2x^2 - 15x + 28 + \dfrac{-3x + 4}{x^2 + 2x + 5}$ **47.** $x + 2$ **49.** $2x + 3$ **51.** $x + 3 + \dfrac{4}{x - 6}$ **53.** $x^2 + 5x + 6 + \dfrac{15}{x - 5}$ **55.** $3x^3 + 4x^2 - 9x - 10 - \dfrac{5}{x - 3}$

57. $x^3 - 6x^2 - 4x + 24 - \dfrac{35}{x + 6}$ **59.** $2x^2 + 8x + 6$ **61. (a)** $\left(\dfrac{f}{g}\right)(x) = x^2 - 2x + 3$ **(b)** 3 **63. (a)** $\left(\dfrac{f}{g}\right)(x) = x + 3$ **(b)** 5

65. (a) $\left(\dfrac{f}{g}\right)(x) = 2x - 1 + \dfrac{2}{x + 3}$ **(b)** $\dfrac{17}{5}$ **67. (a)** $\left(\dfrac{f}{g}\right)(x) = x^2 + 3x - 4$ **(b)** 6 **69. (a)** $\left(\dfrac{f}{g}\right)(x) = x - \dfrac{4}{x - 3}$ **(b)** 6 **71.** -5 **73.** -119

75. 231 **77.** 2 **79.** $x - 2$ is a factor; $f(x) = (x - 2)(x - 1)$ **81.** $x + 2$ is a factor; $f(x) = (x + 2)(2x + 1)$ **83.** $x - 3$ is not a factor **85.** $x + 1$ is

a factor; $f(x) = (x + 1)(4x^2 - 11x + 6)$ **87.** $a^2b - 3a + 6$ **89.** $y + 3$ **91.** $x + 6 + \dfrac{3x + 2}{x^2 + 5}$ **93.** $\dfrac{2}{3}x + \dfrac{1}{2x}$ **95.** $x^2 + 3x - 10 - \dfrac{6}{x + 4}$

97. $x^2 - 8$ **99.** $4x^2 + 6x + 9$ **101.** $f(x) = 3x^2 - 10x - 25$ **103.** $f(x) = x^2 + 5x - 20$ **105.** $(5x + 2)$ ft **107.** $(2x + 5)$ cm

109. (a) $\overline{C}(x) = 0.01x^2 - 0.4x + 13 + \dfrac{400}{x}$ **(b)** \$26 **111.** $a = 2, b = -7, c = -12,$ and $d = -13$, thus, $a + b + c + d = -30$ **113.** The dividend is the polynomial f and has degree n. Remember, (Divisor)(Quotient) + Remainder = Dividend. Since the remainder must be 0 or a polynomial that has lower degree than f, the degree n must be obtained from the product of the divisor and the quotient. The divisor is $x + 4$, which is of degree 1, so the quotient must be of degree $n - 1$. **115.** Yes; $(3x + 4)(2x^2 - 3x + 1) = 6x^3 - x^2 - 9x + 4$.

Putting the Concepts Together (Sections 4.1–4.3) **1.** $5m^4 - 2m^3 + 3m + 8$; Degree: 4 **2.** $-4a^3 + 9a^2 + a - 8$ **3.** $-\dfrac{19}{5}y^2 + 3y - 8$

4. 29 **5.** -25 **6.** $x^2 + 4x + 10$ **7.** $2m^3n^4 - 8m^2n^4 + 12mn^3$ **8.** $9a^2 - 30ab + 25b^2$ **9.** $49n^4 - 9$ **10.** $18a^3 + 6a^2b - ab^2 + 2b^3$

11. $x^3 - 2x^2 + 3x + 22$ **12.** $5z^2 + 3z - 7$ **13.** $2x^2 + 7x - 1 + \dfrac{3}{x + 9}$ **14.** $x^2 + 3x - 1 + \dfrac{4}{x - 1}$ **15.** yes: $f(x) = (x + 5)(3x^2 - 7x + 12)$

Section 4.4 Greatest Common Factor; Factoring by Grouping **1.** factors **2.** prime **3.** greatest common factor **4.** factor **5.** False **6.** 5 **7.** 2z
8. $3xy^3$ **9.** $7z(z - 2)$ **10.** $2y(3y^2 - 7y + 5)$ **11.** $2m^2n^2(m^2 + 4mn^2 - 3n^3)$ **12.** $-5y(y - 2)$ **13.** $-3a(a^2 - 2a + 4)$ **14.** $(a - 3)(4a + 3)$
15. $3(w - 5)(w + 1)$ **16.** $(x + y)(5 + b)$ **17.** $(w - 3)(w^2 + 4)$ **18.** $(2x + 1)(x - 5)$ **19.** $5(a + 7)$ **21.** $-3(y - 7)$ **23.** $7x(2x - 3)$
25. $3z(z^2 - 2z + 6)$ **27.** $-5p^2(p^2 - 2p + 5)$ **29.** $7mn(7m^2 + 12n^2 - 5m^3n)$ **31.** $-2z(9z^2 - 7z - 2)$ **33.** $(3c - 2)(5c - 3)$
35. $2(a - 3)(3a - 2)$ **37.** $(x + y)(5 + a)$ **39.** $(z + 5)(2z^2 - 5)$ **41.** $(w - 5)(w + 3)$ **43.** $2(x - 4)(x - 2)$ **45.** $3x(x + 5)(x - 4)$
47. $(x - y)(2a - b)$ **49.** $5(w - 3)$ **51.** $(2y + 5)(y - 2)$ **53.** $3x^2y(2xy^2 - 7xy + 3)$ **55.** $(x + 1)(x^2 + 3)$ **57.** $2(x - 2)(2x - 3)$
59. $x^2(4x + 1)(20x + 3)$ **61.** $(x + 9)^2(8x + 33)$ **63.** $(x - 5)^2(2x - 1)(10x - 23)$ **65.** $4(x^2 + 1)(4x - 3)^2(7x^2 - 3x + 3)$
67. $x^2(4 - \pi)$ square units **69.** $S = 2\pi r(r + 4)$ square inches **71. (a)** $1.4x$ **(b)** $1.4x - 0.4(1.4x)$ **(c)** $0.84x$ **(d)** No **73. (a)** $x + 0.15x = 1.15x$
(b) $1.15x + 0.1(1.15x)$ **(c)** $1.265x$ **(d)** \$25.30 **75.** $\dfrac{1}{4}(x - 7)$ **77.** $\dfrac{1}{25}b(5b^2 + 8)$ **79.** $x^n(1 + 3x + 6x^n)$ **81.** $2y^{n+2}(2y - 4 + 3y^3)$

Section 4.5 Factoring Trinomials **1.** quadratic trinomial **2.** $m \cdot n; m + n$ **3.** True **4.** $(y + 6)(y + 3)$ **5.** $(p + 2)(p + 12)$
6. $(q - 4)(q - 2)$ **7.** $(x - 2)(x - 6)$ **8.** $(w - 7)(w + 3)$ **9.** $(q - 12)(q + 3)$ **10.** prime **11.** prime **12.** $(q + 9)(q - 5)$
13. $(x + 3y)(x + 5y)$ **14.** $(m + 5n)(m - 4n)$ **15.** False **16.** $2x(x - 9)(x + 3)$ **17.** $-3(z + 2)(z + 5)$ **18.** $12; -13$ **19.** $(2b - 3)(b + 5)$
20. $(2x + 3)(5x + 6)$ **21.** $(2x + 1)(4x + 5)$ **22.** $(6y - 5)(2y + 7)$ **23.** $(6x - y)(5x + 2y)$ **24.** $2(4x + 7y)(x - 3y)$ **25.** $-(6y + 1)(y - 4)$
26. $-(3x + 2y)(3x + 5y)$ **27.** x^2 **28.** $2x - 3$ **29.** $(y^2 + 4)(y^2 - 6)$ **30.** $(x - 1)(4x - 15)$ **31.** $(x + 3)(x + 5)$ **33.** $(p + 6)(p - 3)$
35. $(r + 5)^2$ **37.** $(s + 12)(s - 5)$ **39.** $2(x - 7)(x - 8)$ **41.** $-(w + 6)(w - 4)$ **43.** $(x + 3y)(x + 4y)$ **45.** $p(p - 4q)(p + 6q)$
47. $(2p + 1)(p - 8)$ **49.** $(y - 2)(4y - 3)$ **51.** $(2s - 1)(4s + 3)$ **53.** $(4z - 3)(4z + 5)$ **55.** $-(2y + 1)(9y + 4)$ **57.** $(x + 7y)(2x - 3y)$
59. $3(r - 5s)(4r - 3s)$ **61.** $(3r + 4s)(8r - 3s)$ **63.** $(x^2 + 1)(x^2 + 2)$ **65.** $(mn + 7)(mn - 2)$ **67.** $(x - 7)(x + 3)$ **69.** $3(r - 2)(3r - 5)$
71. $(y + 2)(2y - 3)$ **73.** $(5w + 3)(2w + 7)$ **75.** $2y(8y + 5)$ **77.** $(3x + 2y)(4x + 5y)$ **79.** prime **81.** $(x^2 + 4)(x^2 + 2)$ **83.** $(z^3 + 5)(z^3 + 4)$
85. $(r - 4s)(r - 8s)$ **87.** $(2z + 3)(4z + 3)$ **89.** prime **91.** $2(x - 3)(x + 9)$ **93.** $-3(r - 5)(r - 8)$ **95.** $-2(2m - 5)(4m + 7)$
97. $4(4z + 1)(3z + 7)$ **99.** $3x(x - 10)(x + 8)$ **101.** $4xy^2(x - 7)(2x - 5)$ **103.** $2r^2s(5r - 4)(7r + 2)$ **105. (a)** $V(3) = 1296$; when 3-inch
square corners are cut from the piece of cardboard, the resulting box will have a volume of 1296 cubic inches. **(b)** $V(x) = 4x(x - 15)(x - 12)$

(c) $V(3) = 4(3)(3 - 15)(3 - 12) = 1296$ **(d)** Answers may vary. **107.** $2x - 5$ **109.** $\dfrac{1}{2}(x + 4)(x + 2)$ **111.** $\dfrac{1}{3}(p - 3)(p + 1)$
113. $\dfrac{4}{3}(a - 6)(a + 4)$ **115.** $(2^n - 5)(2^n + 1)$ **117.** $(4^x - 8)(2^x - 2)(2^x + 2)$ **119.** The trial-and-error method is better when ac gets large and there are quite a few factors of the product whose sum must be determined. **121.** There is a common factor of 3 in the factor $3p + 6$ that should have been factored out.

Section 4.6 Factoring Special Products **1.** perfect square trinomial **2.** $(A - B)^2$ **3.** False **4.** $(x - 9)^2$ **5.** $(2x + 5y)^2$ **6.** $2(3p^2 - 7)^2$
7. difference; two; squares **8.** $(P - Q)(P + Q)$ **9.** $(z - 4)(z + 4)$ **10.** $(4m - 9n)(4m + 9n)$ **11.** $(2a - 3b^2)(2a + 3b^2)$ **12.** True
13. $3(b + 2)(b - 2)(b^2 + 4)$ **14.** $(p - 4 - q)(p - 4 + q)$ **15.** $(A + B)(A^2 - AB + B^2)$ **16.** $(A - B)(A^2 + AB + B^2)$
17. $(z + 4)(z^2 - 4z + 16)$ **18.** $(5p - 6q^2)(25p^2 + 30pq^2 + 36q^4)$ **19.** $4(2m + 5n^2)(4m^2 - 10mn^2 + 25n^4)$ **20.** $(-2x + 1)(13x^2 + 5x + 1)$
21. $(x + 2)^2$ **23.** $(w + 6)^2$ **25.** $(2x + 1)^2$ **27.** $(3p - 5)^2$ **29.** $(5a + 9)^2$ **31.** $(3x + 4y)^2$ **33.** $3(w - 5)^2$ **35.** $-5(t + 7)^2$ **37.** $2(4a - 5b)^2$
39. $(z^2 - 3)^2$ **41.** $(x - 3)(x + 3)$ **43.** $(2 - y)(2 + y)$ **45.** $(2z - 3)(2z + 3)$ **47.** $(10m - 9n)(10m + 9n)$ **49.** $(m^2 - 6n^3)(m^2 + 6n^3)$
51. $2(2p - 3q)(2p + 3q)$ **53.** $5r(4p - 7b)(4p + 7b)$ **55.** $(x + y - 3)(x + y + 3)$ **57.** $(x - 2)(x^2 + 2x + 4)$ **59.** $(m + 5)(m^2 - 5m + 25)$
61. $(x^2 - 4y)(x^4 + 4x^2y + 16y^2)$ **63.** $3(2x - 5y)(4x^2 + 10xy + 25y^2)$ **65.** $(p - 2)(p^2 + 5p + 13)$ **67.** $(5y + 1)(7y^2 + 4y + 1)$
69. $(y^2 + z^3)(y^4 - y^2z^3 + z^6)$ **71.** $(y - 1)(y^2 + y + 1)(y^6 + y^3 + 1)$ **73.** $(5x - y)(5x + y)$ **75.** $(2x + 3)(4x^2 - 6x + 9)$ **77.** $(z - 4)^2$

79. $5x(x-2y)(x^2+2xy+4y^2)$ **81.** $(7m-3n)^2$ **83.** $4(x^2+4)$ **85.** $(y-2)^2(y+2)^2$ **87.** $(2ab+3)^2$ **89.** prime **91.** $(n+9)(n+4)$
93. $2(n-m+10)(n+m-10)$ **95.** $3(y-2)(y^2+2y+4)$ **97.** $(4x+3y-10)(4x+3y+10)$ **99.** $(x-3)(x+3)$ square units
101. $(x-4)(x+4)$ square units **103.** $\pi(R-r)(R+r)$ square units **105.** $10a(a-b)(a+b)$ cubic units
107. $A=a\cdot a+a\cdot b+a\cdot b+b\cdot b=a^2+2ab+b^2=(a+b)^2$ **109.** $b=\pm36$; the middle term must equal $\pm$ twice the product of the quantities
that are squared to get the first and last terms. **111.** We need to add 81 to make a perfect square trinomial. **113.** $(b-0.2)^2$
115. $\left(3b-\dfrac{1}{5}\right)\left(3b+\dfrac{1}{5}\right)$ **117.** $\left(\dfrac{x}{3}-\dfrac{y}{5}\right)\left(\dfrac{x}{3}+\dfrac{y}{5}\right)$ **119.** $\left(\dfrac{x}{2}-\dfrac{y}{3}\right)\left(\dfrac{x^2}{4}+\dfrac{xy}{6}+\dfrac{y^2}{9}\right)$

Section 4.7 Factoring: A General Strategy
1. $2q(p+5q)(p-9q)$ **2.** $-3y(3x+1)(5x-9)$ **3.** $(9x-10y)(9x+10y)$
4. $-3n(m-7)(m+7)$ **5.** $(p-8q)^2$ **6.** $5(2x+3)^2$ **7.** $(4y-5)(16y^2+20y+25)$ **8.** $-2(2m+n)(4m^2-2mn+n^2)$ **9.** $5(2z^2-3z+7)$
10. $3x(2y^2+27x^2)$ **11.** $(x^2+2)(2x+5)$ **12.** $3(3x+1)(x-1)(x+1)$ **13.** $(2x+y-9)(2x+y+9)$ **14.** $(4-m-4n)(4+m+4n)$
15. $G(x)=(2x+7)(x-2)$ **16.** $F(p)=3(3p+2)(3p-2)$ **17.** $2(x-12)(x+6)$ **19.** $-3(y-3)(y+3)$ **21.** $(2b+5)^2$
23. $2(2w+y^2)(4w^2-2wy^2+y^4)$ **25.** $-3(z^2-4z+6)$ **27.** $(4y+3)(5y-6)$ **29.** $(x-4)(x^2+5)$ **31.** $2(100x^2+9y^2)$
33. $(x-3)(x+3)(x^2+9)$ **35.** prime **37.** $4q(3q+1)^2$ **39.** $3mn(4m+3)(2m-7)$ **41.** $3r^2(r-2s)(r^2+2rs+4s^2)$
43. $2(x+4)(x-3)(x+3)$ **45.** $(3x^2-1)(3x^2+1)$ **47.** $(w^2+3)(3w^2-5)$ **49.** $2y(2y+1)$ **51.** $(p-6q-5)(p+6q-5)$
53. $(y+2)(y^2-2y+4)(y^3-2)$ **55.** $(p-1)(p+1)(p^2+p+1)(p^2-p+1)$ **57.** $-3(x+5)(x-3)(x+3)$ **59.** $3a(1-3a)(1+3a)$
61. $2t(t^2+4)(2t-3)(2t+3)$ **63.** $2xy(x+5)(x-3)(x+3)$ **65.** $f(x)=(x-9)(x+7)$ **67.** $P(m)=(7m+3)(m+4)$
69. $G(x)=-3(2x+7)(2x-7)$ **71.** $s(t)=-16(t+2)(t-8)$ **73.** $H(a)=(2a+5)(a-4)(a+4)$ **75.** $(x+3)(3x+7)$ square units
77. $(x+y)(x-y)$ square units **79.** $(3x-4)(9x^2+12x+16)$ cubic units **81.** The factors of 9 are: $1,9;3,3;-1,-9;-3,-3$. None of these
factors sum to 0. Therefore, x^2+9 is prime. **83.** $x^{\frac{1}{2}}(x-3)(x+3)$ **85.** $x^{-2}(x+4)(x+2)$ **87.** See pages 364–365.

Section 4.8 Polynomial Equations
1. polynomial equation **2.** $a=0;b=0$ **3.** $\{-7,0\}$ **4.** $\left\{-\dfrac{3}{4},3\right\}$ **5.** quadratic equation **6.** second
7. False **8.** $\{2,3\}$ **9.** $\left\{-\dfrac{1}{3},5\right\}$ **10.** $\left\{-2,-\dfrac{2}{3}\right\}$ **11.** False **12.** $\{-2,-1\}$ **13.** $\{-6,4\}$ **14.** True **15.** $\{-3,1,3\}$ **16.** $\left\{-\dfrac{1}{2},0,2\right\}$
17. (a) -1 and 9; $(-1,12)$ and $(9,12)$ are points on the graph of g. **(b)** 1 and 7; $(1,-4)$ and $(7,-4)$ are points on the graph of g. **18.** Zeros: -4
and $\dfrac{5}{2}$; x-intercepts: $(-4,0)$ and $\left(\dfrac{5}{2},0\right)$ **19.** Zeros: $-\dfrac{2}{5}$ and 1; x-intercepts: $\left(-\dfrac{2}{5},0\right)$ and $(1,0)$ **20.** 9 miles by 15 miles **21.** 60 boxes
22. (a) 4 seconds and 6 seconds after launch **(b)** 10 seconds after launch **23.** $\{-1,3\}$ **25.** $\left\{-\dfrac{4}{3},0\right\}$ **27.** $\{-3,0,5\}$ **29.** $\{0,4\}$ **31.** $\{0,8\}$
33. $\{-5,3\}$ **35.** $\{4,9\}$ **37.** $\{3\}$ **39.** $\left\{-\dfrac{3}{5},1\right\}$ **41.** $\{-3\}$ **43.** $\{-8,5\}$ **45.** $\left\{-6,\dfrac{5}{2}\right\}$ **47.** $\{-6,2\}$ **49.** $\left\{-\dfrac{7}{2},2\right\}$ **51.** $\{-11,3\}$
53. $\{-5,3\}$ **55.** $\left\{-\dfrac{1}{2},0,3\right\}$ **57.** $\left\{-\dfrac{5}{2},-2,2\right\}$ **59.** $\left\{-\dfrac{3}{5},-\dfrac{2}{3},0\right\}$ **61.** $\left\{\dfrac{4}{3},1\right\}$ **63. (a)** $x=-5$ or $x=-2$ **(b)** $x=-8$ or $x=1$
$(-5,2),(-2,2),(-8,20)$, and $(1,20)$ are on the graph of f. **65. (a)** $x=4$ or $x=-1$ **(b)** $x=5$ or $x=-2$ $(-1,3),(4,3),(-2,15)$, and $(5,15)$
are on the graph of g. **67. (a)** $x=0$ or $x=4$ **(b)** $x=5$ or $x=-1$ $(0,5),(4,5),(-1,-10)$, and $(5,-10)$ are on the graph of F.
69. zeros: -7 and -2; x-intercepts: $(-7,0)$ and $(-2,0)$ **71.** zeros: $-\dfrac{1}{3}$ and $\dfrac{9}{2}$; x-intercepts: $\left(-\dfrac{1}{3},0\right)$ and $\left(\dfrac{9}{2},0\right)$ **73.** zeros: -5, 0, and 4; x-intercepts:
$(-5,0)\,(0,0)$, and $(4,0)$ **75.** $\{-4,6\}$ **77.** $\left\{-\dfrac{7}{2},2\right\}$ **79.** $\{0,7\}$ **81.** $\left\{-\dfrac{6}{7}\right\}$ **83.** $\{-5,-2,2\}$ **85.** $\varnothing$ or $\{\ \}$ **87.** $\{-2,1,3\}$
89. $\left\{-2,-\dfrac{1}{2},\dfrac{1}{2},2\right\}$ **91.** $\{-1,0\}$ **93.** $\{x\,|\,x\neq -2,2\}$ **95.** $\left\{x\,|\,x\neq\dfrac{1}{2},1\right\}$ **97.** The width is 16 cm and the length is 8 cm. **99.** The base is 10 ft and the
height is 22 ft. **101.** 8 sides **103.** 40 meters by 20 meters (along the river) or 10 meters by 80 meters (along the river). **105.** 3 ft **107.** The width is
11 in. and the length is 16 in. **109. (a)** $300 **(b)** 20 bicycles **(c)** 15 bicycles **111. (a)** after 2.5 seconds (on the way up), and again after 5 seconds
(on the way down) **(b)** 7.5 seconds after it is hit **113.** $\{-0.61,4.11\}$ **115.** $\{0.60,24.90\}$ **117.** $\{-1.15,1.75\}$

Chapter 4 Review
1. Coefficient: -7; Degree: 4 **2.** Coefficient: $\dfrac{1}{9}$; Degree: 3 **3.** $7x^3-2x^2+x-8$; degree: 3 **4.** y^4-3y^2+2y+3; degree: 4
5. $4x^2+x-11$ **6.** $-x^4+4x^3-5x^2+8x-6$ **7.** $\dfrac{1}{4}x^2-\dfrac{9}{2}x+\dfrac{1}{6}$ **8.** $\dfrac{5}{6}x^2-x+\dfrac{13}{20}$ **9.** $10x^2y^2$ **10.** $-a^2b-6ab^2-4$ **11. (a)** -24 **(b)** -8
(c) -29 **12. (a)** -82 **(b)** -1 **(c)** -7 **13. (a)** x^2+7x-1 **(b)** 29 **14. (a)** $2x^3-2x^2+x-12$ **(b)** -2 **15. (a)** $P(x)=-2.5x^2+280x-3290$
(b) $2147.50 **16. (a)** 6 square units **(b)** 4 square units **17.** $-12x^4y^3$ **18.** $6m^4n^7$ **19.** $-10a^3b^2+5a^2b^3-15a^2b^2$ **20.** $0.85c^3+2.15c^2+4.45c$
21. $x^2-7x-18$ **22.** $-6x^2+26x-8$ **23.** $2m^2-7mn-4n^2$ **24.** $-2a^2-9a+45$ **25.** $3x^3+x^2-9x+2$ **26.** $w^3-3w^2-12w+32$
27. $2m^4+m^3-11m^2+29m-21$ **28.** $2p^3+11p^2q-29pq^2+12q^3$ **29.** $9w^2-1$ **30.** $4x^2-25y^2$ **31.** $36k^2-60k+25$ **32.** $9a^2+12ab+4b^2$
33. x^3+8 **34.** $8x^3-27$ **35. (a)** $18x^2-27x-35$ **(b)** 91 **36. (a)** $3x^3+5x^2-x+2$ **(b)** 270 **37.** $5x^2-30x+53$ **38.** $-2xh+3h-h^2$
39. $4x^2-2x$ **40.** $3w^4-w^2+5w+2$ **41.** $\dfrac{7}{2}y^2+6y-3$ **42.** $\dfrac{m}{2n}+\dfrac{2}{n}-\dfrac{7}{2m}$ **43.** $3x+4$ **44.** $-2x+7+\dfrac{5}{x+5}$ **45.** $3z^2+2-\dfrac{12}{2z+3}$
46. $4k^2+k-2$ **47.** $8x^3+12x^2+18x+27$ **48.** $2x^2-5x+12+\dfrac{2x+7}{x^2-3x+4}$ **49.** $5x+1+\dfrac{6}{x+2}$ **50.** $9a+4$ **51.** $3m^2+2m-11$
52. $n^2+6n-15+\dfrac{7}{n-4}$ **53.** x^3-x^2+7x-7 **54.** $2x^2-4x+13-\dfrac{34}{x+2}$ **55. (a)** x^2+5x-3 **(b)** 11 **56. (a)** $3x+20+\dfrac{9}{3x-2}$ **(b)** $\dfrac{112}{11}$
57. (a) $2x^2+4x-7$ **(b)** -7 **58. (a)** $3x^2-11x+5+\dfrac{2x-3}{x^2-x+5}$ **(b)** $\dfrac{158}{17}$ **59.** 59 **60.** -45 **61.** $x-2$ is a factor; $f(x)=(x-2)(3x+7)$
62. $x+4$ is not a factor **63.** $(5x+1)$ meters **64.** $(2x^2+5x+3)$ square centimeters **65.** $4(z+6)$ **66.** $-7y(y-13)$ **67.** $2xy(7x^2y-4x+y)$
68. $5a^2b(6a^2b^2+3a-5b)$ **69.** $(x+5)(3x-4)$ **70.** $(2c+9)(3-4c)$ **71.** $(x-5y)(6x+5)$ **72.** $(a+7)(2a-b-1)$
73. $(x+6)(x-3)$ **74.** $(c+2)(c-5)$ **75.** $(7z+8)(2z-3)$ **76.** $(3w-4)(7w+2)$ **77.** $2x(x+1)(x-9)$ **78.** $5a^2(2a+3)(a+7)$
79. (a) $\dfrac{1}{2}n(n+1)$ **(b)** 528 **80.** $R(x)=2x(2600-x^2)$ **81.** $(w+2)(w-13)$ **82.** prime **83.** $-1(t-12)(t+6)$ **84.** $(m+3)(m+7)$

85. $(x + 20y)(x - 16y)$ **86.** $(r - 2s)(r - 3s)$ **87.** $(x + 3)(5x - 2)$ **88.** $(2m + 11)(3m + 4)$ **89.** prime **90.** $2(t + 3)(4t - 1)$
91. $(2x - 1)(3x - 5)$ **92.** $(7r + 2s)(3r - s)$ **93.** $(5x - 3y)(4x - 9y)$ **94.** $-2(s - 7)(s + 1)$ **95.** $(x^2 + 1)(x^2 - 11)$ **96.** $(xy + 4)(10xy + 1)$
97. $(a + 7)(a - 8)$ **98.** $w(2w + 7)$ **99.** $(x + 11)^2$ **100.** $(w - 17)^2$ **101.** $(12 - c)^2$ **102.** $(x - 4)^2$ **103.** $(8y + 5)^2$ **104.** $12(z + 2)^2$
105. $(x - 14)(x + 14)$ **106.** $(7 - y)(7 + y)$ **107.** $(t - 15)(t + 15)$ **108.** $(2w - 9)(2w + 9)$ **109.** $(6x^2 - 5y)(6x^2 + 5y)$
110. $20m(2n - 1)(2n + 1)$ **111.** $(x - 7)(x^2 + 7x + 49)$ **112.** $(9 - y)(y^2 + 9y + 81)$ **113.** $(3x - 5y)(9x^2 + 15xy + 25y^2)$
114. $(2m^2 + 3n)(4m^4 - 6m^2n + 9n^2)$ **115.** $2(a - b)(a + b)(a^2 + ab + b^2)(a^2 - ab + b^2)$ **116.** $(y + 3)(y^2 - 6y + 21)$
117. $(x + 1)(x + 6)$ **118.** $-4xy^3(2x - 3)$ **119.** $7x(x - 1)(x - 4)$ **120.** $3(x - 3)(x + 2)$ **121.** $(2z - 15)^2$ **122.** $(3x + 7)(4x - 7)$
123. $(2n - 7)(5n + 1)$ **124.** $-(y - 2)(y + 4)$ **125.** $2(x - 5)(x^2 + 3)$ **126.** $9(h + 2)(3h^2 + 4)$ **127.** $5p(pq - 4)(pq + 4)$
128. $(m - 1)(m + 1)(m - 2)(m + 2)$ **129.** $-2(2m^2 - 7)(4m^4 + 14m^2 + 49)$ **130.** $(h - 1)(h + 1)(h + 2)$
131. $4(3x + y)(9x^2 - 3xy + y^2)$ **132.** $F(c) = (c - 12)^2$ **133.** $f(x) = -3(x - 5)(x + 3)$ **134.** $g(x) = 4(2x + 5)(2x - 5)$
135. $G(y) = 2(2y + 5)(4y^2 - 10y + 25)$ **136.** $f(x) = -4(x^2 + 4)$ **137.** $(x - 5)(x + 5)$ square units **138.** $(4 - \pi)x^2$ square units

139. $\{-5, 13\}$ **140.** $\{-18, -3\}$ **141.** $\{-5, 3\}$ **142.** $\{-4, 0\}$ **143.** $\{-11, 10\}$ **144.** $\left\{\dfrac{2}{5}, -\dfrac{7}{3}\right\}$ **145.** $\{-6, -4\}$ **146.** $\{4, -2\}$
147. $\left\{-\dfrac{5}{2}, -2, 2\right\}$ **148.** $\left\{-3, -\dfrac{1}{5}, 3\right\}$ **149. (a)** $x = -8$ or $x = 3$ **(b)** $x = -7$ or $x = 2$; $(-8, 6), (3, 6), (-7, -4)$, and $(2, -4)$ are on the graph.
150. (a) $x = 0$ or $x = \dfrac{4}{5}$ **(b)** $x = -\dfrac{1}{5}$ or $x = 1$; $(0, 3), \left(\dfrac{4}{5}, 3\right), \left(-\dfrac{1}{5}, 4\right)$, and $(1, 4)$ are on the graph. **151.** zeros: $-4, -2$, and 0; x-intercepts:
$(-4, 0), (-2, 0)$, and $(0, 0)$ **152.** zeros: $-\dfrac{3}{2}$ and 7; x-intercepts: $\left(-\dfrac{3}{2}, 0\right)$ and $(7, 0)$ **153.** after 7 seconds **154. (a)** 0.80 **(b)** 0.90

Chapter 4 Test **1.** $-5x^7 + x^4 + 7x^2 - x + 1$; Degree: 7 **2.** $-\dfrac{5}{3}a^3b^2 + 9a^2b - 5ab - 4$ **3.** 7 **4.** $7x^3 - 4x^2 - 3x + 1$ **5.** $2a^3b^3 - 3a^3b^2 + 4a^2b$
6. $12x^2 + 47x - 17$ **7.** $4m^2 - 4mn + n^2$ **8.** $3z - 7 + \dfrac{-2z + 11}{2z^2 + 1}$ **9.** $5x + 3$ **10.** 2 **11.** 22 **12.** $4ab^2(3a^2 + 2a - 4b)$ **13.** $(2c + 7)(3c - 2)$
14. $(x - 16)(x + 3)$ **15.** $-(2p + 3)(7p - 2)$ **16.** $(z + 3)(5z - 8)$ **17.** $-2(7x - 4)^2$ **18.** $4(2x - 7)(2x + 7)$ **19.** $\left\{1, \dfrac{7}{3}\right\}$
20. 9 meters by 12 meters

Getting Ready for Chapter 5: A Review of Operations on Rational Numbers

1. lowest terms **2.** $\dfrac{5}{6}$ **3.** $\dfrac{20}{3}$ **4.** $-\dfrac{3}{2}$ **5.** $\dfrac{1}{2}$ **6.** $\dfrac{5}{3}$ **7.** $-\dfrac{3}{5}$ **8.** $\dfrac{4}{3}$ **9.** $-\dfrac{5}{9}$ **10.** LCD $= 75$; $\dfrac{3}{25} = \dfrac{9}{75}, \dfrac{2}{15} = \dfrac{10}{75}$ **11.** LCD $= 126$; $\dfrac{5}{18} = \dfrac{35}{126}, -\dfrac{1}{63} = -\dfrac{2}{126}$
12. $\dfrac{19}{20}$ **13.** $\dfrac{17}{60}$ **14.** $-\dfrac{1}{6}$ **15.** $\dfrac{1}{3}$ **17.** $-\dfrac{3}{7}$ **19.** 5 **21.** $\dfrac{5}{3}$ **23.** -3 **25.** $\dfrac{1}{12}$ **27.** $\dfrac{2}{5}$ **29.** $-\dfrac{9}{4}$ **31.** 6 **33.** 2 **35.** $\dfrac{5}{3}$ **37.** $\dfrac{19}{20}$ **39.** $\dfrac{13}{12}$ **41.** $-\dfrac{23}{30}$ **43.** $-\dfrac{41}{90}$
45. $-\dfrac{23}{40}$ **47.** $\dfrac{17}{120}$ **49.** $-\dfrac{41}{45}$ **51.** $-\dfrac{59}{150}$ **53.** To find the least common denominator (LCD) of two rational numbers, factor each denominator as the
product of prime numbers. Copy first the common factors and then the uncommon factors. The product of these factors is the LCD.

Chapter 5 Rational Expressions and Rational Functions

Section 5.1 Multiplying and Dividing Rational Expressions
1. rational expression **2.** numerator; denominator **3.** False
4. $\{x \mid x \neq -6\}$ **5.** $\{z \mid z \neq -7, z \neq 4\}$ **6.** False **7.** $\dfrac{x - 4}{x + 7}$ **8.** $\dfrac{z^2 + 4z + 16}{2z + 5}$ **9.** $-(w + 5)$ **10.** $\dfrac{(3p + 2)(p - 1)}{2(p + 2)}$ **11.** -1 **12.** $-(m + n)$
13. $9a^2b^3$ **14.** $\dfrac{(m - 5)(m + 1)}{2}$ **15.** $\{x \mid x \neq -6, x \neq 5\}$ **16. (a)** $R(x) = x + 1$; $\left\{x \mid x \neq -3, x \neq \dfrac{5}{3}, x \neq 5\right\}$
(b) $H(x) = \dfrac{3x(x - 5)}{4x + 3}$; $\left\{x \mid x \neq -1, x \neq -\dfrac{3}{4}, x \neq 0, x \neq \dfrac{5}{3}\right\}$ **17.** $\{x \mid x \neq -5\}$ **19.** $\{x \mid x \neq -2, x \neq 8\}$ **21.** $\left\{p \mid p \neq -\dfrac{5}{2}, p \neq 2\right\}$
23. $\{x \mid x$ is any real number$\}$ or $(-\infty, \infty)$ **25.** $\{x \mid x \neq 1\}$ **27.** $\dfrac{2}{x - 4}$ **29.** $p + 3$ **31.** $\dfrac{5}{x^2}$ **33.** $\dfrac{q + 3}{q - 2}$ **35.** $\dfrac{y - 4}{y + 5}$ **37.** $-\dfrac{x + 3}{x + 5}$ **39.** $\dfrac{x - 2}{2x(x - 4)}$
41. $\dfrac{x - 3y}{x - 2y}$ **43.** $\dfrac{x^2 + 3}{x - 5}$ **45.** $\dfrac{x^2 - 2x + 4}{x - 7}$ **47.** $\dfrac{1}{4x(x + 3)}$ **49.** $\dfrac{x - 2}{x - 1}$ **51.** $\dfrac{(x + 3)^2}{(x + 5)(x + 7)}$ **53.** $-\dfrac{(2q + 1)(q + 2)}{q + 6}$ **55.** $x - 3$ **57.** 1
59. $\dfrac{9(x + 3)}{8x(x - 4)}$ **61.** $\dfrac{2}{ab}$ **63.** $\dfrac{(p - 5)(p + 3)}{p(2p + 1)}$ **65.** $\dfrac{x^2 + 3x + 1}{3(x + 1)}$ **67.** $\{x \mid x \neq 1\}$ **69.** $\left\{x \mid x \neq -\dfrac{1}{2}, x \neq 4\right\}$ **71.** $\{x \mid x \neq -5, x \neq -1\}$
73. $\left\{x \mid x \neq 2, x \neq \dfrac{5}{2}\right\}$ **75.** $\{x \mid x$ is any real number$\}$ **77. (a)** $\dfrac{(x + 3)(x - 1)}{2x + 3}$; $\left\{x \mid x \neq -6, x \neq -\dfrac{3}{2}, x \neq 5\right\}$
(b) $(x - 5)(3x - 1)$; $\left\{x \mid x \neq -6, x \neq -3, x \neq \dfrac{1}{3}\right\}$ **79. (a)** $\dfrac{(3x + 5)(x + 6)}{(x^2 + x + 1)(2x + 7)}$; $\left\{x \mid x \neq -\dfrac{7}{2}, x \neq 1, x \neq 2\right\}$
(b) $\dfrac{x - 2}{x^2 + x + 1}$; $\left\{x \mid x \neq -\dfrac{5}{3}, x \neq -1, x \neq 1\right\}$ **81.** $\dfrac{z + 3}{2}$ **83.** $\dfrac{m - 2n}{2(m - n)}$ **85.** $\dfrac{4}{w}$ **87.** $\dfrac{2(x + 2)}{(x + 1)(x - 1)}$ **89.** Answers will vary. One possible rational
expression is $\dfrac{x}{x - 3}$. **91.** Answers will vary. One possible rational expression is $\dfrac{7}{(x + 4)(x - 5)} = \dfrac{7}{x^2 - x - 20}$. **93.** $R(x) = \dfrac{4}{x^2 + x - 2}$

95. (a) approximately 9.8208 m/sec² **(b)** approximately 9.8159 m/sec² **(c)** approximately 9.7936 m/sec² **97. (a)** $\{x \mid x \neq 2\}$

(b)

x	3	2.5	2.1	2.01	2.001	2.0001
$f(x)$	1	2	10	100	1000	10,000

f gets larger in the positive direction.

(c)

x	1	1.5	1.9	1.99	1.999	1.9999
$f(x)$	−1	−2	−10	−100	−1000	−10,000

f gets larger in the negative direction.

(d) f gets larger in the positive direction; f gets larger in the negative direction; these results are the same as those in parts (b) and (c).

99. $\dfrac{1}{x(x+1)}$ **101.** A rational expression is simplified provided there are no common factors between the numerator and denominator.

103. It is not a rational expression because $\sqrt{x} = x^{\frac{1}{2}}$. A rational expression is the ratio of two polynomials, and polynomials must have nonnegative integer exponents (1/2 is not a nonnegative integer).

105.   **107.**   **109.**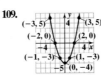

111. Answers will vary. Ideally, the linear functions would be graphed using the properties of a linear function (i.e., slope and intercepts). The functions in problems 109 and 110 are likely graphed using point-plotting, which is less efficient than using properties.

Section 5.2 Adding and Subtracting Rational Expressions

1. $2x - 1$ **2.** $\dfrac{3(x+3)}{x+5}$ **3.** $\dfrac{4x-3}{x-5}$ **4.** least common denominator   **5.** $24x^2y^3$

6. $(x+2)^2(x-7)$ **7.** $\dfrac{9a+8}{30a^2}$ **8.** $-\dfrac{1}{6y}$ **9.** $\dfrac{4x^2+10x-5}{(x-1)(x+2)}$ **10.** $\dfrac{(x-1)(3x+7)}{(x+2)(x+4)(2x+3)}$ **11.** $\dfrac{x^2+15x+5}{(2x-3)(x+2)^2}$ **12.** $\dfrac{-4}{x-2}$ **13.** $\dfrac{3x+5}{x+1}$

15. $\dfrac{2x-1}{2x+5}$ **17.** $\dfrac{1}{x-5}$ **19.** $\dfrac{x-3}{x+3}$ **21.** $\dfrac{3x-1}{x-5}$ **23.** $3x-1$ **25.** LCD $= 8x^3$ **27.** LCD $= 90x^3y^2$ **29.** LCD $= (x-4)(x+2)$

31. LCD $= (x-4)(x+3)(x-5)$ **33.** LCD $= p^2(p+2)(2p-1)$ **35.** $\dfrac{5x+6}{8x^2}$ **37.** $\dfrac{25b-16a}{60a^2b^2}$ **39.** $\dfrac{14(y-1)}{(y-5)(y+3)}$ **41.** $\dfrac{a+4}{a+2}$

43. $\dfrac{-2(x-11)}{(x+3)(x-2)(x+4)}$ **45.** $\dfrac{2x^2-5x+5}{(x+2)(x+1)(x-2)}$ **47.** $\dfrac{-11w+9}{(2w+1)(w+1)(w-3)}$ **49.** $\dfrac{2x^2+xy-5y^2}{(x-3y)^2(x+y)}$ **51.** $\dfrac{1}{(x+1)(x-1)}$ **53.** 0

55. $\dfrac{6}{w(w-2)}$ **57.** $\dfrac{(2p+1)(p-8)}{(p+3)^2(2p-5)}$ **59.** $\dfrac{x+2}{x(x-1)^2}$ **61.** $\dfrac{3}{x^2+3x+9}$ **63.** $x-1$ **65.** $\dfrac{7x+15}{x+3}$ **67.** $\dfrac{-1}{(b-2)(b+4)}$ **69.** $\dfrac{y-3}{y+3}$

71. $\dfrac{7x+9}{(x-2)(x-3)(x+1)}$ **73. (a)** $\dfrac{5x-1}{(x-2)(x+1)}$ **(b)** $\{x \mid x \neq -1, x \neq 2\}$ **75. (a)** $\dfrac{2x+7}{(x-4)(x+3)}$ **(b)** $\{x \mid x \neq -3, x \neq -1, x \neq 4\}$

(c) $\dfrac{-1}{(x-4)(x+3)}$ **(d)** $\{x \mid x \neq -3, x \neq -1, x \neq 4\}$ **77. (a)** $S(x) = \dfrac{2x^3+8000}{x}$ **(b)** $S(10) = 1000$; if the length of the base is 10 inches,

then the surface area of the box will be 1000 square inches.   **79. (a)** $\dfrac{200s+500}{s(s+10)}$ **(b)** $T(50) = 3.5$; if the average speed for the first 50 miles of

the trip is 50 miles per hour, then it will take a total time of 3.5 hours to get to the neighboring university.   **81.** $\dfrac{x+y}{xy}$ **83.** Completely factor each

expression in the denominator of the rational expressions. Copy all common factors that are numerical. Then copy all common variables and use the largest exponent for the LCD. Finally, copy all uncommon factors. The product of these factors represents the LCD.   **85.** $4a^2 - 12a$ **87.** $p^2 - 9$

89. $w^3 - 8$ **91.** $x^2 + 2x + 4$

Section 5.3 Complex Rational Expressions

1. complex rational expression   **2.** True   **3.** $\dfrac{7}{5}$ **4.** $\dfrac{4(z-4)}{z}$ **5.** False   **6.** $\dfrac{1}{4}$ **7.** $\dfrac{7}{5}$ **8.** $\dfrac{4(z-4)}{z}$

9. $\dfrac{-4(x+2)}{x(x+5)}$ **10.** $\dfrac{ab}{3b-a}$ **11.** $\dfrac{4}{9}$ **13.** $\dfrac{x+1}{x-1}$ **15.** $\dfrac{-2a}{(a-2)^2}$ **17.** $\dfrac{1}{(x-1)(x+3)}$ **19.** $-\dfrac{23}{12}$ **21.** $\dfrac{(w-1)(w+1)}{w^2+1}$ **23.** $\dfrac{x+2}{(x-2)(x-7)}$

25. $\dfrac{z^2+4}{4z}$ **27.** $\dfrac{1}{y+3}$ **29.** $2(n-m)$ **31.** $\dfrac{x+3}{x(x-3)}$ **33.** $-x$ **35.** $\dfrac{1}{(x-1)(x-3)}$ **37.** $\dfrac{2(x+1)}{x+4}$ **39.** $-b$ **41.** $\dfrac{3x}{2-x}$ **43.** $\dfrac{x-1}{2(x+1)}$ **45.** $\dfrac{-x+5}{3x+1}$

47. $\dfrac{3xy}{y-x}$ **49.** $\dfrac{mn}{(m+n)^2}$ **51.** $\dfrac{1}{4(b+a)}$ **53. (a)** $\dfrac{R_1R_2}{R_1+R_2}$ **(b)** approximately 2.857 ohms   **55. (a)** $\dfrac{R_1R_2}{(n-1)(R_1+R_2)}$ **(b)** 0.375 meter

57. (a) $\dfrac{2x+1}{x+1}$ **(b)** $\dfrac{3x+2}{2x+1}$ **(c)** $\dfrac{5x+3}{3x+2}$ **(d)** $\dfrac{8x+5}{5x+3}$ **(e)** from part (a): 2, 1, 1; from part (b): 3, 2, 1; from part (c): 5, 3, 2; from part (d): 8, 5, 3;

Each number is the sum of the previous two. The sequence begins 1, 1, 2, 3, 5, 8.   **59.** $-\dfrac{2x+h}{x^2(x+h)^2}$ **61.** $-\dfrac{1}{(x-1)(x+h-1)}$

63. Answers will vary.   **65.** $\{-4\}$ **67.** $\left\{-\dfrac{109}{4}\right\}$ **69.** $\left\{-7, \dfrac{2}{3}\right\}$

Putting the Concepts Together (Sections 5.1–5.3)

1. $\left\{x \mid x \neq -\dfrac{1}{3}, x \neq 6\right\}$ **2.** $\dfrac{-4n}{2n+3}$ **3.** $\dfrac{2p-5q}{3p-4q}$ **4.** $\dfrac{a(a-5)}{2(a+6)}$ **5.** $\dfrac{(x+2)(x-7)}{(3x+1)^2}$

6. $\dfrac{x-5}{x-2}$ **7.** $\dfrac{n}{(n-2)(n-3)}$ **8.** $\dfrac{3y^2-3y-29}{(y-3)(y-4)(y+8)}$ **9.** $\dfrac{3}{(x-7)(2x+1)}$ **10.** $\dfrac{3x^2-14x-1}{(x-4)(x-7)}$ **11.** $\dfrac{n+m}{mn}$ **12.** $\dfrac{z+3}{z-5}$

Section 5.4 Rational Equations

1. rational equation   **2.** $\{-2\}$ **3.** $\left\{\dfrac{2}{3}\right\}$ **4.** $\{-2\}$ **5.** $\left\{-\dfrac{5}{3}\right\}$ **6.** $\{8\}$ **7.** Extraneous solutions   **8.** True

9. $\{\ \}$ or $\varnothing$ **10.** $\{0\}$ **11.** $\left\{-\dfrac{3}{2}, 4\right\}$ **12.** $\{-5\}$ **13.** False   **14.** $\{-1, 5\}$; $(-1, 4)$, $(5, 4)$ **15.** $\left\{-1, \dfrac{3}{2}\right\}$; $(-1, 1)$ and $\left(\dfrac{3}{2}, 1\right)$ **16.** $\dfrac{1}{2}$ hour and 12 hours after ingestion **17.** $\{-4\}$ **19.** $\{-16\}$ **21.** $\varnothing$ or $\{\ \}$ **23.** $\{-1, 5\}$ **25.** $\left\{-\dfrac{1}{3}, \dfrac{3}{2}\right\}$ **27.** $\{1\}$ **29.** $\{4\}$ **31.** $\{-5, 6\}$ **33.** $\{4\}$

35. $\varnothing$ or $\{\ \}$ **37.** $\{-8\}$ **39.** $\left\{\dfrac{1}{4}\right\}$ **41.** $\{-1\}$ **43.** $x = 9$ or $x = 1$; $(1, 10)$ and $(9, 10)$ **45.** $x = -\dfrac{1}{2}$ or $x = -4$; $\left(-\dfrac{1}{2}, -9\right)$ and $(-4, -9)$

47. $x = 6$; $\left(6, \dfrac{9}{2}\right)$ **49.** $x = -5$ or $x = 3$; $(-5, 3)$ and $\left(3, \dfrac{1}{3}\right)$ **51.** $\left\{\dfrac{1}{4}\right\}$ **53.** $\{3\}$ **55.** $\{-19\}$ **57.** $\{2\}$ **59.** $\left\{0, -\dfrac{3}{5}\right\}$ **61.** $-\dfrac{1}{3}$; x-int: $\left(-\dfrac{1}{3}, 0\right)$

63. $-4, \dfrac{3}{2}$; x-int: $(-4, 0), \left(\dfrac{3}{2}, 0\right)$ **65.** $-2, 2$; x-int: $(-2, 0), (2, 0)$ **67.** either 50 or 100 bicycles **69. (a)** 80% **(b)** 90% **71.** 650 walks **73.** Answers will vary. One possibility follows: $\dfrac{2}{x - 1} = \dfrac{3}{x + 1}$ **75.** $\left\{-\dfrac{5}{3}, -1\right\}$ **77.** We find the domain of the rational equation so that we can identify extraneous solutions. **79.** $\left\{-\dfrac{1}{2}, 3\right\}$ **81.** $\dfrac{-2x^2 + 6x + 5}{(x - 2)(x + 1)}$ **83.** $\{-1, 10\}$ **85.** $\{12\}$ **87.** $\{-13\}$ **89.** $\{\ \}$ or $\varnothing$

Section 5.5 Rational Inequalities
1. rational **2.** $\{x | x < -3 \text{ or } x \geq 7\}$; $(-\infty, -3) \cup [7, \infty)$

3. $\{x | -5 < x < 1\}$; $(-5, 1)$ **4.** $\{x | -2 < x < 1\}$; $(-2, 1)$

5. $\{x | x < -1 \text{ or } x > 4\}$ or $(-\infty, -1) \cup (4, \infty)$

7. $\{x | -9 < x < 3\}$ or $(-9, 3)$

9. $\{x | x \leq -10 \text{ or } x > 4\}$ or $(-\infty, -10] \cup (4, \infty)$

11. $\left\{x | x \leq -8 \text{ or } -\dfrac{5}{3} \leq x < 2\right\}$ or $(-\infty, -8] \cup \left[-\dfrac{5}{3}, 2\right)$

13. $\{x | x > -1\}$ or $(-1, \infty)$ **15.** $\left\{x | \dfrac{3}{2} < x < 3\right\}$ or $\left(\dfrac{3}{2}, 3\right)$

17. $\{x | 0 < x \leq 1 \text{ or } x > 4\}$ or $(0, 1] \cup (4, \infty)$

19. $\{x | -5 < x < 2 \text{ or } x \geq 23\}$ or $(-5, 2) \cup [23, \infty)$

21. $\left\{x | -3 < x < \dfrac{1}{2} \text{ or } x > 5\right\}$ or $\left(-3, \dfrac{1}{2}\right) \cup (5, \infty)$

23. $\{x | x > -3\}$; $(-3, \infty)$ **25.** $\{x | -7 \leq x < 8\}$ or $[-7, 8)$

27. $\{x | x \geq 4\}$; $[4, \infty)$ **29.** $\{x | x < -4 \text{ or } x \geq 9\}$ or $(-\infty, -4) \cup [9, \infty)$

31. $\{x | -1 < x \leq 6\}$ or $(-1, 6]$ **33.** $\left\{x | -2 < x < \dfrac{5}{2}\right\}$ or $\left(-2, \dfrac{5}{2}\right)$ **35.** 100 or more bicycles **37.** Answers will vary. One possibility: $\dfrac{1}{x - 2} > 0$

39. Because -1 is not in the domain of the variable x, the solution set is $\{x | -1 < x \leq 4\}$. **41.** $(2, 0)$ **43.** $\left(\dfrac{-7}{2}, 0\right), (2, 0)$ **45.** $\left(\dfrac{2}{3}, 0\right)$

47. $\{x | x \leq -4 \text{ or } x > -1\}$ or $(-\infty, -4] \cup (-1, \infty)$ **49.** $\{x | 7 < x < 26\}$ or $(7, 26)$

Section 5.6 Models Involving Rational Expressions
1. (a) $b = \dfrac{Y - G}{Y}$ **(b)** 0.9 **2.** proportion; similar **3.** 120 feet **4.** $AB = 12$; $DF = 4$

5. approximately 313.5 million people **6.** $\dfrac{40}{3}$ hours or 13 hours, 20 minutes **7.** 33.33 minutes or 33 minutes, 20 seconds **8.** 4 miles per hour

9. $P_1 = \dfrac{V_2 P_2}{V_1}$ **11.** $t = \dfrac{R - r}{R}$ **13.** $x = \dfrac{y - y_1 + mx_1}{m}$ **15.** $v = \dfrac{\omega(I + mr^2)}{rm}$ **17.** $m = \dfrac{MV}{v - V}$ **19.** $AB = 16$; $DF = 5$ **21.** 304,336,135 people

23. approximately \$6.40 **25.** 12.5 pounds **27.** approximately 34.3 minutes **29.** 6 hours **31.** 10 hours **33.** 35 miles per hour **35.** 1.8 feet per second **37.** Gronkowski will be on his own 35-yard line when Urlacher catches up to him. **39.** 8 miles per hour **41.** 45 miles per hour

43. Approximately 2.00 m **45.** a^{15} **47.** $\dfrac{a^4}{b^6}$ **49.** $\dfrac{n^{12}}{9m^4}$

Section 5.7 Variation
1. Variation **2.** $y = kx$

3. (a) $C(g) = 4.2g$
(b) \$19.32
(c)

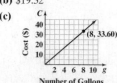

4. $y = \dfrac{k}{x}$ **5. (a)** $y = \dfrac{6}{x}$ **(b)** 1.5 **6. (a)** $V(l) = \dfrac{15,000}{l}$ **(b)** 300 oscillations per second **7.** joint **8.** 1750 joules

9. combined variation **10.** approximately 1.44 ohms **11. (a)** $k = 6$ **(b)** $y = 6x$ **(c)** $y = 42$

13. (a) $k = \dfrac{3}{7}$ **(b)** $y = \dfrac{3}{7}x$ **(c)** $y = 12$ **15. (a)** $k = \dfrac{1}{2}$ **(b)** $y = \dfrac{1}{2}x$ **(c)** $y = 15$ **17. (a)** $k = 20$ **(b)** $y = \dfrac{20}{x}$

(c) $y = 4$ **19. (a)** $k = 21$ **(b)** $y = \dfrac{21}{x}$ **(c)** $y = \dfrac{3}{4}$ **21. (a)** $k = \dfrac{1}{4}$ **(b)** $y = \dfrac{1}{4}xz$ **(c)** $y = 27$

23. (a) $k = \dfrac{13}{10}$ **(b)** $Q = \dfrac{13x}{10y}$ **(c)** $Q = \dfrac{117}{40}$

25. (a) $p(b) = 0.0058358b.$

(b) \$817.01

(c)

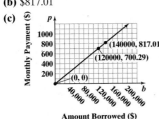

Amount Borrowed (\$)

27. (a) $C(w) = 5.6w$

(b) \$19.60

(c)

29. 96 feet per second   **31. (a)** $D(p) = \dfrac{375}{p}$ **(b)** 125 bags of candy

33. 450 cc   **35.** approximately 119.8 pounds   **37.** 2250 newtons

39. 1.4007×10^{-7} newtons   **41.** 360 pounds   **43. (a)** 314.16

(b) approximately 942.48 meters per minute; approximately 15.708 meters per second   **(c)** $k \approx 0.056$ **(d)** approximately 7.86 newtons

Chapter 5 Review   **1.** $\left\{x \mid x \neq \dfrac{2}{3}\right\}$ **2.** $\{a \mid a \neq -4, a \neq 7\}$ **3.** $\{m \mid m \text{ is any real number}\}$ or $(-\infty, \infty)$ **4.** $\{n \mid n \neq -2, n \neq 4\}$ **5.** $\dfrac{6}{x-5}$

6. $\dfrac{2}{y^3}$ **7.** $\dfrac{w-7}{w+4}$ **8.** $\dfrac{3a+b}{5a+2b}$ **9.** $\dfrac{-1}{3m+1}$ **10.** $\dfrac{n^2+3}{n-4}$ **11.** $\dfrac{p}{2(p-6)}$ **12.** $\dfrac{2q}{3(q-5)}$ **13.** $\dfrac{(x-4)(x+6)}{(x+2)^2}$ **14.** $\dfrac{2(y+5)}{y(y-5)}$ **15.** $\dfrac{3a+b}{2a-b}$ **16.** $\dfrac{3}{m-2}$

17. $\dfrac{9c}{2d^3}$ **18.** $\dfrac{6(z-3)}{7}$ **19.** $\dfrac{x+7}{x-3}$ **20.** $\dfrac{m+4n}{m+5n}$ **21.** $\dfrac{2(p-q)}{p+3q}$ **22.** $\dfrac{2a+5}{9a^2+6a+4}$ **23.** $(x+2)^2; \left\{x \mid x \neq \dfrac{1}{2}, x \neq 5\right\}$

24. $\dfrac{x-5}{x+7}; \left\{x \mid x \neq -7, x \neq -2, x \neq \dfrac{1}{2}\right\}$ **25.** $\dfrac{(x-5)^2}{(2x-1)^2}; \left\{x \mid x \neq -2, x \neq \dfrac{1}{2}, x \neq 5\right\}$ **26.** $\dfrac{(x+2)^2(x+7)}{x-5}; \left\{x \mid x \neq -7, x \neq -2, x \neq \dfrac{1}{2}, x \neq 5\right\}$

27. $\dfrac{4x+3}{x-5}$ **28.** 4   **29.** $\dfrac{a+6}{a-2}$ **30.** $\dfrac{b+5}{2b+3}$ **31.** 3c   **32.** $\dfrac{d+3}{d+1}$ **33.** LCD $= 36x^4$ **34.** LCD $= (y+2)(y-9)$ **35.** LCD $= p^2(2p+5)(p-4)$

36. LCD $= (q+5)(q-1)(q-3)$ **37.** $\dfrac{m^2+4n^2}{m^3n^4}$ **38.** $\dfrac{9x-7y^2}{6x^2y^3}$ **39.** $\dfrac{p^2+q^2}{(p-q)(p+q)}$ **40.** $\dfrac{20x+17}{(x-7)(x-3)(x+4)}$ **41.** $\dfrac{y+8}{(y-1)^2(y+2)}$

42. $\dfrac{11a+7b}{(2a-3b)(2a+3b)}$ **43.** $\dfrac{6x}{x+3}$ **44.** $\dfrac{5}{n^2-5n+25}$ **45.** $\dfrac{2n}{m-7n}$ **46.** $\dfrac{1}{z-3}$ **47.** $\dfrac{3}{y+2}$ **48.** $\dfrac{a-3}{(a+4)(a-2)}$ **49. (a)** $\dfrac{x^2+x+10}{(x-4)(x+2)}$

(b) $\{x \mid x \neq -2, x \neq 4\}$ **50. (a)** $\dfrac{x+8}{(2x-5)(2x+1)}$ **(b)** $\left\{x \mid x \neq -3, x \neq -\dfrac{1}{2}, x \neq \dfrac{5}{2}\right\}$ **51.** $x+1$ **52.** $\dfrac{xy}{x+y}$ **53.** $\dfrac{a-b}{a+b}$ **54.** $\dfrac{-a}{a+3}$ **55.** $\dfrac{3t+4}{5t^2+1}$

56. $\dfrac{1}{a+b}$ **57.** $\dfrac{z+1}{z-1}$ **58.** $\dfrac{x-1}{x+1}$ **59.** $\dfrac{x+y}{x-y}$ **60.** 1   **61.** $\dfrac{2}{x(x-2)}$ **62.** $\dfrac{z-5}{z+5}$ **63.** $\dfrac{n}{m}$ **64.** $\dfrac{2}{(x+1)^2(x-2)}$ **65.** $\dfrac{3y^2-3x^2}{xy}$ **66.** $\dfrac{12d-2c}{c}$ **67.** $\{10\}$

68. $\left\{\dfrac{4}{3}\right\}$ **69.** $\{-2, 7\}$ **70.** $\{9\}$ **71.** $\{2\}$ **72.** $\{3\}$ **73.** $\varnothing$ or $\{\ \}$ **74.** $\{-3\}$ **75.** $\{6\}$ **76.** $\left\{-\dfrac{4}{3}\right\}$ **77.** $\{5\}$ **78.** $\varnothing$ or $\{\ \}$ **79.** $x = 5; (5, 2)$

80. $x = -3$ or $x = 7; (-3, 4)$ and $(7, 4)$ **81.** $\{x \mid x < -2 \text{ or } x \geq 4\}$ or $(-\infty, -2) \cup [4, \infty)$

82. $\{y \mid -4 < y < 5\}$ or $(-4, 5)$

83. $\{z \mid -3 < z < 3\}$ or $(-3, 3)$

84. $\{w \mid w < -7 \text{ or } 2 < w < 4\}$ or $(-\infty, -7) \cup (2, 4)$

85. $\{m \mid -5 < m < 2 \text{ or } m \geq 5\}$ or $(-5, 2) \cup [5, \infty)$

86. $\{n \mid 0 \leq n < 2\}$ or $[0, 2)$ **87.** $\left\{a \mid 2 < a < \dfrac{7}{2}\right\}$ or $\left(2, \dfrac{7}{2}\right)$

88. $\{c \mid c < -6 \text{ or } 0 < c < 2\}$ or $(-\infty, -6) \cup (0, 2)$ **89.** $\left\{x \mid -\dfrac{3}{2} < x < 4\right\}$ or $\left(-\dfrac{3}{2}, 4\right)$

90. $\{x \mid x \leq -5 \text{ or } x > -1\}$ or $(-\infty, -5] \cup (-1, \infty)$ **91.** $C = \dfrac{C_1 C_2}{C_2 + C_1}$ **92.** $T_2 = \dfrac{T_1 P_2 V_2}{P_1 V_1}$ **93.** $G = \dfrac{4\pi^2 a^2}{MT}$ **94.** $x = z \cdot \sigma + \mu$

95. $AB = 24$ and $DF = 10$ **96.** 18.75 feet   **97.** 104 grams   **98.** \$78.00   **99.** 28.8 minutes (or 28 minutes and 48 seconds)   **100.** 2.8 hours (or 2 hours and 48 minutes)   **101.** 21 hours; 28 hours   **102.** 3 minutes   **103.** 60 miles per hour   **104.** 15 miles per hour   **105.** Todd's average walking speed is 3 miles per hour, and his average running speed is 12 miles per hour.   **106.** 60 miles per hour   **107. (a)** $k = 5$ **(b)** $y = 5x$ **(c)** 50   **108. (a)** $k = -6$

(b) $y = -6x$ **(c)** -48 **109. (a)** $k = 60$ **(b)** $y = \dfrac{60}{x}$ **(c)** 12   **110. (a)** $k = \dfrac{3}{4}$ **(b)** $y = \dfrac{3}{4}xz$ **(c)** 42   **111. (a)** $k = 72$ **(b)** $s = \dfrac{72}{t^2}$ **(c)** 8

112. (a) $k = \dfrac{8}{5}$ **(b)** $w = \dfrac{8x}{5z}$ **(c)** $\dfrac{9}{10}$ **113.** 6 inches   **114.** \$352.19   **115.** 1200 kilohertz   **116.** 12 ohms   **117.** 704 cubic cm   **118.** 375 cubic inches

Chapter 5 Test   **1.** $\left\{x \mid x \neq -\dfrac{1}{2}, x \neq 7\right\}$ **2.** $\dfrac{2m-3}{3m-1}$ **3.** $-\dfrac{1}{a+4b}$ **4.** $\dfrac{2x+5}{2x(x-4)}$ **5.** $\dfrac{y+4}{3y+1}$ **6.** $\dfrac{2q}{p-q}$ **7.** $\dfrac{3c-4}{(c-2)(c-1)}$

8. $\dfrac{x-4}{3(x-2)(x-5)}; \{x \mid x \neq -2, x \neq 0, x \neq 2, x \neq 4, x \neq 5\}$ **9.** $\dfrac{3(x^2+2x-4)}{x(x+2)(x-2)}; \{x \mid x \neq -2, x \neq 0, x \neq 2\}$ **10.** $\dfrac{a}{a+1}$ **11.** $\dfrac{4(d-3)}{-3(d+6)}$

12. $\{-1\}$ **13.** $\varnothing$ or $\{\ \}$ **14.** $\left\{x \mid 2 < x \leq \dfrac{11}{2}\right\}$ or $\left(2, \dfrac{11}{2}\right]$ **15.** $k = \dfrac{FD^2}{q_1 q_2}$ **16.** 120 seconds (or 2 minutes)

17. 2.4 hours (or 2 hours and 24 minutes)   **18.** 3 miles per hour   **19.** 20 pounds   **20.** 792 centimeters^2

Cumulative Review Chapters R–5   **1.** 12   **2.** $2x^2 - 2x + 7$ **3.** -3 **4.** $\{-8\}$ **5.** $\left\{x \mid -5 \le x \le \dfrac{1}{3}\right\}; \left[-5, \dfrac{1}{3}\right]$ **6.** $\left\{x \mid x \ne -\dfrac{3}{2}, x \ne 5\right\}$

7. (a) 24   **(b)** $-\dfrac{19}{16}$ **(c)** $x^2 - x - 6$

8.

9. $y = -\dfrac{3}{5}x - 2;\ 3x + 5y = -10$ **10.** $y = 4x - 7;\ 4x - y = 7$ **11.** 2160 mg   **12.** $\begin{cases} 2x - 3y = 7 \\ 5x + 2y = 8 \end{cases}$

13. (a) $\begin{bmatrix} 1 & 1 & 0 & -3 \\ 0 & 2 & -1 & -1 \\ 0 & -5 & 1 & 16 \end{bmatrix}$ **(b)** $\begin{bmatrix} 1 & 1 & 0 & -3 \\ 0 & 1 & -\frac{1}{2} & -\frac{1}{2} \\ 0 & -5 & 1 & 16 \end{bmatrix}$ **14.** $\left(-3, \dfrac{1}{4}\right)$ **15.** -14

16.

17. $8x^2 - 13xy + 9y^2$ **18.** $3x^2 + x + 27$ **19.** $2x^3 - 11x^2 + 24x - 18$ **20.** $2x^2 - 5x + 9$
21. $(x + 5)(x - 2)(x^2 + 2x + 4)$ **22.** $(2x - 3)(3x + 5)$ **23.** $\dfrac{x - 7}{x + 2}$ **24.** $\dfrac{x + 1}{(x - 2)(x - 1)(x + 3)}$ **25.** 48 minutes

Getting Ready for Chapter 6: Square Roots

1. radical sign   **2.** principal square root   **3.** $-4, 4$ **4.** 9   **5.** 30   **6.** $\dfrac{3}{2}$ **7.** 0.4   **8.** 13   **9.** 15   **10.** 10   **11.** 14   **12.** 7   **13.** True   **14.** rational; 20

15. irrational; ≈ 6.32 **16.** not a real number   **17.** rational; -14 **18.** $|a|$ **19.** 14   **20.** $|z|$ **21.** $|2x + 3|$ **22.** $|p - 6|$ **23.** 1   **25.** -10 **27.** $\dfrac{1}{2}$

29. 0.6   **31.** 1.6   **33.** not a real number   **35.** rational; 8   **37.** rational; $\dfrac{1}{4}$ **39.** irrational; 6.63   **41.** irrational; 7.07   **43.** not a real number   **45.** 8   **47.** 19
49. $|r|$ **51.** $|x + 4|$ **53.** $|4x - 3|$ **55.** $|2y + 3|$ **57.** 13   **59.** 17   **61.** not a real number   **63.** 15   **65.** -8 **67.** 6   **69.** not a real number   **71.** -4 **73.** 13

75. The square roots of 36 are -6 and 6; $\sqrt{36} = 6$ **77.** $\dfrac{4\sqrt{13}}{3}$; 4.81   **79.** Answers will vary. One reasonable explanation follows. Because $a^2 \ge 0$, the radicand of $\sqrt{a^2}$ is greater than or equal to 0 (nonnegative). The principal square root of a nonnegative number is nonnegative. The absolute value ensures this result.

Chapter 6 Radicals and Rational Exponents

Section 6.1 nth Roots and Rational Exponents   **1.** index   **2.** 4   **3.** 3   **4.** -6 **5.** not a real number   **6.** $\dfrac{1}{2}$ **7.** 3.68   **8.** 2.99   **9.** 5   **10.** $|z|$ **11.** $3x - 2$ **12.** 2
13. $-\dfrac{2}{3}$ **14.** $\sqrt[n]{a}$ **15.** 5   **16.** -3 **17.** -8 **18.** not a real number   **19.** $\sqrt{b}$ **20.** $(8b)^{\frac{1}{5}}$ **21.** $\left(\dfrac{mn^5}{3}\right)^{\frac{1}{8}}$ **22.** $4y^{\frac{1}{3}}$ **23.** $\sqrt[n]{a^m};\ \left(\sqrt[n]{a}\right)^m$ **24.** 64   **25.** 9   **26.** -8

27. 16   **28.** not a real number   **29.** 13.57   **30.** 1.74   **31.** $a^{\frac{3}{8}}$ **32.** t^3 **33.** $(12ab^3)^{\frac{9}{4}}$ **34.** $\dfrac{1}{9}$ **35.** 4   **36.** $\dfrac{1}{(13x)^{\frac{3}{2}}}$ **37.** 5   **39.** -3 **41.** -5 **43.** $-\dfrac{1}{2}$ **45.** 3

47. 2.92   **49.** 1.86   **51.** 5   **53.** $|m|$ **55.** $x - 3$ **57.** $-|3p + 1|$ **59.** 2   **61.** -6 **63.** 2   **65.** -2 **67.** $\dfrac{2}{5}$ **69.** -5 **71.** not a real number   **73.** $(3x)^{\frac{1}{3}}$

75. $\left(\dfrac{x}{3}\right)^{\frac{1}{4}}$ **77.** 32   **79.** -64 **81.** 16   **83.** 16   **85.** 8   **87.** $\dfrac{1}{12}$ **89.** 125   **91.** 32   **93.** $x^{\frac{3}{4}}$ **95.** $(3x)^{\frac{2}{5}}$ **97.** $\left(\dfrac{5x}{y}\right)^{\frac{3}{2}}$ **99.** $(9ab)^{\frac{4}{3}}$ **101.** 4.47   **103.** 10.08   **105.** 1.26

107. $5x$ **109.** -10 **111.** 8   **113.** 243   **115.** not a real number   **117.** $\dfrac{3}{4}$ **119.** 0.2   **121.** 127   **123.** not a real number   **125.** $3p - 5$ **127.** -18 **129.** 2

131. 8   **133.** 16   **135.** 10; 10   **137. (a)** about $21.25°F$ **(b)** about $17.36°F$ **(c)** about $-15.93°F$ **139. (a)** $\sqrt{\dfrac{8r\rho_w g}{3C\rho}}$ m/s   **(b)** about 7.38 m/s
141. $(-9)^{\frac{1}{2}} = \sqrt{-9}$, but there is no real number whose square is -9. However, $-9^{\frac{1}{2}} = -1 \cdot 9^{\frac{1}{2}} = -1\sqrt{9} = -3$.   **143.** If $\dfrac{m}{n}$, in lowest terms, is positive, then $a^{\frac{m}{n}}$ is a real number provided $\sqrt[n]{a}$ exists. If $\dfrac{m}{n}$, in lowest terms, is negative, then $a^{\frac{m}{n}}$ is a real number provided $a \ne 0$ and $\sqrt[n]{a}$ exists.
145. $(x + 2)(x - 1)^3$ **147.** $\dfrac{z - 6}{z - 1}$

Section 6.2 Simplifying Expressions Using the Laws of Exponents   **1.** $a^r b^r$ **2.** a^{r+s} **3.** $5^{\frac{11}{12}}$ **4.** 8   **5.** 10   **6.** $ab^{\frac{5}{6}}$ **7.** $x^{\frac{2}{3}}$ **8.** $\dfrac{4x^{\frac{1}{2}}}{y^{\frac{2}{3}}}$ **9.** $\dfrac{5x^{\frac{5}{8}}}{y^{\frac{1}{8}}}$ **10.** $\dfrac{200a^{\frac{1}{2}}}{b^{\frac{2}{3}}}$

11. 6   **12.** $2a^2 b^3$ **13.** $\sqrt[12]{x^5}$ **14.** $\sqrt[6]{a}$ **15.** $x^{\frac{1}{2}}(20x + 9)$ **16.** $\dfrac{12x + 1}{x^{\frac{2}{3}}}$ **17.** 25   **19.** 8   **21.** $\dfrac{1}{2^{\frac{7}{6}}}$ **23.** $\dfrac{1}{x^{\frac{7}{12}}}$ **25.** 2   **27.** $\dfrac{125}{8}$ **29.** $x^{\frac{1}{2}}y^{\frac{1}{9}}$ **31.** $\dfrac{x^{\frac{1}{6}}}{y^{\frac{1}{3}}}$ **33.** $\dfrac{2|a|}{b^{\frac{3}{4}}}$ **35.** $\dfrac{x^{\frac{1}{18}}}{2y^{\frac{4}{9}}}$

37. $8x^{\frac{1}{8}}y^{\frac{1}{2}}$ **39.** x^4 **41.** 2   **43.** $2ab^4$ **45.** $\sqrt[4]{x}$ **47.** $\sqrt[6]{x^5}$ **49.** $\sqrt[8]{x^3}$ **51.** $\sqrt[6]{3^7}$ **53.** 1   **55.** $5x^{\frac{1}{2}}(x + 3)$ **57.** $8(x + 2)^{\frac{2}{3}}(3x + 1)$ **59.** $\dfrac{6x + 5}{x^{\frac{1}{2}}}$ **61.** $\dfrac{2(10x - 27)}{(x - 4)^{\frac{1}{3}}}$

63. $5(x^2 + 4)^{\frac{1}{2}}(x^2 + 3x + 4)$ **65.** 2   **67.** 4   **69.** 10   **71.** 5   **73.** 0   **75.** $\dfrac{1}{48}$ **77.** $x^2 - 2x^{\frac{1}{2}}$ **79.** $\dfrac{2}{y^{\frac{1}{3}}} + 6y^{\frac{2}{3}}$ **81.** $4z^3 - 32 = 4(z - 2)(z^2 + 2z + 4)$ **83.** 5   **85.** 3
87. $\sqrt[24]{x}$ **89.** 36   **91.** $\{x \mid x > -1\}$ or $(-1, \infty)$ **93.** $4a^2 - 8a - 6$ **95.** -6

Section 6.3 Simplifying Radical Expressions Using Properties of Radicals   **1.** $\sqrt[n]{ab}$ **2.** $\sqrt{77}$ **3.** $\sqrt[4]{42}$ **4.** $\sqrt{x^2 - 25}$ **5.** $\sqrt{20p^4}$ **6.** 1, 4, 9, 16, 25, 36
7. 1, 8, 27, 64, 125, 216   **8.** $4\sqrt{3}$ **9.** $12\sqrt[3]{2}$ **10.** $10|a|\sqrt{2}$ **11.** Fully simplified   **12.** $2 + \sqrt{5}$ **13.** $\dfrac{-1 + 2\sqrt{2}}{2}$ or $-\dfrac{1}{2} + \sqrt{2}$ **14.** $5a^3\sqrt{3}$ **15.** $3a^2\sqrt{2a}$

16. $4x^2y^3\sqrt[3]{y}$ **17.** $2ab^2\sqrt[4]{ab^3}$ **18.** $4\sqrt{3}$ **19.** $2a^2\sqrt[4]{15}$ **20.** $8ab^3\sqrt[3]{6a}$ **21.** $\dfrac{\sqrt{13}}{7}$ **22.** $\dfrac{3p}{2}$ **23.** $\dfrac{q\sqrt[3]{3}}{2}$ **24.** $2a^2$ **25.** $-2x$ **26.** $\dfrac{5a\sqrt[3]{a}}{b}$ **27.** $\sqrt[12]{10,125}$
28. $2\sqrt[3]{2250}$ **29.** $\sqrt[3]{60}$ **31.** $\sqrt{15ab}$ if $a, b \ge 0$ **33.** $\sqrt{x^2 - 49}$ **35.** $\sqrt{5}$ **37.** $5\sqrt{2}$ **39.** $3\sqrt[3]{2}$ **41.** $4|x|\sqrt{3}$ **43.** $-3x$ **45.** $2|m|\sqrt[4]{2}$ **47.** $2|p|\sqrt{3q}$

49. $9m^2\sqrt{2}$ **51.** $y^6\sqrt{y}$ **53.** $c^2\sqrt[3]{c^2}$ **55.** $5pq^2\sqrt{5p}$ **57.** $-2x^3\sqrt[3]{2}$ **59.** $-m\sqrt[5]{16m^3n^2}$ **61.** $(x-y)\sqrt[4]{x-y}$ **63.** $2\sqrt[3]{x^3-y^3}$ **65.** 5 **67.** $3+\sqrt{2}$

69. $\dfrac{1-\sqrt{2}}{2}$ **71.** 5 **73.** 4 **75.** 2 **77.** $5x\sqrt{3}$ **79.** $2b\sqrt[3]{3b}$ **81.** $18ab^2\sqrt{10}$ **83.** $3pq\sqrt[4]{4p}$ **85.** $-2ab\sqrt[5]{3a}$ **87.** $2(x-y)\sqrt[4]{3(x-y)}$ **89.** $\dfrac{\sqrt{3}}{4}$ **91.** $\dfrac{x\sqrt[4]{5}}{2}$

93. $\dfrac{3y}{5x}$ **95.** $-\dfrac{3x^3}{4y^4}$ **97.** 2 **99.** 4 **101.** $2a\sqrt{2}$ **103.** $\dfrac{2a^2\sqrt{2}}{b}$ **105.** $\dfrac{16a^3}{3b}$ **107.** $a^2\sqrt[3]{26}$ **109.** $\dfrac{3x^3\sqrt{5}}{y}$ **111.** $\sqrt[6]{432}$ **113.** $\sqrt[6]{12}$ **115.** $3\sqrt[6]{12}$ **117.** $\sqrt[3]{18}$

119. $\dfrac{\sqrt[3]{5x}}{2}$ **121.** $\sqrt[3]{45a^2}$ **123.** $6a^2\sqrt{2}$ **125.** $3a\sqrt[3]{2b^2}$ **127.** $\dfrac{-2\sqrt[3]{2}}{a}$ **129.** $-10m\sqrt[3]{4}$ **131.** $3ab^2\sqrt[3]{3ab}$ **133.** 6 **135. (a)**

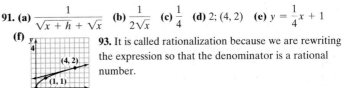

(b) $3\sqrt{5}$ units **137. (a)** roughly \$1,587,000 **(b)** roughly \$2,381,000 **139. (a)** $4x^2$ **(b)** 36

141. $x = -3 - \sqrt{6}$ or $x = -3 + \sqrt{6}$ **143.** $x = \dfrac{-2-\sqrt{7}}{3}$ or $x = \dfrac{-2+\sqrt{7}}{3}$ **145.** The index of each radical must be the same. We can use the Laws of Exponents to rewrite radicals so that they have common indices. **147.** $\left\{\dfrac{5}{2}\right\}$ **149.** $\left\{x \mid x \le \dfrac{130}{3}\right\}$ **151.** Answers will vary.

Section 6.4 Adding, Subtracting, and Multiplying Radical Expressions **1.** like radicals **2.** False **3.** $13\sqrt{13y}$ **4.** $7\sqrt[4]{5}$ **5.** $6\sqrt{2}$ **6.** $-8x\sqrt[3]{2x}$
7. Fully simplified **8.** $(8z+5)\sqrt[3]{z}$ **9.** $2\sqrt{m}$ **10.** $3\sqrt{6}-30$ **11.** $3\sqrt[3]{12}-2\sqrt[3]{3}$ **12.** $-74-27\sqrt{3}$ **13.** False **14.** conjugates **15.** False
16. $9-4\sqrt{5}$ **17.** $25-6\sqrt{14}$ **18.** 1 **19.** $10\sqrt{2}$ **21.** $2\sqrt[3]{x}$ **23.** $14\sqrt{5x}$ **25.** $11\sqrt[3]{5}-11\sqrt{5}$ **27.** $8\sqrt{2}$ **29.** $-2\sqrt[3]{3}$ **31.** $-25\sqrt[3]{2}$ **33.** $9\sqrt{6x}$
35. $4\sqrt{2}+3\sqrt{10}$ **37.** $32x\sqrt{3x}$ **39.** $2x\sqrt{3}-11x\sqrt{2}$ **41.** $(3x-8)\sqrt[3]{2}$ **43.** $5\sqrt{x-1}$ **45.** $3\sqrt{x}$ **47.** $5\sqrt[5]{x}$ **49.** $2\sqrt{3}-3\sqrt{6}$ **51.** $\sqrt{6}+3\sqrt{2}$
53. $\sqrt[3]{12}-2\sqrt[3]{3}$ **55.** $3\sqrt{2x}-2x\sqrt{5}$ **57.** $12+3\sqrt{3}+4\sqrt{2}+\sqrt{6}$ **59.** $12-6\sqrt{7}+2\sqrt{3}-\sqrt{21}$ **61.** $6\sqrt{7}-30$ **63.** $\sqrt{6}-12\sqrt{3}+9\sqrt{2}-4$
65. $31-2\sqrt{15}$ **67.** $4+2\sqrt{3}$ **69.** $7-2\sqrt{10}$ **71.** $x-2\sqrt{2x}+2$ **73.** 1 **75.** -11 **77.** $2x-3y$ **79.** $\sqrt[3]{x^2}+\sqrt[3]{x}-12$ **81.** $\sqrt[3]{4a^2}-25$
83. $\sqrt{15}+5\sqrt{2}$ **85.** $26x^2\sqrt{7x}$ **87.** -13 **89.** $2\sqrt[3]{7}+\sqrt[3]{28}$ **91.** $-4+12\sqrt{2}$ **93.** $20\sqrt{2}$ **95.** $8-4\sqrt{15}$ **97.** $(3x+2)\sqrt[3]{5y}$ **99.** $-7y$
101. $5+x+4\sqrt{x+1}$ **103.** $27+x-10\sqrt{x+2}$ **105.** $\dfrac{11\sqrt{5}}{25}$ **107. (a)** $3\sqrt{3x}$ **(b)** $6\sqrt{3}$ **(c)** $6x$

109. Check $x = -2 + \sqrt{5}$:
$\quad 0 = x^2 + 4x - 1$
$\quad 0 \overset{?}{=} (-2+\sqrt{5})^2 + 4(-2+\sqrt{5}) - 1$
$\quad 0 \overset{?}{=} (-2)^2 + 2(-2)(\sqrt{5}) + (\sqrt{5})^2 - 8 + 4\sqrt{5} - 1$
$\quad 0 \overset{?}{=} 4 - 4\sqrt{5} + \sqrt{25} - 8 + 4\sqrt{5} - 1$
$\quad 0 \overset{?}{=} 9 - 9$
$\quad 0 = 0 \quad$ true
The value is a solution.

Check $x = -2 - \sqrt{5}$:
$\quad 0 = x^2 + 4x - 1$
$\quad 0 \overset{?}{=} (-2-\sqrt{5})^2 + 4(-2-\sqrt{5}) - 1$
$\quad 0 \overset{?}{=} (-2)^2 - 2(-2)(\sqrt{5}) + (\sqrt{5})^2 - 8 - 4\sqrt{5} - 1$
$\quad 0 \overset{?}{=} 4 + 4\sqrt{5} + \sqrt{25} - 8 - 4\sqrt{5} - 1$
$\quad 0 \overset{?}{=} 9 - 9$
$\quad 0 = 0 \quad$ true
The value is a solution.

111. The perimeter is $30\sqrt{2}$ units. The area is 108 square units. **113.** $12\sqrt{6}$ square units **115.** To add or subtract radicals, the indices and radicands must be the same. Then use the Distributive Property "in reverse" to add the "coefficients" of the radicals. **117.** $12a^5b^5$ **119.** $6y^2 + y - 2$
121. $25w^2 - 4$

Section 6.5 Rationalizing Radical Expressions **1.** rationalizing the denominator **2.** $\sqrt{11}$ **3.** $\dfrac{\sqrt{3}}{3}$ **4.** $\dfrac{\sqrt{10}}{4}$ **5.** $\dfrac{\sqrt{10x}}{2x}$ **6.** $\dfrac{4\sqrt[3]{9}}{3}$ **7.** $\dfrac{\sqrt[3]{150}}{10}$ **8.** $\dfrac{3\sqrt[4]{p^3}}{p}$

9. $-2-\sqrt{7}$ **10.** $2(\sqrt{3}-1)$ **11.** $\dfrac{\sqrt{3}+1}{2}$ **12.** $\dfrac{5+\sqrt{10}+4\sqrt{5}+4\sqrt{2}}{3}$ **13.** $\dfrac{\sqrt{2}}{2}$ **15.** $-\dfrac{2\sqrt{3}}{5}$ **17.** $\dfrac{\sqrt{3}}{2}$ **19.** $\dfrac{\sqrt{3}}{3}$ **21.** $\dfrac{\sqrt{2p}}{p}$ **23.** $\dfrac{2\sqrt{2y}}{y^2}$ **25.** $\sqrt[3]{4}$

27. $\dfrac{\sqrt[3]{7q^2}}{q}$ **29.** $-\dfrac{\sqrt[3]{60}}{10}$ **31.** $\dfrac{\sqrt[3]{50y^2}}{5y}$ **33.** $-\dfrac{4\sqrt[4]{27x}}{3x}$ **35.** $\dfrac{12\sqrt[5]{m^2n^3}}{mn}$ **37.** $2(\sqrt{6}+2)$ **39.** $5(\sqrt{5}-2)$ **41.** $2(\sqrt{7}+\sqrt{3})$ **43.** $\dfrac{\sqrt{5}+\sqrt{3}}{2}$

45. $\dfrac{p-\sqrt{pq}}{p-q}$ **47.** $-3(2\sqrt{3}-3\sqrt{2})$ or $3(3\sqrt{2}-2\sqrt{3})$ **49.** $-8-3\sqrt{7}$ or $-(8+3\sqrt{7})$ **51.** $\dfrac{13\sqrt{6}-46}{38}$ **53.** $\dfrac{p+4\sqrt{p}+4}{p-4}$ **55.** $\dfrac{2-3\sqrt{2}}{2}$

57. $\dfrac{4\sqrt{3}}{3}$ **59.** $\dfrac{\sqrt{10}-\sqrt{2}}{2}$ **61.** $\dfrac{22\sqrt{3}}{3}$ **63.** 0 **65.** $\dfrac{1}{2}$ **67.** $\dfrac{\sqrt{2}}{4}$ **69.** $\dfrac{2\sqrt{3}}{3}$ **71.** $\sqrt{3}-2$ **73.** $2(\sqrt{5}-2)$ **75.** 2 **77.** $\dfrac{\sqrt{3}}{3}$ **79.** $\dfrac{\sqrt[3]{18}}{6}$ **81.** $\dfrac{5-\sqrt{3}}{22}$

83. $\dfrac{\sqrt{6}-\sqrt{2}}{4}$ **85.** $\dfrac{1}{3(\sqrt{2}-1)}$ **87.** $\dfrac{x-h}{x+\sqrt{xh}}$

89. $\dfrac{(\sqrt{6})^2 + 2\cdot\sqrt{6}\cdot\sqrt{2} + (\sqrt{2})^2}{4^2} \overset{?}{=} \left(\dfrac{\sqrt{2}+\sqrt{3}}{2}\right)^2$
$\dfrac{6+2\sqrt{12}+2}{16} \overset{?}{=} \dfrac{2+\sqrt{3}}{4}$
$\dfrac{8+2\cdot 2\sqrt{3}}{16} \overset{?}{=} \dfrac{2+\sqrt{3}}{4}$
$\dfrac{8+4\sqrt{3}}{16} \overset{?}{=} \dfrac{2+\sqrt{3}}{4}$
$\dfrac{2+\sqrt{3}}{4} = \dfrac{2+\sqrt{3}}{4}$

91. (a) $\dfrac{1}{\sqrt{x+h}+\sqrt{x}}$ **(b)** $\dfrac{1}{2\sqrt{x}}$ **(c)** $\dfrac{1}{4}$ **(d)** 2; (4, 2) **(e)** $y = \dfrac{1}{4}x + 1$
(f)

93. It is called rationalization because we are rewriting the expression so that the denominator is a rational number.

95.

97.

Putting the Concepts Together (Sections 6.1–6.5)   **1.** -5 **2.** $\dfrac{1}{16}$ **3.** $(3x^3)^{\frac{1}{4}}$ **4.** $7\sqrt[5]{z^4}$ **5.** $2x^{\frac{1}{2}}$ or $2\sqrt{x}$ **6.** $c^2 + c^3$ **7.** $\dfrac{a^2}{b^2}$ **8.** $x^{\frac{5}{8}}$ or $\sqrt[8]{x^5}$ **9.** $\dfrac{x^6}{y}$

10. $\sqrt{30ab}$ **11.** $10m^2n\sqrt{2n}$ **12.** $-2xy$ **13.** $\sqrt{3}$ **14.** $11b\sqrt{2b}$ **15.** $y\sqrt[3]{2y}$ **16.** $12x$ **17.** $7\sqrt{x}$ **18.** $14 - 28\sqrt{2} = 14(1 - 2\sqrt{2})$

19. $41 - 24\sqrt{2}$ **20.** $\dfrac{3\sqrt{2}}{16}$ **21.** $-\dfrac{4(\sqrt{3} + 8)}{61}$

Section 6.6 Functions Involving Radicals   **1.** (a) 4   (b) $2\sqrt{7}$ **2.** (a) -1 (b) 3   **3.** even; odd   **4.** $\{x \mid x \geq -6\}$ or $[-6, \infty)$
5. $\{t \mid t$ is any real number$\}$ or $(-\infty, \infty)$ **6.** $\{m \mid m \leq 2\}$ or $(-\infty, 2]$

7. (a) $\{x \mid x \geq -3\}$ or $[-3, \infty)$ **8.** (a) $\{x \mid x$ is any real number$\}$ or $(-\infty, \infty)$
(b)

9. (a) 3   (b) $\sqrt{14}$ (c) 2   **11.** (a) -5 (b) -1 (c) $-\dfrac{\sqrt{13}}{2}$

13. (a) 4   (b) $4\sqrt{6}$ (c) $\sqrt{6}$ **15.** (a) 2   (b) -2 (c) $-2\sqrt[3]{2}$

17. (a) $\dfrac{\sqrt{5}}{3}$ (b) $\dfrac{\sqrt{2}}{2}$ (c) $\dfrac{\sqrt{6}}{3}$ **19.** (a) 1   (b) $\sqrt[3]{4}$ (c) $\sqrt[3]{3}$

(c) $[0, \infty)$ (c) $(-\infty, \infty)$

21. $\{x \mid x \geq 7\}$ or $[7, \infty)$ **23.** $\left\{x \mid x \geq -\dfrac{7}{2}\right\}$ or $\left[-\dfrac{7}{2}, \infty\right)$

25. $\left\{x \mid x \leq \dfrac{4}{3}\right\}$ or $\left(-\infty, \dfrac{4}{3}\right]$ **27.** $\{z \mid z$ is any real number$\}$ or $(-\infty, \infty)$ **29.** $\left\{p \mid p \geq \dfrac{2}{7}\right\}$ or $\left[\dfrac{2}{7}, \infty\right)$ **31.** $\{x \mid x$ is any real number$\}$ or $(-\infty, \infty)$
33. $\{x \mid x > -5\}$ or $(-5, \infty)$ **35.** $\{x \mid x \leq -3$ or $x > 3\}$ or $(-\infty, -3] \cup (3, \infty)$

37. (a) $\{x \mid x \geq 4\}$ or $[4, \infty)$ **39.** (a) $\{x \mid x \geq -2\}$ or $[-2, \infty)$ **41.** (a) $\{x \mid x \leq 2\}$ or $(-\infty, 2]$ **43.** (a) $\{x \mid x \geq 0\}$ or $[0, \infty)$
(b)

(c) $[0, \infty)$ (c) $[0, \infty)$ (c) $[0, \infty)$ (c) $[3, \infty)$

45. (a) $\{x \mid x \geq 0\}$ or $[0, \infty)$ **47.** (a) $\{x \mid x \geq 0\}$ or $[0, \infty)$ **49.** (a) $\{x \mid x \geq 0\}$ or $[0, \infty)$ **51.** (a) $\{x \mid x \geq 0\}$ or $[0, \infty)$
(b)

(c) $[-4, \infty)$ (c) $[0, \infty)$ (c) $[0, \infty)$ (c) $(-\infty, 0]$

53. (a) $\{x \mid x$ is any real number$\}$ or   **55.** (a) $\{x \mid x$ is any real number$\}$ or   **57.** (a) $\{x \mid x$ is any real number$\}$ **59.** (a) 5 units
$(-\infty, \infty)$ $(-\infty, \infty)$ or $(-\infty, \infty)$ (b) $\sqrt{17} \approx 4.123$ units
(b)   (c) $5\sqrt{17} \approx 20.616$ units
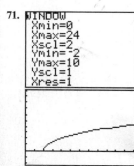
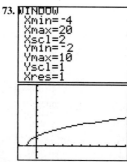
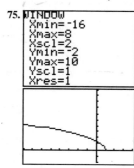

61. (a) $4\sqrt{2} \approx 5.657$ square units
(b) $4\sqrt{5} \approx 8.944$ square units
(c) $2\sqrt{14} \approx 7.483$ square units

(c) $(-\infty, \infty)$ (c) $(-\infty, \infty)$ (c) $(-\infty, \infty)$

63. Shift the graph of $f(x)$ $|c|$ units to the right if $c < 0$ or c units to the left if $c > 0$.   **65.** $\dfrac{5}{6}$ **67.** $\dfrac{4x + 1}{x(x + 1)}$ **69.** $\dfrac{7x + 1}{(x - 1)(x + 1)}$ or $\dfrac{7x + 1}{x^2 - 1}$

71.
```
WINDOW
Xmin=0
Xmax=24
Xscl=2
Ymin=-2
Ymax=10
Yscl=1
Xres=1
```

73.
```
WINDOW
Xmin=-4
Xmax=20
Xscl=2
Ymin=-2
Ymax=10
Yscl=1
Xres=1
```

75.
```
WINDOW
Xmin=-16
Xmax=8
Xscl=2
Ymin=-2
Ymax=10
Yscl=1
Xres=1
```

77.
```
WINDOW
Xmin=-4
Xmax=20
Xscl=2
Ymin=-2
Ymax=10
Yscl=1
Xres=1
```

79.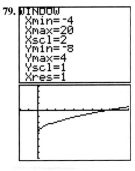

81.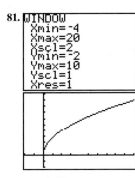

83.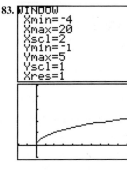

85.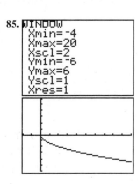

87.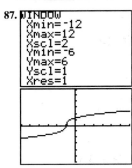

89.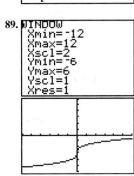

91.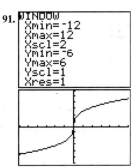

Section 6.7 Radical Equations and Their Applications **1.** radical equation **2.** extraneous **3.** False **4.** {5} **5.** {−5} **6.** ∅ or { } **7.** {4} **8.** {−3} **9.** {15}
10. {−3, 1} **11.** {12} **12. (a)** $L = \dfrac{8T^2}{\pi^2}$ **(b)** 32 feet **13.** {16} **15.** {7} **17.** {11} **19.** ∅ or { } **21.** {2} **23.** {12} **25.** {25} **27.** {11} **29.** {−4} **31.** {1}
33. {−5} **35.** $\left\{0, \dfrac{1}{4}\right\}$ **37.** {3} **39.** {4} **41.** {4} **43.** {6} **45.** {4} **47.** {−3} **49.** {−1, 10} **51.** {0, 4} **53.** {3} **55.** {5} **57.** {2, 10} **59.** {3} **61.** {4} **63.** {6}
65. {−7, 9} **67.** {18} **69.** {3} **71.** {−23, 31} **73.** $r = \dfrac{A^2 - P^2}{P^2}$ **75.** $V = \dfrac{4}{3}\pi r^3$ **77.** $F = \dfrac{q_1 q_2 r^2}{4\pi\varepsilon_0}$ **79.** ∅ or { } **81.** {4} **83.** {2} **85.** {9} **87.** {−3}
89. {−6} **91.** {9} **93.** {5} **95.** {−2}
97. **(a)** {2}; (2, 0) **(b)** {3}; (3, 1) **(c)** {6}; (6, 2) **99. (a)** $y = 5$ or $y = -1$ **101. (a)** after 2 years **(b)** after 16 years
103. (a) in the year 2053 **(b)** in the year 2021
(d) **(e)** The equation $f(x) = -1$ has no solution because the graph of the function does not go below the x-axis. **(b)** **105.** $\left\{-\dfrac{3}{4}, 0\right\}$

107. If the index is even, it is important so that you can identify extraneous solutions. Regardless of the index, it is important to make sure that your answer is correct. **109.** In radical expressions with an even index, the radicand must be greater than or equal to zero. After raising both sides of the equation to an even index, it is possible to obtain solutions that result in a negative radicand. We do not have any restrictions on the value of the radicand when the index is odd, so we don't need to be worried about extraneous solutions. **111.** 0, −4, 12 **113.** $\sqrt{2^3}$, π, $\sqrt[3]{-4}$ **115.** A rational number is a number that can be written as the quotient of two integers, where the denominator is not zero; a rational number is also a number where the decimal either terminates, or does not terminate but repeats. An irrational number has a decimal form that neither terminates nor repeats. The square root of −1 is not real because there is no real number whose square is −1.

117.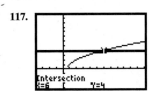

119.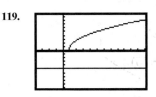

The two graphs do not intersect. Therefore, the equation has no real solution.

Section 6.8 The Complex Number System **1.** imaginary unit **2.** pure imaginary number **3.** $\sqrt{N}i$ **4.** True **5.** $6i$ **6.** $\sqrt{5}i$ **7.** $2\sqrt{3}i$ **8.** $4 + 10i$
9. $-2 - 2\sqrt{2}i$ **10.** $2 - 2\sqrt{2}i$ **11.** $1 + 11i$ **12.** $6 - 9i$ **13.** $-3 + i$ **14.** $12 + 15i$ **15.** $2 + 24i$ **16.** -18 **17.** $38 + 14i$ **18.** $-3 - 5i$ **19.** 73
20. 29 **21.** $\dfrac{1}{3} + \dfrac{4}{3}i$ **22.** $-\dfrac{1}{2} + \dfrac{3}{2}i$ **23.** $-i$ **24.** -1 **25.** $2i$ **27.** $-9i$ **29.** $3\sqrt{5}i$ **31.** $10\sqrt{3}i$ **33.** $\sqrt{7}i$ **35.** $5 + 7i$ **37.** $-2 - 2\sqrt{7}i$ **39.** $2 + i$
41. $\dfrac{1}{3} + \dfrac{\sqrt{2}}{6}i$ **43.** $6 - 2i$ **45.** $-4 + 6i$ **47.** $2 - 5i$ **49.** $3 - 2\sqrt{2}i$ **51.** $24 + 12i$ **53.** $-5 - 2i$ **55.** $5 + 10i$ **57.** $14 - 22i$ **59.** 26 **61.** $-4 + 5\sqrt{2}i$
63. $\dfrac{25}{48} + \dfrac{5}{24}i$ **65.** $5 + 12i$ **67.** $-9 + 40i$ **69.** -6 **71.** $-4\sqrt{5}$ **73.** $84 - 47i$ **75. (a)** $3 - 5i$ **(b)** 34 **77. (a)** $2 + 7i$ **(b)** 53 **79. (a)** $-7 - 2i$
(b) 53 **81.** $\dfrac{1}{3} - \dfrac{1}{3}i$ **83.** $\dfrac{2}{5} + i$ **85.** $\dfrac{6}{5} - \dfrac{3}{5}i$ **87.** $\dfrac{3}{29} - \dfrac{7}{29}i$ **89.** i **91.** $1 + 3i$ **93.** $-\dfrac{1}{5} - \dfrac{7}{5}i$ **95.** i **97.** $-i$ **99.** i **101.** $-i$ **103.** $-15 - 8i$ **105.** $5 + 12i$

107. $\frac{2}{3} + i$ **109.** $\frac{1}{2} - \frac{1}{2}i$ **111.** 12 **113.** $-15 - 20i$ **115.** $-5\sqrt{6}$ **117.** $-\frac{1}{5}i$ **119.** $\frac{2}{5} + \frac{1}{5}i$ **121.** $-\frac{4}{41} - \frac{5}{41}i$ **123. (a)** -1 **(b)** $2i$ **125. (a)** $-7 + 6i$
(b) $4 - 4i$ **127. (a)** $10 - i$ ohms **(b)** 10 ohms **(c)** -1 ohm **129. (a)** 0 **(b)** 0 **131. (a)** 0 **(b)** 0 **(c)** 0 **133.** For a polynomial with real coefficients, the zeros will be real numbers or will occur in conjugate pairs. If the complex number $a + bi$ is a complex zero of the polynomial, then its conjugate $a - bi$ is also a complex zero. **135.** The set of natural numbers is a subset of the set of whole numbers, which is a subset of the set of integers, which is a subset of the set of rational numbers, which is a subset of the set of real numbers, which is a subset of the set of complex numbers. Thus all real numbers are, more generally, complex numbers. **137.** Both methods can rely on the "FOIL" method or the Distributive Property. **139.** $x^3 + 6x^2 + 12x + 8$ **141.** $18 + 26i$
143. Both methods rely on special product formulas to obtain the results. **145.** $-9.9 + 7.2i$ **147.** $-30.86 + 3.59i$ **149.** $-\frac{11}{17} + \frac{27}{17}i$ **151.** $6 - 25i$

Chapter 6 Review **1.** 7 **2.** -5 **3.** $\frac{2}{3}$ **4.** 3 **5.** 3 **6.** 10 **7.** z **8.** $|5p - 3|$ **9.** 9 **10.** not a real number **11.** -2 **12.** 9 **13.** 128 **14.** -9 **15.** -1331 **16.** 6

17. -4.02 **18.** 2.30 **19.** 4.64 **20.** 1.78 **21.** $(5a)^{1/3}$ **22.** $p^{7/5}$ **23.** $(10z)^{3/4}$ **24.** $(2ab)^{5/6}$ **25.** 64 **26.** $\frac{1}{k^{1/4}}$ **27.** $p^2 \cdot q^6$ **28.** $\frac{2b^{1/20}}{a^{3/10}}$
29. $10m^{1/3} + \frac{5}{m}$ **30.** $\frac{1}{2x^{1/3}}$ **31.** $\sqrt[4]{x^3}$ **32.** $11x^2y^5$ **33.** $m^2\sqrt[6]{m}$ **34.** $\frac{1}{\sqrt[3]{c}}$ **35.** $(3m - 1)^{1/4}(3m^2 - 22m + 9)$ **36.** $\frac{(3x + 5)(x - 3)}{(x^2 - 5)^{2/3}}$ **37.** $\sqrt{105}$
38. $\sqrt[4]{12a^3b^3}$ **39.** $4\sqrt{5}$ **40.** $-5\sqrt[3]{4}$ **41.** $3m^2n\sqrt[3]{6n}$ **42.** $p^2|q|\sqrt[4]{50}$ **43.** $8|x^3|\sqrt{y}$ as long as $y \geq 0$ **44.** $(2x + 1)\sqrt{2x + 1}$ as long as $2x + 1 \geq 0$
45. $wz\sqrt{w}$ **46.** $3x^2z\sqrt{5yz}$ **47.** $2a^4b\sqrt[3]{2b^2}$ **48.** $2(x + 1)$ **49.** $3\sqrt{30}$ **50.** $2\sqrt[3]{75}$ **51.** $-2x^2y^3\sqrt[3]{9x}$ **52.** $30xy\sqrt{3xy}$ **53.** $\frac{11}{5}$ **54.** $\frac{a^2\sqrt{5}}{8b}$ **55.** $\frac{\sqrt[3]{6}}{k}$
56. $\frac{-2w^5\sqrt[3]{20}}{7}$ **57.** $2h$ **58.** $\frac{5b^3}{2a}$ **59.** $\frac{-2x^2}{3y}$ **60.** $\frac{2n\sqrt{n}}{m}$ **61.** $\sqrt[6]{500}$ **62.** $2\sqrt[12]{2}$ **63.** $8\sqrt[4]{x}$ **64.** $6\sqrt[3]{4y}$ **65.** $5\sqrt{2} - 4\sqrt{3}$ **66.** $13\sqrt{2}$
67. $\sqrt[3]{2z}$ **68.** $17\sqrt[3]{x^2}$ **69.** $7\sqrt{a}$ **70.** $9x\sqrt{3}$ **71.** $4m\sqrt[3]{4m^2y^2}$ **72.** $(y - 1)\sqrt{y - 4}$ **73.** $\sqrt{15} - 3\sqrt{5}$ **74.** $3\sqrt[3]{5} + \sqrt[3]{20}$ **75.** $7 + \sqrt{5}$
76. $42 + 7\sqrt{2} + 6\sqrt{3} + \sqrt{6}$ **77.** -44 **78.** $9\sqrt[3]{x^2} + 5\sqrt[3]{x} - 4$ **79.** $x - 2\sqrt{5x} + 5$ **80.** $247 + 22\sqrt{10}$ **81.** $2a - b^2$ **82.** $\sqrt[3]{36s^2} - 5\sqrt[3]{6s} - 14$
83. $\frac{\sqrt{6}}{3}$ **84.** $2\sqrt{3}$ **85.** $\frac{4\sqrt{3p}}{p^2}$ **86.** $\frac{5\sqrt{2a}}{2a}$ **87.** $-\frac{\sqrt{6y}}{3y^2}$ **88.** $\frac{3\sqrt[3]{25}}{5}$ **89.** $-\frac{\sqrt[3]{300}}{15}$ **90.** $\frac{27\sqrt[3]{4p^2q}}{2pq}$ **91.** $\frac{42 + 6\sqrt{6}}{43}$ **92.** $-\frac{\sqrt{3} + 9}{26}$
93. $\frac{\sqrt{3}(3 - \sqrt{2})}{7}$ or $\frac{3\sqrt{3} - \sqrt{6}}{7}$ **94.** $\frac{k + \sqrt{km}}{k - m}$ **95.** $\frac{7 + 2\sqrt{10}}{3}$ **96.** $\frac{9 - 6\sqrt{y} + y}{9 - y}$ or $\frac{y - 6\sqrt{y} + 9}{9 - y}$ **97.** $\frac{10\sqrt{2} - 4\sqrt{3}}{19}$
98. $\frac{8\sqrt{2} - 3\sqrt{15}}{7}$ **99.** $\frac{25\sqrt{7}}{21}$ **100.** $\frac{4 + \sqrt{7}}{9}$ **101. (a)** 1 **(b)** 2 **(c)** 3 **102. (a)** 0 **(b)** 2 **(c)** 4 **103. (a)** 1 **(b)** -1 **(c)** 2 **104. (a)** 0 **(b)** 2
(c) $\frac{1}{2}$ **105.** $\left\{x \mid x \geq \frac{5}{3}\right\}$ or $\left[\frac{5}{3}, \infty\right)$ **106.** $\{x \mid x$ is any real number$\}$ or $(-\infty, \infty)$ **107.** $\left\{x \mid x \geq -\frac{1}{6}\right\}$ or $\left[-\frac{1}{6}, \infty\right)$ **108.** $\{x \mid x$ is any real number$\}$ or $(-\infty, \infty)$
109. $\{x \mid x > 2\}$ or $(2, \infty)$ **110.** $\{x \mid x < 0$ or $x \geq 3\}$ or $(-\infty, 0) \cup [3, \infty)$

111. (a) $\{x \mid x \leq 1\}$ or $(-\infty, 1]$
(b)

(c) $[0, \infty)$

112. (a) $\{x \mid x \geq -1\}$ or $[-1, \infty)$
(b)

(c) $[-2, \infty)$

113. (a) $\{x \mid x \geq -3\}$ or $[-3, \infty)$
(b)
(c) $(-\infty, 0]$

114. (a) $\{x \mid x$ is any real number$\}$ or $(-\infty, \infty)$
(b)

(c) $(-\infty, \infty)$

115. $\{169\}$ **116.** $\{-3\}$ **117.** $\left\{\frac{89}{3}\right\}$ **118.** $\varnothing$ or $\{ \}$ **119.** $\{-5\}$ **120.** $\{25\}$ **121.** $\{-512\}$ **122.** $\{2\}$ **123.** $\{5\}$ **124.** $\{2\}$ **125.** $\{11\}$ **126.** $\{6\}$ **127.** $\{9\}$
128. $\varnothing$ or $\{ \}$ **129.** $\left\{\frac{15}{2}\right\}$ **130.** $\{-5, 5\}$ **131.** $h = \frac{3V}{\pi r^2}$ **132.** $v = \frac{30}{f_s^3}$ **133.** $\sqrt{29}i$ **134.** $3\sqrt{6}i$ **135.** $14 - 9\sqrt{2}i$ **136.** $2 + \sqrt{5}i$ **137.** $1 - 2i$
138. $-5 + 10i$ **139.** $5 - 7\sqrt{5}i$ **140.** $-5 + 7i$ **141.** $47 + 13i$ **142.** $8 - \frac{11}{6}i$ **143.** -9 **144.** $67 - 42i$ **145.** 145 **146.** $27 + 38i$ **147.** $\frac{6}{17} - \frac{10}{17}i$
148. $-\frac{21}{53} - \frac{6}{53}i$ **149.** $\frac{4}{29} - \frac{19}{29}i$ **150.** $\frac{1}{2} + \frac{7}{2}i$ **151.** $-i$ **152.** i

Chapter 6 Test **1.** $\frac{1}{7}$ **2.** $6y\sqrt[12]{x^8y^3}$ **3.** $2a^5b^4\sqrt[5]{4a^3b}$ **4.** $\sqrt{39mn}$ **5.** $4x^3y^2\sqrt{2x}$ **6.** $\frac{3a}{2b^2}$ **7.** $(x + 6)\sqrt{5x}$ **8.** $a\sqrt{b}$
9. $33 - 5\sqrt{x} - 2x$ **10.** $-\frac{\sqrt{2}}{18}$ **11.** $5 - 2\sqrt{5}$ **12. (a)** 1 **(b)** 3 **13.** $\left\{x \mid x \leq \frac{5}{3}\right\}$ or $\left(-\infty, \frac{5}{3}\right]$ **14. (a)** $\{x \mid x \geq 0\}$ or $[0, \infty)$ **(b)**

(c) $[-3, \infty)$ **15.** $\{13\}$ **16.** $\{3\}$ **17.** $\{2\}$ **18.** $17 - 13i$ **19.** $29 - 2i$ **20.** $\frac{73}{265} - \frac{89}{265}i$

Chapter 7 Quadratic Equations and Functions

Section 7.1 Solving Quadratic Equations by Completing the Square **1.** $\sqrt{p}; -\sqrt{p}$ **2.** $\{-4\sqrt{3}, 4\sqrt{3}\}$ **3.** $\{-5, 5\}$ **4.** $\{-9, 9\}$ **5.** $\{-6\sqrt{2}i, 6\sqrt{2}i\}$
6. $\{-3i, 3i\}$ **7.** $\{-13, 7\}$ **8.** $\{5 - 4i, 5 + 4i\}$ **9.** 49; $(p + 7)^2$ **10.** $\frac{9}{4}$; $\left(w + \frac{3}{2}\right)^2$ **11.** $\{-4, 2\}$ **12.** $\{4 - \sqrt{7}, 4 + \sqrt{7}\}$ **13.** $\left\{\frac{-3 - \sqrt{11}}{2}, \frac{-3 + \sqrt{11}}{2}\right\}$

14. $\left\{-\dfrac{1}{3}-\dfrac{2\sqrt{5}}{3}i,\ -\dfrac{1}{3}+\dfrac{2\sqrt{5}}{3}i\right\}$ **15.** hypotenuse; legs **16.** False **17.** $c=5$ **18.** Approximately 17.32 miles **19.** $\{-10,10\}$ **21.** $\{-5\sqrt{2},5\sqrt{2}\}$

23. $\{-5i,5i\}$ **25.** $\left\{-\dfrac{\sqrt{5}}{2},\dfrac{\sqrt{5}}{2}\right\}$ **27.** $\{-2\sqrt{2},2\sqrt{2}\}$ **29.** $\{-4,4\}$ **31.** $\left\{-\dfrac{2\sqrt{6}}{3},\dfrac{2\sqrt{6}}{3}\right\}$ **33.** $\{-2i,2i\}$ **35.** $\{1-3\sqrt{2}i,1+3\sqrt{2}i\}$

37. $\{-5-\sqrt{3},-5+\sqrt{3}\}$ **39.** $\left\{-\dfrac{4}{3},\dfrac{2}{3}\right\}$ **41.** $\left\{\dfrac{2}{3}-\dfrac{\sqrt{5}}{3},\dfrac{2}{3}+\dfrac{\sqrt{5}}{3}\right\}$ **43.** $\{-13,5\}$ **45.** $x^2+10x+25;\ (x+5)^2$ **47.** $z^2-18z+81;\ (z-9)^2$

49. $y^2+7y+\dfrac{49}{4};\left(y+\dfrac{7}{2}\right)^2$ **51.** $w^2+\dfrac{1}{2}w+\dfrac{1}{16};\left(w+\dfrac{1}{4}\right)^2$ **53.** $\{-6,2\}$ **55.** $\{2-\sqrt{3},2+\sqrt{3}\}$ **57.** $\{2-i,2+i\}$ **59.** $\left\{-\dfrac{5}{2}-\dfrac{\sqrt{33}}{2},-\dfrac{5}{2}+\dfrac{\sqrt{33}}{2}\right\}$

61. $\{4-\sqrt{19},4+\sqrt{19}\}$ **63.** $\left\{\dfrac{1}{2}-\dfrac{\sqrt{11}}{2}i,\dfrac{1}{2}+\dfrac{\sqrt{11}}{2}i\right\}$ **65.** $\left\{-\dfrac{3}{2},4\right\}$ **67.** $\left\{1-\dfrac{\sqrt{3}}{3},1+\dfrac{\sqrt{3}}{3}\right\}$ **69.** $\left\{\dfrac{5}{4}-\dfrac{\sqrt{17}}{4},\dfrac{5}{4}+\dfrac{\sqrt{17}}{4}\right\}$

71. $\left\{-1-\dfrac{\sqrt{6}}{2}i,-1+\dfrac{\sqrt{6}}{2}i\right\}$ **73.** 10 **75.** 20 **77.** $5\sqrt{2};\ 7.07$ **79.** 2 **81.** $2\sqrt{34};\ 11.66$ **83.** $b=4\sqrt{3}\approx6.93$ **85.** $a=4\sqrt{5}\approx8.94$

87. $\{-3,9\};\ (-3,36),\ (9,36)$ **89.** $\{-2-3\sqrt{2},-2+3\sqrt{2}\};\ (-2-3\sqrt{2},18),(-2+3\sqrt{2},18)$ **91.** $4\sqrt{5}$ units **93.** approximately 104.403 yards **95.** approximately 31.623 feet **97. (a)** approximately 22.913 feet **(b)** 15 feet **99. (a)** 1 second **(b)** approximately 1.732 seconds **(c)** 2 seconds **101.** approximately 9.54% **103.** The triangle is a right triangle; the hypotenuse is 17. **105.** The triangle is not a right triangle.

107. $c^2=(m^2+n^2)^2=m^4+2m^2n^2+n^4$
$a^2+b^2=(m^2-n^2)^2+(2mn)^2$
$\qquad=m^4-2m^2n^2+n^4+4m^2n^2$
$\qquad=m^4+2m^2n^2+n^4$
Because c^2 and a^2+b^2 result in the same expression, a, b, and c are the lengths of the sides of a right triangle.

109. $\{-4,9\}$ **111.** $\left\{-1,\dfrac{1}{2}\right\}$ **113.** In both cases, the simpler equations are linear.

Section 7.2 Solving Quadratic Equations by the Quadratic Formula
1. $\dfrac{-b\pm\sqrt{b^2-4ac}}{2a}$ **2.** $\left\{-\dfrac{3}{2},3\right\}$ **3.** $\left\{-4,\dfrac{1}{2}\right\}$ **4.** False

5. $\left\{1-\dfrac{\sqrt{3}}{2},1+\dfrac{\sqrt{3}}{2}\right\}$ **6.** $\left\{\dfrac{5}{2}\right\}$ **7.** $\left\{2-\dfrac{\sqrt{10}}{2},2+\dfrac{\sqrt{10}}{2}\right\}$ **8.** $\{-1-5i,-1+5i\}$ **9.** discriminant **10.** False **11.** negative **12.** False **13.** True

14. Two complex solutions that are not real **15.** One repeated real solution **16.** Two irrational solutions **17.** True **18.** $\{-3,3\}$

19. $\left\{\dfrac{5}{4}-\dfrac{\sqrt{23}}{4}i,\dfrac{5}{4}+\dfrac{\sqrt{23}}{4}i\right\}$ **20.** $\left\{-\dfrac{5}{3},1\right\}$ **21. (a)** 200 or 600 DVDs **(b)** 400 DVDs **22.** 16 meters by 30 meters **23.** $\{-2,6\}$ **25.** $\left\{-\dfrac{3}{2},\dfrac{5}{3}\right\}$

27. $\left\{1-\dfrac{\sqrt{3}}{2},1+\dfrac{\sqrt{3}}{2}\right\}$ **29.** $\left\{1-\dfrac{2\sqrt{3}}{3},1+\dfrac{2\sqrt{3}}{3}\right\}$ **31.** $\left\{-\dfrac{1}{3}-\dfrac{\sqrt{13}}{3},-\dfrac{1}{3}+\dfrac{\sqrt{13}}{3}\right\}$ **33.** $\{1-\sqrt{6}i,1+\sqrt{6}i\}$ **35.** $\left\{\dfrac{1}{2}-\dfrac{\sqrt{13}}{2}i,\dfrac{1}{2}+\dfrac{\sqrt{13}}{2}i\right\}$

37. $\left\{\dfrac{1}{4}-\dfrac{\sqrt{5}}{4},\dfrac{1}{4}+\dfrac{\sqrt{5}}{4}\right\}$ **39.** $\left\{-\dfrac{2}{3}-\dfrac{\sqrt{7}}{3},-\dfrac{2}{3}+\dfrac{\sqrt{7}}{3}\right\}$ **41.** 21; two irrational solutions **43.** -56; two complex solutions that are not real **45.** 0; one repeated real solution **47.** -8; two complex solutions that are not real **49.** 44; two irrational solutions **51.** $\left\{\dfrac{5}{2}-\dfrac{\sqrt{5}}{2},\dfrac{5}{2}+\dfrac{\sqrt{5}}{2}\right\}$

53. $\left\{-\dfrac{8}{3},1\right\}$ **55.** $\left\{-\dfrac{7}{2},5\right\}$ **57.** $\{-1-\sqrt{7}i,-1+\sqrt{7}i\}$ **59.** $\left\{-\dfrac{3}{2}\right\}$ **61.** $\left\{\dfrac{1}{7}-\dfrac{\sqrt{29}}{7},\dfrac{1}{7}+\dfrac{\sqrt{29}}{7}\right\}$ **63.** $\{-4,4\}$ **65.** $\left\{\dfrac{1}{4}-\dfrac{1}{4}i,\dfrac{1}{4}+\dfrac{1}{4}i\right\}$ **67.** $\left\{-\dfrac{2}{3}\right\}$

69. $\left\{-\dfrac{1}{3}-\dfrac{2\sqrt{7}}{3},-\dfrac{1}{3}+\dfrac{2\sqrt{7}}{3}\right\}$ **71.** $\{2-\sqrt{13},2+\sqrt{13}\}$ **73.** $\{1-\sqrt{5},1+\sqrt{5}\}$ **75.** $\left\{\dfrac{1}{4}-\dfrac{3\sqrt{7}}{4}i,\dfrac{1}{4}+\dfrac{3\sqrt{7}}{4}i\right\}$ **77. (a)** $\{-7,3\}$

(b) $\{-4,0\};\ (-4,-21),\ (0,-21)$ **79. (a)** $\left\{-1-\dfrac{\sqrt{6}}{2},-1+\dfrac{\sqrt{6}}{2}\right\}$ **(b)** $\left\{-1-\dfrac{\sqrt{2}}{2},-1+\dfrac{\sqrt{2}}{2}\right\}$ **81.** $\dfrac{-1-\sqrt{7}}{3},\dfrac{-1+\sqrt{7}}{3}$ **83.** $x=3$; the three sides measure 3, 4, and 5 units **85.** Either $x=1$ and the three sides measure 3, 4, and 5 units, or $x=5$ and the three sides measure 7, 24, and 25 units.

87. $-2+2\sqrt{11}$ inches by $2+2\sqrt{11}$ inches, which is approximately 4.633 inches by 8.633 inches. **89.** The base is $\dfrac{3}{2}+\dfrac{\sqrt{209}}{2}$ inches, which is approximately 8.728 inches; the height is $-\dfrac{3}{2}+\dfrac{\sqrt{209}}{2}$ inches, which is approximately 5.728 inches. **91. (a)** $R(17)=1161.1$; if 17 pairs of sunglasses are sold per week, then the company's revenue will be \$1161.10. $R(25)=1687.5$; if 25 pairs of sunglasses are sold per week, then the company's revenue will be \$1687.50. **(b)** either 200 or 500 pairs of sunglasses **(c)** 350 pairs of sunglasses **93. (a)** after approximately 0.6 second and after approximately 3.8 seconds **(b)** after approximately 1.3 seconds and after approximately 3.0 seconds **(c)** No; the solutions to the equation are complex solutions that are not real. **95.** 12 inches **97. (a)** ages 25 and 68 **(b)** ages 30 and 63 **99.** approximately 4.3 miles per hour **101.** approximately 4.6 hours

103. By the quadratic formula, the solutions of the equation
$ax^2+bx+c=0$ are $x=\dfrac{-b-\sqrt{b^2-4ac}}{2a}$ and
$x=\dfrac{-b+\sqrt{b^2-4ac}}{2a}$.
The sum of these two solutions is
$\dfrac{-b-\sqrt{b^2-4ac}}{2a}+\dfrac{-b+\sqrt{b^2-4ac}}{2a}=\dfrac{-2b}{2a}=-\dfrac{b}{a}$.

105. The solutions of $ax^2+bx+c=0$ are $x=\dfrac{-b\pm\sqrt{b^2-4ac}}{2a}$. The solutions of $ax^2-bx+c=0$ are $x=\dfrac{-(-b)\pm\sqrt{(-b)^2-4ac}}{2a}=\dfrac{b\pm\sqrt{b^2-4ac}}{2a}$.
Now, the negatives of the solutions to $ax^2-bx+c=0$
are $-\left(\dfrac{b\pm\sqrt{b^2-4ac}}{2a}\right)=\dfrac{-b\mp\sqrt{b^2-4ac}}{2a}=\dfrac{-b\pm\sqrt{b^2-4ac}}{2a}$,
which are the solutions to $ax^2+bx+c=0$.

107. Use factoring if the discriminant is a perfect square.

109. (a)

(b) $x = -1$ or $x = -2$

111. (a)

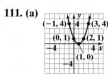

(b) $x = 1$

(c) The x-intercepts of the function $f(x) = x^2 + 3x + 2$ are -2 and -1, which are the same as the solutions of the equation $x^2 + 3x + 2 = 0$.

(c) The x-intercept of the function $g(x) = x^2 - 2x + 1$ is 1, which is the same as the solution of the equation $x^2 - 2x + 1 = 0$.

113. The discriminant is 37; the equation has two irrational solutions. This conclusion based on the discriminant is apparent in the graph because the graph has two x-intercepts.   **115.** The discriminant is -7; the equation has two complex solutions that are not real. This conclusion based on the discriminant is apparent in the graph because the graph has no x-intercepts.

117. (a) $x = -3$ or $x = 8$
  (b) The x-intercepts are -3 and 8.

119. (a) $x = 3$
  (b) The x-intercept is 3.

121. (a) $x = -\dfrac{5}{2} \pm \dfrac{\sqrt{7}}{2}i$

  (b) The graph has no x-intercepts.

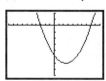

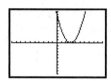

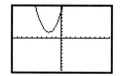

The x-intercepts of $y = x^2 - 5x - 24$ are the same as the solutions of $x^2 - 5x - 24 = 0$.

The x-intercept of $y = x^2 - 6x + 9$ is the same as the solution of $x^2 - 6x + 9 = 0$.

$y = x^2 + 5x + 8$ has no x-intercepts, and the solutions of $x^2 + 5x + 8 = 0$ are not real.

Section 7.3 Solving Equations Quadratic in Form
1. quadratic in form   **2.** $3x + 1$ **3.** True   **4.** $u = \dfrac{1}{x}$ **5.** $\{-3, -2, 2, 3\}$ **6.** $\{-3, 3, \sqrt{2}i, -\sqrt{2}i\}$

7. $\{-3, -2, 2, 3\}$ **8.** $\left\{-\dfrac{\sqrt{3}}{2}, \dfrac{\sqrt{3}}{2}, -i, i\right\}$ **9.** $\left\{\dfrac{4}{9}, 16\right\}$ **10.** $\{25\}$ **11.** $\left\{-\dfrac{5}{2}, -\dfrac{1}{2}\right\}$ **12.** $\{-1, 125\}$ **13.** $\{-2, -1, 1, 2\}$ **15.** $\{-3i, 3i, -2i, 2i\}$

17. $\left\{-2, -\dfrac{1}{2}, \dfrac{1}{2}, 2\right\}$ **19.** $\{-\sqrt{3}, -\sqrt{2}, \sqrt{2}, \sqrt{3}\}$ **21.** $\{2, 10\}$ **23.** $\{-3, -2, 2, 3\}$ **25.** $\{-2i, 2i, -\sqrt{7}i, \sqrt{7}i\}$ **27.** $\{16\}$ **29.** $\varnothing$ or $\{\}$ **31.** $\left\{\dfrac{1}{4}\right\}$ **33.** $\left\{-\dfrac{1}{7}, \dfrac{1}{4}\right\}$

35. $\left\{-\dfrac{2}{3}, \dfrac{5}{2}\right\}$ **37.** $\{-64, 1\}$ **39.** $\{-1, 8\}$ **41.** $\{25\}$ **43.** $\left\{\dfrac{1}{3}, \dfrac{1}{2}\right\}$ **45.** $\left\{-\dfrac{11}{5}, -1\right\}$ **47.** $\left\{1, 3, -\dfrac{1}{2} - \dfrac{\sqrt{3}}{2}i, -\dfrac{1}{2} + \dfrac{\sqrt{3}}{2}i, \dfrac{3}{2} - \dfrac{3\sqrt{3}}{2}i, \dfrac{3}{2} + \dfrac{3\sqrt{3}}{2}i\right\}$

49. $\{-2, 4\}$ **51.** $\{-2i, 2i, -2\sqrt{2}, 2\sqrt{2}\}$ **53.** $\{16\}$ **55.** $\{-3, 3, -2i, 2i\}$ **57.** $\left\{-\dfrac{10}{3}, -2\right\}$ **59.** $\{9, 16\}$ **61.** $\left\{\dfrac{1}{2}, 5\right\}$ **63. (a)** $0, -\sqrt{7}i, \sqrt{7}i$

(b) $-\sqrt{6}i, \sqrt{6}i, -i, i$ **65. (a)** $0, -\sqrt{3}, \sqrt{3}$ **(b)** $-\sqrt{2}i, \sqrt{2}i, -\sqrt{5}, \sqrt{5}$ **67. (a)** $-1, \dfrac{1}{6}$ **(b)** $-\dfrac{1}{2}, \dfrac{1}{7}$ **69.** $-\sqrt{7}i, \sqrt{7}i, -\sqrt{2}i, \sqrt{2}i$ **71.** $\dfrac{81}{4}$ **73.** $-\dfrac{8}{3}, -2$

75. (a) $x = 2$ or $x = 3$ **(b)** $x = 5$ or $x = 6$; comparing these solutions to those in part (a), we note that $5 = 2 + 3$ and $6 = 3 + 3$.   **(c)** $x = 0$ or $x = 1$; comparing these solutions to those in part (a), we note that $0 = 2 - 2$ and $1 = 3 - 2$.   **(d)** $x = 7$ or $x = 8$; comparing these solutions to those in part (a), we note that $7 = 2 + 5$ and $8 = 3 + 5$.   **(e)** The solution set of the equation $(x - a)^2 - 5(x - a) + 6 = 0$ is $\{2 + a, 3 + a\}$.

77. (a) $x = \dfrac{1}{2}$ or $x = 1$ **(b)** $x = \dfrac{5}{2}$ or $x = 3$; comparing these solutions to those in part (a), we note that $\dfrac{5}{2} = \dfrac{1}{2} + 2$ and $3 = 1 + 2$.

(c) $x = \dfrac{11}{2}$ or $x = 6$; comparing these solutions to those in part (a), we note that $\dfrac{11}{2} = \dfrac{1}{2} + 5$ and $6 = 1 + 5$.   **(d)** For $f(x) = 2x^2 - 3x + 1$, the zeros of $f(x - a)$ are $\dfrac{1}{2} + a$ and $1 + a$ **79. (a)** $R(1990) = 3000$; the revenue in 1990 was $3000 thousand (or \$3,000,000)   **(b)** $x = 2000$; in the year 2000, revenue was \$3065 thousand (or \$3,065,000)   **(c)** 2015   **81.** $x = \pm\dfrac{\sqrt{10 + 2\sqrt{17}}}{2}i$ or $x = \pm\dfrac{\sqrt{10 - 2\sqrt{17}}}{2}i$ **83.** $x = \pm\dfrac{3\sqrt{2}}{2}$ **85.** $\{36\}$; Answers will vary.   **87.** Extraneous solutions may result after squaring both sides of the equation.   **89.** $p^{-2} + 4p^{-1} + 9$

91. $5\sqrt[3]{2a} - 4a\sqrt[3]{2a} = (5 - 4a)\sqrt[3]{2a}$

93. Let $Y_1 = x^4 + 5x^2 - 14$.

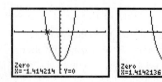

95. Let $Y_1 = 2(x - 2)^2$ and $Y_2 = 5(x - 2) + 1$.

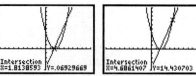

97. Let $Y_1 = x - 5\sqrt{x}$ and $Y_2 = -3$.

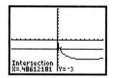

The solution set is approximately $\{-1.41, 1.41\}$.

The solution set is approximately $\{1.81, 4.69\}$.

The solution set is approximately $\{0.49\}$.

99. (a) $Y_1 = x^2 - 5x - 6$

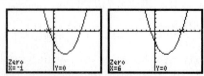

(b) $Y_1 = (x + 2)^2 - 5(x + 2) - 6$

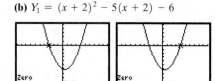

(c) $Y_1 = (x + 5)^2 - 5(x + 5) - 6$

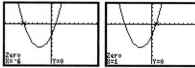

The x-intercepts are -1 and 6.

The x-intercepts are -3 and 4.

The x-intercepts are -6 and 1.

(d) The x-intercepts of the graph of $y = f(x) = x^2 - 5x - 6$ are -1 and 6. The x-intercepts of the graph of $y = f(x + a) = (x + a)^2 - 5(x + a) - 6$ are $-1 - a$ and $6 - a$.

Putting the Concepts Together (Sections 7.1–7.3) **1.** $z^2 + 10z + 25 = (z + 5)^2$ **2.** $x^2 + 7x + \dfrac{49}{4} = \left(x + \dfrac{7}{2}\right)^2$ **3.** $n^2 - \dfrac{1}{4}n + \dfrac{1}{64} = \left(n - \dfrac{1}{8}\right)^2$

4. $\left\{\dfrac{1}{2}, \dfrac{5}{2}\right\}$ **5.** $\left\{-4 - 2\sqrt{3}, -4 + 2\sqrt{3}\right\}$ **6.** $\left\{3 - \sqrt{2}, 3 + \sqrt{2}\right\}$ **7.** $\left\{-\dfrac{4\sqrt{5}}{7}, \dfrac{4\sqrt{5}}{7}\right\}$ **8.** $\left\{4 - \sqrt{10}, 4 + \sqrt{10}\right\}$ **9.** $\left\{-1 - \dfrac{\sqrt{3}}{3}i, -1 + \dfrac{\sqrt{3}}{3}i\right\}$

10. $\left\{-2 - \dfrac{\sqrt{42}}{3}, -2 + \dfrac{\sqrt{42}}{3}\right\}$ **11.** $b^2 - 4ac = 0$; the quadratic equation has one repeated real solution.   **12.** $b^2 - 4ac = 60$; the quadratic equation has two irrational solutions.   **13.** $b^2 - 4ac = -4$; the quadratic equation has two complex solutions that are not real.

14. $c = \sqrt{116} = 2\sqrt{29}$ **15.** $\left\{\dfrac{9}{4}\right\}$ **16.** $\left\{\dfrac{1}{6}, -\dfrac{1}{3}\right\}$ **17.** Revenue will be $12,000 when either 150 microwaves or 200 microwaves are sold.

18. The speed of the wind was approximately 52.9 miles per hour.

Section 7.4 Graphing Quadratic Functions Using Transformations   **1.** quadratic function   **2.** up; down

3.   **4.**   **5.** False   **6.**   **7.** 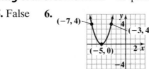   **8.**

9. 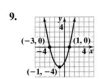   **10.** y; a; vertically stretched; vertically compressed   **11.**   **12.** 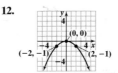   **13.** True

14.   Domain: $\{x \mid x$ is any real number$\}$ or $(-\infty, \infty)$; Range: $\{y \mid y \le 1\}$ or $(-\infty, 1]$

15.   Domain: $\{x \mid x$ is any real number$\}$ or $(-\infty, \infty)$ Range: $\{y \mid y \ge -3\}$ or $[-3, \infty)$ **16.** $f(x) = -(x + 1)^2 + 2$ **17.** (I) (D) (II) (A) (III) (C) (IV) (B)   **19.** shift 10 units to the left   **21.** shift 12 units up   **23.** vertically stretch by a factor of 2 (multiply the y-coordinates by 2) and shift 5 units to the right   **25.** multiply the y-coordinates by -3 (which means it opens down and is stretched vertically by a factor of 3), shift 5 units to the left, and shift up 8 units

27.   **29.**   **31.**   **33.**   **35.**   **37.**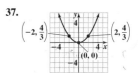

39. 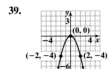   **41.**   **43.** 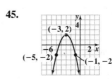   **45.** 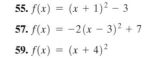   **47.**

49. 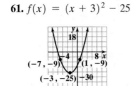   **51.** $f(x) = (x + 1)^2 - 5$ **53.** $g(x) = (x - 2)^2 + 4$ **55.** $f(x) = (x + 1)^2 - 3$

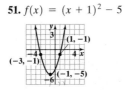   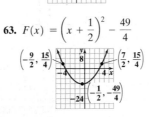   **57.** $f(x) = -2(x - 3)^2 + 7$

59. $f(x) = (x + 4)^2$

61. $f(x) = (x + 3)^2 - 25$

   vertex is $(-3, -25)$; axis of symmetry is $x = -3$; domain is the set of all real numbers or $(-\infty, \infty)$; range is $\{y \mid y \ge -25\}$ or $[-25, \infty)$

63. $F(x) = \left(x + \dfrac{1}{2}\right)^2 - \dfrac{49}{4}$

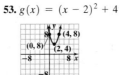   vertex is $\left(-\dfrac{1}{2}, -\dfrac{49}{4}\right)$; axis of symmetry is $x = -\dfrac{1}{2}$ domain is the set of all real numbers or $(-\infty, \infty)$; range is $\left\{y \mid y \ge -\dfrac{49}{4}\right\}$ or $\left[-\dfrac{49}{4}, \infty\right)$

65. $H(x) = 2(x - 1)^2 - 3$

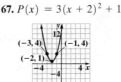   vertex is $(1, -3)$; axis of symmetry is $x = 1$; domain is the set of all real numbers or $(-\infty, \infty)$; range is $\{y \mid y \ge -3\}$ or $[-3, \infty)$ **67.** $P(x) = 3(x + 2)^2 + 1$

vertex is $(-2, 1)$; axis of symmetry is $x = -2$; domain is the set of all real numbers or $(-\infty, \infty)$; range is $\{y \mid y \ge 1\}$ or $[1, \infty)$

69. $F(x) = -(x + 5)^2 + 4$

vertex is $(-5, 4)$; axis of symmetry is $x = -5$; domain is the set of all real numbers or $(-\infty, \infty)$; range is $\{y \,|\, y \le 4\}$ or $(-\infty, 4]$

71. $g(x) = -(x - 3)^2 + 8$

vertex is $(3, 8)$; axis of symmetry is $x = 3$; domain is the set of all real numbers or $(-\infty, \infty)$; range is $\{y \,|\, y \le 8\}$ or $(-\infty, 8]$

73. $H(x) = -2(x - 2)^2 + 4$

vertex is $(2, 4)$; axis of symmetry is $x = 2$; domain is the set of all real numbers or $(-\infty, \infty)$; range is $\{y \,|\, y \le 4\}$ or $(-\infty, 4]$

75. $f(x) = \dfrac{1}{3}(x - 3)^2 + 1$

vertex is $(3, 1)$; axis of symmetry is $x = 3$; domain is the set of all real numbers or $(-\infty, \infty)$; range is $\{y \,|\, y \ge 1\}$ or $[1, \infty)$

77. $G(x) = -12\left(x + \dfrac{1}{2}\right)^2 + 4$

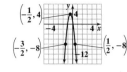

vertex is $\left(-\dfrac{1}{2}, 4\right)$; axis of symmetry is $x = -\dfrac{1}{2}$; domain is the set of all real numbers or $(-\infty, \infty)$; range is $\{y \,|\, y \le 4\}$ or $(-\infty, 4]$

79. Answers may vary. One possibility: $f(x) = (x - 3)^2$ **81.** Answers may vary. One possibility: $f(x) = (x + 3)^2 + 1$ **83.** Answers may vary. One possibility: $f(x) = -(x - 5)^2 - 1$ **85.** Answers may vary. One possibility: $f(x) = 4(x - 9)^2 - 6$ **87.** Answers may vary. One possibility: $f(x) = -\dfrac{1}{3}x^2 + 6$ **89.** The highest or lowest point on a parabola is called the vertex. If $a < 0$, the vertex is the highest point; if $a > 0$, the vertex is the lowest point.   **91.** No. Explanations may vary.   **93.** $29 + \dfrac{1}{12}$ **95.** $x^3 - 5x^2 + 4x - 2 + \dfrac{-2}{2x - 1}$

97.

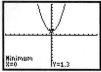

Vertex: $(0, 1.3)$ Axis of symmetry: $x = 0$ Range: $\{y \,|\, y \ge 1.3\}$ or $[1.3, \infty)$

99.

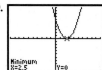

Vertex: $(2.5, 0)$
Axis of symmetry: $x = 2.5$
Range: $\{y \,|\, y \ge 0\}$ or $[0, \infty)$

101.

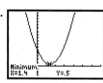

Vertex: $(1.4, 0.5)$
Axis of symmetry: $x = 1.4$
Range: $\{y \,|\, y \ge 0.5\}$ or $[0.5, \infty)$

103.

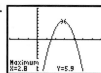

Vertex: $(2.8, 5.9)$
Axis of symmetry: $x = 2.8$
Range: $\{y \,|\, y \le 5.9\}$ or $(-\infty, 5.9]$

Section 7.5 Graphing Quadratic Functions Using Properties   **1.** $-\dfrac{b}{2a}$ **2.** $>$ **3.** 2   **4.** $(-2, -7)$

5.

6.

7.

8.

9. True   **10.** minimum; -7 at $x = 2$ **11.** maximum; 33 at $x = 5$ **12. (a)** The revenue will be maximized at a price of $75.   **(b)** The maximum revenue is $2812.50.   **13.** The maximum area that can be enclosed is 62,500 square feet. The dimensions are 250 feet by 250 feet.

14. The number of boxes that should be sold to maximize revenue is 65, and the maximum revenue is $4225.   **15. (a)** $(3, -25)$ **(b)** The discriminant is positive; there are two x-intercepts: $(-2, 0)$ and $(8, 0)$.   **17. (a)** $(1, -3)$ **(b)** The discriminant is negative; there are no x-intercepts.   **19. (a)** $\left(-\dfrac{1}{2}, 0\right)$

(b) The discriminant is zero; there is one x-intercept: $\left(-\dfrac{1}{2}, 0\right)$.   **21. (a)** $\left(\dfrac{1}{8}, -\dfrac{17}{16}\right)$ **(b)** The discriminant is positive; there are two x-intercepts: $(-0.39, 0)$ and $(0.64, 0)$.

23.

Domain: $\{x \,|\, x$ is any real number$\}$ or $(-\infty, \infty)$
Range: $\{y \,|\, y \ge -9\}$ or $[-9, \infty)$

25.

Domain: $\{x \,|\, x$ is any real number$\}$ or $(-\infty, \infty)$
Range: $\{y \,|\, y \ge -4\}$ or $[-4, \infty)$

27.

Domain: $\{x \,|\, x$ is any real number$\}$ or $(-\infty, \infty)$
Range: $\{y \,|\, y \le 9\}$ or $(-\infty, 9]$

29.

Domain: $\{x \,|\, x$ is any real number$\}$ or $(-\infty, \infty)$
Range: $\{y \,|\, y \ge 0\}$ or $[0, \infty)$

31.

Domain: {x | x is any real number} or (−∞, ∞)
Range: {y | y ≥ 4} or [4, ∞)

33.

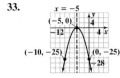

Domain: {x | x is any real number} or (−∞, ∞)
Range: {y | y ≤ 0} or (−∞, 0]

35.

Domain: {x | x is any real number} or (−∞, ∞)
Range: {y | y ≤ −4} or (−∞, −4]

37.

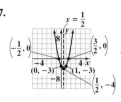

Domain: {x | x is any real number} or (−∞, ∞)
Range: {y | y ≥ −4} or [−4, ∞)

39.

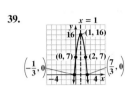

Domain: {x | x is any real number} or (−∞, ∞)
Range: {y | y ≤ 16} or (−∞, 16]

41.

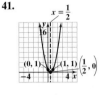

Domain: {x | x is any real number} or (−∞, ∞)
Range: {y | y ≥ 0} or [0, ∞)

43.

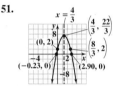

Domain: {x | x is any real number} or (−∞, ∞)
Range: {y | y ≤ 0} or (−∞, 0]

45.

Domain: {x | x is any real number} or (−∞, ∞)
Range: {y | y ≥ 3} or [3, ∞)

47.

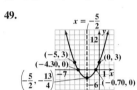

Domain: {x | x is any real number} or (−∞, ∞)
Range: $\left\{ y \mid y \le -\dfrac{3}{4} \right\}$ or $\left(-\infty, -\dfrac{3}{4} \right]$

49.

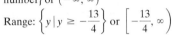

Domain: {x | x is any real number} or (−∞, ∞)
Range: $\left\{ y \mid y \ge -\dfrac{13}{4} \right\}$ or $\left[-\dfrac{13}{4}, \infty \right)$

51.

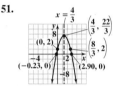

Domain: {x | x is any real number} or (−∞, ∞)
Range: $\left\{ y \mid y \le \dfrac{22}{3} \right\}$ or $\left(-\infty, \dfrac{22}{3} \right]$

53.

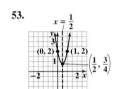

Domain: {x | x is any real number} or (−∞, ∞)
Range: $\left\{ y \mid y \ge \dfrac{3}{4} \right\}$ or $\left[\dfrac{3}{4}, \infty \right)$

55.

Domain: {x | x is any real number} or (−∞, ∞)
Range: {y | y ≤ 3} or (−∞, 3]

57.

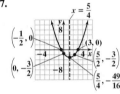

Domain: {x | x is any real number} or (−∞, ∞)
Range: $\left\{ y \mid y \ge -\dfrac{49}{16} \right\}$ or $\left[-\dfrac{49}{16}, \infty \right)$

59.

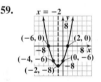

Domain: {x | x is any real number} or (−∞, ∞)
Range: {y | y ≥ −8} or [−8, ∞)

61.

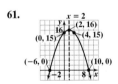

Domain: {x | x is any real number} or (−∞, ∞);
Range: {y | y ≤ 16} or (−∞, 16]

63. minimum; −3 at x = −4   **65.** maximum; 28 at x = −5   **67.** maximum; 23 at x = 3   **69.** minimum; −19 at x = −2   **71.** minimum; $-\dfrac{17}{8}$ at $x = \dfrac{5}{4}$ **73.** maximum; $\dfrac{7}{3}$ at $x = \dfrac{2}{3}$ **75. (a)** $120   **(b)** $36,000

77. 60; $35   **79. (a)** after 7.5 seconds   **(b)** 910 feet   **(c)** about 15.042 seconds   **81. (a)** about 1753.52 feet from the cannon   **(b)** about 886.76 feet   **(c)** about 3517 feet from the cannon   **(d)** The two answers are close. Explanations may vary.   **83. (a)** about 46.5 years   **(b)** $64,661.75   **85.** 18 and 18   **87.** −9 and 9   **89.** 15,625 square yards; 125 yards × 125 yards   **91.** 500,000 square meters; 500 m × 1000 m and the long side is parallel to the river   **93.** 5 inches   **95. (a)** $R = -p^2 + 110p$ **(b)** $55; $3025   **(c)** 55 pairs   **97. (a)** $f(x) = x^2 - 8x + 12$; $f(x) = 2x^2 - 16x + 24$; $f(x) = -2x^2 + 16x - 24$ **(b)** The value of a has no effect on the x-intercepts. **(c)** The value of a has no effect on the axis of symmetry.   **(d)** The x-coordinate of the vertex is 4, which does not depend on a. However, the y-coordinate is −4a, which does depend on a.   **99.** If the discriminant is positive, the equation $ax^2 + bx + c = 0$ will have two distinct real solutions, which means the graph of $f(x) = ax^2 + bx + c$ will have two x-intercepts. If the discriminant is zero, the equation $ax^2 + bx + c = 0$ will have one real solution, which means the graph of $f(x) = ax^2 + bx + c$ will have one x-intercept. If the discriminant is negative, the equation $ax^2 + bx + c = 0$ will have no real solutions, which means the graph of $f(x) = ax^2 + bx + c$ will have no x-intercepts.   **101.** If the price is too high, the quantity demanded will be 0.

103.

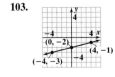

105.

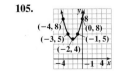

107. Vertex: (3.5, −9.25)

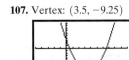

109. Vertex: (3.5, 37.5)

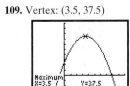

111. Vertex: (−0.3, −20.45)

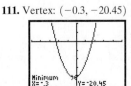

113. Vertex: (0.67, 4.78)   **115.** c is the y-intercept.

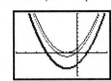

Section 7.6 Polynomial Inequalities   **1.** $\{x \mid x \le -5 \text{ or } x \ge 2\}$; $(-\infty, -5] \cup [2, \infty)$

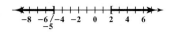

2. $\{x \mid x \le -5 \text{ or } x \ge 2\}$; $(-\infty, -5] \cup [2, \infty)$ **3.** $\{x \mid -6 < x < 4\}$; $(-6, 4)$

4. $\left\{x \mid x < \dfrac{-1 - \sqrt{61}}{6} \text{ or } x > \dfrac{-1 + \sqrt{61}}{6}\right\}$; $\left(-\infty, \dfrac{-1 - \sqrt{61}}{6}\right) \cup \left(\dfrac{-1 + \sqrt{61}}{6}, \infty\right)$

5. $\left\{x \mid x \le -4 \text{ or } \dfrac{3}{2} \le x \le 5\right\}$ or $(-\infty, -4] \cup \left[\dfrac{3}{2}, 5\right]$

6. $\{x \mid -7 < x < -2 \text{ or } x > 2\}$ or $(-7, -2) \cup (2, \infty)$

7. (a) $\{x \mid x < -6 \text{ or } x > 5\}$ or $(-\infty, -6) \cup (5, \infty)$ **(b)** $\{x \mid -6 \le x \le 5\}$ or $[-6, 5]$

9. (a) $\left\{x \mid -6 \le x \le \dfrac{5}{2}\right\}$ or $\left[-6, \dfrac{5}{2}\right]$ **(b)** $\left\{x \mid x < -6 \text{ or } x > \dfrac{5}{2}\right\}$ or $(-\infty, -6) \cup \left(\dfrac{5}{2}, \infty\right)$

11. $\{x \mid x \le -2 \text{ or } x \ge 5\}$ or $(-\infty, -2] \cup [5, \infty)$

13. $\{x \mid -7 < x < -3\}$ or $(-7, -3)$

15. $\{x \mid x < -5 \text{ or } x > 7\}$ or $(-\infty, -5) \cup (7, \infty)$

17. $\left\{n \mid 3 - \sqrt{17} \le n \le 3 + \sqrt{17}\right\}$ or $\left[3 - \sqrt{17}, 3 + \sqrt{17}\right]$

19. $\{m \mid m \le -7 \text{ or } m \ge 2\}$ or $(-\infty, -7] \cup [2, \infty)$

21. $\left\{q \mid q \le -\dfrac{5}{2} \text{ or } q \ge 3\right\}$ or $\left(-\infty, -\dfrac{5}{2}\right] \cup [3, \infty)$

23. $\{x \mid -1 \le x \le 4\}$ or $[-1, 4]$

25. $\{x \mid x < -2 \text{ or } x > 5\}$ or $(-\infty, -2) \cup (5, \infty)$

27. $\left\{x \mid x \le -\dfrac{2}{3} \text{ or } x \ge 4\right\}$ or $\left(-\infty, -\dfrac{2}{3}\right] \cup [4, \infty)$

29. $\left\{x \mid -2 - \sqrt{3} < x < -2 + \sqrt{3}\right\}$ or $\left(-2 - \sqrt{3}, -2 + \sqrt{3}\right)$

31. $\left\{a \mid -\dfrac{1}{2} \le a \le 4\right\}$ or $\left[-\dfrac{1}{2}, 4\right]$

33. $\{z \mid z \text{ is any real number}\}$ or $(-\infty, \infty)$

35. $\varnothing$ or $\{\}$

37. $\{x \mid x \ne 3\}$ or $(-\infty, 3) \cup (3, \infty)$

39. $\{x \mid -1 < x < 2 \text{ or } x > 5\}$ or $(-1, 2) \cup (5, \infty)$

41. $\left\{x \mid x \le -\dfrac{1}{2} \text{ or } 4 \le x \le 9\right\}$ or $\left(-\infty, -\dfrac{1}{2}\right] \cup [4, 9]$

43. $\left\{x \mid -3 \le x \le -\dfrac{1}{2} \text{ or } x \ge 1\right\}$ or $\left[-3, -\dfrac{1}{2}\right] \cup [1, \infty)$

45. $\{x \mid x \le -3 \text{ or } -2 \le x \le 2\}$ or $(-\infty, -3] \cup [-2, 2]$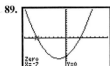

47. $\left\{x \mid -4 < x < -\dfrac{3}{2} \text{ or } x > \dfrac{3}{2}\right\}$ or $\left(-4, -\dfrac{3}{2}\right) \cup \left(\dfrac{3}{2}, \infty\right)$

49. $\{x \mid 0 < x < 5\}$ or $(0, 5)$ **51.** $\{x \mid x \le -4 \text{ or } x \ge 7\}$ or $(-\infty, -4] \cup [7, \infty)$ **53.** $\left\{x \mid x < -\dfrac{5}{2} \text{ or } x > 2\right\}$ or $\left(-\infty, -\dfrac{5}{2}\right) \cup (2, \infty)$

55. $\left\{x \mid x < -\dfrac{7}{4} \text{ or } 0 < x < 2\right\}$ or $\left(-\infty, -\dfrac{7}{4}\right) \cup (0, 2)$ **57.** $\{x \mid x \le -8 \text{ or } x \ge 0\}$ or $(-\infty, -8] \cup [0, \infty)$ **59.** $\{x \mid x \le -5 \text{ or } x \ge 6\}$ or

$(-\infty, -5] \cup [6, \infty)$ **61.** between 2 and 3 seconds after the ball is thrown   **63.** between \$110 and \$130   **65.** $x = -3$; a perfect square cannot be negative. Therefore, the only solution will be where the perfect square expression equals zero, which is -3.   **67.** all real numbers; a perfect square must always be zero or greater. Therefore, it must always be larger than -2.   **69.** Answers may vary. One possibility follows: $x^2 + x - 6 \le 0$

71. $(-\infty, -3) \cup (-1, 2) \cup (4, \infty)$ **73.** $[-5, -2] \cup [2, 5)$ **75.** $\{x \mid -3 < x < -2 \text{ or } x > 2\}$ or $(-3, -2) \cup (2, \infty)$ **77.** Answers will vary. One possibility follows: The inequalities have the same solution set because they are equivalent.   **79.** The square of a real number is greater than or equal to zero. Therefore, $x^2 - 1$ is greater than or equal to -1 for all real numbers.   **81.** No. The inequality $x^2 + 1 > 1$ is true for all real numbers except $x = 0$.   **83.** To solve the quadratic inequality $f(x) > 0$ from the graph of $y = f(x)$, where x is a quadratic function, determine where the graph lies above

the x-axis. This information will give you the solution set.   **85.** $-\dfrac{8m^5}{n^2}$ **87.** $\dfrac{b^{\frac{1}{4}}}{9a^{\frac{7}{9}}}$

89.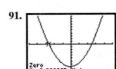

$\{x \mid x < -7 \text{ or } x > 3.5\}$ or $(-\infty, -7) \cup (3.5, \infty)$

91.

$\{x \mid -2.67 \le x \le 2.5\}$ or $[-2.67, 2.5]$

Chapter 7 Review   **1.** $\{-13, 13\}$ **2.** $\{-5\sqrt{3}, 5\sqrt{3}\}$ **3.** $\{-4i, 4i\}$ **4.** $\left\{-\dfrac{2\sqrt{2}}{3}, \dfrac{2\sqrt{2}}{3}\right\}$ **5.** $\{-1, 17\}$ **6.** $\{2 - 5\sqrt{6}, 2 + 5\sqrt{6}\}$ **7.** $\left\{-5, \dfrac{5}{3}\right\}$

8. $\left\{-\dfrac{3\sqrt{14}}{7}, \dfrac{3\sqrt{14}}{7}\right\}$ **9.** $\{-4\sqrt{5}i, 4\sqrt{5}i\}$ **10.** $\left\{-\dfrac{3}{4} - \dfrac{\sqrt{13}}{4}, -\dfrac{3}{4} + \dfrac{\sqrt{13}}{4}\right\}$ **11.** $a^2 + 30a + 225$; $(a + 15)^2$ **12.** $b^2 - 14b + 49$; $(b - 7)^2$

13. $c^2 - 11c + \dfrac{121}{4}$; $\left(c - \dfrac{11}{2}\right)^2$ **14.** $d^2 + 9d + \dfrac{81}{4}$; $\left(d + \dfrac{9}{2}\right)^2$ **15.** $m^2 - \dfrac{1}{4}m + \dfrac{1}{64}$; $\left(m - \dfrac{1}{8}\right)^2$ **16.** $n^2 + \dfrac{6}{7}n + \dfrac{9}{49}$; $\left(n + \dfrac{3}{7}\right)^2$ **17.** $\{2, 8\}$ **18.** $\{-4, 7\}$

19. $\{3 - 2\sqrt{3}, 3 + 2\sqrt{3}\}$ **20.** $\left\{\dfrac{5}{2} - \dfrac{\sqrt{53}}{2}, \dfrac{5}{2} + \dfrac{\sqrt{53}}{2}\right\}$ **21.** $\left\{-\dfrac{1}{2} - \dfrac{3\sqrt{3}}{2}i, -\dfrac{1}{2} + \dfrac{3\sqrt{3}}{2}i\right\}$ **22.** $\{3 - 2\sqrt{2}i, 3 + 2\sqrt{2}i\}$ **23.** $\left\{\dfrac{1}{2}, 3\right\}$

24. $\left\{-\dfrac{1}{2} - \dfrac{3}{2}i, -\dfrac{1}{2} + \dfrac{3}{2}i\right\}$ **25.** $\left\{\dfrac{3}{2} - \dfrac{\sqrt{15}}{6}i, \dfrac{3}{2} + \dfrac{\sqrt{15}}{6}i\right\}$ **26.** $\left\{-\dfrac{2}{3} - \dfrac{\sqrt{10}}{3}, -\dfrac{2}{3} + \dfrac{\sqrt{10}}{3}\right\}$ **27.** $c = 15$ **28.** $c = 8\sqrt{2}$ **29.** $c = 3\sqrt{5}$ **30.** $c = 26$

31. $c = 6$ **32.** $c = 7$ **33.** $b = 3\sqrt{7}$ **34.** $a = 5\sqrt{3}$ **35.** $a = \sqrt{253}$ **36.** approximately 127.3 feet   **37.** $\{-4, 5\}$ **38.** $\left\{-\dfrac{3}{2}, \dfrac{7}{2}\right\}$

39. $\left\{-\dfrac{4}{3} - \dfrac{\sqrt{7}}{3}, -\dfrac{4}{3} + \dfrac{\sqrt{7}}{3}\right\}$ **40.** $\left\{1 - \dfrac{\sqrt{10}}{2}, 1 + \dfrac{\sqrt{10}}{2}\right\}$ **41.** $\left\{-\dfrac{1}{6} - \dfrac{\sqrt{35}}{6}i, -\dfrac{1}{6} + \dfrac{\sqrt{35}}{6}i\right\}$ **42.** $\left\{\dfrac{4}{3}\right\}$ **43.** $\{2 - \sqrt{2}, 2 + \sqrt{2}\}$

44. $\left\{-\dfrac{2}{5} - \dfrac{1}{5}i, -\dfrac{2}{5} + \dfrac{1}{5}i\right\}$ **45.** $\left\{-\dfrac{5}{2} - \dfrac{3\sqrt{3}}{2}i, -\dfrac{5}{2} + \dfrac{3\sqrt{3}}{2}i\right\}$ **46.** $\left\{-\dfrac{3}{2} - \dfrac{\sqrt{5}}{2}i, -\dfrac{3}{2} + \dfrac{\sqrt{5}}{2}i\right\}$ **47.** 57; two irrational solutions   **48.** 0; one repeated

real solution   **49.** -47; two complex solutions that are not real   **50.** -20; two complex solutions that are not real   **51.** 0; one repeated real

solution   **52.** 25; two rational solutions   **53.** $\{-9, 1\}$ **54.** $\left\{-\dfrac{5}{2}, \dfrac{1}{3}\right\}$ **55.** $\{-2 - 3i, -2 + 3i\}$ **56.** $\{-2\sqrt{3}, 2\sqrt{3}\}$ **57.** $\left\{1 - \dfrac{\sqrt{10}}{2}, 1 + \dfrac{\sqrt{10}}{2}\right\}$

58. $\{-4 - 2i, -4 + 2i\}$ **59.** $\{-3, 5\}$ **60.** $\{1 - \sqrt{2}, 1 + \sqrt{2}\}$ **61.** $\left\{-\dfrac{4}{3}, \dfrac{4}{3}\right\}$ **62.** $\{-1 - \sqrt{3}i, -1 + \sqrt{3}i\}$ **63.** $x = 10$; the three sides

measure 5, 12, and 13   **64.** 12 centimeters by 9 centimeters   **65. (a)** either 300 or 600 cellular phones   **(b)** 450 cellular phones   **66. (a)** after approximately 0.5 second and after approximately 2.7 seconds   **(b)** after approximately 4.3 seconds   **(c)** No; the solutions to the equation are complex solutions that are not real. Another explanation: The vertex is (1.5625, 219.0625). Since $a = -16 < 0$, the parabola opens down, so the maximum value of the function is 219.0625.   **67.** approximately 10.8 miles per hour   **68.** approximately 67.8 minutes   **69.** $\{-4i, 4i, -3, 3\}$

70. $\left\{-\dfrac{\sqrt{3}}{2}, \dfrac{\sqrt{3}}{2}, -\sqrt{2}i, \sqrt{2}i\right\}$ **71.** $\left\{-\dfrac{10}{3}, -1\right\}$ **72.** $\{-4, 4, -2\sqrt{2}, 2\sqrt{2}\}$ **73.** $\{16, 81\}$ **74.** $\left\{\dfrac{9}{25}\right\}$ **75.** $\left\{-\dfrac{1}{3}, \dfrac{1}{7}\right\}$ **76.** $\left\{-343, \dfrac{1}{8}\right\}$ **77.** $\{16\}$

78. $\left\{-\dfrac{36}{7}, -\dfrac{19}{4}\right\}$ **79.** $\left\{\dfrac{9}{4}, \dfrac{49}{4}\right\}$ **80.** $\{-2\sqrt{3}, -\sqrt{5}, \sqrt{5}, 2\sqrt{3}\}$

81.

82.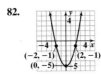

83.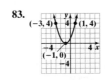

84.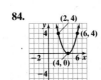

85.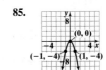

86.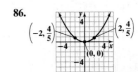

87.

88.

89.

90.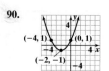

91. vertex is $(3, 1)$; axis of symmetry is $x = 3$; Domain: $\{x \,|\, x \text{ is any real number}\}$ or $(-\infty, \infty)$; Range: $\{y \,|\, y \geq 1\}$ or $[1, \infty)$

92. 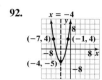 vertex is $(-4, -5)$; axis of symmetry is $x = -4$

93. vertex is $(1, -5)$; axis of symmetry is $x = 1$

94. 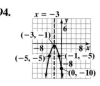 vertex is $(-3, -1)$; axis of symmetry is $x = -3$

Domain: $\{x \,|\, x \text{ is any real number}\}$ or $(-\infty, \infty)$
Range: $\{y \,|\, y \geq -5\}$ or $[-5, \infty)$

Domain: $\{x \,|\, x \text{ is any real number}\}$ or $(-\infty, \infty)$
Range: $\{y \,|\, y \geq -5\}$ or $[-5, \infty)$

Domain: $\{x \,|\, x \text{ is any real number}\}$ or $(-\infty, \infty)$
Range: $\{y \,|\, y \leq -1\}$ or $(-\infty, -1]$

95. vertex is $(2, 4)$; axis of symmetry is $x = 2$
Domain: $\{x \,|\, x \text{ is any real number}\}$ or $(-\infty, \infty)$
Range: $\{y \,|\, y \leq 4\}$ or $(-\infty, 4]$

96. vertex is $(2, 3)$; axis of symmetry is $x = 2$
Domain: $\{x \,|\, x \text{ is any real number}\}$ or $(-\infty, \infty)$ Range: $\{y \,|\, y \geq 3\}$ or $[3, \infty)$

97. $f(x) = 2(x - 2)^2 - 4$ or $f(x) = 2x^2 - 8x + 4$ **98.** $f(x) = -(x - 4)^2 + 3$ or $f(x) = -x^2 + 8x - 13$

99. $f(x) = -\dfrac{1}{2}(x + 2)^2 - 1$ or $f(x) = -\dfrac{1}{2}x^2 - 2x - 3$ **100.** $f(x) = 3(x + 2)^2$ or $f(x) = 3x^2 + 12x + 12$

101.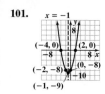

102.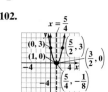

103.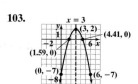

104.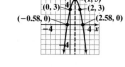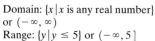

Domain: $\{x \,|\, x \text{ is any real number}\}$ or $(-\infty, \infty)$
Range: $\{y \,|\, y \geq -9\}$ or $[-9, \infty)$

Domain: $\{x \,|\, x \text{ is any real number}\}$ or $(-\infty, \infty)$
Range: $\left\{ y \,|\, y \geq -\dfrac{1}{8} \right\}$ or $\left[-\dfrac{1}{8}, \infty \right)$

Domain: $\{x \,|\, x \text{ is any real number}\}$ or $(-\infty, \infty)$
Range: $\{y \,|\, y \leq 2\}$ or $(-\infty, 2]$

Domain: $\{x \,|\, x \text{ is any real number}\}$ or $(-\infty, \infty)$
Range: $\{y \,|\, y \leq 5\}$ or $(-\infty, 5]$

105.

106.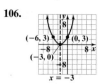

107.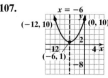

108.

Domain: $\{x \,|\, x \text{ is any real number}\}$ or $(-\infty, \infty)$
Range: $\{y \,|\, y \geq 0\}$ or $[0, \infty)$

Domain: $\{x \,|\, x \text{ is any real number}\}$ or $(-\infty, \infty)$
Range: $\{y \,|\, y \geq 0\}$ or $[0, \infty)$

Domain: $\{x \,|\, x \text{ is any real number}\}$ or $(-\infty, \infty)$
Range: $\{y \,|\, y \geq 1\}$ or $[1, \infty)$

Domain: $\{x \,|\, x \text{ is any real number}\}$ or $(-\infty, \infty)$
Range: $\{y \,|\, y \leq -5\}$ or $(-\infty, -5]$

109. maximum; 22 **110.** minimum; $-\dfrac{11}{8}$ **111.** maximum; 7 **112.** maximum; 5 **113. (a)** $225 **(b)** $16,875 **114. (a)** 3.75 amperes **(b)** 225 watts

115. both numbers are 12 **116. (a)** 3.75 yards by 7.5 yards **(b)** 28.125 square yards **117. (a)** 100 feet **(b)** 50 feet **(c)** 200 feet

118. (a) $R(p) = -0.002p^2 + 60p$ **(b)** $15,000; $450,000 **(c)** 30 automobiles per month **119. (a)** $\{x \,|\, x < -2 \text{ or } x > 3\}$ or $(-\infty, -2) \cup (3, \infty)$

(b) $\{x \,|\, -2 < x < 3\}$ or $(-2, 3)$ **120. (a)** $\left\{ x \,|\, -\dfrac{7}{2} \leq x \leq 1 \right\}$ or $\left[-\dfrac{7}{2}, 1 \right]$ **(b)** $\left\{ x \,|\, x \leq -\dfrac{7}{2} \text{ or } x \geq 1 \right\}$ or $\left(-\infty, -\dfrac{7}{2} \right] \cup [1, \infty)$

121. $\{x \,|\, -4 \leq x \leq 6\}$ or $[-4, 6]$

122. $\{y \,|\, y \leq -8 \text{ or } y \geq 1\}$ or, $(-\infty, -8] \cup [1, \infty)$

123. $\left\{ z \,|\, z < \dfrac{4}{3} \text{ or } z > 5 \right\}$ or $\left(-\infty, \dfrac{4}{3} \right) \cup (5, \infty)$

124. $\left\{ p \,|\, -2 - \sqrt{6} < p < -2 + \sqrt{6} \right\}$ or $\left(-2 - \sqrt{6}, -2 + \sqrt{6} \right)$

125. $\{m \,|\, m \text{ is any real number}\}$ or $(-\infty, \infty)$

126. $\left\{w \mid -\dfrac{1}{3} \le w \le \dfrac{7}{2}\right\}$ or $\left[-\dfrac{1}{3}, \dfrac{7}{2}\right]$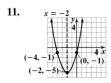

127. $\left\{x \mid x < -1 \text{ or } \dfrac{3}{2} < x < 2\right\}$ or $(-\infty, -1) \cup \left(\dfrac{3}{2}, 2\right)$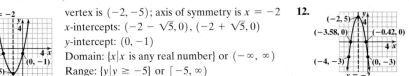

128. $\{x \mid -5 \le x \le -3 \text{ or } x \ge 3\}$ or $[-5, -3] \cup [3, \infty)$

Chapter 7 Test **1.** $x^2 - 3x + \dfrac{9}{4}; \left(x - \dfrac{3}{2}\right)^2$ **2.** $m^2 + \dfrac{2}{5}m + \dfrac{1}{25}; \left(m + \dfrac{1}{5}\right)^2$ **3.** $\left\{-\dfrac{5}{3}, -1\right\}$ **4.** $\{3 - \sqrt{5}, 3 + \sqrt{5}\}$ **5.** $\left\{1 - \dfrac{\sqrt{2}}{2}i, 1 + \dfrac{\sqrt{2}}{2}i\right\}$

6. $\left\{\dfrac{3}{2} - \dfrac{\sqrt{3}}{6}i, \dfrac{3}{2} + \dfrac{\sqrt{3}}{6}i\right\}$ **7.** 57; two irrational solutions **8.** $a = 6\sqrt{2}$ **9.** $\{-3, 3, -2i, 2i\}$ **10.** $\left\{\dfrac{1}{81}\right\}$

11. vertex is $(-2, -5)$; axis of symmetry is $x = -2$ **12.** vertex is $(-2, 5)$; axis of symmetry is $x = -2$

x-intercepts: $(-2 - \sqrt{5}, 0), (-2 + \sqrt{5}, 0)$

y-intercept: $(0, -1)$

Domain: $\{x \mid x$ is any real number$\}$ or $(-\infty, \infty)$

Range: $\{y \mid y \ge -5\}$ or $[-5, \infty)$

x-intercepts: $\left(\dfrac{-4 - \sqrt{10}}{2}, 0\right), \left(\dfrac{-4 + \sqrt{10}}{2}, 0\right)$

y-intercept: $(0, -3)$

Domain: $\{x \mid x$ is any real number$\}$ or $(-\infty, \infty)$

Range: $\{y \mid y \le 5\}$ or $(-\infty, 5]$

13. $f(x) = \dfrac{1}{3}(x + 3)^2 - 5$ or $f(x) = \dfrac{1}{3}x^2 + 2x - 2$ **14.** maximum; 6

15. $\left\{m \mid m < -3 \text{ or } m > \dfrac{5}{2}\right\}$ or $(-\infty, -3) \cup \left(\dfrac{5}{2}, \infty\right)$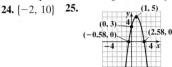

16. $\{x \mid x \le -5 \text{ or } -2 \le x \le 2\}$ or $(-\infty, -5] \cup [-2, 2]$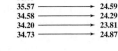

17. 0.4 second and 4.6 seconds **18.** 34.1 hours **19. (a)** $340 **(b)** $28,900 **20. (a)** 12.5 in. by 12.5 in. by 12 in. **(b)** 1875 cubic inches

Cumulative Review Chapters R–7 **1.** 2 **2.** $3c + 11$ **3.** $\{5\}$ **4.** $(-\infty, -3) \cup (1, \infty)$ **5.** $\{x \mid x \ne -4, x \ne 3\}$

6. **7.** $y = -\dfrac{2}{5}x + 1; 2x + 5y = 5$ **8.** $(-2, 5)$ **9.** -4 **10.** **11.** $a^3 - 2a^2 - 5a + 3$

12. $4x^2 + 2x - 3 + \dfrac{4x - 3}{2x^2 - x + 5}$ **13.** $(2m + 3n)(4m^2 - 6mn + 9n^2)$ **14.** $(7y - 5)(y + 4)$ **15.** $\dfrac{w + 5}{w - 2}$ **16.** $\dfrac{k + 23}{(k - 4)(k - 1)(k + 5)}$

17. $\left\{-\dfrac{3}{2}\right\}$ **18.** $(-\infty, -5) \cup [4, \infty)$ or $\{x \mid x < -5 \text{ or } x \ge 4\}$

19. One plane is traveling 126 miles per hour and the other plane is traveling 142 miles per hour. **20.** $7\sqrt{3}$ **21.** $\dfrac{5\sqrt[3]{4}}{4}$ **22.** $\{16\}$ **23.** $x = 11$
24. $\{-2, 10\}$ **25.**

Chapter 8 Exponential and Logarithmic Functions

Section 8.1 Composite Functions and Inverse Functions **1.** composite function **2. (a)** 17 **(b)** 26 **(c)** -63 **3.** False **4.** $g(x) = 4x - 3$
5. (a) $9x^2 + 3x - 1$ **(b)** $3x^2 - 9x + 5$ **(c)** 29 **6.** one-to-one **7.** not one-to-one **8.** one-to-one **9. (a)** not one-to-one **(b)** one-to-one

10.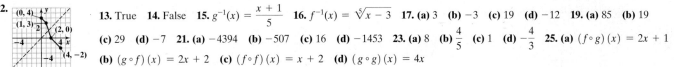

Right Tibia	Right Humerus
36.05	24.80
35.57	24.59
34.58	24.29
34.20	23.81
34.73	24.87

The domain of the inverse function is $\{36.05, 35.57, 34.58, 34.20, 34.73\}$.
The range of the inverse function is $\{24.80, 24.59, 24.29, 23.81, 24.87\}$.

11. $\{(3, -3), (2, -2), (1, -1), (0, 0), (-1, 1)\}$. The domain of the inverse function is $\{3, 2, 1, 0, -1\}$. The range of the inverse function is $\{-3, -2, -1, 0, 1\}$.

12. **13.** True **14.** False **15.** $g^{-1}(x) = \dfrac{x + 1}{5}$ **16.** $f^{-1}(x) = \sqrt[5]{x - 3}$ **17. (a)** 3 **(b)** -3 **(c)** 19 **(d)** -12 **19. (a)** 85 **(b)** 19

(c) 29 **(d)** -7 **21. (a)** -4394 **(b)** -507 **(c)** 16 **(d)** -1453 **23. (a)** 8 **(b)** $\dfrac{4}{5}$ **(c)** 1 **(d)** $-\dfrac{4}{3}$ **25. (a)** $(f \circ g)(x) = 2x + 1$

(b) $(g \circ f)(x) = 2x + 2$ **(c)** $(f \circ f)(x) = x + 2$ **(d)** $(g \circ g)(x) = 4x$

27. (a) $(f \circ g)(x) = -8x + 17$ **(b)** $(g \circ f)(x) = -8x - 23$ **(c)** $(f \circ f)(x) = 4x + 21$ **(d)** $(g \circ g)(x) = 16x - 15$ **29. (a)** $(f \circ g)(x) = x^2 - 6x + 9$
(b) $(g \circ f)(x) = x^2 - 3$ **(c)** $(f \circ f)(x) = x^4$ **(d)** $(g \circ g)(x) = x - 6$ **31. (a)** $(f \circ g)(x) = \sqrt{x + 4}$ **(b)** $(g \circ f)(x) = \sqrt{x} + 4$

(c) $(f \circ f)(x) = \sqrt[4]{x}$ **(d)** $(g \circ g)(x) = x + 8$ **33. (a)** $(f \circ g)(x) = x^2$ **(b)** $(g \circ f)(x) = x^2 + 8x + 12$ **(c)** $(f \circ f)(x) = ||x + 4| + 4|$

(d) $(g \circ g)(x) = x^4 - 8x^2 + 12$ **35. (a)** $(f \circ g)(x) = \dfrac{2x}{x + 1}$, where $x \neq -1, 0$ **(b)** $(g \circ f)(x) = \dfrac{x + 1}{2}$, where $x \neq -1$ **(c)** $(f \circ f)(x) = \dfrac{2(x + 1)}{x + 3}$,

where $x \neq -1, -3$ **(d)** $(g \circ g)(x) = x$, where $x \neq 0$ **37.** one-to-one **39.** not one-to-one **41.** one-to-one **43.** not one-to-one **45.** one-to-one

47. one-to-one **49.** not one-to-one **51.** one-to-one

53.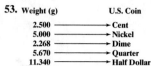

Weight (g)   U.S. Coin

Weight (g)	U.S. Coin
2.500	Cent
5.000	Nickel
2.268	Dime
5.670	Quarter
11.340	Half Dollar
8.100	Dollar

55. $\{(3, 0), (4, 1), (5, 2), (6, 3)\}$ **57.** $\{(3, -2), (1, -1), (-3, 0), (9, 1)\}$

59. 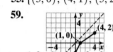 **61.** 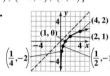 **63.** 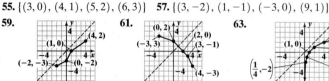 **65.** $f(g(x)) = (x - 5) + 5 = x$
$g(f(x)) = (x + 5) - 5 = x$

67. $f(g(x)) = 5\left(\dfrac{x - 7}{5}\right) + 7 = x - 7 + 7 = x$ **69.** $f(g(x)) = \dfrac{3}{\left(\dfrac{3}{x} + 1\right) - 1} = \dfrac{3}{\dfrac{3}{x}} = 3 \cdot \dfrac{x}{3} = x$

$g(f(x)) = \dfrac{(5x + 7) - 7}{5} = \dfrac{5x}{5} = x$

$g(f(x)) = \dfrac{3}{\dfrac{3}{x - 1}} + 1 = 3 \cdot \dfrac{x - 1}{3} + 1 = x - 1 + 1 = x$

71. $f(g(x)) = \sqrt[3]{(x^3 - 4) + 4} = \sqrt[3]{x^3} = x$ **73.** $f^{-1}(x) = \dfrac{x}{6}$ **75.** $f^{-1}(x) = x - 4$ **77.** $h^{-1}(x) = \dfrac{x + 7}{2}$ **79.** $G^{-1}(x) = \dfrac{2 - x}{5}$

$g(f(x)) = \left(\sqrt[3]{x + 4}\right)^3 - 4 = x + 4 - 4 = x$

81. $g^{-1}(x) = \sqrt[3]{x - 3}$ **83.** $p^{-1}(x) = \dfrac{1}{x} - 3$ **85.** $F^{-1}(x) = 2 - \dfrac{5}{x}$ **87.** $f^{-1}(x) = x^3 + 2$

89. $R^{-1}(x) = \dfrac{2x}{1 - x}$ **91.** $f^{-1}(x) = (x - 4)^3 + 1$ **93.** $A(t) = 400\pi t^2; 3600\pi \approx 11{,}309.73$ sq ft **95. (a)** $C(x) = 2x$ **(b)** \$630 **97.** $f^{-1}(12) = 4$

99. Domain of f^{-1}: $[-5, \infty)$ Range of f^{-1}: $[0, \infty)$ **101.** Domain of g^{-1}: $(-6, 12)$ Range of g^{-1}: $[-4, 10]$ **103.** $x(T) = \dfrac{T + 300}{0.15}$ for $600 \leq T \leq 3960$

105. $\left\{-\dfrac{5}{2}, 3\right\}$ **107.** A function is one-to-one provided no two different inputs correspond to the same output. A function must be one-to-one in order for the inverse to be a function because we interchange the inputs and outputs to obtain the inverse. Thus, if a function is not one-to-one, the inverse will have a single input corresponding to two different outputs. **109.** The domain of f equals the range of f^{-1} because we interchange the roles of the inputs and outputs to obtain the inverse of a function. The same logic explains why the range of f equals the domain of f^{-1}. **111. (a)** 3 **(b)** -3

(c) 19 **(d)** -12 **113. (a)** 85 **(b)** 19 **(c)** 29 **(d)** -7 **115. (a)** -4394 **(b)** -507 **(c)** 16 **(d)** -1453 **117. (a)** 8 **(b)** $\dfrac{4}{5}$ **(c)** 1 **(d)** $-\dfrac{4}{3}$

119. $f(x) = x + 5; g(x) = x - 5$ **121.** $f(x) = 5x + 7; g(x) = \dfrac{x - 7}{5}$

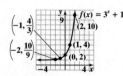

Section 8.2 Exponential Functions **1.** $>; \neq$ **2. (a)** 3.249009585 **(b)** 3.317278183 **(c)** 3.321880096 **(d)** 3.32211036 **(e)** 3.321997085

3. The domain of f is all real numbers or, in interval notation, $(-\infty, \infty)$. The range of f is $\{y \,|\, y > 0\}$ or, in interval notation, $(0, \infty)$.

4. $\left(-1, \dfrac{1}{a}\right)$; $(0, 1)$; $(1, a)$ **5.** True **6.** False

7. The domain of f is all real numbers or, in interval notation, $(-\infty, \infty)$. The range of f is $\{y \,|\, y > 0\}$ or, in interval notation, $(0, \infty)$.

8. The domain of f is all real numbers or, in interval notation, $(-\infty, \infty)$. The range of f is $\{y \,|\, y > 0\}$ or, in interval notation, $(0, \infty)$.

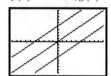

9. The domain of f is all real numbers or, in interval notation, $(-\infty, \infty)$. The range of f is $\{y \,|\, y > 1\}$ or, in interval notation, $(1, \infty)$.

10. 2.71828 **11. (a)** 54.598 **(b)** 0.018 **12.** $\{3\}$ **13.** $\{2\}$ **14.** $\{0, 5\}$ **15.** $\{-1, 3\}$ **16. (a)** 0.918 or 91.8%
(b) 0.998 or 99.8% **17. (a)** approximately 6.91 grams **(b)** 5 grams **(c)** 0.625 gram **(d)** approximately
0.247 gram **18. (a)** \$2102.32 **(b)** \$4227.41 **(c)** \$8935.49 **19. (a)** 11.212 **(b)** 11.587 **(c)** 11.664
(d) 11.665 **(e)** 11.665 **21. (a)** 73.517 **(b)** 77.708 **(c)** 77.924 **(d)** 77.881 **(e)** 77.880 **23.** g **25.** e **27.** f **29.** h

31. 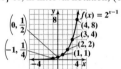 Domain: all real numbers or $(-\infty, \infty)$
Range: $\{y \,|\, y > 0\}$ or $(0, \infty)$

33. 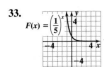 Domain: all real numbers or $(-\infty, \infty)$ Range: $\{y \mid y > 0\}$ or $(0, \infty)$

35. Domain: all real numbers or $(-\infty, \infty)$ Range: $\{y \mid y > 0\}$ or $(0, \infty)$

37. 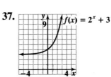 Domain: all real numbers or $(-\infty, \infty)$ Range: $\{y \mid y > 3\}$ or $(3, \infty)$

39. 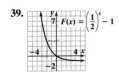 Domain: all real numbers or $(-\infty, \infty)$ Range: $\{y \mid y > -1\}$ or $(-1, \infty)$

41. Domain: all real numbers or $(-\infty, \infty)$ Range: $\{y \mid y > 0\}$ or $(0, \infty)$

43. (a) 21.217 **(b)** 22.472 **(c)** 22.460 **(d)** 22.460 **(e)** 22.459 **45.** 7.389 **47.** 0.135 **49.** 9.974

51. 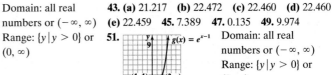 Domain: all real numbers or $(-\infty, \infty)$ Range: $\{y \mid y > 0\}$ or $(0, \infty)$

53. 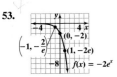 Domain: all real numbers or $(-\infty, \infty)$ Range: $\{y \mid y < 0\}$ or $(-\infty, 0)$

55. $\{5\}$ **57.** $\{-4\}$ **59.** $\{5\}$ **61.** $\{5\}$ **63.** $\left\{\frac{3}{2}\right\}$ **65.** $\{1\}$ **67.** $\{-1, 4\}$ **69.** $\{-4, 2\}$ **71.** $\{9\}$ **73.** $\{-2\}$

75. $\{-2, 2\}$ **77.** $\{-2\}$ **79.** $\{2\}$ **81. (a)** $f(3) = 8$; $(3, 8)$ **(b)** $x = -3$; $\left(-3, \frac{1}{8}\right)$

83. (a) $g(-1) = -\frac{3}{4}$; $\left(-1, -\frac{3}{4}\right)$ **(b)** $x = 2$; $(2, 15)$ **85. (a)** $H(-3) = 24$; $(-3, 24)$ **(b)** $x = 2$; $\left(2, \frac{3}{4}\right)$

87. (a) approximately 323.4 million people **(b)** approximately 474.3 million people **(c)** Answers may vary. One possibility is that the population is not growing exponentially. **89. (a)** \$5308.39 **(b)** \$5983.40 **(c)** \$6744.25 **91. (a)** \$2318.55 **(b)** \$2322.37 **(c)** \$2323.23 **(d)** \$2323.65 **(e)** The future value is higher with more compounding periods. **93. (a)** \$19,841 **(b)** \$15,365 **(c)** \$10,471 **95. (a)** approximately 95.105 grams **(b)** 50 grams **(c)** 25 grams **(d)** approximately 0.661 gram **97. (a)** approximately 300.233°F **(b)** approximately 230.628°F **(c)** yes **99. (a)** approximately 29 words **(b)** approximately 38 words **101. (a)** approximately 0.238 ampere **(b)** approximately 0.475 ampere **103.** $y = 3^x$ **105.** As x increases, the graph increases very rapidly. As x decreases, the graph approaches the x-axis. **107.** Answers may vary. The big difference is that exponential functions are of the form $f(x) = a^x$ (the variable is in the exponent), while polynomial functions are of the form $f(x) = a_n x^n + a_{n-1} x^{n-1} + \cdots + a_1 x + a_0$ (the variable is a base). **109. (a)** 15 **(b)** -5 **111. (a)** undefined **(b)** $\frac{4}{5}$ **113. (a)** 3 **(b)** $3\sqrt{3}$

115. $f(x) = 1.5^x$
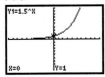
Domain: all real numbers or $(-\infty, \infty)$ Range: $\{y \mid y > 0\}$ or $(0, \infty)$

117. $H(x) = 0.9^x$
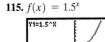
Domain: all real numbers or $(-\infty, \infty)$ Range: $\{y \mid y > 0\}$ or $(0, \infty)$

119. $g(x) = 2.5^x + 3$

Domain: all real numbers or $(-\infty, \infty)$ Range: $\{y \mid y > 3\}$ or $(3, \infty)$

121. $F(x) = 1.6^{x-3}$
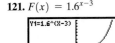
Domain: all real numbers or $(-\infty, \infty)$ Range: $\{y \mid y > 0\}$ or $(0, \infty)$

Section 8.3 Logarithmic Functions **1.** $x = a^y$; $>$; $\neq$ **2.** $3 = \log_4 64$ **3.** $-2 = \log_p 8$ **4.** $2^4 = 16$ **5.** $a^5 = 20$ **6.** $5^{-3} = z$ **7.** 2 **8.** -3 **9.** 2

10. -1 **11.** $\{x \mid x > -3\}$ or $(-3, \infty)$ **12.** $\left\{x \mid x < \frac{5}{2}\right\}$ or $\left(-\infty, \frac{5}{2}\right)$

13. The domain of f is $\{x \mid x > 0\}$ or, in interval notation, $(0, \infty)$. The range of f is all real numbers or, in interval notation, $(-\infty, \infty)$.

14. The domain of f is $\{x \mid x > 0\}$ or, in interval notation, $(0, \infty)$. The range of f is all real numbers or, in interval notation, $(-\infty, \infty)$.

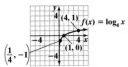

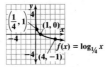

15. 3.146 **16.** 1.569 **17.** -0.523 **18.** $\{16\}$ **19.** $\{4\}$ **20.** $\{e^{-2}\}$ **21.** $\{10,020\}$ **22.** 100 decibels **23.** $3 = \log_4 64$ **25.** $-3 = \log_2\left(\frac{1}{8}\right)$ **27.** $\log_a 19 = 3$

29. $\log_5 c = -6$ **31.** $2^4 = 16$ **33.** $3^{-2} = \frac{1}{9}$ **35.** $5^{-3} = a$ **37.** $a^2 = 4$ **39.** $\left(\frac{1}{2}\right)^y = 12$ **41.** 0 **43.** 3 **45.** -2 **47.** 4 **49.** 4 **51.** $\frac{1}{2}$

53. $\{x \mid x > 4\}$ or $(4, \infty)$ **55.** $\{x \mid x > 0\}$ or $(0, \infty)$ **57.** $\left\{x \mid x > \frac{2}{3}\right\}$ or $\left(\frac{2}{3}, \infty\right)$ **59.** $\left\{x \mid x > -\frac{1}{2}\right\}$ or $\left(-\frac{1}{2}, \infty\right)$ **61.** $\left\{x \mid x < \frac{1}{4}\right\}$ or $\left(-\infty, \frac{1}{4}\right)$

63. 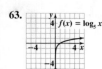   Domain: $\{x \,|\, x > 0\}$ or $(0, \infty)$
Range: all real numbers or $(-\infty, \infty)$

65.   Domain: $\{x \,|\, x > 0\}$ or $(0, \infty)$
Range: all real numbers or $(-\infty, \infty)$

67. 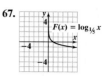   Domain: $\{x \,|\, x > 0\}$ or $(0, \infty)$
Range: all real numbers or $(-\infty, \infty)$

69. $\ln 12 = x$ **71.** $e^4 = x$ **73.** -1 **75.** 3 **77.** 1.826 **79.** 1.686 **81.** -0.456 **83.** -1.609 **85.** 0.097 **87.** -0.981 **89.** $\{4\}$ **91.** $\left\{\dfrac{13}{2}\right\}$ **93.** $\{6\}$

95. $\{3\sqrt{2}\}$ **97.** $\{10\}$ **99.** $\{e^5\}$ **101.** $\left\{\dfrac{11}{20}\right\}$ **103.** $\{-3\}$ **105.** $\{4\}$ **107.** $\{-3, 3\}$ **109. (a)** $f(16) = 4; (16,4)$ **(b)** $x = \dfrac{1}{8}; \left(\dfrac{1}{8}, -3\right)$

111. (a) $G(7) = \dfrac{3}{2}; \left(7, \dfrac{3}{2}\right)$ **(b)** $x = 15; (15, 2)$ **113.** $a = 4$ **115.** $\{x \,|\, x < -1 \text{ or } x > 5\}$ or $(-\infty, -1) \cup (5, \infty)$ **117.** $\{x \,|\, x < -1 \text{ or } x > 4\}$ or
$(-\infty, -1) \cup (4, \infty)$ **119.** $\{x \,|\, x \neq 3\}$ or $(-\infty, 3) \cup (3, \infty)$ **121.** 20 decibels   **123.** 130 decibels   **125.** approximately 7.8 on the Richter scale
127. approximately 794,328   **129. (a)** 12; basic   **(b)** 5; acidic   **(c)** 2; acidic   **(d)** $10^{-7.4}$ mole per liter   **131.** $\{x \,|\, x < -2 \text{ or } x > 5\}$ or $(-\infty, -2) \cup (5, \infty)$
133. $\{x \,|\, x < -1 \text{ or } x > 3\}$ or $(-\infty, -1) \cup (3, \infty)$ **135.** The base of $f(x) = \log_a x$ cannot equal 1 because $y = \log_a x$ is equivalent to $x = a^y$ and a
does not equal 1 in the exponential function. In addition, the graph would be a vertical line $(x = 1)$, which is not a function.   **137.** The domain of
$f(x) = \log_a(x^2 + 1)$ is the set of all real numbers because $x^2 + 1 > 0$ for all x.   **139.** $-3x^2 - 7x + 10$

141. $\dfrac{4x^2 + 2x + 3}{(x + 1)(x - 1)(x + 2)}$ **143.** $5x\sqrt{2x}$

145. $f(x) = \log(x + 1)$

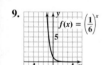   Domain: $\{x \,|\, x > -1\}$ or $(-1, \infty)$
Range: all real number or $(-\infty, \infty)$

147. $G(x) = \ln(x) + 1$

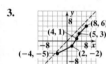   Domain: $\{x \,|\, x > 0\}$ or $(0, \infty)$
Range: all real number or $(-\infty, \infty)$

149. $f(x) = 2\log(x - 3) + 1$

Domain: $\{x \,|\, x > 3\}$ or $(3, \infty)$
Range: all real number or $(-\infty, \infty)$

Putting the Concepts Together (Sections 8.1–8.3)   **1. (a)** $(f \circ g)(x) = 4x^2 - 8x + 3$ **(b)** $(g \circ f)(x) = 8x^2 + 16x + 6$ **(c)** $(f \circ g)(3) = 15$

(d) $(g \circ f)(-2) = 6$ **(e)** $(f \circ f)(1) = 13$ **2. (a)** $f^{-1}(x) = \dfrac{x - 4}{3}$ **(b)** $g^{-1}(x) = \sqrt[3]{x + 4}$

3. 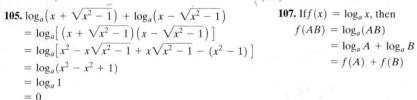   **4. (a)** 14.611   **(b)** 15.206   **(c)** 15.146   **(d)** 15.155   **(e)** 15.154   **5. (a)** $\log_a 6.4 = 4$ **(b)** $\log 278 = x$
6. (a) $2^7 = x$ **(b)** $e^M = 16$ **7. (a)** 4   **(b)** -2 **8.** $\{x \,|\, x > -6\}$ or $(-6, \infty)$

9. 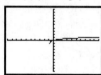   Domain: all real numbers or $(-\infty, \infty)$
Range: $\{y \,|\, y > 0\}$ or $(0, \infty)$

10. 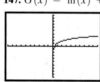   Domain: $\{x \,|\, x > 0\}$ or $(0, \infty)$
Range: all real numbers or $(-\infty, \infty)$

11. $\{-1\}$ **12.** $\{-5\}$ **13.** $\left\{\dfrac{11}{2}\right\}$
14. $\{e^7\}$ **15.** approximately 56 terms

Section 8.4 Properties of Logarithms   **1.** 0   **2.** 0   **3.** 1   **4.** 1   **5.** $\sqrt{2}$ **6.** 0.2   **7.** 1.2   **8.** -4 **9.** False   **10.** $\log_4 9 + \log_4 5$ **11.** $\log 5 + \log w$
12. $\log_7 9 - \log_7 5$ **13.** $\ln p - \ln 3$ **14.** $\log_2 3 + \log_2 m - \log_2 n$ **15.** $\ln q - \ln 3 - \ln p$ **16.** $1.6 \log_2 5$ **17.** $5 \log b$ **18.** $2 \log_4 a + \log_4 b$
19. $2 + 4 \log_3 m - \dfrac{1}{3} \log_3 n$ **20.** 2   **21.** $\log_3\left(\dfrac{x + 4}{x - 1}\right)$ **22.** $\log_5 \dfrac{x}{8}$ **23.** $\log_2 \dfrac{x^2 + 3x + 2}{x^2}$ **24.** 10; 3; 10; 3   **25.** 3.155   **26.** 2.807   **27.** 3   **29.** -7

31. 5   **33.** 2   **35.** 1   **37.** 0   **39.** $a + b$ **41.** $2b$ **43.** $2a + b$ **45.** $\dfrac{1}{2}a$ **47.** $\log a + \log b$ **49.** $4 \log_5 x$ **51.** $\log_2 x + 2 \log_2 y$ **53.** $2 + \log_5 x$

55. $2 - \log_7 y$ **57.** $2 + \ln x$ **59.** $3 + \dfrac{1}{2} \log_3 x$ **61.** $2 \log_5 x + \dfrac{1}{2} \log_5(x^2 + 1)$ **63.** $4 \log x - \dfrac{1}{3} \log(x - 1)$ **65.** $\dfrac{1}{2} \log_7(x + 1) - \dfrac{1}{2} \log_7 x$

67. $\log_2 x + 2 \log_2(x - 1) - \dfrac{1}{2} \log_2(x + 1)$ **69.** 2   **71.** $\log(3x)$ **73.** 2   **75.** 4   **77.** $\log_3 x^3$ **79.** $\log_4\left(\dfrac{x + 1}{x}\right)$ **81.** $\ln(x^2 y^3)$ **83.** $\log_3\left[\sqrt{x}(x - 1)^3\right]$

85. $\log(x^2)$ **87.** $\log(x\sqrt{xy})$ **89.** $\log_8(x - 1)$ **91.** $\log\left(\dfrac{x^{12}}{10}\right)$ **93.** 3.322   **95.** 0.528   **97.** -2.680 **99.** 4.644   **101.** 3   **103.** 1

105. $\log_a\left(x + \sqrt{x^2 - 1}\right) + \log_a\left(x - \sqrt{x^2 - 1}\right)$
$= \log_a\left[\left(x + \sqrt{x^2 - 1}\right)\left(x - \sqrt{x^2 - 1}\right)\right]$
$= \log_a\left[x^2 - x\sqrt{x^2 - 1} + x\sqrt{x^2 - 1} - (x^2 - 1)\right]$
$= \log_a(x^2 - x^2 + 1)$
$= \log_a 1$
$= 0$

107. If $f(x) = \log_a x$, then
$f(AB) = \log_a(AB)$
$= \log_a A + \log_a B$
$= f(A) + f(B)$

109. Answers will vary. One possibility: The logarithm of the product of two expressions equals the sum of the logarithms of the two expressions.
111. Answers will vary. One possibility:
$\log_2(2 + 4) \neq \log_2 2 + \log_2 4$

113. $\left\{\dfrac{5}{2}\right\}$ **115.** $\{-2 - \sqrt{2}, -2 + \sqrt{2}\}$ **117.** $\{47\}$

119. $f(x) = \log_3 x = \dfrac{\log x}{\log 3}$

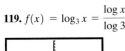

Domain: $\{x \,|\, x > 0\}$ or $(0, \infty)$
Range: all real numbers or $(-\infty, \infty)$

121. $F(x) = \log_{1/2} x = \dfrac{\log x}{\log\left(\frac{1}{2}\right)}$

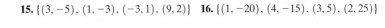

Domain: $\{x \,|\, x > 0\}$ or $(0, \infty)$
Range: all real numbers or $(-\infty, \infty)$

Section 8.5 Exponential and Logarithmic Equations   **1.** $M = N$ **2.** $\{3\}$ **3.** $\{8\}$ **4.** $\{-2\}$ **5.** $\left\{\dfrac{\ln 11}{\ln 2}\right\}$ or $\left\{\dfrac{\log 11}{\log 2}\right\}$; $\{3.459\}$

6. $\left\{\dfrac{\ln 3}{2 \ln 5}\right\}$ or $\left\{\dfrac{\log 3}{2 \log 5}\right\}$; $\{0.341\}$ **7.** $\left\{\dfrac{\ln 5}{2}\right\}$; $\{0.805\}$ **8.** $\left\{-\dfrac{\ln\left(^{20}/_3\right)}{4}\right\}$; $\{-0.474\}$ **9. (a)** approximately 2.85 days   **(b)** approximately

32.52 days   **10. (a)** approximately 6.77 years   **(b)** approximately 11.58 years   **11.** $\{7\}$ **13.** $\{9\}$ **15.** $\{3\}$ **17.** $\{81\}$ **19.** $\{1\}$ **21.** $\{-1\}$ **23.** $\left\{\dfrac{1}{3}\right\}$

25. $\left\{\dfrac{7}{5}\right\}$ **27.** $\{-5\}$ **29.** $\left\{\dfrac{1}{\log 2}\right\} \approx \{3.322\}$ or $\left\{\dfrac{\ln 10}{\ln 2}\right\} \approx \{3.322\}$ **31.** $\left\{\dfrac{\log 20}{\log 5}\right\} \approx \{1.861\}$ or $\left\{\dfrac{\ln 20}{\ln 5}\right\} \approx \{1.861\}$ **33.** $\left\{\dfrac{\log 7}{\log\left(\frac{1}{2}\right)}\right\} \approx \{-2.807\}$ or

$\left\{\dfrac{\ln 7}{\ln\left(\frac{1}{2}\right)}\right\} \approx \{-2.807\}$ **35.** $\{\ln 5\} \approx \{1.609\}$ **37.** $\{\log 5\} \approx \{0.699\}$ **39.** $\left\{\dfrac{\log 13}{2 \log 3}\right\} \approx \{1.167\}$ or $\left\{\dfrac{\ln 13}{2 \ln 3}\right\} \approx \{1.167\}$ **41.** $\left\{\dfrac{\log 3}{4 \log\left(\frac{1}{2}\right)}\right\} \approx \{-0.396\}$ or

$\left\{\dfrac{\ln 3}{4 \ln\left(\frac{1}{2}\right)}\right\} \approx \{-0.396\}$ **43.** $\left\{\dfrac{\log\left(\frac{5}{4}\right)}{\log 2}\right\} \approx \{0.322\}$ or $\left\{\dfrac{\ln\left(\frac{5}{4}\right)}{\ln 2}\right\} \approx \{0.322\}$ **45.** $\{\ln 6\} \approx \{1.792\}$ **47.** $\left\{\dfrac{\log 0.2}{\log 3 - \log 0.2}\right\} \approx \{-0.594\}$ or

$\left\{\dfrac{\ln 0.2}{\ln 3 - \ln 0.2}\right\} \approx \{-0.594\}$ **49.** $\{8\}$ **51.** $\left\{\dfrac{\log 7}{3 \log 5}\right\} \approx \{0.403\}$ or $\left\{\dfrac{\ln 7}{3 \ln 5}\right\} \approx \{0.403\}$ **53.** $\{2\}$ **55.** $\{\ln 15\} \approx \{2.708\}$ **57.** $\left\{-\dfrac{2}{5}\right\}$ **59.** $\{-4, 4\}$ **61.** $\{-2\}$

63. $\{2 + \sqrt{7}\} \approx \{4.646\}$ **65. (a)** 2014   **(b)** 2049   **67. (a)** approximately 5.6 years   **(b)** approximately 11.6 years   **69. (a)** approximately 2.188 years
(b) approximately 10.782 years   **(c)** approximately 23.372 years   **71. (a)** approximately 2.099 seconds   **(b)** 27.62 seconds   **(c)** approximately
45.876 seconds   **73. (a)** approximately 5 minutes   **(b)** approximately 11 minutes   **75. (a)** approximately 82 minutes   **(b)** approximately 396 minutes

(or 6.6 hours)   **77. (a)** approximately 9 years   **(b)** $t = \dfrac{\log 2}{n \log\left(1 + \dfrac{r}{n}\right)}$ **(c)** approximately 8.693 years, which is about the same as the result from the

Rule of 72   **79.** 29 minutes   **81. (a)** 1   **(b)** 11   **(c)** 7   **83. (a)** $-\dfrac{3}{2}$ **(b)** $\dfrac{2}{7}$ **(c)** 0   **85. (a)** 8   **(b)** $\dfrac{1}{4}$ **(c)** 1   **87.** approximately $\{1.06\}$ **89.** approximately
$\{-0.70\}$ **91.** approximately $\{0.05, 1.48\}$ **93.** approximately $\{0.45, 1\}$

Chapter 8 Review   **1. (a)** 32   **(b)** -9 **(c)** 2   **(d)** 13   **2. (a)** 24   **(b)** -28 **(c)** -8 **(d)** 112   **3. (a)** 201   **(b)** 24   **(c)** 163   **(d)** 14   **4. (a)** 23   **(b)** 37
(c) -8 **(d)** 290   **5. (a)** $(f \circ g)(x) = 5x + 1$ **(b)** $(g \circ f)(x) = 5x + 5$ **(c)** $(f \circ f)(x) = x + 2$ **(d)** $(g \circ g)(x) = 25x$ **6. (a)** $(f \circ g)(x) = 2x + 9$
(b) $(g \circ f)(x) = 2x + 3$ **(c)** $(f \circ f)(x) = 4x - 9$ **(d)** $(g \circ g)(x) = x + 12$ **7. (a)** $(f \circ g)(x) = 4x^2 + 4x + 2$ **(b)** $(g \circ f)(x) = 2x^2 + 3$
(c) $(f \circ f)(x) = x^4 + 2x^2 + 2$ **(d)** $(g \circ g)(x) = 4x + 3$ **8. (a)** $(f \circ g)(x) = \dfrac{2x}{x + 1}$, where $x \neq -1, 0$ **(b)** $(g \circ f)(x) = \dfrac{x + 1}{2}$, where $x \neq -1$
(c) $(f \circ f)(x) = \dfrac{2(x + 1)}{x + 3}$, where $x \neq -1, -3$ **(d)** $(g \circ g)(x) = x$, where $x \neq 0$ **9.** not one-to-one   **10.** one-to-one   **11.** one-to-one
12. not one-to-one

13.

Height (inches)	Age
69	24
71	59
72	29
73	81
74	37

14.

Quantity Demanded	Price ($)
112	300
129	200
144	170
161	150
176	130

15. $\{(3, -5), (1, -3), (-3, 1), (9, 2)\}$ **16.** $\{(1, -20), (4, -15), (3, 5), (2, 25)\}$

17.

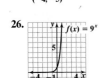

(4, 5)
(2, −2)
(−1, −4)
(−4, −5)

18.

(1, 0)
(5, 1)
$\left(\frac{1}{5}, -1\right)$

19. $f^{-1}(x) = \dfrac{x}{5}$ **20.** $H^{-1}(x) = \dfrac{x - 7}{2}$ **21.** $P^{-1}(x) = \dfrac{4}{x} - 2$ **22.** $g^{-1}(x) = \sqrt[3]{\dfrac{x + 1}{2}}$

23. (a) 27.332   **(b)** 28.975   **(c)** 29.088   **(d)** 29.093   **(e)** 29.091   **24. (a)** 1258.925   **(b)** 1380.384
(c) 1386.756   **(d)** 1385.479   **(e)** 1385.456   **25. (a)** 1.649   **(b)** 0.368   **(c)** 4.482   **(d)** 0.449
(e) 5.885

26.

$f(x) = 9^x$

Domain: all real numbers or $(-\infty, \infty)$
Range: $\{y \,|\, y > 0\}$ or $(0, \infty)$

27.

$g(x) = \left(\dfrac{1}{9}\right)^x$

Domain: all real numbers or $(-\infty, \infty)$
Range: $\{y \,|\, y > 0\}$ or $(0, \infty)$

28.

$H(x) = 4^{x-2}$

Domain: all real numbers or $(-\infty, \infty)$
Range: $\{y \,|\, y > 0\}$ or $(0, \infty)$

29.

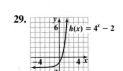

$h(x) = 4^x - 2$

Domain: all real numbers or $(-\infty, \infty)$
Range: $\{y \,|\, y > -2\}$ or $(-2, \infty)$

30. The number e is defined as the number that the expression $\left(1 + \dfrac{1}{n}\right)^n$ approaches as n

increases.   **31.** $\{6\}$ **32.** $\left\{\dfrac{7}{2}\right\}$ **33.** $\{-4, 1\}$ **34.** $\{-2\}$ **35.** $\{9\}$ **36.** $\{-3, 3\}$ **37. (a)** $7513.59
(b) $7652.33   **(c)** $7684.36   **(d)** $7700.00

38. (a) approximately 82.034 grams   **(b)** 50 grams   **(c)** 25 grams   **(d)** approximately 0.263 gram   **39. (a)** approximately 3.883 million people

(b) approximately 5.537 million people   **40. (a)** approximately 151.449°F   **(b)** approximately 94.706°F   **41.** $\log_3 81 = 4$ **42.** $\log_4\left(\dfrac{1}{64}\right) = -3$

43. $\log_b 5 = 3$ **44.** $\log x = 3.74$ **45.** $8^{1/3} = 2$ **46.** $5^r = 18$ **47.** $e^2 = x + 3$ **48.** $10^{-4} = x$ **49.** $\dfrac{7}{3}$ **50.** 0 **51.** -2 **52.** $\dfrac{3}{2}$ **53.** $\{x \mid x > -5\}$ or $(-5, \infty)$

54. $\left\{x \mid x < \dfrac{7}{3}\right\}$ or $\left(-\infty, \dfrac{7}{3}\right)$ **55.** $\{x \mid x > 0\}$ or $(0, \infty)$ **56.** $\left\{x \mid x > -\dfrac{5}{2}\right\}$ or $\left(-\dfrac{5}{2}, \infty\right)$

57.

58.

59. 3.178   **60.** −0.182   **61.** 2.410   **62.** −0.907   **63.** {17}   **64.** {−9, 1}   **65.** $\left\{\dfrac{3}{2}\right\}$ **66.** {6}   **67.** {−142}

68. $\left\{5\sqrt{3}\right\}$ **69.** 80 decibels   **70.** 100,000 millimeters   **71.** 21   **72.** 9.34   **73.** 1   **74.** 0   **75.** 1

76. 16   **77.** $\log_7 x + \log_7 y - \log_7 z$ **78.** $4 - 2\log_3 x$ **79.** $3 + 4\log r$ **80.** $\dfrac{1}{2}\ln(x - 1) - \dfrac{1}{2}\ln x$

81. $\log_3(x^4 y^2)$ **82.** $\ln\left(\dfrac{7\sqrt[4]{x}}{9}\right)$ **83.** −1   **84.** $\log_6(x - 4)$ **85.** 2.183   **86.** 0.606   **87.** −4.419   **88.** 3.723   **89.** {10}   **90.** {−5}   **91.** $\left\{\sqrt{2}\right\} \approx \{1.414\}$

92. {64}   **93.** $\left\{\dfrac{\log 15}{\log 2}\right\} \approx \{3.907\}$ or $\left\{\dfrac{\ln 15}{\ln 2}\right\} \approx \{3.907\}$ **94.** $\left\{\dfrac{\log 27}{3}\right\} \approx \{0.477\}$ **95.** $\left\{\dfrac{\ln 39}{7}\right\} \approx \{0.523\}$ **96.** $\left\{\dfrac{\log 2}{\log 3 - \log 2}\right\} \approx \{1.710\}$ or

$\left\{\dfrac{\ln 2}{\ln 3 - \ln 2}\right\} \approx \{1.710\}$ **97. (a)** after approximately 1.453 days   **(b)** after approximately 23.253 days   **98. (a)** about 2018   **(b)** about 2033

Chapter 8 Test   **1.** not one-to-one   **2.** $f^{-1}(x) = \dfrac{x + 3}{4}$ **3. (a)** 33.360   **(b)** 36.338   **(c)** 36.494   **(d)** 36.463   **(e)** 36.462   **4.** $\log_4 19 = x$

5. $b^y = x$ **6. (a)** −3   **(b)** 4   **7.** $\left\{x \mid x < \dfrac{7}{4}\right\}$ or $\left(-\infty, \dfrac{7}{4}\right)$

8.

Domain: all real numbers or $(-\infty, \infty)$
Range: $\{y \mid y > 0\}$ or $(0, \infty)$

9.

Domain: $\{x \mid x > 0\}$ or $(0, \infty)$
Range: all real numbers or $(-\infty, \infty)$

10. (a) 10   **(b)** 15   **11.** $\dfrac{1}{2}\log_4 x - 3\log_4 y$ **12.** $\log(M^4 N^3)$ **13.** −8.004   **14.** {1}   **15.** {1, 3}   **16.** {4}   **17.** {−17, 17}   **18.** {9}

19. $\left\{\dfrac{\log 17 + \log 3}{\log 3}\right\} \approx \{3.579\}$ or $\left\{\dfrac{\ln 17 + \ln 3}{\ln 3}\right\} \approx \{3.579\}$ **20.** $\left\{2\sqrt{26}\right\} \approx \{10.198\}$ **21. (a)** approximately 35.5 million people

(b) about 2058   **22.** 10 decibels

Chapter 9 Conics

Section 9.1 Distance and Midpoint Formulas   **1.** $d = \sqrt{(x_2 - x_1)^2 + (y_2 - y_1)^2}$ **2.** False   **3.** 5   **4.** $6\sqrt{5} \approx 13.42$

5. (a)

(b) $d(A, B) = 3\sqrt{5}; d(A, C) = 5\sqrt{5}; d(B, C) = 4\sqrt{5}$ **(c)** $[d(A, C)]^2 = [d(A, B)]^2 + [d(B, C)]^2$

(d) 30 square units   **6.** $M = \left(\dfrac{x_1 + x_2}{2}, \dfrac{y_1 + y_2}{2}\right)$ **7.** $\left(\dfrac{3}{2}, 6\right)$ **8.** $\left(1, \dfrac{5}{2}\right)$ **9.** 5   **11.** $4\sqrt{5} \approx 8.94$

13. 5   **15.** 13   **17.** 6   **19.** $3\sqrt{5} \approx 6.71$ **21.** $3\sqrt{7} \approx 7.94$ **23.** $\sqrt{12.56} \approx 3.54$ **25.** (4, 3)   **27.** (3, −1)

29. $\left(-1, \dfrac{7}{2}\right)$ **31.** $\left(-\dfrac{3}{2}, 0\right)$ **33.** $\left(\dfrac{7\sqrt{2}}{2}, \dfrac{5\sqrt{5}}{2}\right)$ **35.** (0.8, −1.6)

37. (a)

(b) $d(A, B) = 2\sqrt{2} \approx 2.83; d(B, C) = 4\sqrt{2} \approx 5.66;$
$d(A, C) = 2\sqrt{10} \approx 6.32$

(c) $[d(A, B)]^2 + [d(B, C)]^2 \overset{?}{=} [d(A, C)]^2$
$(2\sqrt{2})^2 + (4\sqrt{2})^2 \overset{?}{=} (2\sqrt{10})^2$
$4\cdot 2 + 16\cdot 2 \overset{?}{=} 4\cdot 10$
$8 + 32 \overset{?}{=} 40$
$40 = 40 \leftarrow \text{True}$
Therefore, triangle ABC is a right triangle.
(d) 8 square units

39. (a)

(b) $d(A, B) = 5\sqrt{2} \approx 7.07; d(B, C) = 12\sqrt{2} \approx 16.97;$
$d(A, C) = 13\sqrt{2} \approx 18.38$

(c) $[d(A, B)]^2 + [d(B, C)]^2 \overset{?}{=} [d(A, C)]^2$
$(5\sqrt{2})^2 + (12\sqrt{2})^2 \overset{?}{=} (13\sqrt{2})^2$
$25\cdot 2 + 144\cdot 2 \overset{?}{=} 169\cdot 2$
$50 + 288 \overset{?}{=} 338$
$338 = 338 \leftarrow \text{True}$
Therefore, triangle ABC is a right triangle.
(d) 60 square units

41. (2, −3), (2, 5)   **43.** (−6, −3), (10, −3)   **45. (a)** approximately 37.36 blocks   **(b)** approximately 35.13 blocks   **(c)** approximately 71.34 blocks

47. (a) approximately 1.85 seconds **(b)** No. The ball will reach second base (2.65 seconds) before the runner (3.33 seconds). **49.** Answers will vary. Essentially, the Pythagorean Theorem is a theorem that relates the hypotenuse to the lengths of the legs in a right triangle. The hypotenuse in the right triangle is the distance between the two points in the Cartesian plane. **51.** 9; 3 **53.** 81; 3 **55.** If n is a positive integer and $a^n = b$, then

$$\sqrt[n]{b} = \begin{cases} |a| & \text{if } n \text{ is even} \\ a & \text{if } n \text{ is odd.} \end{cases}$$

Section 9.2 Circles
1. circle **2.** radius **3.** $(x - 2)^2 + (y - 4)^2 = 25$ **4.** $(x + 2)^2 + y^2 = 2$ **5.** False **6.** True

7.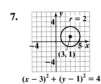
$(x - 3)^2 + (y - 1)^2 = 4$

8.
$(x + 5)^2 + y^2 = 16$

9.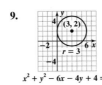
$x^2 + y^2 - 6x - 4y + 4 = 0$

10.
$2x^2 + 2y^2 - 16x + 4y - 38 = 0$

11. $(x - 1)^2 + (y - 2)^2 = 4$
13. $(x - 2)^2 + (y + 1)^2 = 16$
15. $x^2 + y^2 = 9$
17. $(x - 1)^2 + (y - 4)^2 = 4$

19. $(x + 2)^2 + (y - 4)^2 = 36$ **21.** $x^2 + (y - 3)^2 = 16$ **23.** $(x - 5)^2 + (y + 5)^2 = 25$ **25.** $(x - 1)^2 + (y - 2)^2 = 5$

27. $C = (0, 0), r = 6$

29. $C = (4, 1), r = 5$
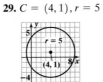

31. $C = (-3, 2), r = 9$
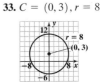

33. $C = (0, 3), r = 8$
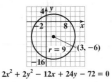

35. $C = (1, -1), r = \dfrac{1}{2}$

37. $C = (3, -1), r = 3$

$x^2 + y^2 - 6x + 2y + 1 = 0$

39. $C = (-5, -2), r = 5$

$x^2 + y^2 + 10x + 4y + 4 = 0$

41. $C = (3, -6), r = 9$

$2x^2 + 2y^2 - 12x + 24y - 72 = 0$

43. $x^2 + y^2 = 20$ **45.** $(x + 3)^2 + (y - 2)^2 = 9$ **47.** $(x + 1)^2 + (y + 1)^2 = 25$ **49.** $A = 64\pi$ square units; $C = 16\pi$ units
51. 32 square units **53.** (a), (e) **55.** The distance formula is used along with the definition of a circle to obtain the equation of the circle.
57. Yes, since it can be written as $x^2 + y^2 = 36$. Center: (0, 0); radius = 6.

59. $f(x) = 4x - 3$

61. $g(x) = x^2 - 4x - 5$

63. $G(x) = -2(x + 3)^2 - 5$

65.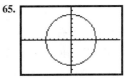
The graph here agrees with that in Problem 27.

67.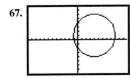
The graph here agrees with that in Problem 29.

69.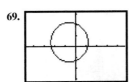
The graph here agrees with that in Problem 31.

71.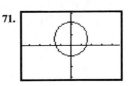
The graph here agrees with that in Problem 33.

73.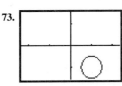
The graph here agrees with that in Problem 35.

Section 9.3 Parabolas
1. parabola **2.** vertex **3.** axis of symmetry

4. $D: x = -2$
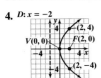

5. $D: x = 5$
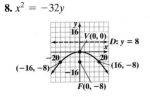

6. $F(0, 1)$
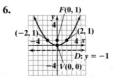

7.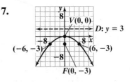

8. $x^2 = -32y$

9. $y^2 = \dfrac{4}{3}x$
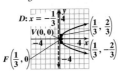

10. $(-3, 2)$

11.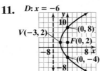

12. The receiver should be located 2 feet from the base of the dish, along its axis of symmetry.

13. c **15.** a **17.** b **19.** e

21. Vertex $(0, 0)$; focus $(0, 6)$; directrix $y = -6$

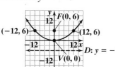

23. Vertex $(0, 0)$; focus $\left(-\dfrac{3}{2}, 0\right)$; directrix $x = \dfrac{3}{2}$

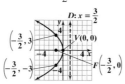

25. Vertex $(0, 0)$; focus $(0, -2)$; directrix $y = 2$

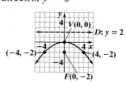

27. $y^2 = 20x$

29. $x^2 = -24y$

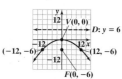

31. $x^2 = 6y$

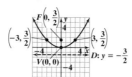

33. $x^2 = -12y$

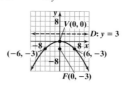

35. $y^2 = -12x$

37. $y^2 = x$

39. Vertex $(2, 4)$; focus $(2, 5)$; directrix $y = 3$

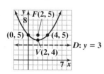

41. Vertex $(-2, -3)$; focus $(-4, -3)$; directrix $x = 0$

43. Vertex $(-5, 1)$; focus $(-5, -4)$; directrix $y = 6$

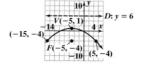

45. Vertex $(-2, -1)$; focus $(-2, -4)$; directrix $y = 2$

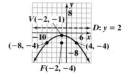

47. Vertex $(1, 4)$; focus $(2, 4)$; directrix $x = 0$

49. Vertex $(-5, 2)$; focus $\left(-5, \dfrac{1}{2}\right)$; directrix $y = \dfrac{7}{2}$

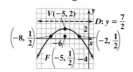

51. 1 inch from the vertex, along the axis of symmetry   **53.** 21.6 feet
55. The height of the bridge is 28.8 feet at a distance of 10 feet from the center, 19.2 feet at a distance of 30 feet from the center, and 0 feet (i.e., ground level) at a distance of 50 feet from the center.
57. $(x - 3)^2 = 4(y + 2)$ **59.** $(y - 3)^2 = -4(x - 2)$

61. (a) Let $x = 4$ and $y = 2$:
$$4^2 \stackrel{?}{=} 8 \cdot 2$$
$$16 = 16 \leftarrow \text{True}$$
Thus, $(4, 2)$ is on the parabola.

(b) The focus of the parabola is $F(0, 2)$, and the directrix is $D: y = -2$.
$$d(F, P) = \sqrt{(0 - 4)^2 + (2 - 2)^2} = \sqrt{16} = 4$$
$$d(P, D) = 2 - (-2) = 4$$
Thus, $d(F, P) = d(P, D) = 4$.
63. 8 units   **65.** See Figure 12.

67. $y = (x + 3)^2$

69. $y = (x + 3)^2$

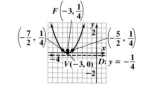

71. $4(y + 2) = (x - 2)^2$

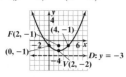

73.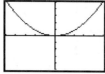

75.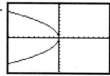

77.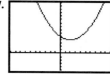

79.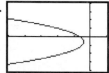

81.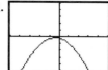

83.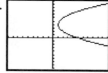

Section 9.4 Ellipses   **1.** ellipse; foci   **2.** major axis   **3.** vertices   **4.** False

5.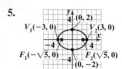

6.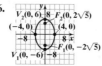

7. $\dfrac{x^2}{40} + \dfrac{y^2}{49} = 1$

8. $(3, -1)$

9.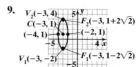

10. $\dfrac{x^2}{2500} + \dfrac{y^2}{1600} = 1$; 40 feet

11. c   **13.** d

15. Foci: $(-3, 0)$ and $(3, 0)$; vertices: $(-5, 0)$ and $(5, 0)$

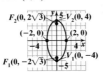

17. Foci: $(0, -8)$ and $(0, 8)$; vertices: $(0, -10)$ and $(0, 10)$

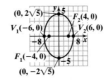

19. Foci: $\left(-3\sqrt{5}, 0\right)$ and $\left(3\sqrt{5}, 0\right)$; Vertices: $(-7, 0)$ and $(7, 0)$

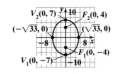

21. Foci: $\left(0, -4\sqrt{3}\right)$ and $\left(0, 4\sqrt{3}\right)$; Vertices: $(0, -7)$ and $(0, 7)$

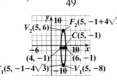

23. Foci: $\left(0, -2\sqrt{3}\right)$ and $\left(0, 2\sqrt{3}\right)$; Vertices: $(0, -4)$ and $(0, 4)$

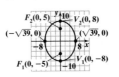

25. $\dfrac{x^2}{36} + \dfrac{y^2}{20} = 1$

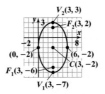

27. $\dfrac{x^2}{33} + \dfrac{y^2}{49} = 1$

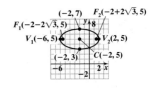

29. $\dfrac{x^2}{100} + \dfrac{y^2}{64} = 1$

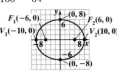

31. $\dfrac{x^2}{39} + \dfrac{y^2}{64} = 1$

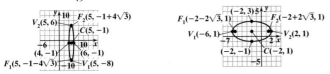

33. $\dfrac{(x-3)^2}{9} + \dfrac{(y+2)^2}{25} = 1$

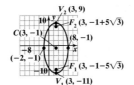

35. $\dfrac{(x+2)^2}{16} + \dfrac{(y-5)^2}{4} = 1$

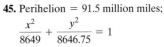

37. $(x-5)^2 + \dfrac{(y+1)^2}{49} = 1$

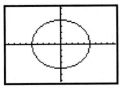

39. $4(x+2)^2 + 16(y-1)^2 = 64$

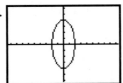

41. $4x^2 + y^2 - 24x + 2y - 63 = 0$

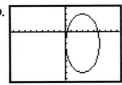

43. (a) $\dfrac{x^2}{225} + \dfrac{y^2}{100} = 1$ (b) Yes

(c) No

45. Perihelion = 91.5 million miles; $\dfrac{x^2}{8649} + \dfrac{y^2}{8646.75} = 1$

47. Perihelion = 460.6 million miles; mean distance = 483.8 million miles; $\dfrac{x^2}{234{,}062.44} + \dfrac{y^2}{233{,}524.2} = 1$ **49.** $\dfrac{(x-1)^2}{16} + \dfrac{(y-2)^2}{9} = 1$

51. $\dfrac{(x-2)^2}{4} + \dfrac{y^2}{16} = 1$

53. Let $a = b$, then

$$\dfrac{x^2}{a^2} + \dfrac{y^2}{b^2} = 1$$
$$\dfrac{x^2}{a^2} + \dfrac{y^2}{a^2} = 1$$
$$a^2\left(\dfrac{x^2}{a^2} + \dfrac{y^2}{a^2}\right) = a^2(1)$$
$$x^2 + y^2 = a^2$$

which is the equation of a circle with center $(0, 0)$ and radius a. $c = 0$; The foci are located at the center point.

55.

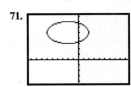

57.

x	5	10	100	1000
$f(x)$	0.71429	0.41667	0.04902	0.00499

59.

x	5	10	100	1000
$f(x)$	5.5	3	2.07216	2.00702

61.

x	5	10	100	1000
$f(x)$	6.83333	11.90909	101.99010	1001.99900
$g(x)$	7	12	102	1002

63. In Problems 57 and 58, the degree of the numerator is less than the degree of the denominator. In Problems 59 and 60, the degree of the numerator and the degree of the denominator are the same.

Conjecture 1: If the degree of the numerator of a rational function is less than the degree of the denominator, then as x increases, the value of the function will approach zero (0).

Conjecture 2: If the degree of the numerator of a rational function equals the degree of the denominator, then as x increases, the value of the function will approach the ratio of the leading coefficients of the numerator and denominator.

65.

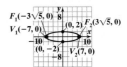

67.

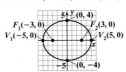

69.

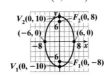

71.

Section 9.5 Hyperbolas   1. hyperbola   2. transverse axis   3. conjugate axis

4.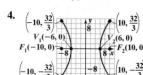

5.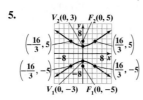

6. False

7. $\dfrac{x^2}{16} - \dfrac{y^2}{20} = 1$

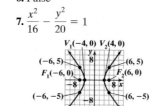

8. $y = -\dfrac{b}{a}x;\; y = \dfrac{b}{a}x$

9.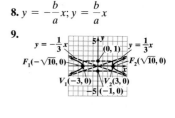

10.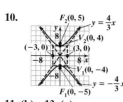

11. (b)   13. (a)

15. $\dfrac{x^2}{4} - \dfrac{y^2}{16} = 1$

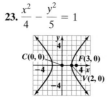

17. $\dfrac{y^2}{25} - \dfrac{x^2}{36} = 1$

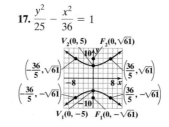

19. $4x^2 - y^2 = 36$

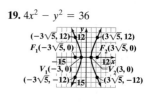

21. $25y^2 - x^2 = 100$

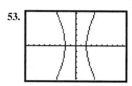

23. $\dfrac{x^2}{4} - \dfrac{y^2}{5} = 1$

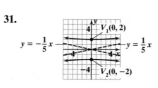

25. $\dfrac{y^2}{25} - \dfrac{x^2}{24} = 1$

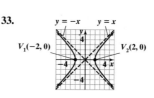

27. $\dfrac{x^2}{49} - \dfrac{y^2}{51} = 1$

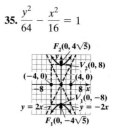

29.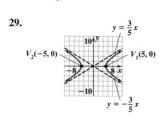

31.

33.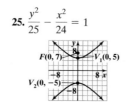

35. $\dfrac{y^2}{64} - \dfrac{x^2}{16} = 1$

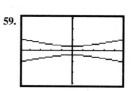

37. $\dfrac{x^2}{4.5} - \dfrac{y^2}{4.5} = 1$

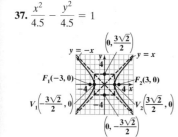

39. $x^2 - y^2 = 1$

41. $\dfrac{y^2}{36} - \dfrac{x^2}{9} = 1$

43. The asymptotes of both hyperbolas are $y = -\dfrac{1}{2}x$ and $y = \dfrac{1}{2}x$. Thus, they are conjugates.

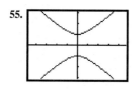

45. Answers will vary.   **47.** $(-3, 1)$

49. $\{(x, y) \mid 2x - 3y = 6\}$

51. $\{(x, y) \mid 6x + 3y = 4\}$

53.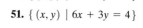

55.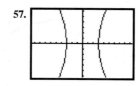

57.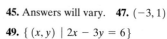

59.

Putting the Concepts Together (Sections 9.1–9.5)   1. $3\sqrt{13}$ 2. $(1, -3)$

3. $C = (-2, 8), r = 6$ **4.** $C = (-3, 2), r = 4$

5. $x^2 + y^2 = 169$ **6.** $(x - 2)^2 + (y - 1)^2 = 25$

7. Vertex: $(-2, 4)$; focus: $(-2, 3)$; directrix: $y = 5$

8. Vertex: $(3, -1)$; focus: $(5, -1)$; directrix: $x = 1$

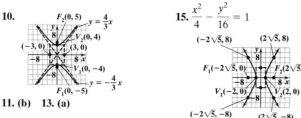

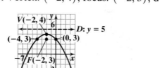

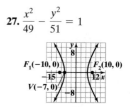

9. $(x + 1)^2 = -12(y + 2)$ **10.** $(y - 3)^2 = 8(x + 3)$

11. Center: $(0, 0)$; foci: $(-6\sqrt{2}, 0)$ and $(6\sqrt{2}, 0)$; vertices: $(-9, 0)$ and $(9, 0)$

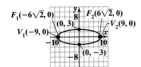

12. Center: $(-1, 2)$; vertices: $(-1, -5)$ and $(-1, 9)$; foci: $(-1, 2 - \sqrt{13})$ and $(-1, 2 + \sqrt{13})$

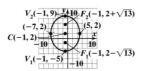

13. $\dfrac{x^2}{45} + \dfrac{y^2}{81} = 1$

14. $\dfrac{(x - 3)^2}{16} + \dfrac{(y + 4)^2}{7} = 1$

15. Vertices: $(0, -9)$ and $(0, 9)$; foci: $\left(0, -3\sqrt{10}\right)$ and $\left(0, 3\sqrt{10}\right)$; asymptotes: $y = 3x$ and $y = -3x$

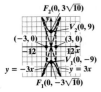

16. Vertices: $(-1, 0)$ and $(1, 0)$; foci: $\left(-\sqrt{26}, 0\right)$ and $\left(\sqrt{26}, 0\right)$; asymptotes: $y = -5x$ and $y = 5x$

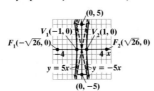

17. $\dfrac{y^2}{4} - \dfrac{x^2}{21} = 1$

18. 6.75 inches from the vertex, along its axis of symmetry

Section 9.6 Systems of Nonlinear Equations   **1.** $(-3, 5)$ and $(1, -3)$ **2.** $\left(\dfrac{11}{5}, -\dfrac{22}{5}\right)$ and $(1, -2)$ **3.** $(0, -4)$, $(-2\sqrt{3}, 2)$, $(2\sqrt{3}, 2)$

4. $\varnothing$ or $\{\ \}$ **5.** $(0, 4)$ and $(1, 5)$ **7.** $(3, 4)$ and $(4, 3)$ **9.** $(0, -2)$, $\left(-\sqrt{3}, 1\right)$, and $\left(\sqrt{3}, 1\right)$ **11.** $(-2, -2)$ and $(2, 2)$

13. $(0, -2)$, $(0, 2)$, $\left(-1, -\sqrt{3}\right)$, and $\left(-1, \sqrt{3}\right)$ **15.** $\varnothing$ **17.** $(0, 0)$, $(-3, 3)$, and $(3, 3)$ **19.** $(-3, 7)$ and $(2, -8)$ **21.** $(-1, 11)$ and $(2, -4)$

23. $(-4, 0)$ and $(4, 0)$ **25.** $(0, 3)$ and $(1, 4)$ **27.** $\left(0, -2 - \sqrt{3}\right)$, $\left(0, -2 + \sqrt{3}\right)$, $(1, -4)$, and $(1, 0)$

29. $\varnothing$ **31.** $(5, 2)$ **33.** $\left(-\dfrac{8}{3}, -\dfrac{2\sqrt{10}}{3}\right)$, $\left(-\dfrac{8}{3}, \dfrac{2\sqrt{10}}{3}\right)$, $\left(\dfrac{8}{3}, -\dfrac{2\sqrt{10}}{3}\right)$, and $\left(\dfrac{8}{3}, \dfrac{2\sqrt{10}}{3}\right)$ **35.** $(0, -5)$, $(3, 4)$, $(4, 3)$ and $(5, 0)$

37. Either -5 and -3, or 3 and 5   **39.** 14 feet by 10 feet   **41.** 19 cm by 10 cm   **43.** $(0, -2)$, $(0, 1)$, and $(2, -1)$ **45.** $(81, 3)$

47. If $r_1 = \dfrac{-b + \sqrt{b^2 - 4ac}}{2a}$, then $r_2 = \dfrac{-b - \sqrt{b^2 - 4ac}}{2a}$; if $r_1 = \dfrac{-b - \sqrt{b^2 - 4ac}}{2a}$, then $r_2 = \dfrac{-b + \sqrt{b^2 - 4ac}}{2a}$.

49. (a) $f(1) = 7$ **(b)** $g(1) = 2$ **51. (a)** $f(3) = 13$ **(b)** $g(3) = 8$ **53. (a)** $f(5) = 19$ **(b)** $g(5) = 32$

55. $(-1, 11)$ and $(2, -4)$ **57.** $(-4, 0)$ and $(4, 0)$ **59.** $(0, 3)$ and $(1, 4)$ **61.** $(0.056, -0.237)$ and $(2.981, -1.727)$ **63.** $\varnothing$

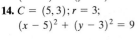

Chapter 9 Review

1. 5   **2.** 10   **3.** $2\sqrt{10} \approx 6.32$ **4.** 6   **5.** $3\sqrt{19} \approx 13.08$ **6.** 2.5   **7.** $(-2, 5)$ **8.** $(6, -2)$ **9.** $\left(-4\sqrt{3}, -3\sqrt{6}\right)$ **10.** $\left(\dfrac{5}{2}, \dfrac{1}{2}\right)$ **11.** $\left(\dfrac{3}{4}, \dfrac{1}{2}\right)$

12. (a)

(b) $d(A, B) = 3\sqrt{2} \approx 4.24$;
$d(B, C) = 2\sqrt{2} \approx 2.83$;
$d(A, C) = \sqrt{26} \approx 5.10$

(c) $[d(A, B)]^2 + [d(B, C)]^2 \overset{?}{=} [d(A, C)]^2$
$\left(3\sqrt{2}\right)^2 + \left(2\sqrt{2}\right)^2 \overset{?}{=} \left(\sqrt{26}\right)^2$
$9 \cdot 2 + 4 \cdot 2 \overset{?}{=} 26$
$18 + 8 \overset{?}{=} 26$
$26 = 26 \leftarrow$ True
Therefore, triangle ABC is a right triangle.

(d) 6 square units

13. $C = (-2, 1)$; $r = 4$; $(x + 2)^2 + (y - 1)^2 = 16$

14. $C = (5, 3)$; $r = 3$; $(x - 5)^2 + (y - 3)^2 = 9$

15. $x^2 + y^2 = 16$

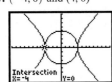

16. $(x + 3)^2 + (y - 1)^2 = 9$

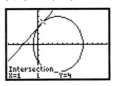

17. $(x - 5)^2 + (y + 2)^2 = 1$

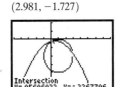

18. $(x - 4)^2 + y^2 = 7$

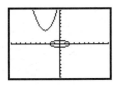

19. $(x - 2)^2 + (y + 1)^2 = 25$
20. $(x + 1)^2 + (y - 3)^2 = 20$

21. $C = (0, 0), r = 5$

22. $C = (1, 2), r = 2$

23. $C = (0, 4), r = 4$
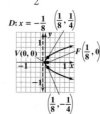

24. $C = (-1, -6), r = 7$
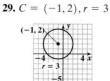

25. $C = \left(-2, \dfrac{3}{2}\right), r = \dfrac{1}{2}$

26. $C = (-3, -3), r = 2$
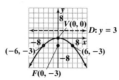

27. $C = (-3, -5), r = 6$

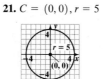

28. $C = (4, -2), r = 2$

29. $C = (-1, 2), r = 3$

30. $C = (5, 1), r = 3$
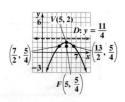

31. $x^2 = -12y$

32. $y^2 = -16x$

33. $y^2 = \dfrac{1}{2}x$

34. $x^2 = 8y$
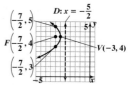

35. Vertex $(0, 0)$; focus $\left(0, \dfrac{1}{2}\right)$; directrix $y = -\dfrac{1}{2}$
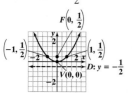

36. Vertex $(0, 0)$; focus $(4, 0)$; directrix $x = -4$

37. Vertex $(-1, 3)$; focus $(-1, 5)$; directrix $y = 1$
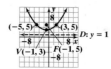

38. Vertex $(-3, 4)$; focus $\left(-\dfrac{7}{2}, 4\right)$; directrix $x = -\dfrac{5}{2}$
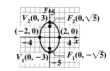
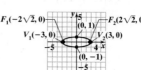

39. Vertex $(5, 2)$; focus $\left(5, \dfrac{5}{4}\right)$; directrix $y = \dfrac{11}{4}$

40. approximately 127.84 feet from the center of the dish, along its axis of symmetry

41. Foci: $\left(-2\sqrt{2}, 0\right)$ and $\left(2\sqrt{2}, 0\right)$; vertices: $(-3, 0)$ and $(3, 0)$
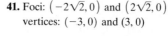

42. Foci: $\left(0, -\sqrt{5}\right)$ and $\left(0, \sqrt{5}\right)$; vertices: $(0, -3)$, and $(0, 3)$

43. $\dfrac{x^2}{16} + \dfrac{y^2}{25} = 1$

44. $\dfrac{x^2}{36} + \dfrac{y^2}{32} = 1$
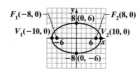

45. $\dfrac{x^2}{100} + \dfrac{y^2}{36} = 1$

46. Center: $(1, -2)$; vertices: $(-6, -2)$ and $(8, -2)$; foci: $\left(1 - 2\sqrt{6}, -2\right)$ and $\left(1 + 2\sqrt{6}, -2\right)$
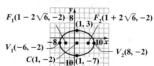

47. Center: $(-3, 4)$; vertices: $(-3, -1)$ and $(-3, 9)$; foci: $(-3, 0)$ and $(-3, 8)$

48. **(a)** $\dfrac{x^2}{900} + \dfrac{y}{256} = 1$
(b) Yes
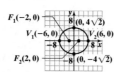

49. Vertices: $(-2, 0)$ and $(2, 0)$; foci: $\left(-\sqrt{13}, 0\right)$ and $\left(\sqrt{13}, 0\right)$

50. Vertices: $(0, -5)$ and $(0, 5)$; foci: $\left(0, -\sqrt{74}\right)$ and $\left(0, \sqrt{74}\right)$
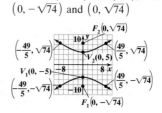
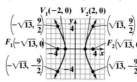

51. Vertices: $(0, -5)$ and $(0, 5)$;
foci: $\left(0, -\sqrt{41}\right)$ and $\left(0, \sqrt{41}\right)$

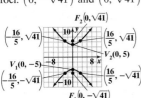

52. Asymptotes: $y = x$ and $y = -x$

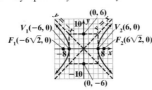

53. Asymptotes: $y = \frac{5}{2}x$ and $y = -\frac{5}{2}x$

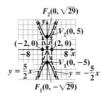

54. $\dfrac{x^2}{9} - \dfrac{y^2}{7} = 1$

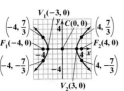

55. $\dfrac{y^2}{9} - \dfrac{x^2}{16} = 1$

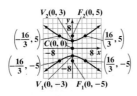

56. $\dfrac{y^2}{16} - \dfrac{x^2}{9} = 1$

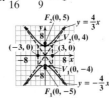

57. $\left(\sqrt{2}, \sqrt{2}\right)$ and $\left(-\sqrt{2}, -\sqrt{2}\right)$

58. $\left(-\dfrac{1}{2}, \dfrac{3}{2}\right)$ and $(1, 3)$

59. $\left(-\dfrac{1}{6}, -6\right)$ and $(1, 1)$

60. $(-5, -1)$, $(-5, 1)$, $(5, -1)$, and $(5, 1)$

61. $\left(2, -2\sqrt{2}\right)$ and $\left(2, 2\sqrt{2}\right)$

62. $(-1, 3)$ and $(1, 3)$

63. $\left(-\sqrt{3}, -\sqrt{5}\right)$, $\left(-\sqrt{3}, \sqrt{5}\right)$, $\left(\sqrt{3}, -\sqrt{5}\right)$, and $\left(\sqrt{3}, \sqrt{5}\right)$ **64.** $(0, 0)$, $\left(5, -\sqrt{15}\right)$, and $\left(5, \sqrt{15}\right)$ **65.** $(2, 4)$ and $(-1, 1)$

66. $(-7, -20)$ and $\left(2, \dfrac{5}{2}\right)$ **67.** $(0, 6)$ and $(-6, 0)$ **68.** $(4, 4)$ and $(1, -2)$ **69.** $\varnothing$ **70.** $(-1, -2)$, $(-1, 2)$, $(1, -2)$, and $(1, 2)$

71. $(-4, 0)$, $(4, 0)$, and $(0, 4)$ **72.** $(4, 0)$ and $(0, -2)$ **73.** 7 and 5 **74.** 12 cm by 5 cm **75.** 72 inches by 30 inches **76.** 12 feet and 9 feet

Chapter 9 Test **1.** $4\sqrt{5}$ **2.** $(-1, 2)$

3. $C = (4, -1)$, $r = 3$

4. $C = (-5, 2)$, $r = 4$

5. $(x + 3)^2 + (y - 7)^2 = 36$

6. $(x + 5)^2 + (y - 8)^2 = 100$

7. Vertex $(1, -2)$; focus $(2, -2)$;
directrix $x = 0$

8. Vertex $(2, 4)$; focus $\left(2, \dfrac{13}{4}\right)$;
directrix $y = \dfrac{19}{4}$

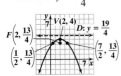

9. $x^2 = -16y$
10. $(y - 4)^2 = 8(x - 1)$

11. Foci: $(-4, 0)$ and $(4, 0)$;
vertices: $(-5, 0)$ and $(5, 0)$

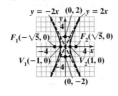

12. Foci: $\left(2, -4 - \sqrt{7}\right)$ and $\left(2, -4 + \sqrt{7}\right)$
vertices: $(2, -8)$ and $(2, 0)$;

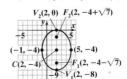

13. $\dfrac{x^2}{9} + \dfrac{y^2}{25} = 1$

14. $\dfrac{(x + 1)^2}{16} + \dfrac{(y - 2)^2}{25} = 1$

15. Vertices: $(-1, 0)$ and $(1, 0)$; foci: $\left(-\sqrt{5}, 0\right)$ and
$\left(\sqrt{5}, 0\right)$; asymptotes: $y = 2x$ and $y = -2x$

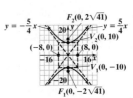

16. Vertices: $(0, -10)$ and $(0, 10)$; foci: $\left(0, -2\sqrt{41}\right)$
and $\left(0, 2\sqrt{41}\right)$; asymptotes: $y = \dfrac{5}{4}x$ and $y = -\dfrac{5}{4}x$

17. $\dfrac{x^2}{9} - \dfrac{y^2}{55} = 1$ **18.** $(-4, 1)$ and $(1, -4)$ **19.** $\left(-\sqrt{5}, -2\right)$, $\left(-\sqrt{5}, 2\right)$, $\left(\sqrt{5}, -2\right)$, and $\left(\sqrt{5}, 2\right)$ **20.** **(a)** $\dfrac{x^2}{225} + \dfrac{y^2}{100} = 1$ **(b)** 6 feet

Cumulative Review Chapters R–9 **1.** $\{x \mid x \neq 7\}$

2.

3. $y = \dfrac{4}{3}x + 2$ or $4x - 3y = -6$ **4.** $(3, -8)$ **5.** -35 **6.** $(2m + n)(4m^2 - 2mn + n^2)$ **7.** $(3p - 4)(3p - 2)$

8. $\dfrac{2a + 1}{a - 7}$ **9.** $\dfrac{1}{(t - 2)(t - 1)}$ **10.** $\{2\}$ **11.** $\{x \mid x < -4 \text{ or } x \geq 5\}$ or $(-\infty, -4) \cup [5, \infty)$ **12.** $9\sqrt{3}$

13. $-3\left(2 - \sqrt{7}\right)$ **14.** $\{5\}$ **15.** $31 + 5i$ **16.** $\left\{\dfrac{2 - \sqrt{10}}{2}, \dfrac{2 + \sqrt{10}}{2}\right\}$

17.   **18.** $\{x \mid -1 \le x \le 4\}$ or $[-1, 4]$ **20.** Domain: $\{x \mid x > 0\}$ or $(0, \infty)$;   **21.** $\dfrac{5}{2}$ **22.** $\{2\}$ **23.** $\{-5, 5\}$

19. Domain: all real numbers or $(-\infty, \infty)$;   Range: all real numbers or $(-\infty, \infty)$ **24.** $(x + 3)^2 + (y - 1)^2 = 10$

Range: $\{y \mid y > 0\}$ or $(0, \infty)$

25. Vertex: $(-3, 3)$; focus: $(-1, 3)$;

      directrix: $x = -5$

Chapter 10 Sequences, Series, and the Binomial Theorem

Section 10.1 Sequences   **1.** sequence   **2.** infinite; finite   **3.** True   **4.** $-1, 1, 3, 5, 7$ **5.** $-4, 8, -12, 16, -20$ **6.** $a_n = 2n + 3$ **7.** $b_n = \dfrac{(-1)^{n+1}}{n + 1}$

8. partial sum   **9.** $3 + 7 + 11 = 21$ **10.** $2 + 9 + 28 + 65 + 126 = 230$ **11.** $\sum_{k=1}^{12} k^2$ **12.** $\sum_{n=1}^{6} \left(\dfrac{1}{2}\right)^{n-1}$ **13.** $8, 11, 14, 17, 20$ **15.** $\dfrac{1}{3}, \dfrac{1}{2}, \dfrac{3}{5}, \dfrac{2}{3}, \dfrac{5}{7}$

17. $-1, 2, -3, 4, -5$ **19.** $3, 5, 9, 17, 33$ **21.** $1, 1, \dfrac{3}{4}, \dfrac{1}{2}, \dfrac{5}{16}$ **23.** $\dfrac{1}{e}, \dfrac{2}{e^2}, \dfrac{3}{e^3}, \dfrac{4}{e^4}, \dfrac{5}{e^5}$ **25.** $a_n = 2n$ **27.** $a_n = \dfrac{n}{n + 1}$ **29.** $a_n = n^2 + 2$ **31.** $a_n = (-1)^n n^2$

33. 54 **35.** $\dfrac{55}{2}$ **37.** 14 **39.** 6 **41.** 50 **43.** 45 **45.** $\sum_{k=1}^{15} k$ **47.** $\sum_{i=1}^{12} \dfrac{1}{i}$ **49.** $\sum_{i=1}^{9} (-1)^{i+1}\left(\dfrac{1}{3^{i-1}}\right)$ **51.** $\sum_{k=1}^{11} (2k + 3)$ **53. (a)** \$12,180   **(b)** \$12,736.36

(c) \$21,768.22   **55. (a)** 323 million   **(b)** 474 million   **57.** $1, 1, 2, 3, 5, 8, 13, 21, 34, 55$ **59.** $10, 10.5, 11.025, 11.57625,$ and 12.1550625 **61.** $8, 10, 13, 17,$ and 22

63. (a) $1; 2; 1.5; 1.\overline{6}; 1.6; 1.625; \dfrac{21}{13} \approx 1.615385; \dfrac{34}{21} \approx 1.619048; \dfrac{55}{34} \approx 1.617647; \dfrac{89}{55} \approx 1.618182$ **(b)** around 1.618   **(c)** $1; 0.5; \dfrac{2}{3} = 0.\overline{6}; 0.6; 0.625;$

$\dfrac{8}{13} \approx 0.615385; \dfrac{13}{21} \approx 0.619048; \dfrac{21}{34} \approx 0.617647; \dfrac{34}{55} \approx 0.618182; \dfrac{55}{89} \approx 0.617978$ **(d)** around 0.618   **65.** The main difference between a function and a sequence is that the domain of a function is based on the real number system, whereas the domain of a sequence is the positive integers.   **67.** The symbol $\sum$ means to add up the terms of a sequence.   **69. (a)** $m = 4$ **(b)** $f(1) = -2; f(2) = 2; f(3) = 6; f(4) = 10$ **71. (a)** $m = -5$

(b) $f(1) = 3; f(2) = -2; f(3) = -7; f(4) = -12$ **73.** $8, 11, 14, 17,$ and 20 **75.** $\dfrac{1}{3}, \dfrac{1}{2}, \dfrac{3}{5}, \dfrac{2}{3},$ and $\dfrac{5}{7}$ **77.** $-1, 2, -3, 4,$ and -5 **79.** $3, 5, 9, 17,$ and 33

81. 54 **83.** $\dfrac{55}{2}$ **85.** 14 **87.** 6

Section 10.2 Arithmetic Sequences   **1.** arithmetic   **2.** Arithmetic $a_1 = -3; d = 2$ **3.** Not arithmetic   **4.** Arithmetic $a_1 = -5; d = 3$ **5.** Not arithmetic

6. Arithmetic $a_1 = 3; d = -2$ **7.** $a_n = a_1 + (n - 1)d$ **8. (a)** $a_n = 6n - 5$ **(b)** 79 **9. (a)** $a_1 = -5; d = 3$ **(b)** $a_n = 3n - 8$ **10.** $10,400$ **11.** 9730

12. 7550 **13.** -1050 **14.** 1470 seats   **15.** $a_n - a_{n-1} = d = 1; a_1 = 6$ **17.** $a_n - a_{n-1} = d = 7; a_1 = 9$ **19.** $a_n - a_{n-1} = d = -3; a_1 = 4$

21. $a_n - a_{n-1} = d = \dfrac{1}{2}; a_1 = \dfrac{11}{2}$ **23.** $a_n = 3n + 1; a_5 = 16$ **25.** $a_n = -5n + 15; a_5 = -10$ **27.** $a_n = \dfrac{1}{3}n + \dfrac{5}{3}; a_5 = \dfrac{10}{3}$ **29.** $a_n = -\dfrac{1}{5}n + \dfrac{26}{5};$

$a_5 = \dfrac{21}{5}$ **31.** $a_n = 5n - 3; a_{20} = 97$ **33.** $a_n = -3n + 15; a_{20} = -45$ **35.** $a_n = \dfrac{1}{4}n + \dfrac{3}{4}; a_{20} = \dfrac{23}{4}$ **37.** $a_1 = 7; d = 5; a_n = 5n + 2$

39. $a_1 = -23; d = 7; a_n = 7n - 30$ **41.** $a_1 = 11; d = -3; a_n = -3n + 14$ **43.** $a_1 = 4; d = -\dfrac{1}{2}; a_n = -\dfrac{1}{2}n + \dfrac{9}{2}$ **45.** $S_{30} = 2670$ **47.** $S_{25} = 700$

49. $S_{40} = -5060$ **51.** $S_{40} = 3160$ **53.** $S_{75} = -9000$ **55.** $S_{30} = 460$ **57.** $x = -\dfrac{3}{2}$ **59.** There are 630 cans in the stack.   **61.** There are 1600 seats in the

auditorium.   **63.** There are 84 terms in the sequence.   **65.** There are 63 terms in the sequence.   **67.** It will take about 15.22 years.   **69.** A sequence is arithmetic if the difference in consecutive terms is constant. Thus, if the actual terms are given, compute the difference between each pair of consecutive terms and determine whether the differences are constant. If a formula is given for the sequence, compute $a_n - a_{n-1}$ and determine whether the result is a constant — if it is, then the sequence is arithmetic.   **71. (a)** 3 **(b)** $3; 9; 27; 81$ **73. (a)** $\dfrac{1}{2}$ **(b)** $5; \dfrac{5}{2}; \dfrac{5}{4}; \dfrac{5}{8}$ **75.** 806.9 **77.** 1427.5

Section 10.3 Geometric Sequences and Series   **1.** geometric   **2.** Geometric; $a_1 = 4, r = 2$ **3.** Not geometric   **4.** Geometric; $a_1 = 9, r = \dfrac{1}{3}$

5. Geometric; $r = 5$ **6.** Not geometric   **7.** Geometric; $r = \dfrac{2}{3}$ **8.** $a_n = 5 \cdot 2^{n-1}; a_9 = 1280$ **9.** $a_n = 50 \cdot \left(\dfrac{1}{2}\right)^{n-1}; a_9 = 0.1953125$ or $a_9 = \dfrac{25}{128}$

10. $S_n = a_1 \cdot \dfrac{1 - r^n}{1 - r}$ **11.** $24,573$ **12.** 7.9921875 **13.** $\dfrac{a_1}{1 - r}$ **14.** $\dfrac{40}{3}$ **15.** $\dfrac{1}{2}$ **16.** $\dfrac{2}{9}$ **17.** The U.S. economy will grow by \$10,000.   **18.** \$244,129.08

19. $\dfrac{a_n}{a_{n-1}} = r = 4; a_1 = 4$ **21.** $\dfrac{a_n}{a_{n-1}} = r = \dfrac{2}{3}; a_1 = \dfrac{2}{3}$ **23.** $\dfrac{a_n}{a_{n-1}} = r = \dfrac{1}{2}; a_1 = \dfrac{3}{2}$ **25.** $\dfrac{a_n}{a_{n-1}} = r = \dfrac{5}{2}; a_1 = \dfrac{1}{2}$ **27. (a)** $a_n = 10 \cdot 2^{n-1}$ **(b)** $a_8 = 1280$

29. (a) $a_n = 100 \cdot \left(\dfrac{1}{2}\right)^{n-1}$ **(b)** $a_8 = \dfrac{25}{32}$ **31. (a)** $a_n = (-3)^{n-1}$ **(b)** $a_8 = -2187$ **33. (a)** $a_n = 100 \cdot (1.05)^{n-1}$ **(b)** $a_8 = 100 \cdot (1.05)^7 \approx 140.71$

35. $a_{10} = 1536$ **37.** $a_{15} = \dfrac{1}{4096}$ **39.** $a_9 = 0.000000005$ **41.** 8190 **43.** 83.3245952 **45.** 6138 **47.** 7.96875 **49.** 2 **51.** 15 **53.** $\dfrac{9}{2}$ **55.** $\dfrac{5}{4}$ **57.** 9

59. $\dfrac{5}{9}$ **61.** $\dfrac{89}{99}$ **63.** arithmetic; $d = 5$ **65.** Neither   **67.** geometric; $r = \dfrac{1}{2}$ **69.** geometric; $r = \dfrac{2}{3}$ **71.** arithmetic; $d = 4$ **73.** Neither   **75.** $x = -4$

77. (a) \$42,000   **(b)** \$62,053   **(c)** \$503,116   **79.** \$13,182   **81. (a)** About 1.891 feet   **(b)** On the 23rd swing   **(c)** About 24.08 feet   **(d)** The pendulum will have swung a total of 60 feet.   **83.** Option A will yield the larger annual salary in the final year of the contract, and option B will yield the larger cumulative salary over the life of the contract.   **85.** The multiplier is 50.   **87.** \$31.14 per share   **89.** \$149,035.94   **91.** \$114,401.52   **93.** \$395.09, or about \$395

95. $0.4\overline{9} = \dfrac{1}{2}$ **97.** 2,147,483,646   **99.** A geometric sequence with $r > 1$ yields faster growth because of the effect of compounding. Answers will vary. However, a reasonable answer is that because geometric growth is faster than arithmetic growth, the population will grow beyond the ability of food supplies to sustain the population, and hunger will ensue.   **101.** A geometric series has a sum provided that the common ratio r is between -1 and 1.
103. 1   **105.** approximately 288.1404315   **107.** approximately 41.66310217

Putting the Concepts Together (Sections 10.1–10.3)   **1.** Geometric with $a_1 = \dfrac{3}{4}$ and common ratio $r = \dfrac{1}{4}$ **2.** Arithmetic with $a_1 = 8$ and $d = 2$
3. Arithmetic with $a_1 = 1$ and $d = \dfrac{7}{9}$ **4.** Neither arithmetic nor geometric   **5.** Geometric with $a_1 = 12$ and $r = 2$

6. Neither arithmetic nor geometric   **7.** 87   **8.** $\displaystyle\sum_{i=1}^{12} \dfrac{1}{2(6+i)}$ **9.** $a_n = 27 - 2n$; 25, 23, 21, 19, and 17   **10.** $a_n = 11n - 35$; $-24, -13, -2$,
9, and 20   **11.** $a_n = 45 \cdot \left(\dfrac{1}{5}\right)^{n-1}$; $45, 9, \dfrac{9}{5}, \dfrac{9}{25}$, and $\dfrac{9}{125}$ **12.** $a_n = 150 \cdot (1.04)^{n-1}$; 150, 156, 162.24, 168.7296, and 175.478784   **13.** $S_{11} = 177{,}146$

14. $S_{20} = 990$ **15.** $\dfrac{10{,}000}{9}$ **16.** A party of 24 people would require 11 tables.

Section 10.4 The Binomial Theorem   **1.** $n(n-1)(n-2) \cdot \cdots \cdot 3 \cdot 2 \cdot 1$ **2.** $1; 1$ **3.** 120   **4.** 840   **5.** False   **6.** True   **7.** 7   **8.** 20

9. $x^4 + 8x^3 + 24x^2 + 32x + 16$ **10.** $32p^5 - 80p^4 + 80p^3 - 40p^2 + 10p - 1$ **11.** 6   **13.** 40,320   **15.** 90   **17.** 336   **19.** 21   **21.** 210
23. $x^5 + 5x^4 + 10x^3 + 10x^2 + 5x + 1$ **25.** $x^4 - 16x^3 + 96x^2 - 256x + 256$ **27.** $81p^4 + 216p^3 + 216p^2 + 96p + 16$
29. $32z^5 - 240z^4 + 720z^3 - 1080z^2 + 810z - 243$ **31.** $x^8 + 8x^6 + 24x^4 + 32x^2 + 16$ **33.** $32p^{15} + 80p^{12} + 80p^9 + 40p^6 + 10p^3 + 1$
35. $x^6 + 12x^5 + 60x^4 + 160x^3 + 240x^2 + 192x + 64$ **37.** $16p^8 - 32p^6q^2 + 24p^4q^4 - 8p^2q^6 + q^8$ **39.** 1.00401   **41.** 0.99004   **43.** $84x^5$
45. $-108{,}864p^3$ **47.** $\dbinom{n}{n-1} = \dfrac{n!}{(n-1)!(n-(n-1))!} = \dfrac{n!}{(n-1)!1!} = \dfrac{n \cdot (n-1)!}{(n-1)!} = n; \dbinom{n}{n} = \dfrac{n!}{n!(n-n)!} = \dfrac{n!}{n!0!} = \dfrac{n!}{n!} = 1$ **49.**
$$\begin{matrix} & & 1 & & \\ & 1 & & 1 & \\ 1 & & 2 & & 1 \\ 1 & 3 & & 3 & 1 \end{matrix}$$
51. The degree of each monomial equals n.   **53.** $a^4 - 8a^3 + 24a^2 - 32a + 16$ **55.** $p^5 + p^4 - 6p^3 - 14p^2 - 11p - 3$

Chapter 10 Review   **1.** $-1, -4, -7, -10$, and -13 **2.** $-\dfrac{1}{5}, 0, \dfrac{1}{7}, \dfrac{1}{4}$, and $\dfrac{1}{3}$ **3.** 6, 26, 126, 626, and 3126   **4.** 3, -6, 9, -12, and 15   **5.** $\dfrac{1}{2}, \dfrac{4}{3}, \dfrac{9}{4}, \dfrac{16}{5}$, and $\dfrac{25}{6}$
6. $\pi, \dfrac{\pi^2}{2}, \dfrac{\pi^3}{3}, \dfrac{\pi^4}{4}$, and $\dfrac{\pi^5}{5}$ **7.** $a_n = -3n$ **8.** $a_n = \dfrac{n}{3}$ **9.** $a_n = 5 \cdot 2^{n-1}$ **10.** $a_n = (-1)^n \cdot \dfrac{n}{2}$ **11.** $a_n = n^2 + 5$ **12.** $a_n = \dfrac{n-1}{n+1}$ **13.** 65   **14.** $\dfrac{33}{2}$ **15.** -30
16. $\dfrac{26}{3}$ **17.** $\displaystyle\sum_{i=1}^{15}(4 + 3i)$ **18.** $\displaystyle\sum_{i=1}^{8}\dfrac{1}{3^i}$ **19.** $\displaystyle\sum_{i=1}^{10}\dfrac{i^3 + 1}{i + 1}$ **20.** $\displaystyle\sum_{i=1}^{7}[(-1)^{i-1} \cdot i^2]$ **21.** arithmetic with $d = 6$ **22.** arithmetic with $d = \dfrac{3}{2}$ **23.** not arithmetic
24. not arithmetic   **25.** arithmetic with $d = 4$ **26.** not arithmetic   **27.** $a_n = 8n - 5$; $a_{25} = 195$ **28.** $a_n = -3n - 1$; $a_{25} = -76$ **29.** $a_n = -\dfrac{1}{3}n + \dfrac{22}{3}$;
$a_{25} = -1$ **30.** $a_n = 6n + 5$; $a_{25} = 155$ **31.** $a_n = \dfrac{18}{5}n - \dfrac{19}{5}$; $a_{25} = \dfrac{431}{5}$ **32.** $a_n = -4n - 4$; $a_{25} = -104$ **33.** 4320   **34.** -2140 **35.** -4080
36. $\dfrac{1875}{4}$, or 468.75   **37.** 2106   **38.** 300 yards   **39.** geometric with $r = 6$ **40.** geometric with $r = -3$ **41.** not geometric   **42.** geometric with $r = \dfrac{2}{3}$
43. geometric with $r = -2$ **44.** not geometric   **45.** $a_n = 4 \cdot 3^{n-1}$; $a_{10} = 78{,}732$ **46.** $a_n = 8 \cdot \left(\dfrac{1}{4}\right)^{n-1}$; $a_{10} = \dfrac{1}{32{,}768}$ **47.** $a_n = 5 \cdot (-2)^{n-1}$;
$a_{10} = -2560$ **48.** $a_n = 1000 \cdot (1.08)^{n-1}$; $a_{10} \approx 1999.005$ **49.** 65,534   **50.** ≈ 45.71428571 **51.** $\dfrac{12{,}285}{4}$ or 3071.25   **52.** $-258{,}280{,}320$ **53.** $\dfrac{20}{3}$ **54.** $\dfrac{100}{3}$
55. $\dfrac{5}{4}$ **56.** $\dfrac{8}{9}$ **57.** After 72 years, there will be 3.125 grams of the tritium remaining.   **58.** After 15 minutes, about 38.15 billion e-mails will have been sent.
59. After 25 years, Scott's 403(b) will be worth $360,114.89.   **60.** The lump sum option would yield more money after 26 years.   **61.** Sheri would need to contribute $534.04, or about $534, each month to reach her goal.   **62.** When Samantha turns 18, the plan will be worth $62,950.79 and will cover about 185 credit hours.   **63.** 120   **64.** 7920   **65.** 5040   **66.** 1716   **67.** 35   **68.** 252   **69.** 1   **70.** 1   **71.** $z^4 + 4z^3 + 6z^2 + 4z + 1$
72. $y^5 - 15y^4 + 90y^3 - 270y^2 + 405y - 243$ **73.** $729y^6 + 5832y^5 + 19{,}440y^4 + 34{,}560y^3 + 34{,}560y^2 + 18{,}432y + 4096$
74. $16x^8 - 96x^6 + 216x^4 - 216x^2 + 81$ **75.** $81p^4 - 216p^3q + 216p^2q^2 - 96pq^3 + 16q^4$ **76.** $a^{15} + 15a^{12}b + 90a^9b^2 + 270a^6b^3 + 405a^3b^4 + 243b^5$
77. $-448x^5$ **78.** $14{,}784x^5$

Chapter 10 Test   **1.** arithmetic with $a_1 = -15$ and $d = 8$ **2.** geometric with $a_1 = -4$ and $r = -4$ **3.** neither arithmetic nor geometric
4. arithmetic with $a_1 = -\dfrac{1}{5}$ and $d = \dfrac{2}{5}$ **5.** neither arithmetic nor geometric   **6.** geometric with $a_1 = 21$ and $r = 3$ **7.** $\dfrac{17{,}269}{1200}$ **8.** $\displaystyle\sum_{i=1}^{8}\dfrac{i + 2}{i + 4}$
9. $a_n = 10n - 4$; 6, 16, 26, 36, and 46   **10.** $a_n = 4 - 4n$; 0, -4, -8, -12, and -16 **11.** $a_n = 10 \cdot 2^{n-1}$; 10, 20, 40, 80, and 160
12. $a_n = (-3)^{n-1}$; 1, -3, 9, -27, and 81   **13.** 720   **14.** $-\dfrac{132{,}860}{9}$ **15.** 324   **16.** 6435   **17.** 792   **18.** $625m^4 - 1000m^3 + 600m^2 - 160m + 16$
19. $6103.11   **20.** 8000 lb

Applications Index

Subject Index

Photo Credits

Working with Radicals (Getting Ready for Chapter 6, Chapter 6)

Simplifying $\sqrt[n]{a^n}$

- If $n \geq 2$ is a positive integer and a is a real number, then

  $\sqrt[n]{a^n} = a$ if $n \geq 3$ is odd

  $\sqrt[n]{a^n} = |a|$ if $n \geq 2$ is even

- If a is a real number and $n \geq 2$ is an integer, then $a^{\frac{1}{n}} = \sqrt[n]{a}$ provided that $\sqrt[n]{a}$ exists.

- If a is a real number, m/n is a rational number in lowest terms with $n \geq 2$, then $a^{\frac{m}{n}} = \sqrt[n]{a^m} = \left(\sqrt[n]{a}\right)^m$ provided that $\sqrt[n]{a}$ exists.

- If m/n is a rational number and if a is a nonzero real number, then $a^{-\frac{m}{n}} = \dfrac{1}{a^{\frac{m}{n}}}$ or $\dfrac{1}{a^{-\frac{m}{n}}} = a^{\frac{m}{n}}$

Product Property of Radicals

If $\sqrt[n]{a}$ and $\sqrt[n]{b}$ are real numbers and $n \geq 2$ is an integer, then $\sqrt[n]{a} \cdot \sqrt[n]{b} = \sqrt[n]{ab}$

Quotient Property of Radicals

If $\sqrt[n]{a}$ and $\sqrt[n]{b}$ are real numbers, $b \neq 0$ and $n \geq 2$ is an integer, then $\dfrac{\sqrt[n]{a}}{\sqrt[n]{b}} = \sqrt[n]{\dfrac{a}{b}}$

Quadratic Equations and Quadratic Functions (Chapter 7)

- **Square Root Property**

  If $x^2 = p$, then $x = \sqrt{p}$ or $x = -\sqrt{p}$

- **Pythagorean Theorem**

  In a right triangle, the square of the length of the hypotenuse is equal to the sum of the squares of the lengths of the legs. That is, $\text{leg}^2 + \text{leg}^2 = \text{hypotenuse}^2$.

- **The Quadratic Formula**

  The solutions to the equation $ax^2 + bx + c = 0$, $a \neq 0$, are given by $x = \dfrac{-b \pm \sqrt{b^2 - 4ac}}{2a}$

- **Discriminant**

  For the quadratic equation $ax^2 + bx + c = 0$, $a \neq 0$:

  - If $b^2 - 4ac > 0$, the equation has two unequal real solutions.
    - If $b^2 - 4ac$ is a perfect square, the equation has two rational solutions.
    - If $b^2 - 4ac$ is not a perfect square, the equation has two irrational solutions.
  - If $b^2 - 4ac = 0$, the equation has a repeated real solution.
  - If $b^2 - 4ac < 0$, the equation has two complex solutions that are not real.

- **Vertex of a Parabola**

  Any quadratic function of the form $f(x) = ax^2 + bx + c$, $a \neq 0$, will have vertex $\left(-\dfrac{b}{2a}, f\left(-\dfrac{b}{2a}\right)\right)$

- **The x-Intercepts of the Graph of a Quadratic Function**

  1. If $b^2 - 4ac > 0$, the graph of $f(x) = ax^2 + bx + c$ has two different x-intercepts.
  2. If $b^2 - 4ac = 0$, the graph of $f(x) = ax^2 + bx + c$ has one x-intercept.
  3. If $b^2 - 4ac < 0$, the graph of $f(x) = ax^2 + bx + c$ has no x-intercepts.

Properties of Logarithms (Chapter 8)

$a^{\log_a M} = M$ $\log_a a^r = r$

$\log_a(MN) = \log_a M + \log_a N$

$\log_a\left(\dfrac{M}{N}\right) = \log_a M - \log_a N$

$\log_a M^r = r \log_a M$

$\log_a M = \dfrac{\log_b M}{\log_b a} = \dfrac{\log M}{\log a} = \dfrac{\ln M}{\ln a}$

Formulas from Chapter 9

The Distance Formula

$d(P_1, P_2) = \sqrt{(x_2 - x_1)^2 + (y_2 - y_1)^2}$

The Midpoint Formula

$M = \left(\dfrac{x_1 + x_2}{2}, \dfrac{y_1 + y_2}{2}\right)$

Standard Form of an Equation of a Circle

$(x - h)^2 + (y - k)^2 = r^2$ with radius r and center (h, k)

General Form of the Equation of a Circle

$x^2 + y^2 + ax + by + c = 0$ when the graph exists

Formulas for Lines and Slope (Chapter 1)

Standard form of a line	$Ax + By = C$
Equation of a vertical line	$x = a$ where a is the x-intercept
Equation of a horizontal line	$y = b$ where b is the y-intercept
Slope of a line	$m = \dfrac{y_2 - y_1}{x_2 - x_1}$, $x_1 \neq x_2$ Slope undefined if $x_1 = x_2$
Point-slope form of a line	$y - y_1 = m(x - x_1)$
Slope-intercept form of a line	$y = mx + b$

Functions (Chapter 2)

- A **function** is a special type of relation where any given input, x, corresponds to only one output y. Functions can be represented through maps, sets of ordered pairs, equations, or graphs.
- **Vertical Line Test:** A set of points in the xy-plane is the graph of a function if and only if every vertical line intersects the graph in at most one point.
- The graph of a function, f, is the set of all ordered pairs $(x, f(x))$.
- When only an equation of a function is given, the **domain** of the function is the largest set of real numbers for which $f(x)$ is a real number.
- The **range** of a function is the set of all outputs of the function.

Steps for Factoring (Chapter 4)

Step 1: Factor out the Greatest Common Factor (GCF), if any exists.

Step 2: Count the number of terms.

Step 3: (a) 2 terms
- Is it the difference of two squares? If so, $A^2 - B^2 = (A - B)(A + B)$
- Is it the difference of two cubes? If so, $A^3 - B^3 = (A - B)(A^2 + AB + B^2)$
- Is it the sum of two cubes? If so, $A^3 + B^3 = (A + B)(A^2 - AB + B^2)$

(b) 3 terms
- Is it a perfect square trinomial? If so, $A^2 + 2AB + B^2 = (A + B)^2$ or $A^2 - 2AB + B^2 = (A - B)^2$
- Is the coefficient of the square term 1? If so, $x^2 + bx + c = (x + m)(x + n)$ where $mn = c$ and $m + n = b$
- Is the coefficient of the square term different from 1? If so,
  - **a.** Use factoring by grouping
  - **b.** Use trial and error

(c) 4 terms
- Use factoring by grouping

Step 4: Check your work by multiplying out the factored form.

The Rules of Exponents (Getting Ready for Chapter 4, Chapter 6)

If a and b are real numbers and if r and s are rational numbers, then assuming the expression is defined,

Zero Exponent Rule:	$a^0 = 1$	if $a \neq 0$
Negative Exponent Rule:	$a^{-r} = \dfrac{1}{a^r}$	if $a \neq 0$
Product Rule:	$a^r \cdot a^s = a^{r+s}$	
Quotient Rule:	$\dfrac{a^r}{a^s} = a^{r-s} = \dfrac{1}{a^{s-r}}$	if $a \neq 0$
Power Rule:	$(a^r)^s = a^{r \cdot s}$	
Product to Power Rule:	$(a \cdot b)^r = a^r \cdot b^r$	
Quotient to Power Rule:	$\left(\dfrac{a}{b}\right)^r = \dfrac{a^r}{b^r}$	if $b \neq 0$
Quotient to a Negative Power Rule:	$\left(\dfrac{a}{b}\right)^{-r} = \left(\dfrac{b}{a}\right)^r$	if $a \neq 0, b \neq 0$

Working with Rational Expressions (Chapter 5)

Multiplying Rational Expressions	$\dfrac{a}{b} \cdot \dfrac{c}{d} = \dfrac{ac}{bd}$	$b \neq 0, d \neq 0$
Adding Rational Expressions	$\dfrac{a}{c} + \dfrac{b}{c} = \dfrac{a + b}{c}$	$c \neq 0$
Subtracting Rational Expressions	$\dfrac{a}{c} - \dfrac{b}{c} = \dfrac{a - b}{c}$	$c \neq 0$
Dividing Rational Expressions	$\dfrac{a}{b} \div \dfrac{c}{d} = \dfrac{\frac{a}{b}}{\frac{c}{d}} = \dfrac{a}{b} \cdot \dfrac{d}{c} = \dfrac{ad}{bc}$	$b \neq 0, c \neq 0,$ $d \neq 0$